Trial by Judge or Jury

Should a person who must go on trial choose to be tried by a judge or by a jury? Statistics based on a classic study will reveal the answer.

Jane Blalock and the LPGA

When Jane Blalock, star professional golfer, was banned from participating in tournaments by the Ladies' Professional Golf Association, she sued the organization under the federal antitrust laws. How much money would she have earned? Her court battle was based on probability theory and the Case Study gives the surprising result of her suit.

Lenin as Statistician

V. I. Lenin is well known as the leader of the Russian revolution. It is not so well known that in his formative years he used comparative statistics to support the argument that the plight of Russian peasants had worsened under the Czar.

The Federalist Papers

The 13 colonies in America joined forces from 1776 to 1783 to fight for their independence from Great Britain, but then a family squabble broke out as the independent colonies wrestled with their own self-governance. The most instrumental documents supporting the ratification of the Constitution were The Federalist papers, but for almost 200 years the identity of the authors of some of The Federalist papers remained unknown. Conclusive evidence was first published in 1964 using statistical methods.

Allowance for Doubtful Accounts

In business an unpaid bill is called a bad debt, and some bad debts are never collected. Large retail firms must account for a percentage of money due that they never will collect. The Case Study shows how accountants use Markov chains to give firms a structured means of predicting the amount of bad debts to account for.

Game Theory and the Normandy Invasion

Taken from a classic paper written by Colonel O. G. Haywood, Jr., in 1954, the Case Study shows that game theory can be used to analyze military decisions by centering on a key battle in World War II.

The Norway Cement Company Merger

A very stormy merger of Norway's three cement corporations took place in the 1960s. Even though three groups were involved in the negotiation, the conflict was resolved by using a two-person-game-theoretic approach.

Finite Mathematics with Calculus:
An Applied Approach

Finite Mathematics with Calculus: An Applied Approach

Second Edition

David E. Zitarelli
Raymond F. Coughlin

Temple University

Saunders College Publishing

Harcourt Brace Jovanovich College Publishers

Forth Worth   Philadelphia   San Diego   New York
Orlando   Austin   San Antonio   Toronto
Montreal   London   Sydney   Tokyo

Text Typeface: Times Roman
Compositor: General Graphic Services
Acquisitions Editor: Robert Stern
Developmental Editor: Richard Koreto
Managing Editor: Carol Field
Copy Editors: Elaine Honig and Merry Post
Manager of Art and Design: Carol Bleistine
Art Assistant: Caroline McGowan
Text Designer: William Boehme
Cover Designer: Lawrence R. Didona
Text Artwork: Vantage Art, Inc.
Director of EDP: Tim Frelick
Production Manager: Charlene Squibb
Product Manager: Monica Wilson

Cover Credit: The Wente Torus, a surface of constant mean curvature clipped by a sphere and with color derived from surface parameters. Mathematical programming by Yi Fang. Graphics by James T. Hoffman.

Printed in the United States of America

FINITE MATHEMATICS WITH CALCULUS: AN APPLIED APPROACH, Second edition

0-03-055849-2

Library of Congress Catalog Card Number: 91-058103

234   039   98765432

Preface

In the summer of 1991 a committee of the National Research Council, which is part of the National Academy of Sciences, issued a report called *Moving Beyond Myths*.* The report describes many serious problems in undergraduate mathematics education and provides an action plan for attacking them. One of the myths that this committee identified (see page 12) is

Myth: *Only scientists and engineers need to study mathematics.*

Reality: Mathematics is a science of patterns that is useful in many areas. Indeed, the most rapid areas of growth in applications of mathematics have been in the social, biological, and behavioral sciences. Financial analysts, legal scholars, political pollsters, and sales managers all rely on sophisticated mathematical models to analyze data and make projections. Even artists and musicians use mathematically based computer programs to aid in their work. No longer just a tool for the physical sciences, mathematics is a language for all disciplines.

The goal of this book is to reverse this myth. Although the table of contents is standard, the means of achieving the goal differs from other texts in two notable ways. First, in every section the Referenced Exercises refer to areas outside mathematics and are accompanied by footnotes so the student can pursue them in greater depth or the instructor can assign them as part of special projects. Because such applications deal with real-life situations, calculators are frequently necessary to carry out the computations.

We hope to show students through these exercises that mathematics is everywhere. No matter what major a student chooses, mathematics can play a key role in solving interesting problems in that discipline. Here is a representative list of selected Referenced Exercises, along with the section where each can be found.

Application	Source	Section
stereo speakers	Teledyne Acoustic Research	1–1
archaeological deposits	*American Antiquity*	2–1
crop rotations	*American Journal of Agricultural Economics*	3–3
dairy farming	*Journal of the Operational Research Society*	4–1
athletes' contracts	text *Financial Management*	5–2

*William E. Kirwan (Editor), *Moving Beyond Myths: Revitalizing Undergraduate Mathematics,* Washington D.C., National Academy Press, 1991.

Application	Source	Section
pitch classes	*Journal of Music Theory*	6–1
homicide and deterrence	*Southern Economics Journal*	7–5
economic forecasts	*Management Science*	8–3
urban status	*American Journal of Sociology*	9–1
auditing procedures	*The Accounting Review*	10–2
stocks and bonds	*Forbes Magazine*	11–3
income tax	*Wall Street Journal*	12–3
telecommunications	*Forecasting Public Utilities*	13–2
regeneration of trees	*Ecology*	14–5
learning curves	*Decision Sciences*	15–2
pollution control	*American Journal of Agricultural Economics*	16–4
gifted students	*Journal of Mathematical Psychology*	17–1
cancer treatment	*Physics in Medicine and Biology*	18–2

In addition, every chapter has at least one Case Study that treats a particular application in greater detail than the Referenced Exercises. The topics have been chosen not only to illustrate areas where mathematics has played a crucial role in solving an important problem, but to highlight areas of particular human interest.

Even if the instructor does not have time to cover all the Referenced Exercises or Case Studies, it is hoped that their relevance and wealth (in both diversity and number) will make a lasting impression.

Changes in the Second Edition

This second edition of the book differs from the first edition in several ways.

1. The exercises are divided into three sets. The first set consists of the standard assortment of problems, usually numbering between 50 and 70. The first 20 or so generally reflect the examples in the text and are routine, while the later problems are a bit more challenging. The second set, titled ''Referenced Exercises,'' contains problems from areas outside mathematics; these exercises are accompanied by complete references to the literature. The third set, ''Cumulative Exercises,'' contains problems whose solutions call on material from the preceding sections in that chapter. These problems often require different skills than do the usual problems. The authors have found this feature very useful in their own classes as a way to continually help students review old material as they study the current section. In total, there are over 4,000 exercises in this book, about 20% of them new to this edition. Although the book continues to contain application exercises associated with many subjects, many simple drill problems have been added as well to help students master concepts before applying them.
2. There are many new and updated Referenced Exercises.
3. There are two new Case Studies, one on ''Trial by Jury'' and the other on ''Lenin as Statistician.''

4. The normal exercise sets and the cumulative exercise sets include two types of problems that several users of the first edition requested: those that are stated in words and require the translation from the ordinary language into English, and those that require a geometric interpretation.

5. At the end of each chapter is a set of exercises for Programmable Calculators. An Appendix discusses such calculators. The programmable, or graphing, calculator is an important new tool in the teaching of applied mathematics. However, instructors who do not use this calculator in class can skip the programmable calculator material without any loss of continuity. This edition, like the first, also includes references to ordinary scientific calculators in key places.

6. Many instructors who used the first edition felt that four important topics should have their own sections. Section 6–5 is now exclusively devoted to the binomial theorem, Section 7–8 to expected value, Section 12–4 to continuous and differentiable functions, and Section 18–6 to volume.

Organization

The text covers two distinct areas, finite mathematics (Chapters 1 through 10) and calculus (Chapters 11 through 18). Chapter R presents a brief review of those topics whose mastery is necessary beforehand. The first four sections of Chapter R present background material for finite mathematics, and the remaining four sections for calculus.

The material on finite mathematics can be divided into three parts: linear mathematics (Chapters 1 through 4), probability and statistics (Chapters 6, 7, and 8), and applications (Chapters 5, 9 and 10).

Linear Mathematics

Chapter 1 introduces the Cartesian coordinate system and linear equations. Systems of two equations in two variables are presented separately from systems of three equations in three variables for those instructors who want to proceed directly to the geometric solution of linear programming problems in Chapter 3.

Chapter 2 introduces matrices and their connection to systems of linear equations. Section 1–4 on Gaussian elimination is a prerequisite for Section 2–3 on the Gauss-Jordan method. The matrix inverse method is applied to input-output analysis.

Chapters 3 and 4 deal with linear programming. Chapter 3 illustrates the geometric method of solving linear programming problems with two variables. In Chapter 4 the simplex method is used for solving linear programming problems with more than two variables.

Probability and Statistics

Chapter 6 lays the foundation for this part. It reviews the basic aspects of set theory, including Venn diagrams and tree diagrams, and presents counting techniques, permutations, and combinations. In the second edition we have expanded our coverage of the Binomial Theorem by devoting a new section to it. In Chapter 7 the basic aspects of probability are motivated by a study of auto accidents on a major highway in Boston. After the set-theoretic foundations are described, laws governing the addition, subtraction, multiplication, and division of probabilities are derived. We have also expanded our coverage of the expected value and binomial experiments by devoting a section to each in this edition. Chapter 8 presents the rudiments of statistics, including graphical methods, measures of central tendency, measures of spread from the central tendency, the normal curve, and binomial experiments.

Applications

Chapter 5 is independent of the rest of the book. It deals primarily with the mathematics of finance, but it also treats inflation, cost of living, and population growth.

Chapters 9 and 10 apply matrices and probability to Markov chains and game theory. The last section in the book ties this material to linear programming, and the epilog discusses the historical meeting that produced the confluence of these seemingly unrelated fields.

The second part of the text can also be divided into three parts: differential calculus (Chapters 11 through 15), integral calculus (Chapters 16 and 17), and functions of several variables (Chapter 18).

Differential Calculus

Chapter 11 covers the algebraic techniques needed for the subsequent material. Functions are defined and studied, and various properties of polynomial functions and rational functions are illustrated. Chapter 12 explains the limit of a function by discussing velocity, rate of change of a moving object, and the tangent to a graph. The definition of the derivative is used to calculate the derivatives of several functions.

Chapter 13 presents various techniques for computing the derivatives of functions. Chapter 14 applies these techniques in a wide variety of problems.

Chapter 15 introduces exponential and logarithmic functions. Their derivatives are obtained by calculator experiments aimed at suggesting general rules. Those rules are then proved rigorously.

Integral Calculus

Chapter 16 introduces integration via antidifferentiation. After indefinite integrals have been defined, the Fundamental Theorem of Calculus relates the derivative to the integral. Then the integral is used to compute areas bounded by curves.

Chapter 17 presents two techniques of integration, the use of tables of integrals, and numerical integration.

Functions of Several Variables

Chapter 18 extends the definition of the derivative and the definition of the integral to functions of more than one variable. Partial derivatives are defined and applied to the sketching of surfaces, while double integrals are evaluated for functions of two variables and used to compute volumes.

Format

We have tried to keep the length of each section to what can be covered in a typical 50-minute class. Sometimes, however, the topic has dictated more extensive coverage. Each numbered section has been partitioned into subsections to help the instructor prepare lectures and to help the student organize the material.

There are two kinds of examples. One explains a new skill which is being encountered for the first time. It is labeled simply EXAMPLE. The other illustrates a skill which was explained beforehand. It uses the following format:

E X A M P L E

Problem

Solution

The student should be able to make a good attempt at solving the problem independently before reading its solution.

Each chapter ends with a review of the terms, notation, and formulas that were introduced in the chapter. It also contains review problems that can be used to review for tests.

Supplements

A number of supplements are available for use with this text.

The **Instructor's Manual**, by Richard Shores, Lynchburg College, contains the partially worked-out solutions to all exercises in the text. (Answers to the odd-numbered exercises are also in the back of the text.) Also contained are additional teaching hints and aids for the instructor.

A **Student Solutions Manual and Study Guide**, also by Richard Shores, contains fully worked-out solutions to every other odd-numbered problem. This manual is designed to help students with their problem-solving skills: these solutions can be used as models in solving similar problems.

The software package **MathPath** by George Bergeman, Northern Virginia Community College, is available free to users of this text. MathPath has been revised for this edition. It supplies graphical and computational support for many of the important topics in each chapter. The software requires an IBM or IBM-compatible computer with at least 512K. Prospective users might want to consult an article that compared the software packages that accompany six books: Joan Wyzkoski Weiss, "IBM Software for Finite Mathematics, Part I," *The College Mathematics Journal* **22**(3) (May, 1991), 248–254.

The test bank, by Robert Kurtz and Pao-sheng Hsu, of the University of

Maine at Orono, contains about 1,800 questions, all new to this edition. "Writing Across the Curriculum" problems, which apply mathematics to different fields, make this manual unique. Even instructors who use their own test banks will find this one useful.

The test bank is also available in a computerized format for IBM and IBM-compatible computers. This computerized version allows an instructor to custom-design tests and to sort the questions by several different categories. It requires 256K and two disk drives or 384K and a hard drive, graphics card and monitor, and a printer capable of handling graphs. Full instructions are included.

Acknowledgments

It is a pleasure to acknowledge the help we received from many people. Our biggest support has continued to come from our wives, Anita and Judy, and children, Paul and Nicole, and Virginia, Christina, and Sara.

Several reviewers deserve special thanks for their constructive criticism of the entire manuscript.

Second Edition Reviewers

Carol Benson, Illinois State University
Priscilla Chaffe-Stengel, California State University, Fresno
Charles C. Clever, South Dakota State University
Lynne Doty, Marist College
Keith Ferland, Plymouth State College
Nancy Fisher, University of Alabama
Jerry Goldman, DePaul University
Frances Gulick, University of Maryland, College Park
Denise Hennicke, Collin County Community College
David Jones, Glendale Community College
Robert Kurtz, University of Maine, Orono
Sarah L. Mabrouk, Boston University
Claire C. McAndrew, Fitchburg State College
Richard J. McGovern, Marist College
Richard J. Shores, Lynchburg College
Lowell Stultz, Kalamazoo Valley Community College
William R. Trott, University of Mississippi
Melvin F. Tuscher, West Valley College
P.L. Waterman, Northern Illinois University

We would like to thank again the very helpful reviewers who read the manuscript of the first edition at various stages of development.

First Edition Reviewers

Gail Broome, Providence College
Garret Etgen, University of Houston

Lou Hoelzle, Bucks County Community College
Ron Jeppson, Moorhead State University
Lawrence Maher, North Texas State University
Giles Maloof, Boise State University
Gordon Schilling, University of Texas at Arlington
Daniel Symancyk, Anne Arundel Junior College

A very important part of the reviewing process is the checking of errors. For their aid in this process, we would like to thank Bob Martin, Tarrant County Junior College, and Paul Allen, University of Alabama, who performed the arduous task of checking galleys and page proofs for mathematical accuracy.

We extend special gratitude to our friend Vincent Damiano, who is a Senior Project Engineer for Chevron USA, Inc. He read the entire manuscript and suggested several realistic linear programming problems.

The staff at Saunders has been a pleasure to work with. We especially thank our editors, Robert Stern and Richard Koreto, for the interest, encouragement, guidance, and persistence they have bestowed upon this project. We also thank the project editor, Carol Field, and the art director, Carol Bleistine, not only for their professional expertise but for their patience and humor as well. Our appreciation also goes to Elaine Honig and Merry Post for their excellent copyediting.

David E. Zitarelli
Raymond F. Coughlin
Philadelphia, PA
December 1991

Contents

7   **Probability** / 317

8   **Statistics** / 406

9   **Markov Processes** / 469

10   **Game Theory** / 510

11 Functions / 558

12 The Derivative / 612

13 Differentiation Rules / 655

14 Applications of the Derivative / 693

15 Exponential and Logarithmic Functions / 761

16   The Integral / 817

17   Techniques of Integration / 865

18   Functions of Several Variables / 896

Appendix A: The Programmable Calculator / A.1

Appendix B: Tables / A.6

Index of Applications

Business and Economics

Science and Technology

Social Sciences

Algebra Review

Mathematics and high finance: the New York Stock Exchange. (H. Armstrong Roberts)

Chapter Overview

The material in this chapter is a review in the sense that you have learned most of it previously. Some of it will be fresh in your mind, but other concepts will have to be learned anew. Mastering this material makes it much easier to understand the rest of the text.

The first four sections can be regarded as a review of algebraic topics that are used in finite mathematics. The last four sections cover some precalculus techniques.

The first two sections deal with equations and inequalities that contain only one variable. Later in the text these ideas are expanded to include equations and inequalities with more than one variable. There are many uses of exponents in finite mathematics and calculus. They are reviewed in Section R–3. The absolute value of numbers and expressions containing absolute values are covered in Section R–4.

Polynomials are building blocks of many branches of mathematics, especially calculus. They are introduced in Section R–5 by studying those polynomials with one variable. Factoring polynomials is an important skill that is reviewed in Section R–6. Solving equations is necessary for a thorough understanding of many concepts in calculus as well as in finite mathematics. In Section R–7 we cover the solution of quadratic equations. The last section deals with rational expressions that are ratios of polynomials.

R–1   Equalities in One Variable

One of the fundamental building blocks of mathematics is the study of equations. An equation states that one expression is equal to another, such as

$$4 + 6 = 10 \qquad x + 6 = 10 \qquad 4 + 2K = 10$$

The first example is always true while the other two equations are true for some values of the **variable,** which is what the unknown quantity is called. Although most of the time the variables used are x and y, any letter will suffice. The second equation uses x while the third uses K.

An equation that can be written in the form $ax + b = c$, for real numbers a, b, and c with $a \neq 0$, is called a **linear equation in one variable.** The variable can be represented by any letter. For example, $x + 6 = 10$ and $4 + 2K = 10$ are linear equations in x and K, respectively. Examples of equations that are **nonlinear** (not linear) are $x^2 = 1$ and $x^{1/2} = 4$.

The equation $x + 6 = 10$ is true when $x = 4$ and is false for any other value of x. We say that $x = 4$ is the **solution of the equation.** To solve the equation $4 + 2K = 10$, isolate K on the left-hand side of the equation by first subtracting 4 from each side of the equation and then dividing the resulting equation by 2.

$$4 + 2K = 10$$
$$4 + 2K - 4 = 10 - 4$$
$$2K + 4 - 4 = 10 - 4$$
$$2K = 6$$
$$K = 3$$

This means that the solution of the equation is $K = 3$. The solution should always be checked by substituting it into the original equation. In this case substitute $K = 3$ into the equation $4 + 2K = 10$ and check whether the resulting statement is true.

$$4 + 2(3) \overset{?}{=} 10$$

The left-hand side equals the right-hand side, meaning that the solution checks.

The following properties of real numbers are used to solve equations. A variable represents a real number so that it also obeys the properties.

Properties of Real Numbers Used in Solving Equations

Let a, b, and c represent real numbers or expressions that represent real numbers.

1. If $a = b$, then $a + c = b + c$
2. If $a = b$, then $ac = bc$
3. $a + b = b + a$ **(the commutative law of addition)**
4. $a(b + c) = ab + ac$, $(a + b)c = ac + bc$ **(the distributive laws)**

Sometimes an equation will have more than one term containing the variable, such as $2x + 4x$. In such a case use property 4, the distributive rule, to "gather like terms." That is, $2x + 4x = (2 + 4)x = 6x$. Example 1 demonstrates how to solve a linear equation in one variable when more than one term contains the variable.

EXAMPLE 1

Problem

Solve (a) $6x + 9 = 2x - 3$ (b) $2(3p - 1) + 2p = 3$

Solution (a) First isolate x on the left-hand side by subtracting $2x$ and 9 from each side, and then gathering like terms and dividing by the appropriate number; that is, use rule 2 with $c = \frac{1}{4}$.

$$6x + 9 = 2x - 3$$
$$6x + 9 - 2x - 9 = 2x - 3 - 2x - 9$$
$$4x = -12$$
$$x = -3$$

The solution is $x = -3$.

(b) First multiply 2 times the expression in the parentheses according to property 4 and then proceed as in part (a).

$$2(3p - 1) + 2p = 3$$
$$6p - 2 + 2p = 3$$
$$8p = 5$$
$$p = \tfrac{5}{8}$$

The solution is $p = \tfrac{5}{8}$.

Example 2 shows how to solve a linear equation in one variable when some additional arithmetic is needed.

EXAMPLE 2

Problem

Solve $\dfrac{6x - 1}{5} = \dfrac{3x + 10}{4}$

Solution First eliminate the fractions by multiplying through the equation by the least common multiple of the denominators, 20. This yields

$$20\,\frac{(6x - 1)}{5} = 20\,\frac{(3x + 10)}{4}$$

$$4(6x - 1) = 5(3x + 10)$$

Use property 4.

$$24x - 4 = 15x + 50$$

Isolate x on the left-hand side by subtracting $15x$ and adding 4 to each side, and then gathering like terms and dividing by the appropriate number.

$$24x - 4 = 15x + 50$$
$$24x - 4 - 15x + 4 = 15x + 50 - 15x + 4$$
$$9x = 54$$
$$x = 6$$

These techniques can be used to solve equations that are not linear.

EXAMPLE 3

Problem

Solve $7 - 3/x = -2$.

Solution First multiply each side of the equation by x.

$$(7 - 3/x)x = -2x$$
$$7x - 3 = -2x$$

Isolate x on the left-hand side by adding $2x + 3$ to each side, and then gathering like terms and dividing by the appropriate number:

$$7x + 2x = 3$$
$$9x = 3$$
$$x = \tfrac{1}{3}$$

EXERCISE SET R–1

In Problems 1 to 24 solve the linear equation.

1. $3x + 5 = 4x - 2$
2. $4y - 3 = 5 - 2y$
3. $3 + 4x = 4 - 2x$
4. $1 - 3z = 5z - 2$

5. $3(s - 5) + 3 = 2s$

6. $4r - 5(3 - 2r) = r + 3$

7. $3q - 4(23 - 21q) = 15q + 30$

8. $2 - 4(2.5 - 5.1t) = 2(1.5t + 2.2)$

9. $3(x - 1) + 3 = 2(1 + x)$

10. $4y - 5(3 - y) = 2(2y + 3)$

11. $2x/5 - 3/8 = 5/8 - 7x/20$

12. $3y/4 - 1/8 = 5y/12 - 7y/8$

13. $1 + 2(1.5 - 1.1y) = 2(0.5 + 1.2y)$

14. $2 + 3(1.5 - x) = 4(1.5 + 0.5x)$

15. $t/5 - 3t/2 = 1/10 - t/20$

16. $y/2 + y/8 = 5y/2 - 7$

17. $x/2 - 4x/3 = 3/10 - 7x/20$

18. $y/4 + 3/8 = 5y/8 - 7y/4$

19. $\dfrac{x - 1}{5} = \dfrac{x + 3}{7}$

20. $\dfrac{x + 2}{5} = \dfrac{x + 3}{6}$

21. $\dfrac{3x + 2}{4} = \dfrac{4x + 3}{5}$

22. $\dfrac{2x - 1}{3} = \dfrac{5x - 2}{4}$

23. $\dfrac{7x + 2}{3} = \dfrac{2x + 8}{5} + 1$

24. $\dfrac{2x - 2}{3} = \dfrac{3x - 1}{6} - 1$

In Problems 25 to 44 solve the equation.

25. $3/x + 1 = 2$

26. $2 - 3/x = -1$

27. $4/x + 1 = 10$

28. $5 - 1/x = 1$

29. $4/x - 3/x = 1$

30. $5/x + 7/x = 19$

31. $2/3x + 1 = 1/x$

32. $3 - 3/2x = 1/x$

33. $2(1 - 1/x) = 1$

34. $3(1 + 1/x) = 6$

35. $4(2/x - 1) = 4$

36. $3(3/x + 1) = 12$

37. $1 + 2(1 - 2/x) = 2$

38. $3 - 2(2 + 4/x) = 7$

39. $1/x = 2(1 + 1/x)$

40. $2/3x = 4(1 - 1/x)$

41. $\dfrac{1}{(x - 1)} = 1$

42. $\dfrac{2}{(3 - x)} = 1$

43. $1 + \dfrac{3}{(x + 1)} = 2$

44. $2 - \dfrac{5}{(1 + 2x)} = 1$

R–2 Linear Inequalities in One Variable

An **inequality** states that one expression is unequal to another. Sometimes an equals sign is included with the **inequality sign** to indicate that the expressions can be equal. There are four inequality signs.

$<$ represents "is less than"
$>$ represents "is greater than"
$\leq$ represents "is less than or equal to"
$\geq$ represents "is greater than or equal to"

Examples of inequalities without a variable are

$$3 < 4 \qquad 3 \leq 5 \qquad -1 > -2 \qquad -3 \leq -3$$

These represent statements about real numbers. They are all true but a linear inequality can also be false, such as $3 < 0$, and $-1 < -2$. If a variable is present, the inequality is sometimes true and sometimes false, depending on the value assigned to the variable. Examples of inequalities in one variable are

$$3x < 4 \qquad t + 3 \leq 5 \qquad 3x + 4 > 10 \qquad 6 + 2y \geq 10$$

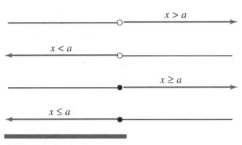

FIGURE 1

A **linear inequality in one variable** is one that can be expressed as $ax < b$, where a and b are real numbers, or expressed with any of the other three inequality signs and any letter for the variable. Thus $3x < 4$ is a linear inequality with $a = 3$ and $b = 4$. The properties of real numbers used to solve inequalities are similar to those properties used to solve equations. The important difference is when an inequality is multiplied by a negative number, property 3 here.

Properties of Real Numbers Used in Solving Inequalities

Let a, b, and c represent real numbers or expressions that represent real numbers. The properties are stated for "less than" ($<$), but they also hold for the other three inequality signs.

1. If $a < b$, then $a + c < b + c$
2. If $a < b$, and $c > 0$, then $ac < bc$
3. If $a < b$, and $c < 0$, then $ac > bc$

Property 3 states that the sign of the inequality changes if the inequality is multiplied by a negative number. For example, $3 > 2$ but $3(-4) < 2(-4)$ because $-12 < -8$. Dividing by a negative number also changes the sign of the inequality because division is the same as multiplying by the multiplicative inverse of the number. For example, $6 > 2$ but $6(-1/2) < 2(-1/2)$ because $-3 < -1$.

The **solution of a linear inequality** is expressed as $x < a$, $x > a$, $x \leq a$, or $x \geq a$. These expressions represent the set of all real numbers less than a or greater than a, where a is not included in the set for the first two expressions but a is included in the set for the other two expressions. The graph of these expressions is shown in Figure 1. The graphs use a shaded line to indicate the solution set with an open circle to indicate that the endpoint is not included and a shaded circle if it is included.

Example 1 illustrates how to solve a linear inequality in one variable.

EXAMPLE 1

Problem

Solve $3x - 5 < 7$

FIGURE 2

Solution Isolate x on the left-hand side of the inequality and then divide by 3.

$$3x - 5 < 7$$
$$3x - 5 + 5 < 7 + 5$$
$$3x < 12$$
$$x < 4$$

The graph of the solution is in Figure 2. The solution is not one number; it consists of infinitely many numbers.

Example 2 uses more properties to solve a linear inequality.

EXAMPLE 2

Problem

Solve $3(1 - 2x) > 3x + 15$

Solution First use the distributive law and then proceed as in Example 1.

$$3 - 6x > 3x + 15$$
$$3 - 6x - 3 - 3x > 3x + 15 - 3 - 3x$$
$$-9x > 12$$
$$x < -\tfrac{4}{3}$$

The last step used property 3 for inequalities with $c = -\tfrac{1}{9}$ so that the sense of the inequality is reversed. The solution is graphed in Figure 3.

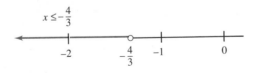

FIGURE 3

Example 3 solves an inequality that includes an equal sign in the inequality.

EXAMPLE 3

Problem

Solve $2(3 + 5x) \le 6(x - 3)$

FIGURE 4

Solution First use the distributive law and then proceed as in Example 1.

$$6 + 10x \leq 6x - 18$$
$$6 + 10x - 6x - 6 \leq 6x - 18 - 6x - 6$$
$$4x \leq -24$$
$$x \leq -6$$

The solution is graphed in Figure 4.

EXERCISE SET R–2

Solve the linear inequality.

1. $3x - 4 < 5$

2. $5x - 3 < 7$

3. $4 + 3x > 10$

4. $6 + 2x > 10$

5. $4 < x + 5$

6. $7 > 4x - 5$

7. $5x - 2 \leq 8$

8. $7x - 3 \leq 11$

9. $3(x + 1) > 9$

10. $5(x - 1) > 10$

11. $2(1 - 3x) < 14$

12. $3(3 - 2x) < 3$

13. $2x + 1 < 3x - 2$

14. $3y - 5 < 1 - 2y$

15. $5x + 2 > x - 2$

16. $2y - 3 > 1 - 4y$

17. $x - 2 \geq 3 - 2x$

18. $4y - 3 \leq 2 - y$

19. $4(2x - 3) > 7x$

20. $3(4x + 1) > 11x$

21. $3x < 2 + 4(x + 1)$

22. $4x > 8 - 3(2x - 4)$

23. $2x < 6 + 4(1 - 2x)$

24. $-3x > 2(3 - x) - 5$

25. $2(x + 3) > 3(x + 1)$

26. $4(x - 5) > 5(1 + x)$

27. $2(3x - 1) < 4(3x + 7)$

28. $6(2x + 1) > 2(x + 13)$

29. $2(t - 5) + 1 \geq 3t$

30. $6r - 2(3 - r) \geq r + 1$

31. $3x - 2(2 - 2x) \leq 5x + 3$

32. $2 - 3(2.1 - 5.2x) \leq 3(1.5x + 2.5)$

33. $2x/5 - 3/4 > 1/4 - 3x/10$

34. $y/3 - 1/4 > 5(y/2 - 7) - y/4$

35. $\dfrac{x - 1}{2} > \dfrac{x + 3}{3}$

36. $\dfrac{x + 2}{3} > \dfrac{x + 3}{2}$

37. $\dfrac{3x + 2}{4} < \dfrac{x + 3}{2}$

38. $\dfrac{2x - 3}{3} < \dfrac{x - 2}{4}$

39. $\dfrac{5x + 2}{3} \leq \dfrac{2x + 8}{2} + 1$

40. $\dfrac{2x - 2}{3} \geq \dfrac{3x - 1}{-2} - 1$

R–3   Exponents

In this section we review the basic formulas that govern the use of **exponents**. If n is a positive integer, then a^n means $a \cdot a \cdots a \cdot a$ with n factors of a. That is,

$$a^n = \underbrace{a \cdot a \cdots a \cdot a}_{n \text{ times}}$$

For example,

$$2^3 = 2 \cdot 2 \cdot 2 = 8$$
$$3^4 = 3 \cdot 3 \cdot 3 \cdot 3 = 81$$
$$2^5 = 2 \cdot 2 \cdot 2 \cdot 2 \cdot 2 = 32$$

An exponent can also be negative. It means that the reciprocal is to be taken. For example,

$$2^{-3} = \frac{1}{2^3} = \frac{1}{8}$$

$$3^{-4} = \frac{1}{3^4} = \frac{1}{81}$$

In general,

$$a^0 = 1$$

$$a^{-n} = \frac{1}{a^n}$$

The basic properties of exponents are presented here. The examples given so far use only **integer exponents** but the properties are true for any real numbers.

Let $a > 0$ and n be a real number.

1. $a^0 = 1$

2. $a^{-1} = \dfrac{1}{a}$

3. $a^{-n} = \dfrac{1}{a^n}$

4. $(a^n)^m = a^{nm}$

5. $a^n a^m = a^{n+m}$

6. $\dfrac{a^n}{a^m} = a^{n-m}$

7. $a^n b^n = (ab)^n$

8. $\dfrac{a^n}{b^n} = \left(\dfrac{a}{b}\right)^n$

EXAMPLE 1

Problem

Simplify
(a) $(2^3)^2 2^2$
(b) $(10^{-3} 10^2)^3$

(c) $\dfrac{3^{-3}}{3^{-2}}$

Solution
(a) $(2^3)^2 2^2 = 2^6 2^2 = 2^8 = 256$

(b) $(10^{-3} 10^2)^3 = (10^{-3+2})^3 = (10^{-1})^3 = 10^{-3}$

$$= \frac{1}{10^3} = 0.001$$

(c) $\dfrac{3^{-3}}{3^{-2}} = 3^{-3}3^2 = 3^{-1} = \dfrac{1}{3}$

The properties also hold for any letter as the base since it represents any real number. In the next example assume the letters stand for positive real numbers.

EXAMPLE 2

Problem

Simplify
(a) $(x^{1/2})^6 x^2$
(b) $(t^{-1/2}s^2)^4$

(c) $\dfrac{y^{7/3}}{y^{2/3}}$

Solution
(a) $(x^{1/2})^6 x^2 = x^3 x^2 = x^5$

(b) $(t^{-1/2}s^2)^4 = (t^{-1/2})^4(s^2)^4 = t^{-2}s^8 = \dfrac{s^8}{t^2}$

(c) $\dfrac{y^{7/3}}{y^{2/3}} = y^{7/3}y^{-2/3} = y^{5/3}$

The properties listed for integer exponents also hold for rational exponents, provided that the base a is nonnegative when n is even. The expression $a^{1/n}$, for n a positive integer, represents the positive **nth root** of a. It is that number which when raised to the nth power equals a. In other words, $a^{1/n}$ is that number b such that $b^n = a$. For instance, $9^{1/2}$ is the positive 2nd root, or **square root,** of 9, which is 3; thus $9^{1/2} = 3$ because 3 raised to the 2nd power is 9 (or simply $3^2 = 9$). Also, $8^{1/3}$ is the positive 3rd root, or the **cube root,** of 8, which is 2. That is, $8^{1/3} = 2$ because $2^3 = 8$. We also write

$$\sqrt{9} = 3 \qquad \text{and} \qquad \sqrt[3]{8} = 2$$

Sometimes the nth root of a number cannot be expressed as an integer. For instance, $2^{1/2} = \sqrt{2}$ is an irrational number, meaning that it cannot be expressed as a rational number, so it cannot be written as an integer.

If n is even, $a^{1/n}$ is not defined for a negative number a. If n is odd, then a can be negative and $a^{1/n}$ is also negative.

EXAMPLE 3

Problem

Compute the numbers.

(a) $16^{1/2}$ (b) $25^{1/2}$ (c) $64^{1/3}$ (d) $\left(-\dfrac{1}{8}\right)^{1/3}$

Solution (a) $16^{1/2} = 4$ because $4^2 = 16$.
(b) $25^{1/2} = 5$ because $5^2 = 25$.
(c) $64^{1/3} = 4$ because $4^3 = 64$.

(d) $\left(-\dfrac{1}{8}\right)^{1/3} = -\dfrac{1}{2}$ because $\left(-\dfrac{1}{2}\right)^3 = -\dfrac{1}{8}$

A common use of exponents is to evaluate expressions in x, or some other variable, which are raised to a power.

EXAMPLE 4

Problem

Evaluate the expressions at the given value of x.
(a) x^2 at $x = 2$
(b) x^{-2} at $x = 3$
(c) $x^{-1/2}$ at $x = 16$
(d) $x^{-2/3}$ at $x = 27$

Solution (a) Evaluate x^2 at $x = 2$ means compute $2^2 = 4$.

(b) Evaluate x^{-2} at $x = 3$ means compute $3^{-2} = \dfrac{1}{3^2} = \dfrac{1}{9}$.

(c) Evaluate $x^{-1/2}$ at $x = 16$ means compute $16^{-1/2} = \dfrac{1}{16^{1/2}} = \dfrac{1}{4}$.

(d) Evaluate $x^{-2/3}$ at $x = 27$ means compute $27^{-2/3} = \dfrac{1}{27^{2/3}} = \left(\dfrac{1}{27^{1/3}}\right)^2 =$

$\dfrac{1}{3^2} = \dfrac{1}{9}$.

EXERCISE SET R–3

Assume all bases are positive.
In Problems 1 to 20 simplify the expression.

1. $(3^2)^2 3^2$

2. $(4^2)^2 4^2$

3. $(10^{-2} 10^3)^2$

4. $(10^{-4} 10^3)^2$

5. $\dfrac{2^{-4}}{2^{-5}}$

6. $\dfrac{4^{-5}}{3^{-4}}$

7. $27^{1/3} 8^{1/3}$

8. $64^{1/6} 64^{1/3}$

9. $100^{1/2} 81^{1/2}$

10. $1000^{1/3} 241^{1/3}$

11. $(x^2)^2 x^6$

12. $(y^3)^4 y^3$

13. $(x^{-2} y^2)^3$

14. $(t^{-4} r^3)^3$

15. $\dfrac{x^{1/2}}{x^{5/2}}$

16. $\dfrac{t^{5/4}}{t^{-3/4}}$

17. $A^{1/3} A^{-4/3} A^{-1}$

18. $B^{1/6} B^{1/3} B^{1/2}$

19. $(x^{1/2} x^{-3/2})^2$

20. $(y^{1/3} y^{5/3})^3$

In Problems 21 to 44 compute the numbers.

21. $49^{1/2}$

22. $64^{1/2}$

23. $81^{1/2}$

24. $144^{1/2}$

25. $36^{-1/2}$

26. $144^{-1/2}$

27. $(1/49)^{1/2}$

28. $(1/100)^{1/2}$

29. $(4/49)^{1/2}$

30. $(81/100)^{1/2}$

31. $27^{1/3}$

32. $125^{1/3}$

33. $(8/27)^{1/3}$

34. $(27/125)^{1/3}$

35. $64^{-1/3}$

36. $27^{-1/3}$

37. $64^{2/3}$

38. $27^{2/3}$

39. $8^{-5/3}$

40. $27^{-4/3}$

41. $(-8)^{1/3}$

42. $(-27)^{1/3}$

43. $(-8)^{-1/3}$

44. $(-27)^{-1/3}$

49. $x^{1/2}$ at $x = 121$

50. $x^{1/2}$ at $x = 1/4$

51. $x^{-1/3}$ at $x = 64$

52. $x^{-2/3}$ at $x = 64$

53. $x^{1/5}$ at $x = 32$

54. $x^{3/4}$ at $x = 81$

55. $x^{1/5}$ at $x = -32$

56. $x^{-3/4}$ at $x = -81$

In Problems 45 to 56 evaluate the expression at the given value of x.

45. x^2 at $x = 5$

46. x^{-2} at $x = 4$

47. x^{-3} at $x = 1$

48. x^{-4} at $x = 2$

In Problems 57 to 60 compute the numbers.

57. $2(49^{1/2})(27^{1/3})$

58. $3(64^{1/2})(8^{1/3})$

59. $3^2(81^{1/2})(8^{-1/3})$

60. $4^2(144^{1/2})(36^{-1/2})$

R–4 Absolute Value

The absolute value of a number measures the **distance** that the number is from 0. If the number is positive, then the distance is equal to the number itself, so the absolute value of a positive number is the number itself. For example, the absolute value of 4 is 4 because it is four units from 0. Since -4 is also four units from 0, the absolute value of -4 is also 4. To get 4 from -4 via a mathematical operation, we multiply by -1. That is, $-(-4) = 4$. This leads to the formal definition of absolute value.

DEFINITION

If a is any real number, the **absolute value** of a, written $|a|$, is defined by the following:

1. If $a \geq 0$, then $|a| = a$
2. If $a < 0$, then $|a| = -a$

The absolute value of a number is always greater than or equal to 0. Part (2) of the definition might seem to indicate that $|a|$ is negative because there is a negative sign in the formula; but remember that in part (2) a is negative, and so $-a$ is positive. Example 1 clarifies the definition.

EXAMPLE 1

Problem

Evaluate each expression.
(a) $|-4|$,   (b) $|0|$,   (c) $-|-5|$,   (d) $-|4 - 9|$

Solution (a) $|-4| = 4$ by part (2) of the definition.
(b) $|0| = 0$ by part (1) of the definition.
(c) $-|-5| = -(5) = -5$ since $|-5| = 5$ by part (2) of the definition.
(d) $-|4 - 9| = -|-5| = -5$

Sometimes it will be necessary to evaluate a variable expression with absolute values at a particular value of x. For example, let us evaluate the following expression at $x = 2$:

$$x + |3x - 9| \quad \text{evaluated at } x = 2 \text{ is equal to}$$
$$2 + |3 \cdot 2 - 9| = 2 + |6 - 9| = 2 + |-3| = 2 + 3 = 5$$

The following examples show how to evaluate such expressions.

EXAMPLE 2

Problem

Evaluate the expression

$$2x - |5 - 4x|$$

at (a) $x = 1$ and (b) $x = 2$.

Solution (a) Substituting $x = 1$ into the expression yields

$$2 - |5 - 4| = 2 - |1| = 2 - 1 = 1$$

(b) Substituting $x = 2$ into the expression yields

$$4 - |5 - 4 \cdot 2| = 4 - |5 - 8| = 4 - |-3| = 4 - 3 = 1$$

EXAMPLE 3

Problem

Evaluate the expression

$$|2x + 1| - 2|1 - 4x|$$

at (a) $x = 0$ and (b) $x = -2$.

Solution (a) Substitute $x = 0$ into the expression.

$$|2(0) + 1| - 2|1 - 4(0)| = |1| - 2|1| = 1 - 2 = -1$$

(b) Substituting $x = -2$ into the expression yields

$$|2(-2) + 1| - 2|1 - 4(-2)|$$
$$= |-4 + 1| - 2|1 + 8| = |-3| - 2|9|$$
$$= 3 - 2(9) = -15$$

EXAMPLE 4

Problem

Evaluate the expressions at $x = -1$.

(a) $\dfrac{1}{|2x + 1|}$

(b) $|1 - 4x^2|$

Solution Substituting $x = -1$ into the expressions yields

(a) $\dfrac{1}{|2(-1) + 1|} = \dfrac{1}{|-2 + 1|} = \dfrac{1}{|-1|} = \dfrac{1}{1} = 1$

(b) $|1 - 4(-1)^2| = |1 - 4(1)| = |1 - 4| = |-3| = 3$

Sometimes it is necessary to solve equations and inequalities with absolute values.

EXAMPLE 5

Problem

Solve the equation

$$|2x + 1| = 7$$

Solution There are two cases, when (a) $2x + 1 \geq 0$ or (b) $2x + 1 < 0$.
(a) Suppose $2x + 1 \geq 0$. Then

$$|2x + 1| = 2x + 1 = 7$$
$$2x = 6$$
$$x = 3$$

(b) Suppose $2x + 1 < 0$. Then

$$|2x + 1| = -(2x + 1) = 7$$
$$-2x - 1 = 7$$
$$-2x = 8$$
$$x = -4$$

The solutions are $x = 3, -4$.

EXAMPLE 6

Problem

Solve the inequalities
(a) $|2 - 3x| < 7$
(b) $|2 - 3x| > 7$

Solution (a) There are two cases, $2 - 3x \geq 0$ or $2 - 3x < 0$. They can be handled in the same step by writing

$$-7 < 2 - 3x < 7$$
$$-7 - 2 < -3x < 7 - 2$$
$$-9 < -3x < 5$$
$$3 > x > -\tfrac{5}{3}$$
$$-\tfrac{5}{3} < x < 3$$

(b) The same two cases hold as in part (a), but they cannot be treated simultaneously as previously. This is because the expression cannot be greater than 7 and less than -7 for the same value of x. Each case must be handled individually.

(i) Suppose $2 - 3x \geq 0$. Then $|2 - 3x| = 2 - 3x$. The inequality becomes

$$|2 - 3x| = 2 - 3x > 7$$
$$-3x > 5$$
$$x < -\tfrac{5}{3}$$

(ii) Suppose $2 - 3x < 0$. Then $|2 - 3x| = -(2 - 3x)$. The inequality becomes

$$|2 - 3x| = -2 + 3x > 7$$
$$3x > 9$$
$$x > 3$$

The solution is $x < -\tfrac{5}{3}$ or $x > 3$. As expected, these are the numbers that are not mentioned in the solution of part (a), except for $x = -\tfrac{5}{3}$ and $x = 3$, which are not solutions to either inequality.

EXERCISE SET R–4

In Problems 1 to 10 evaluate each expression.

1. $|-6|$

2. $|-(-9)|$

3. $-|-15|$

4. $-\left|-\tfrac{3}{2}\right|$

5. $-|6 - 7|$

6. $-|12 - 3 - 10|$

7. $-5 + |-5|$

8. $-|-6| - |-7|$

9. $-|-9| + 2|-5|$

10. $-3|-6| - 4|-7|$

In Problems 11 to 14 evaluate the expression at $x = 2$.

11. $3x + |7 - 4x|$

12. $-3x + |9 - 3x|$

13. $|x - 6| + |10 - 5x|$

14. $|-3x - 1| - 2|8 - 5x|$

In Problems 15 to 18 evaluate the expression at $x = -2$.

15. $2x - |7x - 4|$

16. $-5x - |9x + 8|$

17. $|2 - x| + 3|2 - 5x|$

18. $|-4x - 1| - 2|3x - 15|$

In Problems 19 to 26 evaluate the expression at $x = -1$.

19. $\dfrac{1}{|4 - x|}$

20. $\dfrac{1}{|x - 4|}$

21. $\dfrac{2}{|2x - 3|}$

22. $\dfrac{4}{|4 + 5x|}$

23. $x|3 - 2x|$

24. $x|1 + 5x|$

25. $x + 2x|3x - 4|$

26. $3x - 2x|7 - 2x|$

In Problems 27 to 34 evaluate the expression at $x = 3$.

27. $\dfrac{x}{|1 - x|}$

28. $\dfrac{x}{|x - 5|}$

29. $\dfrac{2x}{|2x - 7|}$

30. $\dfrac{4x}{|9 - 5x|}$

31. $\dfrac{x - 4}{|1 - 2x|}$

32. $\dfrac{1 - x}{|2x - 5|}$

33. $\dfrac{|2x - 7|}{|x - 7|}$

34. $\dfrac{|4 - x|}{|9 - 2x|}$

In Problems 35 to 42 evaluate the expression at $x = -1$.

35. $|4 - 3x^2|$

36. $|x^2 - 4|$

37. $\dfrac{2}{|2x^2 - 3|}$

38. $\dfrac{4}{|4 - 5x^2|}$

39. $2x|1 - 2x^3| - \dfrac{1}{|7x^2 - 6|}$

40. $-5x|4x^3| - \dfrac{x^2}{|8x + 9|}$

41. $4x|3 - 5x^2| - \dfrac{1}{|8x^3 - 1|}$

42. $-2x^2|4x^3 - x - 5| - \dfrac{2x}{|3x^2 + 1|}$

In Problems 43 to 46 solve the equation.

43. $|7 - 4x| = 11$

44. $|9 - 3x| = 12$

45. $-6 + |10 + 2x| = 5$

46. $|3x - 1| - 2 = 0$

In Problems 47 to 54 solve the inequality.

47. $|7 - 4x| < 11$

48. $|9 - 3x| < 12$

49. $|10 + 3x| > 5$

50. $|3 - 2x| > 2$

51. $|7x - 4| < 11$

52. $|9x - 5| < 14$

53. $|7x - 4| > 11$

54. $|9x - 5| > 14$

R–5   Polynomials in One Variable

A polynomial in one variable is an expression of the form

$$c_n x^n + c_{n-1} x^{n-1} + \cdots + c_1 x + c_0$$

where the c_i are real numbers, n is a natural number, and x is the variable. Some examples of polynomials in the variable x are

$$x + 4 \qquad 3x - 10 \qquad x^2 - 3x + 5 \qquad 4x^4 - 5x$$

Any letter can be used for the variable. For instance, a polynomial with K as the variable is $6K^2 - 5K$. It has $n = 2$, $c_2 = 6$, $c_1 = -5$, and $c_0 = 0$.

The polynomial $3x^2 - 5$ has two **terms**, $3x^2$ and -5. In the term $3x^2$, 3 is the **coefficient** and 2 is the **exponent**. A **linear polynomial** is a polynomial that can be expressed as $ax + b$, while a **quadratic polynomial** can be expressed as $ax^2 + bx + c$.

Two or more polynomials can be combined by using the properties of real numbers since the variables represent real numbers. The most common property used is the distributive property. It is used to **combine like terms**; that is, add terms with the same exponent and multiply out parentheses. Example 1 illustrates how to proceed.

E X A M P L E 1

Problem

Simplify the expression

$$(3x^2 - 2x + 5) + (5x^2 + 6)$$

Solution Combine like terms.

$$(3x^2 - 2x + 5) + (5x^2 + 6) = (3x^2 + 5x^2) - 2x + (5 + 6)$$
$$= 8x^2 - 2x + 11$$

Example 2 shows how to multiply two polynomials by using the distributive law.

EXAMPLE 2

Problem

Simplify the expression

$$(2x^4 + 3x^2 + 5x) - 2x(3x^3 + 6x^2 - x + 3)$$

Solution First multiply out the parentheses and then combine like terms.

$$(2x^4 + 3x^2 + 5x) - 2x(3x^3 + 6x^2 - x + 3)$$
$$= (2x^4 + 3x^2 + 5x) - (6x^4 + 12x^3 - 2x^2 + 6x)$$
$$= (2x^4 + 3x^2 + 5x) - 6x^4 - 12x^3 + 2x^2 - 6x$$
$$= (2x^4 - 6x^4) - 12x^3 + (3x^2 + 2x^2) + (5x - 6x)$$
$$= -4x^4 - 12x^3 + 5x^2 - x$$

Certain products occur frequently, so we single out their statements:

1. $(a + b)(a - b) = a^2 - b^2$
2. $(a + b)^2 = a^2 + 2ab + b^2$
3. $(a - b)^2 = a^2 - 2ab + b^2$
4. $(ax + b)(cx + d) = acx^2 + (ad + bc)x + bd$
5. $(a + b)^3 = a^3 + 3a^2b + 3ab^2 + b^3$
6. $(a - b)^3 = a^3 - 3a^2b + 3ab^2 - b^3$

EXAMPLE 3

Problem

Express the following as polynomials.
(a) $(x + 3)(x - 3)$
(b) $(5 + x)^2$
(c) $(y - 4)^2$
(d) $(2x + 3)(4x - 1)$
(e) $(y - 2)^3$

Solution (a) In the first formula let $a = x$ and $b = 3$. Then

$$(x + 3)(x - 3) = x^2 - 3^2 = x^2 - 9$$

(b) In the second formula let $a = 5$ and $b = x$. Then

$$(5 + x)^2 = 5^2 + 2(5)x + x^2 = 25 + 10x + x^2$$

(c) In the third formula let $a = y$ and $b = 2$. Then

$$(y - 4)^2 = y^2 - 2y(4) + 4^2 = y^2 - 8y + 16$$

(d) In the fourth formula let $a = 2$, $b = 3$, $c = 4$, and $d = -1$. Then

$$(2x + 3)(4x - 1) = 8x^2 + (-2 + 12)x + 3(-1)$$
$$= 8x^2 + 10x - 3$$

(e) In the sixth formula let $a = y$ and $b = 2$. Then

$$(y - 2)^3 = y^3 - 3y^2(2) + 3y(2)^2 - 2^3$$
$$= y^3 - 6y^2 + 12y - 8$$

EXERCISE SET R–5

In Problems 1 to 20 simplify the polynomial expression.

1. $(2x^3 - 2x^2 + 5) + (5x^3 + 6x^2 - 2)$

2. $(4x^3 - 3x^2 - 6x - 1) + (2x^3 + 6x + 7)$

3. $(5x^4 + 3x^3 - 5x^2) - (2x^4 + 7x^3 - 2x^2)$

4. $(7x^4 - 10x^3 - 5x^2 - x) -$
   $(2x^4 + 7^3 - 2x^2 + 1)$

5. $x(x^4 + 3x^2 + x + 20)$

6. $3x(x^5 + 4x^3 - 2x^2 + 1)$

7. $3x^2(x^4 + x^3 - x^2 + 2)$

8. $2x^3(x^3 - 5x^2 - 6x)$

9. $x^4(2x^3 + 3x + 5)$

10. $3x^5(x^4 + 3x^3 + 2x - 2)$

11. $(x + 1)(x^2 + 2x)$

12. $(x^2 - 1)(x + 7)$

13. $(2x + 5)(x^2 + x)$

14. $(3x^2 - 2)(3x + 5)$

15. $x(x^3 + x^2 + x + 2) +$
    $3(x^4 + 6x^3 - x^2 + 2)$

16. $2x(x^3 - 3x^2 - 6) + x^2(2x^2 + 3x + 1)$

17. $x^2(x^2 + 2x) - (x^2 - 1)(x + 7)$

18. $(x + 1)(x^2 + 3) - x^2(3x + 7)$

19. $(3x - 4)(x^2 - 1) - (x^2 - 2)(x^2 + 3)$

20. $(2x + 3)(x^3 - 2) - (x^2 + 1)(x^2 + 1)$

In Problems 21 to 58 write the expression as a polynomial.

21. $(x + 1)(x - 1)$

22. $(x + 4)(x - 4)$

23. $(x - 5)(x + 5)$

24. $(x - 7)(x + 7)$

25. $(1 + y)(1 - y)$

26. $(4 + t)(4 - t)$

27. $(x + 3)^2$

28. $(x + 7)^2$

29. $(3 + x)^2$

30. $(7 + x)^2$

31. $(t + 8)^2$

32. $(y + 2)^2$

33. $(y - 4)^2$

34. $(y - 7)^2$

35. $(t - 8)^2$

36. $(y - 2)^2$

37. $(3 - x)^2$

38. $(7 - x)^2$

39. $(3x + 2)(5x + 4)$

40. $(3x + 2)(5x - 4)$

41. $(3x - 2)(5x - 4)$

42. $(3x - 2)(5x + 4)$

43. $(2 - 3y)(4 + 3y)$

44. $(7 + 2z)(3 - 5z)$

45. $(4 + 3x)(5 - 3x)$

46. $(4 + 3x)(7 - 3x)$

47. $(4 + 3x)(4 - 3x)$

48. $(5 + 3x)(5 - 3x)$

49. $(3x + 2)^2$

50. $(5x - 4)^2$

51. $(3x + y)^2$

52. $(5x - y)^2$

53. $(3x - y)(5x + y)$

54. $(2x - 3y)(4x + 3y)$

55. $(3x + 1)^3$

56. $(2x + 3)^3$

57. $(2x - y)^3$

58. $(4x - 3z)^3$

R–6 Factoring

The reverse process of multiplication of polynomials is called **factoring,** breaking down into **factors.** For example, the product of the two polynomials x and $x + 3$ is $x(x + 3) = x^2 + 3x$. The reverse process starts with the polynomial $x^2 + 3x$

and expresses it as the product of the two terms x and $x + 3$. It is written $x^2 + 3x = x(x + 3)$. The term x is said to be **factored out** of $x^2 + 3x$. The process of factoring starts with one polynomial and expresses it as the product of other polynomials.

EXAMPLE 1

Problem

Factor the polynomials
(a) $x^3 + 5x^2$ and (b) $12y^3 - 3y^2 + 6y$.

Solution (a) Look for a factor that is common to each term. Since x^2 is such a factor, find the other factor of each term: express the first term as $x^3 = x^2(x)$ and the second as $5x^2 = x^2(5)$. Then write

$$x^3 + 5x^2 = x^2(x) + x^2(5) = x^2(x + 5)$$

(b) The common factor is $3y$: express $12y^3 = 3y(4y^2)$, $3y^2 = 3y(y)$, and $6y = 3y(2)$. Then write

$$12y^3 - 3y^2 + 6y = 3y(4y^2 - y + 2)$$

A polynomial may not have a factor common to each term, other than 1 and -1, and still be factorable. For instance, $x^2 - 1$ does not have a **common factor** (other than 1 and -1), but it can be expressed as $(x + 1)(x - 1)$. To check that this is correct, multiply the two polynomials and get $x^2 - 1$. This latter step is the key to another type of factoring. Consider the multiplication formula

$$(a + b)(a - b) = a^2 - b^2$$

It can also be used to factor a polynomial of the form $x^2 - b^2$ by letting $a = x$. In the preceding polynomial let $a = x$ and $b = 1$. Then the formula yields

$$(x + 1)(x - 1) = x^2 - 1^2 = x^2 - 1$$

EXAMPLE 2

Problem

Factor the polynomials.
(a) $x^2 - 25$ and (b) $2y^3 - 18y$.

Solution (a) In the formula let $a = x$ and $b = 5$. Then

$$x^2 - 25 = (x + 5)(x - 5)$$

(b) First factor out the common factor $2y$.

$$2y^3 - 18y = 2y(y^2 - 9)$$

Now let $a = y$ and $b = 3$.

$$2y^3 - 18y = 2y(y + 3)(y - 3)$$

To multiply two linear expressions of the form $x + a$ and $x + b$, use the formula

$$(x + a)(x + b) = x^2 + (a + b)x + ab$$

This same formula helps factor a quadratic expression of the form $x^2 + cx + d$ by expressing it as the product of two linear factors. It is necessary to express c, the coefficient of x, as the sum of two numbers, $a + b$, and d, the constant term, as the product of the same two numbers, ab. For example, to factor $x^2 + 4x + 3$, set

$$a + b = 4 \qquad \text{and} \qquad ab = 3$$

The only integral factors of 3 are 3 and 1, and their sum is 4, so let $a = 3$ and $b = 1$. This yields

$$x^2 + 4x + 3 = (x + 3)(x + 1)$$

Factoring in this way is a trial-and-error method that sometimes does not work. The previous illustration was easy because there is only one set of factors for 3. If there is more than one set of factors, each set must be inspected to see if the sum is the coefficient of the variable. Example 3 describes the method.

E X A M P L E 3

Problem

Factor $x^2 - 5x + 6$.

Solution Let

$$ab = 6 \qquad \text{and} \qquad a + b = -5$$

The factors of 6 are	Their sums are
6 and 1	$6 + 1 = 7$
-6 and -1	$-6 + (-1) = -7$
3 and 2	$3 + 2 = 5$
-3 and -2	$-3 + (-2) = -5$

The last set of factors, -3 and -2, works. Hence

$$x^2 - 5x + 6 = [x + (-3)][x + (-2)]$$
$$= (x - 3)(x - 2)$$

Thus far the coefficient of x has always been 1. If it is another number, then a more general formula must be used.

$$(ax + c)(bx + d) = abx^2 + (ad + cb)x + cd$$

E X A M P L E 4

Problem

Factor $6x^2 + 7x + 2$.

Solution Set

$$ab = 6 \qquad ad + cb = 7 \qquad cd = 2$$

The factors of 6 are 6 and 1; -6 and -1; 3 and 2; -3 and -2. The factors of 2 are 2 and 1; -2 and -1. These possibilities are tested until a combination is found such that $ad + cb = 7$. If we choose $a = 6$, $b = 1$, $c = 1$, and $d = 2$, then $ad + cb = 6 \cdot 2 + 1 = 13$, which is not equal to 7. The correct choice is $a = 2$, $b = 3$, $c = 1$, and $d = 2$ since then $ad + cb = 7$. (An equivalent choice would be $a = 3$, $b = 2$, $c = 2$, and $d = 1$.) Therefore

$$6x^2 + 7x + 2 = (2x + 1)(3x + 2)$$

Sometimes it is necessary to factor more than once.

EXAMPLE 5

Problem

Factor $6x^4 + 7x^3 + 2x^2$.

Solution First factor x^2 from each term to get

$$6x^4 + 7x^3 + 2x^2 = x^2(6x^2 + 7x + 2)$$

Then factor the remaining quadratic polynomial. This was done in Example 4.

$$6x^2 + 7x + 2 = (2x + 1)(3x + 2)$$

Therefore

$$\begin{aligned} 6x^4 + 7x^3 + 2x^2 &= x^2(6x^2 + 7x + 2) \\ &= x^2(2x + 1)(3x + 2) \end{aligned}$$

EXERCISE SET R–6

Factor the polynomial.

1. $x^2 - 6x$
2. $x^2 + 16x$
3. $2t^2 - 60t$
4. $4y^2 + 16y$
5. $x^3 - 8x^2 + x$
6. $x^3 + 10x^2 - 5x$
7. $4y^3 - 8y^2 + 12y$
8. $5p^3 + 10p^2 - 25p$
9. $x^2 - 36$
10. $x^2 - 121$
11. $49 - t^2$
12. $100 - p^2$
13. $1 - 4x^2$
14. $1 - 9p^2$
15. $9 - 4x^2$
16. $16 - 25t^2$
17. $2x^2 - 162$
18. $4x^2 - 36$
19. $2x^3 - 18x$
20. $5x^3 - 20x$
21. $x - 4x^3$
22. $y - 9y^3$
23. $x^4 - 16x^2$
24. $y^4 - 36y^2$
25. $6x^4 - 24x^2$
26. $5x^4 - 45x^2$
27. $x^2 + 4x + 4$
28. $x^2 + 6x + 5$
29. $x^2 - x - 2$
30. $x^2 - 8x + 15$
31. $2x^2 + 3x + 1$
32. $3x^2 + 4x + 1$
33. $4x^2 + 5x + 1$
34. $6x^2 - 11x - 10$
35. $2y^2 - 9y - 5$
36. $2t^2 - 7t - 15$
37. $4t^2 + 8t + 3$
38. $4s^2 + 9s + 5$
39. $6x^2 - 7x - 3$
40. $6x^2 - 11x - 2$
41. $5 - 4x - x^2$
42. $7 - 6y - x^2$
43. $6 - x - x^2$
44. $16 + 4x - 2x^2$
45. $2x^3 + 5x^2 + 2x$
46. $3x^3 + 4x^2 + x$

R–7 Quadratic Equations

The **general form of a quadratic equation** is

$$ax^2 + bx + c = 0$$

where $a \neq 0$. It is an equation that can be put into the form of a quadratic polynomial set equal to 0. Some examples of quadratic equations are

$$x^2 + 2x + 1 = 0 \qquad 3x^2 - 3x - 5 = 0$$
$$x^2 = 5 - 2x \qquad x = 4x^2$$

The last two examples can be arranged so that the equation is in general form. The left equation becomes $x^2 + 2x - 5 = 0$, and the right one becomes $4x^2 - x = 0$, where the constant term is 0. When solving a quadratic equation, it is best to put it into general form.

One way to solve a quadratic equation is to factor the polynomial and set each factor equal to 0. This method uses the property of real numbers: if a and b are real numbers such that $ab = 0$, then $a = 0$, $b = 0$, or both a and b equal 0. To demonstrate how to apply this property to equations, consider the linear equation

$$4(x - 1) = 0$$

This expression equals 0 when either $4 = 0$ or $x - 1 = 0$. Of course, 4 is not equal to 0, so the solution is obtained by solving $x - 1 = 0$, which yields $x = 1$. The procedure is similar for a quadratic equation. Consider the **quadratic equation in factored form**

$$(x - 1)(x - 4) = 0$$

To solve the equation, set each factor equal to 0. The equation equals 0 only if $x - 1 = 0$ or $x - 4 = 0$. Solve each linear equation.

$$x - 1 = 0 \qquad x - 4 = 0$$
$$x = 1 \qquad\quad x = 4$$

Thus the solutions of the quadratic equation $(x - 1)(x - 4) = 0$ are $x = 1$ and $x = 4$. The solutions can always be checked by substituting them into the original equation.

EXAMPLE 1

Problem

Solve the equations (a) $(x - 4)(x + 3) = 0$ and (b) $x(2x + 3) = 0$.

Solution (a) Let each linear factor equal 0 and solve for x.

$$x - 4 = 0 \qquad x + 3 = 0$$
$$x = 4 \qquad\quad x = -3$$

The solutions are $x = 4$ and $x = -3$.

(b) Let each linear factor equal 0 and solve for x.

$$x = 0 \qquad 2x + 3 = 0$$
$$2x = -3$$
$$x = -\tfrac{3}{2}$$

The solutions are $x = 0$ and $x = -\tfrac{3}{2}$.

If the polynomial in the equation is not in factored form, factor it, if possible, and proceed as in Example 1.

E X A M P L E 2

Problem

Solve $x^2 + 2x - 3 = 0$.

Solution First factor the polynomial.

$$x^2 + 2x - 3 = (x - 1)(x + 3)$$

Set each factor equal to zero.

$$x - 1 = 0 \qquad x + 3 = 0$$
$$x = 1 \qquad\qquad x = -3$$

The solutions are $x = 1$ and $x = -3$.

Sometimes the equation must be rearranged so that it is in general form to be solved.

E X A M P L E 3

Problem

Solve the equation $x^2 = 1$.

Solution Express the equation in general form.

$$x^2 - 1 = 0$$

Factor the polynomial on the left-hand side of the equation.

$$(x + 1)(x - 1) = 0$$

The solutions are $x = -1$ and $x = 1$.

If the quadratic equation cannot be solved directly by factoring, then the **quadratic formula** is used. The derivation of the quadratic formula can be found in most algebra texts, so it will not be given here.

If a, b, and c are real numbers with $a \neq 0$, then the solutions of the quadratic equation $ax^2 + bx + c = 0$ are

$$x = \frac{-b \pm \sqrt{b^2 - 4ac}}{2a}$$

EXAMPLE 4

Problem

Solve $5x^2 = 3x + 1$.

Solution First put the equation in general form. It becomes

$$5x^2 - 3x - 1 = 0$$

Use the quadratic formula with $a = 5$, $b = -3$, and $c = -1$. The solutions are

$$x = \frac{-(-3) \pm \sqrt{(-3)^2 - 4(5)(-1)}}{2(5)}$$

$$= \frac{3 \pm \sqrt{9 + 20}}{10}$$

$$= \frac{3 \pm \sqrt{29}}{10}$$

EXERCISE SET R–7

In Problems 1 to 8 solve the equation.

1. $(x - 6)(x - 9) = 0$

2. $(x - 7)(x + 2) = 0$

3. $(2 - x)(x + 2) = 0$

4. $(4 - x)(6 + x) = 0$

5. $(2x - 5)(x - 6) = 0$

6. $(x - 5)(7x - 6) = 0$

7. $x(3x + 16) = 0$ 8. $x(2 - 5x) = 0$

In Problems 9 to 32 solve by factoring.

9. $x^2 + 4x + 4 = 0$ 10. $x^2 + 6x + 5 = 0$

11. $x^2 - x - 2 = 0$

12. $x^2 - 8x + 15 = 0$

13. $2x^2 + 3x + 1 = 0$

14. $3x^2 = -4x - 1$

15. $x^2 - 4 = 0$ 16. $x^2 - 9 = 0$

17. $y^2 = 25$ 18. $y^2 = 144$

19. $36 - x^2 = 0$ 20. $81 - x^2 = 0$

21. $9x^2 - 4 = 0$ 22. $4x^2 - 49 = 0$

23. $1 - 9x^2 = 0$ 24. $16 - 9x^2 = 0$

25. $x^3 - 16x = 0$ 26. $y^3 = 36y$

27. $9x = 25x^3$ 28. $36x - 121x^3 = 0$

29. $x^4 - 9x^2 = 0$ 30. $18x^3 - 8x = 0$

31. $x^4 = 25x^2$ 32. $36x^3 = 4x$

In Problems 33 to 42 solve by the quadratic formula.

33. $x^2 - 5x + 1 = 0$ 34. $x^2 - 4x - 2 = 0$

35. $x^2 + 5x - 1 = 0$ 36. $x^2 + 4x + 2 = 0$

37. $2x^2 + 5x - 1 = 0$ 38. $3x^2 + 4x + 2 = 0$ 41. $2x^3 + 5x^2 = 2x$ 42. $3x^3 + 4x^2 = x$

39. $-2x^2 + 3x = 1$ 40. $-3x^2 - x = -1$

R–8   Rational Expressions

A **rational expression** is an algebraic expression that can be expressed as a quotient of two polynomials. Some examples of rational expressions are

$$\frac{x}{2x - 1} \qquad \frac{2x^2 - 5x}{x^2 - 3} \qquad \frac{x^3 - x^2 + 4}{4x^5 + 5}$$

Often it is necessary to simplify a rational expression or to express it in an alternate form. The properties used to perform arithmetic operations with rational expressions are given here.

> Let A, B, C, and D represent rational expressions, none of which is 0, so they can be used in the denominator of a fraction. The following properties are valid:
>
> 1. $\dfrac{A}{B} + \dfrac{C}{D} = \dfrac{AD}{BD} + \dfrac{BC}{BD}$ **(addition rule)**
>
> 2. $\dfrac{A}{B} \cdot \dfrac{C}{D} = \dfrac{AC}{BD}$ **(multiplication rule)**
>
> 3. $\dfrac{A}{B} - \dfrac{C}{D} = \dfrac{AD}{BD} - \dfrac{BC}{BD}$ **(subtraction rule)**
>
> 4. $\dfrac{A}{B} \div \dfrac{C}{D} = \dfrac{A}{B} \cdot \dfrac{D}{C} = \dfrac{AD}{BC}$ **(division rule)**
>
> 5. $\dfrac{AB}{CB} = \dfrac{A}{C}$ **(cancellation rule)**

We emphasize that none of the expressions in the denominator can equal 0. This is because division by 0 is not defined.

EXAMPLE 1

Problem

Express each rational expression in lowest terms.

(a) $\dfrac{3}{3x - 6}$ (b) $\dfrac{15x}{x(10x - 5)}$

Solution (a) First factor 3 from the denominator and then divide the numerator and denominator by the common factor 3.

$$\frac{3}{3x - 6} = \frac{3}{3(x - 2)} = \frac{1}{x - 2}$$

(b) Factor 5 from the denominator and get

$$\frac{15x}{x(10x - 5)} = \frac{15x}{5x(2x - 1)}$$

Divide the numerator and denominator by the common factor 5.

$$\frac{15x}{x(10x - 5)} = \frac{3(5x)}{5x(2x - 1)} = \frac{3}{2x - 1}$$

EXAMPLE 2

Problem

Express each rational expression in lowest terms.

(a) $\dfrac{x^2 - x}{x}$ (b) $\dfrac{x^3 - x}{x^2 - x}$ (c) $\dfrac{x^2 - x - 2}{x^3 + x^2}$

Solution (a) First factor $x^2 - x = x(x - 1)$. Then use property 5 with $A = x - 1$, $B = x$, and $C = 1$.

$$\frac{x(x - 1)}{x} = \frac{x - 1}{1} = x - 1$$

(b) Factor the numerator and denominator: $x^3 - x = x(x^2 - 1) = x(x + 1)(x - 1)$, and $x^2 - x = x(x - 1)$. Then use property 5.

$$\frac{x^3 - x}{x^2 - x} = \frac{x(x + 1)(x - 1)}{x(x - 1)} = x + 1$$

(c) Factor the numerator and denominator: $x^2 - x - 2 = (x - 2)(x + 1)$ and $x^3 + x^2 = x^2(x + 1)$. Then use property 5.

$$\frac{x^2 - x - 2}{x^3 + x^2} = \frac{(x - 2)(x + 1)}{x^2(x + 1)} = \frac{x - 2}{x^2}$$

EXAMPLE 3

Problem

Perform the indicated operation on the rational expressions.

(a) $\dfrac{2x + 5}{x - 1} + \dfrac{x}{x + 1}$

(b) $\dfrac{2x + 5}{x - 1} \div \dfrac{x}{x + 1}$

Solution (a) The denominators are different, so their least common multiple must be found, which will be the common denominator. It is $(x - 1)(x + 1)$. Multiply the first fraction by $(x + 1)/(x + 1)$ and the second by $(x - 1)/(x - 1)$ so that each fraction has the same denominator. This yields

$$\frac{(2x + 5)(x + 1)}{(x - 1)(x + 1)} + \frac{x(x - 1)}{(x + 1)(x - 1)}$$

$$= \frac{(2x + 5)(x + 1) + x(x - 1)}{(x - 1)(x + 1)}$$

$$= \frac{(2x^2 + 7x + 5) + x^2 - x}{(x - 1)(x + 1)}$$

$$= \frac{3x^2 + 6x + 5}{x^2 - 1}$$

(b) Use property 4 to invert and multiply. Then use property 2.

$$\frac{2x + 5}{x - 1} \div \frac{x}{x + 1} = \frac{2x + 5}{x - 1} \cdot \frac{x + 1}{x}$$

$$= \frac{(2x + 5)(x + 1)}{(x - 1)x} = \frac{2x^2 + 7x + 5}{x^2 - x}$$

EXERCISE SET R–8

In Problems 1 to 26 express the rational expression in lowest terms.

1. $\dfrac{6}{6x - 3}$

2. $\dfrac{4}{8x - 12}$

3. $\dfrac{10}{15x - 20}$

4. $\dfrac{8}{10x - 14}$

5. $\dfrac{x}{x^2 - x}$

6. $\dfrac{3x}{x^2 + x}$

7. $\dfrac{3x}{3x^2 + 6x}$

8. $\dfrac{6x}{3x^2 + 9x}$

9. $\dfrac{x^2 - 4x}{x}$

10. $\dfrac{2x^2 + 6x}{2x}$

11. $\dfrac{2x^3 + 6x^2}{x^2}$

12. $\dfrac{x^3 - 2x^2}{x^2}$

13. $\dfrac{x^3 - 3x}{x^2 - 2x}$

14. $\dfrac{2x^3 + 5x}{5x^2 - 2x}$

15. $\dfrac{x^2 - 9}{x - 3}$

16. $\dfrac{x^2 - 16}{x - 4}$

17. $\dfrac{4x^2 - 16}{x + 2}$

18. $\dfrac{2x^2 - 50}{x + 5}$

19. $\dfrac{x + 3}{9 - x^2}$

20. $\dfrac{x + 10}{100 - x^2}$

21. $\dfrac{x^4 + x^2}{x^3 - 2x^2}$

22. $\dfrac{2x^4 + 5x^2}{5x^3 - 2x^2}$

23. $\dfrac{x^2 - x - 6}{x^3 + 2x^2}$

24. $\dfrac{x^2 - x - 12}{x^3 + 3x^2}$

25. $\dfrac{x^2 + x - 6}{x^3 - 2x^2}$

26. $\dfrac{x^2 + x - 12}{x^3 - 3x^2}$

In Problems 27 to 40 perform the indicated operation and reduce the answer to lowest terms.

27. $\dfrac{2x + 1}{x - 1} + \dfrac{3x}{x + 1}$

28. $\dfrac{3x - 4}{x - 1} + \dfrac{x - 1}{x + 1}$

29. $\dfrac{2x + 1}{x} - \dfrac{3x}{x - 2}$

30. $\dfrac{2x - 5}{x - 2} - \dfrac{x + 1}{x}$

40. $\dfrac{2x - 5}{x - 2} - \dfrac{x + 1}{x} + \dfrac{1}{x^2}$

31. $\dfrac{3x + 6}{x - 1} \cdot \dfrac{x}{x + 2}$

32. $\dfrac{2x + 7}{x} \cdot \dfrac{x}{x - 2}$

In Problems 41 to 46 simplify the expression.

33. $\dfrac{2x - 6}{x - 1} \div \dfrac{x - 3}{x + 2}$

34. $\dfrac{2x - 4}{x} \div \dfrac{x}{x - 2}$

41. $\dfrac{(x + h)^2 + 2 - (x^2 + 2)}{h}$

35. $\dfrac{x^2 - 1}{x^3 + x^2} + \dfrac{x}{x^2(x - 1)}$

42. $\dfrac{(x + h)^2 (x + h) + 2 - (x^2 + x + 2)}{h}$

43. $(x^2 - 1)[-(3 - 5x)^{-2}(-5)] + (3 - 5x)^{-1}(2x)$

36. $\dfrac{x^2 - x}{x^3 - x^2} - \dfrac{x^2 + x - 2}{x^2(x - 1)}$

44. $(x^2 - 1)[-2(3 - x^3)^{-3}(-3x^2)]$
$+ (3 - x^3)^{-2}(2x)$

37. $\dfrac{2x + 1}{x - 1} + \dfrac{3x}{x + 1} - \dfrac{1}{x}$

45. $\dfrac{4(3x^2 - 2) - (4x - 1)(6x)}{(3x^2 - 2)}$

38. $\dfrac{3x - 4}{x - 1} + \dfrac{x - 1}{x + 1} - \dfrac{2}{x}$

46. $\dfrac{8x(3x^2 - 2x) - (4x^2 - 1)(6x - 2)}{(3x^2 - 2x)}$

39. $\dfrac{2x + 1}{x} - \dfrac{3x}{x - 2} - \dfrac{1}{x^2}$

CHAPTER REVIEW

Key Terms

R–1   Equalities in One Variable
Variable
Equation in One Variable
Linear Equation in One Variable

Solution of an Equation
Nonlinear Equations

R–2   Linear Inequalities in One Variable
Inequality
Inequality Sign

Linear Inequality in One Variable
Solution of a Linear Inequality

R–3   Exponents
Exponent
Integer Exponents
nth Root

Square Root
Cube Root

R–4   Absolute Value
Distance
Absolute Value

R–5   Polynomials in One Variable
Terms
Coefficients
Linear Polynomial

Quadratic Polynomial
Combine Like Terms

R–6   Factoring

Factoring   Factored Out
Factors   Common Factor

R–7   Quadratic Equations

General Form of a Quadratic Equation   Quadratic Formula
Quadratic Equation in Factored Form

R–8   Rational Expressions

Rational Expression   Division Rule
Addition Rule   Cancellation Rule
Multiplication Rule   Lowest Terms
Subtraction Rule

Summary of Important Concepts

Properties of Real Numbers Used in Solving Equations

Let a, b, and c represent real numbers or expressions that represent real numbers.

1. If $a = b$, then $a + c = b + c$
2. If $a = b$, then $ac = bc$
3. $a + b = b + a$ **(the commutative law of addition)**
4. $a(b + c) = ab + ac$, $(a + b)c = ac + bc$ **(the distributive laws)**

Properties of Real Numbers Used in Solving Equations

Let a, b, and c represent real numbers or expressions that represent real numbers. The properties are stated for "less than" ($<$), but they also hold for the other three inequality signs.

1. If $a = b$, then $a + c = b + c$
2. If $a = b$, then $ac = bc$
3. $a + b = b + a$ **(the commutative law of addition)**
4. $a(b + c) = ab + ac$, $(a + b)c = ac + bc$ **(the distributive laws)**

Quadratic Formula

If a, b, and c are real numbers with $a \neq 0$, then the solutions of the quadratic equation $ax^2 + bx + c = 0$ are

$$x = \frac{-b \pm \sqrt{b^2 - 4ac}}{2a}$$

REVIEW PROBLEMS

In Problems 1 to 4 solve the equation.

1. $2x + 3(x - 1) = x - 5$

2. $x/2 - 5/2 = 4x$

3. $\dfrac{x - 4}{3} = 2 + \dfrac{x + 1}{6}$

4. $2/x + 1 = 3/x$

In Problems 5 to 8 solve the linear inequality.

5. $8 - 5x < 1 + 2x$

6. $2x - 4(x + 1) < 8$

7. $2(3x - 4) > 4(1 - 3x)$

8. $\dfrac{x - 4}{3} \geq \dfrac{2x + 1}{2}$

In Problems 9 and 10 simplify the expression.

9. $x^{-2}(x^3)^{-1}$

10. $\dfrac{(x^{1/3})^6}{x^{-5}}$

In Problems 11 and 12 compute the numbers.

11. $(1/144)^{1/2}$

12. $(-8)^{-1/3}$

In Problems 13 and 14 express each as a number.

13. $-3|4 - 7| + \dfrac{|5 - 6|}{|-1|}$

14. $\dfrac{-|-4|}{|4|} - \dfrac{4|-1 - 5|}{3}$

In Problems 15 and 16 evaluate the expression at $x = -2$.

15. $2x + |2x| - \dfrac{6}{|4x + 2|}$

16. $\dfrac{3x}{|3x + 5|} + x|4 - 3x|$

In Problems 17 and 18 simplify the polynomial.

17. $x(x^2 - 4x + 1) - x^2(3x^2 + 2x - 1)$

18. $x^3(x^3 - x + 5) - 2x^2(x^4 + 2x^2 - 1)$

In Problems 19 and 20 write the expression as a polynomial.

19. $(x - 4)(x^2 - 4)$

20. $(3x + 5)^2$

In Problems 21 to 24 factor the polynomial.

21. $5x^3 - 10x^2$

22. $4x^4 + 16x^3$

23. $6x^2 - 11x - 2$

24. $5 - 4x - x^2$

In Problems 25 and 26 solve the equation by factoring.

25. $6x^2 - 11x = 2$

26. $5 - 4x - x^2 = 0$

In Problems 27 and 28 solve the equation by the quadratic formula.

27. $2x^2 - 4x + 1 = 0$

28. $2 - 3x - x^2 = 0$

In Problems 29 and 30 express the rational expression in lowest terms.

29. $\dfrac{x^3 - 3x^2}{x^2}$

30. $\dfrac{2x^2 - 32}{x - 4}$

In Problems 31 and 32 perform the indicated operation and reduce the answer to lowest terms.

31. $\dfrac{2x - 1}{x - 1} + \dfrac{5x}{x + 1}$

32. $\dfrac{2x + 2}{x(x - 1)} \div \dfrac{x + 1}{2x}$

Linear Equations

1

The four-minute mile is the subject of a case study in this chapter.
(H. Armstrong Roberts)

1

Chapter Overview

A linear equation in two variables is an equation whose graph is a straight line. The association of a curve (like a line) with an equation (like a linear equation) was a revolutionary idea that occurred during the early part of the 17th century. The first two sections discuss the correspondence between curves in two dimensions and equations in two variables. Section 1–1 introduces the Cartesian coordinate system and sketches the graphs of various equations. Section 1–2 discusses three standard forms of linear equations.

The next two sections deal with systems of linear equations. Section 1–3 presents several methods for solving systems of two linear equations in two variables. It emphasizes the method of elimination and discusses its advantages. Section 1–4 introduces Gaussian elimination to solve systems of three equations in three variables.

Section 1–5, which can be covered after Section 1–3, discusses least squares. It investigates voting trends in recent presidential elections and lays the foundation for the Case Study.

CASE STUDY PREVIEW

Records for the mile run have been kept since 1864. Walter George set the world record four times between 1880 and 1884. His strategy revolutionized the way that milers run the race. Since then the world has experienced many drastic changes, yet the record for the mile run has been lowered with remarkable regularity.

The Case Study examines this surprising phenomenon. It derives an equation that projects the year in which the world record will be lowered to 3 minutes. The equation also predicts what the world record will be in any given year. You can test its validity for the year you are reading this book.

1–1 The Cartesian Coordinate System

About 1630 two Frenchmen, René Descartes and Pierre Fermat, independently revolutionized the way of expressing scientific laws. The idea they hit upon has come to be called "analytic geometry" or "algebraic geometry," because it is a blend of algebra and geometry. Descartes is usually given credit for being the founder of this field, and the term "Cartesian coordinate system" reflects humankind's debt to him, but Fermat deserves at least as much credit for his pioneering work.* Sections 1–1 and 1–2 are devoted to their discovery.

Unit distance

(a)

The Graph of an Equation

(b)

Consider the formula $y = 3x + 1$. It is an algebraic expression that describes the relationship between the variables x and y, but it can also be viewed as a geometric

*For the historical details, see Section 17.11 (pp. 386–388) of Carl B. Boyer and Uta C. Merzbach, *A History of Mathematics* (second edition), New York, John Wiley & Sons, 1989.

FIGURE 1

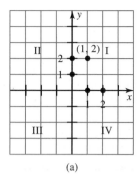

(a)

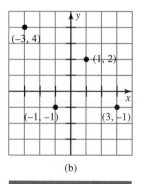

(b)

FIGURE 2

object, namely, a line in the plane. To accomplish this both Descartes and Fermat introduced a *coordinate system,* which we will describe briefly.

We start with the **real line.** It is a straight line with one point marked as the **origin,** corresponding to the number 0 (see Figure 1a). A point to the right of 0 is marked and it corresponds to the number 1. The distance from 0 to 1 is called a **unit distance.** Points that are located at one-unit intervals to the right of 1 are marked 2, 3, and so on. Negative numbers are located to the left of 0. In this way, each real number a corresponds to the point that is $|a|$ units to the right of 0 if a is positive or to the left if a is negative. Additional points, corresponding to some rational and irrational numbers, are marked off in Figure 1b.

Next construct a vertical line perpendicular to the horizontal real line at the origin, and mark off a unit length on it. The positive direction is upward. Usually the horizontal line is called the **x-axis** and the vertical line the **y-axis.** Every point in the plane is represented by an ordered pair of numbers. The ordered pair (1, 2) corresponds to the point that is one unit to the right of the y-axis and two units above the x-axis. (See Figure 2a.) In the ordered pair (1, 2), 1 is called the **x-coordinate** and 2 the **y-coordinate.**

In Figure 2b additional points are plotted in the coordinate system. The numbers in the ordered pairs are also called the first and second **coordinates.** This coordinate system is called the **Cartesian coordinate system** after Descartes. The choice of the letters x and y is arbitrary; we will change them when a particular problem dictates. The Cartesian coordinate system divides the plane into four distinct **quadrants** denoted by the Roman numerals shown in Figure 2a.

Consider the equation $y = 3x + 1$. If $x = 1$, then $y = 4$. These values are treated as the point (1, 4) in the Cartesian coordinate system. The point (1, 4) is said to "satisfy" the equation $y = 3x + 1$. The set of all points that satisfy an equation is called the **graph** of the equation. The graph is obtained from a table of values of x and the corresponding values of y. Example 1 illustrates the procedure for drawing the graph of an equation.

EXAMPLE 1

Consider the equation $y = 3x + 1$. To find its graph, make a table of some values of x and the corresponding values of y. If $x = -2$, then $y = 3x + 1 = 3(-2) + 1 = -5$. This means that the point $(-2, -5)$ lies on the graph. The table lists four additional points corresponding to $x = -1, 0, 1,$ and 2.

x	-2	-1	0	1	2
y	-5	-2	1	4	7
point	$(-2, -5)$	$(-1, -2)$	$(0, 1)$	$(1, 4)$	$(2, 7)$

Next plot the five points $(-2, -5)$, $(-1, -2)$, $(0, 1)$, $(1, 4)$, and $(2, 7)$ on a Cartesian coordinate system. When these points are joined the resulting graph appears to be the line drawn in Figure 3.

How can we be sure that the graph of $y = 3x + 1$ is a line? One way is to plot additional points. By letting x be *any* real number in the equation

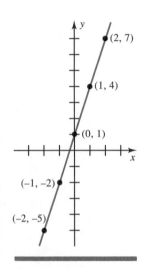

FIGURE 3

$y = 3x + 1$ there will be an infinite number of points of the form $(x, 3x + 1)$ on the graph. These points constitute the graph of $y = 3x + 1$ shown in Figure 3.

The five values that we chose for x in the table were selected arbitrarily. The graph of $y = 3x + 1$ will be the same line even if other values of x are selected.

Example 1 illustrates how algebra and geometry were molded into one mathematical discipline called analytic geometry. The essence is that equations in two variables correspond to curves in two dimensions. The procedure for sketching the graph of an equation is called the **plotting method.** It consists of three steps:

1. Make a table of values of x and y.
2. Treat the values as ordered pairs on a coordinate system and plot them.
3. Connect the points with a smooth curve.

The graph of the equation is the curve that results from step 3. In Section 1–2 we will see how to proceed in the opposite direction, from the graph to the equation.

This simple idea allows algebraic equations to be pictured geometrically, reflecting the old adage that "a picture is worth a thousand words." The next example illustrates it.

EXAMPLE 2

Problem

Sketch the graph of the equation $4x + 2y = 1$ by the plotting method.

Solution Construct a table for the equation $4x + 2y = 1$ by choosing a value for x and substituting it into the equation to get the corresponding value of y. If $x = 1$, then $4(1) + 2y = 1$, so $y = -\frac{3}{2}$. Therefore $(1, -\frac{3}{2})$ is a point on the graph of $4x + 2y = 1$. The table provides four additional points.

x	-2	-1	0	1	2
y	$\frac{9}{2}$	$\frac{5}{2}$	$\frac{1}{2}$	$-\frac{3}{2}$	$-\frac{7}{2}$
point	$(-2, \frac{9}{2})$	$(-1, \frac{5}{2})$	$(0, \frac{1}{2})$	$(1, -\frac{3}{2})$	$(2, -\frac{7}{2})$

Plot these points and draw the curve through them. The graph is shown in Figure 4.

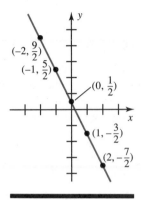

FIGURE 4

Notice that the graph of the equation $4x + 2y = 1$ shown in Figure 4 is a line. Plotting additional points will confirm this fact. It is possible to prove the following statement.

> The graph of an equation of the form $Ax + By = C$ is a line when A, B, and C are real numbers such that A and B are not both zero.

An equation in two variables whose graph is a line is called a **linear equation.** Examples 1 and 2 show that $y = 3x + 1$ and $4x + 2y = 1$ are linear equations.

Applications

For many businesses the total cost C (in dollars) of producing a commodity and the number x of units produced can be written as a linear equation. For instance, let

$$C = 500 + 4x$$

If no units are produced, then $x = 0$, so $C = \$500$. This amount is called the fixed cost. If $x = 25$ units are produced, then the total cost is $C = \$600$.

Often the total revenue R (in dollars) derived from selling the commodity can be written as a linear equation. For instance, let

$$R = 8x$$

If 25 units are produced, then $R = \$200$, in which case the revenue is \$400 less than the cost. The business will have to produce more units in order to turn a profit. In Section 1–3 we will explain how to compute the number of units needed to derive a profit from the sales, but for now we lay the foundation for the solution.

EXAMPLE 3

Problem

Sketch the graphs of the cost equation and the revenue equation on the same coordinate axis.

Solution Use the plotting method to sketch the graph of the cost $C = 500 + 4x$. A calculator can be quite helpful in making the table. For instance, if $x = 150$, then the value of C can be obtained from the following sequence of key strokes.

| 4 | × | 150 | = | + | 500 | = |

x	0	50	100	150
C	500	700	900	1100
point	(0, 500)	(50, 700)	(100, 900)	(150, 1100)

Sketch the graph of $R = 8x$ for the same values of x.

x	0	50	100	150
R	0	400	800	1200
point	(0, 0)	(50, 400)	(100, 800)	(150, 1200)

The graphs of the two equations are shown in Figure 5.

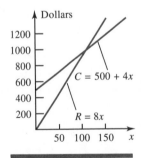

FIGURE 5

Most large firms have a planning group which constructs sales forecasts. These forecasts are critical for production scheduling, plant design, and financial planning. Graphs are often used for viewing the relationship between sales growth and external financial requirements. Example 4 examines the Arinson Products Company, whose owner, John Arinson, sought funds to maintain exclusive control of the company.†

EXAMPLE 4

Let x represent the sales growth rate of the Arinson Products Company and let y represent the external funds required (in thousands of dollars), where

$$y = 494x - 16$$

Problem

(a) Draw the graph of the equation. (b) If the sales growth rate is 5%, what external funding is required? (c) If \$107,500 is available for external funding, what is the sales growth rate?

Solution (a) If $x = 0.1$ (corresponding to a 10% sales growth rate), then $y = 494(0.1) - 16 = 33.4$. The table lists several other values of x and the corresponding values of y.

x	−0.2	−0.1	0	0.1	0.2
y	−114.8	−65.4	−16	33.4	82.8

The graph is sketched in Figure 6.

(b) For a sales growth rate of 5% set $x = 0.05$. Then $y = 494(0.05) - 16 = 8.7$. Since the units for y represent thousands of dollars, the amount of external funds required is \$8700.

(c) If \$107,500 is available for external funding, set $y = 107.5$. The sales growth rate is obtained by substituting $y = 107.5$ into the equation and solving for x.

$$494x - 16 = y$$
$$494x - 16 = 107.5$$
$$x = \frac{107.5 + 16}{494} = 0.25$$

The sales growth rate is $x = 25\%$.

†B. Campsey and E. Brigham, *Introduction to Financial Management*, Hinsdale, Ill., The Dryden Press, 1985, pp. 190–204.

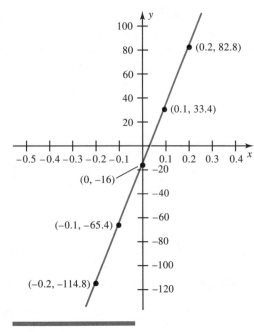

FIGURE 6

EXAMPLE 5

Based on figures supplied by the U.S. Census Bureau, the population of California in the decade 1980–1990 can be described by the equation $C = 617,134.8x + 23,667,902$, where x is the number of years after 1980.

Problem

(a) What was the population of California in 1980? (b) What was the population of California in 1990? (c) What was the increase in California's population in the 1980s?

Solution (a) Since x is the number of years after 1980, the population in 1980 can be found by setting $x = 0$. Then $C = 23,667,902$, which was California's population in 1980.
(b) For the year 1990 set $x = 10$. Then $C = 6,171,348 + 23,667,902 = 29,839,250$. This was California's population in 1990.
(c) The increase in population in the 1980s is found by subtracting the population in 1980 from the population in 1990. Hence the increase was $29,839,250 - 23,667,902 = 6,171,348$.

Notice that the answer to Example 5(c) is ten times the coefficient of x in the equation $C = 617,134.8x + 23,667,902$. This is not an accident. The exercises in this section and the next two sections will explore the matter further.

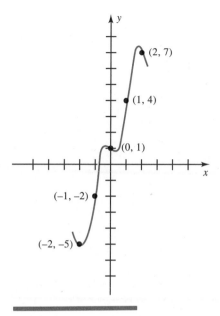

FIGURE 7

What's in a Name?

All of the graphs in this section have been lines. Of course, these are not the only graphs of algebraic equations according to the method of Descartes and Fermat; there are also circles, parabolas, and many others. It is useful to be able to picture a graph when its equation is given.

So far, graphs have been drawn by the plotting method, but this method has its shortcomings. Given the table of five points for the equation $y = 3x + 1$ in Example 1, it is conceivable that its graph could be the curve in Figure 7. In Section 1–2 we will see why it must be a straight line.

EXERCISE SET 1–1

In Problems 1 to 4 fill in the table for the given values.
Supply your own values of x in Problems 3 and 4.

1. $y = 2x - 3$

x	-2	-1	0	1	2	3
y						
point						

2. $y = x - 2$

x	-2	-1	0	1	2	3
y						
point						

3. $y = 4 - x$

x	
y	
point	

4. $3x - y = 2$

x	
y	
point	

In Problems 5 to 10 use the plotting method to sketch the graph of the equation.

5. $y = 2x - 3$ 6. $y = x - 2$

7. $y = 4 - x$ 8. $3x - y = 2$

9. $2x + y = 6$ 10. $4x + y = 7$

In Problems 11 to 14 the graphs of the given equations are straight lines. Plot two points and then sketch the graph.

11. $x - y = 3$ 12. $2x + y = 3$

13. $3x - y = -1$ 14. $3x + y = 1$

In Problems 15 to 18 sketch the graphs of the total cost C and the total revenue R on the same coordinate axis.

15. $C = 25 + 3x$ 16. $C = 40 + x$
     $R = 8x$ $R = 5x$

17. $C = 100 + 3x$ 18. $C = 300 + x$
     $R = 13x$ $R = 31x$

Problems 19 and 20 refer to the equation $y = 494x - 16$ given in Example 4.

19. (a) If the sales growth rate is 15%, what external funding is required? (b) If $33,400 is available for external funding, what is the sales growth rate?

20. (a) If the sales growth rate is -5%, what external funding is required? (b) If $82,800 is available for external funding, what is the sales growth rate?

In Problems 21 to 24 use the plotting method to sketch the graph of the equation.

21. $x + 4y = 7$ 22. $x + 3y = 3$

23. $3x + 4y = -12$ 24. $3x - 4y = 6$

In Problems 25 to 34 the equations are nonlinear. Sketch the graph using the plotting method.

25. $y = x^2 - 3$ 26. $y = x^2 + 1$

27. $y = -x^2$ 28. $y = -2x^2$

29. $y = 1 - 2x^2$ 30. $y = 1 - x^2$

31. $y = x^3$ 32. $y = x^3 - 8$

33. $y = \sqrt{x}$ 34. $y = \sqrt{x} - 1$

In Problems 35 and 36 sketch the graphs of the two equations on the same coordinate axis.

35. $x + y = 120$
      $100x + 40y = 6000$

36. $45x + 9y = 45$
     $10x + 6y = 20$

Problems 37 to 40 refer to the equation $5F - 9C = 160$, where F is degrees Fahrenheit and C is degrees Celsius.

37. Sketch the graph of the equation.

38. Find the degrees Celsius when $F = 68$ degrees.

39. Find the degrees Celsius when $F = 98.6$ degrees.

40. Find the degrees Fahrenheit when $C = 20$ degrees.

Problems 41 and 42 refer to a company whose total cost is $C = 0.5x + 20,000$, where x is the amount of sales and the fixed cost is $20,000.

41. If the company sells $60,000 worth of goods, what is the total cost?

42. If the company sells $30,000 worth of goods, what is the total cost?

43. A person's height can be estimated by the length of the tibia bone, which runs from the knee to the ankle. The formula for the height y [in centimeters (cm)] in terms of the length x (in cm) of the tibia is

$$3y - 8x = 211$$

(a) Draw the graph of the equation.
(b) How tall is a person whose tibia is 37 cm long?
(c) How long is the tibia if the person is 185 cm tall?

44. A person's height can be estimated by the length of the radius bone, which runs from the wrist to the elbow. The formula for the height y (in cm) in terms of the length x (in cm) of the radius is

$$2y - 7x = 166$$

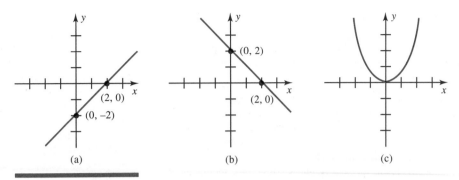

FIGURE 8 Problems 51 and 52.

(a) Draw the graph of the equation.

(b) How tall is a person whose radius bone is 24 cm long?

(c) How long is the radius bone if the person is 174 cm tall?

45. What is the meaning of the value of x at which the cost line and the revenue line meet in Figure 5?

Problems 46 to 48 refer to Figure 5 in the text.

46. What is the meaning of the value x at which the cost line and the revenue line meet?

47. For what numbers of items produced is there a profit?

48. How many items produce a revenue of $800,000?

Problems 49 and 50 are based on figures supplied by the U.S. Census Bureau.

49. The population of Michigan in the decade 1980–1990 is given by the equation $M = 6,670.6x + 9,262,078$, where x is the number of years after 1980. What was the total increase in Michigan's population in the 1980s?

50. The population of Illinois in the decade 1980–1990 is given by the equation $I = 4,016.4x + 11,426,518$, where x is the number of years after 1980. What was the total increase in Illinois's population in the 1980s?

In Problems 51 and 52 match the graph in Figure 8 to the given equation.

51. $y = x - 2$

52. $y = x^2$

Referenced Exercise Set 1–1

1. The percentage share of the national radio audience for FM radio stations (contrasted with AM radio stations) in any year since 1972 can be approximated by the formula

$$S = 3.514286Y - 226.2953$$

where* S is a percent and Y is the last two digits in the year* (For instance, for the year 1989 $Y = 89$.)

(a) What was the approximate percentage share in 1985?

(b) In what year was the percentage share about 50%?

2. The amplifier in a stereo system must be strong enough to produce loud peak volume levels (106 decibels) in the speakers. The strength needed depends on the character of the room. For instance, a "live" room is one in which a hand clap will cause a fluttering echo. For Acoustic Research loudspeakers,† the linear relationship between the power y (in watts) and the volume x of a "live" room (in cubic feet) is given by

$$y = \frac{x}{180} + \frac{80}{9}$$

(a) What power is needed for a room measuring 2000 cubic feet?

(b) What power is needed for a room measuring 6500 cubic feet?

(c) If a system produces 40 watts, what size room will allow it to produce loud peak levels?

VCRs have invaded American homes in the last few years. Most VHS systems have a speed switch that allows the

---

*This equation was derived from data supplied by Statistical Research, Incorporated.

†Source: Teledyne Acoustic Research of Canton, Massachusetts. This company manufactures the Acoustic Research loudspeakers.

viewer to record in three different speeds: 2-hour standard play (SP), 4-hour long play (LP), and 6-hour super long standard play (SLP). Problems 3 and 4 involve equations that allow a viewer to get the highest quality from a recording.‡

3. Suppose a viewer wishes to record a program that lasts between 2 and 4 hours. Let x be the length of the program in hours. Then the highest quality recording using the entire T-120 videocassette can be obtained by recording in LP mode for $y = 2x - 4$ hours and in SP mode for $4 - y$ hours. How long should the viewer record in LP mode if the program lasts (a) 3 hours? (b) $3\frac{1}{2}$ hours? (c) 2 hours and 40 minutes?

4. Many people see little difference in quality between LP and SLP modes. Suppose a viewer wishes to record a program that lasts between 2 and 6 hours using LP and SLP modes. Let x be the length of the program in hours. Then the highest quality recording using the entire T-120 videocassette can be obtained by recording in SLP mode for $y = 1.5x - 3$ hours and in SP mode for $4 - y$ hours. How long should the viewer record in SLP mode if the program lasts (a) 4 hours? (b) $3\frac{1}{2}$ hours? (c) 4 hours and 40 minutes?

5. The first federal income tax went into effect in 1862, one year after President Abraham Lincoln signed it into legislation. Phil Lapsansky, the curator of the Afro-American collection at the Library Company of Philadelphia, recently discovered an original tax form in the stacks of that library, which Benjamin Franklin founded in 1731.§ Computing taxes was a lot easier then; the tax rate was 3% on income from $600 to $10,000 and 5% beyond that amount.
   (a) Write an equation for the tax T of a person whose income in 1862 was x dollars, where x is between $600 and $10,000.
   (b) Write an equation for the tax T of a person whose income in 1862 was x dollars, where x is more than $10,000.

6. Consider the statement: "There are six times as many students as professors at this university." Let S stand for the number of students and P stand for the number of professors.
   (a) If $P = 1200$, what is S?
   (b) If $S = 1200$, what is P?
   (c) Write an equation using the variables S and P to represent the given statement.‖
   (d) Substitute the values you obtained in parts (a) and (b) into the equation in part (c). If they do not satisfy the equation, revise the equation and repeat this part.

---

‡G. N. Fiore, "An Application of Linear Equations to the VCR," *Mathematics Teacher*, October 1988, pp. 570–572.

§Leonard W. Boasberg, "The Tax That Was Due for 1862," *The Philadelphia Inquirer*, April 12, 1990, pp. 1-D, 4-D.

‖It has been reported that only 63% of engineering students and 43% of social science students get this equation correct. Sources for the study addressing this issue can be found in the article by Annie A. Selden and John Selden, Jr., entitled "Do You Know the Students-and-Professors Problem?" which appeared in a newsletter of the Mathematical Association of America called *UME Trends*, May 1989, p. 6.

1–2 Lines

There are two basic aspects to the Descartes–Fermat invention of analytic geometry. One is a fixed frame of reference, called the Cartesian coordinate system. The other is the relationship between equations and curves. Section 1–1 showed how to begin with a linear equation in two variables and to draw its graph. This section illustrates the opposite direction: given the graph of a line in some form, construct its equation.

General Form

In Section 1–1 we stated that a linear equation in two variables is an equation whose graph is a line. We examine three forms of linear equations.

Recall from Section 1–1 that an equation in two variables x and y is linear if and only if it can be written in the form $Ax + By = C$. This form of a linear equation is called the *general form*. Therefore the equation $4x + 3y = 12$ is linear, but the equations $x^2 - y = 0$ and $\sqrt{x} - y = 0$ are nonlinear.

DEFINITION

The **general form** of a linear equation is

$$Ax + By = C$$

where A, B, and C are real numbers, with A and B not both equal to 0.

Example 1 illustrates a method of graphing linear equations in general form that is more effective than the plotting method shown in Example 2 of Section 1–1. It makes use of the intercepts of a line. The **x-intercept** of a line is the point where the line crosses the x-axis, and the **y-intercept** is the point where it crosses the y-axis. Thus an x-intercept must have a y-coordinate equal to 0 and a y-intercept must have an x-coordinate equal to 0.

EXAMPLE 1

Problem

Sketch the graph of the linear equation

$$3x + 2y = 4$$

Solution Find the y-intercept by setting $x = 0$. Then $y = 2$, so the y-intercept is $(0, 2)$. Find the x-intercept by setting $y = 0$. Then $x = \frac{4}{3}$, so the x-intercept is $(\frac{4}{3}, 0)$. Draw the line through the two intercepts. This is shown in Figure 1.

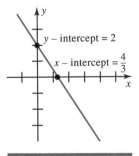

FIGURE 1 Graph of $3x + 2y = 4$.

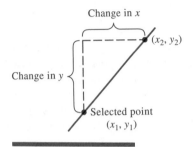

FIGURE 2 Slope.

Slope-Intercept Form

Every equation of the form $y = mx + b$, where m and b are numbers, is linear since it can be written in the general form $mx - y = -b$. Therefore the graph of the equation $y = 3x + 1$ is a line with $m = 3$ and $b = 1$; its graph cannot be the curve shown in Figure 7 of Section 1–1. The expression "the line $y = mx + b$" refers to the graph of the linear equation $y = mx + b$.

The numbers m and b in the linear equation $y = mx + b$ play an important role. The property measured by m is intuitively described as "steepness." It is called the *slope* of the line. To measure the slope, start at some point (x_1, y_1) on the line, proceed in the vertical direction some distance, then proceed in the horizontal direction until the line is intersected at a point (x_2, y_2). (See Figure 2.) The vertical distance is called the "change in y" (or the "rise") and the horizontal distance is called the "change in x" (or the "run"). The slope is then defined to be the ratio of the change in y to the change in x (or the ratio of the rise to the run).

DEFINITION

Let (x_1, y_1) and (x_2, y_2) be distinct points on a line with $x_1 \neq x_2$. The **slope** of the line is the number

$$m = \frac{\text{change in } y}{\text{change in } x} = \frac{y_2 - y_1}{x_2 - x_1}$$

The points $(1, -1)$ and $(4, 5)$ lie on the line $y = 2x - 3$. (See Figure 3.) The slope of the line is

$$m = \frac{\text{change in } y}{\text{change in } x} = \frac{5 - (-1)}{4 - 1} = \frac{6}{3} = 2$$

Figure 4 shows the graphs of several lines. As the values of x increase from left to right, those with positive slope tend upward (or increase) while those with negative slope tend downward (or decrease).

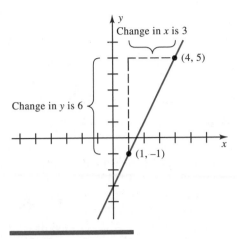

FIGURE 3 Slope of line $y = 2x - 3$.

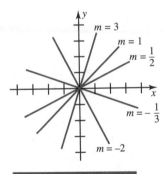

FIGURE 4 Lines with various slopes.

Example 2 investigates the slope determined by two distinct pairs of points on the same line.

EXAMPLE 2

Problem

Find the slope of the line $y = 3x + 1$ through these pairs of points.
(a) $(0, 1)$ and $(3, 10)$
(b) $(-1, -2)$ and $(4, 13)$

Solution (a) The slope of the line through the points $(0, 1)$ and $(3, 10)$ is

$$m = \frac{10 - 1}{3 - 0} = \frac{9}{3} = 3$$

(b) The slope of the line through the points $(-1, -2)$ and $(4, 13)$ is

$$m = \frac{13 - (-2)}{4 - (-1)} = \frac{13 + 2}{4 + 1} = \frac{15}{5} = 3$$

In general the slope of a line is independent of the two points chosen on the line. Example 2 illustrates this fact. Another way of viewing the slope can be seen from a table of points.

x	-2	-1	0	1	2
y	-5	-2	1	4	7

Each time the value of x increases by 1, the value of y increases by the slope 3.

The number b in the linear equation $y = mx + b$ can be obtained by setting $x = 0$. Then $y = m \cdot 0 + b = b$, so the line passes through the point $(0, b)$. The number b is referred to as the *y-intercept* of the line $y = mx + b$ even though formally the y-intercept is the point $(0, b)$.

The linear equation $y = mx + b$ is called the *slope-intercept form* of a line because m is the slope and b is the y-intercept.

DEFINITION

The **slope-intercept form** of a line is

$$y = mx + b$$

where m is the slope and b is the y-intercept.

EXAMPLE 3

Problem

Sketch the graph of the equation $y = -2x + 3$.

Solution The y-intercept is $b = 3$, so the line passes through $(0, 3)$. The slope is $m = -2$. Since m is negative, the line slopes downward (from left to right). The graph is sketched in Figure 5.

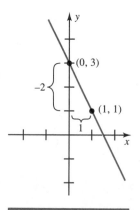

FIGURE 5 Graph of $y = -2x + 3$.

Parallel Lines

Two lines are **parallel** if they do not meet in either direction when extended indefinitely. Since parallel lines have the same steepness, it follows that two lines are parallel if and only if they have the same slope.

EXAMPLE 4

Problem

Write the equation of the line that is parallel to the line $y = -2x + 2$ and passes through the origin.

Solution The graph of the linear equation $y = -2x + 2$ is a line with slope $m = -2$. The new line will have the same slope $m = -2$. Since the new line passes through $(0, 0)$, its y-intercept is $b = 0$. Substituting the values $m = -2$ and $b = 0$ into the slope-intercept form $y = mx + b$ yields $y = -2x$. This is the equation of the line parallel to $y = -2x + 2$ and passing through the origin. The lines are sketched in Figure 6.

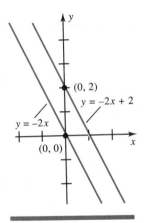

FIGURE 6 Parallel lines.

There are two special kinds of lines. **Horizontal lines** are parallel to the x-axis and **vertical lines** are parallel to the y-axis. The slope of a horizontal line can be determined by choosing any two points $(x_1, 0)$ and $(x_2, 0)$ on the x-axis. By definition the slope of the line through these points is

$$m = \frac{0 - 0}{x_2 - x_1} = 0$$

Thus the slope of the x-axis is 0. Since parallel lines have the same slope, it follows that *the slope of any horizontal line is zero.*

Next consider vertical lines. If $(0, y_1)$ and $(0, y_2)$ are two points on the y-axis, the slope is

$$m = \frac{y_2 - y_1}{0 - 0}$$

Since division by 0 is not permitted, the slope of the y-axis is not defined. It follows that *the slope of any vertical line is not defined.*

EXAMPLE 5

Problem

Write the equations of the horizontal and vertical lines that pass through the point (2, 3).

Solution The slope of the horizontal line is $m = 0$. Since the horizontal line passing through (2, 3) is parallel to the x-axis, its y-intercept is $b = 3$. Substituting $m = 0$ and $b = 3$ into the equation $y = mx + b$ yields $y = 3$.

The equations of vertical lines cannot be obtained from the linear equation $y = mx + b$ because m is not defined. The vertical line through (2, 3) is located two units to the right of the y-axis. Some additional points on it are (2, 2), (2, 1), (2, 0), and (2, −1). The value of x is always 2. Hence the equation of the line is $x = 2$.

Both lines are sketched in Figure 7.

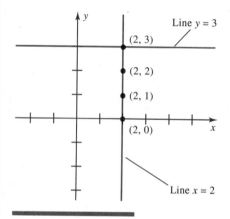

FIGURE 7 Horizontal and vertical lines.

Table 1

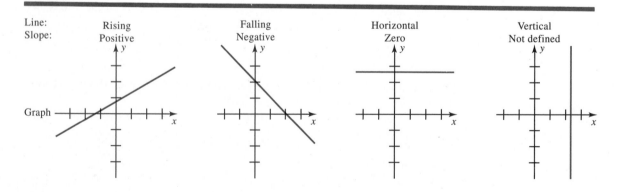

Line:	Rising	Falling	Horizontal	Vertical
Slope:	Positive	Negative	Zero	Not defined
Graph				

FIGURE 8   Mathematicians always run from left to right.

Example 5 suggests the following standard forms for all horizontal and vertical lines:

horizontal lines: $y = c$,   a number

vertical lines: $x = k$,   a number

Table 1 summarizes the results on the slopes of lines, while Figure 8 presents the same concepts from a roadrunner's perspective.

Point-Slope Form

The equation of a line that is neither vertical nor horizontal can be expressed in several forms. We have already indicated that the slope-intercept form gives the equation of a line when its slope and y-intercept are known. But what if the slope m and another point (x_1, y_1) on the line are known? Then an arbitrary point (x, y)

on the line can be used with (x_1, y_1) to compute the slope $m = (y - y_1)/(x - x_1)$. So $y - y_1 = m(x - x_1)$. This form is aptly named the *point-slope form*.

DEFINITION

The **point-slope form** of a line is

$$y - y_1 = m(x - x_1)$$

where m is the slope and (x_1, y_1) is the given point on the line.

Example 6 shows how to convert from the point-slope form to the slope-intercept form.

EXAMPLE 6

Problem

Find the point-slope form and the slope-intercept form of the line that passes through the point $(3, 2)$ and has slope 4.

Solution We have $x_1 = 3$, $y_1 = 2$, and $m = 4$. The equation of the line in point-slope form is

$$y - 2 = 4(x - 3)$$

Use the properties of algebra indicated on the right to write the equation of the line in slope-intercept form.

$$y - 2 = 4(x - 3) \qquad \text{(point-slope form)}$$
$$y = 4(x - 3) + 2 \qquad \text{(add 2 to each side)}$$
$$y = 4x - 12 + 2 \qquad \text{(distributive law)}$$
$$y = 4x - 10 \qquad \text{(add } -12 \text{ and 2)}$$

Therefore $m = 4$ and $b = -10$. The graph is sketched in Figure 9.

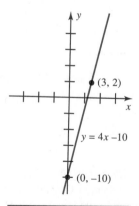

FIGURE 9 Line through $(3, 2)$ with $m = 4$.

Example 7 shows how to determine the equation of a line when two points on the line are known.

EXAMPLE 7

Problem

Find an equation of the line that passes through the points $(1, 2)$ and $(-1, 5)$.

Solution The slope of the line is

$$m = \frac{5 - 2}{-1 - 1} = \frac{3}{-2} = -\frac{3}{2}$$

Using the point-slope form with the point $(1, 2)$ [we could have chosen the point $(-1, 5)$ instead] the equation is

$$y - 2 = -\frac{3}{2}(x - 1)$$

The equation in slope-intercept form is found by solving for y.

$$y = -\frac{3}{2}x + \frac{7}{2}$$

Linear Depreciation

The tax law requires businesses to determine the current value of every asset. Most assets lose value as they get older. It is said that they *depreciate* in value. There are several ways to compute the depreciation of an asset. One method, called **linear depreciation,** assumes that the current value y of the asset is related to the age x of the asset by a linear equation. This equation usually takes the form $y = b - mx$, where b is the purchase price of the asset, or the value when $x = 0$, and m is the rate of depreciation.

EXAMPLE 8

A firm purchases a machine for $50,000 and the machine has a useful life (determined by tax laws) of ten years. If the machine depreciates $4,000 per year, the value of the machine after x years is

$$y = 50,000 - 4,000x$$

The salvage value of an asset is its value when the asset reaches its useful life.

Problem

Determine the salvage value of the machine.

Solution Find the value of y when $x = 10$.

$$y = 50,000 - 4,000(10) = 50,000 - 40,000$$

$$= 10,000$$

The salvage value of the machine is $10,000.

Here is a summary of the main items in this section.

1. The slope is $m = \dfrac{y_2 - y_1}{x_2 - x_1}$ (unless $x_1 = x_2$).

2. The slope measures steepness. As x increases from left to right, y increases if the slope is positive and y decreases if the slope is negative.

3. Horizontal lines have form $y = c$; their slope is 0. Vertical lines have form $x = k$; their slope is undefined.

4. Three principal forms for the equation of a line follow.

Name of Form	Equation	Variables
general	$Ax + By = C$	A, B, C real numbers A, B not both 0
slope-intercept	$y = mx + b$	$m = $ slope $b = y$-intercept
point-slope	$y - y_1 = m(x - x_1)$	$m = $ slope $(x_1, y_1) = $ point

EXERCISE SET 1–2

In Problems 1 to 4 sketch the graph of the line.

1. $5x - 2y = 10$
2. $5x + 3y = 15$
3. $3x + 4y = 12$
4. $3x - y = 6$

In Problems 5 to 10 find the slope of the line passing through the two points.

5. $(1, 2), (2, 4)$
6. $(3, 1), (4, 3)$
7. $(-1, 2), (3, -4)$
8. $(-3, 8), (3, -8)$
9. $(1.6, 4.1), (2.5, 5.9)$
10. $(\frac{3}{2}, 5), (-1, -\frac{3}{2})$

In Problems 11 to 14 sketch the graph of the line.

11. $y = 4x - 5$
12. $y = 2x + 1$
13. $y = -2x + 3$
14. $y = -3x - 2$

In Problems 15 and 16 determine if the lines are parallel.

15. Line 1 passes through $(0, 1)$ and $(1, 2)$; line 2 passes through $(-1, 6)$ and $(6, -1)$.

16. Line 1 passes through $(1, 3)$ and $(5, -1)$; line 2 passes through $(1, 1)$ and $(2, 0)$.

In Problems 17 and 18 determine the slope of the line.

17. (a) $x = 2$ (b) $y = -7$

18. (a) $x = 15$ (b) $y = 2$

In Problems 19 and 20 find the slope-intercept form of the line passing through the given points.

19. $(2, 1), (3, -6)$
20. $(2, -1), (-3, 6)$

In Problems 21 to 34 find an equation of the line from the information given. Sketch the graph of the line.

21. $m = 2, b = 7$
22. $m = -2, b = 7$

23. The horizontal line that passes through the point $(-5, -3)$.

24. The vertical line that passes through $(-5, -3)$.

25. The slope is undefined and the x-intercept is -2.

26. The slope is 0 and $b = -4$.

27. $m = -2$; the line passes through $(6, -1)$.

28. $m = 1$; the line passes through $(6, -1)$.

29. $b = -1$ and the line is parallel to $y = 2x + 1$.

30. The line passes through $(1, 1)$ and is parallel to the line $x - 3y = 4$.

31. The line passes through $(-2, 0)$ and $(0, 2)$.

32. The line passes through $(2, 0)$ and $(0, -2)$.

33. The line passes through $(-2, 0)$ and $(-2, -2)$.

34. The line passes through $(0, 4)$ and $(2, 4)$.

In Problems 35 and 36 determine if the slope of the line is positive, negative, zero, or undefined.

35.

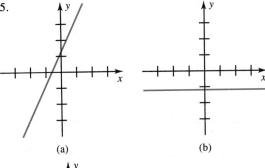

(a)   (b)

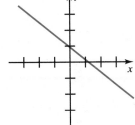

(c)

36.

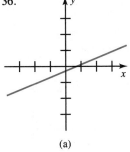

(a)   (b)

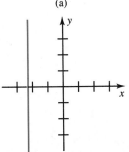

(c)

37. (a) Put the equation $3x + 2y = 12$ in slope-intercept form. (b) What is the slope? (c) What is the y-intercept?

38. (a) Put the equation $5x + 3y = 15$ in slope-intercept form. (b) What is the slope? (c) What is the y-intercept?

In Problems 39 and 40 determine the equation of the line passing through the two intercepts.

39.

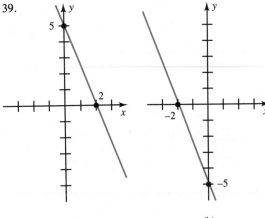

(a)   (b)

40.

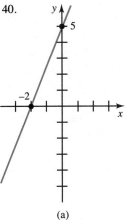

(a)   (b)

In Problems 41 and 42 write the equation (in point-slope form) of the line that has the given table.

41.

x	-1	0	1
y	5	1	-3

42.

x	-2	0	2
y	5	1	-3

43. For tax purposes some firms assume that the current value y of an asset is related to the asset's age x by a linear equation. Suppose a firm buys a work station with ten microcomputers and assumes that after x years the work station's worth is given by

$$y = 60{,}000 - 15{,}000x$$

(a) Draw the graph of the equation.
(b) What is the work station's worth after three years?
(c) What economic interpretation can be given to the y-intercept?

Problems 44 to 47 are based on figures supplied by the U.S. Census Bureau.

44. The population of Texas was 14,229,191 in 1980. The average increase for each year during the 1980s was 283,061.4. (a) Write an equation to describe the population of Texas during that decade. (b) Use this equation to determine the population of Texas in 1990.

45. The population of New York was 17,558,072 in 1980. The average increase for each year during the 1980s was 48,643.3. (a) Write an equation to describe the population of New York during the decade. (b) Use this equation to determine the population of New York in 1990.

46. In Section 1–1 the equation for the population of California in the 1980s was given as $C = 617{,}134.8x + 23{,}667{,}902$. Explain the meaning of the slope of this equation.

47. The equation for the population of Iowa in the 1980s is $I = -12{,}638.4x + 2{,}913{,}808$. Explain the meaning of the negative slope.

In Problems 48 to 51 match the graphs in Figures a to d to the given linear equations.

48. $2x + 5y = 20$

49. $3x + 4y = 24$

50. $y = 5x + 4$

51. $y = 2x + 6$

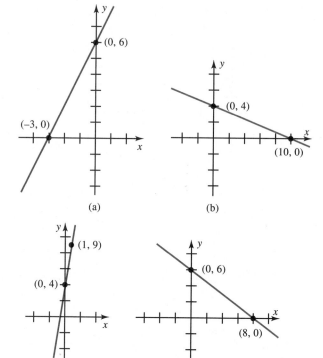

(a)   (b)

(c)   (d)

52. Prove that the slope of the line $y = mx + b$ is m. *Hint:* Let (x_1, y_1) and (x_2, y_2) be points on the line with $x_1 \neq x_2$. Show that

$$\frac{(y_2 - y_1)}{(x_2 - x_1)} = m$$

53. Is it true that every equation in the variables x and y in which the highest power of x is 1 and the highest power of y is 1 must be linear?

Referenced Exercise Set 1–2

1. The petroleum industry uses "decline curve analysis" to estimate reserves of oil and gas wells, even though more sophisticated techniques are available. The reason for the continued emphasis on this method is that the required data are almost always available. Recently two petroleum engineers developed a simple graphing technique for evaluating production data from a linear equation.* Let y represent the loss ratio (a technical term

*D. A. Rowland and C. Lin, "New Linear Method Gives Constants of Hyperbolic Decline," *Oil and Gas Journal*, January 14, 1985, pp. 86–90.

measured here in months) and x represent the time (also in months). The line which the engineers discovered for a particular set of data has slope 0.5 and y-intercept 7.5.
(a) What is the equation of the line? (b) Sketch the line on the interval of 0 months to 72 months. (c) What is the loss ratio after 6 years?

2. This problem deals with four "rules of thumb" relating the height and weight of adults to those of children.† Write each relationship as an equation. (This requires the introduction of appropriate variables.) (a) The height of a male adult is double the height at age 2. (b) The height of a female adult is double the height at age $1\frac{1}{2}$. (c) The weight of a male adult is five times the weight at age 2. (d) The weight of a female adult is five times the weight at age $1\frac{1}{2}$.

3. Figure 10 shows the power necessary to achieve loud peak volume levels (106 decibels) for a pair of Acoustic Research loudspeakers.‡ The three different types of rooms ("dead," "average," and "live") require amplifiers of different sizes to produce loud peak volume levels. Let y be the power (in watts) and x be the volume of the room (in cubic feet). For each type of room, express y in terms of x in slope-intercept form.

4. A carpenter asked University of Delaware Professor Richard J. Crouse the following question:§ "I am building roofs that are 30 feet (ft.) long at the base. The pitch of each roof rises 4 inches (in.) for every horizontal foot. I have to install vertical supports every 16 in. (See Figure 11.) This forces me to climb the ladder, measure 16 in. horizontally, measure the vertical height at that point, climb down the ladder, saw the support, and climb back up to put it in place. Is there a formula that I can use to determine the lengths of the supports in advance?"
(a) What are the coordinates of points A and B in Figure 11?
(b) Write the equation of the line through points A and B, where x is the horizontal distance and y is the height.
(c) Use the equation in part (b) to find the length of the support at 16 in.
(d) Use the equation in (b) to find the length of the support at 8 ft.

Problems 5 and 6 deal with the percentage of energy that is supplied to the body by fats and carbohydrates during treadmill running.‖ Figure 12 shows that each energy source is a linear function of time.

5. (a) Interpret the meaning of the fact that during the first few minutes of running the slope of the carbohydrate function is positive.

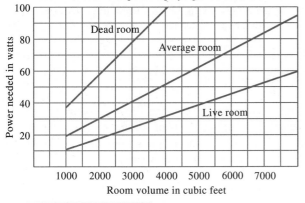

Power per channel required for a pair of acoustic, research loudspeakers playing at 106dB S.P.L.

FIGURE 10   Problem 3.

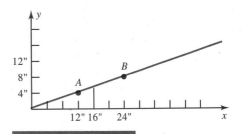

FIGURE 11   Problem 4.

†Wilton M. Krogman, *Child Growth,* Ann Arbor, Mich., University of Michigan Press, 1972, p. 42.

‡Source: Teledyne Acoustic Research of Canton, Massachusetts.

§R. J. Crouse, "Linear Equation Saves Carpenter's Time," *Mathematics Teacher,* May 1990, pp. 400–401.

‖Trevor Smith, "Chemistry, Exercise, and Weight Control," *Today's Chemist,* February 1990, pp. 10–11, 21.

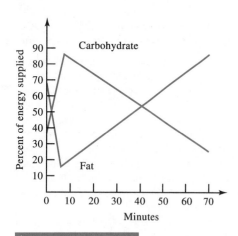

FIGURE 12 Problems 5 and 6.

(b) Interpret the meaning of the fact that after a few minutes of running the slope of the carbohydrate function is negative.

6. (a) Interpret the meaning of the fact that during the first few minutes of running the slope of the fat function is negative.

(b) Interpret the meaning of the fact that after a few minutes of running the slope of the fat function is positive.

Cumulative Exercise Set 1–2

In Problems 1 to 3 use the plotting method to sketch the graph of the equation.

1. $y = 2x - 3$

2. $x + 4y = 4$

3. $y = -x^2 + 1$

4. Sketch the graphs of the total cost function C and the total revenue function R.

$$C = 50 + 2x \qquad R = 12x$$

5. Find the slope-intercept form of the line passing through the points $(-1, 4)$ and $(1, -4)$.

6. Find the equations of the vertical line and the horizontal line passing through the point $(2, 5)$.

In Problems 7 and 8 find an equation of the line from the given information.

7. The line has y-intercept 5 and is parallel to the line $2x - 4y = 1$.

8. The line has x-intercept 7 and the slope is undefined.

9. A company has total cost $C = 0.4x + 15,000$, where x is the number of sales and the fixed cost is $15,000. If the company sells $30,000 worth of goods, what is the total cost? Sketch the graph of the equation.

1–3 Intersection of Lines

Many applications of mathematics involve more than one linear equation at the same time. In this section we describe several methods of solving systems of two linear equations in two variables. The next section extends these techniques to systems having more than two equations or more than two variables.

The Solution of a System

Consider the two linear equations

$$x + y = 7$$

$$x - y = 3$$

Taken together they form a **system of two linear equations in two variables.** A **solution** is an ordered pair (x, y) of numbers that satisfy both equations at the same

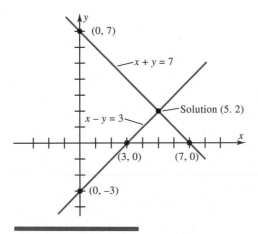

FIGURE 1 Intersection of lines.

time, so (5, 2) is a solution because

$$5 + 2 = 7$$
$$5 - 2 = 3$$

However, the pair (6, 1) is not a solution because it does not satisfy the second equation, even though it does satisfy the first equation.

The solution (5, 2) was obtained by inspection; it is fairly obvious. Another way of solving the system is to draw the graphs of both equations on the same coordinate system. Since both graphs are lines, the solution of the system is the point of intersection of the two lines because the point (5, 2) satisfies both equations. (See Figure 1.) Example 1 illustrates the graphing method.

EXAMPLE 1

Problem

Solve this system of linear equations by graphing.

$$x + 3y = 5$$
$$x - \ y = 1$$

Solution The graph of each equation is a line. The two lines are drawn on the same coordinate system in Figure 2. It appears from the figure that the lines intersect at the point (2, 1). To verify this, substitute $x = 2$ and $y = 1$ into the equations

$$2 + 3(1) = 5$$
$$2 - \ 1 \ = 1$$

Since the two equations are satisfied, (2, 1) is the solution of the system.

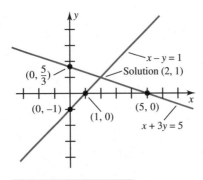

FIGURE 2 Graphing method.

The graphical method is valuable for approximating a solution to a system of two equations, but often it is not precise enough to produce exact solutions.

Elimination

The method of elimination is a very efficient method of solving systems of linear equations. It consists of eliminating one variable and then solving for the other. We illustrate it on this system of linear equations.

$$2x + 3y = 4$$
$$3x - 2y = 6$$

Choose a variable to eliminate. Sometimes it is easier to eliminate one variable than the other, but here it does not matter so we choose y. Multiply the first equation by 2 and the second by 3.

$$4x + 6y = 8$$
$$9x - 6y = 18$$

The coefficients of y are negatives of each other. To eliminate y, add the two equations.

$$13x = 26$$

A solution to this equation is also a solution to the original system of equations. Solve for x.

$$x = 2$$

Substitute $x = 2$ into either one of the original equations to get y. From the first equation

$$2x + 3y = 4$$

$$2(2) + 3y = 4$$

$$y = 0$$

Since $x = 2$ and $y = 0$, the solution to the system is $(2, 0)$.

The variable y was eliminated by producing the numbers 6 and -6 as coefficients of y and then adding the two equations. This two-step procedure is the essence of the method of elimination.

Two Steps in Eliminating a Variable

1. Multiply the equations by appropriate numbers to produce coefficients of a variable such that one is the negative of the other.
2. Add the two resulting equations.

EXAMPLE 2

Problem

Solve this system of linear equations by elimination.

$$3x - 4y = 1$$

$$4x - 2y = 3$$

Solution Eliminating y can be accomplished with only one multiplication. Multiply the second equation by -2 and add the resulting equations.

$$\begin{aligned} 3x - 4y &= 1 \\ -8x + 4y &= -6 \\ \hline -5x &= -5 \\ x &= 1 \end{aligned}$$

Substitute $x = 1$ into the first equation of the given system to get $y = \frac{1}{2}$. The solution is $(1, \frac{1}{2})$.

The next example applies elimination to a system in which some variable has the same coefficient in both equations.

EXAMPLE 3

Problem

Solve this system of linear equations.

$$3x + 2y = 2$$

$$3x - y = 5$$

Solution Multiply the second equation by -1 and add it to the first equation.

$$\begin{array}{rcr} 3x + 2y = & 2 \\ -3x + y = & -5 \\ \hline 3y = & -3 \end{array}$$

Solve for y.

$$y = -1$$

Substitute $y = -1$ into the first equation to get $x = \frac{4}{3}$. The solution is $(\frac{4}{3}, -1)$.

Substitution

The method of substitution involves solving one equation for one variable and then substituting it into the other equation. This method also leads to one equation in one variable.

EXAMPLE 4

Problem

Solve this system of linear equations.

$$2x + 4y = 1$$
$$x - 5y = -2$$

Solution Solve the second equation for x.

$$x = 5y - 2$$

Substitute this expression into the first equation.

$$2x + 4y = 1$$
$$2(5y - 2) + 4y = 1$$
$$14y - 4 = 1$$
$$y = \frac{5}{14}$$

To find x, substitute $y = \frac{5}{14}$ to get $x = 5y - 2 = -\frac{3}{14}$. The solution is $(-\frac{3}{14}, \frac{5}{14})$.

Systems Without Unique Solutions

Every system that we have encountered so far contains exactly one solution. This is not always the case, however, as can be seen by examining the geometric meaning of a system of linear equations in two unknowns. The graph of each equation is a

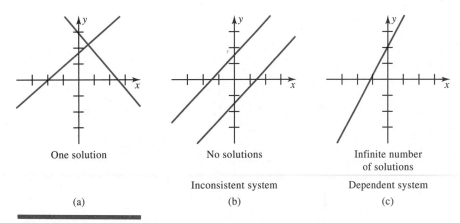

One solution   No solutions   Infinite number
                                                       of solutions

                         Inconsistent system   Dependent system
(a)   (b)   (c)

F I G U R E 3   Systems of two equations.

line, so the system consists of two lines, l_1 and l_2. Figure 3 shows that such a system can have one solution, no solutions, or infinitely many solutions.

A system with no solutions is called **inconsistent.** Figure 3b shows that the graph of such a system consists of two parallel lines. Example 5 shows how to detect such a system algebraically.

E X A M P L E 5

Problem

Solve this system of linear equations.

$$x - y = 1$$
$$2x - 2y = 3$$

Solution To solve this system by elimination, multiply the first equation by -2 and add it to the second equation.

$$-2x + 2y = -2$$
$$\underline{2x - 2y = 3}$$
$$0 = 1$$

If an ordered pair (x, y) were to satisfy the system, then $0 = 1$, so a contradiction would result. Therefore the original system cannot have any solutions.

This can be seen geometrically by writing each equation in slope-intercept form: $y = x - 1$ and $y = x - (\frac{3}{2})$. The lines have equal slopes but different y-intercepts, so the equations represent distinct parallel lines.

A system with infinitely many solutions is called **dependent.** Dependent systems occur when the two equations represent the same line. (See Figure 3c.) Example 6 shows how to uncover a dependent system algebraically.

E X A M P L E 6

Problem

Solve this system of linear equations.

$$x - y = 1$$
$$2x - 2y = 2$$

Solution To solve this system by elimination, multiply the first equation by -2 and add it to the second equation.

$$-2x + 2y = -2$$
$$\underline{2x - 2y = 2}$$
$$0 = 0$$

This is true for *all* ordered pairs (x, y) that satisfy the first equation. The reason is that the second equation has the same set of solutions as the first.

Breakeven Analysis

In financial management the **breakeven point** is the level of operations where neither profits nor losses are incurred. Above this point, each unit sold contributes to profits; below it, each unit sold contributes to losses. Therefore the breakeven point is the intersection of the sales revenue and the total cost. The revenue is derived from the sales of the units. The total cost usually involves a fixed cost (like salaries and rental expenses) and a variable cost (like sales commissions and advertising). Example 7 illustrates these terms.

E X A M P L E 7

A candy company specializes in making mints. Each box of mints sells for $2 and has a variable cost of $1.20. The fixed cost is $40,000.

Problem

Determine the company's breakeven point.

Solution The sales revenue is derived from the $2 price for each box of mints. Let x be the number of boxes sold. If R is the sales revenue, then $R = 2x$. The variable cost is $1.20 per box, so it is $1.2x$. The total cost C, which includes the $40,000 fixed cost, is given by $C = 1.2x + 40{,}000$.

The breakeven point occurs at a value of x where $R = C$. Therefore, at this point the equation for total cost can be written $R = 1.2x + 40{,}000$. The breakeven point is found by solving the system of linear equations

$$R = 2x$$

$$R = 1.2x + 40{,}000$$

The solution is $x = 50{,}000$ and $R = 100{,}000$. Therefore the breakeven point occurs when 50,000 boxes of mints are sold.

The company breaks even when it sells 50,000 boxes of mints. The graphs of $R = 2x$ and $C = 1.2x + 40,000$ are given in Figure 4. The breakeven point (50,000, 100,000) is the point of intersection of these two lines. If the company sells more than 50,000 boxes it makes a profit, while if it sells less than 50,000 boxes it suffers a loss.

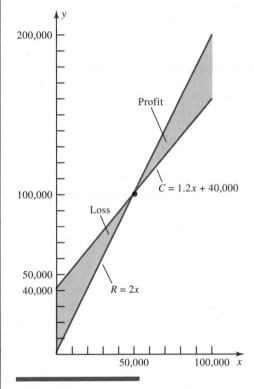

FIGURE 4

EXERCISE SET 1–3

In Problems 1 and 2 solve the system by inspection. (Think, do not write.)

1. $x + y = 5$
   $x + 2y = 6$

2. $x + y = 2$
   $x - y = 2$

In Problems 3 and 4 solve the system by graphing.

3. $2x + y = 4$
   $x - y = 2$

4. $x + 3y = 1$
   $2x - y = 9$

In Problems 5 to 10 solve the system by elimination.

5. $2x + y = 3$
   $3x + 4y = 7$

6. $2x + 5y = 2$
   $3x - 2y = 3$

7. $5x + 2y = 1$
   $2x - 3y = 8$

8. $8x - 3y = 1$
   $3x + 4y = 9$

9. $2x - 3y = 1$
   $3x - 2y = 3$

10. $-4x + 2y = -3$
    $3x - 2y = 3$

In Problems 11 to 14 solve the system by substitution.

11. $x + y = -2$
    $x - y = 4$

12. $2x - 3y = 2$
    $3x - y = 3$

13. $2x - 4y = 1$
    $x - y = 2$

14. $x + 3y = 4$
    $2x - 6y = 1$

In Problems 15 to 18 determine whether the system is inconsistent or dependent.

15. $3x - y = 1$
    $6x - 2y = 5$

16. $8x + y = 2$
    $8x + y = 3$

17. $2x - 4y = 2$
    $3x - 6y = 3$

18. $12x - 6y = 3$
    $8x - 4y = 2$

In Problems 19 and 20 determine the breakeven point of a candy company that specializes in making mints.

19. Each box of mints sells for $3 and has a variable cost of $1.50. The fixed cost is $30,000.

20. Each box of mints sells for $5 and has a variable cost of $2.00. The fixed cost is $30,000.

In Problems 21 to 36 solve the system by any method. If there is not a unique solution, state whether there are no solutions or infinitely many solutions.

21. $2x - y = 3$
    $4x + y = 1$

22. $-x + 2y = 3$
    $2x + 6y = 1$

23. $4x + y = 18$
    $4x - 3y = 10$

24. $2x + y = 6$
    $x - 2y = 3$

25. $x + y = 120$
    $100x + 40y = 6000$

26. $45x + 9y = 45$
    $10x + 6y = 20$

27. $2x + 3y = 4$
    $3x - 2y = 6$

28. $x + 2y = 2$
    $x - y = 5$

29. $25x - 50y = 1500$
    $30x - 15y = 2700$

30. $1.2x - 3.5y = 23$
    $8.5x + 7y = 1.5$

31. $x + 1.5y = 0$
    $y = 8.5x$

32. $15x - 110y = 4000$
    $10x + 55y = 1000$

33. $x + y = 5$
    $2x + 2y = 8$

34. $x - y = 3$
    $3x - 3y = 9$

35. $2x + 5y = 40$
    $2x + y = 24$

36. $2x + 4y = 40$
    $50x + 200y = 1600$

37. A travel agency is planning charter flights to Hawaii for exactly 1400 passengers aboard two types of jets. Each 747 jet holds 400 passengers and each 707 holds 150 passengers. The rental fee is $15,000 for each 747 and $9000 for each 707. If exactly $66,000 must be used for renting the jets, how many of each type should be ordered?

38. A dietician is preparing a meal of chicken and corn that must provide exactly 63 grams (g) of protein and 21 milligrams (mg) of iron. Each serving of chicken contains 45 g of protein and 9 mg of iron; each serving of corn contains 9 g of protein and 6 mg of iron. How many servings of each ingredient should be prepared?

39. Suppose that $5000 is invested in two securities, school bonds and public utility bonds. School bonds pay 6% interest and public utility bonds pay 8% interest. If an interest of $370 is required, how much money should be invested in each security?

40. Suppose that regular gasoline costs $1 per gallon and premium gasoline costs $1.20 per gallon. How many gallons of each gasoline should be blended to produce 400 gallons of mixture that is worth $1.05 per gallon?

41. In the 1980 census Ohio was the sixth highest state in population while Florida was seventh. Ten years later Ohio had dropped to seventh while Florida had jumped to fourth. The equations for the populations in these states in the 1980s are

    $$F = 325,703.8x + 9,746,324$$

    $$O = 8,969.5x + 10,797,638$$

    where x is the number of years after 1980. In what month of what year did Florida's population overcome Ohio's?

42. In 1980 Alaska was the least populated state and Wyoming was next. The equations for the populations in these states in the 1980s are

    $$A = 15,009.6x + 401,851$$

    $$W = -1,358.2x + 469,557$$

    where x is the number of years after 1980. In what month of what year did Alaska lose the distinction of being the least populated state?

In Problems 43 and 44 match the graph in Figure 5 to a system of equations having the stated number of solutions.

43. No solutions.

44. Infinitely many solutions.

45. When Vince and Anne Marie took their two children to the movies, the cost of the four tickets was $15. Peg took her three children to the same movie and the cost of the four tickets was $12.50. How much did each adult's ticket cost? How much did each child's ticket cost?

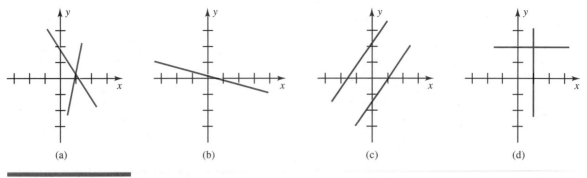

(a)   (b)   (c)   (d)

FIGURE 5 Problems 43 and 44.

Referenced Exercise Set 1–3

Problems 1 and 2 are adapted from an article on the ability of the Big Three American automakers to reduce their breakeven point substantially to become more competitive with foreign competition.* The variable x represents the number of cars sold (in millions of cars) and y represents the sales revenue and the total cost (in millions of dollars).

1. (a) Determine the breakeven point if the sales revenue is $y = 1.5x$ and the total cost is $y = x + 6.1$.
   (b) Determine the breakeven point if the sales revenue is $y = 2x$ and the total cost is $y = 0.5x + 13.35$.
   (c) The solution in (a) was the breakeven point for the Big Three in 1979. The solution in (b) was the breakeven point in 1983. By how many cars did the Big Three drop the breakeven point in four years?

2. (a) Determine the breakeven point if the sales revenue is $y = 0.5x$ and the total cost is $y = 0.25x + 0.6$.
   (b) Determine the breakeven point if the sales revenue is $y = 0.8x$ and the total cost is $y = 0.6x + 0.28$.
   (c) The solution in (a) was the breakeven point for Chrysler in 1981. The solution in (b) was the breakeven point in 1983. By how many cars did Chrysler drop the breakeven point in two years?

3. Figure 6 shows the relationship between short-term interest rates and the volume of funds to be raised in credit markets. The graph is due to Peter Crawford, who is in charge of financial analysis for Citicorp Information Services of New York, NY.† (a) At what rate of interest did the supply of credit equal the demand for credit in 1985? (b) At what rate of interest did the supply of credit equal the demand for credit in 1983?

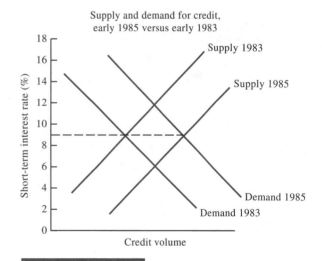

FIGURE 6 Problem 3.

---

*Adapted from Lynn Adkins, "Detroit Gets Lean and Mean," *Dun's Business Month,* January 1983, pp. 56–58.

†Peter Crawford, "Interest Rate Outlook: Focus on Credit Supply," *Business Economics,* July 1985, pp. 34–39.

4. In 1939 L. F. Richardson proposed a mathematical model of an arms race that is still valid today.‡ The model is based upon equations from calculus called differential equations, but we have adapted it to a system of linear equations. Let X and Y be two countries with arms expenditures x and y, respectively. Denote each change in expenditure by Δx and Δy. Then

$$\Delta x = \quad ay - \quad bx + \quad c \quad (a > 0, b > 0)$$

$$\Delta y = \quad dx - \quad ey + \quad f \quad (d > 0, e > 0)$$

$$\underbrace{\text{spurring}}_{\text{term}} \quad \underbrace{\text{breaking}}_{\text{term}} \quad \overbrace{\text{grievance}}^{}\\ \text{term}$$

A state of equilibrium occurs when there is no change in expenditures, that is, when $\Delta x = \Delta y = 0$. Find the equilibrium point for $a = 0.01$, $b = 0.02$, $c = 2$, $d = 0.0125$, $e = 0.1$, and $f = 6.25$.

Cumulative Exercise Set 1–3

1. Let $x + 2y = 5$. What is the value of x when $y = 3$?

2. The total cost C (in dollars) for a company is $C = 0.4x + \$15,000$, where x is the amount of sales. If the company sells $68,500 worth of goods, what is the total cost?

3. Sketch the graph of the line $100x + 40y = 6000$.

4. What is the equation of the vertical line whose x-intercept is -3?

5. Find the slope-intercept form of the line passing through the points $(-1, 4)$ and $(3, -8)$.

6. What is the slope of the line $3x - 6y = 8$?

7. What is the equation of the line parallel to the line $y = 3x + 2$ and passing through the point $(0, -7)$?

8. Write the slope-intercept form of the line whose table is

x	0	1
y	-1	-3

9. Solve the system of linear equations by any method.

$$3x - 2y = -16$$

$$2x + 4y = 16$$

10. Find the point of intersection of the lines $4x + 2y = 10$ and $y = x + 1$.

11. Determine whether the system of linear equations is inconsistent or dependent.

$$6x - 3y = 1$$

$$8x - 4y = 0$$

12. BikeLine manufactures 10-speed and 12-speed bicycles. Each 10-speed bike requires 2 minutes to assemble the frame and 8 minutes to install the gears. Each 12-speed bike requires 5 minutes to assemble the frame and 19 minutes to install the gears. How many bicycles of each speed can be made per day if exactly 60 minutes must be used for assembling frames and 232 minutes for installing gears?

13. Draw a line having points in the first, second, and third quadrants. (a) Is the slope positive or negative? (b) Is the y-intercept positive or negative? (c) Is the x-intercept positive or negative?

14. A farmer estimates that his seasonal farming costs are $9700 plus an additional $130 per acre of corn that he plants.
(a) Write an equation for the cost C in terms of acres of corn planted.
(b) Sketch the graph of the equation.
(c) What is the cost of planting 10 acres of corn?

15. A daytime telephone call from New York City to Los Angeles costs $.23 for the first minute and $.24 for each additional minute. For nights and weekends the rate is a flat $.13 per minute.
(a) Write an equation for the cost D of a daytime call in terms of minutes x.
(b) Write an equation for the amount of money saved by calling during a night or weekend instead of calling during the daytime.
(c) Interpret the meaning of the slope of the line in part (b).

‡L. F. Richardson, "Generalized Foreign Politics," *British Journal of Psychology,* Monograph Supplement XXIII, Cambridge, England, Cambridge University Press, 1939. A BASIC program for solving this system is given in the article by Leonard M. Wapner, "Modeling with Difference Equations: Two Examples," *The Mathematics Teacher,* February 1984, pp. 136–140.

1–4   Systems of Linear Equations

Around 1800 the German mathematician Karl Friedrich Gauss (1777–1855) introduced an effective method for solving systems of linear equations by eliminating the variables one at a time. This method was known in China 500 years earlier, but it remained unknown in the West, where it is now called Gaussian elimination. This section applies Gaussian elimination to systems of three linear equations in three variables.

Gaussian Elimination

A **linear equation in three variables** is an equation of the form $ax + by + cz = k$, where a, b, c, and k are constants such that a, b, and c are not all 0. A similar definition applies to linear equations in four or more variables. The equation $2x - 3y + 0.5z = 4$ is a linear equation in three variables. Its graph is a plane in three-dimensional space.

Consider this system of three linear equations in three variables.

$$x + y + z = 10$$
$$x - y \quad\ = 2$$
$$y - z = 1$$

A solution is the ordered triple of numbers $(5, 3, 2)$ because all three equations are satisfied when the values $x = 5$, $y = 3$, and $z = 2$ are substituted into the three equations. However, $(6, 4, 0)$ is not a solution because it does not satisfy the third equation, even though it does satisfy the first two equations.

It is possible to use the methods of inspection, graphing, and substitution to solve systems with three or more variables, but in general these methods are not very practical. However, Gaussian elimination can be performed easily on a system of three equations in three variables, extended to systems of other sizes, and can be applied to whatever numbers are involved in the system (fractions or decimals). Although only a slight variation of elimination, it is very efficient. We will describe it in detail on the following system, whose equations are denoted by E1, E2, and E3.

$$x + \ y + 2z = 9 \qquad\qquad\text{(E1)}$$
$$2x + 4y - 3z = 1 \qquad\qquad\text{(E2)}$$
$$3x + 6y - 5z = 0 \qquad\qquad\text{(E3)}$$

The first step is to eliminate the variable x from E2 by multiplying E1 by -2, adding it to E2, and replacing E2 by this new equation. The number -2 is called the **multiplier** of E2. The notation

$$(-2)(\text{E1}) + (\text{E2})$$

represents this procedure. The multiplier for the third equation is -3. The original system is replaced by a simpler system with the new equations in place of the original, second, and third equations.

$$x + \ y + \ 2z = \quad 9$$
$$2y - \ 7z = -17 \qquad (-2)(E1) + (E2)$$
$$3y - 11z = -27 \qquad (-3)(E1) + (E3)$$

Notice that the variable x has been eliminated from the second and third equations.

The set of solutions of a system of equations is called the **solution set.** Two systems of linear equations are called **equivalent** if their solution sets are equal. The new system is equivalent to the original system because multiplying an equality by a number produces another equality, and adding two equalities also produces an equality. The next step eliminates the variable y from the third equation.

$$x + \ y + 2z = \quad 9$$
$$2y - 7z = -17$$
$$- \tfrac{1}{2}z = -\tfrac{3}{2} \qquad (-\tfrac{3}{2})(E2) + (E3)$$

This system of equations is called the **reduced system** of the original system.

The purpose of Gaussian elimination is to convert the original system of equations to an equivalent, reduced system, then to solve the reduced system. The first part, called **elimination,** reduces the original system to an equivalent, reduced system. The second part, called **back-substitution,** solves the reduced system of equations by working from the bottom equation up. The solution to the third equation is $z = 3$. Substitute it into the second equation and solve for y.

$$2y - 7(3) = -17$$
$$y \qquad = \quad 2$$

Finally, the values $y = 2$ and $z = 3$ are substituted into the first equation.

$$x + 2 + 2(3) = 9$$
$$x = 1$$

The reduced system has one solution: $x = 1, y = 2, z = 3$. Since the original system has the same solution set, the triple $(1, 2, 3)$ is the only solution of the original system too. It should be checked by substituting it into the original system.

Check:

$$1 + \ \ 2 + 2(3) = 9$$
$$2(1) + 4(2) - 3(3) = 1$$
$$3(1) + 6(2) - 5(3) = 0$$

The accompanying display describes the entire method for solving systems of three linear equations in three variables x, y, and z.

Gaussian Elimination

Elimination

1. Eliminate x by replacing E2 by E2 plus its multiplier times E1, and by replacing E3 by E3 plus its multiplier times E1.
2. Eliminate y from the resulting bottom equation by replacing E3 with the sum of E3 and its multiplier times E2.

Back-Substitution

3. Solve the third equation for z.
4. Substitute the value of z into the second equation and solve for y.
5. Substitute the values of y and z into the first equation and solve for x.

Check
6. Substitute the triple (x, y, z) into the original system.

Example 1 illustrates the method of Gaussian elimination by solving the system that began this section.

EXAMPLE 1

Problem

Solve this system of linear equations.

$$x + y + z = 10$$
$$x - y \quad = 2$$
$$y - z = 1$$

Solution   In step 1 the multiplier for equation 2 is -1. The variable x is already eliminated from equation 3, so no operation is necessary. The system becomes

$$x + y + z = 10$$
$$\quad -2y - z = -8 \qquad (-1)(E1) + (E2)$$
$$\quad y - z = 1$$

In step 2 the multiplier for equation 3 is $\frac{1}{2}$. Thus

$$x + y + z = 10$$
$$\quad -2y - z = -8$$
$$\quad -(\tfrac{3}{2})z = -3 \qquad (\tfrac{1}{2})(E2) + (E3)$$

Now apply back-substitution. Step 3 yields $z = 2$. By step 4

$$-2y - z = -8$$
$$-2y - 2 = -8$$
$$y = 3$$

Apply step 5 with $y = 3$ and $z = 2$.

$$x + y + z = 10$$
$$x + 3 + 2 = 10$$
$$x = 5$$

The solution is $(5, 3, 2)$, which confirms the solution noted earlier.

Simplifications

The efficiency of Gaussian elimination is due to the fact that it uses only one operation repeatedly: multiply an equation by a number, add it to a second equation, and replace the second equation by the new equation. Gaussian elimination is the method that is used on almost all computers to solve systems of linear equations because it is fast and accurate. However, when we humans solve such a system it is advantageous to simplify some of the computations. There are two additional operations on equations that are easy to carry out by hand and do not change the solution set: interchange equations and multiply an equation by a nonzero number.

Interchanging two equations is sometimes helpful in avoiding fractions (which often cause numerical errors), but at other times it is absolutely necessary. For instance, consider this system.

$$y - z = 1 \tag{E1}$$
$$x - y \phantom{{}+z} = 2 \tag{E2}$$
$$x + y + z = 10 \tag{E3}$$

It is impossible to find a multiplier for x in E1 because the coefficient of x is 0. The operation E1 $\leftrightarrow$ E3 leads to the system solved in Example 1. The operation E1 $\leftrightarrow$ E2 could have been used instead.

Here is a list of the three operations that preserve the solution set of a system of linear equations.

Elementary Operations on a System of Linear Equations

1. Interchange two equations.
2. Multiply each number in an equation by a nonzero number.
3. Replace an equation by the sum of that equation and a multiple of another equation.

EXAMPLE 2

Problem

Solve the system of equations

$$-2x \qquad + 3z = -5$$
$$2x + 3y - z = 9$$
$$x - y + 2z = -3$$

Solution It is helpful to interchange the top and bottom equations so the coefficient of x is 1.

$$x - y + 2z = -3 \qquad \text{E1} \leftrightarrow \text{E3}$$
$$2x + 3y - z = 9$$
$$-2x \qquad + 3z = -5$$

The multiplier for equation 2 is -2 and the multiplier for equation 3 is 2.

$$x - y + 2z = -3$$
$$5y - 5z = 15 \qquad (-2)(\text{E1}) + (\text{E2})$$
$$-2y + 7z = -11 \qquad (2)(\text{E1}) + (\text{E3})$$

To avoid fractions, reduce the second equation.

$$x - y + 2z = -3$$
$$y - z = 3 \qquad (\tfrac{1}{5})(\text{E2})$$
$$-2y + 7z = -11$$

The multiplier for equation 3 is 2.

$$x - y + 2z = -3$$
$$y - z = 3$$
$$5z = -5 \qquad (2)(\text{E2}) + (\text{E3})$$

By back-substitution, the solution is $(1, 2, -1)$. It checks when substituted into the original system.

Example 3 illustrates these operations in a practical setting.

EXAMPLE 3

A software firm makes disks in three sizes: $3\tfrac{1}{2}''$, $5\tfrac{1}{4}''$, and $8''$. They are tested in three phases: A, B, and C. The $3\tfrac{1}{2}''$ disks require exactly 1 second in each of the three phases. The $5\tfrac{1}{4}''$ disks require 1 second in phase A, 2 seconds in phase B, and 3 seconds in phase C. The $8''$ disks require 4 seconds in phase A, 8 seconds in phase B, and 8 seconds in phase C. Phase A is available for exactly 27 seconds, phase B for 49 seconds, and phase C for 63 seconds.

Problem

How many disks of each size can be tested?

Solution Let

x = the number of $3\frac{1}{2}''$ disks

y = the number of $5\frac{1}{4}''$ disks

z = the number of $8''$ disks

In phase A the $3\frac{1}{2}''$ disks require 1 second, the $5\frac{1}{4}''$ disks require 1 second, and the $8''$ disks require 4 seconds. There are exactly 27 seconds available, so

$$x + y + 4z = 27$$

Similarly, the conditions on phases B and C lead to the equations

$$x + 2y + 8z = 49$$

$$x + 3y + 8z = 63$$

This system reduces to the equivalent system

$$x + y + 4z = 27$$
$$y + 4z = 22$$
$$-4z = -8$$

By back-substitution, the solution is $x = 5$, $y = 14$, and $z = 2$. Therefore five $3\frac{1}{2}''$ disks, fourteen $5\frac{1}{4}''$ disks, and two $8''$ disks can be tested.

Number of Solutions

All of the 3×3 systems in this section have had precisely one solution. However, some 3×3 systems have infinitely many solutions and some have no solution, just like the 2×2 case. Are there any other possibilities? For instance, is it possible for a system to have exactly two solutions? Or between five and ten solutions? To answer these questions, we turn to the geometrical interpretation of a linear system.

Equations of the form $ax + by + cz = d$ represent planes in space. Three such equations correspond to three planes, and their simultaneous solution corresponds to the intersection of planes. Three distinct planes can intersect in one point (view the floor, front wall, and a side wall of the room you are sitting in); no points (view the floor, the ceiling, and the front wall); or infinitely many points (the front wall, a side wall, and an imaginary wall that runs through the line where the front and side walls meet). There are no other possibilities. Thus every system of three linear equations in three variables has either no solution, a unique solution, or infinitely many solutions.

EXERCISE SET 1–4

In Problems 1 to 4 a system of equations and three ordered triples are given. Determine whether any of the triples are solutions.

1. $x + y + z = 3$
   $x - y - 2z = -2$
   $2x + y - z = 2$
   $(1, 0, 2), (3, 1, -1), (1, 1, 1)$

2. $2x + y - z = 2$
   $3x - 2y + z = 2$
   $x - 3y + 2z = 0$
   $(1, 1, 1), (1, 0, 1), (0, 4, -6)$

3. $2x - y = 3$
   $x - y + z = 3$
   $3x - 8z = 3$
   $(3, 1, 2), (0, -3, 0), (2, -1, 0)$

4. $3x - y = 0$
   $y - 2z = -2$
   $x - y = 0$
   $(1, 3, 1), (0, 0, 1), (0, 0, 0)$

The systems in Problems 5 to 8 have unique solutions. Find them by inspection. (Think, don't write.)

5. $x + y + z = 3$
   $x \quad\quad - z = 0$
   $\quad\quad y - z = 0$

6. $x + y + z = 0$
   $x \quad\quad - z = 0$
   $\quad\quad y - z = 0$

7. $x + y + z = 0$
   $x + y \quad\quad = 0$
   $x \quad\quad + z = 0$

8. $x + y + z = 6$
   $x + y \quad\quad = 3$
   $x \quad\quad + z = 4$

In Problems 9 to 16 use Gaussian elimination to find the unique solution. Make sure to check your answers.

9. $x + y - z = 2$
   $2x + 3y - z = 6$
   $3x + 2y - z = 7$

10. $x - 2y - z = -3$
    $2x - 3y = -2$
    $-x + y + z = 1$

11. $2x - y + 3z = 5$
    $x - y - z = 2$
    $-3x - 2z = -9$

12. $2y - z = 3$
    $3x + 5y - z = 6$
    $x + 2y - 3z = 0$

13. $2x + 4y - 2z = -2$
    $3x + 5y - z = 3$
    $-2x + 3y - z = -7$

14. $3x - 6y + 9z = -24$
    $4x + 2y = -2$
    $-x + y - 3z = 5$

15. $-x + y + 2z = 5$
    $x + y + z = 3$
    $2x + 3y + z = 5$

16. $2x + 3y + z = 5$
    $-x + y + 2z = 0$
    $x + y + z = 1$

17. A software firm tests small, medium, and large disks. The small disks require 1 second in phase A, 2 seconds in phase B, and 3 seconds in phase C. The medium disks require 1 second in phase A, 3 seconds in phase B, and 4 seconds in phase C. The large disks require 2 seconds in phase A, 6 seconds in phase B, and 9 seconds in phase C. Phase A is available for exactly 14 seconds, phase B for 37 seconds, and phase C for 55 seconds. How many disks of each size can be tested?

18. A software firm tests small, medium, and large disks. The small disks require 1 second in phase A, 4 seconds in phase B, and 3 seconds in phase C. The medium disks require 2 seconds in phase A, 9 seconds in phase B, and 6 seconds in phase C. The large disks require 1 second in phase A, 3 seconds in phase B, and 2 seconds in phase C. Phase A is available for exactly 21 seconds, phase B for 89 seconds, and phase C for 60 seconds. How many disks of each size can be tested?

In Problems 19 to 28 solve each system of equations by finding the unique solution when it exists. Otherwise, state that there is not a unique solution.

19. $x + y - z = 4$
    $2x + 3y - z = 9$
    $3x + 2y - z = 8$

20. $x - 2y - z = -5$
    $2x - 3y = -4$
    $-x + y + z = 3$

21. $2x - y + 3z = 5$

$$x - y - z = 5$$
$$-3x \qquad - 2z = -10$$

22.
$$2y - z = 1$$
$$3x + 5y - z = -2$$
$$x + 2y - 3z = 4$$

23.
$$2x + 4y - 2z = -2$$
$$3x + 5y - z = 3$$
$$-2x + 3y - z = -7$$

24.
$$2x + 4y - 2z = 0$$
$$3x + 5y - z = 0$$
$$-2x + 3y - z = 0$$

25.
$$x - 2y + z = 1$$
$$3x - y + 2z = 2$$
$$5x - 5y + 4z = 4$$

26.
$$2x - y + z = 1$$
$$3x + 3y - z = 11$$
$$x + 4y - 2z = 10$$

27.
$$2x + y - 2z = 2$$
$$x + 3y + 2z = 6$$
$$x + 2y - z = 2$$

28.
$$x + y + 2z = 1$$
$$2x + y + 3z = 2$$
$$4x + 2y - z = 9$$

29. A sound studio uses special equipment for mixing and dubbing the recording of solo artists and groups. The equipment is available exactly 16 hours a day for mixing and 13 hours a day for dubbing. Each solo artist requires 2 hours for mixing and 5 hours for dubbing, while each group requires 7 hours for mixing and 4 hours for dubbing. How many solo artists and groups can be recorded each day?

30. A meat manufacturer mixes beef and pork into two types of sausage links, regular and all-beef. Each pound (lb) of regular links contains 0.6 lb of beef and 0.4 lb of pork, while each pound of all-beef links contains 0.8 lb of beef and 0.2 lb of pork. There are exactly 24 lb of beef and 11 lb of pork in stock. How many pounds of each type of sausage links can be made?

31. A hospital nutritionist prepares a meal whose primary ingredients are fish, corn, and rice. Each ounce of fish contains 100 calories (cal), 2 grams (g) of fat, and 3 milligrams (mg) of iron; each ounce of corn contains 200 cal, 5 g of fat, and 7 mg of iron; and each ounce of rice contains 200 cal, 6 g of fat, and 9 mg of iron. The meal must provide exactly 1000 cal, 24 g of fat, and 35 mg of iron. How many ounces of each ingredient should be used?

32. A shoe company makes special shoes for runners, hikers, and bikers on three machines. Each running shoe requires 1 hour on each machine; each hiking shoe requires 2 hours on machine A, 1 hour on machine B, and 2 hours on machine C; each biking shoe requires 2 hours on machine A, 1 hour on machine B, and 1 hour on machine C. Machine A is available exactly 26 hours a week, machine B 21 hours a week, and machine C 23 hours a week. How many shoes of each type can be made each week?

In Problems 33 and 34 a word problem and a system of linear equations is given. The system of equations is the translation of the word problem. The equations represent the constraints of the word problem, which are the number of tables manufactured per week, the manufacturing cost, and the shipping cost. The specific costs for each type of table have been left out of the statement of the word problem. In each exercise restate the word problem, inserting the specific manufacturing cost and the shipping cost for each type of table and then solve the system of equations.

33. A manufacturer makes small, medium, and large tables. How many of each type of table should be made per week if the manufacturer must make a total of 100 tables per week, the total manufacturing cost per week is $1700, and the total shipping cost per week is $860?

$$x + y + z = 100$$
$$10x + 15y + 20z = 1700$$
$$5x + 8y + 10z = 860$$

34. A manufacturer makes small, medium, and large tables. How many of each type of table should be made per week if the manufacturer must make a total of 200 tables per week; the total manufacturing cost per week is $3000 and the total shipping cost per week is $1550?

$$x + y + z = 200$$
$$10x + 15y + 20z = 3000$$
$$5x + 8y + 10z = 1550$$

A system of linear equations is called **homogeneous** if all of the numbers on the right-hand side are equal to zero. In Problems 35 and 36 solve the stated problems with the same coefficients on the left-hand side but with all zeros on the right-hand side.

35. Problems 19, 25, and 27.

36. Problems 20, 26, and 28.

In Problems 37 to 40 determine whether the statement is true or false, using Problems 35 and 36 as a guide.

37. Every homogeneous system of linear equations has exactly one solution.

38. Every homogeneous system of linear equations has at least one solution.

39. The triple $(0, 0, 0)$ is always a solution to a homogeneous system of linear equations.

40. The triple $(0, 0, 0)$ is the unique solution to every homogeneous system of linear equations.

41. Apply Gaussian elimination to solve a general system of two equations in two variables.

$$ax + by = e$$

$$cx + dy = f$$

where a, b, c, d, e, and f are real numbers with $ad - bc \neq 0$.

Referenced Exercise Set 1–4

1. In a pari-mutuel wagering system such as horse racing, *dutching* means to wager on more than one horse to win in the same race. This problem deals with dutching three horses.* (Two problems in Sections 2–3 and 2–5 will consider other aspects of dutching five horses and two horses, respectively.) If a bettor seeks to win $\$w$ on three horses whose odds of winning are $a:1$ (read "a to 1"), $b:1$, and $c:1$, then the amounts which should be bet on each horse are x, y, and z, respectively, where

$$ax - y - z = w$$

$$-x + by - z = w$$

$$-x - y + cz = w$$

If the odds on three particular horses are $3:1$, $3:1$, and $6:1$, how much money should be bet on each horse in order to win \$100?

2. The Chinese were the first society to solve systems of linear equations. Their method, which is strikingly similar to Gaussian elimination, was based on computing board techniques that used red rods for positive numbers and black rods for negative numbers. This explains the origin of the financial phrases "to be in the red" and "to be in the black." Solve the following problem from the book *Chiu chang suan shu* (*Nine Chapters on the Mathematical Art*), which was written about 200 B.C.†

Of three classes of cereal plants, 3 bundles of the first, 2 of the second, and 1 of the third will produce 39 *tou* of corn after threshing; 2 bundles of the first, 3 of the second, and 1 of the third will produce 34 *tou*; while 1 of the first, 2 of the second, and 3 of the third will produce 26 *tou*. Find the measure of corn contained in 1 bundle of each class. (1 *tou* = 10.3 liters)

Cumulative Exercise Set 1–4

In Problems 1 and 2 use the plotting method to sketch the graph of the equation.

1. $y = 2 - 3x$

2. $x^2 + y = 4$

3. Find the slope-intercept form of the line passing through the points $(-1, -2)$ and $(1, -4)$.

4. Find an equation of the line that has x-intercept 5 and is parallel to $6x - 2y = 3$.

---

*See, for example, William L. Scott, *Investing at the Race Track*, New York, Simon & Schuster, 1981. The article "Some Mathematical Applications of Pari-mutuel Wagering" by Frank P. Soler (which appeared in *The Mathematics Teacher*, May 1987, pp. 394–399) presents a method in which the bettor can use a hand-held calculator to carry out the computations on dutching three horses in 15 seconds.

†The Chinese solution to this problem is described on page 14 of the article by Frank Swetz, titled "The Evolution of Mathematics in Ancient China," which appeared in the January 1979 issue of the journal *Mathematics Magazine*.

5. Solve the system by inspection.

$$x + y = 4$$

$$x - y = 0$$

In Problems 6 to 8 solve the system by elimination.

6. $x + y = 4$
    $2x - y = 11$

7. $x - 3y = 8$
    $2x + 3y = -2$

8. $4x - 6y = 8$
    $2x - 3y = -2$

In Problems 9 and 10 use Gaussian elimination to find the unique solution.

9. $x + y - 2z = 1$
    $2x - y \quad\;\; = 3$
    $4x - y - 3z = 4$

10. $2x + y - 3z = 2$
    $\;\;x - y \quad\;\; = -2$
    $3x + y - 2z = 4$

11. A company makes small, medium, and large tents. Each small tent requires 1 hour to cut, 2 hours to sew, and 1 hour to assemble. Each medium tent requires 1 hour to cut, 1 hour to sew, and 1 hour to assemble. Each large tent requires 1 hour to cut, 3 hours to sew, and 2 hours to assemble. The sewing department is available for exactly 25 hours, the cutting department is available for exactly 65 hours, and the assembling department is available for exactly 45 hours. How many tents of each type can be made?

12. A tour director has sold 450 tickets for a weekend excursion which includes a bus trip. The director can lease three types of buses. Type 1 buses can carry 50 passengers, type 2 can carry 75 passengers, and type 3 can carry 80 passengers. Each type 1 bus will cost $900, each type 2 bus will cost $1,000, and each type 3 bus will cost $1,200. The cost for the buses is to be $54,000. The director will lease eight buses. How many buses of each type should be leased?

1–5 Least Squares

Many applications of mathematics are concerned with the relationship between one quantity x and another quantity y. For instance, x can be a person's height and y can be that person's shoe size. If you record the height and shoe size of ten people and then plot the ten ordered pairs of points (x, y) on a coordinate system, you will most likely find that the points do not lie on a straight line. However, they will appear to be distributed along a line, in the sense that a line can be drawn to pass very near to all of them. Of all the lines that come close to the ten points the one that comes closest is called the *line of best fit*, and the method of least squares produces its equation.

This section shows how to construct the equation for the line of best fit. It is applied to recent voting trends in presidential elections in the United States.

The Line of Best Fit

Table 1

x	y
2	1
3	6
4	9
5	13
6	15

Table 1 lists five points in the plane. Regard them as ordered pairs (x, y) and graph them on a coordinate system. (See Figure 1.) The points do not lie on the same line but seem to be spaced in a linear fashion. The method of least squares produces the line of best fit for these points, where by "best fit" we mean that the sum of squares of the distances from the points to the line are minimal. Since distances are measured in terms of squares and least is synonymous with minimal, the method is called "least squares."

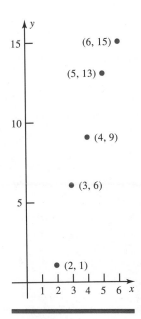

FIGURE 1
Five points.

The method of least squares involves two equations in two unknowns, m (slope) and b (y-intercept). The Greek letter sigma, Σ, denotes the "sum."

$$Pb + (\Sigma\, x)m \; = \; \Sigma\, y \tag{LS1}$$

$$(\Sigma\, x)b + (\Sigma\, x^2)m = \Sigma\, xy \tag{LS2}$$

where

P = the number of points

$\Sigma\, x$ = the sum of the x values

$\Sigma\, y$ = the sum of the y values

$\Sigma\, x^2$ = the sum of the squares x^2

$\Sigma\, xy$ = the sum of the products xy

We refer to the equations as LS1 and LS2. Their derivation involves calculus so we will illustrate them without deriving them.

For the five points in Table 1, set $P = 5$. To obtain the four sums, expand Table 1 by two more columns. Place a column for x^2 to the left of the x column and a column for xy to the right of the y column. This is shown in Table 2, where the row of totals yields $\Sigma\, x = 20$, $\Sigma\, y = 44$, $\Sigma\, x^2 = 90$, and $\Sigma\, xy = 211$. The least squares equations are

$$5b + 20m = \; 44$$

$$20b + 90m = 211$$

This system can be solved by elimination. Multiply the top equation by -4, then add the two equations.

$$\begin{aligned} -20b - 80m &= -176 \\ \underline{20b + 90m} &= \underline{211} \\ 10m &= 35 \end{aligned}$$

The slope of the line of best fit is $m = 3.5$. Substitute this value into the equation

$$5b + 20m = 44$$

Table 2

	x^2	x	y	xy
	4	2	1	2
	9	3	6	18
	16	4	9	36
	25	5	13	65
	36	6	15	90
Totals	90	20	44	211

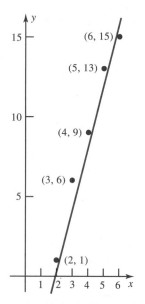

FIGURE 2
Line of best fit.

Then

$$5b + 20(3.5) = 44$$
$$b = -5.2$$

The y-intercept is $b = -5.2$. The line of best fit is uniquely determined by its slope and y-intercept.

$$y = 3.5x - 5.2$$

The line and the five points are graphed in Figure 2.

The Method of Least Squares
For a given set of data points (x, y) the **line of best fit** is

$$y = mx + b$$

where m and b are obtained by solving the system of **least squares equations**

$$Pb + (\Sigma\ x)m = \Sigma\ y \tag{LS1}$$
$$(\Sigma\ x)b + (\Sigma\ x^2)m = \Sigma\ xy \tag{LS2}$$

where

$$P\ =\ \text{number of points}$$
$$\Sigma\ x\ =\ \text{sum of the } x \text{ values}$$
$$\Sigma\ y\ =\ \text{sum of the } y \text{ values}$$
$$\Sigma\ x^2\ =\ \text{sum of the squares } x^2$$
$$\Sigma\ xy\ =\ \text{sum of the products } xy$$

EXAMPLE 1

Problem

Find the line of best fit for these points.

$$(1, 1), \quad (2, 2), \quad (4, 3), \quad (5, 6)$$

Solution Write a vertical table of x values and y values. Place a column of the squares x^2 to the left of the x values and a column of the products xy to the right of the y values.

	x^2	x	y	xy
	1	1	1	1
	4	2	2	4
	16	4	3	12
	25	5	6	30
Totals	46	12	12	47

There are four points, so $P = 4$. The row of totals shows that

$$\Sigma\ x^2 = 46, \qquad \Sigma\ x = 12, \qquad \Sigma\ y = 12, \qquad \Sigma\ xy = 47$$

The least squares equations are

$$4b + 12m = 12 \qquad \text{(LS1)}$$

$$12b + 46m = 47 \qquad \text{(LS2)}$$

Multiply LS1 by -3, then add it to LS2. The resulting equation is $10m = 11$, so $m = 1.1$. Substituting $m = 1.1$ into LS1 yields $b = -0.3$, so the line of best fit is

$$y = 1.1x - 0.3$$

It is graphed in Figure 3.

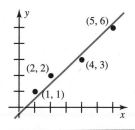

FIGURE 3 Line of best fit $y = 1.1x - 0.3$.

Voting Trends

Example 2 applies the method of least squares to presidential elections. We will then discuss one of the drawbacks to basing predictions of future outcomes on historical results.

EXAMPLE 2

Table 3 lists the percentage of eligible voters who have voted in the presidential elections from 1960 to 1980.

Table 3

Problem

Find the equation of the line of best fit.

Solution Set $x =$ the year of the presidential election and $y =$ the percentage of voters who voted. Write the years without the prefix 19 so that $x = 60$ stands for 1960. Expand Table 3 to include the two additional columns.

Year	Percentage
1960	62.8
1964	61.9
1968	60.9
1972	55.5
1976	54.4
1980	52.3

	x^2	x	y	xy
	3600	60	62.8	3768.0
	4096	64	61.9	3961.6
	4624	68	60.9	4141.2
	5184	72	55.5	3996.0
	5776	76	54.4	4134.4
	6400	80	52.3	4184.0
Totals	29,680	420	347.8	24,185.2

The system of least squares equations is

$$6b + 420m = 347.8 \qquad \text{(LS1)}$$

$$420b + 29680m = 24185.2 \qquad \text{(LS2)}$$

Multiply LS1 by -70, then add it to LS2.

$$
\begin{array}{r}
-420b - 29400m = -24{,}346.0 \\
420b + 29680m = 24{,}185.2 \\
\hline
280m = -160.8
\end{array}
$$

The slope is $m = -1608/2800 = -201/350$. Substituting m into LS1 yields the y-intercept b.

$$6b + 420\left(\frac{-201}{350}\right) = 347.8$$

$$6b = 347.8 + \left(\frac{84{,}420}{350}\right)$$

$$6b = \frac{206{,}150}{350} = 589$$

$$b = \frac{589}{6}$$

Thus the line of best fit is $y = -(201/350)x + (589/6)$.

The primary importance of the line of best fit is for projecting future values. To obtain the projection for 1984, substitute $x = 84$ and solve for y.

$$y = -(201/350)(84) + (589/6) \approx 49.93$$

The projection was that slightly less than 50% of the voters would vote in the 1984 election. However, 53.3% voted. Does this reveal a limitation on basing projections on historical evidence, or is the 1984 figure an exception to the rule of continuing decline?

EXERCISE SET 1–5

In Problems 1 to 14 find the equation of the line of best fit for the given points.

1. $(1, 3), (2, 1), (3, 4), (4, 3)$

2. $(1, 3), (2, 7), (4, 9), (5, 12)$

3. $(1, 3), (2, 4), (3, 1), (4, 2)$

4. $(1, 12), (2, 9), (4, 7), (5, 3)$

5. $(-2, -2), (-1, 2), (1, 4), (2, 7)$

6. $(-1, 0), (0, -2), (1, 1), (2, 0)$

7. $(2, 15), (3, 13), (4, 9), (5, 6), (6, 1)$

8. $(-1, -5), (0, 0), (1, 3), (2, 7), (3, 9)$

9. $(0.3, 0.4), (0.4, 0.3), (0.1, 0.2), (0.3, 0.1)$

10. $(0.1, 0.5), (0.9, 0.4), (0.7, 0.2), (0.3, 0.1)$

11. $(-3, 0), (-2, 2.5), (-1, 4.5), (0, 6), (1, 7.5), (2, 10), (3, 13)$

12. $(-3, -4)$, $(-2, -4)$, $(-1, -3)$, $(0, -3)$, $(1, -2.5)$, $(2, -2)$, $(3, -1)$

13. $(-3, 1)$, $(-2, 0)$, $(-1, 0)$, $(0, -0.5)$, $(1, -1)$, $(2, -1.5)$, $(3, -2)$

14. $(-3, 9)$, $(-2, 4.5)$, $(-1, 0)$, $(0, -3.5)$, $(1, -8)$, $(2, -12)$, $(3, -16)$

Problems 15 to 18 refer to figures compiled by the Department of Health and Human Services. The average 5-year-old boy is 43 inches (in.) tall, the average 10-year-old boy is 54 in. tall, and the average 15-year-old boy is 67 in. tall.

15. (a) Plot these values on a coordinate system with age along the x-axis and height along the y-axis.
    (b) Find the equation for the line of best fit that links age to height.

16. Use the equation for the line of best fit in Problem 15 to approximate the height of an average 8-year-old boy.

17. Use the equation for the line of best fit in Problem 15 to approximate the height of an average 17-year-old boy.

18. Use the equation for the line of best fit in Problem 15 to approximate the height of an average 12-year-old boy.

In Problems 19 to 22 a figure shows a set of data points. Determine whether the points are spaced linearly or nonlinearly.

19.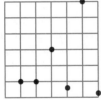

20.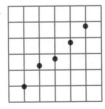

21.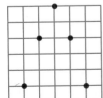

22.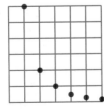

Referenced Exercise Set 1–5

1. Samuel Messick, an official of the Educational Testing Service, admitted that "coaching" for the SATs could be effective.* He revealed that a typical student will increase the SAT-Math score by 10 points after 8 hours of coaching, 20 points after 19 hours, 30 points after 45 hours, and 40 points after 107 hours.
   (a) Plot these points with hours on the x-axis and points on the y-axis.
   (b) Do the points appear to be spaced linearly?
   (c) What are the two equations for the method of least squares?
   (d) What is the equation for the line of best fit?
   (e) Use part (d) to project how many points the score will be raised with 200 hours of coaching.
   (f) Use part (d) to project how many hours of coaching will be required to raise the score 100 points.

2. Use Table 4 to find the equation of the line of best fit between U.S. shoe sizes and European shoe sizes† (a) for women (b) for men.

Table 4 Shoe Sizes

Women		Men	
U.S.	Europe	U.S.	Europe
$4\frac{1}{2}$	35	$9\frac{1}{2}$	43
6	37	10	44
$6\frac{1}{2}$	37.5	$10\frac{1}{2}$	44.5
7	38	11	45
$7\frac{1}{2}$	38.5	$11\frac{1}{2}$	46
8	39	12	46.5
$8\frac{1}{2}$	40	$13\frac{1}{2}$	48
9	40.5	14	49
$9\frac{1}{2}$	41		
10	42		

3. Table 5 has been brought to you in part by Delta Air Lines. Draw a coordinate system with the altitude (ft) as the x-axis and the temperature (degrees F) as the y-axis. Plot the points to see if they are linear. Then compute the line of best fit.

*Edward B. Fiske, "Kicking and Boosting the S.A.T. Scores," *New York Times,* December 23, 1980, p. C4.

†L. Diane Miller and Jim Miller, "Metric Week—The Capitol Way," *Mathematics Teacher,* September 1989, pp. 454–458.

4. Use Problem 3 to project
   (a) the temperature at 40,000 ft.
   (b) the temperature at 4,000 ft.
   (c) the altitude at 32 degrees.
   (d) the altitude at -30 degrees.

Table 5 How Cold Is Up?

There you are, bareheaded and with no gloves and in the lightest of summer clothes, eating a cold salad at 69° below ($-56°C$) zero!

It's routine on Delta jets, summer or any other season, for you to travel in the upper levels of air far from surface heat. It's a case of the higher, the lower. Perhaps you'd like to know exactly how the temperature changes with altitude, how it can be 104°F (40°C) in the sun of the parking lot and well below zero just four miles from the lot, straight up!

The seasons influence lower level temperatures, of course, but on a "standard" day (59°F) (15°C) at sea level, the mercury would drop this way as you went up:

Feet	Meter	F	C
1,000	300	56°	13°
5,000	1,500	41°	5°
10,000	3,000	23°	$-5°$
15,000	4,500	5°	$-15°$
20,000	6,000	$-15°$	$-26°$
30,000	9,000	$-47°$	$-44°$
36,087	10,826	$-69°$	$-56°$
	(Stratosphere)		

Source: Courtesy of Delta Air Lines.

5. In financial management an area called "trend analysis" deals with the construction of forecasting models to predict a firm's cash flow (expenditures and receipts). Figure 4 is taken from a study of a particular firm's cash receipts, where it is called Exhibit 3.‡ The answers to parts (a) and (b) can be obtained directly from the

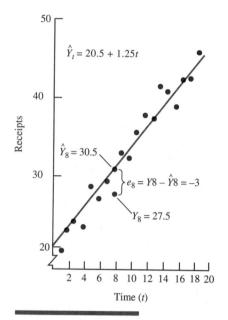

FIGURE 4 Graph of the least squares forecasting model.

figure. (a) What kind of mathematical method is being used for the forecast model? (b) What is the equation of the line of best fit? (c) How many receipts does this model forecast after $t = 20$ time periods? (d) How many time periods does this model forecast for 33 receipts?

6. Figure 5 depicts the relationship between fat intake and death from breast cancer for the inhabitants of many countries around the world.§ Does the relationship appear to be linear?

Problems 7 and 8 refer to Table 6, which describes the doctorates awarded in mathematics to U.S. citizens over the past two decades.‖

7. (a) Find the equation of the line of best fit for the percentage of women who received doctorates. Set $x = 1973$ for the year 1972–1973.
   (b) Use the line to project the percentage in the year 2000–2001.

‡James L. Pappas and George P. Huber, "Probabilistic Short-term Financial Planning," *Financial Management,* Autumn 1973, pp. 36–44.

§Leonard A. Cohen, "Diet and cancer," *Scientific American,* November 1987, pp. 42–48.

‖Donald E. McClure, "Report on the 1990 Survey of New Doctorates," *Notices of the American Mathematical Society,* November 1990, pp. 1217–1222.

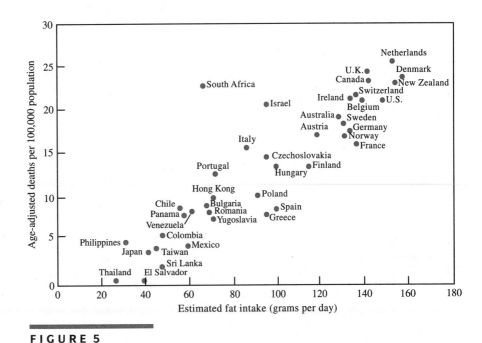

FIGURE 5

Table 6 U.S. Citizen Doctorates, Male and Female

	Doctorates Who Are U.S. Citizens	Male	Female	% Female
1972–1973	774	696	78	10
1973–1974	677	618	59	9
1974–1975	741	658	83	11
1975–1976	722	636	86	12
1976–1977	689	602	87	13
1977–1978	634	545	89	14
1978–1979	596	503	93	16
1979–1980	578	491	87	15
1980–1981	567	465	102	18
1981–1982	519	431	88	17
1982–1983	455	366	89	20
1983–1984	433	346	87	20
1984–1985	396	315	81	20
1985–1986	386	304	82	21
1986–1987	362	289	73	20
1987–1988	363	287	76	21
1988–1989	411	313	98	24
1989–1990	401	312	89	22

8. (a) Does the total number of doctorates awarded in mathematics to U.S. citizens seem to be linear over this period?
   (b) Find the equation of the line of best fit for the total number of U.S. citizens who received doctorates. Set $x = 1973$ for the year 1972–1973.
   (c) Use the line to project the total number in the year 2000–2001.

Cumulative Exercise Set 1–5

1. Let $3x - 2y = 8$. What is the value of y when $x = -2$?

2. Sketch the graph of the equation $2x - 5y = 0$.

3. Find an equation of the line that passes through the origin and has slope $m = 3$.

4. Find the slope of the line passing through the points $(-1, 4)$ and $(1, -1)$.

5. What is the y-intercept of the line $3x - 6y = 8$?

6. Write an equation of the line whose table is

x	1	2
y	-1	-3

7. Solve the system of linear equations:

   $x - y = 4$

   $x - 2y = 12$

8. Find the point of intersection of the lines $y = 3x + 5$ and $y = 5x + 3$.

9. Which triple listed below is a solution of the stated system of equations?

   $(2, -4, 1), \quad (0, 0, 0), \quad (1, -2, 1)$

   $2x + y + 4z = 4$

   $3x + 4y + 5z = 0$

   $-2x + 4y + z = -9$

10. Solve the system of linear equations:

    $x + 2y + 5z = -8$

    $-4y + 2z = -6$

    $-6x + y + 7z = 24$

11. Does the system of linear equations below have any solutions besides $x = 0$ and $y = 0$? Explain your answer geometrically.

    $3x - y = 0$

    $2x + y = 0$

    $x - 3y = 0$

12. Determine the equation of the line of best fit for the points $(1, 5.2)$, $(1.5, 4.7)$, $(1.8, 4.4)$, $(2, 3.8)$, $(2.2, 3.6)$.

13. Is the slope of a line positive or negative when
    (a) the x-intercept is positive and the y-intercept is negative?
    (b) the x-intercept is negative and the y-intercept is positive?
    (c) both intercepts are positive?

14. A taxi in Chicago charges $3 plus $.25 for each tenth of a mile. (a) Write an equation for the cost C in terms of miles. (b) Sketch the graph of the equation. (c) What is the cost of a 2.6-mile ride?

CASE STUDY   The Mile Run

Of all track-and-field events, the mile run is perhaps the most international. It is the only nonmetric event that is still contested. The first world record for the mile run was recorded in 1864 with a time of 4:56. Within a year that standard was lowered to 4:36.5.

Walter George of Great Britain broke the record four times between 1880 and 1884. His 1884 record of 4:18.4 lasted ten years. However, George's greatest gift to running is his race strategy, which consists of a fast first quarter-mile (when the runner is fresh), a slower second quarter, a still slower third quarter, and then as fast a final quarter as possible.

Most milers adopted George's strategy, and by 1945 the world record was lowered to 4:01.4. Then the "barrier" set in. It was felt that the human body was incapable of running a sub-4-minute mile. In spite of the increased popularity after World War II of track-and-field sports throughout the world, no miler was able to better Gunder Haegg's time of 4:01.4.

Then on May 6, 1954 an English medical student named Roger Bannister set out to smash the barrier in a dual meet at Oxford University. Following George's strategy, he ran the first three quarters in 57.5, 60.7, and 62.3 seconds, respectively. His time going into the final quarter was 3:00.5. Bannister was known as a "kick runner," one who preferred to use a great finishing kick to win a race in the stretch. But could he use his great kick to defeat the clock instead of another runner? Yes. Bannister ran the last quarter in 58.9 seconds. His world record of 3:59.4 proved that the 4-minute-mile barrier was psychological, not physiological. Further proof was obtained less than two months later when John Landy ran the mile in 3:58.

The world record has been lowered with surprising regularity since then, and as of this writing (June 1991), it is 3:46.31. Take a look at Table 1, which lists the world record for the mile run since Roger Bannister's historic race. Of course, the times get lower and lower. Is it possible to say more? Is there a pattern to the decrease in times that is obscured by the apparent chaos of facts?

One way to answer these questions is to take a geometric approach. Figure 1 shows the graph of the world record for the mile run when the x-axis is the year and the y-axis is the time of the record in minutes and seconds. For instance, when $x = 1975$, the value $y = 3:50$ reflects Filbert Bayi's world record of 3 minutes 50 seconds.

Figure 1 reveals a remarkable pattern: the points seem to lie along a line. This means that the world record for the mile run is approximately linear, and if this is true, then the future of the event can be predicted with some degree of confidence. It means that we can project what the mile record will be in 1995. And in the year 2000. In fact, we can project when the mile will be run in 3 minutes flat. Beyond that, we can project when it will be run in 0 seconds flat.

Table 1 World Record for the Mile Run

Name (country)	Year	Time
Roger Bannister (Great Britain)	1954	3:59.4
John Landy (Australia)	1954	3:58
Derek Ibbotson (Great Britain)	1957	3:57.2
Herb Elliott (Australia)	1958	3:54.5
Peter Snell (New Zealand)	1962	3:54.4
Peter Snell (New Zealand)	1964	3:54.1
Michel Jazy (France)	1965	3:53.6
Jim Ryun (United States)	1966	3:51.3
Jim Ryun (United States)	1967	3:51.1
Filbert Bayi (Tanzania)	1975	3:50
John Walker (New Zealand)	1975	3:49.4
Sebastian Coe (Great Britain)	1979	3:49.1
Steve Ovett (Great Britain)	1980	3:48.8
Sebastian Coe (Great Britain)	1981	3:48.53
Steve Ovett (Great Britain)	1981	3:48.4
Sebastian Coe (Great Britain)	1981	3:47.33
Steve Cram (Great Britain)	1985	3:46.31

Zero seconds flat! No way.

Let us investigate these intriguing possibilities.

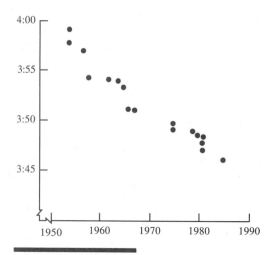

FIGURE 1   World records in the mile run.

EXAMPLE 1

Problem

Determine the line of best fit for the mile run based on the world records for the years 1954–1985.

Solution   Revise Table 1 by writing the years without the prefix 19 so that 54 stands for 1954, and by converting the time to seconds so that 3:59.4 is written as 239.4 seconds. See the top portion of Table 2.

The record was set 17 times during this period, so $P = 17$. The sums of the columns are

$$\Sigma\,x\ = 1184 \qquad \Sigma\,y\ = 3941.47$$

$$\Sigma\,x^2 = 84{,}254 \qquad \Sigma\,xy = 273{,}869$$

The system of linear equations is

$$17b +\ \ 1{,}184m =\ \ \ 3{,}941.47$$

$$1184b + 84{,}254m = 273{,}869.00$$

The solution to this system can be obtained by elimination.

$$b = 256.835 \qquad m = -0.358725$$

The line of best fit is

$$y = -0.358725x + 256.835$$

Consider the meaning of the slope $m = -0.358725$ in Example 1. The negative sign means that the record is being lowered, that the time of the event is decreasing. The number 0.358725 is about $\frac{1}{3}$, so the record will drop about $\frac{1}{3}$ second per year, or about 1 second every three years. Since the record in 1985

Table 2

x^2	x	y	xy
2916	54	239.40	12927.6
2916	54	238.00	12852.0
3249	57	237.20	13520.4
3364	58	234.50	13601.0
3844	62	234.40	14532.8
4096	64	234.10	14982.4
4225	65	233.60	15184.0
4356	66	231.30	15265.8
4489	67	231.10	15483.7
5625	75	230.00	17250.0
5625	75	229.40	17205.0
6241	79	229.10	18098.9
6400	80	228.80	18304.0
6561	81	228.53	18510.9
6561	81	228.40	18500.4
6561	81	227.33	18413.7
7225	85	226.31	19236.4
84254	1184	3941.47	273869.0

Year	Record	Year	Record
1980	3:48.14	1991	3:44.19
1981	3:47.78	1992	3:43.83
1982	3:47.42	1993	3:43.48
1983	3:47.06	1994	3:43.12
1984	3:46.70	1995	3:42.76
1985	3:46.34	1996	3:42.40
1986	3:45.99	1997	3:42.04
1987	3:45.63	1998	3:41.68
1988	3:45.27	1999	3:41.32
1989	3:44.91	2000	3:40.96
1990	3:44.55	2001	3:40.61

was 3:46.31, it should be about 3:43.31 in 1994. It will not be long before we see how accurate this projection is.

Example 2 improves these projections.

EXAMPLE 2

Problem

Use the line of best fit to project what the world record will be in the years 1989 and 2000.

Solution To find the exact times that the line of best fit projects for the future, substitute the desired year into the equation $y = -0.358725x + 256.835$. Remember that the values for x omit the prefix 19. For the year 1989 set $x = 89$. Then $y = -0.358725(89) + 256.835 = 224.91$, so if the world record had been set in 1989 it would have been 3:44.91.

To find the projection for the world record in the year 2000, set $x = 100$. Then $y = 220.96$, so the projected time is 3:40.96.

Table 2 gives the projected times for the years 1980–2001 based on the line of best fit in Example 1. Note that the projected figure for 1985 was just 0.03 seconds from the world mark.

Sometimes it is necessary to determine when a performance will reach a certain level, for instance, when the mile will first be run in 3 minutes flat. To do this, substitute the appropriate time for the value of y and then solve for x. Since the time is given in terms of seconds, $y = 180$. Then

$$y = -0.358725x + 256.835$$

$$180 = -0.358725x + 256.835$$

$$x = 214.2$$

Since x denotes the number of years after 1900, the year is 2114. Thus the mile will first be run in 3 minutes flat in the year 2114. Probably none of us will witness this event.

Let us extend this result to what now seems an inhuman feat. For $y = 120$ the line of best fit yields $x = 381$, so the mile will be run in 2 minutes flat in 2281. It will be run in 1 minute flat in 2448. Carrying these projections to an extreme reveals that the mile will be run in 0 seconds flat sometime during the year 2616. This is patently absurd. Let us discuss the ramifications.

The least squares method can be carried out on any set of data, but it is most effective when the points appear to lie along a straight line. If the points are not linear, then the method of least squares will not provide reasonable projections, so curves other than lines have to be used. And that is what will eventually occur with the mile run. Since the world record for that event was set in 1865, progress has been remarkably linear. We have isolated the years since 1954 to illustrate this fact. At some time in the future the world record will not continue to follow a linear path, and at that time we will have to apply another curve to model the situation.

For now, though, it provides us with a fascinating glimpse into the near future. And it will allow you to see just how accurate our projections will be.

Case Study Exercises

1. Use Table 1 to make a table of the world records for the mile in the period 1967–1980. What are the two equations for the method of least squares?
2. Use Table 1 to make a table of the world records for the mile in the period 1957–1965. What are the two equations for the method of least squares?
3. Use Problem 1 to determine the line of best fit for the mile run based on the years 1967–1980.
4. Use Problem 2 to determine the line of best fit for the mile run based on the years 1957–1965.

5. Use Problem 3 to project the world record for the mile run in the year 2000 based on the period 1967–1980.
6. Use Problem 4 to project the world record for the mile run in the year 2000 based on the period 1957–1965.
7. Use Problem 3 to project the first year in which the mile will be run in 3 minutes flat, based on the years 1967–1980.
8. Use Problem 4 to project the first year in which the mile will be run in 3 minutes flat, based on the years 1957–1965.
9. The line of best fit for the mile run based on the period 1975–1985 is roughly

$$y = -0.32x + 254$$

Use this line to project
(a) what the world record will be in the year 2000.
(b) what year the world record will be 3:40.
10. Repeat Problem 9 for the period 1954–1964. The line of best fit is roughly

$$y = -0.5x + 264$$

11. Problem 9 describes the line of best fit for the years 1975–1985.
(a) What is the slope of the line?
(b) How many seconds will the record improve each year?
(c) How many seconds will the record improve in a century?
12. Repeat Problem 11, using the line of best fit described in Problem 10.

CHAPTER REVIEW

Key Terms

1–1 The Cartesian Coordinate System
Real Line
Origin
Unit Distance
x-Axis; y-Axis
x-Coordinate; y-Coordinate
Coordinates

Cartesian Coordinate System
Quadrants
Graph
The Plotting Method
Linear Equation

1–2 Lines
General Form
x-Intercept
y-Intercept
Slope
Slope-Intercept Form

Parallel Lines
Horizontal Lines
Vertical Lines
Point-Slope Form
Linear Depreciation

1–3 Intersection of Lines
System of Equations
Solution of a System
Inconsistent System

Dependent System
Breakeven Point

1–4 Systems of Linear Equations

Linear Equation in
   Three Variables
Multiplier
Solution Set
Equivalent Systems

Reduced System
Elimination
Back-Substitution
Gaussian Elimination
Homogeneous System

1–5 Least Squares

Line of Best Fit

Least Squares Equations

Summary of Important Concepts

The slope of a line is $m = \dfrac{y_2 - y_1}{x_2 - x_1}$ (unless $x_1 = x_2$).

Horizontal lines have form $y = c$; their slope is 0. Vertical lines have form $x = k$; their slope is undefined.

Three principal forms for the equation of a line are

Name of Form	Equation
general	$Ax + By = C$
slope-intercept	$y = mx + b$
point-slope	$y - y_1 = m(x - x_1)$

The Method of Least Squares

For a given set of data points (x, y) the line of best fit is

$$y = mx + b$$

where m and b are obtained by solving the system of least squares equations

$$Pb + (\Sigma\, x)m = \Sigma\, y$$
$$(\Sigma\, x)b + (\Sigma\, x^2)m = \Sigma\, xy$$

REVIEW PROBLEMS

1. For the equation $y = 3 - 2x$ find the values of y for $x = -2, -1, 0, 1,$ and 2. Plot them and sketch the graph of the equation.

2. Find the slope of the line passing through the points $(-2, 3)$ and $(1, -3)$.

In Problems 3 to 5 sketch the graph of the stated line.

3. $y = -x - 5$

4. $y - 2 = 3(x + 1)$

5. $2x - 4y = 8$

6. Find the slope-intercept form of the line that is parallel to the line $x + y = 0$ and whose x-intercept is equal to -2.

7. Write an equation of the line whose y-intercept is -3 and whose slope is 3.

8. Write the equation of the horizontal line that passes through the point $(-2, -5)$.

9. Write the equation for one day's car rental cost (C) in terms of miles driven (x) if the charges are \$25 per day and 22 cents for each mile the car is driven.

In Problems 10 and 11 determine whether the system has

a solution. If it does, write the solution. If it does not, tell whether it is inconsistent or dependent.

10. $15x - 5y = 20$
       $-12x + 4y = -16$

11. $x + 2y = 1$
    $x - y = -3$

12. Determine the point of intersection of the lines
    $y = 2x - 1$ and $y = -x + 5$.

In Problems 13 and 14 determine how many solutions the system has. If it has one, state it; if there are more than one, state two of them.

13. $x - y = 2$
       $2x - 2y = 4$

14. $2x + 4y = -6$
       $-3x - 5y = 12$

15. Use Gaussian elimination to find the unique solution to this system of equations.

    $x - 2y - 3z = 5$

    $2x + y - z = -5$

    $-x - 3y + 2z = 2$

16. Computer chips are manufactured for micro-, mini-, and mainframe computers. Suppose that each chip for

a microcomputer requires 1 A-transistor, 1 B-transistor, and 2 C-transistors; each chip for a minicomputer requires 1 A-transistor, 3 B-transistors, and 7 C-transistors; and each chip for a mainframe computer requires 2 A-transistors, 7 B-transistors, and 13 C-transistors. How many computer chips of each kind can be manufactured if there are 29 A-transistors, 59 B-transistors, and 126 C-transistors available?

17. Find the equation of the line of best fit for the points $(-1, 5)$, $(0, 3)$, $(1, 6)$, and $(2, 5)$.

18. (a) Plot the data from the table on a coordinate system in which the altitude (in thousands of meters) is the x-axis and the temperature (in degrees Celsius) is the y-axis.

altitude	1	2	3	4	6
temperature	50	35	23	10	-15

   (b) Write the least squares equations.
   (c) Find the equation of the line of best fit and graph it on the same coordinate system.
   (d) What will the temperature be at 5000 meters?
   (e) At what altitude will the temperature be 0 degrees?

 # PROGRAMMABLE CALCULATOR EXPERIMENTS

1. The formula for the area of a rectangle is $A = xy$, where x and y are the length and width, respectively. Write a program that graphs the lines $A = xy$ for the following fixed values of the width y: (a) $y = 1$, (b) $y = 2$, (c) $y = 2.5$, (c) $y = 3$, (d) $y = 3.5$. What is the slope and y-intercept of each line? What do these lines have in common?

2. The formula for the perimeter of a rectangle is $P = 2x + 2y$, where x and y are the length and width, respectively. Write a program that graphs the lines $P = 2x + 2y$ for the following fixed values of the width y: (a) $y = 1$, (b) $y = 2$, (c) $y = 3$, (c) $y = 5$, (d) $y = 10$. What is the slope and y-intercept of each line? What do these lines have in common?

3. Write a program that graphs the rectangles that have the following fixed perimeters: (a) $P = 10$, (b) $P = 20$, (c) $P = 30$, (d) $P = 40$. What is the slope and y-intercept of each line? What do these lines have in common?

4. Graph the three lines l_1, l_2, and l_3 in the standard viewing rectangle:

   l_1: $x + 2y = 12$

   l_2: $x + 3y = 15$

   l_3: $x + y = 9$

   Determine the solution of the system from the graph. Check your answer algebraically.

5. In Problem 4 change l_3 to $x + y = 9.1$ and graph the three lines in the same viewing rectangle. Is there a solution? Illustrate the answer to the last question by using the zoom-in utility.

6. Plot the following points in the standard viewing rectangle: $(0, 9)$, $(1, 8)$, $(2, 6)$, $(3, 8)$, $(4, 4)$, $(5, 5)$, $(6, 3)$, $(7, 1)$. Draw a line in the viewing rectangle that approximates the line of best fit. Find the equation of this line. Calculate the line of best fit algebraically. How close was your approximation?

Matrices

2

Manufacturers find matrices invaluable. (H. Armstrong Roberts)

Chapter Overview

This chapter defines the fundamental concept of a matrix and applies it to various settings. Sections 2–1 and 2–2 introduce the three basic operations on matrices: addition, scalar multiplication, and matrix multiplication.

The next three sections return to the theme of solving systems of linear equations. Section 2–3 explains the Gauss–Jordan method, which is used in Section 2–4 to compute the inverse of a matrix. Section 2–5 solves systems of equations by the matrix inverse method, contrasting it with the methods presented earlier.

Section 2–6 applies matrices to a branch of economics called input-output analysis. It sets the background for the Case Study.

CASE STUDY PREVIEW

Wassily W. Leontief entered college at age 15 and graduated with a doctorate at age 22. While a young faculty member at Harvard University he single-handedly introduced matrix methods into economics. He was later awarded the Nobel prize in economics for his pioneering work.

Leontief's theory correctly predicted that World War II would not be followed by massive unemployment and a depression. His theory also helped avert a crisis in that war by forewarning the government that a massive infusion of silver would be required to build 50,000 badly needed airplanes. The Case Study examines the background of this eminent octogenarian and describes one of his articles on an input-output model of the U.S. economy.

2–1 Addition and Scalar Multiplication

In this section we introduce matrices, define the two basic operations stated in the section title, and strengthen the link between matrices and systems of equations.

Notation

The Pressman-Gutman Company is a fabric-finishing firm that deals primarily with two fabrics, cotton and synthetics. The company buys fabric from a mill, colors it, and gives it a smooth finish. Then the company sells the fabric to a manufacturing company whose end product is a finished garment, like a pair of jeans.

The Pressman-Gutman Company uses finishing plants throughout the country, with three main plants in Fall River (MA), Passaic (NJ), and Los Angeles (CA). The company stores its records in a computer as a matrix or an array. A **matrix** is a rectangular array of numbers with a specific number of rows and columns. The numbers in a matrix are called **entries.** Consider the following matrix.

$$
\begin{array}{c}
\quad\quad\quad\text{cotton}\quad\text{synthetics}\\
\begin{array}{c}
\text{Fall River}\\
\text{Passaic}\\
\text{Los Angeles}
\end{array}
\begin{bmatrix}
2.0 & 4.5\\
2.5 & 4.0\\
3.5 & 1.5
\end{bmatrix}
\end{array}
$$

Each entry represents the number of linear yards of material that was shipped to that plant in the month of January 1988. The units are in millions of yards. Look at the first column of numbers. They show that 2 million yards of cotton were shipped to the Fall River plant, 2.5 million yards of cotton to the Passaic plant, and 3.5 million yards of cotton to the Los Angeles (LA) plant. The second column shows that 4.5 million yards of synthetics were shipped to Fall River, 4 million yards to Passaic, and 1.5 million to LA. By eliminating the labels from the rows and columns, we arrive at the matrix P.

$$P = \begin{bmatrix} 2.0 & 4.5 \\ 2.5 & 4.0 \\ 3.5 & 1.5 \end{bmatrix}$$

We generally use uppercase letters to denote matrices. Consider this matrix.

$$M = \begin{bmatrix} 1 & -1 & 2 & -3 \\ 2 & 3 & -1 & 9 \\ -2 & 0 & 3 & -5 \end{bmatrix}$$

This notation means that M is the name of the matrix whose entries are the numbers inside the brackets. We also say that M is a 3×4 matrix (read "3 by 4") because M consists of 3 rows and 4 columns. In general, an $m \times n$ matrix has m rows and n columns. This is referred to as the **size** of the matrix. If $m = n$, the matrix is a **square matrix.**

The entries of a matrix are denoted by the corresponding lowercase letter with two subscripts. The first subscript refers to the row that the entry lies in and the second subscript refers to the column. For the matrix M, m_{23} refers to the entry in row 2 and column 3, so $m_{23} = -1$. (The notation m_{23} is read "m sub 2, 3.") Sometimes m_{23} is called the "(2, 3) entry."

The general entry in a matrix M is written m_{ij}. This denotes the number that is located in row i and column j.

A matrix with one row is called a **row matrix** and a matrix with one column is called a **column matrix.** The matrix R is a 1×4 row matrix, while C is a 3×1 column matrix.

$$R = \begin{bmatrix} -3 & 4 & 6 & 8 \end{bmatrix} \qquad C = \begin{bmatrix} 2 \\ 9 \\ -1 \end{bmatrix}$$

Addition

Two matrices are **equal** if they have the same size and their corresponding entries are equal. For instance, let

$$R = \begin{bmatrix} 1 & 2 \\ 0 & -1 \end{bmatrix} \qquad S = \begin{bmatrix} 1 & 2 \\ (-3)1 + 3 & (-3)2 + 5 \end{bmatrix}$$

$$T = \begin{bmatrix} 1 & 2 & 0 \\ 0 & -1 & 0 \end{bmatrix}$$

Then $R = S$ because all four entries are equal. However, $R \neq T$ since they have different sizes, even though the four entries of R are equal to the corresponding entries in T.

Consider the two matrices E and F.

$$E = \begin{bmatrix} -5 & 2 & 3 \\ 6 & z & 8 \end{bmatrix} \qquad F = \begin{bmatrix} x & 2 & 3 \\ 6 & -1 & y \end{bmatrix}$$

If $E = F$, then $x = -5$, $y = 8$, and $z = -1$.

The *sum* of two matrices M and N of the same size is the matrix $M + N$ of the same size and whose entries are the sums of the corresponding entries in M and N. The operation is called **addition.** Matrices of different sizes cannot be added.

DEFINITION

Let A and B be $m \times n$ matrices. The **sum** of A and B, $A + B$, is the $m \times n$ matrix M whose entry in the ith row, jth column, m_{ij}, is

$$m_{ij} = a_{ij} + b_{ij}$$

For instance, consider the matrices A, B, and C.

$$A = \begin{bmatrix} 2 & -1 \\ 3 & -6 \end{bmatrix} \qquad B = \begin{bmatrix} 4 & -1 \\ -3 & 8 \end{bmatrix} \qquad C = \begin{bmatrix} 4 & -1 & 0 \\ -3 & 8 & 0 \end{bmatrix}$$

Then

$$A + B = \begin{bmatrix} 2 + 4 & -1 + (-1) \\ 3 + (-3) & -6 + 8 \end{bmatrix} = \begin{bmatrix} 6 & -2 \\ 0 & 2 \end{bmatrix}$$

However, the sum $A + C$ cannot be formed because A and C have different sizes.

The Pressman-Gutman example shows how addition arises in applications. Recall that P represents the amounts of fabrics that were shipped to the three finishing plants in January 1988. Let Q represent the corresponding shipments in February 1988.

$$P = \begin{bmatrix} 2.0 & 4.5 \\ 2.5 & 4.0 \\ 3.5 & 1.5 \end{bmatrix} \qquad Q = \begin{bmatrix} 3.0 & 5.0 \\ 2.5 & 4.5 \\ 4.5 & 1.0 \end{bmatrix}$$

The sum $P + Q$ gives the total amounts of material shipped in January and February.

$$P + Q = \begin{bmatrix} 2.0 + 3.0 & 4.5 + 5.0 \\ 2.5 + 2.5 & 4.0 + 4.5 \\ 3.5 + 4.5 & 1.5 + 1.0 \end{bmatrix} = \begin{bmatrix} 5.0 & 9.5 \\ 5.0 & 8.5 \\ 8.0 & 2.5 \end{bmatrix}$$

The $(1, 1)$ entries show that since 2 million yards of cotton were shipped to Fall River in January and 3 million yards were shipped in February, 5 million yards were shipped in both months.

Example 1 combines equality and addition.

E X A M P L E 1

Problem

Find values for u, v, w, x, y, and z such that

$$\begin{bmatrix} x & 1 \\ -2 & 3 \\ 4 & y \end{bmatrix} + \begin{bmatrix} 4 & 1 \\ 2 & -3 \\ -3 & 6 \end{bmatrix} = \begin{bmatrix} 8 & u \\ v & 0 \\ 1 & 1 \end{bmatrix}$$

Solution First add the two matrices on the left-hand side of the equation. Then set the sum equal to the right-hand side.

$$\begin{bmatrix} x + 4 & 2 \\ 0 & 0 \\ 1 & y + 6 \end{bmatrix} = \begin{bmatrix} 8 & u \\ v & 0 \\ 1 & 1 \end{bmatrix}$$

By the definition of equality

$$x + 4 = 8 \qquad\qquad 2 = u$$

$$0 = v \qquad y + 6 = 1$$

Then $u = 2$ and $v = 0$. The remaining values are obtained by solving the respective equations algebraically.

$$x + 4 = 8 \qquad \text{implies} \qquad x = 4$$

$$y + 6 = 1 \qquad \text{implies} \qquad y = -5$$

Scalar Multiplication

The matrix P represents the linear yards of cotton and synthetics shipped to three finishing plants by the Pressman-Gutman Company in January 1988. To determine the number of linear feet of material shipped, multiply each number in P by 3. For example, the 2 million yards of cotton that were shipped to the Fall River plant convert to 6 million feet. The matrix obtained by multiplying each entry in P by 3 is called the *scalar product* and is denoted by $3P$.

$$3P = \begin{bmatrix} 3(2.0) & 3(4.5) \\ 3(2.5) & 3(4.0) \\ 3(3.5) & 3(1.5) \end{bmatrix} = \begin{bmatrix} 6.0 & 13.5 \\ 7.5 & 12.0 \\ 10.5 & 4.5 \end{bmatrix}$$

Since numbers are sometimes called scalars, the term ''scalar multiplication'' refers to multiplication of a matrix by a number.

DEFINITION

Let A be an $m \times n$ matrix and let r be a real number. The **scalar product** of A and r, rA, is the $m \times n$ matrix M whose entry in the ith row, jth column, m_{ij}, is

$$m_{ij} = ra_{ij}$$

The next example shows how addition and scalar multiplication together lead to an equation whose constant terms are matrices.

EXAMPLE 2

Problem

Find values of x and y such that

$$x\begin{bmatrix} 2 \\ 5 \end{bmatrix} + y\begin{bmatrix} 1 \\ 3 \end{bmatrix} = \begin{bmatrix} 1 \\ -3 \end{bmatrix}$$

Solution Here x and y are unknown scalars. The left-hand side is

$$x\begin{bmatrix} 2 \\ 5 \end{bmatrix} + y\begin{bmatrix} 1 \\ 3 \end{bmatrix} = \begin{bmatrix} 2x \\ 5x \end{bmatrix} + \begin{bmatrix} y \\ 3y \end{bmatrix} = \begin{bmatrix} 2x + y \\ 5x + 3y \end{bmatrix}$$

Set the left-hand side equal to the right-hand side.

$$\begin{bmatrix} 2x + y \\ 5x + 3y \end{bmatrix} = \begin{bmatrix} 1 \\ -3 \end{bmatrix}$$

The entries are equal, so

$$2x + y = 1$$
$$5x + 3y = -3$$

The solution to this system is $x = 6$ and $y = -11$.

It is advisable to check the answer.

$$6\begin{bmatrix} 2 \\ 5 \end{bmatrix} + (-11)\begin{bmatrix} 1 \\ 3 \end{bmatrix} = \begin{bmatrix} 12 \\ 30 \end{bmatrix} + \begin{bmatrix} -11 \\ -33 \end{bmatrix} = \begin{bmatrix} 1 \\ -3 \end{bmatrix}$$

Example 2 solved the matrix equation $Ax + By = C$, where

$$A = \begin{bmatrix} 2 \\ 5 \end{bmatrix} \quad B = \begin{bmatrix} 1 \\ 3 \end{bmatrix} \quad \text{and} \quad C = \begin{bmatrix} 1 \\ -3 \end{bmatrix}$$

This is the matrix analog of the general form of a linear equation introduced in Section 1–2.

Properties

We examine some properties of addition and scalar multiplication on the set of all 2×3 matrices. These properties hold for all $m \times n$ matrices.

The set is **closed** under addition because if A and B are 2×3 matrices, then so is $A + B$. This follows from the definition of addition.

Matrix addition is **associative** because

$$(A + B) + C = A + (B + C)$$

Here the parentheses indicate the particular order in which addition is carried out. The left-hand side means that the matrix $A + B$ is formed first and then added to C. On the right-hand side the matrix $B + C$ is computed first and then added to A. Example 3 verifies the associative law but it does not constitute a general proof.

EXAMPLE 3

Problem

Verify that $(A + B) + C = A + (B + C)$ for

$$A = \begin{bmatrix} -1 & 2 & 4 \\ 3 & 8 & 9 \end{bmatrix} \quad B = \begin{bmatrix} -6 & 4 & 2 \\ 3 & 0 & -2 \end{bmatrix} \quad \text{and} \quad C = \begin{bmatrix} 0 & 0 & 1 \\ 1 & 0 & 0 \end{bmatrix}$$

Solution Compute both sides of the equality separately and then show that they are equal. The left-hand side is $(A + B) + C$. First form $A + B$.

$$A + B = \begin{bmatrix} -7 & 6 & 6 \\ 6 & 8 & 7 \end{bmatrix}$$

Then add $A + B$ to C.

$$(A + B) + C = \begin{bmatrix} -7 & 6 & 6 \\ 6 & 8 & 7 \end{bmatrix} + \begin{bmatrix} 0 & 0 & 1 \\ 1 & 0 & 0 \end{bmatrix} = \begin{bmatrix} -7 & 6 & 7 \\ 7 & 8 & 7 \end{bmatrix}$$

Next evaluate the right-hand side $A + (B + C)$. First form $B + C$.

$$B + C = \begin{bmatrix} -6 & 4 & 3 \\ 4 & 0 & -2 \end{bmatrix}$$

Then add A to $B + C$.

$$A + (B + C) = \begin{bmatrix} -1 & 2 & 4 \\ 3 & 8 & 9 \end{bmatrix} + \begin{bmatrix} -6 & 4 & 3 \\ 4 & 0 & -2 \end{bmatrix} = \begin{bmatrix} -7 & 6 & 7 \\ 7 & 8 & 7 \end{bmatrix}$$

All six entries in $A + (B + C)$ are equal to the corresponding six entries in $(A + B) + C$. Thus $(A + B) + C = A + (B + C)$

The associative law dispenses with the need for parentheses. Therefore the expression $A + B + C$ is meaningful because it can be evaluated as either $(A + B) + C$ or $A + (B + C)$. Addition is also **commutative:** $A + B = B + A$. Both the associative and commutative properties hold because they hold for real numbers.

There is a special matrix called the **zero matrix.**

$$Z = \begin{bmatrix} 0 & 0 & 0 \\ 0 & 0 & 0 \end{bmatrix}$$

It behaves like the number zero, in that

$$M + Z = M \qquad \text{and} \qquad Z + M = M$$

for every 2×3 matrix M.

For any matrix A we write $-A$ for $(-1)A$. The matrix $-A$ satisfies

$$A + (-A) = Z \qquad \text{and} \qquad (-A) + A = Z$$

Therefore $-A$ is called the **additive inverse** of A. Subtraction is defined by $X - Y = X + (-Y)$.

There are two properties linking scalar multiplication to addition

$$s(A + B) = (sA) + (sB)$$

$$(r + s)A = rA + sA$$

where r and s are scalars and A and B are matrices.

EXERCISE SET 2–1

Problems 1 to 6 refer to these matrices.

$$A = \begin{bmatrix} 1 & -6 & 7 \\ -2 & 4 & 3 \end{bmatrix} \qquad B = \begin{bmatrix} 2 & -6 \\ 5 & 0 \\ -1 & 6 \end{bmatrix}$$

$$C = \begin{bmatrix} 1 \\ 9 \\ 7 \end{bmatrix} \qquad D = \begin{bmatrix} 1 & 2 & 3 \\ 4 & 5 & 6 \\ 7 & 8 & 9 \end{bmatrix}$$

$$E = \begin{bmatrix} 1 & 2 \\ 3 & 4 \end{bmatrix} \qquad F = [1 \quad 0 \quad 1 \quad 2]$$

1. What are the sizes of A, C, and E?

2. What are the sizes of B, D, and F?

3. Find (a) a_{22} (b) b_{31} (c) d_{13} (d) e_{11}.

4. Find (a) d_{33} (b) e_{12} (c) b_{22} (d) a_{21}.

5. What notation describes these entries in the matrix A?
   (a) -6 (b) 3 (c) 4

6. What notation describes these entries in the matrix B?
   (a) 5 (b) 6 (c) 0

In Problems 7 and 8 find values for x, y, and z such that $A = B$.

7. $A = \begin{bmatrix} x & 2 \\ 3 & y \\ 4 & 7 \end{bmatrix} \qquad B = \begin{bmatrix} 5 & 2 \\ 3 & 6 \\ 4 & z \end{bmatrix}$

8. $A = \begin{bmatrix} 0 & 9 \\ x & 3 \\ 5 & y \end{bmatrix} \qquad B = \begin{bmatrix} 0 & z \\ 7 & 3 \\ 5 & 8 \end{bmatrix}$

Problems 9 to 16 refer to the matrices R, S, and T.

$$R = \begin{bmatrix} 1 & -1 \\ 2 & -2 \\ 3 & -3 \end{bmatrix} \qquad S = \begin{bmatrix} -3 & 6 \\ 10 & 8 \\ 4 & 0 \end{bmatrix}$$

$$T = \begin{bmatrix} 11 & -4 \\ -3 & 10 \\ -5 & -1 \end{bmatrix}$$

9. Compute (a) $R + S$ (b) $(R + S) + T$ (c) $R + S + T$.

10. Compute (a) $S + R$ (b) $T + (S + R)$ (c) $T + S + R$.

11. Compute (a) $3R$ (b) $(-2)S$ (c) $-T$.

12. Compute (a) $2R$ (b) $(-3)T$ (c) $-S$.

13. Compute (a) $2R + 3S$ (b) $2R - 3S$.

14. Compute (a) $3S + T$ (b) $3S - T$.

15. Verify the equality $R + S = S + R$.

16. Verify the equality $3(R + S) = 3R + 3S$.

In Problems 17 to 20 the matrices A, B, D, and E are defined in Problems 1 to 6 and Z is a zero matrix.

17. What is the size of Z if (a) $A + Z = A$?
    (b) $D + Z = D$?

18. What is the size of Z if (a) $B + Z = B$?
    (b) $E + Z = E$?

19. Find a matrix X such that $A + X = Z$.

20. Find a matrix X such that $B + X = Z$.

21. Suppose that in March the Pressman-Gutman Company shipped 3 million yards of cotton and 1.5 million yards of synthetics to the Fall River plant, 0.5 million yards of cotton and 3 million yards of synthetics to the Passaic plant, and 5 million yards of cotton and 2 million yards of synthetics to the LA plant. Write the shipment matrix M for this month.

22. Suppose that in April the Pressman-Gutman Company shipped 5 million yards of cotton and 2 million yards of synthetics to the Fall River plant, 3.5 million yards of cotton and 2 million yards of synthetics to the Passaic plant, and 1.5 million yards of cotton and 2.5 million yards of synthetics to the LA plant. Write the shipment matrix A for this month.

23. Write the shipment matrix for the period of March through April. (Refer to Problems 21 and 22.)

24. If the shipments in May double those in April, what is the shipment matrix for May?

In Problems 25 and 26 find values for x and y such that $E = F$.

25. $E = \begin{bmatrix} -2 & 2x \\ 35 & y - 2 \end{bmatrix}$ $F = \begin{bmatrix} -2 & 12 \\ 35 & 2 \end{bmatrix}$

26. $E = \begin{bmatrix} 21 & 5 \\ -3 & 9 \end{bmatrix}$ $F = \begin{bmatrix} 21 & x - 2 \\ -3 & -3y \end{bmatrix}$

Problems 27 to 36 refer to the matrices R, S, and T defined in Problems 9 to 16. The matrix U is

$$U = \begin{bmatrix} u + 2 & x \\ v & y + 1 \\ w - 3 & z + 5 \end{bmatrix}$$

Find values of x, y, and z so the equality holds.

27. $R = U$ 28. $T = U$

Find values of x, y, and z so the equality holds.

29. $R - S = T + U$ 30. $R + S = T - U$

31. $(-2)R = U$ 32. $(-3)S = U$

Compute the stated operations.

33. (a) $R - (S + T)$ (b) $(R - S) - T$

34. (a) $R - (S - T)$ (b) $(R - S) + T$

35. What do you conclude from Problem 33?

36. What do you conclude from Problem 34?

In Problems 37 to 40 determine whether it is possible to find values of x and y that satisfy the equality. If so, find x and y.

37. $x\begin{bmatrix} 1 \\ 1 \end{bmatrix} + y\begin{bmatrix} 2 \\ -1 \end{bmatrix} = \begin{bmatrix} 5 \\ 2 \end{bmatrix}$

38. $x\begin{bmatrix} 1 \\ 1 \end{bmatrix} + y\begin{bmatrix} 1 \\ -1 \end{bmatrix} = \begin{bmatrix} 10 \\ 2 \end{bmatrix}$

39. $x\begin{bmatrix} 1 \\ 2 \end{bmatrix} + y\begin{bmatrix} 2 \\ 4 \end{bmatrix} = \begin{bmatrix} 0 \\ 1 \end{bmatrix}$

40. $x\begin{bmatrix} 2 \\ 4 \end{bmatrix} + y\begin{bmatrix} 3 \\ 6 \end{bmatrix} = \begin{bmatrix} 1 \\ 0 \end{bmatrix}$

41. The Clone Corporation makes three computer models (PC, XT, and AT) that are manufactured in Korea and assembled in the United States. The total costs (in dollars) are described by the matrices K and U, which stand for Korea and the United States, respectively.
    (a) Write the matrix of the total costs.
    (b) Write the matrix of the difference in costs between Korea and the United States.
    (c) Write the matrix of the situation where the costs in Korea are cut in half while the costs in the United States are doubled.

$$\begin{array}{c} \\ \text{PC} \\ K = \text{XT} \\ \text{AT} \end{array} \begin{bmatrix} \overset{\text{manufacture}}{150} & \overset{\text{ship}}{12} \\ 325 & 15 \\ 700 & 18 \end{bmatrix}$$

$$\begin{array}{c} \\ \text{PC} \\ U = \text{XT} \\ \text{AT} \end{array} \begin{bmatrix} \overset{\text{assemble}}{15} & \overset{\text{ship}}{6} \\ 20 & 8 \\ 30 & 10 \end{bmatrix}$$

42. A company owns two branch stores, Barney's and Carol's. The amounts of the retail sales and discount sales (in $1000) in three divisions (furniture, appliances, and clothing) for each branch store are given by the matrices B and C, respectively.
    (a) Write the matrix of the total sales for the stores.
    (b) If management wants Barney's to double its sales and Carol's to triple its sales, what matrix will represent the total sales?

$$B = \begin{array}{ccc} \text{furniture} & \text{appliances} & \text{clothing} \\ \begin{bmatrix} 38 & 47 & 19 \\ 62 & 153 & 23 \end{bmatrix} & & \begin{array}{l} \text{retail} \\ \text{discount} \end{array} \end{array}$$

$$C = \begin{bmatrix} 17 & 18 & 67 \\ 25 & 22 & 75 \end{bmatrix} \begin{array}{l} \text{retail} \\ \text{discount} \end{array}$$

43. Let M be an $m \times n$ matrix and let Z be the $m \times n$ zero matrix. Prove that there always exists an $m \times n$ matrix X such that $M + X = Z$.

44. Let A and B be $m \times n$ matrices. Prove that $A + B = B + A$.

Referenced Exercise Set 2–1

1. The matrix gives the shortest navigable distances in nautical miles between the ports of San Francisco, Vancouver, and Panama City.* Write the corresponding matrix in statute miles, where 1 nautical mile = 1.15 statute miles.

$$\begin{array}{c c c c} & \text{SF} & \text{V} & \text{PC} \\ \begin{array}{c} \text{SF} \\ \text{V} \\ \text{PC} \end{array} & \begin{bmatrix} 0 & 812 & 3246 \\ 812 & 0 & 4021 \\ 3246 & 4021 & 0 \end{bmatrix} \end{array}$$

2. An important part of an archaeologist's work is to assign dates to deposits and finds. The Brainerd-Robinson study investigated eight stratified deposits taken from three trenches.† The pieces of pottery found in the deposits were sorted into eight types. The 8×8 matrix B is defined by $b_{ij} = 1$ if deposit i contains an artifact of type j, and $b_{ij} = 0$ otherwise.

$$B = \begin{bmatrix} 0 & 0 & 1 & 1 & 1 & 1 & 1 & 0 \\ 1 & 1 & 1 & 1 & 1 & 1 & 1 & 1 \\ 1 & 0 & 1 & 0 & 0 & 1 & 0 & 1 \\ 1 & 0 & 1 & 1 & 1 & 1 & 1 & 0 \\ 1 & 0 & 1 & 1 & 1 & 1 & 1 & 1 \\ 1 & 0 & 1 & 0 & 0 & 0 & 1 & 1 \\ 1 & 1 & 0 & 0 & 0 & 0 & 1 & 0 \\ 1 & 1 & 1 & 0 & 0 & 1 & 0 & 1 \end{bmatrix}$$

(a) Explain the meaning of the entry b_{24}.
(b) Explain the meaning of the entry b_{73}.

3. Correspondence analysis is a technique for graphing tables for marketing purposes. It has many applications in marketing research and involves the scaling of the rows and columns of a data matrix. This problem illustrates its use with an example representing soft drink consumption.‡ The matrix in Table 1 shows the result of a survey of the beverage preferences of 34 MBA students at Columbia. It is a "binary matrix" because each entry is either 1 (the beverage was consumed at least once over a two-week period) or 0.
    (a) What is the size of the matrix?
    (b) Find the "correspondence matrix" P, where $P = (1/99)X$ and X is the binary matrix in Table 1.
    (c) Write the 8×8 matrix D_c, where d_{jj} is equal to the sum of the entries in the jth column of P and all other entries are equal to 0.
    (d) Write the 34×34 matrix D_r, where d_{jj} is equal to the sum of the entries in the jth row of P and all other entries are equal to 0.

*Source: "Distances Between Ports," Defense Mapping Agency, Hydrographic/Topographic Center, 1990.

†This work was done jointly, but published in two separate articles: W. S. Robinson, "A Method for Chronologically Ordering Archaeological Deposits," *American Antiquity*, Vol. 16, 1951, pp. 293–301; and G. W. Brainerd, "The Place of Chronological Ordering in Archaeological Analysis," *idem*, pp. 301–313. A very readable account of the work can be found in the paper by Alan Shuchat, "Matrix and Network Models in Archaeology," *Mathematics Magazine*, January 1984, pp. 3–14.

‡Donna L. Hoffman and George R. Franke, "Correspondence Analysis: Graphical Representation of Categorical Data in Marketing Research," *Journal of Marketing Research*, August 1986, pp. 213–227.

Table 1 The 34 × 8 Binary Indicator Matrix of Beverage Purchase and Consumption

Individual				Soft Drink				
	Coke	Diet Coke	Diet Pepsi	Diet 7Up	Pepsi	Sprite	Tab	7Up
1	1	0	0	0	1	1	0	1
2	1	0	0	0	1	0	0	0
3	1	0	0	0	1	0	0	0
4	0	1	0	1	0	0	1	0
5	1	0	0	0	1	0	0	0
6	1	0	0	0	1	1	0	0
7	0	1	1	1	0	0	1	0
8	1	1	0	0	1	1	0	1
9	1	1	0	0	0	1	1	1
10	1	0	0	0	1	0	0	1
11	1	0	0	0	1	1	0	0
12	0	1	0	0	0	0	1	0
13	0	0	1	1	0	1	0	1
14	1	0	0	0	0	1	0	0
15	0	1	1	0	0	0	1	0
16	0	0	0	0	1	1	0	0
17	0	1	0	0	0	1	0	0
18	1	1	0	0	1	0	0	0
19	1	0	0	0	0	0	0	1
20	1	1	1	0	1	0	0	0
21	1	0	0	0	1	0	0	0
22	1	0	0	0	1	0	0	0
23	0	1	0	1	0	0	1	0
24	1	1	0	0	1	0	0	0
25	0	1	1	1	0	0	0	0
26	0	1	0	1	0	0	1	0
27	0	1	0	0	0	0	1	0
28	1	0	0	0	0	1	0	1
29	1	0	0	0	0	1	0	0
30	0	1	1	0	0	0	1	0
31	1	0	0	0	1	0	0	1
32	0	1	1	0	0	0	1	0
33	1	0	0	0	1	0	0	1
34	0	1	1	1	0	0	1	0

2–2 Matrix Multiplication

Matrix multiplication is quite different from addition and scalar multiplication. It might seem artificial and contrived at first, yet it is one of the most useful operations in all of mathematics. In this section we define matrix multiplication and illustrate its properties. The link between systems of equations and matrices shows why the definition of multiplication is the one that is commonly used.

Row-by-Column Multiplication

Recall that the Pressman-Gutman Company is a fabric-finishing firm that buys cotton and synthetics from a mill and finishes them into smooth, colored products. Three of its plants are located in Fall River, Passaic, and Los Angeles. The matrix P represents the number of linear yards of each material that was shipped to each plant in January 1988.

$$
\begin{array}{c}
\quad\text{cotton}\quad\text{synthetics} \\
\begin{array}{l}
\text{Fall River} \\
\text{Passaic} \\
\text{Los Angeles}
\end{array}
\begin{bmatrix}
2.0 & 4.5 \\
2.5 & 4.0 \\
3.5 & 1.5
\end{bmatrix}
\end{array}
$$

Suppose that it costs \$2 to ship each linear yard of unfinished cotton and \$1 to ship each linear yard of unfinished synthetics. The cost involved in shipping these fabrics to each of the finishing plants is computed by multiplying each number in the first column of P by \$2 and each number in the second column by \$1 and then adding these products. For instance, the cost of shipping the fabrics to Fall River is

$$(2.0)(\$2) + (4.5)(\$1) = \$8.5$$

Since the numbers in P are in terms of millions of yards, the cost of shipping the fabrics to Fall River is \$8.5 million, or \$8,500,000.

There is a convenient way of organizing the arithmetic in this problem by putting the two costs in a column matrix.

$$
C = \begin{bmatrix} 2 \\ 1 \end{bmatrix}
$$

Regard the first row of P as a row matrix. We define "row-by-column multiplication" to model the Pressman-Gutman problem.

$$
[2.0 \quad 4.5] \begin{bmatrix} 2 \\ 1 \end{bmatrix} = [(2.0)(2) + (4.5)(1)] = [8.5]
$$

This process can be extended to the multiplication of an entire matrix times a column matrix by including the remaining rows of P.

$$
PC = \begin{bmatrix} 2.0 & 4.5 \\ 2.5 & 4.0 \\ 3.5 & 1.5 \end{bmatrix} \begin{bmatrix} 2 \\ 1 \end{bmatrix} = \begin{bmatrix} (2.0)(2) + (4.5)(1) \\ (2.5)(2) + (4.0)(1) \\ (3.5)(2) + (1.5)(1) \end{bmatrix} = \begin{bmatrix} 8.5 \\ 9.0 \\ 8.5 \end{bmatrix}
$$

Notice that the top entry in PC is equal to 8.5, which corresponds to the \$8.5 million cost of shipping to the Fall River plant.

It is not necessary for matrix multiplication to have a physical interpretation like the Pressman-Gutman example, nor for one matrix to be a column matrix. The display gives the formal definition, which is illustrated in Figure 1.

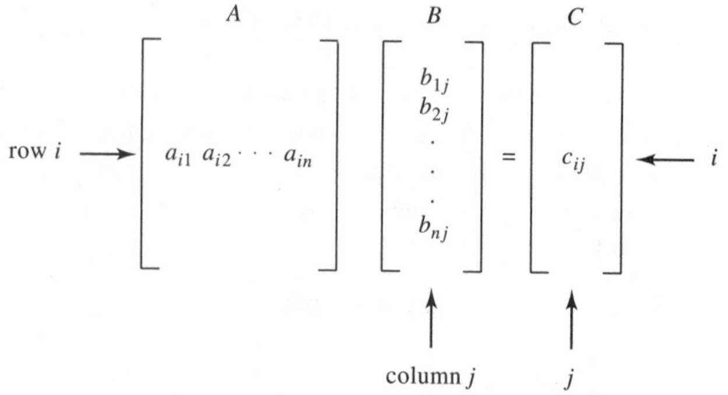

where $c_{ij} = a_{i1}b_{1j} + a_{i2}b_{2j} + \ldots + a_{in}b_{nj}$.

FIGURE 1

DEFINITION

Let A be an $m \times n$ matrix and B an $n \times p$ matrix. The **product** of A and B is the $m \times p$ matrix C whose entry in the ith row, jth column, c_{ij}, is

$$c_{ij} = a_{i1}b_{1j} + a_{i2}b_{2j} + \cdots + a_{in}b_{nj}$$

Example 1 carries out the details of this definition. The operation is called **matrix multiplication.** Sometimes it is called row-by-column multiplication.

EXAMPLE 1

Problem

Compute the product AB, where

$$A = \begin{bmatrix} 7 & 0 \\ 8 & 1 \\ 9 & 2 \end{bmatrix} \quad \text{and} \quad B = \begin{bmatrix} 1 & 2 & 3 \\ 4 & 5 & 6 \end{bmatrix}$$

Solution Multiply each row of A by each column of B. The first row of A times the first column of B is

$$[7 \quad 0] \begin{bmatrix} 1 \\ 4 \end{bmatrix} = [(7)(1) + (0)(4)] = [7]$$

The first row of A times the second column of B is

$$[7 \quad 0] \begin{bmatrix} 2 \\ 5 \end{bmatrix} = [(7)(2) + (0)(5)] = [14]$$

The first row of A times the third column is equal to 21. This completes the first row of the product AB. The details of the entire multiplication are shown here.

$$AB = \begin{bmatrix} 7 & 0 \\ 8 & 1 \\ 9 & 2 \end{bmatrix} \begin{bmatrix} 1 & 2 & 3 \\ 4 & 5 & 6 \end{bmatrix}$$

$$= \begin{bmatrix} (7 \cdot 1) + (0 \cdot 4) & (7 \cdot 2) + (0 \cdot 5) & (7 \cdot 3) + (0 \cdot 6) \\ (8 \cdot 1) + (1 \cdot 4) & (8 \cdot 2) + (1 \cdot 5) & (8 \cdot 3) + (1 \cdot 6) \\ (9 \cdot 1) + (2 \cdot 4) & (9 \cdot 2) + (2 \cdot 5) & (9 \cdot 3) + (2 \cdot 6) \end{bmatrix}$$

$$= \begin{bmatrix} 7 & 14 & 21 \\ 12 & 21 & 30 \\ 17 & 28 & 39 \end{bmatrix}$$

It is not always possible to complete row-by-column multiplication. For instance, let

$$A = \begin{bmatrix} 1 & 3 & 4 \\ 4 & 5 & 6 \end{bmatrix} \quad \text{and} \quad B = \begin{bmatrix} 7 & 8 & 9 \\ 0 & 1 & 2 \end{bmatrix}$$

The product AB cannot be formed because row 1 of A has three entries but column 1 of B has only two. We say that the product AB is **undefined.** In general, if A is an $m \times n$ matrix and B is a $q \times p$ matrix, then AB is an $m \times p$ matrix if $n = q$ and AB is undefined otherwise. In brief,

$$\begin{array}{ccc} (m \times n) & (n \times p) & = (m \times p) \\ \underbrace{(m \times n)}_{\text{size of } A} & \underbrace{(q \times p)}_{\text{size of } B} & \underbrace{\text{is undefined if } n \neq q}_{\substack{\text{size of} \\ \text{product } AB}} \end{array}$$

Example 2 shows why caution must be exercised when extending properties of the multiplication of numbers to matrix multiplication.

EXAMPLE 2

Problem

Compute BC and CB, where

$$B = \begin{bmatrix} -1 & 0 \\ 6 & 8 \end{bmatrix} \quad \text{and} \quad C = \begin{bmatrix} 0 & 4 & 1 \\ 6 & -1 & 0 \end{bmatrix}$$

Solution The product BC is equal to

$$BC = \begin{bmatrix} -1 & 0 \\ 6 & 8 \end{bmatrix} \begin{bmatrix} 0 & 4 & 1 \\ 6 & -1 & 0 \end{bmatrix}$$

$$= \begin{bmatrix} -1 \cdot 0 + 0 \cdot 6 & -1 \cdot 4 + 0 \cdot -1 & -1 \cdot 1 + 0 \cdot 0 \\ 6 \cdot 0 + 8 \cdot 6 & 6 \cdot 4 + 8 \cdot -1 & 6 \cdot 1 + 8 \cdot 0 \end{bmatrix}$$

$$= \begin{bmatrix} 0 & -4 & -1 \\ 48 & 16 & 6 \end{bmatrix}$$

However, CB is undefined because C has three columns and B has two rows.

Systems of Equations

Suppose the Pressman-Gutman Company determines that to break even the Fall River plant must produce $15,000,000 worth of material, the Passaic plant must produce $15,500,000, and the Los Angeles plant $13,500,000. The company wants to find the price per yard of cotton and of synthetic material that it must charge in order to break even. To solve the problem, let x be the price per yard of cotton, y be the price per yard of synthetic material, and

$$X = \begin{bmatrix} x \\ y \end{bmatrix} \quad \text{be the price matrix.}$$

Then PX represents the income from the material finished at each plant. In order to break even, x and y must satisfy

$$\begin{bmatrix} 2.0 & 4.5 \\ 2.5 & 4.0 \\ 3.5 & 1.5 \end{bmatrix} \begin{bmatrix} x \\ y \end{bmatrix} = \begin{bmatrix} 15.0 \\ 15.5 \\ 13.5 \end{bmatrix}$$

This matrix equation is solved by multiplying the matrices on the left-hand side and setting the product equal to the right-hand side. Thus

$$\begin{bmatrix} 2.0x + 4.5y \\ 2.5x + 4.0y \\ 3.5x + 1.5y \end{bmatrix} = \begin{bmatrix} 15.0 \\ 15.5 \\ 13.5 \end{bmatrix}$$

The unknown quantities x and y must satisfy the system of equations

$$2.0x + 4.5y = 15.0$$

$$2.5x + 4.0y = 15.5$$

$$3.5x + 1.5y = 13.5$$

By Gaussian elimination the solution is $x = 3$ and $y = 2$. It can be written as the column matrix $\begin{bmatrix} 3 \\ 2 \end{bmatrix}$.

This illustrates why matrix multiplication is defined the way it is. The **coefficient matrix** of the system is the Pressman-Gutman matrix P. The **variable matrix** is the column matrix X. The **constant matrix** is the column matrix of numbers on the right-hand side.

$$C = \begin{bmatrix} 15.0 \\ 15.5 \\ 13.5 \end{bmatrix}$$

The product PX becomes

$$PX = \begin{bmatrix} 2.0 & 4.5 \\ 2.5 & 4.0 \\ 3.5 & 1.5 \end{bmatrix} \begin{bmatrix} x \\ y \end{bmatrix} = \begin{bmatrix} 2.0x + 4.5y \\ 2.5x + 4.0y \\ 3.5x + 1.5y \end{bmatrix}$$

Its entries are the left-hand side of the given system of equations. Thus the entire system can be written as the matrix equation

$$PX = C$$

In this equation P and C are given matrices (that is, their entries are numbers) and X is the unknown (or variable) matrix that is sought. In this sense the matrix equation $PX = C$ resembles the linear equation $ax = b$ for numbers.

> Every system of linear equations can be written as a matrix equation of the form $AX = C$. Conversely, every such matrix equation describes a system of linear equations.

EXAMPLE 3

Problem

Write this system of equations as a matrix equation.

$$\begin{aligned} x - 2y &= 8 \\ 4y - 3z &= -7 \end{aligned}$$

Solution The coefficient matrix is

$$A = \begin{bmatrix} 1 & -2 & 0 \\ 0 & 4 & -3 \end{bmatrix}$$

Be sure to include the zeros. Since the system has three variables, the variable matrix is

$$X = \begin{bmatrix} x \\ y \\ z \end{bmatrix}$$

The constant matrix has two entries.

$$C = \begin{bmatrix} 8 \\ -7 \end{bmatrix}$$

The matrix equation is then $AX = C$.

Sections 2–3 and 2–5 will further develop the link between systems of equations and matrices.

Algebra of Square Matrices

The set of 2×3 matrices is not closed under matrix multiplication because of the rule

$$(2 \times \underbrace{3) \quad (2}_{} \times 3)$$

not defined because $3 \neq 2$

For square matrices the situation is different. If A and B are $n \times n$ matrices, then so is AB because

$$(n \times n)(n \times n) = (n \times n)$$

equal

The set of $n \times n$ matrices is closed under multiplication.

The multiplication of real numbers and the multiplication of matrices have some properties in common. For instance, the associative law holds.

$$(AB)C = A(BC)$$

whenever the sizes of the matrices permit them to be multiplied. But there are several differences that show that care must be exercised when dealing with matrices. For instance, the **cancellation law** does not hold, meaning that $AB = AC$ does not necessarily imply that $B = C$. Example 4 shows that the set of $n \times n$ matrices is not commutative under multiplication.

EXAMPLE 4

Let

$$A = \begin{bmatrix} 1 & 2 & 3 \\ 4 & 5 & 6 \\ 7 & 8 & 7 \end{bmatrix} \quad \text{and} \quad B = \begin{bmatrix} 1 & 4 & 7 \\ 2 & 5 & 8 \\ 3 & 6 & 9 \end{bmatrix}$$

Problem

Determine whether $AB = BA$.

Solution The definition of the product AB shows that the $(1, 1)$ entry is equal to 14. Similarly, the $(1, 1)$ entry of BA is equal to 66. In order for AB to be equal to BA their entries must all be equal, respectively. Since the $(1, 1)$ entries differ, they are not equal. Therefore $AB \neq BA$.

The **main diagonal** of an $n \times n$ matrix A is the diagonal of numbers running from the $(1, 1)$ entry to (n, n). The main diagonal of this matrix is shaded.

$$A = \begin{bmatrix} 5 & 4 & 0 \\ 6 & -3 & 7 \\ 9 & -1 & 4 \end{bmatrix}$$

The $n \times n$ matrix with 1's down the main diagonal and zeros elsewhere is called the **identity matrix** and is denoted by I_n. The 3×3 identity matrix is

$$I_3 = \begin{bmatrix} 1 & 0 & 0 \\ 0 & 1 & 0 \\ 0 & 0 & 1 \end{bmatrix}$$

$AI_n = A$ and $I_nA = A$ for every $n \times n$ matrix A.

The operations of addition and multiplication of matrices are tied together by the same property that ties them together for numbers, the **distributive laws**

$$A(B + C) = AB + AC$$

and

$$(D + E)F = DF + EF$$

EXERCISE SET 2–2

In Problems 1 to 6 find the product of the two matrices whenever it is defined.

$$A = \begin{bmatrix} 1 & -1 & 0 \\ 0 & 10 & 0 \\ 0 & 1 & 1 \end{bmatrix} \qquad B = \begin{bmatrix} -2 & 1 & 0 \\ 0 & 6 & 3 \\ 0 & 0 & -2 \end{bmatrix}$$

$$C = \begin{bmatrix} 1 & 0 & -1 & 6 \\ 2 & 5 & 2 & 3 \\ 4 & 1 & 3 & 1 \end{bmatrix} \qquad D = \begin{bmatrix} 0 \\ -1 \\ -2 \end{bmatrix}$$

$$E = \begin{bmatrix} 6 & 3 & 1 & 4 \\ 5 & 7 & 0 & -4 \end{bmatrix}$$

$$F = \begin{bmatrix} -1 & 10 \\ 6 & -7 \end{bmatrix} \qquad G = \begin{bmatrix} 16 & 3 \\ -2 & 0 \end{bmatrix}$$

1. AB

2. FG

3. AC

4. FE

5. AD

6. DA

In Problems 7 to 10 compute the stated entry in the product $J = KL$, where

$$K = \begin{bmatrix} -1 & 6 & 7 \\ 2 & 4 & 1 \\ 3 & 0 & 6 \\ 0 & 0 & 0 \\ -6 & 3 & 0 \end{bmatrix}$$

$$L = \begin{bmatrix} 1 & 6 & 0 & -1 \\ 6 & -2 & 0 & -3 \\ 0 & 6 & 0 & 6 \end{bmatrix}$$

7. j_{12}

8. j_{42}

9. j_{22}

10. j_{33}

In Problems 11 and 12 determine whether the product AB is defined. If it is, state its size.

11. (a) A is 4×6, B is 6×3
    (b) A is 4×3, B is 4×3

12. (a) A is 2×4, B is 2×4
    (b) A is 1×4, B is 4×9

In Problems 13 and 14 a system of equations is given. Write the coefficient matrix A, the variable matrix X, and the constant matrix C for each system.

13. $-x + 4y - 3z = 11$
    $2x - 2y + 7z = -4$

14. $4x - 6y = 8$
    $-x + 6y = 0$
    $2x - 3y = 4$

In Problems 15 to 18 write the matrix equation $AX = Z$ as a system of equations, where

$$X = \begin{bmatrix} x \\ y \\ z \end{bmatrix} \quad \text{and} \quad Z = \begin{bmatrix} 0 \\ 0 \\ 0 \end{bmatrix}$$

15. $A = \begin{bmatrix} 1 & 0 & 6 \\ -2 & 3 & 4 \\ 1 & 2 & 0 \end{bmatrix}$

16. $A = \begin{bmatrix} 2 & 6 & 7 \\ 0 & 3 & -1 \\ 0 & 0 & 4 \end{bmatrix}$

17. $A = \begin{bmatrix} 2 & 6 & 7 \\ 3 & 5 & -2 \end{bmatrix}$

18. $A = \begin{bmatrix} 1 & 1 & -1 \\ 2 & 5 & 4 \end{bmatrix}$

In Problems 19 to 22 determine whether $AB = BA$, where

$$B = \begin{bmatrix} 10 & 2 \\ -1 & 4 \end{bmatrix}$$

19. $A = \begin{bmatrix} 1 & 2 \\ 3 & 4 \end{bmatrix}$

20. $A = \begin{bmatrix} -1 & 0 \\ 0 & 3 \end{bmatrix}$

21. $A = \begin{bmatrix} 3 & 0 \\ 0 & -2 \end{bmatrix}$

22. $A = \begin{bmatrix} 0 & 3 \\ -2 & 0 \end{bmatrix}$

Problems 23 and 24 refer to this Pressman-Gutman matrix.

$$
\begin{array}{c}
\\
\text{Fall River} \\
\text{Passaic} \\
\text{Los Angeles}
\end{array}
\begin{array}{cc}
\text{cotton} & \text{synthetics} \\
\begin{bmatrix}
2.0 & 4.5 \\
2.5 & 4.0 \\
3.5 & 1.5
\end{bmatrix}
\end{array}
$$

23. If it costs $1 to ship each linear yard of cotton and $1.50 to ship each linear yard of synthetics, what matrix represents the cost of shipping these fabrics to each of the finishing plants?

24. If it costs $4 to ship each linear yard of cotton and $3.50 to ship each linear yard of synthetics, what matrix represents the cost of shipping these fabrics to each of the finishing plants?

Problems 25 to 28 refer to these matrices.

$$
A = \begin{bmatrix} -1 & 0 \\ 2 & 3 \end{bmatrix} \qquad
B = \begin{bmatrix} 4 & -1 \\ 0 & 2 \end{bmatrix}
$$

$$
C = \begin{bmatrix} 6 & -9 \\ -4 & 5 \end{bmatrix}
$$

25. (a) Compute $(AB)C$.
    (b) Compute $A(BC)$.
    (c) What do you conclude from parts (a) and (b)?
    (d) Compute ABC.

26. (a) Compute $AA - BB$.
    (b) Compute $(A + B)(A - B)$.
    (c) What do you conclude from parts (a) and (b)?

27. (a) Compute $A(B + C)$.
    (b) Compute $AB + AC$.
    (c) What do you conclude from parts (a) and (b)?

28. (a) Compute $(A + B)C$.
    (b) Compute $AC + BC$.
    (c) What do you conclude from parts (a) and (b)?

In Problems 29 and 30 let E_{ij} be the 3×3 matrix with the number 1 in the (i, j) entry and 0's elsewhere. Let

$$
A = \begin{bmatrix} 1 & 2 & 3 \\ 4 & 5 & 6 \\ 7 & 8 & 9 \end{bmatrix}
$$

29. (a) Compute $E_{12}A$.
    (b) Compute $E_{23}A$.
    (c) What do you conclude from parts (a) and (b)?

30. (a) Compute AE_{12}.
    (b) Compute AE_{23}.
    (c) What do you conclude from parts (a) and (b)?

31. Let A be a 3×3 matrix whose bottom row consists entirely of 0's and let B be any 3×4 matrix. What can you conclude about the product AB?

32. Let B be a 3×3 matrix whose middle column consists entirely of 0's and let A be any 4×3 matrix. What can you conclude about the product AB?

33. A researcher for a dairy company is experimenting with blends of low-fat milk and skim milk. The number of grams per cup of protein, carbohydrate, and fat contained in each kind of milk is given by the matrix

$$
\begin{array}{c}
\\
A =
\end{array}
\begin{array}{cc}
\text{skim} & \text{low-fat} \\
\begin{bmatrix}
9 & 9 \\
13 & 13 \\
1 & 5
\end{bmatrix}
\begin{array}{l}
\text{protein} \\
\text{carbohydrate} \\
\text{fat}
\end{array}
\end{array}
$$

The number of cups of milk fed to people in three age categories is given by the matrix

$$
\begin{array}{c}
\\
B =
\end{array}
\begin{array}{ccc}
\text{children} & \text{adolescents} & \text{adults} \\
\begin{bmatrix}
3 & 1 & 0 \\
2 & 1 & 1
\end{bmatrix}
\begin{array}{l}
\text{skim} \\
\text{low-fat}
\end{array}
\end{array}
$$

What do the entries in the matrix AB tell the researcher?

34. A company makes records, tapes, and compact disks in Los Angeles and Nashville. Each item must be manufactured and quality tested. The time (in hours) for each operation is given by the matrix A.

$$
\begin{array}{c}
\\
A =
\end{array}
\begin{array}{cc}
\text{manufacture} & \text{test} \\
\begin{bmatrix}
1 & \frac{1}{3} \\
3 & \frac{1}{2} \\
8 & 1
\end{bmatrix}
\begin{array}{l}
\text{record} \\
\text{tape} \\
\text{compact disk}
\end{array}
\end{array}
$$

The labor costs (in dollars per hour) in each city for each operation is given by the matrix B.

$$
\begin{array}{c}
\\
B =
\end{array}
\begin{array}{cc}
\text{Los Angeles} & \text{Nashville} \\
\begin{bmatrix}
7 & 6 \\
10 & 5
\end{bmatrix}
\begin{array}{l}
\text{manufacture} \\
\text{test}
\end{array}
\end{array}
$$

What do the entries in the matrix AB mean?

35. An automobile dealership has three franchises (F1, F2, and F3) in Florida and the same three in

Georgia. The October sales volume, in cars sold, is shown in the matrix C.

$$C = \begin{array}{c} \\ F1 \\ F2 \\ F3 \end{array} \begin{array}{c} \text{FL} \quad \text{GA} \\ \begin{bmatrix} 126 & 92 \\ 76 & 87 \\ 45 & 78 \end{bmatrix} \end{array}$$

Each car sold for $17,000 in Florida and $16,200 in Georgia. Use matrix multiplication to determine the total sales for each franchise in October.

36. An automobile dealership had two franchises in each state: Texas, Oklahoma, and Nebraska. The November sales volume, in cars sold, is shown in the matrix C.

$$C = \begin{array}{c} \\ F1 \\ F2 \end{array} \begin{array}{c} \text{TX} \quad \text{OK} \quad \text{NB} \\ \begin{bmatrix} 423 & 68 & 52 \\ 186 & 45 & 47 \end{bmatrix} \end{array}$$

Each car sold for $14,500 in Texas, $15,000 in Oklahoma, and $16,200 in Nebraska. Use matrix multiplication to determine the total sales for each franchise in November.

37. A bank pays interest rates of 10, 7, and 5% on certificates of deposit (CDs) of $5000, $2000, and $1000, respectively. Use row-by-column matrix multiplication to determine the total interest earned on the CDs.

38. A bank pays interest rates of 12, 10, 7, and 5% on certificates of deposit (CDs) of $25,000, $5000, $2000, and $1000, respectively. Use row-by-column matrix multiplication to determine the total interest earned on the CDs.

39. Two juices (A, B) are blended from four ingredients (W, X, Y, Z). The numbers of gallons (gal) of each ingredient in a 20-gal container of each juice are given by the matrix J.

$$J = \begin{array}{c} \\ A \\ B \end{array} \begin{array}{c} W \quad X \quad Y \quad Z \\ \begin{bmatrix} 8 & 6 & 4 & 2 \\ 9 & 7 & 3 & 1 \end{bmatrix} \end{array}$$

Use matrix multiplication to find the total cost of making 20 gal of each juice if W, X, Y, and Z cost $0.25, $0.50, $1, and $2 per gallon, respectively.

40. Use Problem 39 to compute the cost of making 60 gal of juice A and 80 gal of juice B under the same conditions given there.

41. A developer is planning to build a high-rise building with apartments and condominiums in 1-bedroom, 2-bedroom, and 3-bedroom models. The matrix D shows the number of each unit that will be built, while the matrix C lists the costs per unit. What is the total cost of building all 100 units?

$$D = \begin{array}{c} \\ \\ \end{array} \begin{array}{c} 1 \quad\ 2 \quad\ 3 \\ \begin{bmatrix} 25 & 10 & 5 \\ 20 & 30 & 10 \end{bmatrix} \end{array} \begin{array}{c} \text{apartments} \\ \text{condominiums} \end{array}$$

$$C = \begin{bmatrix} 40,000 \\ 55,000 \\ 65,000 \end{bmatrix} \begin{array}{l} \text{1-bedroom} \\ \text{2-bedroom} \\ \text{3-bedroom} \end{array}$$

42. Write the stated entry in the identity matrix I_3.
(a) The (3, 2) entry
(b) The (2, 2) entry

43. Let Y be the matrix defined below. If X is a matrix such that $XY = 0$, must X be the zero matrix?

$$Y = \begin{bmatrix} 0 & 2 \\ 3 & 0 \end{bmatrix}$$

44. If a system of two linear equations has four unknowns, what is the size of the corresponding variable matrix X in the matrix equation $AX = C$?

45. Let A be an arbitrary 2×2 matrix and let S be a scalar matrix, where s, a, b, c, and d are any numbers.

$$A = \begin{bmatrix} a & b \\ c & d \end{bmatrix} \qquad S = \begin{bmatrix} s & 0 \\ 0 & s \end{bmatrix}$$

(a) Verify that $SA = AS$.
(b) Prove that if B satisfies $AB = BA$ for all 2×2 matrices A, then $B = S$ for some number s.

46. Find 2×2 matrices A, B, and C, with C not the zero matrix, such that $AC = BC$ but $A \neq B$.

47. Find two 2×2 matrices A and B, different from the zero matrix Z, such that $AB = Z$.

48. Find a 2×2 matrix A, different from the zero matrix Z, such that $AA = Z$.

Referenced Exercise Set 2–2

Decision theory makes use of matrices to organize and examine data that can be helpful to individuals when making personal decisions. Problems 1 and 2 apply decision theory to two unrelated situations. In both problems a matrix X describes the relative preferences of each option with respect to each criterion. The absolute preferences are obtained from the rows of the product XY.

1. A study was conducted for various universities in the Province of Ontario after the government ended its practice of administering universal exams (called "departmentals") for admission into the universities, because a period of grade inflation ensued at some of the high schools.* There were three options ($G =$ to continue to accept high school grades, $C =$ to calibrate high school grades, $E =$ to construct new admission exams) and four criteria ($F =$ fairness, $P =$ predictability, $L =$ low cost, and $A =$ acceptability).

$$X = \begin{array}{c} \\ \\ \\ \end{array} \begin{array}{cccc} F & P & L & A \\ \left[\begin{array}{cccc} .105 & .297 & .644 & .674 \\ .258 & .540 & .271 & .101 \\ .637 & .163 & .085 & .225 \end{array}\right] & & & \end{array} \begin{array}{c} G \\ C \\ E \end{array}$$

$$Y = \begin{bmatrix} .543 \\ .085 \\ .213 \\ .159 \end{bmatrix}$$

(a) What is the sum of the entries in each column of X? Why?
(b) Which option was most preferable?
(c) Which option was least preferable?

2. A plant pathologist observes a leaf wilt in her neighbor's tobacco plant; the wilt can be caused by either virus A or virus B.† Virus A cannot be treated, so that if the plant has this virus it will die, causing a loss of $15. There is a $5 treatment that will save the plant if it has virus B. If the plant has virus B and is not treated it will die. There are two options ($T =$ treat the plant and $N =$ do not treat the plant) and two criteria ($H =$ healthy plants and $L =$ low cost).

$$X = \begin{array}{c} \\ \\ \end{array} \begin{array}{cc} H & L \\ \left[\begin{array}{cc} \frac{4}{5} & \frac{1}{3} \\ \frac{1}{5} & \frac{2}{3} \end{array}\right] & \end{array} \begin{array}{c} T \\ N \end{array} \qquad Y = \begin{bmatrix} \frac{1}{4} \\ \frac{3}{4} \end{bmatrix}$$

Which option is more preferable?

In the 1970s Wendy's International, Inc. grew from a single store to a conglomerate of 1800 stores in the United States, Europe, and Japan. Today Wendy's is the third largest fast-food hamburger chain in the United States. The top managers at Wendy's faced three important strategy decisions: (1) whether to emphasize expansion of its domestic market or its foreign market, (2) whether to continue to appeal to the 82% of its customers who were over 25 years of age, and (3) whether to utilize stock or debt to finance new construction.

Problems 3 to 5 introduce a technique for dealing with these strategies called QSPM (Quantitative Strategic Planning Matrix).‡ The "sum total attractiveness scores" are the products of a rating matrix r and a matrix of the values for various factors. The best strategy is the option with the highest sum total attractiveness score.

3. Where should Wendy's concentrate its markets, domestic D or foreign F? The rating matrix is

$$r = [3 \quad 1 \quad 2 \quad 2 \quad 1 \quad 2 \quad 4 \quad 1 \quad 1]$$

and the value matrices are

$$D = \begin{bmatrix} 4 \\ 4 \\ 2 \\ 3 \\ 4 \\ 2 \\ 4 \\ 4 \\ 3 \end{bmatrix} \qquad F = \begin{bmatrix} 3 \\ 2 \\ 1 \\ 2 \\ 1 \\ 3 \\ 2 \\ 1 \\ 2 \end{bmatrix}$$

4. Which age group should Wendy's appeal to, $U =$ 25 or under, or $O =$ over 25? The rating matrix is

$$r = [4 \quad 1 \quad 2 \quad 2 \quad 4 \quad 1 \quad 2 \quad 4 \quad 1 \quad 1]$$

---

*Ed Barbeau, "Perron's Result and a Decision on Admissions Tests," *Mathematics Magazine,* February 1986, pp. 12–22.

†JoAnne Growney, "Personal Decisions/Medical Decisions: Consider the Analytic Hierarchy Process," *UMAP Journal,* Winter 1985, pp. 7–35.

‡Fred R. David, "The Strategic Planning Matrix—A Quantitative Approach," *Long Range Planning* (*The International Journal of Strategic Management and Corporate Planning*), October 1986, pp. 102–107.

and the value matrices are

$$U = \begin{bmatrix} 3 \\ 4 \\ 1 \\ 3 \\ 3 \\ 2 \\ 4 \\ 2 \\ 2 \\ 2 \end{bmatrix} \qquad O = \begin{bmatrix} 4 \\ 2 \\ 4 \\ 2 \\ 2 \\ 4 \\ 2 \\ 4 \\ 3 \\ 3 \end{bmatrix}$$

5. How should Wendy's finance its construction, by debt D or common stock S? The rating matrix is

$$r = [2 \quad 1 \quad 1 \quad 1]$$

and the value matrices are

$$D = \begin{bmatrix} 2 \\ 1 \\ 2 \\ 3 \end{bmatrix} \qquad S = \begin{bmatrix} 4 \\ 4 \\ 4 \\ 2 \end{bmatrix}$$

6. The product of the matrices X and Y below can be used to multiply two 12-digit numbers on a calculator that displays only 8 digits.[§]

$$X = \begin{bmatrix} a & b & c \\ 0 & a & b \\ 0 & 0 & a \end{bmatrix} \qquad Y = \begin{bmatrix} f & 0 & 0 \\ e & f & 0 \\ d & e & f \end{bmatrix}$$

What is the (1, 2) entry of the product XY?

[§]R. P. Boas, "Multiplying long numbers," *Mathematics Magazine*, June 1989, pp. 173–174.

Cumulative Exercise Set 2–2

Problems 1 to 6 refer to the following matrices.

$$A = \begin{bmatrix} 1 & -1 \\ -3 & 0 \end{bmatrix} \qquad B = \begin{bmatrix} 4 & 0 \\ -1 & 6 \\ 5 & 7 \end{bmatrix}$$

$$C = \begin{bmatrix} -7 & 6 \\ 3 & -5 \\ 0 & 0 \end{bmatrix} \qquad D = \begin{bmatrix} 5 & 0 \\ 0 & -2 \end{bmatrix}$$

1. (a) What is the size of B?
   (b) Find b_{21}.

2. What notation describes the entry -1 in A?

3. Perform, where possible, the following matrix operations:
   (a) $A + B$ (b) $B + C$ (c) $3B - C$

4. Perform, where possible, the following matrix operations:
   (a) AB (b) BA (c) $3BA$

5. Compute the entry m_{22} where $M = CA$.

6. Is is true that $AD = DA$?

7. Find values of x and y, if possible, such that $x \neq 0$, $y \neq 0$, and

$$x\begin{bmatrix} 1 \\ 1 \end{bmatrix} + y\begin{bmatrix} 2 \\ -1 \end{bmatrix} = \begin{bmatrix} 0 \\ 0 \end{bmatrix}$$

8. Find the coefficient matrix U, the variable matrix X, and the constant matrix V for the following system of equations.

$$-x - 2y + 2z = 4$$
$$2x + 3y - 4z = 7$$

2–3 The Gauss–Jordan Method

The word *echelon* in French means "a rung of a ladder." In English, echelon refers to a formation whose units are arranged in steplike fashion, like the formation of airplanes at the top of page 84.

The mathematical use of the word describes a matrix in a form resembling an echelon formation. In this section we describe how the echelon form of a matrix is used to solve systems of linear equations. The procedure is called the Gauss–Jordan

method because it is a refinement of Gaussian elimination due to Jordan.* We present this method for three reasons: (1) the Gauss–Jordan method is somewhat easier than Gaussian elimination when solving a system by hand, (2) it is indispensable for computing the inverse of a matrix in Sections 2–4 and 2–5, and (3) it forms an integral part of the simplex method (Chapter 4).

Elementary Operations

In solving a system of linear equations, the coefficients are the essential ingredient and matrices are used as a device for keeping the coefficients aligned. By definition the **coefficient matrix** is the matrix of coefficients of the variables, and the **augmented matrix** consists of the coefficient matrix and the column of constants. The notation (R_) is used in place of (E_) because the operations are performed on rows instead of equations.

The permissible operations on the augmented matrix of a system of linear equations are called **elementary row operations.** They correspond to the permissible operations on the system of equations.

Elementary Row Operations

1. Interchange two rows.
2. Multiply each number in a row by a nonzero number.
3. Replace a row by the sum of that row and a multiple of another row.

---

*Traditionally this refinement has been ascribed to the French mathematician Camille Jordan (1838–1922), but recent evidence suggests that the refinement is due to the German geodesist Wilhelm Jordan (1842–1899) instead. Renate McLaughlin presented the discovery (due to Steven Althoen and Renate McLaughlin) in a paper titled "Will the Real Jordan Please Stand Up?" at the International Congress of Mathematicians at Berkeley in 1986. Their results were published in the February 1987 issue of the *American Mathematical Monthly,* pp. 130–142.

We describe how to use the Gauss–Jordan method to solve a system of equations that was solved by Gaussian elimination in Section 1–4.

$$x + y + z = 10 \qquad \text{(E1)}$$
$$x - y \quad\;\; = 2 \qquad \text{(E2)}$$
$$y - z = 1 \qquad \text{(E3)}$$

First form the augmented matrix. Write each variable atop the column of coefficients that it refers to.

$$
\begin{array}{ccc}
x & y & z
\end{array}
$$
$$
\left[\begin{array}{ccc|c}
1 & 1 & 1 & 10 \\
1 & -1 & 0 & 2 \\
0 & 1 & -1 & 1
\end{array}\right]
\begin{array}{l}
\text{(R1)} \\
\text{(R2)} \\
\text{(R3)}
\end{array}
$$

The Gauss–Jordan method moves from the first column to the second column to the third column. For each column the aim is to produce a 1 as the first nonzero number of a row and to produce 0's above and below that 1. For the preceding system of equations we begin with the first column in the same way that we proceeded with Gaussian elimination.

$$
\left[\begin{array}{ccc|c}
1 & 1 & 1 & 10 \\
0 & -2 & -1 & -8 \\
0 & 1 & -1 & 1
\end{array}\right]
\qquad (-1)(\text{R1}) + (\text{R2})
$$

Move to the second column to produce a 1 in the second row and 0's elsewhere. The 1 can be obtained by interchanging R2 and R3. Do *not* interchange R2 and R1 because the first column would no longer be in proper form.

$$
\left[\begin{array}{ccc|c}
1 & 1 & 1 & 10 \\
0 & 1 & -1 & 1 \\
0 & -2 & -1 & -8
\end{array}\right]
\qquad (\text{R2}) \leftrightarrow (\text{R3})
$$

$$
\left[\begin{array}{ccc|c}
1 & 0 & 2 & 9 \\
0 & 1 & -1 & 1 \\
0 & 0 & -3 & -6
\end{array}\right]
\begin{array}{l}
(-1)(\text{R2}) + (\text{R1}) \\[4pt]
2(\text{R2}) + (\text{R3})
\end{array}
$$

The final step is to transform the third column so that its third entry is 1 and the other entries are 0.

$$
\begin{array}{ccc}
x & y & z
\end{array}
$$
$$
\left[\begin{array}{ccc|c}
1 & 0 & 2 & 9 \\
0 & 1 & -1 & 1 \\
0 & 0 & 1 & 2
\end{array}\right]
\qquad (-\tfrac{1}{3})(\text{R3})
$$

$$
\left[\begin{array}{ccc|c}
1 & 0 & 0 & 5 \\
0 & 1 & 0 & 3 \\
0 & 0 & 1 & 2
\end{array}\right]
\begin{array}{l}
(-2)(\text{R3}) + (\text{R1}) \\
(\text{R3}) + (\text{R2})
\end{array}
$$

Now the augmented matrix is in a form that yields the solution by inspection: $x = 5$, $y = 3$, $z = 2$.

EXAMPLE 1

Problem

Solve the following system of equations by the Gauss–Jordan method.

$$x + 2y + 3z = 1$$
$$2x + 5y + 7z = 2$$
$$3x + 5y + 7z = 4$$

Solution Form the augmented matrix of the system

$$\begin{array}{ccc} x & y & z \end{array}$$
$$\left[\begin{array}{ccc|c} 1 & 2 & 3 & 1 \\ 2 & 5 & 7 & 2 \\ 3 & 5 & 7 & 4 \end{array}\right]$$

Convert the first column to the desired form.

$$\left[\begin{array}{ccc|c} 1 & 2 & 3 & 1 \\ 0 & 1 & 1 & 0 \\ 0 & -1 & -2 & 1 \end{array}\right] \quad \begin{array}{l} (-2)(R1) + (R2) \\ (-3)(R1) + (R3) \end{array}$$

The desired second column is obtained as follows:

$$\left[\begin{array}{ccc|c} 1 & 0 & 1 & 1 \\ 0 & 1 & 1 & 0 \\ 0 & 0 & -1 & 1 \end{array}\right] \quad \begin{array}{l} (-2)(R2) + (R1) \\ \\ (R2) + (R3) \end{array}$$

Next produce a 1 in the third row, third column.

$$\left[\begin{array}{ccc|c} 1 & 0 & 1 & 1 \\ 0 & 1 & 1 & 0 \\ 0 & 0 & 1 & -1 \end{array}\right] \quad (-1)(R3)$$

Convert the third column to the desired form.

$$\begin{array}{ccc} x & y & z \end{array}$$
$$\left[\begin{array}{ccc|c} 1 & 0 & 0 & 2 \\ 0 & 1 & 0 & 1 \\ 0 & 0 & 1 & -1 \end{array}\right] \quad \begin{array}{l} (-1)(R3) + (R1) \\ (-1)(R3) + (R2) \end{array}$$

The solution can be read directly.

$$x = 2 \qquad y = 1 \qquad z = -1$$

The next example shows what can happen when a column cannot be converted to the desired form. It introduces the general solution of a system that has more than one solution and shows how to obtain particular solutions.

E X A M P L E 2

Problem

Solve this system of linear equations by the Gauss–Jordan method.

$$\begin{aligned} x - 3y &= 4 \\ -3x + 8y - z &= -10 \\ 2x - 7y - z &= 10 \end{aligned}$$

Solution Form the augmented matrix.

$$\left[\begin{array}{ccc|c} 1 & -3 & 0 & 4 \\ -3 & 8 & -1 & -10 \\ 2 & -7 & -1 & 10 \end{array}\right]$$

Convert the first column to the desired form.

$$\left[\begin{array}{ccc|c} 1 & -3 & 0 & 4 \\ 0 & -1 & -1 & 2 \\ 0 & -1 & -1 & 2 \end{array}\right] \quad \begin{aligned} &3(R1) + (R2) \\ &(-2)(R1) + (R3) \end{aligned}$$

Put the second column in proper form.

$$\left[\begin{array}{ccc|c} 1 & 0 & 3 & -2 \\ 0 & 1 & 1 & -2 \\ 0 & 0 & 0 & 0 \end{array}\right] \quad \begin{aligned} &-3(R2) + (R1) \\ &(-1)(R2) \\ &(-1)(R2) + (R3) \end{aligned}$$

Notice that the third column cannot be converted to the desired form. The corresponding system of equations is

$$\begin{aligned} x + 3z &= -2 \\ y + z &= -2 \end{aligned}$$

Since this system of equations has infinitely many solutions, the original system does too. The **general solution** is written in the form

$$\begin{aligned} x &= -2 - 3z \\ y &= -2 - z \\ z &= \text{any number} \end{aligned}$$

A **particular solution** is obtained from the general solution by choosing a particular number for z and then substituting to find values of x and y. Here one particular solution has $z = 0$, in which case we get $x = -2$, $y = -2$, and $z = 0$. This solution is written $(-2, -2, 0)$. Another particular solution, which has $z = 1$, is $(-5, -3, 1)$.

The remainder of this section shows how to form the general solution, and some particular solutions, to systems of equations like the one in Example 2.

Echelon Form

A matrix is said to be in **echelon form** if it satisfies these conditions:

1. If a row has a nonzero number, then the leftmost nonzero number in the row is 1. It is called the **leading 1.**
2. Rows with all zeros lie beneath the rows having a nonzero number.
3. When two rows have a leading 1, the leading 1 in the upper row is to the left of the leading 1 in the lower row.
4. Each column that contains a leading 1 has zeros everywhere else.

Echelon form resembles the airplane formation that began this section, with the leading 1's being the airplanes. The reduced matrices in Examples 1 and 2 are in echelon form.

$$
\begin{bmatrix}
1 & 0 & 0 & \bigm| & 2 \\
0 & 1 & 0 & \bigm| & 1 \\
0 & 0 & 1 & \bigm| & -1
\end{bmatrix}
\qquad
\begin{bmatrix}
1 & 0 & 3 & \bigm| & -2 \\
0 & 1 & 1 & \bigm| & -2 \\
0 & 0 & 0 & \bigm| & 0
\end{bmatrix}
$$

A variable is called a **free variable** if it can be any number and a **basic variable** if it depends on a free variable. Basic variables correspond to columns with leading 1's in echelon form. In Example 2, x and y are basic variables and z is a free variable.

According to property 4 of the definition of echelon form, a column with a leading 1 has 0's in every other entry of the column. The term "basic" is used because it is straightforward to determine a solution by assigning to each basic variable the value in the last column of its row. Example 3 illustrates how to do this.

EXAMPLE 3

Let the augmented matrix of a system of linear equations be

$$
\begin{array}{cccc}
x & y & u & v
\end{array}
$$
$$
M = \begin{bmatrix}
1 & 0 & 1 & 2 & \bigm| & 1 \\
0 & 1 & 0 & 3 & \bigm| & 0
\end{bmatrix}
$$

Problem

Determine which variables are free and which are basic. Derive the general solution to the system of equations and find two particular solutions.

Solution The matrix M is already in echelon form, so x and y are basic variables while u and v are free. The system of equations is

$$
\begin{aligned}
x + \quad u + 2v &= 1 \\
y + \quad\quad 3v &= 0
\end{aligned}
$$

Solving them for the basic variables gives $x = 1 - u - 2v$ and $y = -3v$. The general solution is written

$$x = 1 - u - 2v$$

$$y = -3v$$

$$u = \text{any number}$$

$$v = \text{any number}$$

Particular solutions are obtained by choosing specific numbers for u and v. Two particular solutions are

$$x = 1 \qquad y = 0 \qquad u = 0 \qquad v = 0$$

and

$$x = 0 \qquad y = 0 \qquad u = 1 \qquad v = 0$$

EXAMPLE 4

Problem

Solve the system of linear equations

$$\begin{aligned} x + 4y - 5z &= 2 \\ 3x + 12y - 11z &= 2 \\ x + 4y - 3z &= 0 \end{aligned}$$

Solution Form the augmented matrix

$$\begin{array}{ccc} x & y & z \end{array}$$
$$\left[\begin{array}{ccc|c} 1 & 4 & -5 & 2 \\ 3 & 12 & -11 & 2 \\ 1 & 4 & -3 & 0 \end{array}\right]$$

Then

$$\left[\begin{array}{ccc|c} 1 & 4 & -5 & 2 \\ 0 & 0 & 4 & -4 \\ 0 & 0 & 2 & -2 \end{array}\right] \quad \begin{array}{l} (-3)(\text{R1}) + (\text{R2}) \\ (-1)(\text{R1}) + (\text{R3}) \end{array}$$

$$\left[\begin{array}{ccc|c} 1 & 4 & -5 & 2 \\ 0 & 0 & 1 & -1 \\ 0 & 0 & 1 & -1 \end{array}\right] \quad \begin{array}{l} (\frac{1}{4})(\text{R2}) \\ (\frac{1}{2})(\text{R3}) \end{array}$$

$$\left[\begin{array}{ccc|c} 1 & 4 & 0 & -3 \\ 0 & 0 & 1 & -1 \\ 0 & 0 & 0 & 0 \end{array}\right] \quad \begin{array}{l} 5(\text{R2}) + (\text{R1}) \\ \\ (\text{R3}) - (\text{R2}) \end{array}$$

This matrix is in echelon form. The variable y is free; x and z are basic. The top row gives $x + 4y = -3$ and the second row gives $z = -1$, so the general solution is

$$x = -3 - 4y$$

$$y = \text{any number}$$

$$z = -1$$

For $y = 0$ a particular solution is $(-3, 0, -1)$, and for $y = 1$ a particular solution is $(-7, 1, -1)$.

The discussion at the conclusion of Section 1–4 used the intersection of planes to show that a system of three equations in three variables will have exactly one solution, infinitely many solutions, or no solutions. Example 5 illustrates one such possibility.

EXAMPLE 5

Problem

Solve the system of linear equations

$$
\begin{array}{rcrcrcr}
x & - & 2y & + & z & = & 4 \\
3x & + & 4y & - & 2z & = & 2 \\
2x & + & 6y & - & 3z & = & -4
\end{array}
$$

Solution

$$
\begin{bmatrix}
1 & -2 & 1 & \bigm| & 4 \\
3 & 4 & -2 & \bigm| & 2 \\
2 & 6 & -3 & \bigm| & -4
\end{bmatrix}
$$

$$
\begin{bmatrix}
1 & -2 & 1 & \bigm| & 4 \\
0 & 10 & -5 & \bigm| & -10 \\
0 & 10 & -5 & \bigm| & -12
\end{bmatrix}
$$

$$
\begin{bmatrix}
1 & -2 & 1 & \bigm| & 4 \\
0 & 1 & -\frac{1}{2} & \bigm| & -1 \\
0 & 0 & 0 & \bigm| & 2
\end{bmatrix}
$$

We stop the Gauss–Jordan method at this point because the equation corresponding to the bottom row is $0 = 2$, which is impossible. Therefore the system has no solution. Geometrically, this means that the three planes do not intersect because at least two of them are parallel.

EXAMPLE 6

A hospital nutritionist prepares a menu for patients on a low-fat, high-protein diet. The primary ingredients are chicken and rice. The meal must contain exactly 6 grams (g) of fat, 16.5 milligrams (mg) of iron, and 550 calories (cal). Each 3-ounce serving of chicken contains 4 g of fat, 9 mg of iron, and 500 cal. Each 1-cup serving of rice contains 2 g of fat, 6 mg of iron, and 150 cal.

Problem

How many servings of chicken and rice should be included in the meal?

Solution Let x be the number of servings of chicken and y be the number of servings of rice. Then

$$
\begin{aligned}
4x + 2y &= 6 &&(\text{fat}) \\
9x + 6y &= 16.5 &&(\text{iron}) \\
500x + 150y &= 550 &&(\text{calories})
\end{aligned}
$$

The augmented matrix is

$$
\left[\begin{array}{cc|c}
4 & 2 & 6 \\
9 & 6 & 16.5 \\
500 & 150 & 550
\end{array}\right]
$$

The steps in the reduction to echelon form are

$$
\left[\begin{array}{cc|c}
1 & 0.5 & 1.5 \\
9 & 6 & 16.5 \\
10 & 3 & 11
\end{array}\right]
\qquad
\begin{aligned}
&(\tfrac{1}{4})(\text{R1}) \\
&\\
&(\tfrac{1}{50})(\text{R3})
\end{aligned}
$$

$$
\left[\begin{array}{cc|c}
1 & 0.5 & 1.5 \\
0 & 1.5 & 3 \\
0 & -2 & -4
\end{array}\right]
\qquad
\begin{aligned}
&(-9)(\text{R1}) + (\text{R2}) \\
&(-10)(\text{R1}) + (\text{R3})
\end{aligned}
$$

$$
\left[\begin{array}{cc|c}
1 & 0.5 & 1.5 \\
0 & 1 & 2 \\
0 & 1 & 2
\end{array}\right]
\qquad
\begin{aligned}
&(\tfrac{2}{3})(\text{R2}) \\
&(-\tfrac{1}{2})(\text{R3})
\end{aligned}
$$

$$
\left[\begin{array}{cc|c}
1 & 0 & 0.5 \\
0 & 1 & 2 \\
0 & 0 & 0
\end{array}\right]
\qquad
\begin{aligned}
&(-0.5)(\text{R2}) + (\text{R1}) \\
&\\
&(\text{R2}) - (\text{R3})
\end{aligned}
$$

Then $x = 0.5$ and $y = 2$. The meal should consist of a half serving of chicken and two servings of rice.

EXERCISE SET 2–3

In Problems 1 to 6 use the Gauss–Jordan method to find the unique solution. Make sure to check your answers.

1. $\begin{aligned}
x + y - z &= 2 \\
2x + 3y - z &= 6 \\
3x + 2y - z &= 7
\end{aligned}$

2. $\begin{aligned}
x - 2y - z &= -3 \\
2x - 3y &= -2 \\
-x + y + z &= 1
\end{aligned}$

3. $\begin{aligned}
2x - y + 3z &= 5 \\
x - y - z &= 2 \\
-3x - 2z &= -9
\end{aligned}$

4. $\begin{aligned}
2x + 4y - 2z &= -2 \\
3x + 5y - z &= 3 \\
-2x + 3y - z &= -7
\end{aligned}$

5. $\begin{aligned}
3x - 6y + 9z &= -24 \\
4x + 2y &= -2 \\
-x + y - 3z &= 5
\end{aligned}$

6. $\begin{aligned}
-x + y + 2z &= 5 \\
x + y + z &= 3 \\
2x + 3y + z &= 5
\end{aligned}$

In Problems 7 to 10 the solution of the corresponding system with variables x, y, and z can be determined directly from the given augmented matrix. Find the solution by thinking, not writing.

7. $\begin{bmatrix} 1 & 0 & 0 & | & 3 \\ 0 & 1 & 0 & | & -1 \\ 0 & 0 & 1 & | & 6 \end{bmatrix}$

8. $\begin{bmatrix} 1 & 0 & 0 & | & -2 \\ 0 & 0 & 1 & | & 3 \\ 0 & 1 & 0 & | & 0 \end{bmatrix}$

9. $\begin{bmatrix} 1 & 0 & 0 & | & 6 \\ 0 & 0 & 1 & | & 4 \\ 0 & 0 & 0 & | & 0 \end{bmatrix}$

10. $\begin{bmatrix} 1 & 0 & -2 & | & 1 \\ 0 & 1 & 17 & | & 2 \\ 0 & 0 & 1 & | & 0 \end{bmatrix}$

In Problems 11 to 14 find the general solution to the corresponding system whose variables are x, y, and z. List two particular solutions. Note that Problem 11 completes Example 2.

11. $\begin{bmatrix} 1 & 0 & 3 & | & -2 \\ 0 & 1 & 1 & | & -2 \\ 0 & 0 & 0 & | & 0 \end{bmatrix}$

12. $\begin{bmatrix} 1 & 2 & 0 & | & 5 \\ 0 & 0 & 1 & | & 4 \\ 0 & 0 & 0 & | & 0 \end{bmatrix}$

13. $\begin{bmatrix} 1 & 0 & 0 & | & 4 \\ 0 & 1 & -1 & | & 2 \\ 0 & 0 & 0 & | & 0 \end{bmatrix}$

14. $\begin{bmatrix} 1 & 2 & 3 & | & 12 \\ 0 & 0 & 0 & | & 0 \\ 0 & 0 & 0 & | & 0 \end{bmatrix}$

In Problems 15 to 18 find the general solution to the corresponding system whose variables are x, y, u, and v. Determine which variables are basic and which are free.

15. $\begin{bmatrix} 1 & 1 & 0 & 2 & | & 1 \\ 0 & 0 & 1 & 3 & | & 0 \end{bmatrix}$

16. $\begin{bmatrix} 1 & 1 & 0 & 0 & | & 4 \\ 0 & 0 & 1 & 0 & | & 3 \\ 0 & 0 & 0 & 1 & | & 2 \end{bmatrix}$

17. $\begin{bmatrix} 1 & 0 & 0 & 1 & | & 0 \\ 0 & 1 & 0 & 2 & | & 5 \\ 0 & 0 & 1 & 3 & | & 4 \end{bmatrix}$

18. $\begin{bmatrix} 1 & 2 & 3 & 0 & | & 0 \\ 0 & 0 & 0 & 1 & | & 5 \end{bmatrix}$

In Problems 19 to 22 determine whether each system has either no solution or infinitely many solutions. If there are infinitely many solutions, derive the general form and list the basic variables.

19. $\begin{aligned} x - y + 2z &= 4 \\ 2x - y - 2z &= -1 \\ 8x - 6y + 4z &= 14 \end{aligned}$

20. $\begin{aligned} x - y + 2z &= 4 \\ 2x - y - 2z &= -1 \\ 8x - 6y + 4z &= 0 \end{aligned}$

21. $\begin{aligned} -x - 2y + 2z &= 4 \\ 2x + 3y - 4z &= 7 \\ -x - 3y + 2z &= 19 \end{aligned}$

22. $\begin{aligned} -x - 2y + 2z &= 4 \\ 2x + 3y - 4z &= 7 \\ -x - 3y + 2z &= 0 \end{aligned}$

23. A hospital nutritionist prepares a menu of chicken and rice for patients on a low-fat, high-protein diet. It must contain exactly 12 grams (g) of fat, 9 milligrams (mg) of iron, and 1100 calories (cal). Each 3-ounce serving of chicken contains 3 g of fat, 2 mg of iron, and 200 cal. Each 1-cup serving of rice contains 6 g of fat, 5 mg of iron, and 700 cal. How many servings of chicken and rice should be included in the menu?

24. A hospital nutritionist prepares a menu of chicken and rice for patients on a low-fat, high-protein diet. It must contain exactly 16 g of fat, 44 mg of iron, and 1000 cal. Each 3-ounce serving of chicken contains 2 g of fat, 5 mg of iron, and 100 cal. Each 1-cup serving of rice contains 6 g of fat, 17 mg of iron, and 400 cal. How many servings of chicken and rice should be included in the menu?

Solve the systems of equations in Problems 25 to 34 by the Gauss–Jordan method. If there are infinitely many solutions, find the general form and two particular solutions.

25. $\begin{aligned} 2x + y - 2z &= 10 \\ 3x + 2y + 2z &= 1 \\ 5x + 4y + 3z &= 4 \end{aligned}$

26. $\begin{aligned} 2x + y - 2z &= 0 \\ 3x + 2y + 2z &= 0 \\ 5x + 4y + 3z &= 0 \end{aligned}$

27. $\begin{aligned} x - 3y + 5z &= -17 \\ 3x - 8y + 13z &= -47 \\ -2x + 7y - 12z &= 38 \end{aligned}$

28. $\begin{aligned} 2x + 5y - z &= 8 \\ x + 2y - z &= 3 \\ -3x - 5y + 4z &= -7 \end{aligned}$

29. $\begin{aligned} 2x + y - 3z &= 5 \\ 3x - 2y + 2z &= 5 \\ 5x - 3y - z &= 16 \end{aligned}$

30. $\begin{aligned} 3x - 3y - 5z &= 9 \\ x - y - 2z &= 4 \\ -2x + 2y + z &= 1 \end{aligned}$

31. $\begin{aligned} 4x - 3y + 8z &= 8 \\ x - y + 2z &= 1 \\ 4x - 5y + 8z &= 0 \end{aligned}$

32. $\begin{aligned} 2x + 4y + 6z &= 2 \\ x - 2y - 5z &= 4 \\ x + 2y + 3z &= 0 \end{aligned}$

33. $3x - 6y - 9z = 12$
     $x + 2y - 5z = 4$
     $x - 2y - 3z = 0$

34. $-2x - 3y + 11z = -8$
     $-x - 4y + 18z = -4$
     $x + 2y - 8z = 4$

35. A hospital nutritionist prepares a menu of chicken and rice for patients on a low-fat, high-protein diet. It must contain exactly 5 g of fat, 11 mg of iron, and 1400 cal. Each 3-ounce serving of chicken contains 1 g of fat, 2 mg of iron, and 200 cal. Each 1-cup serving of rice contains 3 g of fat, 7 mg of iron, and 1000 cal. How many servings of chicken and rice should be included in the menu?

36. A hospital nutritionist prepares a menu of chicken and rice for patients on a low-fat, high-protein diet. It must contain exactly 7 g of fat, 30 mg of iron, and 1500 cal. Each 3-ounce serving of chicken contains 1 g of fat, 4 mg of iron, and 100 cal. Each 1-cup serving of rice contains 2 g of fat, 9 mg of iron, and 600 cal. How many servings of chicken and rice should be included in the menu?

37. A computer company makes three kinds of disk drives: hard, 1.2MB floppy, and 360K floppy. Each drive must be manufactured and tested. The hard drives require 3 hours of manufacturing and 1 hour of testing, 1.2MB drives require 6 hours of manufacturing and 2 hours of testing, and the 360K drives require 7 hours of manufacturing and 2 hours of testing. Manufacturing is possible exactly 67 hours per week and testing 20 hours per week. How many drives can be made per week?

38. An automobile plant makes cars in three price ranges: low (L), medium (M), and high (H). L-cars require 1 minute on the assembly line and 2 minutes of finishing; M-cars require 2 minutes on the assembly line and 3 minutes of finishing, and H-cars require 3 minutes on the assembly line and 8 minutes of finishing. The assembly line is available for exactly 63 minutes and finishing is available for 130 minutes. How many cars in each price range can be made?

In Problems 39 to 42 match the graphs to the system of equations. Tell from the graph whether the system has a solution.

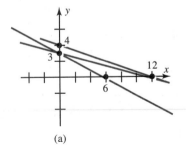

(a)

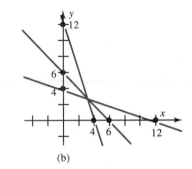

(b)

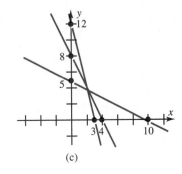

(c)

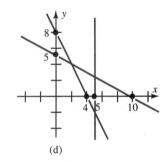

(d)

39. $x + y = 6$
     $x + 3y = 12$
     $3x + y = 12$

40. $x + 2y = 6$
     $x + 3y = 12$
     $x + 4y = 12$

41. $2x + y = 8$
     $x + 2y = 10$
     $x \phantom{{}+ 2y} = 5$

42. $2x + y = 8$
     $x + 2y = 10$
     $4x + y = 12$

43. Let a, b, c, and d be real numbers with $ad - bc \neq 0$. Reduce this matrix to echelon form.

$$\begin{bmatrix} a & b & 1 & 0 \\ c & d & 0 & 1 \end{bmatrix}$$

44. In Problem 37, what is the largest number of drives that can be made per week? What is the smallest number? (Assume that fractional parts of a drive cannot be made.)

Referenced Exercise Set 2–3

1. Cost accounting is concerned with determining the cost of each department in a firm so that the firm can carry out a policy of setting prices, hiring or firing personnel, adding or eliminating departments, and preparing external reports.* Consider a firm with four departments: A = accounting, M = maintenance, P = production, and F = finishing. Find the total costs x, y, z, w for each department, respectively, where

   $x = 0.1y + 9{,}000$
   $y = 0.2x \phantom{{}+ 0.1y} + 6{,}000$
   $z = 0.4x + 0.4y + 20{,}000$
   $w = 0.4x + 0.5y + 16{,}000$

2. If a bettor in a pari-mutuel wagering system (such as horse racing or dog racing) seeks to win \$$w$ by betting

on five horses whose odds of winning are $a{:}1$, $b{:}1$, $c{:}1$, $d{:}1$, and $e{:}1$, then the amounts of money that should be bet on each horse are solutions to this system of equations.†

$$
\begin{aligned}
ax - y - z - r - s &= w \\
-x + by - z - r - s &= w \\
-x - y + cz - r - s &= w \\
-x - y - z + dr - s &= w \\
-x - y - z - r + es &= w
\end{aligned}
$$

If the odds on five particular horses are $2{:}1$, $4{:}1$, $7{:}1$, $10{:}1$, and $12{:}1$, how much money should be bet on each horse in order to win \$100?

Cumulative Exercise Set 2–3

Problems 1 to 3 refer to matrices R, S, and T.

$$R = \begin{bmatrix} 2 & 1 & 4 \\ 1 & 6 & 2 \\ 1 & 3 & 1 \\ -3 & 1 & -2 \end{bmatrix}$$

$$S = \begin{bmatrix} 1 & 3 & -1 \\ 0 & 1 & 3 \\ 2 & 0 & -2 \\ 0 & 0 & 0 \end{bmatrix}$$

$$T = \begin{bmatrix} 0 & 3 & 7 \\ 0 & 1 & 0 \\ 0 & 2 & 2 \\ 1 & 5 & 5 \end{bmatrix}$$

1. Compute (a) $R + S$ and (b) $R + S + T$.

2. Compute (a) $2R - 3S$ and (b) $2R - 3(S + T)$.

3. Find values of x, y, and z so that the equality $U = R$ holds for the matrix

$$U = \begin{bmatrix} x & 1 & 4 \\ 1 & 3y & 2 \\ 1 & 3 & 1+z \\ -3 & 1 & -2 \end{bmatrix}$$

*Adapted from the article by Roland Minch and Enrico Petri, "Matrix Models of Reciprocal Service Cost Allocation," *The Accounting Review*, July 1972, pp. 576–580.

†The terminology and appropriate reference were given in Problem 1 in Referenced Exercise Set 1–4.

4. Find the product if possible: (a) AB (b) BC.

$$A = \begin{bmatrix} 0 & 2 & 3 \\ 1 & 2 & 1 \\ 1 & 3 & 4 \\ -1 & 1 & -2 \end{bmatrix} \quad B = \begin{bmatrix} 1 & 3 \\ 0 & 1 \\ 2 & 0 \end{bmatrix}$$

$$C = \begin{bmatrix} 2 & 3 & 7 \\ 3 & 1 & 0 \\ 1 & 0 & 2 \\ 1 & 2 & -1 \end{bmatrix}$$

5. Compute ABC.

$$A = \begin{bmatrix} 2 & -2 \\ 0 & 3 \end{bmatrix} \quad B = \begin{bmatrix} 4 & 1 \\ 1 & 3 \end{bmatrix}$$

$$C = \begin{bmatrix} -3 & 1 \\ 1 & 2 \end{bmatrix}$$

6. A firm manufactures three types of coats, small, medium, and large. Each small coat takes 2 hours to cut and 3 hours to assemble. Each medium coat takes 2 hours to cut and 2 hours to assemble. Each large coat takes 4 hours to cut and 1 hour to assemble. In one day the firm produces 1000 small coats, 1500 medium coats, and 2000 large coats. Use matrix multiplication to determine how much time must be allotted for cutting and for assembling to produce these coats.

7. Use the Gauss–Jordan method to find the unique solution.

$$5x + 2y + z = 11$$
$$2x + y + z = 6$$
$$x + 2y + z = 3$$

In Problems 8 and 9 the variables are x, y, and z.

8. Find the solution by thinking, not writing.

$$\begin{bmatrix} 1 & 0 & 0 & | & 2 \\ 0 & 1 & 0 & | & 5 \\ 0 & 0 & 1 & | & -3 \end{bmatrix}$$

9. Find the general solution to the system.

$$\begin{bmatrix} 1 & 0 & 1 & | & 2 \\ 0 & 1 & 2 & | & 3 \\ 0 & 0 & 0 & | & 0 \end{bmatrix}$$

10. Find the general solution to the system whose variables are x, y, u, and v.

$$\begin{bmatrix} 1 & 0 & 1 & 2 & 0 & | & 1 \\ 0 & 1 & 0 & 3 & 0 & | & 0 \\ 0 & 0 & 0 & 0 & 1 & | & 3 \end{bmatrix}$$

11. Solve the system of equations.

$$x - 2y + 2z = 1$$
$$2x + y - 3z = 6$$
$$4x - 3y + z = 8$$

In Problems 12 and 13 express the word problem as a system of equations and solve the system by the Gauss–Jordan method.

12. A firm manufactures three types of coats, small, medium, and large. Each small coat takes 2 hours to cut, 1 hour to trim, and 3 hours to assemble. Each medium coat takes 2 hours to cut, 1 hour to trim, and 2 hours to assemble. Each large coat takes 4 hours to cut, 1 hour to trim, and 1 hour to assemble. If the firm uses 1200 hours to cut, 400 hours to trim, and 700 hours to assemble, how many coats of each size can be made?

13. A hospital nutritionist prepares a menu of meat, corn, and rice. It must contain exactly 8 grams (g) of fat, 25 milligrams (mg) of iron, and 1700 calories (cal). Each 3-ounce serving of meat contains 1 g of fat, 4 mg of iron, and 800 cal. Each 1-cup serving of corn contains 3 g of fat, 10 mg of iron, and 400 cal. Each 1-cup serving of rice contains 1 g of fat, 1 mg of iron, and 100 cal. How many servings of each food should be contained in the menu?

2–4 The Inverse

If you drive from Los Angeles to San Francisco and then return by retracing your route, the return trip is called the inverse route. More generally, any time you perform a procedure and then reverse it in such a way that you retrace the steps and end where you had begun, the reverse procedure is called the inverse. This section introduces the inverse of a matrix and describes a procedure for computing it. *All matrices in this section will be square*.

Invertible Matrices

Consider these 3×3 matrices.

$$A = \begin{bmatrix} 1 & 0 & 8 \\ 1 & 2 & 3 \\ 2 & 5 & 3 \end{bmatrix} \qquad B = \begin{bmatrix} 9 & -40 & 16 \\ -3 & 13 & -5 \\ -1 & 5 & -2 \end{bmatrix}$$

They satisfy $AB = I_3$ and $BA = I_3$, so B is called the *inverse* of A. Also, A is the inverse of B. In short, A and B are inverses of each other. The display contains the basic definitions.

DEFINITION

An $n \times n$ matrix A is **invertible** (or **nonsingular**) if there is an $n \times n$ matrix B such that

$$AB = I_n \qquad \text{and} \qquad BA = I_n$$

B is called the **inverse** of A and is denoted by A^{-1} (read "A inverse"). Thus

$$AA^{-1} = A^{-1}A = I_n$$

If A does not have an inverse, then A is called **noninvertible** (or **singular**).

An invertible matrix has only one inverse. Equivalently, *if a matrix has an inverse, then the inverse is unique*. The proof of this assertion is left to the exercises. The preceding 3×3 matrix A is invertible and its inverse is $A^{-1} = B$.

It is not usually easy to pick out which matrix is the inverse of another, even when you know the inverse exists. Example 1 illustrates this point. It also reinforces the definition of the inverse of a matrix.

EXAMPLE 1

Consider the matrix

$$A = \begin{bmatrix} 1 & 2 \\ 3 & 4 \end{bmatrix}$$

Problem

Which of these matrices is the inverse of A?

$$R = \begin{bmatrix} 4 & -2 \\ -3 & 1 \end{bmatrix} \qquad S = \begin{bmatrix} 1 & -2 \\ 0 & 1 \end{bmatrix} \qquad T = \begin{bmatrix} -2 & 1 \\ \frac{3}{2} & -\frac{1}{2} \end{bmatrix}$$

Solution Form the matrix product AR.

$$AR = \begin{bmatrix} -2 & 0 \\ 0 & -2 \end{bmatrix}$$

Thus R is not the inverse of A.

The initial computations for AS seem to indicate that S is the inverse of A because

$$AS = \begin{bmatrix} 1 & 0 \\ * & * \end{bmatrix}$$

However, the $(2, 1)$ entry is equal to 3, not 0, so S is not the inverse of A.

The matrix T is the inverse of A because $AT = I$ and $TA = I$.

In Example 1 two equalities were needed to verify that T is the inverse of A: $AT = I$ and $TA = I$. However, only one of these equalities needs to hold because if X and Y are $n \times n$ matrices with $XY = I_n$, then it follows that $YX = I_n$. The proof of this fact is beyond the scope of the book.

Gauss–Jordan Elimination

The definition of the inverse of a matrix does not indicate how to construct it. The procedure can be seen by examining a specific matrix A.

$$A = \begin{bmatrix} 1 & 0 & 8 \\ 1 & 2 & 3 \\ 2 & 5 & 3 \end{bmatrix}$$

Let the inverse of A be

$$A^{-1} = \begin{bmatrix} a & b & c \\ d & e & f \\ g & h & i \end{bmatrix}$$

By definition $AA^{-1} = I$, so

$$\begin{bmatrix} 1 & 0 & 8 \\ 1 & 2 & 3 \\ 2 & 5 & 3 \end{bmatrix} \begin{bmatrix} a & b & c \\ d & e & f \\ g & h & i \end{bmatrix} = \begin{bmatrix} 1 & 0 & 0 \\ 0 & 1 & 0 \\ 0 & 0 & 1 \end{bmatrix}$$

This equality leads to three systems of three linear equations in three variables.

$$
\begin{array}{lll}
a \qquad\quad + 8g = 1 & b \qquad\quad + 8h = 0 & c \qquad\quad + 8i = 0 \\
a + 2d + 3g = 0 & b + 2e + 3h = 1 & c + 2f + 3i = 0 \\
2a + 5d + 3g = 0 & 2b + 5e + 3h = 0 & 2c + 5f + 3i = 1
\end{array}
$$

Since all three systems have the same coefficient matrix, they can be solved simultaneously by augmenting the coefficient matrix with the identity matrix I_3 to form a 3×6 matrix. We state the procedure for finding the inverse of a matrix, then complete the details in Example 2.

Procedure for Finding the Inverse of an $n \times n$ Matrix M

1. Form the $n \times 2n$ matrix $[M \mid I]$.
2. Reduce it to echelon form $[X \mid Y]$, making sure to perform the same row operations on the augmented matrix I.
3. If X is the identity matrix, then Y is the inverse of M. If X is not the identity matrix, then M is not invertible.

EXAMPLE 2

Problem

Find the inverse of the matrix A, where

$$
A = \begin{bmatrix} 1 & 0 & 8 \\ 1 & 2 & 3 \\ 2 & 5 & 3 \end{bmatrix}
$$

Solution First form the 3×6 matrix

$$
\begin{bmatrix}
1 & 0 & 8 & 1 & 0 & 0 \\
1 & 2 & 3 & 0 & 1 & 0 \\
2 & 5 & 3 & 0 & 0 & 1
\end{bmatrix}
$$

The vertical bar separates the entries of A from the identity matrix. Next perform a series of row operations to reduce A to the identity matrix while carrying out the same operations on the right-hand side.

$$
\begin{bmatrix}
1 & 0 & 8 & 1 & 0 & 0 \\
0 & 2 & -5 & -1 & 1 & 0 \\
0 & 5 & -13 & -2 & 0 & 1
\end{bmatrix}
\qquad
\begin{array}{l}
(-1)(R1) + (R2) \\
(-2)(R1) + (R3)
\end{array}
$$

$$
\begin{bmatrix}
1 & 0 & 8 & 1 & 0 & 0 \\
0 & 1 & -\frac{5}{2} & -\frac{1}{2} & \frac{1}{2} & 0 \\
0 & 5 & -13 & -2 & 0 & 1
\end{bmatrix}
\qquad
(\tfrac{1}{2})(R2)
$$

$$
\begin{bmatrix}
1 & 0 & 8 & 1 & 0 & 0 \\
0 & 1 & -\frac{5}{2} & -\frac{1}{2} & \frac{1}{2} & 0 \\
0 & 0 & -\frac{1}{2} & \frac{1}{2} & -\frac{5}{2} & 1
\end{bmatrix}
\qquad
(-5)(R2) + (R3)
$$

$$\begin{bmatrix} 1 & 0 & 8 \\ 0 & 1 & -\frac{5}{2} \\ 0 & 0 & 1 \end{bmatrix} \begin{array}{|ccc} 1 & 0 & 0 \\ -\frac{1}{2} & \frac{1}{2} & 0 \\ -1 & 5 & -2 \end{array} \qquad (-2)(R3)$$

$$\begin{bmatrix} 1 & 0 & 0 \\ 0 & 1 & 0 \\ 0 & 0 & 1 \end{bmatrix} \begin{array}{|rrr} 9 & -40 & 16 \\ -3 & 13 & -5 \\ -1 & 5 & -2 \end{array} \qquad \begin{array}{l} (-8)(R3) + (R1) \\ (\frac{5}{2})(R3) + (R2) \end{array}$$

The inverse of A is

$$A^{-1} = \begin{bmatrix} 9 & -40 & 16 \\ -3 & 13 & -5 \\ -1 & 5 & -2 \end{bmatrix}$$

Example 3 shows that the Gauss–Jordan method of constructing an inverse also detects noninvertible matrices. There are several other tests for invertibility, but this one produces the inverse at the same time, if it exists.

EXAMPLE 3

Problem

Find the inverse, if it exists, of the matrix M.

$$M = \begin{bmatrix} 1 & 2 & 3 \\ 4 & 5 & 6 \\ 7 & 8 & 9 \end{bmatrix}$$

Solution

$$\begin{bmatrix} 1 & 2 & 3 \\ 4 & 5 & 6 \\ 7 & 8 & 9 \end{bmatrix} \begin{array}{|ccc} 1 & 0 & 0 \\ 0 & 1 & 0 \\ 0 & 0 & 1 \end{array}$$

$$\begin{bmatrix} 1 & 2 & 3 \\ 0 & -3 & -6 \\ 0 & -6 & -12 \end{bmatrix} \begin{array}{|rrr} 1 & 0 & 0 \\ -4 & 1 & 0 \\ -7 & 0 & 1 \end{array} \qquad \begin{array}{l} (-4)(R1) + (R2) \\ (-7)(R1) + (R3) \end{array}$$

$$\begin{bmatrix} 1 & 2 & 3 \\ 0 & -3 & -6 \\ 0 & 0 & 0 \end{bmatrix} \begin{array}{|rrr} 1 & 0 & 0 \\ -4 & 1 & 0 \\ 1 & -2 & 1 \end{array} \qquad (-2)(R2) + (R3)$$

The matrix M cannot be reduced to I on the left-hand side of the vertical bar because of the row of 0's. Consequently, M does not have an inverse.

The 2 × 2 Case

We single out a procedure for forming the inverse of a 2 × 2 matrix because it is needed frequently. The derivation applies the Gauss–Jordan method to a general 2 × 2 matrix; it is left as an exercise. Unfortunately, the corresponding procedure for larger matrices is much more complicated.

Let M be a general 2×2 matrix.

$$M = \begin{bmatrix} a & b \\ c & d \end{bmatrix}$$

The inverse of M involves a number D, called the **determinant** of M, defined by $D = ad - bc$. For the matrix A from Example 1

$$A = \begin{bmatrix} 1 & 2 \\ 3 & 4 \end{bmatrix}$$

the determinant is $D = 1 \cdot 4 - 3 \cdot 2 = -2$.

Procedure for Forming the Inverse of $M = \begin{bmatrix} a & b \\ c & d \end{bmatrix}$

1. Let $D = ad - bc$. If $D = 0$, then M is not invertible. If $D \neq 0$, then perform steps 2, 3, and 4.
2. Interchange a and d and negate b and c.
3. Multiply this matrix by the number $1/D$.
4. The inverse of M is

$$M^{-1} = \frac{1}{D} \begin{bmatrix} d & -b \\ -c & a \end{bmatrix}$$

EXAMPLE 4

Problem

What is the inverse of $M = \begin{bmatrix} -2 & -4 \\ 3 & 5 \end{bmatrix}$?

Solution

$$M^{-1} = \frac{1}{(-2)(5) - (3)(-4)} \begin{bmatrix} 5 & 4 \\ -3 & -2 \end{bmatrix} = \frac{1}{2} \begin{bmatrix} 5 & 4 \\ -3 & -2 \end{bmatrix}$$

$$= \begin{bmatrix} \frac{5}{2} & 2 \\ -\frac{3}{2} & -1 \end{bmatrix}$$

The Inverse of a Product

If A and B are $n \times n$ matrices, then so is the product AB. Suppose that A and B are invertible. Then $(AB)(B^{-1}A^{-1}) = A(BB^{-1})A^{-1} = AI_nA^{-1} = AA^{-1} = I_n$. This means that AB is invertible. Since the inverse is unique, the inverse of AB is given by the formula

$$(AB)^{-1} = B^{-1}A^{-1}$$

Example 5 shows how to verify this formula. It will serve as a model for verifying other properties of the inverse.

EXAMPLE 5

Problem

Verify that $(AB)^{-1} = B^{-1}A^{-1}$, where

$$A = \begin{bmatrix} 1 & 2 \\ 3 & 4 \end{bmatrix} \qquad \text{and} \qquad B = \begin{bmatrix} -2 & -4 \\ 3 & 5 \end{bmatrix}$$

Solution Form AB by matrix multiplication.

$$AB = \begin{bmatrix} 4 & 6 \\ 6 & 8 \end{bmatrix}$$

The inverse of AB is

$$(AB)^{-1} = -\tfrac{1}{4}\begin{bmatrix} 8 & -6 \\ -6 & 4 \end{bmatrix}$$

A^{-1} and B^{-1} were computed in Examples 1 and 4.

$$B^{-1} = \begin{bmatrix} \frac{5}{2} & 2 \\ -\frac{3}{2} & -1 \end{bmatrix} \qquad \text{and} \qquad A^{-1} = \begin{bmatrix} -2 & 1 \\ \frac{3}{2} & -\frac{1}{2} \end{bmatrix}$$

Multiply B^{-1} times A^{-1} in that order.

$$B^{-1}A^{-1} = \begin{bmatrix} -2 & \frac{3}{2} \\ \frac{3}{2} & -1 \end{bmatrix}$$

Note that $(AB)^{-1} = B^{-1}A^{-1}$.

A common mistake is to write $(AB)^{-1} = A^{-1}B^{-1}$. From Example 5 we can see that this equality does not always hold since

$$A^{-1}B^{-1} = \begin{bmatrix} -\frac{13}{2} & -5 \\ \frac{9}{2} & \frac{7}{2} \end{bmatrix}$$

Note that $(AB)^{-1} \neq A^{-1}B^{-1}$.

EXERCISE SET 2–4

In Problems 1 to 4 a matrix A is given. Determine which of the matrices X, Y, or Z is the inverse of A.

1. $A = \begin{bmatrix} 1 & 2 \\ 3 & 4 \end{bmatrix} \qquad X = \begin{bmatrix} 1 & -2 \\ 0 & 1 \end{bmatrix}$

$Y = \begin{bmatrix} 4 & -2 \\ -3 & 1 \end{bmatrix} \qquad Z = \begin{bmatrix} -2 & 1 \\ \frac{3}{2} & -\frac{1}{2} \end{bmatrix}$

2. $A = \begin{bmatrix} 4 & 3 \\ 3 & 2 \end{bmatrix} \qquad X = \begin{bmatrix} 1 & 3 \\ -1 & -4 \end{bmatrix}$

$Y = \begin{bmatrix} 2 & -3 \\ -3 & 4 \end{bmatrix} \qquad Z = \begin{bmatrix} -2 & 3 \\ 3 & -4 \end{bmatrix}$

3. $A = \begin{bmatrix} 1 & 0 & 0 \\ 0 & 2 & 0 \\ 0 & 0 & 3 \end{bmatrix} \qquad X = \begin{bmatrix} 1 & 1 & 2 \\ 0 & -1 & 0 \\ 0 & 0 & -1 \end{bmatrix}$

$Y = \begin{bmatrix} 1 & 0 & 0 \\ 0 & \frac{1}{2} & 0 \\ 0 & 0 & \frac{1}{3} \end{bmatrix} \qquad Z = \begin{bmatrix} 1 & 0 & 0 \\ 0 & \frac{1}{2} & 0 \\ 1 & 0 & \frac{1}{3} \end{bmatrix}$

4. $A = \begin{bmatrix} 1 & 0 & 1 \\ 0 & 1 & 0 \\ 0 & 0 & 1 \end{bmatrix} \qquad X = \begin{bmatrix} 1 & 0 & 1 \\ 0 & 1 & 0 \\ 0 & 0 & -1 \end{bmatrix}$

$Y = \begin{bmatrix} 1 & 0 & -1 \\ 0 & 1 & 0 \\ 0 & 0 & 1 \end{bmatrix} \qquad Z = \begin{bmatrix} 1 & 0 & 1 \\ 0 & 1 & 0 \\ -1 & 0 & -1 \end{bmatrix}$

In Problems 5 to 10 determine whether the given matrix is invertible and find its inverse if it is.

5. $\begin{bmatrix} 3 & 1 \\ 7 & 2 \end{bmatrix}$

6. $\begin{bmatrix} 8 & 4 \\ 4 & 2 \end{bmatrix}$

7. $\begin{bmatrix} 2 & 3 \\ 4 & 6 \end{bmatrix}$

8. $\begin{bmatrix} 5 & 4 \\ 4 & 3 \end{bmatrix}$

9. $\begin{bmatrix} 1 & -2 \\ -2 & 8 \end{bmatrix}$

10. $\begin{bmatrix} 1 & -2 \\ 3 & 8 \end{bmatrix}$

In Problems 11 to 18 determine whether the given matrix is invertible. If it is, compute its inverse.

11. $\begin{bmatrix} 1 & 1 & 1 \\ 2 & 1 & 1 \\ 2 & 2 & 3 \end{bmatrix}$

12. $\begin{bmatrix} 1 & -1 & 0 \\ 0 & 1 & 1 \\ 2 & 0 & 1 \end{bmatrix}$

13. $\begin{bmatrix} 1 & 3 & 2 \\ 1 & 3 & 3 \\ 2 & 7 & 8 \end{bmatrix}$

14. $\begin{bmatrix} 1 & 4 & 2 \\ -2 & -8 & -3 \\ 0 & 1 & 1 \end{bmatrix}$

15. $\begin{bmatrix} 1 & -2 & -3 \\ 0 & 1 & 4 \\ 0 & 0 & 1 \end{bmatrix}$

16. $\begin{bmatrix} 1 & -5 & 12 \\ 0 & 1 & -2 \\ 0 & 0 & 1 \end{bmatrix}$

17. $\begin{bmatrix} 0 & -3 & 0 \\ 2 & 0 & 0 \\ 0 & 0 & 4 \end{bmatrix}$

18. $\begin{bmatrix} 0 & 0 & 2 \\ 0 & 1 & 0 \\ -1 & 0 & 1 \end{bmatrix}$

In Problems 19 and 20 two matrices A and B are given. Verify that $(AB)^{-1} = B^{-1}A^{-1}$.

19. $A = \begin{bmatrix} 5 & -2 \\ -3 & 1 \end{bmatrix}$ $B = \begin{bmatrix} 3 & 1 \\ 7 & 3 \end{bmatrix}$

20. $A = \begin{bmatrix} 7 & 3 \\ 5 & 2 \end{bmatrix}$ $B = \begin{bmatrix} 5 & -2 \\ -4 & 2 \end{bmatrix}$

In Problems 21 to 23 determine whether the given matrix is invertible, and find its inverse if it is.

21. $\begin{bmatrix} 0 & -1 \\ 1 & 4 \end{bmatrix}$ 22. $\begin{bmatrix} 2 & 0 \\ 0 & -4 \end{bmatrix}$ 23. $\begin{bmatrix} 1 & 3 \\ 0 & 0 \end{bmatrix}$

24. Let A be a matrix that has a row with all entries equal to zero. Can A be invertible? (*Hint:* See Problem 23.)

25. Compute the inverse of this matrix, if possible.

$\begin{bmatrix} 1 & 0 & 0 \\ 0 & 3 & 0 \\ 0 & 0 & -1 \end{bmatrix}$

26. Use the result of Problem 25 to compute the inverse of the following matrix, where $a \neq 0$, $b \neq 0$, and $c \neq 0$:

$\begin{bmatrix} a & 0 & 0 \\ 0 & b & 0 \\ 0 & 0 & c \end{bmatrix}$

In Problems 27 and 28 find the inverse of the 4×4 matrix by the method for 3×3 matrices shown in Example 2.

27. $\begin{bmatrix} 1 & -2 & 3 & 1 \\ 0 & 1 & 2 & -4 \\ 0 & 0 & 1 & 0 \\ 0 & 0 & 0 & 1 \end{bmatrix}$

28. $\begin{bmatrix} 1 & 2 & 5 & -8 \\ 0 & 1 & 3 & 0 \\ 0 & 0 & 1 & 2 \\ 0 & 0 & 0 & 1 \end{bmatrix}$

Let A be an invertible matrix. Then the inverse of A^{-1} is denoted by $(A^{-1})^{-1}$. In Problems 29 to 32 find A^{-1} and $(A^{-1})^{-1}$.

29. $A = \begin{bmatrix} 3 & 1 \\ 7 & 2 \end{bmatrix}$

30. $A = \begin{bmatrix} 5 & 4 \\ 4 & 3 \end{bmatrix}$

31. $A = \begin{bmatrix} 1 & -2 \\ -2 & 8 \end{bmatrix}$

32. $A = \begin{bmatrix} 1 & -2 \\ 3 & 8 \end{bmatrix}$

33. Use Problems 29 to 32 to discover a formula for $(A^{-1})^{-1}$.

34. Compute (a) $(A^{-1}B)^{-1}$ (b) $(AB^{-1}C)^{-1}$.

35. Find $(AB)^{-1}$ and AB, where

$A^{-1} = \begin{bmatrix} 5 & -2 \\ -3 & 1 \end{bmatrix}$ and $B^{-1} = \begin{bmatrix} 3 & 1 \\ 7 & 3 \end{bmatrix}$

36. Find A, where

$A^{-1} = \begin{bmatrix} 1 & 0 & 3 \\ 0 & 1 & 0 \\ 0 & 0 & 1 \end{bmatrix}$

37. Let A and B be the following matrices. (a) Is B the inverse of A? (b) If not, what is?

$A = \begin{bmatrix} -3 & 1 \\ 2 & 0 \end{bmatrix}$ $B = \begin{bmatrix} 0 & 1 \\ 2 & 3 \end{bmatrix}$

38. Let A and B be the following matrices. (a) Is B the inverse of A? (b) If not, what is?

$$A = \begin{bmatrix} 4 & -6 & 2 \\ 6 & 3 & -3 \\ -14 & 3 & 5 \end{bmatrix}$$

$$B = \begin{bmatrix} 2 & 3 & 1 \\ 1 & 4 & 2 \\ 5 & 6 & 4 \end{bmatrix}$$

In Problems 39 and 40 the figure shows the graphs of two lines. Determine directly from the figure whether matrix A is invertible.

$$A = \begin{bmatrix} a & b \\ c & d \end{bmatrix}$$

39.

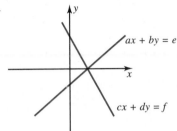

40.

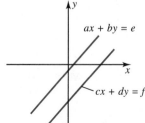

41. Determine whether each matrix is invertible.

$$A = \begin{bmatrix} 1 & 2 & 3 \\ 4 & 5 & 6 \\ 7 & 8 & 9 \end{bmatrix} \qquad B = \begin{bmatrix} 1 & 2 & 3 \\ 4 & 5 & 6 \\ 7 & 8 & 9.001 \end{bmatrix}$$

42. Let A, P, and D be the following matrices.

$$A = \begin{bmatrix} 22 & 20 \\ -25 & -23 \end{bmatrix} \quad P = \begin{bmatrix} 1 & -4 \\ -1 & 5 \end{bmatrix}$$

$$D = \begin{bmatrix} 2 & 0 \\ 0 & -3 \end{bmatrix}$$

(a) Show that $A = PDP^{-1}$.
(b) Find A^3.
(c) Find PD^3P^{-1}.
(d) Compare the results of parts (b) and (c).

43. Prove that an invertible matrix A has exactly one inverse. (*Hint:* Assume there are two inverses, B and C. Substitute $CA = I$ into $B = IB$ and substitute $AB = I$ into $C = CI$ to show that $B = C$.)

44. Use the Gauss–Jordan method to compute the inverse of an arbitrary 2×2 matrix

$$A = \begin{bmatrix} a & b \\ c & d \end{bmatrix}$$

where $a \neq 0$ and $ad - bc \neq 0$.

45. Let A be the task of "putting on a sock" and B be the task of "putting on a shoe."
(a) Interpret the meanings of A^{-1} and B^{-1}.
(b) Interpret the meanings of AB and $(AB)^{-1}$.
(c) Explain why the right-hand side of the equation $(AB)^{-1} = B^{-1}A^{-1}$ is in the proper order.

Referenced Exercise Set 2–4

1. In accounting, matrices are used to describe the financial interdependence of the departments within a firm. The inverse of a matrix is needed to compute the total cost of each department. Compute the inverse of the matrix M, which appeared in a classic paper in accounting.*

$$M = \begin{bmatrix} 1 & 0 & -0.05 & -0.10 & -0.20 \\ 0 & 1 & -0.10 & -0.05 & -0.20 \\ -0.10 & -0.10 & 1 & -0.05 & -0.20 \\ -0.05 & 0 & -0.10 & 1 & -0.20 \\ -0.10 & -0.10 & -0.05 & 0 & 1 \end{bmatrix}$$

Cumulative Exercise Set 2–4

Problems 1 to 4 refer to the following matrices.

$$A = \begin{bmatrix} 1 & -1 \\ -3 & 0 \end{bmatrix} \qquad B = \begin{bmatrix} 4 & 0 \\ -1 & 6 \\ 5 & 7 \end{bmatrix}$$

$$C = \begin{bmatrix} -7 & 6 \\ 3 & -5 \\ 0 & 0 \end{bmatrix} \qquad D = \begin{bmatrix} 5 & 0 \\ 0 & -2 \end{bmatrix}$$

*Thomas H. Williams and Charles H. Griffin, "Matrix Theory and Cost Allocation," *The Accounting Review*, July 1964, pp. 671–678.

1. (a) What is the size of A?
   (b) Find c_{22}.

2. Perform, where possible, the following matrix operations.
   (a) $A + D$ (b) $C + D$ (c) $3A - D$

3. Perform, where possible, the following matrix operations.
   (a) AC (b) CA (c) CAD

4. Write the matrix equation $BX = K$ as a system of equations, where

   $$X = \begin{bmatrix} x \\ y \end{bmatrix} \qquad K = \begin{bmatrix} 5 \\ 0 \\ 2 \end{bmatrix}$$

5. Find values of x and y such that

   $$x \begin{bmatrix} 1 \\ 1 \end{bmatrix} + y \begin{bmatrix} 2 \\ -1 \end{bmatrix} = \begin{bmatrix} 0 \\ 0 \end{bmatrix}$$

6. Solve the following system of linear equations by the Gauss–Jordan method.

   $$\begin{aligned} x + 3y - z &= 5 \\ 2x + 7y + z &= 12 \\ -x + y - 9z &= 3 \end{aligned}$$

7. Find the general solution to the corresponding system of linear equations below with variables x, y, u, and v. Determine which variables are basic and which are free.

   $$\left[\begin{array}{cccc|c} 1 & -1 & 0 & -1 & 4 \\ 1 & 0 & 1 & -1 & 3 \end{array} \right]$$

8. Solve the following system of linear equations by the Gauss–Jordan method. If there are infinitely many solutions, find the general form and two particular solutions.

   $$\begin{aligned} x + 3y - z &= 5 \\ 2x + 7y + z &= 12 \\ -x - 4y - 2z &= -7 \end{aligned}$$

9. Determine whether each of the following matrices is invertible. If it is, compute its inverse.

   (a) $\begin{bmatrix} 1 & 3 \\ 3 & 11 \end{bmatrix}$ (b) $\begin{bmatrix} 4 & 2 \\ 10 & 5 \end{bmatrix}$

10. Determine whether the matrix below is invertible. If it is, compute its inverse.

    $$\begin{bmatrix} 1 & 1 & -2 \\ -1 & 0 & 1 \\ 1 & -1 & 1 \end{bmatrix}$$

11. Show that $A = MDM^{-1}$ and then find A^{10}.

    $$A = \begin{bmatrix} 0 & -2 & 1 \\ 1 & 3 & -1 \\ 0 & 0 & 1 \end{bmatrix} \qquad M = \begin{bmatrix} -2 & 1 & -1 \\ 1 & 0 & 1 \\ 0 & 1 & 0 \end{bmatrix}$$

    $$D = \begin{bmatrix} 1 & 0 & 0 \\ 0 & 1 & 0 \\ 0 & 0 & 2 \end{bmatrix}$$

12. If A and B are invertible 3×3 matrices, is it always true that $A^{-1}BA = B$? Justify your answer.

2–5 Systems of Equations

We have discussed two procedures for solving systems of linear equations, Gaussian elimination and the Gauss–Jordan method. This section describes an entirely different procedure, called the "inverse matrix method." Unlike the two elimination methods, the inverse matrix method applies only to square matrices. Therefore *all matrices in this section will be square matrices.*

The Matrix Equation

Let us return to the Pressman-Gutman example to show how matrix equations arise and how they can be solved. We need only consider the Fall River and Passaic finishing plants. Suppose the company determines that to break even the Fall River plant must produce $15 (= $15,000,000) worth of material and the Passaic plant must produce $15.5 (= $15,500,000). The problem is to find the prices to charge for cotton and synthetic material in order to break even. Let x be the price per yard of cotton and y the price per yard of synthetic material. To break even, x and y must satisfy the linear equations

$$2.0x + 4.5y = 15.0$$
$$2.5x + 4.0y = 15.5$$

Let A be the coefficient matrix, X the matrix of variables, and C the matrix of constants.

$$A = \begin{bmatrix} 2.0 & 4.5 \\ 2.5 & 4.0 \end{bmatrix} \qquad X = \begin{bmatrix} x \\ y \end{bmatrix} \qquad C = \begin{bmatrix} 15.0 \\ 15.5 \end{bmatrix}$$

Then the system of linear equations can be written as the **matrix equation**

$$AX = C$$

This matrix equation is similar to the numerical equation $ax = c$, where a and c are specified numbers and x is the number that is sought. Take a particular case.

$$2x = 5$$

It is solved by multiplying both sides by $\frac{1}{2}$ because $\frac{1}{2} = 2^{-1}$ is the multiplicative inverse of 2.

$$(\tfrac{1}{2}) \cdot 2x = (\tfrac{1}{2}) \cdot 5 = 2^{-1} \cdot 5$$
$$x = \tfrac{5}{2}$$

The solution to the matrix equation $AX = C$ is obtained in a similar way when the coefficient matrix A is invertible. The reasons are supplied in parentheses.

$$AX = C$$
$$A^{-1}(AX) = A^{-1}C \qquad \text{(multiply on the left by } A^{-1})$$
$$(A^{-1}A)X = A^{-1}C \qquad \text{(associative law)}$$
$$I_n X = A^{-1}C \qquad \text{(definition of } A^{-1})$$
$$X = A^{-1}C \qquad \text{(definition of } I_n)$$

If A is an invertible matrix, then the solution to the matrix equation

$$AX = C$$

is

$$X = A^{-1}C$$

If the matrix A is not invertible, then the equation $AX = C$ has either no solution or infinitely many solutions.

The solution by the matrix method involves finding the inverse and performing matrix multiplication. The multiplication is straightforward, but the procedure for computing the inverse is cumbersome for larger systems. Example 1 illustrates the matrix method when A^{-1} is given. Example 2 demonstrates the entire procedure from beginning to end.

EXAMPLE 1

Problem

Solve the Pressman-Gutman system of equations $AX = C$, where the inverse of the coefficient matrix A is

$$A^{-1} = \frac{-1}{3.25} \begin{bmatrix} 4.0 & -4.5 \\ -2.5 & 2.0 \end{bmatrix}$$

Solution The solution to the matrix equation $AX = C$ is

$$X = A^{-1}C$$

$$\begin{bmatrix} x \\ y \end{bmatrix} = \frac{-1}{3.25} \begin{bmatrix} 4.0 & -4.5 \\ -2.5 & 2.0 \end{bmatrix} \begin{bmatrix} 15.0 \\ 15.5 \end{bmatrix}$$

$$\begin{bmatrix} x \\ y \end{bmatrix} = \begin{bmatrix} 3 \\ 2 \end{bmatrix}$$

Then $x = 3$ and $y = 2$. This solution can be checked by substituting directly into the original system of equations.

EXAMPLE 2

Problem

Solve the system of equations

$$\begin{aligned} x + 2y &= 8 \\ 3x - 4y &= -26 \end{aligned}$$

Solution Write the system as a matrix equation.

$$\begin{bmatrix} 1 & 2 \\ 3 & -4 \end{bmatrix} \begin{bmatrix} x \\ y \end{bmatrix} = \begin{bmatrix} 8 \\ -26 \end{bmatrix}$$

The solution is

$$\begin{bmatrix} x \\ y \end{bmatrix} = \begin{bmatrix} 1 & 2 \\ 3 & -4 \end{bmatrix}^{-1} \begin{bmatrix} 8 \\ -26 \end{bmatrix}$$

$$\begin{bmatrix} x \\ y \end{bmatrix} = \left(\frac{1}{-10} \right) \begin{bmatrix} -4 & -2 \\ -3 & 1 \end{bmatrix} \begin{bmatrix} 8 \\ -26 \end{bmatrix}$$

$$\begin{bmatrix} x \\ y \end{bmatrix} = \left(\frac{1}{-10} \right) \begin{bmatrix} 20 \\ -50 \end{bmatrix}$$

$$\begin{bmatrix} x \\ y \end{bmatrix} = \begin{bmatrix} -2 \\ 5 \end{bmatrix}$$

Therefore $x = -2$ and $y = 5$.

The display assembles the four steps of the **inverse matrix method.**

Matrix Method for Solving a System of Equations

1. Translate the system into $AX = C$, where A is the coefficient matrix, X the matrix of variables, and C the matrix of constants.
2. Compute A^{-1} if A is invertible.
3. Construct the solution $X = A^{-1}C$.

Example 3 shows how to solve a system of three linear equations in three variables by the matrix inverse method.

EXAMPLE 3

Problem

Solve this system by the matrix method.

$$\begin{aligned} x + y + 2z &= 4 \\ 2x + y + 2z &= 3 \\ 2x \quad\;\; + z &= 2 \end{aligned}$$

Solution Translate the system into $AX = C$, where

$$A = \begin{bmatrix} 1 & 1 & 2 \\ 2 & 1 & 2 \\ 2 & 0 & 1 \end{bmatrix} \qquad X = \begin{bmatrix} x \\ y \\ z \end{bmatrix} \qquad \text{and} \qquad C = \begin{bmatrix} 4 \\ 3 \\ 2 \end{bmatrix}$$

Compute A^{-1} by forming the 3×6 matrix $[A \mid I]$ and reducing A to the identity matrix while carrying out the same operations on the right-hand side.

$$\left[\begin{array}{ccc|ccc} 1 & 1 & 2 & 1 & 0 & 0 \\ 2 & 1 & 2 & 0 & 1 & 0 \\ 2 & 0 & 1 & 0 & 0 & 1 \end{array}\right]$$

$$\left[\begin{array}{ccc|ccc} 1 & 1 & 2 & 1 & 0 & 0 \\ 0 & -1 & -2 & -2 & 1 & 0 \\ 0 & -2 & -3 & -2 & 0 & 1 \end{array}\right] \qquad \begin{array}{l}(-2)(R1) + (R2) \\ (-2)(R1) + (R3)\end{array}$$

$$\left[\begin{array}{ccc|ccc} 1 & 0 & 0 & -1 & 1 & 0 \\ 0 & 1 & 2 & 2 & -1 & 0 \\ 0 & 0 & 1 & 2 & -2 & 1 \end{array}\right] \qquad \begin{array}{l}(R1) + (R2) \\ (-1)(R2) \\ (-2)(R2) + (R3)\end{array}$$

$$\left[\begin{array}{ccc|ccc} 1 & 0 & 0 & -1 & 1 & 0 \\ 0 & 1 & 0 & -2 & 3 & -2 \\ 0 & 0 & 1 & 2 & -2 & 1 \end{array}\right] \qquad (-2)(R3) + (R2)$$

The 3×3 matrix on the right-hand side is A^{-1}.

The solution to the original system of equations is $X = A^{-1}C$, so

$$\begin{bmatrix} x \\ y \\ z \end{bmatrix} = \begin{bmatrix} -1 & 1 & 0 \\ -2 & 3 & -2 \\ 2 & -2 & 1 \end{bmatrix} \begin{bmatrix} 4 \\ 3 \\ 2 \end{bmatrix} = \begin{bmatrix} -1 \\ -3 \\ 4 \end{bmatrix}$$

The solution is $x = -1$, $y = -3$, and $z = 4$.

Contrast in Methods

The two major ways to solve a system of n linear equations in n variables are elimination (Gaussian or Gauss–Jordan) and the inverse matrix method. Elimination can be applied to all systems, but the inverse matrix method applies only to square systems with an invertible coefficient matrix. Nonetheless, there are times when the inverse matrix method is preferable.

Example 4 contrasts Gauss–Jordan elimination with the inverse matrix method. The coefficient matrix is very important in quantum mechanics and heat transfer.*

EXAMPLE 4

Problem

Solve the system of equations

$$\begin{array}{rcl} 2x - y & = & -7 \\ -x + 2y - z & = & 9 \\ -y + 2z & = & -5 \end{array}$$

Solution (1) *Elimination.* Apply the Gauss–Jordan method by forming the augmented matrix and reducing it to echelon form.

$$\begin{bmatrix} 2 & -1 & 0 & | & -7 \\ -1 & 2 & -1 & | & 9 \\ 0 & -1 & 2 & | & -5 \end{bmatrix}$$

$$\begin{bmatrix} -1 & 2 & -1 & | & 9 \\ 2 & -1 & 0 & | & -7 \\ 0 & -1 & 2 & | & -5 \end{bmatrix} \qquad \text{(R1)} \leftrightarrow \text{(R2)}$$

$$\begin{bmatrix} 1 & -2 & 1 & | & -9 \\ 0 & 3 & -2 & | & 11 \\ 0 & -1 & 2 & | & -5 \end{bmatrix} \qquad \begin{array}{l} (-1)(\text{R1}) \\ 2(\text{R1}) + (\text{R2}) \end{array}$$

$$\begin{bmatrix} 1 & -2 & 1 & | & -9 \\ 0 & -1 & 2 & | & -5 \\ 0 & 3 & -2 & | & 11 \end{bmatrix} \qquad \text{(R2)} \leftrightarrow \text{(R3)}$$

$$\begin{bmatrix} 1 & 0 & -3 & | & 1 \\ 0 & 1 & -2 & | & 5 \\ 0 & 0 & 4 & | & -4 \end{bmatrix} \qquad \begin{array}{l} (-2)(\text{R2}) + (\text{R1}) \\ (-1)(\text{R2}) \\ 3(\text{R2}) + (\text{R3}) \end{array}$$

*See the article "What Is Quantum Field Theory?" by David C. Brydges in the *Bulletin of the American Mathematical Society,* January 1983, pp. 31–40. The coefficient matrix is described on page 34.

$$\begin{bmatrix} 1 & 0 & -3 & | & 1 \\ 0 & 1 & -2 & | & 5 \\ 0 & 0 & 1 & | & -1 \end{bmatrix} \quad (\tfrac{1}{4})(R3)$$

$$\begin{bmatrix} 1 & 0 & 0 & | & -2 \\ 0 & 1 & 0 & | & 3 \\ 0 & 0 & 1 & | & -1 \end{bmatrix} \quad \begin{array}{l} 3(R3) + (R1) \\ 2(R3) + (R2) \end{array}$$

The solution is $x = -2$, $y = 3$, and $z = -1$.

(2) *Inverse Matrix Method.* The first step is

$$\begin{bmatrix} x \\ y \\ z \end{bmatrix} = \begin{bmatrix} 2 & -1 & 0 \\ -1 & 2 & -1 \\ 0 & -1 & 2 \end{bmatrix}^{-1} \begin{bmatrix} -7 \\ 9 \\ -5 \end{bmatrix}$$

Next perform steps similar to those in part (1) to find the inverse of the coefficient matrix. The solution is

$$\begin{bmatrix} x \\ y \\ z \end{bmatrix} = \left(\frac{1}{4}\right) \begin{bmatrix} 3 & 2 & 1 \\ 2 & 4 & 2 \\ 1 & 2 & 3 \end{bmatrix} \begin{bmatrix} -7 \\ 9 \\ -5 \end{bmatrix}$$

$$= \left(\frac{1}{4}\right) \begin{bmatrix} -8 \\ 12 \\ -4 \end{bmatrix} = \begin{bmatrix} -2 \\ 3 \\ -1 \end{bmatrix}$$

Once again, the solution is $x = -2$, $y = 3$, and $z = -1$.

In general, if one system is being solved it is better to use elimination. However, the inverse matrix method is preferable when there are many systems of equations having the same coefficients but different constants (on the right-hand side of the equations). For instance, the breakeven point for the Fall River and Passaic finishing plants is the solution to this system.

$$\begin{bmatrix} 2.0 & 4.5 \\ 2.5 & 4.0 \end{bmatrix} \begin{bmatrix} x \\ y \end{bmatrix} = \begin{bmatrix} 15.0 \\ 15.5 \end{bmatrix}$$

The numbers 15 and 15.5 represent the amounts of material that the two plants must produce. If a different amount of material has to be produced during the same month, then another matrix equation with the same coefficient matrix has to be solved. Example 5 illustrates how to handle this situation.

EXAMPLE 5

Problem

Solve the system of equations

$$3x - 2y = a$$
$$-5x + 4y = b$$

for each of these pairs of values of a and b.

(1) $a = -1$ (2) $a = -7$ (3) $a = 12$
     $b = 3$ $b = 13$ $b = -20$

Solution Rather than solve the three systems separately, apply the matrix method once. The solution to the general system is

$$\begin{bmatrix} x \\ y \end{bmatrix} = \begin{bmatrix} 3 & -2 \\ -5 & 4 \end{bmatrix}^{-1} \begin{bmatrix} a \\ b \end{bmatrix}$$

$$\begin{bmatrix} x \\ y \end{bmatrix} = \begin{bmatrix} 2 & 1 \\ \frac{5}{2} & \frac{3}{2} \end{bmatrix} \begin{bmatrix} a \\ b \end{bmatrix}$$

$$\begin{bmatrix} x \\ y \end{bmatrix} = \begin{bmatrix} 2a + b \\ (\frac{5}{2})a + (\frac{3}{2})b \end{bmatrix}$$

To obtain the solutions to the three specific systems, substitute their given values for a and b and set the corresponding entries equal. (1) In the first system $a = -1$ and $b = 3$, so

$$\begin{bmatrix} x \\ y \end{bmatrix} = \begin{bmatrix} 2(-1) + 3 \\ (\frac{5}{2})(-1) + (\frac{3}{2}) \cdot 3 \end{bmatrix} = \begin{bmatrix} 1 \\ 2 \end{bmatrix}$$

The solution is $x = 1$ and $y = 2$. (2) Similarly, the pair $a = -7$ and $b = 13$ yields $x = -1$ and $y = 2$, while (3) $a = 12$ and $b = -20$ yields $x = 4$ and $y = 0$.

EXERCISE SET 2–5

In Problems 1 to 4 solve the system of linear equations by the inverse matrix method. Use the fact that the inverse of the coefficient matrix is given in Example 2 of Section 2–4.

1. $x + + 8z = -6$
   $x + 2y + 3z = -3$
   $2x + 5y + 3z = -6$

2. $x + + 8z = 20$
   $x + 2y + 3z = 9$
   $2x + 5y + 3z = 11$

3. $x + + 8z = 27$
   $x + 2y + 3z = 9$
   $2x + 5y + 3z = 7$

4. $x + + 8z = 6$
   $x + 2y + 3z = 3$
   $2x + 5y + 3z = 6$

In Problems 5 to 10 solve the system of linear equations by the inverse matrix method. Make use of the rule for computing the inverse of a 2 × 2 matrix described in Section 2–4.

5. $2x + y = 3$
   $3x + 4y = 7$

6. $2x + 5y = 2$
   $3x - 2y = 3$

7. $5x + 2y = 1$
   $2x - 3y = 8$

8. $8x - 3y = 1$
   $3x + 4y = 9$

9. $2x - 3y = 1$
   $3x - 2y = 3$

10. $-4x + 2y = -3$
    $3x - 2y = 3$

In Problems 11 to 14 follow the method shown in Example 3 to solve the system of linear equations.

11. $x + y + z = -2$
    $2x + y + z = 3$
    $2x + 2y + 3z = -8$

12. $x - y = 8$
    $y + z = -7$
    $2x + z = 6$

13. $x + 3y + 2z = 9$
    $x + 3y + 3z = 5$
    $2x + 7y + 8z = 7$

14. $x + 4y + 2z = 19$
    $-2x - 8y - 3z = -29$
    $y + z = 10$

In Problems 15 to 18 solve the systems of linear equations for each pair of values of a and b.

15. $5x + 2y = a$
    $7x + 3y = b$
    (a) $a = 13$ (b) $a = 5$ (c) $a = 6$
    $\quad b = 19$ $\quad\quad b = 6$ $\quad\quad b = 9$

16. $7x + 2y = a$
    $3x + y = b$
    (a) $a = 1$ (b) $a = 8$ (c) $a = -4$
    $\quad b = 0$ $\quad\quad b = 4$ $\quad\quad b = -1$

17. $5x + 3y = a$
    $4x + 2y = b$
    (a) $a = -7$ (b) $a = 17$ (c) $a = 21$
    $\quad b = -6$ $\quad\quad b = 12$ $\quad\quad b = 12$

18. $3x + 5y = a$
    $2x + 2y = b$
    (a) $a = -11$ (b) $a = -10$ (c) $a = 11$
    $\quad b = -2$ $\quad\quad b = 0$ $\quad\quad b = 10$

In Problems 19 and 20 solve the systems of linear equations for each set of values of a, b, and c.

$2x - y \quad\quad = a$
$-x + 2y - z = b$
$\quad - y + 2z = c$

19. (a) $a = 6, b = 3, c = -4$
    (b) $a = 11, b = 4, c = -17$
    (c) $a = 2, b = 0, c = -2$

20. (a) $a = 8, b = -9, c = 10$
    (b) $a = -11, b = 0, c = 13$
    (c) $a = -1, b = 0, c = -1$

In Problems 21 to 28 solve the system by the inverse matrix method if there is a unique solution.

21. $x + y + z = 3$
    $-x - 2y - 2z = -4$
    $4x + 3y + 4z = 9$

22. $2x \quad\quad + 4z = 0$
    $3x - 4y \quad\quad = 1$
    $\quad - 2y - 3z = 0$

23. $x + 3y + 2z = 20$
    $2x + 6y + 5z = 44$
    $x + 4y + 5z = 37$

24. $x - y \quad\quad = 12$
    $x \quad\quad + z = 13$
    $2x \quad\quad + z = 18$

25. $x - y \quad\quad = -3$
    $\quad y + z = 7$
    $2x \quad\quad - z = 11$

26. $x + 4y + 2z = 11$
    $2x + 9y + 4z = 23$
    $-x - 4y - z = -6$

27. $2x \quad\quad + 4z = 1$
    $3x - 4y \quad\quad = 0$
    $\quad - 2y - 3z = 0$

28. $x + y + z = 1$
    $-2x - y - z = 0$
    $\quad y + z = 2$

In Problems 29 and 30 use the results of Problems 27 and 28, respectively, of Section 2–4 to solve the system of four equations in four unknowns.

29. $x - 2y + 3z + t = -42$
    $\quad y + 2z - 4t = 0$
    $\quad\quad z \quad\quad = -4$
    $\quad\quad\quad t = 2$

30. $x + 2y + 5z - 8t = 8$
    $\quad y + 3z \quad\quad = 9$
    $\quad\quad z + 2t = 1$
    $\quad\quad\quad t = -1$

31. A company makes 120 suits of two types, wool and blend. It costs the company $100 to make each wool suit and $40 to make each blend suit. If the company allots exactly $6000 to make the suits, how many can it make of each type?

32. A company makes 450 pairs of denim and corduroy slacks. It costs the company $10 to make each pair of denim slacks and $25 to make each pair of corduroy slacks. If the company allots exactly $6750 to make the slacks, how many can it make of each type?

33. A person takes two daily vitamin supplements, C and E, via colored pills. Each red pill yields 45 units of C and 10 units of E. Each blue pill yields 9 units of C and 6 units of E. How many pills of each color must be taken daily to achieve 72 units of C and 28 units of E?

34. A person takes two daily vitamin supplements, C and E, via colored pills. Each red pill yields 90 units of C and 20 units of E. Each blue pill yields 36 units of C and 24 units of E. How many pills of each color must be taken daily to achieve 126 units of C and 44 units of E?

35. A transit company offers a discount on the first class fare, called an "executive" fare, to those passengers on a train who have no luggage. The train holds 190 passengers. The company is limited to a cost of $2250 per train, with the cost being $15 for each first class passenger, $15 for each executive passenger, and $5 for each economy class passenger. The company must reserve 10 cubic feet (cu ft) of storage for luggage for each first class passenger and 5 cu ft of storage for each economy class passenger; 1350 cu ft of storage must be used on each train. How many tickets of each type should be sold?

36. Repeat Problem 35 for a bus in which the costs and storage space remain the same, but the bus holds 45 passengers, the cost is $425 per bus, and the storage is 125 cu ft.

In Problems 37 to 40 solve the system equations.

37. $x + y + 2z + t = 6$
$\phantom{x + {}} y + t = 3$
$\phantom{x + y + {}} z - t = 3$
$\phantom{x + y + 2z + {}} t = 3$

38. $x + 2z + t = 6$
$\phantom{x + {}} y + t = 4$
$\phantom{x + y + {}} z - t = 0$
$\phantom{x + y + 2z + {}} t = 2$

39. $x \phantom{+ y + {}} + t + w = 6$
$\phantom{x + {}} y \phantom{+ {}} + t = 3$
$\phantom{x + y + {}} z - t + w = 3$
$\phantom{x + y + z + {}} t = 3$
$\phantom{x + y + z + t + {}} w = 1$

40. $x + 2y + z + t + w = 11$
$\phantom{x + {}} y - z + t = 3$
$\phantom{x + 2y + {}} z - t + w = 3$
$\phantom{x + 2y + z + {}} t - w = -2$
$\phantom{x + 2y + z + t + {}} w = 2$

Referenced Exercise Set 2–5

1. Problem 1 in Referenced Exercise Set 2–3 introduced a cost accounting procedure for determining the total costs for each department in a firm with several interdependent departments. Solve that problem for x, y, z, w using matrix methods:*

$$\begin{bmatrix} x \\ y \end{bmatrix} = \begin{bmatrix} 0 & 0.1 \\ 0.2 & 0 \end{bmatrix} \begin{bmatrix} x \\ y \end{bmatrix} + \begin{bmatrix} 9000 \\ 6000 \end{bmatrix}$$

$$\begin{bmatrix} z \\ w \end{bmatrix} = \begin{bmatrix} 0.4 & 0.4 \\ 0.4 & 0.5 \end{bmatrix} \begin{bmatrix} x \\ y \end{bmatrix} + \begin{bmatrix} 20,000 \\ 16,000 \end{bmatrix}$$

2. Problem 1 in Referenced Exercise Set 2–4 asked for the inverse of the coefficient matrix of a system of linear equations derived from the system below. Use that result to solve the system, where each variable represents the total cost of one of the five service departments in a firm.†

$x = \phantom{0.05z +{}} 0.05z + 0.1u + 0.2v + 8000$
$y = \phantom{0.05z +{}} 0.1z + 0.05u + 0.2v + 12000$
$z = 0.1x + 0.1y + \phantom{0.05z +{}} 0.05u + 0.2v + 6000$
$u = 0.05x + \phantom{0.1y +{}} 0.1z + \phantom{0.05u +{}} 0.2v + 11000$
$v = 0.1x + 0.1y + 0.05z + 13000$

3. A paper in accounting proposed an approximation method to the traditional reciprocal method of determining cost.‡ This new technique is much faster, and hence cheaper, to carry out. The matrices A and b are taken from that paper.

$$A = \begin{bmatrix} 0.05 & 0.15 & 0.15 \\ 0.10 & 0.10 & 0.15 \\ 0.20 & 0.20 & 0.05 \end{bmatrix} \quad b = \begin{bmatrix} 30,000 \\ 60,000 \\ 40,000 \end{bmatrix}$$

(a) Solve the matrix equation $(I_3 - A)x = b$ by the matrix inverse method (called the reciprocal method in accounting).

(b) Solve the matrix equation $(I_3 - A)x = b$ by the approximation method: $\hat{x} = 70{,}270y + b$, where the number 70,270 is derived from the entries in A and b, and y is a column matrix whose entries are the sums of the rows of A.

(c) Find the relative error in each component of $\hat{x}$ as a percentage of each component of x.

4. Suppose that a bettor in a pari-mutuel wagering system seeks to win w by dutching on two horses A and

*Roland Minch and Enrico Petri, "Matrix Models of Reciprocal Service Cost Allocation," *The Accounting Review*, July 1972, pp. 576–580.

†Thomas H. Williams and Charles H. Griffin, "Matrix Theory and Cost Allocation," *The Accounting Review*, July 1964, pp. 671–678.

‡Frederick H. Jacobs and Ronald M. Marshall, "A Reciprocal Service Cost Approximation," *The Accounting Review*, January 1987, pp. 67–78.

§The terminology and appropriate references were given in Problem 1 in Referenced Exercise Set 1–4.

B, whose odds of winning are $a:1$ and $b:1$, respectively.§ Let $\$x$ be the amount of money which the bettor should bet on A and $\$y$ be the amount of money which the bettor should bet on B. Then x and y can be found by solving the matrix equation

$$\begin{bmatrix} a & -1 \\ -1 & b \end{bmatrix} \begin{bmatrix} x \\ y \end{bmatrix} = \begin{bmatrix} w \\ w \end{bmatrix}$$

If the odds on two particular horses are $2:1$ and $8:1$, how much money should be bet on each horse in order to win each of these amounts?
(a) \$50   (b) \$25   (c) \$1000   (d) \$5000

Cumulative Exercise Set 2–5

1. Compute (a) $R + S$ and (b) RS for the following matrices.

$$R = \begin{bmatrix} 0 & 2 & -1 \\ 1 & 2 & -2 \\ 1 & 4 & 1 \end{bmatrix} \quad S = \begin{bmatrix} 2 & 0 & -1 \\ 0 & 3 & -3 \\ 1 & -1 & 0 \end{bmatrix}$$

2. Use the Gauss–Jordan method to find the unique solution.

$$\begin{aligned} x + 3y - z &= 1 \\ 2x - y + z &= 6 \\ x + 2y - z &= 2 \end{aligned}$$

3. Find the general solution to the system whose variables are x, y, and z, and list two particular solutions.

$$\begin{bmatrix} 1 & 0 & -1 & 4 \\ 0 & 1 & 1 & 5 \\ 0 & 0 & 0 & 0 \end{bmatrix}$$

4. Solve the system of equations by the Gauss–Jordan method.

$$\begin{aligned} 3x - y + 3z &= 4 \\ x + y - z &= 6 \\ 5x - 3y + 7z &= 2 \end{aligned}$$

5. Determine whether the matrix is invertible and find its inverse if it is.

$$\begin{bmatrix} 1 & 0 & -2 \\ 0 & 1 & 1 \\ 1 & 3 & 0 \end{bmatrix}$$

6. Verify that $(AB)^{-1} = B^{-1}A^{-1}$ for the given matrices.

$$A = \begin{bmatrix} 1 & 2 \\ 1 & 4 \end{bmatrix} \quad B = \begin{bmatrix} 0 & 3 \\ 1 & -1 \end{bmatrix}$$

7. Find A^{-1} and $(A^{-1})^{-1}$ for

$$A = \begin{bmatrix} 1 & -2 \\ 2 & 3 \end{bmatrix}$$

8. Find A^{-1} for

$$A = \begin{bmatrix} 1 & 0 & -1 & 0 \\ 0 & 1 & 3 & 1 \\ 0 & 0 & 1 & 2 \\ 0 & 0 & 0 & 1 \end{bmatrix}$$

In Problems 9 and 10 solve the system of equations by the inverse matrix method.

9. $\begin{aligned} 3x - y &= 5 \\ x + y &= 3 \end{aligned}$

10. $\begin{aligned} x + \quad\ + 3z &= 7 \\ x + y \qquad &= 1 \\ 2x - 2y + z &= 4 \end{aligned}$

11. Solve the system of equations for each pair of values of a and b.

$$\begin{aligned} 2x - 3y &= a \\ x + y &= b \end{aligned}$$

(a) $a = -1$ (b) $a = 1$ (c) $a = 6$
  $\ \ b = 2$ $b = 3$ $b = 3$

12. Solve the system of four equations in four unknowns.

$$\begin{aligned} x + y + 2z + t &= 6 \\ y \qquad\quad + t &= 4 \\ z - t &= 2 \\ t &= 1 \end{aligned}$$

13. A person takes three daily vitamin supplements, B, C, and E, via three types of pills, regular, extra-strength, and fortified. Each regular pill contains 10 units of B, 20 units of C, and 20 units of E. Each extra-strength pill contains 20 units of B, 60 units of C, and 10 units of E. Each fortified pill contains 10 units of B, 20 units of C, and 40 units of E. How many pills of each type should be taken in order to achieve 50 units of B, 120 units of C, and 90 units of E?

2—6 Input-Output Analysis

Shortly after World War II broke out President Franklin Delano Roosevelt submitted an order for 50,000 airplanes, which required the country to produce more aluminum. But the potlines used to make aluminum required massive bus bars through which copper conducts electricity, and an unforeseen copper shortage threatened the entire production schedule. The government averted the crisis by substituting silver for copper. Where did they get all the silver? They borrowed it from Fort Knox.

The immediate result was that the 50,000 airplanes were built, but the end result was much more pervasive. It led to the construction of detailed tables analyzing the interdependence of various industries within the economy and to a new area of economics called "input-output analysis."

Input-output analysis has become one of the cornerstones of modern economics because it provides a picture of various interlocking parts of a given economy. The general pattern of VCRs, stereos, cars, soft drinks, insurance, and other items produced and consumed by Americans can be followed across the page of an input-output table. Such tables are of great use in economic forecasting. An input-output table constructed by the Department of Labor during World War II predicted correctly that, contrary to widespread opinion, there would not be a depression and massive unemployment at war's end. Today industrial giants like ITT and General Motors, many smaller businesses, and some universities base their corporate policy on input-output analysis.

Input-Output Tables

At the heart of input-output analysis is an **input-output table,** which measures how much each industry depends on other industries for its production requirements. It is like a road map that allows you to read off distances from one city to another. In place of cities you have industries. And in place of miles between cities you have numbers giving the share of one industry's sales that are purchased by the other industries.

Table 1 describes a simplified economy consisting of a utility company that produces electricity (E) and water (W). The basic unit is of primary importance. Here we make it one dollar ($1). Thus the number 0.8 corresponds to $0.80 since $0.8 \times \$1 = \$0.80$.

Table 1 is like an economic crossword puzzle in which each entry can be read horizontally as the output of one industry or vertically as the input of another. The input columns explain the meaning of the entries. Consider the column under E. To produce each dollar worth of electricity, the utility company must spend $0.20 on electricity and $0.40 on water.

The rows of an input-output table tell what amounts of goods must be produced to meet **internal demands.** For Table 1 let

x = dollar amount of electricity produced

y = dollar amount of water produced

Table 1

		Input (purchase)	
		E	W
Output	E	0.2	0.3
(sale)	W	0.4	0.1

where each amount represents the production per household per month. The first row shows that in order to produce x dollars of electricity, the utility company must use $0.2x$ dollars of electricity and $0.3y$ dollars of water. Thus the internal demand for electricity can be expressed as

$$0.2x + 0.3y$$

The internal demand for water is determined from the second row of Table 1.

$$0.4x + 0.1y$$

In addition to internal demands there are also **external demands** by customers outside the company. Suppose the utility company determines that it must produce $20 of electricity and $5 of water per household each month. The total demands, both internal and external, for electricity and water become

electricity $0.2x + 0.3y + 20$
water $0.4x + 0.1y + \ \ 5$

$$\underbrace{\hphantom{0.4x + 0.1y}}_{\text{internal}}\ \underbrace{\hphantom{+5}}_{\text{external}}$$

The amounts produced, x for electricity and y for water, must equal the demand. This leads to two equations

$$x = 0.2x + 0.3y + 20$$
$$y = 0.4x + 0.1y + \ \ 5$$

This system can easily be solved by elimination. But in practice the numbers in an input-output table remain the same while only the external demand changes. Matrix methods present the most efficient means of solving such systems.

The Leontief Formula

Let X be the matrix of unknowns and D be the constant matrix.

$$X = \begin{bmatrix} x \\ y \end{bmatrix} \qquad D = \begin{bmatrix} 20 \\ 5 \end{bmatrix}$$

The constant matrix D is called the **bill of demands.** Let T be the coefficient matrix of the right-hand side of the system of equations. It is precisely Table 1 and is called the **technology matrix** (or the *consumption matrix* or the *input-output matrix*). The **Leontief Formula** is the equation

$$X = TX + D$$

It corresponds to the equality

production = (internal demand) + (external demand)

The solution to the Leontief Formula is obtained as follows:

$$X = TX + D$$
$$IX - TX = D$$
$$(I - T)X = D$$
$$X = (I - T)^{-1}D$$

Procedure for Solving the Leontief Formula

Given
T and D.

Problem
Solve for X, where $X = TX + D$.

Steps

1. Form $I - T$.
2. Compute $(I - T)^{-1}$.
3. Multiply to get $X = (I - T)^{-1}D$.
4. Interpret the entries of X.

EXAMPLE 1

Consider an input-output analysis of a utility company that produces electricity (E) and water (W), where the technology matrix and the bill of demands are given by

$$T = \begin{matrix} \\ E \\ W \end{matrix} \begin{matrix} E & W \\ \begin{bmatrix} 0.2 & 0.3 \\ 0.4 & 0.1 \end{bmatrix} \end{matrix} \qquad D = \begin{bmatrix} 20 \\ 5 \end{bmatrix}$$

Problem

Find the production schedule X that meets this demand.

Solution Let x be the production schedule for E and let y be the production schedule for W. The unknown production schedule is

$$X = \begin{bmatrix} x \\ y \end{bmatrix}$$

The aim is to solve for X, where $X = TX + D$. By step 1

$$I - T = \begin{bmatrix} 1 & 0 \\ 0 & 1 \end{bmatrix} - \begin{bmatrix} 0.2 & 0.3 \\ 0.4 & 0.1 \end{bmatrix} = \begin{bmatrix} 0.8 & -0.3 \\ -0.4 & 0.9 \end{bmatrix}$$

By step 2 the inverse of $I - T$ is

$$(I - T)^{-1} = \begin{bmatrix} \frac{3}{2} & \frac{1}{2} \\ \frac{2}{3} & \frac{4}{3} \end{bmatrix}$$

Step 3 is to multiply the inverse by D.

$$X = (I - T)^{-1}D = \begin{bmatrix} \frac{3}{2} & \frac{1}{2} \\ \frac{2}{3} & \frac{4}{3} \end{bmatrix} \begin{bmatrix} 20 \\ 5 \end{bmatrix} = \begin{bmatrix} 32.5 \\ 20 \end{bmatrix}$$

By step 4 the utility company must produce \$32.50 worth of electricity and \$20 worth of water to meet the internal and external demands.

The solution to the system of equations in Example 1 is

$$32.5 = (0.2)(32.5) + (0.3)(20) + 20$$

$$20 \ = (0.4)(32.5) + (0.1)(20) + \ \ 5$$

The first equation means that each month the utility company should produce \$32.50 worth of electricity per household, with \$6.50 going for the internal demand for electricity (since \$6.50 = 0.2 × \$32.50), \$6 going for water (since \$6 = 0.3 × \$20), and the remaining \$20 going to meet external demands. The utility company should produce \$20 worth of water, with \$13 going for electricity, \$2 for the internal use of water, and \$5 to meet external demands.

Example 2 shows why matrix methods are preferred to elimination for solving input-output problems.

EXAMPLE 2

Problem

Let T be the technology matrix from Example 1. Find the production schedules for each of these demands.

$$\text{(a)} \quad D = \begin{bmatrix} 6 \\ 12 \end{bmatrix} \qquad \text{(b)} \quad D = \begin{bmatrix} 12 \\ 6 \end{bmatrix}$$

Solution Example 1 carried out steps 1 and 2 of the Leontief procedure. Therefore finding the production schedule involves only matrix multiplication by the inverse of $I - T$, which was computed in Example 1.

$$\text{(a)} \quad X = (I - T)^{-1}D = \begin{bmatrix} \frac{3}{2} & \frac{1}{2} \\ \frac{2}{3} & \frac{4}{3} \end{bmatrix} \begin{bmatrix} 6 \\ 12 \end{bmatrix} = \begin{bmatrix} 15 \\ 20 \end{bmatrix}$$

$$\text{(b)} \quad X = (I - T)^{-1}D = \begin{bmatrix} \frac{3}{2} & \frac{1}{2} \\ \frac{2}{3} & \frac{4}{3} \end{bmatrix} \begin{bmatrix} 12 \\ 6 \end{bmatrix} = \begin{bmatrix} 21 \\ 16 \end{bmatrix}$$

The matrix method is used to solve input-output problems because the bill of demands changes frequently but the technology matrix remains the same. Example 2 shows that production schedules are easy to compute once one production schedule has been computed.

Example 3 examines a three-sector economy.

EXAMPLE 3

Problem

Find the production schedule for an economy with three industries, A, B, and C, where the technology matrix T and bill of demands D are given by

$$
\begin{array}{c}
\quad\quad A \quad B \quad C \\
T = \begin{array}{c} A \\ B \\ C \end{array}
\begin{bmatrix}
0.8 & 0.1 & 0 \\
0.1 & 0.8 & 0.1 \\
0 & 0.1 & 0.8
\end{bmatrix}
\end{array}
\qquad
D = \begin{bmatrix} 8 \\ 6 \\ 12 \end{bmatrix}
$$

Solution By step 1

$$
I - T = \begin{bmatrix} 1 & 0 & 0 \\ 0 & 1 & 0 \\ 0 & 0 & 1 \end{bmatrix} - \begin{bmatrix} 0.8 & 0.1 & 0 \\ 0.1 & 0.8 & 0.1 \\ 0 & 0.1 & 0.8 \end{bmatrix} = \begin{bmatrix} 0.2 & -0.1 & 0 \\ -0.1 & 0.2 & -0.1 \\ 0 & -0.1 & 0.2 \end{bmatrix}
$$

The inverse of $I - T$ is

$$
(I - T)^{-1} = 10 \begin{bmatrix} \frac{3}{4} & \frac{1}{2} & \frac{1}{4} \\ \frac{1}{2} & 1 & \frac{1}{2} \\ \frac{1}{4} & \frac{1}{2} & \frac{3}{4} \end{bmatrix}
$$

The solution is

$$
X = (I - T)^{-1}D = 10 \begin{bmatrix} \frac{3}{4} & \frac{1}{2} & \frac{1}{4} \\ \frac{1}{2} & 1 & \frac{1}{2} \\ \frac{1}{4} & \frac{1}{2} & \frac{3}{4} \end{bmatrix} \begin{bmatrix} 8 \\ 6 \\ 12 \end{bmatrix}
$$

$$
= 10 \begin{bmatrix} 12 \\ 16 \\ 14 \end{bmatrix} = \begin{bmatrix} 120 \\ 160 \\ 140 \end{bmatrix}
$$

This means that A must produce 120 units, B 160 units, and C 140 units to meet the internal and external demands placed on them.

Input-output analysis did not really blossom until after World War II, when the advent of computers made the computation of matrix inverses quick and reliable. The inverse of the matrix $I - T$ in Example 3 is computed in Exercise 13.

EXERCISE SET 2–6

In Problems 1 to 4 use the technology matrix from Example 1 to compute the production schedule for the given bill of demands.

1. $D = \begin{bmatrix} 40 \\ 10 \end{bmatrix}$

2. $D = \begin{bmatrix} 4 \\ 16 \end{bmatrix}$

3. $D = \begin{bmatrix} 20 \\ 25 \end{bmatrix}$

4. $D = \begin{bmatrix} 12 \\ 5 \end{bmatrix}$

In Problems 5 to 8 let the technology matrix be

$$
T = \begin{bmatrix} 0.5 & 0.4 \\ 0.75 & 0.2 \end{bmatrix}
$$

Compute the production schedule to meet the given bill of demands.

5. $D = \begin{bmatrix} 20 \\ 10 \end{bmatrix}$

6. $D = \begin{bmatrix} 4 \\ 2 \end{bmatrix}$

7. $D = \begin{bmatrix} 4 \\ 6 \end{bmatrix}$

8. $D = \begin{bmatrix} 40 \\ 60 \end{bmatrix}$

In Problems 9 and 10 use the technology matrix from Example 3 to compute the production schedule for the given bill of demands.

9. $D = \begin{bmatrix} 12 \\ 6 \\ 8 \end{bmatrix}$

10. $D = \begin{bmatrix} 20 \\ 2 \\ 28 \end{bmatrix}$

11. A company makes hard and soft contact lenses. For each \$1 output of hard lenses the company must spend \$0.60 on hard lenses and \$0.10 on soft lenses. For each \$1 of output on soft lenses the company must spend \$0.20 on hard lenses and \$0.70 on soft lenses. Find the production schedule that satisfies a bill of demands of \$60,000 for hard lenses and \$50,000 for soft lenses.

12. Repeat Problem 11 for a bill of demands of \$100,000 for hard lenses and \$750,000 for soft lenses.

13. Verify that the matrices A and B are inverses, where

$$A = \begin{bmatrix} 0.2 & -0.1 & 0 \\ -0.1 & 0.2 & -0.1 \\ 0 & -0.1 & 0.2 \end{bmatrix}$$

$$B = \begin{bmatrix} 7.5 & 5 & 2.5 \\ 5 & 10 & 5 \\ 2.5 & 5 & 7.5 \end{bmatrix}$$

14. Show that if $A = (0.1)B$, then $A^{-1} = (10)B^{-1}$.

15. If $A = kB$ for some number $k \neq 0$, how are A^{-1} and B^{-1} related?

Cumulative Exercise Set 2–6

1. If I is the 3×3 identity matrix, and A and B are 3×3 matrices satisfying $AB - I = 0$, how are the matrices A and B related?

2. Compute $AB - BA$ for the matrices

$$A = \begin{bmatrix} 1 & -1 \\ 2 & 5 \end{bmatrix} \quad B = \begin{bmatrix} 1 & 0 \\ -1 & 1 \end{bmatrix}$$

3. Find the general solution to the system of linear equations, with variables x, y, and z defined by the augmented matrix below. Determine which variables are basic and which are free.

$$A = \begin{bmatrix} 1 & 0 & -1 & | & -1 \\ 0 & 1 & -2 & | & 12 \end{bmatrix}$$

4. Solve the following system of linear equations by the Gauss–Jordan method. If there are infinitely many solutions, find the general form and two particular solutions.

$$\begin{aligned} x - 2y + 2z &= 3 \\ 3x + 4y - z &= 8 \\ -x \quad\quad + z &= 2 \end{aligned}$$

5. Determine whether the matrix below is invertible. If it is, compute its inverse.

$$\begin{bmatrix} 1 & -2 & 4 \\ 0 & 3 & 5 \\ 0 & 0 & -2 \end{bmatrix}$$

6. Suppose A and B are 2×2 matrices such that

$$AB = \begin{bmatrix} 2 & 0 \\ 0 & 2 \end{bmatrix}$$

How are the entries of A^{-1} related to the entries of B?

7. Solve the system of linear equations

$$\begin{aligned} -2x + y - z &= 3 \\ x \quad\quad + z &= 0 \\ y \quad\quad &= 2 \end{aligned}$$

using the fact that the inverse of the coefficient matrix is

$$\begin{bmatrix} -1 & -1 & 1 \\ 0 & 0 & 1 \\ 1 & 2 & -1 \end{bmatrix}$$

8. Solve the following system of linear equations by the inverse matrix method.

$$\begin{aligned} x - 2y &= 1 \\ 5x - 9y &= -33 \end{aligned}$$

9. Solve the systems of linear equations for each pair of values of a and b.

$$\begin{aligned} x - 2y &= a \\ 5x - 9y &= b \end{aligned}$$

(a) $a = -1$ (b) $a = -2$ (c) $a = 0$
     $b = 1$ $b = 5$ $b = 0$

10. Use the inverse matrix method to determine if the system of linear equations below has a unique solution. If it does, find the solution.

$$x - 2y + 3z = 1$$
$$x + 2y + z = 0$$
$$2x - 4y + 6z = 1$$

11. Compute the production schedule where the technology matrix and the bill of demands are

$$T = \begin{bmatrix} 0.4 & 0.3 \\ 0.2 & 0.1 \end{bmatrix} \qquad D = \begin{bmatrix} 96 \\ 120 \end{bmatrix}$$

12. Compute the production schedule where the technology matrix and the bill of demands are

$$T = \begin{bmatrix} 0.25 & 0.5 & 0.1 \\ 0.25 & 0.5 & 0 \\ 0 & 0.65 & 0.55 \end{bmatrix} \qquad D = \begin{bmatrix} 8{,}000{,}000 \\ 6{,}000{,}000 \\ 5{,}000{,}000 \end{bmatrix}$$

using the fact that

$$(I - T)^{-1} = \begin{bmatrix} 1.43096 & 0.29288 & 0.31799 \\ 0.37656 & 1.12970 & 0.08368 \\ 0.54393 & 1.63179 & 2.34309 \end{bmatrix}$$

13. A financial adviser offers low-risk investments that pay 10% per year and high-risk investments that pay 15% per year to five customers: V, W, X, Y, and Z. The matrix A shows each customer's investment and desired return. How should each investor combine the two options to obtain the desired return?

$$A = \begin{bmatrix} V & W & X & Y & Z & \\ 100{,}000 & 120{,}000 & 200{,}000 & 500{,}000 & 1{,}000{,}000 & \text{invest-} \\ 12{,}000 & 14{,}000 & 25{,}000 & 65{,}000 & 140{,}000 & \text{desired} \end{bmatrix} \begin{matrix} \text{ment} \\ \text{return} \end{matrix}$$

14. Write the following system of equations as a matrix equation:

$$2x - 3y + 4z = 5, \quad 9x - z = 0.$$

CASE STUDY   **The Leontief Model of the Economy**

Economics and mathematics have not always shared mutual interests. In fact, Adam Smith's classic work on economics, *Wealth of Nations,* written in 1776, contains almost no mathematics. The real intrusion of mathematical methods into economics occurred a century later in the work of the 19th-century economist Alfred Marshall, who has been dubbed a "closet mathematician" because of his habit of removing all traces of the underlying mathematical principles from his published work in favor of real-life examples that illustrate his theory. However, one of Marshall's students at Cambridge, John Maynard Keynes, gained notoriety and world renown by applying mathematical methods to economic principles. Keynes not only took his degree in mathematics, but he published a book entitled *A Treatise on Probability.*

Of course not all economists today share the Keynesian view of mathematics. Harvard Professor John Kenneth Galbraith, himself a living disciple of Keynes, regards mathematics with a considerable degree of skepticism. In his *Economics, Peace, and Laughter* he wrote, "There can be no question that prolonged commitment to mathematical exercises in economics can be damaging. It leads to the atrophy of judgment and intuition."

In spite of Galbraith's viewpoint, a brief look at the Nobel prize laureates in economics attests to the close relationship that economics and mathematics have enjoyed.* These prizes were first awarded in 1969 by the National Bank of

*Two excellent articles that describe the relationship between economics and mathematics are "Theoretical Models: Mathematical Form and Economic Content," by Gerard Debreu, the 1985 Nobel prize laureate in economics (the November 1986 issue of *Econometrica,* pp. 1259–1270), and "Mathematical Methods of Economics" by Joel Franklin, based on a lecture given at a meeting of the Society for Industrial and Applied Mathematics (printed in the April 1983 issue of the *American Mathematical Monthly,* pp. 229–246).

Sweden to commemorate its 300th anniversary. The first winners, Ragnar Frisch and Jan Tinbergen, were cited for developing "the mathematical expression of economic theory." In 1975 the mathematician Leonid Kantorovich (1912–1986) shared the prize for developing the mathematical theory of linear programming and for applying it to economic problems in optimizing the allocation of resources. (Chapters 3 and 4 are devoted to linear programming.) Altogether 14 of the 22 prizes in the years 1969–1990 have been awarded for work that is heavily mathematical.

This Case Study examines the work of the economist Wassily W. Leontief, who was awarded the 1973 Nobel prize in economics "for the development of the input-output method and for its application to important economic problems." Before describing his work, we will survey Leontief's accomplishments.†

Leontief was born in Leningrad in 1905. At the age of 15 he entered Leningrad University, graduating five years later with the title of "Learned Economist." It took him just two more years to obtain his Ph.D. from the University of Berlin. From there he traveled to China as an economic adviser and then to New York as a research associate at the National Bureau of Economic Research. In 1931 he began a 35-year tenure on the faculty of Harvard University. Since 1966 he has directed the Institute for Economic Analysis in New York. He lives in a Greenwich Village apartment with his wife Estelle, who is a poet.

Input-output analysis began in the 1920s when the federal government organized data to get a clearer picture of the workings of its economy. Leontief, then a young professor at Harvard, developed the primary methods of input-output analysis that later were instrumental in the policies of the important Controlled Materials Plan during World War II. After World War II the Air Force and the Department of Labor funded a project, called Project Scoop, which provided a detailed analysis of the economy. Since then input-output analysis has been regarded as one of the cornerstones of economic theory.

In 1965 Leontief published an account of the U.S. economy based on statistics from 1958.‡ His analysis examined the interdependence among 81 industrial sectors. We will present a more modest version in which the industries are lumped together

---

†See the July 1984 issue of *Omni* for an insightful interview with Leontief (pp. 78–82 and 110–118).
‡See Wassily W. Leontief, "The Structure of the U.S. Economy," *Scientific American*, April 1965, pp. 25–35.

Table 1

Sector	Abbreviation	Industries
Final nonmetal	FN	Apparel, food, drugs, furniture
Final metal	FM	Motor vehicles, aircraft, communications
Basic metal	BM	Iron, steel, heating, plumbing
Basic nonmetal	BN	Livestock, agriculture products, fabrics, lumber
Energy	E	Electricity, gas, water, coal, petroleum
Services	S	Transportation, insurance, real estate, finance

Table 2

		Input					
		FN	FM	BM	BN	E	S
	FN	0.170	0.004	0.000	0.029	0.000	0.008
	FM	0.003	0.295	0.018	0.002	0.004	0.016
Output	BM	0.025	0.173	0.460	0.007	0.011	0.017
	BN	0.348	0.037	0.021	0.403	0.011	0.048
	E	0.007	0.001	0.039	0.025	0.358	0.025
	S	0.120	0.074	0.104	0.123	0.173	0.234

into six sectors. Table 1 lists the sectors and some representative industries from each. Notice that it includes most of the articles involved in our daily life.

Table 2 gives Leontief's input-output table for the six-sector economy of the United States for 1958. The basic unit is one million dollars. Thus the number 0.004 in the (1, 2) entry of the table corresponds to $4000, since

$$0.004 \times \$1,000,000 = \$4000.$$

To solve the Leontief formula set

r = amount of FN goods produced in a year

s = amount of FM goods produced in a year

t = amount of BM goods produced in a year

x = amount of BN goods produced in a year

y = amount of E goods produced in a year

z = amount of S goods produced in a year

By including the external demands, Table 2 leads to the following system of six linear equations in six unknowns.

$$r = 0.170r + 0.004s \qquad\qquad + 0.029x \qquad\qquad + 0.008z + \quad 99,640$$

$$s = 0.003r + 0.295s + 0.018t + 0.002x + 0.004y + 0.016z + \quad 75,548$$

$$t = 0.025r + 0.173s + 0.460t + 0.007x + 0.011y + 0.007z + \quad 14,444$$

$$x = 0.348r + 0.037s + 0.021t + 0.403x + 0.011y + 0.048z + \quad 33,501$$

$$y = 0.007r + 0.001s + 0.039t + 0.025x + 0.358y + 0.025z + \quad 23,527$$

$$z = 0.120r + 0.074s + 0.104t + 0.123x + 0.173y + 0.234z + 263,985$$

Example 1 solves this system and interprets its meaning for the U.S. economy.

EXAMPLE 1

Problem

Find the production schedule to run the 1958 U.S. economy.

Solution Let X be the unknown vector consisting of the production schedules for the six sectors of the economy and let D be the bill of demands.

$$X = \begin{bmatrix} r \\ s \\ t \\ x \\ y \\ z \end{bmatrix} \qquad D = \begin{bmatrix} 99,640 \\ 75,548 \\ 14,444 \\ 33,501 \\ 23,527 \\ 263,985 \end{bmatrix}$$

The technology matrix T is precisely Table 2. The Leontief Formula is $X = TX + D$, and its solution is $X = (I - T)^{-1}D$.

We used the software package MATLAB to obtain the inverse of $I - T$, which is the following 6×6 matrix (rounded to four decimal places).

$$\begin{bmatrix} 1.2345 & 0.0139 & 0.0068 & 0.0640 & 0.0060 & 0.0175 \\ 0.0176 & 1.4365 & 0.0561 & 0.0139 & 0.0190 & 0.0329 \\ 0.0845 & 0.4710 & 1.8838 & 0.0416 & 0.0512 & 0.0568 \\ 0.7522 & 0.1335 & 0.1008 & 1.7416 & 0.0659 & 0.1242 \\ 0.0612 & 0.0454 & 0.1305 & 0.0834 & 1.5781 & 0.0612 \\ 0.3412 & 0.2366 & 0.3079 & 0.3155 & 0.3767 & 1.3529 \end{bmatrix}$$

The solution vector X is obtained by multiplying this matrix by the bill of demands D. The result is

$$X = \begin{bmatrix} 131,059 \\ 120,686 \\ 88,805 \\ 179,174 \\ 67,491 \\ 432,897 \end{bmatrix}$$

This means that in order to run the economy, the FN sector should produce $131,059 worth of goods, the FM sector should produce $120,686, the BM sector $88,805, the BN sector $179,174, the E sector $67,491, and the S sector $432,897.

Case Study Exercises

In Problems 1 to 4 find the production schedule to run the 1958 U.S. economy for the given bill of demands.

$$1.\ D = \begin{bmatrix} 100,000 \\ 600,000 \\ 500,000 \\ 800,000 \\ 200,000 \\ 100,000 \end{bmatrix} \qquad 2.\ D = \begin{bmatrix} 30,000 \\ 20,000 \\ 40,000 \\ 10,000 \\ 10,000 \\ 90,000 \end{bmatrix}$$

$$3.\ D = \begin{bmatrix} 300{,}000 \\ 200{,}000 \\ 500{,}000 \\ 100{,}000 \\ 100{,}000 \\ 900{,}000 \end{bmatrix} \qquad 4.\ D = \begin{bmatrix} 10{,}000 \\ 60{,}000 \\ 50{,}000 \\ 80{,}000 \\ 20{,}000 \\ 10{,}000 \end{bmatrix}$$

Problems 5 and 6 refer to the technology matrix T for the 1947 U.S. economy, where the three sectors are agriculture, manufacturing, and household.

$$T = \begin{bmatrix} 0.245 & 0.102 & 0.051 \\ 0.099 & 0.291 & 0.279 \\ 0.433 & 0.376 & 0.011 \end{bmatrix}$$

Find the production schedule to run the 1947 U.S. economy to meet the given bill of demands, using the fact that

$$(I - T)^{-1} = \begin{bmatrix} 1.454 & 0.291 & 0.157 \\ 0.533 & 1.763 & 0.525 \\ 0.837 & 0.791 & 1.278 \end{bmatrix}$$

$$5.\ D = \begin{bmatrix} 2{,}000{,}000{,}000 \\ 1{,}000{,}000{,}000 \\ 4{,}000{,}000{,}000 \end{bmatrix} \qquad 6.\ D = \begin{bmatrix} 1{,}000{,}000{,}000 \\ 3{,}000{,}000{,}000 \\ 100{,}000{,}000 \end{bmatrix}$$

CHAPTER REVIEW

Key Terms

2–1   Addition and Scalar Multiplication

Matrix	Addition of Matrices
Entries	Scalar Product
Size	Closed
Square Matrix	Associative
Row Matrix	Commutative
Column Matrix	Zero Matrix
Equal Matrices	Additive Inverse
Sum	

2–2   Matrix Multiplication

Matrix Multiplication	Constant Matrix
Product	Cancellation Law
Product Undefined	Main Diagonal
Coefficient Matrix	Identity Matrix
Variable Matrix	Distributive Laws

2–3 The Gauss–Jordan Method

Coefficient Matrix

Augmented Matrix

Elementary Row Operations

General Solution

Particular Solution

Echelon Form

Leading 1

Free Variable

Basic Variable

2–4 The Inverse

Invertible Matrix

Nonsingular Matrix

Inverse

Noninvertible Matrix

Singular Matrix

Determinant

2–5 Systems of Equations

Matrix Equation

Inverse Matrix Method

2–6 Input-Output Analysis

Input-Output Table

Internal Demands

External Demands

Bill of Demands

Leontief Formula

Technology Matrix

Summary of Important Concepts

The **sum** of A and B, $A + B$, is the matrix M whose entry in the ith row, jth column, m_{ij}, is

$$m_{ij} = a_{ij} + b_{ij}$$

The **scalar product** of A and r, rA, is the matrix M whose entry in the ith row, jth column, m_{ij}, is

$$m_{ij} = ra_{ij}$$

The **product** of A and B is the matrix C whose entry in the ith row, jth column, c_{ij}, is

$$c_{ij} = a_{i1}b_{1j} + a_{i2}b_{2j} + \cdots + a_{in}b_{nj}$$

Elementary Row Operations

1. Interchange two rows.
2. Multiply each number in a row by a nonzero number.
3. Replace a row by the sum of that row and a multiple of another row.

Matrix Method for Solving a System of Equations

1. Translate the system into $AX = C$, where A is the coefficient matrix, X the matrix of variables, and C the matrix of constants.
2. Compute A^{-1} if A is invertible.
3. Construct the solution $X = A^{-1}C$.

Procedure for Solving the Leontief Formula

Solve for X where $X = TX + D$

1. Form $I - T$.
2. Compute $(I - T)^{-1}$.
3. Multiply to get $X = (I - T)^{-1}D$.
4. Interpret the entries of X.

REVIEW PROBLEMS

1. (a) What is the size of M?
   (b) Find m_{42}.
   (c) What notation describes the entry 7?

$$M = \begin{bmatrix} 1 & -6 & 9 \\ -2 & 4 & 3 \\ 1 & 2 & 3 \\ 4 & 5 & 6 \\ 2 & 7 & 8 \end{bmatrix}$$

2. Compute $3A + 2B - C$, where

$$A = \begin{bmatrix} 1 & -1 & 4 \\ 3 & -3 & 7 \end{bmatrix} \quad B = \begin{bmatrix} -3 & 6 & 5 \\ 4 & 0 & 1 \end{bmatrix}$$

$$C = \begin{bmatrix} 5 & 3 & 7 \\ 1 & 2 & 6 \end{bmatrix}$$

3. Determine whether it is possible to find values of x and y that satisfy the equality. If so, find x and y.

$$x \begin{bmatrix} 2 \\ 5 \end{bmatrix} + y \begin{bmatrix} 3 \\ 7 \end{bmatrix} = \begin{bmatrix} 7 \\ 16 \end{bmatrix}$$

4. Compute AB, where

$$A = \begin{bmatrix} 1 & 4 & 5 \\ 0 & 0 & 0 \\ 3 & 11 & 2 \end{bmatrix}$$

$$B = \begin{bmatrix} -2 & 1 & 0 \\ 0 & 6 & 3 \\ 0 & 0 & -2 \end{bmatrix}$$

5. Compute XY, where

$$X = \begin{bmatrix} 1 & 4 & 5 \\ 0 & 0 & 0 \\ 3 & 11 & 2 \end{bmatrix}$$

$$Y = \begin{bmatrix} 1 & 0 & -1 & 6 \\ 2 & 5 & 2 & 3 \\ 4 & 1 & 3 & 1 \end{bmatrix}$$

Problems 6 to 8 refer to the following matrices.

$$A = \begin{bmatrix} 0 & -1 \\ 0 & 3 \end{bmatrix} \quad B = \begin{bmatrix} 4 & -5 \\ 6 & -3 \end{bmatrix}$$

$$C = \begin{bmatrix} 5 & -4 \\ 6 & -3 \end{bmatrix}$$

6. Compute AB.

7. Compute AC.

8. What do you conclude from Problems 6 and 7?

Problems 9 and 10 deal with powers of a matrix A. By definition, $A^2 = A \cdot A$, $A^3 = A \cdot A^2$, and so on.

9. Compute A^3 for the matrix $A = \begin{bmatrix} 1 & 2 \\ 3 & 4 \end{bmatrix}$

10. Compute A^{10} for the matrix $A = \begin{bmatrix} 1 & 0 \\ -1 & 2 \end{bmatrix}$

11. Write the coefficient matrix A, the variable matrix X, and the constant matrix C for the system of equations

$$\begin{aligned} w + x - 5y &= -1 \\ 3x + y + 7z &= 8 \end{aligned}$$

12. Use the Gauss–Jordan method to find the unique solution to this system of equations.

$$\begin{aligned} x + y - z &= 4 \\ 2x - y + 3z &= 5 \\ 5x - 3y - 5z &= 12 \end{aligned}$$

13. Find the general solution to the system. List two particular solutions. The variables are x, y, and z.

$$\begin{bmatrix} 1 & 0 & 1 & | & -2 \\ 0 & 1 & -1 & | & 4 \\ 0 & 0 & 0 & | & 0 \end{bmatrix}$$

14. Find the general solution to the system. List two particular solutions. The variables are x, y, and z.

$$\left[\begin{array}{ccc|c} 1 & -2 & 0 & -2 \\ 0 & 0 & 1 & 4 \\ 0 & 0 & 0 & 0 \end{array}\right]$$

15. Determine whether this system has either no solution or infinitely many solutions. If there are infinitely many solutions, derive the general form and list the basic variables.

$$-x + 2y - 3z = 19$$
$$-x + 2y - 2z = 4$$
$$2x - 4y + 3z = 7$$

16. Compute the inverse of this matrix if possible.

$$\left[\begin{array}{ccc} 1 & 1 & 1 \\ 2 & 1 & 1 \\ 0 & 1 & 1 \end{array}\right]$$

17. Compute the inverse of this matrix if possible.

$$\left[\begin{array}{ccc} 1 & -1 & 0 \\ 0 & 1 & 1 \\ 2 & 0 & -1 \end{array}\right]$$

18. The matrix $A = \left[\begin{array}{cc} 1 & -2 \\ -2 & 8 \end{array}\right]$ is invertible. The square of A^{-1} is denoted by A^{-2}.
    (a) Compute A^{-2}.
    (b) Compute A^2 and $(A^2)^{-1}$.
    (c) How are A^{-2} and $(A^2)^{-1}$ related?

19. Compute (a) $(AB^{-1})^{-1}$ (b) $(ABC^{-1})^{-1}$.

20. Find A where $A^{-1} = \left[\begin{array}{cc} 1 & 3 \\ 2 & 5 \end{array}\right]$.

21. Derive A^{-1} by forming AB for the following matrices.

$$A = \left[\begin{array}{ccc} 3 & 2 & -1 \\ 1 & -1 & 0 \\ 2 & 0 & -1 \end{array}\right]$$

$$B = \left[\begin{array}{ccc} 1 & 2 & -1 \\ 1 & -1 & -1 \\ 2 & 4 & -5 \end{array}\right]$$

In Problems 22 and 23 solve the system by the inverse matrix method if possible.

22.
$$x + y + z = 1$$
$$-2x - y - z = 0$$
$$y + z = 0$$

23.
$$2x \qquad + 4z = 16$$
$$- 2y - 3z = 1$$
$$3x - 4y + z = 29$$

24. Solve the systems of linear equations for each pair of values of a and b.

$$2x + 3y = a$$
$$5x + 7y = b$$

   (a) $a = 3$ (b) $a = 4$ (c) $a = -6$
       $b = 9$ $b = 0$ $b = 3$

25. Compute the production schedule where the technology matrix and the bill of demands are

$$T = \left[\begin{array}{cc} 0.4 & 0.5 \\ 0.2 & 0.75 \end{array}\right] \qquad D = \left[\begin{array}{c} 100 \\ 80 \end{array}\right]$$

PROGRAMMABLE CALCULATOR EXPERIMENTS

1. Write a program to calculate the determinant of an arbitrary 2×2 matrix.

2. Write a program to calculate the inverse of an arbitrary 2×2 matrix.

3. Write a program to calculate the determinant of an arbitrary 3×3 matrix.

4. Write a program to calculate the inverse of an arbitrary 3×3 matrix.

5. Use the programs in Problems 1 and 2 to calculate X for the given bill of demands matrix D and technology matrix T.

$$D = \left[\begin{array}{c} 30 \\ 10 \end{array}\right] \qquad T = \left[\begin{array}{cc} 0.1 & 0.4 \\ 0.3 & 0.2 \end{array}\right]$$

6. For the technology matrix given in Problem 5 calculate X for each of the given bill of demand matrices.

   (a) $D = \left[\begin{array}{c} 30 \\ 10 \end{array}\right]$ (b) $D = \left[\begin{array}{c} 10 \\ 30 \end{array}\right]$

   (c) $D = \left[\begin{array}{c} 25 \\ 15 \end{array}\right]$ (d) $D = \left[\begin{array}{c} 20 \\ 20 \end{array}\right]$

Linear Programming: Geometric Approach

3

Linear programming has affected business and political history. (C.P. Cushing/H. Armstrong Roberts)

Chapter Overview

Linear programming is a method of solving problems in which a particular quantity is to be minimized or maximized. The problem could be to maximize profit, minimize cost, or minimize caloric intake, or some similar optimization problem. The quantity to be optimized is limited in some ways. Profit is dependent on production capacity while cost depends on personnel. The solution of these problems involves linear inequalities and linear equations, and in Chapter 4 we will use matrices to solve more complicated problems.

Section 3–1 introduces linear inequalities. In Section 3–2 linear programming problems are translated into the language of mathematics. Then in Section 3–3 we use the skills developed in Section 3–1 to solve the linear programming problems using a geometric approach.

CASE STUDY PREVIEW

In 1947 George Dantzig was a young mathematician looking for a method of solving very complicated optimization problems. The problems he considered were real-world problems originating in the military, business, and economics. He was employed by the military but it was his friends who were economists and mathematicians, not to mention the engineers who were working on building a new type of machine called a ''computer,'' who led him on the path to success.

Many of the problems facing Dantzig involved systems of equations with 50 or even 100 equations and as many unknowns. The method he created, called the *simplex method,* described in Chapter 4, became an instant success. One of the first problems that he applied the method to was a transportation problem involving the large steel industry. These companies had built a virtual monopoly by claiming that the transportation cost of steel was too complicated to compute, so all steel companies, large and small, had to use the same simplified formula that favored the large companies. The simplex method showed them how to compute the cost, thereby allowing smaller companies to use their proximity to the user to cut cost and remain competitive.

After Section 3–2, we present a Case Study that shows how modern businesses use linear programming to solve major problems. It tells about Federal Express Corporation's decision to shift their operations from small aircraft to larger capacity ones.

The second Case Study, after Section 3–3, is based on the Berlin airlift, one of the first and most significant applications of linear programming. The problem was to supply Berlin and break the Russian blockade after World War II.

3–1 Linear Inequalities

The general form of a **linear inequality** in two variables is

$$ax + by < c$$

where a, b, and c are constants. The inequality sign $<$ can be replaced by $\leq$, $>$, or $\geq$, so there are actually four versions of the general form of a linear inequality.

Recall that $a \leq b$ means that $a < b$ or $a = b$. To review the notation for inequalities, remember that $6 < 8$, $6 \leq 6$, $-1 > -3$, and $-3 \geq -3$. Some elementary properties of inequalities are

Property 1. If $a < b$ and c is any number, then $a + c < b + c$.
Property 2. If $a < b$ and $c > 0$, then $ac < bc$.
Property 3. If $a < b$ and $c < 0$, then $ac > bc$.

A **solution** to a linear inequality is a pair of numbers that when substituted into the linear inequality produces a true statement. For instance, a solution to $3x + 5y < 10$ is $(0, 1)$ because $3 \cdot 0 + 5 \cdot 1 = 5$, which is less than 10. But $(2, 1)$ is not a solution because $3 \cdot 2 + 5 \cdot 1 = 11$, which is not less than 10.

Inequalities in two variables are graphed in the plane. The set of points that solve the inequality are **half-planes.** For example, to graph $y > 2x + 1$, first graph the set of points on the line $y = 2x + 1$. (See Figure 1.) They are not part of the solution to the inequality because y is not strictly greater than $2x + 1$. Select a point on the line, say, $(1, 3)$. Consider any point directly above $(1, 3)$, say, $(1, 5)$. Since $5 > 3$ and 3 equals $2x + 1$ when $x = 1$, we have $5 > 2x + 1$ when $x = 1$. This says that $(1, 5)$ is a solution to the inequality because it lies directly above a point on the line. Similarly all points that lie above the line are solutions. Let us summarize.

1. The set of points above the line satisfy $y > 2x + 1$.
2. The set of points below the line satisfy $y < 2x + 1$.
3. The set of points on the line satisfy $y = 2x + 1$.

In general, the solution to a linear inequality is determined by the corresponding equality. The solution to the inequality is one of the two half-planes into which the line divides the plane. All that is necessary is to test one of the points in the plane that is not on the line to determine if it is a solution to the inequality. If it is a solution, then the half-plane containing the point is the graph of the inequality. If it is not a solution, then the other half-plane is the graph.

Usually, the easiest point to choose is $(0, 0)$ because it is the simplest to substitute into the inequality. As long as $(0, 0)$ does not lie on the line we will select it as the test point.

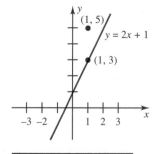

FIGURE 1

Procedure for Graphing Linear Inequalities

1. Replace the inequality sign by an equals sign and graph the corresponding straight line.
2. If the inequality sign is $\leq$ or $\geq$, then the line itself is part of the graph. Otherwise the line is not part of the graph.
3. Choose a test point that is not on the line and substitute the coordinates of the point into the inequality to determine if it satisfies the inequality. [Choose (0, 0) if it does not lie on the line.]
4. If the test point satisfies the inequality, then the half-plane containing the point is the graph of the inequality. If the test point does not satisfy the inequality, then the half-plane not containing the point is the graph.

We adopt the following two conventions when graphing inequalities.

1. If the line is part of the graph, it will be drawn as a solid line. If it is not part of the graph, it will be drawn as a dotted line.
2. The graph of the inequality will be the unshaded half-plane. In other words, think of the shaded part as being ''thrown away'' and the unshaded part, the graph of the inequality, as kept.

The second convention might seem strange at first, but it will become clear why it is done this way when systems of inequalities are solved. Think of shading points as ''erasing'' them from the graph. The points that remain unshaded become the solution set.

The next two examples illustrate the procedure.

EXAMPLE 1

Problem

Graph the inequalities.

(a) $2x + 3y \leq 6$ (b) $2x + 3y > 6$

Solution (a) Replace the inequality sign with an equality sign. Graph the equation $2x + 3y = 6$. (See Figure 2.) Choose a convenient point that is not on the

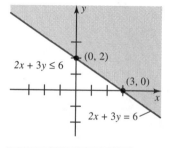

2x + 3y ≤ 6

(0, 2)

(3, 0)

2x + 3y = 6

FIGURE 2

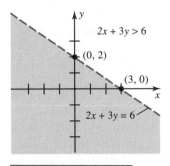

FIGURE 3

line; (0, 0) is usually the easiest choice. Test (0, 0) and get $2 \cdot 0 + 3 \cdot 0 = 0 \le 6$, which is true. Thus (0, 0) is a solution. From Figure 2, (0, 0) lies below the line so that region is the graph. The region above the line is then shaded. Since the "equals" part of the inequality is included, the line is drawn as a solid line.
(b) The only difference between parts (a) and (b) is the different inequality sign. The solutions differ in two respects. Since the equals part of the inequality sign is not included, the line is graphed as a dotted line. Also, when we test (0, 0) we get $2 \cdot 0 + 3 \cdot 0 = 0 > 6$, which is not true. Hence (0, 0) is not a solution and the graph is the region not containing (0, 0), the half-plane above the line. (See Figure 3.)

Thus far when the inequality was $>$ the region above the line was the solution. It might seem that this will always be the case; that $>$ means above the line and $<$ means below the line. It is not true, as Example 2 shows.

EXAMPLE 2

Problem

Graph the inequality $2x - 5y \le 10$.

Solution Follow the steps in the procedure.

1. Graph the line $2x - 5y = 10$. (See Figure 4.)
2. Use a solid line because the line is part of the graph.

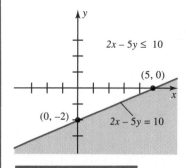

FIGURE 4

3. Test $(0, 0)$; $2 \cdot 0 - 5 \cdot 0 = 0 \le 10$, which is true so $(0, 0)$ is a solution and is part of the graph.
4. Since $(0, 0)$ is in the half-plane above the line, that region is the graph. Shade the region below the line.

Notice that the inequality in Example 2 has a $\le$ sign but the graph is the half-plane above the line.

Systems of Linear Inequalities

Often in applications there is more than one inequality. A set with more than one inequality is called a **system of inequalities.** A *solution* to the system of inequalities is the set of points that simultaneously satisfy all the inequalities. For instance, in Figure 5 we graph the linear inequalities from Examples 1(b) and 2. They comprise the system of linear inequalities

$$2x + 3y > 6$$
$$2x - 5y \le 10$$

The graph of the system is simply the two graphs superimposed on the same coordinate system. The two lines separate the plane into four regions; the region above both lines is the solution to the system. Because we have shaded the half-planes that are not the solutions, the unshaded region in Figure 5 is immediately seen to be the graph of the system.

EXAMPLE 3

Problem

Graph the system of inequalities.

$$x + y \le 4$$
$$2x - y \ge -4$$
$$y \ge 0$$

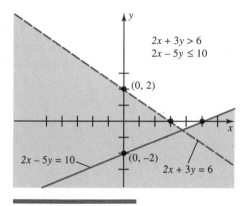

FIGURE 5

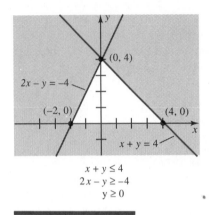

$$x + y \leq 4$$
$$2x - y \geq -4$$
$$y \geq 0$$

FIGURE 6

Solution Graph each line: $x + y = 4$, $2x - y = -4$, and $y = 0$. (See Figure 6.) Test $(0, 0)$ in the first two inequalities; it is a solution to both. Since $(0, 0)$ lies below each line, the partial solution is the region below both lines. Since $(0, 0)$ is on $y = 0$, choose another point to test, say, $(0, 1)$; it is a solution and lies above the line, so shade the region below the line. The solution to the system is the unshaded triangular region in Figure 6.

The next example shows how many real-world problems can be expressed with linear inequalities. It is typical of the type of problem in the next two sections.

EXAMPLE 4

Problem

A sausage manufacturer makes two types of sausages, regular and deluxe. Each regular sausage contains 2 pounds (lb) of filler and 1 lb of beef while each deluxe sausage contains 1 lb of filler and 2 lb of beef. How many sausages of each kind can be made with no more than 8 lb of filler and no more than 10 lb of beef?

Solution Express the data in a table where the rows are labeled "filler" and "beef" and the columns are labeled "regular," "deluxe," and "limiting value."

	Regular	Deluxe	Limiting Value
Filler	2	1	8
Beef	1	2	10

Let x represent the number of regular sausages and y the number of deluxe sausages. Then $2x$ represents the amount of filler used in regular sausages and y represents the number of pounds of filler used in deluxe sausages. Since no more than 8 lb of filler can be used, we get

$$2x + y \leq 8$$

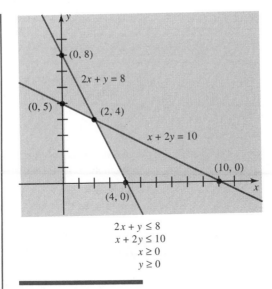

$$2x + y \le 8$$
$$x + 2y \le 10$$
$$x \ge 0$$
$$y \ge 0$$

FIGURE 7

Similarly, x is the amount of beef used in regular sausages and $2y$ is the amount of beef used in deluxe sausages. Since no more than 10 lb of beef can be used, we get

$$x + 2y \le 10$$

It makes no sense to let x and y take on negative values, so we get two additional inequalities:

$$x \ge 0 \qquad \text{and} \qquad y \ge 0$$

The system of inequalities that represents the problem is

$$2x + \ y \le \ 8$$
$$x + 2y \le 10$$
$$x \ge 0, \ y \ge 0$$

The solution is obtained by graphing the lines $2x + y = 8$, $x + 2y = 10$, $x = 0$, and $y = 0$ and shading the appropriate regions. The last two inequalities limit the graph to the first quadrant. The solution to each of the other regions is the region below the lines because $(0, 0)$ is a solution to each inequality. (See Figure 7.)

EXERCISE SET 3–1

In Problems 1 to 4 determine whether the points satisfy the given inequality.

1. $x + y > 4$, $(1, 2)$, $(2, 2)$

2. $2x - y < 0$, $(3, -1)$, $(-2, 0)$

3. $-x + 7y \ge 8$, $(-1, 1)$, $(0, 0)$

4. $3x - y \le 3$, $(1, 0)$, $(0, 1)$

In Problems 5 to 8 determine whether the points lie above, below, or on the line.

5. $x + y = 4$, $(0, 0)$, $(1, 7)$

6. $2x - y = 0$, $(0, 0)$, $(2, -1)$

7. $x - 3y = 1$, $(1, -1)$, $(-1, -1)$

8. $2x + 3y = 6$, $(3, 0)$, $(0, 0)$

In Problems 9 to 18 sketch the graph of the linear inequality by shading those points that do not satisfy the inequality.

9. $x + y > 2$

10. $x + y < -3$

11. $2x + 3y \leq 12$

12. $2x + 3y \geq 12$

13. $x + 3y > -6$

14. $-x - 3y \leq 6$

15. $4x + 3y \geq 12$

16. $4x - 3y \geq 12$

17. $y \leq 0$

18. $x \leq 0$

In Problems 19 to 30 graph the system of linear inequalities.

19. $x + 3y \leq 6$
    $x - 2y > 4$

20. $x + 2y \leq 4$
    $x - 3y \leq 6$

21. $2x + y \geq 4$
    $x - y < 5$

22. $x + y < -2$
    $2x - 5y \geq 10$

23. $x + y \leq 8$
    $2x + y \leq 14$

24. $x + y \leq 4$
    $x + 2y \leq 6$

25. $x + 3y \leq 6$
    $x - 2y > 4$
    $x \geq 0, y \geq 0$

26. $x + 2y \leq 4$
    $x - 3y \leq 6$
    $x \geq 0, y \geq 0$

27. $2x + y \geq 4$
    $x + y < 5$
    $x \geq 0, y \geq 0$

28. $x + y < -2$
    $2x + 5y \geq 10$
    $x \geq 0, y \geq 0$

29. $2x + y \leq 6$
    $x + y \leq 4$
    $x \geq 0, y \geq 0$

30. $x + 3y \leq 6$
    $x + 2y \leq 4$
    $x \geq 0, y \geq 0$

In Problems 31 to 34 express the word problem as an inequality.

31. A manufacturer makes two types of jackets, a wool jacket and a cotton jacket. It costs $200 to make each wool jacket and $130 to make each cotton jacket. The manufacturer can budget no more than $2600 to produce the jackets.

32. The manufacturer in Problem 31 can make no more than 100 jackets.

33. A travel agent has sold 400 tickets for a tour of Mexico. The agent can use two types of airplanes for the charter flights. Type 1 can carry at most 40 passengers and type 2 can carry at most 50 passengers.

34. The travel agent in Problem 33 is limited to a total air cost of $145,000. Each flight of a type 1 aircraft will cost $9000 and each type 2 flight will cost $15,000.

In Problems 35 to 38 express the word problem as a system of inequalities and graph its solution.

35. A manufacturer makes large and small toasters. It costs $20 to make each small toaster and $40 to make each large toaster. The manufacturer can budget no more than $2400 to make the toasters and can make no more than 100 toasters.

36. A garden store sells two types of lawnmowers, gas and electric. The store wants to sell at least 45 mowers per week. The profit on each gas mower is $50 and the profit on each electric mower is $40. The store wants to earn at least $2000 per week on the sale of lawnmowers.

37. A hospital nutritionist prepares a menu consisting of beef and rice that must contain at least 45 grams (g) of protein and at least 15 milligrams (mg) of iron. Each serving of beef contains 45 g of protein and 9 mg of iron. Each serving of rice contains 9 g of protein, and 6 mg of iron.

38. A meat manufacturer mixes beef and pork in sausage links. Two types of links are made, regular and spicy. Each pound of regular sausage meat contains 0.3 pound (lb) of beef and 0.1 lb of pork, while each pound of spicy sausage meat contains 0.2 lb of beef and 0.4 lb of pork. There are 120 lb of beef and 60 lb of pork in stock.

In Problems 39 to 45 graph the system of linear inequalities.

39. $x + 3y \leq 6$
    $x + 4y > 4$
    $x \geq 1, y \geq 1$

40. $x + 3y \leq 3$
    $x + 2y \leq 6$
    $x \geq 1, y \geq 2$

41. $4x + y \geq 8$
    $x + y \leq 5$
    $6x + y \leq 6$
    $x \geq 0, y \geq 0$

42. $x + y \leq 8$
    $x + 2y \leq 10$
    $2x + 5y \geq 10$
    $x \geq 0, y \geq 0$

43. $x + y \leq 10$
    $x + 2y \leq 12$
    $2x + y \leq 12$
    $x \geq 0, y \geq 0$

44. $x + y \leq 10$
    $x + 2y \geq 12$
    $2x + y \leq 12$
    $x \geq 0, y \geq 0$

45. $x + y \leq 10$
    $x + 2y \geq 12$
    $2x + y \geq 12$
    $x \geq 0, y \geq 0$

Referenced Exercise Set 3–1

1. The transportation problem was one of the first types of practical problems to which linear programming was applied in the early 1950s. In a more recent application Choypeng applied the procedure to problems in Thailand.* In one part of the study two types of ships were considered, carriers and transport ships. The total number of ships available was 100. Each carrier required 30 administrative personnel and each transport required 10 administrative personnel. There was a total of 2500 administrative personnel available. Express each constraint as an inequality.

2. In an article describing how to apply linear programming to credit investigation and credit granting procedures, John Stowe used the following analysis for customers with different types of credit ratings.† The net present value N of the loan, assuming credit was granted to the customer, was related to the size of the credit request x. For two types of customers the two variables were related by

$$N \geq -6.1 + .26x$$
$$N \geq -2.4 + .23x$$

Sketch the graph of the solution set of the system of inequalities.

3. In an article studying the effects of learning on resource allocation and scheduling, Joyce Chen used as an example a firm producing two products, type 1 and type 2.‡ To make each type 1 product requires 19 units of material and to make each type 2 product requires 20 units of material. No more than 14,440 units of material are to be used, and the total number of products made is not to exceed 750. Express each constraint as an inequality.

4. This problem shows how various expressions can be represented by inequalities. It is taken from an article by David Dishneau, a business reporter for the Associated Press.§ Let x be the number of Boeing airplanes ordered and write an inequality for each expression.
   (a) "United Airlines yesterday announced orders for as many as 370 Boeing aircraft valued at $15.74 billion."
   (b) "United orders up to 370 planes from Boeing."

---

*P. Choypeng, et al., "Optimal Ship Routing and Personnel Assignment for Naval Recruitment in Thailand," *Interfaces,* Vol. 16, 1986, pp. 47–52.

†John D. Stowe, "An Integer Programming Solution for the Optimal Credit Investigation/Credit Granting Sequence," *Financial Management,* Vol. 14, 1985, pp. 66–76.

‡Joyce Chen, "Modeling Learning Curve and Learning Complementarity for Resource Allocation and Production Scheduling," *Decision Sciences,* Vol. 14, 1983, pp. 170–186.

§David Dishneau, "United Orders Up to 370 Planes from Boeing," Associated Press, April 27, 1989.

3–2 Linear Programming Problems

A linear programming problem involves finding the largest or smallest value of a variable given that certain restrictions are met. For instance, a firm wants to maximize profit subject to limitations on production capacity and cost.

In this section we present three examples of typical linear programming problems. They are analogous to the ones that originated with George Dantzig's research in this area. The aim is to develop a method for translating a given problem into mathematical terms. In the next section we will show how to solve the problems once the translation has been carried out.

We first state a typical problem and then use it to illustrate the first few steps in expressing linear programming problems in mathematical terms. It is a type of problem that is frequently found in the business world.

An Example: The Manufacturing Problem

Manufacturing problem A clothing manufacturer makes two types of suits, wool and a blend of wool and polyester. The wool suit sells to the retail stores for $160 each and the blend for $90 each. The cost for manufacturing each wool suit is $100 and the cost of each blend suit is $40. If the manufacturer makes no more than 120 suits a week and budgets no more than $6000 per week, how many of each type should be made to maximize profit?

To express a linear programming problem in mathematical language, we must identify certain parts of the problem.

1. The variables, especially the one that is to be maximized or minimized, must be labeled. Assign a letter to each quantity and be careful about the units that are used.
2. Determine the equation that governs the quantity to be maximized or minimized. This equation is called the **objective function.**
3. Each problem contains restrictions that are called **constraints.** They are translated into linear inequalities. For each constraint the terms must have the same units.

Begin by rereading the problem once or twice. Often it is convenient to organize the data in a table. Let us demonstrate the technique by translating the manufacturing problem.

EXAMPLE 1

There are two types of suits, wool and blend. The problem entails finding the number of each type of suit that maximizes profit. Therefore the two variables are seen to be

x = the number of wool suits manufactured per week

y = the number of blend suits manufactured per week

One constraint is that the total number of suits manufactured cannot exceed 120. This translates into the inequality

$x + y \leq 120$

Generally, the coefficients of the variables are entered into the table, but for this constraint it seems to be clearer if the variables themselves, x and y, are put into the table. Hence x and y are entered in the first row together with the limiting value, 120. The next constraint is cost, so we enter the cost of each type of suit, $100 and $40, respectively, together with the limiting value, $6000. From the last sentence in the problem the objective function entails maximizing profit, so one of the rows in the table consists of profit. The cost of each wool suit is $100 and the revenue is $160, so the profit of each wool suit is $60. The profit on each blend suit is $90 − $40 = $50. These values are entered in the last row of the table. The problem can be summarized in Table 1.

Table 1

	Wool Suits	Blend Suits	Limiting Value
Number manufactured	x	y	120
Cost ($)	100	40	6000
Profit ($)	60	50	

In the third row there is no limiting value because the object is to find the largest profit under the given constraints. Let P represent profit and then the third row is interpreted as the equation

$$P = 60x + 50y$$

Notice that each term represents dollars. P is dollars of profit, and since 60 represents dollars per wool suit and x is the number of wool suits, $60x$ is

(dollars/wool suit)(number of wool suits) = dollars

Similarly, $50y$ is

(dollars/blend suit)(number of blend suits) = dollars

The second row of Table 1 represents cost, and since no more than $6000 is to be spent per week, it is interpreted as

$$100x + 40y \leq 6000$$

Two additional constraints are implicitly included in the problem. It makes no sense to allow x or y to take on negative values, and thus we must include the two inequalities $x \geq 0$ and $y \geq 0$ in the list of constraints. These constraints are easy to overlook because the problem does not explicitly mention them, but they are a part of almost every linear programming problem.

The problem is stated in mathematical language by the following.

Maximize $P = 60x + 50y$ objective function
Subject to $100x + 40y \leq 6000$
$\ \ \ \ x + \ \ y \leq \ 120$ constraints
$\ \ \ \ x \geq 0, y \geq 0$

The objective function is separated from the constraints because they play different roles in the solution to the linear programming problem.

Definition of a Linear Programming Problem

A linear programming problem originates as a word problem. When it is translated into mathematics it is expressed in a similar way to the problem in Example 1. While each problem has specific characteristics, it is possible to describe in general the formulation of linear programming problems in two variables. In the next chapter we will study linear programming problems in more than two variables.

DEFINITION

A **linear programming problem** in two variables is a problem that entails maximizing or minimizing a linear expression $z = ax + by$, called the objective function, subject to several constraints. The constraints are expressed as linear inequalities such as $ax + by \geq c$ or $ax + by \leq c$. The object is to find values of x and y that satisfy the constraints and yield the largest (or smallest) value of z.

Key Steps in Translating the Problem

Example 1 illustrates the key steps in translating a linear programming problem into mathematical language. We record them for easy reference.

Key Steps in Translating a Linear Programming Problem into Mathematical Language

1. Read and reread the problem to assess what is known and what is to be determined.
2. Organize the data, preferably into a table.
3. Identify unknown quantities and assign variables to them.
4. Determine the objective function.
5. Translate the constraints into linear inequalities.

Steps 2, 3, and 4 were described earlier in more detail. Notice that their order was different. This is to emphasize that the objective function can be determined before or after the constraints; just be sure to keep it separate and express it as an equation rather than an inequality like the constraints.

Remember that most linear programming problems are complicated and require a second or third reading to get started. Also, when organizing the data in a table, check to ensure that the numbers in each row and column agree with its description.

In the remainder of this section we present two more examples of linear programming problems. One is from the health field and the other is a transportation problem.

EXAMPLE 2

A hospital nutritionist prepares a menu for patients on low-fat diets. The primary ingredients are chicken and corn. The meal must contain enough iron and protein to meet at least the U.S. recommended daily allowance (RDA) for each nutrient. The RDA for protein is 45 grams (g) and for iron it is 20 milligrams (mg). Each 3-ounce serving of chicken contains 45 g of protein, 10 mg of iron, and 4 g of fat. Each 1-cup serving of corn contains 9 g of protein, 6 mg of iron, and 2 g of fat. How much chicken and corn should be included in the meal in order to minimize the amount of fat?

TABLE 2

	Chicken (3-oz serving)	Corn (1-cup serving)	Limiting Value
Protein (g)	45	9	45
Iron (mg)	10	6	20
Fat (g)	4	2	

Problem

Translate the problem into mathematical language.

Solution There are two types of food: chicken and corn. The limiting values are at least 45 g of protein and at least 20 mg of iron. Hence the constraints are protein and iron, respectively. They become the rows of the table while chicken and corn are the columns. The objective function is to minimize the amount of fat. It becomes the third row. (See Table 2.)

The variables are

x = the number of servings of chicken

y = the number of servings of corn

To determine the objective function, let F represent the amount of fat in the meal (in grams). From the third row it becomes

$$F = 4x + 2y$$

The first two rows are the explicit constraints. In each case the limiting value is prefaced by the term "at least," which translates into $\geq$. Thus the two constraints are $45x + 9y \geq 45$ and $10x + 6y \geq 20$. As before, it makes no sense to let x and y take on negative values, so the implicit constraints are $x \geq 0$ and $y \geq 0$. The problem can be stated as follows:

Minimize $F = 4x + 2y$ objective function

Subject to $\left.\begin{array}{l} 45x + 9y \geq 45 \\ 10x + 6y \geq 20 \\ x \geq 0, \ y \geq 0 \end{array}\right\}$ constraints

The next example is a common type of problem facing many businesses. It is called a transportation problem. We have chosen to illustrate this type of problem by maximizing profit subject to time and cost. In actual practice sometimes a firm needs to minimize time or cost subject to other limitations.

EXAMPLE 3   A bicycle manufacturer owns a retail outlet and, in addition, sells some bikes to a large chain store. One truck makes all deliveries from the manufacturing warehouse to the retail outlet and to the chain store distribution center. It takes 2 hours to make a round trip to the retail outlet and 4 hours to make a round trip to the distribution center. The cost of a round trip to the retail outlet is $50 and the cost of a round trip to the distribution center is $200. The profit of each truckload sold through the retail outlet is $210, and the profit is $400 for each truckload sold to the chain store. How many shipments per week should be scheduled to each store in order to maximize profit if the manufacturer is limited to no more than 40 hours of travel time per week and a cost of $1600?

Problem

Translate the problem into mathematical language.

Solution The variables can be determined from the last sentence of the problem.

x = the number of trips per week to the retail outlet

y = the number of trips per week to the chain store

P = profit

Since the constraints involve time and cost, they are the labels of the rows. The data can then be put into a table. (See Table 3.)

Table 3

	Retail Outlet	Chain Store	Limiting Value
Time for a round trip (hours)	2	4	40
Cost of a round trip ($)	50	200	1600
Profit ($)	210	400	

The first two rows represent the constraints on time and cost. Since each uses the phrase "no more than," the inequalities involve $\leq$. They are $2x + 4y \leq 40$ and $50x + 200y \leq 1600$. The third row, representing profit, is the objective function, $P = 210x + 400y$. The problem can be stated as follows:

$$
\begin{aligned}
\text{Maximize} \quad & P = 210x + 400y \quad && \text{objective function} \\
\text{Subject to} \quad & 2x + 4y \leq 40 \\
& 50x + 200y \leq 1600 \quad && \text{constraints} \\
& x \geq 0, \ y \geq 0
\end{aligned}
$$

The method of solving linear programming problems involves two distinct operations. First the word problem must be expressed in mathematics and then the mathematical formulation must be solved. This section concentrates on the first skill, translating the word problem into mathematics. In the next section the method of solving the mathematical formulation of the linear programming problem will be described.

When doing the exercises it is important to distinguish the objective function from the constraints. This is because the solution in the next section treats them separately. The first step in the solution is to graph the system of linear inequalities formed by the constraints. This graph produces a region in the plane. It is called the region of feasible solutions because the points in the region are the only ones that are allowed to be substituted into the objective function. Any point that is not in this region is not a possible solution. Then the points of intersection of the boundaries of the region are located. These points are called corner points. The corner points are then substituted into the objective function to find the solution. In the next section we will see why this method produces the solution.

EXERCISE SET 3–2

In Problems 1 to 26 translate the linear programming problem into mathematical language.

1. A clothing manufacturer makes two types of outerwear, a wool coat and a polyester all-weather coat. The wool coat sells to the retail stores for $190 each and the polyester coat for $130 each. The cost for manufacturing each wool coat is $150 and the cost of each polyester coat is $100. If the manufacturer makes no more than 100 coats a week and budgets no more than $12,000 per week, how many of each type should be made to maximize profit?

2. A clothing manufacturer makes two types of jogging attire, regular and designer. The attire sells to the retail stores for $10 each for the regular suits and $40 each for the designer suits. The cost for manufacturing each regular suit is $5 and it is $10 for a designer suit. If the manufacturer makes no more than 200 suits a week and budgets no more than $1500 per week, how many of each type should be made to maximize profit?

3. A hospital nutritionist prepares a menu for patients on low-fat diets. The primary ingredients are chicken and corn. The meal must contain at least 45 grams (g) of protein and at least 15 mg of iron. Each 3-ounce serving of chicken contains 45 g of protein, 9 mg of iron, and 4 g of fat. Each 1-cup serving of corn contains 9 g of protein, 6 mg of iron, and 2 g of fat. How much chicken and corn should be included in the meal in order to minimize the amount of fat?

4. Work Problem 3 with the constraint that the nutritionist wants the meal to contain at least the full U.S. RDA for protein but requires only that at least one-half the U.S. RDA for iron be met.

5. The computer center director at a large university wants to staff consulting stations with two types of shifts: type 1 will contain 2 senior programmers and 1 student worker and type 2 will consist of 1 senior programmer and 5 student workers. The director wants to use no more than 36 individuals. There will be at least 24 hours to be filled during the week, with a type 1 shift serving for 3 hours and a type 2 shift serving for 2 hours. The cost of a type 1 shift is $44 per hour and $40 per hour for a type 2 shift. Determine the number of shifts of each type in order to minimize cost.

6. Work Problem 5 if a type 2 shift is to consist of 1 senior programmer and 7 student workers, so that the cost of a type 2 shift is $48 and all other figures remain the same.

7. A bicycle manufacturer has a retail outlet and, in addition, sells some bikes to a large chain store. One truck makes all deliveries, and it takes 3 hours to make a round trip to the retail store and 4 hours to make a round trip to the chain store. The cost of a round trip to the retail store is $100 and the cost of a round trip to the chain store is $200. The profit of each truckload sold through the retail store is $500; it is $800 for each truckload sold to the chain store. How many shipments per week should be scheduled to each store in order to maximize profit if the manufacturer is limited to no more than 48 hours per week and a cost of no more than $2000?

8. Work Problem 7 if the profit on each truckload sold through the retail store is $600 and it is $800 for a truckload sold to the chain store.

9. A meat manufacturer mixes beef and pork in sausage links. Two types of links are made, regular and "all-beef." Each pound of regular sausage meat contains 0.3 pound (lb) of beef and 0.2 lb of pork, while each pound of all-beef sausage meat contains 0.4 lb of beef and 0.1 lb of pork. There are 120 lb of beef and 60 lb of pork in stock. If the profit for regular links is 50 cents per pound and the profit for all-beef links is

70 cents per pound, how many pounds of each type link should be made to maximize profit?

10. A consortium of travel agents has sold 1200 tickets to the Super Bowl. The weekend package includes air fare, and they have a choice of two types of airplanes for the charter flights. Type 1 can carry 100 passengers and type 2 can carry 150 passengers. Each flight of a type 1 aircraft will cost $9000 and each type 2 flight will cost $15,000. The consortium is allowed to lease no more than ten planes. How many airplanes of each type should be leased in order to minimize cost?

11. The administration of a money market fund wants to invest up to $15 million in two types of investments. The funds are to be divided between short-term bank notes and Treasury notes. The current yield for bank notes is 15% and for Treasury notes it is 12%. Because the yield for bank notes fluctuates unpredictably, it has been decided that at least twice as much money is to be invested in Treasury notes as in bank notes. At least $2 million must be allocated in each type of investment. How much money should be invested in each type of investment to produce the largest gain?

12. A window manufacturer produces windows in two styles, regular and thermopane. It costs $100 to make each regular window, which the manufacturer sells for $150. It costs $120 to make each thermopane window and each sells for $175. The daily production capacity is 110 windows and the daily cost cannot exceed $12,000. How many windows of each type should be made per day to maximize profit?

13. An automobile leasing company buys two types of cars, compacts and midsized. The company wants to purchase at most 1000 cars. The cost of each compact is $8000 and of each midsized car is $10,000. The company has decided to allocate no more than $9 million for the purchase of the cars. It must purchase at least 200 of each type of car. If the anticipated return from the sale of the cars at the end of the company's three-year leasing period is $1000 for a compact and $1200 for a midsized car, how many of each type should be purchased in order to maximize return?

14. Officials of the state high school basketball tournament must plan accommodations for at least 660 players. They have two types of rooms available, dormitory rooms that sleep three people and motel rooms that sleep two people. For meals they have budgeted $10 daily per person for those in dorms and $20 daily per person for those in motels, and they must not spend more than $12,000 in meal money per day. They have available at most 200 dormitory rooms and 150 motel rooms. If the daily cost of a room is $20 for a dormitory room and $30 for a motel room, how many rooms of each kind must they schedule in order to minimize cost?

15. A heating oil delivery firm has two processing plants, plant A and plant B. Plant A processes daily 150 barrels of high-grade oil for commercial use and 50 barrels of low-grade for residential use. Plant B processes daily 100 barrels of high-grade oil and 50 barrels of low-grade oil. The daily cost of operation is $20,000 for plant A and $15,000 for plant B. An order is placed for 1000 barrels of high-grade oil and 450 barrels for low-grade oil. Find the number of days that each plant should operate to fill the order and minimize the cost?

16. An automobile leasing company has warehouses in Los Angeles and San Francisco. A company in Las Vegas sends an order to lease 25 cars and a firm in Fresno orders 15 cars. The Los Angeles warehouse has 30 cars and the San Francisco warehouse has 20 cars. It costs $100 to ship a car from Los Angeles to Las Vegas, $90 to ship a car from Los Angeles to Fresno, $150 to ship a car from San Francisco to Las Vegas, and $80 to ship a car from San Francisco to Fresno. Find the number of cars to be shipped from each warehouse to each city in order to minimize cost.

17. A consortium of travel agents has sold 1800 tickets to the Super Bowl. The weekend package includes air fare, and there is a choice of two types of airplanes for the charter flights. Type 1 can carry 100 passengers and type 2 can carry 300 passengers. Each flight of a type 1 aircraft will cost $9000 and each type 2 flight will cost $15,000. The consortium is allowed to lease no more than 12 planes. Each type 1 flight requires 2 stewards and each type 2 flight requires 3 stewards. No more than 30 stewards are available. How many airplanes of each type should be leased in order to minimize cost?

18. A manufacturer makes two types of sausage meat. Each pound of type 1 sausage meat contains 3 ounces (oz) of beef, 1 oz of pork, and 2 oz of spices. Each pound of type 2 sausage meat contains 6 oz of beef, 3 oz of pork, and 1 oz of spices. There is 225 lb of

beef, 100 lb of pork, and 200 lb of spices in stock. If the profit for type 1 sausage meat is 90 cents per pound and the profit for type 2 sausage meat is 70 cents per pound, how many pounds of each type should be made to maximize profit?

19. A farmer plants two crops, wheat and corn. The profit per acre for wheat is $600 and for corn it is $900. The cost per acre for fertilizer is $8 for wheat and $16 for corn. The cost per acre for maintenance is $6 for wheat and $3 for corn. No more than 12 acres will be planted with wheat and corn. No more than $160 is to be spent for fertilizer and no more than $66 is to be spent for maintenance. How many acres of each crop should be planted to maximize profit?

20. A farmer plants two crops, wheat and corn. The profit per acre for wheat is $500 and for corn it is $600. The cost per acre for maintenance is $6 for wheat and $3 for corn. Time spent per acre per week for maintenance is one hour for wheat and two hours for corn. No more than 12 acres will be planted with wheat and corn. No more than $66 is to be spent for maintenance. Time spent for maintenance per week must be at least 20 hours. How many acres of each crop should be planted to maximize profit?

21. A lamp manufacturer makes two types of lamps: deluxe and regular. The cost of manufacturing each type of lamp is $60 for a deluxe model and $30 for a regular model. It costs $10 to ship each deluxe model and $15 to ship each regular model. The maximum weekly costs are $3600 for manufacturing and $1200 for shipping. No more than 100 lamps can be manufactured per week. How many lamps of each type should be made to maximize profit if the profit is $140 on each deluxe model? $90 on each regular model?

22. A consulting firm has two types of consulting teams to handle different types of business: type I consists of 2 architects and 2 salespersons; type II consists of 1 architect and 3 salespersons. There are 12 architects available and 24 salespersons available to make up the teams. They must form at least ten teams. If the average weekly revenue from each type of team is $4000 for type I and $5000 for type II, how many teams of each type should be made to maximize profit?

23. An animal shelter can house no more than 100 cats and dogs. The weekly cost for feeding each dog is $15 and the weekly cost for feeding each cat is $10. The shelter has a weekly allowance of $1050 for feed-

ing the animals. If the city government budgets $10 to house each dog and $8 to house each cat, how many animals of each type can be sheltered to minimize the housing cost?

24. A pet store owner specializes in bunnies and puppies. Each bunny costs $5.50 and sells for $11.95. Each puppy costs $13 and sells for $19.95. How many animals of each type should the store maintain to maximize profit if the costs cannot exceed $200 and the store cannot house more than 25 animals?

25. A car rental agency must rent at least 100 midsized cars and compact cars per day. It costs $15 per day to maintain a midsized car and $12 to maintain a compact car. Personnel costs related to the cars are $1.75 for each midsized car and $.85 for each compact car. A maximum of $3000 per day is available for maintenance and $200 for personnel costs. If the profit on each midsized car is $7 per day and the profit on each compact car is $5 per day, how many cars of each type should be rented each day to maximize profit?

26. An electronics firm trucks TV sets from an assembly plant in Atlanta to distribution outlets in New York (N.Y.) and Los Angeles (L.A.). The plant can assemble at most 3500 TV sets a week and there are 60 trucks available. Each truck to the N.Y. outlet carries 50 sets and each truck to the L.A. outlet carries 70 sets. It costs $400 for each truck to N.Y. and $500 for each truck to L.A. How many trucks should be dispatched each week to each distribution outlet to minimize cost?

In Problems 27 to 30 match Tables a to d with the given word problem.

27. A manufacturing firm makes steel tubing. Each tube is shaped and cut on machine I and then is buffed on machine II. Each large tube requires 4 minutes of time on machine I and 5 minutes of time on machine II. Each small tube requires 1 minute of time on machine I and 2 minutes of time on machine II. On a given shift machine I can operate no more than 70 minutes and machine II can operate no more than 80 minutes. If the profit on each large tube is $30 and it is $20 on each small tube, how many of each type should be made to maximize profit?

28. A manufacturing firm makes steel tubing. Each tube is shaped and cut on machine I and then is buffed on machine II. Each large tube requires 4 minutes of

Table a

		Limiting Value
2	1	20
4	5	30
Objective Function 80	70	

Table b

		Limiting Value
4	1	70
5	2	80
Objective Function 30	20	

Table c

		Limiting Value
2	4	80
1	5	70
Objective Function 20	30	

Table d

		Limiting Value
4	5	30
1	2	20
Objective Function 70	80	

time on machine I and 1 minute of time on machine II. Each small tube requires 5 minutes of time on machine I and 2 minutes of time on machine II. On a given shift machine I can operate no more than 30 minutes and machine II can operate no more than 20 minutes. If the profit on each large tube is $70 and it is $80 on each small tube, how many of each type should be made to maximize profit?

29. A manufacturer makes shirts. Each shirt is cut on machine I and then is sewn on machine II. On machine I each long sleeve shirt requires 2 minutes of time

and each short sleeve shirt requires 1 minute of time. On machine II each long sleeve shirt requires 4 minutes of time and each short sleeve shirt requires 5 minutes of time. On a given shift machine I can operate no more than 20 minutes and machine II can operate no more than 30 minutes. If the profit on each short sleeve shirt is $70 and it is $80 on each long sleeve shirt, how many of each type should be made to maximize profit?

30. A manufacturer makes shirts. Each shirt is cut on machine I and then is sewn on machine II. On machine I each long sleeve shirt requires 2 minutes of time and each short sleeve shirt requires 4 minutes of time. On machine II each long sleeve shirt requires 1 minute of time and each short sleeve shirt requires 5 minutes of time. On a given shift machine I can operate no more than 80 minutes and machine II can operate no more than 70 minutes. If the profit on each short sleeve shirt is $30 and it is $20 on each long sleeve shirt, how many of each type should be made to maximize profit?

Referenced Exercise Set 3–2

1. In 1967 the Missouri Air Conservation Commission adopted air pollution regulations for metropolitan St. Louis that created heated controversy. Coal burning accounted for more than 30% of the pollutants, and a major coal producer sought an injunction to prohibit enforcement of the regulations, claiming, among other things, that it was economically infeasible. A great deal of study ensued, thereby establishing precedent on the legality of pollution standards. A study by Robert Kohn applied linear programming to determine the economic impact of the regulations on the coal industry.* One problem investigated by Kohn dealt with two methods of controlling the pollutant ammonia nitrate. The two control methods were the use of a wet scrubber and a cyclone collector. The regulation required that there be a reduction of at least 165 million lb of sulfur dioxide. Another constraint was that if the two controls operated at peak efficiency, the largest reduction of particulates would be 230 million lb. For each ton of ammonia nitrate that was controlled by the wet scrubber, there was a reduction of sulfur dioxide by 9.2 lb and particulates by 7.6 lb. For each ton of ammonia nitrate

*Robert Kohn, "Application of Linear Programming to a Controversy on Air Pollution Control," *Management Science*, Vol. 17, 1971, pp. 609–621.

that was controlled by the cyclone collector, there was a reduction of sulfur dioxide by 4.6 lb and particulates by 3.1 lb. If the cost per ton was $.07 to operate the wet scrubber and $.03 to operate the cyclone collector, find the number of tons of ammonia nitrate production controlled by each type of control method in order to minimize cost.

2. The allocation of space is often a complicated problem for large enterprises. Large school systems and universities especially have a major task assigning classroom space. Gosselin and Truchon devised a linear programming model to help solve this type of allocation of space problem. The procedure has been applied at various locales, including the Université Laval.† As an example of their procedure, consider a segment of a university that uses rooms with 20 seats, called preferred rooms, and rooms with 40 seats, called regular rooms. The total number of these rooms available is 50. There are 1600 students to accommodate. The objective is to minimize the number of times a request for a particular type of room is denied. This is called a penalty. Suppose 10% of requests for preferred rooms are usually denied and 5% of requests for regular rooms are usually denied. Find the number of each type of room that should be allocated to this segment of the university to minimize the penalty function.

3. In his book on linear programming G. H. Symonds describes one of the first practical applications of the subject, which centered on the solution of oil refinery production problems.‡ The following problem is adapted from this book. An oil refinery produces at most 2000 gallons of fuel oil per day. Two types are produced, commercial fuel oil and residential fuel oil. Blending requirements are that the number of gallons of commercial fuel oil be at least three times as many as the number of gallons of residential fuel oil. If the profit per gallon is $.34 for commercial fuel oil and $.22 for

residential fuel oil, how many gallons of each type should be produced to maximize profit?

Cumulative Exercise Set 3–2

1. Determine whether each point lies above, below, or on the line $2x - 3y = 6$.
   (a) $(1, 0)$ (b) $(0, -2)$ (c) $(1, -2)$

2. Sketch the graph of the linear inequality $x - 3y \leq 9$.

In Problems 3 to 6 sketch the graph of the system of linear inequalities.

3. $2x + y \geq 3$
   $3x + 4y \leq 7$

4. $5x + 2y \leq 1$
   $2x - 3y \geq 8$

5. $x + y \geq -2$
   $x - y \leq 4$
   $3x + y \geq 0$

6. $3x + 4y \geq 12$
   $x \geq 0$
   $y \geq 0$

In Problems 7 and 8 a linear programming problem is stated. For each problem (a) write the objective function, (b) write the constraints, and (c) sketch the graphs of the constraints.

7. A travel agent has sold 3000 tickets to Hawaii via two types of airplanes. Type I planes can carry 150 passengers and type II planes can carry 300 passengers. Each flight on type I costs the agent $12,000, while each flight on type II costs the agent $15,000. The agent is allowed to lease no more than 15 planes. How many airplanes of each type should be leased in order to minimize cost?

8. A door manufacturer produces doors in two styles, regular and deluxe. It costs $100 to make each regular door, which the manufacturer sells for $175. It costs $120 to make each deluxe door, each of which sells for $250. The daily production capacity is 110 doors, and the daily cost cannot exceed $12,000. How many doors of each type should be made per day to maximize profit?

---

†K. Gosselin and M. Truchon, "Allocation of Classrooms by Linear Programming," *Journal of the Operational Research Society*, Vol. 37, 1986, pp. 561–569.

‡G. H. Symonds, *Linear Programming: The Solution of Refinery Problems*, New York, Esso Standard Oil Co., 1955.

CASE STUDY

Federal Express Corporation*

Seldom does a research paper in a college political science course amount to more than simply a grade for the writer. But Mr. Frederick W. Smith's 1964 paper, written while he was a student at Yale University, describing a new type of business venture, was put into practice in 1972. It was then that Mr. Smith founded a company he called Federal Express Corporation. His concept was to airlift small packages overnight with door-to-door pickup and delivery.

Just to open for business Mr. Smith needed a great deal of capital and planning. All airplanes, delivery trucks, offices, personnel, and government licenses had to be in place. In the eight years since writing his political science paper Mr. Smith firmed up his ideas while running a very successful airplane company. After leaving the Marines he purchased a company that serviced airplanes and gradually changed the business to buying and selling used aircraft. Over a two-year period Mr. Smith increased sales from $1 million to $9 million and turned its profit situation from a $40,000 loss to a $250,000 profit.

To verify the profit potential of his overnight delivery concept, Mr. Smith hired two business consultant firms. Both were enthusiastic. In fact, the president of one firm, Mr. Arthur Bass, joined Mr. Smith as a Senior Vice-President for Planning in 1972. It was Mr. Bass who decided to use analytical methods to help in making the complex decisions Federal Express would have to face in the near future. According to Mr. Bass, it was his job to translate Mr. Smith's far-reaching ideas into reality, or to turn "Mr. Smith's view of Federal Express in 1990 into a profit in the 1970s." It was this view to the future, their ability to visualize and anticipate future trends, that provided them with almost limitless potential and distinguished them from the competition. In 1978, reflecting on the meteoric growth of Federal Express, Mr. Smith said:

> In five years we have gone through phases of growth which other companies normally experience over a period of twelve to fifteen years. From June 1977 to June 1978 alone our revenues have jumped from $108 million to $160 million. In order to understand where we are, we have to look not only at the past but also at the future.

The past achievements of Federal Express and potential growth in a market they defined were summarized in an article that appeared in *The South Magazine,* March–April 1975, which stated that:

> If anyone wants a refresher course in the free enterprise system, in the basic premise that creativity, dedication and sweat adds up to better, in the assumption that a work force can be inspired by the challenge of its mission, and that free give-and-take competition equals progress, then go to Federal Express. It is undoubtedly one of America's great adventures.

---

*Adapted from John C. Camillus, "Federal Express Corporation: History and Concept of Business," in *Business Policy and Strategy Management,* by William Glueck and Laurence Jauck (eds.), New York, McGraw-Hill Book Co., 1980.

The key ingredient in Federal Express's strategy was that in a developed economy such as in the United States, with advanced technology and high capital investment, the need for swift transportation of parts and supplies is inevitable and frequent. While the needs of any individual customer might be unpredictable, from a statistical point of view, the timing and transportation requirements between large population centers would be fairly constant.

Federal Express's management believed that they could meet these demands and win 100% of the market if they could satisfy six primary requirements of their customers. They were: (1) coverage—they had to service all major population centers, (2) timing—they had to deliver before noon of the following day, (3) reliability—they would guarantee delivery, (4) cost—equitable rates were needed, (5) billing—it must be timely and accurate, (6) service—door-to-door pickup and delivery were essential.

Federal Express's most important consideration was the type of aircraft to use. The initial decision was guided by the Civil Aeronautics Board (CAB) ruling that distinguished between "air taxi" operators and cargo firms. The former could fly anywhere in the United States without the need to obtain a certificate of "public necessity and convenience." Operators not falling into this classification had to obtain a certificate for each city-pair route, of which Federal Express had 5000. No new certificate had been granted in over 20 years, as existing carriers' objections had to be overruled, with the burden of proof resting on the applicant. Federal Express's strategy to start up overnight with a multitude of city-pair routes made it impossible to classify their business as air cargo. They had to fall into the air taxi classification. This meant that the capacity of their planes was limited to 7500 pounds (lb). They chose the French-built Dassault Fanjet Falcon 20, and later added some Learjets.

These served as their only aircraft until 1978 when Federal Express's Washington lobbying resulted in an amendment to the CAB ruling, which essentially exempted cargo carriers from CAB control. This meant Federal Express could purchase larger aircraft. The obstacle was that their delivery system, based on the swift economy of small aircraft, was working efficiently and profitably. Would even a partial shift to large planes enhance operating efficiency? And if so, what mix would be most desirable?

United Airlines was retiring 20 Boeing 727s and Federal Express's management decided to analyze the feasibility of purchasing some or all of them. A comparison of the two types of planes yielded the following data:

	Boeing 727	Falcon 20
Direct operating cost	$1400/hr	$500/hr
Payload	42,000 lb	6000 lb

Hourly operating cost was limited to $35,000 and total payload had to be at least 672,000 lb. These facts translate into two constraints, but a third constraint was that only twenty 727s were available. The objective was to maximize the number of aircraft given these constraints.

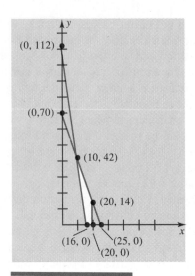

FIGURE 1

To solve the problem, let

x = the number of 727s

y = the number of Falcon 20s

Then the problem becomes

Maximize $N = x + y$ subject to
$$1,400x + 500y \leq 35,000$$
$$42,000x + 6,000y \geq 672,000$$
$$x \geq 0, y \geq 0, x \leq 20$$

In the next section we will see how to find the solution to this type of problem. The solution is (10, 42). That is, to maximize the number of aircraft under the given constraints, Federal Express should purchase ten 727s and 42 Falcon 20s. (See Figure 1.)

In 1978, following the removal of the CAB regulation, Federal Express purchased 10 Boeing 727s from United Airlines. The management of Federal Express was sure that the choice of the Fanjet Falcons and the Learjets played a vital role in the firm's early success. But as competition grew, both from new private companies and the U.S. Post Office, the larger aircraft would provide more flexibility and a wider range of cargo capability. In fact, as they were purchasing the 727s, Federal Express management was studying the feasibility of using 747s.

By the way, Mr. Smith received a grade of C on his paper.

Case Study Exercises

In Problems 1 to 4 restate the linear programming problem in mathematical language with the given changes in the Federal Express word problem.

1. The direct operating costs were $1200 and $600 for the 727 planes and the Falcons, respectively.
2. The direct operating costs were $1500 and $700 for the 727 planes and the Falcons, respectively.
3. The payload had to be at least 40,000 pounds (lb) for the 727s and 550 lb for the Falcons, respectively.
4. The payload had to be at least 43,000 lb for the 727s and 650 lb for the Falcons, respectively.

3–3 Graphical Solutions to Linear Programming Problems

Once a linear programming problem in two variables has been translated into mathematical language, then the graphical solution can be applied. In this section we demonstrate this geometric approach to the solution. It is effective in handling problems in two unknowns, like those in Section 3–2. For problems with more variables the simplex method is used, which is covered in Chapter 4.

The Manufacturing Problem (Revisited)

Let us demonstrate the steps of the graphical method by looking at the manufacturing problem of Section 3–2. A clothing manufacturer made two types of suits. The problem was to maximize profit subject to constraints of cost and capacity. We start with the mathematical formulation of the problem.

$$\text{Maximize } P = 60x + 50y \text{ subject to}$$
$$100x + 40y \leq 6000$$
$$x + y \leq 120$$
$$x \geq 0, y \geq 0$$

The task is to find the values of x and y that simultaneously satisfy the inequalities and yield the largest value of P. The method is separated into three steps: The first step is to graph the inequalities. The second is to find the **corner points,** which are the points of intersection of the boundaries of the region of solutions of the inequalities. These steps have already been covered in Sections 3–1 and 1–3, respectively. The only new idea is the definition of "corner point." The third step

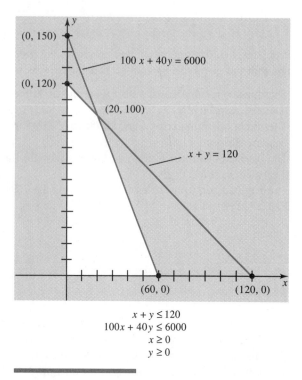

$$x + y \leq 120$$
$$100x + 40y \leq 6000$$
$$x \geq 0$$
$$y \geq 0$$

FIGURE 1

is to find which corner point yields the largest number when substituted into the objective function. We will solve the manufacturing problem and then show why the method works.

1. *Graph the region defined by the inequalities.* As in Section 3–1, replace the inequality signs with equal signs. Then graph the lines by plotting the intercepts and drawing a line through them. For each line shade the region that is not part of the solution. The unshaded region is the solution of the inequalities. (See Figure 1.)

2. *Find the corner points.* The boundary of the solution of the unshaded region consists of the line segments of the four lines $x + y = 120$, $100x + 40y = 6000$, $x = 0$, and $y = 0$. From Figure 1 the boundary consists of four line segments, and thus there are four points of intersection. Three of them were determined when we graphed the lines: $(0, 0)$, $(60, 0)$, and $(0, 120)$. The fourth is the point of intersection of $x + y = 120$ and $100x + 40y = 6000$. By the methods of Section 1–3, multiply the first equation by -40 and add the two equations. The point of intersection is $(20, 100)$. Hence the four corner points are $(0, 0)$, $(60, 0)$, $(0, 120)$, and $(20, 100)$.

3. *Evaluate the objective function at the corner points.* Put the computations in the following table:

Corner Point	Value of $60x + 50y$	$= P$
(0, 0)	$60 \cdot 0 + 50 \cdot 0$	$= 0$
(60, 0)	$60 \cdot 60 + 50 \cdot 0$	$= 3600$
(0, 120)	$60 \cdot 0 + 50 \cdot 120$	$= 6000$
(20, 100)	$60 \cdot 20 + 50 \cdot 100$	$= 6200$

The largest value of P is 6200. Therefore the solution is (20, 100); it occurs when $x = 20$ and $y = 100$. Referring to the meaning of the description of the manufacturing problem, the largest profit is \$6200 when the manufacturer makes $x = 20$ wool suits and $y = 100$ blend suits.

The Graphical Method of Solution

Before presenting the general method we need some terminology. The set of points that are solutions of the constraints, the unshaded region in Figure 1, is called the **region of feasible solutions.** The feasible solutions are the only points in the plane that have a chance to be solutions of the linear programming problem. The vertices of the boundaries are called **corner points** and the **optimal solution** is the point (or points) that yields the largest or smallest value of the objective function, depending on whether it is to be maximized or minimized.

It is possible that a linear programming problem does not have a solution. For instance, if the region of feasible solutions is unbounded, then the objective function $z = x + y$ cannot be maximized. This means that if the region is bounded, the solution occurs at a corner point. If the region is unbounded, first check whether a solution exists. If it does, it will occur at a corner point.

The Graphical Method for Solving a Linear Programming Problem Expressed in Mathematical Language

1. Graph the constraints and find the region of feasible solutions.
2. Find the corner points of the region of feasible solutions. They are the points of intersection of the boundary line segments.
3. Evaluate the objective function at each corner point. The optimum solution, if it exists, is that point that yields
   (a) the largest value if the objective function is to be maximized.
   (b) the smallest value if the objective function is to be minimized.

Justification of the Method

The natural question to ask is why the method works. There is an infinite number of points in the region of feasible solutions, each of which could possibly be a point that maximizes P. How can we be sure that the correct point is among the corner points? Let us answer this question for the manufacturing problem; the essence of the argument can be used to show that the method works for any linear programming problem.

The problem is to find values of x and y that are in the region of feasible solutions and yield the largest value of $P = 60x + 50y$. If P takes on a value, say, $P = 600$, then this equation becomes $600 = 60x + 50y$, which is a linear equation in x and y. The slope-intercept form of this equation is found by solving for y; so $50y = 600 - 60x$ and dividing by 50 yields the slope-intercept form

$$y = -(\tfrac{6}{5})x + 12$$

This line has slope $-\tfrac{6}{5}$ and y-intercept 12. Now let P take on another value, say, $P = 6200$. This produces another linear equation in x and y. Its slope-intercept form is

$$y = -(\tfrac{6}{5})x + 124$$

This line has the same slope as the previous line. In general, if P takes on a value, say, $P = P_0$, then this expression is a linear equation in x and y whose slope is $-\tfrac{6}{5}$ and whose y-intercept is $P_0/50$. This means that all of these equations are

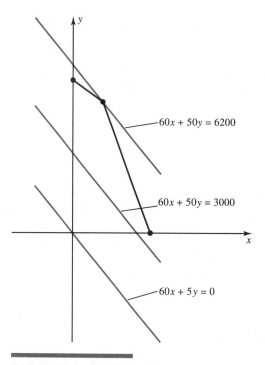

FIGURE 2 Some examples of the lines $60x + 50y = P_0$, the lines with slope $-\tfrac{6}{5}$.

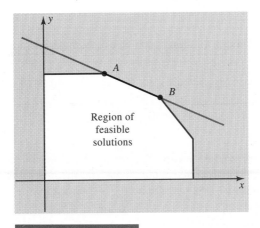

FIGURE 3

parallel and they differ only in their y-intercept. Figure 2 gives some examples of the lines $60x + 50y = P_0$.

The optimal solution is a point in the region of feasible solutions that is also substituted into the objective function, $P = 60x + 50y$. Graphically, this means we are looking for a point in the region of feasible solutions that is also on one of the lines $60x + 50y = P_0$. We need the point that yields the largest value of P_0. Making P_0 as large as possible means choosing the line with the largest y-intercept. Therefore we are searching for the line with the largest y-intercept that also touches the region of feasible solutions. From Figure 2 that line is the line that goes through $(20, 100)$. Any line below it has a smaller y-intercept and any line above it does not touch the region of feasible solutions. This is why $(20, 100)$ is the optimal solution.

This argument shows that the line with the largest y-intercept (or smallest if the objective function is to be minimized) that also touches the region of feasible solutions must be a corner point. Thus the corner points are the only ones that must be evaluated; the optimal solution must be one of them.

We have implied that there may be more than one optimum solution. This can happen if the objective function has the same slope as one of the boundaries. In Figure 3 not only are points A and B optimum solutions, but all points on the line segment between A and B are also optimum solutions. Since they lie on the same line they yield the same value of the objective function.

Applying the Procedure

We can now apply the graphical method to solve the problems that were translated into mathematical language in Section 3–2.

EXAMPLE 1

Problem

Find the optimum solution to the diet problem in Example 2 of Section 3–2.

$$\text{Minimize}\quad F = 4x + 2y$$
$$\text{Subject to}\quad 45x + 9y \geq 45$$
$$10x + 6y \geq 20$$
$$x \geq 0,\ y \geq 0$$

Solution

1. Graph the constraints and find the region of feasible solutions. The inequality signs are replaced by equality signs and the intercepts are plotted. The graph is in Figure 4; it is the infinite region above the line segments.
2. The corner points are $(2, 0)$, $(0, 5)$, and $(0.5, 2.5)$. The latter is the point of intersection of the two lines $45x + 9y = 45$ and $10x + 6y = 20$.
3. Evaluate the objective function at each corner point. The computations are in the following table:

Corner Point	Value of $4x + 2y$	$= F$
$(2, 0)$	$4\cdot2 + 2\cdot0$	$= 8$
$(0, 5)$	$4\cdot0 + 2\cdot5$	$= 10$
$(0.5, 2.5)$	$4\cdot0.5 + 2\cdot2.5$	$= 7$

The smallest value of F is 7. Therefore the solution is $(0.5, 2.5)$; it occurs when $x = 0.5$ and $y = 2.5$. Referring to the meaning of the description of the diet

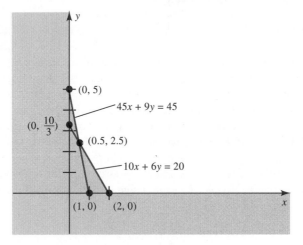

$$45x + 9y \geq 45$$
$$10x + 6y \geq 20$$
$$x \geq 0$$
$$y \geq 0$$

FIGURE 4

problem, the optimum meal with the smallest amount of fat has 0.5 serving of chicken (1.5 ounces) and 2.5 servings of corn (2.5 cups).

If the problem in Example 1 called for maximizing the objective function, there would be no solution because the region of feasible solutions is unbounded. The objective function, $F = 4x + 2y$, can be made as large as can be by choosing large enough values of x and y. Since the region is unbounded, such values can always be found in the region. This would be an example of the situation where you must check to see if the problem has a solution before applying the method.

EXAMPLE 2

Problem

Find the optimum solution to the transportation problem in Example 3 of Section 3–2.

Maximize $\quad P = 210x + 400y$
Subject to $\quad 2x + 4y \le 40$
$\quad\quad\quad\quad 50x + 200y \le 1600$
$\quad\quad\quad\quad\quad x \ge 0, y \ge 0$

Solution

1. Graph the constraints and find the region of feasible solutions. (See Figure 5.)

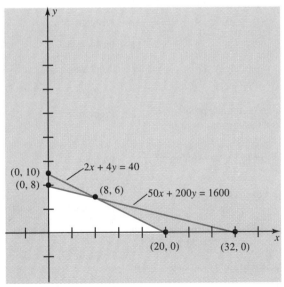

$2x + 4y \le 40$
$50x + 200y \le 1600$
$x \ge 0$
$y \ge 0$

FIGURE 5

2. The corner points are (20, 0), (0, 8), and (8, 6).

3. Evaluate the objective function at each corner point.

Corner Point	Value of $210x + 400y = P$
(20, 0)	$210 \cdot 20 + 400 \cdot 0 = 4200$
(0, 8)	$210 \cdot 0 + 400 \cdot 8 = 3200$
(8, 6)	$210 \cdot 8 + 400 \cdot 6 = 4080$

The optimum solution is (20, 0). The manufacturer should schedule 20 trips to the retail outlet and no trips to the chain store each week to maximize profit.

EXERCISE SET 3–3

In Problems 1 to 6 draw the graph of the set of feasible solutions and list the corner points.

1. $x + 4y \le 12$
$x + 3y \le 10$
$x \ge 0, y \ge 0$

2. $x + y \le 10$
$3x + 2y \le 12$
$x \ge 0, y \ge 0$

3. $x + 6y \le 18$
$2x + 3y \ge 12$
$x \ge 0, y \ge 0$

4. $x + 3y \le 12$
$2x + 5y \le 10$
$x \ge 0, y \ge 0$

5. $2x + 3y \le 10$
$4x + 7y \ge 28$
$x \ge 0, y \ge 0$

6. $3x + 2y \ge 18$
$x + 4y \le 12$
$x \ge 0, y \ge 0$

In Problems 7 to 12 solve the linear programming problem with the given objective function and the constraints from the stated problem in this section's exercises.

7. Maximize $z = 4x + 10y$ subject to the constraints in Problem 1.

8. Minimize $z = 3x + 4y$ subject to the constraints in Problem 1.

9. Minimize $z = 20x + 18y$ subject to the constraints in Problem 3.

10. Maximize $z = 10x + 12y$ subject to the constraints in Problem 2.

11. Minimize $z = 20x + 18y$ subject to the constraints in Problem 5.

12. Maximize $z = 10x + 12y$ subject to the constraints in Problem 6.

In Problems 13 to 30 solve the indicated problem from the exercise set in Section 3–2.

13. Problem 1

14. Problem 2

15. Problem 3

16. Problem 4

17. Problem 5

18. Problem 6

19. Problem 7

20. Problem 8

21. Problem 9

22. Problem 10

23. Problem 11

24. Problem 12

25. Problem 13

26. Problem 14

27. Problem 15

28. Problem 16

29. Problem 17

30. Problem 18

In Problems 31 to 38 find the values of x and y that maximize $P = 2x + 3y$ subject to the given constraints.

31. $x + 3y \le 6$
     $x + 4y \ge 4$
     $x \ge 1, y \ge 1$

32. $x + 3y \le 3$
     $x + 2y \le 6$
     $x \ge 1, y \ge 2$

33. $4x + y \le 8$
     $x + y \ge 5$
     $x + 3y \ge 9$
     $x \ge 0, y \ge 0$

34. $x + y \le 8$
     $x + 2y \le 10$
     $2x + 5y \ge 10$
     $x \ge 0, y \ge 0$

35. $x + y \le 10$
     $x + 2y \le 12$
     $2x + y \le 12$
     $x \ge 0, y \ge 0$

36. $x + y \le 10$
     $x + 2y \ge 12$
     $2x + y \le 12$
     $x \ge 0, y \ge 0$

37. $x + y \le 10$
     $x + 2y \ge 12$
     $2x + y \ge 12$
     $x \ge 0, y \ge 0$

38. $x + y \le 20$
     $x + 3y \ge 30$
     $y \le 25$
     $x \ge 5, y \ge 0$

39. A firm manufactures two types of lamps, gold-plated and chrome-plated. Each gold-plated lamp requires 2 hours of plating, 1 hour of finishing, and 3 hours of assembling. Each chrome-plated lamp requires 2 hours of plating, 2 hours of finishing, and $1\frac{1}{2}$ hours of assembling. The capacity of the plating department is 200 hours, the capacity of the finishing department is 120 hours, and the capacity of the assembly department is 180 hours. If the profit of each gold-plated lamp is $50 and the profit of each chrome-plated lamp is $30, find the number of each type of lamp that should be manufactured to maximize profit.

40. A hospital nutritionist prepares a menu consisting of chicken and rice that must contain at least 40 units of protein, at least 90 units of carbohydrates, and at least 50 units of iron. Each serving of chicken contains 4 units of protein, 1 unit of carbohydrates, 1 unit of iron, and 4 units of fat. Each serving of rice contains 1 unit of protein, 1 unit of carbohydrates, 3 units of iron, and 5 units of fat. How many servings of chicken and rice should the menu contain to minimize the amount of fat?

41. A meat manufacturer mixes beef and pork in sausage links. Two types of links are made, regular and spicy.

Each pound of regular sausage meat contains 0.1 pound (lb) of beef, 0.1 lb of pork, and 0.1 lb of spices. Each pound of spicy sausage meat contains 0.3 lb of beef, 0.1 lb of pork, and 0.2 lb of spices. The manufacturer can use no more than 120 lb of beef and 80 lb of pork, but at least 100 lb of spices must be used. The profit of each pound of regular sausage meat is $.40, while the profit of each pound of spicy sausage meat is $.60. Find the number of pounds of each type of sausage meat that should be made to maximize profit.

42. Solve Problem 41 if the manufacturer is required to manufacture at least 100 lb of regular sausage meat but no more than 700 lb of spicy sausage meat.

Referenced Exercise Set 3–3

1. Many farming strategies require crop rotations. In a model that was tested on a northeastern Oregon farm, researchers studied crop rotations of four crops: corn, potatoes, wheat, and alfalfa.* In one part of the study the number of acres planted with corn and the number of acres planted with potatoes were investigated. The total number of acres planted was limited to at most 100. Each acre of corn required 1 lb of nitrogen while each acre of potatoes required 3 lb of nitrogen. No more than 270 lb of nitrogen was to be used. Find the number of acres of each type of crop that should be planted to maximize profit if the profit per acre was $126 for corn and $1600 for potatoes.

2. In an article exploring new techniques in the application of linear programming to various aspects of planning and management, Robert Bland studied a small brewery whose ale and beer are in high demand but whose production is limited by materials that are in short supply.† Ale and beer are made from different proportions of corn, hops, and malt. Each barrel of ale requires 5 lb of corn, 4 oz of hops, and 35 lb of malt. Each barrel of beer requires 15 lb of corn, 4 oz of hops, and 20 lb of malt. The brewery has on hand 480 lb of corn, 160 oz of hops, and 1190 lb of malt. If the profit per barrel is $13 for ale and $23 for beer, how many barrels of each type should be produced to maximize profit?

*Talaat El-Nazer and Bruce McCarl, "The Choice of Crop Rotation: A Modeling Approach and Case Study," *American Journal of Agricultural Economics,* Vol. 68, 1983, pp. 127–136.

†Robert G. Bland, "The Allocation of Resources by Linear Programming," *Scientific American,* June 1983, pp. 126–144.

3. An integrated crop and intensive beef production firm is a farm enterprise that grows much of its own foodstuffs for its cattle. To operate with a minimum cost goal requires detailed long-term planning. In an article applying linear programming to the solution of this type of problem, J. J. Glen considered various foodstuff rations for the livestock.[‡] In one part of the study two variables were considered, the number of cattle fed primarily enterprise-grown wheat and the number of cattle fed primarily enterprise-grown hay. The number of cattle was between 200 and 500. Each animal fed primarily enterprise-grown wheat consumed 0.2 hectare of foodstuff, while each animal fed primarily enterprise-grown hay consumed 0.7 hectare of foodstuff. The number of hectares for these animals was limited to 140. Find the number of animals that should be given each type of ration to minimize cost if the cost is $35 to feed an animal the wheat ration diet and the cost is $30 to feed an animal the hay ration diet.

Problems 4 to 6 reveal an unexpected application of the fundamental concepts of linear programming. They show why the two leagues in major league baseball play the type of schedule that they do.[§] In Problems 4 and 5 assume that each team plays 162 games a year.

4. The National League is split into two divisions, each with six teams. Suppose a team plays each of the other five teams in its division x times and each of the six teams outside its division y times. Then

$$5x + 6y = 162$$

Because it is desirable for a team to play more games against teams in its own division than teams outside the division, it is reasonable to assume that

$$x > y$$

Moreover, because the baseball season is divided into series, and each series consists of at least two games in each of the two parks, it is reasonable to assume that

$$y \geq 4$$

(a) Find the region of feasible solutions of the two inequalities.

(b) Find the two pairs (x, y) of whole numbers in the region of feasible solutions satisfying $5x + 6y = 162$.

(c) Why do you suppose the National League adopted one solution instead of the other?

5. The American League is split into two divisions, each with seven teams. Suppose a team plays each of the other six teams in its division x times and each of the seven teams outside its division y times. Then

$$6x + 7y = 162$$

The same two constraints as in Problem 4 still apply.

$$x > y$$
$$y \geq 4$$

(a) Find the region of feasible solutions of the two inequalities.

(b) Find the two pairs (x, y) of whole numbers in the region of feasible solutions satisfying $6x + 7y = 162$.

(c) Why do you suppose the American League adopted one solution instead of the other?

6. For many years the teams in both divisions played 154 games a year.

(a) In the present alignment of the National League (see Problem 4), how many games would each team play against teams in its own division during a 154-game season?

(b) In the present alignment of the American League (see Problem 5), how many games would each team play against teams in its own division during a 154-game season?

Cumulative Exercise Set 3–3

1. Sketch the graph of the linear inequality by shading those points that do not satisfy the inequality.

$$2x + 3y \leq 24$$

In Problems 2 and 3 graph the system of linear inequalities.

2. $x + 2y \leq 6$
   $x - 2y > 4$

[‡]J. J. Glen, "A Linear Programming Model for an Integrated Crop and Intensive Beef Production Enterprise," *Journal Operational Research Society*, Vol. 37, 1986, pp. 487–494.

[§]John F. Kurtzke, "The Baseball Schedule: A Modest Proposal," *Mathematics Teacher*, May 1990, pp. 346–350.

3. $2x + 3y \leq 12$
   $x + 2y \geq 7$
   $x \geq 0, y \geq 0$

4. Express the word problem as a system of inequalities and graph its solution.
   A manufacturer makes two types of jackets—a wool jacket and a cotton jacket. The manufacturer sells the wool jacket to the retail stores for $250 each and sells the cotton jacket for $125 each. The manufacturer can budget no more than $2000 per week to produce the coats and can make no more than 100 jackets a week.

In Problems 5 to 8 translate the linear programming problem into mathematical language.

5. A tour director has sold 450 tickets for a weekend excursion, which includes a bus trip. The director can lease two types of buses. Type 1 can carry 50 passengers and type 2 can carry 75 passengers. Each type 1 bus will cost $900 and each type 2 bus will cost $1000. The director will lease no more than eight buses. How many buses of each type should be leased in order to minimize cost?

6. A door manufacturer produces doors in two styles, regular and ornate. It costs $500 to make each regular door, which the manufacturer sells for $650. It costs $700 to make each ornate door and each sells for $975. The daily production capacity is 100 doors and the daily cost cannot exceed $60,000. How many doors of each type should be made per day to maximize profit?

7. An automobile leasing company buys two types of cars, compacts and midsized. The company wants to purchase at most 2000 cars. The cost of each compact is $10,000 and of each midsized car is $12,000. The company has decided to allocate no more than $21 million for the purchase of the cars. It must purchase at least 200 of each type of car. If the anticipated return from the sale of the cars at the end of the company's 3-year leasing period is $1000 for a compact and $1200 for a midsized car, how many of each type should be purchased in order to maximize return?

8. An oil delivery firm has two processing plants, plant A and plant B. Plant A processes daily 100 barrels of high-grade oil for commercial use and 200 barrels of low-grade for residential use. Plant B processes daily 150 barrels of high-grade oil and 500 barrels of low-grade oil. The daily cost of operation is $20,000 for plant A and $35,000 for plant B. An order is placed for 500 barrels of high-grade oil and 1400 barrels for low-grade oil. Find the number of days that each plant should operate to fill the order and minimize the cost.

In Problems 9 to 12 solve the linear programming problem referred to in this cumulative exercise set.

9. Problem 5

10. Problem 6

11. Problem 7

12. Problem 8

CASE STUDY   **The Berlin Airlift**

One of the first concrete problems that George Dantzig's linear programming helped solve was the massive supply to West Berlin shortly after World War II.

Berlin surrendered to the Russian army on May 2, 1945. During the following weeks the Russians shipped most of the city's industrial goods to Russia. The American, British, and French troops did not arrive until July 1945. Even before the war ended the Allies decided to divide the city into four sectors, each country occupying one. Berlin lay deep in the Russian sector of the country, but the Western powers assumed that the Soviets would allow them free access to the city. However, on June 24, 1948 the Soviets blocked all land and water routes through East Germany to Berlin. They hoped to drive the Western powers from East Germany. The problem facing the Western Allies was how to keep West Berlin supplied during the Russian blockade.

The problem was turned over to the Planning Research Division of the U.S. Air Force. Their staff had been working on similar programming problems, most of which were theoretical. But now they were asked to apply their new methods to a very practical problem. Their solution helped shatter Soviet hopes of using the blockade to win total occupation of Berlin. A gigantic airlift was organized to supply more than 2 million people in West Berlin. It was a large-scale program that required intricate planning. To break the blockade, hundreds of American and British planes delivered massive quantities of food, clothing, coal, petroleum, and other supplies. At its peak a plane landed in West Berlin every 45 seconds.

The number of variables in the formulation of the problem exceeded 50. They included the number of planes, crew capacity, runways, supplies in Berlin, supplies in West Germany, and money. A simplified version of the problem is to maximize the cargo capacity of the planes subject to the restrictions: (1) The number of planes is limited to 44; (2) the American planes are larger than the British planes and therefore need twice the number of personnel per flight, so if we let one "crew" represent the number of flight personnel required for a British plane, and thus two "crews" are required for an American flight, then the restriction is that the number of crews available is 64; and (3) the cost of an American flight is $9000 and the cost of a British flight is $5000 and the total cost per week is limited to $300,000. The cargo capacity of an American plane is 30,000 cubic feet and 20,000 cubic feet for a British plane.

These data can be arranged in the following table:

	American Planes	British Planes	Limiting Value
Number of planes	x	y	44
Crew capacity	2	1	64
Cost of flights	9,000	5,000	300,000
Capacity	30,000	20,000	

The corresponding linear programming problem is

Maximize $C = 30,000x + 20,000y$ subject to
$$x + y \le 44$$
$$2x + y \le 64$$
$$9,000x + 5,000y \le 300,000$$

The solution is $x = 20$ and $y = 24$. Thus 20 American and 24 British planes should be used to maximize cargo capacity.

Case Study Exercises

In Problems 1 to 4 solve the linear programming problem in the case study for the given change in the objective function.

1. Maximize $C = 20,000x + 30,000y$
2. Maximize $C = 15,000x + 40,000y$
3. Maximize $C = 40,000x + 15,000y$
4. Maximize $C = 45,000x + 10,000y$

CHAPTER REVIEW

Key Terms

3–1 Linear Inequalities

Linear Inequality

Solution

Half-Plane

System of Linear Inequalities

3–2 Linear Programming Problems

Linear Programming Problems

Objective Function

Constraints

3–3 Graphical Solutions to Linear Programming Problems

Region of Feasible Solutions

Corner Point

Optimal Solution

Summary of Important Concepts

Key Steps in Translating Linear Programming Problems into Mathematical Language

1. Read and reread the problem.
2. Organize the data.
3. Identify the variable.
4. Determine the objective function.
5. Translate the constraints into linear inequalities.

The Graphical Method of Solution

1. Graph the constraints and find the region of feasible solution.
2. Find the corner points.
3. Evaluate the objective function at the corner points:
   a. If it is a maximum problem, select the largest value.
   b. If it is a minimum problem, select the smallest value.

REVIEW PROBLEMS

In Problems 1 to 3 sketch the graph of the linear inequality by shading those points that do not satisfy the inequality.

1. $2x + 3y \geq 12$

2. $3x - 5y \leq 30$

3. $5x + 2y < 20$

In Problems 4 to 8 graph the system of linear inequalities.

4. $x + 2y \geq 6$
   $x + y \geq 4$

5. $x + 3y \leq 6$
   $x - 2y > 4$

6. $5x + 3y \leq 30$
   $y \geq 5$

7. $x + 3y \leq 6$
   $x - 2y > 4$
   $x \geq 0, y \geq 0$

8. $2x + 3y \leq 12$
   $x + 3y \geq 9$
   $x \geq 0, y \geq 0$

9. Express the word problem as an inequality and graph its solution. A manufacturer makes two types of stoves,

gas and electric. The cost of a gas stove is $280 and the cost of an electric stove is $330. The manufacturer can budget no more than $4800 per week to produce the stoves.

10. Express the word problem as a system of inequalities and graph its solution. A manufacturer makes two types of lamps, large and small. The small lamp sells for $20 and the large lamp sells for $40. The manufacturer can budget no more than $2400 per week to produce the lamps and can make no more than 100 lamps per week.

In Problems 11 to 14 translate the linear programming problem into mathematical language.

11. A consortium of travel agents has sold 2500 tickets to the world's fair. The weekend package includes air fare, and there is a choice of two types of airplanes for the charter flights. Type 1 can carry 150 passengers and type 2 can carry 200 passengers. Each flight of a type 1 aircraft will cost $12,000 and each type 2 flight will cost $15,000. The consortium is allowed to lease no more than 15 planes. How many airplanes of each type should be leased in order to minimize cost?

12. A door manufacturer produces doors in two styles, regular and deluxe. It costs $100 to make each regular door, which the manufacturer sells for $175. It costs $120 to make each deluxe door and each sells for $250. The daily production capacity is 110 doors and the daily cost cannot exceed $12,000. How many doors of each type should be made per day to maximize profit?

13. Officials of a convention must plan accommodations for at least 800 participants. They have two types of rooms available, motel rooms that sleep three people and hotel rooms that sleep two people. For meals, they have budgeted $20 daily per person for those in motel rooms and $40 daily per person for those in hotels, and they must not spend more than $24,000 in meal money per day. If the daily cost of a room is $40 for a motel room and $60 for a hotel room, how many rooms of each kind must they schedule in order to minimize cost?

14. A manufacturer makes two types of hot dog meat. Each pound of type 1 hot dog meat contains 3 ounces (oz) of beef, 1 oz of pork, and 1 oz of ham. Each pound of type 2 hot dog meat contains 4 oz of beef, 2 oz of pork, and 1 oz of ham. There is 120 pounds (lb) of beef, 50 lb of pork, and 30 lb of ham in stock. If the profit for type 1 hot dog meat is $.30 per pound and the profit for type 2 hot dog meat is $.40 per pound, how many pounds of each type should be made to maximize profit?

15. Find the region of feasible solutions, the corner points, and the point that maximizes $z = 4x + 5y$ subject to

$$2x + 3y \geq 12$$
$$3x + \ y \leq 11$$
$$x, y \geq 0$$

16. Find the region of feasible solutions, the corner points, and the point that minimizes $z = x + 8y$ subject to

$$x + 3y \geq 24$$
$$x + 2y \leq 20$$
$$x, y \geq 0$$

In Problems 17 to 20 solve the linear programming problem for the given problem in this set of exercises.

17. Problem 11

18. Problem 12

19. Problem 13

20. Problem 14

PROGRAMMABLE CALCULATOR EXPERIMENTS

1. Write a program to find the four intercepts of the system of linear inequalities.

$$4x + 7y \geq 28$$
$$3x + 8y \leq 24$$
$$x \geq 0, y \geq 0$$

2. Write a program to find the four intercepts of a system of linear inequalities of the form

$$ax + by \leq c$$
$$dx + ey \leq f$$
$$x \geq 0, y \geq 0$$

3. Write a program to graph the system of linear inequalities given in Problem 1.

4. Write a program to calculate the point of intersection of the lines in the system of linear inequalities given in Problem 1.

5. Write a program to calculate the point of intersection of the lines in the system of linear inequalities given in Problem 2. First have the program test to determine whether the point of intersection exists.

6. Write a program to find the intercepts and the points of intersection of a system of linear inequalities of the form

$$ax + by \leq c$$
$$dx + ey \leq f$$
$$gx + hy \leq k$$
$$x \geq 0, y \geq 0$$

7. Consider the following linear programming problem.

Maximize $P = px + qy$ subject to
$$ax + by \leq c$$
$$dx + ey \leq f$$
$$x \geq 0, y \geq 0$$

Write a program that calculates the corner points of the region of feasible solutions and determines which is the solution of the problem.

8. Approximate the point of intersection of the following two lines graphically using the zoom utility of your calculator. Then solve the system algebraically. Which is easier? Which is more accurate?

$$4x + 7y = 28$$
$$3x + 8y = 24$$

Linear Programming: The Simplex Method

4

A case study shows how an American oil giant makes use of linear programming.
(M. Roessler/H. Armstrong Roberts)

Chapter Overview

In the previous chapter we considered linear programming problems with two variables. The approach was geometric and we could "see" the solution emerging in a figure because it could be viewed graphically. Unfortunately, this method does not extend to problems with more than two variables because it is much more difficult, if not impossible, to get a graphical depiction.

The simplex method can handle problems with many variables so it is applied to a wide variety of settings. However, unlike the geometric approach there is little intuition to guide you; thus it will appear to be a "cookbook" approach by contrast.

The first section lays the groundwork by defining an appropriate matrix and a special solution associated with it. The next section shows how to move to the next matrix if the present one does not yield the final solution. Section 4–3 finalizes the method for standard problems and Section 4–4 handles nonstandard problems. Section 4–5 studies a general type of problem.

CASE STUDY PREVIEW

The problems facing George Dantzig and his contemporaries in the late 1940s were monumental. They arose in the military, management, shipping, and economics. The most perplexing part was that most of the problems they faced could be expressed only in systems of equations and inequalities that contained far too many variables—often one hundred or more. Linear programming was created in 1947 when Dantzig invented the simplex method—a way to skirt most of the seemingly necessary calculations.

Today most large corporations have a department called something similar to "Operations Research" or "Industrial Management." They use mathematical analysis to solve problems facing the firm. One of the primary tools of these departments is linear programming. They solve problems in various areas, such as transportation cost, task assignment, production levels, and needs analysis of personnel and machinery.

The Case Study deals with the Shell Oil Company's method of handling the distribution of their oil. Since 20% of Shell's revenue is spent on transportation of oil, an efficient distribution is essential. Shell uses linear programming to determine how many thousands of barrels of oil to ship from their refineries to their storage and reshipment terminals. The Case Study investigates the solution of Shell's oil transportation problem in the Chicago region.

4–1 Slack Variables and the Simplex Tableau

At first we solve linear programming problems that are expressed in a specific form. In Section 4–4 we consider other types of linear programming problems and express them in this form. This shows that the method applies in all cases.

Standard Form

Consider the following two linear programming problems.

<table>
<tr><td align="center">**Problem 1**</td><td align="center">**Problem 2**</td></tr>
</table>

Maximize $m = 6x + 8y$ subject to   Minimize $m = 2x + 3y$ subject to

$$x + y \le 10 \qquad\qquad\qquad x + 2y \ge 15$$
$$2x + 3y \le 12 \qquad\qquad\qquad 4x + 3y \le 12$$
$$x \ge 0 \qquad\qquad\qquad\qquad x > -10$$
$$y \ge 0 \qquad\qquad\qquad\qquad y \ge 0$$

We say that Problem 1 is in *standard form* because it has three special properties.

D E F I N I T I O N

═══════════════════════════

A linear programming problem is in **standard form** if:

i. The objective function is to be maximized.
ii. The variables are all nonnegative; that is, they all are greater than or equal to 0.
iii. In the other constraints the linear expressions involving the variables are less than or equal to a nonnegative constant; that is, they are all in the form

(linear polynomial) $\le a$

where a is a positive number or 0.

Problem 2 is not in standard form for three reasons: because the objective function is to be minimized, violating condition (i); because x can be negative, violating condition (ii); and because the first constraint is in the form greater than or equal 15, which violates condition (iii).

The First Step: Slack Variables

From this point on until Section 4–4 we assume that each linear programming problem is in standard form. In particular, throughout the entire chapter we assume that all variables are greater than or equal to 0, so we will not continue to write this stipulation.

The first step in the simplex method is to translate each inequality into an equation. To illustrate, consider the inequality

$$2x + y \le 10$$

The point $(1, 3)$ satisfies the inequality because

$$2(1) + (3) = 5 \le 10$$

If $2x + y$ is less than 10, then let s_1 represent the difference between these two numbers. Hence

$2x + y + s_1 = 10$

Note that $s_1 \geq 0$; it is a positive number if $2x + y < 10$ and it is 0 if $2x + y = 10$. For the point $(1, 3)$, $s_1 = 5$ because

$2(1) + 3 + 5 = 10$

$x \qquad y \quad s_1$

Another way of saying that s_1 is the difference between $2x + y$ and 10 is that s_1 "takes up the slack" between them. For this reason s_1 is called a **slack variable.** Slack variables are used in this way to transform the inequalities of a linear programming problem into equalities. The subscripts indicate the inequality into which the slack variable was introduced. Example 1 clarifies these ideas.

EXAMPLE 1

Problem

Use slack variables to transform the linear programming problem with inequalities into one with equalities.

Maximize $m = 6x + 8y$ subject to
$x + \quad y \leq 10$
$2x + 3y \leq 12$

Solution Use s_1 with the first inequality to get the equation $x + y + s_1 = 10$. For the second inequality use s_2 and get $2x + 3y + s_2 = 12$. Hence the modified linear programming problem is

Maximize $m = 6x + 8y$ subject to
$x + \quad y + s_1 \qquad = 10$
$2x + 3y + \quad + s_2 = 12$

In Example 1 two new variables, s_1 and s_2, were introduced, so the modified system has four variables. Since x and y are greater than or equal to 0, so must s_1 and s_2 be greater than or equal to 0. Thus all four variables satisfy condition (ii) of the definition of standard form.

As in Chapter 2, each variable will be kept in a single column. The constraints are expressed as equations to take advantage of the techniques for solving systems of equations and matrices that were covered in Chapters 1 and 2. Another advantage is that the objective function and the constraints can be treated together in the same system of equations rather than separately.

The Proper Form of the Objective Function

The objective function is not in the same form of an equation as the constraints. It does not have a nonnegative constant to the right of the equals sign. By rearranging terms the objective function can be expressed in proper form. For example, the objective function in Problem 1 can be written as

$$-6x - 8y + m = 0$$

Using this form of the objective function, we can express Problem 1 in terms of linear equations as follows:

Maximize m such that

$$\left.\begin{array}{l} x + y + s_1 \qquad\qquad = 10 \\ 2x + 3y \qquad + s_2 \qquad = 12 \end{array}\right\} \text{ constraints}$$

$$-6x - 8y \qquad\qquad + m = 0 \qquad \text{objective function}$$

Hence the constraints and the objective function form a system of linear equations. The augmented matrix of the system is the matrix A.

$$A = \begin{array}{ccccc} x & y & s_1 & s_2 & m \\ \left[\begin{array}{ccccc|c} 1 & 1 & 1 & 0 & 0 & 10 \\ 2 & 3 & 0 & 1 & 0 & 12 \\ \hline -6 & -8 & 0 & 0 & 1 & 0 \end{array}\right] \end{array}$$

The columns are labeled to emphasize the variables to which each corresponds. We also separate the last column corresponding to the constant terms. In addition, the last row corresponding to the objective function must be distinguished from the others.

The Simplex Tableau

The matrix of the linear programming problem (with the constraints and the objective function as the corresponding equations), is called the **initial simplex tableau.** The term "tableau" is synonymous here with "matrix." The term "simplex" refers to the method of solution employed. "Initial" implies that we will derive other tableaus in the simplex method.

A simplex tableau has more columns than rows so there is an infinite number of solutions to the system. Each is a feasible solution to the linear programming problem.

Which feasible solution is best? The key is the form of the initial simplex tableau. In matrix A the third, fourth, and fifth columns each have exactly one entry equal to 1 and all others equal to 0. This type of column is in **basic form** and the variable corresponding to it is called a **basic variable.** Hence the basic variables for matrix A are s_1, s_2, and m. Columns 1 and 2 are not in basic form, so their variables, x and y, are called **nonbasic variables.**

Basic Feasible Solutions

The particular solution derived from the simplex tableau is found by setting the nonbasic variables equal to 0. Then the remainder of the solution is apparent. For example, in matrix A we set the nonbasic variables x and y equal to 0, so the corresponding equations become

$$
\begin{aligned}
s_1 \quad &= 10 \\
s_2 \quad &= 12 \\
m &= \ 0
\end{aligned}
$$

This type of solution is called a **basic feasible solution** of the linear programming problem. Hence the basic feasible solution for matrix A is $x = y = 0$, $s_1 = 10$, $s_2 = 12$, $m = 0$. We first encountered solutions to systems like this in Section 2–3. The general procedure can easily be extended to linear programming problems with more variables and constraints.

Procedure for Determining a Basic Feasible Solution

Let matrix M be a simplex tableau having some columns in basic form and some not in basic form.

1. Set each nonbasic variable of M equal to 0.
2. The resulting system of equations has one and only one basic variable in each equation. Each equation yields a value in the last column for a basic variable.

The basic feasible solution of a simplex tableau is a corner point of the region of feasible solutions. For the initial tableau it is the origin, which is almost never the optimum solution. The simplex method consists of constructing a sequence of tableaus and basic feasible solutions, which are also corner points of the region of feasible solutions. This sequence ultimately yields an optimal solution to the original linear programming problem. Section 4–2 will describe the construction of this sequence.

Example 2 shows how to apply the procedure for determining a basic feasible solution to tableaus that are not initial ones.

EXAMPLE 2

Consider the following two examples of simplex tableaus:

(a)

x	y	s_1	s_2	m	
-1	0	1	-2	0	2
1	1	0	1	0	5
1	0	0	3	1	15

(b)

x	y	z	s_1	s_2	s_3	m	
1	1	0	0	1	0	0	2
0	1	0	1	-1	1	0	5
0	-1	1	-2	3	0	0	10
0	2	0	0	2	0	1	15

Problem

Identify the basic and nonbasic variables and then compute the basic feasible solution for each matrix.

Solution (a) The columns in basic form are the second, third, and fifth columns, so the basic variables are y, s_1, and m. Thus the nonbasic variables are x and s_2. To compute the basic feasible solution, let $x = s_2 = 0$, and so the equations corresponding to the matrix are

$$
\begin{aligned}
s_1 &= 2 \\
y &= 5 \\
m &= 15
\end{aligned}
$$

The basic feasible solution is $s_1 = 2$, $y = 5$, $m = 15$, and $x = s_2 = 0$.

(b) The basic variables are x, z, s_3, and m and the nonbasic variables are y, s_1, and s_2. Let $y = s_1 = s_2 = 0$, so the equations corresponding to the matrix are

$$
\begin{aligned}
x &= 2 \\
s_3 &= 5 \\
z &= 10 \\
m &= 15
\end{aligned}
$$

The basic feasible solution is $x = 2$, $z = 10$, $s_3 = 5$, $m = 15$, $y = s_1 = s_2 = 0$.

In this section we showed both how to express a linear programming problem so it can be handled by the simplex method and how to compute the basic feasible solution. Example 3 combines these procedures.

EXAMPLE 3

Problem

Find the initial simplex tableau for the linear programming problem and compute its basic feasible solution.

Maximize $m = 2x + 3y + 4z$ subject to
$$
\begin{aligned}
x + 2y + 3z &\le 12 \\
2x + 5z &\le 10 \\
3x + y + z &\le 6
\end{aligned}
$$

Solution Add slack variables and put the objective function in proper form. The equations are

$$
\begin{aligned}
x + 2y + 3z + s_1 &= 12 \\
2x + 5z + s_2 &= 10 \\
3x + y + z + s_3 &= 6 \\
-2x - 3y - 4z + m &= 0
\end{aligned}
$$

The initial simplex tableau is

$$
\begin{array}{ccccccc}
x & y & z & s_1 & s_2 & s_3 & m \\
\end{array}
$$

$$
\left[
\begin{array}{ccccccc|c}
1 & 2 & 3 & 1 & 0 & 0 & 0 & 12 \\
2 & 0 & 5 & 0 & 1 & 0 & 0 & 10 \\
3 & 1 & 1 & 0 & 0 & 1 & 0 & 6 \\
\hline
-2 & -3 & -4 & 0 & 0 & 0 & 1 & 0 \\
\end{array}
\right]
$$

The basic variables are s_1, s_2, s_3, and m. The nonbasic variables are x, y, and z. To determine the basic feasible solution, set $x = y = z = 0$. The resulting equations are

$$
\begin{array}{rl}
s_1 & = 12 \\
s_2 & = 10 \\
s_3 & = 6 \\
m & = 0 \\
\end{array}
$$

Hence the basic feasible solution is $x = y = z = 0$, $s_1 = 12$, $s_2 = 10$, $s_3 = 6$, $m = 0$.

To summarize, we give the procedure for translating a linear programming problem in standard form into the mathematical form for solving it by the simplex method.

Procedure for Translating Linear Programming Problems in Standard Form into Mathematical Language

1. Add a slack variable to each constraint that is expressed as an inequality in order to transform it into an equation.
2. Put the objective function in proper form and include it with the constraints as a system of linear equations.
3. Form the augmented matrix of the system. Label the columns with the corresponding variables. Separate the last row corresponding to the objective function and the last column corresponding to the constant terms.

EXERCISE SET 4–1

In Problems 1 to 4 determine whether the linear programming problem is in standard form.

1. Maximize $m = 3x + 4y$ subject to

$$
\begin{aligned}
x + y &\leq 15 \\
2x - y &\leq 6 \\
x \geq 0, \; y &\geq 0
\end{aligned}
$$

2. Minimize $m = 3x + 5y$ subject to

$$
\begin{aligned}
2x + 3y &\leq 10 \\
x + 2y &\geq 5 \\
x \geq 0, \; y &\geq 0
\end{aligned}
$$

3. Maximize $m = 5x - y$ subject to
$$x + 3y \leq 0$$
$$3x + y \leq 8$$
$$x \geq 0, y \geq 0$$

4. Maximize $m = 2x + y + z$ subject to
$$x + 3z \leq 5$$
$$x + y \leq 8$$
$$x + 2y \leq 5$$
$$x \geq 0, y \geq 0, z \geq 0$$

In Problems 5 and 6 add slack variables to the inequalities to translate them into equations.

5. $x + 2y \leq 4$ 6. $x - y + z \leq 8$
   $x - y \leq 5$ $x + y - z \leq -2$

In Problems 7 and 8 put the objective function in proper form.

7. $m = 3x + 4y$ 8. $m = 2x + 8y + 3z$

Throughout the rest of the exercises assume the problem is in standard form.

In Problems 9 to 14 find the initial simplex tableau for the linear programming problem.

9. Maximize $m = 3x + 4y$ subject to
$$x + 2y \leq 10$$
$$3x + y \leq 8$$

10. Maximize $m = 4x + 9y$ subject to
$$3x + 4y \leq 12$$
$$2x + 5y \leq 10$$

11. Maximize $m = x + 5y$ subject to
$$3x + 8y \leq 1$$
$$4x - 5y \leq 4$$
$$2x + 7y \leq 6$$

12. Maximize $m = x + 2y + 4z$ subject to
$$x + 2y + 8z \leq 16$$
$$x + y + 3z \leq 12$$
$$x + z \leq 5$$

13. Maximize $m = 3x + y + 5z$ subject to
$$x + z \leq 15$$
$$x + 2y \leq 8$$
$$y + 5z \leq 10$$

14. Maximize $m = x - z$ subject to
$$x + 4y + 5z \leq 10$$
$$5x + 2y - z \leq 20$$
$$x + 3y + 6z \leq 12$$

In Problems 15 to 20 determine the basic variables and the nonbasic variables corresponding to the simplex tableau.

15.
x	y	s_1	s_2	m	
2	1	0	2	0	3
3	0	1	1	0	1
-1	0	0	3	1	6

16.
x	y	s_1	s_2	m	
0	3	2	1	0	-1
1	1	0	0	0	2
0	-2	0	0	1	0

17.
x	y	s_1	s_2	m	
1	0	2	0	0	3
0	1	3	1	0	4
0	-1	-6	0	1	8

18.
x	y	s_1	s_2	m	
-2	0	1	0	0	3
0	1	0	-1	0	4
0	0	0	5	1	8

19.
x	y	z	s_1	s_2	s_3	m	
1	0	1	0	0	1	0	1
2	0	0	0	1	2	0	3
0	1	0	3	0	1	0	4
0	0	0	-4	0	3	1	6

20.
x	y	z	s_1	s_2	s_3	m	
0	0	0	2	0	1	1	2
1	7	0	-1	1	2	0	5
0	1	1	0	2	1	0	14
0	0	0	0	-2	3	0	22

In Problems 21 to 26 compute the basic feasible solution to the simplex tableaus.

21. Problem 15.

22. Problem 16.

23. Problem 17.

24. Problem 18.

25. Problem 19.

26. Problem 20.

In Problems 27 to 30 construct the original linear programming problem (without slack variables) assuming that the matrix is the initial simplex tableau.

27.
1	2	1	0	0	3
4	5	0	1	0	6
-2	-1	0	0	1	0

28.
$$\left[\begin{array}{ccccc|c} 3 & 1 & 1 & 0 & 0 & 3 \\ 4 & 0 & 0 & 1 & 0 & 6 \\ \hline -1 & 0 & 0 & 0 & 1 & 0 \end{array}\right]$$

29.
$$\left[\begin{array}{ccccccc|c} 1 & 2 & 1 & 1 & 0 & 0 & 0 & 23 \\ 0 & 3 & 2 & 0 & 1 & 0 & 0 & 30 \\ 3 & 4 & 7 & 0 & 0 & 1 & 0 & 41 \\ \hline -1 & -2 & -5 & 0 & 0 & 0 & 1 & 0 \end{array}\right]$$

30.
$$\left[\begin{array}{ccccccc|c} 1 & 3 & 0 & 1 & 0 & 0 & 0 & 3 \\ 1 & 4 & 1 & 0 & 1 & 0 & 0 & 0 \\ 0 & 4 & 2 & 0 & 0 & 1 & 0 & 1 \\ \hline -9 & -2 & 0 & 0 & 0 & 0 & 1 & 0 \end{array}\right]$$

In Problems 31 to 34 consider the two inequalities

$x + 2y \le 10$
$2x + 5y \le 20$

Let s_1 and s_2 be slack variables that turn these inequalities into equalities so that

$x + 2y + s_1 = 10$
$2x + 5y + s_2 = 20$

Find the values of s_1 and s_2 that satisfy the equations for the given values of x and y.

31. $x = 3$, $y = 2$ 32. $x = 2$, $y = 1$

33. $x = 8$, $y = 0$ 34. $x = 1$, $y = 3$

In Problems 35 and 36 find the initial simplex tableau for the linear programming problems. Assume the problem is in standard form.

35. Maximize $m = x + 3y + 5z + 6w$ subject to

$$\begin{aligned} x \quad\;\; + 3z + \;\; w &\le 10 \\ y + \;\; z + 2w &\le 13 \\ y + 2z + 3w &\le 8 \\ y + 3z + 2w &\le 12 \end{aligned}$$

36. Maximize $m = x + z + 2w$ subject to

$$\begin{aligned} x + 4y + 5z \quad\quad\; &\le 18 \\ 5x + 2y \quad\;\; + 3w &\le 29 \\ x + 3y \quad\;\; + 6w &\le 22 \\ 2y + 3z + \;\; w &\le 20 \end{aligned}$$

In Problems 37 and 38 determine the basic variables, the nonbasic variables, and the basic feasible solution of the simplex tableau.

37.
x	y	z	s_1	s_2	s_3	m	
0	0	0	2	1	1	0	5
1	0	0	1	0	0	0	8
0	6	1	3	0	0	0	2
0	−1	0	−5	3	0	1	9

38.
x	y	z	s_1	s_2	s_3	m	
1	1	0	0	0	0	0	1
0	0	0	0	1	1	0	4
0	0	1	1	0	0	0	6
0	−1	0	2	1	0	1	2

39. Show that the number of variables of a linear programming problem with slack variables is greater than the number of equations.

In Problems 40 to 42 find the initial simplex tableau for the linear programming problem.

40. A firm manufactures three types of tents, small, medium, and large. Each small tent takes 2 hours to cut and 4 hours to sew. Each medium tent takes 3 hours to cut and 5 hours to sew. Each large tent takes 4 hours to cut and 7 hours to sew. There is a total of 310 hours available to cut and 500 hours available to sew the tents. No more than 100 tents are to be made. How many of each type of tent should be made to maximize profit if the profit is $30 on each small tent, $50 on each medium tent, and $70 on each large tent?

41. A soft-drink company makes three types of fruit juice drinks by mixing lemonade, limeade, and orange juice. Each gallon of lemon-lime drink uses 0.5 gallon (gal) of lemonade and 0.5 gal of limeade. Each gallon of orangeade drink uses 0.5 gal of orange juice, 0.25 gal of lemonade, and 0.25 gal of limeade. Each gallon of fruit punch drink uses 0.10 gal of orange juice, 0.5 gal of lemonade, and 0.4 gal of limeade. The company will use no more than 30 gal of orange juice, 40 gal of lemonade, and 25 gal of limeade. If the profit per gallon is $2 for lemon-lime, $1 for orangeade, and $2.50 for fruit punch drink, how many gallons of each type of drink should be made to maximize profit?

42. A farmer has 100 acres of land and wants to plant corn, beans, and tomatoes. Each acre of corn requires 4 days of labor, $140 in seed and fertilizer, and $50 to harvest. Each acre of beans requires 5 days of labor, $100 in seed and fertilizer, and $80 to harvest. Each acre of tomatoes requires 2 days of labor, $130 in seed and fertilizer, and $90 to harvest. The farmer has available 120 total days of labor, $10,500 to invest in seed and fertilizer, and $6800 to spend in harvesting. If the profit per acre is $320 for corn, $400 for beans, and $500 for tomatoes, how many acres of each plant should the farmer plant to maximize profit?

Referenced Exercise Set 4–1

In Problems 1 to 3 a linear programming problem adapted from a research study is given. Express the problem mathematically, find the initial simplex tableau, and determine whether it is in standard form.

1. The following problem is from an article studying the dairy industry of New Zealand.* The production of three types of products is studied: whole milk, cream, and buttermilk. The total number of liters of the three products produced per day is no more than 12 million. The number of liters of cream and buttermilk produced is less than one-half the number of liters of whole milk produced. If the profit per liter is $.54 for whole milk, $.44 for cream, and $.37 for buttermilk, how many millions of liters of each type of product should be produced to maximize profit?

2. This problem is from an article studying stock location problems, with a specific example of a library with limited storage capacity.† A new library is divided into three general storage areas: literature, science, and fiction. The total number of books stored is no more than 1 million. The number of nonfiction books is no more than 700,000. The number of literature books is at least 200,000. Find how many of each type of book should be stored to minimize cost if the cost is $25 for each literature book, $30 for each science book, and $22 for each fiction book.

3. Cleanup capability is a necessary evil in the oil industry. In an article studying various aspects of cleanup of oil spills, researchers set up the following type of problem.‡ Three types of cleanup equipment were considered. The transportation cost (in dollars per gallon of oil spilled per mile to the site) to a particular potential spill site is $.0015 for type 1 equipment, $.0015 for type 2 equipment, and $.00015 for type 3 equipment. The response time (in hours to reach the site) is 3 hours for type 1 equipment, 4 hours for type 2 equipment, and 5 hours for type 3 equipment. The operating cost (in dollars per gallon of oil spilled) is $5 for type 1 equipment, $7 for type 2 equipment, and $1 for type 3 equipment. If the total transportation cost is limited to $.012/gal/mi and the total response time is limited to 30 hours, find the number of each type of equipment that should be used to minimize the operating cost.

*B. R. Benseman, "Production Planning in the New Zealand Dairy Industry," *Journal of the Operational Research Society,* Vol. 37, 1986, pp. 747–754.

†James R. Evans, "The Factored Transportation Problem," *Management Science,* Vol. 30, 1984, pp. 1021–1024.

‡Harilaos N. Psaraftis, et al., "Optimal Response to Oil Spills: The Strategic Decision Case," *Operations Research,* Vol. 34, 1986, pp. 203–217.

4–2   The Maximality Test and Pivoting

The simplex method begins by expressing the linear programming problem as a system of equations. Then from the initial simplex tableau a sequence of simplex tableaus is computed to obtain a succession of solutions. The procedure ensures that the value of the objective function increases at each step. We continue the process until an optimal solution is found or we find that there is no solution.

Test for a Maximal Solution

For each new tableau in the procedure we must test its basic feasible solution to determine if it is the final optimal solution of the linear programming problem. Since the problem is in standard form, the objective function must be maximized.

The test itself is easy to apply, but understanding why it works is another matter. Suppose the last row of a simplex tableau is

$$\begin{array}{ccccc} x & y & s_1 & s_2 & m \\ [2 & -1 & 0 & 0 & 1 & 70] \end{array}$$

This row corresponds to the objective function. Its corresponding equation is

$$2x - y + m = 70$$

The numbers in the row are the coefficients of the variables in the equation. The object of the problem is to increase m as much as possible. This is done by increasing the other variables, in this case x and y. The central idea is the sign of the coefficients: x has a positive coefficient and y has a negative one. If $x = y = 0$, then $m = 70$. If x increases to 10 and $y = 0$, then m decreases to 50. If y increases to 10 and $x = 0$, then m increases to 80. Thus positive coefficients cause a decrease in m while negative coefficients cause an increase in m. This means that m cannot be increased further when there are no negative coefficents in the objective function. Since these coefficients are the numbers in the last row, we get the following test.

> ## Maximality Test
>
> A simplex tableau will yield an optimal solution if and only if the last row, corresponding to the objective function, contains no negative entries.

Remember that this test is applied to each tableau. If there are no negative entries in the last row, the solution is maximal. If the test fails, we compute another simplex tableau using a method that will be described shortly. Example 1 demonstrates how to use the maximality test.

EXAMPLE 1

Suppose A and B are the last rows of two simplex tableaus.

$$\begin{array}{ccccccc} & x & y & z & s_1 & s_2 & s_3 & m \\ A[& 1 & 0 & 2 & -2 & 1 & 0 & 0 & 6] \end{array}$$

$$B[\; 0 \quad 0 \quad 1 \quad\;\; 2 \quad 12 \quad 0 \quad 1 \quad 32 \;]$$

Problem

Apply the maximality test to each row.

Solution There is a negative number in row A so the basic feasible solution is not maximal. Another simplex tableau must be computed. In row B there are no negative numbers so the basic feasible solution is maximal. Therefore the maximum value of m is 32.

Suppose the Test Fails: The Entering Variable

If there are negative entries in the last row of the simplex tableau, then the maximality test says that the basic feasible solution is not maximal, so we must compute another tableau. This is done by **pivoting,** that is, by selecting one entry to be the pivot entry and using elementary row operations to change it to 1 and the other entries in its column to 0. This will be done in such a way that m increases by the largest possible amount.

The pivot operation uses elementary row operations to transform a prescribed column into a column with a 1 in a certain entry, called the *pivot entry,* and zeros elsewhere. There are three distinct procedures here: select the pivot column, select the pivot row (this identifies the pivot entry), and then perform the pivot operation. In the remainder of this section we explain how to select the pivot column and how to perform the pivot operation. In the next section we will show how to select the pivot row, which is a bit more complicated than the other two operations.

After the pivot operation is completed the variable corresponding to its column becomes a basic variable because its column will be in basic form. Thus pivoting turns a nonbasic variable into a basic one. It is easy to determine the proper variable, called the *entering variable* (because it "enters" the set of basic variables). If there is only one negative entry in the last row of the tableau, then the variable corresponding to the column containing that negative entry is the entering variable and that column is called the *pivot column.* Suppose, however, there are two or more negative entries. Then the one that is most negative will produce the largest increase in m and its variable is chosen as the entering variable. This is the next step in the procedure.

Selecting the Pivot Column

The column that has the most negative entry in the last row is the **pivot column.** The variable corresponding to its column is the **entering variable.**

If there are two or more entries that are equal and the most negative, any one of them can be chosen for the pivot column. Example 2 will help clarify how to select the pivot column.

EXAMPLE 2

Suppose A and B are the last rows of two simplex tableaus.

$$A \begin{bmatrix} x & y & z & s_1 & s_2 & s_3 & m \\ 3 & 0 & -4 & 0 & 0 & -1 & 1 & 84 \end{bmatrix}$$

$$B \begin{bmatrix} -8 & -1 & 0 & -8 & 0 & 0 & 1 & 12 \end{bmatrix}$$

Problem

Identify the pivot column and the entering variable for each tableau.

Solution For matrix A the most negative entry is -4, so the pivot column is column 3 and the entering variable is z. For matrix B the most negative entry is -8. Since it appears twice, the pivot column is either column 1 or column 4. The entering variable is either x or s_1.

Once the pivot column has been selected, elementary row operations are used to transform it into basic form. That is, make one entry 1 and the others 0. This is what is meant by the pivot process.

The Pivot Operation

To *pivot* a matrix about a nonzero entry, called the **pivot entry**, use elementary row operations to:

1. Transform the pivot entry into a 1.
2. Transform all other entries in the same column into 0's.

The next example shows how to pivot.

EXAMPLE 3

Consider tableau A.

$$A = \begin{array}{c} \begin{array}{ccccc} x & y & s_1 & s_2 & m \end{array} \\ \left[\begin{array}{ccccc|c} 1 & 2 & 0 & 3 & 0 & 26 \\ 0 & ① & 1 & 1 & 0 & 10 \\ 0 & -2 & 0 & 10 & 1 & 12 \end{array} \right] \\ \uparrow \end{array}$$

The maximality test fails because there is a negative entry in the last row. There is only one negative entry in the last row, so the second column is the pivot column and y is the entering variable. The $(2, 2)$ entry (the row 2, column 2 entry), is circled because it is the pivot entry (we will explain how it was chosen in the next section). To transform the second column into basic form, insert a 0 in the first row (in place of 2) and 0 in the third row (in place of -2). This is done by the two operations: replace the first row by -2 times row 2 plus row 1, R1 + (-2)R2, and replace row 3 by 2 times row 2 plus row 3, R3 + 2R2. This yields the new tableau B.

$$B = \begin{array}{c} \begin{array}{ccccc} x & y & s_1 & s_2 & m \end{array} \\ \left[\begin{array}{ccccc|c} 1 & 0 & -2 & 1 & 0 & 6 \\ 0 & 1 & 1 & 1 & 0 & 10 \\ 0 & 0 & 2 & 12 & 1 & 32 \end{array} \right] \end{array} \quad \begin{array}{l} \text{R1} + (-2)\text{(R2)} \\[1.5em] \text{R3} + 2\text{(R2)} \end{array}$$

We have pivoted about the $(2, 2)$ entry, thus letting y "enter" the set of basic variables.

There are two important observations about Example 3. First, elementary row operations are used in the pivot process to compute tableau B from tableau A. Hence the two matrices are *equivalent,* meaning that they have identical solution sets. Therefore the solutions to tableau B are the solutions of the original linear programming problem. This ensures that when we compute a new simplex tableau using the pivot process, the same solution set is retained. Also, the test for a maximal solution is applied whenever a new simplex tableau is computed. The matrix B passes the maximality test so its basic feasible solution is the maximal solution for the original linear programming problem.

When a new tableau is computed, the set of basic variables changes: the variable corresponding to the pivot column becomes a new basic variable and so one of the former basic variables becomes a nonbasic variable. In Example 3 the new basic variable is y, and s_1 changes from a basic variable to a nonbasic variable.

The next example illustrates the pivot process. It begins with a linear programming problem and uses three pivots to obtain the maximal solution. Recall that we are assuming that each linear programming problem in this section and the next is in standard form, so, in particular, all variables are greater than or equal to 0.

EXAMPLE 4

Problem

Maximize $m = 3x + 2y + z$ subject to

$$
\begin{aligned}
x + \quad z &\le 200 \\
y + 2z &\le 400 \\
2x + \quad y &\le 600
\end{aligned}
$$

Solution First use slack variables to express the problem as a system of equations while expressing the objective function in proper form.

$$
\begin{aligned}
x \quad\quad + \quad z + s_1 \quad\quad\quad\quad\quad &= 200 \\
y + 2z \quad\quad + s_2 \quad\quad\quad &= 400 \\
2x + \quad y \quad\quad\quad\quad\quad + s_3 \quad &= 600 \\
-3x - 2y - \quad z \quad\quad\quad\quad\quad\quad + m &= 0
\end{aligned}
$$

The initial simplex tableau is

x	y	z	s_1	s_2	s_3	m	
①	0	1	1	0	0	0	200
0	1	2	0	1	0	0	400
2	1	0	0	0	1	0	600
−3	−2	−1	0	0	0	1	0

The maximality test fails because there are negative numbers in the last row. The pivot column is the first column since −3 is the most negative number in the last row. Suppose the pivot entry is the (1, 1) entry. We get

$$
\begin{array}{ccccccc}
x & y & z & s_1 & s_2 & s_3 & m \\
\end{array}
$$

$$
\left[\begin{array}{ccccccc|c}
1 & 0 & 1 & 1 & 0 & 0 & 0 & 200 \\
0 & 1 & 2 & 0 & 1 & 0 & 0 & 400 \\
0 & 1 & -2 & -2 & 0 & 1 & 0 & 200 \\
0 & -2 & 2 & 3 & 0 & 0 & 1 & 600
\end{array}\right]
\quad
\begin{array}{l}
\\
\\
R3 + (-2)R1 \\
R4 + 3(R1)
\end{array}
$$

The maximality test fails because there is a negative entry in the last row. The pivot column is column 2. Let us pivot about the (3, 2) entry. We get

$$
\begin{array}{ccccccc}
x & y & z & s_1 & s_2 & s_3 & m \\
\end{array}
$$

$$
\left[\begin{array}{ccccccc|c}
1 & 0 & 1 & 1 & 0 & 0 & 0 & 200 \\
0 & 0 & 4 & 2 & 1 & -1 & 0 & 200 \\
0 & 1 & -2 & -2 & 0 & 1 & 0 & 200 \\
0 & 0 & -2 & -1 & 0 & 2 & 1 & 1000
\end{array}\right]
\quad
\begin{array}{l}
\\
R2 + (-1)R3 \\
\\
R4 + 2(R3)
\end{array}
$$

The maximality test fails again. The pivot column is column 3. Let us pivot about the (2, 3) entry. First perform $\frac{1}{4}$(R2) to make the pivot entry 1.

$$
\begin{array}{cccccccc}
x & y & z & s_1 & s_2 & s_3 & m \\
\end{array}
$$

$$
\left[\begin{array}{ccccccc|c}
1 & 0 & 0 & \frac{1}{2} & -\frac{1}{4} & \frac{1}{4} & 0 & 150 \\
0 & 0 & 1 & \frac{1}{2} & \frac{1}{4} & -\frac{1}{4} & 0 & 50 \\
0 & 1 & 0 & -1 & \frac{1}{2} & \frac{1}{2} & 0 & 300 \\
0 & 0 & 0 & 0 & \frac{1}{2} & \frac{3}{2} & 1 & 1100
\end{array}\right]
\quad
\begin{array}{l}
R1 + (-1)(R2) \\
\frac{1}{4}(R2) \\
R3 + 2(R2) \\
R4 + 2(R2)
\end{array}
$$

The maximality test passes and the basic feasible solution is maximal. It is

$$
x = 150, \quad y = 300, \quad z = 50, \quad s_1 = s_2 = s_3 = 0, \quad m = 1100
$$

Hence the solution to the original problem is $m = 1100$ when $x = 150$, $y = 300$, and $z = 50$.

EXERCISE SET 4–2

In Problems 1 to 6 assume that each row matrix is the last row of a simplex tableau. Determine whether it passes the maximality test.

$$
\begin{array}{ccccc}
x & y & s_1 & s_2 & m \\
\end{array}
$$
1. $\begin{bmatrix} 2 & 0 & 0 & -1 & 1 & 10 \end{bmatrix}$

$$
\begin{array}{ccccc}
x & y & s_1 & s_2 & m \\
\end{array}
$$
2. $\begin{bmatrix} 0 & 3 & -8 & 0 & 1 & 0 \end{bmatrix}$

$$
\begin{array}{ccccc}
x & y & s_1 & s_2 & m \\
\end{array}
$$
3. $\begin{bmatrix} 0 & 0 & 3 & 0 & 1 & 0 \end{bmatrix}$

$$
\begin{array}{cccccc}
x & y & z & s_1 & s_2 & s_3 & m \\
\end{array}
$$
4. $\begin{bmatrix} 2 & 1 & 0 & 0 & 3 & 0 & 1 & 10 \end{bmatrix}$

$$
\begin{array}{cccccc}
x & y & z & s_1 & s_2 & s_3 & m \\
\end{array}
$$
5. $\begin{bmatrix} -2 & 1 & 0 & 0 & -5 & 1 & 6 & 8 \end{bmatrix}$

$$
\begin{array}{ccccccc}
x & y & z & s_1 & s_2 & s_3 & m \\
\end{array}
$$
6. $\begin{bmatrix} 3 & 0 & 1 & 0 & 8 & 0 & 0 & 0 \end{bmatrix}$

In Problems 7 to 10 find the basic feasible solution of the simplex tableau and then apply the maximality test.

$$
\begin{array}{ccccc}
x & y & s_1 & s_2 & m \\
\end{array}
$$
7. $\left[\begin{array}{ccccc|c}
0 & 3 & 1 & 2 & 0 & 10 \\
1 & -2 & 0 & 1 & 0 & 15 \\
0 & -1 & 0 & 0 & 1 & 20
\end{array}\right]$

$$
\begin{array}{ccccc}
x & y & s_1 & s_2 & m \\
\end{array}
$$
8. $\left[\begin{array}{ccccc|c}
1 & 3 & 0 & 2 & 0 & 12 \\
0 & -2 & 1 & 1 & 0 & 25 \\
0 & 1 & 0 & 0 & 1 & 50
\end{array}\right]$

9.
$$\begin{array}{c} \quad x \quad y \quad z \quad s_1 \quad s_2 \quad\quad s_3 \quad m \\ \left[\begin{array}{ccccccc|c} 0 & 0 & 1 & 1 & 1 & 1 & 0 & 10 \\ 1 & 0 & 0 & 2 & 3 & -7 & 0 & 28 \\ 0 & 1 & 0 & 4 & 4 & 4 & 0 & 34 \\ \hline 0 & 0 & 0 & 0 & 0 & 31 & 1 & 22 \end{array}\right] \end{array}$$

10.
$$\begin{array}{c} \quad x \quad y \quad z \quad s_1 \quad s_2 \quad\quad s_3 \quad m \\ \left[\begin{array}{ccccccc|c} 2 & 0 & 1 & 1 & 0 & 3 & 0 & 10 \\ 1 & 0 & 0 & 2 & 1 & -1 & 0 & 28 \\ 8 & 1 & 0 & 4 & 0 & 2 & 0 & 34 \\ \hline -3 & 0 & 1 & 0 & 0 & -1 & 1 & 22 \end{array}\right] \end{array}$$

In Problems 11 to 16 select the pivot column and the entering variable.

11.
$$\begin{array}{c} \quad x \quad\quad y \quad\quad s_1 \quad s_2 \quad m \\ \left[\begin{array}{ccccc|c} 0 & 0 & 0 & 1 & 0 & 1 \\ 1 & 1 & 1 & 0 & 0 & 3 \\ \hline 0 & -3 & -1 & 0 & 1 & 8 \end{array}\right] \end{array}$$

12.
$$\begin{array}{c} \quad x \quad y \quad s_1 \quad\quad s_2 \quad m \\ \left[\begin{array}{ccccc|c} 2 & 1 & 0 & 1 & 0 & 1 \\ 1 & 0 & 1 & 0 & 0 & 7 \\ \hline -8 & 0 & 0 & -3 & 1 & 5 \end{array}\right] \end{array}$$

13.
$$\begin{array}{c} \quad x \quad y \quad z \quad\quad s_1 \quad\quad s_2 \quad s_3 \quad m \\ \left[\begin{array}{ccccccc|c} 1 & 1 & 0 & 0 & 2 & 0 & 0 & 3 \\ 0 & 0 & 0 & 1 & 1 & 1 & 0 & 8 \\ 0 & 0 & 1 & 3 & 0 & 0 & 0 & 9 \\ \hline 0 & 2 & 0 & -1 & -2 & 0 & 1 & 3 \end{array}\right] \end{array}$$

14.
$$\begin{array}{c} \quad x \quad y \quad z \quad s_1 \quad\quad s_2 \quad s_3 \quad m \\ \left[\begin{array}{ccccccc|c} 0 & 0 & 1 & 0 & 1 & 0 & 0 & 0 \\ 1 & 0 & 0 & 1 & 2 & 1 & 0 & 4 \\ 2 & 1 & 0 & 0 & 0 & 1 & 0 & 2 \\ \hline -3 & 0 & 0 & 0 & -1 & 1 & 1 & 5 \end{array}\right] \end{array}$$

15.
$$\begin{array}{c} \quad x \quad y \quad\quad z \quad s_1 \quad\quad s_2 \quad s_3 \quad m \\ \left[\begin{array}{ccccccc|c} 0 & 0 & 1 & 0 & -4 & 0 & 0 & 10 \\ 1 & 0 & 3 & 1 & 2 & 1 & 0 & 9 \\ 0 & 1 & -1 & 0 & 1 & 1 & 0 & 2 \\ \hline 0 & 0 & -9 & 0 & -5 & 3 & 1 & 45 \end{array}\right] \end{array}$$

16.
$$\begin{array}{c} \quad x \quad\quad y \quad z \quad s_1 \quad\quad s_2 \quad s_3 \quad m \\ \left[\begin{array}{ccccccc|c} 2 & -1 & 1 & 0 & -4 & 0 & 0 & 10 \\ 1 & 0 & 0 & 1 & 2 & 0 & 0 & 18 \\ 0 & 1 & 0 & 0 & 0 & 1 & 0 & 2 \\ \hline 4 & -1 & 0 & 0 & -2 & 0 & 1 & 45 \end{array}\right] \end{array}$$

In Problems 17 to 28 perform the pivot process on the simplex tableau in the indicated problem using the given entry as the pivot entry.

17. Problem 11, the (2, 2) entry.

18. Problem 12, the (1, 1) entry.

19. Problem 13, the (1, 5) entry.

20. Problem 14, the (3, 1) entry.

21. Problem 15, the (2, 3) entry.

22. Problem 16, the (2, 5) entry.

23. Problem 15, the (1, 3) entry.

24. Problem 16, the (1, 5) entry.

25. Problem 15, the (3, 5) entry.

26. Problem 16, the (3, 2) entry.

27. Problem 15, the (3, 2) entry.

28. Problem 16, the (1, 1) entry.

In Problems 29 to 34 find the initial simplex tableau and form a new tableau by pivoting about the indicated entry.

29. Maximize $m = 5x + 7y$ subject to
$$x + 3y \le 12$$
$$2x + 3y \le 18$$
Pivot about the (1, 2) entry.

30. Maximize $m = 8x + 5y$ subject to
$$4x + 3y \le 24$$
$$9x + 2y \le 18$$
Pivot about the (2, 1) entry.

31. Maximize $m = 4x + 7y$ subject to
$$3x + 5y \le 15$$
$$x + 5y \le 10$$
Pivot about the (2, 2) entry.

32. Maximize $m = x + 2y + 3z$ subject to
$$x + 2y + z \le 40$$
$$x + y + z \le 30$$
Pivot about the (2, 3) entry.

33. Maximize $m = x + 3y + 4z$ subject to
$$x + 3y \qquad \le 300$$
$$x \qquad + z \le 150$$
$$x + 2y + z \le 200$$
Pivot about the (2, 3) entry.

34. Maximize $m = 4x + 6y + 5z$ subject to
$$2x + 3y + z \le 900$$
$$3x + y + z \le 350$$
$$4x + 2y + z \le 400$$
Pivot about the (3, 2) entry.

In Problems 35 to 40 translate the linear programming problem into mathematics. Then find the initial simplex tableau and form a new tableau by pivoting about the (1, 1) entry.

35. A clothing manufacturer makes three types of jogging attire, discount, regular, and designer. The manufacturer sells the attire to the retail stores for $5 each for the discount suits, $10 each for the regular suits, and $40 each for the designer suits. The cost for manufacturing each discount suit is $3, for each regular suit $5, and for each designer suit $10. The number of designer suits produced must be no more than the sum of discount and regular suits made. If the manufacturer makes no more than 200 suits a week and budgets no more than $1500 per week, how many of each type should be made to maximize profit?

36. A hospital nutritionist prepares a menu for patients on low-fat, high-protein diets. The primary ingredients are chicken, corn, and rice. The meal must contain no more than 1000 calories (cal), 5 grams (g) of fat, and 15 milligrams (mg) of iron. Each 3-ounce serving of chicken contains 45 g of protein, 500 cal, 9 mg of iron, and 4 g of fat. Each 1-cup serving of corn contains 9 g of protein, 150 cal, 6 mg of iron, and 2 g of fat. Each 1-cup serving of rice contains 7 g of protein, 100 cal, 5 mg of iron, and 2 g of fat. How much chicken, corn, and rice should be included in the meal in order to maximize the amount of protein?

37. The computer center director at a large university wants to staff consulting sections with three types of shifts: type 1 will contain 1 senior programmer and 2 student workers, type 2 will consist of 1 senior programmer and 5 student workers, and type 3 will consist of 1 senior programmer and 3 student workers. The director wants to use no more than 50 individuals. The number of senior programmers is limited to 10. There can be no more than 40 hours filled during the week, with a type 1 shift serving for 3 hours, a type 2 shift serving for 2 hours, and a type 3 shift serving for 1 hour. A type 1 shift processes 22 students per hour, a type 2 shift processes 20 students per hour, and a type 3 shift processes 15 students per hour. Determine the number of shifts of each type in order to maximize the number of students processed per hour.

38. A bicycle manufacturer has two retail outlets and, in addition, sells some bikes to a large chain store. One truck makes all deliveries, and it takes 3 hours to make a round trip to the first retail store, 2 hours to the second retail store, and 4 hours to the chain store.

The cost of a round trip to the first retail store is $100, to the second retail store $75, and to the chain store $200. The profit of each truckload sold through the first retail store is $500, for the second retail store $400, and for the chain store $800. How many shipments per week should be scheduled to each store in order to maximize profit if the manufacturer is limited to no more than 48 hours per week and a cost of $3000?

39. An investment manager wants to invest at most $25,000 in four types of mutual funds. From past experience it is assumed that in the short run type A fund will yield 8%, type B fund will yield 10%, type C fund will yield 12%, and type D fund will yield 11%. The manager wants to invest no more than $10,000 in type A and type B funds together; no more than $15,000 in type A, type B, and type C funds together; and no more than $20,000 in type A and type D funds together. How much should be invested in each type of fund to maximize yield?

40. A firm manufactures four types of tents, small, medium, large, and deluxe. Each small tent takes 2 hours to cut and 4 hours to sew; each medium tent takes 3 hours to cut and 5 hours to sew; each large tent takes 4 hours to cut and 7 hours to sew; and each deluxe tent takes 6 hours to cut and 9 hours to sew. There is a total of 310 hours available to cut and 500 hours available to sew the tents. No more than 100 tents are to be made. How many of each type of tent should be made to maximize profit if the profit is $90 on each small tent, $50 on each medium tent, $70 on each large tent, and $80 on each deluxe tent?

Referenced Exercise Set 4–2

1. Under Title III of the 1973 Older Americans Act over 500 Area Agencies on Aging (AAAs) have been established to coordinate services to the elderly. In an article analyzing how AAAs can most efficiently operate their centers, researchers based their recommendations on a preference index.* Persons served by the centers were divided into one of three categories: type 1 consisted of persons with no impairment, type 2 persons had one impairment, and type 3 persons had more than one impairment. Each center could serve all types of persons, but some were equipped to serve a larger proportion of one type than another. The preference index

*Richard M. Burton, and David C. Dellinger, ''Making the Area Agencies on Aging Work: The Role of Information,'' *Socio-Economic Planning Science*, Vol. 14, 1980, pp. 1–9.

measured the proportion of each type of person a center was best equipped to serve. For example, one center had the preference index $P = 5x + 6y + 3z$, where x is the number of type 1 persons it could serve, y is the number of type 2 persons, and z is the number of type 3 persons. The object was to maximize the preference index. This means the center could best serve twice as many type 2 persons as type 3 persons, and just a few less type 1 than type 2. For this center the number of persons it could serve was 60. Its rehabilitation department, which served types 2 and 3 persons, could serve at most 40 persons. The remedial training department, which served types 1 and 2 persons, could serve at most 50 persons. Write the initial simplex tableau.

2. The management of school districts is big business. Many large districts have budgets exceeding $25 million. Most corporations in the private sector with budgets this high have large accounting departments to manage their fiscal operations. In an article analyzing financial management for school districts, the primary technique is linear programming.[†] The most important consideration is cash flow. School districts have large revenues that are usually paid in lump sums, such as property taxes and state aid. Most expenditures are also made in several lump sums. Revenues must be invested wisely before an expenditure is due to optimize financial management. Suppose a district keeps its capital in three types of investments: liquid accounts (such as savings accounts and money markets), one-month certificates of deposit (CDs), and six-month CDs. The liquid accounts earn 6% interest, the one-month CDs earn 8% interest, and the six-month CDs earn 9%. The sum of the amounts in the three accounts is limited to $2.5 million. At least $1 million must be kept in the liquid accounts, meaning that the sum of the amounts in the CDs can be no more than $1.5 million. State law requires that the sum of 80% of the liquid accounts and 30% of the one-month CDs must be no more than the district's short-term debt, which is $1.8 million. Write the initial simplex tableau.

3. The following problem is adapted from an article studying how to maximize profit from a large-scale assembly line.[‡] The problem is in standard form. The initial simplex tableau is given. Express the problem as a word problem, that is, describe the objective function and the constraints. The variables represent the capacity of the four machines in the assembly line; that is, they are the number of items that the machine can treat per day. The numbers in the last row represent the profit per item per machine.

x	y	z	w	s_1	s_2	s_3	m	
1	1	1	1	1	0	0	0	300
1	1	0	0	0	1	0	0	100
0	1	1	1	0	0	1	0	230
−2	−3	−4	0	0	0	0	1	0

Cumulative Exercise Set 4–2

1. Find the initial simplex tableau for the linear programming problem. Assume the problem is in standard form.

   Maximize $m = 6x + 7y$ subject to
   $$x + 4y \le 10$$
   $$2x + y \le 6$$

2. Construct the original linear programming problem (without slack variables) assuming that the matrix is the initial simplex tableau.

0	4	5	1	0	0	0	7
1	6	2	0	1	0	0	0
1	3	1	0	0	1	0	2
−3	−5	−2	0	0	0	1	55

3. Find the initial simplex tableau for the linear programming problem. Assume the problem is in standard form.

   Maximize $m = 2x + y + 4z + 3w$ subject to
   $$x + \quad\;\; 3z + 4w \le 20$$
   $$x + 5y + \qquad\quad w \le 6$$
   $$y + 2z + 3w \le 20$$
   $$x + \qquad z + 3w \le 15$$

4. Determine the basic variables, the nonbasic variables, and the basic feasible solution of the simplex tableau.

x	y	z	s_1	s_2	s_3	m	
0	0	0	2	1	3	0	13
1	0	0	1	0	4	0	22
0	7	1	2	0	0	0	32
0	−6	0	−9	0	0	1	50

[†]Frederick L. Dembowski, "An Integer Programming Approach to School District Financial Management," *Socio-Economic Planning Science,* Vol. 14, 1980, pp. 147–153.

[‡]R. Chandrasekaran, "Production Planning in Assembly Line Systems," *Management Science,* Vol. 30, 1984, pp. 713–719.

In Problems 5 to 8 assume the problems are in standard form.

5. Find the basic feasible solution of the simplex tableau and then apply the maximality test.

$$
\begin{array}{ccccc}
x & y & s_1 & s_2 & m \\
\left[\begin{array}{ccccc|c}
0 & 5 & 1 & 7 & 0 & 20 \\
1 & -6 & 0 & 8 & 0 & 25 \\
0 & -2 & 0 & 0 & 1 & 60
\end{array}\right]
\end{array}
$$

6. Select the pivot column and the entering variable.

$$
\begin{array}{ccccc}
x & y & s_1 & s_2 & m \\
\left[\begin{array}{ccccc|c}
0 & 1 & 5 & 1 & 0 & 10 \\
1 & 4 & 2 & 0 & 0 & 24 \\
0 & -3 & -4 & 0 & 1 & 98
\end{array}\right]
\end{array}
$$

7. Perform the pivot process on the simplex tableau using the (1, 1) entry as the pivot entry.

$$
\begin{array}{ccccccc}
x & y & z & s_1 & s_2 & s_3 & m \\
\left[\begin{array}{ccccccc|c}
1 & -2 & 1 & 0 & -4 & 0 & 0 & 20 \\
2 & 1 & 0 & 1 & 1 & 0 & 0 & 90 \\
1 & 0 & 0 & 0 & 0 & 1 & 0 & 25 \\
-6 & -1 & 0 & 0 & -1 & 0 & 1 & 85
\end{array}\right]
\end{array}
$$

8. Find the initial simplex tableau and form a new tableau by pivoting about the (3, 2) entry.

Maximize $m = 5x + 8y + 7z$ subject to
$$
\begin{aligned}
5x + 2y + z &\leq 900 \\
2x + y + z &\leq 350 \\
x + 2y + z &\leq 600 \\
x, y, z &\geq 0
\end{aligned}
$$

4–3 The Simplex Method for Problems in Standard Form

There is one more skill in the simplex method. In Section 4–2 we saw that the simplex method uses elementary row operations to compute a sequence of tableaus that ends when the maximality test passes. Example 4 illustrated such a sequence, but at each step the pivot entry was given. In this section the rule to select the pivot entry is presented. Then all the steps in the simplex method are reviewed.

Choosing the Pivot Row and the Pivot Entry

Once the pivot column is identified, the pivot entry is chosen by selecting the correct row—the **pivot row.** In keeping with the goal of finding a maximal solution, we choose the pivot entry so that, after pivoting, the new basic feasible solution will give the largest increase in m. We must also avoid making any variable negative since the problem must remain in standard form.

The key to choosing the correct pivot entry is the last column because that is where the values of the basic variables in the basic feasible solution are found. The main objective is to avoid generating negative numbers in the last column when pivoting because that would violate the stipulation that the variables be nonnegative. So let us explain how a negative number could be produced.

Suppose a tableau has the following two rows where the first column is the pivot column and all its entries are 1.

$$
\begin{array}{cc}
\text{pivot column} & \text{constant} \\
\end{array}
$$

$$
\begin{array}{c}
\text{row 1} \\
\text{row 2}
\end{array}
\begin{bmatrix}
1 & \cdots & 3 \\
1 & \cdots & 1 \\
\;\cdot & & \\
\;\cdot & & \\
\;\cdot & &
\end{bmatrix}
$$

Look at the consequence of choosing row 1 as the pivot row. To pivot, perform $R2 + (-1)(R1)$ to put a 0 in the (2, 1) entry. In the last column of the second row this results in $1 + (-1)3 = -2$, a negative number—exactly what we must avoid. It is produced because the entry in the pivot row, 3, was larger than the entry in the other row, 1. This means that if all the entries in the pivot column are 1, then the pivot row must be that row having the smallest nonnegative number in the last column.

Suppose the numbers in the pivot column are not all 1. Then the pivot row is selected by forming quotients. Each entry in the pivot column is the denominator of a quotient, unless that entry is 0 or a negative number, in which case no quotient is formed. The numerator of each quotient is the entry from the constant column of the corresponding row. No quotient is formed for the objective function, the bottom row. Hence quotients are formed only for rows above the bottom row with positive numbers in the pivot column. The pivot row is then the row with the smallest quotient. This is the criterion.

Choosing the Pivot Row

For each row, except the last one, with a positive entry in the pivot column, compute the ratio of the number in the last column to the number in the pivot column. The pivot row is the row with the smallest nonnegative ratio.

If there are two or more rows that have the same ratio, any of these rows can be selected. The next example shows how to apply the criterion.

EXAMPLE 1

Problem

For each tableau find the pivot column and the pivot row. Then identify the pivot entry.

$$
A = \begin{array}{c}
\\
\\
\\
\end{array}
\begin{array}{ccccc}
x & y & s_1 & s_2 & m \\
\end{array}
\left[
\begin{array}{ccccc|c}
3 & -1 & 1 & 0 & 0 & 6 \\
5 & 2 & 0 & 1 & 0 & 15 \\
\hline
-4 & 1 & 0 & 0 & 1 & 10
\end{array}
\right]
$$

$$
B = \begin{array}{c} \\ \\ \\ \\ \end{array}
\begin{array}{c}
x \quad\ y \ \ z \quad\ s_1 \ \ s_2 \quad\ s_3 \ \ m
\end{array}
$$

$$
B = \left[\begin{array}{rrrrrrr|r}
1 & 3 & 0 & -10 & 0 & 3 & 0 & 10 \\
0 & -1 & 0 & 5 & 1 & 8 & 0 & 50 \\
0 & 2 & 1 & 6 & 0 & -2 & 0 & 30 \\
\hline
0 & -3 & 0 & -5 & 0 & 6 & 1 & 10
\end{array}\right]
$$

Solution For tableau A the first column is the pivot column because -4 is the only negative number in the last row. The two ratios are

for row 1 the ratio is $\frac{6}{3}$ $= 2$

for row 2 the ratio is $\frac{15}{5}$ $= 3$

Row 1 has the smaller nonnegative ratio and is therefore the pivot row. Thus the $(1, 1)$ entry, 3, is the pivot entry.

For tableau B column 4 is the pivot column because -5 is the most negative. Do not consider the ratio for row 1 because its entry in the pivot column, -10, is negative. The other ratios are

for row 2 the ratio is $\frac{50}{5}$ $= 10$

for row 3 the ratio is $\frac{30}{6}$ $= 5$

Row 3 is the pivot row. Thus the pivot entry is the $(3, 4)$ entry, 6.

There is a reason for stipulating that only nonnegative ratios be considered. It might seem superfluous because all the constants are nonnegative so the ratios must be nonnegative. We inserted the stipulation because in the next section constraints whose constant term is negative are considered and we want the same procedure to hold.

All the major steps in the simplex method have been covered. They are assembled in the following procedure.

The Simplex Method for Problems in Standard Form

1. State the problem in terms of a system of linear equations by inserting slack variables into the constraints and expressing the objective function in proper form.
2. Construct the initial simplex tableau.
3. Apply the maximality test. If the basic feasible solution is maximal, the problem is solved. If not, go to step 4.
4. Compute a new simplex tableau using the following procedure:
   (a) Choose the pivot column.
   (b) Choose the pivot row.
   (c) Pivot about the pivot entry.
5. Go to step 3.

It might be necessary to use steps 3, 4, and 5 several times before the test for a maximal solution passes. It is possible to determine when a linear programming problem does not have a solution by looking at the ratios used in step 4, part (b). If all the ratios are negative numbers, the problem has no solution. Example 2 illustrates the procedure.

EXAMPLE 2

Problem

Maximize $m = 15x + 30y$ subject to
$$x + 2y \leq 40$$
$$5x + 2y \leq 80$$

Solution

1. State the problem in terms of linear equations.

$$
\begin{array}{rcl}
x + 2y + s_1 & = & 40 \\
5x + 2y + s_2 & = & 80 \\
-15x - 30y + m & = & 0
\end{array}
$$

2. Construct the initial simplex tableau.

$$
\begin{array}{ccccc}
x & y & s_1 & s_2 & m \\
\left[\begin{array}{ccccc|c}
1 & 2 & 1 & 0 & 0 & 40 \\
5 & 2 & 0 & 1 & 0 & 80 \\
\hline
-15 & -30 & 0 & 0 & 1 & 0
\end{array}\right]
\end{array}
$$

3. The maximality test fails because the last row has negative numbers.
4. The pivot column is column 2 and the ratios are $\frac{40}{2} = 20$ for row 1 and $\frac{80}{2} = 40$ for row 2. Hence row 1 is the pivot row so the pivot entry is the $(1, 2)$ entry. Pivoting produces the tableau

$$
\begin{array}{ccccc}
x & y & s_1 & s_2 & m \\
\left[\begin{array}{ccccc|c}
\frac{1}{2} & 1 & \frac{1}{2} & 0 & 0 & 20 \\
4 & 0 & -1 & 1 & 0 & 40 \\
0 & 0 & 15 & 0 & 1 & 600
\end{array}\right]
\end{array}
\quad
\begin{array}{l}
(\frac{1}{2})R1 \\
R2 + (-2)R1 \\
R3 + (30)R1
\end{array}
$$

5. The maximality test passes and therefore the basic feasible solution is the maximal solution: it is $x = s_1 = 0$, $y = 20$, $s_2 = 40$, $m = 600$.

Hence the solution to the original problem is $m = 600$ when $x = 0$ and $y = 20$.

The next example requires three pivots to obtain the optimal solution.

EXAMPLE 3

Problem

Maximize $m = 3x + 5y + 4z$ subject to
$$x + y \qquad \leq 200$$
$$2x \qquad + z \leq 350$$
$$2y + z \leq 650$$

Solution First use slack variables to express the problem as a system of equations while expressing the objective function in proper form.

$$
\begin{array}{rcl}
x + y & + s_1 & = 200 \\
2x & + z & + s_2 & = 350 \\
2y + z & + s_3 & = 650 \\
-3x - 5y - 4z & + m = & 0
\end{array}
$$

Then find the initial simplex tableau. It is

$$
\begin{array}{ccccccc|c}
x & y & z & s_1 & s_2 & s_3 & m & \\
1 & 1 & 0 & 1 & 0 & 0 & 0 & 200 \\
2 & 0 & 1 & 0 & 1 & 0 & 0 & 350 \\
0 & 2 & 1 & 0 & 0 & 1 & 0 & 650 \\
\hline
-3 & -5 & -4 & 0 & 0 & 0 & 1 & 0
\end{array}
$$

The maximality test fails because there are negative numbers in the last row. The pivot column is the second column since -5 is the most negative number in the last row. There are two positive numbers in the second column, so we compute the two ratios: for row 1 the ratio is $200/1 = 200$; for row 3 the ratio is $650/2 = 325$. So row 1 is the pivot row. Hence the pivot entry is the $(1, 2)$ entry, 1. Pivoting yields the new tableau

$$
\begin{array}{ccccccc|c}
x & y & z & s_1 & s_2 & s_3 & m & \\
1 & 1 & 0 & 1 & 0 & 0 & 0 & 200 \\
2 & 0 & 1 & 0 & 1 & 0 & 0 & 350 \\
-2 & 0 & 1 & -2 & 0 & 1 & 0 & 250 \\
2 & 0 & -4 & 5 & 0 & 0 & 1 & 1000
\end{array}
\qquad
\begin{array}{l}
\\
\\
R3 + (-2)R1 \\
R4 + 5R1
\end{array}
$$

The maximality test again fails because there is still a negative entry in the last row. The pivot column is column 3. There are two positive numbers in column 3, so we compute the two ratios: for row 2 the ratio is $350/1 = 350$; for row 3 the ratio is $250/1 = 250$. So row 3 is the pivot row. Hence the pivot entry is the $(3, 3)$ entry, 1. Pivoting yields the new tableau

$$
\begin{array}{ccccccc|c}
x & y & z & s_1 & s_2 & s_3 & m & \\
1 & 1 & 0 & 1 & 0 & 0 & 0 & 200 \\
4 & 0 & 0 & 2 & 1 & -1 & 0 & 100 \\
-2 & 0 & 1 & -2 & 0 & 1 & 0 & 250 \\
-6 & 0 & 0 & -3 & 0 & 4 & 1 & 2000
\end{array}
\qquad
\begin{array}{l}
\\
R2 + (-1)R3 \\
\\
R4 + 4R3
\end{array}
$$

The maximality test fails again. The pivot column is column 1. There are two positive numbers in the pivot column, so we compute the two ratios: for row 1 the ratio is $200/1 = 200$; for row 2 the ratio is $100/4 = 25$. So the pivot row is row 2. Hence the pivot entry is the $(2, 1)$ entry, 4. First perform $(\frac{1}{4})$R2 and then the remaining operations. The new tableau is

$$
\begin{array}{cccccccc|c}
x & y & z & s_1 & s_2 & s_3 & m & \\
0 & 1 & 0 & \frac{1}{2} & -\frac{1}{4} & \frac{1}{4} & 0 & 175 \\
1 & 0 & 0 & \frac{1}{2} & \frac{1}{4} & -\frac{1}{4} & 0 & 25 \\
0 & 0 & 1 & -1 & \frac{1}{2} & \frac{1}{2} & 0 & 300 \\
0 & 0 & 0 & 0 & \frac{3}{2} & \frac{5}{2} & 1 & 2150
\end{array}
\qquad
\begin{array}{l}
\text{R1} + (-1)\text{R2} \\
(\frac{1}{4})\text{R2} \\
\text{R3} + 2\text{R2} \\
\text{R4} + 6\text{R2}
\end{array}
$$

The maximality test passes and the basic feasible solution is maximal. It is

$$x = 25, \quad y = 175, \quad z = 300, \quad s_1 = s_2 = s_3 = 0, \quad m = 2150$$

Hence the solution to the original problem is $m = 2150$ when $x = 25$, $y = 175$, and $z = 300$.

The next example shows how to determine when a linear programming problem has no solution.

EXAMPLE 4

Problem

Maximize $m = x + 2y$ subject to
$3x - 2y \le 18$
$3x - y \le 24$

Solution Construct the initial simplex tableau

$$
\begin{array}{ccccc|c}
x & y & s_1 & s_2 & m & \\
3 & -2 & 1 & 0 & 0 & 18 \\
3 & -1 & 0 & 1 & 0 & 24 \\
-1 & -2 & 0 & 0 & 1 & 0
\end{array}
$$

The most negative number in the last row is -2 in the second column. But both entries above -2 are negative, which means there is no number that qualifies as the pivot. Hence there is no solution. The graph of the region of feasible solutions is an unbounded region. (See Figure 1.) The objective function, $m = x + 2y$, can be made as large as desired by choosing large enough values of x and y in the region of feasible solutions. Therefore m does not have a maximum value.

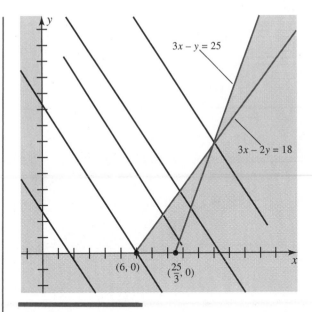

FIGURE 1 Various lines $m = x + 2y$. There is no largest value for $m = x + 2y$.

To compare the simplex method with the graphical method of solution of linear programming problems covered in the previous chapter, let us solve a problem using both methods. Consider the transportation problem in Example 2 of Section 3–3.

Maximize $P = 210x + 400y$
Subject to $x + 2y \leq 20$
$x + 4y \leq 32$
$x \geq 0, y \geq 0$

The constraints are graphed in Figure 2. The corner points are (0, 8), (20, 0) and (8, 6). In the graphical method each corner point is substituted into the objective function and the largest value is chosen.

Corner Point	Value
(20, 0)	4200
(0, 8)	3200
(8, 6)	4080

The optimum solution is (20, 0) corresponding to the maximum value of the objective function 4200.

The simplex method starts with the initial simplex tableau

$$\begin{bmatrix} 1 & 2 & 1 & 0 & 0 & | & 20 \\ 1 & 4 & 0 & 1 & 0 & | & 32 \\ \hline -210 & -400 & 0 & 0 & 1 & | & 0 \end{bmatrix}$$

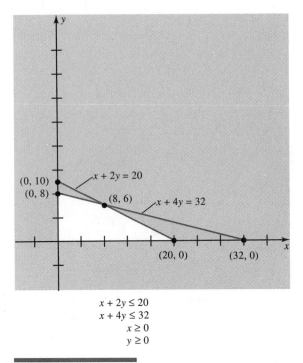

$$x + 2y \le 20$$
$$x + 4y \le 32$$
$$x \ge 0$$
$$y \ge 0$$

FIGURE 2

The first pivot entry is the (2, 2) entry. The next tableau is

$$\begin{bmatrix} \frac{1}{2} & 0 & 1 & -\frac{1}{2} & 0 & 4 \\ \frac{1}{4} & 1 & 0 & \frac{1}{4} & 0 & 8 \\ \hline -110 & 0 & 0 & 100 & 1 & 3200 \end{bmatrix}$$

The next pivot entry is the (1, 1) entry. The next tableau is

$$\begin{bmatrix} 1 & 0 & 2 & -1 & 0 & 8 \\ 0 & 1 & -\frac{1}{2} & \frac{1}{2} & 0 & 6 \\ \hline 0 & 0 & 220 & -10 & 1 & 4080 \end{bmatrix}$$

The next pivot entry is the (2, 4) entry. The next tableau is

$$\begin{bmatrix} 1 & 2 & 1 & 0 & 0 & 20 \\ 0 & 2 & -1 & 1 & 0 & 12 \\ \hline 0 & 20 & 210 & 0 & 1 & 4200 \end{bmatrix}$$

This tableau yields the optimum solution of 4200, which occurs at (20, 0). Look at each tableau. Each basic feasible solution corresponds to a corner point, and the value in the (3, 6) entry is the value obtained by substituting that corner point into the objective function.

In general, the simplex method searches through the corner points of the region of feasible solution, always ensuring that the value of the objective function increases when a new tableau is computed. When all the numbers in the last row are nonnegative the objective function cannot be increased, so the corresponding basic feasible solution is the optimal solution.

Just as the graphical method investigates the corner points in search of the maximum value of the objective function, so too the simplex method looks at the corner points as basic feasible solutions to simplex tableaus.

The graphical method may seem easier in this problem because there are only two variables and only three corner points. The simplex method is superior to the graphical method when there are many constraints or more than two variables. In fact, if there are four or more variables, it makes no sense even to consider the graph of the region of feasible solutions.

EXERCISE SET 4–3

Assume all problems are in standard form.

In Problems 1 to 6 two columns of a simplex tableau are given. Assume the one on the left is the pivot column and the other is the last column. Compute the appropriate ratios and select the pivot entry.

1.
$$\begin{bmatrix} 2 & \cdots & 10 \\ 1 & \cdots & 6 \\ -2 & \cdots & 10 \end{bmatrix}$$

2.
$$\begin{bmatrix} 3 & \cdots & 8 \\ 6 & \cdots & 12 \\ -1 & \cdots & 10 \end{bmatrix}$$

3.
$$\begin{bmatrix} -1 & \cdots & 4 \\ 2 & \cdots & 10 \\ -5 & \cdots & 10 \end{bmatrix}$$

4.
$$\begin{bmatrix} 3 & \cdots & 9 \\ 3 & \cdots & 8 \\ -1 & \cdots & 1 \end{bmatrix}$$

5.
$$\begin{bmatrix} 3 & \cdots & 11 \\ -2 & \cdots & 2 \\ 6 & \cdots & 12 \\ -3 & \cdots & 8 \end{bmatrix}$$

6.
$$\begin{bmatrix} 0 & \cdots & 1 \\ -3 & \cdots & 3 \\ 1 & \cdots & 10 \\ -8 & \cdots & 10 \end{bmatrix}$$

In Problems 7 to 12 find the pivot entry and perform the pivot operation.

7.
$$\begin{array}{ccccc} x & y & s_1 & s_2 & m \\ \end{array}$$
$$\left[\begin{array}{ccccc|c} 0 & 4 & 1 & 2 & 0 & 70 \\ 1 & 1 & 0 & 1 & 0 & 15 \\ 0 & -1 & 0 & 0 & 1 & 20 \end{array}\right]$$

8.
$$\begin{array}{ccccc} x & y & s_1 & s_2 & m \\ \end{array}$$
$$\left[\begin{array}{ccccc|c} 0 & 2 & 0 & 1 & 0 & 2 \\ 1 & 1 & 1 & 0 & 0 & 3 \\ 0 & -4 & -1 & 0 & 1 & 8 \end{array}\right]$$

9.
$$\begin{array}{ccccc} x & y & s_1 & s_2 & m \\ \end{array}$$
$$\left[\begin{array}{ccccc|c} 2 & 1 & 0 & 1 & 0 & 10 \\ 1 & 0 & 1 & 0 & 0 & 8 \\ -8 & 0 & 0 & -4 & 1 & 20 \end{array}\right]$$

10.
$$\begin{array}{ccccccc} x & y & z & s_1 & s_2 & s_3 & m \\ \end{array}$$
$$\left[\begin{array}{ccccccc|c} 2 & 0 & 1 & 1 & 0 & 3 & 0 & 18 \\ 2 & 0 & 0 & 2 & 1 & -1 & 0 & 28 \\ 4 & 1 & 0 & 5 & 0 & 1 & 0 & 34 \\ -4 & 0 & 2 & 0 & 0 & -3 & 1 & 22 \end{array}\right]$$

11.
$$\begin{array}{ccccccc} x & y & z & s_1 & s_2 & s_3 & m \\ \end{array}$$
$$\left[\begin{array}{ccccccc|c} 1 & 3 & 0 & 0 & 2 & 0 & 0 & 20 \\ 0 & 1 & 0 & 1 & 0 & 1 & 0 & 2 \\ 0 & 1 & 1 & 3 & 1 & 0 & 0 & 9 \\ 0 & 1 & 0 & -1 & -3 & 0 & 1 & 3 \end{array}\right]$$

12.
$$\begin{array}{ccccccc} x & y & z & s_1 & s_2 & s_3 & m \\ \end{array}$$
$$\left[\begin{array}{ccccccc|c} 2 & 0 & 1 & 0 & 1 & 0 & 0 & 20 \\ 1 & 0 & 0 & 1 & 2 & 1 & 0 & 30 \\ 2 & 1 & 0 & 0 & 0 & 1 & 0 & 10 \\ -3 & 0 & 0 & 0 & -1 & 1 & 1 & 5 \end{array}\right]$$

In Problems 13 to 32 solve the linear programming problem by the simplex method.

13. Maximize $m = 5x + 7y$ subject to
$$x + 3y \le 12$$
$$2x + 3y \le 18$$

14. Maximize $m = 8x + 5y$ subject to
$$4x + 3y \le 24$$
$$9x + 2y \le 18$$

15. Maximize $m = 4x + 7y$ subject to
$$3x + 5y \le 15$$
$$x + 5y \le 10$$

16. Maximize $m = 3x + 7y$ subject to
$$3x + y \le 15$$
$$4x + 5y \le 20$$

17. Maximize $m = x + 2y + 3z$ subject to
$$x + 2y + z \le 40$$
$$x + y + z \le 30$$

18. Maximize $m = 3x + 7y + 2z$ subject to
$$3x + 5y + 3z \le 15$$
$$4x + 5y + 2z \le 20$$

19. Maximize $m = 3x + 4y + 2z$ subject to
$$x + 5y + 3z \le 15$$
$$x + 2y + 2z \le 20$$

20. Maximize $m = 4x + 7y + z$ subject to
$$3x + y + z \le 30$$
$$4x + y + 2z \le 45$$

21. Maximize $m = 5x + 7y$ subject to
$$3x + 5y \le 30$$
$$x + 5y \le 20$$
$$x \le 8$$

22. Maximize $m = 12x + 7y$ subject to
$$3x + 5y \le 30$$
$$x + 5y \le 20$$
$$x \le 8$$

23. Maximize $m = 10x + 7y$ subject to
$$3x + y \le 48$$
$$2x + 5y \le 84$$
$$y \le 14$$

24. Maximize $m = 12x + 7y$ subject to
$$3x + y \le 30$$
$$2x + 3y \le 55$$
$$y \le 16$$

25. Maximize $m = 6x + 8y$ subject to
$$x + y \le 10$$
$$2x + 3y \le 24$$
$$3x + 2y \le 24$$

26. Maximize $m = 7x + 5y$ subject to
$$x + y \le 20$$
$$x + 2y \le 20$$
$$2x + 3y \le 20$$

27. Maximize $m = x + 3y + 4z$ subject to
$$x + 3y \le 300$$
$$x + z \le 150$$
$$x + 2y + z \le 200$$

28. Maximize $m = 4x + 6y + 5z$ subject to
$$2x + 3y + z \le 550$$
$$3x + y + z \le 350$$
$$4x + 2y + z \le 400$$

29. Maximize $m = 2x + 3y + 5z + 2w$ subject to
$$x + 3y + 2w \le 8$$
$$x + 4y + 2z \le 6$$
$$x + y + w \le 2$$
$$y + 3w \le 9$$

30. Maximize $m = 4x + 7y + 9z + 6w$ subject to
$$x + 3z + w \le 200$$
$$x + 5y \le 650$$
$$x + w \le 90$$
$$x + 4y + z \le 350$$

31. Maximize $m = 8x + 7y + 6w$ subject to
$$x + z + 2w \le 30$$
$$x + 2y + 2w \le 30$$
$$x + z + w \le 40$$
$$y + z \le 20$$

32. Maximize $m = x + 2y + 3z$ subject to
$$x + 2y + z \le 30$$
$$x + z + w \le 60$$
$$x + 3y + z + 4w \le 50$$
$$y + z + 2w \le 80$$

33. An airline offers a discount on their first class fare to those who have no luggage to check in. The new class is called an "executive" fare. The plane on which it is offered has 190 seats available. The price of a first class ticket is $310, the price of an executive ticket is $270, and the price of an economy class ticket is $140. The airline's cost is $150 for a first class ticket, $150 for an executive class ticket, and $50 for an economy class ticket. The airline must reserve 10 cubic feet (cu ft) of storage for luggage for each first class passenger and 5 cu ft of storage for each economy class passenger. It is decided that the cost of each flight will not exceed $22,500 and the maximum storage space available is 1350 cu ft. How many tickets of each type should be sold to maximize profit?

34. In one part of the farm a farmer plants three crops, tomatoes, potatoes, and corn. The profit per acre for tomatoes is $1200, for potatoes $700, and for corn $800. The cost per acre for fertilizer is $8 for tomatoes, $3 for potatoes, and $6 for corn. The cost per acre for maintenance is $8 for tomatoes, $9 for potatoes, and $6 for corn. Time spent per acre per week for maintenance is 2 hours for tomatoes, 1 hour for potatoes, and 1 hour for corn. No more than $780 is to be spent for fertilizer and no more than $804 is to be spent for maintenance. Time spent for maintenance is to be limited to no more than 150 hours per week.

How many acres of each crop should be planted to maximize profit?

35. A lamp manufacturer makes three types of lamps: a deluxe model that is sold in the firm's own store, a regular model that is shipped to a nearby department store, and an economy model that is shipped to a discount store across town. The cost of manufacturing each type of lamp is $80 for a deluxe model, $60 for a regular model, and $30 for an economy model. It costs $20 to ship each regular model and $50 to ship each economy model. The maximum weekly costs are $7000 for manufacturing and $2200 for shipping. It takes 40 minutes to assemble each deluxe model and 20 minutes to assemble each regular and each economy model. The maximum amount of time available for assembly each week is 500 machine-hours; that is, ten machines are available for 50 hours per week. How many lamps of each type should be made to maximize profit if the profit is $140 on each deluxe model, $110 on each regular model, and $90 on each economy model?

36. A city planning/consulting firm has three types of consulting teams to handle different types of business: type I consists of 1 engineer, 1 architect, and 3 salespersons; type II consists of 3 engineers, 1 architect, and 3 salespersons; and type III consists of 4 engineers, 2 architects, and 2 salespersons. There are 100 engineers available, 40 architects available, and 100 salespersons available to make up the teams. If the average weekly revenue from each type of team is $4000 for type I, $7000 for type II, and $8000 for type III, how many teams of each type should be made to maximize profit?

37. A firm manufactures three types of tool sheds, small, medium, and large. Each small shed takes 2 hours to cut and 2 hours to assemble, each medium shed takes 1 hour to cut and 3 hours to assemble, and each large shed takes 1 hour to cut and 4 hours to assemble. There is a total of 320 hours available to cut and 640 hours available to assemble the sheds. No more than 200 sheds are to be made. How many of each type of shed should be made to maximize profit if the profit is $90 on each small shed, $60 on each medium shed, and $30 on each large shed?

38. A paint manufacturer runs a sale on three colonial colors of indoor trim paint. The color is mixed by blending yellow, red, and blue pigment in base white paint. Each gallon of gun-metal blue has 0.1 ounce (oz) of yellow, 0.2 oz of red, and 0.4 oz of blue. Each gallon of Williamsburg orange has 0.3 oz of yellow, 0.2 oz of red, and 0.1 oz of blue. Each gallon of Saratoga rust has 0.2 oz of yellow, 0.4 oz of red, and 0.1 oz of blue. The company will use no more than 32 oz of yellow, 30 oz of red, and 10 oz of blue. If the profit per gallon is $1 for gun-metal blue, $2 for Williamsburg orange, and $3 for Saratoga rust, how many gallons of each type of paint should be made to maximize profit?

In Problems 39 to 42 match the given graph (Figures a to d) with a linear programming problem in Problems 21 to 24.

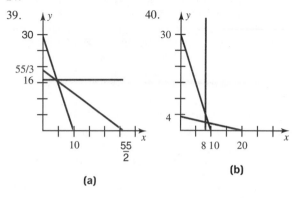

(a)

(b)

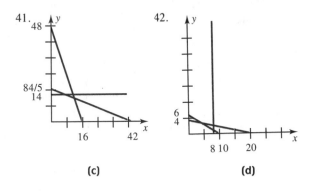

(c)

(d)

Referenced Exercise Set 4–3

1. In 1983 the Dominican Republic began planning an emergency medical service for the capital city of Santo Domingo. One component of the plan used the simplex method to solve a linear programming problem that described how to allocate ambulance locations effi-

ciently.* The following problem is adapted from this study. The city is divided into zones, each covered by an ambulance location. There are three types of ambulances: type 1 is designed to cover 2 zones, type 2 covers 3 zones, and type 3 covers 4 zones. The total number of ambulances available in this part of the study is 72. The average response time is 5 minutes for a type 1 ambulance, 7 minutes for a type 2 ambulance, and 10 minutes for a type 3 ambulance. The total response time is to be no greater than 800 minutes. Find the number of each type of ambulance needed to maximize the number of zones covered.

2. Large meat processing firms usually blend sausage meats using computer-based linear programming models.† A typical problem would entail maximizing profit while the constraints include the number of pounds of meat and spice available, the amount of protein content, and the amount of fat in the blend. Suppose the amount of the blend of beef, pork, and spice is to be no more than 400 lb. Each pound of beef contains 90 grams (g) of protein and each pound of pork contains 60 g of protein while the spice contains no protein. The total amount of protein in the blend is limited to 15,000 g. Each pound of beef contains 10 g of fat, each pound of pork contains 5 g of fat, and each pound of spice contains 2.5 g of fat. The total amount of fat in the blend is limited to 1400 g. The profit per pound is $1 for beef, $2 for pork, and $.50 for spice. Find the number of pounds of each ingredient that maximizes profit.

3. Banks for cooperatives (BCs) are specialty banks that service specific types of firms, such as farm cooperatives. Their loan policies differ from traditional lending institutions in that they tailor payment and interest schedules to the individual customer. They are also governed by different tax and accounting regulations. In an article analyzing lending policies of BCs, linear programming was used to maximize the profit, which is called the *present value of the loan,* subject to constraints, including variable interest rates.‡ For example, consider a loan that is to be paid back in three variable payments, p_1, p_2, and p_3. The sum of the payments is to be no more than $50,000. The variable interest rate is 9% on the first payment, 8% on the second payment, and 6% on the third payment. The total interest is to be no more than $4000. The first payment is to be no more than $20,000. The present value of the loan is $.9p_1 + .8p_2 + .7p_3$. Find the amount of each payment that maximizes the present value.

Cumulative Exercise Set 4–3

1. Determine whether the following linear programming problem is in standard form.

   Maximize $m = 4x + y + 6z$ subject to
   $$2x + 4y + 3z \leq 660$$
   $$x + 3y + 7z \leq 445$$
   $$y + z \geq 87$$
   $$x \geq 0, y \geq 0, z \geq 0$$

2. Find the initial simplex tableau for the following linear programming problem.

   Maximize $m = 14x + 11y$ subject to
   $$3x + 2y + 17z \leq 489$$
   $$7x + 8y + 11z \leq 112$$
   $$x + 2y + 20z \leq 334$$
   $$x \geq 0, y \geq 0, z \geq 0$$

3. Assuming that the matrix below is the initial simplex tableau for a linear programming problem in standard form, construct the original linear programming problem (without slack variables).

x	y	z	s_1	s_2	s_3	s_4	m	
3	1	2	1	0	0	0	0	41
4	7	8	0	1	0	0	0	23
7	9	6	0	0	1	0	0	57
1	5	7	0	0	0	1	0	12
−4	−7	−1	0	0	0	0	0	0

4. Determine the basic variables, the nonbasic variables, and the basic feasible solution for the following simplex tableau.

*David J. Eaton, et al., ''Determining Ambulance Deployment in Santo Domingo, Dominican Republic,'' *Journal of the Operational Research Society,* Vol. 37, 1986, pp. 113–126.

†Ralph E. Steuer, ''Sausage Blending Using Multiple Objective Linear Programming,'' *Management Science,* Vol. 30, 1984, pp. 1376–1384.

‡Ismail Ahmad, et al., ''Analyzing Equity Capital Programs of Banks for Cooperatives,'' *American Journal of Agricultural Economics,* Vol. 68, 1986, pp. 849–856.

$$\begin{array}{ccccccc} x & y & z & s_1 & s_2 & s_3 & m \\ \begin{bmatrix} 0 & 1 & 0 & 2 & -4 & 4 & 0 \\ 0 & 0 & 1 & -2 & 1 & 1 & 0 \\ 1 & 0 & 0 & 2 & 4 & -4 & 0 \\ 0 & 0 & 0 & 0 & 3 & 5 & 1 \end{bmatrix} & \begin{matrix} 86 \\ 30 \\ 55 \\ 1641 \end{matrix} \end{array}$$

5. Determine whether the row below, which is the bottom row of a simplex tableau, passes the maximality test.

$$\begin{array}{ccccccc} x & y & z & s_1 & s_2 & s_3 & m \\ \begin{bmatrix} 0 & 0 & 1 & 0 & 0 & -1 & 1 \end{bmatrix} & 20 \end{array}$$

6. Find the basic feasible solution of the following simplex tableau and apply the maximality test.

$$\begin{array}{ccccc} x & y & s_1 & s_2 & m \\ \begin{bmatrix} 2 & 1 & 2 & 0 & 0 \\ 4 & 0 & -3 & 1 & 0 \\ 0 & 0 & 9 & 0 & 1 \end{bmatrix} & \begin{matrix} 12 \\ 15 \\ 800 \end{matrix} \end{array}$$

7. Select the pivot column and the entering variable of the following simplex tableau, then pivot around the (3, 4) entry.

$$\begin{array}{ccccccc} x & y & z & s_1 & s_2 & s_3 & m \\ \begin{bmatrix} \frac{1}{2} & \frac{1}{2} & 0 & 0 & 1 & 0 & 0 \\ -\frac{1}{2} & -\frac{1}{2} & 0 & 1 & 0 & 1 & 0 \\ 0 & 0 & 1 & 3 & 0 & 0 & 0 \\ 1 & 3 & 0 & -1 & 0 & 0 & 1 \end{bmatrix} & \begin{matrix} \frac{3}{2} \\ \frac{13}{2} \\ 9 \\ 6 \end{matrix} \end{array}$$

8. Write the initial simplex tableau, then form a new tableau by pivoting about the (1, 2) entry.

Maximize $m = 4x + 5y$ subject to
$$x + y \le 350$$
$$3x + y \le 900$$
$$2x + y \le 400$$
$$x \ge 0, y \ge 0$$

9. Select the pivot entry in the simplex tableau below, where the column on the left is the pivot column.

$$\begin{bmatrix} 5 & \cdots & 6 \\ 3 & \cdots & -3 \\ 1 & \cdots & 2 \\ 6 & \cdots & 7 \\ -4 & \cdots & 15 \end{bmatrix}$$

10. Find the pivot entry and perform the pivot operation.

$$\begin{array}{ccccccc} x & y & z & s_1 & s_2 & s_3 & m \\ \begin{bmatrix} 2 & 3 & 1 & 1 & 0 & 0 & 0 \\ 7 & 0 & 2 & -1 & 3 & 0 & 0 \\ 8 & 0 & 1 & -2 & 0 & 3 & 0 \\ 0 & 0 & -3 & 2 & 0 & 0 & 1 \end{bmatrix} & \begin{matrix} 550 \\ 500 \\ 100 \\ 1100 \end{matrix} \end{array}$$

In Problems 11 and 12 solve the linear programming problem by the simplex method.

11. Maximize $m = 4x + 5y$ subject to
$$2x + 3y \le 12$$
$$3x + y \le 11$$
$$x \ge 0, y \ge 0$$

12. An automobile leasing company leases compact, midsized, and luxury cars. To get a compact car ready for a customer requires 25 minutes for paper work, 10 minutes to wash the car, and 15 minutes to inspect it. A midsized car requires 22 minutes for paper work, 12 minutes to wash, and 18 minutes to inspect, while a luxury car requires 25 minutes for paper work, 10 minutes to wash, and 15 minutes to inspect. The profits on compact, midsized, and luxury cars are $50, $55, $60, respectively. If in a given day 200 minutes are available for paper work, 75 minutes for washing, and 60 minutes for inspecting, how many cars of each model should be leased to maximize profit?

4–4 Nonstandard Problems

Thus far we have restricted ourselves to linear programming problems in standard form. In this section we apply the simplex method to other types of problems. Recall that a problem in standard form satisfies three properties:

  i. The objective function is to be maximized.
  ii. The variables are all nonnegative; that is, they are greater than or equal to 0.

iii. The other constraints must be of the form

(linear polynomial) $\leq a$

where a is nonnegative.

In this section we consider problems on which these restrictions are relaxed. Property (ii) is still assumed, so we still assume that all the variables are greater than or equal to 0. Since property (ii) still holds, a problem that is not in standard form requires that the objective function either be minimized or have constraints with greater than ($>$) instead of less than ($<$). We first study the type of nonstandard problem with the objective function to be minimized and then handle the second type.

Minimizing the Objective Function

Suppose the problem requires the objective function to be minimized, so property (i) is not satisfied. We need a fact about numbers. Let S be a finite set of numbers and let $-S$ be the set of negatives of all the numbers in S. If p is the smallest number in S, then $-p$ is the largest number in $-S$. This is true because $p \leq x$ for all x in S implies that $-p \geq -x$ for all $-x$ in $-S$. This means that if the problem is to minimize m, then we can solve the corresponding problem that maximizes $-m$. Example 1 makes use of this property.

EXAMPLE 1

Suppose a linear programming problem is to minimize

$m = 10x + 30y$

Suppose there are four feasible solutions: (2, 2), (3, 1), (4, 3), and (0, 4). Let us compute m and $-m$ for each point.

Feasible Solutions	$10x + 30y = m$	$-10x - 30y = -m$
(2, 2)	$20 + 60 = 80$	$-20 - 60 = -80$
(3, 1)	$30 + 30 = 60$	$-30 - 30 = -60$
(4, 3)	$40 + 90 = 130$	$-40 - 90 = -130$
(0, 4)	$0 + 120 = 120$	$0 - 120 = -120$

The smallest value of m is $m = 60$, which occurs at (3, 1). The largest value of $-m$ is $-m = -60$, also at (3, 1). Hence the same point that minimizes m also maximizes $-m$.

Example 1 shows that minimum problems can be solved via the simplex method by multiplying the objective function m by -1 and maximizing $-m$. The problem then satisfies property (i). The optimal solution that maximizes $-m$ will also minimize m.

For example, suppose the problem is to

Minimize $m = 5x + 7y$ subject to
$x + 5y \leq 10$
$x + 3y \leq 12$

This problem is not in standard form because the objective function is to be minimized. To put it into standard form express it as

Maximize $-m = -5x - 7y$ subject to
$x + 5y \leq 10$
$x + 3y \leq 12$

The preceding discussion shows that the optimal solution of this problem is simultaneously the optimal solution of the original problem.

Inequalities in Nonstandard Form

Now we deal with the second type of nonstandard problem, one that violates property (iii), that is, one in which an inequality has greater than or equal to ($\geq$) instead of less than or equal to ($\leq$). For instance, consider

$x - 3y \geq 5$

To write the inequality using $\leq$ instead of $\geq$, multiply it by -1 and get

$-x + 3y \leq -5$

However, this violates the part of property (iii) that requires the constant term to be nonnegative. To alleviate that difficulty, we form the simplex tableau and perform an appropriate pivot.

In order to see how to handle a negative constant, consider the following row of a simplex tableau and assume that column 4 is in basic form.

$$\begin{array}{ccccc} x & y & s_1 & s_2 & m \\ \left[\begin{array}{ccccc} -1 & 5 & 0 & 1 & 0 & -50 \end{array}\right] \end{array}$$

The difficulty is the negative number in the last column. When we calculate the basic feasible solution corresponding to this tableau, we get $s_2 = -50$. This violates property (ii), which requires all the variables to be nonnegative. In order to apply the simplex method, we must change the sign of any number in the last column that is negative, except the last one. This is done by **pivoting by a negative number** that is in the same row as the negative number in the last column. In the preceding tableau -50 appears in the last column. Since all the variables are greater than or equal to 0, the only way a number in the last column can be negative is for at least one of the numbers in its row to be negative. Otherwise the equation corresponding to that row would have all the positive coefficients and variables on the left-hand side of the equation set equal to a negative constant, which is the number in the last column. This cannot happen since it implies a negative number is set equal to a positive number.

Refer again to the preceding row with -50 in the last column. Locate a negative number in that row. There is only one, -1, in the first column. Use it as the pivot column. Pivoting in this manner will remove the negative number from that entry, as is described in Example 2. This procedure might have to be repeated several times because another negative number might appear or might have been in the last column to begin with.

Remember that we are not concerned with negative numbers in the bottom entry of the last row. It corresponds to the objective function and a negative number in that spot does not violate property (iii).

EXAMPLE 2

Consider the simplex tableau

$$\begin{bmatrix} 1 & 2 & 1 & 0 & 0 & | & 40 \\ -1 & 3 & 0 & 1 & 0 & | & -20 \\ 1 & -4 & 0 & 0 & 1 & | & 0 \end{bmatrix}$$

Problem

Remove the negative number in the last column.

Solution Look for a negative number in the same row as the negative number in the last column. The $(2, 1)$ entry is negative, so choose its column as the pivot column. As before, compute the ratios and select the pivot entry by identifying the smaller positive ratio. In this case the ratios are

for the first row the ratio is $\frac{40}{1} = 40$

for the second row the ratio is $\frac{-20}{-1} = 20$

The second row has the smaller ratio, and thus it is the pivot row. Hence the pivot entry is the $(2, 1)$ entry, -1. When we pivot about it we get the new tableau

$$\begin{bmatrix} 0 & 5 & 1 & 1 & 0 & | & 20 \\ 1 & -3 & 0 & -1 & 0 & | & 20 \\ 0 & -1 & 0 & 1 & 1 & | & -20 \end{bmatrix}$$

Notice that all the entries in the last column (except the last one) are nonnegative. (It might be necessary to pivot more than once to achieve this.) Thus the basic feasible solution has all the variables nonnegative. Now the simplex method can be applied since the tableau is in standard form.

We now summarize the simplex method for problems that are not in standard form.

The Simplex Method for Problems in Nonstandard Form

1. If the problem is to minimize m, then maximize $-m$.
2. If any constraint is in the form

   (linear polynomial) $\geq a$

then multiply the inequality by -1 to get it in the form

(linear polynomial) $\leq -a$

Then form the initial simplex tableau.

3. If no negative number appears in the last column (except the last entry), go to step 6.
4. If a negative number appears in the last column (except the last one), change it to a positive number by pivoting as follows:
   (a) Select a negative entry in the same row as the negative number in the last column. Its column becomes the pivot column.
   (b) Compute all ratios, including those corresponding to negative entries in the pivot column. The pivot row is the one with the smallest positive ratio.
   (c) Perform the pivot about the pivot entry.
5. Go to step 3.
6. Now the tableau is in standard form. Use the simplex method for problems in standard form as described in Section 4–3.

In the next example the method is applied twice, and each time a negative entry appears in the last column. It also shows how to handle the case when two negative numbers appear.

EXAMPLE 3

Problem

Minimize $m = x + y$ subject to
$$2x - y \geq 30$$
$$-x + y \geq 50$$

Solution

1. Multiply the objective function by -1, which yields

   Maximize $-m = -x - y$

2. Both constraints are in the form (linear polynomial) greater than or equal to a, so multiply each by -1, which yields

$$-2x + y \leq -30$$
$$x - y \leq -50$$

The initial simplex tableau is

$$\begin{bmatrix} -2 & 1 & 1 & 0 & 0 & -30 \\ 1 & -1 & 0 & 1 & 0 & -50 \\ \hline 1 & 1 & 0 & 0 & 1 & 0 \end{bmatrix}$$

3. There are two negative numbers in the last column, so we go to step 4.
4. Either number can be chosen. We arbitrarily choose -50 in the second row. There is one other negative entry in the second row, -1, in the second column. The two ratios are $-30/1 = -30$ and $-50/-1 = 50$. Since 50 is the only

positive ratio, choose the (2, 2) entry as the pivot entry. The resulting simplex tableau is

$$\begin{bmatrix} -1 & 0 & 1 & 1 & 0 & | & -80 \\ -1 & 1 & 0 & -1 & 0 & | & 50 \\ \hline 2 & 0 & 0 & 1 & 1 & | & -50 \end{bmatrix}$$

The number -80 is the only negative number in the last column above the last entry. The only other negative entry in the first row is -1 in the first column. The two ratios are $-80/-1 = 80$ and $50/-1 = -50$. The only positive ratio is 80, so the (1, 1) entry, -1, is the pivot entry. The resulting simplex tableau is

$$\begin{bmatrix} 1 & 0 & -1 & -1 & 0 & | & 80 \\ 0 & 1 & -1 & -2 & 0 & | & 130 \\ \hline 0 & 0 & 2 & 3 & 1 & | & -210 \end{bmatrix}$$

There is no negative number in the last row to the left of the last entry, so the maximality test passes and hence the basic feasible solution for this tableau yields the optimal solution. The basic feasible solution is $x = 80$, $y = 130$, $s_1 = s_2 = 0$, $-m = -210$. Hence the values $x = 80$ and $y = 130$ yield the minimum value of $m = 210$.

Dantzig's Breakthrough

The simplex method for solving linear programming problems is more complex than the geometric approach used in Chapter 3. There are two reasons that the simplex method is superior. First, it can handle many equations and variables while the geometric method can deal with only two variables. Second, it is very mechanical and does not rely on visual or geometric interpretations of data, and so it is much easier to program a computer to perform the calculations.

In fact, in 1947 when George Dantzig was searching for a method to solve these problems, he was able to consider the simplex method only because the invention of the computer was imminent. His simplex method reduced the number of calculations drastically, but there were still too many for the method to be useful. Once the computer was invented, the simplex method became a way of life for economists and big business. The Case Study at the end of this chapter describes an example in detail.

EXERCISE SET 4–4

In Problems 1 to 6 determine which of the three defining properties of a linear programming problem in standard form is (or are) violated.

1. Maximize $m = x + y$ subject to
$$2x - y \le 3$$
$$x + y \ge 4$$

2. Maximize $m = -x - y$ subject to
$$3x + y \ge -1$$
$$2x - y \ge 4$$

3. Minimize $m = x + y - 3z$ subject to
$$x - z \le 2$$
$$y + z \ge 4$$

4. Minimize $m = x + 3z$ subject to
$$x + y \quad\ \ \geq 1$$
$$x \quad\ \ - z \leq 2$$

5. Minimize $m = x + 3y + 3z$ subject to
$$x + y - z \leq 12$$
$$x - y + z \geq 4$$
$$x + y + z \geq 3$$

6. Minimize $m = x + y + z$ subject to
$$x + y \quad\ \ \geq -1$$
$$y - z \leq \quad 0$$

In Problems 7 to 12 convert the objective function to one in standard form and then write the last row of the initial simplex tableau, assuming that there is the given number of variables and constraints in the problem.

7. Minimize $m = x + y$, two variables, two constraints.

8. Minimize $m = 2x + 5y$, two variables, two constraints.

9. Minimize $m = x + 3y$, two variables, three constraints.

10. Minimize $m = 2x + 3y$, two variables, three constraints.

11. Minimize $m = x - y + 2z$, three variables, three constraints.

12. Minimize $m = 2x + z$, three variables, four constraints.

In Problems 13 to 18 convert the linear programming problem into one in standard form.

13. Minimize $m = x + y$ subject to
$$x + 2y \leq 4$$
$$x + 3y \geq 1$$

14. Minimize $m = 3x + y$ subject to
$$3x + 2y \leq 14$$
$$4x + 3y \geq 11$$

15. Minimize $m = 2x - y$ subject to
$$2x + y \geq 2$$
$$x - y \geq 4$$

16. Minimize $m = x - 5y$ subject to
$$2x + 3y \geq 42$$
$$5x - 3y \geq 54$$

17. Minimize $m = 3x + y + z$ subject to
$$x \quad\ \ - z \leq -2$$
$$x + y \quad\ \ \geq \quad 4$$

18. Minimize $m = x - 3y$ subject to
$$x + y - 2z \geq \quad 0$$
$$x \quad\ \ - 4z \leq -2$$
$$2x + 3y \quad\ \ \geq -4$$

In Problems 19 and 20 suppose the solution of a minimum problem yields the given matrix as the final simplex tableau. What is the minimum value that is obtained by the optimal solution to the original problem?

19. $\begin{bmatrix} 1 & 0 & -1 & 0 & 0 & | & 12 \\ 0 & 1 & 2 & 1 & 0 & | & 15 \\ 0 & 0 & 1 & 1 & 1 & | & -60 \end{bmatrix}$

20. $\begin{bmatrix} 1 & 2 & 0 & 3 & 0 & | & 10 \\ 0 & 1 & 1 & 1 & 0 & | & 6 \\ 0 & 1 & 0 & 0 & 1 & | & -110 \end{bmatrix}$

In Problems 21 to 33 solve the linear programming problems by the simplex method.

21. Maximize $m = -x + 3y$ subject to
$$5x + 2y \leq 100$$
$$y \geq 10$$
$$x \geq \quad 0$$

22. Maximize $m = x + 4y$ subject to
$$6x + 9y \leq 18$$
$$x \geq 2$$
$$y \geq 0$$

23. Minimize $m = 4x + 13y$ subject to
$$x + 2y \geq 11$$
$$x + y \geq 10$$

24. Minimize $m = 2x + 7y$ subject to
$$x - y \geq 10$$
$$x + 2y \geq 6$$

25. Minimize $m = 2x - 4y$ subject to
$$2x + 6y \leq 18$$
$$-2x + 3y \geq 6$$

26. Minimize $m = x - 2y$ subject to
$$3x + y \geq 18$$
$$x + y \leq 9$$

27. Minimize $m = x + z$ subject to
$$x + 2y \quad\ \ \leq 8$$
$$x \quad\ \ - z \leq 4$$
$$y + 4z \geq 4$$

28. Minimize $m = x + y$ subject to
$$x + y \quad\ \ \geq 10$$
$$x \quad\ \ - z \leq 14$$
$$y + 2z \geq 6$$

29. Minimize $m = x + 2y + z$ subject to
$$2x + y \quad\quad \leq 10$$
$$3x + y \quad\quad \leq 12$$
$$\quad\quad y + 2z \geq 8$$

30. Minimize $m = 6x + 7y + 3z$ subject to
$$2x + 5y \quad\quad \leq 20$$
$$2x + y \quad\quad \leq 8$$
$$\quad\quad y + z \leq 8$$

31. Minimize $m = x + 2y + z$ subject to
$$3x + 2y + z \leq 12$$
$$x \quad\quad + z \leq 8$$
$$2x + 5y \quad\quad \leq 10$$

32. Minimize $m = x + 3y + 2z$ subject to
$$x + 5y + z \leq 20$$
$$x \quad\quad + z \leq 10$$
$$2x + 5y + z \leq 15$$

33. Minimize $m = 2x + y + 3z + u$ subject to
$$x + y \quad\quad\quad \leq 100$$
$$x + y + z \quad\quad \leq 150$$
$$\quad y + z \quad\quad \leq 100$$
$$x \quad\quad\quad + u \leq 200$$

34. Show that if S is a finite set of numbers and p is the smallest number in S, then $-p$ is the largest number in $-S$, the set of all negatives of numbers in S.

35. A television service store manager must purchase at least 200 VCR tapes. There are two types of tapes to choose from, high-quality and regular. The high-quality tapes play for 2 hours and cost $8 each and the regular tapes play for 1 hour and cost $6. The tapes must produce at least 100 hours of playing time. How many of each type should the manager purchase to minimize cost?

36. A fuel oil manufacturing firm has two processing plants, plant I and plant II. Plant I processes daily 250 barrels of high-grade oil for commercial use and 100 barrels of low-grade oil for residential use. Plant II processes daily 100 barrels of high-grade oil and 100 barrels of low-grade oil. The daily cost of operation is $50,000 for plant I and $30,000 for plant II. An order is placed for 1500 barrels of high-grade oil and 1200 barrels for low-grade oil. Find the number of days that each plant should operate to fill the order and minimize the cost.

37. A candy manufacturer makes three types of 1-pound (lb) chocolate bars. The chocolate-nut bar contains $\frac{3}{4}$ lb chocolate and $\frac{1}{4}$ lb nuts, the chocolate-peanut butter bar contains $\frac{3}{4}$ lb chocolate and $\frac{1}{4}$ lb peanut butter, and the chocolate-cluster bar contains $\frac{1}{2}$ lb chocolate, $\frac{1}{4}$ lb nuts, and $\frac{1}{4}$ lb peanut butter. Each day the manufacturer must use at least 10 lb of nuts, no more than 40 lb of chocolate, and no more than $7\frac{1}{2}$ lb of peanut butter. If it costs $4 to make each chocolate-nut bar, $2 to make each chocolate-peanut butter bar, and $3 to make each chocolate-cluster bar, how many bars of each type should the manufacturer make each week to minimize cost?

38. An overnight delivery service has daily flights to primary depots in four cities, Boston, Chicago, Dallas, and Los Angeles. The total daily flights to Boston and Chicago must be at least 10; the total daily flights to Chicago and Dallas must be at least 12; there must be no more than 15 total daily flights to Boston and Los Angeles; there must be no more than 10 total daily flights to Dallas and Los Angeles. The cost of each flight is $8000 to Boston, $20,000 to Chicago, $10,000 to Dallas, and $6000 to Los Angeles. How many flights to each city should be scheduled to minimize cost?

Referenced Exercise Set 4–4

1. Linear programming has been used in the petroleum industry for over 25 years to minimize the cost of blending fuel. In an article analyzing linear programming techniques used at Exxon, blending gasoline in processing tanks was studied.* As an example, suppose that at one point of the blending process there are three tanks. The total amount of gasoline that passes through the three tanks is no more than 5 million gallons, and no more than 1 million gallons pass through tank 3. One goal of the blending process is to extract lead from the gasoline. From each gallon in tank 1, .2 grams (g) of lead is extracted; from each gallon in tank 2, .5 g of lead is extracted; and from each gallon in tank 3, .1 g of lead is extracted. At least 1.5 million g of lead must be extracted. The cost per gallon is $.10 for tank 1, $.15 for tank 2, and $.20 for tank 3. Find the number of gallons that should be processed by each tank to minimize cost.

---

*Thomas E. Baker, and Leon S. Lasdon, "Successive Linear Programming at Exxon," *Management Science*, Vol. 31, 1985, pp. 264–274.

2. The cost of transportation is an important consideration for many enterprises. School systems especially feel the pressure to hold down busing costs while maintaining adequate service. Swersey and Ballard applied linear programming to a bus scheduling problem in the school system of New Haven, Connecticut.† A typical problem in the study involves a busing schedule utilizing three types of routes that service a school with 2000 students. The object is to minimize the number of buses. The first type of route uses two buses, and the second and third types of routes use three buses each. Each bus on a type 1 route has 50 seats, while each bus on a type 2 or 3 route has 60 seats. The operating cost per year is $15,000 for each bus used on a type 1 route, $20,000 for each bus used on a type 2 route, and $25,000 for each bus used on a type 3 route. If the cost per year to maintain the fleet is limited to $600,000, find the number of each type of route that should be used to minimize the number of buses.

Cumulative Exercise Set 4–4

1. Construct the original linear programming problem (without slack variables) assuming that the matrix is the initial simplex tableau.

$$\begin{bmatrix} 1 & 0 & 2 & 1 & 0 & 0 & 0 & 20 \\ 3 & 8 & 3 & 0 & 1 & 0 & 0 & 35 \\ 4 & 2 & 7 & 0 & 0 & 1 & 0 & 60 \\ -8 & -9 & -3 & 0 & 0 & 0 & 1 & 100 \end{bmatrix}$$

2. Determine the basic variables, the nonbasic variables, and the basic feasible solution of the simplex tableau.

x	y	z	s_1	s_2	s_3	m	
1	3	0	0	0	1	0	113
0	4	1	0	0	1	0	100
0	0	0	6	1	0	0	320
0	-4	0	-3	0	0	1	500

3. Find the basic feasible solution of the simplex tableau and then apply the maximality test. Assume the problem is in standard form.

x	y	z	s_1	s_2	s_3	m	
0	2	1	0	0	1	0	250
0	9	0	1	9	5	0	490
1	0	0	0	1	2	0	250
0	-8	0	0	0	-2	1	1300

4. Perform the pivot process on the simplex tableau using the (1, 2) entry as the pivot entry.

x	y	z	s_1	s_2	s_3	m	
1	2	0	1	0	1	0	4
1	1	0	0	1	2	0	3
3	0	1	0	0	1	0	8
-2	-7	0	0	0	-1	1	35

5. Find the initial simplex tableau and form a new tableau by pivoting about the (3, 3) entry.

Maximize $m = 2x + 5y + 6z$ subject to
$$5x + 2y + 3z \le 90$$
$$2x + y + z \le 35$$
$$x + 2y + 3z \le 60$$
$$x, y, z \ge 0$$

In Problems 6 and 7 find the pivot entry and perform the pivot operation.

6.

x	y	s_1	s_2	m	
0	2	1	4	0	60
1	1	0	2	0	20
0	-1	0	-3	1	80

7.

x	y	z	s_1	s_2	s_3	m	
1	0	1	0	1	0	0	20
2	0	0	1	2	1	0	50
3	1	0	0	0	1	0	70
-4	0	0	0	-3	0	1	5

In Problems 8 and 9 solve the linear programming problem by the simplex method.

8. Maximize $m = 3x + 8y$ subject to
$$x + 3y \le 24$$
$$2x + 3y \le 36$$

9. Maximize $m = 5x + 4y + 6z$ subject to
$$x + 2y + 3z \le 110$$
$$x + 3y + z \le 70$$
$$x + 4y + 2z \le 80$$

10. A firm manufactures three types of tents, small, medium, and large. Each small tent takes 2 hours to cut and 1.5 hours to assemble. Each medium tent takes 2 hours to cut and 2 hours to assemble. Each large tent takes 4 hours to cut and 1 hour to assemble.

†Arthur J. Swersey, and Wilson Ballard, "Scheduling School Buses," *Management Science,* Vol. 30, 1984, pp. 844–853.

There is a total of 16 hours available to cut and 80 hours available to assemble the tents. No more than 50 tents are to be made. How many of each type of tent should be made to maximize profit if the profit is $30 on each small tent, $20 on each medium tent, and $10 on each large tent?

11. Minimize $m = 3x + 3y$ subject to
$$x + y \quad\quad \geq 40$$
$$x - \quad\quad z \leq 56$$
$$y + 2z \geq 24$$

12. Minimize $m = 3x + 3y + 6z$ subject to
$$x + 3y + 2z \leq 36$$
$$x + y \quad\quad \leq 24$$
$$2y + 5z \leq 310$$

4–5 The Dual Problem

This section describes the dual method for solving a nonstandard problem whose objective function is minimized and whose constraints are all in the form greater than or equal to ($\geq$). Such a problem can be solved by the methods of Section 4–4, but they require numerous pivots. The **dual method,** however, converts the original problem to a problem that is in standard form and whose solution contains the solution to the original problem.

Matrix Inequalities

Let M and N be two matrices. Recall that $M = N$ if they have the same size and their corresponding entries are equal. Similarly, we define $M \leq N$ if they have the same size and $m_{ij} \leq n_{ij}$ for all i,j.

Linear programming problems can be written in terms of matrices. Consider the diet problem from Section 3–2.

$$\text{Minimize} \quad m = 4x + 2y$$
$$\text{Subject to} \quad 5x + y \geq 5$$
$$5x + 3y \geq 10$$
$$x \geq 0, y \geq 0$$

Let A be the coefficient matrix of the first two constraints, C be the constant matrix, R be the row matrix of coefficients of m, and X be the column matrix of variables.

$$A = \begin{bmatrix} 5 & 1 \\ 5 & 3 \end{bmatrix} \quad C = \begin{bmatrix} 5 \\ 10 \end{bmatrix} \quad R = \begin{bmatrix} 4 & 2 \end{bmatrix} \quad X = \begin{bmatrix} x \\ y \end{bmatrix}$$

The diet problem becomes

$$\text{Minimize} \quad m = RX$$
$$\text{Subject to} \quad AX \geq C$$
$$X \geq 0$$

In Example 1 another linear programming problem is converted to matrix notation.

EXAMPLE 1

Problem

Write this linear programming problem in terms of matrices.

Minimize $m = 7x + 5y$
Subject to $x + 3y \geq 2$
$2x + y \geq 4$
$x + y \geq 6$
$x \geq 0, y \geq 0$

Solution Let A be the coefficient matrix of the first three constraints, C be the constant matrix, R be the row matrix of coefficients of m, and X be the column matrix of variables.

$$A = \begin{bmatrix} 1 & 3 \\ 2 & 1 \\ 1 & 1 \end{bmatrix} \quad C = \begin{bmatrix} 2 \\ 4 \\ 6 \end{bmatrix} \quad R = [7 \quad 5] \quad X = \begin{bmatrix} x \\ y \end{bmatrix}$$

The linear programming problem can be written in terms of matrices as

Minimize $m = RX$
Subject to $AX \geq C$
$X \geq 0$

Notice in Example 1 that A and C must have the same number of rows and that R and X must have the same number of entries.

The Transpose

The **transpose** of a matrix A, denoted by A', is formed by switching the rows and columns of A so that the first row becomes the first column, the second row becomes the second column, and so on. For instance,

$$\text{if} \quad A = \begin{bmatrix} 1 & 3 & 0 \\ 2 & 7 & 4 \end{bmatrix} \quad \text{then} \quad A' = \begin{bmatrix} 1 & 2 \\ 3 & 7 \\ 0 & 4 \end{bmatrix}$$

In general, the transpose of an $m \times n$ matrix is an $n \times m$ matrix. Thus the transpose of a row matrix is a column matrix, and vice versa. The central role of the transpose in the dual problem is described in the display.

The Dual Problem

If a linear programming problem is

Minimize $m = RX$
Subject to $AX \geq C$
$X \geq 0$

then its *dual problem* is

$$\text{Maximize} \quad m = C'U$$
$$\text{Subject to} \quad A'U \leq R'$$
$$U \geq 0$$

where U is a column matrix whose size is the same size as C.

The dual problem is in standard form so it can be solved by standard means. Its objective function m is the same as the objective function m of the original problem. The proof of this fact is beyond the scope of this book.

There is a more efficient way to express the dual problem. Let us illustrate it on the diet problem.

$$\text{Minimize} \quad m = 4x + 2y$$
$$\text{Subject to} \quad 5x + y \geq 5$$
$$5x + 3y \geq 10$$
$$x \geq 0, \, y \geq 0$$

First form the matrix T, called the **primal matrix,** whose bottom row comes from the objective function and whose other rows come from the explicit constraints.

$$T = \begin{bmatrix} 5 & 1 & 5 \\ 5 & 3 & 10 \\ 4 & 2 & 0 \end{bmatrix}$$

Its transpose is

$$T' = \begin{bmatrix} 5 & 5 & 4 \\ 1 & 3 & 2 \\ 5 & 10 & 0 \end{bmatrix}$$

The dual problem can be read from T', with the objective function again coming from the bottom row.

$$\text{Maximize} \quad m = 5u + 10v$$
$$\text{Subject to} \quad 5u + 5v \leq 4$$
$$u + 3v \leq 2$$
$$u \geq 0, \, v \geq 0$$

EXAMPLE 2

Problem

Write the dual problem of the linear programming problem in Example 1.

Solution Form the primal matrix T.

$$T = \begin{bmatrix} 1 & 3 & 2 \\ 2 & 1 & 4 \\ 1 & 1 & 6 \\ 7 & 5 & 0 \end{bmatrix}$$

Then form the transpose T'.

$$T' = \begin{bmatrix} 1 & 2 & 1 & 7 \\ 3 & 1 & 1 & 5 \\ 2 & 4 & 6 & 0 \end{bmatrix}$$

The dual problem of the original problem is

Maximize $m = 2u + 4v + 6w$
Subject to $u + 2v + w \le 7$
             $3u + v + w \le 5$
             $u \ge 0, v \ge 0, w \ge 0$

The Solution

The dual problem is in standard form, so it is solved by the usual means. The final tableau yields the solution to the original problem because the variable m in the dual problem is the same as the variable m in the original problem. The values of the variables x and y are obtained from the final tableau in a distinctive way. Consider the diet problem and its dual together.

Diet Problem	**Dual Problem**
Minimize $m = 4x + 2y$	Maximize $m = 5u + 10v$
Subject to $5x + y \ge 5$	Subject to $5u + 5v \le 4$
$5x + 3y \ge 10$	$u + 3v \le 2$
$x \ge 0, y \ge 0$	$u \ge 0, v \ge 0$

The initial tableau of the dual problem is

$$\begin{array}{ccccc} u & v & s_1 & s_2 & m \\ \left[\begin{array}{ccccc|c} 5 & 5 & 1 & 0 & 0 & 4 \\ 1 & 3 & 0 & 1 & 0 & 2 \\ \hline -5 & -10 & 0 & 0 & 1 & 0 \end{array}\right] \end{array}$$

It reduces to the final tableau

$$\begin{array}{ccccc} u & v & s_1 & s_2 & m \\ \left[\begin{array}{ccccc|c} 1 & 0 & \frac{3}{10} & -\frac{1}{2} & 0 & \frac{1}{5} \\ 0 & 1 & -\frac{1}{10} & \frac{1}{2} & 0 & \frac{3}{5} \\ \hline 0 & 0 & \frac{1}{2} & \frac{5}{2} & 1 & 7 \end{array}\right] \end{array}$$

The solution to the dual problem is $u = \frac{1}{5}$, $v = \frac{3}{5}$, and $m = 7$. The value $m = 7$ is the maximum value of the objective function of the dual problem and the minimum value of the objective function of the original problem. The values for x and y lie in the bottom row of the final tableau in the columns of the slack variables, $x = \frac{1}{2}$ and $y = \frac{5}{2}$. (The proof of this fact is beyond the scope of this book.) This solution corresponds to the point $(\frac{1}{2}, \frac{5}{2})$ derived geometrically in Example 1 of Section 3–3.

Example 3 continues Examples 1 and 2.

EXAMPLE 3

Problem

Solve the problem in Example 1.

Solution From Example 2 the dual problem is

$$\begin{aligned}
\text{Maximize} \quad & m = 2u + 4v + 6w \\
\text{Subject to} \quad & u + 2v + w \le 7 \\
& 3u + v + w \le 5 \\
& u \ge 0, v \ge 0, w \ge 0
\end{aligned}$$

The initial tableau is

$$\begin{array}{ccccccc}
u & v & w & s_1 & s_2 & m & \\
\left[\begin{array}{cccccc|c}
1 & 2 & 1 & 1 & 0 & 0 & 7 \\
3 & 1 & 1 & 0 & 1 & 0 & 5 \\
\hline
-2 & -4 & -6 & 0 & 0 & 1 & 0
\end{array}\right]
\end{array}$$

Pivot about the (2, 3) entry.

$$\begin{array}{cccccc}
u & v & w & s_1 & s_2 & m \\
\left[\begin{array}{cccccc|c}
-2 & 1 & 0 & 1 & -1 & 0 & 2 \\
3 & 1 & 1 & 0 & 1 & 0 & 5 \\
\hline
16 & 2 & 0 & 0 & 6 & 1 & 30
\end{array}\right]
\end{array}$$

The values for x and y lie in the bottom row under the slack variables. The solution is $x = 0$, $y = 6$, $m = 30$.

The dual method consists of three steps: (1) convert the original problem to its dual problem, (2) solve the dual problem by standard techniques, and (3) interpret the final tableau of the dual problem to obtain the solution to the original problem. Example 4 illustrates the entire method.

EXAMPLE 4

Problem

Solve by the dual method.

$$\begin{aligned}
\text{Minimize} \quad & m = 12x + 3y + 4z \\
\text{Subject to} \quad & 2x + 3y + 4z \ge 4 \\
& 3x + y + z \ge 6 \\
& x \ge 0, y \ge 0, z \ge 0
\end{aligned}$$

Solution To convert the original problem to its dual, form the primal matrix T and its transpose.

$$T = \begin{bmatrix} 2 & 3 & 4 & 4 \\ 3 & 1 & 1 & 6 \\ 12 & 3 & 4 & 0 \end{bmatrix} \qquad T' = \begin{bmatrix} 2 & 3 & 12 \\ 3 & 1 & 3 \\ 4 & 1 & 4 \\ 4 & 6 & 0 \end{bmatrix}$$

The dual problem is

Maximize $m = 4u + 6v$
Subject to $2u + 3v \leq 12$
$\qquad\qquad 3u + v \leq 3$
$\qquad\qquad 4u + v \leq 4$
$\qquad\qquad u \geq 0, v \geq 0$

The initial tableau is

$$
\begin{array}{cccccc|c}
u & v & s_1 & s_2 & s_3 & m & \\
2 & 3 & 1 & 0 & 0 & 0 & 12 \\
3 & 1 & 0 & 1 & 0 & 0 & 3 \\
4 & 1 & 0 & 0 & 1 & 0 & 4 \\
\hline
-4 & -6 & 0 & 0 & 0 & 1 & 0
\end{array}
$$

Pivot about the (2, 2) entry.

$$
\begin{array}{cccccc|c}
u & v & s_1 & s_2 & s_3 & m & \\
-7 & 0 & 1 & -3 & 0 & 0 & 3 \\
3 & 1 & 0 & 1 & 0 & 0 & 3 \\
1 & 0 & 0 & -1 & 1 & 0 & 1 \\
\hline
14 & 0 & 0 & 6 & 0 & 1 & 18
\end{array}
$$

This passes the maximality test. The solution to the original problem is $x = 0$, $y = 6$, $z = 0$, $m = 18$.

EXERCISE SET 4–5

In Problems 1 to 4 write the linear programming problem in terms of matrices. In all problems $x \geq 0$, $y \geq 0$, and $z \geq 0$.

1. Minimize $m = 4x + 2y$
   Subject to $3x + 2y \geq 4$
   $\qquad\qquad 4x - 5y \geq 9$

2. Minimize $m = 4x + y + z$
   Subject to $x + 3y - 2z \geq 6$
   $\qquad\qquad x - 2y + 3z \geq 6$

3. Minimize $m = 2x + 4y$
   Subject to $3x + 5y \geq 8$
   $\qquad\qquad -x + 3y \geq 4$
   $\qquad\qquad 2x - 6y \geq 7$

4. Minimize $m = 4x + y + z$
   Subject to $x + 3y - 2z \geq 6$
   $\qquad\qquad x - 2y + 3z \geq 6$
   $\qquad\qquad -x + 4y + 5z \geq 9$

In Problems 5 to 10 form the transpose of the given matrix.

5. $\begin{bmatrix} 5 & -1 \\ 2 & -3 \end{bmatrix}$

6. $\begin{bmatrix} 4 & 6 \\ -5 & 8 \\ 10 & 7 \end{bmatrix}$

7. $\begin{bmatrix} -4 \\ 2 \end{bmatrix}$

8. $\begin{bmatrix} 4 & 0 & -3 \\ 2 & -1 & 8 \end{bmatrix}$

9. $\begin{bmatrix} 1 & -4 & 11 \\ 6 & 15 & 0 \\ -2 & 1 & -2 \end{bmatrix}$

10. $\begin{bmatrix} 3 & 0 & -6 \end{bmatrix}$

In Problems 11 to 14 write the dual problem of the given linear programming problem.

11. Problem 1.

12. Problem 2.

13. Problem 3.

14. Problem 4.

In Problems 15 to 18 solve each linear programming problem by the dual method. In all problems $x \geq 0$, $y \geq 0$, and $z \geq 0$.

15. Minimize $m = x + 2y$
    Subject to $2x + y \geq 10$
    $3x + y \geq 12$

16. Minimize $m = 6x + 7y$
    Subject to $2x + 5y \geq 20$
    $2x + y \geq 8$

17. Minimize $m = 4x + 13y$
    Subject to $x + 2y \geq 11$
    $x + y \geq 10$

18. Minimize $m = 2x + 7y$
    Subject to $x - y \geq 10$
    $x + 2y \geq 6$

19. A dietician is preparing a meal of chicken and rice so that each meal contains at least 5 grams (g) of protein and 5 milligrams (mg) of iron. Each serving of chicken contains 5 g of protein, 3 mg of iron, and 4 g of fat. Each serving of rice contains 1 g of protein, 2 mg of iron, and 2 g of fat. How many servings of chicken and rice should be prepared to minimize the amount of fat?

20. A heating oil firm has two processing plants. Plant A processes daily 3000 barrels of No. 2 fuel for residential use and 1000 barrels of No. 6 fuel for industrial use. Plant B processes daily 2000 barrels of No. 2 fuel and 1000 barrels of No. 6 fuel. The daily cost of operation is $20,000 for plant A and $15,000 for plant B. An order is placed for 20,000 barrels of No. 2 fuel and 9000 barrels of No. 6 fuel. Find the number of days that each plant should be operated to fill the order and minimize the cost.

In Problems 21 to 28 solve each linear programming problem by converting it to the dual problem. Assume that $x \geq 0$, $y \geq 0$, $z \geq 0$, and $w \geq 0$.

21. Minimize $m = x + 2y + z$
    Subject to $3x + 2y + z \geq 12$
    $x \qquad + z \geq 8$
    $2x \qquad + 5z \geq 10$

22. Minimize $m = 2x + y + 3z + w$
    Subject to $x + y \qquad\qquad \geq 100$
    $x + y + z \qquad \geq 150$
    $y + z \qquad \geq 100$
    $x \qquad\qquad + w \geq 0$

23. Minimize $m = 3x + 4y + 2z + 3w$
    Subject to $x + y \qquad + w \geq 8$
    $2x \qquad + z \qquad \geq 7$
    $y + z + w \geq 9$
    $2x + y \qquad + 2w \geq 6$

24. Minimize $m = 5x + 6y + 8z + 3w$
    Subject to $x + y \qquad + w \geq 1$
    $3x \qquad + z + 2w \geq 2$
    $x + y + z + w \geq 0$
    $4x + y + 2z \qquad \geq 3$

25. Minimize $m = x + 2y + z$
    Subject to $3x + 2y + z \geq 12$
    $x \qquad + z \geq 8$

26. Minimize $m = 2x + y + 3z + w$
    Subject to $x + y + z \qquad \geq 15$
    $y + z \qquad \geq 10$
    $x \qquad + w \geq 0$

27. Minimize $m = 9x + 21y$
    Subject to $x + 3y \geq 16$
    $x + y \geq 12$

28. Minimize $m = 3x + y$
    Subject to $10x + 2y \geq 84$
    $8x + 4y \geq 120$

29. A dietician is preparing a menu of chicken, rice, and spinach so that each meal contains at least 30 g of protein, 50 mg of iron, and 40 mg of starch. Each serving of chicken contains 1 g of protein, 1 mg of iron, and 20 g of fat. Each serving of rice contains 2 g of protein, 1 mg of starch, and 35 g of fat. Each serving of spinach contains 2 g of protein, 1 mg of starch, and 65 g of fat. How many servings of chicken, rice, and spinach should be prepared to minimize the amount of fat?

30. A heating oil firm has three processing plants. The amount of No. 2 fuel oil processed per hour is 1000 barrels at plant A and 3000 barrels at plant B. The amount of No. 6 fuel processed per hour is 1000 barrels at plant A and 1000 barrels at plant C. The amount of blended oil processed per hour is 1000 barrels at plant A, 2000 barrels at plant B, and 1000 barrels at plant C. The daily cost of operation is $100,000 for plant A, $300,000 for plant B, and $400,000 for plant C. An order is placed for 300,000 barrels of No. 2 fuel oil, 150,000 barrels of No. 6 fuel, and 200,000 barrels of blended oil. Find the number of hours that each plant should be operated to fill the order and minimize the cost.

31. In a certain Nintendo game you need to gather swords and axes. Each sword kills 18 ogres, 30 slimes, and 10 pirates, while each ax kills 9 ogres, 20 slimes, and 30 pirates. To beat the game you must kill at least 270 ogres, 540 slimes, and 3900 pirates, but you must sacrifice 30 friends for each sword and 10 friends for each ax. How many swords and axes should you use to beat the game while minimizing the loss of friends?

32. A hospital patient is required to take two different vitamins, each composed of substances X and Y. Each vitamin A pill consists of 1 gram (g) of X and 10 g of Y, while each vitamin B pill consists of 4 g of X and 8 g of Y. The patient must take at least 24 g of substance X and 120 g of substance Y. If each gram of substance X causes the patient to lose 1 square centimeter (cm^2) of hair and each gram of substance Y causes the patient to lose 3 cm^2 of hair, how many vitamin pills of each type should the patient take to minimize the loss of hair?

Referenced Exercise Set 4–5

1. One of the most widely used applications of linear programming is in scheduling problems. In an article describing plant design and scheduling problems in the food industry, Regina Beneviste analyzed efficient ways to mix powdered ingredients in weighing machines, called hoppers.* A specific plant used three hoppers in mixing batches of several recipes. One type of problem involved minimizing the amount of time spent weighing the ingredients subject to the number of times certain recipes were required. Recipe A used hopper 1 and hopper 2. Recipe B used hopper 2 and hopper 3. Recipe C used hopper 1 and hopper 3. The batch requires recipe A be made at least 100 times, recipe B be made at least 240 times, and recipe C be made at least 140 times. It takes .4 minutes to use hopper 1, .8 minutes to use hopper 2, and .7 minutes to use hopper 3. Find the number of times each hopper should be scheduled to minimize the amount of weighing time.

2. Dotmar Packaging Ltd. is a firm with a corrugated cardboard box factory that operates four finishing machines in the corrugation process. Boxes are made by cutting a continuous strip of corrugated board into customer-specified dimensions. Cost is affected by the amount of waste left from the cutting process. Dotmar hired consultants to develop a production schedule that would minimize cutting cost while satisfying customer demand.† A typical problem involved four cutting patterns performed on the four finishing machines. Pattern A uses machines 1 and 2. Pattern B uses machines 3 and 4. Pattern C uses machines 1 and 3. Pattern D uses machines 2 and 4. The demand on a particular day requires 75 runs of pattern A, 180 runs of pattern B, 105 runs of pattern C, and 150 runs of pattern D. If the cost per run is $1.00 for machine 1, $1.80 for machine 2, $1.60 for machine 3, and $.90 for machine 4, find the number of times each machine should be used to minimize cost while satisfying demand.

Cumulative Exercise Set 4–5

1. Determine whether each linear programming problem is in standard form. If it is, write the initial simplex tableau. If it is not, convert it to standard form and then write the initial simplex tableau.
    (a) Minimize $m = 3x - y + z$ subject to
    $$2x + 3y + z \leq 6$$
    $$x - 4y - z \leq 5$$
    $$x \geq 0, y \geq 0$$
    (b) Maximize $m = 2x + 3y + 2z$ subject to
    $$x + 4y + 2z \leq 9$$
    $$2x + 3y + z \leq 7$$
    $$x \geq 0, y \geq 0, z \geq 0$$

2. Let the row below be the bottom row of the initial simplex tableau of a linear programming problem.
    (a) What is the objective function?
    (b) How many constraints are there?

$$\begin{bmatrix} x & y & z & t & s_1 & s_2 & s_3 & m \\ -2 & -1 & 0 & -5 & 0 & 0 & 0 & 1 \end{bmatrix} | \, 0 \,]$$

3. Suppose the objective function of a linear programming problem is to minimize $m = 2x + 3y + 5z$.
    (a) Convert the objective function to standard form.
    (b) Write the bottom row of the initial simplex tableau assuming there are three variables and four constraints.

*Regina Beneviste, "An Integrated Plant Design and Scheduling Problem in the Food Industry," *Journal of the Operational Research Society,* Vol. 37, 1986, pp. 435–461.

†James H. Bookbinder, and James K. Higginson, "Customer Service vs. Trim Waste in Corrugated Box Manufacture," *Journal of the Operational Research Society,* Vol. 37, 1986, pp. 1061–1071.

4. (a) Convert the following linear programming problem into one in standard form.
   (b) Write the initial simplex tableau.

   Minimize $m = 7x - y - 2z$ subject to
   $$2x + 3y \quad\quad \leq 11$$
   $$x \quad\quad - 4z \leq 12$$
   $$y + 2z \geq 8$$
   $$x \geq 0, y \geq 0, z \geq 0$$

5. (a) Write the following linear programming problem in terms of matrices.

   (b) Write the objective function and constraints of the dual of the linear programming problem.
   Minimize $m = 2x + 3y + 4z + t$ subject to
   $$x - 2y + z + t \geq 6$$
   $$3x + 3y - z + 2t \geq 12$$
   $$4x - y \quad\quad + 4t \geq 15$$
   $$x \geq 0, y \geq 0, z \geq 0, t \geq 0$$

6. (a) Write the primal matrix T of the linear programming problem in Problem 5.
   (b) Form the transpose T'.
   (c) Write the dual problem using T'.

7. Solve the following linear programming problems by any appropriate method.
   (a) Maximize $m = 4x + 5y$ subject to
   $$2x + 3y \leq 12$$
   $$3x + y \geq 11$$
   $$x \geq 0, y \geq 0$$
   (b) Maximize $m = 4x + 8y + 5z$ subject to
   $$x + 4y + z \leq 6$$
   $$2x + 6y + 4z \leq 15$$
   $$x + 2y + 3z \leq 18$$
   $$x \geq 0, y \geq 0, z \geq 0$$

8. Solve the following linear programming problems by the simplex method or the dual method.
   (a) Minimize $m = 3x + 4y$ subject to
   $$x + 4y \geq 8$$
   $$2x + 3y \geq 12$$
   $$2x + y \geq 6$$
   $$x \geq 0, y \geq 0$$
   (b) Minimize $m = 2x + 3y$ subject to
   $$3x + 4y \geq 5$$
   $$x + 2y \geq 2$$
   $$5x + 3y \geq 7$$
   $$x \geq 0, y \geq 0$$

Solve Problems 9 to 12 by any appropriate method.

9. The Nintendo hotline center employs expert and novice game counselors in two shifts. The day shift employs 2 experts and 1 novice, while the night shift employs 1 expert and 5 novices. Nintendo wants to use no more than 36 counselors altogether. There will be at least 24 hours to be filled during the week, with the day shift working for 3 hours and the night shift working for 2 hours. The cost is $44 per hour for the day shift and $40 per hour for the night shift. How many shifts of each type should be employed to minimize the cost?

10. A truck makes deliveries of Commodore computers to a retail outlet and to a chain store. The time of a round trip to the retail outlet is 3 hours and the time of a round trip to the chain store is 4 hours. It costs $100 for a round trip to the retail outlet and $200 for a round trip to the chain store. The profits are $500 for each truckload to the retail outlet and $800 for each truckload to the chain store. How many truckloads per week should be scheduled to each destination to maximize profit if the truck is limited to 48 hours per week and Commodore's cost cannot exceed $2000?

11. A summer camp concocts a breakfast cereal that is mixed from packages of Cheerios and Wheaties. Each package of Cheerios adds 2 units of protein, 1 unit of iron, and 1 unit of thiamine to the mix, while each package of Wheaties adds 1 unit of protein, 1 unit of iron, and 3 units of thiamine. Suppose the cost of a Cheerios package is $.30 and the cost of a Wheaties package is $.40. How many packages of each type of cereal should be used in the mix to minimize the cost if the mix must provide at least 12 units of protein, 9 units of iron, and 15 units of thiamine?

12. A nutritionist prepares a menu for people whose diet requires at least 5 grams (g) of protein, 15 milligrams (mg) of iron, and 1000 calories. The primary ingredients are chicken, beans, and potatoes. Each serving of chicken contains 45 g of protein, 9 mg of iron, 500 calories, and 4 g of fat; each serving of beans contains 9 g of protein, 6 mg of iron, 150 calories, and 2 g of fat; and each serving of potatoes contains 7 g of protein, 5 mg of iron, 100 calories, and 2 g of fat. How many servings of each ingredient should the menu include in order to minimize the cost?

CASE STUDY   ## Shell Oil Company

One of the most important applications that George Dantzig's simplex method was used for was the solution of the transportation problem. Virtually the same methods are used today in industry to solve similar problems. All large firms, and many smaller ones, have very complex schemes and formulas to determine the cost of transporting their merchandise. Usually the procedure entails the use of a computer.

As an example we will look at the Shell Oil Company and their distribution problems.* Since 20% of Shell's revenue is spent on transportation of oil, an efficient distribution is critical. Shell divides the country into regions and each region is divided into subregions. We will center our attention on the Chicago area subregion. This area has two primary refineries where the oil is turned from crude oil to the many varied grades of commercial grade oil. They are located in East Chicago, Indiana, and Hammond, Indiana. The two major storage and reshipment terminals are located in Des Plaines, Illinois, and Niles, Michigan. Figure 1 illustrates Shell's distribution scheme. In actual practice, the problem is more complex than we can consider here. It involves over 1200 variables and 800 constraints. This is because there are more complex decisions to be made such as which mode of transportation to use (including pipelines, barges, trucks, and tankers). The typical problem faced each day is solved on a computer. It takes about $\frac{1}{2}$ hour to run and costs about $100. Such a report will usually generate around ten optional reports because there are various goals and different types of individuals using the same data.

Let us now concentrate on a typical problem encountered in the Chicago area subregion. On a given day the East Chicago refinery will produce at least 50 thousand barrels of commercial grade oil; the Hammond refinery will produce

*T. K. Zierer, et al., "Practical Applications of Linear Programming to Shell's Distribution Problems," *Interfaces,* Vol. 6, No. 4, August 1976.

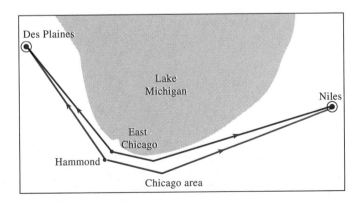

FIGURE 1

at least 120 thousand barrels; the Niles terminal requires at least 70 thousand barrels; and the Des Plaines terminal requires at least 100 thousand barrels. The average transportation cost of shipping from the East Chicago refinery is $8 (per mile per 1000 barrels) to Niles and $16 to Des Plaines, while the average cost from the Hammond refinery is $14 to Niles and $7 to Des Plaines. The problem is to determine how many barrels should be shipped from each refinery to each terminal in order to minimize the transportation cost subject to the given constraints. To solve the problem we can arrange the data in Table 1.

Table 1

From		East Chicago		Hammond		Limiting
	To	Niles	Des Plaines	Niles	Des Plaines	Value
Produced E		1	1	0	0	50
by H		0	0	1	1	120
Received N		1	0	1	0	70
by D		0	1	0	1	100
Cost		8	16	14	7	

We let the variables represent the number of barrels (measured in 1000s) shipped per day from a refinery to a terminal, so that

x = the number from East Chicago to Niles

y = the number from East Chicago to Des Plaines

z = the number from Hammond to Niles

t = the number from Hammond to Des Plaines

c = the total cost

Thus the problem becomes

Minimize $c = 8x + 16y + 14z + 7t$ subject to

$$x + y \qquad\qquad \geq 50$$
$$z + t \geq 120$$
$$x + \quad z \quad \geq 70$$
$$y \quad + t \geq 100$$
$$x, y, z, t \geq 0$$

Note that the problem is not in standard form because it is a minimum problem and each constraint is in the greater than or equal to ($\geq$) form. Therefore if we express the problem in matrix form, we can solve the dual problem. In matrix form the problem is

Minimize $c = RX$
Subject to $AX \geq C$

where

$$R = \begin{bmatrix} 8 & 16 & 14 & 7 \end{bmatrix} \quad X = \begin{bmatrix} x \\ y \\ z \\ t \end{bmatrix} \quad A = \begin{bmatrix} 1 & 1 & 0 & 0 \\ 0 & 0 & 1 & 1 \\ 1 & 0 & 1 & 0 \\ 0 & 1 & 0 & 1 \end{bmatrix} \quad C = \begin{bmatrix} 50 \\ 120 \\ 70 \\ 100 \end{bmatrix}$$

In order to form the dual problem, we compute the transpose of each matrix. We get

$$R' = \begin{bmatrix} 8 \\ 16 \\ 14 \\ 7 \end{bmatrix}, \quad U = \begin{bmatrix} u \\ v \\ w \\ r \end{bmatrix}, \quad A' = \begin{bmatrix} 1 & 0 & 1 & 0 \\ 1 & 0 & 0 & 1 \\ 0 & 1 & 1 & 0 \\ 0 & 1 & 0 & 1 \end{bmatrix}, \quad C' = \begin{bmatrix} 50 & 120 & 70 & 100 \end{bmatrix}$$

Thus the dual problem is

Maximize $c = C'U$ subject to
$A'U \geq R'$

The corresponding simplex tableau is

$$\begin{bmatrix}
1 & 0 & 1 & 0 & 1 & 0 & 0 & 0 & 0 & 8 \\
1 & 0 & 0 & 1 & 0 & 1 & 0 & 0 & 0 & 16 \\
0 & 1 & 1 & 0 & 0 & 0 & 1 & 0 & 0 & 14 \\
0 & 1 & 0 & 1 & 0 & 0 & 0 & 1 & 0 & 7 \\
-50 & -120 & -70 & -100 & 0 & 0 & 0 & 0 & 1 & 0
\end{bmatrix}$$

The entry in the last row that is most negative is -120 in the second column. The two ratios are $\frac{7}{1} = 7$ and $\frac{14}{1} = 14$, so the pivot entry is 1, the (4, 2) entry. The new tableau after pivoting is

$$\begin{bmatrix}
1 & 0 & 1 & 0 & 1 & 0 & 0 & 0 & 0 & 8 \\
1 & 0 & 0 & 1 & 0 & 1 & 0 & 0 & 0 & 16 \\
0 & 0 & 1 & -1 & 0 & 0 & 1 & -1 & 0 & 7 \\
0 & 1 & 0 & 1 & 0 & 0 & 0 & 1 & 0 & 7 \\
-50 & 0 & -70 & 20 & 0 & 0 & 0 & 120 & 1 & 840
\end{bmatrix}$$

The most negative entry in the last row is -70 in the third column. The ratios are $\frac{7}{1} = 7$ and $\frac{8}{1} = 8$ so the pivot entry is 1, in the third row. The new tableau is

$$\begin{bmatrix}
1 & 0 & 0 & 1 & 1 & 0 & -1 & 1 & 0 & 1 \\
1 & 0 & 0 & 1 & 0 & 1 & 0 & 0 & 0 & 16 \\
0 & 0 & 1 & -1 & 0 & 0 & 1 & -1 & 0 & 7 \\
0 & 1 & 0 & 1 & 0 & 0 & 0 & 1 & 0 & 7 \\
-50 & 0 & 0 & -50 & 0 & 0 & 70 & 50 & 1 & 1330
\end{bmatrix}$$

The most negative entry in the last row is -50, and it appears in the first and fourth columns so we can use either one. We arbitrarily choose the first row. The ratios are $\frac{1}{1} = 1$ and $\frac{16}{1} = 16$, so the pivot entry is 1 in the first row. The new tableau is

$$\left[\begin{array}{ccccccccc|c}
1 & 0 & 0 & 1 & 1 & 0 & -1 & 1 & 0 & 1 \\
0 & 0 & 0 & 0 & -1 & 1 & 1 & -1 & 0 & 15 \\
0 & 0 & 1 & -1 & 0 & 0 & 1 & -1 & 0 & 7 \\
0 & 1 & 0 & 1 & 0 & 0 & 0 & 1 & 0 & 7 \\
\hline
0 & 0 & 0 & 0 & 50 & 0 & 20 & 100 & 1 & 1380
\end{array}\right]$$

The maximality test passes, and thus the solution is $c = 1380$. Therefore the minimum cost for Shell under the given constraints is $1380 (per mile per 1000 barrels). The values of the variables are found in the last row in the columns headed by the slack variables. Hence we have $x = 50$, $y = 0$, $z = 20$, and $t = 100$. This means that Shell will achieve its minimum cost by shipping 50 thousand barrels from East Chicago to Niles, none from East Chicago to Des Plaines, 20 thousand barrels from Hammond to Niles, and 100 thousand barrels from Hammond to Des Plaines.

Case Study Exercises

1. Suppose the average transportation cost of shipping from the East Chicago refinery is $16 (per mile per 1000 barrels) to Niles and $8 to Des Plaines, while the average cost from the Hammond refinery is $7 to Niles and $14 to Des Plaines. Determine how many barrels should be shipped from each refinery to each terminal in order to minimize the transportation cost.
2. Suppose all the transportation costs are $10. Determine how many barrels should be shipped from each refinery to each terminal in order to minimize the transportation cost.

CHAPTER REVIEW

Key Terms

4–1   Slack Variables and the Simplex Tableau

Standard Form	Basic Form
Slack Variables	Basic Variable
Simplex Tableau	Nonbasic Variable
Initial Simplex Tableau	Basic Feasible Solution

4–2   The Maximality Test and Pivoting

Maximality Test	Pivot Column
Pivoting	Entering Variable
Pivot Entry	

4–3   The Simplex Method for Problems in Standard Form
Pivot Row

4–4   Nonstandard Problems
Pivoting by a Negative Number

4–5   Dual Problem

Dual Method   Primal Matrix
Transpose

Summary of Important Concepts

Procedure for Determining the Basic Feasible Solution

1. Set each nonbasic variable equal to 0.
2. Each resulting equation has one basic variable and the last column contains its value.

Procedure for Translating Linear Programming Problems in Standard Form into Mathematical Language

1. Add a slack variable to each constraint to make it an equation.
2. Put the objective function into proper form.
3. Form the augmented matrix.

Maximality Test
A simplex tableau will yield an optimal solution if and only if the last row contains no negative entries.

Selecting the Pivot Column
The column that has the most negative entry in the last row is the pivot column.

Choosing the Pivot Row
The row with the smallest nonnegative ratio.

The Simplex Method for Problems in Standard Form

1. State the problem as a system of equations with slack variables.
2. Construct the initial simplex tableau.
3. Apply the maximality test. If the solution is maximal, the problem is solved. If not, go to step 4.
4. Compute a new simplex tableau by pivoting.
5. Go to step 3.

The Simplex Method for Problems in Nonstandard Form

1. If it is a minimum problem, maximize $-m$.
2. State every inequality in terms of $\leq$.
3. If no negative number appears in the last column (except the last entry), go to step 6.
4. If a negative number appears in the last column (except the last entry), change it to a positive number by pivoting.
5. Go to step 3.
6. The problem is now in standard form.

REVIEW PROBLEMS

In Problems 1 and 2 determine whether the linear programming problem is in standard form.

1. Minimize $m = 3x + 5y$ subject to
$$2x + 3y \leq 10$$
$$x + 2y \leq 5$$
$$x \geq 0, y \geq 0$$

2. Maximize $m = 7x + 8y$ subject to
$$3x + 7y \leq 50$$
$$2x + 5y \geq 45$$
$$x \geq 0, y \geq 0$$

3. Add slack variables to the inequalities and express them as equations.
$$4x + 3y \leq 30$$
$$2x + 7y \leq 65$$

In Problems 4 and 5 find the initial simplex tableau for the linear programming problem. Assume the problem is in standard form.

4. Maximize $m = 2x + 5y$ subject to
$$2x + 3y \leq 10$$
$$3x + 4y \leq 12$$

5. Maximize $m = x + 4y + 2z$ subject to
$$x + 2y + 5z \leq 30$$
$$2x + 3y + z \leq 40$$
$$4x + 3y \leq 20$$

6. Determine the basic variables, the nonbasic variables, and the basic feasible solution of the simplex tableau.

x	y	z	s_1	s_2	s_3	m	
0	0	0	1	1	1	0	15
1	0	0	2	0	0	0	18
0	4	1	4	1	0	0	22
0	-1	0	-5	3	0	1	49

7. Apply the maximality test to the simplex tableau.

x	y	z	s_1	s_2	s_3	m	
0	0	0	1	1	0	0	22
1	3	-3	2	0	0	0	35
0	4	2	4	0	1	0	28
0	1	0	-5	0	0	1	82

8. Select the pivot column and the entering variable.

x	y	z	s_1	s_2	s_3	m	
2	0	1	0	4	0	0	20
1	0	0	1	2	1	0	10
1	1	0	0	0	1	0	12
-3	0	0	0	3	1	1	55

9. For the simplex tableau in Problem 8 perform the pivot process using the (2, 1) entry as the pivot entry.

10. Find the initial simplex tableau and form a new tableau by pivoting about the indicated entry, then pivot about the (1, 2) entry.

Maximize $m = 2x + 3y$ subject to
$$x + 3y \leq 15$$
$$2x + 5y \leq 10$$
$$x \geq 0, y \geq 0$$

11. Translate the linear programming problem into mathematics. Then find the initial simplex tableau and form a new tableau by pivoting about the (1, 1) entry.

A hospital nutritionist prepares a menu for patients on low-fat, high-protein diets. The primary ingredients are chicken, beans, and potatoes. The meal must contain no more than 1000 calories (cal), 15 grams (g) of fat, and 15 milligrams (mg) of iron. Each 3-ounce serving of chicken contains 45 g of protein, 500 cal, 5 mg of iron, and 4 g of fat. Each serving of beans contains 9 g of protein, 150 cal, 6 mg of iron, and 2 g of fat. Each serving of potatoes contains 7 g of protein, 100 cal, 5 mg of iron, and 2 g of fat. How much chicken, beans, and potatoes should be included in the meal in order to maximize the amount of protein?

In Problems 12 and 13 find the pivot entry and perform the pivot operation.

12.
x	y	s_1	s_2	m	
2	0	1	1	0	20
1	1	0	0	0	16
-8	0	0	-4	1	40

13.
x	y	z	s_1	s_2	s_3	m	
1	0	1	1	0	2	0	12
5	0	0	2	1	-1	0	24
4	1	0	0	1	0	0	25
-4	0	3	0	0	-2	1	42

In Problems 14 to 16 solve the linear programming problem by the simplex method.

14. Maximize $m = 8x + 7y$ subject to
$$2x + 3y \leq 21$$
$$x + 3y \leq 18$$
$$x \geq 0, y \geq 0$$

15. Maximize $m = 4x + 5y + 3z$ subject to
$$x + 5y + 3z \leq 15$$
$$x + 2y + 2z \leq 20$$
$$x \geq 0, y \geq 0, z \geq 0$$

16. A firm manufactures three types of sails, small, medium, and large. Each small sail takes 3 hours to cut and 2 hours to assemble; each medium sail takes 4 hours to cut and 2 hours to assemble; each large sail takes 2 hours to cut and 4 hours to assemble. There is a total of 80 hours available to cut and 320 hours available to assemble the sails. No more than 50 sails are to be made. How many of each type of sail should be made to maximize profit if the profit is $100 on each small sail, $150 on each medium sail, and $50 on each large sail?

17. Convert the linear programming problem into one in standard form.

Minimize $m = 2x + 5y$ subject to
$$x + 4y \leq 4$$
$$x + 2y \geq 7$$
$$x \geq 0, y \geq 0$$

18. Solve the linear programming problem by the simplex method.

Minimize $m = x + y + 2z$ subject to
$$x + 4y \qquad \leq 8$$
$$x - y \qquad \leq 6$$
$$y + 4z \geq 4$$
$$x \geq 0, y \geq 0, z \geq 0$$

In Problems 19 and 20 solve the linear programming problem by the dual method.

19. Minimize $m = 12x + 6y$ subject to
$$x + y \geq 2$$
$$4x + y \geq 3$$
$$x \geq 0, y \geq 0$$

20. Minimize $m = 40x + 30y$
$$x + y \geq 1$$
$$2x + y \geq 2$$
$$x + y \geq 3$$
$$x \geq 0, y \geq 0$$

PROGRAMMABLE CALCULATOR EXPERIMENTS

Problems 1 to 3 refer to the following linear programming problem.

Maximize $P = px + qy$ subject to
$$ax + by \leq c$$
$$dx + ey \leq f$$
$$x \geq 0, y \geq 0$$

1. Write a program that computes the initial simplex tableau of the linear programming problem.

2. Write a program that computes the pivot column of the initial simplex tableau in Problem 1.

3. Write a program that computes the pivot row of the initial simplex tableau in Problem 1.

4. Write a program that pivots about the pivot entry of the initial simplex tableau in Problem 1.

5. Modify the programs in Problems 1 to 4 to linear programming problems with three equations in three unknowns.

Finance

5

Finite math helps manage personal accounts and billion-dollar portfolios. (R. Kord/H. Armstrong Roberts)

Chapter Overview

In this chapter we deal primarily with the mathematics of finance, but we also discuss inflation and population growth. In Section 5–1 we introduce simple interest on savings and loans and investigate why the "simple discount method" is really no discount at all. In Section 5–2 the interest is compounded over different periods, with the emphasis on the customary banking practice of interest compounded daily. We also show how much money should be invested at a given time to produce a desired amount at some later date. This kind of problem is a particular concern to college students who must plan tuition payments years ahead of time.

In Section 5–3 we apply the exponential function to inflation, cost of living, and population growth.

In Sections 5–4 and 5–5 we describe transactions in which the payments are made periodically. Both sections rely on the formula for the sum of a geometric series, which is derived in Section 5–4. In Section 5–5 we discuss the kinds of loans on cars and houses that most of us have to take out during our lifetime.

Since tables cannot possibly contain all of the necessary figures, a calculator is required. We introduce estimations before some of the solutions to encourage students to think about the reasonableness of answers. This is particularly important when a calculator is used.

CASE STUDY PREVIEW

"Tastes great!!!" "Less filling!!!!"

The commercials based on these two claims have become marketing classics. The Miller Brewing Company developed the advertisements for product identification in 1974. At the time Miller held onto a distant fifth place share of the brewing market in the United States. Within five years, however, the advertisements vaulted Miller into a close second place behind Anheuser-Busch, and the two breweries have dominated the domestic market ever since. The Case Study examines Miller's rise to the top.

5–1 Simple Interest and Discount

In this section we describe simple interest and introduce the concept of present value. We also examine the so-called "discount interest method." A good working knowledge of percents is required. The solutions to all financial problems are expressed in a dollars-and-cents form, so a calculator might be necessary for some computations.

A Matter of Interest

Suppose that you invest $1000 at a 6% rate of interest. The amount of money invested is called the **principal,** denoted by P. Here $P = \$1000$. The **interest** I is the amount of money earned on the investment, and the **rate of interest** r is the percentage that is earned. Here $r = 6\%$. In all calculations we will write r in decimal form, so $r = 0.06$.

The rate of interest can be a simple rate or a rate that is compounded over various periods. In **simple interest** the rate of interest applies only to the principal; in **compound interest** the rate of interest applies to the previously accumulated interest as well as to the principal. The formula for simple interest is given in the display.

Simple Interest
The simple interest I earned on a principal P invested at $r = p\%$ annual rate of interest for t years is

$$I = Prt$$

For the $1000 invested at a 6% rate of interest the simple interest in one year is $I = (\$1000)(0.06)(1) = \$60$. When interest accrues over 5 years, the simple interest is $I = (\$1000)(0.06)(5) = \$300$.

Newspaper advertisements for banks frequently tout interest rates that grab our attention, but the fine print usually reveals conditions like ''minimum deposit $10,000.'' Example 1 illustrates the simple interest formula on a $10,000 investment over fractional parts of a year. It assumes an **exact interest year** of 365 days (as opposed to an **ordinary interest year** of 360 days). In order to follow banking industry standards, we will assume 365 days in a year on savings and 360 days in a year on loans. When interest on savings is calculated in terms of days, $t = \cdot\dfrac{d}{365}$, where d is the number of days; when it is calculated in terms of months, $t = \dfrac{m}{12}$, where m is the number of months.

EXAMPLE 1

Suppose that a $10,000 certificate is deposited at a 10.75% rate of simple interest.

Problem

(a) What is the simple interest for $2\frac{1}{2}$ years? (b) What is the simple interest from March 8 to April 15?

Estimation 10% of $10,000 yields $1000 per year, so the interest will be about $2500 for $2\frac{1}{2}$ years and $100 for 5 weeks (since 5 weeks is about $\frac{1}{10}$ of a year).

Solution From the information given, $P = 10,000$ and $r = 10.75\%$. Write $r = 0.1075$. (a) Here $t = 2.5$, so $I = Prt = (10,000)(0.1075)(2.5) = 2687.5$. The simple interest is $2687.50. (b) There are 38 days from March 8 to April 15, so the time is $t = 38/365 = 0.10411$. Then $I = Prt = (\$10,000)(0.1075)(0.10411) = \$111.92$.

The examples so far have emphasized savings, but in reality most students are concerned with loans. Aside from a different viewpoint and a year based on 360 days, however, the basic concepts remain the same. Example 2 illustrates this fact.

EXAMPLE 2

Problem

What interest must be paid on a loan of $2500 for 90 days at 14% simple interest?

Solution Set $P = 2500$, $r = 0.14$, and $t = 90/360 = 0.25$. Then $I = Prt = (2500)(0.14)(0.25) = 87.5$, so the interest on the loan is $87.50.

Almost all loans of a year or less are calculated as simple interest. However, most savings earn compound interest.

Present Value

We have seen that the simple interest on a principal $P = \$1000$ that is invested at $r = 6\%$ simple interest for $t = 5$ years is $I = \$300$. The sum of $\$1300 = \$1000 + \$300$ is called the **future amount.** It is denoted by A and given by the formula $A = P + I$. By substituting for I and factoring the common term P, we get $A = P + Prt = P(1 + rt)$. This is called the Future Amount Formula.

Future Amount Formula

$$A = P(1 + rt)$$

EXAMPLE 3

Problem

If a $10,000 savings certificate is purchased at a $6\frac{1}{2}\%$ rate of simple interest, what is the future amount in (a) 60 days, (b) 30 months, (c) 20 years?

Solution In all three cases $P = 10,000$ and $r = 0.065$. The remaining variable in the Future Amount Formula is t. (a) For 60 days the time is $t = 60/365 = 0.16438$. Then $A = P(1 + rt) = 10,000(1 + 0.065 \cdot 0.16438) = 10,106.85$. Thus the future amount is $10,106.85. (b) For 30 months $t = 30/12 = 2.5$. Then the future amount is $A = P(1 + rt) = \$10,000(1 + 0.065 \cdot 2.5) = \$11,625$. (c) For 20 years $t = 20$, so the future amount is $A = \$10,000(1 + 0.065 \cdot 20) = \$23,000$.

In the Future Amount Formula the principal P is called the **present value.** There is a reciprocal relationship between A and P: A is the future amount of P, and P is the present value of A. In Example 3 the present value P was given and the future amount A was computed. The reverse case occurs when some future amount is needed and money can be invested at a given rate. The present value is obtained by dividing both sides of the Future Amount Formula by $(1 + rt)$.

Present Value Formula

$$P = \frac{A}{1 + rt}$$

Example 4 considers a resourceful parent who wants to invest a lump sum to help pay for a newborn child's future college education.

EXAMPLE 4

Problem

How much money should be invested at 12% simple interest to yield $20,000 in 18 years?

Estimation Since 12% of $10,000 yields $1200 per year, 18 years will yield more than $20,000. Therefore the answer must be less than $10,000.

Solution The future amount is $A = 20,000$, the rate of interest is $r = 0.12$, and the time period is $t = 18$ years. By the Present Value Formula

$$P = \frac{20,000}{1 + (0.12)(18)} = \frac{20,000}{3.16} = 6,329.11$$

Therefore $6329.11 should be invested now in order to produce $20,000 in 18 years.

Simple Discount

Some lending institutions loan money by the **discount interest method.** This method involves deducting the interest on the loan before giving the balance to the borrower. The deducted money is called the **discount** and the amount the borrower actually receives is called the **proceeds.**

Proceeds Formula

$$PR = P - I$$

For instance, consider a loan of $1000 for 9 months at 8% simple interest using the discount interest method. Then $P = 1000$, $r = 0.08$, and $t = \frac{9}{12} = 0.75$. The discount corresponds to simple interest.

$$I = Prt = (1000)(0.08)(0.75) = 60$$

Thus the discount is $60. The proceeds PR are defined by the formula $PR = P - I$, so $PR = \$940$.

Here is how the discount interest method works. The $60 is deducted from the $1000 loan first. Then the borrower is handed the proceeds of $940. This is really the amount the borrower is borrowing. In effect, the borrower is paying $60 interest on only $940. Example 5 will compute the real interest being charged.

There is another way to view discount interest. It involves the true annual interest rate, which is called the **effective rate** and is computed from the formula for simple interest in the form $I = (PR)(er)t$. Solving for the variable er yields

$$er = \frac{I}{(PR)t}$$

The effective rate allows the consumer to compare the discount rate with the rate that is actually being charged. Example 5 carries out such a comparison.

EXAMPLE 5

Suppose that $1000 is borrowed for 9 months at 8% simple interest using the discount interest method.

Problem

What is the effective rate?

Solution The term is 9 months, so $t = \frac{9}{12} = 0.75$. We have already calculated the discount $I = \$60$ and the proceeds $PR = \$940$. The effective rate is

$$er = \frac{I}{(PR)t} = \frac{60}{(940)(0.75)} = 0.0851$$

The effective rate is 8.51%. Compare this with the stated rate of 8%.

The "Truth in Lending Law" requires that the effective rate of interest appear on all contracts. Example 5 should convince you that it was in the consumer's interest to enact this law. After all, the term "discount interest method" is a real misnomer!

EXERCISE SET 5–1

In Problems 1 to 4 find the simple interest on each of the following savings. Assume 365 days in a year.

1. $1000 at $5\frac{1}{2}$% interest for
   (a) 7 years   (b) 7 months   (c) 1 week

2. $2500 at 6% interest for
   (a) 3 years   (b) 3 months   (c) 30 days

3. $1988 at 14.2% interest for
   (a) 13 years   (b) 21 months   (c) 58 days

4. $1988 at 14.1% interest for
   (a) 13 years   (b) 21 months   (c) 58 days

In Problems 5 and 6 find the simple interest on each of the following loans. Assume 360 days in a year.

5. $1580 at 11% interest for
   (a) 5 years   (b) 40 days

6. $1480 at 12% interest for
   (a) 4 years   (b) 50 days

In Problems 7 to 10 find the future amount of each of the following savings. Assume 365 days in a year.

7. $1000 at $5\frac{1}{2}$% interest for
   (a) 7 years   (b) 7 months   (c) 1 week

8. $2500 at 6% interest for
   (a) 3 years   (b) 3 months   (c) 30 days

9. $1988 at 14.2% interest for
   (a) 13 years   (b) 21 months   (c) 58 days

10. $1988 at 14.1% interest for
    (a) 13 years   (b) 21 months   (c) 58 days

In Problems 11 and 12 find the total amount that must be repaid on each of the following loans. Assume 360 days in a year.

11. $1580 at 11% interest for
    (a) 4 years   (b) 4 months   (c) 4 days

12. $1480 at 12% interest for
    (a) 4 years   (b) 4 months   (c) 4 days

In Problems 13 and 14 find the present value of the given future amount in the time specified. Assume 365 days in a year.

13. $5000 at 10% interest for
    (a) 10 years   (b) 10 months   (c) 10 days

14. $1000 at 5% interest for
    (a) 5 years   (b) 5 months   (c) 5 days

In Problems 15 to 18 find the discount, proceeds, and effective rate. Assume 360 days in a year.

15. $2000 at 10% interest for 3 years.

16. $1000 at 5% interest for 8 months.

17. $176 at 11.2% interest for 60 days.

18. $1776 at 15.1% interest for March 1 through April 15.

19. Suppose a loan at 9% simple interest is necessary to pay for a semester's tuition of $2998. How much interest will have to be paid in 4 years?

20. Suppose that a parent refuses to pay the senior year's tuition because of the liberal life-style the student has adopted. How much interest does the student have to pay on a loan to cover the $3200 tuition at $6\frac{1}{2}$% interest for 8 months?

In Problems 21 and 22 find the simple interest on each of the following loans. Assume 360 days in a year.

21. $10,000 at 12.3% interest made on September 10 and due on (a) February 10   (b) February 28

22. $12,345 at 10% interest made on April 3 and due on (a) November 3   (b) November 26

In Problems 23 and 24 find the total amount that must be repaid on these loans. Assume 360 days in a year.

23. $10,000 at 12.3% interest made on September 10 and due on (a) February 10   (b) February 28

24. $12,345 at 10% interest made on April 3 and due on (a) November 3   (b) November 26

25. A family wants to deposit a certain amount of money in an account that pays 14.2% simple interest so their first offspring can have $100,000 at age 18 for 4 years' tuition at an Ivy League school. How much should they deposit now?

26. A family wants to send its newborn child to a state university when the child turns 18 years old. How much should they invest now in an account that pays 13.4% simple interest to ensure that their child will have $40,000 for 4 years' tuition?

In Problems 27 to 30 the principal P, interest I, and time t of an investment are given. Find the simple interest rate.

27. $P = \$1000$, $I = \$48.75$, $t = 9$ months.

28. $P = \$2500$, $I = \$131.25$, $t = 1$ year.

29. $P = \$500$, $I = \$31.25$, $t = 1$ year.

30. $P = \$1000$, $I = \$28.75$, $t = 6$ months.

In Problems 31 to 34 the future amount A, principal P, and rate of simple interest r are given. Find the time t of the investment.

31. $A = \$2000$, $P = \$1000$, $r = 12\%$.

32. $A = \$1500$, $P = \$1000$, $r = 8\%$.

33. $A = \$500$, $P = \$375$, $r = 5.5\%$.

34. $A = \$500$, $P = \$250$, $r = 9.8\%$.

In Problems 35 to 38 the future amount A, principal P, and time t are given for an investment that earns simple interest. Find the interest rate r as a percent.

35. $A = \$2000$, $P = \$1000$, $t = 39$ months.

36. $A = \$1500$, $P = \$1000$, $t = 4$ years.

37. $A = \$500$, $P = \$375$, $t = 3$ years.

38. $A = \$500$, $P = \$425$, $t = 30$ months.

Referenced Exercise Set 5–1

Problems 1 and 2 consider a surprising view of the historic tax bill that was passed by the Senate in 1986. John D. Paulus, the managing director of Morgan Stanley's economics department, contended that interest rates would drop when the cost of borrowing rose. This paradox ran counter to the prevailing theory of finance.*

1. The Senate tax bill lowered the tax rate for someone in the highest tax bracket from 50 to 27%. Suppose such a person borrowed $100,000.
   (a) What is the interest if the money is borrowed for 1 year at 10% simple interest?
   (b) Compare the costs of borrowing the money if 50% of the interest is tax deductible versus if 27% of the interest is tax deductible.
   (c) What percent did the after-tax cost rise while the tax rate dropped from 50 to 27%?

2. The Senate tax bill also affected people in lower tax brackets. This problem concerns a person who borrowed $10,000 and whose tax bracket was reduced from 30 to 15%.
   (a) What is the interest if the money is borrowed for 1 year at 10% simple interest?
   (b) Compare the costs of borrowing the money if 30% of the interest is tax deductible versus if 15% of the interest is tax deductible.
   (c) What percent did the after-tax cost rise while the tax rate dropped from 30 to 15%?

*J. A. Livingston, "Business Outlook: Taking Interest in a Paradox," *The Philadelphia Inquirer*, June 11, 1986.

5–2 Compound Interest

In this section we develop the "Compound Interest Formula" for finding the principal on investments and loans whose interest rates are compounded annually, quarterly, monthly, and daily. The emphasis is on interest rates that are compounded daily because most banks use this method.

The solutions to the problems are expressed in a dollars-and-cents form, which can be obtained by means of a calculator or from one of three tables. Some expressions, however, cannot be evaluated by the tables, so they will require a calculator with an exponential key. Because calculators round their answers in different ways, your results might differ slightly from ours. The tables were compiled on a microcomputer; thus they too are subject to slight rounding errors.

Annual Rates

When money is invested at a simple rate of interest the interest applies only to the principal. When money is invested at a compound rate of interest the interest applies to the principal and the previously earned interest. The expression **interest compounded annually** refers to an interest rate that is applied once a year.

We compare simple interest and compound interest on a principal of $P = \$1000$ that is deposited at an annual interest rate of $r = 6\%$ for $t = 3$ years. The simple interest is

$$I = Prt = (\$1000)(0.06)(3) = \$180$$

An amount of $60 is paid at the end of each year.

For compound interest the $60 interest that is earned the first year is added to the $1000 principal at the end of the year. The second year's interest is earned on the new principal of $1060. We change the notation to reflect the change in the principal. Let P_t be the principal after t years. In particular, P_1 is the principal after 1 year.

$$P_1 = P + P(0.06) = P(1.06) = \$1000(1.06) = \$1060$$

The principal after 2 years is

$$P_2 = P_1 + P_1(0.06) = P_1(1.06)$$

$$= (\$1060)(1.06) = \$1123.60$$

The principal after 3 years is

$$P_3 = P_2(1.06) = (\$1123.60)(1.06) = \$1191.02$$

Therefore the compound interest after 3 years is $\$1191.02 - \$1000 = \$191.02$, compared to the simple interest of $180.

This method of calculating interest compounded annually leads to a derivation of a formula for computing the principal in any year directly without having to compute the interests for the intervening years. Recall that the Future Amount Formula is $A = P(1 + rt)$. Therefore the principal after $t = 1$ year is

$$P_1 = P(1 + r)$$

When interest is compounded annually the principal after 2 years is

$$P_2 = P_1(1 + r) = P(1 + r)(1 + r)$$

$$= P(1 + r)^2$$

The principal after 3 years is

$$P_3 = P_2(1 + r) = P(1 + r)^2(1 + r)$$

$$= P(1 + r)^3$$

By continuing this process, we find the formula for the principal after t years is

$$P_t = P(1 + r)^t$$

This is the formula for the principal when interest is compounded annually. We refer to it as "the formula for interest compounded annually." It is the analog of the Future Amount Formula for simple interest. The term $(1 + r)^t$ is called the **growth factor.**

For the preceding example a principal of $1000 that is invested for 3 years at 6% interest compounded annually is obtained directly by substituting $P = 1000$, $r = 0.06$, and $t = 3$.

$$P_3 = \$1000(1 + 0.06)^3 = \$1000(1.06)^3$$

This expression can be converted into dollars and cents by evaluating the growth factor $(1.06)^3$ by a calculator or by the Compound Interest Table in the Appendix. If the calculator has an exponential key $\boxed{y^x}$, the usual method for evaluating $(1.06)^3$ is to use the key sequence

$$1.06 \boxed{y^x} 3 \boxed{=}$$

Otherwise ordinary multiplication can be used.

$$(1.06)^3 = (1.06)(1.06)(1.06) = 1.191016$$

After rounding, the principal is

$$P_3 = \$1000(1.06)^3 = \$1000(1.19102) = \$1191.02$$

The growth factor $(1.06)^3$ is located in the Compound Interest Table at $n = 3$ and $r = 6\%$. The entry 1.19102 agrees with the figure obtained by calculator.

Example 1 applies the formula for interest compounded annually to a "loan problem."

EXAMPLE 1

Problem

What is the interest on a $2500 loan that is borrowed at 10% interest compounded annually for 5 years?

Estimation The answer must exceed $(5)(10\%)(\$2500) = \$1250$.

Solution It is given that $P = 2500$, $r = 0.1$, and $t = 5$. Therefore

$$P_5 = 2500(1 + 0.1)^5 = 2500(1.1)^5$$

By calculator or the Compound Interest Table (at $n = 5$ and $r = 10\%$)

$$P_5 = 2500(1.1)^5 = 2500(1.61051) = 4026.275$$

Thus the principal is $4026.28. The interest is equal to

$$P_5 - P = \$4026.28 - \$2500 = \$1526.28$$

The Compound Interest Formula

We introduce interest rates that are compounded over various periods by examining a principal of $1000 that is invested for 2 years at 8% interest compounded quarterly. The word "quarterly" means that the interest is added to the principal four times a year. Since $\dfrac{12 \text{ months}}{4} = 3$ months, the interest is added every 3 months. This is called the **conversion period** (or just the **period**). Moreover, the rate of interest that is paid every 3 months is $\dfrac{0.08}{4} = 0.02$. It is called the **periodic interest.** Thus the expression "8% compounded quarterly" means that 2% interest is earned every 3 months. The interest earned after the first quarter is 2% of $1000 (or $20), so the principal increases to $1020. It can be obtained from the formula

$$Q_1 = P + P(0.02) = P(1.02)$$

The principal after two quarters is

$$Q_2 = Q_1(1.02) = P(1.02)(1.02) = P(1.02)^2$$

The principal after three quarters is

$$Q_3 = Q_2(1.02) = P(1.02)^2(1.02) = P(1.02)^3$$

Similarly, the principal after four quarters, or 1 year, is

$$Q_4 = P(1.02)^4$$

and the principal after 2 years is

$$Q_8 = P(1.02)^8$$

This formula is a particular case of the **Compound Interest Formula,** which determines the principal on interest rates that are compounded over given time periods. Its derivation is analogous to this specific case.

Compound Interest Formula

$$A = P(1 + i)^n$$

where

    A is the compound amount

    P is the principal

    $i = r/k$ is the periodic interest, where

        r is the annual rate of interest

        k is the number of periods per year the interest is compounded

    $n = tk$ is the total number of periods, where

        t is the number of years

By the Compound Interest Formula the principal after 2 years on $1000 invested at 8% interest compounded quarterly is obtained by substituting $r = 0.08$, $k = 4$, $t = 2$, and $P = 1000$. Then $i = 0.08/4 = 0.02$ and $n = 8$, so

$$A = 1000(1 + 0.02)^{2 \cdot 4} = 1000(1.02)^8$$

The growth factor $(1.02)^8$ can be obtained by calculator or from the Compound Interest Table at $n = 8$ (not $n = 2$) and $r = 2\%$ (not $r = 8\%$) as 1.17116. Then

$$A = 1000(1.02)^8 = (1000)(1.17166) = 1171.66$$

Therefore the compound amount after 2 years is $1171.66.

EXAMPLE 2

Problem

Compute the compound amount on a $1000 certificate that is deposited for 5 years in an account that pays 16% interest compounded quarterly.

Estimation The answer must exceed $1000 + (5)($160) = $1800.

Solution We have $P = \$1000$, $r = 0.16$, and $t = 5$. Since the interest is compounded quarterly, $k = 4$. Then $i = 0.16/4 = 0.04$ and $n = 20$, so $A = 1000(1.04)^{20}$. The growth factor in the table at $n = 20$ and $r = 4\%$ is 2.19112, so $(1.04)^{20} = 2.19112$. The compound amount is $A = \$1000(2.19112) = \$2191.12$.

Example 3 considers a loan at an interest rate that is compounded monthly. It shows that a calculator must be used when the available tables do not contain the given rate.

EXAMPLE 3

Problem

How much interest is paid on a 3-year loan of $2500 at 10% interest compounded monthly?

Solution First, we find the compound amount after 3 years. Set $P = 2500$, $r = 0.1$, $t = 3$, and $k = 12$. Then

$$i = \frac{0.1}{12} \text{ and } n = 36$$

so

$$A = 2500\left(1 + \frac{0.1}{12}\right)^{36}$$

The following calculator key sequence can be used to evaluate the expression for A:

.1 $\boxed{\div}$ 12 $\boxed{=}$ $\boxed{+}$ 1 $\boxed{=}$ $\boxed{y^x}$ 36 $\boxed{=}$ $\boxed{\times}$ 2500 $\boxed{=}$

It yields 3370.4546, so the compound amount after 3 years is $A = \$3370.45$. To obtain the interest, subtract the borrowed amount of $2500 from the compound amount A. Then $I = A - P = \$3370.45 - \$2500 = \$870.45$.

Annual Yield

The most common interest rates are those that are compounded daily. We assume a 365-day year.

EXAMPLE 4

Problem

What is the interest on a $1000 investment for 1 year at 8% interest compounded daily?

Solution We are given $P = \$1000$, $t = 1$, $r = 0.08$, and $k = 365$. Then $i = 0.08/365 = 0.0002191$ and $n = tk = 365$. Substitute these values into the Compound Interest Formula and use a calculator.

$$A = \$1000(1 + 0.0002191)^{365} = \$1083.28$$

The interest is $\$1083.28 - \$1000.00 = \$83.28$.

In Example 4 the stated rate of 8% is called the **nominal rate.** It is equivalent to a simple interest rate of 8.328%, which is called the **effective rate.** The effective rate provides a common unit for comparing various nominal rates. The Truth in Lending Law provides the consumer with this common unit of measure.

Most banks today use the term **annual yield** for the effective rate of a nominal rate that is compounded daily. Example 5 shows how to find the annual yield.

EXAMPLE 5

Problem

What is the annual yield on an account that pays 7.7% compounded daily?

Solution The nominal rate is 7.7%. The compound interest in 1 year from an initial investment of $1 is obtained by substituting $P = 1$, $i = 0.077/365 = 0.000210958$, and $n = 365$ into the Compound Interest Formula.

$$A = 1(1.000210958)^{365} = 1.08003$$

HIGH YIELD CERTIFICATES

Annual Yield	Annual Rate	Term
8.00%	**7.70%**	*30 Months*
7.88%	**7.58%**	*18 Months*
7.63%	**7.35%**	*6 Months*
7.30%	**7.05%**	*91 Days*

Effective rate — Rate of compound interest

FIGURE 1

This means that every dollar will grow to $1.08003. The annual yield is obtained by converting $A - 1$ into a percent. Since $A - 1 = 0.08003$, the annual yield is 8.003%.

The annual yield of 8.003% in Example 5 is customarily stated as 8%. See the top row in Figure 1 for a newspaper advertisement listing these rates. The general method for converting from nominal rates to annual yields is to calculate A and then convert $A - 1$ to a percent.

Present Value

The Compound Interest Formula can also be used to calculate the amount that should be invested to produce a desired amount at some later date. For instance, suppose that $1000 will be needed in 3 years from an investment that earns 6% compounded annually. The problem is to determine the present value P from the Compound Interest Formula.

$$A = P(1 + i)^n$$

$$A = P(1 + 0.06)^3$$

$$1000 = P(1.06)^3$$

Multiply both sides by $(1.06)^{-3}$ to get

$$1000(1.06)^{-3} = P$$

The value of $(1.06)^{-3}$ can be found in the Present Value I table in the Appendix or by a calculator. The entry in the table at $n = 3$ and $r = 6\%$ is 0.839619, meaning that $(1.06)^{-3} = 0.839619$. For a calculator with an exponential key $\boxed{y^x}$, use the key sequence

$$1.06 \boxed{y^x} - 3 \boxed{=}$$

Sometimes -3 is obtained by the sequence $\boxed{3}\boxed{\pm}$. The present value is $P = 1000(0.839619) = 839.619$. Therefore an investment of $839.62 will mature to $1000 if it is invested for 3 years at an interest rate of 6% compounded annually.

E X A M P L E   6

Problem

How much money should be invested to produce $1500 in 2 years at 8% interest compounded annually?

Estimation The answer must be less than $1500.

Solution We are given that $t = 2$, $A = 1500$, $r = 0.08$, and $k = 1$. Then $i = r/k = 0.08$ and $n = tk = 2$. Substitute these values into the Compound Interest Formula.

$$A = P(1 + i)^n$$

$$1500 = P(1.08)^2$$

Solve for P.

$$P = 1500(1.08)^{-2}$$

By the Present Value I table, the entry for $n = 2$ and $r = 8\%$ is 0.857339. Thus $P = 1500(0.857339) = 1286.0085$. The present value is \$1286.01.

The next example reconsiders an example from Section 5–1 at an interest rate that is compounded daily. Its solution makes use of the Present Value II table in the Appendix. The Present Value I table is used in problems whose interest rates are compounded annually while the Present Value II table is used in problems whose interest rates are compounded daily.

EXAMPLE 7

A parent of a newborn child wants to buy a certificate that will mature to \$20,000 when the child turns 18. A bank advertises "All-Savers" certificates that pay 12% interest compounded daily.

Problem

How much should be invested in the certificate?

Solution From the information given $t = 18$, $A = 20,000$, $r = 0.12$, and $k = 365$. Substitute these values into the Compound Interest Formula.

$$A = P(1 + i)^n$$

$$20,000 = P\left(1 + \frac{0.12}{365}\right)^{18 \cdot 365}$$

Solve for P.

$$P = 20,000\left(1 + \frac{0.12}{365}\right)^{-18 \cdot 365}$$

This form can be evaluated by a calculator with an exponential key or from the Present Value II table. The entry at $r = 12\%$ and $n = 18$ is 0.115376, so

$$\left(1 + \frac{0.12}{365}\right)^{-18 \cdot 365} = 0.115376$$

The present value is

$$P = 20,000(0.115376) = 2307.52$$

Therefore an investment of \$2307.52 at 12% interest compounded daily will mature to \$20,000 in 18 years.

Contrast Example 7 with Example 4 of Section 5–1, which required \$6329.11 at 12% simple interest.

EXERCISE SET 5–2

In Problems 1 to 6 compute the interest on each savings or loan, where the rate is compounded annually.

1. $500 for 3 years at 6%.

2. $1000 for 10 years at 10%.

3. $1000 for 10 years at 8.15%.

4. $1500 for 3 years at 5.75%.

5. $1994 for 10 years at 13.056%.

6. $1984 for 10 years at 13.678%.

In Problems 7 to 10 compute the compound amount on the following investments, where the interest is compounded annually for 30 months.

7. $2500 at 10%   8. $500 at 6%

9. $2500 at 9.67%   10. $2500 at 10.16%

In Problems 11 and 12 determine the compound amount on the savings, where the interest rates are compounded quarterly.

11. $1000 for 10 years at 8%.

12. $2500 for 3 years at 12%.

In Problems 13 and 14 determine the interest on loans at interest rates that are compounded quarterly.

13. $2500 for 3 years at 12%.

14. $5000 for 2 years at 16%.

In Problems 15 and 16 determine the interest on loans at interest rates that are compounded monthly.

15. $1000 for 5 years at 10.77%.

16. $500 for 10 years at 5.5%.

In Problems 17 and 18 determine the interest on savings accounts at interest rates that are compounded daily.

17. $1000 for 10 years at 8%.

18. $2500 for 3 years at 12%.

In Problems 19 and 20 determine the interest on loans at interest rates that are compounded daily.

19. $2500 for 3 years at 12%.

20. $5000 for 2 years at 16%.

In Problems 21 and 22 determine the annual yield on the following nominal rates. Assume 365 days in a year.

21. 5.5%   22. 13.8%

23. Determine the amount of money that should be invested now to return $10,000 in 5 years at an interest rate of 6% compounded annually.

24. Determine the amount of money that should be invested now to return $10,000 in 10 years at an interest rate of 6% compounded annually.

25. Determine the amount of money that should be invested now to return $10,000 in 30 years at an interest rate of 10% compounded annually.

26. Determine the amount of money that should be invested now to return $10,000 in 50 years at an interest rate of 2% compounded annually.

27. Determine the amount of money that should be invested now to return $10,000 in 5 years at an interest rate of 6% compounded daily.

28. Determine the amount of money that should be invested now to return $10,000 in 50 years at an interest rate of 2% compounded daily.

In Problems 29 and 30 determine what size certificate should be purchased now in order to return the stated dividend when the interest is compounded daily.

29. $4000 in 3 years at 12% interest.

30. $8000 in 4 years at 10% interest.

31. How much interest is owed on a 3-year $2500 loan at 10% interest compounded in the following ways?
    (a) quarterly   (b) monthly   (c) daily

32. How much interest is owed on a $2500 loan at 14% interest compounded daily for the following periods?
    (a) 60 days   (b) 9 months   (c) 2 years

33. Verify that the annual yield of 7.3% corresponds to the nominal rate of 7.05% in Figure 1.

34. Would you prefer to have interest compounded daily at 5% for the first 3 years and at 6% for the next 3 years, or vice versa?

35. Use the Present Value I table to find the present value, in dollars-and-cents form, of $10,000 in 3 years at 8% compounded semiannually.

36. Use the Present Value I table to find the present value, in dollars-and-cents form, of $10,000 in 3 years at 8% compounded quarterly.

37. Which is worth more, a lump sum of $10,000 or a

DIME 30-MONTH SAVINGS CERTIFICATES
IN EFFECT THRU JAN. 18th GIVE YOU:

15.25% effective annual yield on 14.00% a year interest rate

MINIMUM DEPOSIT $500

GIFTS AROUND THE CIRCLE OR $20 CASH
When you renew or buy a $10,000 6-Month Certificate OR When you buy a new All Savers Tax-Free or a 30-Month Certificate for $5,000 or more.

DIME 6-MONTH SAVINGS CERTIFICATES IN EFFECT THRU
JANUARY 18th GIVE YOU:

13.678% effective annual yield on 13.056% a year interest rate

ALL SAVERS 1-YEAR CERTIFICATES NOW GIVE YOU:

10.16% effective annual yield on 9.67% a year interest rate

UP TO $2,000 INTEREST TAX-EXEMPT
Any existing 6-Month Certificate can be converted to an All Savers Certificate without penalty.

FIGURE 2

$5000 certificate that was purchased 9 years ago at an 8% rate of interest compounded annually?

38. Repeat Problem 37 for an interest rate that is compounded daily.

39. Determine the annual yield on a nominal rate of interest of 4.69%. Assume 365 days in a year.

40. Determine the amount of money that should be invested now to return $2995 in 3 years at an interest rate of 6.25% compounded annually.

41. Determine the amount of money that should be invested now to return $14,995 in 5 years at an interest rate of 6.57% compounded daily.

42. Determine the amount of money that should be invested now to return $1,000,000 in 50 years at an interest rate of 8% compounded daily.

18-MONTH IRA RATES

CURRENT VARIABLE RATE	GUARANTEED FIXED RATE
13.69%	13.67%
COMPOUNDED DAILY FOR AN EFFECTIVE ANNUAL YIELD OF	COMPOUNDED DAILY FOR AN EFFECTIVE ANNUAL YIELD OF
14.67%	14.60%

FIGURE 3

Referenced Exercise Set 5–2

1. Figure 2 displays an advertisement from the *New York Times* for the Dime Savings Bank offering various rates of interest on savings certificates. Do any of the three annual yields correspond to the stated nominal rates?

2. Figure 3 shows part of a newspaper advertisement for 18-month IRA rates. Verify that each annual yield in the bottom row corresponds to the nominal rate above it.

3. Find the annual yield on the money market funds listed in Figure 4, which displays the "Consumer Rates" chart from the *New York Times* of April 17, 1988.

4. Experiment with the Compound Interest Formula to determine about how long it takes a $1000 investment to double if the 8% interest is compounded daily.

Hershel Walker signed a $6 million, 4-year contract to play football in the fledgling USFL. Steve Young signed a $40 million, 43-year contract at the same time to play in the same league. Most sources feel that Hershel Walker got the better financial deal.* Problems 5 to 8 present a simplified analysis of the comparison using a guaranteed rate of interest of 10% compounded annually.

5. Steve Young was paid $9 million in 1990. What was the present value in 1984, the first year of the contract?

6. Steve Young will be paid $30 million in the year 2027. What was the present value in 1990, when those payments began?

7. Hershel Walker was paid $1.5 million in 1984, 1985, 1986, and 1987. What were the present values in 1984 of each of the last 3 years?

8. Whose contract contained the higher present value in 1984, Steve Young's or Hershel Walker's?

Cumulative Exercise Set 5–2

1. Find the simple interest on a $3500 savings account that earns $6\frac{1}{4}\%$ interest for the stated time period. Assume 365 days in a year.
   (a) 4 years   (b) 90 days

*B. Campsey and E. Brigham, "Reenlisting with the Generals," which appeared in Chapter 14 of the book *Introduction to Financial Management,* Hinsdale, IL, The Dryden Press, 1985, pp. 355–378.

CONSUMER RATES
in percent

Tax-Exempt Bonds Bond Buyer 20-Bond Index	7.81
Money Market Funds Donoghue's Money Fund 7-Day Average Yield	6.28
Bank Money Market Accounts Bank Rate Monitor	5.67
Home Mortgage Federal Home Loan Bank, national average	10.34
Unsecured Personal Loans Minimum rate, 3-year maturity	
Chase Manhattan, New York	13.90
Continental Illinois, Chicago	14.00
Bank of America, San Francisco	17.25

FIGURE 4

2. Find the interest on a $2500 loan at 12% simple interest for the stated time period. Assume 360 days in a year.
   (a) 120 days   (b) 42 months

3. Find the future amount of a $7000 savings account at 8.23% simple interest for the stated time period.
   (a) 3 years   (b) 30 months

4. A college student wants to deposit a certain amount of money in an account that pays 8% simple interest for 4 years in order to have $1000 to throw a graduation party. How much money should be deposited now?

5. A person wishes to put $5000 into a savings account for 30 months. Bank A offers an annual rate of 7.7% interest and Bank B offers an annual rate of 7.58%. How much more money will be saved in Bank A than in Bank B?

6. What is the compound amount of a $700 savings account that pays 6.25% interest compounded annually for 4 years?

7. Determine the amount of money that must be deposited in order to have $80,000 at the end of 4 years if the interest rate is 8% compounded (a) annually and (b) daily.

8. What is the annual yield if the nominal rate is 7.05%?

9. A physician needs $12,350 to purchase computer equipment for her new office. If she takes out a loan at a simple interest rate of 16% per year, how much interest must she pay if she pays off the loan in 9 months?

10. Suppose Mellon Bank East pays 5.62% interest compounded monthly on savings accounts and Mellon Bank West pays 5.6% interest compounded daily on savings accounts. Which bank offers the better interest?

CASE STUDY   The Advertising Coup of the Miller Brewing Company*

The modern era of the brewing industry began after World War II when several firms decided to sell beer nationally instead of operating from a single plant. The number of breweries decreased from 404 in 1947 to 110 in 1970 as the large national firms cut into the sales of the smaller local companies. Even though some firms increased sales dramatically, the industry as a whole showed little growth. This was due to demographic factors. According to *Brewers Almanac 1976* (p. 82), the age group 21–45 accounts for better than 70% of the market. This age group had little growth from 1948 to 1959. While the total consumption of beer declined from 1959 to 1970, the number of firms with capacity of greater than 4 billion barrels increased from 2 to 11. By 1974 the top five breweries accounted for over 60% of the market. Their barrel shipments and market share in 1974 are given in Table 1.

*William Glueck and Lawrence Jauck, *Business and Policy Management*, 4th ed., New York, McGraw-Hill, 1984, pp. 622–640.

Table 1

	1974	
	Barrels (billions)	Market Share (%)
Anheuser-Busch	34.1	23.2
Schlitz	22.7	15.4
Pabst	14.3	9.7
Coors	12.3	8.4
Miller	9.1	6.2

In order to keep their market share, the national companies used advertising, especially television commercials, to increase product identification. While competition was fierce in the 1960s, in 1970 a new rivalry developed when Philip Morris Inc., a large multiproduct company, bought Miller Brewing. The assets and sales of Philip Morris were larger than those of the other top four breweries combined, and Philip Morris was an experienced, aggressive advertiser. In 1975 Miller launched two campaigns that changed the market dramatically. Miller devised a brilliant strategy to promote a low calorie beer, Lite, which they had bought from Meister Brau Inc. of Chicago in 1972. Low calorie beers had been advertised heavily in the past with little success. Miller reasoned that a large segment of the market was comprised of young sports fans who dreamed of athletic prowess. People were developing a passion for exercise. Miller used retired athletes in appealing commercials to spread the message that you could drink plenty of good-tasting Lite and still retain athletic skills. Lite was not just for weight-watchers anymore.

In an equally successful promotion Miller's ads for their premium beer were directed at young people having a good time at boisterous parties. The competition came out with comparatively bland messages. For instance, Schlitz commercials had a gritty cowboy pointing a gun at the screen and threatening the viewer to buy its beer.

In 1975 Miller's market share increased from 6.2 to 8.5%. In the same period Schlitz's market share increased from 15.4 to 15.5%. Example 1 determines the percentage increase for each company.

EXAMPLE 1

Problem

What is the percentage increase when a firm's market share increases (a) from 6.2 to 8.5% (b) from 15.4 to 15.5%?

Solution (a) The actual increase is $8.5\% - 6.2\% = 2.3\% = 0.023$. The percentage increase is $0.023/0.062 \approx 0.37$, or about 37%. (b) The actual increase is $15.5\% - 15.4\% = 0.1\% = 0.001$. The percentage increase is $0.001/0.154 \approx 0.006$, or about 0.6%.

In the next 3 years Miller's market share increased to 12.0% in 1976, 15.2% in 1977, and 18.9% in 1978. Meanwhile Schlitz's market share was 15.8% in 1976, 13.9% in 1977, and 11.8% in 1978. Example 2 determines the percentage increase (or decrease) for each year and each company.

EXAMPLE 2

Problem

Find the percentage increase (or decrease) of Miller's and Schlitz's market share in the years 1976 to 1978.

Solution The percentages can be found by dividing the present year's share by the previous year's share and then subtracting 1 from that number. For Miller the dividends are 12.0/8.5 ≈ 1.41, 15.2/12.0 ≈ 1.27, and 18.9/15.2 ≈ 1.24. Therefore Miller's market share increased by 41, 27, and 24% in the period from 1976 to 1978. Schlitz's dividends are 15.8/15.5 ≈ 1.02, 13.9/15.8 ≈ 0.88, and 11.8/13.9 ≈ 0.85. Therefore Schlitz's market share increased by 2% in 1976, decreased by 12% in 1977 (because $0.88 - 1 = -0.12$), and decreased by 15% in 1978 (because $0.85 - 1 = -0.15$).

Compounded Annual Growth Rate   Marketing analysts use *compounded annual growth rate* instead of simple growth rate as a measure of a firm's performance over a period of years. This is because compounded growth rate takes into account not only the actual annual percentage increase or decrease, but also the rate of increase on the previous year's increase. That is, just as compound interest includes interest on interest, compound growth rate includes the percent of growth on the previous year's growth rate.

In the brewing industry marketing analysts use barrel shipments as a measure of market share. Table 2 gives the increased barrel shipments, the market share percentage change, and the compounded annual shipment growth in the period 1974–1978 for the five market leaders.

Miller's aggressive ad campaigns paid huge dividends. The company moved from a distant fifth in market share in 1974 to a close second in 1978, as its market share increased from 6.2 to 18.9%. The Anheuser-Busch market share went from 23.2% in 1974 to 25.1% in 1978. Schlitz went from 15.4% in 1974 to 11.8% in 1978.

While the change in market share is significant, marketing analysts point to the compounded annual shipment growth rate when assessing Miller's phenomenal success in this time span. Compare Miller's increase to that of Anheuser-Busch. The change in market share is about six times greater for Miller, but the compounded annual shipment growth rate is better than seven times larger.

To quote August Busch III, Chairman of the Board of Anheuser-Busch, "This business is now a two-horse race."

So who says advertising doesn't pay?

Table 2

	1974–1978		
	Increased Barrels (billions)	Change in Market Share (%)	Compounded Annual Shipment Growth Rate (%)
Anheuser-Busch	7.5	+ 1.9	5.1
Miller	22.2	+12.7	36.2
Schlitz	− 3.1	− 3.6	−3.3
Pabst	1.1	− 0.4	1.9
Coors	0.3	− 0.8	0.4

Case Study Exercises

1. Miller's market share increased from 6.2% in 1974 to 18.9% in 1978. Find the percentage increase.
2. Schlitz's market share decreased from 15.4% in 1974 to 11.8% in 1978. Find the percent of decrease.
3. The actual number of barrel shipments by Miller from 1974 to 1978 were (in millions) 9.1, 12.8, 18.4, 24.2, and 31.3. Find the percentage increase for each year and show that the compounded annual shipment growth was 36.2%.
4. The actual number of barrel shipments by all other breweries (all except the top five) from 1974 to 1978 were (in millions) 53.0, 49.7, 48.2, 45.2, and 41.7. Find the percentage increase for each year and show that the compounded annual shipment growth was −4.9%.
5. Show that the two methods for determining percentage increase in Examples 1 and 2 are equivalent; that is, let a and b represent the two percentages and show that $(b - a)/a = b/a - 1$.

5–3 **Inflation and Population Growth**

The Compound Interest Formula can be applied to many areas outside banking. In this section we discuss inflation, cost of living, and population growth.

Inflation

The Compound Interest Formula provides a model of economic inflation when the compound amount A is interpreted as the cost in t years of an item that costs P dollars today. However, it must be assumed that the annual rate of inflation remains fixed from year to year. Then $k = 1$, so $n = tk = t$ and $i = r/k = r$.

Take, for instance, a record album that cost $9 in 1988. If the annual rate of inflation is 10%, how much will it cost in 1994? This is equivalent to finding the future amount of a $9 account that is deposited for 6 years at 10% interest compounded annually. Substitute $n = 6$ and $i = 0.1$ into the Compound Interest Formula.

$$A = P(1 + i)^n = 9(1.1)^6$$

From the Compound Interest Table or by calculator

$$A = 9(1.77156) = 15.944$$

Therefore the album will cost $15.94 in 1994.

Example 1 considers how salary increases affect base salaries.

EXAMPLE 1

Problem

If a base salary was $15,600 in 1985 and a 6% pay raise is received in each of the next 32 years, what will the base salary be in 2017?

Solution From the information given, $P = 15,600$, $i = 0.06$, and $n = 32$. The salary in the year 2017 will be the compound amount in the Compound Interest Formula.

$$A = \$15,600(1.06)^{32} = \$15,600(6.45338) = \$100,672.83$$

Thus the salary will be more than $100,000 a year.

The next example applies the Compound Interest Formula to a present value problem. It determines what today's millionaire was worth 32 years ago.

EXAMPLE 2

Problem

If the rate of inflation remained a constant 4% over the past 32 years, what amount would have grown to $1,000,000 today?

Solution Substitute $r = 0.04$, $n = 32$, and $A = 1,000,000$ into the Compound Interest Formula.

$$A = P(1 + i)^n$$

$$1,000,000 = P(1.04)^{32}$$

Solve for P.

$$P = 1,000,000(1.04)^{-32}$$

By calculator or the Present Value I table (with $n = 32$ and $r = 4\%$) the present value is

$$P = 1,000,000(0.285059) = 285,059$$

This means that someone who was worth $285,059 thirty-two years ago is equivalent to a millionaire today.

Table 1 Effects of a 10% Annual Rate of Inflation

Item	1988 Cost ($)	1994 Cost ($)	2020 Cost ($)	2088 Cost ($)
Nothing	.05	.09	1.06	689
Album	9	15.94	190	124,025
Bicycle	100	177	2,111	1,378,061
Automobile	8,000	14,172	168,910	110,244,899
House	50,000	88,578	1,055,689	689,030,617

The effects of high inflation become more apparent when presented in practical terms. Take a look at Table 1. We have already calculated the cost of an album in 1994. Notice that a bicycle that cost $100 in 1988 will cost over a million dollars a century later!

Cost of Living

Cost of living problems can also be solved by the Compound Interest Formula. Consider this news item.

> The cost of living rose eight-tenths of one percent in January. This amounts to an increase of over 10% for the year.

How can such a conclusion be warranted? After all, a year has 12 months and $(12)(0.8\%) = 9.6\%$ is less than 10%.

Example 3 will answer the question by considering an item that costs $1 on January 1. It computes the cost 1 year later under the assumption that the price is determined solely by the cost of living.

EXAMPLE 3

Problem

If the cost of living rises 0.8% each month for 1 year, what is the cost of living over the entire year?

Solution Special care is needed to put this problem in a form that can be handled by the Compound Interest Formula. It is equivalent to finding A for $i = 0.8\%$ and $n = 12$. Let $P = \$1$ serve as a standard unit. Substitute into the Compound Interest Formula.

$$A = \$1(1.008)^{12} = \$1.10034$$

This means that a $1 item will be worth over $1.10 a year later. The rise in the cost of living over a year is $A - 1 = 0.10034$, which is 10.034%. This verifies the conclusion "over 10% for the year."

Population Growth

The Compound Interest Formula also serves as a model of population growth under certain assumptions. Example 4 demonstrates one aspect.

EXAMPLE 4

Problem

The world's population has been increasing at a fairly constant rate of 2% per year. How much larger will the population be in 100 years if it continues to increase at this rate?

Solution Assume that the principal P is 1. Substitute $i = 0.02$ and $n = 100$ into the Compound Interest Formula.

$$A = 1(1.02)^{100} = 7.24$$

Thus the population will be 7.24 times as large in 100 years as it is today.

Example 4 contains a dire implication in spite of the seemingly small population growth. It indicates that the world population will increase sixfold in the next century. (Doubling is regarded as a onefold increase.) In different terms, it means that the 170×170 yards of land and water now occupied by each of us 5 billion earthlings will reduce to 24×24 yards in 100 years. The commandment "love thy neighbor" will, of necessity, attain even greater significance.

EXERCISE SET 5–3

In Problems 1 to 4 assume a constant rate of inflation during the stated time period.

1. What will a $9 album cost in 20 years at 10% inflation?

2. What will an $8000 car cost in 10 years at 10% inflation?

3. What will a $20,000 salary be in 10 years with an 8% pay raise each year?

4. What will a $100,000 salary be in 5 years with a 12% pay raise each year?

In Problems 5 and 6 assume a constant 2% increase each year in the population over the given time period.

5. The population of the United States in 1990 was 249,632,692. What will it be in the year 2000?

6. In 1980 the population of China, including Tibet, was approximately 971 million. Wow! What will it be in the year 2000?

7. If the cost of living rises 0.6% per month for a year, how much will it rise over the entire year?

8. If the cost of living rises 0.2% per month for a year, how much will it rise over the entire year?

9. Suppose the rate of inflation had remained a constant 4% for the past 50 years. What amount of money then would be worth $1,000,000 today?

10. Repeat Problem 9 for a period of 10 years.

In Problems 11 to 14 assume a constant rate of inflation of 6% from 1968 to 1988. How much would each item have cost in 1968 if its price in 1988 is the one stated?

11. $18 ticket to a concert

12. $100 bicycle

13. $12,000 automobile

14. $80,000 house

Problems 15 and 16 list the 1990 population for certain states and the average rate that the population changed per year from 1980 to 1990. Find each state's population in the year 2000, assuming the same change throughout the 1990s.

15.	State	1990 Population	Percent Change
(a)	California	29,839,250	2.61
(b)	Arizona	3,677,985	3.53
(c)	Oklahoma	3,157,604	0.44
(d)	North Carolina	6,657,630	1.32
(e)	Illinois	11,466,682	0.04
(f)	New York	18,044,505	0.28

16.	State	1990 Population	Percent Change
(a)	Alaska	551,947	3.74
(b)	Colorado	3,307,912	1.45
(c)	Texas	17,059,805	1.99
(d)	South Carolina	3,505,707	1.23
(e)	Ohio	10,887,325	0.08
(f)	Rhode Island	1,005,984	0.62

17. In 1985 the rate of inflation in Brazil was 63.7%. If this rate remains constant, what will a 2500-cruzeiro item (about $1) cost in 1990?

18. In 1980 the rate of inflation in Israel was 132.9%. If this incredible rate continued for a decade, what would a 25-cent bagel have cost in 1990?

19. Use the Compound Interest Table to determine how many years it takes for prices to double if the annual rate of inflation is 6%.

20. In 1982 the rate of inflation in Great Britain was 14%. If this rate continues, what would a 120,000-pound Rolls Royce have cost in 1990?

21. Rare coins minted before 1933 have increased in value at an annual appreciation rate of 28.7%. How much would a 1928 quarter be worth in 1998?

22. (a) Pieter Minuit, a Dutch colonist, bought Manhattan Island from the Indians in 1626 for about $24. If instead he had invested the $24 in a bank paying 5% interest per year, what would the principal have been when the United States celebrated its bicentennial in 1976? (b) If Minuit had put his $24 in 1626 in a savings account that paid 5% interest compounded daily, how much would have been in his account in 1976?

23. In January 1982 the cost of living in Argentina rose 9%. If this increase continued each month throughout the year, how much would a 100-peso item cost in January 1983?

24. If the cost of living rises 0.75% each month for a year, will the annual cost of living exceed 10%?

25. Gerald Baumann, a numismatist at Manfred, Tordella & Brooks, stated, "People have become very aware of U.S. coins as an investment vehicle. They no longer have any qualms about paying more than $50,000 for a single coin." Rare coins increased their value from 1968 to 1986 at an average rate of 12.7% per year. If this growth continued, how much would a $50,000 coin purchased in 1986 be worth in the year 2000?

26. What will a $15-compact disc cost in 10 years at 8% inflation?

27. Assume a 6% annual rate of inflation from 1980 to 1990. How much would a television set have cost in 1980 if it cost $400 in 1990?

28. If a diamond has appreciated in value at an annual rate of 15% since 1960, how much would the diamond have been worth in 1990 if it was worth $500 in 1960?

29. There has been a recent boom market in celebrity autographs, especially sports figures. Suppose a baseball signed by Pete Rose has risen in value at a rate of 22% per year in the last 3 years. If this growth continues and the signed ball is worth $15 in 1992, how much will it be worth in 1995?

Referenced Exercise Set 5–3

The facts stated in Problems 1 and 2 are based on a newspaper article which appeared in 1987, the year that the world's population reached 5 billion.*

1. It is predicted that Mexico City's population will soar from 15 million in 1987 to 26 million by the year 2000. Substitute various values for i into the Compound Interest Formula to approximate the annual growth rate of Mexico City's population between 1987 and 2000.

2. The Population Crisis Committee predicted that the population of the world would grow by an additional billion by the end of the century. Substitute various values for i into the Compound Interest Formula to approximate the annual growth rate of the world's population between 1987 and the year 2000.

Cumulative Exercise Set 5–3

1. Find the simple interest on a loan of $1680 for 3 years at 11% interest. Assume 360 days.

2. Find the future amount of the savings of $2500 at the simple interest rate of 9% for 10 years.

3. Find the total amount that must be repaid on a loan of $6800 at the simple interest rate of 12% for 4 years.

4. Find the time t of an investment with the principal of $5000, the future amount of $10,000, and the simple interest rate of 8%.

5. Compute the interest on a loan of $1600 for 10 years at 10% compounded annually.

6. Determine the compound amount on the savings of $5000 for 5 years at 9% compounded quarterly.

7. How much interest is owed on a 4-year $8000 loan at 10% interest compounded (a) quarterly, (b) monthly, (c) daily?

*Dan Shannon, "World Population at 5 Billion," *Los Angeles Times*, July 12, 1987.

8. Which is worth more, a lump sum of $20,000 or a $7000 certificate purchased 12 years ago at a 10% rate of interest compounded quarterly?

9. What will a $10 tape cost in 15 years at 7% inflation?

10. Assume a 6% annual rate of inflation from 1970 to 1990. How much would a book have cost in 1970 if it cost $20 in 1990?

11. If certain types of rare stamps have increased in value at an annual appreciation rate of 25% since 1960, how much would such a stamp be worth in 1990 if it was worth $2 in 1960?

12. Use the Compound Interest Table to determine how many years it takes for prices to triple if the annual rate of inflation is 8%.

13. In January 1982 the cost of living in Bolivia rose 5%. If this increase continued each month throughout the year, how much would a 1000-peso item cost in January 1983?

14. Many people believe that the next boom market in collectibles will be in baseball trivia. A recent article stated that a baseball card of Hank Aaron in his rookie year has increased in value at a rate of 25% per year in the last 2 years. If this growth continues and the card is worth $10 in 1992, how much will it be worth in 1995?

5–4 Annuities and Sinking Funds

In this section each investment consists of a sequence of payments instead of one lump sum. We rely on the Compound Interest Formula from Section 5–2. As in earlier sections, all answers are written in a dollars-and-cents form that is obtained by either a calculator or a table.

Annuities

A family is planning a special vacation 2 years in advance by depositing $500 at the end of every 3 months in a savings account that earns 8% interest compounded quarterly. After 2 years (eight deposits) the amount deposited will total $4000 altogether. How much interest will accrue from the deposits?

This scenario is a typical annuity problem, where an **annuity** is a sequence of equal payments made over equal time periods. It is an **ordinary annuity** because the time periods correspond to the period of the compound rate of interest and the payments are made at the end of the time period. The sum of the payments and the interest is called the **future value** (or the **amount**). We will calculate the future value of the family's annuity.

The Compound Interest Formula $A = P(1 + i)^n$ gives the compound amount on each payment $P = \$500$. The first payment, which is deposited at the end of the first quarter, earns 8% interest compounded quarterly for seven quarters, so $r = 0.08$, $k = 4$, and $t = 7/4$. Then $i = r/k = 0.08/4 = 0.02$, and $n = tk = 7$; thus by the Compound Interest Formula the compound amount is $500(1.02)^7$. Figure 1 shows a time line with the compound amount of each payment. The top row lists the compound amount on the first payment. The other compound amounts are also listed. The future value A of the annuity is the sum of these compound amounts.

$$A = 500(1.02)^7 + 500(1.02)^6 + \cdots + 500(1.02) + 500$$

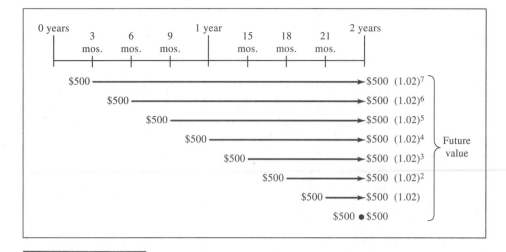

FIGURE 1

Each term can be evaluated by the Compound Interest Table or a calculator. However, it is more efficient to derive a formula that gives the future value without having to evaluate each of these expressions. Set $a = 1.02$. Then

$$A = 500a^7 + 500a^6 + \cdots + 500a + 500$$

$$= 500(a^7 + a^6 + \cdots + a + 1)$$

The sum in the parentheses is called a **geometric series.** A geometric series can be evaluated as follows:

$$a^n + a^{n-1} + \cdots + a + 1 = \frac{(a^n + a^{n-1} + \cdots + a + 1)(a - 1)}{(a - 1)}$$

$$= \frac{a^{n+1} - 1}{a - 1}$$

The general result is stated in the display.

Sum of a Geometric Series

$$a^n + \cdots + a + 1 = \frac{a^{n+1} - 1}{a - 1}$$

In particular, $a^7 + \cdots + a + 1 = \dfrac{a^8 - 1}{a - 1}$, so

$$A = 500 \frac{a^8 - 1}{a - 1} = 500 \frac{(1.02)^8 - 1}{0.02}$$

The last expression is a particular case of the Annuity Formula, whose derivation is similar. It is stated in the display.

Annuity Formula

$$A = R \frac{(1 + i)^n - 1}{i}$$

where
  A is the future value (or amount)
  R is the periodic payment
  $i = r/k$ is the periodic interest rate
  $n = tk$ is the number of periods
(The payments are made at the end of each period.)

The fractional multiple of R in the Annuity Formula is commonly denoted by $s_{\overline{n}|i}$ (read "s angle n at i").

$$s_{\overline{n}|i} = \frac{(1 + i)^n - 1}{i}$$

The Annuity Table in the Appendix provides some values of $s_{\overline{n}|i}$. For $n = 8$ and $i = 2\%$, $s_{\overline{n}|i} = 8.58297$. Therefore the future value of the family's annuity is $A = \$500(8.58297) = \$4291.48$. The value of $s_{\overline{8}|0.02}$ can also be obtained by a calculator with an exponential key $\boxed{y^x}$ using this key sequence.

$$1.02 \boxed{y^x}\ 8 \boxed{=} \boxed{-}\ 1 \boxed{=} \boxed{\div}\ .02 \boxed{=}$$

The display should approximate 8.58297.

Examples 1 and 2 demonstrate the Annuity Formula.

EXAMPLE 1

Problem

An annuity consists of payments of $1000 made at the end of each year for 12 years at 6% compounded annually. What is the future value of the annuity?

Solution We are given that $R = 1000$, $i = 0.06$, and $n = 12$. By the Annuity Formula, and either a calculator or the Annuity Table,

$$A = 1000 \frac{(1.06)^{12} - 1}{0.06} = 1000(16.86994) = 16,869.94$$

Therefore the future value of the annuity is $16,869.94.

EXAMPLE 2

Problem

What is the future value of an annuity consisting of payments of $100 made at the end of each month for 3 years at 18% compounded monthly?

Estimation The answer must exceed $(3)(12)(\$100) = \$3600$.

Solution From the information given $R = 100$, $i = r/k = 0.18/12 = 0.015$, and $n = tk = 3(12) = 36$. The future value of the annuity is

$$A = \$100 \frac{(1.015)^{36} - 1}{0.015} = \$100(47.27597) = \$4727.60$$

A calculator has to be used to obtain the final form because the Annuity Table in the Appendix does not provide entries for $s_{\overline{n}|i}$ when $i = 1.5\%$.

Suppose the family that was planning its vacation 2 years in advance decided to deposit the payments at the beginning of each quarter instead of at the end. The future value of this annuity is calculated in Example 3. When the payments on an annuity are paid at the beginning of the period it is called an **annuity due.**

EXAMPLE 3

Problem

What is the future value of an annuity consisting of payments of $500 made at the beginning of each quarter for 2 years at 8% compounded quarterly?

Solution Refer to Figure 2. The time line divides the 2 years into eight quarters. The exponent in the first row under the line is 8 because the first payment

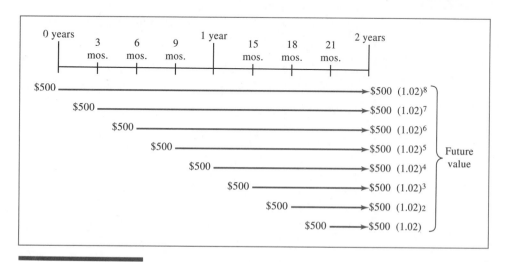

FIGURE 2

earns interest for eight periods (quarters). The future value of the annuity is the sum of all the compound amounts.

$$A = 500(1.02)^8 + \cdots + 500(1.02)$$

Notice that $A + 500$ corresponds to an annuity for nine periods. By the Annuity Formula

$$A + 500 = 500 \frac{(1.02)^9 - 1}{0.02} = 500(9.75463) = 4877.31$$

Then $A = 4877.31 - 500 = 4377.31$, so the future value is $4377.31.

From Example 3 the future value of the family's annuity is $4377.31 if payments are made at the beginning of each quarter. Compare this with the future value of the annuity (calculated earlier) if payments are made at the end of each quarter: $4291.48.

Sinking Funds

The Annuity Formula expresses a relationship between the future value A and the payments R. So far we have been given R and asked to solve for A. Example 4 illustrates the reverse situation for a problem we encountered in Sections 5–1 and 5–2.

EXAMPLE 4

A parent wants to present a child with $20,000 when the child turns 18. The parent wants to make a deposit on each birthday, from the first to the 18th, and present it on the 18th birthday.

Problem

How large should the deposits be if the account earns 6% compounded annually?

Solution According to the Annuity Formula

$$20,000 = R \frac{(1.06)^{18} - 1}{0.06}$$

By calculator or the Annuity Table

$$20,000 = R(30.90564)$$

Solve for R.

$$R = \frac{20,000}{30.90564} = 647.13$$

Each payment will be $647.13.

In Example 4 the annual payment of $647.13 seems small compared to the amount of $20,000. It attests to the power of the exponential function y^x. The annuity in this example is called a **sinking fund,** which is an account established to receive periodic payments to build some future amount. Example 5 provides another illustration.

EXAMPLE 5

Problem

Determine the quarterly payments of a sinking fund that will provide $4000 in 5 years from an account earning 12% compounded quarterly.

Estimation Each payment must be less than ($4000/5)/4 = $200.

Solution We have $A = 4000$, $k = 4$, $t = 5$, and $r = 0.12$. Then $i = 0.03$ and $n = 20$. By calculator,

$$4000 = R \frac{(1.03)^{20} - 1}{0.03} = R(26.87037)$$

Therefore the quarterly payments are

$$R = \frac{\$4000}{26.87037} = \$148.86$$

EXERCISE SET 5–4

In Problems 1 to 4 determine the future value of the stated annuity, where the payments are made at the end of each year and the interest is compounded annually.

1. The payments are $1000 per year for 8 years at 6%.

2. The payments are $1000 per year for 12 years at 8%.

3. The payments are $2000 per year for 5 years at 12%.

4. The payments are $5000 per year for 3 years at 10%.

In Problems 5 to 10 determine the future value of the stated annuity where the payments are made at the end of each period.

5. The payments are $500 per quarter for 2 years at 8% compounded quarterly.

6. The payments are $1000 per quarter for 5 years at 16% compounded quarterly.

7. The payments are $100 per month for 2 years at 12% compounded monthly.

8. The payments are $100 per month for 3 years at 18% compounded monthly.

9. The payments are $300 per quarter for 3 years at 16% compounded quarterly.

10. The payments are $200 per quarter for 4 years at 6% compounded quarterly.

In Problems 11 to 14 determine the future value of the stated annuity, where the payments are made at the beginning of each period.

11. The payments are $500 per quarter for 2 years at 8% compounded quarterly.

12. The payments are $1000 per quarter for 5 years at 16% compounded quarterly.

13. The payments are $100 per month for 2 years at 12% compounded monthly.

14. The payments are $100 per month for 3 years at 18% compounded monthly.

In Problems 15 to 18 a parent wants to present a child with a cash gift by making periodic deposits in an account paying the stated interest compounded annually. How large should the deposits be?

15. $10,000 gift after 5 years at 6%.

16. $5000 gift after 3 years at 8%.

17. $2000 gift after 2 years at 10%.

18. $800 gift after 4 years at 12%.

In Problems 19 to 22 determine the quarterly payments of a sinking fund that is set up to satisfy the given conditions.

19. $3000 in 3 years at 8% compounded quarterly.

20. $3000 in 3 years at 16% compounded quarterly.

21. $4000 in 4 years at 6% compounded quarterly.

22. $4000 in 4 years at 10% compounded quarterly.

23. Suppose that a professional athlete deposits $25,000 at the end of each year for 10 years in an account paying 8% compounded annually. How much money will be in the account at that time?

24. Suppose that a TV news anchor deposits $2000 every 3 months for the duration of a 3-year contract in an account paying 8% compounded quarterly. How much money will be in the account at the end of the contract?

25. Tax laws allow all individuals to deposit $2000 per year in IRAs (individual retirement accounts). If at age 30 someone deposits $2000 a year in an account paying 8% compounded annually, how much will be in the account upon retirement at age 67?

26. Tax laws allow self-employed individuals to open a Keogh retirement plan, which is not taxable until the money is withdrawn. If $5000 is deposited every year for 30 years at 10% compounded annually, how much will be in the account upon retirement?

27. For the first 5 years of their marriage a couple deposits $400 at the end of each month in an account paying 8% compounded monthly. How much will be in the account after 5 years?

28. The couple in Problem 27 could have invested $5000 at the end of each year in an account paying 9% compounded annually. Did they make a wise choice?

29. A family wants to purchase a $1600 camcorder in 6 months. How much money should they deposit at the end of each month in an account paying 8% compounded monthly?

30. A 14-year-old boy dreams of saving his earnings from his newspaper route to buy a $22,000 Corvette when he turns 18. How much money should he deposit at the end of each month in an account paying 6% compounded monthly to realize his dream?

31. A 17-year-old dreamer wants to become a millionaire at the age of 40 by making monthly deposits in an account paying 12% compounded monthly. How big must the payments be for the dream to come true?

32. A newly married couple wants to be able to put down $30,000 on a new house in 3 years. How much money should they deposit at the end of each month in an account paying 9% compounded monthly?

33. If you deposit $10 a day in an account paying 6% compounded daily, how much will be in the account after 1 year?

34. If you deposit $1 a day in an account paying 8% compounded daily, how much will be in the account after 1 year?

35. A firm sets up a sinking fund to produce $400,000 in equipment in 5 years at 16% compounded quarterly. How big are the quarterly payments?

36. A city establishes a sinking fund to produce $4 million revenue in 20 years at 7% compounded annually. How big are the annual payments?

37. What daily deposits have to be made in an account paying 8% compounded daily to have $1000 in 1 year?

38. What daily deposits have to be made in an account paying 6% compounded daily to have $1000 in 1 year?

39. The physicist Albert Bartlett, concerned about the energy crisis, wrote, ''Many people find it hard to believe that when the rate of consumption is growing a mere 7% per year, the consumption in one decade exceeds the total of all of the previous consumption.'' Use the assumption that the rate of consumption is 7% and the fact that 1000 trillion barrels of oil were consumed in 1986 to project how many trillion will be consumed in (a) 1987 (b) 1988 (c) 1996 (d) the sum of all of the years from 1987 through 1996.

40. Determine the future amount of an annuity in which a payment of $1500 is made at the end of every quarter for 5 years at 8% compounded quarterly.

41. Determine the future amount of an annuity in which a payment of $1500 is made at the beginning of every quarter for 5 years at 8% compounded quarterly.

Table 1 The Nation's Top Markets (in millions of dollars)

Metro Area	1985	1990 Estimate	Percent Increase 1985–1990
1. Los Angeles–Long Beach	$50,985	$76,557	50.2
2. New York	41,390	59,971	44.9
3. Chicago	35,628	52,260	46.7
4. Philadelphia	28,394	42,106	48.3
5. Boston-Lawrence-Salem-Lowell-Brockton	27,704	40,438	56.0
6. Washington, D.C.	25,220	38,781	53.8
7. Detroit	26,267	37,800	43.9
8. Houston	21,643	31,823	47.0
9. Nassau-Suffolk, N.Y.	19,423	28,819	48.4
10. Atlanta	17,419	27,593	58.4

Source: Sales and Marketing Management.

42. Determine the quarterly payments of a sinking fund that must produce $25,000 in 5 years at 8% compounded quarterly.

43. Determine the quarterly payments of a sinking fund that must produce $250,000 in 15 years at 8% compounded quarterly.

Referenced Exercise Set 5–4

Problems 1 and 2 are based on Table 1, which lists the ten metropolitan areas in the United States which had the biggest markets in 1985.

1. If, for every year from 1985 to 2000, Atlanta experiences an 11.6% increase and Detroit experiences an 8.7% increase, will Atlanta's market exceed Detroit's by the year 2000?

2. If the Los Angeles–Long Beach area experiences a 10% increase for every year from 1985 to 2000, what will the market be in the year 2000?

Cumulative Exercise Set 5–4

1. Find the total amount that must be repaid on a loan of $20,000 at 9% simple interest for 30 months.

2. Find the present value of a $7000 future amount at 7.56% simple interest for the stated time period.
   (a) 3 years   (b) 30 months

3. Determine the interest on a $5000 loan for 42 months at 10% interest compounded annaully.

4. Determine the interest on a $2000 loan for 4 years at an interest rate of 16% compounded quarterly.

5. What amount of money can be invested now into an account that pays 12% interest compounded daily so that it pays $40,000 in 20 years?

6. If the cost of compact discs rises 8% a year for the next 10 years, what will the price of a $12 disc be then?

7. If the cost of living rises 0.5% each month for a year, what amount does it rise in a year?

8. If a city had a population of 1,400,000 in 1980 and the population decreases by 1.2% each year until 2000, will the population exceed a million in the year 2000?

9. If someone has $1,000,000 today, what was the equivalent amount 25 years ago under the assumption that the rate of inflation remained a constant 6% over that time period?

10. Determine the future amount of an annuity in which a payment of $250 is made at the beginning of every quarter for 3 years at 8% compounded quarterly.

11. Determine the quarterly payments of a sinking fund that must produce $2500 in 5 years at 12% compounded quarterly.

12. If a person puts $10 a day for 1 year into a savings account that pays 6.5% interest compounded daily, what will the balance be after 1 year?

13. A man has to make an alimony payment of $2500 in 9 months. What lump sum can he invest now at 12% interest compounded monthly in order to fulfill his obligation?

14. The U.S. Census Bureau reported that New Jersey's population in 1990 was 7,748,734. The report stated that in the period 1980–1990 the "change in people" was +383,811 and the "percent change" was +5.2%.

(a) What kind of "percent change" is being used?
(b) What compounded annual percentage increase could also be used to describe this population change?

5–5 Amortization

In this section we make use of the present value concept from Section 5–2 to develop a formula for the payments on loans that many of us take out at one time or another. Here are two typical amortization problems that we will solve in this section.

Problem A   A new car costs $17,000. The dealer offers $5000 for a trade-in. The down payment is $2000. What will the monthly installments be if the $10,000 loan is taken out for 4 years at 18% annual interest compounded monthly?

Problem B   A house costs $90,000. The terms of sale are 10% down and the remainder to be paid in monthly payments for 25 years at an annual interest rate of 9% compounded monthly. What are the monthly payments?

Present Value

Consider an investment that pays 8% annual interest compounded quarterly. How much money should be invested as a lump sum today in order to remove $1500 every 3 months for 2 years? The answer must be less than $12,000. How much less? To answer these questions, regard each of the $1500 dividends separately. A certain amount of money A_1 must produce the first dividend of $1500 in 3 months. The quarterly rate of interest is $\dfrac{0.08}{4} = 0.02$, so by the Compound Interest Formula $1500 = A_1(1.02)$. Then $A_1 = \dfrac{1500}{1.02}$. Recall that A_1 is called the present value of $1500. Similarly, the present value A_2 of the $1500 dividend for two periods (6 months) is $A_2 = \dfrac{1500}{1.02^2}$. This process continues through eight periods (2 years), with the present value of the final dividend being $A_8 = \dfrac{1500}{1.02^8}$. The eight present values are shown below the time line in Figure 1.

Since the sequence of payments is an annuity, the total amount P that will produce the annuity is called the **present value of the annuity.** P is the sum of

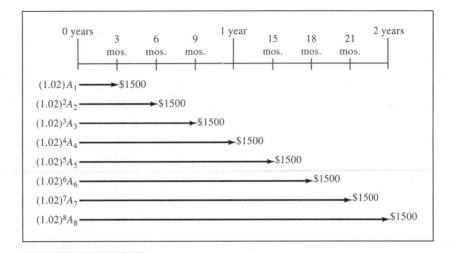

FIGURE 1

the individual present values. Each one could be evaluated separately but, like the last section, it is more efficient to derive a general formula.

$$P = A_1 + A_2 + \cdots + A_8$$

$$= \frac{1500}{1.02} + \frac{1500}{1.02^2} + \cdots + \frac{1500}{1.02^8}$$

Set $a = 1.02$, factor the common term 1500, and simplify.

$$P = 1500 \left(\frac{1}{a} + \frac{1}{a^2} + \cdots + \frac{1}{a^8} \right)$$

$$= \frac{1500}{a^8} \left(\frac{a^8}{a} + \frac{a^8}{a^2} + \cdots + \frac{a^8}{a^8} \right)$$

$$= \frac{1500}{a^8} (a^7 + a^6 + \cdots + 1)$$

Now form the sum of the geometric series and simplify.

$$P = \frac{1500}{a^8} \cdot \frac{a^8 - 1}{a - 1}$$

$$= 1500 \cdot \frac{1 - \dfrac{1}{a^8}}{a - 1}$$

$$= 1500 \cdot \frac{1 - \dfrac{1}{1.02^8}}{0.02}$$

The last expression is a particular case of the formula for the present value of an annuity. Its derivation is similar to the one derived here. It is summarized in the display.

Present Value of an Annuity

$$P = R \cdot \frac{1 - \dfrac{1}{(1 + i)^n}}{i}$$

where
- P = the present value of all payments
- R = the periodic payment
- i = r/k, the periodic interest rate
- n = tk, the number of periods

The fractional multiplier of R is commonly denoted by $a_{\overline{n}|i}$ (read "a angle n at i").

$$a_{\overline{n}|i} = \frac{1 - \dfrac{1}{(1 + i)^n}}{i}$$

The Amortization Table in the Appendix provides some values of $a_{\overline{n}|i}$. For $n = 8$ and $i = 2\%$, $a_{\overline{n}|i} = 7.32548$. Therefore the present value for the stated investment is $P = (\$1500)(7.32548) = \$10{,}988.22$. The value of $a_{\overline{8}|0.02}$ can also be obtained by a calculator with an exponential key and a reciprocal key using this procedure: calculate $(1 + i)^n$, take its reciprocal, subtract from 1, and divide by i. The present value is the product of the number displayed (here 7.32548) and the periodic payment R (here \$1500).

EXAMPLE 1

Problem

What is the present value of an annuity that pays \$50 a month for 4 years at 12% annual interest compounded monthly?

Estimation The answer must be less than $(4)(12)(\$50) = \$2400$.

Solution From the information given, $r = 0.12$, $k = 12$, and $t = 4$, so the monthly interest is $i = 0.01$ and the number of periods is $n = 48$. By the formula for the present value of an annuity

$$P = 50 \cdot \frac{1 - \dfrac{1}{1.01^{48}}}{0.01}$$

By calculator, $a_{\overline{48}|0.01} = 37.97396$. Therefore the present value is $P = (\$50)(37.97396) = \$1898.70$.

Amortization

The root of the term "amortize" is the French word *mort* meaning death. Thus the expression "to **amortize** a debt" means to pay it off (literally to "kill" it). Problems A and B from the beginning of this section are amortization problems that ask for the monthly payments on loans. We will develop the formula for solving these problems by solving a similar problem whose figures are more tractable.

Consider a $4000 loan that must be repaid in five yearly payments at an interest rate of 8% compounded annually. What will the size of the payments R be? This is an annuity problem in which the present value is known and the periodic payment must be solved for. Each payment must exceed $4000/5 = $800. To determine the exact payment, substitute $P = 4000$, $i = 0.08$, and $n = 5$ into the formula for the present value of an annuity.

$$4000 = R \cdot \frac{1 - \dfrac{1}{(1.08)^5}}{0.08} = R \cdot a_{\overline{5}|0.08}$$

By the Amortization Table or calculator $a_{\overline{5}|0.08} = 3.99271$, so

$$4000 = R \cdot 3.99271$$

Then

$$R = \frac{4000}{3.99271} = 1001.83$$

Therefore the payments that amortize the loan are $1,001.83.

In general, suppose a loan of P must be repaid in equal periodic payments of R at an interest rate that is compounded over the same period of time. The Amortization Formula determines the size of the payments. Its derivation is similar to the preceding particular case.

Amortization Formula

$$R = P \cdot \frac{i}{1 - \dfrac{1}{(1 + i)^n}} = P \cdot \frac{1}{a_{\overline{n}|i}}$$

where
   $P =$ the present value of all payments
   $R =$ the periodic payment
   $i = r/k$, the periodic interest rate
   $n = tk$, the number of periods

The Amortization Formula allows us to solve the problems stated at the beginning of this section. Example 2 solves Problem A.

EXAMPLE 2

Suppose that a new car costs $17,000. The dealer offers $5000 for a trade-in with a down payment of $2000.

Problem

What will the monthly installments be if the $10,000 loan is for 4 years at 18% annual interest compounded monthly?

Estimation The answer must exceed $10,000/48 \approx $200.

Solution We have $P = 10,000$. There are 4 years of monthly payments, so $n = 48$. The monthly interest rate is $i = 0.18/12 = 0.015$. By the Amortization Formula

$$R = 10,000 \cdot \frac{1}{a_{\overline{48}|0.015}}$$

By calculator $a_{\overline{48}|0.015} = 34.04255$, so

$$R = \frac{10,000}{34.04255} = 293.75$$

Therefore the monthly payments are $293.75.

Let us reflect on Example 2. The total amount of money paid over the life of the loan is ($293.75)(48) = $14,100. Therefore the amount of interest paid on the $10,000 loan is $4100. A *payment schedule* (also called an **amortization schedule**) lists all payments of a loan and breaks down each payment into the principal and interest. Let us illustrate this with an example having fewer periods.

Reconsider the $4000 loan at 8% interest compounded annually. It was repaid in five yearly payments of $R = $1001.83. For the first payment the interest applies to the entire loan, so the interest is 8% of $4000, or $320. The principal that is applied to the loan is the difference between the payment and the interest, or $1001.83 − $320 = $681.83. Then the remaining balance of the loan after one payment is $4000 − $681.83 = $3318.17. This is shown in the row for payment 1 in the accompanying payment schedule. The figures for payment 2 are obtained in a similar manner. The interest is 8% of the remaining balance, or (0.08)(3318.17) = 265.45. The principal that is applied to the loan is $1001.83 − $265.45 = $736.38, so the remaining balance of the loan after two payments is $3318.17 − $736.38 = $2581.79. The other payments are obtained similarly and are shown in the payment schedule.

Payment	Interest	Principal	Balance
0			$4000.00
1	$320.00	$681.83	3318.17
2	265.45	736.38	2581.79
3	206.54	795.29	1786.50
4	142.92	858.91	927.59
5	74.21	927.62	−0.03

Payment schedule for $4000 loan at 8% interest for 5 years

Sometimes an adjustment has to be made in the final payment. Here the final payment would be reduced by three cents.

Example 3 solves Problem B stated at the beginning of the section.

EXAMPLE 3

A house costs $90,000. The terms of sale are 10% down and the remainder to be paid in monthly payments for 25 years. The loan is taken at an annual interest rate of 9% compounded monthly.

Problem

What are the monthly payments? Construct the payment schedule for the first four payments.

Solution Initially 10% of $90,000 (= $9000) must be paid, so the amount of the loan is $P = \$81,000$. There are $n = (25)(12) = 300$ monthly periods and the monthly interest rate is $i = 0.09/12 = 0.0075$. By the Amortization Formula the monthly payments are

$$R = P \cdot \frac{1}{a_{\overline{300}|0.0075}}$$

The value of $a_{\overline{300}|0.0075}$ is not provided in the Amortization Table. By calculator $a_{\overline{300}|0.0075} = 119.16162$, so $R = 81,000/119.16162 = 679.75$. Therefore the monthly payments are $679.75.

Here is the payment schedule for the first four payments.

Payment	Interest	Principal	Balance
0			$81,000.00
1	$607.50	$72.25	80,927.75
2	$606.96	$72.79	80,854.96
3	$606.41	$73.34	80,781.62
4	$605.86	$73.89	80,707.73

Payment schedule for $81,000 loan at 0.75%
per month for 300 months

The total amount paid for the house in Example 3 is ($679.75)(300) = $203,925. Therefore the amount of interest paid is $203,925 − $81,000 = $122,925, which is one and one-half times the amount of the loan. The abbreviated payment schedule supplies some evidence for this fact because after four payments of $679.75 each, the principal of the loan has been lowered from $81,000 to only $80,707.73. Most homeowners have experienced the anguish that accompanies this realization.

EXERCISE SET 5–5

In Problems 1 to 4 determine the amount of money that should be invested as a lump sum today to receive the stated dividend.

1. A dividend of $400 every 3 months for 2 years in an account that pays 8% compounded quarterly.

2. A dividend of $400 every 3 months for 1 year in an account that pays 8% compounded quarterly.

3. A dividend of $100 every 3 months for 1 year in an account that pays 6% compounded quarterly.

4. A dividend of $100 every 3 months for 4 years in an account that pays 6% compounded quarterly.

In Problems 5 to 8 determine the present value of the stated annuity.

5. Pays $100 a month for 2 years at 24% annual interest compounded monthly?

6. Pays $150 a month for 1 year at 24% annual interest compounded monthly?

7. Pays $250 a month for 2 years at 12% annual interest compounded monthly?

8. Pays $20 a month for 4 years at 12% annual interest compounded monthly?

In Problems 9 and 10 determine the size of the payments for a $4000 loan that must be repaid in five yearly payments at the stated annual interest rate.

9. 6%   10. 10%

In Problems 11 and 12 determine the size of the payments for a $3500 loan that must be repaid in four yearly payments at the stated annual interest rate.

11. 4%   12. 12%

In Problems 13 to 16 determine the monthly car installments on the stated loan at the annual interest rate compounded monthly.

13. $2000 loan for 4 years at 18%.

14. $1000 loan for 2 years at 18%.

15. $2500 loan for 2 years at 6%.

16. $1200 loan for 3 years at 6%.

In Problems 17 to 20 determine the monthly house payments on the stated loan at the annual interest rate compounded monthly.

17. $60,000 loan for 20 years at 9%.

18. $40,000 loan for 25 years at 8%.

19. $120,000 loan for 30 years at 8%.

20. $75,000 loan for 25 years at 9%.

21. Construct the payment schedule for the loan in Problem 9.

22. Construct the payment schedule for the loan in Problem 11.

23. Construct the payment schedule for the first four payments of the loan in Problem 17.

24. Construct the payment schedule for the first four payments of the loan in Problem 18.

In Problems 25 to 28 determine the monthly payments on a $20,000 loan under the given repayment conditions.

25. 10 years at 10% compounded quarterly.

26. 10 years at 10% compounded monthly.

27. 15 years at 10% compounded quarterly.

28. 15 years at 10% compounded monthly.

29. Suppose that a new car costs $13,000. The dealer offers $4000 for a trade-in with a down payment of $1000. What will the monthly installments be if the loan is for 4 years at 12% annual interest compounded monthly?

30. Suppose that a new car costs $22,000. The dealer offers $8000 for a trade-in. What will the monthly installments be if there is no down payment and the loan is for 4 years at 12% annual interest compounded monthly?

31. A house costs $800,000. The terms of sale are 25% down and the remainder to be paid in monthly payments for 25 years. The loan is taken at an annual interest rate of 9% compounded monthly.
    (a) What are the monthly payments?
    (b) Construct the payment schedule for the first four payments.

32. A house costs $65,000. The terms of sale are 20% down and the remainder to be paid in monthly payments for 30 years. The loan is taken at an annual interest rate of 9% compounded monthly.
    (a) What are the monthly payments?
    (b) Construct the payment schedule for the first four payments.

33. Construct the entire payment schedule for the loan in Problem 15.

34. Construct the entire payment schedule for the loan in Problem 14.

35. A movie star signs a "10 Million Dollar Contract," which pays $1,000,000 a year for 10 years. If the studio can invest its money at 10% interest com-

pounded annually, what is the present value of the contract?

36. Two donors agree to give the same amount of money to their alma mater. One pledges $1000 a year for 10 years. The other prefers to donate one lump sum now. What size should the lump sum be if the money earns 10% interest compounded annually?

37. A student wants to have $200 each month for spending money for the next 9 months at school. How much money has to be set aside from summer earnings to do this if the money is deposited in an account that pays 6% interest compounded monthly?

38. If the monthly payments for a car loan were $150 for 4 years at 9% interest compounded monthly, what was the amount of the loan?

39. An art museum benefactor donated $500,000 for acquisitions, with the understanding that the money was to be withdrawn quarterly for 5 years. If the account paid 8% annual interest compounded quarterly, how much money could the art museum withdraw each quarter?

40. What are the monthly payments on a $3000 dining room set that is paid for in 24 monthly installments at a monthly interest rate of 1.5%?

41. A small business buys a local network of microcomputers costing $30,000. The business amortizes the purchase with monthly payments over 6 years at 12% interest compounded monthly. What is the size of each payment?

42. Determine the monthly payments on a $10,000 loan under each of these repayment conditions.
    (a) 5 years at 12% interest compounded monthly.
    (b) 3 years at 9% interest compounded monthly.

43. If a house was purchased with a down payment of $10,000 and a 30-year mortgage at 9.5% interest compounded monthly with monthly payments of $400, what was the price of the house?

Cumulative Exercise Set 5–5

1. Find the simple interest on a loan of $5580 for 4 years at 10% interest. Assume 365 days.

2. Find the total amount that must be repaid on a loan of $8500 at 15% compounded annually for 5 years.

3. Find the simple interest rate as a percent of an investment with the principal of $1500, the future amount of $2000, and the time of 2 years.

4. Compute the interest on a loan of $15,000 for 20 years at 10% compounded annually.

5. Determine the compound amount on the savings of $10,000 for 8 years at 7% compounded quarterly.

6. Determine the amount of money that should be invested now to return $10,000 in 10 years at 8% compounded annually.

7. Determine the future value of an annuity where payments are $2000 per year for 10 years at 6% compounded annually.

8. Determine the quarterly payments of a sinking fund that will provide $4500 in 5 years from an account earning 10% compounded annually.

9. A person deposits $10,000 at the end of each year for 5 years in an account paying 7% interest compounded annually. How much money will be in the account at that time?

10. For 5 years a couple deposits $200 at the end of each month in an account paying 9% compounded monthly. How much money will be in the account at the end of 5 years?

11. Determine the monthly car installments on a $5000 loan for 5 years at 12% interest compounded monthly.

12. Determine the monthly house payments on an $80,000 loan for 20 years at 10% interest compounded monthly.

13. If a house was purchased with a down payment of $20,000 and a 15-year mortgage at 10% interest compounded monthly with monthly payments of $600, what was the price of the house?

14. If a house was purchased for $200,000 with a down payment of $30,000 and a 30-year mortgage at 10% interest compounded monthly, calculate the monthly payments.

CHAPTER REVIEW

Key Terms

5–1 Simple Interest and Discount

Principal

Interest

Rate of Interest

Simple Interest

Compound Interest

Exact Interest Year

Ordinary Interest Year

Future Amount

Present Value

Discount Interest Method

Discount

Proceeds

Effective Rate

5–2 Compound Interest

Interest Compounded:

    Annually

    Quarterly

    Monthly

    Daily

Growth Factor

Period

Conversion Period

Periodic Interest

Compound Interest Formula

Nominal Rate

Effective Rate

Annual Yield

5–4 Annuities and Sinking Funds

Annuity

Ordinary Annuity

Future Value

Geometric Series

Annuity Due

Sinking Fund

5–5 Amortization

Present Value of an Annuity

Amortize

Amortization Schedule

Summary of Important Concepts

Formulas

Simple interest

$$I = Prt$$

Future amount

$$A = P(1 + rt)$$

Interest compounded annually

$$P_t = P(1 + r)^t$$

Compound Interest Formula

$$A = P(1 + i)^n$$

Sum of a geometric series

$$a^n + \cdots + a + 1 = \frac{a^{n+1} - 1}{a - 1}$$

Annuity Formula

$$A = R \cdot \frac{(1 + i)^n - 1}{i} = R \cdot s_{\overline{n}|i}$$

Present value of an annuity

$$P = R \cdot \frac{1 - \dfrac{1}{(1 + i)^n}}{i} = R \cdot a_{\overline{n}|i}$$

Amortization Formula

$$R = P \cdot \frac{i}{1 - \dfrac{1}{(1 + i)^n}} = P \cdot \frac{1}{a_{\overline{n}|i}}$$

REVIEW PROBLEMS

1. Find the simple interest on a $2500 savings account that earns $5\frac{1}{4}$% interest for 2 years.

2. Find the simple interest on a $4000 loan for 90 days at 18% interest.

3. Determine the interest on a $500 loan for 3 years at 15.5% interest compounded annually.

4. After 3 years, what is the future value of a savings account with $700 that pays 5.25% interest compounded daily?

5. Determine the present value of $10,000 over 3 years at 8% compounded (a) annually (b) daily.

6. What is the annual yield if the nominal rate is 7.58%?

7. If the cost of a ticket to a Broadway play rises 8% a year for the next 20 years, what will the price of a $40 ticket be in 20 years?

8. In 1985 Mexico's rate of inflation was 63.7%. If a burrito cost 1000 pesos in 1985, and that inflation continued for a decade, what would the burrito cost in 1995?

9. In May of 1986 Mexico's rate of inflation was 5.6%. If that rate continued for an entire year, what would the annual rate of inflation be?

10. If the world population increases 2% each year for the next 25 years, how much larger will the population be?

11. Determine the future value of an annuity in which a payment of $250 is made at the end of every quarter for 3 years at 8% compounded quarterly.

12. Determine the quarterly payments of a sinking fund that must produce $12,000 in 4 years at 8% compounded quarterly.

13. If an individual deposits $2000 a year in a retirement account paying 6% compounded annually, how much will be in the account after 25 years?

14. How much money should be invested as a lump sum today to receive a dividend of $500 every 3 months for 4 years in an account that pays 8% compounded quarterly?

15. What is the present value of an annuity that pays $250 a month for 4 years at 9% annual interest compounded monthly?

16. Construct the payment schedule for a $5000 loan that must be repaid in four yearly payments at an interest rate of 10% compounded annually.

17. A house costs $80,000. The terms of sale are 25% down and the remainder to be paid in monthly payments for 30 years. The loan is taken at an annual interest rate of 12% compounded monthly.

    (a) What are the monthly payments?
    (b) Construct the payment schedule for the first four payments.

PROGRAMMABLE CALCULATOR EXPERIMENTS

1. Write a program that computes the sum of a geometric series given the values of a and r.

2. Write a program that computes the annuity given the values of R and i.

3. Write a program that computes the present value given the values of R and i.

4. Write a program that computes the amortization given the values of P and i.

5. Write a program that constructs a payment schedule for a loan in the amount of A that must be repaid in n years at an interest rate of i% compounded (a) annually, (b) quarterly, (c) monthly, (d) daily.

Sets and Counting

6

How many different nuts go with how many different bolts? (H. Armstrong Roberts)

Chapter Overview

This chapter lays the foundation for the next two chapters on probability and statistics. It begins with a review of the basic terms and notation from set theory, where various sets of real numbers are introduced. Since several meanings are attached to the verb "count," a discussion of the mathematical meaning is given. It is followed by a development of two fundamental counting techniques, the Inclusion-Exclusion Principle, and the Multiplication Principle. Venn diagrams and tree diagrams are introduced as aids in visualizing certain types of situations, like the scheme concocted by a gang of crooks to rig the Pennsylvania lottery. The last concepts described, permutations and combinations, are applied to the binomial theorem, which itself has applications in probability, statistics, and calculus.

CASE STUDY PREVIEW

Write down the license plate number of your car (or a friend's car if you do not own one). Most likely it is not a "number" at all but, rather, consists of three letters followed by three digits. The reason for this assertion is that more than half of the states currently use this so-called standard format.

California has more passenger vehicles than any other state; so many that the standard format cannot account for all of the license plates. A second format was introduced, and the combined formats accounted for all of the state's motoring needs. But then California introduced yet a third format. Is this an instance of irrational behavior in the Golden State, or is there a mathematical justification? The Case Study will reveal the answer.

6–1 Sets

In mathematics the little word "set" plays a very big role. Set theory was developed to organize, synthesize, and unify all of mathematics.

This section introduces some basic aspects of set theory and describes various sets of real numbers. We have already used sets intuitively in Chapters 3 and 4, where sets of feasible solutions to linear programming problems were encountered.

Notation

There are several ways to describe a **set.** The most direct way is the *descriptive method,* in which the phrase "the set of all . . . ." precedes a description of the set. The set must be *well defined* in the sense that it is possible to tell whether or not a given object is in the set. For instance, the "set of all large numbers" is not well defined because the word "large" is ambiguous.

EXAMPLE 1

Problem

Determine whether each of the following collections is a well-defined set:
(a) good senators; (b) nonnegative numbers.

Solution (a) Each senator is regarded as good by some people and bad by others. Since the collection is not well defined, it is not a set.

(b) The nonnegative numbers consist of zero and all positive numbers. This collection is well defined, so it is a set.

Usually capital letters stand for sets. *Throughout this chapter the letter S will denote the set of all license plates in standard format,* where **standard format** means three letters followed by three digits.

The objects in a set are called **elements.** The symbol $\in$ means "is an element of" while the symbol $\notin$ means "is not an element of." Thus AOK 007 $\in$ S but 007 AOK $\notin$ S.

Another way to specify a set is to use *braces* to enclose the elements. The notation $A = \{4, 5, 6\}$ means that A is the name of the set whose elements are 4, 5, and 6. It is read "A is equal to the set consisting of 4, 5, and 6." Braces can also define sets implicitly. The set

$$F = \{1, 2, 3, \ldots, 50\}$$

refers to the whole numbers from 1 to 50. The sets A and F are finite. Two examples of infinite sets are the **natural numbers**

$$N = \{1, 2, 3, \ldots\}$$

and the **integers**

$$Z = \{\ldots, -3, -2, -1, 0, 1, 2, 3, \ldots\}$$

A third way of describing sets is to use *set-builder notation* in the form $\{\_ : \_\_\_\}$. This is read "the set of all $\_$ such that $\_\_\_$." The set of **rational numbers** Q, which is sometimes called the set of fractions, can be defined by

$$Q = \left\{ \frac{a}{b} : a, b \in Z, \quad b \neq 0 \right\}$$

In words, Q consists of all quantities that can be expressed in the form a/b, where a and b are integers such that $b \neq 0$. Some examples of rational numbers are $\frac{8}{5}$, $4 (= \frac{4}{1})$, $0 (= \frac{0}{1})$, $1.25 (= \frac{5}{4})$, and $0.333\ldots (= \frac{1}{3})$.

Subsets

Let X and Y be arbitrary sets. Then X is a **subset** of Y if every element in X also belongs to Y. Equivalently, X is a subset of Y if and only if $x \in X$ implies that $x \in Y$. It is written $X \subset Y$. If $X \subset Y$, then Y *contains* X. If X is not a subset of Y, we write $X \not\subset Y$. This occurs when some element is in the set X but not in the set Y. Since every natural number is an integer, it follows that for the infinite sets N and Z described earlier, $N \subset Z$.

EXAMPLE 2

Problem

Is the set $D = \{0, \frac{1}{2}, 1\}$ a subset of N? Is $D \subset Z$? Is $D \subset Q$?

Solution The number 0 belongs to D but not to N, so $D \not\subset N$. Similarly, $D \not\subset Z$ since $\frac{1}{2} \in D$ but $\frac{1}{2} \notin Z$. Each element of D is a rational number, so $D \subset Q$.

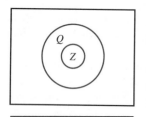

FIGURE 1

Since the set of rational numbers contains all of the natural numbers, all of their negatives, and 0, it follows that $Z \subset Q$. A **Venn diagram** depicts this in Figure 1. The innermost circle represents Z and is contained in the larger circle representing Q.

Two sets are **equal** if they contain the same elements. Thus $\{i, o, u\} = \{u, o, i\}$. This means that when elements are listed in a set the order in which the elements are written is not important. However, it is customary to write the elements in alphabetical or numerical order.

Two sets are not equal if one of the sets contains at least one element that is not contained in the other set. Thus $\{i, o, u\} \neq \{i, o, x\}$.

Operations on Sets

The **complement** of a set X is the set of elements that do not belong to X. It is written X', read "X prime." The definition raises the issue of how far to look outside X. Consider, for instance, the set Q of rational numbers. Is the kitchen sink "not in Q"? Yes, it is not in Q. However, we usually limit the search to a set called the **universal set,** denoted by U. The universal set for Q in this book will be the set R of real numbers.

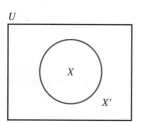

FIGURE 2

The Venn diagram in Figure 2 shows a set and its complement. The universal set U is the area inside the rectangle, X is the area inside the circle, and X' is the area that is inside the rectangle U but outside the circle X.

Usually the universal set is understood from context but sometimes it must be specified. Example 3 addresses this issue.

EXAMPLE 3

Problem

Find A' for the following choices of a set A and a universal set U.
(a) $A = \{1, 2, 3\}$, $U = \{1, 2, 3, 4, 5, 6\}$
(b) $A = \{1, 2, 3\}$, $U = N$
(c) A is the set of even positive integers, $U = N$

Solution (a) $A' = \{4, 5, 6\}$ (b) $A' = \{4, 5, 6, 7, \ldots\}$ (c) A' is the set of odd positive integers.

Two additional operations on sets are union and intersection. Let X and Y be arbitrary sets. The **intersection** of X and Y, written $X \cap Y$, is the set of all elements that are in X and Y simultaneously. The **union** of the set X with the set Y, written $X \cup Y$, is the set of all elements that are in X or in Y (or in their intersection).

EXAMPLE 4

Problem

Consider the sets $A = \{0, 1, 2, 3, 4\}$, $B = \{1, 3, 5\}$, and natural numbers N. Find (a) $A \cap B$, (b) $A \cap N$, (c) $A \cup B$, (d) $A \cup N$.

Solution (a) $A \cap B = \{1, 3\}$ (b) $A \cap N = \{1, 2, 3, 4\}$
(c) $A \cup B = \{0, 1, 2, 3, 4, 5\}$ (d) $A \cup N = \{0, 1, 2, 3, \ldots\}$

The number $\sqrt{2}$ cannot be written as the ratio of any pair of integers, so it is called an **irrational number.** (We will not prove this fact in this book.) The set Q' consists of all irrational numbers. Since no number can be both rational and irrational simultaneously, $Q \cap Q'$ has no elements. The set with no elements is called the **empty** (or **null**) **set,** denoted by $\varnothing$. Hence $Q \cap Q' = \varnothing$.

Two sets X and Y having no elements in common are called **disjoint.** Every set X and its complement X' are disjoint because $X \cap X' = \varnothing$. In particular, Q and Q' are disjoint.

For any set X and universal set U we have $X \cup X' = U$. What about the union $Q \cup Q'$? It is equal to the set R of all **real numbers.** Therefore R is the union of the set of rational numbers and the set of irrational numbers so that $R = Q \cup Q'$, where Q and Q' are disjoint sets. The Venn diagram in Figure 3 shows the set inclusions

$$N \subset Z \subset Q \subset R$$

Example 5 applies set theory to a kind of problem that is encountered frequently.

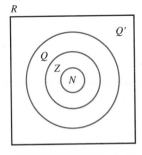

FIGURE 3

EXAMPLE 5

In 1987 the American Chamber of Commerce Researchers Association conducted a study to see how much it costs to maintain a middle-management life-style in 242 cities. Table 1 shows some parts of the study for 10 major cities. Let G be the set of cities whose grocery items cost less than \$100, and L be the set of cities whose utilities cost less than \$100.

Table 1 Relative Costs of Living in Ten Major Cities[a]

City	All Items	Grocery Items	Housing	Utilities	Transportation	Health Care	Misc. Goods & Services
New York	136.3	107.6	179.8	172.9	117.6	164.3	108.9
San Francisco	127.2	109.9	189.9	59.5	119.9	151.6	113.4
San Diego	116.8	95.5	162.1	82.6	119.7	122.6	105.6
Philadelphia	116.4	106.8	117.8	152.4	113.1	125.2	107.2
Los Angeles	114.1	95.4	145.8	65.2	113.1	130.5	115.9
San Jose, Calif.	113.0	102.4	158.6	57.7	109.9	130.4	103.2
Dallas	112.0	110.5	119.6	120.7	111.3	108.8	105.2
Denver	107.6	105.8	134.9	75.1	110.1	111.7	98.7
Phoenix	106.3	99.2	118.6	102.5	106.9	119.0	99.6
Columbus, Ohio	105.2	106.8	98.2	120.3	107.7	104.5	103.1

[a]Cost-of-living index for major items; national average = \$100
Source: American Chamber of Commerce Researchers Association.

Problem

Determine the elements in the sets G, L, $G \cap L$, and $G \cup L$.

Solution The sets G and L can be read from the third and fifth columns in Table 1.

$$G = \{\text{San Diego, Los Angeles, Phoenix}\}$$

$$L = \{\text{San Francisco, San Diego, Los Angeles, San Jose, Denver}\}$$

The definitions of intersection and union lead to

$$G \cap L = \{\text{San Diego, Los Angeles}\}$$
$$G \cup L = \{\text{San Francisco, San Diego, Los Angeles, San Jose, Denver, Phoenix}\}$$

EXERCISE SET 6–1

In Problems 1 to 4 determine if the collection is well defined.

1. The set of all large dogs.

2. The set of all tall people.

3. The set of all positive integers.

4. The set of rational numbers between 1 and 2 inclusive.

In Problems 5 to 8 list the elements in the set.

5. The set of positive integer divisors of 12.

6. The set of integers between 0 and 5 inclusive.

7. The set of integers between -3 and 0 inclusive.

8. The set of even integers between -10 and -4 inclusive.

In Problems 9 to 12 determine whether A or B is a subset of X.

9. $X = \{1, 2, 3, 4\}$, $A = \{0, 1, 2\}$, $B = \{2, 3, 4\}$

10. $X = \{-1, 0, 1, 2, 3, 4\}$, $A = \{0, 1, 4\}$, $B = \{1, 3, 4\}$

11. $X = Z$, $A = \{-2, 0, 2, 4\}$, $B = \{2, 4, 6, 8\}$

12. $X = Z$, $A = \{0\}$, $B = \{-\frac{1}{2}, 1, 2\}$

In Problems 13 to 16 find X' for the given choice of U.

13. $X = \{1, 2, 3\}$, $U = \{1, 2, 3, 4, 5, 6\}$

14. $X = \{1, 2, 3\}$, $U = \{0, 1, 2, 3, 4, 5, 6, 7\}$

15. $X = \{1, 2, 3\}$, $U = N$

16. $X = \{2, 4, 6, 8, 10\}$, $U = N$

In Problems 17 to 20 find $X \cup Y$.

17. $X = \{1, 2, 3\}$, $Y = \{2, 3, 4\}$

18. $X = \{-1, 0, 1\}$, $Y = \{-1, 1\}$

19. $X = Q$, $Y = \varnothing$

20. $X = Q$, $Y = Z$

In Problems 21 to 24 find $X \cap Y$.

21. $X = \{1, 2, 3, 4\}$, $Y = \{3, 4, 5\}$

22. $X = \{1, 2, 3, 4\}$, $Y = \{4, 5\}$

23. $X = \{-2, -1, 0, 1, 2\}$, $Y = N$

24. $X = \{\frac{2}{2}, \frac{3}{2}, \frac{4}{2}, \frac{5}{2}\}$, $Y = Z$

In Problems 25 to 28 determine whether A or B is a subset of X.

25. $X = Q$, $A = \{x : x^2 = 4\}$, $B = \{x : x^2 = 2\}$

26. $X = Q$, $A = \{x : x^2 = -1\}$, $B = \{x : x^2 = 1\}$

27. $X = Q$, $A = \{\ldots, -3, -2, -1\}$, $B = \{x \in Z : x < 0\}$

28. $X = Q$, $A = \{5, 6, 7, \ldots\}$, $B = \{x \in Z : x < -4\}$

In Problems 29 and 30 find X' for the given choice of U.

29. $X = \{1, 3, 5, \ldots\}$, $U = \{1, 2, 3, \ldots\}$

30. $X = \{4, 8, 12, \ldots\}$, $U = \{2, 4, 6, \ldots\}$

In Problems 31 to 34 find $X \cup Y$.

31. $X = \{-2, -1, 0, 1\}$, $Y = N$

32. $X = \{\sqrt{2}, 2, \sqrt{3}, 3\}$, $Y = Q$

33. $X = N$, $Y = \{x \in Z : x < 0\}$

34. $X = N$, $Y = \{x \in Z : x \geq 0\}$

In Problems 35 to 38 find $X \cap Y$.

35. $X = Q$, $Y = \{\sqrt{2}, \sqrt{3}, \sqrt{4}, \sqrt{5}, \ldots\}$

36. $X = Q'$, $Y = \{\sqrt{2}, \sqrt{3}, \sqrt{4}, \sqrt{5}, \ldots\}$

37. $X = N$, $Y = \{x \in Z : x < 4\}$

38. $X = \{x \in Z : x < 7\}$, $Y = \{x \in Z : x < 3\}$

39. Describe $F \cap X$ where $F = \{4, 8, 12, 16, \ldots\}$ and $X = \{6, 12, 18, 24, \ldots\}$.

40. Describe $M \cap K$ where M consists of all multiples of m and K consists of all multiples of k, where m and k are given positive integers.

41. Let U be the set of letters in the alphabet, let V be the set of vowels, and let M be the set of letters in "mathematics." Find (a) $V \cap M$ (b) $V \cup M$ (c) $V' \cap M$.

42. Let U be the set of letters in the alphabet, $F = \{a, b, c, d, e\}$ and $V = \{a, e, i, o, u\}$. Find (a) $F \cap V$ (b) $F \cup V$ (c) $F \cap V'$.

43. Let $U = \{1, 2, 3, \ldots, 100\}$, let E be the set of even numbers in U, and let T be the first ten numbers in U. Find (a) $E \cap T$ (b) $E \cup T$ (c) $E' \cap T$.

44. Let $U = \{1, 2, 3, \ldots, 20\}$, E be the set of even numbers in U, and $T = \{3, 6, 9, 12, 15, 18\}$. Find (a) $E \cap T$ (b) $E \cup T$ (c) $E' \cap T$.

45. Let N be the set of natural numbers and $A = \{-2, -1, 0, 1, 2\}$. Find $N \cup A$.

46. Let N be the set of natural numbers and $A = \{-1, 0, 1, 2, 3, 4\}$. Find $N \cup A$.

47. Let $B = \{x : x$ is an integer and $x^2 < 16\}$, and $C = \{3\}$. Answer each of the following as true or false. (a) $3 \in C$ (b) $\{-2\} \in B$ (c) $C \subset B$

48. List all of the elements in the set $S = \{x : x$ is a positive even integer and $x < 10\}$

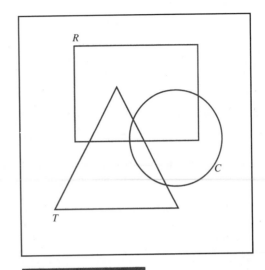

FIGURE 4

49. Shade the set $R \cap C'$ in Figure 4.

50. Shade the set $T \cup R$ in Figure 4.

Referenced Exercise Set 6–1

Problems 1 to 4 refer to Table 1 in Example 5.

1. List the set of cities whose cost of housing or miscellaneous goods and services is less than $100.

2. List the set of cities whose cost of housing and utilities exceeds $100.

3. Let $U = 10$ major cities, $H = $ cost of housing that is more than $100, and $M = $ cost of miscellaneous goods and services that is more than $100. Describe in words the fact that $H \cup M = U$.

4. Let $U = 10$ major cities and $T = $ cost of transportation that is more than $100. Describe verbally the fact that $U = T$.

Problems 5 and 6 refer to Table 2, which lists certain countries and the percentage of their population for two age groups as of 1986.

Table 2

Country	Brazil	Canada	China	Ireland	U.S.	USSR
Age 15–29	28.6	28.9	29.1	24.9	22.4	26.6
Age 30–44	16.4	20.0	17.5	17.2	19.1	19.1

Source: 1986 Britannica Book of the Year, pp. 826–831.

Let F be the three countries with the highest percentages aged 15–29 and let T be the three countries with the highest percentages aged 30–44.

5. Find (a) F (b) T.

6. Find (a) $F \cap T$ (b) $F \cup T$.

7. What is the difference between $\emptyset$ and $\{\emptyset\}$?

8. In music theory "the diatonic universe of seven pitch classes" consists of the notes A, B, C, D, E, F, and G, irrespective of inflection by sharps and flats.* (a) Use set notation to describe this universal set. (b) Use set notation to describe the set in which each note can also be sharp or flat (refer to a piano keyboard; that is, G, G-sharp, and G-flat are regarded as different note classes).

9. In 1988 the Office Network, a consortium of real estate firms, conducted a study of rental rates in the central business districts (CBDs) of several cities. Table 3 shows part of the results of the survey. (a) List the cities in the set A of those cities having the five highest CBD rates. (b) List the cities in the set B of those cities having the five highest rates outside CBDs. (c) What would it mean for A ∩ B = A? (d) For this study, is it true that A = B?

Table 3 Office Rental Costs
Typical annual rate, in dollars per square foot, for rent and operating expenses in space that is new or competitive with new space. Rates are for space inside central business district or outside CBD.

City	CBD	Outside
Baltimore	$23.38	$17.50
Boston	35.00	15.50
Chicago	28.95	19.56
Houston	12.00	9.50
Los Angeles	34.00	22.00
Miami	29.00	19.00
New York	40.00	35.00
Philadelphia	22.10	14.85
San Francisco	23.65	22.25
Washington	32.00	22.50
Wilmington	18.00	16.10

N.A. = Not applicable.
Source: The Office Network

*John Clough, "Aspects of Diatonic Sets," *Journal of Music Theory,* Spring 1979, pp. 45–62.

6–2 Venn Diagrams and Counting

The word **count** usually means "one, two, three, four, ..." In mathematics, however, it has a wider usage. Perhaps a more appropriate word is "enumerate," meaning "to ascertain the number of." In mathematics the expression "count the number of . . ." does not mean to count the objects individually but, rather, to determine how many objects there are altogether. Sometimes enumerating a large set can be a complicated task. Indeed, the root of the word enumerate is the Latin *numerus,* meaning "nimble."

For instance, consider a room that has 12 rows of chairs with 10 chairs in each row. If all but 6 of the chairs are occupied, how many people are sitting in the chairs? A counting technique for answering this question is to multiply 12 by 10, then subtract 6, to get 114. The chairs did not have to be counted individually to arrive at this answer.

This section presents a basic counting principle that is explained and illustrated by Venn diagrams.

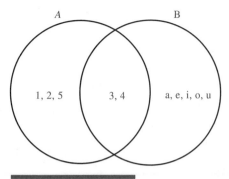

FIGURE 1

Number of Elements

Let $n(X)$ denote the number of distinct elements in a finite set X. The expression "count the number of elements in X" means to determine $n(X)$ by some method. Counting techniques are efficient methods for computing $n(X)$ directly.

Let $A = \{1, 2, 3, 4, 5\}$ and $B = \{a, e, i, o, u, 3, 4\}$. Then $n(A) = 5$ and $n(B) = 7$. Since $A \cap B = \{3, 4\}$ it follows that $n(A \cap B) = 2$. The Venn diagram in Figure 1 shows the distribution of the elements in A and B. It also shows that $n(A \cup B) = 10$. Notice that the formula $n(A \cup B) = n(A) + n(B) - n(A \cap B)$ holds because $10 = 5 + 7 - 2$. This is a particular case of a basic counting technique called the Inclusion-Exclusion Principle. It holds in this case because when counting the elements in $A \cup B$ by first counting the elements in A and then counting the elements in B, the two elements in $A \cap B$ are counted twice, so they must be subtracted once in order to compensate.

Inclusion-Exclusion Principle
If X and Y are any sets, then

$$n(X \cup Y) = n(X) + n(Y) - n(X \cap Y)$$

Example 1 illustrates the reasoning behind this principle.

EXAMPLE 1

Suppose a survey of 100 people asks

Have you eaten at McDonald's this year?   Y   N
Have you eaten at Wendy's this year?   Y   N

The respondents circle Y or N for each question.

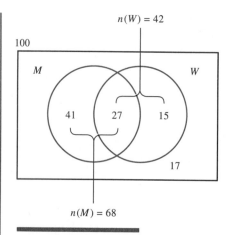

FIGURE 2

Problem

If 68 people circled Y for McDonald's, 42 circled Y for Wendy's, and 27 circled Y for both questions, how many respondents circled N for both questions?

Solution Let M be the set of respondents who circled Y for McDonald's and let W be the set of respondents who circled Y for Wendy's. Then $n(M) = 68$ and $n(W) = 42$. Since 27 circled both Y, $n(M \cap W) = 27$. By the Inclusion-Exclusion Principle

$$n(M \cup W) = n(M) + n(W) - n(M \cap W)$$

$$= 68 + 42 - 27$$

$$= 83$$

This means that 83 respondents circled Y for either McDonald's or Wendy's (or both). Since there were 100 respondents altogether, 17 respondents circled N for both questions. The Venn diagram is shown in Figure 2.

Recall that license plates in standard format S consist of three letters followed by three digits. License plates in *reverse format R* consist of three digits followed by three letters. Example 2 counts the number of license plates in either one of these formats, where the numbers of license plates in each format will be derived in the Case Study.

EXAMPLE 2

Problem

If $n(S) = 14{,}866{,}000$ and $n(R) = 16{,}208{,}000$, how many license plates are in either the standard or reverse format?

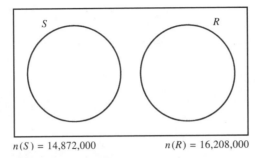

$n(S) = 14,872,000$ $n(R) = 16,208,000$

FIGURE 3

Solution The problem reduces to finding $n(S \cup R)$ because $S \cup R$ represents all license plates in either the standard format or the reverse format. Each license plate in S begins with a letter and each license plate in R begins with a number, so there are no license plates in both formats. Thus $n(S \cap R) = 0$. By the Inclusion-Exclusion Principle

$$n(S \cup R) = n(S) + n(R) - n(S \cap R)$$
$$= 14,866,000 + 16,208,000 - 0$$
$$= 31,074,000$$

This means that over 31 million license plates can be made using either the standard format or the reverse format.

The sets S and R in Example 2 are disjoint. They are shown in Figure 3. The formula for $n(S \cup R)$ is a particular case of the Inclusion-Exclusion Principle.

If X and Y are disjoint sets, then

$$n(X \cup Y) = n(X) + n(Y)$$

Three Circles

One subset divides a universal set U into two regions. (See Figure 2 in Section 6–1.) Two subsets that are not disjoint divide U into four regions. (See Figure 2 in this section.) Three subsets in which no two subsets are disjoint divide U into eight regions, as Example 3 will show. The example refers to Table 1, which lists stock activity (sales, last price, and net change) during the week of April 11–15, 1988.

Table 1 Most Active Stocks Week Ended April 15, 1988 (Consolidated)

Company	Sales	Last	Net Chng
US West o	24,346,100	$51\frac{5}{8}$	$-1\frac{1}{4}$
NYSEG	13,857,000	$22\frac{1}{8}$	$-\frac{1}{8}$
IBM	11,068,100	$114\frac{1}{8}$	$+2\frac{1}{2}$
Varity	10,662,100	$3\frac{1}{2}$	$-\frac{1}{8}$
Gen El	10,515,100	40	$-2\frac{1}{2}$
Exxon	10,238,100	$43\frac{5}{8}$	-1
Texaco	10,083,400	$48\frac{5}{8}$	$-\frac{5}{8}$
AT&T	9,417,400	$26\frac{5}{8}$	$-1\frac{1}{2}$
Ford M	7,694,300	$45\frac{1}{4}$	$-1\frac{3}{8}$
Zayre	7,160,500	24	$+\frac{1}{8}$
Phil Pet	7,058,400	$17\frac{5}{8}$	$-\frac{7}{8}$
USX	7,019,100	32	$+\frac{5}{8}$
INCO	6,802,400	$26\frac{7}{8}$	$+1\frac{1}{4}$
Navistr	6,801,300	$6\frac{3}{8}$	$-\frac{3}{8}$
Am Exp	6,664,000	$24\frac{3}{4}$	-1

Source: The New York Times, April 17, 1988.

EXAMPLE 3

Define these three sets of companies:

S = sales over $10,000,000

L = last price over $30

C = net change greater than $0

Problem

(a) Which companies are in all three sets? (b) Which companies are in C but in neither S nor L?

Solution Refer to Figure 4, which partitions the companies into the three sets. From Table 1, using the first letter of each company, except X = AmExp, K = INCO, Y = NYSEG, and W = USWest o, we get

S = {W, Y, I, V, G, E, T}

L = {W, I, G, E, T, F, U}

C = {I, Z, U, K}

The set $S \cap L \cap C$ represents those companies in all three sets, so $S \cap L \cap C$ = { I }. This answers part (a).

To answer (b), note that $S \cap C$ = {I} and $L \cap C$ = {I, U}. The companies that are in C = {I, Z, U, K} but are in neither S nor L are obtained by removing from C those companies listed in either $S \cap C$ or $L \cap C$. This set is {I, U}, so the remaining set is {Z, K}, which answers part (b).

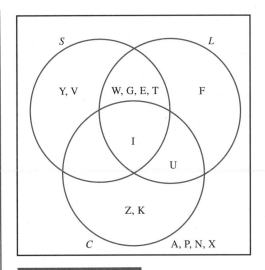

FIGURE 4

Example 4 is typical of problems involving the partition of a set by three of its subsets. It builds on the concepts introduced in Example 3.

EXAMPLE 4

A survey asked 200 people which sports they played regularly. It revealed that 30 played tennis, 25 played golf, 40 played softball, 10 played tennis and golf, 20 played tennis and softball, 15 played golf and softball, and 8 played all three sports.

Problem

How many of those surveyed played (a) golf and softball but not tennis (b) only golf (c) none of these sports?

Solution A Venn diagram provides a convenient method for tabulating these data. Let T = tennis, G = golf, and S = softball. Three circles are needed to present these sets. [See Figure 5(a).] Since 8 people played all three sports, we have $n(T \cap G \cap S) = 8$. This is shown in Figure 5(b).

Consider the fact that 15 people played golf and softball: $n(G \cap S) = 15$. Restrict your attention to the G circle and the S circle in Figure 5(b). We have already accounted for 8 of the 15 people in $G \cap S$. The remaining 7 lie in the $G \cap S$ region but outside $T \cap G \cap S$. Figure 5(c) shows their proper alignment. Notice that these 7 played golf and softball but not tennis, answering part (a).

The same kind of reasoning will place the numbers 2 and 12 in their respective areas in Figure 5(d).

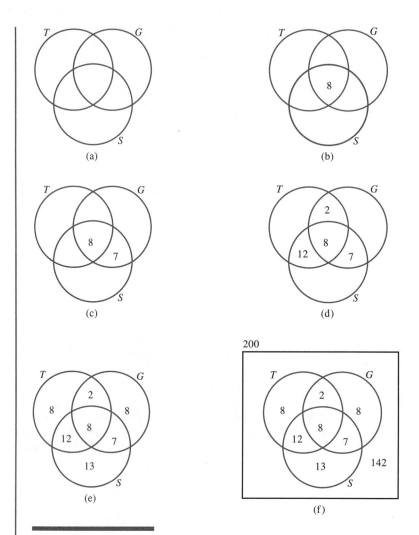

FIGURE 5

Consider only the G circle in Figure 5(d). Adding 2, 8, and 7 accounts for 17 people who played golf. Since $n(G) = 25$, 8 people in the G circle are unaccounted for. These 8 are the people who played only golf, which answers part (b).

Similar reasoning shows that 8 people played only tennis and 13 played only softball. These numbers are shown in Figure 5(e). The sum of all of the numbers in Figure 5(e) is 58. The remaining people did not play any of these sports. Since 200 people were surveyed altogether, 142 people did not play these sports. This answers part (c), which is shown in Figure 5(f).

EXERCISE SET 6-2

In Problems 1 to 6 let $U = \{a, b, c, \ldots, z\}$, $V = \{a, e, i, o, u\}$, and $H = \{o, u, c, h\}$. Find the number of elements in the following sets:

1. V 2. H 3. $V \cup H$

4. $V \cup H'$ 5. $V \cap H'$ 6. $V \cap H$

In Problems 7 and 8 find $n(A \cup B)$.

7. $n(A) = 16$, $n(B) = 9$, and $n(A \cap B) = 5$

8. $n(A) = 6$, $n(B) = 12$, and $n(A \cap B) = 3$

In Problems 9 and 10 find $n(A \cap B)$.

9. $n(A) = 16$, $n(B) = 9$, and $n(A \cup B) = 21$

10. $n(A) = 6$, $n(B) = 12$, and $n(A \cup B) = 14$

In Problems 11 and 12 find $n(A \cup B \cup C)$.

11. $n(A) = 40$, $n(B) = 20$, $n(C) = 30$, $n(A \cap B) = 9$, $n(B \cap C) = 11$, $n(A \cap C) = 8$, and $n(A \cap B \cap C) = 3$

12. $n(A) = 60$, $n(B) = 50$, $n(C) = 70$, $n(A \cap B) = 19$, $n(B \cap C) = 20$, $n(A \cap C) = 13$, and $n(A \cap B \cap C) = 11$

13. The numbers of possible license plates in standard format and in reverse format (without regard to any restrictions) are 17,576,000 each. How many license plates can be made in one or the other of these formats?

14. An English class consists of 5 freshmen, 8 sophomores, and 22 upperclassmen (juniors and seniors). How many students are in the course altogether?

15. The combined membership of the Mathematical Association of America (MAA) and the American Mathematical Society (AMS) is approximately 45,000, of whom 27,000 belong to the MAA and 23,000 belong to the AMS. How many people belong to both organizations?

16. A history class consists of 35 students, of whom 20 are women. Two of the women and five of the men are veterans. How many of the men are not veterans?

In Problems 17 and 18 draw a Venn diagram to answer the questions.

17. A survey of 36 students revealed the following information:
14 had studied French

11 had studied Spanish
14 had studied German
4 had studied French and Spanish
3 had studied Spanish and German
5 had studied French and German
1 had studied all three languages
(a) How many had studied German and Spanish but not French?
(b) How many had studied only French?
(c) How many had not studied any of these languages?

18. A survey of 200 people revealed that 30 played tennis, 25 played golf, 40 played baseball, 10 played tennis and golf, 15 played golf and baseball, 20 played tennis and baseball, and 8 played all three sports.
(a) How many played baseball and golf but not tennis?
(b) How many played only tennis?
(c) How many did not play any of these sports?

In Problems 19 and 20 A and B are disjoint sets. Find $n(A)$.

19. $n(A \cup B) = 10$ and $n(A) = n(B) + 2$

20. $n(A \cup B) = 8$ and $n(A) + 2 = n(B)$

In Problems 21 and 22 find $n(A \cap B)$.

21. $n(A \cup B) = 12$, $n(A) = 7$, and $n(B) = 8$

22. $n(A \cup B) = 4$, $n(A) = 3$, and $n(B) = 3$

In Problems 23 and 24 find $n(A \cup B \cup C)$ where A and B are disjoint sets.

23. $n(A) = 10$, $n(B) = 5$, $n(C) = 9$, $n(A \cap C) = 1$, and $n(B \cap C) = 3$

24. $n(A) = 8$, $n(B) = 9$, $n(C) = 12$, $n(A \cap C) = 7$, and $n(B \cap C) = 1$

In Problems 25 and 26 draw a Venn diagram of the universal set U and subsets A and B. Use the given information to fill in the number of elements for each region.

25. $n(U) = 20$, $n(A') = 11$, $n(B') = 15$, and $n(A \cap B) = 2$

26. $n(U) = 20$, $n(A) = 4$, $n(B') = 8$, and $n(A \cup B) = 13$

In Problems 27 and 28 find $n(B)$.

27. $n(A \cup B) = 22$, $n(A \cap B) = 6$, and $n(A \cap B') = 8$

28. $n(A \cup B) = 16$, $n(A \cap B) = 6$, and $n(A \cap B') = 8$

29. Is it possible for $n(A \cap B) = n(A \cup B)$?

30. Is it possible for $n(A \cap B) > n(A \cup B)$?

31. A business section article on the stock market revealed that of the 100 stocks studied 1 week, 55 stocks advanced during the week, 24 stocks had a net change of more than 10%, 18 stocks had split within the last 12 months, 18 of the stocks that advanced during the week also had a net change of more than 10%, 10 of the stocks that advanced during the week also had split within the last 12 months, 8 stocks that had a net change of more than 10% also had split within the last 12 months, 3 stocks that advanced and also had a net change of more than 10% had also split within the last 12 months. How many stocks (a) advanced but neither split nor had a net change of more than 10%? (b) had a net change of more than 10% but neither advanced nor split?

32. Work the previous problem if the number of stocks that had all three properties was zero instead of 3.

33. An article on the banking industry revealed that of the 200 banks surveyed 1 month, 62 banks lowered their prime rate, 54 banks lowered their mortgage rate, 38 banks lowered their CD rate, 30 of the banks that lowered their prime rate also lowered their mortgage rate, 24 of the banks that lowered their prime rate also lowered their CD rate, 20 of the banks that lowered their mortgage rate also lowered their CD rate, 18 banks lowered all three rates. How many of the banks lowered none of the rates?

34. Work the previous problem if the number of banks that lowered all three rates was 10 instead of 18.

35. If $n(X) = 3$, how many subsets does X have?

36. If $n(X) = k$, how many subsets does X have?

37. Is there a difference between $n(\emptyset)$ and $n(\{\emptyset\})$?

38. Is there a difference between $\emptyset$ and $\{\emptyset\}$?

Referenced Exercise Set 6–2

Problems 1 to 4 refer to a study in sociology on the economic mobility of the poor.* The accompanying table presents the status of the one year poor, temporarily poor (1 to 2 years), and persistently poor (7 to 8 years) from that study. Find $n(X)$ where X is the set of

1. White people who are persistently poor.

2. Black people who are persistently poor.

3. Men who are one year poor.

4. Women who are one year poor.

	One Year	Temporarily Poor	Persistently Poor	Poor
Women				
White	494	112	327	933
Black	562	307	122	991
Men				
White	472	40	1102	1614
Black	247	198	225	670
Total	1775	657	1776	4208

In 1985 New Zealand's prime minister, David Lange, proposed that his country withdraw from the Anzus pact instead of permitting U.S. nuclear submarines to land in its ports. A national opinion poll† revealed that 70 percent favored remaining in the alliance while 60 percent approved Mr. Lange's course. Problems 5 and 6 refer to the poll.

5. What percent agreed with both positions?

6. What percent favored Mr. Lange's proposal but did not want to remain in the alliance?

7. The three primary colors in the visible spectrum are red, blue, and green.‡ Draw a Venn diagram with

*Greg J. Duncan, et al., *Years of Poverty, Years of Plenty*, Ann Arbor, Survey Research Center, Institute for Social Research, 1984, p. 49.

†*The New York Times*, October 6, 1985, p. 24.

‡Raymond A. Serway, *College Physics*, Philadelphia, Saunders College Publishing, 1985, pp. 660–663.

three overlapping circles, labeled R, B, and G, and color each circle with a primary color. Describe the color of each of these regions: (a) R ∩ G (b) R ∩ B (c) B ∩ G (d) R ∩ G ∩ B (e) (R ∪ G ∪ B)′.

8. In the 1984–1985 National Hockey League season, there were 234 games in which at least one team scored exactly two goals.§ The nonlosing team scored two goals in 38 games and the nonwinning team scored two goals in 217 games. How many games ended in a tie?

9. Draw a Venn diagram with six sets.‖ How many regions are there?

10. The language problem that has dogged Canada since its founding some 125 years ago reached a peak in the summer of 1990. The figures stated here are taken from an analysis of that situation by well-known political commentator John Burns.#
   (a) In Quebec 6,035,316 people speak French, 2,595,821 speak English, and 2,226,756 speak both languages. How many people in Quebec speak either French or English?
   (b) In the rest of Canada 18,187,255 people speak English, 1,978,585 speak French, and 1,829,415 speak both languages. How many Canadians outside Quebec speak either French or English?
   (c) How many Canadians altogether speak either French or English?

Cumulative Exercise Set 6–2

1. List the elements in the set of integers between 0 and 8 inclusive.

2. Determine whether A is a subset of B.
   (a) $A = \{1, 2, 3, 4, 5\}$ $B = \{2, 3, 4\}$
   (b) $A = \{a, b, c\}$ $B = \{1, 2, a, b, c, d\}$

3. Find $A \cup B$ and $A \cap B'$.
   (a) $A = \{1, 2, 3, 4\}$ $B = \{2, 3, 4, 5, 6\}$
      $U = \{0, 1, 2, 3, 4, 5, 6, 7, 8\}$

   (b) $A = \{a, b, c\}$ $B = \{b, c, d, g\}$
      $U = \{a, b, c, d, e, f, g, h\}$

4. List the elements in $A \cap B'$, where A is the set of divisors of 12, B is the set of integers from 2 to 5 inclusive, and $U = \{0, 1, 2, \ldots, 14\}$.

5. Find $n(A \cup B)$ when $n(A) = 20$, $n(B) = 15$, and $n(A \cap B) = 3$.

6. A club has 40 members, 21 of whom are female and 18 of whom are business majors. If 5 of the males are business majors, how many females are not business majors?

7. A survey of 80 restaurants revealed that 36 served wine, 30 served pizza, 32 served hamburgers, 13 served wine and pizza, 19 served pizza and hamburgers, 8 served wine and hamburgers, 6 served all three. How many served none of the foods?

8. Find $n(A \cup B)$ when $n(U) = 34$, $n(A') = 21$, $n(B) = 20$, and $n(A \cap B) = 3$.

9. An article in *Investment Dealer's Digest* on the trading of recent stock issues revealed that of the 30 stocks studied, 7 were initial price offerings (IPOs), 19 had a high price offer (HPOs) of more than $10, 11 had a high percent change (HPCs) of greater than 10% from the previous month, 2 IPOs were also HPOs, 10 HPCs were also HPOs, none of the IPOs was an HPC. How many of the HPOs that were not IPOs did not have an HPC?

10. An article in the *New York Times* (12-30-90) on the performance of mutual funds listed 40 funds and categorized them according to whether they were a no-load (NL) fund, meaning they charged no sales tax, a growth (G) fund, or a high yield (HY) fund, meaning its return over the last 5 years was greater than 90%. There were 17 NL funds, 30 G funds and 9 HY funds. There were 13 funds that were NL and G, 6 funds that were G and HY, 2 funds that were NL and HY, and there were no funds with all three properties. How many funds had none of these properties?

§Benny Ercolani and Jeff Boyle, *NHL Data, 1984–85 Season*, Montreal, Canada, National Hockey League Communications Department, 1985.

‖The Venn diagram is drawn in J. Chris Fisher, E. Koh, and Branko Grünbaum, "Diagrams Venn and How," *Mathematics Magazine*, February 1988, pp. 36–40.

#John Burns, "Canada's Crucial Seven Day Countdown," *New York Times*, June 17, 1990, Sec. 4 (Week in Review), p. 3.

6-3 Tree Diagrams and the Multiplication Principle

Hundreds of thousands of people plunk down a dollar or more daily in hopes of "hitting the big one" in the state lottery. Each day, as their ticket chunks out of the computer, visions of expensive cars, wall-to-wall stereo systems, and VCRs dance in their heads. The odds against winning are monumental, but on they play.

If you *knew* what the number was going to be, actually knew it by some foolproof method, would you only bet $1? Of course not! You would gather all available cash and purchase tickets in a flurry. This is unlikely, however, because state governments have taken great precautions to prevent even the slightest chance of cheating.

Why then did a white Cadillac pull up in front of the Dew Drop Inn in South Philadelphia in 1980 and the driver purchase 904 lottery tickets? That was just the beginning. By the end of the day the driver purchased over 10,000 tickets, all on numbers that consisted of 4's and 6's. Throughout the state accomplices purchased another 10,000 tickets.

The Pennsylvania daily lottery is run by selecting a Ping-Pong ball from each of three urns with ten balls numbered 0 to 9. Violet Lowery, a senior citizen, drew the number 666, making many of those 20,000 tickets big winners. The gang held tickets worth over $1.75 million.

The next day many sellers of the tickets, most of whom are local shop owners who also pay out the winnings for a fee from the state, complained that their payoffs were unusually high. The number 666 was very suspicious. The Bible calls it "the number of the beast." But how suspicious was it?

Here are three questions raised by the possible rigging of the "daily number." What are the chances of the number 666 being drawn? Is it less likely, the same, or more likely to be drawn than any other three-digit number? What facts would the state need in order to launch a full-scale investigation to determine if the fix was on? All three questions will be answered in the epilogue to this section.

Tree Diagrams

A **tree diagram** is a convenient way to visualize and enumerate the number of outcomes in an event that occurs in two or more stages. It begins with the possible outcomes of the first stage joined by lines to one common point. These lines form the branches of the tree. Each of these outcomes is joined to the outcomes of the second stage by another set of branches. The tree, which is lying on its side, grows from left to right with each successive stage.

For instance, consider a dinner menu that offers these options:

Salad	Entrée	Dessert
House (h)	Turkey (t)	Pie (p)
Spinach (s)	Veal (v)	Cake (c)
	Fish (f)	Ice cream (i)
	Burger (b)	

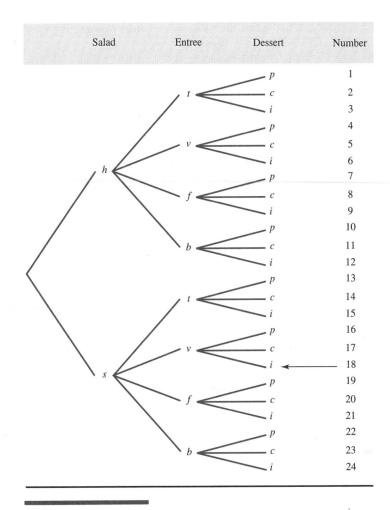

Salad	Entree	Dessert	Number

FIGURE 1

If the restaurant wants its customers to order their meals by number, how many numbers are necessary? The tree diagram in Figure 1 portrays all of the possibilities. At the first stage there are two possibilities, the house salad and a spinach salad. This stage has two branches emanating from the starting point. Each of the salads can be ordered with one of the four entrées, resulting in eight selections for these two courses. There are three desserts that can be ordered for each of these eight selections, so altogether 24 complete meals can be ordered. They are numbered in Figure 1. Note that each meal is defined by three branches so that meal 18 consists of a spinach salad, veal, and ice cream.

Word processors play a central role in business and education today. Example 1 deals with macros in the word processor "WordPerfect." A "macro" is a combination of word processing commands that can be accessed with just a few keystrokes. For instance, deleting a sentence in WordPerfect requires three steps:

1. Hold down the Control key while pressing the F4 key.
2. Press 1 for "Sentence."
3. Press 3 for "Delete."

Instead, a macro with the filename SD can perform these three steps by typing only SD.

EXAMPLE 1

Consider those macros that can be applied to three blocks of text (Words, Sentences, and Paragraphs) by one of the following 19 editing commands, where capital letters denote that block or command.*

Bold	Copy	Double underline
Erase	eXtra large type	Fine type
Goto start	Italics	lArge type
Lower case	Move	Redline
Small type	sPell check	striKeout
Transpose	uNderline	Upper case
Very large type		

Problem

How many two-letter names for macros can be formed in which the first letter denotes the block of text and the second denotes the editing command?

Solution View each macro as consisting of two stages in a tree diagram, the first stage being the block of text and the second stage being the editing command. The first stage will have three branches, one for each block of text. Each of these branches will have 19 branches corresponding to the second stage. Thus the total number of macros will be 57 since $3 \cdot 19 = 57$. The tree diagram is shown in Figure 2.

Trimmed Trees

Suppose a daily number is formed by drawing three balls from an urn containing ten balls numbered 0 to 9. The second and third balls are drawn from the urn without replacing those that have already been drawn. This is called **selection without replacement.** A tree diagram could present all the possibilities but it would be very unwieldy. We modify it by drawing only one line between the different stages and by writing above that line the number of branches that would have connected the two stages. The trimmed tree for the lottery is

$$\text{start} \underset{}{\overset{10}{\rule{2cm}{0.4pt}}} \text{1st draw} \underset{}{\overset{9}{\rule{2cm}{0.4pt}}} \text{2nd draw} \underset{}{\overset{8}{\rule{2cm}{0.4pt}}} \text{3rd draw}$$

*See Table 1 in the article by Susan Neff, "Mnemonic Macros," *WordPerfect: The Magazine,* September 1990, pp. 19–23.

Block	Edit command	Macro filename
Word	Bold	WB
	Copy	WC
	Double underline	WD
	Erase	WE
	eXtra large type	WX
	Fine type	WF
	Goto start	WG
	Italics	WI
	lArge type	WA
	Lower case	WL
	Move	WM
	Redline	WR
	Small type	WS
	sPell check	WP
	striKeout	WK
	Transpose	WT
	uNderline	WN
	Upper case	WU
	Very large type	WV
Sentence	Bold	SB
	Copy	SC
	Double underline	SD
	Erase	SE
	eXtra large type	SX
	Fine type	SF
	Goto start	SG
	Italics	SI
	lArge type	SA
	Lower case	SL
	Move	SM
	Redline	SR
	Small type	SS
	sPell check	SP
	striKeout	SK
	Transpose	ST
	uNderline	SN
	Upper case	SU
	Very large type	SV
Paragraph	Bold	PB
	Copy	PC
	Double underline	PD
	Erase	PE
	eXtra large type	PX
	Fine type	PF
	Goto start	PG
	Italics	PI
	lArge type	PA
	Lower case	PL
	Move	PM
	Redline	PR
	Small type	PS
	sPell check	PP
	striKeout	PK
	Transpose	PT
	uNderline	PN
	Upper case	PU
	Very large type	PV

FIGURE 2

The 10 represents the ten branches corresponding to the ten balls in the urn. Each of these branches is connected to nine branches leading to stage 2, and each of these is connected to eight branches leading to stage 3. Altogether there are $10 \cdot 9 \cdot 8 = 720$ ways of selecting the daily number without replacement.

Suppose the daily number is formed by selecting three balls from the urn in such a way that a ball is selected, its number is recorded, and then it is returned to the urn for another selection. This is called **selection with replacement.** The trimmed tree can be written with only the numbers and not the branches.

$$10 \quad \cdot \quad 10 \quad \cdot \quad 10 \; = \; 1000$$

first   second   third
stage   stage   stage

There are 1000 such daily numbers altogether.

The Multiplication Principle

The **Multiplication Principle** follows from the preceding examples, which were visualized in stages of two or more selections each.

> ### Multiplication Principle
>
> If a counting problem can be divided into two stages such that at stage 1 there are n_1 ways to make a selection and at stage 2 there are n_2 ways, then the total number of ways to make the two selections is
>
> $$n_1 \cdot n_2$$
>
> Similarly, if the problem requires a third stage in which there are n_3 ways to make the third selection, then the total number of ways to make the three selections is
>
> $$n_1 \cdot n_2 \cdot n_3$$
>
> In general, if there are m stages and at the ith stage there are n_i ways to make a selection, then the total number of ways to make the m selections is
>
> $$n_1 \cdot n_2 \cdot n_3 \cdots n_m$$

Example 2 applies the Multiplication Principle to license plates in standard format (three letters followed by three digits without regard to restrictions.)

EXAMPLE 2

Problem

How many license plates are in standard format?

Solution A license plate in standard format can be viewed as a tree with six stages: three letters (L) and three digits (D). There are 26 possible letters for each of the first three stages and ten possible digits for each of the last three stages. The trimmed tree is

$$\underline{\quad 26 \quad}_{L} \underline{\quad 26 \quad}_{L} \underline{\quad 26 \quad}_{L} \underline{\quad 10 \quad}_{D} \underline{\quad 10 \quad}_{D} \underline{\quad 10 \quad}_{D}$$

Thus there are 17,576,000 license plates in standard format.

In 1982 California had more than 18 million licensed vehicles. Example 2 shows that they could not all have license plates in standard format. So California added the reverse format (three digits followed by three letters). By the Multiplication Principle the reverse format also covers 17,576,000 license plates, allowing over 35 million license plates in these two formats. Why, then, did California adopt yet a third format? The answer lies in this chapter's Case Study.

The results of Example 3 will be needed repeatedly in Chapters 7 and 8.

EXAMPLE 3

Problem

(a) If a coin is flipped two times, in how many ways can it land? (b) If a coin is flipped three times, in how many ways can it land? (c) If a coin is flipped ten times, in how many ways can it land?

Solution (a) There are two choices for the first flip of the coin, heads (H) or tails (T). There are also two choices for the second flip. By the Multiplication Principle there are $2 \cdot 2 = 4$ ways that the coin can land. These ways are listed in the following tree diagram:

1st flip	2nd flip	Possibilities
H	H	HH
	T	HT
T	H	TH
	T	TT

(b) By the Multiplication Principle there are $2 \cdot 2 \cdot 2 = 2^3$ ways the coin can land when it is flipped three times. Thus there are eight ways altogether.
(c) By the Multiplication Principle there are $2^{10} = 1024$ ways the coin can land when it is flipped ten times. Notice that the exponent of 2 is equal to the number of times the coin is flipped.

Epilogue: The Lottery Fix Revisited

We can now answer the questions posed in the introduction to this section. When the number 666 was drawn there should not have been any alarm that the lottery was fixed because 666 is simply one of the 1000 possibilities.

At first, state officials vehemently denied that foul play could have been associated with the drawing. They had made large payoffs before and there are certain numbers that people choose more often than others. The payoff for 666 was extremely large but it could be attributed to its unusual character.

Within a few days a number of investigative reporters were on the scent. They discovered that all of the numbers consisting of 4's and 6's were bet heavily under very similar wagering patterns. How many such numbers are there? By the Multiplication Principle there are three stages with two selections (4 or 6) at each stage. So there are $2 \cdot 2 \cdot 2 = 8$ ways of arranging the two numbers in a three-digit number.

The gamblers were confident that one of these eight numbers would be drawn. They placed heavy bets on all eight numbers. This provided the authorities with good evidence of a conspiracy but it was still not known how they pulled it off.

Pennsylvania lost $1 million. Did state authorities recognize immediately that the fix was on? Are you kidding? Of course not! In fact State Revenue Secretary Howard Cohen, whose department was responsible for running the lottery, stated emphatically for months that it was "impossible" to rig the lottery. Six months later when the facts were uncovered, he was forced to resign.

Can you guess how the lottery was rigged? In each urn they weighted those balls numbered 4 and 6, ensuring that the tubes that pull one ball from each urn would pull a 4 or a 6. The nonweighted balls continued to whirl around, thus giving the appearance that the lottery was fair.

In 1982 two men were sentenced to prison terms of up to 7 years for their roles in the lottery fix. Ironically, the day on which they went to prison Delaware's daily number came up 555. There was no unusually heavy betting on it, however, and no investigation was begun.

EXERCISE SET 6—3

1. A restaurant offers a breakfast menu with 2 juices (orange, apple), 4 eggs (scrambled, poached, soft-boiled, sunny), and 2 drinks (coffee, tea). Draw a tree diagram to list the possibilities. How many outcomes are there?

2. A study classifies human beings according to sex (male, female), height (tall, medium, short), and weight (heavy, normal, light). Draw a tree diagram to list the possibilities. How many outcomes are there?

3. Suppose that one coin is selected from each of two pockets, with each pocket containing a dime and a penny. Draw a tree diagram to list the possible ways of selecting the coins. How many selections are there?

4. Toss two dice. (This is equivalent to tossing one die two times.) Draw a tree diagram to list the possible outcomes. How many outcomes are there?

5. A competitor in a triathlon must choose from one of 3 swim suits (Jantzen, Laguna, Catalina), 2 bicycles (Schwinn, Raleigh), and 4 running shoes (New Balance, Nike, Adidas, Puma). Draw a tree diagram to list the possible choices. How many choices are there?

6. A coin is flipped and then a die is tossed. Draw a tree diagram to list the possible outcomes. How many outcomes are there?

7. Suppose that a family has five people in it. In how many ways is it possible for their birthdays to be in five different months?

8. The computer test bank for this chapter contains 12 problems on Section 6–1, 15 on Section 6–2, 20 each on Sections 6–3 and 6–4, and 10 on Section 6–5. If a test has one problem from each section, how many different tests are possible?

9. A slot machine has the 12 zodiac signs on one wheel, 5 fruit on another wheel, and the 7 days of the week on the third. How many groups of three figures can occur?

10. A menu in a Chinese restaurant lists ten dinners in column A and seven dinners in column B. How many selections can be made with one from column A and one from column B?

11. An urn contains 26 balls, each with a letter of the alphabet printed on it. How many three-letter "words" can be formed by selecting three balls without replacement from the urn?

12. A daily "word" is selected by drawing 1 ball from each of three urns. Each urn contains 26 balls with the letters of the alphabet printed on them. How many three-letter "words" can be formed?

13. License plates in Hawaii consist of two letters followed by three digits. How many such license plates are possible?

14. License plates in Louisiana consist of three digits, followed by one letter, followed by three digits. How many such license plates are possible?

15. A combination lock has a dial with ten digits. Three numbers are chosen to form the combination that opens the lock. (For example, the combination 3-5-3 means to turn the dial to the right to 3, then to the left to 5, and then to the right to 3.) How many combinations can be formed if the digits can be repeated?

16. In most locks whose combination has three digits the first and second digits cannot be the same but the second and third can. Given this restriction, how many combinations are possible if the dial has ten digits?

17. If the dial of a lock has 30 numbers and repetition of the digits is allowed, how many combinations with 3 numbers are possible?

18. If the dial of a lock has 30 numbers, how many combinations with 3 numbers are possible if the first and second numbers cannot be the same but the second and third can?

19. A bag contains numerous $1 and $5 bills. You reach in and grab one bill at a time. You stop whenever you grab a $5 bill or whenever three bills have been grabbed. Draw a tree diagram to list the possible grabs. Write the amount of money grabbed beside each possibility.

20. The Miller's Lite teams (T = tastes great, L = less filling) decide to end their feud with a series in which the first team to win two games is the winner. Draw a tree diagram to list the possible series outcomes. How many possible outcomes are there?

21. If three dice are tossed, in how many ways can they land? (See Problem 4.)

22. If four dice are tossed, in how many ways can they land?

23. If ten dice are tossed, in how many ways can they land?

24. If n dice are tossed, in how many ways can they land?

In Problems 25 to 28 at each intersection the driver of the car can make either a left-hand turn or a right-hand turn or go straight ahead.

25. How many different routes are possible if two intersections are encountered?

26. How many different routes are possible if three intersections are encountered?

27. How many different routes are possible if ten intersections are encountered?

28. How many different routes are possible if n intersections are encountered?

29. License plates in several states consist of 6 characters, where a character is one of the 26 letters or one of the 10 digits. How many such license plates are possible?

30. License plates in Washington, D.C., consist of six digits. How many such license plates are possible if the first digit cannot be 0 and all six digits cannot be the same?

31. How many seven-digit telephone numbers can be formed if the first digit can be neither 0 nor 1?

32. How many seven-digit telephone numbers can be formed if the first three digits cannot be 800?

33. In how many ways can the letters of the name JACK be arranged?

34. In how many ways can the letters of the name VINCE be arranged?

35. Make a list of the different ways in which the letters of the name JILL can be arranged. How many ways are there?

36. Make a list of the different ways in which the letters of the name ANITA can be arranged. How many ways are there?

37. A contest consists of pairs of cards. Each pair is formed by selecting one card with a letter *A*, *B*, *C*, or *D* on it, and the other card with a digit 6, 7, 8, or 9 on it. How many pairs of cards are there in the contest?

38. A contest has five cards with the letters *A*, *B*, *C*, *D*, *E* and five cards with the digits 5, 6, 7, 8, 9. The object is to match one letter with one digit. How many different matchings are there altogether?

39. How many different arrangements of four characters are there in which the first two characters are letters and the last two characters are digits?

40. How many different arrangements of four characters are there in which the first two characters are distinct letters and the last two characters are distinct digits?

41. Suppose that a hand consists of 2 cards drawn (without replacement) from an ordinary deck of 52 cards. How many hands can be drawn?

42. Suppose that a hand consists of 3 cards drawn (without replacement) from an ordinary deck of 52 cards. How many hands can be drawn?

43. A bag contains numerous $1 and $5 bills. You reach in and grab one bill at a time. You stop whenever you grab at least $6 or whenever three bills have been grabbed. Draw a tree diagram to list the possible grabs.

44. A bag contains numerous $1 and $5 bills. You reach in and grab one bill at a time until you have grabbed at least $5. Draw a tree diagram to list the possible grabs.

45. There are 20 students in a class and each student must receive a grade of A, B, C, D, or F. How many different ways are there for the grades to be assigned?

46. A developer is building a development with three models of homes: Cape Cod, split-level, and colonial. All models come with a one-car or two-car garage. The Cape Cods come with one or two bedrooms, the split-level with two, three, or four bedrooms, and the colonial with three, four, or five bedrooms. If the builder wants to construct a sample house of each possible type, how many houses must be constructed?

47. Suppose that a primary ballot lists the names of four candidates for governor, three for lieutenant governor, and six candidates for state treasurer. How many different ways can the offices be filled if one person is chosen to run for each position?

48. How many license plates consist of two letters followed by two digits, if the first letter cannot be *I* or *O* and the second letter cannot be *I*, *O*, *Q*, or *Z*?

49. How many license plates consist of three letters followed by two digits, if the third letter cannot be *I*, *O*, *Q*, or *Z* and the first digit cannot be 1 or 0?

50. A man has 5 shirts, 4 pairs of pants, and 3 pairs of shoes. How many different outfits can he wear to work if an outfit consists of a shirt, a pair of pants, and a pair of shoes?

51. How many arrangements of four letters begin with one consonant, end with a different consonant, and have two distinct vowels in between? [The vowels are *a*, *e*, *i*, *o*, *u*. All other letters are consonants.]

52. How many different ways of choosing three initials are there that have exactly two initials alike?

53. How many different ways of choosing three initials are there that have two or three initials alike?

Referenced Exercise Set 6–3

1. Over 1000 bombings take place in the United States each year. It is now possible for federal authorities to trace the explosives used in the bombs by identifying microscopic chips of layered paints called ''taggants.''[†] A typical identification taggant consists of ten different colored layers. How many different taggants can be made from ten different colors?

2. Maalox Plus is an antacid that is packaged with six tablets on a cardboard card.[‡] Each tablet is white on one side and yellow on the other. Two possible arrangements on a card are

 and

†Richard D. Lyons, ''Tagging Bombs, Trapping Bombers,'' *New York Times*, June 24, 1979.

‡Bonnie Litwiller and David Duncan, ''Maalox Lottery: A Novel Probability Problem,'' *The Mathematics Teacher*, September 1987, pp. 455–456.

How many different arrangements are possible on a card?

3. The IBM disk operating system for the PC-AT microcomputer makes use of *directories* to organize computer programs (called *files*) into groupings.§ Draw a tree diagram which has one root directory that leads to four subdirectories (WP, DATA, SPREAD, and UTILITIES), where WP has four files (WordPerfect, Multimate, Microsoft Word, Word Star), DATA has three files (dBase, DataEase, Oracle), SPREAD has three files (Lotus, Excel, Quattro), and UTILITIES has two files (Norton, PC Tools).

4. Arthur Geoffrion introduced the concept of "structured modeling" in business studies because of his belief that advances in the discipline of modeling paled in comparison with the disciplines of analyzing and solving models once they had been brought into being. Figure 3, which refers to a Feedmix Model, is taken from that study.‖ How many possibilities does the tree diagram represent if each possibility consists of Feedmix with one or more other ingredients?

5. Max Wertheimer, the founder of Gestalt psychology, proposed the following brain-teaser to his friend, Albert Einstein.# Answer the question using the Multiplication Principle.

An amoeba propagates by simple division; it takes 3 minutes for each split. I put such an amoeba into a glass container with a nutrient fluid; the rate of multiplication rises, of course; it takes 1 hour until the vessel is full of amoebas. Question: How long would it take to fill the vessel likewise if I start not with one, but with two amoebas?

Cumulative Exercise Set 6–3

1. List the elements in the set $S = \{x \in N : x$ is even and $x < 10\}$.

2. Find X' for each choice of U.
   (a) U is the set of days in a week and X consists of those days that begin with the letter T.
   (b) $U = Z$ (the integers) and $X = \{x \in Z : x \leq 0\}$.

3. Let the universal set U consist of the months of the year. Let A be the set of months having days of summer and B be the set of months having exactly 30 days.
   (a) List the elements in U.
   (b) List the elements in A.
   (c) Find B'.
   (d) Express the set B' in words.
   (e) Find $A \cap B$.
   (f) Find $A \cup B$.

4. Find $n(A \cup B)$ where
   (a) $n(A) = 3$, $n(B) = 4$, and A and B are disjoint sets.
   (b) $n(A) = 3$, $n(B) = 4$, and $n(A \cap B) = 2$.

§IBM, *Reference Manual: Disk Operating System Version 3.20*, Boca Raton, FL, 1986.

‖See page 558 of Geoffrion's article "An Introduction to Structured Modeling" in the May 1987 issue of *Management Science*.

#Edith H. and Abraham S. Luchins, "The Einstein-Wertheimer Correspondence on Geometric Proofs and Mathematical Puzzles," *The Mathematical Intelligencer*, Vol. 12, No. 2, 1990, pp. 35–43. See the English translation of the postscript to the letter on page 40.

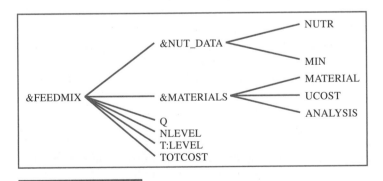

FIGURE 3

(c) $n(A' \cap B) = 3$, $n(A \cap B') = 4$, and $n(A \cap B) = 2$.

5. If $n(A) = 8$, $n(B) = 4$, and $n(A \cap B) = 4$, how are the sets A and B related?

6. A club has 40 members, of whom 22 are married, 5 are married women, and 8 are unmarried men. How many women are unmarried?

7. A ballot has four candidates for president and 2 candidates for secretary. In how many different ways can the two offices be filled?

8. In how many ways can the grades of A, B, C, D, or F be assigned to a class of six students?

9. Of the 110 students in class, 60 are freshmen, 7 are veterans, 65 are enrolled in business, 2 are freshmen who are veterans, 5 are veterans who are enrolled in business, 32 are freshmen who are enrolled in business, and 2 are freshmen who are veterans and enrolled in business.
   (a) How many are freshmen and veterans but are not enrolled in business?
   (b) How many are freshmen but neither veterans nor enrolled in business?
   (c) How many are neither freshmen nor veterans nor enrolled in business?

10. Draw a tree diagram to list the possible outcomes when a die is tossed and then two coins are flipped.

11. Define a set of initials to consist of exactly three letters.
   (a) How many sets of initials are there?
   (b) Is the outcome JFK different from the outcome FJK?
   (c) How many sets of initials have all three letters different?

12. If a penny, nickel, dime, and quarter are flipped, in how many ways can they land so that
   (a) three of the coins land with heads facing up?
   (b) at least one coin lands with heads facing up?

13. A builder can build homes with two, three, or four bedrooms, with four possible exteriors, with three different floor plans, and with or without a basement. If the builder would like to construct a sample house of each possible type, how many houses must be constructed?

6—4 Permutations and Combinations

Many useful counting techniques involve permutations and combinations. This section explains the meanings of these terms and develops rules for computing them. In Section 6–5 these concepts will be applied to the Binomial Theorem. Chapter 7 contains additional applications to probability problems.

Order

Suppose that a scholarship committee is going to award a total of $1200 to two students and that the committee has narrowed its list to five candidates. Suppose also that the committee is undecided whether to grant two $600 awards or to give one award of $800 and another award of $400. Examples 1 and 2 address these two cases.

EXAMPLE 1

Problem

In how many ways can the scholarship committee select two of the five students so that one student is awarded $800 and the other student $400?

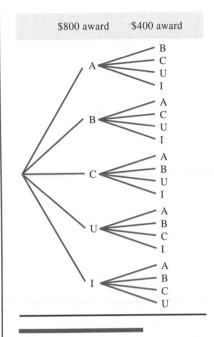

$800 award $400 award

FIGURE 1

Solution Use the Multiplication Principle with two stages. The first stage has five choices for the $800 scholarship while the second stage has four choices for the $400 scholarship. No student can receive both scholarships. Figure 1 shows the $5 \cdot 4 = 20$ ways for students named A, B, C, U, and I. Note the difference between U getting $800 and I getting $400 versus I getting $800 and U getting $400.

EXAMPLE 2

Problem

In how many ways can the scholarship committee select two of the five students so that each one wins $600?

Solution This problem cannot be solved by means of the Multiplication Principle alone. Certain pairs of students in Figure 1 reduce to one possibility here. For instance, the U-I and I-U pairs refer to separate awards, but here the order of the students is not relevant because each award is $600, so the two pairs reduce to the one set $\{U, I\}$. Originally, Figure 1 displayed 20 pairs, with each pair of students listed twice. But when each pair is listed once, there are ten possibilities, as shown by the tree diagram in Figure 2.

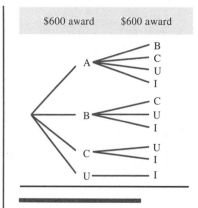

FIGURE 2

The concept of order distinguishes Example 1 from Example 2. In Example 1 it mattered which student came first because one award was $800 and the other was $400. In Example 2 the order of the students was immaterial because both awards were for $600. The following definition distinguishes these two concepts for objects that are selected without replacement:

DEFINITIONS

A list of objects in a specific order is called a **permutation.** A list of objects in which the order is not considered is called a **combination.**

The 20 lists in Example 1 are permutations while the 10 lists in Example 2 are combinations.

Factorials

The expression $n!$ is read "n **factorial.**" It is defined by

$$n! = n(n - 1)(n - 2) \cdots 3 \cdot 2 \cdot 1$$

with $0! = 1$. Thus $5! = 120$ because

$$5! = 5 \cdot 4 \cdot 3 \cdot 2 \cdot 1$$

Many calculators have a key for computing factorials, so that pressing $\boxed{9}\;\boxed{n!}$ yields 362,880. Factorials play a central role in permutations and combinations, so it is important to be able to evaluate them in various forms. For instance, $(7 - 2)! = 5! = 120$. Example 3 shows how to evaluate two other forms.

EXAMPLE 3

Problem

Evaluate (a) $\dfrac{8!}{(8-5)!}$ (b) $\dfrac{8!}{(8-5)!\cdot 5!}$

Solution Expand each factorial and cancel the common factors.

(a) $\dfrac{8!}{(8-5)!} = \dfrac{8!}{3!} = \dfrac{8\cdot7\cdot6\cdot5\cdot4\cdot3\cdot2\cdot1}{3\cdot2\cdot1} = 8\cdot7\cdot6\cdot5\cdot4 = 6720$

(b) $\dfrac{8!}{(8-5)!\cdot 5!} = \dfrac{8!}{3!\cdot5!} = \dfrac{8\cdot7\cdot6\cdot5\cdot4\cdot3\cdot2\cdot1}{3\cdot2\cdot1\cdot5\cdot4\cdot3\cdot2\cdot1} = \dfrac{8\cdot7\cdot6}{3\cdot2\cdot1} = 56$

Formulas for Permutations and Combinations

The notation $P(n, r)$ denotes the number of permutations of n objects taken r at a time. Thus $P(10, 4)$ is the number of permutations of ten objects taken four at a time. By the Multiplication Principle, $P(10, 4) = 10\cdot9\cdot8\cdot7$. This number can be written in terms of factorials.

$$P(10, 4) = 10\cdot9\cdot8\cdot7 = \dfrac{10!}{6!} = \dfrac{10!}{(10-4)!}$$

The final form is a specific case of the general formula for evaluating permutations. The derivation is similar.

$$P(n, r) = n\cdot(n-1)\cdots(n-r+1) = \dfrac{n!}{(n-r)!}$$

The number of ways of permuting n distinct objects taken r at a time is

$$P(n, r) = \dfrac{n!}{(n-r)!}$$

EXAMPLE 4

Problem

How many permutations of three letters can be selected from the alphabet?

Solution Since the alphabet has 26 letters, $n = 26$. Here $r = 3$. The number of permutations is

$$P(26, 3) = \dfrac{26!}{(26-3)!} = \dfrac{26!}{23!} = 26\cdot25\cdot24 = 15{,}600$$

A particular case of the formula for $P(n, r)$ occurs when $n = r$. Then

$$P(n, n) = \frac{n!}{(n - n)!} = \frac{n!}{0!} = \frac{n!}{1} = n!$$

Therefore the number of permutations of n objects is $n!$. Example 5 uses this fact. It also provides the reasoning for the way in which the formula for combinations is derived from the formula for permutations.

EXAMPLE 5

Problem

(a) How many governing boards (president, vice-president, and secretary) can be formed from a law firm with eight partners?
(b) How many three-member investigative committees can be formed from a law firm with eight partners?

Solution (a) The order in which partners are selected to the governing board is important because the three positions are distinguishable. Therefore the problem deals with permutations. There are eight partners so $n = 8$. There are three officers so $r = 3$. The number of governing boards is

$$P(8, 3) = \frac{8!}{(8 - 3)!} = 8 \cdot 7 \cdot 6 = 336$$

Thus 336 different governing boards can be formed.

(b) In forming three-member committees the order of selection is not important because the positions are not distinguishable. The problem reduces to finding the number of combinations of eight partners taken three at a time. We derive this number from $P(8, 3)$.

Consider a particular committee consisting of Adams (A), Buren (B), and Carter (C). The committee can be regarded as the set $\{A, B, C\}$. By the Multiplication Principle there are six different arrangements (permutations) of the partners in the set $\{A, B, C\}$.

A-B-C A-C-B B-A-C B-C-A C-A-B C-B-A

They are distinct arrangements; for instance, the arrangement A-B-C (Adams is president, Buren vice-president, and Carter secretary) is different from A-C-B (Adams is president, Carter vice-president, and Buren secretary). But as a three-member committee in which each member has equal standing, they reduce to the one set $\{A, B, C\}$ because, for instance, $\{A, B, C\} = \{A, C, B\}$. It follows that for every six permutations there is one combination, so the number of combinations is one-sixth the number of permutations.

$$\frac{P(8, 3)}{6} = \frac{336}{6} = 56$$

Altogether 56 committees of three members can be formed.

The notation $C(n, r)$ denotes the number of combinations of n objects taken r at a time. Another commonly used notation is $\binom{n}{r}$. Part (b) of Example 5 showed that $C(8, 3)$ was derived as follows:

$$C(8, 3) = \frac{P(8, 3)}{6} = \frac{P(8, 3)}{3!} = \frac{8!}{(8 - 3)! \cdot 3!}$$

To evaluate $C(n, r)$ for any n and r (where $r \leq n$) means to count the number of subsets containing r elements of a set with n elements. There are $P(n, r)$ arrangements that consist of r elements. However, because the order of the elements in a subset does not matter, each subset of r elements corresponds to $r!$ arrangements. Therefore the number of subsets is $P(n, r)$ divided by $r!$. This can be expressed in a more useful form as follows:

$$C(n, r) = \frac{P(n, r)}{r!} = \frac{P(n, r)}{1} \cdot \frac{1}{r!} = \frac{n!}{(n - r)!} \cdot \frac{1}{r!}$$

The number of combinations of n distinct objects taken r at a time is

$$C(n, r) = \frac{n!}{(n - r)! \cdot r!}$$

It is important to be able to distinguish between permutations and combinations because in many problems the essential ingredient is the proper identification of the concept being used. Keep in mind that a permutation is an "arrangement" that implies order while a combination is a "collection" without regard to order.

Example 6 applies the formulas for permutations and combinations to wagering. A "trifecta" refers to three objects that finish first, second, and third in a prescribed order. A "boxed trifecta" refers to three objects that finish first, second, and third in any order.

EXAMPLE 6

Problem

How many trifecta and boxed trifecta tickets are possible in a horse race involving ten horses?

Solution Since there are ten horses and three winners, $n = 10$ and $r = 3$. Trifecta tickets are permutations because the order must be exact. There are $P(10, 3) = 720$ trifecta tickets.

Boxed trifecta tickets are combinations because only the names of the three winners must be listed, not their order of finish. Thus there are $C(10, 3) = 120$ boxed trifecta tickets.

	Red Cards		Black Cards	
	Hearts	Diamonds	Spades	Clubs
Ace	A♥	A♦	A♠	A♣
King	K♥	K♦	K♠	K♣
Queen	Q♥	Q♦	Q♠	Q♣
Jack	J♥	J♦	J♠	J♣
10	10♥	10♦	10♠	10♣
9	9♥	9♦	9♠	9♣
8	8♥	8♦	8♠	8♣
7	7♥	7♦	7♠	7♣
6	6♥	6♦	6♠	6♣
5	5♥	5♦	5♠	5♣
4	4♥	4♦	4♠	4♣
3	3♥	3♦	3♠	3♣
2	2♥	2♦	2♠	2♣

FIGURE 3

Example 7 combines the three counting techniques that were studied in the last two sections: the Multiplication Principle, permutations, and combinations. Since many problems deal with decks of ordinary cards, it might be helpful to refer to Figure 3, which indicates three ways to partition the 52 cards. The top row of Figure 3 shows the partition into 26 red cards and 26 black cards. The second row shows the partition into four suits (hearts, diamonds, spades, and clubs). The first column shows the partition into 13 types of cards. There are 12 picture cards: 4 kings, 4 queens, and 4 jacks.

EXAMPLE 7

A poker hand consists of 5 cards drawn from an ordinary deck of 52 cards.

Problem

How many poker hands consist of 3 aces and 2 kings?

Solution The procedure will be to find the number of ways of selecting 3 aces, then to find the number of ways of selecting 2 kings. Each desired hand consists of 3 aces *and* 2 kings. By the Multiplication Principle the total number of such hands is obtained by multiplying these two numbers.

There are 4 aces, so there are $C(4, 3) = 4$ ways to select 3 aces from them. Similarly, there are $C(4, 2) = 6$ ways to select 2 kings from the 4 kings. By the Multiplication Principle there are $6 \cdot 4 = 24$ hands consisting of 3 aces and 2 kings.

EXERCISE SET 6–4

In Problems 1 to 8 evaluate the given expression.

1. (a) $P(5, 2)$
   (b) $C(5, 2)$

2. (a) $P(5, 3)$
   (b) $C(5, 3)$

3. (a) $P(6, 2)$
   (b) $C(10, 2)$

4. (a) $P(10, 5)$
   (b) $C(10, 7)$

5. (a) $P(5, 0)$
   (b) $C(5, 0)$

6. (a) $P(5, 5)$
   (b) $C(5, 5)$

7. (a) $0!$
   (b) $(8 - 3)!$

8. (a) $1!$
   (b) $52!/(50! \cdot 2!)$

In Problems 9 to 26 determine whether the solution involves the Multiplication Principle, permutations, or combinations, then use the appropriate formula to answer the question.

9. In how many ways can a club with ten members elect a president and a vice-president?

10. In how many ways can four people line up in a bakery?

11. If the call letters of a radio station must begin with the letter W, how many different stations could be designated by using four letters, such as WIND, without using repetitions?

12. The Coast Guard uses flags raised on a flagpole to signal ships concerning matters such as weather conditions. If three different flags are available, how many signals can be sent using all three flags if the order of the arrangement is important?

13. The number of signals available in Problem 12 can be increased by using signals of two or three flags. How many such signals are possible if the order of the flags is important?

14. In how many ways can ten volumes of a set of encyclopedias be arranged on a shelf?

15. In how many ways can the gold, silver, and bronze medals be awarded if ten women compete in a marathon?

16. In how many ways can four students be given assigned seats in a classroom with ten seats?

17. An agricultural expert has designed an experiment to test the effects of two types of insecticide and four types of fertilizer on six different types of crops. Each plot of land uses one insecticide, one fertilizer, and one crop. How many plots are needed?

18. A veterinarian wishes to test three different levels of dosage of a vaccine on four types of animals at two different age levels of the animals. How many classes of vaccinations are possible?

19. There are 20 candidates for an all-star basketball game and 5 will be selected for the first team. How many different first teams can be chosen?

20. Twelve people work as programmers in a computer center, with teams of three serving as consultants. In how many ways can teams be chosen?

21. Seven astronauts have reached the final training level for a particular space flight. (a) How many two-person crews can be chosen from the seven? (b) How many two-person crews will have one person designated captain and the other designated copilot?

22. How many 2-card hands can be dealt from an ordinary deck of 52 cards?

23. How many poker hands consist of three queens and two jacks?

24. How many poker hands consist of four queens and one jack?

25. How many poker hands consist of two queens, two jacks, and one king?

26. How many poker hands consist of one queen, three jacks, and one king?

27. Evaluate the expression

$$C(6, 6) + C(6, 5) + C(6, 4) + C(6, 3)$$
$$+ C(6, 2) + C(6, 1) + C(6, 0)$$

28. Evaluate the expression

$$C(4, 4) + C(4, 3) + C(4, 2) + C(4, 1) + C(4, 0)$$

29. A basketball team has 5 starters selected from the 12 players. (a) How many starting teams are possible? (b) How many starting teams are possible if two of the players' contracts specify that they must start every game?

30. (a) How many poker hands can be drawn? (b) How many poker hands contain all four aces?

31. In how many ways can three initials be chosen so that two initials are the same but not all three?

32. If a penny, nickel, and dime are tossed in a fountain, in how many ways can they land so that two of them are heads?

33. How many poker hands consist of three red cards and two black cards?

34. How many poker hands consist of three clubs and two spades?

35. How many three-digit numbers can be chosen if repetition is allowed? (For example, 989 is permitted.)

36. How many three-digit numbers can be chosen if no repetitions are permitted?

37. How many four-digit numbers can be chosen if repetition is allowed? (For example, 1989 is permitted.)

38. How many four-digit numbers can be chosen if no repetitions are permitted?

39. How many boxed three-digit numbers can be formed?

40. How many three-digit numbers have exactly two of the digits the same?

41. In how many ways can four walls be painted if four colors are available and (a) repetition is allowed, (b) repetition is not allowed?

42. In how many ways can a sorority with 20 members elect (a) a president and vice-president, (b) a three-member committee?

43. There are six finalists for two awards. In how many ways can the awards be given if (a) the awards are equal and any finalist can win two, one, or no award, (b) the awards are equal but no finalist can win more than one award, (c) one award is twice as large as the other and no finalist can win more than one award?

44. A pharmaceutical salesperson wants to hand out $1000 worth of samples to ten doctors. In how many ways can the samples be given if (a) all ten get an equal amount worth $100, (b) five doctors are chosen at random and are given equal amounts worth $200, (c) two doctors are chosen at random and one is given $600 worth of samples and the other is given $400 worth?

45. In wagering, an "exacta" refers to two objects that finish first and second in a specified order. If a dog race contains nine greyhounds, how many exacta tickets can be purchased?

46. Many harness races consist of exactly nine horses. (a) How many different trifecta tickets can be purchased? (b) How many different boxed trifecta tickets can be purchased?

47. (a) Why is it true that $C(14, 5) = C(14, 9)$?
(b) Why is it true that $C(n, r) = C(n, n - r)$?

48. (Pigs-get-fat-hogs-get-slaughtered department) It has been conjectured that the Pennsylvania lottery scam team could have milked the state for much more money by placing about one-half the number of bets and by repeating the procedure often. Why?

Referenced Exercise Set 6–4

1. Campanology is the art of ringing bells.* Since modern bells often weigh more than a ton, it takes about 2 seconds to advance or retard each bell. Consequently, a "ring" of bells is an orderly sequence of permutations on the bells instead of a melody. An *extent* is the maximum number of rings so that the extent of six bells (called a Minor) is 6! = 720. What is the extent of a Major, which consists of eight bells? (Plain Bob Major, which was played in 1963 and 1977, required over 18 hours to ring!)

2. In music theory, chords and lines are basic terms.† A chord refers to a subset of the white keys on a piano, while a line is a sequence of distinct white keys. (a) Which concept refers to a permutation? (b) Which concept refers to a combination?

Problems 3 and 4 require a knowledge of the programming language BASIC. They are based on an article that describes how computer programs can be used to introduce permutations and combinations.‡

3. What is the output of this program?

```
10 FOR I = 1 TO 4
20     FOR J = 1 TO 4
30         IF I ≥ J THEN 50
40         PRINT I;J
50     NEXT J
60 NEXT I
```

4. What is the output of the program in Problem 3 if the inequality sign in line 30 is replaced by =?

Cumulative Exercise Set 6–4

1. Determine whether B is a subset of A.
(a) $A = \{1, 2, 3, \ldots, 15\}$ $B = \{2, 3, 4\}$
(b) $A = \{a, b, c, d, e\}$ $B = \{1, a, b, c, d\}$

*Arthur T. White, "Ringing the Cosets," *American Mathematical Monthly*, October 1987, pp. 721–746.

†John Clough and Gerald Myerson, "Musical Scales and the Generalized Circle of Fifths," *American Mathematical Monthly*, November 1986, pp. 695–701.

‡Clark Kimberling, "Permutations and Combinations," *The Mathematics Teacher*, May 1987, pp. 403–407.

2. Find $n(A \cup B)$ when $n(A) = 51$, $n(B) = 35$, and $n(A \cap B) = 15$.

3. A class has 32 students, 12 of whom are seniors and 18 of whom are business majors. If 7 of the seniors are business majors, how many students who are not business majors are also not seniors?

4. A poll of 100 voters revealed that 36 will vote for Proposition 1, 42 will vote for Proposition 2, 33 will vote for Proposition 3, 21 will vote for Propositions 1 and 2, 15 will vote for Propositions 1 and 3, 16 will vote for Propositions 2 and 3, and 7 will vote for all three propositions. How many will vote for at least one proposition?

5. A coin is tossed and heads or tails is recorded and then a die is tossed and the number on the top face is recorded. Draw a tree diagram and list the possible outcomes.

6. A person must make selections in three categories for an insurance policy. The person must choose from three liability options (a $10,000, $50,000, or $100,000 maximum), two uninsured motorist options (yes or no), and four deductible options ($50, $100, $200, or $500). How many different policies are there with these options?

7. Three dice are tossed. How many ways can they land if you record (a) the sum of the top faces, (b) the numbers on each of the top faces (for example, one

such outcome would be 1-1-1 and another would be 2-4-6).

8. How many three-digit numbers have exactly two digits the same?

9. How many seven-digit telephone numbers can be formed if the first digit must be 2 or 3 and the last digit must be even?

10. How many different arrangements of four characters are there if the characters are chosen from the set $\{a, b, c, d, e\}$ and (a) repetition is allowed, (b) repetition is not allowed?

11. In how many ways can a club with 12 members elect (a) a president, a vice-president, and a secretary; (b) a three-member committee?

12. There are ten finalists for two awards. In how many ways can the awards be given if (a) the awards are equal and any finalist can win two, one, or no award, (b) the awards are equal but no finalist can win more than one award, (c) one award is twice as large as the other and no finalist can win more than one award?

13. A company has $500 to award in bonuses to a department of ten people. In how many ways can the bonuses be given if (a) all ten get an equal award of $50, (b) five people are chosen at random and are given equal awards of $100, (c) two people are chosen at random, one is given $400 and the other is given $100, while the remaining eight people get nothing?

6–5 The Binomial Theorem

The expression $(x + y)^n$ is often seen in mathematics in various forms. Sometimes x and y are replaced by numbers or different letters, as in the expression $(2x + 5)^3$. They are especially important in the study of probability in the next chapter. When these expressions are multiplied out, they are called *expansions*. For example, the first few expansions are

$(x + y)^1 = x + y$

$(x + y)^2 = x^2 + 2xy + y^2$

$(x + y)^3 = x^3 + 3x^2y + 3xy^2 + y^3$

$(x + y)^4 = x^4 + 4x^3y + 6x^2y^2 + 4xy^3 + y^3$

Inspection of these expressions shows that they follow a pattern. Can you guess what the expansion of $(x + y)^5$ looks like?

Binomial Coefficients

Consider the expression $(x + y)^6$. It is called a "binomial form" because it contains two variables. The expression can be expanded by multiplying $(x + y)$ by itself six times.

$$(x + y)^6 = (x + y)(x + y)(x + y)(x + y)(x + y)(x + y)$$

This leads to the equality

$$(x + y)^6 = x^6 + \_x^5y + \_x^4y^2 + \_x^3y^3 + \_x^2y^4 + \_xy^5 + y^6$$

The underlined spaces represent coefficients called **binomial coefficients,** which must be filled in with numbers.

Instead of multiplying the six factors one at a time to determine the binomial coefficients, it is instructive to compute one of them, say, $\_x^4y^2$. The term x^4y^2 arises every time the $(x + y)$ terms are multiplied to produce four x's and two y's. One such way is

$$(\underline{x} + y)(\underline{x} + y)(\underline{x} + y)(\underline{x} + y)(x + \underline{y})(x + \underline{y})$$

The appropriate variables have been underlined. Another way is

$$(\underline{x} + y)(\underline{x} + y)(\underline{x} + y)(x + \underline{y})(\underline{x} + y)(x + \underline{y})$$

The total number of ways to obtain four x's and two y's is equal to the number of ways of selecting four objects from a set of six objects. This is $C(6, 4)$. We use the conventional notation $\binom{6}{4}$ to denote this binomial coefficient. Therefore the appropriate term is $\binom{6}{4}x^4y^2$. Similarly, the next-to-last term in the expansion is $\binom{6}{1}x^1y^5$. The entire expansion can be written in the form

$$(x + y)^6 = \binom{6}{6}x^6 + \binom{6}{5}x^5y + \binom{6}{4}x^4y^2$$
$$+ \binom{6}{3}x^3y^3 + \binom{6}{2}x^2y^4 + \binom{6}{1}xy^5 + \binom{6}{0}y^6$$

Therefore

$$(x + y)^6 = x^6 + 6x^5y + 15x^4y^2 + 20x^3y^3 + 15x^2y^4 + 6xy^5 + y^6$$

The reasoning used to derive the binomial coefficients $\binom{n}{r}$ will be used again when we discuss binomial trials in the chapters on probability and statistics. For now we state the general result, called the **Binomial Theorem.**

Binomial Theorem

$$(x + y)^n = \binom{n}{n} x^n + \binom{n}{n-1} x^{n-1}y + \binom{n}{n-2} x^{n-2}y^2$$

$$+ \cdots + \binom{n}{2} x^2 y^{n-2} + \binom{n}{1} xy^{n-1} + \binom{n}{0} y^n$$

EXAMPLE 1

Problem

What is the coefficient of the x^2y^6 term in the expansion of $(x + y)^8$?

Solution The coefficient is $\binom{8}{2}$. The top number 8 comes from the exponent of $(x + y)^8$. The bottom number 2 comes from the exponent of x in x^2y^6. Since $\binom{8}{2} = \dfrac{8!}{2! \cdot 6!} = \dfrac{8 \cdot 7}{2} = 28$, the coefficient is 28.

The next example is central to the proof that a set with n elements has 2^n subsets. (See Exercise 28.)

EXAMPLE 2

Problem

Evaluate

$$\binom{5}{5} + \binom{5}{4} + \binom{5}{3} + \binom{5}{2} + \binom{5}{1} + \binom{5}{0}$$

Solution One approach is to evaluate each binomial coefficient and then add them. Alternately, by the binomial theorem

$$(x + y)^5 = \binom{5}{5} x^5 + \binom{5}{4} x^4 y + \binom{5}{3} x^3 y^2 + \binom{5}{2} x^2 y^3 + \binom{5}{1} xy^4 + \binom{5}{0} y^5$$

Let $x = 1$ and $y = 1$. Then the right-hand side of the equation is equal to the desired expression. The left-hand side equals $(1 + 1)^5 = 32$. Therefore

$$\binom{5}{5} + \binom{5}{4} + \binom{5}{3} + \binom{5}{2} + \binom{5}{1} + \binom{5}{0} = 32$$

Pascal's Triangle

Another method to compute the binomial coefficients is to use the array of numbers given in Figure 1, called **Pascal's triangle.** If we refer to the first number 1 at the

top of the triangle as the 0th row so that the next row, consisting of two 1's, "1   1"
is the first row, then the nth row in the array gives the coefficients of $(x + y)^n$.
For example, consider the fourth row. Compare the numbers in this row,
"1   4   6   4   1," with the coefficients of the expansion

$$(x + y)^4 = x^4 + 4x^3y + 6x^2y^2 + 4xy^3 + y^4$$

The triangle has an easy pattern: (1) start with 1 at the top of the array, (2) start
and end each row with 1, (3) every other number is the sum of the two numbers
in the previous row directly above it, one to its left and one to its right. For instance,
the 6 in the fourth row is the sum of 3 and 3.

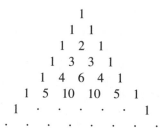

A disadvantage of this method is that in order to find the coefficients of an expansion
the entire triangle must be produced down to the row that corresponds to the desired
expansion.

EXAMPLE 3

Problem

Use Pascal's triangle to compute $(a + b)^5$.

Solution The first term is a^5 and each succeeding term has a raised to one less
power times b to one greater power. The remaining terms contain a^4b, a^3b^2,
a^2b^3, a^4b, and b^5. The coefficients are the numbers in the 5th row of Pascal's
triangle, "1   5   10   10   5   1." Therefore

$$(a + b)^5 = a^5 + 5a^4b + 10a^3b^2 + 10a^2b^3 + 5a^4b + b^5$$

EXERCISE SET 6–5

In Problems 1 to 8 write out the binomial expansion.

1. $(x + y)^5$
2. $(x + y)^7$
3. $(a + b)^4$
4. $(s + t)^4$
5. $(a + b)^6$
6. $(s + t)^7$
7. $(2x + y)^3$
8. $(x + 3y)^5$

In Problems 9 to 12 write the first four terms of the binomial expansion.

9. $(x + y)^9$
10. $(x + y)^{13}$
11. $(a + b)^{10}$
12. $(s + t)^{14}$

In Problems 13 to 16 find the given coefficient.

13. The x^2y^7 coefficient in $(x + y)^9$.
14. The x^3y^6 coefficient in $(x + y)^9$.
15. The x^5y^6 coefficient in $(x + y)^{11}$.
16. The x^4y^7 coefficient in $(x + y)^{11}$.

In Problems 17 to 20 evaluate the expression.

17. $\binom{4}{4} + \binom{4}{3} + \binom{4}{2} + \binom{4}{1} + \binom{4}{0}$

18. $\binom{6}{6} + \binom{6}{5} + \binom{6}{4} + \binom{6}{3} + \binom{6}{2} + \binom{6}{1}$
$+ \binom{6}{0}$

19. $\binom{7}{6} + \binom{7}{5} + \binom{7}{4} + \binom{7}{3} + \binom{7}{2} + \binom{7}{1}$
$+ \binom{7}{0}$

20. $\binom{7}{6} + \binom{7}{5} + \binom{7}{4} + \binom{7}{3} + \binom{7}{2} + \binom{7}{1}$

In Problems 21 to 26 write out the binomial expansion.

21. $(2x - 3y)^4$
22. $(4x - 3y)^5$

23. $(a^2 + b)^4$
24. $(s + 2t^3)^5$

25. $(x^2 + 3)^6$
26. $(5 - 2c^2)^5$

27. Use the Binomial Theorem and Example 2 to prove that a set with five elements has 2^5 subsets, including the empty set and the set itself.

28. Use the Binomial Theorem and Example 2 to prove that a set with n elements has 2^n subsets including the empty set and the set itself.

In Problems 29 to 32 use the result in Problem 28 to determine how many subsets there are of a set with the given number of elements.

29. 6 elements
30. 10 elements

31. 12 elements
32. 20 elements

33. How many different tips could you leave in a restaurant if you had a quarter, a half-dollar, a dollar bill, and five-dollar bill and you used all four denominations?

34. A salad bar offers six toppings that can be added to the salad: tomatoes, peppers, bacon bits, mushrooms, cheese, and onions. How many different salads can be made?

35. A customer is invited to sample six different types of cookies while waiting in line at a bakery. In how many ways can a person sample the cookies if the person selects either none or one of each type of cookie?

36. How many different committees can be selected from seven people given that each committee can have as few as one person or as many as seven people?

37. Use the Binomial Theorem to evaluate the expression

$$\left(\frac{1}{2}\right)^6 \left[\binom{6}{6} + \binom{6}{5} + \binom{6}{4} + \cdots + \binom{6}{0} \right]$$

38. Use the Binomial Theorem to evaluate the expression

$$\left(\frac{1}{3}\right)^4 + \binom{4}{3}\left(\frac{1}{3}\right)^3\left(\frac{2}{3}\right) + \binom{4}{2}\left(\frac{1}{3}\right)^2\left(\frac{2}{3}\right)^2$$
$$+ \binom{4}{1}\left(\frac{1}{3}\right)\left(\frac{2}{3}\right)^3 + \left(\frac{2}{3}\right)^4$$

Cumulative Exercise Set 6–5

1. Determine whether B is a subset of A.
   (a) $A = \{1, 2, 3, \ldots, 15\}$ $B = \{1, 2, \ldots, 20\}$
   (b) $A = \{a, b, d, e\}$ $B = \{a, b, c, d, e, f\}$

2. Find $n(A \cap B)$ when $n(A) = 6$, $n(B) = 15$ and $n(A \cup B) = 19$.

3. An accountant has 26 clients, 12 of which are retailers and 10 of which have more than 100 employees. If 7 of the retailers also have more than 100 employees, how many of the retailers also have 100 or fewer employees?

4. A poll of 1000 voters revealed that 580 will vote for the Republican mayoral candidate, 420 will vote for the Republican senatorial candidate, 533 will vote for the Republican gubinatorial candidate, 280 will vote for both the Republican mayoral candidate and the Republican senatorial candidate, 320 will vote for both the Republican mayoral candidate and the Republican gubinatorial candidate, 110 will vote for both the Republican senatorial candidate and the Republican gubinatorial candidate, 60 will vote for all three Republican candidates. How many will vote for exactly one Republican candidate?

5. Two coins are tossed and the number of heads is recorded, then a die is tossed and the number on the top face is recorded. List the possible outcomes.

6. A person must make selections in three categories for a job placement questionnaire. The person must choose from two location options—city or suburbs—four career path options—data base entry, word processing, communications, or sales—and two salary options—high,

medium, low. How many different ways are there to fill out the questionnaire with these options?

7. Four dice are tossed. How many ways can they land if you record (a) the sum of the top faces, (b) the numbers on each of the top faces (for example, one such outcome would be 1-1-1-1 and another would be 1-2-4-6).

8. How many five-digit telephone numbers can be formed if the first digit must be 1 or 2?

9. How many different arrangements of four characters are there if the characters are vowels {*a, e, i, o, u*} and (a) repetition is allowed, (b) repetition is not allowed?

10. There are six finalists for two awards. In how many ways can the awards be given if (a) the awards are equal and any finalist can win two, one, or no award, (b) the awards are equal but no finalist can win more than one award.

CASE STUDY   License Plates*

Figure 1 shows the license plates for passenger vehicles in each of the 50 states, the territories, and the Armed Forces of the United States.† Most of the states have adopted the *standard format,* which consists of three letters followed by three digits. But some states, such as New Jersey, have adopted the *reverse format* as well. It consists of three digits followed by three letters.

There are over 20 million licensed vehicles in California, more than in any other state. The Golden State decided not only to use the standard format and the reverse format for its license plates but also to use a third format. It consists of a single digit followed by three letters and three digits; we will call it the *7-Cal format.*

How many usable license plates do these three formats allow? We will answer this question by defining the term "usable license plate" and by applying the counting techniques we have developed in this chapter.

Usable License Plates   A convenient way to view the standard format is *LLLDDD,* where *L* stands for the 26 letters and *D* for the 10 digits. By the Multiplication Principle there are 17,576,000 possible license plates in standard format.

$$\underline{\quad 26 \quad}_L \underline{\quad 26 \quad}_L \underline{\quad 26 \quad}_L \underline{\quad 10 \quad}_D \underline{\quad 10 \quad}_D \underline{\quad 10 \quad}_D$$

This is the *theoretical number* of license plates because not all license plates are usable. For instance, JKI123 is not usable because the third letter can easily be confused with the first digit. This is an example of a license plate being restricted

*Adapted from Edwin G. Landauer, "Counting Using License Plates and Phone Numbers: A Familiar Experience," *The Mathematics Teacher,* March 1984, pp. 183–187, 234.

†We are grateful to Carolyn Edwards and David R. McElhaney of the Office of Highway Information Management for supplying Figure 1 and other relevant information.

for reasons of "clarity." Other license plates are reserved for disabled persons and disabled veterans. Two examples are DPW123 and VET789.

The set of *usable license plates* refers to those license plates that are available to the general public. This set excludes license plates that are restricted for clarity or reserved for disabled persons and disabled veterans. Example 1 counts the number of usable license plates in standard format.

EXAMPLE 1

The number of theoretical license plates in standard format *LLLDDD* is 17,576,000. The number of usable license plates is $17,576,000 - n(C \cup D)$ where

C = license plates restricted for clarity

D = license plates reserved for the disabled

The clarity rule is as follows:

1. The letters I, O, Q, and Z are not allowed in the third-letter position. Since four letters cannot be used in the third position

$$n(C) = 26 \cdot 26 \cdot 4 \cdot 10 \cdot 10 \cdot 10 = 2,704,000$$

There are two rules for reserved license plates.

2. The combinations DPW, DPX, DPY, and DPZ are designated for vehicles of disabled persons.
3. The combinations VET, VTN, and VTR are designated for vehicles of disabled veterans.

By the Multiplication Principle there are 1000 license plates of the form DPW_ _ _ because each blank can have ten possible digits. Similarly, each combination DPX, DPY, and DPZ eliminates 1000 license plates, making 4000 license plates that are reserved for disabled persons by rule 2. Three such combinations are also reserved for disabled veterans by rule 3, so $n(D) = 7000$.

It follows from rules 1 and 2 that a license plate lies in both C and D only if the letters are DPZ. There are 1000 license plates with the letters DPZ, so

$$n(C \cap D) = 1000$$

By the Inclusion-Exclusion Principle

$$n(C \cup D) = 2,704,000 + 7000 - 1000 = 2,710,000$$

Thus the number of usable license plates is

$$17,576,000 - 2,710,000 = 14,866,000$$

The theoretical number of license plates in reverse format is the same as the theoretical number in standard format. But the number of restrictions is different. Exercises 1 to 3 count the number of usable license plates in reverse format.

(a)

FIGURE 1

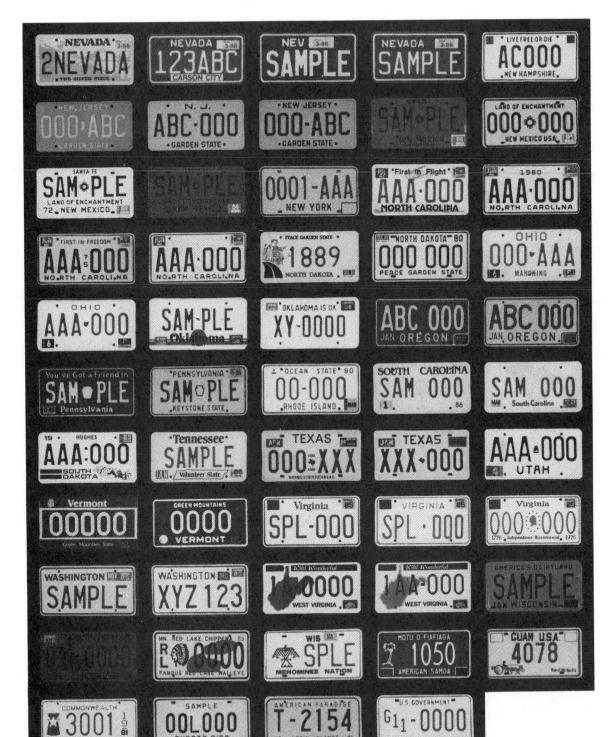

(b)

The 7-Cal Format   The 7-Cal format consists of a digit from 1 to 9 followed by the standard format. It can be written *ELLLDDD,* where *E* represents one of the digits from 1 to 9. The theoretical number of license plates in 7-Cal format is

$$9 \cdot 26 \cdot 26 \cdot 26 \cdot 10 \cdot 10 \cdot 10 = 158,184,000$$

This number is over half the population of the entire United States! It stands to reason that the 7-Cal format can accommodate all of the licensed passenger vehicles in the state of California.

However, the 7-Cal format has four restrictions.

1. The letters I and O are not allowed in the first-letter position.
2. The letters I, O, Q, and Z are not allowed in the third-letter position.
3. The three-character combination 1DP is reserved for vehicles of disabled persons.
4. The three-character combination 1VT and the four-character combination 1VET are reserved for vehicles of disabled veterans.

Example 2 counts the number of usable license plates in the 7-Cal format by an approach that is quite different from the approach in Example 1.

EXAMPLE 2

By the Multiplication Principle the number of usable license plates in the 7-Cal format *ELLLDDD* is

$$9 \cdot 24 \cdot 26 \cdot 22 \cdot 10 \cdot 10 \cdot 10 = 123,552,000$$

The number 24 comes from rule 1 and the number 22 from rule 2. This total includes the restrictions for clarity but not the reservations for disabled persons.

By rule 3 the characters 1DP are followed by one letter and three digits. The letter cannot be Z because of rule 2, so the number of license plates reserved for disabled persons is

$$25 \cdot 10 \cdot 10 \cdot 10 = 25,000$$

Next consider rule 4. There are 25,000 license plates beginning with 1VT. Also, 1000 license plates begin with 1VET. Thus 26,000 license plates are reserved for disabled veterans. The sets of license plates that are reserved for disabled persons and reserved for disabled veterans are disjoint. Hence the number of reserved license plates is $25,000 + 26,000 = 51,000$.

Altogether there are $123,552,000 - 51,000 = 123,501,000$ usable license plates in the 7-Cal format.

Table 1 summarizes the number of usable license plates in California.

Future Needs   More than 31 million usable license plates are available in either the standard or the reverse format. Yet in 1980 California, with less than 18 million licensed vehicles, began issuing license plates in the 7-Cal format. Why did the Golden State adopt a third format?

Table 1

Format	Number of Usable License Plates
Standard	14,866,000
Reverse	16,208,000
7-Cal	123,501,000
Total	154,575,000

For two reasons: First, California does not reissue regular license plates, except for vanity plates. This reason alone creates the need for a new format, but it is indigenous to California. Second, all states have lists of unacceptable combinations of letters and numbers. Most of them are obscene, so we will not list them here (for reasons of taste as well as litigation).

Will California ever have the need for a fourth format? We doubt it. After all, the population of the state is less than 30 million, so the 150 million usable license plates allow five vehicles per resident. However, if its population should happen to increase dramatically, or its residents take up the hobby of collecting cars, California can always adopt a new format of a single letter followed by the reverse format. Exercise 4 discusses this possibility.

What format does your state use? How many usable license plates are available? The exercises examine Louisiana, Montana, and California.

Case Study Exercises

1. For reasons of clarity the letters I and O are not allowed in the first-letter position of license plates in the reverse format *DDDLLL*. How many license plates does this restriction eliminate?

2. Two sets of three-letter combinations are reserved for license plates in the reverse format *DDDLLL*. Those designated for vehicles of disabled persons are DPW, DPX, DPY, DPZ, RDP, SDP, TDP, UDP, VDP, WDP, XDP, YDP, and ZDP. Those designated for vehicles of disabled veterans are VET, VTN, and VTR. How many license plates do these reservations eliminate?

3. How many usable license plates are in the reverse format *DDDLLL?* (You must make use of the answers to Problems 1 and 2.)

4. How many usable license plates consist of a letter followed by three digits and three letters, subject to the restrictions stated in Problems 1 and 2?

5. License plates in Louisiana consist of three digits, a letter, and three digits. (See Figure 1.) The letter designates the state police troop area that issued the plate (A, B, C, D, E, F, G, H, I, K, and L). The letter X means that the plate was issued by mail. How many usable license plates does this format allow?

6. Can you think of any restrictions in Louisiana's format that are due to reasons of clarity?

7. Louisiana currently has about 2,700,000 licensed vehicles. Will a new format be needed soon?

8. Montana's license plate format consists of one or two digits, followed by a dot, and then up to five more digits. (See Figure 1.) The first group of digits runs from 1 to 56; each one designates the county in which the owner resides. The second group runs from 1 to the number of cars licensed in that county. How many usable license plates does this format allow?

CHAPTER REVIEW

Key Terms

6–1 Sets

Set	Complement
Standard Format	Universal Set
Element	Intersection
Natural Number	Union
Integer	Irrational Number
Rational Number	Real Number
Subset	Empty (Null) Set
Venn Diagram	Disjoint Sets
Equal Sets	

6–2 Venn Diagrams and Counting
"Count"

6–3 Tree Diagrams and the Multiplication Principle

Tree Diagram	Selection With Replacement
Selection Without Replacement	The Multiplication Principle

6–4 Permutations and Combinations

Permutations	Factorial
Combinations	

6–5 The Binomial Theorem
Binomial Coefficients
Binomial Theorem
Pascal's Triangle

Summary of Important Concepts

Notation

$\{ \quad \}$	$\not\subset$
$\{ \ : \ \}$	$\varnothing$
$\in$	$n(X)$
$\notin$	$n!$
$\subset$	

Formulas

Inclusion-Exclusion Formula

$$n(A \cup B) = n(A) + n(B) - n(A \cap B)$$

Multiplication Principle

$$n_1 \cdot n_2 \cdots n_r$$

Permutations

$$P(n, r) = \frac{n!}{(n - r)!}$$

Combinations

$$C(n, r) = \frac{n!}{(n - r)!r!}$$

Binomial Theorem

$$(x + y)^n = \binom{n}{n}x^n + \binom{n}{n - 1}x^{n-1}y + \binom{n}{n - 2}x^{n-2}y^2 + \cdots + \binom{n}{2}x^2y^{n-2}$$

$$+ \binom{n}{1}xy^{n-1} + \binom{n}{0}y^n$$

REVIEW PROBLEMS

1. Which of these collections is well defined?
   (a) nice numbers.
   (b) U.S. cities whose population in the 1990 census exceeded 100,000.

In Problems 2 and 3 let $U = \{a, b, c, \ldots, z\}$, $V = \{a, e, i, o, u\}$, and $A = \{a, b, c, d, e\}$.

2. Find (a) $V \cup A$ (b) $V \cap A$.

3. Find (a) A' (b) $V' \cap A'$ (c) $(V \cup A)'$.

4. Determine whether A or B is a subset of the set of rational numbers Q where $A = \{x \in Z : x^2 = -1\}$ and $B = \{x \in Z : x^2 > 9\}$.

5. Find $n(A \cup B)$ where $n(A) = 9$, $n(B) = 5$, and $n(A \cap B) = 4$.

6. Find $n(A \cup B)$ where $n(A \cap B') = 9$, $n(A' \cap B) = 5$, and $n(A \cap B) = 4$.

7. What is X' if X is the set of rational numbers and the universal set is the set of real numbers?

8. The 30,000 members of the MAA can subscribe to three journals: *Monthly* (M), *Math Magazine* (G), and *College Math Journal* (C). In 1988 the numbers of subscribers to the journals (in thousands) were $n(M \cap G \cap C) = 1$, $n(M \cap G) = 5$, $n(M \cap C) = 7$, $n(G \cap C) = 1$, $n(M) = 18$, $n(G) = 7$, and $n(C) = 11$. How many members subscribed to
   (a) Both *Math Magazine* and the *College Math Journal?*
   (b) The *College Math Journal* only?
   (c) None of the three journals?

9. Refer to the figure to determine the number of elements in each set.
   (a) $A \cap B$ (b) $A \cup B$ (c) A' (d) U

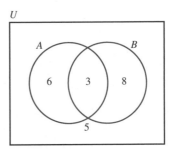

10. In how many different ways can four people be seated in four seats in a row?

11. In the game of "Clue" a suggestion consists of a suspect, a weapon, and a room. There are 6 suspects, 6 weapons, and 9 rooms. How many suggestions are possible?

12. How many arrangements of four letters begin with one consonant, end with a different consonant, and have two distinct vowels in between?

13. Four people enter the elevator in a building on the ground floor. The elevator can stop at floors 1 through 12.
    (a) In how many different ways can the four people exit?
    (b) In how many different ways can the four people exit if no two exit on the same floor?

14. Each ticket in the game of "Lotto" has six different numbers punched from the numbers 1 to 40. How many different tickets are possible if order does not matter?

15. In how many different ways can the letters of the word MAJORITY be arranged?

In Problems 16 to 18 a tournament with four prizes starts with 64 teams. Each losing team is eliminated and a winning team continues to play until it loses a game.

16. How many orderings of the first through the fourth place teams are possible?

17. How many combinations of teams can make the final four?

18. The final four teams in the 1991 NCAA basketball tournament were Duke, University of Nevada at Las Vegas, North Carolina, and Kansas. In how many different ways could these four teams have finished first through fourth?

19. Twenty dancers audition for roles in *A Chorus Line*.
    (a) How many different selections of 3 of them can be made for the first, second, and third leading roles?
    (b) How many different selections of 3 of the remaining 17 dancers can be made for 3 indistinguishable understudy roles?

20. Two dice (one red and the other green) are rolled and their sum is recorded. In how many different ways can the dice land so that the sum is equal to 10?

21. An urn contains 26 balls, each with a letter of the alphabet printed on it. How many three-letter "words" can be formed by drawing 3 balls from the urn without replacement?

22. What is the coefficient of the x^2y^5 term in the expansion of $(x + y)^7$?

PROGRAMMABLE CALCULATOR EXPERIMENTS

1. Write a program that computes $n!$ for all n less than a given number.

2. Write a program that computes $P(n, r)$ for a given n and for all r less than n.

3. Write a program that computes $C(n, r)$ for a given n and for all r less than n.

4. Write a program that computes the binomial coefficients for a given n.

Probability

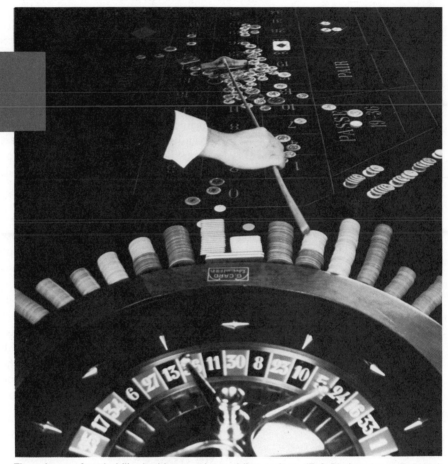

The science of probability had its start in gambling casinos. (Zefa/H. Armstrong Roberts)

Chapter Overview

In many aspects of mathematics a formula or process is used to predict the future. Sometimes the prediction is exact, but often events take place randomly and it is not possible to predict exactly what will happen. A model must be constructed that takes this random nature of outcomes into account. For example, a meteorologist cannot tell precisely when, or even if, it will rain, but instead, assigns probabilities to the possible types of weather to expect. Sections 7–1 and 7–2 show how to construct probability models and also illustrate their fundamental properties. The later sections cover more specific topics that govern when we are able to add, subtract, multiply, and divide probabilities. We also calculate probabilities using the counting techniques of Chapter 6.

Throughout the chapter games of chance are used as examples because they are easy to explain and understand. We also mention several significant applications of probability theory in the real world. In fact, the first example is from a study of automobile accidents on a major Boston thoroughfare that used probability theory in its analysis. The two case studies describe applications of probability more thoroughly.

CASE STUDY PREVIEWS

Suppose you face a trial and your lawyer says it is a toss-up whether to select a trial by jury or a trial by judge alone. Which would you choose? Two political scientists researched this question, and their results, based on probabilistic reasoning, are given in the first of two case studies in this chapter.

When Jane Blalock, star professional golfer, was banned from participating in tournaments by the Ladies' Professional Golf Association, she sued the organization under the federal antitrust laws. The court reinstated her but the decision came too late for her to participate in the tournament. She then sued for damages. But how much money would she have earned? She might have won the tournament; however, she just as well might have finished out of the money. Her court battle was based on probability theory. The second case study gives the surprising result of her suit.

7–1   Sample Spaces

Many events in the real world seem to happen by chance. By studying the behavior of some of these seemingly unpredictable events, we can often determine patterns. The theory of probability is the branch of mathematics that helps find these patterns. In order to describe the basic ideas of probability, let us look at a concrete example of such a study.

Introduction: An Example

In the early 1970s the Massachusetts Department of Public Works conducted a study of speed limits in an effort to improve motor vehicle safety. Highway accident data were analyzed on a suburban stretch of Route 2, a major road entering Boston, Massachusetts, from the west. In an article about the study William B. Fairley

Table 1 Accidents on Route 2, 1970–1972

Quarter	Quarterly Count	Relative Frequencies
1	302	0.24
2	244	0.19
3	283	0.22
4	<u>440</u>	0.35
Total	1269	

Source: William B. Fairley, ''Accidents on Route 2: Two-Way Structures for Data,'' *Statistics and Public Policy,* by Fairley and Mosteller, eds., Reading, MA, Addison-Wesley, pp. 23–50, 1977.

raised the critical question, What season of the year had the most accidents? The year was divided into four quarters, the first being January to March, then April to June, and so forth. Data were gathered for each quarter from 1970 to 1972. Table 1 gives the quarterly count, called the **frequency,** and the average for each quarter. For example, during the three years of the study there was a total of 302 accidents in the first quarter: 138 in 1970, 74 in 1971, and 90 in 1972. The averages are computed by dividing each quarterly count by the total number of accidents in all quarters, 1269. Thus the quarterly average for quarter 1 is $302/1269 = 0.24$ rounded to two decimal places.

By looking at the averages, we can easily determine that not only were there more accidents in the fourth quarter, but there were almost twice as many in quarter 4 as in quarter 2. The percentages are called **relative frequencies.** They arise from the fraction, m/N, where m is the number of occurrences of accidents in a quarter and N is the total number of accidents. If we select an accident at random from the total of 1269, we say that the **probability** that it occurred in quarter 4 is .35.

To say that the probability of this event is .35 means that since it has occurred 35% of the time in the past, one can reasonably predict that the event will happen 35% of the time in the future. In this way probability is used to predict future trends. That is, under the same conditions 35% of the accidents on Route 2 will occur in quarter 4.

The phrase ''under the same conditions'' spells the difference between studies of real-world phenomena and games of chance, such as tossing coins or rolling dice, or playing cards or roulette. Games are repeated exactly, but it is much more difficult to repeat real-world experiments under the same conditions. Therefore games are used to illustrate the concepts of probability because they are familiar and easy to demonstrate, and on each play they are repeated identically. In this chapter we refer to specific experiments using simple descriptive models. In the next chapter we will study experiments in statistics that involve data analysis.

Sample Spaces

The basic idea in probability is to study outcomes of experiments. By an **experiment** we mean an action whose possible outcomes can be determined and recorded. For example, four simple experiments are:

Experiment *A*: Toss a coin and observe heads or tails.
Experiment *B*: Roll a die and record the number on the top face.
Experiment *C*: Select a ball from an urn containing balls numbered 0, 1, . . ., 9.
Experiment *D*: Choose an accident along Route 2 and record the quarter in which it occurs.

Each time an experiment is repeated an outcome is observed and recorded. The set of possible outcomes is called the *sample space* for the experiment. For instance, Experiment *A* has two outcomes: heads and tails. If we let "*H*" represent heads and "*T*" represent tails, then $S = \{H, T\}$ is the sample space for Experiment *A*.

EXAMPLE 1

Problem

Find the sample space for Experiments *B*, *C*, and *D*.

Solution Experiment *B* has six possible outcomes: 1, 2, 3, 4, 5, 6. Hence the sample space for Experiment *B* is

$$S = \{1, 2, 3, 4, 5, 6\}$$

Experiment *C* has ten outcomes. Its sample space is

$$S = \{0, 1, 2, 3, 4, 5, 6, 7, 8, 9\}$$

Experiment *D* has four outcomes. Its sample space is

$$S = \{\text{quarter 1, quarter 2, quarter 3, quarter 4}\}$$

DEFINITION

The **sample space** of an experiment is the set of all possible outcomes of the experiment. Each time the experiment is conducted exactly one outcome is recorded.

Thus an experiment is described by its sample space. The next two examples illustrate how to determine the sample space. An experiment can be viewed as consisting of two steps, an action followed by recording something about the action. The possible quantities that can be recorded are the **outcomes**. The set of outcomes is the sample space. In Experiment *D* the action is to select an accident at random, and then the quarter in which it occurred is recorded. Example 2 shows that even if the same action is performed, a new experiment is defined if the quantity recorded is changed.

EXAMPLE 2

Problem

An experiment consists of selecting at random an accident from the Fairley study on Route 2 and recording the month in which it occurred. What is the sample space of the experiment?

Solution The outcomes of the experiment are the 12 months of the year. Thus the sample space S is

$S = \{$Jan., Feb., Mar., Apr., May, Jun., Jul., Aug., Sep., Oct., Nov., Dec.$\}$

The next example shows that the same action can generate two different experiments by recording different outcomes, thus producing two separate sample spaces.

EXAMPLE 3

Toss two coins. Record (a) the sequence of H's and T's and (b) the number of H's.

Problem

Find the sample space for each experiment.

Solution (a) There are four outcomes for the experiment: HH, HT, TH, and TT. Thus the sample space is

$S = \{HH, HT, TH, TT\}$

(b) There are three outcomes: 0, 1, and 2. Hence the sample space is

$S = \{0, 1, 2\}$

The significance of the distinction between the two experiments in Example 3 will be explained in Section 7–2.

Events

Often we are interested in looking at the possibility of more than one outcome being observed. For example, in Experiment D suppose we wanted to determine how many accidents occurred in the first six months of the year. The corresponding outcomes would be quarter 1 and quarter 2. In mathematical language this is the set {quarter 1, quarter 2}, which is a subset of the sample space. In general, a subset of the sample space is called an *event*. The next two examples describe events associated with Experiments B and C.

EXAMPLE 4

Problem

Describe in terms of sets the two events E and F associated with Experiment B, in which a die is rolled and the number on its top face is recorded, and where

  E: The event that an odd number occurs.
  F: The event that a number less than 5 occurs.

Solution Event E consists of the three outcomes, 1, 3, and 5. Thus

  $E = \{1, 3, 5\}$

Event F consists of the outcomes 1, 2, 3, and 4. Thus

$F = \{1, 2, 3, 4\}$

EXAMPLE 5

Problem

Describe in terms of sets the three events E, F, and G that are associated with Experiment C whose sample space is $\{0, 1, 2, 3, 4, 5, 6, 7, 8, 9\}$.

E: An odd number occurs.
F: A number less than 5 occurs.
G: Either an odd number or a number less than 5 occurs.

Solution Event E consists of the five outcomes: 1, 3, 5, 7, and 9. Thus

$E = \{1, 3, 5, 7, 9\}$

Event F consists of the outcomes 0, 1, 2, 3, and 4. Thus

$F = \{0, 1, 2, 3, 4\}$

Event G consists of the odd numbers 1, 3, 5, 7, and 9, together with the numbers less than 5, which are 0, 1, 2, 3, and 4. Notice that 1 and 3 are mentioned twice, but it makes sense to write them only once in the set G. Thus

$G = \{0, 1, 2, 3, 4, 5, 7, 9\}$

DEFINITION

An **event** of an experiment is a subset of the sample space. The event E *occurs* if the outcome of the experiment is an element of the subset E.

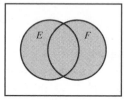

$E \cup F$

FIGURE 1

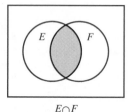

$E \cap F$

FIGURE 2

Since events are defined in terms of sets, we can use set theory to define new events from other events. For instance, in Example 5, event G is the union of E and F because $E \cup F$ occurs when E or F (or both) occurs. Thus $G = E \cup F$. We can also define the event $E \cap F$, which is the event that occurs when both E and F occur at the same time. In Example 5, $E \cap F = \{1, 3\}$.

Often in an experiment an event is defined by a description rather than a list of its elements. In this case the word "or" generally represents the union of two events while "and" refers to the intersection. For example, consider the experiment to roll a die and observe the number on the top face. Let E be "observe an odd number" and F be "observe a prime number." Then the event "observe a number that is odd and a prime" is the intersection of the two events, $E = \{1, 3, 5\}$ and $F = \{2, 3, 5\}$. It is $E \cap F = \{3, 5\}$. The event "observe a number that is odd or a prime" is the union of E and F. It is $E \cup F = \{1, 2, 3, 5\}$.

Figures 1 and 2 give the Venn diagrams for $E \cup F$ and $E \cap F$.

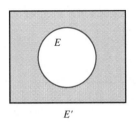

E'

FIGURE 3

In a probability experiment the universal set is the entire sample space. In Experiment *C*, whose sample space is given in Example 5, consider the event "a number not less than 5 occurs." This set is the **complement** of *F*.

$$F' = \{5, 6, 7, 8, 9\}$$

Figure 3 gives the Venn diagram for the complement, *E'*, of an event *E*. Thus the three set operations of union, intersection, and complement allow us to build new events from given events.

DEFINITION

If *E* and *F* are two events of an experiment, then

1. $E \cup F$ occurs when *E* or *F*, or both, occur.
2. $E \cap F$ occurs when both *E* and *F* occur.
3. *E'* occurs when *E* does not occur.

Two events that cannot both occur at the same time are called **mutually exclusive events.** For example, when a die is tossed, the events "an even number" and "an odd number" cannot both occur, so they are mutually exclusive. In terms of sets, two events *E* and *F* are mutually exclusive if $E \cap F = \varnothing$. Figure 4 gives the Venn diagram for mutually exclusive events. Let us look at two examples to illustrate these points.

EXAMPLE 6

Let the sample space for an experiment be $S = \{a, b, c, d, e\}$. Define the events

$$E = \{a, b\} \quad F = \{b, c, d\} \quad G = \{a, b, c, e\}$$

Problem

Find (a) $E \cup F$, (b) $E \cap F$, (c) G', (d) $E \cap G'$.

Solution

(a) $E \cup F = \{a, b, c, d\}$
(b) $E \cap F = \{b\}$
(c) $G' = \{d\}$
(d) $E \cap G' = \varnothing$, so *E* and *G'* are mutually exclusive.

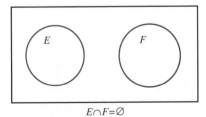

$E \cap F = \varnothing$

FIGURE 4

In the next example we use the notation $n(A)$ to represent the number of elements in the set A. This notation was introduced in Chapter 6.

EXAMPLE 7

The second part of the Fairley study of accidents on Route 2 centered on the incidence of accidents per month to see if there was a finer trend to be discovered. The data are as follows:

Month	Jan.	Feb.	Mar.	Apr.	May	Jun.
Number of accidents	120	85	97	81	88	75
Month	Jul.	Aug.	Sep.	Oct.	Nov.	Dec.
Number of accidents	95	84	104	149	143	148

The experiment is to select an accident at random and to record its month. Define the events

J: The accident occurred in January.

F: The accident occurred in February.

A: The accident occurred in quarter 2 (Apr., May, Jun.).

S: The accident occurred in Jun., Jul., or Aug.

Problem

Determine the sample space and find how many accidents are associated with each of the following events: (a) $J \cap F$, (b) $J \cup F$, (c) $A \cap S$, (d) J'.

Solution   The sample space consists of the 12 months, so it is

$S = \{$Jan., Feb., Mar., Apr., May, Jun., Jul., Aug., Sep., Oct., Nov., Dec.$\}$

(a) J and F are mutually exclusive since $J \cap F = \emptyset$, so $n(J \cap F) = 0$.
(b) There were 120 accidents in January and 85 in February, so $n(J) = 120$ and $n(F) = 85$. Since $J \cap F = \emptyset$, $n(J \cup F) = 120 + 85 = 205$.
(c) $A \cap S = \{$Jun.$\}$, so $n(A \cap S) = n($Jun.$) = 75$.
(d) There were 1269 accidents altogether and 120 occurred in January, so $n(J') = 1269 - 120 = 1149$.

EXERCISE SET 7–1

In Problems 1 to 14 find the sample space of the experiment.

1. Roll a die and record whether the number on the top face is even or odd.

2. Roll a die and record whether the number on the top face is less than 3, equal to 3, or greater than 3.

3. Select a letter from the word "probability."

4. Select a letter from the word "dzo."

5. Select a card from an ordinary deck and record its color.

6. An ordinary deck of cards is shuffled and a card is selected at random and its suit is recorded.

7. A survey of voters is taken to determine their affiliation. They are classified as Democratic, Republican, or independent. A voter is selected at random and the voter's affiliation is recorded.

8. A person is selected and his or her birth month is recorded.

9. At a weekly meeting of Weight Watchers, members record whether they gained weight, lost weight, or stayed the same. Two members are selected at random and their weight changes are recorded.

10. A manufacturer chooses a sample of four items from a shipment and records the number of defective items in the sample.

11. A financial analyst keeps a record of whether the closing price of a particular stock increases, decreases, or is unchanged from the previous day's close. The record is kept for four days.

12. The analyst in Problem 11 records the number of days of the four days that the closing price of the stock increases.

13. A farmer records over a three-year period whether corn, potatoes, or alfalfa is planted in a particular field in a given year. Each crop is planted once in the field in the three-year period.

14. The farmer in Problem 13 records whether the same three crops were planted in a field. In this field no crop was planted in successive years.

In Problems 15 to 24 find $E \cup F$, $E \cap F$, and E'.

15. A die is rolled and the number on the top face is recorded.

$E = \{1, 2, 3\}, \qquad F = \{2, 3, 4\}$

16. A die is rolled and the number on the top face is recorded.

$E = \{1, 2, 3, 4\}, \qquad F = \{2, 4, 6\}$

17. A letter is selected from the word "probability."

$E =$ a vowel, $\qquad F =$ a consonant

18. A letter is selected at random from the word "dzo."

$E = \{d, z\}, \qquad F = \{z, o\}$

19. An ordinary deck of cards is shuffled and a card is selected and its suit is recorded.

$E =$ heart, $\qquad F =$ red card

20. The 12 picture cards of an ordinary deck of cards are shuffled and one of them is selected at random.

$E =$ heart, $\qquad F =$ king

21. The sample space is $\{2, 3, 4, 5, 6, 7, 8, 9, 10, 11, 12\}$.

$E = \{2, 3, 4\}, \qquad F = \{3, 4, 5, 6, 7\}$

22. The sample space is $\{-3, -2, -1, 0, 1, 2, 3, 4\}$.

$E = \{-3, -2, 2, 3\}, \qquad F = \{2, 3, 4\}$

23. A coin is tossed two times and the number of heads is recorded.

$E =$ at most one H
$F =$ at least one H

24. A coin is tossed two times and the number of heads is recorded.

$E =$ at most one H
$F =$ exactly one H

In Problems 25 to 28 let S be the sample space of an experiment. Determine whether E and F are mutually exclusive events.

25. $S = \{a, b, c, d\}, \quad E = \{a, c\}, \quad F = \{b, d\}$

26. $S = \{a, b, c, d, e, f\}, \qquad E = \{a, b, c\},$
$F = \{b, d, f\}$

27. $S = \{2, 3, 4, 5, 6, 7, 8, 9, 10, 11, 12\},$
$E = \{2, 3, 4\}, \quad F = \{4, 5, 6, 7\}$

28. $S = \{-3, -2, -1, 0, 1, 2, 3, 4\},$
$E = \{-3, -2, 0, 1\}, \quad F = \{2, 3, 4\}$

29. An experiment consists of choosing a committee of two individuals from a list of six people whose last initials are A, B, C, R, S, T. Determine the sample space of the experiment and list the elements of the events E, F, and G.

E: A, B, or C is on the committee.

F: Neither A nor B is on the committee.

G: Either C or S is on the committee.

30. The experiment consists of selecting a marble from a bag containing red, white, and blue marbles and recording its color. Define the events

E: It is not red.
F: It is red or blue.

(a) Determine the sample space S of the experiment. Find (b) $E \cup F$, (c) $E \cap F$, (d) E'.

31. A survey asks people whether they (1) smoke at least a pack of cigarettes a day, (2) take vitamin C regularly, and (3) have more than two colds in a year. The experiment consists of selecting a person and recording the number of questions answered "yes." Define the events

    E: At least two questions were answered "Yes."
    F: At most two questions were answered "Yes."

    (a) Determine the sample space S of the experiment. Find (b) E ∪ F, (c) E ∩ F, (d) E'.

32. Two bags contain red, white, and blue marbles. The experiment consists of first selecting a bag and then a marble from the bag and recording its color. Define the events

    E: It is not red.
    F: It is red or blue.

    (a) Determine the sample space S of the experiment. Find (b) E ∪ F, (c) E ∩ F, (d) E'.

33. Let E and F be events. The following events are described in words; express them in terms of ∪, ∩, and ( )'.
    (a) Both E and F occur.
    (b) E occurs but F does not.
    (c) E does not occur.
    (d) E or F occurs, perhaps both.

34. Let E, F, and G be events. The following events are described in words; express them in terms of ∪, ∩, and ( )'.
    (a) All three occur.
    (b) E occurs but F and G do not.
    (c) Neither E nor F occurs.
    (d) At least one of the three occurs.

35. An experiment consists of tossing a coin three times and recording the number of heads. (a) Find the sample space and determine the elements in the two events

    E: More than one heads
    F: An even number of heads

    Find (b) E ∪ F, (c) E ∩ F, (d) E'.

36. An experiment consists of tossing a coin five times and recording the number of heads. (a) Find the sample space and determine the elements in the two events

    E: More than one heads
    F: An even number of heads

    Find (b) E ∪ F, (c) E ∩ F, (d) E'.

37. For any event E, is it true that E and E' are mutually exclusive? Why?

38. Let E and F be mutually exclusive events.
    (a) Must E' and F' be mutually exclusive?
    (b) Must E ∩ F and (E ∪ F)' be mutually exclusive? Why?

39. Determine all possible events that are associated with the sample space S where (a) S = {1, 2, 3}, (b) S = {1, 2, . . . , n}.

Referenced Exercise Set 7–1

1. In a medical experiment* designed to study the effect of Reye's syndrome on the inner ear, patients were asked whether they had contracted an influenza infection, had taken aspirin during their illness, and had an arsenine-deficient diet. Define the events I, A, and D as follows:

    I: The patient had influenza.
    A: The patient took aspirin.
    D: The patient had an arsenine-deficient diet.

    Define the following events in words:
    (a) I ∪ A, (b) A ∪ D, (c) I ∩ D, (d) A ∩ D, (e) I', (f) I' ∩ A', (g) I ∩ A ∩ D.

2. In a study† designed to explore the possibility of a link between air pollution and ill-health the investigators discussed the effects of their research on mortality rates characterized by age, race, and sex. They defined the following events:

    I: Infant (under 1 year)
    C: Child (from 1–14 years old)
    A: Adult (over 14 years old)
    N: Nonwhite   W: White
    F: Female   M: Male

---

*K. E. Rarey, et al., "Effects of Influenza Infection, Aspirin, and an Arsenine-Deficient Diet on the Inner Ear in Reye's Syndrome," *Annals of Otology, Rhinology, and Laryngology,* Vol. 93, pp. 551–557, November–December 1984.

†John W. Pratt (ed.), *Statistical and Mathematical Aspects of Pollution Problems,* New York, Marcel Dekker, Inc., pp. 223–247, 1974.

Define the following events in words.
(a) $I \cup A$, (b) $A \cup F$, (c) $C \cap M$, (d) $N \cap F$,
(e) F', (f) $A' \cap W'$, (g) $I \cap N \cap M$

3. Farmers classify damage to crops in three categories: damage due to insects, disease, and weather. In an article illustrating methods farmers use to estimate damage probabilities, Pingali and Carlson gathered data from 47 apple orchards in Henderson County, North Carolina.‡ Farmers were asked to estimate the percentage of the total damage done to the crop by each type of damage. (a) Find the sample space of the experiment; that is, determine the outcomes the farmers were asked to consider. (b) Suppose the farmers estimate that the percentages of each type of damage will be equal. Find the fraction of the total damage assigned to each individual type of damage. (c) Suppose the farmers estimate that the damage due to insects and disease will be equal but there will be twice as much damage due to weather as to insects. Find the fraction of the total damage assigned to each type of damage. (d) Suppose the farmers estimate that the damage due to insects will be twice as high as the damage due to disease and the damage due to disease will be twice as much as the damage due to weather. Find the fraction assigned to each type of damage.

‡Prabhu L. Pingali and Gerald A. Carlson, "Human Capital, Adjustments in Subjective Probabilities, and the Demand for Pest Controls," *American Journal of Agricultural Economics*, Vol. 67, pp. 853–861, 1985.

7–2   Fundamental Probability Principles

In this section we assign probabilities to the outcomes of a sample space. For example, when a coin is tossed we assume that heads and tails are equally likely to occur so each will happen half of the time. If H represents heads and T represents tails, then we say that the *probability of the event H is* $\frac{1}{2}$, which we express as $Pr(H) = \frac{1}{2}$. Also, $Pr(T) = \frac{1}{2}$. This means that if we conducted the experiment many times, we would expect H to occur half of the time and T to occur half of the time.

If the outcomes of the sample space are **equally likely,** the probability of an event can be determined by the following fundamental principle.

DEFINITION

Suppose a sample space S of an experiment has N equally likely outcomes. Then the probability of each outcome is $1/N$. If the event E consists of m distinct outcomes, then the **probability of E**, written $Pr(E)$, is

$$Pr(E) = \frac{m}{N} = \frac{n(E)}{n(S)}$$

When a die is rolled and the number on the top face is recorded the six outcomes are equally likely, so each has the probability $\frac{1}{6}$. For instance, we write $Pr(2) = \frac{1}{6}$. The next example shows how to compute the probability of events that consist of more than one outcome for this experiment.

EXAMPLE 1

The experiment is to toss a die and record the number on the top face. Define the events E, F, and G as follows:

$$E = \{3, 4\}$$

$$F = \{1, 3, 4, 5, 6\}$$

G: An even number occurs

Problem

Find (a) $Pr(E)$, (b) $Pr(F)$, (c) $Pr(G)$.

Solution (a) In the formula $m = n(E) = 2$ and $N = n(S) = 6$, so $Pr(E) = \frac{2}{6} = \frac{1}{3}$.
(b) Here $m = 5$ and $N = 6$, so $Pr(F) = \frac{5}{6}$.
(c) There are three outcomes that are even, so $G = \{2, 4, 6\}$. Thus $m = 3$ and so $Pr(G) = \frac{3}{6} = \frac{1}{2}$.

Probability Distributions

It is sometimes helpful to list the outcomes together with their associated probabilities in a table, called a **probability distribution.** For example, the probability distributions for the experiment of tossing a coin and rolling a die are as follows:

Toss a Coin		Roll a Die	
Outcome	Probability	Outcome	Probability
H	$\frac{1}{2}$	1	$\frac{1}{6}$
T	$\frac{1}{2}$	2	$\frac{1}{6}$
		3	$\frac{1}{6}$
		4	$\frac{1}{6}$
		5	$\frac{1}{6}$
		6	$\frac{1}{6}$

Sometimes the sample space does not have equally likely outcomes. In this case we usually have to picture the experiment in a different setting, one in which the outcomes are equally likely. For example, the experiment of tossing two coins and recording the number of heads has three outcomes; 0, 1, 2. They are not equally likely, as seen in the next example.

EXAMPLE 2

Two coins are tossed. There are two natural sample spaces: we can record (a) the sequence of heads and tails or (b) the number of heads that occur.

Problem

Assign probabilities to the elements of each sample space.

Solution (a) To see the difference between the sample spaces, we need to distinguish the coins. Suppose the first coin is a nickel and the second is a quarter. The sample space is {*HH, HT, TH, TT*}, where *HH* represents *H* on both coins, *HT* represents *H* on the nickel and *T* on the quarter, and so on. The following tree diagram gives a pictorial view of the sample space:

From experience we know that these outcomes are equally likely, so the definition says that each has a probability $\frac{1}{4}$. The probability distribution is given as follows:

Toss Two Coins

Record the *H*'s and *T*'s	
Outcome	Probability
HH	$\frac{1}{4}$
HT	$\frac{1}{4}$
TH	$\frac{1}{4}$
TT	$\frac{1}{4}$

(b) When the number of heads is recorded the sample space is {0, 1, 2}. The event "0," meaning "0 heads," corresponds to the event "*TT*" in the previous sample space, so the probability of "0" is the same as the probability of "*TT*." Therefore $Pr(0) = \frac{1}{4}$. Similarly, the event "2" corresponds to "*HH*," so $Pr(2) = \frac{1}{4}$. For the event "1" there are two ways to have exactly one heads occur; *H* on the nickel and *T* on the quarter, or *T* on the nickel and *H* on the quarter. Thus the event "1" corresponds to "*TH* or *HT*," so $Pr(1) = \frac{1}{2}$. The probability distribution for this sample space is given here, alongside the one in part (a).

Toss Two Coins			Toss Two Coins	
			Record the Number of Heads	
Record the *H*'s and *T*'s				
Outcome	Probability		Outcome	Probability
TT	$\frac{1}{4}$		0	$\frac{1}{4}$
HT	$\frac{1}{4}$		1	$\frac{1}{2}$
TH	$\frac{1}{4}$			
HH	$\frac{1}{4}$		2	$\frac{1}{4}$

Let us look at another example of how to assign probabilities when the outcomes are not equally likely. This example has a little twist to the familiar experiment of tossing a die.

EXAMPLE 3

A fair die has three sides colored red, two sides colored white, and one side colored blue. The experiment consists of tossing the die and recording the color of the top face.

Problem

Find the sample space and assign probabilities to the outcomes.

Solution There are three possible outcomes—red, white, and blue. So the sample space is {red, white, blue}. Of the six equally likely faces, only one is colored blue, so $Pr(\text{blue}) = \frac{1}{6}$. Two of the faces are white, so $Pr(\text{white}) = \frac{2}{6} = \frac{1}{3}$. Three of the faces are red, so $Pr(\text{red}) = \frac{3}{6} = \frac{1}{2}$. The assignment of probabilities can be easily read from the following table:

Outcome	Probability
Red	$\frac{3}{6} = \frac{1}{2}$
White	$\frac{2}{6} = \frac{1}{3}$
Blue	$\frac{1}{6}$

Basic Properties of Distributions

Probability distributions have several important basic properties. When we assign a probability to an outcome, it must be a number between 0 and 1, inclusive. An event that is certain to happen, called a "**certain event,**" has probability 1, while an "**impossible event**" has probability 0. For example, if a die is rolled, a certain event would be "observe a number less than 7" and an impossible event would be "observe a number greater than 6."

The sample space must exhaust all possible outcomes. This means that the sum of the probabilities of the outcomes in the sample space must add up to 1. Finally, the outcomes in the sample space must be mutually exclusive.

Basic Properties of Probability Distributions
Let S be the sample space for an experiment and let the probabilities assigned to the outcomes in S be $p_1, p_2, \ldots, p_n$. Then

1. For all $i = 1, \ldots, n,$ $0 \le p_i \le 1.$
2. $p_1 + p_2 + \cdots + p_n = 1.$
3. If E is a certain event, then $Pr(E) = 1.$
4. If E is an impossible event, then $Pr(E) = 0.$

Examples 2 and 3 demonstrate that it is sometimes necessary to examine an experiment from a different perspective to assign probabilities. As another example that also illustrates the properties of sample spaces, consider again the study of accident rates on Route 2 in the next example.

EXAMPLE 4

In Section 7–1 the idea of a probability distribution was introduced by looking at the Fairley study of accidents on Route 2, a major road into Boston. The number of accidents in a three-year span was gathered per quarter. Recall that the quarterly average is m/N, where m is the number of accidents that occurred in the quarter and N is the total number of accidents, 1269. For instance, in quarter 1, $m = 302$, and so $m/N = 302/1269 = .24$. Table 1 contains the probability distribution.

Table 1 Accidents on Route 2, 1970–1972

Quarter	Quarterly Count	Relative Frequency
1	302	0.24
2	244	0.19
3	283	0.22
4	440	0.35
Total	1269	

Source: William B. Fairley, "Accidents on Route 2: Two-Way Structures for Data," *Statistics and Public Policy,* by Fairley and Mosteller, eds., Reading, MA, Addison-Wesley, pp. 23–50, 1977.

This study can be expressed in terms of probabilities by regarding it as an experiment whose sample space is {quarter 1, quarter 2, quarter 3, quarter 4}. The relative frequencies are probabilities. Thus the probabilities of the four outcomes are, respectively, .24, .19, .22, and .35. That is, if we select an accident at random, then the probability that it occurred in quarter 1 is Pr(quarter 1) = .24, while the probability that it occurred in quarter 2 is Pr(quarter 2) = .19. The four outcomes are not equally likely.

To picture the experiment with equally likely events, think of it in this way. Let the sample space consist of the 1269 individual accidents. If we select one at random, each is as likely to be chosen as another, so each has probability 1/1269. With this sample space the event "quarter 1" has 302 outcomes, each with probability 1/1269. To get the probability of quarter 1, we add them to get 302/1269. We are allowed to add the probabilities because they are mutually exclusive.

Let us compare the two sample spaces by taking a closer look at the event quarter 1 in each sample space. The first sample space has 4 outcomes that are not equally likely. The second sample space has 1269 outcomes that are equally likely. In the first sample space quarter 1 is a single outcome. In the second sample space quarter 1 consists of 302 distinct outcomes. In each sample space its probability is the same, 302/1269.

Example 4 suggests that if two or more events are mutually exclusive, the probability of their union is the sum of their probabilities. This is expressed in the next principle.

Addition Principle for Mutually Exclusive Events
If two events E and F are mutually exclusive, that is, $E \cap F = \emptyset$, then

$$Pr(E \cup F) = Pr(E) + Pr(F)$$

The next two examples illustrate this principle.

EXAMPLE 5

Two coins are tossed and the number of heads is recorded.

Problem

Find the probability of observing at most one heads.

Solution Let E represent "observing at most one heads." If we let F and G represent the two events "exactly 0 heads" and "exactly 1 heads," respectively, then $E = F \cup G$ and $F \cap G = \emptyset$. Hence by the Addition Principle, $Pr(E) = Pr(F) + Pr(G) = \frac{1}{4} + \frac{1}{2} = \frac{3}{4}$.

In Example 6 remember that an ace is not considered a picture card, which is also referred to as a face card. The picture cards are the kings, queens, and jacks.

EXAMPLE 6

The experiment is to draw a card from an ordinary deck. The sample space consists of the 52 cards, each assigned the probability $\frac{1}{52}$. Define the events

E: A ten
F: A red ace
G: A picture card

Problem

Find $Pr(E \cup F)$, $Pr(E \cup G)$, and $Pr(F \cup G)$.

Solution The events are mutually exclusive since no two can occur at the same time. Since there are 4 tens, 2 red aces, and 12 picture cards, we have $Pr(E) = \frac{4}{52}$, $Pr(F) = \frac{2}{52}$ and $Pr(G) = \frac{12}{52}$. Then

$$Pr(E \cup F) = Pr(E) + Pr(F) = \frac{4}{52} + \frac{2}{52} = \frac{6}{52} = \frac{3}{26}$$

$$Pr(E \cup G) = Pr(E) + Pr(G) = \frac{4}{52} + \frac{12}{52} = \frac{16}{52} = \frac{4}{13}$$

$$Pr(F \cup G) = Pr(F) + Pr(G) = \frac{2}{52} + \frac{12}{52} = \frac{14}{52} = \frac{7}{26}$$

Complements

Suppose an urn contains 52 marbles, 4 of which are red and the rest are various other colors. If a marble is drawn at random from the urn, what is the probability that it is not red? There are 4 red marbles, so there are $52 - 4 = 48$ marbles that are not red, implying that the probability is $48/52 = 12/13$. This idea can be expressed in terms of complements. Let E represent "the marble is red." Then the desired event, "the marble is not red," is the complement, E'. In terms of probabilities, the last computation can be expressed as

$$Pr(E') = \frac{52 - 4}{52} = \frac{48}{52}$$

Still another way to write it is

$$Pr(E') = \frac{52}{52} - \frac{4}{52} = 1 - \frac{4}{52} = 1 - Pr(E)$$

This generalizes into the **rule for complements.**

Rule for Complements
If E is any event and E' is its complement, then

$$Pr(E') = 1 - Pr(E)$$

Since E is the complement of E', this rule can also be written as $Pr(E) = 1 - Pr(E')$.

Verification of the Rule for Complements
Since $E \cap E' = \emptyset$, the addition rule for mutually exclusive events yields

$$Pr(E \cup E') = Pr(E) + Pr(E')$$

By definition of E', $E \cup E' = S$, and since S is the certain event, we have

$$Pr(E \cup E') = Pr(S) = 1$$

Hence, putting these two equations together, we get

$$Pr(E) + Pr(E') = 1$$

Therefore

$$Pr(E') = 1 - Pr(E)$$

The next example shows how to apply the rule for complements.

E X A M P L E 7

Problem

In the experiment of Example 4, find the probability that a particular accident occurred in neither quarter 1 nor quarter 2.

Solution Let E be the desired event.

  E: The accident occurred in neither quarter 1 nor quarter 2.

Then E' is "the accident occurred in quarter 1 or quarter 2," and so $Pr(E') = .24 + .19 = .43$. By the rule for complements

$$Pr(E) = 1 - Pr(E') = 1 - .43 = .57$$

EXERCISE SET 7–2

In Problems 1 to 6 a die is rolled and the number on the top face is recorded. Find the probability of the event.

1. 6.

2. 5 or 6.

3. An odd number.

4. A number greater than 4.

5. A number between 2 and 5 inclusive.

6. A number less than 5.

In Problems 7 to 10 two coins are tossed and the number of heads is recorded. Find the probability of the events.

7. One heads.

8. Two heads.

9. Less than two heads.

10. An odd number of heads.

In Problems 11 to 14 the experiment is the Fairley study of accidents on Route 2. Find the probability of the events.

11. The accident occurred in quarter 1 or quarter 2.

12. The accident occurred in quarter 2, quarter 3, or quarter 4.

13. The accident occurred in the first three quarters.

14. The accident occurred in the last two quarters.

In Problems 15 to 18 a card is drawn from an ordinary deck. Define the events

E: A red card A: The ace of spades

F: A spade B: A red 2

G: A club C: A picture card

Find the probability of the events.

15. $E \cup F$.

16. $E \cup G$.

17. $A \cup B$.

18. $B \cup C$.

In Problems 19 and 20 a card is drawn from an ordinary deck. Find the probability of the given event in two ways, first directly and then by using the rule for complements.

19. The card is not a black ace.

20. The card is not the king of hearts.

In Problems 21 to 24 an experiment has a sample space $S = \{a, b, c, d, e, f\}$. Is the table a possible probability distribution for the experiment? If not, state why.

21. | outcomes | a | b | c | d | e | f |
|---|---|---|---|---|---|---|
| probability | .1 | .1 | .2 | .3 | .2 | .1 |

22. | outcomes | a | b | c | d | e | f |
|---|---|---|---|---|---|---|
| probability | .2 | .3 | .1 | .1 | .2 | .1 |

23. | outcomes | a | b | c | d | e | f |
|---|---|---|---|---|---|---|
| probability | .4 | .1 | .2 | .1 | .2 | .1 |

24. | outcomes | a | b | c | d | e | f |
|---|---|---|---|---|---|---|
| probability | .11 | .1 | .38 | .1 | .21 | .1 |

25. Three coins are tossed and the number of heads is recorded. Find the probability of the events (a) less than two heads, (b) an odd number of heads.

26. Four coins are tossed and the number of heads is recorded. Find the probability of the events (a) at least three heads, (b) no more than one heads.

27. American roulette is a game where a ball is rolled on a spinning wheel and it lands in one of 38 slots, 18 of which are red, 18 are black, and 2 are green. What is the probability that the ball lands on a red number?

28. Select a ball from an urn containing ten balls numbered 0, 1, . . . , 9. What is the probability that the number selected is (a) even, (b) greater than 7, (c) less than 3 or greater than 7?

29. A marble is chosen from an urn containing 3 red and 7 white marbles. What is the probability that it is red?

30. A marble is chosen from an urn containing 3 red, 2 blue, and 7 white marbles. What is the probability that it is (a) red, (b) blue, (c) red or blue?

31. A card is drawn from an ordinary deck. Find the probability that the card is (a) a king or a jack, (b) a red ace or a black king, (c) a club or a diamond.

In Problems 32 to 34 an experiment has the sample space $S = \{a, b, c, d\}$ with the following probability distribution:

Outcome	Probability
a	.32
b	.51
c	.07
d	p

Define the events $E = \{a, b\}$ and $F = \{c, d\}$.

32. Find p.

33. Find $Pr(E)$ and $Pr(F)$.

34. Find $Pr(E \cup F)$.

In Problems 35 to 38 an experiment has the sample space $S = \{a, b, c, d, e, f\}$ with the following probability distribution:

Outcome	Probability
a	.25
b	.20
c	.15
d	.05
e	.10
f	.25

Define the events $E = \{a, b\}$, $F = \{c, d\}$, and $G = \{e, f\}$.

35. Find $Pr(E)$, $Pr(F)$, and $Pr(G)$.

36. Find $Pr(E \cup F)$ and $Pr(E \cup G)$.

37. Find $Pr(E')$ and $Pr(F')$.

38. Find $Pr[(E \cup F)']$.

In Problems 39 and 40 let E and F be events such that $Pr(E) = .3$, $Pr(F) = .4$, and $Pr(E \cap F) = 0$.

39. Find $Pr(E \cup F)$.

40. Find $Pr[(E \cup F)']$.

41. In an introductory business course students were asked if they were or were not accounting majors. The survey also recorded the sex of the student. The results are given in the following table.

	Accounting Major	Not Accounting Major
Men	10	18
Women	6	16

The experiment is to select a student at random. Find the probability of each event.
(a) The applicant was male and was an accounting major.
(b) The applicant was female and was not an accounting major.
(c) The applicant was female or was not an accounting major.

42. In an introductory political science course students were asked if they were registered as a Democrat or a Republican and whether or not they were in favor of a business tax increase. The results are given in the following table:

	Business Tax Increase	
	In Favor	Not in Favor
Democrat	13	7
Republican	8	12

The experiment is to select a student at random. Find the probability of each event.
(a) The applicant was a Democrat and in favor of the increase.
(b) The applicant was a Republican and in favor of the increase.
(c) The applicant was a Republican or the applicant was in favor of the increase.

43. A decimal die has ten faces, each with a number from 1 to 10 on it. If a decimal die is tossed, what is the probability that the top face is at least 3?

44. Assume that the eight possible sex distributions in a three-child family are equally likely. Compute the probability that more than one child is male?

45. A die has two faces with 3 spots, two faces with 4 spots, and two faces with 5 spots. If the die is thrown once, what is the probability that the top face has (a) 3 spots? (b) an odd number of spots?

46. A die has two faces with 3 spots, one face with 4 spots, and three faces with 5 spots. If the die is thrown once, what is the probability that the top face has (a) 5 spots? (b) at least 4 spots?

47. Three coins are tossed and the number of heads is recorded. What is the probability of recording less than two heads?

48. A die is rolled and the number on the top face is recorded. What is the probability that the number is at least 5?

49. Suppose an experiment has sample space $S = \{a, b, c\}$ such that $Pr(a) = 2Pr(b) = 3Pr(c)$. Find the probability distribution.

50. Suppose an experiment has sample space $S = \{a, b, c\}$ such that $Pr(a) = .1 + Pr(b)$. Define the event $E = \{a, b\}$. If $Pr(E') = .7$, find the probability distribution.

Referenced Exercise Set 7–2

1. An important issue in our society is determining whether discrimination because of sex or ethnic identity is being practiced. In an article by Peter Bickel et al. the following data were gathered while studying the graduate admissions at the University of California, Berkeley, for the fall quarter, 1973.

Applicants	Outcome	
	Admit	Deny
Men	3738	4704
Women	1494	2827

Source: From Peter Bickel et al., "Sex Bias in Graduate Admissions: Data from Berkeley," *Science,* Vol. 187, pp. 398–404, February 7, 1975.

The question studied was whether the decision to admit or deny was influenced by the sex of the applicant.

Define the events

E: The applicant was a man and was admitted.

F: The applicant was a man and was denied admittance.

G: The applicant was a woman and was admitted.

H: The applicant was a woman and was denied admittance.

Select an applicant at random and find (a) $Pr(E \cup F)$, (b) $Pr(G \cup H)$, (c) $Pr(E \cup G)$, (d) $Pr(F \cup H)$.

2. In an article on the possible underestimating of catastrophic accident probabilities, William B. Fairley reviewed data supplied to the Federal Power Commission by importers of liquefied natural gas. A large spill in a port could cause a catastrophic fire, resulting in the loss of lives and property similar to a major accident at a nuclear power plant. Fairley defined the following probability distribution based on natural gas shipments to Staten Island and Providence in a ten-year period.

Outcome	Probability
Accident occurred based on officially reported data	8/10,000
Accident occurred based on unreported accounts	32/10,000
No accident occurred	

Source: From William Fairley, "Evaluating the 'Small' Probability of a Catastrophic Accident from the Marine Transportation of Liquefied Natural Gas," *Statistics and Public Policy* by Fairley and Mosteller, eds., Reading, MA, Addison-Wesley, pp. 331–353, 1977.

Determine (a) $Pr($No accident occurred.$)$, (b) $Pr($An accident occurred.$)$. (c) What is the factor by which Fairley multiplied the historical probability of an accident in order to obtain the probability of an unreported accident; that is, how much larger is the latter than the former?

3. Probability theory is often used to assess the potential outcomes when certain risks are taken. Farming is a profession where risks are taken as a matter of course. In an article investigating the practical selection of which types of risks to take, Atwood studied Philippine rice yields. The table gives the yield in pounds per acre and the corresponding probability of attaining that yield. (a) Find the probability of attaining at most 40

Yield	0–40	40–50	50–60	60–70	70–80	80–90	>90
Probability	.03	.05	.09	.13	.18	.23	

Source: From Joseph Atwood, ''Demonstration of the Use of Lower Partial Moments to Improve Safety-first Probability Limits,'' *American Journal of Agricultural Economics,* Vol. 67, pp. 787–795, 1985.

pounds per acre. (b) Find the probability of attaining more than 90 pounds per acre.

Cumulative Exercise Set 7–2

1. Find the sample space of each experiment.
   (a) Select a day of the week.
   (b) Select a month of the year.
   (c) Interview four voters and record the number who voted for the incumbent in the last election.

2. Consider the experiment of rolling a die and recording the number on the top face. Let $E = \{1, 3, 5\}$ and $F = \{4, 5, 6\}$. Find $E \cup F$, $E \cap F$, and E'.

3. If a nickel and a dime are flipped, what is the probability that one will land heads and the other tails?

4. Assume that in a three-child family the eight possible sex distributions of the children are equally likely. Compute the probability that exactly one child is female.

5. If a card is selected at random from an ordinary deck of cards, what is the probability that it is a red card or a black king?

6. If a customer selects an ice-cream cone from a list of flavors limited to vanilla, chocolate, and strawberry, what is the probability that the flavor is not strawberry?

7. If a cookie is selected from a jar that contains four vanilla cookies and six chocolate cookies, what is the probability that it is chocolate?

8. If one die is cast and the value of the top face is recorded, what is the probability that the value is at least 3?

7–3   The Addition Principle

In the previous section some elementary rules allowed us to compute probabilities of various events. In all cases the events were mutually exclusive. In this section we consider events that have a nonempty intersection.

For example, select a card from a deck and define the two events

E: The card is a spade.

F: The card is a king.

To find $Pr(E \cup F)$, we count the number of outcomes favorable to the event $E \cup F$. There are 13 spades and 4 kings; but the king of spades was counted twice, once as a spade and once as a king. Thus in $E \cup F$ there are 13 spades and only 3 additional kings, the king of spades having already been counted. Therefore

$$Pr(E \cup F) = \frac{16}{52} = \frac{4}{13}$$

The Venn diagram in Figure 1 helps to explain where the king of spades causes the difficulty—in $E \cap F$. Hence to find the probability of the union of two events that are not mutually exclusive, the intersection of the events is critical. This discussion leads to the general result called the **Addition Principle.**

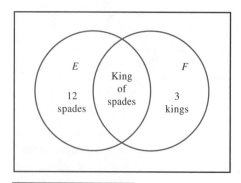

FIGURE 1

The Addition Principle
If E and F are any two events in a sample space, then

$$Pr(E \cup F) = Pr(E) + Pr(F) - Pr(E \cap F)$$

The Addition Principle for Mutually Exclusive Events is a special case of the Addition Principle because it is used when $E \cap F$ is the empty set. The Addition Principle can be used in all cases.

To illustrate how to use the Addition Principle, we consider the preceding discussion and find $Pr(E \cup F)$. We have $Pr(E) = 13/52$, $Pr(F) = 4/52$, and $Pr(E \cap F) = 1/52$. Thus

$$Pr(E \cup F) = \frac{13}{52} + \frac{4}{52} - \frac{1}{52} = \frac{16}{52}$$

It is necessary to subtract $Pr(E \cap F)$ because it was accounted for twice, once in $Pr(E)$ and again in $Pr(F)$.

The fraction 16/52 can be reduced to 4/13 as was done earlier. However, sometimes it is advantageous to leave the fraction in its unreduced form. In this case the fraction 16/52 illustrates that 16 of the 52 cards are in $E \cup F$. In general, we reduce a fraction unless it is more instructive not to. For instance, in the next two examples that further illustrate the Addition Principle, some fractions are left in their unreduced form because they will be added to other fractions and their least common denominator is easier to find in this form.

EXAMPLE 1

The experiment is to draw a card from a deck. Define

E: A heart $\quad$ F: A picture card $\quad$ G: A red card

Problem

Find (a) $Pr(E \cup F)$, (b) $Pr(F \cup G)$.

Solution There are 13 hearts, 12 picture cards, and 26 red cards, so $Pr(E) = 13/52$, $Pr(F) = 12/52$, and $Pr(G) = 26/52$. (a) There are 3 hearts that are picture cards, so $Pr(E \cap F) = 3/52$. Therefore

$$Pr(E \cup F) = \frac{13}{52} + \frac{12}{52} - \frac{3}{52} = \frac{22}{52} = \frac{11}{26}$$

(b) There are 6 red cards that are picture cards, so $Pr(F \cap G) = 6/52$. Therefore

$$Pr(F \cup G) = \frac{12}{52} + \frac{26}{52} - \frac{6}{52} = \frac{32}{52} = \frac{8}{13}$$

The next example deals with the experiment of tossing two dice and recording the numbers on the top two faces. What is the sample space, S? It might seem as though there are 11 elements in S, the numbers 2, 3, 4, 5, 6, 7, 8, 9, 10, 11, 12. These numbers, however, are the sums of the top faces. The experiment is to record the numbers on each face. It is helpful to distinguish between the two dice. Let one be white and the other black (see Figure 2a), or let one be the first die and the other be the second (see Figure 2b). In any case the sample space consists of 36 elements. Thus "5 on the white die and 6 on the black" is represented by the ordered pair (5, 6). Note that $(5, 6) \neq (6, 5)$. The 36 outcomes are equally likely, so the probability of each is $\frac{1}{36}$.

In these examples it is assumed that the dice are "fair," meaning that they are well balanced and each face is equally likely to occur. If a die is "loaded," meaning that one or more sides have extra weight so that one number is more likely to occur, the probabilities would change. Casinos take great precaution to emphasize to their customers that their dice are fair. After a set amount of plays the dice are replaced and the old ones are destroyed.

If the experiment is to toss two dice and record the sum of the top faces, the sample space consists of the 11 sums, 2 through 12. These outcomes are not equally likely. A 2 comes up much less frequently than a 7. To find the probability of each outcome, we determine how many ordered pairs are associated with each sum. For instance, a sum of 2 can occur in only one way, that is, observing (1, 1). So $Pr(2) = \frac{1}{36}$. But a 5 can occur in four ways; (1, 4), (2, 3), (3, 2), or (4, 1), so $Pr(5) = \frac{4}{36} = \frac{1}{9}$. The probability distribution is as follows. Notice the symmetry of the distribution; 2 and 12 have the same probability, as do 3 and 11, and so forth.

outcome	2	3	4	5	6	7	8	9	10	11	12
probability	$\frac{1}{36}$	$\frac{2}{36}$	$\frac{3}{36}$	$\frac{4}{36}$	$\frac{5}{36}$	$\frac{6}{36}$	$\frac{5}{36}$	$\frac{4}{36}$	$\frac{3}{36}$	$\frac{2}{36}$	$\frac{1}{36}$
(reduced)		$\frac{1}{18}$	$\frac{1}{12}$	$\frac{1}{9}$		$\frac{1}{6}$		$\frac{1}{9}$	$\frac{1}{12}$	$\frac{1}{18}$	

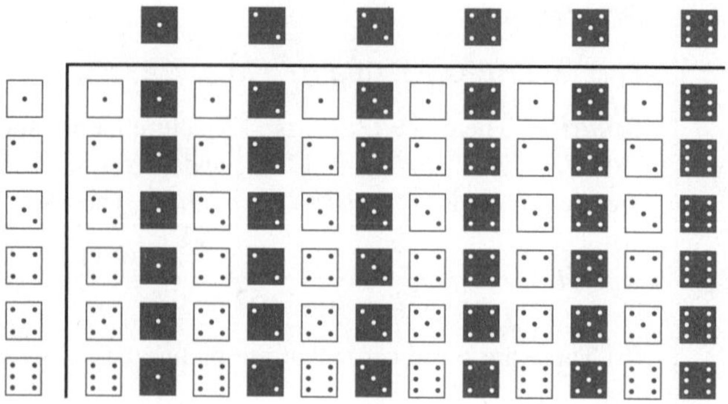

(a)

Second die

First die

		(1, 1)	(1, 2)	(1, 3)	(1, 4)	(1, 5)	(1, 6)
		(2, 1)	(2, 2)	(2, 3)	(2, 4)	(2, 5)	(2, 6)
		(3, 1)	(3, 2)	(3, 3)	(3, 4)	(3, 5)	(3, 6)
		(4, 1)	(4, 2)	(4, 3)	(4, 4)	(4, 5)	(4, 6)
		(5, 1)	(5, 2)	(5, 3)	(5, 4)	(5, 5)	(5, 6)
		(6, 1)	(6, 2)	(6, 3)	(6, 4)	(6, 5)	(6, 6)

(b)

FIGURE 2

EXAMPLE 2

Toss two dice and record the sum of the top faces. Define the events

E: A sum of 3, 4, or 5 occurs.

F: A sum less then 7 occurs.

G: An odd sum occurs.

Problem

Find (a) $Pr(E)$, $Pr(F)$, and $Pr(G)$; (b) $Pr(E \cup G)$.

Solution (a) The events 3, 4, and 5 are mutually exclusive, so $Pr(E)$ is the sum of their respective probabilities. Hence,

$$Pr(E) = Pr(3) + Pr(4) + Pr(5)$$
$$= \tfrac{2}{36} + \tfrac{3}{36} + \tfrac{4}{36} = \tfrac{9}{36} = \tfrac{1}{4}$$

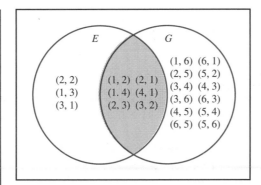

FIGURE 3

Similarly,

$$Pr(F) = Pr(2, 3, 4, 5, 6) = Pr(2) + Pr(3) + Pr(4) + Pr(5) + Pr(6)$$
$$= \tfrac{1}{36} + \tfrac{2}{36} + \tfrac{3}{36} + \tfrac{4}{36} + \tfrac{5}{36} = \tfrac{15}{36} = \tfrac{5}{12}$$

Also

$$Pr(G) = Pr(3, 5, 7, 9, 11) = Pr(3) + Pr(5) + Pr(7) + Pr(9) + Pr(11)$$
$$= \tfrac{2}{36} + \tfrac{4}{36} + \tfrac{6}{36} + \tfrac{4}{36} + \tfrac{2}{36} = \tfrac{18}{36} = \tfrac{1}{2}.$$

(b) There are two outcomes, 3 and 5, in the intersection of E and G, so $Pr(E \cap G) = Pr(3, 5) = Pr(3) + Pr(5) = \tfrac{2}{36} + \tfrac{4}{36} = \tfrac{6}{36} = \tfrac{1}{6}$. In Figure 3 the ordered pairs corresponding to E and G are indicated with $E \cap G$ shaded. From the Addition Principle

$$Pr(E \cup G) = \frac{9}{36} + \frac{18}{36} - \frac{6}{36} = \frac{21}{36} = \frac{7}{12}$$

Verification of the Addition Principle

Recall that if two events A and B are mutually exclusive, then $Pr(A \cup B) = Pr(A) + Pr(B)$. Consider the Venn diagram in Figure 4. The set $E \cup F$ is the union of three disjoint sets:

$$E \cup F = (E \cap F') \cup (E \cap F) \cup (E' \cap F)$$

Because these sets are disjoint, we write

$$Pr(E \cup F) = Pr(E \cap F') + Pr(E \cap F) + Pr(E' \cap F)$$

Also, from Figure 4 notice that

$$E = (E \cap F') \cup (E \cap F) \qquad \text{and} \qquad F = (F \cap E') \cup (F \cap E)$$

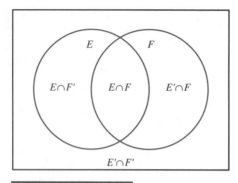

FIGURE 4

In each equality the sets are disjoint, so

$$Pr(E) = Pr(E \cap F') + Pr(E \cap F)$$

$$Pr(F) = Pr(F \cap E') + Pr(F \cap E)$$

Solving for $Pr(E \cap F')$ and $Pr(F \cap E')$, respectively, yields

$$Pr(E \cap F') = Pr(E) - Pr(E \cap F)$$

$$Pr(F \cap E') = Pr(F) - Pr(F \cap E)$$

Substituting these equations into the original statement for $Pr(E \cup F)$ gives us

$$Pr(E \cup F) = [Pr(E) - Pr(E \cap F)] + Pr(E \cap F)$$
$$+ [Pr(F) - Pr(E \cap F)]$$

Two of the three terms with $Pr(E \cap F)$ cancel, and we get

$$Pr(E \cup F) = Pr(E) + Pr(F) - Pr(E \cap F)$$

An Application to Social Science

The following application from political science provides another example of the Addition Rule. There is a more thorough description of this study in the Case Study after Section 7–5. In an article* that studied many uses of probability in the law,

*Hans Zeisel and Harry Kalvern, "Parking Tickets and Missing Women: Statistics and the Law," *Statistics: A Guide to the Unknown,* by J. A. Tanur, et al., eds., San Francisco, Holden-Day, Inc., pp. 102–111, 1972. The authors also investigated the 1968 trial of the pediatrician-author Dr. Benjamin Spock for conspiracy to violate the Selective Service Act by encouraging resistance to the war in Vietnam. They asked the question, "What is the probability that 'this total absence of women jurors was an accident'?" They computed the probability that it was an accident to be 1/1,000,000,000,000,000,000.

Zeisel and Kalvern investigated the perennial debate over the merits of the jury system. The question they raised is, Would there be a difference between verdicts reached by a jury and verdicts that the judge would have reached alone if there were no jury? They studied 3576 jury trials in which the judge reported how he or she would have ruled if there were no jury. The data are as follows: in 17% of the cases the judge would have acquitted the defendant, in 33% of the cases the jury acquitted the defendant, and in 14% of the cases they both reached the verdict of acquittal.

EXAMPLE 3

Problem

A case from the Zeisel and Kalvern data is chosen at random. What is the probability that

(a) Either the jury acquitted or the judge would have acquitted the defendant?

(b) The jury convicted the defendant and the judge would have acquitted the defendant?

Solution Define the events

E: The judge would have acquitted.

F: The jury acquitted.

Then $Pr(E) = .17$, $Pr(F) = .33$, and $Pr(E \cap F) = .14$. See the Venn diagram in Figure 5.

(a) The desired event is $E \cup F$. By the Addition Principle

$$Pr(E \cup F) = .17 + .33 - .14 = .36$$

(b) The desired event is $E \cap F'$. From the Venn diagram in Figure 5 we see that $E = [E \cap F'] \cup [E \cap F]$. Since these sets are disjoint, we have

$$Pr(E) = Pr(E \cap F') + Pr(E \cap F)$$

Solving for $Pr(E \cap F')$ yields

$$Pr(E \cap F') = Pr(E) - Pr(E \cap F)$$

$$= .17 - .14 = .03$$

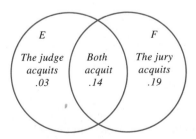

FIGURE 5

A qualified conclusion can be reached from Example 3. If all other factors are insignificant (which is a significant assumption, of course) a defendant has a better chance for acquittal with a jury trial. In only 3% of the cases would a judge have acquitted the defendant when the jury found the defendant guilty. However, in 19% of the cases the jury reached acquittal when the judge would have found the defendant guilty.

Odds

When probabilities are assigned in empirical studies they are usually based on past experience. Theoretically, if the probability $\frac{1}{4}$ is assigned to the event E, it means that if the exact experiment were performed many times, E would occur about one-fourth of the time. But often the experiment will be conducted only once, as when a weather forecaster predicts the likelihood that a hurricane will hit land, or an economist predicts the likelihood that the Dow-Jones average will increase by at least 20%. Frequently this type of probability is expressed in terms of "odds." Consider the statements

The odds against a hurricane reaching land are 3 to 1.
The odds in favor of a 20% or more increase in the Dow-Jones average are 5 to 3.

The first statement means that under exactly the same weather conditions three times the hurricane will not reach land and once it will. Let E be the event that the hurricane will not reach land. Then E will occur three out of four times, so $Pr(E) = 3/4$. Notice that the first statement mentions the "odds against" the event occurring, while the second statement mentions the "odds in favor." One is the complement of the other, so when computing odds we must be careful which odds are desired. If the odds against an event are 3 to 1, then the odds in favor of the event are 1 to 3. We generalize these remarks as follows:

DEFINITION

If the **odds in favor of an event E** are m to n, then $Pr(E) = \dfrac{m}{m + n}$.

If $Pr(E) = m/N$, then the odds in favor of E are m to $N - m$.
If the odds in favor of E are m to n, then the odds against E are n to m.

EXAMPLE 4

Problem

The odds in favor of a general strike are 6 to 5.
(a) What is the probability that a strike will take place? (b) What are the odds against a strike?

Solution (a) Let E be the event that a strike will take place. Then $m = 6$ and $n = 5$ in the formula, so $Pr(E) = 6/(6 + 5) = 6/11$. (b) The probability that a strike will not occur is $1 - 6/11 = 5/11$ so the odds against a strike are 5 to 6.

An alternate formula for odds in favor of event E is

$$\text{Odds in favor of } E = \frac{Pr(E)}{Pr(E')} = \frac{Pr(E)}{1 - Pr(E)}$$

This definition is used if $Pr(E)$ is given and the odds are to be computed. The original definition is used when the odds are given and $Pr(E)$ is to be computed. The next example illustrates that the two definitions are equivalent.

EXAMPLE 5

Problem

The odds in favor of an increase tomorrow in the Dow-Jones average are 5 to 3. Let E represent the event that an increase will take place tomorrow.
(a) Use the original formula to compute $Pr(E)$.
(b) Use $Pr(E)$ computed in (a) in the alternate formula to compute the odds in favor of E.

Solution (a) From the original formula $m = 5$ and $n = 3$, so $Pr(E) = 5/(5 + 3) = 5/8$.
(b) Since the odds are given as 5 to 3, the answer should be 5/3 if the definitions are equivalent. Substituting $Pr(E) = 5/8$ and $Pr(E') = 1 - Pr(E) = 1 - 5/8 = 3/8$ into the alternate formula yields

$$\text{Odds in favor of } E = \frac{5/8}{3/8} = \frac{5}{3}$$

Theoretical Versus Empirical Probability

There are several ways to assign probabilities to events. Everyone uses probabilistic reasoning in making daily decisions. If you are in a hurry to cross a busy intersection, you weigh the chance of being injured with the consequence of being late. If time is running short when studying for an exam, you make a decision on what to concentrate based on the likelihood that the material will be on the exam. These are called **subjective probabilities** because there is no concrete formula for assigning a probability to the event. Decisions are made subjectively based on experience.

Another way to assign probabilities is to rely on concrete data. This is called empirical probability. Most experiments in this text are theoretical rather than empirical. Let us explain the difference between these two types of probabilities.

In Section 7–1 we mentioned the difference between games of chance and studies of real-world phenomena. In assigning probabilities to the outcomes of games, we have **theoretical probabilities** because we compute them by theoretical reasoning rather than by gathering data. For example, when we say the probability of observing a 2 when tossing a die is $\frac{1}{6}$, we recognize that there are six equally likely outcomes. Experience at tossing dice bears this out, but the number $\frac{1}{6}$ came from theoretical reasoning rather than actual data. These types of experiments can be repeated exactly any number of times.

However, most studies of real-world phenomena cannot be repeated under exactly the same conditions. In this case the only way to assign probabilities is to look at past experience. The probabilities are then relative frequencies, often called **empirical probabilities.** Each view of probability has advantages in particular situations.

One of the difficulties with empirical probability is that two studies might yield significantly different results. The science of statistics establishes safeguards so that separate studies will produce as similar results as possible.

EXERCISE SET 7–3

In Problems 1 to 4 two dice are tossed and the sum of the top faces is recorded. Find $Pr(E \cup F)$ for the given events E and F.

1. E: A sum of at most 4 F: A sum of 4 or 5

2. E: A sum of at most 7 F: A sum of 6, 7, or 8

3. E: An even sum F: A sum divisible by 3

4. E: A sum less than 7 F: An odd sum

In Problems 5 to 8 a card is chosen from an ordinary deck. Find the probability that the card is

5. a picture card or a black card.

6. a red picture card or a king.

7. an ace or a queen or a picture card.

8. a jack or a red ace or a card that is not a king.

In Problems 9 and 10 a marble is selected at random from an urn with 3 red, 4 white, 5 blue, and 6 green marbles. Find $Pr(E \cup F)$ for the given events E and F.

9. E: A red or white marble
   F: A blue or red marble

10. E: A red or blue marble
    F: A green or blue marble

In Problems 11 to 14 an experiment is conducted with the following probability distribution:

Outcome	Probability
a	.25
b	.30
c	.10
d	.35

Define the events

$$E = \{a, b, c\} \quad F = \{b, c, d\} \quad G = \{b, d\}$$

Find the probability of the event.

11. $E \cup F$.

12. $E \cup G$.

13. $E' \cup F$.

14. $F' \cup G'$.

In Problems 15 to 18 suppose $Pr(E) = .16$, $Pr(F) = .38$, and $Pr(E \cap F) = .05$. Use a Venn diagram to find the probability of the event.

15. $E' \cap F$.

16. $E' \cup F$.

17. $E \cap F'$.

18. $E \cup F'$.

In Problems 19 to 22 find $Pr(E)$.

19. The odds in favor of E are 4 to 7.

20. The odds in favor of E are 8 to 3.

21. The odds against E are 5 to 2.

22. The odds against E are 3 to 7.

In Problems 23 to 26 find the odds in favor of E and the odds against F.

23. $Pr(E) = 2/9$, $Pr(F) = 3/7$

24. $Pr(E) = 3/4$, $Pr(F) = 1/8$

25. $Pr(E) = .26$, $Pr(F) = .38$

26. $Pr(E) = .42$, $Pr(F) = .05$

27. Five coins are tossed and the number of heads is recorded. Find the probability of observing at least three heads or at most four heads.

28. Suppose $Pr(E) = .45$, $Pr(F) = .31$, and $Pr(E \cap F) = .22$. Find (a) $Pr(E' \cap F')$ (b) $Pr(E' \cup F')$.

29. Suppose $Pr(E \cap F) = .15$, $Pr(E \cup F) = .77$, and $Pr(E \cap F') = .30$. Find (a) $Pr(E' \cap F)$ (b) $Pr(E \cap F')$.

30. In a lottery a three-digit number is selected with replacement. What is the probability that it is (a) less than 100, (b) greater than 50, (c) greater than 50 or less than 100?

31. An American roulette wheel has 38 slots. Two slots are green and labeled 0 and 00. The other slots are numbers 1 to 36 and are evenly divided between red and black. If the ball lands in a slot at random, what is the probability that the slot is (a) red, (b) black, (c) red or black, (d) red or green?

32. In American roulette, what is the probability that the slot in which the ball lands is (a) red or a number greater than 30,   (b) black or a number less than 5.

33. Suppose $Pr(E) = .35$ and $Pr(F) = .65$. Find (a) the odds in favor of E, (b) the odds against F.

34. The odds in favor of the first proposal passing are 7 to 6, the odds in favor of the second proposal passing are 9 to 4, and the odds in favor that both will pass are 5 to 8. What are the odds in favor that at least one will pass?

Use the following table in Problems 35 to 40, which gives the results of a survey that measured how many people from three levels of income regularly invest in the stock market:

Regularly Invest in the Stock Market

		Yes	No
	High	18	6
Income Level	Medium	21	16
	Low	15	24

A person in the survey is chosen at random. Find the probability of each event.

35. The person has a high income or has a low income.

36. The person does not have a high income or has a low income.

37. The person has a low income or regularly invests in the stock market.

38. The person does not have a low income or regularly invests in the stock market.

In Problems 39 and 40 define the events

H: The person has a high income.
L: The person has a low income.
Y: The person regularly invests in the stock market.
N: The person does not regularly invest in the stock market.

39. Find $(H \cup L) \cap N$

40. Find $(H' \cup L) \cap Y$

Problems 41 to 44 refer to the following word problem.

An article on the banking industry revealed that of the 200 banks surveyed one month, 80 banks lowered their prime rate, 65 banks lowered their mortgage rate, 55 banks lowered their certificate of deposit (CD) rate, 40 of the banks that lowered their prime rate also lowered their mortgage rate, 25 of the banks that lowered their prime rate also lowered their CD rate, 30 of the banks that lowered their mortgage rate also lowered their CD rate, 10 banks lowered all three rates. A bank is chosen at random. In Problems 41 to 44 find the probability of the given event.

41. The bank did not lower any of the three rates.

42. The bank lowered exactly one of the three rates.

43. The bank lowered exactly two of the three rates.

44. The bank lowered exactly one or two of the three rates.

45. Suppose E, F, and G are events. Show that

$$Pr(E \cup F \cup G) = Pr(E)$$
$$+ Pr(F) + Pr(G) - Pr(E \cap F)$$
$$- Pr(E \cap G) - Pr(F \cap G)$$
$$+ Pr(E \cap F \cap G)$$

46. Show that the odds in favor of E can be expressed as $Pr(E)/Pr(E')$, that is, the ratio of "$Pr(E)$ to $Pr(E')$."

Referenced Exercise Set 7–3

1. An article by Colton and Buxbaum* on state-mandated vehicle inspection studied the effectiveness (as mea-

*Theodore Colton and Robert Buxbaum, "Motor Vehicle Inspection and Motor Vehicle Accident Mortality," *American Journal of Public Health*, Vol. 58, No. 6, pp. 1090–1099, 1968.

sured by loss of life, personal injury, and property damage) of those states that had mandatory vehicle inspection versus those that did not. One measure they considered was the percentage of mortality rates (rate per 100,000 population) by sex. They found that 58% of the mortalities were in noninspection states, 24% were women, and 14% were women in noninspection states. Define the events

E: The accident occurred in a noninspection state.
F: The victim was a woman.

Find $Pr(E \cup F)$ and $Pr(E \cap F')$.

2. The admission process at many universities is a very complex operation, especially in the face of considerable uncertainty of student performance. In an article by Holtzman and Johnson† designed to use probabilistic models to help colleges choose the right admissions policy for a particular situation, computer simulation was used to compare student performance of those students who were accepted at a major university versus the predicted performance of those students who were rejected. Define the events

E: The applicant was accepted.
F: The applicant's GPA (grade-point average) was (or was predicted to be) greater than or equal to 2.0.

The event *F* includes the first year performance of those students who were accepted as well as the predicted GPA of those who were rejected. The data showed that $Pr(E) = .64$, $Pr(F) = .60$, and $Pr(E \cap F) = .50$.
(a) Express these three events in words and then find
(b) $Pr(E \cup F)$, (c) $Pr(E' \cap F)$, (d) $Pr(E \cap F')$.

3. Probability theory assigns numbers from 0 to 1 to events in order to measure the likelihood of the event occurring. English words, such as "seldom" and "toss-up," also are used to describe the chance of an event occurring. In an article studying whether individuals agree on the meaning of such words, Budescu and Wallsten asked subjects to assign numbers from 0 to 1 to 19 words and phrases that connote a probabilistic meaning.‡ Thirty-two subjects were divided into two groups,

and each person was asked to give a rank ordering to the probability phrases. For one phrase, 94% of the individuals in group 1 agreed with its ordering while 81% of group 2 agreed with its ordering. Find the probability that an individual selected at random from either group would agree with its ordering.

Problems 4 and 5 provide two instances where proper reasoning leads to unexpected conclusions.§

4. Players *A* and *B* each toss a pair of dice two times and add the numbers on the top faces each time. Player *A* wins if the sum is 7 both times; player *B* wins if a sum of 6 appears once and a sum of 8 appears once (in either order).
(a) Since the probability of throwing 7 is greater than the probability of throwing either 6 or 8, is it correct to conclude that *A* has a greater chance of winning than *B*?
(b) What is the probability that *A* will win the game?
(c) What is the probability that *B* will win the game?
(d) Is your answer to part (a) surprising in light of your answers to parts (b) and (c)?

5. Each of three players has a spinner that can point to two equally likely numbers. (See the accompanying figure.) Player *A* can spin 3 or 5, player *B* can spin 1

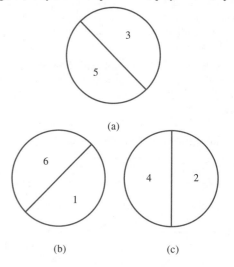

(a)

(b)   (c)

†Albert G. Holzman and Donald B. Johnson, "A Simulation Model of the College Admission Process," *Interfaces*, Vol. 5, No. 3, pp. 34–43, May 1975.

‡David Budescu and Thomas Wallsten, "Consistency in Interpretation of Probabilistic Phrases," *Organizational Behavior and Human Decision Processes*, Vol. 36, pp. 391–405, 1985.

§H. Shultz and B. Leonard, "Probability and Intuition," *Mathematics Teacher*, pp. 52–53, January 1989.

or 6, and player C can spin 2 or 4. When two players compete the winner is the one whose spinner points to the higher number.
(a) When A and B compete, do both players have the same chance of winning?
(b) When B and C compete, do both players have the same chance of winning?
(c) Based on the results of parts (a) and (b), does it seem correct to conclude that players A and C have the same chance of winning?
(d) When A and C compete, do both players have the same chance of winning?
(e) Is your answer to part (d) surprising in light of your answer to part (c)?

Cumulative Exercise Set 7–3

1. Find the sample space of the experiment of rolling a die and recording the number on the top face and then tossing a coin and recording heads or tails.

2. An experiment consists of choosing a committee of two individuals from a list of four people whose last initials are A, B, C, and D. Determine the sample space of the experiment and list the elements of the events

E: A, B, or C is on the committee.
F: Neither A nor B is on the committee.
G: Either C or S is on the committee.

3. The experiment consists of tossing a coin three times and recording the number of heads. Define the two events

E: More than one heads
F: An even number of heads

Find (a) $E \cup F$, (b) $E \cap F$, (c) E'.

4. Find the probability of the events in Problem 3 in parts (a), (b), and (c).

5. An experiment has the sample space, $S = \{a, b, c, d, e, f\}$ with the following probability distribution:

outcomes	a	b	c	d	e	f
probability	.1	.1	.2	.3	.2	.1

Define the events

$E = \{a, c\}$, $F = \{c, d, e\}$, $G = \{e, f\}$

Find (a) $Pr(E \cup F)$, (b) $Pr(E \cup G)$, (c) $Pr(F')$, (d) $Pr(E' \cup G')$.

6. In an introductory business course students were asked if they were or were not marketing majors. The survey also recorded the sex of the student. The results are given in the table.

	Marketing Major	Not Marketing Major
Men	12	20
Women	8	15

A person is chosen at random. Find the probability that
(a) the applicant was male and was an accounting major.
(b) the applicant was female and was not an accounting major.
(c) the applicant was female or was not an accounting major.

7. Two dice are tossed and the sum of the top faces is recorded. Find the probability that
(a) the sum is at most 4.
(b) the sum is at least 9.
(c) the sum is odd.

8. The experiment consists of selecting a marble from an urn containing 2 red, 3 white, and 5 blue marbles and recording its color. Find the probability that the marble
(a) is not red.
(b) is red or blue
(c) is red or white.

9. An experiment is conducted with the following probability distribution:

Outcome	Probability
a	.52
b	.16
c	.03
d	p

Define the events

$E = \{a, b\}$ and $F = \{c, d\}$.
(a) Find p.
(b) Find $Pr(E)$ and $Pr(F)$.
(c) Find $Pr(E \cup F)$.

10. Suppose $Pr(E) = .35$, $Pr(F) = .21$, and $Pr(E \cap F) = .12$. Find (a) $Pr(E' \cap F')$ and (b) $Pr(E' \cup F')$.

11. The following table gives the results of a survey that measured how many people from three levels of income regularly invest in real estate:

		Regularly Invest in Real Estate	
		Yes	No
	High	18	6
Income Level	Medium	21	16
	Low	15	24

A person in the survey is chosen at random. Define the events

H: The person has a high income.

L: The person has a low income.

Y: The person regularly invests in the stock market.

Find the probability of the event (a) $H \cup L$, (b) $H' \cup Y$.

12. The odds in favor of the first proposal passing are 8 to 5, the odds of the second proposal passing are 9 to 2, and the odds that both will pass are 4 to 5. What are the odds that at least one of the proposals will pass?

7–4 Counting Techniques

In this section we use the counting techniques introduced in Chapter 6 to solve more complex problems in probability. Their solutions depend on the equally likely outcomes principle, first covered in Section 7–2. Let us recall it here.

If a sample space S contains N equally likely outcomes and an event E contains m outcomes, then

$$Pr(E) = \frac{m}{N}$$

In other words, if $n(E)$ is the number of outcomes in E and $n(S)$ is the number of outcomes in S, then

$$Pr(E) = \frac{n(E)}{n(S)}$$

We make use of the Multiplication Principle and the formulas for permutations and combinations, $P(n, r)$ and $C(n, r)$, respectively. The first example uses two of these formulas.

EXAMPLE 1

Five coins are tossed and the number of heads is recorded.

Problem

What is the probability of observing (a) exactly two heads, (b) at most two heads?

Solution The sample space is $S = \{0, 1, 2, 3, 4, 5\}$, but these outcomes are not equally likely. To get equally likely outcomes, distinguish the coins and consider the sequence of H's and T's.

We first find the denominator, $n(S)$, in each problem. It is the total number of ways that five coins can land. There are two choices, H and T, for each of the five coins. By the Multiplication Principle presented in Chapter 6, there are $2^5 = 32$ outcomes or sequences of H's and T's. Thus the denominator for each problem is $n(S) = 32$.

(a) From Chapter 6 we see that the number of ways of getting exactly two heads with five coins is $C(5, 2)$. Thus

$$Pr(2H) = \frac{C(5, 2)}{2^5} = \frac{10}{32} = \frac{5}{16}$$

(b) The event "at most two heads" is the union of the three mutually exclusive events: "exactly 0 heads," "exactly one heads" and "exactly two heads." The denominator for the probability of each event is 32 and their numerators are $C(5, 0)$, $C(5, 1)$, and $C(5, 2)$, respectively. Thus

$$Pr(\text{at most } 2 \ H\text{'s}) = Pr(0 \ H) + Pr(1 \ H) + Pr(2 \ H)$$

$$= \frac{C(5, 0)}{32} + \frac{C(5, 1)}{32} + \frac{C(5, 2)}{32}$$

$$= \frac{1}{32} + \frac{5}{32} + \frac{10}{32} = \frac{16}{32} = \frac{1}{2}$$

Many problems in probability can be reduced to urn problems since an experiment usually consists of selecting m objects from a total of N objects. For instance, the coin tossing problem in Example 1 can be thought of as an urn problem in two ways. Think of the experiment as consisting of 5 urns, each with 2 marbles labeled H and T (or red and white). The experiment is to select a marble from each urn. Equivalently, consider the experiment as consisting of 1 urn with 2 marbles labeled H and T, and then select 1 marble five times with replacement. The sample space in each of these urn problems is identical with the sample space in Example 1. The Multiplication Principle is used for the denominator because the selection is done with replacement. The numerators use combinations because we are selecting m of the urns to have the outcome H, so the remaining $5 - m$ urns will have the outcome T.

In previous sections we selected only one object from an urn. But now counting techniques will enable us to tackle more complex situations.

EXAMPLE 2

An urn contains 3 red marbles and 9 white marbles. Three marbles are selected at random without replacement.

Problem

Find the probability of observing (a) 3 red marbles, (b) 3 white marbles, (c) 2 red and 1 white marbles (in any order).

Solution There are 12 marbles in all, so the denominator for each probability is $C(12, 3)$. To determine the numerators, think of the marbles as distinguished, say, each has a different number. The order in which they are selected is not important.

(a) There is one way to select all 3 red marbles. By formula $C(3, 3) = 1$. Hence

$$Pr(3 \text{ red}) = \frac{C(3, 3)}{C(12, 3)} = \frac{1}{220}$$

(b) The number of ways of selecting 3 white marbles is $C(9, 3)$. Hence

$$Pr(3 \text{ white}) = \frac{C(9, 3)}{C(12, 3)} = \frac{84}{220} = \frac{21}{55}$$

(c) There are $C(3, 2)$ ways to select 2 of the 3 red marbles and $C(9, 1)$ ways to select 1 of the 9 white marbles. By the Multiplication Principle there are $C(3, 2) \cdot C(9, 1)$ ways to select 2 red and 1 white marble. Hence

$$Pr(2 \text{ red and 1 white}) = \frac{C(3, 2) \cdot C(9, 1)}{C(12, 3)}$$

$$= \frac{3 \cdot 9}{220} = \frac{27}{220}$$

Poker Hands

When the big loser inevitably screams ''Wait! What beats what?'' the player is demanding to know which poker hand is better than another. One hand beats another when the probability of its occurring is smaller. In fact, the rules of poker are based on probability. To illustrate we consider three hands: a pair (2 matching cards with 3 remaining unmatched cards), a full house (3 of a kind with a pair), and a flush (all the cards the same suit). The order in which the cards appear is immaterial.

EXAMPLE 3

Problem

Compute the probability of being dealt the following 5-card poker hands: (a) a flush, (b) a full house, (c) a pair.

Solution The number of possible hands with 5 cards is the denominator. Since we need 5 of the 52 cards, the denominator will be $C(52, 5)$.

(a) For a flush we need to draw 5 of the 13 cards in one of the four suits. Thus for each suit there are $C(13, 5)$ ways to get a flush.* Since there are four distinct suits, use the Multiplication Principle to show that there are $4 \cdot C(13, 5)$ ways to

---

*To be precise we should subtract the 40 ways of getting a straight flush. That is, in each suit there are 10 ways the cards can also be a straight, like 2 through 6, and a straight flush beats every other hand.

draw a flush. Therefore

$$Pr(\text{a flush}) = \frac{4 \cdot C(13, 5)}{C(52, 5)} \approx .002$$

(b) To get 3 of a kind, we must draw 3 of the 4 cards of a particular type, and there are 13 different types of cards. Thus there are $13 \cdot C(4, 3)$ ways of getting 3 of a kind. But we must also fill out the hand with another pair. There are 12 remaining types of cards. Each type has 4 cards and we must draw 2 of them. Thus there are $12 \cdot C(4, 2)$ ways to fill out the hand. Therefore

$$Pr(\text{full house}) = \frac{13 \cdot C(4, 3) \cdot 12 \cdot C(4, 2)}{C(52, 5)} \approx .0014$$

(c) There are 13 different types of cards, each in a separate suit. To get a pair, we must draw 2 of the 4, which can be done in $C(4, 2)$ ways. Since this can be done in 13 ways, there are $13 \cdot C(4, 2)$ ways to get a pair. The other 3 cards must not match. In each suit there are $C(12, 3)$ ways to select 3 different types from the remaining 12 types of cards. The 3 cards can be in any suit, so there are 4^3 ways of selecting each of the $C(12, 3)$ possibilities. Hence there are $C(12, 3) \cdot 4^3$ ways to fill out the hand. Therefore

$$Pr(\text{a pair}) = \frac{13 \cdot C(4, 2) \cdot C(12, 3) \cdot 4^3}{C(52, 5)} \approx .42$$

Thus a full house beats a flush, which in turn beats a pair.

One of the most useful applications of probability theory is sampling. When an entire set of data cannot be inspected, usually because of time or cost considerations, a sample of the data is selected and studied. Then an inference about the whole set is made from the sample; for example, when a firm is shipped goods often a sample is chosen at random and tested for defectives. If more than a prescribed number of defectives is found, the entire shipment is rejected. The next example illustrates how probability and sampling are used in accounting. (See Referenced Exercise 2 for another application with more realistic data.)

EXAMPLE 4

An auditor generally does not study all the accounts of a firm, but rather, selects a representative sample. In a certain category the auditor chooses a sample of 4 from a total of 20 accounts to see how many are delinquent. Suppose that, unknown to the auditor, 5 of the 20 are delinquent. If the auditor finds that 1 out of the 4 in the sample is delinquent, the sample will have exactly the same percentage of delinquencies as the entire set—25%. If the sample is off by 1, so that 0 or 2 delinquencies are found, the sample will not be too far off. But if 3 or 4 are discovered, then the sample will be a very poor representation of the data and the auditor's conclusions will be grossly in error.

Problem

Given that 5 out of 20 accounts are delinquent, if a sample of 4 is selected, what is the probability that 3 or 4 of the accounts are delinquent?

Solution There are $C(20, 4)$ ways to select 4 accounts at random, which is the denominator. There are $C(5, 4)$ ways to select 4 of the 5 delinquent accounts. So $Pr(4 \text{ delinquencies}) = C(5, 4)/C(20, 4)$. There are $C(5, 3)$ ways to select 3 of the 5 delinquencies, and for each of these there are $C(15, 1)$ ways to select 1 of the other 15 accounts. Hence $Pr(3 \text{ delinquencies}) = C(5, 3) \cdot C(15, 1)/C(20, 4)$. Therefore

$$Pr(3 \text{ or } 4 \text{ delinquencies}) = Pr(4) + Pr(3)$$

$$= \frac{C(5, 4)}{C(20, 4)} + \frac{C(5, 3) \cdot C(15, 1)}{C(20, 4)}$$

$$= \frac{5}{4845} + \frac{30}{969} = \frac{31}{969} \approx .03$$

Thus even with a small sample of 4 out of 20, there is only a 3% chance that the sample will yield a large discrepancy from the whole set of data.

The next example demonstrates how to use the complement principle to simplify some computations. Recall that if E' is the complement of E, then $Pr(E) = 1 - Pr(E')$.

EXAMPLE 5

An urn contains 10 marbles numbered 1 to 10. Select 3 marbles from the urn with replacement and record the numbers.

Problem

What is the probability that at least 2 will have the same number?

Solution Let E represent the event that at least 2 will have the same number. Then E' represents the event that the 3 marbles do not have the same number. Think of the experiment in three stages; at each stage select a marble, record its number, and then replace it. For E' to occur the three numbers must be different—at the first stage there are ten possible numbers, at the second stage there are nine numbers, and eight at the third stage. So the number of ways of selecting 3 marbles with different numbers is $P(10, 3) = 10 \cdot 9 \cdot 8 = 720$. Permutations are used because the order of the digits is important; a different order can produce a different number, such as 123 versus 321. There are 10^3 total ways to select three numbers. Hence

$$Pr(E') = Pr(\text{three different numbers})$$

$$= \frac{P(10, 3)}{10^3}$$

$$= \frac{720}{1000} = \frac{18}{25}$$

Therefore the probability of getting at least two numbers the same is

$Pr(E) = Pr$(at least two numbers the same)

$\qquad = 1 - Pr(E') = 1 - Pr$(three different numbers)

$\qquad = 1 - \dfrac{18}{25} = \dfrac{7}{25}$

Most people find this result surprising. It seems as if the probability would be smaller. Here is another result that is even more perplexing. How many people would you have to select so that the probability of two of them having the same birthday is greater than 50%? After the next example we answer this question.

EXAMPLE 6

Problem

Four people are selected at random. What is the probability that at least two of them have the same birthday? ·

Solution This problem is similar to Example 5. Here we have four stages instead of three and at each stage there are 365 possibilities instead of 10. Let E be the event that at least two of them have the same birthday. Then E' is the event that all four have different birthdays. Think of the possibilities at each of the four stages for E' to occur.

stage	1	2	3	4
possibilities	365	364	363	362

Hence there are $P(365, 4)$ ways that the four can have different birthdays. There are 365^4 total ways of selecting four birthdays. Thus

$Pr(E') = Pr($ four different birthdays$)$

$\qquad = \dfrac{P(365, 4)}{365^4}$

$\qquad \approx .984$

Therefore the probability of at least two people having the same birthday is

$Pr(E) = Pr$(at least two birthdays the same)

$\qquad = 1 - Pr(E') = 1 - Pr($ four different birthdays$)$

$\qquad = 1 - .984 = .016$

Example 6 says that if 4 people are chosen at random, the probability that at least 2 will have the same birthday is just 1.6%. As expected, this is a small number. The same process can be used to compute the probability that at least 2 people have the same birthday if 10, 20, or any number of people were selected. What is the smallest number of people that we can choose so that the probability is greater than 50%? Suppose n people are selected. Let E be the event that at least 2 of the n people have the same birthday. Then the formula for $Pr(E)$ is

$$Pr(E) = 1 - \frac{P(365, n)}{365^n}$$

The following table gives the values for various choices of n:

n	5	10	15	20	22	23	25	30	40	50
$Pr(E)$	.027	.117	.253	.411	.476	.507	.569	.706	.891	.970

The probability for 22 people is less than .5, but for 23 people the probability is .507, greater than 50%. Many people guess that the number needed for the probability to exceed .5 is closer to 50 or more, but notice that if 50 people are selected, it is almost certain that at least 2 will have the same birthday. It is not difficult to conduct this experiment, and a large group of people is not needed. Simply write the numbers from 1 to 365 on slips of paper and select them from a hat with replacement until the same number appears. The number of draws will most likely be close to 23.

EXERCISE SET 7–4

In Problems 1 to 6 a coin is tossed n times and the number of heads is recorded. Find the probability of observing the given event E.

1. $n = 6$, E is "exactly one heads."

2. $n = 6$, E is "exactly two heads."

3. $n = 6$, E is "at most two heads."

4. $n = 6$, E is "at least two heads."

5. $n = 7$, E is "at least two heads."

6. $n = 7$, E is "at most two heads."

In Problems 7 to 10 an urn contains 7 red and 5 white marbles. Four marbles are selected without replacement. Find the probability of the event.

7. Exactly 4 are red.

8. Three are red and 1 is white.

9. At least 3 are red.

10. At most 1 is white.

11. A manufacturer ships a package of $n = 12$ automobile batteries. The customer selects $m = 4$ of them and

tests for defectives. Suppose that the package has two defectives. Find the probability that the sample will contain at most one defective.

In Problems 12 to 14 solve Problem 11 with the different choices of n and m.

12. $n = 20$, $m = 4$.

13. $n = 20$, $m = 6$.

14. $n = 30$, $m = 5$.

15. An auditor selects a sample of 5 accounts from a total of $n = 20$ accounts and examines them for delinquency. Suppose there are $m = 5$ delinquent accounts. Find the probability that the sample will have 4 or 5 delinquent accounts.

In Problems 16 to 18 solve Problem 15 with the different choices of n and m.

16. $n = 20$, $m = 6$.

17. $n = 50$, $m = 5$.

18. $n = 100$, $m = 6$.

In Problems 19 to 22 a three-card poker hand is dealt. Find the probability of observing the described hand.

19. All three cards are the same color.

20. Two of a kind.

21. All three cards are the same suit.

22. Three of a kind.

In Problems 23 to 26 an urn contains n marbles numbered 1 to n. Select m marbles with replacement and record the number. Find the probability that at least 2 of them are the same number.

23. $n = 10$, $m = 4$.

24. $n = 15$, $m = 4$.

25. $n = 20$, $m = 5$.

26. $n = 50$, $m = 4$.

27. Ten coins are tossed and the number of heads is recorded. Find the probability of observing (a) exactly three heads, (b) at most three heads, (c) at least three heads.

28. A true-false test consists of ten questions. If the student guesses at each answer, find the probability that the number of correct answers is at least seven.

29. An urn contains 10 red, 6 white, and 4 blue marbles. Four marbles are selected without replacement. Find the probability that (a) all 4 are red, (b) at least 3 are red.

30. In the game of Lotto the player selects six numbers from 1 to 40 inclusive, without repetition. The order in which the numbers are selected is not important. What is the probability of winning?

31. In poker, what is the probability of drawing (a) 4 of a kind, (b) a straight (the 5 cards are in sequence, but not necessarily dealt in sequence, so ace-2-3-4-5 in any order is a straight).

32. For a group of ten people compute the probability that at least two of them have the same birthday.

33. There are 50 members in a union and 10 agree with the president on whether to take a pay cut. The president selects 4 at random and asks their opinion. What is the probability that at least 1 will agree with the president?

34. Seven letters are selected from the alphabet with replacement to form a code. Find the probability that the code has at least one letter repeated.

35. There are 15 banks in a city and 8 of them offer students a discount on checking accounts. A student calls 6 banks at random and asks if they offer a discount. Find the probability that at least 3 of the banks questioned offer a discount.

36. Thirty people are called for jury duty, 18 of whom are women. Find the probability that at most 1 woman is selected for the 12-person jury.

37. A shipment of 100 engines contains five defectives. The quality control inspector selects 3 engines at random to be tested. Find the probability that none of the engines tested is defective.

38. A batch of 50 computers contains five defectives. The quality control inspector selects 4 computers at random to be tested. Find the probability that at least 1 of the computers tested is defective.

39. Assume that the possible sex distributions in a four-child family are equally likely. Compute the probability that exactly two of the children are girls.

40. If a fair coin is tossed ten times, what is the probability that it will land heads exactly five times?

41. Two cookies are grabbed (without replacement) from a jar containing 4 vanilla cookies and 4 chocolate cookies. What is the probability that the cookies will be different flavors?

42. A box of ornaments contains 7 good ornaments and 3 defective ornaments. If a sample of 2 ornaments is selected, what is the probability that exactly one ornament will be defective?

43. Suppose there are two used-car lots, a small lot with 14 goods cars and 2 lemons, and a big lot with 57 good cars and 7 lemons.
    (a) If a car is selected from the big lot, what is the probability that it is a lemon?
    (b) If a lot is selected and one car is bought from it, what is the probability that the car is good?
    (c) If a lot is selected and two cars are bought from it, what is the probability that both cars are good?

44. A jar contains a dozen cookies: 8 chocolate and 4 vanilla. If you reach into the jar and grab 3 cookies, one at a time and without replacement, what is the probability that all 3 cookies are chocolate?

45. A true-false test consists of 20 questions. If the student guesses at each answer, find the probability that the number of correct answers is at least 14.

46. An urn contains 20 red, 12 white, and 8 blue marbles. Five marbles are selected without replacement. Find the probability that (a) all 5 are red, (b) at least 3 are red, (c) 2 are red and at least 2 are white.

Referenced Exercise Set 7–4

1. A study of how a bank should best improve waiting time in its drive-in facility found that bank officials define service as poor if a customer must wait more than 4 minutes.* This is because people base their impatience on the movement of the minute hand between two marks, which represents 5 minutes on a clock. The average service time was just less than 2 minutes, and out of every 14 customers, 3 would require more than $2\frac{1}{2}$ minutes. Poor service would result if 2 of these people entered the same line. Assume that a bank has 7 teller stations and 14 people are in line, two at each teller station. Suppose 3 of the 14 require more than $2\frac{1}{2}$ minutes. Find the probability that 2 of these 3 are in the same line.

2. In an experiment, reported in an article by Neter,† to test the accuracy of their sampling method of accounting, executives of the Chesapeake & Ohio Railroad compared the actual amount of revenue due them on interline freight shipments versus the amount computed by sampling. Sampling techniques have many checks to ensure that the sample contains percentages similar to the actual data. This is where probability is used. In the six-month period of the study there were 23,000 total bills, so they divided them into categories according to the size of the bill. In category 3, $11 to $20, the sample consisted of 20% of the bills. There were 8000 total bills in category 3, so 1600 bills were in the sample. If too many bills in the sample were greater than $18, the sample would give a much larger gross figure than the actual amount. There were 800 bills in category 3 that were greater than $18. Set up the formula for the probability that the sample had at most 200 bills that were greater than $18.

3. How much better are experts than nonexperts at making subjective probabilities? This is the subject of an article studying ways to measure how well experts make probability assessments about events in their field of expertise.‡ The researchers questioned 22 casino dealers and 24 others on various situations in the card game blackjack. They posed the same 19 questions to each group. An example of a question is, "Out of 1000 cases in which the dealer's first card is a 4, how often will the dealer finally bust (meaning get a score of more than 21)?" Of the 46 individuals, five answered this question correctly, where "correctly" means their answer came within 100 cases of the correct answer. Given that each person has the same probability of answering correctly, find the probability that all 5 who answered the question correctly were dealers.

4. The language problem in Canada reached a peak in the summer of 1990. An article by John Burns,§ a well-known political commentator, revealed the following statistics about the ability of Canadians to speak either French or English:

	English Only	French Only	Both
Quebec	369,065	3,808,560	2,226,756
Rest of Canada	16,357,840	149,170	1,829,415

(a) If a person is chosen at random in Quebec, what is the probability that the person speaks French only?

(b) If a person is chosen at random outside Quebec, what is the probability that the person speaks English only?

(c) If a person is chosen at random in Canada, what is the probability that the person speaks both languages?

*B. L. Foote, "A Queuing Case Study of Drive-In Banking," *Interfaces,* Vol. 6, No. 4, August 1976.

†John Neter, "How Accountants Save Money by Sampling," *Statistics: A Guide to the Unknown,* by J. A. Tanur et al., eds., San Francisco, Holden-Day, Inc., 1976.

‡W. A. Wagenaar and G. B. Keren, "Calibration of Probability Assessments by Professional Blackjack Dealers, Statistical Experts and Lay People," *Organizational Behavior and Human Decision Processes,* Vol. 36, pp. 406–416, 1985.

§John Burns, "Canada's Crucial Seven Day Countdown," *New York Times,* June 17, 1990, Sec. 4 (Week in Review), p. 3.

Cumulative Exercise Set 7–4

1. Find the sample space of each experiment.
   (a) Select two cards from an ordinary deck and record the number of queens.
   (b) Roll a die and record whether the number on the top face is at least 2.
   (c) Administer a drug to three patients and record the number of patients who were cured by the drug.

2. Select a card from an ordinary deck of cards. Let E be the event of drawing a queen and F the event of drawing a heart. Find $E \cup F$, $E \cap F$, and E'.

3. If a penny, nickel, dime, and quarter are flipped, what is the probability that two coins will land heads and two coins will land tails?

4. What is the probability that a family with five children will have three girls and two boys?

5. If a card is selected at random from an ordinary deck of cards, what is the probability that it is a red ace or a black king?

6. The enrollment in a certain math class can be described by the following table:

	Commuters	Dormies
Men	10	35
Women	15	45

Let W = women and C = commuters. Find the following probabilities when a person is selected at random from the class.
(a) $Pr(W)$, (b) $Pr(C')$, (c) $Pr(W \cup C)$, (d) $Pr(W \cap C)$.

7. If two dice are tossed, what is the probability that the sum is at least 3?

8. Choose a card from an ordinary deck of cards. Find $Pr(E \cup F)$ for the events E and F defined by

   E: A black picture card F: A jack

9. A hat checker hangs 12 London Fog raincoats and 4 Gucci raincoats. If three raincoats are chosen at random without replacement, what is the probability that 2 are London Fogs and 1 is a Gucci?

10. If five coins are flipped what is the probability that
    (a) exactly one of them will land heads?
    (b) at most two of them will land heads?

11. A box of holiday lights contains 6 red lights and 4 green lights. Two of the lights are selected from the box without replacement. Compute each probability.
    (a) Both are red.
    (b) One is red and the other is green.
    (c) Neither is red.

12. Three hats are placed on a shelf. Each hat contains slips of paper numbered 1 through 20. If one slip is selected from each hat, what is the probability that at least two of the slips have the same number?

13. Suppose 10 people are waiting in the lobby of a hotel for an elevator. Suppose further that the hotel has 20 identical floors and that the 10 people were randomly assigned to 10 different rooms.
    (a) What is the probability that at least 2 people have rooms on the same floor? (Does it seem that the answer should be $\frac{1}{2}$?)
    (b) If you are one of the 10 people, what is the probability that at least one other person is on the same floor as you?

7–5 Conditional Probability

Probability theory is most often used to make predictions about the future, using past performance as a guide. Sometimes the likelihood of an event occurring is altered by new information. We often get a different perspective of a problem by looking at it from another angle. In this section we see how probability handles the situation when a new or different piece of information is included in the problem.

To illustrate, let us consider two problems. The experiment is to draw a card from a deck.

Problem A: Find the probability of observing the king of clubs.

Problem B: Suppose the card is known to be a picture card, find the probability that it is the king of clubs.

To solve Problem A note that there are 52 cards, they are equally likely to occur, and there is one king of clubs, so the probability is 1/52. To solve Problem B note that there is a restricted sample space. Since we know the card is a picture card, we are restricted to those 12 cards. One of them is the king of clubs so the probability is 1/12.

The solution to Problem B is called a *conditional probability*. If E and F are two events, then we write $E \mid F$ to represent the event "E given that F has occurred" or simply "E given F" or "E after F." For example, in Problem B if we let

E: The king of clubs   and F: A picture card

then $Pr(E \mid F) = 1/12$. This expression is called a conditional probability and is read "the probability of E given F is 1/12."

A conditional probability can be expressed in terms of the intersection of the two events. To demonstrate, consider Problem B again and note that $Pr(E \cap F) = 1/52$ and

$$Pr(E \mid F) = \frac{Pr(E \cap F)}{Pr(F)} = \frac{1/52}{12/52} = \frac{1}{12}$$

This is true for any two events E and F. Let us generalize this result for events in a sample space S with N equally likely outcomes. Refer to the Venn diagram in Figure 1. Suppose F has n outcomes and $E \cap F$ has m outcomes. Thus $Pr(F) = n/N$ and $Pr(E \cap F) = m/N$. One way to find $Pr(E \mid F)$ is to consider the restricted sample space of F. That is, since we know that F has occurred, one of the n outcomes in F must occur; there are m outcomes that are also in E, so $Pr(E \mid F) = m/n$. The preceding expression gives the same answer.

$$Pr(E \mid F) = \frac{Pr(E \cap F)}{Pr(F)} = \frac{m/N}{n/N} = \frac{m}{n}$$

This leads us to the definition of $Pr(E \mid F)$.

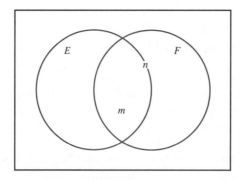

FIGURE 1

DEFINITION

If E and F are two events with $Pr(F) \neq 0$, then the **conditional probability** *of E given F is*

$$Pr(E \mid F) = \frac{Pr(E \cap F)}{Pr(F)}$$

The discussion preceding the definition implies that there is an alternate way to view conditional probabilities in a sample space with equally likely outcomes. In this case compute $Pr(E \mid F)$ by dividing the number of outcomes in $E \cap F$ by the number of outcomes in F. Thus *for equally likely outcomes*

$$Pr(E \mid F) = \frac{n(E \cap F)}{n(F)}$$

In the first example we use the definition because the sample space does not have equally likely outcomes. The second example demonstrates the use of the alternate formula.

EXAMPLE 1

Roll two dice and record the sum of the top faces. Define the two events.

E: Observe an even sum.

F: Observe a sum of 5, 6, or 7.

Problem

Find (a) $Pr(E \mid F)$, (b) $Pr(F \mid E)$.

Solution We need to compute $Pr(E)$, $Pr(F)$, and $Pr(E \cap F)$. We have

$$Pr(E) = Pr(2, 4, 6, 8, 10, 12) = \frac{1 + 3 + 5 + 5 + 3 + 1}{36} = \frac{18}{36} = \frac{1}{2}$$

$$Pr(F) = Pr(5, 6, 7) = \frac{4 + 5 + 6}{36} = \frac{15}{36} = \frac{5}{12}$$

$$Pr(E \cap F) = Pr(6) = \frac{5}{36}$$

(a) By the definition

$$Pr(E \mid F) = \frac{Pr(E \cap F)}{Pr(F)} = \frac{5/36}{15/36} = \frac{5}{15} = \frac{1}{3}$$

(b) Since $E \cap F = F \cap E$, by the definition we get

$$Pr(F \mid E) = \frac{Pr(E \cap F)}{Pr(E)} = \frac{5/36}{18/36} = \frac{5}{18}$$

EXAMPLE 2

A composition course has 25 students of whom 14 are male and 9 are sophomores. In addition, 5 of the sophomores are men.

Problem

A student is chosen at random. Find the probability that the student is
(a) a sophomore given that the student is a man.
(b) a woman given that the student is a sophomore.

Solution Define the events

   M: The student is a man.

   S: The student is a sophomore.

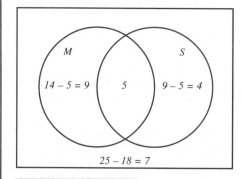

FIGURE 2

Then we must compute (a) $Pr(S \mid M)$ and (b) $Pr(M' \mid S)$. Refer to the Venn diagram in Figure 2.
(a) We have $n(S \cap M) = 5$ and $n(M) = 14$. Hence

$$Pr(S \mid M) = \frac{n(S \cap M)}{n(M)} = \frac{5}{14}$$

(b) We have $n(M' \cap S) = 4$ and $n(S) = 9$. Hence

$$Pr(M' \mid S) = \frac{n(M' \cap S)}{n(S)} = \frac{4}{9}$$

Product Rule

In applications sometimes a conditional probability is easy to compute but the probability of the intersection is not. In this case the formula in the definition of conditional probability can be altered to give the probability of the intersection. If we multiply each side of the equation in the definition by $Pr(F)$, we obtain the **Product Rule.**

> **Product Rule**
> If E and F are any two events, then
> $$Pr(E \cap F) = Pr(F) \cdot Pr(E \mid F)$$

The Product Rule is especially effective when solving problems with tree diagrams. The next example shows how tree diagrams can help solve probability problems that occur in stages.

EXAMPLE 3

A utility company submits a rate increase proposal to the state utility commission. The proposal will go to one of two commission members, either Ms. Allen or Mr. Barnes. In the past Ms. Allen has accepted 60% of the company's proposals, while Mr. Barnes has accepted 75%. There is a 70% chance that Ms. Allen will receive the proposal.

Problem

Determine the probability that the proposal will be accepted.

Solution Consider the tree diagram in Figure 3, where A represents the result that Ms. Allen receives the proposal and B represents Mr. Barnes receiving it. Let S represent success and F failure of the proposal. On the branch of the tree to A we put $Pr(A) = .70$, the probability that A gets the proposal. On the branch to B we put $Pr(B) = .30$. On the branch from A to S we put $Pr(S \mid A) = .60$, so on the branch from A to F we put $Pr(F \mid A) = .40$. On the branches from B to S and from B to F we put $Pr(S \mid B) = .75$ and $Pr(F \mid B) = .25$, respectively. There are two paths that correspond to success; (1) from the start to A and then to S and (2) from the start to B and then to S. They are the events $A \cap S$ and $B \cap S$. The Product Rule is used to compute their probabilities by multiplying the numbers on the corresponding branches. Hence

$$Pr(A \cap S) = Pr(S \mid A) \cdot Pr(A) = (.60)(.70) = .42$$

$$Pr(B \cap S) = Pr(S \mid B) \cdot Pr(B) = (.75)(.30) = .225$$

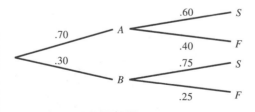

FIGURE 3

The event S can occur in one of the two mutually exclusive ways, $S \cap A$ and $S \cap B$. Therefore

$$Pr(S) = Pr(S \cap A) + Pr(S \cap B) = .42 + .225 = .645$$

Hence there is a 64.5% chance of the proposal being accepted.

The next example shows how conditional probability is used in medicine. The article by Gilbert et al. concerns one of the first attempts by the medical profession to distinguish well-designed experiments from poor ones. By well-designed is meant a **controlled experiment,** that is, one in which the patients who get the treatment are compared with some who do not. In particular, does a controlled experiment give more reliable results?

The investigators reviewed 53 studies of a particular type of operation, the portacaval shunt, which treats bleeding in the esophagus. They measured the degree of control of the experiment versus the enthusiasm of the physician. They measured whether the study had a good control group, a poor control group, or none at all. They compared that measure with whether the physician was very enthusiastic, had moderate enthusiasm, or showed no enthusiasm for the operation.

EXAMPLE 4

The following table contains the data of the study of the portacaval shunt operation.

	Degree of Enthusiasm			
Degree of Control	Marked	Moderate	None	Totals
Well-controlled	0	3	3	6
Poorly controlled	10	3	2	15
Uncontrolled	24	7	1	32
Totals	34	13	6	53

Source: John P. Gilbert et al., ''Assessing Social Innovations: An Empirical Base for Policy,'' *Statistics and Public Policy,* by Fairley and Mosteller, eds., Reading, MA, Addison-Wesley, pp. 185–241, 1977.

Define the events

   W: The experiment was well controlled.

   P: The experiment was poorly controlled or uncontrolled.

   M: The physician showed marked enthusiasm.

   N: The physician showed no enthusiasm.

Problem

Find (a) $Pr(M \mid W)$, (b) $Pr(N \mid W)$, (c) $Pr(W \mid N)$, (d) $Pr(P \mid M)$

Solution For the probabilities of W, M, and N we look at the totals and compute $Pr(W) = 6/53$, $Pr(M) = 34/53$, $Pr(N) = 6/53$. To calculate the probabilities of the intersections we look at the intersection of the column and row described. We get $Pr(W \cap M) = 0/53 = 0$, $Pr(W \cap N) = 3/53$, $Pr(P \cap M) = (10 + 24)/53 = 34/53$. Now we compute the conditional probabilities.

(a) $Pr(M \mid W) = Pr(W \cap M)/Pr(W) = 0/(6/53) = 0$

(b) $Pr(N \mid W) = Pr(W \cap N)/Pr(W) = (3/53)/(6/53) = 3/6 = 1/2$

(c) $Pr(W \mid N) = Pr(W \cap N)/Pr(N) = (3/53)/(6/53) = 3/6 = 1/2$

(d) $Pr(P \mid M) = Pr(P \cap M)/Pr(M) = (34/53)/(34/53) = 34/34 = 1$

The significance of this study lies in the fact that the investigators found that after four years the survival rates of the patients in the control groups were the same as those who elected to have the operation. Because $Pr(P \mid M) = 1$, the study shows that uncontrolled and poorly controlled experiments led to far greater enthusiasm than well-controlled experiments. Later investigation showed, however, that this enthusiasm was unwarranted. On the other hand, since $Pr(M \mid W) = 0$, the experimenters that used well-controlled studies had a more accurate opinion of effectiveness of the operation. This shows that performing poorly controlled experiments wastes time and can lead to false conclusions.

Independent Events

Each spin of a roulette wheel is a new experiment where the probabilities remain the same. But try to convince gamblers of that! If red has come up ten straight times, the amount of money placed on black is enormous. Yet the probability that black occurs is still $18/38 = 9/19$. Each spin of the wheel is said to be independent of the others; that is, what happens on one spin does not affect the probability of any other spin. Intuitively two events E and F are said to be independent if the likelihood of one occurring does not affect the probability of the other.

DEFINITION

Two events E and F are **independent** if

$$Pr(E \mid F) = Pr(E)$$

EXAMPLE 5

Two dice are rolled and the sum of the top faces is recorded. Define the three events

E: A sum of 2, 7, or 11

F: An odd number

G: A sum of 4, 7, or 10

Problem

Determine whether (a) E and F are independent, (b) F and G are independent.

Solution We compute $Pr(E) = 9/36 = 1/4$, $Pr(F) = 18/36 = 1/2$, $Pr(G) = 12/36 = 1/3$, $Pr(E \cap F) = Pr(7 \text{ or } 11) = 8/36 = 2/9$, and $Pr(F \cap G) = Pr(7) = 6/36 = 1/6$.

(a) We must compare $Pr(E)$ and $Pr(E \mid F)$. By definition

$$Pr(E \mid F) = \frac{8/36}{18/36} = \frac{8}{18} = \frac{4}{9}$$

Since $Pr(E) = 1/4 \neq 4/9$, E and F are not independent.

(b) We must compare $Pr(F)$ and $Pr(F \mid G)$. By definition

$$Pr(F \mid G) = \frac{6/36}{12/36} = \frac{6}{12} = \frac{1}{2}$$

Since $Pr(F) = 1/2 = Pr(F \mid G)$, F and G are independent.

In part (b) of Example 5 we could just as well have shown that $Pr(G) = Pr(G \mid F)$. That is, in the definition of independent events one equality implies the other. (See Problem 35.) This shows that the expression "independent events" is not a one-sided relationship. That is, E and F are independent if $Pr(E \mid F) = Pr(E)$, which also implies that $Pr(F \mid E) = Pr(F)$.

The definition of independence can be coupled with the Product Rule to form another useful rule. It governs when we can multiply the probabilities of independent events. Suppose E and F are independent. Then $Pr(E \mid F) = Pr(E)$, and so the Product Rule yields

$$Pr(E \cap F) = Pr(E \mid F) \cdot Pr(F) = Pr(E) \cdot Pr(F)$$

The same steps in reverse show that if

$$Pr(E) \cdot Pr(F) = Pr(E \cap F)$$

then E and F are independent. This is an important rule, so we single it out.

The Product Rule for Independent Events

Two events E and F are independent if and only if

$$Pr(E) \cdot Pr(F) = Pr(E \cap F)$$

For instance, in Example 5 we have

$$Pr(F) \cdot Pr(G) = \left(\frac{1}{2}\right) \cdot \left(\frac{1}{3}\right) = \frac{1}{6} = Pr(F \cap G)$$

Thus the Product Rule confirms that F and G are independent events.

The rule can be used in the opposite direction also. That is, when two events E and F are known to be independent, $Pr(E \cap F)$ can be computed by multiplying $Pr(E)$ and $Pr(F)$. An important application of this idea is in reliability. The **reli-**

ability of a machine or process is defined as the probability that it will not fail during an operation.*

For example, suppose the probability of a computer failing is .05. Then its reliability is the probability that it will not fail, which is $1 - .05 = .95$. Suppose this computer operates independently of another computer; that is, they both perform the same operation but independently. If the probability of the second failing is .02, then the probability of both failing is the intersection of the two independent events that each fails individually. Since we assume that they are independent, to find the probability that both fail we multiply the probabilities and get $(.05)(.02) = .001$.

This demonstrates the need for what is often called a *backup* system. One computer backs up another in the sense that it can handle the load if one fails. A concrete example of such a system is the computer facility aboard the space shuttle. Each shuttle is equipped with four computers that operate independently. Each has its own tasks but they also serve as backups for the others, especially for essential computations. If two do not agree with a calculation, the third and fourth are consulted. The historic first flight of the space shuttle *Columbia* was delayed for several days when the countdown was halted in the last few minutes because the reliability of the computer system was too low.

*Hans Zeisel and Harry Kalvern, Jr., "Parking Tickets and Missing Women: Statistics and the Law," *Statistics: A Guide to the Unknown,* by J.A. Tanur et al., eds., San Francisco, Holden-Day, Inc., pp. 102–111, 1972.

EXERCISE SET 7–5

In Problems 1 to 4 a card is chosen from a deck. Find the conditional probability for the following events:

E: An ace or a king
F: A picture card
G: A club or a red card

1. $Pr(E \mid F)$.

2. $Pr(F \mid E)$.

3. $Pr(E \mid G)$.

4. $Pr(F \mid G)$.

In Problems 5 to 8 a club has 45 members, 20 of whom are women, 25 are less than 20 years old, and 7 women are less than 20 years old. A member is chosen at random. Find the conditional probability.

5. *Pr*(the person is a woman given that the person is less than 20).

6. *Pr*(the person is less than 20 given that the person is a woman).

7. *Pr*(the person is a man given that the person is 20 or more).

8. *Pr*(the person is less than 20 given that the person is a man).

In Problems 9 and 10 a company sends a package by an overnight courier. The package will be picked up by one of three delivery vans, depending on which is nearest. The chances are equal for any of the vans to make the pickup. The problem gives the past record for each van's success in delivering packages on time. Find the probability that the package will be delivered on time.

9. Pr(success for van 1) $= .95$.
   Pr(success for van 2) $= .96$.
   Pr(success for van 3) $= .90$.

10. Pr(success for van 1) $= .90$.
    Pr(success for van 2) $= .93$.
    Pr(success for van 3) $= .99$.

In Problems 11 and 12 urn 1 and urn 2 have the indicated number of white and red marbles. The experiment consists of first selecting an urn and then selecting a marble from the urn and recording its color. For the given number of marbles in each urn, find the probability of observing a red marble.

11. Urn 1 has 5 white and 7 red marbles; urn 2 has 3 white and 9 red marbles.

12. Urn 1 has 9 white and 8 red marbles; urn 2 has 2 white and 8 red marbles.

Problems 13 to 16 refer to the table whose entries are the data from the Zeisel and Kalvern study of criminal jury trials, first mentioned in Section 7–3. After the jury gave its verdict the judge was asked what verdict he or she would have given if there had been no jury. The percentages of agreement and disagreement are in the table. One way to analyze the data is by conditional probabilities.

| | Jury Verdict | |
Judge Verdict	Acquitted	Convicted
Acquitted	14	3
Convicted	19	64

Source: Data from Hans Zeisel and Harry Kalvern, "Parking Tickets and Missing Women: Statistics and the Law," *Statistics: A Guide to the Unknown,* by J. A. Tanur et al., eds., San Francisco, Holden-Day, Inc., pp. 102–111, 1972.

In Problems 13 to 16 find the indicated conditional probability.

13. $Pr($judge would have convicted $|$ jury convicted$)$.

14. $Pr($judge would have acquitted $|$ jury convicted$)$.

15. $Pr($jury acquitted $|$ judge would have convicted$)$.

16. $Pr($jury convicted $|$ judge would have acquitted$)$.

In Problems 17 to 20 two dice are rolled and the sum of the top faces is recorded. For the given events E and F find $Pr(E \mid F)$ and determine if E and F are independent.

17. E: A sum of 2, 3, 4, or 5
    F: A sum of 4, 5, 6, or 7.

18. E: A sum of 2, 6, 8, or 12
    F: A sum of 2, 3, 11, or 12.

19. E: An odd sum
    F: A sum of 2, 4, or 5.

20. E: A sum greater than 5
    F: A sum less than 6.

In Problems 21 and 22 two machines operate independently and have reliability .90 and .95, respectively.

21. Find the probability that they both fail.

22. Find the probability that neither fails.

23. Let E and F be events with $Pr(E) = .35$, $Pr(F) = .67$, and $Pr(E \cap F) = .24$. Find $Pr(E \mid F')$ and $Pr(E' \mid F')$.

24. An economics class has 40 students, of whom 23 are seniors, 12 are juniors, and 5 of the juniors and 3 of the seniors are marketing majors. A student is selected at random. Find the probability that the student is a marketing major given that the student is a senior.

25. There are 3 urns: urn 1 has 1 red and 2 white marbles, urn 2 has 1 red and 3 white marbles, urn 3 has 3 red and 2 white marbles. A die is rolled: if 1 occurs, then a marble is selected from urn 1; if 2 or 3 occurs, then a marble is selected from urn 2; if 4, 5, or 6 occurs, then a marble is selected from urn 3. Find the probability that the marble is red.

26. For the experiment in Problem 25, if a red marble is selected, then it is replaced and a second marble is selected from urn 1, and if the first marble is white, a second marble is selected from urn 2. Find the probability that the second marble is red.

27. An urn contains 6 red and 9 white marbles. Two marbles are selected without replacement. Find the probability that (a) both are red, (b) both are white, (c) the second is red given that the first is white.

28. A factory has three assembly lines; line 1, line 2, and line 3. The reliability of line 1 is .90, meaning that 90% of its products are acceptable, 10% are defective. The reliability of line 2 is .85 and the reliability of line 3 is .95. If line 1 makes 60%; line 2, 30%; and line 3, 10% of the products, find the probability that a product selected at random is defective.

29. A survey shows that 20% of a population are left-handed, 60% of the left-handed people are men while 55% of the right-handed people are women. What percentage of the people are (a) left-handed women, (b) right-handed men?

Problems 30 to 33 refer to the table whose entries are the data from a large metropolitan university. Each entry is the percentage of students enrolled in the university that is described by the category heading of its row and column.

The categories are full-time and part-time students, and students in the College of Business and students in the College of Arts and Sciences.

College	Full-Time	Part-Time
Business	25	18
Arts and Sciences	45	12

A student is selected at random. In Problems 30 to 33 find the indicated conditional probability.

30. Pr(student is full-time | student is a business major).

31. Pr(student is part-time | student is a business major).

32. Pr(student is full-time | student is an arts and science major).

33. Pr(student is part-time | student is an arts and science major).

34. Suppose E and F are mutually exclusive events and $Pr(F)$ is nonzero. Find $Pr(E \mid F)$.

35. By definition if E and F are independent, then $Pr(E \mid F) = Pr(E)$. Show that it is also true that $Pr(F \mid E) = Pr(F)$.

36. For any two events E and F, show that $Pr(F' \mid E) = 1 - Pr(F \mid E)$.

37. Let E and F be events with nonzero probabilities. Show that, in general, only one of the following statements is true:
   (a) $Pr(E \mid F) + Pr(E' \mid F') = 1$.
   (b) $Pr(E \mid F) + Pr(E' \mid F) = 1$.
   (c) $Pr(E \mid F) + Pr(E \mid F') = 1$.

38. Suppose E and F are independent. Determine whether E' and F' are independent.

39. Suppose E and F are independent. Show that $Pr(E \cup F) = 1 - Pr(E') \cdot Pr(F')$.

40. A family has two children. Suppose that each child is equally likely to be a boy or a girl. Assume that the conditional probability that both children are boys given that the older child is a boy is $\frac{1}{2}$. What is the conditional probability that both are boys given that at least one of the children is a boy?

Problems 41 to 44 refer to the following word problem. Find the probability of the given event.

An article on the stock market revealed that of the 100 stocks studied one week, 40 stocks advanced during the week, 20 stocks had a net change of more than 10%, 20 stocks had split within the last 12 months, 15 of the stocks that advanced during the week also had a net change of more than 10%, 12 of the stocks that advanced during the week also had split within the last 12 months, 8 stocks that had a net change of more than 10% also had split within the last 12 months, and 5 stocks that advanced and also had a net change of more than 10% had also split within the last 12 months. A stock is chosen at random.

41. The stock split given that the stock advanced.

42. The stock advanced given that the stock split.

43. The stock advanced given that the stock had a net change of more than 10%.

44. The stock split given that the stock had a net change of more than 10%.

Referenced Exercise Set 7–5

1. Brightman and Haslanger studied whether past performance of portfolio managers for mutual funds is a good indication of their ability to perform well in the future. They collected data on 52 portfolio managers over a six-year period. They ranked each manager's performance according to which quartile they were in (i.e., a ranking in the first quartile means the manager was ranked in the top quarter, or top 13). They first measured the performance in the 1971–1973 period and then compared that with the performance in the 1974–1976 period. The data are given in the following table:

		1971–1973 Performance Quartile				
		1	2	3	4	Total
1974–1976	1	3	1	1	8	13
Performance	2	2	3	5	3	13
Quartile	3	4	3	5	1	13
	4	4	6	2	1	13
Total		13	13	13	13	52

Source: Brightman and Haslanger, "Past Investment Performance: Seductive but Deceptive," *Journal of Portfolio Management*, Vol. 6, pp. 43–45, 1980.

Find the probability that a manager was ranked
(a) in the first quartile in the 1974–1976 period given

that the manager was ranked in the first quartile in the 1971–1973 period.

(b) in the first quartile in the 1974–1976 period given that the manager was ranked in the first or second quartile in the 1971–1973 period.

(c) in the third or fourth quartile in the 1974–1976 period given that the manager was ranked in the first or second quartile in the 1971–1973 period.

2. In one of the first studies of the side effects of oral contraceptive users, Jick et al. compared the incidence of thromboembolic disease with the user's blood type. A control group of healthy women was studied for comparison. The data are in the following table:

Blood Group	Women with Thromboembolism	Healthy Women
A	32	51
B	8	19
AB	6	5
O	9	70

Source: Jick et al., "Venous Thromboembolic Disease and ABO Blood Types," *Lancet*, Vol. 1, No. 1, p. 539, 1969.

If a woman is selected at random, find the probability that she

(a) had thromboembolism given that she had blood group A.

(b) had thromboembolism given that she had blood group O.

(c) had blood group A given that she had thromboembolism.

3. One of the most perplexing and agonizing problems in our society is whether capital punishment deters homicide. In an article studying data from Vital Statistics, Stephen K. Laysons† analyzed arrest, conviction and execution frequencies based on conditional probability. Let C represent the event "conviction" and E represent "execution." Then $Pr(C) = .0046$ and $Pr(E \mid C) = .00016$. Find $Pr(C \cap E)$ from the formula for conditional probability. Use the fact that $C \cap E = E$ to determine whether C and E are independent.

4. One of the games on the popular TV show "The Price Is Right" is played as follows.‡ The contestant is presented with four items, each carrying an incorrect price. The contestant is asked if the actual price of each item is higher or lower than the stated price. The total number N of correct responses is recorded, then the contestant is presented with a table with four shells on it. One of the shells covers a small ball. The contestant is allowed to choose N of the shells, and if one of them is covering the ball, the contestant wins the grand prize.

(a) Do you think that a contestant who guesses every price has a good chance of winning the grand prize?

(b) If a contestant guesses every price, what is the probability of winning the grand prize?

5. Many of us who live in the continental United States think of Anchorage, Alaska, as a sunny city. However, in a given year Anchorage experiences 64 clear days, 67 partly cloudy days, and 234 cloudy days.§

(a) What is the probability that a day of the year is cloudy?

(b) Assuming that the weather on one day is independent of the weather on the next day, what is the probability that Anchoraginians will experience three consecutive days of cloudy weather?

Cumulative Exercise Set 7–5

1. The experiment consists of tossing a coin five times and recording the number of heads. Find the sample space and determine the elements in the two events

   E: More than three heads

   F: An even number of heads

   Find (a) $E \cup F$, (b) $E \cap F$, (c) E'.

2. Let E and F be events with $Pr(E) = .65$, $Pr(F) = .47$, and $Pr(E \cap F) = .14$. Find $Pr(E \cap F')$ and $Pr(E' \cap F')$.

---

†Stephen K. Layson, "Homicide and Deterrence: A Reexamination of the United States Time-Series Evidence," *Southern Economics Journal*, Vol. 52, pp. 68–77, 1985.

‡D. Turner, D. Young, and V. Marco, "Maybe the Price Doesn't Have to Be Right: Analysis of a Popular TV Game Show," *College Mathematics Journal*, pp. 419–421, October 1988.

§*Source:* Rand McNally Places Rated Almanac.

3. A survey of students enrolled in a university revealed the following information about full-time and part-time students, and students in the College of Business and students in the College of Arts and Sciences:

College	Full-Time	Part-Time
Business	30	10
Arts and Sciences	45	15

A student is selected at random. Find
(a) Pr(the student is full-time and is a business major).
(b) Pr(the student is part-time or is a business major).

4. An urn contains n marbles numbered 1 to 10. Select five marbles with replacement and record the number. Find the probability that at least two of them are the same number.

5. Forty people are called for jury duty, 22 of whom are men. Find the probability that at most 1 woman is selected for the six-person jury.

6. A shipment of 50 engines contains 5 defectives. The quality control inspector selects 3 engines at random to be tested. Find the probability that none of the engines tested is defective.

7. A card is chosen from a deck. Find the conditional probability for the events

E: A queen or a king

F: A club or a red card

(a) $Pr(E \mid F)$, (b) $Pr(F \mid E)$

8. The given table has entries that are the data from a study of criminal jury trials.

Judge Verdict	Jury Verdict	
	Acquitted	Convicted
Acquitted	15	5
Convicted	20	60

Find the indicated conditional probability.
(a) Pr(Judge would have convicted | jury acquitted).
(b) Pr(Judge would have acquitted | jury convicted).

9. Two dice are rolled and the sum of the top faces is recorded. For the given events E and F, find $Pr(E \mid F)$ and determine if E and F are independent.

E: A sum of 3, 4, 5, or 6

F: A sum of 4, 5, 6, or 11

10. Two machines operate independently and have reliability .90 and .85, respectively. Find the probability that (a) they both fail, (b) neither fails.

11. An accounting class has 50 students, 13 of whom are seniors, 10 of whom are juniors, and 5 of the juniors and 3 of the seniors are marketing majors. A student is selected at random. Find the probability that the student is a marketing major given that the student is a senior.

12. A factory has three assembly lines, line 1, line 2, and line 3. The reliability of line 1 is .90, meaning that 90% of its products are acceptable, 10% are defective. The reliability of line 2 is .80 and the reliability of line 3 is .85. If line 1 makes 50%; line 2, 10%; and line 3, 40% of the products, find the probability that a product selected at random is defective.

CASE STUDY   **Trial by Judge or Jury?**

As if Jon didn't have problems enough. He was being tried for a crime he didn't commit, which was bad enough, but then his lawyer asked him if he would rather be tried by judge or by jury. Isn't that what you hire a lawyer for, to make those decisions? His lawyer did point out the advantages and disadvantages of each. If your case depends on some technical legal issues that would be difficult for a jury to follow, it is best to select a trial by judge. But if a major part of the defense is emotional, a trial by jury is preferable. Jon's case fit neither description, so it was a toss-up.

The basic question facing Jon and his lawyer is, given that there is no clear reason for choosing one type of trial over the other, which is better for the defendant? And how do you even answer such a question?

Two political scientists, Hans Zeisel and Harry Kalvern,* Jr., of the University of Chicago, raised the same issue in an article investigating various ways probability and statistics are used in the law. In one section of the article Zeisel and Kalvern investigated the perennial debate over the merits of the jury system. The question they raised is: Would there be a difference between verdicts reached by a jury and verdicts by a judge if there were no jury? The argument is: If the judge and jury hardly ever differed, the jury would be a wasteful institution; if they differed too often, serious questions would be raised about the rationality of the system, including juries, judges, and the very meaning of justice itself. They studied 3576 jury trials in which the judge reported how he or she would have ruled if there were no jury. The percentages of agreement and disagreement are in the following table:

	Judge Verdict	
Jury Verdict	Acquitted	Convicted
Acquitted	14	19
Convicted	3	64

View the study as a probability experiment where one of the cases is chosen at random and the decisions of the jury and of the judge are recorded. It is a two-stage experiment where the first stage is the jury's verdict (Jur) and the second stage is the judge's verdict (Jud). There are two possibilities at each stage of the experiment, acquit (*A*) or convict (*C*). Thus there are four outcomes to the experiment which can be illustrated in a tree diagram.

First stage   Second stage
   Jur   Jud   Outcome

                 A   Jur*A* and Jud*A*

  A

                 C   Jur*A* and Jud*C*

                 A   Jur*C* and Jud*A*

  C

                 C   Jur*C* and Jud*C*

Each outcome can also be viewed as an intersection. For example, the outcome "Jur*A* and Jud*C*" can be viewed as the intersection of the event "The jury acquits the defendant." and the event "The judge convicts the defendant." Then the table consists of four numbers (percentages) that represent the probability that a case

*Hans Zeisel and Harry Kalvern, Jr., "Parking Tickets and Missing Women: Statistics and the Law," *Statistics: A Guide to the Unknown*, by J.A. Tanur et al., eds., San Francisco, Holden-Day, Inc., pp. 102–111, 1972.

chosen at random had the given two decisions. Another way to look at the table is as follows:

Jury Verdict	Judge Verdict	
	Acquitted	Convicted
Acquitted	Jur$A$ and Jud$A$ 14	Jur$A$ and Jud$C$ 19
Convicted	Jur$C$ and Jud$A$ 3	Jur$C$ and Jud$C$ 64

The question facing Jon is whether the data yield a definitive answer to his question: Which type of trial is better for him? The answer is not obvious. It requires some critical analysis. The jury and the judge agree in 78% of the cases, in 14% they both acquit and in 64% they both convict, meaning that in most cases the evidence was compelling. The more pertinent data are the cases where they disagree. Jon is most interested in when he might make a mistake. Sometimes the key step in critical thinking is to ask the right question. In Jon's dilemma here is the key question: What is the likelihood of choosing the wrong type of trial? Jon obviously wants to choose the type of trial that has the smaller margin for error. More specifically, Jon makes a mistake if he chooses one type of trial and is convicted, but he would have been acquitted if he had chosen the other type. This can happen in two ways:

1. He is convicted by the jury but would have been acquitted by the judge.
2. He is convicted by the judge but would have been acquitted by the jury.

These are conditional events. Expressing them in the language of probability, we get:

Event I. He would have been acquitted by the judge given that he was convicted by the jury?

Event II. He would have been acquitted by the jury given that he was convicted by the judge?

Now we compute the probability of each event. If the probability of event I is larger, he would make a mistake by choosing a trial by jury. We now compute these two probabilities:

$$Pr(\text{event I}) = Pr(\text{Jud}A \mid \text{Jur}C) = \frac{Pr(\text{Jud}A \cap \text{Jur}C)}{Pr(\text{Jur}C)} = \frac{3}{3 + 64} \simeq .04$$

$$Pr(\text{event II}) = Pr(\text{Jur}A \mid \text{Jud}C) = \frac{Pr(\text{Jur}A \cap \text{Jud}C)}{Pr(\text{Jud}C)} = \frac{19}{19 + 64} \simeq .23$$

This means defendants make a mistake by choosing a trial by jury 4% of the time and they make a mistake by choosing a trial by judge 23% of the time. In other words, there is almost a six times greater chance of making a mistake by choosing a trial by judge.

Jon chose a trial by jury.

Case Study Exercises

Compute the conditional probabilities from the table.

1. $Pr(\text{Jud}C \mid \text{Jur}A)$.
2. $Pr(\text{Jur}C \mid \text{Jud}A)$.
3. $Pr(\text{Jud}A \mid \text{Jur}A)$.
4. $Pr(\text{Jud}C \mid \text{Jur}C)$.
5. $Pr(\text{Jur}A \mid \text{Jud}A)$.
6. $Pr(\text{Jur}C \mid \text{Jud}C)$.

Problems 7 to 12 refer to data similar to that in the Zeisel and Kalvern study. From this table compute the indicated probabilities.

	Judge Verdict	
Jury Verdict	Acquitted	Convicted
Acquitted	10	20
Convicted	5	65

7. $Pr(\text{Jud}A \mid \text{Jur}C)$.
8. $Pr(\text{Jur}A \mid \text{Jud}C)$.
9. $Pr(\text{Jud}C \mid \text{Jur}A)$.
10. $Pr(\text{Jud}C \mid \text{Jur}C)$.
11. $Pr(\text{Jur}A \mid \text{Jud}A)$.
12. $Pr(\text{Jur}C \mid \text{Jud}C)$.

7–6 Bayes' Theorem

The previous section dealt with conditional probability, the probability of an event E occurring given that event F has occurred. In this section the process is reversed. We use conditional probability to determine the probability that an earlier event F occurred, given that a later one E has happened. In other words, if we know $Pr(E \mid F)$, can we find $Pr(F \mid E)$?

For example, suppose a medical experiment separates patients into two groups, the treatment group and the control group. Then the treatment is administered to the former group and a placebo is given to the control group. A typical question answered in the last section would be, "What is the probability that a patient was cured given that the patient was in the control group?" In this section a typical question would be, "What is the probability that a patient was in the control group given that the patient was cured?"

The experiments in this section consist of two stages. Tree diagrams will be useful. Let us look at a simple two-stage problem to illustrate the type of problem and its solution.

EXAMPLE 1

The experiment is to choose a marble from either an urn or a hat. The urn contains 1 red and 2 white marbles and the hat contains 3 red and 2 white marbles. (See Figure 1.)

Problem

Find the probability that the marble came from the urn given that it is a red marble.

Solution The tree diagram in Figure 1 shows that the experiment consists of two stages, where the first stage is to choose either the urn or the hat and the second is to select a marble. Let U, H, R, and W represent selecting the urn, the hat, a red marble, and white marble, respectively. Then we assign the following probabilities: $Pr(U) = Pr(H) = \frac{1}{2}$, $Pr(R \mid U) = \frac{1}{3}$, $Pr(R \mid H) = \frac{3}{5}$, $Pr(W \mid U) = \frac{2}{3}$, and $Pr(W \mid H) = \frac{2}{5}$. The problem is to find $Pr(U \mid R)$.

Notice that R can occur in two ways, from U or from H. The corresponding branches are U-R and H-R. In terms of sets this can be expressed as $R = (U \cap R) \cup (H \cap R)$. Because these two sets are mutually exclusive, we have

$$Pr(R) = Pr(U \cap R) + Pr(H \cap R) \tag{1}$$

From the definition of conditional probability we have

$$Pr(U \mid R) = \frac{Pr(U \cap R)}{Pr(R)}$$

Substitute equation 1 into this equation to get

$$Pr(U \mid R) = \frac{Pr(U \cap R)}{Pr(U \cap R) + Pr(H \cap R)} \tag{2}$$

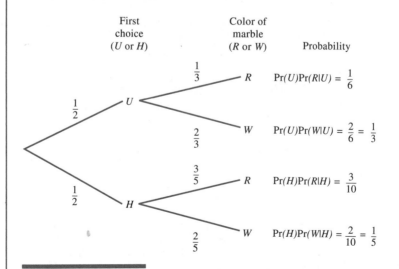

First choice (U or H)	Color of marble (R or W)	Probability

$Pr(U)Pr(R|U) = \frac{1}{6}$

$Pr(U)Pr(W|U) = \frac{2}{6} = \frac{1}{3}$

$Pr(H)Pr(R|H) = \frac{3}{10}$

$Pr(H)Pr(W|H) = \frac{2}{10} = \frac{1}{5}$

FIGURE 1

These probabilities can be computed from the tree diagram by using the Product Rule. Thus

$$Pr(U \cap R) = Pr(U) \cdot Pr(R \mid U) = \frac{1}{2} \cdot \frac{1}{3} = \frac{1}{6}$$

$$Pr(H \cap R) = Pr(H) \cdot Pr(R \mid H) = \frac{1}{2} \cdot \frac{3}{5} = \frac{3}{10}$$

Substituting these numbers into equation 2 yields

$$Pr(U \mid R) = \frac{1/6}{1/6 + 3/10} = \frac{1/6}{28/60} = \frac{5}{14}$$

Therefore the probability that the marble came from the urn given that it is red is 5/14.

The key step in Example 1 is equation 2. It expresses the conditional probability of an earlier event given a later event in terms of known probabilities. A Venn diagram helps to get a handle on the meaning of equation 2. In Figure 2 event R intersects U and H and $U \cap R$ and $H \cap R$ are mutually exclusive sets. This shows that equation 1 holds:

$$Pr(R) = Pr(U \cap R) + Pr(H \cap R) \tag{1}$$

Now suppose that R intersects three mutually exclusive sets, say, U_1, U_2, and U_3. The counterpart to equation 1 for three sets is

$$Pr(R) = Pr(U_1 \cap R) + Pr(U_2 \cap R) + Pr(U_3 \cap R) \tag{3}$$

(See Figure 3.) Just as we substituted equation 1 into the Product Rule to get equation 2, so too we can substitute equation 3 into the Product Rule to find $Pr(U_1 \mid R)$.

$$Pr(U_1 \mid R) = \frac{Pr(U_1 \cap R)}{Pr(R)} = \frac{Pr(U_1 \cap R)}{Pr(U_1 \cap R) + Pr(U_2 \cap R) + Pr(U_3 \cap R)}$$

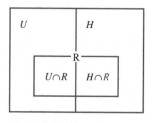

FIGURE 2

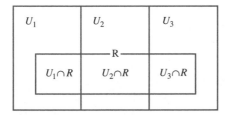

FIGURE 3

In the same way we can extend this result to any finite number of mutually exclusive sets. The theorem is called **Bayes' Formula,** after the Presbyterian minister and part-time mathematician, Reverend Thomas Bayes (1702–1736).*

Bayes' Formula

Suppose $U_1, U_2, \ldots, U_n$ are mutually exclusive events whose union is the entire sample space. Let E be an event with nonzero probability. Then

$$Pr(U_i \mid E) = \frac{Pr(U_i \cap E)}{Pr(E)}$$

$$= \frac{Pr(U_i \cap E)}{Pr(U_1 \cap E) + Pr(U_2 \cap E) + \cdots + Pr(U_n \cap E)}$$

$$= \frac{Pr(U_i)Pr(E \mid U_i)}{Pr(U_1)Pr(E \mid U_1) + \cdots + Pr(U_n)Pr(E \mid U_n)}$$

The next example demonstrates the use of Bayes' Formula in an experiment with three possibilities at the first stage.

EXAMPLE 2

A department store classifies credit card customers according to the maximum amount of credit allowed per month. Preferred customers are allowed over $500 credit per month, regular customers are allowed a maximum of $500 per month, and basic customers are allowed a maximum of $50 per month. A minimum monthly charge is assessed each account. If the minimum is not paid, the account is frozen and no new charges are allowed. Past experience shows that

*Recent evidence indicates that Bayes' Formula was known before Bayes was born. See Stephen H. Stigler, "Who Discovered Bayes' Theorem?" *The American Statistician*, pp. 290–296, November 1983.

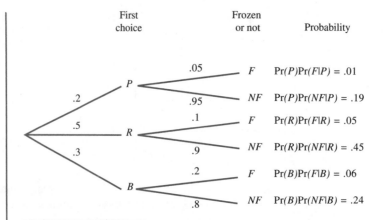

First choice		Frozen or not	Probability

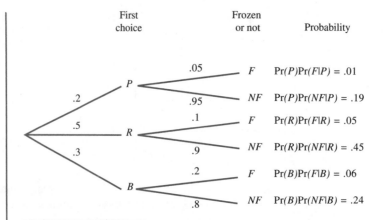

FIGURE 4

5% of the preferred accounts are frozen, 10% of the regular accounts are frozen, and 20% of the basic accounts are frozen each month. Suppose 20% of the accounts are preferred, 50% are regular accounts, and 30% are basic accounts.

Problem

Find the probability that a particular frozen account is a preferred account.

Solution The first step is to construct a tree diagram. (See Figure 4.) At the first stage let P, R, and B represent preferred, regular, and basic accounts, respectively. At the second stage let F and NF represent frozen and not frozen, respectively. Label each branch with its corresponding probability. The probability that a frozen account is a preferred account is $Pr(P \mid F)$. Using Bayes' Formula yields

$$Pr(P \mid F) = \frac{Pr(F \cap P)}{Pr(F \cap P) + Pr(F \cap R) + Pr(F \cap B)}$$

The probabilities on the tree diagram can be used to get

$$Pr(P \mid F) = \frac{(.20)(.05)}{(.20)(.05) + (.5)(.1) + (.3)(.2)}$$

$$= \frac{.01}{.01 + .05 + .06} = \frac{.01}{.12}$$

$$= \frac{1}{12} \approx .0833$$

Therefore the probability that an account selected at random is preferred given that it is frozen is $\frac{1}{12}$, or about 8.3% of the frozen accounts are preferred.

The next example illustrates how to use Bayes' Formula when there are more than three possibilities at the first stage.

EXAMPLE 3

A television manufacturer purchases cabinets from five different suppliers. The following table gives the percentage of the manufacturer's business that each supplier has as well as the percent of defective cabinets from each supplier.

Supplier	Percent of Business	Percent Defectives	Product
1	.35	.02	.007
2	.25	.04	.010
3	.20	.06	.012
4	.10	.07	.007
5	.10	.09	.009
Total			.045

Problem

Find the probability that a defective cabinet was manufactured by the first supplier.

Solution Let S_i be the event that the cabinet was manufactured by supplier i. Let D be the event that the cabinet is defective. We want to compute $Pr(S_1 \mid D)$. In Bayes' Formula let $n = 5$ and replace U_i by S_i and E by D. The denominator is the sum of the products of the two columns of percentages. Hence

$$Pr(S_1 \mid D) = \frac{Pr(S_1 \cap D)}{Pr(D)}$$

$$= \frac{Pr(S_1)Pr(D \mid S_1)}{Pr(S_1)Pr(D \mid S_1) + \cdots + Pr(S_5)Pr(D \mid S_5)}$$

$$= \frac{(.35)(.02)}{(.35)(.02) + (.25)(.04) + (.2)(.06) + (.1)(.07) + (.1)(.09)}$$

$$= \frac{.007}{.007 + .01 + .012 + .007 + .009} = \frac{.007}{.045}$$

$$\approx .1555$$

Thus about 15.6% of the defectives are made by supplier 1.

The problem in Example 3 can also be solved by a tree diagram. (See Figure 5.) The first stage corresponds to the percent of business of each supplier. The second stage is the percent of defectives for each supplier. The product of the numbers on the branches of each path is then formed and the sum of the products is the solution.

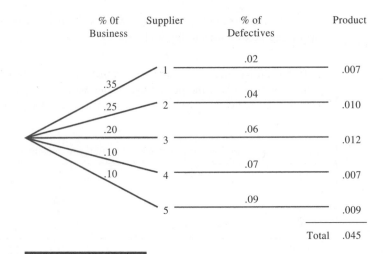

% Of Business	Supplier	% of Defectives	Product

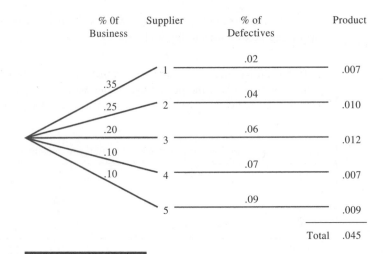

FIGURE 5

EXERCISE SET 7–6

In Problems 1 to 4 two mutually exclusive events U_1 and U_2 and their probabilities are given together with the conditional probabilities $Pr(E \mid U_1)$ and $Pr(E \mid U_2)$. Find $Pr(U_1 \mid E)$.

1. $Pr(U_1) = .6$, $Pr(U_2) = .4$
   $Pr(E \mid U_1) = .3$, $Pr(E \mid U_2) = .2$

2. $Pr(U_1) = .35$, $Pr(U_2) = .65$
   $Pr(E \mid U_1) = .1$, $Pr(E \mid U_2) = .2$

3. $Pr(U_1) = .4$, $Pr(U_2) = .6$
   $Pr(E \mid U_1) = .1$, $Pr(E \mid U_2) = .7$

4. $Pr(U_1) = .51$, $Pr(U_2) = .46$
   $Pr(E \mid U_1) = .3$, $Pr(E \mid U_2) = .1$

In Problems 5 and 6 three mutually exclusive events U_1, U_2, and U_3 and their probabilities are given together with the conditional probabilities $Pr(E \mid U_1)$, $Pr(E \mid U_2)$, and $Pr(E \mid U_3)$. Find $Pr(U_1 \mid E)$.

5. $Pr(U_1) = .2$, $Pr(U_2) = .5$, $Pr(U_3) = .3$
   $Pr(E \mid U_1) = .1$, $Pr(E \mid U_2) = .2$,
   $Pr(E \mid U_3) = .05$

6. $Pr(U_1) = .35$, $Pr(U_2) = .45$, $Pr(U_3) = .20$
   $Pr(E \mid U_1) = .2$, $Pr(E \mid U_2) = .1$,
   $Pr(E \mid U_3) = .25$

In Problems 7 to 10 find the indicated probabilities by referring to the accompanying tree diagram:

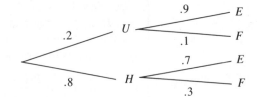

7. $Pr(U \cap E)$, $Pr(H \cap E)$

8. $Pr(U \cap F)$, $Pr(H \cap F)$

9. $Pr(U \mid E)$, $Pr(H \mid E)$

10. $Pr(U \mid F)$, $Pr(H \mid F)$

In Problems 11 to 18 find the indicated probabilities by referring to the accompanying tree diagram:

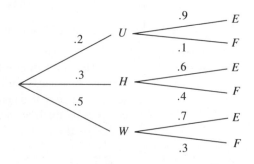

11. $Pr(U \cap E)$, $Pr(H \cap E)$, $Pr(W \cap E)$

12. $Pr(U \cap F)$, $Pr(H \cap F)$, $Pr(W \cap F)$

13. $Pr(U \mid E)$ 15. $Pr(W \mid E)$ 17. $Pr(H \mid F)$

14. $Pr(H \mid E)$ 16. $Pr(U \mid F)$ 18. $Pr(W \mid F)$

In Problems 19 to 22 refer to the Venn diagram. The number of elements in each set is given. For example, $n(U) = 15$ and $n(U \cap E) = 5$. Find the indicated probabilities.

U	10	H	4	W	13
		E			
	5		6		2

19. $Pr(U \mid E)$.

20. $Pr(H \mid E)$.

21. $Pr(W \mid E)$.

22. $Pr(U \mid E')$.

In Problems 23 and 24 the experiment is first to choose at random either an urn or a hat, each containing some red and white marbles, and then to select a marble. The urn has 4 red and 6 white marbles and the hat has 7 red and 5 white marbles. Find the probability of each event.

23. The urn was chosen given that the marble is red.

24. The hat was chosen given that the marble is white.

In Problems 25 to 27 the experiment is first to choose at random one of 3 urns, each containing some red and white marbles, and then to select a marble. Urn 1 has 4 red and 8 white marbles, urn 2 has 3 red and 7 white marbles, and urn 3 has 7 red and 5 white marbles. Find the probability of each event.

25. Urn 1 was chosen given that the marble is red.

26. Urn 2 was chosen given that the marble is white.

27. Urn 3 was chosen given that the marble is white.

28. A firm that specializes in filling out income tax forms has two types of accounts—large (gross income greater than $50,000) and small (gross income less than $50,000). They find that 5% of the large accounts and 2% of the small accounts are audited by the IRS. If 40% of their accounts are large accounts, and an account is audited, what is the probability that it is large?

29. A medical research company conducted a survey of a new ointment that claims to reduce swelling if applied immediately after an injury. They choose 100 people to be in the treatment group and 50 in the control group. Those in the control group were given a similar looking ointment that is known to be ineffective in reducing swelling. Of those in the treatment group, 80 said that the swelling was reduced significantly. Of those in the control group, 30 reported that the swelling was reduced significantly. Find the probability that a person chosen at random was in the control group given that the person reported that the swelling was reduced significantly.

Problems 30 and 31 refer to the accompanying Venn diagram, where U, V, W, A, and B are mutually exclusive sets whose union is the sample space:

U	10	V	5	W	3	A	1	B	4
				E					
	5		6		2		1		1

30. Compute $Pr(U \mid E)$.

31. Compute $Pr(V \mid E)$.

32. The experiment consists of choosing at random 1 of 5 urns, each containing some red and white marbles, and then selecting a marble. Urn i has i red marbles and $1 + i$ white marbles. (Thus urn 1 has 1 red and $1 + 1 = 2$ white marbles; urn 2 has 2 red and $1 + 2 = 3$ white marbles, and so on.) Find the probability that urn 1 was chosen given that the marble is red.

33. The following table gives the distribution of family income, the proportion of the families with that income, and the proportion of families with two wage earners. If a family with two wage earners is selected at ran-

dom, what is the probability that the family income is at least $30,000?

Family Income (per year)	Proportion of Families	Proportion with Two Wage Earners
<10,000	.08	.1
10,000–20,000	.20	.2
20,001–30,000	.40	.3
30,001–40,000	.20	.2
>40,000	.12	.2

34. In a survey of skiing accidents at five hospitals, the incidence of severe fractures (a fracture that requires surgery) was as follows: hospital 1, 22% of the fractures were severe; hospital 2, 12%; hospital 3, 5%; hospital 4, 8%; hospital 5, 3%. Assuming that there were the same number of accidents at each hospital, find the probability that a severe fracture occurred at hospital 1 or hospital 2.

35. In order to measure the degree of visibility of an ad campaign, interviewers often ask consumers if they can remember seeing certain ads. An advertising firm questioned 1000 TV viewers and asked if they recalled seeing a particular commercial. They were then asked if they had bought the product recently. Of the 110 people who definitely remembered the commercial, 60 had bought the product recently. Of the 330 who vaguely remembered the commercial, 66 had bought the product recently. Of the 560 who did not remember seeing the commercial, 40 had bought the product recently. Find the probability that a person definitely remembered seeing the ad given that the person bought the product recently.

36. Red-green color-blind people see these two colors the same. The probability that this type of color blindness occurs is .049. The proportion of men who are red-green color-blind is .085. Suppose a person is selected at random. Assume the probability that the person is a man is .5. Find the probability that the person is a man given that the person is red-green color-blind.

37. Two cookie jars are on a shelf. The red jar contains 5 vanilla cookies and 3 chocolate cookies; the green jar contains 4 vanilla cookies and 8 chocolate cookies. A jar is grabbed at random and one cookie is selected from it. If the selected cookie is chocolate, what is the probability that the green jar was selected?

38. A shelf holds two boxes of ornaments. The blue box contains 7 good ornaments and 3 defective ornaments; the white box contains 4 good ornaments and 1 defec-

tive ornament. A box is selected at random and one ornament is chosen from it. If the ornament is good, what is the probability that the blue box was chosen?

39. Suppose there are three used-car lots, a small lot with 14 good cars and 2 lemons, a medium lot with 28 good cars and 4 lemons, and a big lot with 57 good cars and 7 lemons. If Joe bought a lemon from one of these lots, what is the probability that he bought it from the big lot?

40. There are three jars on a shelf, each containing a dozen cookies. Jar A has 8 chocolate and 4 vanilla cookies, jar B has 6 chocolate and 6 vanilla cookies, jar C has 3 chocolate and 9 vanilla cookies. A jar is selected at random and one cookie is selected from it. If the cookie is vanilla, what is the probability that it was selected from jar C?

41. Suppose $A, B, C,$ and D are mutually exclusive events in a sample space S where $S = A \cup B \cup C \cup D$. Let E be any event. Show that

$$Pr(E) = Pr(E \cap A) + Pr(E \cap B)$$
$$+ Pr(E \cap C) + Pr(E \cap D)$$

42. Suppose $A, B,$ and C are mutually exclusive events in a sample space S where $S = A \cup B \cup C$. Let E be any event. Show that

$$Pr(E) = Pr(E \mid A)Pr(A) + Pr(E \mid B)Pr(B)$$
$$+ Pr(E \mid C)Pr(C)$$

Referenced Exercise Set 7–6

1. In a study of 1405 students, participation in religious activities was compared with the answer to the question, "How happy has your home life been?" C. L. Stone found the following:

Religious Participation	Answer to Question	
	Happy	Not Happy
Not at all	183	25
Very little	406	30
Somewhat	521	24
Very much	203	13

Adapted from C. L. Stone, "Church Participation and Social Adjustment of High School and College Youth," *Rural Sociology Series on Youth*, No. 12, Bulletin 550, Agricultural Research Center, Washington State University, 1954.

Find the probability that a person who did not have a happy home life participated very little or not at all in religious activities.

2. In an article that studied the effects of seeding hurricanes Howard et al. used Bayes' Formula extensively.[†] The possibility of mitigating the destructive force of hurricanes by seeding them with silver oxide was suggested as early as 1961, but strong evidence was not obtained until 1969 when "Hurricane Debbie" was seeded. The researchers defined the events: H_1—seeding reduces wind speed; H_2—seeding has no effect; H_3—seeding increases wind speed. They determined the theoretical probabilities $Pr(H_1) = .49 = Pr(H_2)$, $Pr(H_3) = .02$. They defined the event "Debbie" to be that wind reduction of at least 31% was observed. They computed $Pr(\text{Debbie} \mid H_1) = .6$, $Pr(\text{Debbie} \mid H_2) = .25$, and $Pr(\text{Debbie} \mid H_3) = .4$. Find the probability that H_1 occurred given that Debbie occurred.

3. The Equal Employment Opportunity Act of 1972 extended Title VII of the Civil Rights Act protection against sex discrimination in employment to academic personnel. In a paper studying the extent to which the salary structure of a large public university has changed since 1972 the researchers applied their findings to the University of Arizona.[‡] In one part of the study the percentage of female faculty in the College of Liberal Arts was compared with the percentage of female faculty in other colleges in 1982. These data are given in the table below. Use Bayes' Formula to find the probability that a faculty member selected at random is in the College of Liberal Arts given that the person is female.

	College	
	Liberal Arts	Other
Female	51	106
Male	408	572

Cumulative Exercise Set 7–6

1. Find the sample space for each of the following experiments:
   (a) Roll three dice and record the number of 6's that face up.

(b) Time four rats in a maze and record whether each one exits the maze in less than one minute.

2. What is the probability that in a family with six children there will be three girls and three boys?

3. If a card is selected at random from an ordinary deck of cards, what is the probability that it is an ace or a picture card?

4. Toss two dice. Define the events E and F by

   E: The sum is at most 3.

   F: The sum is 3 or 5.

   What is $Pr(E \cup F)$?

5. Let $Pr(E) = .6$, $Pr(F) = .5$, and $Pr(E \cap F) = .3$. Find (a) $Pr(E \cup F)$, (b) $Pr(E \cap F')$.

6. There are two cookie jars on a shelf, each containing 6 vanilla and 4 chocolate cookies. One cookie selected from each jar. Compute each of the following probabilities:
   (a) Both are vanilla.
   (b) Neither is vanilla.
   (c) One is vanilla and the other is chocolate.

7. In a group of 20 people 15 are right-handed and 5 are left-handed. If 3 people are selected at random from the group, what is the probability that all 3 are right-handed given that the first 2 are right-handed?

8. At a high school five-year reunion, half of the 168 graduates were women and 89 of the graduates had gone to college; 46 of those 89 were women.
   (a) If a man is selected at random, what is the probability that he had gone to college?
   (b) If a person who did not go to college is selected at random, what is the probability that the person is a man?

9. Each entry in the following table is the number of different investments that a broker sold during a given period:

	Stocks	Bonds
Taxable	44	6
Tax-free	2	28

[†]Ronald A. Howard et al., "The Decision to Seed Hurricanes," *Science,* Vol. 176, pp. 1191–1202, June 1972.

[‡]Sharon B. Megdal and Michael R. Ransom, "Longitudinal Changes in Salary at a Large Public University: What Response to Equal Pay Legislation?" *American Economics Review,* Vol. 75, pp. 271–274, 1985.

If an investment is chosen at random, what is the probability that

(a) it is tax-free given that it is a stock?
(b) it is a bond given that it is taxable?

10. Find the indicated probabilities from the accompanying tree diagram.

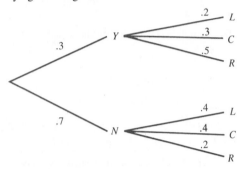

(a) $Pr(N \cap C)$
(b) $Pr(Y \mid L)$

11. There are two cookie jars on a shelf, a red jar with 4 vanilla cookies and 3 chocolate cookies and a blue jar with 5 vanilla cookies and 6 chocolate cookies. A jar is selected at random and a cookie is selected from it. If the cookie is chocolate, what is the probability that the red jar was selected? (Notice that there are 9 cookies of each flavor. Does this mean that the answer to the question is .5?)

12. Three professors administer a common exam to their classes of 40 students each. Suppose that 32 of the students in class A, 30 of the students in class B, and 30 of the students in class C pass the exam. If a student who has passed the exam is chosen at random, what is the probability that the student is in class A?

13. Of 60 people who were asked their source of news, 35 said they watched television, 20 listened to radio, and 5 both watched TV and listened to the radio. None of the 10 people who said they relied on print media (primarily newspapers and weekly magazines) watched TV or listened to the radio for news content. If one of these 60 people is selected at random, what is the probability that the person did not rely on any of these three media for the source of news?

14. Each team in the NFL plays 8 away games during the regular season. The start of each game is determined by a coin flip, with the visiting team guessing the outcome. What is the probability that a team will guess the outcome correctly on exactly 4 of its 8 away games? [Does it seem reasonable that the answer should be $\frac{1}{2}$?]

15. Suppose that a small used-car lot has 4 good cars and 2 lemons and that a big used-car lot has 5 good cars and 7 lemons.

(a) If a car is selected from the small lot, what is the probability that it is a lemon?
(b) If a lot is selected and one car is bought from it, what is the probability that the car is a lemon?
(c) If a lot is selected and two cars are bought from it, what is the probability that both cars are lemons?

7–7   Bernoulli Trials

A common type of problem in probability is one in which an experiment is repeated several times. To illustrate, consider the following problems:

1. Toss a coin five times and record H or T each time.
2. Roll a die four times and record the number on the top face of each roll.
3. Select ten patients and determine if each is cured.

Each repetition is called a **trial.** In Problem 1 the trial is tossing a coin and is repeated five times. In Problem 3 the trial is to select a patient and is repeated ten times. Each problem has specific characteristics: a physical experiment or trial is repeated several times and the trials are identical. In Problems 1 and 3 each trial has the same two outcomes. The outcomes are usually called success and failure, or simply S and F. In most experiments the connotation of "success" is arbitrary. Thus S can be assigned to either outcome and then F is its complementary event. Therefore if $Pr(S) = p$, then $Pr(F) = 1 - p$.

Bernoulli Trials and Binomial Experiments

Experiments of this kind are called *binomial experiments* and the trials themselves are called *Bernoulli trials*. Thus a binomial experiment consists of the repetition of several Bernoulli trials. In Problem 2 the Bernoulli trial is to roll a die, and the binomial experiment is the repetition of the trial four times. Let us summarize.

DEFINITION

A **binomial experiment** is a probability problem consisting of one experiment repeated several times. Each repetition is called a **Bernoulli trial.** The experiment has the following properties:

1. The experiment in each trial remains the same when it is repeated.
2. There are two outcomes, called success and failure, or simply S and F.
3. The trials are independent.
4. The probability of the outcomes remains the same for each trial.

Sometimes a binomial experiment is called a **Bernoulli process.**

We considered experiments like this in earlier sections, but we had to solve each problem from scratch. Example 1 illustrates the derivation of a formula that will help solve this common type of problem.

EXAMPLE 1

Problem

Roll a die four times. Find the probability of observing (a) exactly four 1's, (b) exactly three 1's.

Solution Let us assign S to the event of observing 1. For each trial $Pr(S) = 1/6$ and $Pr(F) = 5/6$. Hence
(a) $Pr(\text{exactly four 1's}) = Pr(SSSS) = (1/6)(1/6)(1/6)(1/6) = 1/1296$.
(b) There are $C(4, 3) = 4$ ways to observe exactly three 1's: $SSSF$, $SSFS$, $SFSS$, and $FSSS$. The probability of each is the product of three $(1/6)$'s and one $5/6$. Hence

$$Pr(\text{exactly three 1's}) = Pr(\text{exactly three } S\text{'s})$$

$$= C(4, 3) \left(\frac{1}{6}\right)^3 \left(\frac{5}{6}\right)$$

$$= 4\left(\frac{1}{216}\right)\left(\frac{5}{6}\right) = \frac{20}{1296} = \frac{5}{324}$$

The second part of Example 1 provides a model for the general case. Suppose a binomial experiment consists of n repetitions of a Bernoulli trial with $Pr(S) = p$. The problem is to find the probability of observing exactly x successes. One way to observe x successes is for the S's to come first followed by the F's. There would be x S's and $n - x$ F's.

$$\underbrace{SS \cdots SS}_{x \text{ times}}\underbrace{FF \cdots FF}_{n - x \text{ times}}$$

Since $Pr(S) = p$, we have $Pr(F) = q = 1 - p$. Hence

$$Pr(\underbrace{SS \cdots SS}_{x \text{ times}}\underbrace{FF \cdots FF}_{n - x \text{ times}}) = \underbrace{Pr(S)Pr(S) \cdots Pr(S)}_{x \text{ times}}\underbrace{Pr(F) \cdots Pr(F)}_{n - x \text{ times}}$$

$$= \underbrace{pp \cdots pp}_{x \text{ times}}\underbrace{(1 - p) \cdots (1 - p)}_{n - x \text{ times}}$$

$$= p^x (1 - p)^{n-x}$$

The event "exactly x S's" in n trials occurs with any arrangement of the x S's and $n - x$ F's, and there are $C(n, x)$ such arrangements. This leads to the following general result, where we substitute q for $1 - p$.

Probabilities of Bernoulli Trials

In a binomial experiment consisting of n repetitions of a Bernoulli trial with $Pr(S) = p$ and $Pr(F) = 1 - p = q$, the probability of x successes (implying there are $n - x$ failures) is

$$Pr(\text{exactly } x \text{ } S\text{'s}) = C(n, x)p^x q^{n-x}$$

The next two examples illustrate the use of the formula.

EXAMPLE 2

An urn contains 3 red and 7 white marbles. Five marbles are drawn with replacement.

Problem

Find the probability of observing (a) exactly 3 red marbles, (b) at least 3 red marbles.

Solution This is a binomial experiment with five repetitions of the Bernoulli trial of drawing a marble. Let S represent drawing a red marble. Then $p = Pr(S) = \frac{3}{10} = .3$ and $q = Pr(F) = 1 - \frac{3}{10} = \frac{7}{10} = .7$.
(a) We need to compute "exactly three successes." From the formula with $n = 5$ and $x = 3$ we have

$$Pr(x = 3) = C(5, 3)(.3)^3(.7)^{5-3}$$

$$= 10(.027)(.49) = .1323$$

(b) The event "at least 3 red marbles" consists of the three mutually exclusive events: "exactly 3," "exactly 4," and "exactly 5" marbles. As in part (a), we find the latter two probabilities from the formula with $n = 5$ and $x = 4$ and 5, respectively:

$$Pr(x = 4) = C(5, 4)(.3)^4(.7)^1 = .02835$$

$$Pr(x = 5) = C(5, 5)(.3)^5(.7)^0 = .00243$$

Therefore if we let "at least three S's" be represented by "$x \geq 3$,"

$$Pr(x \geq 3) = Pr(3) + Pr(4) + Pr(5)$$

$$= .1323 + .02835 + .00243 = .16308$$

EXAMPLE 3

Food companies use what is known as a "triangle taste test" to judge the significance of the taste of one product over another. For example, when the Coca-Cola Co. was preparing to introduce a new brand of Coke they used this test to evaluate the difference in taste between the old brand, now called "Coke Classic," against the new version. In the triangle taste test the taster is given three samples, two of which are alike, and asked to choose the odd food by tasting. Suppose the taster does not have a well-developed sense of taste for this product and chooses at random.

Problem

If the taster makes six identical taste tests of the product, find the probability that the taster will make (a) no correct choices, (b) exactly one correct choice, (c) at least three correct choices.

Solution This is a binomial experiment with six repetitions and with $Pr(S) = 1/3$ where S represents choosing the odd food. In the formula let $n = 6$ and $x = 0$, $x = 1$, and $x \geq 3$, respectively. We have
(a) $Pr(x = 0) = C(6, 0)(1/3)^0(2/3)^6 = 64/729 \approx .09$
(b) $Pr(x = 1) = C(6, 1)(1/3)^1(2/3)^5 = 192/729 \approx .26$
(c) We need to compute

$$Pr(x \geq 3) = 1 - [Pr(0) + Pr(1) + Pr(2)]$$

Hence we need to calculate

$$Pr(x = 2) = C(6, 2)\left(\frac{1}{3}\right)^2\left(\frac{2}{3}\right)^4 = \frac{240}{729} \approx .33$$

Therefore

$$Pr(x \geq 3) = 1 - [Pr(0) + Pr(1) + Pr(2)]$$

$$\approx 1 - [.09 + .26 + .33] = 1 - .68 = .32$$

Thus there is only about a 32% chance that the taster will get at least three of six choices correct.

EXERCISE SET 7–7

In Problems 1 to 4 the experiment is to toss a coin six times and to record the sequence of heads and tails. Find the probability of the event.

1. Exactly four heads.

2. Exactly five heads.

3. At least four heads.

4. At least five heads.

In Problems 5 and 6 the experiment is to toss a coin eight times and to record the sequence of heads and tails. Find the probability of the event.

5. Exactly two heads.

6. At least seven heads.

In Problems 7 to 10 an urn contains 5 red and 15 white marbles. Select n marbles with replacement and find the probability of the event.

7. $n = 5$, exactly 4 red marbles.

8. $n = 5$, at least 4 red marbles.

9. $n = 6$, at least 5 red marbles.

10. $n = 7$, at most 2 red marbles.

In Problems 11 and 12 an urn contains 5 red and 20 white marbles. Select n marbles with replacement and find the probability of the event.

11. $n = 5$, exactly 4 red marbles.

12. $n = 5$, at least 4 red marbles.

In Problems 13 to 20 a taste test described in Example 3 is conducted. The taster makes the given number n of taste tests of the product. Find the probability of the taster making the given number m of correct choices.

13. $n = 5$, $m = 2$.

14. $n = 5$, $m = 3$.

15. $n = 7$, $m = 2$.

16. $n = 8$, $m = 2$.

17. $n = 5$, at least four correct choices.

18. $n = 5$, at least three correct choices.

19. $n = 8$, at most two correct choices.

20. $n = 8$, at most three correct choices.

In Problems 21 to 26 a survey of n voters is conducted in an election where it is assumed that 40% of the voters favor a candidate. Find the probability that at least half of those surveyed will vote for the candidate.

21. $n = 5$.

22. $n = 7$.

23. $n = 8$.

24. $n = 9$.

25. $n = 11$.

26. $n = 12$.

27. It is known that 80% of the population in a ward favor an upcoming bond issue. Eight people are selected. Find the probability that (a) exactly six of them will favor the bond issue, (b) at least six of them will favor the bond issue.

28. It is known that 75% of the population in a ward favor an upcoming bond issue. Ten people are selected. Find the probability that (a) exactly eight of them will favor the bond issue, (b) at most eight of them will favor the bond issue.

29. A student takes a ten-question true-false test and guesses at the answers. What is the probability of getting (a) exactly seven correct, (b) at least seven correct?

30. A student takes a 20-question true-false test and guesses at the answers. What is the probability of getting (a) exactly 15 correct, (b) at least 18 correct?

31. The past performance of a machine shows that it produces 5% defectives. What is the probability that in a run of ten items (a) exactly eight will be defective, (b) at least eight will be defective?

32. A die is rolled eight times. What is the probability of observing (a) exactly five 6's, (b) at most five 6's?

33. A die is rolled ten times. What is the probability of observing (a) exactly eight 6's, (b) at least eight 6's?

34. A part of a final exam consists of ten questions that the student has not studied. If the questions are multiple choice, each with five answers, and the student guesses, find the probability that the student gets (a) exactly three correct, (b) at least three correct.

35. A part of a final exam consists of five questions that the student has not studied. If the questions are multiple choice, each with four answers, and the student

guesses, find the probability that the student gets (a) exactly three correct, (b) at most three correct.

36. If a baseball player has a batting average of .300 and gets to bat five times in the next game, find the probability that the player will get (a) exactly three hits, (b) at most three hits.

37. If a basketball player has a shooting percentage of 60%, find the probability that the player will make at least two of her next six shots.

38. Each year a firm selects several undergraduates to participate in a co-op program. In the past the firm has hired 40% of the co-op workers to full-time employment after graduation. If ten students participate this year, find the probability that five will be hired after graduation.

39. In a small city it is expected that 30% of the population will contract the latest strain of influenza. If a company has 15 employees, find the probability that at least 3 employees will contract the disease.

40. A company advertises that its new toothpaste will prevent tartar buildup. It claims that 40% of people will have less tartar after three months of using the toothpaste. To justify the claim a clinical experiment is run with 20 people using the toothpaste for three months. Find the probability that at least 4 of them will have less tartar.

41. A die is rolled ten times. What is the probability of observing (a) exactly two 6's, (b) at most two 6's?

42. Each month a company selects several employees to participate in an advertising campaign. In the past the firm has invited back the next month 20% of the employees for the following month. If ten employees participate this month, find the probability that five will participate the following month.

43. In a city it is expected that 5% of the population will have a serious car accident. If a company has 20 employees, find the probability that at least 3 employees will have a serious car accident.

44. A fraternity has 20% of its members on the Dean's list. If ten members are chosen at random, find the probability that at least four of them will be on the Dean's list.

Referenced Exercise Set 7–7

1. In an attempt to determine if performance on ESP tests is affected by the person's belief in ESP, Ryzl* scored people according to their relative ability on an ESP test. Forty percent of the people performed well on the ESP test. Suppose ten are selected at random. Find the probability that at least nine of them performed well.

Cumulative Exercise Set 7–7

1. The experiment consists of tossing two dice and recording the sum of the top faces. Find the sample space and determine the elements in the two events
   E: The sum is greater than 9.
   F: The sum is even.
   Find (a) $Pr(E \cup F)$, (b) $Pr(E \cap F')$.

2. There are 50 members of a state legislature and 30 will vote to pass the next bill. If three are selected at random and asked how they will vote, what is the probability that at least one will vote to pass the bill?

3. Urn 1 has 3 blue and 7 red marbles and urn 2 has 2 blue and 8 red marbles. The experiment is to select an urn and then to select a marble and record its color. Find the probability of observing a red marble.

4. There are 100 patients in a hospital, 40 of whom are in outpatient care, 35 are being treated for influenza, and 10 of the outpatients are being treated for influenza. If a patient is selected at random, find the probability that the patient is an outpatient given that the patient is not being treated for influenza.

5. In a survey of students' attitudes in five different classes the percentage of students who rated the professor the best they ever had was as follows: class 1, 20%; class 2, 15%, class 3, 10%; class 4, 25%; and class 5, 5%. Assuming that there were the same number of students in each class, find the probability that a student selected at random from those who rated the professor the best they ever had was selected from class 1 or class 2.

6. A bank has three types of checking accounts—regular, deluxe, and executive class. Past experience shows that 15% of regular accounts are assessed an overdraft charge, 10% of deluxe accounts are charged an over-

*M. Ryzl, ''Precognition Scoring and Attitude Toward ESP,'' *Journal of Parapsychology,* Vol. 32, pp. 1–8, 1968. Ryzl found that 96% of the people who scored well on the test also believed in ESP. Do you think there is a connection between scoring well on the ESP test and believing in ESP?

draft charge, and 5% of executive accounts are charged an overdraft charge. Suppose 60% of the accounts are regular, 30% are deluxe, and 10% are executive class. Find the probability that an account that was charged for an overdraft was an executive class account.

7. A die is rolled ten times. What is the probability of observing (a) exactly seven 6's, (b) at most seven 6's?

8. It is known that 60% of the population in a ward favor an upcoming bond issue. Ten people are selected. Find the probability that (a) exactly eight of them will favor the bond issue, (b) at least eight of them will favor the bond issue.

9. A student takes a ten-question multiple-choice test with four choices per question and guesses at the answers. What is the probability of getting (a) exactly one correct, (b) at most two correct?

10. A firm is going to introduce a new product and wants to predict how well it will do. Past records show that

similar products at least break even 70% of the time. If there are ten sales regions, find the probability that the product will break even in at least eight regions.

11. There are 2 used-car lots. The small lot has 6 overpriced cars and 4 underpriced cars and the large lot has 15 overpriced cars and 5 underpriced cars.
    (a) If a car is selected at random from the small lot, what is the probability that it is underpriced?
    (b) If a lot is selected at random and a car is selected from it at random, what is the probability that it is underpriced?

12. There are 2 used-car lots. The small lot has 20 overpriced cars and 5 underpriced cars and the large lot has 35 overpriced cars and 15 underpriced cars.
    (a) If a lot is selected at random and three cars are selected from it at random, what is the probability that all three are underpriced?
    (b) If a car is selected at random and found to be underpriced, what is the probability that it was selected from the small lot?

7–8 Expected Value

An important topic in many applications of probability is that of the "expected value" of an experiment. The expected value is very similar to the concept of "average value." Intuitively, the average value is used to measure the center of a set of numbers, whereas the expected value measures the center of a probability distribution. In the next chapter we will show the close connection between the average value and the expected value.

For example, consider the experiment of tossing a coin 100 times. How many times would we expect heads to appear? While we might not get exactly 50 heads each time the experiment is run, we would certainly expect the number of heads to be close to 50 since $Pr(H) = \frac{1}{2}$ and $(\frac{1}{2}) \cdot 100 = 50$. If a die were rolled 600 times, about how many 2's would appear? Since $Pr(2) = \frac{1}{6}$, we would expect about $(\frac{1}{6}) \cdot 600 = 100$ 2's. Therefore for the binomial experiment of rolling a die 600 times and recording the number of 2's, the expected value is 100.

These examples suggest the following result concerning the expected value of a binomial experiment.

If a binomial experiment has n repetitions with $Pr(S) = p$, then the expected value E is

$$E = np$$

The expected value is a useful property for any probability distribution whose outcomes are numbers. If the outcomes are not numbers, such as the experiment to toss a coin and record H or T, it makes no sense to speak of the "average" outcome. Therefore in this section all probability distributions have numerical outcomes.

To illustrate how to compute the expected value, look at a simple experiment whose outcomes are not equally likely. An urn contains ten marbles. One marble has the number 1 marked on it, three marbles have the number 2, and six have the number 3. A marble is selected from the urn and its number is recorded. What is the expected value? If the experiment were done many times, the number 1 would be expected to appear one-tenth of the time, 2 would appear three-tenths of the time, and 3 would appear six-tenths of the time. The probability distribution is

x	1	2	3
Pr	$\frac{1}{10}$	$\frac{3}{10}$	$\frac{6}{10}$

This means that if the experiment were repeated 100 times, we would expect to observe about 10 1's, 30 2's, and 60 3's. The average of these numbers is

$$\frac{1(10) + 2(30) + 3(60)}{100} = \frac{250}{100} = 2.5$$

Another way to view this computation is to divide each term in the numerator by 100 to get

$$1\left(\frac{10}{100}\right) + 2\left(\frac{30}{100}\right) + 3\left(\frac{60}{100}\right) = 2.5$$

Dividing each numerator and denominator by 10 gives

$$1\left(\frac{1}{10}\right) + 2\left(\frac{3}{10}\right) + 3\left(\frac{6}{10}\right) = \frac{25}{10} = 2.5$$

This is what we call the expected value. It is the sum of the products of the outcomes and their respective probabilities. It is independent of how many times the experiment is actually repeated. If the experiment were done 1000 times, we would expect the average of the 1000 outcomes, that is, their sum divided by 1000, to be the same value, 2.5.

This expression demonstrates that the expected value is the sum of the products of the outcomes and their corresponding probabilities. We can now formally state the definition.

DEFINITION

If an experiment has numerical outcomes and the probability distribution is

x_i	x_1	x_2	$\ldots$	x_n
p_i	p_1	p_2	$\ldots$	p_n

then the expected value, E, of the experiment is

$$E = x_1 p_1 + x_2 p_2 + \ldots + x_n p_n$$

The following examples illustrate the use of this definition. Example 1 discusses the concept of a "fair game," meaning one in which each person has an equal chance to win. That is, if the game were played many times, each player would expect to win about the same amount of money. This means that a game is fair if its expected value is zero.

EXAMPLE 1

Two players, A and B, play the following game. A draws a card from a shuffled deck. If a picture card is drawn, A wins \$3. If any other kind of card is drawn, B wins \$1.

Problem

Is the game fair?

Solution We compute the expected value and then interpret its meaning. There are two outcomes: A either wins \$3 or loses \$1, so the two outcomes are 3 and -1. The outcome 3 occurs if one of the picture cards is drawn, so the corresponding probability is $\frac{3}{13}$. Hence $Pr(3) = \frac{3}{13}$. On the other hand, $Pr(-1) = 1 - \frac{3}{13} = \frac{10}{13}$, which means that there are 40 of the 52 cards that are not picture cards. Using the notation in the definition, we have $x_1 = 3$, $p_1 = \frac{3}{13}$, $x_2 = -1$, and $p_2 = \frac{10}{13}$. Thus the probability distribution is

x_i	3	-1
p_i	$\frac{3}{13}$	$\frac{10}{13}$

Hence the expected value E is

$$E = x_1 p_1 + x_2 p_2$$
$$= 3 \cdot (\tfrac{3}{13}) + (-1) \cdot \tfrac{10}{13}$$
$$= \tfrac{9}{13} - \tfrac{10}{13} = -\tfrac{1}{13}$$

Therefore the game is not fair because A will likely lose \$1 every 13 plays in the long run.

EXAMPLE 2

A game consists of rolling a fair die and the player receives an amount of dollars equal to the number turning up on the top face if the number 1, 2, 3, or 4 occurs. If 5 or 6 occurs, the player loses $5 or $6, respectively.

Problem

Is the game fair?

Solution We compute the expected value by first finding the probability distribution. There are six outcomes: 1, 2, 3, 4, -5, and -6. Each has probability $\frac{1}{6}$ because they are equally likely. Thus the probability distribution is

x_i	1	2	3	4	-5	-6
p_i	$\frac{1}{6}$	$\frac{1}{6}$	$\frac{1}{6}$	$\frac{1}{6}$	$\frac{1}{6}$	$\frac{1}{6}$

Hence the expected value E is

$$E = 1\cdot(\tfrac{1}{6}) + 2\cdot(\tfrac{1}{6}) + 3\cdot(\tfrac{1}{6}) + 4\cdot(\tfrac{1}{6}) + (-5)\cdot(\tfrac{1}{6}) + (-6)\cdot(\tfrac{1}{6})$$

$$= -\tfrac{1}{6}$$

Therefore the game is not fair because the player will likely lose an average of one-sixth of a dollar, or about $0.17, per roll. In other words, the player can expect to lose about $1 every six plays in the long run.

EXAMPLE 3

Problem

In the game described in Example 2, change the value that the player wins when 1 is rolled in order to make the game fair.

Solution A fair game has expected value zero. We must replace the outcome 1 by a variable outcome, say, a, and compute the expected value of this probability distribution, and then set it equal to 0 and solve for a. Thus the probability distribution is

x_i	a	2	3	4	-5	-6
p_i	$\frac{1}{6}$	$\frac{1}{6}$	$\frac{1}{6}$	$\frac{1}{6}$	$\frac{1}{6}$	$\frac{1}{6}$

Hence the expected value E is

$$E = a\cdot\left(\frac{1}{6}\right) + 2\cdot\left(\frac{1}{6}\right) + 3\cdot\left(\frac{1}{6}\right) + 4\cdot\left(\frac{1}{6}\right) + (-5)\cdot\left(\frac{1}{6}\right) + (-6)\cdot\left(\frac{1}{6}\right)$$

$$= \frac{a}{6} - \frac{2}{6} = \frac{(a-2)}{6}$$

Now let $E = 0$ and solve for a.

$$0 = \frac{(a - 2)}{6}$$

$$0 = a - 2$$

$$a = 2$$

Therefore the game is fair if the player wins $2 when 1 occurs.

The next two examples show how the expected value is often used in business applications. When starting a new venture, one must have a good knowledge of the prospects for success. Often the expected value can give a good indication of the likelihood for success.

EXAMPLE 4   A person has applied for a donut shop franchise. Two locations look profitable. Each site has a good traffic pattern, indicating that a heavy volume of customers is possible. Site 1 has most of its traffic in the morning and afternoon: 50% of the customers will come in the morning, 40% in the afternoon, and 10% in the evening. Site 2 has a more homogeneous clientele: 30% will come in the morning, 30% in the afternoon, and 40% in the evening. Each site has the same projected volume—about 1000 customers per day. The parent company has determined that the average customer spends $5 in the morning, $4 in the afternoon, and $3 in the evening.

Problem

Which site has the larger expected profit?

Solution The problem consists of finding the expected value for two experiments—site 1 and site 2. Here are the probability distributions for each experiment.

Period of Day	Revenue per Customer, x_i	Percentage of Volume Site 1, p_i	Percentage of Volume Site 2, p_i
Morning	$5	.5	.3
Afternoon	$4	.4	.3
Evening	$3	.1	.4

The expected value for site 1 is

$$E = 5 \cdot (.5) + 4 \cdot (.4) + 3 \cdot (.1) = 2.5 + 1.6 + .3 = 4.4$$

The expected value for site 2 is

$$E = 5 \cdot (.3) + 4 \cdot (.3) + 3 \cdot (.4) = 1.5 + 1.2 + 1.2 = 3.9$$

Hence site 1 has a better expected value of $4.40 per customer than site 2, which has the expected value of $3.90 per customer. To interpret the meaning of these expected values, assume there is a volume of 1000 customers per day, then the owner can expect a revenue of about $4400 per day from site 1 and about $3900 per day from site 2.

EXAMPLE 5

A real estate company is bidding on rights to develop two properties, property A and property B. The company estimates that a profit of $600,000 will be made on property A while property B will realize a profit of $400,000. Legal and other costs of bidding amount to $9000 for property A and $5000 for property B. The probability of being awarded the contract if the company bids for property A is .4 while it is .7 for property B.

Problem

For which property should the company bid?

Solution The problem consists of finding the expected value for two experiments: property A and property B. Here are the probability distributions for each experiment.

	Property A	Property B
Profit	$600,000	$400,000
Cost of bidding	$9000	$5000
Probability of success	.4	.7

The expected value for property A is

$$E = 600,000 \cdot (.4) + (-9000) \cdot (.6)$$

$$= 234,600$$

The expected value for property B is

$$E = 400,000 \cdot (.7) + (-5000) \cdot (.3)$$

$$= 278,500$$

Hence property B has a higher expected value.

EXERCISE SET 7–8

In Problems 1 to 4 find the expected value of the binomial experiment.

1. Toss a coin 50 times and record the number of heads.

2. Toss two coins 200 times and record the number of times two heads occur.

3. Roll a die 12 times and record the number of times 6 occurs on the top face.

4. Roll two dice 10 times and record the number of times the sum of 2 occurs.

In Problems 5 and 6 find the expected value of the probability experiment when a marble is selected from the urn and its number is recorded.

5. An urn contains 10 marbles. One marble has the number 1 marked on it, 4 marbles have the number 2, and 5 marbles have the number 3.

6. An urn contains 20 marbles. One marble has the number 2 marked on it, 8 marbles have the number 4, and 11 have the number 6.

In Problems 7 to 12 find the expected value for the games. Is the game fair? (A game is fair if the expected value is 0.)

7. Player A chooses a card from a deck of 52 cards and wins $4 if a picture card appears and loses $1 if it is not a picture card.

8. Player A chooses a card from a deck of 52 cards and wins $13 if a picture card appears and loses $3 if it is not a picture card.

9. Player A rolls two dice and records the sum of the top faces. A wins $9 if 7 or 11 appears and loses $2 otherwise.

10. Player A rolls 2 dice and records the sum of the top faces. A wins $3 if 2, 7, or 11 appears and loses $1 otherwise.

11. A game consists of rolling a fair die and the player receives an amount of dollars equal to twice the number turning up on the top face if the number 1, 2, 3, or 4 occurs. If 5 or 6 occurs, the player loses $9.

12. A game consists of rolling a fair die and the player receives an amount of dollars equal to twice the number turning up on the top face if the number 1, 2, 3, or 4 occurs. If 5 or 6 occurs, the player loses $10.

In Problems 13 and 14 consider the game described in Problem 11. Change the value that the player wins when the given number is rolled in order to make the game fair.

13. Change the value the player wins when 2 is rolled.

14. Change the value the player wins when 4 is rolled.

In Problems 15 and 16 determine which site is predicted to produce the larger revenue for a proposed franchise.

15.

Revenue		Site 1	Site 2
		Percentage	
Morning	$5	.5	.6
Afternoon	$3	.1	.1
Evening	$6	.2	.2
Late night	$2	.2	.1

16.

Revenue		Site 1	Site 2
		Percentage	
Morning	$4	.4	.3
Afternoon	$3	.3	.4
Evening	$5	.1	.2
Late night	$2	.2	.1

In Problems 17 to 20 a real estate company is bidding on rights to develop two properties, property A and property B. The estimated profit for each property is given along with the costs of bidding and the probabilities of being awarded the contract if the company bids for property A and for property B. For which property should the company bid?

17.

	Property A	Property B
Profit	$200,000	$100,000
Cost of bidding	$8000	$3000
Probability of success	.4	.5

18.

	Property A	Property B
Profit	$400,000	$500,000
Cost of bidding	$8000	$9000
Probability of success	.7	.6

19.

	Property A	Property B
Profit	$300,000	$500,000
Cost of bidding	$12,000	$9000
Probability of success	.3	.4

20.

	Property A	Property B
Profit	$500,000	$500,000
Cost of bidding	$10,000	$9000
Probability of success	.3	.4

In Problems 21 to 25 find the expected value of the binomial experiment.

21. It is known that 70% of the population in a ward favor an upcoming bond issue. Five people are selected at random and each is asked whether the person is in favor.

22. A student takes a 20-question true-false test and guesses at the answers. It is recorded whether the student is right or wrong on each question.

23. The past performance of a machine shows that it produces 2% defectives. Ten items are selected at

random and it is recorded whether each item is defective.

24. A die is rolled eight times and it is recorded whether the number is a 6 on each roll.

25. A part of a final exam consists of five questions that the student has not studied. The questions are multiple choice, each with four answers, the student guesses each answer, and the number of correct responses is recorded.

26. Find the expected value for the following probability distribution:

x	2	3	4	5	6
Pr	.05	.25	.35	.15	.20

27. An urn contains 3 red and 7 white marbles. Ten marbles are selected with replacement and the number of red marbles is recorded. Find the expected value of the experiment.

28. A raffle offers a first prize of $5000, two second prizes of $1000, four third prizes of $500, and ten prizes of $100. One thousand tickets are sold at $20 each. Find the expected value for the person who buys one ticket.

29. A lottery consists of a player choosing a three-digit number. Repetitions are allowed. Each ticket costs $1 and it pays $500 to win. Compute the expected earnings of the player who purchases one ticket.

30. Compute the expected earnings of the lottery player in Problem 29 if repetitions are not allowed.

31. In a lottery a player purchases a ticket with a three-digit number for $2. Repetitions are allowed. If the number is drawn, the player wins $400. If the number drawn is not the player's but the digits are the same except for the order, the player wins $200. Find the expected earnings.

32. The annual premium for a $6000 insurance policy against the loss of jewelry is $100. If the probability that the jewelry will be lost is .005, what is the expected gain or loss on the insurance policy?

33. A common way that salespeople assess their potential is to assign probabilities to their success at selling certain volumes of business. For the following month a salesperson determines from past records that the probability of selling (a) less than $10,000 worth of

business next month is .10, (b) between $10,000 and $20,000 is .45, (c) between $20,000 and $30,000 is .40, and (d) greater than $30,000 is .05. Find the salesperson's expected earnings.

34. A professional tennis player is asked what the player expects to win each week. Of course, it is impossible to determine how the player will perform in any given week, but it is possible to calculate the player's expected winnings from past performances. Here is a table that gives the likelihood that the tennis player finishes the next tournament in one of the first five places or in a lower position. It also gives the average prize for that place. Find the player's expected winnings for the next tournament.

Place	1	2	3	4	5	>5
Probability	.02	.04	.08	.12	.20	.54
Prize ($1000)	50	25	12	5	2.5	1

In Problems 35 to 38 we consider games that are similar to those encountered in Chapter 10, Game Theory. The outcomes in the games are given along with their probabilities. Find the expected value.

35.

x_i	−1	0	2	3
p_i	.3	.1	.5	.1

36.

x_i	−1	−5	1	10
p_i	.2	.3	.4	.1

37.

x_i	−10	2	3	3
p_i	.15	.75	.08	.02

38.

x_i	−10	1	2	3
p_i	.15	.7	.1	.05

39. A firm has five portfolios. Portfolio A has 50 accounts, 22 of which are executive accounts. Portfolio B has 70 accounts, 15 of which are executive accounts. Portfolio C has 40 accounts, 25 of which are executive accounts. Portfolio D has 90 accounts, 20 of which are executive accounts. Portfolio E has 75 accounts, 35 of which are executive accounts. If an account is selected at random, find the probability that the account is an executive account.

40. In a survey of skiing accidents at six hospitals the incidence of fractures was as follows: hospital 1, 20% of the injuries were fractures; hospital 2, 10%; hospital 3, 5%; hospital 4, 8%; hospital 5, 2%; hospital 6, 15%. Assume that there were the same number of accidents at each hospital and suppose an accident is

selected at random. Find the probability that it was a fracture.

41. For the previous problem answer the same question given that the number of accidents at each hospital was as follows: hospital 1, 100 accidents; hospital 2, 50 accidents; hospital 3, 60 accidents; hospital 4, 50 accidents; hospital 5, 100 accidents; hospital 6, 80 accidents.

42. Find the expected value of the binomial experiment of tossing two coins 500 times and recording the number of times two heads occur.

Referenced Exercise Set 7–8

1. There is a difference between the game of roulette played in America versus Europe. The American roulette wheel has 38 slots, numbered 1 to 36 with one labeled 0 and another 00. Of the 36 number slots, 18 are red and 18 are black; the other two are green. The European wheel does not have the 00 slot and the betting differs. The differences can be compared via the expected value to see which version gives the player the better advantage. Consider first the version played in America. If you bet $1 on red, you have two possibilities: you may win $1 (in addition to your initial bet of $1) with probability $\frac{18}{38} = \frac{9}{19}$, or you may lose your bet of $1 with probability $\frac{20}{38} = \frac{10}{19}$. Here is the probability distribution.

American Roulette

Outcomes, x_i	Probability, p_i	$x_i p_i$
1	$\frac{9}{19}$	$\frac{9}{19}$
-1	$\frac{10}{19}$	$-\frac{10}{19}$

In European roulette there is a third outcome besides the same two for the American version—if the ball lands on green, you do not necessarily lose your bet. Your bet is carried to the next spin where there are again three possibilities: if the ball lands on red, you get your $1 bet back, so you break even; if it lands on black, you lose $1; and if it lands on green again, you get back $.50, which means you lose $.50 of your initial bet. There are four outcomes: 1 (or win $1), 0 (or break even), $-.5$ (or lose $.50), and -1 (or lose $1). Construct the probability distribution for European roulette and find the expected value for each game.

2. Land allocation for farmers is always done under risk. They are guided by return on their crops, but various conditions force decisions to be made whose outcomes are unknown: What will the weather conditions be during the season? How much fertilizer will produce the best harvest? In an article weighing various options for farmers R. N. Collender and D. Zilberman use the expected value to predict the return for several different strategies. They center their attention on farmland in the lower Mississippi drainage area. The land can be used to grow corn or cotton. Two types of plantings of each crop are to be made. Type 1 corn and cotton do well under good weather conditions while type 2 corn and cotton do not do well under good weather conditions. Certain percentages of the available land are planted with each type of crop. There are constraints dictated by the marketplace; for instance, no more than 60% of the land can be used for cotton, and each crop must be planted on at least 10% of the land. The table gives the return per acre expected from each type of crop and two strategies for the percent of land used for each type of crop. Which strategy will give the highest return?

Type of Crop	Return ($/acre)	Percent Land Planted	
		Strategy A	Strategy B
Type 1 corn	53	10	10
Type 2 corn	77	30	30
Type 1 cotton	244	20	30
Type 2 cotton	351	40	30

Source: R. N. Collender and D. Zilberman, "Land Allocation Under Uncertainty for Alternative Specifications of Return Distributions," *American Journal Agricultural Economics*, Vol. 67, 1985.

3. Probabilistic inferences and forecasts are an integral part of many decision-making situations. In an article evaluating such forecasts Keith Levi studies the decision-making process of a patient who has found a lump under the skin and wants two specialists to determine whether the tumor is malignant or benign. To evaluate the accuracy of the two physicians, data from past diagnoses are studied. There are four possibilities for each case: the diagnosis was that the tumor was benign and it was actually either benign or malignant, or the diagnosis was that it was malignant and it was actually benign or malignant. The best case is when it is benign and it is diagnosed as such. This is referred to as event *A*. The next best case, event *B*, is when the tumor is malignant and it is diagnosed as such. An incorrect diagnosis is made otherwise. The worst error, event

C, would be that the tumor was malignant and was diagnosed as benign. The remaining possibility is event *D*, when the tumor is benign but diagnosed as malignant. The next step in the analysis is to assign a weight, called a *utility*, to each possibility. The worst case, event *C*, is arbitrarily assigned 0, and 1 is assigned to the best case, event *A*. Then the relative merits, or

demerits, of the remaining cases are determined. The patient in the study decided to assign .7 to event *B* and .2 to event *D*. The percentage of cases handled in each category by each physician was then determined, using 180 cases for each physician. The data are given in the table. Compute the expected value for each physician to evaluate the relative performance of each.

Diagnosis of Tumors: Physician Performance

Diagnosis: Actual:	Benign Benign	Malignant Malignant	Benign Malignant	Malignant Benign
Event	A	B	C	D
Utility	1	.7	0	.2
Past performance				
Physician 1	.54	.09	.01	.36
Physician 2	.29	.35	.27	.09

Source: Keith Levi, "A Signal Detection Framework for the Evaluation of Probabilistic Forecasts," *Organizational Behavior and Human Decision Processes*, Vol. 36, pp. 143–166, 1985.

Cumulative Exercise Set 7–8

1. The experiment consists of tossing two dice and recording the sum of the top faces. Find the sample space and determine the elements in the two events

   E: The sum is less than 6.
   F: The sum is 4, 5, or 6.

   Find (a) $Pr(E \cup F)$, (b) $Pr(E' \cap F)$.

2. Work Problem 1 if three dice are tossed.

3. There are 40 members in a union and 10 agree to take a pay cut. If three are selected at random and asked their opinion, what is the probability that at least one will agree to take a pay cut?

4. Urn 1 has 5 white and 7 red marbles and urn 2 has 4 white and 8 red marbles. The experiment is to select an urn and then select a marble and record its color. Find the probability of observing a red marble.

5. A portfolio has 50 accounts, 22 of which are executive accounts and there are 10 delinquent accounts, while 5 of the executive accounts are delinquent. If an account is selected at random, find the probability that the account is an executive account given that it is not delinquent.

6. In a survey of biking accidents at five hospitals the incidence of fractures was as follows: hospital 1, 20%

of the injuries were fractures; hospital 2, 10%; hospital 3, 5%; hospital 4, 8%; hospital 5, 2%. Assuming that there was the same number of biking accidents at each hospital, find the probability that a fracture occurred at hospital 1 or hospital 2.

7. Find the expected value of the binomial experiment of tossing two coins 200 times and recording the number of times two heads occur.

8. Determine which site is predicted to produce the larger revenue for a proposed franchise.

Expenditure	Percentage	
	Site 1	Site 2
Morning $4	.4	.6
Afternoon $2	.3	.1
Evening $7	.2	.2
Late night $3	.1	.1

9. A student takes a ten-question true-false test and guesses at the answers. What is the probability of getting (a) exactly eight correct, (b) at least six correct?

10. The annual premium for a $9000 insurance policy against the loss of jewelry is $80. If the probability that the jewelry will be lost is .005, what is the expected gain or loss on the insurance policy?

11. A common way that salespeople assess their potential is to assign probabilities to their success at selling certain volumes of business. For the following month a salesperson determines from past records that the probability of selling (a) less than $10,000 worth of business next month is .15, (b) between $10,000 and $20,000 is .55, (c) between $20,000 and $30,000 is .25, and (d) greater than $30,000 is .05. Find the salesperson's expected earnings.

12. The outcomes of a game are given along with their probabilities. Find the expected value of the game.

x_i	-1	0	2	3
p_i	.2	.1	.4	.3

13. Traffic police monitor cars parked in a two-hour parking zone by marking on the tires with chalk the hour of the day rounded to the next largest hour. Thus there are 12 possible marks corresponding to the 12 hours on a clock face. For example, if it is 11:45, they would put a chalk mark at the top of the tire to indicate 12:00. Suppose a car drives away and returns to the same parking spot.
   (a) What is the probability that the mark will be in the same spot on the tire?
   (b) Suppose the officer marks both curbside tires and suppose the tires rotate at different speeds, making the events of the hour that the chalk mark on the front tire independent of the chalk mark on the back tire. What is the probability that the marks will come to rest in the original locations?

14. A city has three tests it administers to drivers to test for alcohol intoxication: a breathalizer, walk a straight line, and touch index fingers by swinging them from an outstretched sideways position. A type I error is made when the driver is judged intoxicated when the driver is actually sober. Suppose the probability of a type I error for the breathalizer is .01, for the walk it is .1, and for the finger test it is .15. Suppose for a sober person the tests are independent. What is the probability that a type I error is made on all three tests?

CASE STUDY   ## Jane Blalock and the LPGA

In a celebrated court case in 1974 Ms. Jane Blalock sued a group of her peers for damages, claiming they had deprived her of her means of income. Ms. Blalock joined the Ladies Professional Golf Association (LPGA) in 1969 and soon became a star performer. By May 1972 she had become the year's leading money winner on the tour. Then during the final round of the Bluegrass Invitational Tournament she was disqualified from the tournament for an alleged infraction of the rules. A committee of her competitors, selected from the LPGA, also suspended her from the next tournament, the Ladies Carling Open. Blalock sought relief in court under the Sherman Antitrust Act, which holds that individuals cannot be suspended from their profession by their peers because it would tend to lessen competition. This legal precedent appeared justified because her peers would clearly benefit by her absence from the tournament. Blalock received an injunction enabling her to play in the Carling Open, but it was issued too late for her to join the tournament.

Besides being denied the chance to earn prize money in the Carling Open, Blalock had incurred a large legal fee, so she again sued the LPGA for damages under the Sherman Act. The only legal point the court could rule on was her loss of potential earnings from being denied the right to participate in the tournament. This is a very sticky issue in most antitrust cases because the court must decide what the earnings of the claimant, if any, would have been had the competitors not interfered, in this case by suspending the claimant. Thus her problem was to convince the court that, had she been allowed to play, she would most likely have scored low enough to win prize money—and how much (remember that in golf the lower the score the better). Therefore she had to estimate her score in a tournament in which she never played—and most important, her estimate and her means of obtaining it, had to convince a jury that it was valid.

One of the key issues for Blalock was to use a measure that would not only give a clear indication of her most probable score, but would also be easy for a jury to understand. She used the expected value.

A wide range of statistics was gathered and her legal team decided to use the nine tournaments before the disqualification to compare Blalock's performance with that of her competitors. Some fairly complicated statistical measures were used to measure the probability of Blalock's attaining a particular score. These probabilities were obtained by comparing Blalock's scores in the previous nine tournaments with those of her main competitors. We summarize the results in Table 1. The possible scores, or outcomes, are the integers from 209 (that being the lowest since it was the winning score in the tournament) to 232 (that being the highest because anyone with a higher score did not win any money). The data become more manageable if we limit the 24 outcomes to 8, so we let the outcome 210 represent the three possible scores, 209, 210, or 211. The corresponding probability, .07, is therefore the probability that she would score either 209, 210, or 211. The probability that she would score between 212 and 214, represented by the outcome 213, is .16. In Table 1 the possible outcomes are the x_i's and the probabilities are represented by p_i's. In the third column we compute $x_i p_i$. Then the sum of the $x_i p_i$'s is the expected value.

Table 1

Possible Outcome x_i	Probability p_i	$x_i p_i$
210	.07	14.70
213	.16	34.08
216	.23	49.68
219	.24	52.56
222	.17	37.74
225	.09	20.25
228	.03	6.84
231	.01	2.31
Total		218.19

Source: Adapted from Ferdinand K. Levy, "Anti-trust and the Links—Estimating a Golfer's Tournament Score," *Interfaces*, Vol. 6, pp. 6–17, 1976.

Thus the expected value is 218.19, or a score of about 218. Blalock reasoned that this was the best estimate of her score for the tournament in which she was disqualified, and thus she should be awarded a judgment equal to the prize money given to the person(s) who scored 218 in the tournament. Three women shot 218 and tied for fifth place, each earning $1427.50. If the jury ruled in her favor, these damages would be tripled and her legal fees included under the provisions of the Sherman Act.

This argument was so overwhelmingly convincing to the jury that they decided that not only were Blalock's chances of winning the prescribed amount very good, but the likelihood of her winning the tournament and garnering the first place prize was great enough that they awarded her the first place money— $4500. This was then tripled and her legal expenses were included.

Sometimes it pays to know statistics.

Case Study Exercises

In Problems 1 to 4 refer to Table 1 and find the probability of Jane Blalock's attaining the described score.

1. A score less than or equal to 220.
2. A score less than or equal to 226.
3. A score between 218 and 226 inclusively.
4. A score between 221 and 229 inclusively.

In Problems 5 and 6 assume that Jane Blalock's probability of attaining a particular outcome in a tournament is given in the table. As in Table 1, assume that the outcomes represent the probability of her getting that score, or the one lower or the one higher. Thus the outcome of 222 represents the probability of her getting 221, 222, or 223. Find the expected value.

5. Outcome	Probability		6. Outcome	Probability
210	.05		210	.10
213	.20		213	.15
216	.25		216	.20
219	.22		219	.25
222	.15		222	.22
225	.08		225	.04
228	.03		228	.03
231	.02		231	.01

CHAPTER REVIEW

In Section 6–1 we said that set theory forms the basis for probability. In the list of key terms for this chapter we put the corresponding terminology from Chapter 6 in parentheses.

Key Terms

7–1 Sample Spaces
Frequency
Relative Frequency

Probability
Experiment
Sample Space (universal set)
Outcome (element)
Event (subset)
One or the Other Occurs
    (union)
Both Occur (intersection)
Mutually Exclusive Events
    (disjoint subsets)
Does Not Occur (complement)
    Rule for Complements

7–2   Fundamental Probability Principles
Equally Likely Events
Probability Distribution
Certain Event
Impossible Event
Addition Principle for
    Mutually Exclusive Events
Rule for Complements

7–3   The Addition Principle
Addition Principle
Odds in Favor of an Event E
Subjective Probability
Theoretical Probability
Empirical Probability

7–5   Conditional Probability
Conditional Probability
Product Rule
Controlled Experiment
Independent Events
Reliability

7–6   Bayes' Formula
Bayes' Formula

7–7   Bernoulli Trials
Trial
Binomial Experiment
Bernoulli Trial
Bernoulli Process

7–8   Expected Value
Expected Value
Average Value

Summary of Important Concepts

Notation

Probability	$Pr(\ )$
Does Not Occur (Complement)	E'
Conditional Probability	$Pr(E \mid F)$

Formulas

Addition Principle for Mutually Exclusive Events	$Pr(E \cup F) = Pr(E) + Pr(F)$
Certain Event	$Pr(E) = 1$
Impossible Event	$Pr(E) = 0$
Addition Principle	$Pr(E \cup F) = Pr(E) + Pr(F) - Pr(E \cap F)$
Odds in Favor of an Event E	$Pr(E)/Pr(E')$
	$Pr(E) = m/(m + n)$
Conditional Probability	$Pr(E \mid F) = Pr(E \cap F)/Pr(F)$
Product Rule	$Pr(E \cap F) = Pr(E|F){\cdot}Pr(F)$
Independent Events	$Pr(E \mid F) = Pr(E)$
Bayes' Formula	

$$Pr(U_i|E) = Pr(U_i \cap E)/Pr(E) = \frac{Pr(U_i)Pr(E|U_i)}{[Pr(U_1)Pr(E \mid U_1) + \cdots + Pr(U_n)Pr(E \mid U_n)]}$$

Bernoulli Trial	$Pr(x\ S\text{'s in } n \text{ trials}) = C(n, x)p^x q^{n-x}$
Binomial Experiment	$E = p_1 x_1 + \cdots + p_n x_n$
Expected Value	

REVIEW PROBLEMS

In Problems 1 and 2 find the sample space of the experiment.

1. Select a letter in the word "mathematics."

2. Three coins are tossed and the number of heads is recorded.

3. Three urns contain red and blue marbles. The experiment consists of first selecting an urn and then a marble from the urn and recording its color. Define the events

   E: It is not red.

   F: It is red or blue.

   Determine the sample space S of the experiment. Find (a) $E \cup F$, (b) $E \cap F$, (c) E'.

4. A marble is selected from an urn containing 4 red, 5 white, and 6 blue marbles, and its color is recorded. Find (a) $Pr(\text{red})$, (b) $Pr(\text{red or blue})$.

5. A card is drawn from a deck. Define the events $E = \{king\}$, $F = \{black\ ace\}$. Find (a) $Pr(E)$, (b) $Pr(F)$, (c) $Pr(E \cup F)$.

6. A die is rolled and the number on the top face is recorded. Define the two events $E = \{2, 3\}$ and $F = \{3, 4, 5\}$. Find $Pr(E \cup F)$.

In Problems 7 and 8 a sample space has the following probability distribution.

outcome	a	b	c	d
probability	.6	.1	.2	p

7. (a) Find $Pr(d)$, (b) Define the two events, $E = \{a, c\}$, and $D = \{c, d\}$. Find $Pr(E \cup F)$.

8. Define the events

   $E = \{a, b, c\} \qquad F = \{b, c, d\}$

   Find $Pr(E \cup F')$.

9. If $Pr(E) = .27$, $Pr(F) = .65$, and $Pr(E \cup F) = .78$, find $Pr(E \cap F)$ and $Pr(E' \cup F')$.

10. Suppose $Pr(E) = .38$, $Pr(F) = .68$, and $Pr(E \cup F) = .84$. Find $Pr(E' \cup F')$.

11. If the odds in favor of E are 6 to 7, find the probability of E occurring.

12. If a three-card hand is dealt, find the probability that all of the cards are picture cards.

13. If 12 coins are tossed, find the probability that at least 10 are heads.

14. An auditor selects a sample of 5 accounts from a total of 20 accounts and examines them for delinquency. Suppose there are 4 delinquent accounts. Find the probability that the sample will have 3 or 4 delinquent accounts.

15. A club has 40 members, 25 of whom are seniors and 18 of whom are women. If 12 of the women are seniors, find the probability that a person selected is a man given that the person is a senior.

16. Urn 1 has 4 red marbles and 8 white marbles and urn 2 has 6 red marbles and 4 white marbles. The experiment consists of selecting one of the two urns and then drawing a marble from the urn and recording its color. Find the probability of selecting a red marble.

17. Two dice are rolled and the sum of the top faces is recorded. Define the two events $E = \{\text{odd sum}\}$ and $F = \{\text{a sum less than 7}\}$. Are E and F independent?

18. Refer to the accompanying Venn diagram. The number of elements in each set is given. For example, $n(A) = 12$ and $n(A \cap E) = 3$. Find $Pr(A \cap E)$, $Pr(H \cap E)$, $Pr(W \cap E)$, $Pr(A \mid E)$.

A	12	H	8	W	1
			E		
	3		7		2

19. Choose 1 of 3 urns, each containing some red and white marbles, and then select a marble. Urn 1 has 1 red and 9 white marbles, urn 2 has 4 red and 6 white marbles, and urn 3 has 2 red and 8 white marbles. Find the probability that urn 1 was chosen given that the marble is red.

20. The three mutually exclusive events A, B, and C have probabilities $Pr(A) = .4$, $Pr(B) = .1$, and $Pr(C) = .5$, and $Pr(E \mid A) = .1$, $Pr(E \mid B) = .2$, and $Pr(E \mid C) = .1$. Find $Pr(A \mid E)$.

21. Find the expected value of the experiment of tossing two coins 300 times and recording the number of times two heads occur.

22. A raffle offers a first prize of $15,000, 2 prizes of $10,000, 4 prizes of $5000, and 20 prizes of $500. Two thousand tickets are sold at $50 each. Find the expected value for the person who buys one ticket.

PROGRAMMABLE CALCULATOR EXPERIMENTS

1. Write a program that computes the probability of observing "exactly no heads" for the experiment of tossing n coins for (a) $n = 3$, (b) $n = 4$, (c) $n = 10$.

2. Write a program that computes the probability of observing "exactly one heads" for the experiment of tossing n coins for $n = 2, 3, \ldots, 10$.

3. Write a program that computes the probability of observing "at most two heads" for the experiment of tossing n coins for (a) $n = 5$, (b) $n = 6$, (c) $n = 10$.

4. Write a program that computes the probability of observing "at most two heads" for the experiment of tossing n coins for $n = 1, 2, 3, \ldots, 10$.

5. Write a program that computes the probability that at least two people will have the same birthday when n people are selected at random for $n = 10, 20, 21, 22, 23, 24, 25, 26$.

6. Write a program that computes the probability of observing "exactly three 1's" for the experiment of rolling n dice for $n = 2, 3, 4, 5, 6$.

Statistics

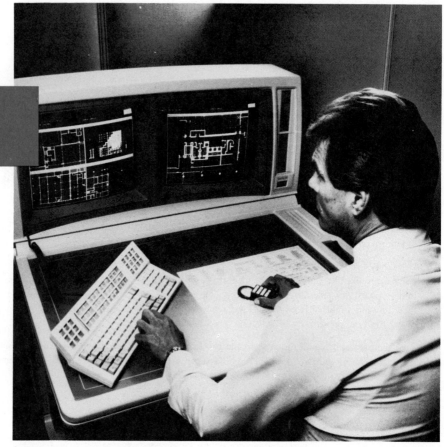

Sophisticated computers handle the collection and organization of statistics. (Group 4 Production/H. Armstrong Roberts)

8

Chapter Overview

Statistics is concerned with presenting, describing, and interpreting sets of data. This chapter begins by placing probability distributions in a statistical setting. It shows how histograms present various kinds of distributions in a graphic manner. Next, five units of measure are introduced. The mean, median, and expected value describe the ''average'' of a set of numbers, while the variance and standard deviation describe the spread from the average. The normal curve is introduced to interpret data from the binomial experiments discussed in Section 7–7.

Statistics is one of the most useful areas of mathematics. This chapter considers applications to diverse fields such as marketing, law, medicine, sociology, ecology, agriculture, industrial accidents, gambling, drug studies, and the overbooking of reservations by airlines.

CASE STUDY PREVIEW

The two Case Studies examine two entirely different issues in two entirely different revolutions in two entirely different countries. The first Case Study takes place in the U.S.S.R.

The Russian revolutionary leader V. I. Lenin is well known for his political savvy and cunning. A recent study concludes that he was an accomplished statistician as well. Lenin's early studies, in which comparative statistics support his argument about the worsening plight of Russian peasants, reflect his command of the subject.

The second Case Study takes place in the U.S.A.

The 13 colonies in America joined forces from 1776 to 1783 to fight for their independence from Great Britain. But then a family squabble broke out as the independent colonies wrestled with their own self-governance. The writing of a constitution was finally completed in September of 1787. The next step was its ratification by the colonies, and one of the most instrumental documents supporting it was The Federalist papers.

As we all know, the Constitution was ratified unanimously. However, for almost 200 years the identity of the authors of some of The Federalist papers remained unknown. Conclusive evidence was first published in 1964. The statistical methods described in this chapter will enable you to understand the results of that historic study.

8–1 Probability Distributions

This section bridges the gap between probability and statistics. First we present some topics from Chapter 7 in a statistical setting. Then it is shown how certain graphs called histograms represent data geometrically. Finally, we demonstrate how histograms are related to probability.

Histograms can be seen in most daily newspapers, especially in the business section. In addition, when major manufacturers of personal computers market a new model they frequently picture it with a full-color display of a histogram.

Triskaidekaphobia

Many superstitions are associated with Friday the 13th. Hollywood has exploited a fear of this date by filming several horror movies with the same title, yet the 13th of a month is more likely to fall on a Friday than any other day. Let us explain.

The calendar repeats itself every 400 years. There are 4800 months in this time period since every year has 12 months. Table 1 lists the days of the week that the 13th falls on for all 4800 months. It shows that the 13th occurs on a Friday more frequently than on any other day.*

Table 1 is called a **frequency distribution.** The sample space consists of the seven days of the week. Since numerical outcomes are more convenient to handle, we assign integers 1 through 7 to the days of the week. They are denoted by the letter X. The number of months that the 13th falls on each day is called the **frequency.** The frequencies are listed in the column under f.

A **random variable** is a rule that assigns a real number to each outcome in the sample space. The term "random variable" refers to the unpredictable nature of the outcomes. In the previous example, the variable X is a random variable.

In Table 1 the frequency of $X = 6$ is 688. This means that the 13th of a month falls on a Friday 688 of the 4800 months. The proportion of these occurrences is called the **relative frequency.** The relative frequency for $X = 6$, rounded to four decimal places, is $688/4800 = 0.1433$. Since the relative frequency can be interpreted as a probability, we write

$$Pr(X = 6) = .1433$$

In general, $Pr(X = x)$ stands for the probability that the random variable X is equal to x. Table 2 is formed from Table 1 by omitting the names of the days and by

*For an interesting history of the role mathematics played in the calendar, see Jacques Dutka, "On the Gregorian Revision of the Julian Calendar," *The Mathematical Intelligencer,* Vol. 7, No. 1, pp. 56–64, 1985.

Table 1

Day	x	f
Sunday	1	687
Monday	2	685
Tuesday	3	685
Wednesday	4	687
Thursday	5	684
Friday	6	688
Saturday	7	684
Total		4800

Source: From V. Frederick Rickey, "Mathematics of the Gregorian Calendar," *The Mathematical Intelligencer,* Vol. 7, No. 1, pp. 53–56, 1985.

Table 2

x	f	p
1	687	.1431
2	685	.1427
3	685	.1427
4	687	.1431
5	684	.1425
6	688	.1433
7	684	.1425
Total	4800	.9999

adding a column of relative frequencies under the letter p (for probability). It is called a **probability distribution.**

   Example 1 shows how to find the probability that a value of X falls between two numbers a and b. The notation for this is $Pr(a \leq X \leq b)$.

EXAMPLE 1

Problem

What is the probability that the 13th of a month falls on a weekday?

Solution Weekdays are assigned integers between 2 and 6 inclusive. Thus the problem is to find $Pr(2 \leq X \leq 6)$. It is obtained by summing the relative frequencies.

$$Pr(2 \leq X \leq 6) = Pr(X = 2) + Pr(X = 3) + Pr(X = 4)$$
$$+ Pr(X = 5) + Pr(X = 6)$$
$$= .1427 + .1427 + .1431 + .1425 + .1433$$
$$= .7143$$

The probability of the 13th of a month falling on a weekday is .7143.

   Example 2 applies these concepts to coin-tossing experiments that will occur repeatedly in this chapter. The notation $Pr(X \geq c)$ refers to the probability that X is greater than or equal to the number c.

EXAMPLE 2

Toss three coins and let X represent the number of heads that occur.

Problem

Construct the probability distribution and find $Pr(X \geq 1)$.

Solution The random variable X assumes the values 0, 1, 2, and 3 for the number of heads that can occur. From Chapter 7 the probability distribution is

x	f	p
0	1	$\frac{1}{8}$
1	3	$\frac{3}{8}$
2	3	$\frac{3}{8}$
3	1	$\frac{1}{8}$

The notation $Pr(X \geq 1)$ stands for the probability of tossing at least one head. It can be obtained directly.

$$Pr(X \geq 1) = Pr(X = 1) + Pr(X = 2) + Pr(X = 3)$$
$$= \tfrac{3}{8} + \tfrac{3}{8} + \tfrac{1}{8} = \tfrac{7}{8}$$

Alternately, it can be obtained from the rule for complementary probabilities:
$$Pr(X \geq 1) = 1 - Pr(X < 1) = 1 - Pr(X = 0) = 1 - \tfrac{1}{8} = \tfrac{7}{8}$$

Histograms

Statistics makes vital use of various graphing techniques to analyze data. A **histogram** represents data by means of a bar graph. Figure 1 shows the histogram for the days on which the 13th of a month occurs. The horizontal axis represents the days of the week and the vertical axis represents the frequencies. The jagged line indicates that an interval of data has been omitted, resulting in a graph whose true height is distorted.

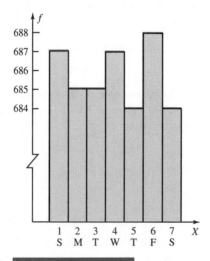

FIGURE 1

A set of data can have a histogram for its frequency distribution and a histogram for its probability distribution. In both cases the units along the horizontal axis are the same. However, the units along the vertical axis are either the frequencies or the relative frequencies. The shapes of the two histograms are identical, but the histogram with relative frequencies has a direct connection with probability. When the widths of the bars are equal to one unit and the heights of the bars are the relative frequencies, the probability of each outcome is equal to the area of that bar. In this way the probability of an event E is equal to the sum of the areas of the bars of all outcomes in E. Example 3 illustrates this property.

EXAMPLE 3

Problem

Construct a histogram for the probability distribution in Example 2. Which bars represent the probability of tossing at least one head?

Solution The histogram for the probability distribution is shown in Figure 2.

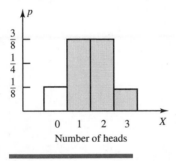

Number of heads

FIGURE 2

The shaded portion represents the fact that $Pr(X \geq 1) = \frac{7}{8}$ because the total area of the shaded rectangles equals $(\frac{3}{8})(1) + (\frac{3}{8})(1) + (\frac{1}{8})(1) = \frac{7}{8}$.

Histograms are often used to compare different sets of frequencies of the same sample space. The Contraceptive Drug Study, which examined the possible side effects of oral contraceptives on blood pressure, made use of such a comparison. Table 3 presents the results of the study for women aged 17 to 24 who were neither

Table 3

Nonusers		Users	
x	f	x	f
100	14	100	9
110	26	110	21
120	33	120	27
130	18	130	25
140	7	140	14
150	2	150	4

pregnant nor taking hormonal medication. It gives one frequency distribution for women who used the Pill and another frequency distribution for women who did not. The random variable X represents systolic blood pressure (in millimeters) and the relative frequency represents the percentage of women with the given blood pressure.

EXAMPLE 4

Problem

Draw the histograms for each frequency distribution in Table 3. Shade the portion that represents blood pressures that are 130 or lower.

Solution The histograms are drawn in Figure 3. They show graphically that women aged 17 to 24 who used the Pill had a much higher probability of having a blood pressure over 130 than women who did not use the Pill.

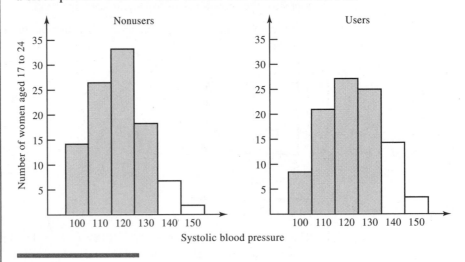

FIGURE 3

Table 4

Weight	Frequency
170–179	4
180–189	5
190–199	3
200–209	2
210–219	3
220–229	5
230–239	5
240–249	3
250–259	1
260–269	2
270–279	6
280–289	1
290–299	3

Intervals

Sometimes intervals of numbers are tabulated instead of the numbers themselves. For statistical purposes the entries become the midpoints of the intervals. Table 4 gives the weights of the 1988 Super Bowl champion team, the Washington Redskins. Example 5 shows how to draw the histogram for these data.

EXAMPLE 5

Problem

Construct the frequency distribution and draw the histogram for the weights of the Washington Redskins.

Solution The random variable X assumes values that are midpoints of the intervals. The interval 170–179 includes all players who weigh between 170 and 180, so the midpoint is 175. The remaining entries are 185, 195, and so on. This leads to the accompanying frequency distribution. The histogram is drawn in Figure 4.

x	f
175	4
185	5
195	3
205	2
215	3
225	5
235	5
245	3
255	1
265	2
275	6
285	1
295	3

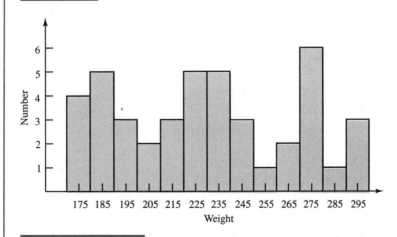

FIGURE 4

The histogram in Figure 4 is nearly level. It is close to what is called a **uniform distribution.** Contrast it with the histogram in Figure 2, which is close to what is called a **normal distribution** (or a **bell-shaped curve**). For the general population the distribution of weight is a normal distribution. It is somewhat surprising that the distribution of weights on an NFL team is not a normal distribution. We certainly expect the weights to be higher than the general population, but we do not expect them to be so uniform. It is because players are selected to perform specific types of tasks, so it is not a random selection.

The distribution of income is an example of a skewed distribution. It will be illustrated in the exercises.

EXERCISE SET 8–1

Problems 1 to 4 refer to Example 1. Find

1. Pr(the 13th of a month occurs on a weekend).

2. Pr(the 13th of a month does not occur on a Friday).

3. $Pr(X \geq 5)$.   4. $Pr(X \leq 6)$.

In Problems 5 to 8 toss three coins and let X be the number of heads that appear. Find

5. $Pr(X = 1)$.   6. $Pr(X = 3)$.

7. $Pr(X \geq 2)$.   8. $Pr(X \leq 2)$.

In Problems 9 to 12 draw the histogram for the experiment of tossing three coins and shade the stated portion.

9. exactly one head.   10. exactly three heads.

11. at least two heads.   12. at most two heads.

In Problems 13 and 14 toss two coins and let X be the number of heads that appear.

13. Construct the frequency distribution and draw its histogram.

14. Construct the probability distribution and draw its histogram.

Problems 15 to 18 refer to the Contraceptive Drug Study tabulated in Table 3.

15. Draw the two histograms and shade the portions for blood pressures that are 120 or less.

16. Draw the two histograms and shade the portions for blood pressures that are 110, 120, or 130.

17. What conclusions can be drawn about blood pressures that are 120 or less? (Use Problem 15.)

18. What conclusions can be drawn about blood pressures that are 110, 120, or 130? (Use Problem 16.)

19. Draw the histogram for the ages of students in a marketing course from this table.

age	17–19	20–22	23–25	26–28	29–31
frequency	10	14	7	2	1

20. Draw the histogram for the scores on an exam as tabulated in this table.

score	50–59	60–69	70–79	80–89	90–99
frequency	4	5	6	5	5

In Problems 21 to 26 construct the probability distribution, draw its histogram, and find the stated probability.

21. Toss four coins and let X be the number of heads that occur. Find $Pr(X \geq 1)$.

22. A jar contains four vanilla cookies and three chocolate cookies. Two cookies are selected at random without replacement. Let X be the number of vanilla cookies. Find $Pr(X \geq 1)$.

23. Toss two dice and let X be the sum of the top faces. Find $Pr(6 \leq X \leq 8)$.

24. Toss two dice and let X be the sum of the top faces. Find $Pr(7 \leq X \leq 10)$.

25. Select 3 cards from an ordinary deck without replacement and let X be the number of hearts. Find $Pr(X \leq 2)$.

26. Select 2 cards from an ordinary deck without replacement and let X be the number of kings. Find $Pr(X \geq 1)$.

Problems 27 to 32 refer to this table of scores on two quizzes for a class of 24 students.

score	10	9	8	7	6	5
quiz 1	7	7	5	1	1	3
quiz 2	2	3	8	6	1	4

27. Find $Pr(X \geq 8)$ for quiz 1.

28. Find $Pr(X \geq 8)$ for quiz 2.

29. Find $Pr(6 \leq X \leq 8)$ for quiz 1.

30. Find $Pr(6 \leq X \leq 8)$ for quiz 2.

31. Draw the histogram for each quiz.

32. Which quiz scores are higher? Do the histograms drawn in Problem 31 convey this visually?

Problems 33 to 36 refer to the histogram shown.

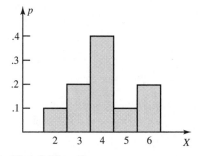

33. Find $Pr(X \geq 4)$.

34. Find $Pr(2 \leq X \leq 4)$.

35. Find $Pr(3 \leq X \leq 5)$.

36. Find $Pr(X \geq 5)$.

Problems 37 to 40 refer to the histogram shown.

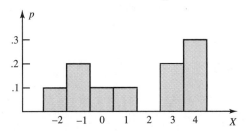

37. Find $Pr(X \leq 2)$.

38. Find $Pr(3 \leq X \leq 4)$.

39. Find $Pr(1 \leq X \leq 3)$.

40. Find $Pr(X \leq 0)$.

Referenced Exercise Set 8–1

1. This table lists the total attendance (in millions) and average ticket price (in dollars) for the Broadway theater over a 12-year span. Draw the histograms for total attendance and ticket price. What conclusion can be drawn from comparing the histograms?

year	74–75	75–76	76–77	77–78	78–79	79–80
attendance	6.6	7.3	8.8	9.6	9.6	9.6
$	8.63	9.72	10.56	11.87	13.95	15.20

year	80–81	81–82	82–83	83–84	84–85	85–86
attendance	11.0	10.1	8.4	7.9	7.4	6.5
$	17.90	22.07	24.88	28.73	28.78	28.96

Source: From The League of American Theatres and Producers.

2. Draw the histogram for Table 5, which lists the distribution of income of families in the United States. What kind of distribution is it?

Table 5 Distribution of Income of Families in the United States, 1979

Income Level ($)	Percent
0– 5,000	4
5,000–10,000	13
10,000–15,000	17
15,000–20,000	20
20,000–25,000	16
25,000–30,000	12
30,000–35,000	7
35,000–40,000	4
40,000–45,000	3
45,000–50,000	2
50,000–55,000	1

Source: From The Current Population Survey, 1980.

3. Table 6 lists the height of men and women in the United States. Draw the histogram for each distribution. What kind of distributions are they?

Table 6 Height of U.S. Men and Women, 1979

Men		Women	
Height (in.)	Number (in thousands)	Height (in.)	Number (in thousands)
60	1	53	0.5
61	5	54	0.5
62	9	55	0.2
63	17	56	0.2
64	37	57	10
65	35	58	13
66	70	59	38
67	63	60	45
68	94	61	83
69	54	62	104
70	63	63	73
71	32	64	90
72	28	65	47
73	11	66	44
74	6	67	14
75	1	68	12
		69	1
		70	1

Source: From The Current Population Survey, 1980.

4. Table 7 traces the size of Hershey's standard milk chocolate bar from its introduction in 1950 to its discontinuation in 1974. Draw its histogram with the horizontal axis for the years from 1950 to 1974.

Table 7 Prices of Hershey's Standard Milk Chocolate Bar

Common Retail Price: 10¢	Ounces
1950	2
July 1951	1⅞
February 1954	1⅝
August 1954	1½
July 1955	1¾
February 1958	1½
June 1960	2
August 1963	1¾
August 1965	2
June 1966	1¼
February 1968	1½
March 1970	1⅜
January 1973	1.26
Discontinued 1-1-74	

Source: From H. J. Maidenberg, "Misadventures of Cocoa Trade," *New York Times*, February 25, 1979, and Hershey Foods Corp.

5. What does the word "triskaidekaphobia" mean?

Problems 6 and 7 refer to Table 8, which lists the weights, in intervals, of the Denver Broncos, the losing team in the 1988 Super Bowl.

6. (a) Construct a frequency distribution with the midpoints of the intervals as entries.
   (b) Draw the histogram.

7. Does the histogram resemble the uniform curve drawn in Figure 4?

Table 8

Weight	Frequency
170–179	4
180–189	5
190–199	9
200–209	2
210–219	2
220–229	3
230–239	5
240–249	3
250–259	2
260–269	8
270–279	2

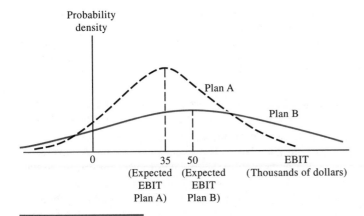

FIGURE 5 EBIT probability distribution.

8. This problem describes the way in which probability distributions are used in financial management to assess a firm's operating leverage. Figure 5 shows two typical EBIT (earnings before interest and taxes) distributions for Industrial Technology, Inc.* The vertical axis represents the probability of achieving profits under Plans A and B.
   (a) What is the EBIT at Plan A's highest probability?
   (b) What is the EBIT at Plan B's highest probability?
   (c) Is it possible to conclude from Figure 5 which plan is more volatile, A or B? If so, which one?

9. The histogram in Figure 6 is taken from a study that contrasted several forecasting models for estimating the expected cash receipts of a company.† Does it represent a probability distribution? Why?

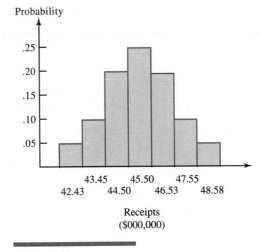

FIGURE 6

*B. J. Campsey and Eugene F. Brigham, *Introduction to Financial Management*, Hinsdale, IL, The Dryden Press, pp. 226–230, 1985.

†James L. Pappas and George P. Huber, "Probabilistic Short-term Financial Planning," *Financial Management*, pp. 36–44, August 1973.

8–2 Measures of Central Tendency

Section 8–1 showed that a histogram provides a quick overall impression of a set of data, giving it a form and structure that is sometimes missing from the mere listing of the individual entries. A histogram reveals other characteristics of the data as well, like the average. This section examines three meanings of the term "average": mean, median, and expected value.

Average

The League of American Theatres and Producers reported that the average price for a ticket to a Broadway show during the 1985–1986 season was $28.96. By the term "average" the League meant the mean, which was obtained by dividing the total ticket revenue by the total attendance. However, another type of "average" is the median ticket price that is directly in the middle of all ticket prices, so half the tickets cost more and half the tickets cost less than the median price.

The mean and median are numerical values that give information about an entire collection of data. Since averages measure the center of a set of data, they are called "measures of central tendency." The discussion of Broadway theater ticket prices illustrates these two measures of central tendency, the mean and the median. It is important to state precisely which one is being used. The referenced exercises examine a federal court ruling in a civil rights discrimination suit that revolved around the distinction between these two terms.

The *mean* of a set of data, denoted by $\bar{x}$, is defined in the display.

DEFINITION

The **mean** of a set of n numbers $x_1, x_2, \ldots, x_n$, is

$$\bar{x} = \frac{x_1 + x_2 + \cdots + x_n}{n}$$

Calculators with a $\bar{x}$ key can find the mean of any set of data directly. The stated average ticket price of $28.96 to a Broadway show was, more precisely, the mean price. As another example, the mean of five test grades 82, 84, 78, 76, 95 is

$$\frac{82 + 84 + 78 + 76 + 95}{5} = 83$$

The **median** of a set of data is the middle number when the entries are arranged in order. Curiously, there is no special notation for this concept. To find the median, first determine if there is an odd or even number of elements in the set. If it is odd, the median is the middle number. If it is even, the median is the mean of the middle two numbers. Thus the median grade on tests 82, 84, 78, 76, 95 is 82 because when these grades are listed in order, 82 is the middle grade.

$$76, \quad 78, \quad 82, \quad 84, \quad 95$$
$$\uparrow$$
median

Notice that for this set of grades the mean is 83 and the median is 82. The median grade on the first four tests is equal to the mean of 78 and 82, which is 80.

76, 78, 82, 84

↖↗

median = 80

For this set of four grades the mean and median are equal.

Example 1 shows how to compute the mean and median for sets with repeated entries.

EXAMPLE 1

A branch of a bank has 8 tellers who earn $15,000 per year, 2 managers who earn $35,000, a vice-president who earns $70,000, and a president who earns $100,000.

Problem

What are the mean and median salaries?

Solution The mean salary is equal to the total salaries divided by the number of employees.

$$\text{total salary} = 15 + 15 + 15 + 15 + 15 + 15 + 15 + 15$$
$$\text{(in 1000's)} \quad + 35 + 35 + 70 + 100$$
$$= 8(15) + 2(35) + 1(70) + 1(100) = 360$$

$$\text{number of employees} = 8 + 2 + 1 + 1 = 12$$

Thus

$$\bar{x} = \frac{8(15) + 2(35) + 1(70) + 1(100)}{8 + 2 + 1 + 1} = \frac{360}{12} = 30$$

The mean salary is $30,000.

Since there are 12 employees, the median salary is the mean of the sixth and seventh numbers.

15, 15, 15, 15, 15, 15, 15, 15, 35, 35, 70, 100

↖↗

median

Thus the median salary is $15,000.

Example 1 suggests the derivation of an alternate formula for the mean in terms of frequencies.

Alternate Formula for the Mean

If a set has f_1 entries equal to x_1, f_2 entries equal to x_2, . . . , and f_n entries equal to x_n, then

$$\bar{x} = \frac{f_1 x_1 + f_2 x_2 + \cdots + f_n x_n}{f_1 + f_2 + \cdots + f_n}$$

A convenient way to evaluate $\bar{x}$ from this expression is to make a frequency table with a third column for the products of each entry x and its frequency f.

x	f	xf
15	8	120
35	2	70
70	1	70
100	1	100
Totals	12	360

$$\bar{x} = \frac{360}{12} = 30$$

EXAMPLE 2

A psychology professor enters the 100 scores on a ten-question test into a grade book. To find the mean, the grades are grouped as follows:

number correct	10	9	8	7	6	5	4	3	2	1	0
number of students	11	16	20	15	13	10	10	4	1	0	0

Problem

What is the mean?

Solution Write the entries x and their frequencies f in a table, with a third column of the products xf.

x	f	xf
10	11	110
9	16	144
8	20	160
7	15	105
6	13	78
5	10	50
4	10	40
3	4	12
2	1	2
Total	100	701

$$\bar{x} = \frac{701}{100} = 7.01$$

The mean number of correct answers is slightly more than 7.

Expected Value

The alternate formula for the mean leads directly to its analogue for probability distributions. In the alternate formula for $\bar{x}$ set $N = f_1 + f_2 + \cdots + f_n$. Then

$$\bar{x} = \frac{f_1 x_1 + f_2 x_2 + \cdots + f_n x_n}{f_1 + f_2 + \cdots + f_n}$$

$$= \frac{f_1 x_1 + f_2 x_2 + \cdots + f_n x_n}{N}$$

$$= \left(\frac{f_1}{N}\right) x_1 + \left(\frac{f_2}{N}\right) x_2 + \cdots + \left(\frac{f_n}{N}\right) x_n$$

Each fractional part f_i/N is a relative frequency, so it can be interpreted as a probability. The mean of a probability distribution is called the *expected value*, denoted by $E(X)$. By setting $p_i = f_i/N$, the formula for the expected value becomes

$$E(X) = p_1 x_1 + p_2 x_2 + \cdots + p_n x_n$$

We encountered expected value in Chapter 7, where it referred to the average number of outcomes that occur when a probability experiment is repeated many times. To compute it, extend the probability distribution of a random variable X to a third column containing the products Xp.

x	p	xp
x_1	p_1	$x_1 p_1$
x_2	p_2	$x_2 p_2$
$\vdots$	$\vdots$	$\vdots$
x_n	p_n	$x_n p_n$
		Total $= E(X)$

The formula for $E(X)$ shows that it is equal to the sum of the numbers in the Xp column.

> The **expected value** of the random variable X is
>
> $$E(X) = x_1 p_1 + x_2 p_2 + \cdots + x_n p_n$$

Both the mean and the expected value describe the average of a set of numbers, but the mean is computed from a frequency distribution, whereas the expected value is computed from a probability distribution. Example 3 finds the expected value for the sum of two dice.

EXAMPLE 3

Toss two dice and let X be the sum of the faces.

Problem

Find $E(X)$.

Solution The probability distribution given here is taken from the table before Example 2 in Section 7–3.

x	p	xp
2	1/36	2/36
3	2/36	6/36
4	3/36	12/36
5	4/36	20/36
6	5/36	30/36
7	6/36	42/36
8	5/36	40/36
9	4/36	36/36
10	3/36	30/36
11	2/36	22/36
12	1/36	12/36
		Total = 252/36

The expected value is $E(X) = 252/36 = 7$.

Example 3 means that if you toss two dice many times and form the sum each time, the mean of all the sums will be very close to 7. Example 4 computes an expected value directly from the probability distribution.

EXAMPLE 4

Suppose that the New Jersey Turnpike Commission reports the number of accidents during the Friday evening rush hour to be 0, 1, 2, or 3 with probabilities .92, .03, .02, and .03, respectively.

Problem

What is the expected number of accidents during the Friday evening rush hour?

Solution The term "expected number" refers to the expected value $E(X)$, where the random variable X represents the number of accidents. Form the expanded probability distribution and total the Xp column.

x	p	xp
0	.92	.00
1	.03	.03
2	.02	.04
3	.03	.09
		Total = .16

Thus $E(X) = .16$. Since the expected number of accidents is approximately $\frac{1}{6}$ per Friday evening rush hour, one accident will occur about every six weeks.

EXERCISE SET 8–2

In Problems 1 to 8 find the mean and median.

1. 2, 5, 7, 9, 2, 1, 13, 12, 50, 60

2. 2, 5, 7, 9, 2, 1, 13, 12, −50, −60

3. .06, .07, .02, .10, .05, .02

4. 5.2, 6.1, 10.8, 11.2, 6.3, .5, 8.7, 7.2

5.

value	0	1	2	3
frequency	2	5	3	4

6.

value	1	2	3	4
frequency	8	2	1	5

7.

value	3	5	7	9
frequency	2	1	0	4

8.

value	11	12	13	14	15
frequency	3	5	2	0	1

In Problems 9 to 12 find the expected value from the probability distribution.

9.

x	p
2	$\frac{1}{8}$
3	$\frac{1}{2}$
5	$\frac{1}{4}$
7	$\frac{1}{8}$

10.

x	p
2	$\frac{1}{2}$
4	$\frac{1}{3}$
6	$\frac{1}{6}$

11.

x	p
2	.2
3	.3
4	.5

12.

x	p
1	.2
3	.4
5	.1
7	.3

In Problems 13 and 14 find the mean from the given histogram.

13.

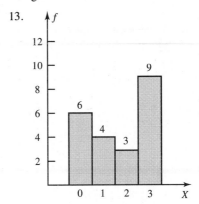

14.

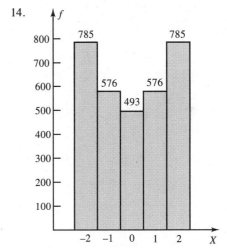

In Problems 15 and 16 find the expected value from the given histogram.

15.

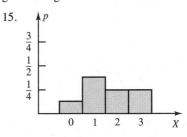

16.

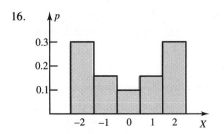

In Problems 17 to 22 find the mean and median.

17. $-3, 2, -3, 1, -3, 1, -1, 0, 0, 1$

18. $3, -3, 2, 2, 2, -3, -1, 0, 0, -2$

19.
value	4.4	5.1	6.4	7.3	8.2
frequency	3	4	1	5	7

20.
value	0.5	0.6	0.7	0.9
frequency	9	5	2	4

21.
value	-3	-2	-1	0	2	3
frequency	2	1	1	2	3	1

22.
value	-3	-1	0	1	2
frequency	3	1	2	3	1

In Problems 23 to 26 find the expected value from the probability distribution.

23.
x	p
-3	.2
-1	.2
0	.1
1	.4
3	.1

24.
x	p
-2	.2
-1	.1
0	.4
1	.2
2	.1

25.
x	p
-10	.2
-5	.4
0	.1
5	.2
10	.1

26.
x	p
-5	.1
-3	.3
0	.1
1	.2
4	.3

27. Toss three coins and let X be the number of heads that occur. Find $E(X)$.

28. Toss four coins and let X be the number of heads that occur. Find $E(X)$.

29. Find the mean and median weights of the Washington Redskins from the table in Example 5 of Section 8–1.

30. What is the median grade on the psychology test described in Example 2?

In Problems 31 and 32 the data are tabulated in intervals. Find the mean using the midpoints of the intervals as the data.

31.
interval	0–2	2–4	4–6	6–8	8–10
frequency	4	5	3	0	1

32.
interval	0–5	5–10	10–15	15–20	20–25	25–30
frequency	16	14	10	8	2	1

33. A set of data has five entries. Each entry is either 0, 1, or 2.
 (a) What is the set if the mean is 0?
 (b) What is the set if the mean is 2?
 (c) Can the mean be 3?

34. A set of data has five entries. Each entry is either 10, 11, or 12. What is the set if the mean is 10.8 and the median is 10?

35. A psychologist studies whether reward or punishment is a more effective method of inducing learning by running 11 mice through two sets of mazes. The differences (in seconds) between the rewarded trials and the punished trials are $-1.1, 2.0, 1.3, -0.9, -1.9,$ $0.8, 1.2, -1.2, 4.0, 0.2, -0.1.$
 (a) Find the mean of these rates.
 (b) Which method is more effective?

36. An entymologist studying the morphological variation in a species of mosquito selects ten samples and measures their body lengths (in centimeters): 1.2, 1.4, 1.3, 1.6, 1.0, 1.5, 1.7, 1.1, 1.2, 1.3. What are the mean and median body lengths?

In Problems 37 and 38 find the mean and the median of the data in the distribution.

37.
value	0	2	4	6	8
frequency	4	8	10	8	10

38.
value	10	15	20	25	30
frequency	5	4	6	3	2

39. Find the expected value.

x	-1	0	1	2	3	4
p	.2	.1	.1	.3	.1	.2

40. Find the mean.

interval	0-2	3-5	6-8	9-11	12-14
frequency	8	18	10	10	4

Referenced Exercise Set 8-2

1. The 1964 Civil Rights Act prohibits discrimination by race, sex, or national origin in hiring and promotions. In 1981, 20 people filed suit that they were discriminated against when they were not hired in 1975 by the Federal Reserve Bank. The court agreed. The attorneys for the two sides disagreed on the amount of money the bank should pay each plaintiff. A sample of 319 employee records revealed this pertinent information about the job turnover rate, where M denotes the number of months worked before termination and E denotes the number of employees. (See Table 1.)
   (a) Find the mean number of months worked.
   (b) Find the median number of months worked.
   (c) Which side argued for which average?
   (d) Each of the 20 plaintiffs was to be compensated at a rate of seven (years) times $9500 (the yearly salary), divided by the average number of months worked. What is the difference in the amount of money the bank would have to pay according to whether the average is interpreted as the mean or the median?

2. J. E. Kerrich described a coin-tossing experiment while interned during World War II.* He tossed a coin 1000 times and recorded the number of heads. He conducted the experiment ten times, obtaining these results.

   502, 511, 497, 529, 504, 476, 507, 528, 504, 529

   What are the mean and median number of heads?

3. Sociologist David Phillips designed a study to determine whether some individuals have a will that is strong enough to prolong life. He recorded the number of months between the birthday and day of death of 1251 famous people, reasoning that famous people anticipate being honored on their birthday and might ward off death for a period of time. Table 2 lists the results, where B is the birth month and D is the number of deaths. What is the expected death month?

4. What is the mean income of families in the United States according to Table 5 in Section 8-1?† What is the median income?

5. Table 6 in Section 8-1 lists the height of women in the United States.† What are the mean and median heights?

*J. E. Kerrich, *An Experimental Introduction to the Theory of Probability,* Copenhagen, Belgisk Import Co.

†Source: The Current Population Survey, 1980.

Table 1

M	0.5	1	2	3	4	5	8	10	11	14	18	20	24	30	35	40	60
E	15	21	27	33	42	9	17	32	18	12	22	26	11	9	6	8	11

Source: Data from EEOC vs. *Federal Reserve Bank* 30 FEP Cases 1138-1169.

Table 2

B	-6	-5	-4	-3	-2	-1	0	1	2	3	4	5
D	90	100	87	96	101	86	119	118	121	114	113	106

Source: Data from J. M. Tanur et al. (eds.), *Statistics: A Guide to the Unknown,* San Francisco, Holden-Day Inc., pp. 71-85, 1978.

Table 3

x	0	1	2	3	4	5	6	7	8	10	15
$P(X = x)$	.154	.026	.179	.179	.179	.077	.102	.026	.026	.026	.026

Source: Joseph N. Kane, *Facts About the Presidents,* 4th ed., New York, H. W. Wilson Co., pp. 348–349, 1981.

6. Toss *n* coins and let *X* be the number of heads that occur. Find $E(X)$.

7. The probability distribution for the number of games played in each World Series from 1922 to 1987 is given below. What is the expected number of games played?

x	4	5	6	7
$P(X = x)$	0.167	0.197	0.197	0.439

Source: Craig Carter (ed.), *The Complete Baseball Record Book,* St. Louis, The Sporting News Publishing Co., p. 323, 1988.

8. The probability distribution for the number *x* of children of the presidents of the United States from George Washington to Ronald Reagan is given in Table 3. What is the expected number of children of a president of the United States based on this information?

9. Professors at the University of Wisconsin devised a five-question multiple-choice vocabulary test whose results were expected to be close to what would happen if the students guessed each question.‡ There are five possible answers for each question on the test.
   (a) What is the expected number of correct answers if a student guesses on each question?
   (b) What is the expected number of correct answers based on the accompanying table, which shows the results from one class that took the test? Here *x* is the number of correct answers and *f* is the frequency.

x	0	1	2	3	4	5
f	6	11	6	1	1	0

   (c) Do the answers to parts (a) and (b) suggest that the students in the class probably knew some of the answers?

‡Ken Kundert, "Student-Generated Data in Elementary Statistics," *Mathematics Teacher,* pp. 322–325, April 1990.

Cumulative Exercise Set 8–2

1. Draw the histogram for the experiment of tossing three coins where *X* is the number of heads, and shade the portion corresponding to $Pr(X = 2)$.

2. Draw the histogram for the ages of the students in a business course from this table.

age	17–18	19–20	21–22	23–24	25–26
frequency	13	10	8	7	2

3. For the experiment of tossing two dice with *X* the sum of the top faces, construct the probability distribution, draw the histogram, and find $Pr(6 \leq X \leq 9)$.

4. For the experiment of tossing five coins where *X* is the number of heads, construct the probability distribution, draw the histogram, and find $Pr(0 \leq X \leq 2)$.

For Problems 5 and 6, find the mean and the median of the data in the distribution.

5.
value	0	2	4	6	8
frequency	5	6	3	4	2

6.
value	20	25	30	35	40	50
frequency	5	4	7	8	4	2

7. Find the expected value.

x	−2	−1	0	1	2	3	4
p	.2	.1	.1	.3	.1	.15	.05

8. Find the mean.

interval	0–2	3–5	6–8	9–11	12–14
frequency	12	8	15	5	10

CASE STUDY   ## Lenin as Statistician

Vladimir Ilich Lenin (1870–1924) is well known throughout the world for his pivotal role in the revolution that toppled the Czarist regime in Russia and transformed it into a Communist state. It is not widely known that Lenin was an accomplished statistician as well as a highly successful political leader. His role in statistics was recently evaluated by two prominent statisticians.*

Lenin's use of statistics is part of the cult of his personality in the Soviet Union. Each April numerous articles about his life appear in Soviet publications to commemorate his birth. Many such articles citing his contributions to statistics have appeared in *Vestnik Statistiki*, the official journal of the Central Statistical Association of the Soviet Union. These articles reflect Lenin's standing with the leadership of the country at the time of their writing.

Thus there was less praise for Lenin's work at the height of Stalin's regime after World War II (1946–1953) than during the Khrushchev era and the early years of Brezhnev's era (1954–1970). The adulation diminished slightly during the peak period of Brezhnev's consolidation of power (the later 1970s and early 1980s). According to Kotz and Seneta, "At present, in the first years of Gorbachev's era Lenin's [standing] remains intact and his pronouncements have not as yet been challenged even in the avant-garde, pro-reform and pro-democratic, liberal press in the Soviet Union."

In 1894 Lenin wrote an article called "What Are 'the Friends of the People' and how do they fight with the Social Democrats?" The aim of the article was largely to attack "anti-Marxist liberal *narodniks*," whom Lenin scathingly accused of giving an erroneous picture of the peasant economy. To provide a more accurate picture, he analyzed the budgets of 24 peasant households within a county in Voronezh. It was crucial for Lenin's argument that his sample space be typical of peasant households in that entire region since otherwise his findings would not be taken seriously. To achieve this, the 24-year-old revolutionary compared four measures with the corresponding measures from an extensive study that had been gathered, analyzed, and published by the eminent Russian statistician, F. A. Shcherbina, in 1887. Lenin called Shcherbina's study "extraordinarily complete." Others regarded it as the definitive study of economics in that area.

Table 1 contains the items from Shcherbina's study that Lenin used for comparison. Shcherbina's original table lumped together households with two or three draft animals into one category and lumped together households with four or more draft animals into another category. We have adjusted his figures to be consistent with his findings. The corresponding entries from Lenin's study are given in Table 2.

Notice that the strata classes in the two studies are different. Shcherbina's study was based on the legal classification of draft animals, which were mostly

*Samuel Kotz and Eugene Seneta, "Lenin as a Statistician: A Non-Soviet View," *Journal of the Royal Statistical Society* Vol. A 153 (Part 1), pp. 73–94, 1990. We are much indebted to Professor Kotz for sending us a copy of this fascinating paper.

Table 1 Shcherbina's Study of Peasant Households

Draft Animals	Households	Land Allotment	Rented Land	Family Size	Cattle
0	8,728	6.2	0.2	4.6	0.7
1	10,510	9.4	1.3	5.7	3.0
2	6,763	13.9	2.1	7.7	5.1
3	4,428	14.0	3.5	8.5	6.9
4	3,152	21.3	12.3	11.2	14.3

Table 2 Lenin's Study of Peasant Budgets

Strata Class	Households	Land Allotment	Rented Land	Family Size	Cattle
Farm laborers	2	7.2	0.0	4.5	0.5
Poor peasants	5	8.7	3.9	5.6	2.8
Middle peasants	11	9.2	7.7	8.3	8.1
Prosperous	6	22.1	8.8	7.8	13.5

horses. Lenin chose an entirely different classification. His aim was to dramatize the deteriorating condition of Russian peasants, so his strata were organized along socioeconomic lines. It was important, however, that the averages (more specifically, the means) of each column in the studies be as close as possible. We will carry out two of these calculations in Example 1.

EXAMPLE 1

The land allotment was measured in *dessiatina*. Each *dessiatina* is equal to roughly 2.7 acres.

Problem

Calculate the mean number of *dessiatinas* that were allotted per household in both the Lenin and Shcherbina studies.

Solution Consider Shcherbina's study first. (See Table 1.) The random variable is the number of *dessiatina* in each strata class. The frequency is the number of households in each class. Thus the relevant table is as follows.

x	f	xf
6.2	8,728	54,113.6
9.4	10,510	98,794.0
13.9	6,763	94,005.7
14.0	4,428	61,992.0
21.3	3,152	67,137.6
Totals	33,581	376,042.9

The mean is $\bar{x} = \dfrac{376,042.9}{33,581} \simeq 11.20$. The relevant table for the Lenin study (see Table 2) is

x	f	xf
7.2	2	14.4
8.7	5	43.5
9.2	11	101.2
22.1	6	132.6
Totals	24	291.7

The mean is $\bar{x} = \dfrac{291.9}{24} \simeq 12.15$.

Example 1 shows that the land allotments were remarkably similar for the two groups of households. In the exercises you will be asked to verify that the mean family sizes per household were just as remarkably similar and that the mean rented land per household and the mean number of cattle owned per household were very similar.

Kotz and Seneta wrote, "We can imagine Lenin's elation on seeing the agreement." Although as modern statisticians they view the agreement as an impressive coincidence and *not* a statistically significant fact, mainly due to the small sample size within each of Lenin's strata classes, they admit that he derived great benefit from his data.

Lenin used his study for political ends. He concluded that the legal classification used by Shcherbina concealed huge socioeconomic differences that his own analysis revealed. He went on to describe how the plight of the peasant in the Soviet Union had worsened under the Czar, a conviction that remained one of his fundamental tenets for the rest of his life.

This example shows that Lenin was certainly not beyond turning statistical analysis to political ends.

Case Study Exercises

1. Compare the mean of the family sizes per household in the Lenin study with the mean of the family sizes per household in the Shcherbina study.
2. Compare the mean amount of rented land (in *dessiatina*) per household in the Lenin study with the mean amount of rented land per household in the Shcherbina study.
3. Compare the mean number of cattle owned per household in the Lenin study with the mean number of cattle owned per household in the Shcherbina study.

Table 3 Shcherbina's 1889 Study of 66 Peasant Households

Draft Animals	Households	Land Allotment	Rented Land	Family Size	Cattle
0	12	5.9	0.0	4.08	0.8
1	18	7.4	1.5	4.94	2.6
2	17	12.7	3.0	8.23	4.9
3	9	18.5	5.6	13.0	9.1
4	5	22.9	14.2	14.2	12.8
5	5	23.5	27.4	16.0	19.5

4. Based on the results obtained in Example 1 and Problems 1, 2, and 3, do you agree that the 24 households in Lenin's study of peasant budgets are typical of the peasant households in Shcherbina's study? Explain your answer.

In 1889 Shcherbina published a detailed study of peasant budgets based on 66 households. The relevant data are given in Table 3. As before, Shcherbina's numbers have been altered slightly so that the final computations agree with his figures. Problems 5 and 6 refer to Table 3.

5. (a) Calculate the mean land allotment (in *dessiatina*) per household in this study.
   (b) Calculate the mean rented land (in *dessiatina*) per household.
   (c) Calculate the mean family size per household.
   (d) Calculate the mean number of cattle owned per household
6. Based on the results obtained in Example 1 and Problems 1, 2, 3, and 5, do you agree that the 24 households in Lenin's study of peasant budgets are typical of the peasant households in Shcherbina's study of 66 households? Explain your answer.

8–3 Standard Deviation

The mean and expected value are numerical values that describe the central tendency. The Greek letter μ (mu) is the conventional symbol for this measure, so

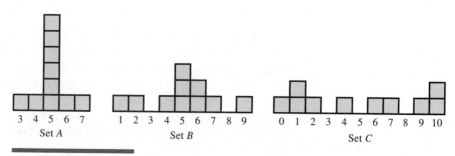

FIGURE 1

$\mu = \bar{x}$ for a frequency distribution and $\mu = E(X)$ for a probability distribution. We refer to it as "the mean μ."

In this section we discuss two values, the variance and standard deviation, which measure the spread of the data from the mean μ. A knowledge of a set's mean and standard deviation enables many conclusions about a study to be drawn.

Measures of Dispersion

Consider these three sets of data.

set A: 6, 5, 5, 4, 5, 3, 5, 7, 5, 5

set B: 6, 9, 7, 6, 5, 5, 2, 1, 4, 5

set C: 6, 2, 10, 9, 4, 1, 0, 1, 7, 10

Their histograms are shown in Figure 1. The mean of each set is 5, but their spreads are quite different. The entries in set C are more dispersed than the entries in set B, and the entries in set B are more dispersed than the entries in set A. The variance and standard deviation are measures that describe these dispersions quantitatively.

In Table 1 we list the entries of set B and find $\mu = 5$. To calculate the deviation of each entry x from the mean, we compute $x - \mu$ and list it in the second column. Notice that the sum of this column is 0; the definition of μ ensures that it will be 0 for every set of data. To compensate for the negative numbers canceling out the positive numbers, we compute the square of each deviation, $(x - \mu)^2$, and list it in the third column. The *variance*, denoted by $V(X)$, is the mean of the numbers in this column. The name "variance" refers to the variability of the data. For set B, $V(X) = \frac{48}{10} = 4.8$. The *standard deviation*, which is denoted by the Greek letter σ (sigma), is equal to the square root of the variance. Here $\sigma = \sqrt{4.8} \approx 2.2$.

Table 1 Set B

x	$x - \mu$	$(x - \mu)^2$
6	1	1
9	4	16
7	2	4
6	1	1
5	0	0
5	0	0
2	-3	9
1	-4	16
4	-1	1
5	0	0

Total　　50　　　　0　　　　48

$\mu = \frac{50}{10} = 5$　　$V(X) = \frac{48}{10} = 4.8$　　$\sigma = \sqrt{4.8} \approx 2.2$

DEFINITION

For a set of data with n entries $x_1, x_2, \ldots, x_n$ and with mean μ the **variance** is

$$V(X) = \frac{(x_1 - \mu)^2 + (x_2 - \mu)^2 + \cdots + (x_n - \mu)^2}{n}$$

The **standard deviation** is $\sigma = \sqrt{V(X)}$

Some calculators have a σ key for computing the standard deviation. However, there are two types of standard deviation, the population standard deviation, which we are using, and the sample standard deviation. If you are using a calculator, you should check your manual on the set B to see which standard deviation is displayed.

Formulas

The formulas for the variance and standard deviation are appropriate for describing their meanings as measures of the spread of data from the mean. However, another approach is more convenient for computations. It can be obtained by simplifying the formula for $V(X)$.

Computational Formula for Variance

$$V(X) = \frac{x_1^2 + x_2^2 + \cdots + x_n^2}{n} - \mu^2$$

Example 1 contrasts the two methods of computing $V(X)$. In both cases $\sigma = \sqrt{V(X)}$.

EXAMPLE 1

Problem

Find the standard deviation of set C.

6, 2, 10, 9, 4, 1, 0, 1, 7, 10

Solution The tables list the data for the two formulas. The table on the left contains the calculation by definition while the table on the right contains the calculation by formula.

x	$x - \mu$	$(x - \mu)^2$		x	x^2
6	1	1		6	36
2	−3	9		2	4
10	5	25		10	100
9	4	16		9	81
4	−1	1		4	16
1	−4	16		1	1
0	−5	25		0	0
1	−4	16		1	1
7	2	4		7	49
10	5	25		10	100
Totals 50	0	138		Totals 50	388

$$\mu = \frac{50}{10} = 5$$

$$V(X) = \frac{138}{10} = 13.8$$

$$\mu = \frac{50}{10} = 5$$

$$V(X) = \frac{388}{10} - (5)^2 = 13.8$$

Both formulas yield $V(X) = 13.8$, so $\sigma = \sqrt{13.8} \approx 3.7$.

The frequency formula for the variance is used when data are given by a frequency distribution.

Frequency Formula for the Variance

If the entries $x_1, x_2, \ldots, x_n$ occur with frequencies $f_1, f_2, \ldots, f_n$, respectively, then

$$V(X) = \frac{f_1 x_1^2 + f_2 x_2^2 + \cdots + f_n x_n^2}{f_1 + f_2 + \cdots + f_n} - \mu^2$$

EXAMPLE 2

An ecologist investigates the reproductive behavior of certain water fowl to determine whether DDT is causing infertile eggs. The species could become endangered if the eggs are infertile. A frequency distribution gives the number of nests that hold between zero and five eggs.

eggs in nest	0	1	2	3	4	5
frequency	42	44	32	19	11	2

Problem

Compute the mean and standard deviation.

Solution Table 2 contains the essential ingredients.

Table 2

x	f	fx	x^2	fx^2
0	42	0	0	0
1	44	44	1	44
2	32	64	4	128
3	19	57	9	171
4	11	44	16	176
5	2	10	25	50
Totals	150	219		569

$$\mu = \frac{219}{150} \approx 1.46$$

$$V(X) = \frac{569}{150} - \left(\frac{219}{150}\right)^2 \approx 1.66$$

$$\sigma = \sqrt{1.66} \approx 1.3$$

Thus the mean number of bird eggs in a nest is 1.46 with a standard deviation of 1.3.

Probability Distributions

The Frequency Formula for $V(X)$ leads to an interpretation of the variance for probability distributions. The procedure is analogous to the derivation of the formula for $E(X)$ from the formula for $\bar{x}$ that was carried out in Section 8–2. The Frequency Formula is

$$V(X) = \frac{f_1x_1^2 + f_2x_2^2 + \cdots + f_nx_n^2}{N} - \mu^2$$

where $N = f_1 + f_2 + \cdots + f_n$. Then

$$V(X) = \left(\frac{f_1}{N}\right)x_1^2 + \left(\frac{f_2}{N}\right)x_2^2 + \cdots + \left(\frac{f_n}{N}\right)x_n^2 - \mu^2$$

Therefore

$$V(X) = p_1x_1^2 + p_2x_2^2 + \cdots + p_nx_n^2 - \mu^2$$

This is the desired computational form of the variance for probability distributions.

Variance for Probability Distributions
If the probabilities of the outcomes $x_1, x_2, \ldots, x_n$ are $p_1, p_2, \ldots, p_n$, respectively, and μ is the mean, then

$$V(X) = p_1 x_1^2 + p_2 x_2^2 + \cdots + p_n x_n^2 - \mu^2$$

EXAMPLE 3

Problem

Find the standard deviation for the probability distribution given in this table.

x	p
4	.1
5	.3
6	.2
7	.4

Solution The procedure is to compute $V(X)$ and to set $\sigma = \sqrt{V(X)}$. Expand the table to include columns for xp, x^2, and x^2p. The total of the xp column gives the mean $\mu = 5.9$. The total of the x^2p column is 35.9. By the formula for probability distributions, $V(X) = 35.9 - (5.9)^2 = 1.09$. Then $\sigma = \sqrt{1.09} \approx 1.04$.

x	p	xp	x^2	x^2p
4	.1	0.4	16	1.6
5	.3	1.5	25	7.5
6	.2	1.2	36	7.2
7	.4	2.8	49	19.6
Totals		5.9		35.9

Example 4 uses the expected value and standard deviation to compare two sets of data.

EXAMPLE 4

A farmer must determine which of two hybrid grains, A or B, has the higher yield and smaller variation. The following distributions are known, where the units for x are in thousands of bushels.

Type A

x	p
1	.2
2	.1
3	.4
4	.2
5	.1

Type B

x	p
2	.4
3	.4
4	.1
5	.1

Problem

Determine the expected value and standard deviation for each type of hybrid grain. Which type should the farmer plant?

Solution Expand the tables for both types of hybrid grain.

Type A

x	p	xp	x^2	x^2p
1	.2	.2	1	.2
2	.1	.2	4	.4
3	.4	1.2	9	3.6
4	.2	.8	16	3.2
5	.1	.5	25	2.5
Totals		2.9		9.9

Type B

x	p	xp	x^2	x^2p
2	.4	.8	4	1.6
3	.4	1.2	9	3.6
4	.1	.4	16	1.6
5	.1	.5	25	2.5
Totals		2.9		9.3

The two grains have the same expected value $\mu = 2.9$, so they produce the same yield. For type A, $V(X) = 9.9 - (2.9)^2 = 1.49$ and $\sigma = \sqrt{1.49} \approx 1.22$. For type B, $V(X) = 9.3 - (2.9)^2 = 0.89$ and $\sigma = \sqrt{0.89} \approx 0.94$. Since the variation in type B's yield is smaller than the variation in type A's yield, the farmer should plant type B.

Chebyshev's Theorem

The expression "within k standard deviations of the mean" refers to those entries whose distance to the mean μ is less than or equal to $k\sigma$. We write "within $k\sigma$" for short. The entries can lie to the left or right of μ, as Figure 2 shows.

Recall set B from the beginning of this section.

set B: 6, 9, 7, 6, 5, 5, 2, 1, 4, 5

We showed that $\mu = 5$ and $\sigma \approx 2.2$. An entry lies within 2σ of the mean if its value is between $\mu - 2\sigma \approx 5 - 4.4 = 0.6$ and $\mu + 2\sigma \approx 5 + 4.4 = 9.4$. All ten entries of B lie within this interval.

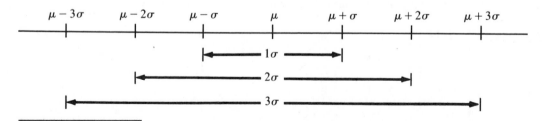

FIGURE 2

Since the standard deviation measures the spread of data, it seems reasonable that as k increases, the number of entries within $k\sigma$ of the mean will also increase. This fact, which was proved by the Russian mathematician Pafnuti Chebyshev (1821–1894), is called **Chebyshev's Theorem.***

Chebyshev's Theorem

In any experiment the probability that an entry lies within k standard deviations of the mean is at least $1 - (1/k)^2$.

For $k = 2$ Chebyshev's Theorem asserts that the proportion of entries within 2σ of the mean is at least $1 - (\frac{1}{2})^2 = \frac{3}{4}$, or 75%. Indeed, for set B, 100% of the entries lie in this interval. The importance of Chebyshev's result is that it refers to *any* distribution. Later we will improve the approximation for data with a normal distribution.

EXAMPLE 5

When a coin is tossed 100 times, the expected number of heads is $\mu = 50$. It will be shown in Section 8–5 that $\sigma = 5$.

Problem

What is the lower bound for the probability of tossing between 35 and 65 heads?

Solution We must first determine the number of standard deviations from the mean for $x = 35$ and $x = 65$. Set $x = 65$. Then $x - \mu = 65 - 50 = 15$. Since $\sigma = 5$ and $k\sigma = 15$, we get $k = \frac{15}{5} = 3$. This means that $x = 65$ lies 3σ to the right of the mean. (See Figure 3.) Similarly, $x = 35$ lies 3σ to the left of the mean. By Chebyshev's Theorem the probability of an entry lying within 3σ of the mean is at least $1 - (\frac{1}{3})^2 = 1 - (\frac{1}{9}) = \frac{8}{9}$.

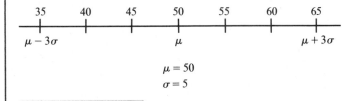

$\mu = 50$
$\sigma = 5$

FIGURE 3

---

*Stephen M. Stigler has discovered that the French mathematician Bienaymé "stated and proved the Chebyshev inequality a decade before and in greater generality than Chebyshev's first work on the topic." See S. M. Stigler, "Stigler's Law of Eponymy," *Transactions of the New York Academy of Sciences,* p. 148, April 1980.

EXERCISE SET 8–3

In Problems 1 to 10 calculate the variance and standard deviation.

1. 0, 1, 3, 1, 2, 0, 6, 2, 7, 3

2. −1, 0, −3, 1, 4, 0, −2, 5, 1, −3

3. 10, 15, 18, 11, 12, 15, 20, 10, 12, 18

4. 25, 20, 18, 15, 24, 22, 16, 20, 17, 23

5. 22, 18, 32, 40, 20, 28, 38, 42, 15, 45

6. 40, 48, 60, 55, 58, 65, 42, 45, 52, 35

7.

x	1	3	6	9	11
f	3	7	5	10	5

8.

x	2	6	10	14	18
f	15	18	20	28	20

9.

x	11	13	15	17	19	21
f	1	8	10	5	15	3

10.

x	1	2	3	4	5	6
f	10	8	7	8	6	3

In Problems 11 to 14 find the variance and standard deviation of the probability distribution. The expected values were calculated in Problems 9 to 12 in Section 8–2.

11.

x	p
2	$\frac{1}{8}$
3	$\frac{1}{2}$
5	$\frac{1}{4}$
7	$\frac{1}{8}$

12.

x	p
2	$\frac{1}{2}$
4	$\frac{1}{3}$
6	$\frac{1}{6}$

13.

x	p
2	.2
3	.3
4	.5

14.

x	p
1	.2
3	.4
5	.1
7	.3

15. An automobile dealer has four types of cars selling for $21,000, $17,000, $14,000, and $9000. The probability that a shopper will buy each type of car is, respectively, .1, .2, .4, and .3. What is the expected sales price? What is the standard deviation?

16. A student taking an economics exam has a probability of .75 of scoring 80%, .15 of scoring 85%, .05 of scoring 90%, and .05 of scoring 100%. What is the expected score? What is the standard deviation?

In Problems 17 and 18 a coin is tossed 100 times.

17. What is the lower bound for the probability that it will land heads between 40 and 60 times?

18. What is the lower bound for the probability that it will land heads between 20 and 80 times?

In Problems 19 to 24 calculate the variance and standard deviation.

19. 2.5, 2.0, 1.8, 1.5, 2.4, 2.2, 1.6, 2.0, 1.7, 2.3

20. 1.0, 1.5, 1.8, 1.1, 1.2, 1.5, 2.0, 1.0, 1.2, 1.8

21. 4.0, 4.8, 6.0, 5.5, 5.8, 6.5, 4.2, 4.5, 5.2, 3.5

22. 2.2, 1.8, 3.2, 4.0, 2.0, 2.8, 3.8, 4.2, 1.5, 4.5

23.

x	20	60	100	140	180
f	15	18	20	28	20

24.

x	110	130	150	170	190	210
f	1	8	10	5	15	3

In Problems 25 to 28 find the variance and standard deviation of the probability distribution. The expected values were calculated in Problems 23 to 26 in Section 8–2.

25.

x	p
−3	.2
−1	.2
0	.1
1	.4
3	.1

26.

x	p
−2	.2
−1	.1
0	.4
1	.2
2	.1

27.

x	p
−10	.2
−5	.4
0	.1
5	.2
10	.1

28.

x	p
−5	.1
−3	.3
0	.1
1	.2
4	.3

In Problems 29 to 32 find the standard deviation from the given histogram. The mean or expected value was computed in Problems 13 to 16 of Section 8–2

29.

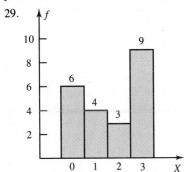

30.

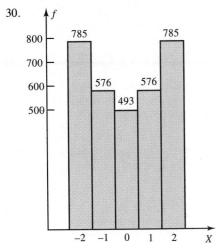

31.

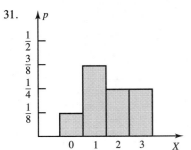

32.

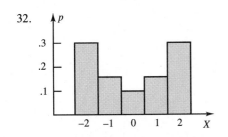

33. Toss a coin three times and let X be the number of heads. In Problem 27 of Section 8–2 it was found that $E(X) = \frac{3}{2}$. Find the standard deviation.

34. Two cookies are selected without replacement from a jar with four vanilla cookies and three chocolate cookies. Find the expected number of vanilla cookies and the standard deviation.

35. A biologist investigating airborne bacteria places a petri plate containing glucose under a ventilation duct. The number of bacteria colonies that appear per square millimeter is tabulated, as follows:

colonies per cell	0	1	2	3	4	5
cells	5	7	10	6	2	1

(a) What is the average number of colonies per cell?
(b) What is the standard deviation?

36. An economist determines trends in the stock market by comparing the average price per share of ten selected stocks from month to month. The prices for the first month are $15.75, $17, $6.25, $7.25, $47, $32, $21.50, $19, $40, and $12.88. What is the average price per share, and how much variation is there?

37. (a) Find the standard deviation of the numbers 1, 2, 3, 4, 5, 6, 7.
    (b) Find the standard deviation of the numbers 2, 3, 4, 5, 6, 7, 8.
    (c) Find the standard deviation of another set of seven consecutive integers.
    (d) Based on parts (a), (b), and (c), what do you conclude about the standard deviation of seven consecutive integers?

38. (a) Which distribution in the accompanying figure has the largest variance? (b) Which has the smallest?

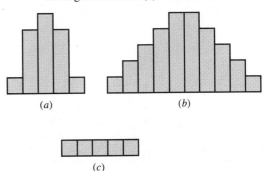

39. Toss a fair coin four times. (a) What is the expected number of tails that appears? (b) What is the variance?

40. Toss a fair coin five times. (a) What is the expected number of tails that appears? (b) What is the variance?

In Problems 41 and 42 a coin is flipped 100 times. The expected number of heads is 50 and the standard deviation is 5.

41. What is a lower bound for the probability of tossing between 40 and 60 heads?

42. What is a lower bound for the probability of tossing between 30 and 70 heads?

43. Prove that the formula for variance by definition and the computational formula are equivalent for $n = 2$. (*Hint:* Prove this equality.)

$$\frac{(x_1 - \mu)^2 + (x_2 - \mu)^2}{2} = \frac{x_1{}^2 + x_2{}^2}{2} - \mu^2$$

44. State the equality for the fact that the variance by definition is equal to the computational formula for the variance for the case $n = 3$.

45. If a coin is tossed n times, the expected number of heads is $n/2$. What is the standard deviation?

Referenced Exercise Set 8–3

1. The expected birth month for the Phillips study on strong wills prolonging life was found in Problem 3 of Referenced Exercise Set 8–2. Calculate the standard deviation.

2. Problem 2 in Referenced Exercise Set 8–2 computed the mean number of heads in J. E. Kerrich's coin-tossing experiment from these data: 502, 511, 497, 529, 504, 476, 507, 528, 504, 529. Find the standard deviation.

3. Table 3 presents three methods that were used to forecast the consumption of gasoline during 1982–1983. The actual gasoline consumption is listed in the last column. Compute the mean and standard deviation for each forecast and for the actual consumption. Which method of forecasting is most accurate?

4. The national speed limit was reduced to 55 miles per hour in 1974. Table 4 would seem to confirm the reasonableness of this policy, yet some critics object to it. Arnold Barnett, of the Sloan School of Management at MIT, used elementary statistics to rebut the critics.
   (a) Compute the mean number of road fatalities and the standard deviation for the years 1970–1973.
   (b) Within how many standard deviations of the mean does the number of deaths in 1974 lie?

5. For assets which are not held in a diversified portfolio, it is commonplace in business to identify the standard deviation σ with the measure of risk. Eastern Com-

Table 3

	Econometric Forecast	Time Series Forecast	Composite Forecast	Actual Gasoline Consumption
	(million barrels per month)			
1982				
4	195.93	194.38	200.30	206.7
5	210.36	204.03	202.92	206.15
6	207.12	206.96	202.06	204.36
7	217.03	206.36	205.01	210.77
8	220.03	209.09	206.14	206.31
9	202.20	199.11	196.79	195.21
10	211.54	207.64	201.19	198.12
11	200.70	194.63	194.54	196.77
12	213.96	208.96	203.06	202.99
1983				
1	193.56	191.02	192.46	185.41
2	175.85	180.26	182.02	174.64
3	200.60	201.18	196.56	212.13

Source: From Anthony E. Bopp, "On Combining Forecasts: Some Extensions and Results," *Management Science*, pp. 1492–1498, December 1985.

Table 4

Year	Deaths
1970	54,800
1971	54,700
1972	56,600
1973	55,700
1974	46,600

Source: From Arnold Barnett, "Misapplications Reviews: Speed Kills?" *Interfaces,* pp. 63–68, March–April 1986.

Table 5 Projected Rates of Return for the Standard and Designer Phone Lines

State of the Economy	Probability	Designer Line Rate of Return (%)	Standard Line Rate of Return (%)
Boom	0.3	100	20
Normal	0.4	15	15
Recession	0.3	−70	10

Source: From B. J. Campsey and Eugene F. Brigham, *Introduction to Financial Management,* Hinsdale, Ill. The Dryden Press, pp. 426–432, 1985.

munications, Inc., a manufacturer and retailer of business and consumer telephones, projected its rate of return on two new lines, Standard and Designer, over a wide range of states of the economy. Table 5 summarizes the projections. (a) Compute σ (as a percent) for the Standard Phone line. (b) Compute σ (as a percent) for the Designer Phone line. (c) Which line had the smaller chance of significant loss from investing in that product line?

6. Table 6 lists the expected EPS (earnings-per-share) and the standard deviation of the EPS for various D/A (debts/assets) ratios for Santa Clara Industries.
   (a) For which D/A ratios is there a negative EPS within one standard deviation of the expected EPS?
   (b) For which D/A ratios is there a negative EPS within 1.5 standard deviations of the expected EPS?

Table 6

D/A Ratio (%)	Expected EPS	Standard Deviation of EPS
0	$2.40	$1.52
10	$2.56	$1.69
20	$2.75	$1.90
30	$2.97	$2.17
40	$3.20	$2.53
50	$3.36	$3.04
60	$3.30	$3.79

Source: From Campsey and Brigham, idem, p. 539.

7. The well-known writer Stephen Jay Gould made use of statistics to support his claim that oldtime baseball players, several of whom hit .400, were no better than the best players of today.* Gould wrote, "As (standard) variation shrinks around a constant mean batting average, .400 hitting disappears. It is, I think, as simple as that." Are you convinced by this argument?

8. Many studies use statistical analyses, based on the mean and standard deviation, to draw conclusions. One such study conducted by Paula Cook asked whether teaching geometry to general students in high school using computers (the Logo group) would improve student achievement over the standard pencil-and-paper approach (the traditional group).† This problem asks you to derive conclusions from Dr. Cook's findings.
   (a) An achievement test was administered to determine if the two groups of students were equivalent at the beginning of the study. The relevant results are tabulated in Table 7. Do you think that the groups were equivalent?
   (b) The students took both a pretest and a posttest on achievement in geometry. The relevant data are

Table 7

Group	Mean number of correct answers
Logo	29.28
Traditional	27.47

*Stephen Jay Gould, "Entropic Homogeneity Isn't Why No One Hits .400 Any More," *Discover,* pp. 60–66, August 1986.

†Paula J. Cook, "The Effects of an Instructional Unit Utilizing Logo and the Computer on Achievement in Geometry and Attitude Toward Mathematics of Selected High School General Mathematics Students," doctoral dissertation, Temple University, Philadelphia, PA, 1988.

tabulated in Table 8. Do you think that the use of the computer improved the scores over the standard approach?

Table 8

Group	Pretest $\bar{x}$	σ	Posttest $\bar{x}$	σ
Logo	10.47	2.51	17.03	4.40
Traditional	9.57	3.33	16.27	3.38

(c) The students took both a pretest and a posttest on their attitude toward mathematics. The relevant data are tabulated in Table 9. Do you think that the use of a computer changed their attitudes significantly toward mathematics?

Table 9

Group	Pretest $\bar{x}$	σ	Posttest $\bar{x}$	σ
Logo	55.00	20.07	60.31	17.93
Traditional	67.37	13.24	61.87	19.11

9. S. Yeshurun wrote a provocative article on cognitive psychology that dealt with left-brain/right-brain differences and with information processing approaches to learning. The author revealed that the mean number of items that the brain's short-term memory can hold is 7, with standard deviation 2.‡ How many standard deviations from the mean is a brain whose short-term memory holds between 3 and 11 items?

Cumulative Exercise Set 8–3

1. Toss a coin four times and record the sequence of heads and tails. Let X be the number of heads recorded.
   (a) Construct the frequency distribution.
   (b) Draw the histogram.

2. Draw the histogram for the cost of 1991 automobiles from the following table.

cost (in $000)	5–11	11–17	17–23	23–29	29–35
frequency	8	8	14	6	5

3. Draw the histogram for the number of heads when a penny is tossed five times from the probability distribution below.

x	0	1	2	3	4	5
p	$\dfrac{1}{32}$	$\dfrac{5}{32}$	$\dfrac{10}{32}$	$\dfrac{10}{32}$	$\dfrac{5}{53}$	$\dfrac{1}{32}$

4. Find the mean for the data 5, 5, 6, 6, 6, 6, 9, 10, 10.

5. Find the (a) mean and (b) median for the data in the table.

x	10	9	8	7	6	5
f	7	7	5	1	1	3

6. The number of new accounts opened by a new salesperson during her first ten days on the job is 5, 4, 4, 6, 3, 5, 5, 4, 7, 6. Find the mean number of accounts opened per day.

7. Find the (a) mean and (b) median for the data in the table.

height (in inches)	56–58	58–60	60–62	62–64	64–66
frequency	400	800	1000	700	500

8. The department of transportation reported that the probability of no accident occurring per day on Route 2 near Boston is .28, the probability of one accident is .12, the probability of two accidents is .25, and the probability of three accidents is .35. What is the expected number of accidents per day?

9. A set of data has five entries. Each entry is either 6, 7, or 8. What is the set if the mean is 6.2 and the median is 6?

10. Find the expected value of the random variable X from the probability distribution table.

x	-2	-1	2	4
p	.1	.4	.2	.3

‡S. Yeshurun, "A First Look at Psychomathematics and Neuropsychomathematics," *International Journal of Mathematical Education in Science and Technology*, pp. 229–251, March–April 1987.

11. Calculate the standard deviation for the data 3, 4, 5, 6, 7, 8, 9.

12. Calculate the standard deviation for the data in the table.

x	2	3	4	5
p	.4	.4	.1	.1

8–4 The Normal Curve

Suppose an expert archer and a complete novice shoot at a target on an archery range. Although the expert's shots will be more accurate, the novice's shots will land in a surprisingly similar manner. This does not mean that the outcome of each single shot can be predicted. Rather, if each archer shoots 1000 arrows and the distance between the hits and the target is measured, then each histogram of the distances will assume a definite shape that does not depend on their relative abilities.

The shape of each histogram will approximate an ideal distribution known as the normal curve. The more shots that are made, the closer the histogram will be to this curve. Because the normal curve resembles a bell, it is sometimes called a "bell-shaped curve."

The normal curve describes many mental and physical characteristics of human beings. This is fortunate because many important facts are known about the normal curve, including its mean and standard deviation. This section examines these facts and shows how to use tables of values to answer questions involving heights and IQs (intelligence quotient).

Fundamental Properties

Consider a probability experiment of tossing six coins and recording the number of heads. Its histogram is shown in Figure 1. We have superimposed upon it a curve that passes through the midpoints at the top of the rectangles. It resembles a curve known as the **normal curve.**

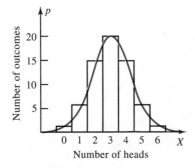

FIGURE 1

The normal curve has an unwieldy looking equation.

$$y = \frac{1}{\sqrt{2\pi}} e^{-z^2/2}$$

where $\pi \approx 3.1416$ and $e \approx 2.7183$. The graph of the normal curve is given in Figure 2. The values on the z-axis are standard deviations. The units on the y-axis correspond to the relative frequencies on a histogram, and therefore a normal curve can be used to approximate the histogram.

There are four basic properties of the normal curve:

1. The mean corresponds to the value $z = 0$.
2. The curve is symmetric about the mean.
3. The curve always lies above the z-axis.
4. The area under the curve and above the z-axis is exactly 1.

We make use of these properties to find areas that lie under the curve and above the z-axis. The reason for the emphasis on areas is that the area under a normal curve corresponds to the probability of an event. There are three fundamental areas:

The area between $z = -1$ and $z = 1$ is about .68.
The area between $z = -2$ and $z = 2$ is about .95.
The area between $z = -3$ and $z = 3$ is about .99.

These areas refer to the proportion of entries that lie within 1σ, 2σ, and 3σ of the mean. Thus for a normal curve about two-thirds of the entries will lie within 1σ of the mean, about 95% will lie within 2σ, and virtually all will lie within 3σ. These properties are shown in Figure 2.

According to Chebyshev's Theorem for $k = 2$, at least three-fourths of the entries in any distribution lie within 2σ of the mean. For data with a normal distribution this approximation can be improved to 95%. For 3σ Chebyshev's Theorem gives a figure of at least $\frac{8}{9} = .89$, while the corresponding figure for the normal curve is known to be .99.

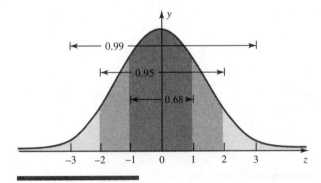

FIGURE 2

Examples 1 and 2 combine the four basic properties of the normal curve with the three designated areas to derive other areas under the normal curve.

EXAMPLE 1

Problem

What is the area under the normal curve between $z = 0$ and $z = 1$?

Solution The area under the normal curve between $z = -1$ and $z = 1$ is about .68. Since the normal curve is symmetric about the mean, the mean divides this area into two equal parts. (See Figure 3.) The area between $z = 0$ and $z = 1$ is one-half the area between $z = -1$ and $z = 1$, so its area is about $(.5)(.68) = .34$.

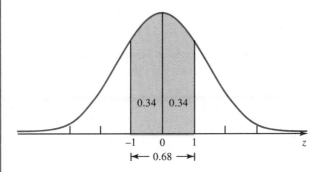

FIGURE 3

EXAMPLE 2

Problem

What is the area under the normal curve to the right of $z = 2$?

Solution The area under the normal curve to the right of $z = 2$ is derived from the fact that the area between $z = -2$ and $z = 2$ is about .95. There is about .05 remaining. There are two regions under the curve that comprise this area,

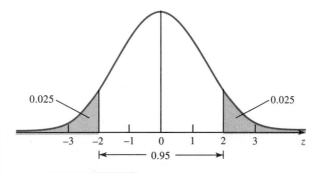

FIGURE 4

the region to the right of $z = 2$ and the region to the left of $z = -2$. These regions have equal areas because of the symmetry of the normal curve. Hence the area under the normal curve to the right of $z = 2$ is one-half of .05, or .025. (See Figure 4.)

The area under the normal curve to the left of $z = 2$ can be obtained from Example 2 by using property 4. Since the total area is 1 and the area to the right is about .025, the area to the left is approximately $1 - .025 = .975$. An alternate way to obtain this area is to look it up in a table.

Using a Table

The values we have supplied for areas under the normal curve are very limited; they are rough approximations that apply only to integral multiples of σ. Table 1 contains several more values that are approximated to four decimal places. A more complete table is printed in the Appendix. For each value of z, called the **z-score,** $A(z)$ gives the area under the normal curve to the left of z. The area $A(z)$ is shaded in Figure 5.

There are three kinds of areas under the normal curve:

L: The area to the left of z
R: The area to the right of z
B: The area between two values of z

Table 1 accounts for L-areas. For instance, the area to the left of $z = 2$ is $A(2) = .9773$. This is a better approximation than the value .975 obtained after Example 2. Example 3 shows how to compute R-areas.

Table 1

z	$A(z)$
2.00	.9773
1.50	.9332
1.00	.8413
0.50	.6915
0.00	.5000
−0.50	.3085
−1.00	.1587
−1.50	.0668
−2.00	.0227

E X A M P L E   3

Problem

Find the area under the normal curve to the right of $z = 1$.

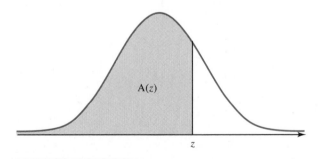

$A(z)$

z

FIGURE 5

Solution From Table 1 the area to the left of $z = 1$ is $A(1) = .8413$. Since the total area under the normal curve is 1, the area to the right of $z = 1$ is $1 - .8413 = .1587$. See Figure 6.

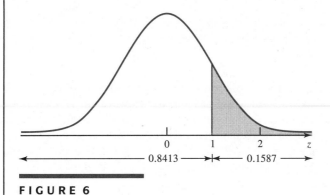

FIGURE 6

The following rule emerges from Example 3. It shows how to compute an R-area.

The area under the normal curve to the right of z is $1 - A(z)$.

EXAMPLE 4

Problem

Find the area under the normal curve between $z = -0.5$ and $z = 0.5$.

Solution First set $z_1 = 0.5$, then $A(0.5) = .6915$ is the area to the left of $z_1 = 0.5$. Similarly, $A(-0.5) = .3085$ is the area to the left of $z_2 = -0.5$. The desired area is the difference between these two areas. See Figure 7 for a geometric view of why this is true. Thus the area between $z_1 = -0.5$ and $z_2 = 0.5$ is $A(0.5) - A(-0.5) = .6915 - .3085 = .3830$.

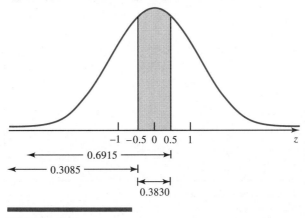

FIGURE 7

The following rule emerges from Example 4. It shows how to compute a B-area.

> The area under the normal curve between z_1 and z_2, where $z_1 < z_2$, is $A(z_2) - A(z_1)$.

The normal curve is a mathematical abstraction that is not derived from any specific real-world phenomena. However, it is very useful for describing any set whose histogram resembles the normal curve. The main difference between the normal curve and such a set is that the normal curve is *continuous,* while data sets are necessarily *discrete.* In spite of this difference the normal curve's utility is vast. Let us turn to its applications.

Applications

So far we have referred to *the* normal curve. Actually, it is called the **standard normal curve.** Any list of data that is assumed to be normally distributed will have a histogram, and hence a normal curve, that resembles the standard normal curve. The **normal approximation** is a procedure for finding the area under any normal curve by adjusting the units of the horizontal and vertical axes so that the resulting curve resembles the standard normal curve. Let us explain this procedure.

The Health Examination Survey, conducted by the Department of Health and Human Services, found that the average height of women in the United States forms a normal curve, with $\mu = 64''$ and $\sigma = 2.5''$. *What percentage of women are less than $67\frac{3}{4}''$ tall?* To answer this question, draw an x-axis and mark the heights at μ, $\mu \pm 1\sigma$ (66.5" and 61.5"), and $\mu \pm 2\sigma$ (69" and 59"). Draw a z-axis below it, marking the corresponding number of standard deviations. (See Figure 8.) The

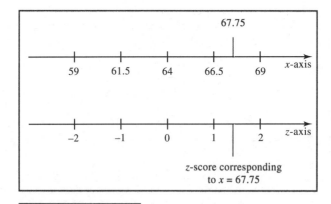

FIGURE 8

value $x = 67.75''$ on the upper line lies between 66.5'' and 69'', so the corresponding z-score on the lower line is between 1 and 2. Since $\mu = 64''$ and $\sigma = 2.5$, the precise z-score is obtained from this formula.

$$z = \frac{\text{distance between } x \text{ and } \mu}{\text{standard deviation}} = \frac{67.75 - 64}{2.5} = \frac{3.75}{2.5} = 1.5$$

This means that $x = 67.5''$ is $z = 1.5$ standard deviations from the mean $\mu = 64''$. The formula for the z-score is described in the display.

If a set of data is normally distributed with mean μ and standard deviation σ, the z-score corresponding to a value of x is

$$z = \frac{x - \mu}{\sigma}$$

The question about the percentage of women whose height is less than $x = 67.75''$ translates into z-scores less than $z = 1.5$. This is an L-area problem. From Table 1 the area to the left of $z = 1.5$ is $A(1.5) = .9332$. Thus about 93.32% of American women are less than 67.75'' tall.

In summary, the normal approximation is a three-part procedure. First the x-scores are converted to z-scores. Then the problem is translated into one of the fundamental areas: L, R, or B. Finally, a table of z-scores for the standard normal curve provides the area, which is interpreted as either a probability or a percentage. Example 5 illustrates this procedure for a B-area.

EXAMPLE 5

Assume that the heights of women in the United States are normally distributed with $\mu = 64''$ and $\sigma = 2.5''$.

Problem

What percentage of heights are between 5' and 5'2''?

Solution Since $5' = 60''$, set $x_1 = 60$. The corresponding z-score is $z_1 = (60 - 64)/2.5 = -1.6$. From the Appendix, $A(-1.6) = .0548$. Next, set $x_2 = 62$. Then $z_2 = (62 - 64)/2.5 = -0.8$ and $A(-0.8) = .2119$.

The percentage of heights between $x_1 = 60''$ and $x_2 = 62''$ translates to the area between $z_1 = -1.6$ and $z_2 = -0.8$. This area is $A(z_2) - A(z_1) = A(-0.8) - A(-1.6) = .2119 - .0548 = .1571$. Consequently, 15.71% of American women are between 5' and 5'2''. This percentage is roughly 1 in 6.

Example 6 applies the normal approximation procedure to IQ (intelligence quotient) tests under the usual assumption that IQ scores are normally distributed. It illustrates the method for R-areas.

E X A M P L E 6

Many IQ tests are designed so that the mean score is 100 and the standard deviation is 20.

Problem

If a person is selected at random, what is the probability that the IQ score is greater than 130?

Solution The z-score corresponding to $x = 130$ is $z = (130 - 100)/20 = 1.5$. From Table 1, $A(1.5) = .9332$. Scores above 130 translate to the area to the right of $z = 1.5$. This is an R-area, which is equal to $1 - A(1.5) = 1 - .9332 = .0668$. Thus the probability is .0668.

In many public schools the cutoff point for a student to be classified as "gifted" is an IQ score of 130. Example 6 shows that about 7% of the population is gifted under the assumption that $\mu = 100$ and $\sigma = 20$.

EXERCISE SET 8–4

In Problems 1 to 8 find the area under the normal curve within the given limits, using the four basic properties and three fundamental areas of the normal curve.

1. between $z = 0$ and $z = 2$.

2. between $z = 0$ and $z = 3$.

3. between $z = -1$ and $z = 0$.

4. between $z = -2$ and $z = 0$.

5. to the right of $z = 1$.

6. to the right of $z = 3$.

7. to the left of $z = -2$.

8. to the left of $z = -1$.

In Problems 9 to 12 use Table 1 to find the area under the normal curve within the given limits.

9. to the right of $z = 1.5$.

10. to the right of $z = -0.5$.

11. between $z = -1$ and $z = 1$.

12. between $z = -0.5$ and $z = 1.5$.

In Problems 13 to 16 assume that the heights of women in the United States are normally distributed with $\mu = 64''$ and $\sigma = 2.5''$. Determine the percentage of heights that are in the range

13. between 5'3" and 5'6".

14. between 5'1" and 5'5".

15. taller than 5'7".

16. shorter than 5'.

In Problems 17 to 20 the mean IQ is 100 and the standard deviation is 20. Determine the percentage of IQs that are in the given range.

17. between 110 and 130.

18. between 90 and 115.

19. greater than 95.

20. less than 110.

In Problems 21 to 26 find the area under the normal curve within the given limits, using the four basic properties and three basic areas of the normal curve.

21. $z = 1$ and $z = 2$.

22. $z = 2$ and $z = 3$.

23. $z = -1$ and $z = 2$.

24. $z = -2$ and $z = 1$.

25. to the right of $z = -1$.

26. to the left of $z = 3$.

In Problems 27 to 30 use the table in the Appendix to find the area under the normal curve within the given limits.

27. to the left of $z = 0.38$.

28. to the left of $z = -1.05$.

29. between $z = -1.74$ and $z = 0.51$.

30. between $z = -0.56$ and $z = 2.15$.

31. If the mean IQ is 100 with standard deviation 20, what percentage of IQ scores is between 102 and 118?

32. If the mean on an exam is 475 with standard deviation 100, what percentage of scores is between 465 and 527?

33. If the mean of a set of data that is normally distributed is 4.1 with standard deviation 0.1, what percentage of data is greater than 4.21?

34. If the mean of a set of data that is normally distributed is 1.1 with standard deviation 0.5, what percentage of data is less than 1.2?

Referenced Exercise Set 8–4

The mean and median scores on the SAT exams vary slightly from year to year.* In Problems 1 to 4 assume that they form a normal distribution and that, for the mathematics section, $\mu = 470$ and $\sigma = 100$.

*William H. Angoff, "How We Calibrate College Board Scores," *The College Board Review*, No. 68, Summer 1968.

1. What percentage of students scored above 500?

2. What percentage of students scored between 400 and 600?

3. If a student is selected at random, what is the probability that the student scored between 430 and 530?

4. If a student is selected at random, what is the probability that the student scored above 440?

Problems 5 to 8 refer to the Stanford study on the effects of long-distance running on the joints. Table 2, which is taken from that study, tabulates the effects on the knee joint using three measures.

5. For nonrunners the mean sclerosis measure is $\mu = 5.2$ with $\sigma = 0.3$. What percentage of the nonrunners had a sclerosis measure greater than 5.9, which is the mean sclerosis measure for runners?

6. For nonrunners the mean spur measure is $\mu = 5.7$ with $\sigma = 0.7$. What percentage of the nonrunners had a spur measure greater than 7.3, which is the mean spur measure for runners?

Table 2 Knee Joint Roentgenograms (Mean Values)

| | All Subjects | | | | Matched Pairs | | | | | |
| | | | | | All Pairs | | Female Pairs | | Male Pairs | |
	Runners $(n = 65)$	Nonrunners $(n = 33)$	50+ Runners $(n = 41)$	Controls $(n = 57)$	50+ Runners $(n = 41)$	Controls $(n = 41)$	50+ Runners $(n = 18)$	Controls $(n = 18)$	50+ Runners $(n = 23)$	Controls $(n = 23)$
Sclerosis										
Mean	5.9	5.2	5.9	5.5	6.0	5.3	6.7	5.1	5.5	5.5
SE	0.3	0.3	0.3	0.3	0.3	0.3	0.5	0.3	0.4	0.5
					(0.44)		(0.63)		(0.56)	
Spurs										
Mean	7.3	5.7	7.4	6.3	7.5	6.0	8.4	5.1	6.7	6.8
SE	0.5	0.7	0.6	0.5	0.6	0.6	0.9	0.7	0.8	0.9
					(0.81)		(0.96)		(1.16)	
Narrowing (mm)										
Mean	7.0	6.9	7.0	6.9	7.0	6.7	6.6	6.3	7.4	7.1
SE	0.2	0.3	0.3	0.2	0.3	0.3	0.3	0.4	0.4	0.4
					(0.32)		(0.57)		(0.36)	

Source: From Nancy E. Lane et al., "Long-Distance Running, Bone Density, and Osteoarthritis," *Journal of the American Medical Association*, pp. 1148–1151, March 7, 1986.

Table 3 Sample Data

	Proportion of Wins	Mean Win (chips)	Standard Deviation (chips)	Mean Player Edge (%)	Largest Player Deficit (chips)
Las Vegas	0.87	2084	1786	1.35	4961
Atlantic City	0.79	1108	1471	0.91	3621

Source: From Martin H. Millman, ''A Statistical Analysis of Casino Blackjack,'' *American Mathematical Monthly,* pp. 431–436, August–September 1983.

7. For runners the sclerosis measure is $\mu = 5.9$ and $\sigma = 0.3$.
   (a) What percentage of runners had a sclerosis measure less than 5.5?
   (b) What percentage of nonrunners had a sclerosis measure less than 5.5?

8. Refer to Table 2 for the mean and standard deviation of the joint space narrowing.
   (a) What percentage of runners had a joint space narrowing between 6.8 and 7.2 mm (millimeters)?
   (b) What percentage of nonrunners had a joint space narrowing between 6.8 and 7.2 mm?

Problems 9 and 10 refer to Table 3, which lists the data from a computer simulation of 20,000 games of blackjack in Atlantic City and Las Vegas. (The actual playing of 20,000 hands would consume some 300 to 400 human hours.)

9. What is the probability of a player winning at least 300 chips in (a) Las Vegas and (b) Atlantic City?

10. What is the probability of a player winning less than 100 chips in (a) Las Vegas and (b) Atlantic City?

Cumulative Exercise Set 8–4

1. For the experiment of tossing six coins where X is the number of heads, construct the probability distribution, draw the histogram, and find $Pr(0 \le X \le 2)$.

2. Find the mean and the median of the data in the distribution.

value	1	3	5	7	9
frequency	8	4	2	4	2

3. Find the expected value.

x	2	3	4	8	9	10
p	.2	.1	.1	.3	.1	.2

4. Calculate the variance and the standard deviation.

   10, 12, 14, 10, 11, 12, 12, 10, 11, 10

5. Find the standard deviation.

value	1	3	5	7	9	11
frequency	10	4	2	6	2	1

6. Find the standard deviation.

x	2	3	4	5	6	7
p	.1	.2	.1	.3	.1	.2

7. A student taking a biology exam has probability .70 scoring 80%, .15 scoring 85%, .10 scoring 90%, and .05 scoring 95%. Find the expected score and the standard deviation.

8. Find the area under the normal curve
   (a) to the left of $z = 1$.
   (b) to the right of $z = 1$.

9. Find the area under the normal curve
   (a) to the right of $z = -1.72$.
   (b) to the left of $z = 2.08$.
   (c) between $z = -1.72$ and $z = 2.08$.

10. If the mean IQ is 100 with standard deviation 20, what percentage of IQ scores is between 106 and 127?

11. If the mean on an exam is 470 with standard deviation 100, what percentage of scores is between 460 and 567?

12. If the mean of a set of data that is normally distributed is 5.7 with standard deviation 0.1, what percentage of data is greater than 5.81?

8–5 The Normal Approximation to Binomial Experiments

When you apply to a college the admissions office either accepts or rejects your application. Each decision from a college represents a **Bernoulli trial** because it consists of two possible outcomes. Some other examples of Bernoulli trials are the effectiveness of a medication (the patient is or is not cured), the movement of the Dow-Jones average (up or down), and the firing of a missile (hit or miss).

A **binomial experiment** consists of n Bernoulli trials. In Section 7–7 we saw that if the probability of success is p and the probability of failure is $q = 1 - p$, then the probability of obtaining exactly x successes in n trials is

$$Pr(X = x) = C(n, x)p^x q^{n-x}$$

This formula is useful when the value of n is relatively small, but the computations become tedious when n is large. Fortunately, the normal curve provides a very close approximation to a binomial experiment in such a case. This section shows how to carry out and apply the normal approximation procedure.

Coin Tossing

Coin tossing serves as a convenient model for binomial experiments. As an example, toss a coin six times and let the random variable X represent the number of heads that appear. Then $n = 6$ and $p = \frac{1}{2}$. The probability of tossing exactly four heads is

$$Pr(X = 4) = C(6, 4)\left(\frac{1}{2}\right)^4 \left(\frac{1}{2}\right)^2 = \frac{15}{64}$$

The entire probability distribution, including the particular case $X = 4$, is given in Table 1. Notice the values in the row of totals. The column of probabilities p adds up to 1, which serves as a check on the individual probabilities. The sum of the xp column is the expected value $\mu = 192/64 = 3$. Common sense confirms that six tosses of a coin should result in three heads on the average. Recall from

Table 1

x	p	xp	$x^2 p$
0	1/64	0/64	0/64
1	6/64	6/64	6/64
2	15/64	30/64	60/64
3	20/64	60/64	180/64
4	15/64	60/64	240/64
5	6/64	30/64	150/64
6	1/64	6/64	36/64
Totals	1	192/64	672/64

Section 7–7 that another way of obtaining this value is from the formula $\mu = np$. Here $\mu = (6)(1/2) = 3$.

Recall that the variance $V(X)$ is equal to the total of the x^2p column of Table 1 minus the square of μ.

$$V(X) = \frac{672}{64} - 3^2 = \frac{21}{2} - 9 = \frac{3}{2}$$

For a binomial experiment it is possible to compute this value from the formula $V(X) = npq$. In this example $V(X) = 6(1/2)(1/2) = 3/2$, which confirms the value derived from Table 1. The appropriate formulas are given in the display.

If a binomial experiment consists of n Bernoulli trials, each with probability of success p and probability of failure q, then

The mean is $\mu = np$.

The variance is $V(X) = npq$.

The standard deviation is $\sigma = \sqrt{npq}$.

EXAMPLE 1

Toss a coin 100 times and record the number of heads that appear.

Problem

What are the mean and standard deviation?

Solution We have $n = 100$ and $p = \frac{1}{2}$ so $\mu = np = 50$. Since $p = \frac{1}{2}$, it follows that $q = 1 - p = \frac{1}{2}$, so $\sigma = \sqrt{npq} = \sqrt{25} = 5$.

Figure 1 shows the histogram for the probability distribution of tossing six coins and recording the number of heads. The histogram can be approximated by the normal curve that is superimposed upon it. The normal curve has $\mu = 3$ and $\sigma = \sqrt{3/2} \approx 1.225$. Let us see how to use this normal curve to approximate probabilities.

Consider the value of $X = 4$ heads. Its probability is equal to the area of the shaded bar above 4 in the histogram. This area can be approximated by the area under the normal curve between $x = 3.5$ and $x = 4.5$. The procedure for finding this area, as explained in Section 8–4, uses the formula $z = (x - \mu)/\sigma$.

for $x_1 = 3.5$, $\quad z_1 = \dfrac{(3.5 - 3)}{1.225} \approx 0.41$

for $x_2 = 4.5$, $\quad z_2 = \dfrac{(4.5 - 3)}{1.225} \approx 1.22$

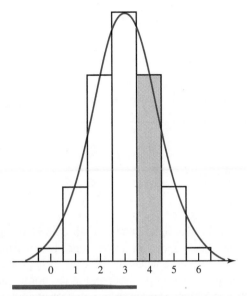

FIGURE 1

From the table in the Appendix, $A(z_1) = .6591$ and $A(z_2) = .8888$. The area under the normal curve between 3.5 and 4.5 is approximately equal to $A(z_2) - A(z_1) = .8888 - .6591 = .2297$. This value compares favorably with the theoretical probability of $15/64 \approx .2344$ listed in Table 1.

The probability of tossing at least four heads in six tosses is given by $Pr(4) + Pr(5) + Pr(6) = 22/64 \approx .3438$. This value can be approximated by the area under the normal curve to the right of $x = 3.5$ since the area of the histogram in Figure 1 corresponds to this probability. For $x = 3.5$ we found that $z = 0.41$ and $A(0.41) = .6591$. The desired area, which lies to the right of $z = 0.41$, is $1 - A(z) = 1 - .6591 = .3409$. This value compares favorably with the theoretical probability of $.3438$.

Example 2 applies a similar approximation to the question of whether a coin is fair. Do you think a coin that lands heads 60 times in 100 tosses is fair?

EXAMPLE 2

Problem

What is the probability of a coin landing heads 60 or more times in 100 tosses?

Solution From Example 1, $\mu = 50$ and $\sigma = 5$. The problem is to find $Pr(X \geq 60)$. The normal curve is shown in Figure 2, where the desired area is shaded. The value $x = 59.5$ is used for $X = 60$ because the bar corresponding to 60 runs from 59.5 to 60.5. Then $z = (59.5 - 50)/5 = 9.5/5 = 1.9$. From the table in the Appendix, the area to the left of $z = 1.9$ is $A(1.9) = .9713$. The area to the right is $1 - .9713 = .0287$. This is the desired probability.

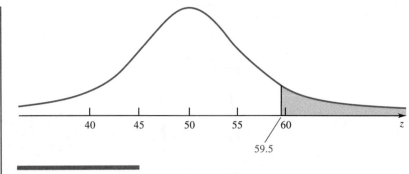

FIGURE 2

It follows from Example 2 that if a fair coin is tossed 100 times, it will land heads 60 or more times about 3% of the time. It is reasonable to conclude that if a coin produces 60 or more heads in 100 tries, it is probably not fair.

The display lists the steps for applying the normal approximation procedure to a binomial experiment. To appreciate its power, notice that the alternative to Example 2 is to carry out these computations.

$$Pr(X \geq 60) = Pr(X = 60) + Pr(X = 61) + \cdots + Pr(X = 99) + Pr(X = 100)$$

$$= C(100, 60)(\tfrac{1}{2})^{60}(\tfrac{1}{2})^{40} + C(100, 61)(\tfrac{1}{2})^{61}(\tfrac{1}{2})^{39} + \cdots + C(100, 99)(\tfrac{1}{2})^{99}(\tfrac{1}{2})^{1} + C(100, 100)(\tfrac{1}{2})^{100}(\tfrac{1}{2})^{0}$$

Procedure for Approximating a Binomial Experiment by the Normal Curve

1. Compute $\mu = np$ and $\sigma = \sqrt{npq}$.
2. Sketch the normal curve and shade the desired area.
3. Convert X to x, then x to z using $z = (x - \mu)/\sigma$.
4. Use a table to determine the area.

EXAMPLE 3

Problem

If a pair of dice is tossed 2880 times, what is the probability of obtaining between 450 and 500 doubles inclusive? (In this section, "between" is meant to be inclusive of the first and last numbers.)

Solution We have $n = 2880$. The probability of tossing doubles is 6/36, so $p = 1/6$ and $q = 5/6$. Then $\mu = 480$ and $\sigma = \sqrt{npq} = 20$. The problem is to find $Pr(450 \leq X \leq 500)$. The desired area is sketched in Figure 3. Set $x_1 = 449.5$ and $x_2 = 500.5$. Then $z_1 = (449.5 - 480)/20 = -1.525$ and $z_2 = (500.5 - 480)/20 = 1.025$. The area between z_1 and z_2 is $A(z_2) - A(z_1) = .8461 - .0643 = .7818$. Thus the probability of obtaining between 450 and 500 doubles is approximately .7818.

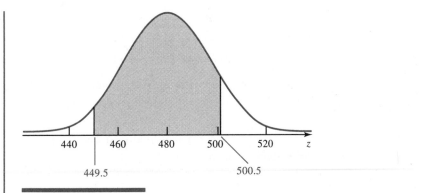

FIGURE 3

Many situations involve either one outcome or another. We label each outcome a "success" or "failure" even though the meanings of these terms need not apply. Consider the cause of accidents. Fatigue causes a certain percent p of accidents depending on the particular occupation. An accident is either due or not due to fatigue, so it represents a Bernoulli trial. If an accident is due to fatigue, we label it a success and assign it the probability p, even though in the human arena an accident caused by fatigue could hardly be called a success.

Example 4 considers industrial accidents.

EXAMPLE 4

Fatigue is the cause of 20% of all industrial accidents. Last year there were 2500 accidents in the steel industry.

Problem

What is the probability that at least 525 of the accidents were caused by fatigue?

Solution Since $p = 0.2$ and $n = 2500$, $\mu = (2500)(0.2) = 500$ and $\sigma = \sqrt{(2500)(0.2)(0.8)} = \sqrt{400} = 20$. The problem is to find $Pr(X \geq 525)$. The desired area is sketched in Figure 4. Set $x = 524.5$. Then

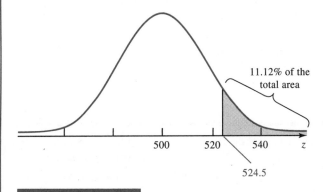

FIGURE 4

$z = (524.5 - 500)/20 = 1.225$. From the table in the Appendix, $A(1.22) = .8888$. The desired area lies to the right of $z = 1.225$, so

$$Pr(X \geq 525) \approx 1 - .8888 = .1112$$

Thus the probability of at least 525 of the accidents being caused by fatigue is about 11%.

Example 5 shows how statistical studies allow some professions to prepare for certain events beforehand.

EXAMPLE 5

Next year 62,500 people will use a drug that has caused severe side effects (loss of weight, rash, depression) in 10% of all people who have already taken it.

Problem

What is the probability that at least 6200 people will suffer side effects?

Solution We have $p = 0.1$ and $n = 62,500$. Then $\mu = 6250$ and $\sigma = \sqrt{npq} = \sqrt{5625} = 75$. The problem is to compute $Pr(X \geq 6200)$. The desired area is sketched in Figure 5. Set $x = 6199.5$. Then

$$z = \frac{6199.5 - 6250}{75} = \frac{-50.5}{75} \approx -.67$$

The table in the Appendix gives $A(-0.67) = .2514$. Therefore the probability is approximately equal to $1 - .2514 = .7486$, which is about 75%.

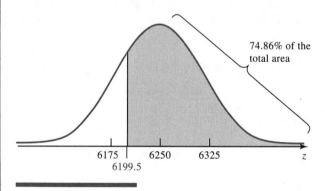

74.86% of the total area

6175   6250   6325 z
6199.5

FIGURE 5

EXERCISE SET 8–5

In Problems 1 to 4 toss n coins and record the number of heads. Find μ and σ.

1. $n = 144$ 2. $n = 36$

3. $n = 81$ 4. $n = 121$

In Problems 5 to 8 a coin is tossed 100 times. Find the probability of each event.

5. at least 55 heads.

6. at least 45 heads.

7. between 45 and 50 heads.

8. at most 60 heads.

In Problems 9 to 12 a coin is tossed 36 times. Find the probability of each event.

9. less than 15 heads.

10. at least 20 heads.

11. between 15 and 20 heads.

12. more than 20 heads.

In Problems 13 to 16 a pair of dice is tossed 2880 times. Find the probability of each event.

13. at least 450 doubles.

14. at least 510 doubles.

15. between 500 and 540 doubles.

16. at most 490 doubles.

In Problems 17 to 20 a pair of dice is tossed 180 times. Find the probability of each event.

17. at most 25 doubles.

18. at most 35 doubles.

19. between 25 and 35 doubles.

20. at least 20 doubles.

Problems 21 and 22 refer to an industry where 20% of the accidents are caused by fatigue. Suppose that 900 accidents occur. Compute the probability of each event.

21. at least 150 accidents were caused by fatigue.

22. between 162 and 192 accidents were caused by fatigue.

Problems 23 and 24 refer to a drug that causes severe side effects in 10% of the people who take it. Suppose that 100 people take the drug. Compute the probability of each event.

23. Between 5 and 15 people suffer severe side effects.

24. At least 12 people suffer severe side effects.

In Problems 25 to 28 a coin is tossed 400 times. Find the probability of each event.

25. at most 205 heads.

26. at most 195 heads.

27. between 195 and 220 heads.

28. at least 210 heads.

In Problems 29 to 34 a survey of 1200 families with exactly two children is conducted. Find the probability of each event.

29. At most 315 of the families have two boys.

30. At most 290 of the families have two girls.

31. At least 310 of the families have two girls.

32. At least 285 of the families have two boys.

33. Between 270 and 330 of the families have two girls.

34. Between 300 and 345 of the families have two girls.

35. About 2% of a dentist's patients cancel their appointments. What is the probability that out of 10,000 patients, at least 180 appointments will be canceled?

36. A clairvoyant draws a card from an ordinary deck of 52 cards and announces the suit. This is performed 48 times. What is the probability that the clairvoyant will name the correct suit at least 18 times?

37. A jar contains 3 red jelly beans and 9 black jelly beans. Select 75 jelly beans with replacement. What is the probability that between 15 and 20 are red?

38. A fluoride treatment is rated 98% effective. If it is applied in 100 cases, what is the probability that between 90 and 98 are effective?

39. A student known as a "partier" must guess at every answer on a 50-question multiple-choice test, which has five possible answers to every question. What is the probability that the student will obtain a grade of 40% or better?

40. A manufacturer of disks has found that only 1% of the disks is defective. What is the probability that a shipment of 1100 disks will contain less than 12 defective disks?

41. The Dow-Jones average will rise on a given day about 55% of the time. What is the probability that it will rise on at least 210 days during a regular year?

42. Of 1000 two-children families, what is the probability that 600 or more of them will have one boy and one girl?

43. Toss a pair of dice 180 times and record the number of times that doubles will land. What is the probability that doubles will land less than 27 times?

44. Toss one die 180 times and record the number of times that the number 3 turns up. What is the probability that a 3 will turn up between 24 and 32 times?

45. If 100 people select a Ping-Pong ball from 100 containers of Ping-Pong balls and each container holds 10 Ping-Pong balls numbered 0 through 9, what is the probability that the number of people who select the number 7 is between 9 and 11?

46. If 100 people select a Ping-Pong ball from 100 containers of Ping-Pong balls and each container holds 10 Ping-Pong balls numbered 0 through 9, what is the probability that at least 12 of them will select the number 7?

Referenced Exercise Set 8–5

1. USAir knows from experience that 15% of the reservations will not be honored.*
   (a) The airline sells 160 tickets for a flight that has 148 seats. What is the probability that more than 148 people will show up?
   (b) What is the maximum number of tickets the airline should sell so that there is a 95% chance that everyone who shows up will get a seat?

2. In Dade County, Florida, the number of malpractice lawsuits against doctors doubled from the year 1975 to the year 1985. At present, 25% of all physicians in the state of Florida are sued annually.† If a hospital has 48 physicians on staff, what is the probability that at least 15 of them will be sued during a year?

3. When Hewlett-Packard Company (H-P) entered into a joint venture with Yokogawa Corporation in 1983 to produce circuit boards for computers, it found that boards produced at the Tokyo plant contained eight defects per million connections compared with the rate of six defects per thousand connections at the H-P plant in Cupertino, California.‡
   (a) If a batch of 50 million connections is produced at the Tokyo plant, what is the probability that less than 460 connections will be defective?
   (b) If a batch of 50 million connections is produced at the Cupertino plant, what is the probability that less than 298,360 connections will be defective?

4. In American jury trials, prospective jurors can be "peremptorily" challenged by the counsel for either side. Although the state of Florida ordinarily allows six peremptory challenges for each side, the judge in the Theodore Bundy trial permitted 20 because of the national publicity surrounding Bundy's brutal murder of two sorority sisters. A pretrial survey conducted by the Public Defender's office revealed that 40 percent of the eligible jurors felt that Bundy was guilty.§ If 30 people were questioned, what is the probability that at least 12 of them did not already feel that Bundy was guilty?

Cumulative Exercise Set 8–5

1. Toss a coin three times and record the sequence of heads and tails. Let X be the number of heads recorded.
   (a) Construct the probability distribution.
   (b) Draw the histogram.

2. A set of data has five entries. Each entry is either 6, 7, or 8. What is the set if the mean is 7.8 and the median is 8?

3. A player rolls two dice and wins $5 if doubles occur and loses $1 if not. What is the expected value?

4. Calculate the standard deviation for the data in the table.

x	1	2	3	4	5
p	.2	.1	.4	.2	.1

5. A fair coin is tossed three times. The random variable X is the number of heads that appears.
   (a) What is the expected number of heads?
   (b) What is the variance?

6. Approximate the area under the normal curve between $z = 0$ and $z = 2$ without using a table.

7. What is the area under the normal curve between $z = -0.5$ and $z = 1.5$?

---

*We are grateful to Dennis Dombroski of USAir for supplying us with these figures.

†William F. Buckley, Jr., "Malpractice Brings a Truce, of Sorts," Nationally Syndicated Column, April 4, 1988.

‡John Schneidawind, "Computer Company Capitalizes on Japanese Methods," Knight-Ridder News Service, May 13, 1988.

§LeRoy A. Franklin, "Bayes' Theorem, Binomial Probabilities, and Fair Numbers of Peremptory Challenges in Jury Trials," *College Mathematics Journal,* September 1987, pp. 291–299.

8. Suppose the SAT scores form a normal distribution with a mean of 920 and a standard deviation of 50. What is the percentage of scores that are at least 820?

9. Roll a pair of dice 180 times and record the number of times they land doubles.
   (a) What is the mean for this experiment?
   (b) What is the standard deviation for this experiment?

10. If a quarter is flipped 100 times, what is the probability that it will land heads at most 52 times?

11. If a pair of dice is tossed 180 times, what is the probability that the sum of 7 will occur at least 35 times?

12. A shelf holds 25 cookie jars, each of which contains 8 chocolate cookies and 2 vanilla cookies. If one cookie is selected from each jar, what is the probability that the number of vanilla cookies selected is between 4 and 6?

C A S E   S T U D Y   ## The Federalist Papers

Let us recall some events from the American Revolution. In the 1760s the British government adopted the policy of taxing the American colonies. The most famous and most important of these measures was the Stamp Act, which was later repealed and replaced by the Townshend Acts, both of which provided taxes to maintain the British army in America. This led to the Boston Tea Party, and the decision by King George III that the colonies should be subdued. Revolt followed. Hostilities broke out in April 1775, and on July 4 of the following year the delegates to the second Continental Congress approved the declaration that ''These United Colonies are, and of right ought to be, free and independent states.'' The war came to an official end in 1783 with the Treaty of Paris, in which Britain recognized the independence of the United States.

That sequence of events answered one of the two great questions from the revolutionary era: How should power be distributed between Britain and America? But another question remained unresolved: How should power be divided among the Americans? It is instructive to think not of *the* revolution, but of two revolutions, one external and the other internal. We just outlined the external revolution; now let us describe the internal revolution.

The Continental Congress functioned as a national government until 1781, when the Articles of Confederation, the first national constitution, were adopted. However, the Articles contained one major defect: they did not provide for a strong central government. This is understandable; American heads of state had no intention of replacing one form of despotism by another. Consequently, the central government's lack of power to regulate commerce caused grave commercial problems, and its lack of power to tax led to a financial disaster. The situation was so serious that historians have come to call the period from 1783 to 1787 not *a* critical period but *the* critical period in American history.

A constitutional convention was convened in hot, steamy Philadelphia from May to September of 1787. There the delegates hammered out a document that would prove to be the cornerstone of our government—the Constitution. The 13 colonies were much less united than the 50 states are today, yet politicians from all ideologies recognized the importance of the passage of the Constitution. The Federalist papers were published by Alexander Hamilton, John Jay, and James

Madison to persuade the citizens of New York State to vote for ratification. They have earned an important niche in political philosophy.

A problem arose afterward because The Federalist papers had been published anonymously over the signature "Publius." Of the 85 papers there was general agreement that Hamilton wrote 51, Madison wrote 14, and Jay wrote 5. Three were of joint authorship. The remaining 12 were written by either Madison or Hamilton, but it was not clear which of these distinguished statesmen wrote them.

Madison or Hamilton—who was the author of the disputed Federalist papers? Even though they held opposing political views, neither would claim the honor in deference to the greater good of convincing the New York electorate to vote for ratification of the new Constitution. The usual means for determining authorship by political content and writing style failed to solve the problem. Besides, Madison and Hamilton consciously adopted and perfected a common legal writing style. Thus the usual linguistic methods of studying sentence length and word usage shed no further light on the problem. The papers became known as the "disputed papers."

The first conclusive evidence was provided nearly two centuries later in a historic study by Mosteller and Wallace.* It was based on certain words, called *marker words,* which are exceptionally common to one author's style and not the other's. However, the use of marker words presents two big difficulties: finding them and proving that the difference in their usage is significant. The nagging

*Frederick Mosteller and David L. Wallace, *Inference and Disputed Authorship: The Federalist Papers*, Reading, MA, Addison-Wesley, 1964. This study was updated in a second edition of the book under the title *Applied Bayesian and Classical Inference: The Case of the Federalist Papers*, New York, Springer-Verlag, 1984.

Table 1 Frequency Distribution of Rate per 1000 Words in 48 Hamilton Papers for the Word "Upon"

Rate per 1000 Words		Number of			
Interval	Midpoint x_i	Papers f_i	$f_i x_i$	x_i^2	$f_i x_i^2$
0.5–1.5	1	5	5	1	5
1.5–2.5	2	10	20	4	40
2.5–3.5	3	12	36	9	108
3.5–4.5	4	10	40	16	160
4.5–5.5	5	8	40	25	200
5.5–6.5	6	2	12	36	72
6.5–7.5	7	1	7	49	49
		48	160	140	634

$$\bar{x} = \frac{f_1 x_1 + \cdots + f_7 x_7}{f_1 + \cdots + f_7} = \frac{160}{48} \approx 3.33$$

$$\sigma = \sqrt{\frac{f_1 x_1^2 + \cdots + f_7 x_7^2}{f_1 + \cdots + f_7} - \bar{x}^2} = \sqrt{\frac{634}{48} - \left(\frac{160}{48}\right)^2} \approx 1.45$$

Table 2 Frequency Distribution of Rate per 1000 Words in 50 Madison Papers for the Word "Upon"

Rate per 1000 Words		Number of			
Interval	Midpoint x_i	Papers f_i	$f_i x_i$	x_i^2	$f_i x_i^2$
0.0–0.4	0.2	41	8.2	0.04	1.64
0.4–0.8	0.6	2	1.2	0.36	0.72
0.8–1.2	1.0	4	4.0	1.00	4.00
1.2–1.6	1.4	1	1.4	1.96	1.96
1.6–2.0	1.8	2	3.6	3.24	6.48
		50	18.4	6.60	14.80

$$\bar{x} = \frac{18.4}{50} \approx 0.37$$

$$\sigma = \sqrt{\frac{14.8}{50} - \left(\frac{18.4}{50}\right)^2} \approx 0.40$$

uncertainty never disappears, as perfect markers can never be found, but sometimes a nearly conclusive case can be made.

The question of the authorship of The Federalist papers forged an important link between statistics and history. The standard deviation was the key element. We present the data gathered by Mosteller and Wallace for the marker word "upon" from 50 essays written by Madison, 48 essays written by Hamilton, and the 12 disputed papers. Tables 1 and 2 present the distribution for the essays by Hamilton and Madison, respectively, while Table 3 gives the distribution for the 12 disputed papers.

The means and standard deviations are summarized in Table 4. They show that the usage of the marker word "upon" in the disputed papers fits the usage in the Madison essays much more closely than it fits the Hamilton essays. The

Table 3 Frequency Distribution of Rate per 1000 Words in 12 Disputed Federalist Papers for the Word "Upon"

Rate per 1000 Words		Number of			
Interval	Midpoint x_i	Papers f_i	$f_i x_i$	x_i^2	$f_i x_i^2$
0.0–0.4	0.2	11	2.2	0.04	0.44
0.4–0.8	0.6	0	0.0	0.36	0.00
0.8–1.2	1.0	0	0.0	1.00	0.00
1.2–1.6	1.4	1	1.4	1.96	1.96
		12	3.6	3.36	2.40

$$\bar{x} = \frac{3.6}{12} = 0.3$$

$$\sigma = \sqrt{\frac{2.4}{12} - \left(\frac{3.6}{12}\right)^2} \approx 0.33$$

Table 4 Mean and Standard Deviation for 48
Hamilton, 50 Madison, and 12 Disputed Papers for
the Word "Upon"

	Hamilton	Madison	Disputed
Mean	3.33	0.37	0.30
σ	1.45	0.40	0.33

same conclusion was obtained for many other marker words. The inference from the Mosteller and Wallace study is clear: Madison is the author of the disputed Federalist papers.

Case Study Exercises

Problems 1 to 10 provide the data for ten marker words in The Federalist papers. The problems are adapted from Table 4.8–3 of the 1984 edition of the Mosteller and Wallace book, pp. 167–172. H stands for Hamilton, M for Madison, and D for disputed. The intervals refer to the rate per 1000 words, and the numbers in the other columns give the number of papers having the stated rate. (There were 48 papers of Hamilton, 50 of Madison, and 12 disputed.) For each marker word, determine whether the results are strong enough to assign the authorship of the disputed papers.

1. Marker Word: "also"

Interval	H	M	D
0 –0.4	28	18	0
0.4–0.8	13	11	8
0.8–1.2	5	16	3
1.2–1.6	2	1	0
1.6–2.0	0	3	1
3.2–3.6	0	1	0

2. Marker Word: "an"

Interval	H	M	D
0– 2	1	1	2
2– 4	7	20	5
4– 6	20	19	4
6– 8	15	8	1
8–10	4	2	0
10–12	0	0	0
12–14	1	0	0

3. Marker Word: "by"

Interval	H	M	D
0– 2	1	0	0
2– 4	3	0	8
4– 6	14	4	3
6– 8	12	2	0
8–10	12	11	1
10–12	6	11	0
12–14	0	12	0
14–16	0	4	0
16–18	0	4	0
18–20	0	2	0

4. Marker Word: "of"

Interval	H	M	D
32–36	0	1	0
44–48	0	1	0
48–52	2	4	0
52–56	2	10	3
56–60	13	16	7
60–64	10	8	0
64–68	8	7	1
68–72	3	3	1
72–76	6	0	0
76–80	4	0	0

5. Marker Word: "on"

Interval	H	M	D
0– 2	9	0	0
2– 4	23	2	1
4– 6	13	14	1
6– 8	2	14	2
8–10	1	9	4
10–12	0	8	4
12–14	0	2	0
14–16	0	1	0

6. Marker Word: "there"

Interval	H	M	D
0– 1	2	24	6
1– 2	8	16	3
2– 3	12	5	3
3– 4	10	4	0
4– 5	10	1	0
5– 6	4	0	0
6– 7	1	0	0
9–10	1	0	0

7. Marker Word: "this"

Interval	H	M	D
0– 2	1	2	0
2– 4	1	8	1
4– 6	8	13	3
6– 8	14	18	5
8–10	19	9	2
10–12	5	0	1

8. Marker Word: "to"

Interval	H	M	D
23–26	0	3	3
26–29	2	2	2
29–32	2	11	1
32–35	4	11	3
35–38	6	7	2
38–41	10	7	1
41–44	8	6	0
44–47	10	1	0
47–50	3	2	0
50–53	1	0	0
53–56	1	0	0
56–59	1	0	0

9. Marker Word: "both"

Interval	H	M	D
0 –0.4	23	10	3
0.4–0.8	10	12	4
0.8–1.2	11	10	2
1.2–1.6	3	6	3
1.6–2.0	1	6	0
2.0–2.4	0	1	0
2.4–2.8	0	2	0
2.8–3.2	0	1	0
3.2–3.6	0	2	0

10. Marker Word: "though"

Interval	H	M	D
0 –0.4	9	21	8
0.4–0.8	13	22	3
0.8–1.2	12	5	1
1.2–1.6	6	1	0
1.6–2.0	5	0	0
2.0–2.4	1	0	0
2.4–2.8	2	0	0
2.8–3.2	0	0	0
3.2–3.6	0	1	0

CHAPTER REVIEW

Key Terms

8–1 Probability Distributions
Frequency Distribution
Frequency

Random Variable
Relative Frequency

Probability Distribution   Uniform Distribution
Histogram   Normal Distribution

8–2   Measures of Central Tendency
Mean   Expected Value
Median

8–3   Standard Deviation
Variance   Chebyshev's Theorem
Standard Deviation

8–4   The Normal Curve
Normal Curve   Standard Normal Curve
z-Score   Normal Approximation

8–5   The Normal Approximation to Binomial Experiments
Bernoulli Trial   Binomial Experiment

Summary of Important Concepts

Formulas

Mean: $\bar{x} = \dfrac{x_1 + x_2 + \cdots + x_n}{n}$

Mean: $\bar{x} = \dfrac{f_1 x_1 + f_2 x_2 + \cdots + f_n x_n}{f_1 + f_2 + \cdots + f_n}$

Expected value: $E(X) = x_1 p_1 + x_2 p_2 + \cdots + x_n p_n$

Variance: $V(X) = \dfrac{x_1^2 + x_2^2 + \cdots + x_n^2}{n} - \mu^2$

$$V(X) = \dfrac{f_1 x_1^2 + f_2 x_2^2 + \cdots + f_n x_n^2}{f_1 + f_2 + \cdots + f_n} - \mu^2$$

$$V(X) = p_1 x_1^2 + p_2 x_2^2 + \cdots + p_n x_n^2 - \mu^2$$

Standard deviation: $\sigma = \sqrt{V(X)}$

For a normally distributed set of data:

Area within 1σ of mean is approximately .68.

Area within 2σ of mean is approximately .95.

Area within 3σ of mean is approximately .99.

Normal approximation: $z = \dfrac{x - \mu}{\sigma}$

Binomial experiments

Mean $\mu = np$

Variance $V(X) = npq$

Standard deviation $\sigma = \sqrt{npq}$

REVIEW PROBLEMS

1. Draw the histogram for the experiment of tossing four coins and recording the number of heads. Shade the portion corresponding to less than three heads.

Problems 2 to 4 refer to the probability experiment of tossing two dice and letting X be the sum of the top faces.

2. Construct the probability distribution.

3. Draw the histogram.

4. Find $Pr(3 \le X \le 7)$.

Problems 5 to 7 refer to the probability experiment of choosing 3 cards from an ordinary deck without replacement and letting X be the number of clubs selected.

5. Construct the probability distribution.

6. Draw the histogram.

7. Find $Pr(0 \le X \le 2)$.

8. Find the (a) mean and (b) median using the midpoints of the intervals as the data given in the table.

interval	0–2	2–4	4–6	6–8	8–10
frequency	3	1	3	2	1

Problems 9 and 10 refer to the frequency distribution in the table.

value	1	2	3	4	5
frequency	6	7	2	3	2

9. Find the (a) mean and (b) median.

10. Find the (a) variance and (b) standard deviation.

Problems 11 and 12 refer to the probability distribution in the table.

x	p
-2	$\frac{1}{8}$
-1	$\frac{1}{2}$
1	$\frac{1}{4}$
2	$\frac{1}{8}$

11. Find the expected value.

12. Find (a) the variance and (b) the standard deviation.

13. Ten employees in a company have a mean salary of $15,000.
    (a) If another person is hired at a salary of $15,000, what is the mean salary of the 11 employees?
    (b) If a twelfth person is hired at a salary of $50,000, what is the mean salary of the 12 employees?

14. A leap year has 4 months with 30 days, 7 months with 31 days, and February with 29 days.
    (a) Find the mean number of days in a month.
    (b) Find the standard deviation.

15. A college wishes to estimate the average cost of books per student during the year. Compute the average cost and the standard deviation from the table.

class	freshmen	sophomore	junior	senior
percent	35%	25%	21%	19%
cost of books	$100	$160	$180	$190

In Problems 16 and 17 find the area under the normal curve within the given limits.

16. (a) between $z = -2$ and $z = 2$.
    (b) to the right of $z = 2$.

17. (a) between $z = -0.73$ and $z = 1.44$.
    (b) to the right of $z = -1.05$.

In Problems 18 and 19 suppose that the SAT-Math scores form a normal distribution with $\mu = 470$ and $\sigma = 20$.

18. What percentage of scores are between 430 and 480?

19. If a student is selected at random, what is the probability that the SAT-Math score is at least 500?

20. If a pair of dice is tossed 180 times, what is the probability of obtaining a sum of 7 at least 25 times?

Problems 21 to 24 refer to a binomial experiment consisting of 18 trials, each with a probability of success equal to $\frac{2}{3}$.

21. Compute the probability of obtaining exactly 14 successes.

22. What is the mean number of successes?

23. What is the standard deviation?

24. Compute the probability of obtaining at least 14 successes.

Problems 25 to 27 refer to the probability experiment of drawing 1 card from each of 4800 decks of ordinary playing cards and recording the number of hearts drawn.

25. What is the expected value?

26. What is the standard deviation?

27. What is the probability that more than 1150 hearts will be drawn?

PROGRAMMABLE CALCULATOR EXPERIMENTS

1. Write a program that computes the standard deviation of a set of data with ten entries and their given frequencies using the frequency formula for the variance.

2. Write a program that computes the standard deviation of a set of data with ten entries and their given probability of occurrence using the formula for the variance of a probability distribution.

3. Write a program that computes the probability of observing "at least 70 heads" for the experiment of tossing n coins for $n = 100$ directly from the formula for probabilities of Bernoulli trials in Chapter 7, and then compute the same probability using the normal approximation of a Bernoulli experiment.

4. Do problem 3 for $n = 120, 150, 200,$ and 250.

Markov
Processes

9

Markov processes are a useful tool in the management of stocks and bonds.
(K. Harriger/H. Armstrong Roberts)

Chapter Overview

Many real-world situations are processes that can be viewed as a sequence of experiments. Some examples are the administering of a vaccine to a large population, the voting patterns of a particular populace over a series of elections, and the managing of a portfolio of stocks by a stockbroker. The outcomes of these experiments depend on probabilities so that probability theory can be used effectively to predict the future.

In this chapter we are interested in certain types of processes called ''Markov processes'' or ''Markov chains.'' A Markov process is a sequence of identical experiments in which the outcomes of the current experiment depend only on the outcomes of the previous experiment. The first section is an introduction to the terminology and elementary skills. Section 9–2 deals with the most common type of Markov chain, regular chains. The last section studies the long-term trend of Markov chains by looking at absorbing chains.

One of the primary purposes of Markov processes is to predict the future. Political analysts study the voting trends of certain types of people to help them predict the outcome of future elections; investment counselors study the trends of the value of stocks to improve the total value of their customers' portfolios; and advertisers monitor the effectiveness of their ads in order to achieve the best impact. Therefore, when we view these real-world phenomena as Markov processes we see that long-term trends of the probabilities are important. This is our ultimate goal in this chapter.

CASE STUDY PREVIEW

What happens when someone, for whatever reason, fails to pay you for something you did? You try various means to collect the money due. In business an unpaid bill is called a bad debt. Some bad debts are never collected. Large retail firms must account for a percentage of money due that they never will collect. The case study shows how accountants use Markov chains to give firms a structured means of predicting the amount of bad debts to account for in the future based on past records.

9–1 Basic Properties

Political scientists have learned that people with a strong religious affiliation tend to support one political party and rarely split their vote. Thus if they voted Democratic in one election, it is very likely that they will vote Democratic in the next election. In a study* of voting trends in Elmira, New York, researchers found that the Christian community's vote, as a whole, depended primarily on the previous election's results. The sequence of elections forms a Markov process. A **Markov process** is a probability experiment consisting of a sequence of identical trials or stages. The outcome of the current stage of the process depends only on the outcome of the previous stage. The outcomes at each stage of the preceding process are

*P. Lazarsfeld, B. Berelson, and W. McPhee, *Voting: A Study of Opinion Formation in a Presidential Election*, Chicago, IL, University of Chicago Press, 1954.

"Democratic" and "Republican." The probabilities of each outcome, which we study in Example 1, are the percentages of people who voted Democratic and Republican, respectively.

Definitions

Each trial of a Markov process is called a **stage** of the process; the present stage is called the **current stage** and the following one is the **next stage.** The outcomes are called **states.** The first time the experiment is performed it is called the **initial stage.** As the process goes from the current stage to the next stage, the state either changes or it stays the same. This is called the **transition** from state to state. We will see how it can be described by tree diagrams and matrices.

EXAMPLE 1

In the Elmira example researchers found that of the people who voted Democratic in a given election, 90% voted Democratic in the next election and 10% voted Republican. Of those who voted Republican in the current election, 70% voted Republican in the next election and 30% voted Democratic. The tree diagrams in Figure 1 graphically depict these numbers. There are two states for this experiment: the first state is "voted Democratic," represented by "Dem.," and the second is "voted Republican," represented by "Rep." These numbers form a matrix called the **transition matrix.** We label the matrix T and the rows and columns of T with the two states, "Dem." and "Rep."

$$
\begin{array}{c}
\text{Current Stage} \\
\overbrace{\quad\quad\quad\quad\quad}
\end{array}
$$

$$
\begin{array}{cc}
\text{state 1} & \text{state 2} \\
\text{Dem.} & \text{Rep.}
\end{array}
$$

$$
\begin{matrix}
\text{Next} \\
\text{Stage}
\end{matrix}
\begin{cases}
\text{state 1} \mid \text{Dem.} \\
\text{state 2} \mid \text{Rep.}
\end{cases}
\begin{bmatrix}
.9 & .3 \\
.1 & .7
\end{bmatrix} = T
$$

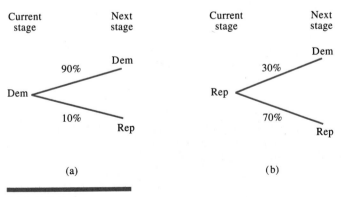

(a)   (b)

FIGURE 1

To see how to interpret the transition matrix, view the tree diagram from a different perspective.

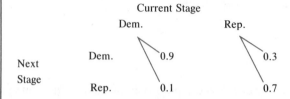

The branches of the tree diagram become the columns of the transition matrix. To construct the matrix T, delete the lines and move the column headings to the right.

$$T = \begin{bmatrix} .9 & .3 \\ .1 & .7 \end{bmatrix}$$

In the following sections we drop the column and row headings and simply refer to the matrix as T.

Each entry in the transition matrix is a conditional probability. For example, the probability that a person will vote Democratic in the next election given that the person voted Democratic in the current election is .9, and the probability that a person will vote Republican in the next election given that the person voted Democratic in the current election is .1.

Let the event "will vote Democratic in the next election" be represented by "D-next" and the event "voted Democratic in the current election" be represented by "D-current." Define "R-next" and "R-current" similarly. With this notation the event "will vote Democratic in the next election given that the person voted Democratic in the current election" is "D-next | D-current." Thus Pr(D-next | D-current) = .9. Each of the entries in T can be represented as

$$T = \begin{bmatrix} .9 & .3 \\ .1 & .7 \end{bmatrix} = \begin{bmatrix} Pr(\text{D-next} \mid \text{D-current}) & Pr(\text{D-next} \mid \text{R-current}) \\ Pr(\text{R-next} \mid \text{D-current}) & Pr(\text{R-next} \mid \text{R-current}) \end{bmatrix}$$

The columns sum to 1 because they represent all possible outcomes in a sample space. Hence a transition matrix satisfies the following properties.

DEFINITION

The defining properties for a **transition matrix** are:

1. It is a square matrix since the rows and columns represent the same states. That is, there are the same number of states at each stage, so there are the same number of rows and columns.
2. Its entries are greater than or equal to 0.
3. The entries in each column sum to 1.

EXAMPLE 2

Problem

Which of the following matrices could be transition matrices for a Markov process?

(a) $\begin{bmatrix} .2 & .3 & .6 \\ .8 & .7 & .4 \end{bmatrix}$ (b) $\begin{bmatrix} .1 & .4 \\ .9 & .7 \end{bmatrix}$ (c) $\begin{bmatrix} .2 & .9 & .5 \\ .6 & .0 & .2 \\ .2 & .1 & .3 \end{bmatrix}$

Solution (a) The matrix is not square, so it is not a transition matrix. Note that if it were a transition matrix, the current stage would have three states corresponding to the three columns, while the next stage would have two states corresponding to the two rows. This contradicts the assumption of a Markov process, which requires the states to be the same from one stage to the next.
(b) The second column does not sum to 1, so this matrix is not a transition matrix.
(c) This matrix satisfies all three properties, so it is a transition matrix.

    The next example illustrates how to construct a transition matrix with three states.

EXAMPLE 3

The manager of an investment fund has a conservative portfolio containing only blue chip stocks that are intended for future growth. Because it is very stable, the daily management decision depends primarily on the previous day's performance. Each day the manager makes one of three decisions for each stock in the portfolio: buy 5% new shares, sell 5% of the shares, or make no transition. The manager employs the following strategy:

1. If a stock issue was purchased in one day's trading, then 10% of the time it will be purchased the next day, 70% of the time it will be sold, and 20% of the time it will be left unchanged.
2. If a stock issue was sold in one day's trading, then 60% of the time it will be purchased the next day, 10% of the time it will be sold, and 30% of the time it will be left unchanged.
3. If a stock issue was left unchanged in one day's trading, then 40% of the time it will be purchased the next day, 40% of the time it will be sold, and 20% of the time it will be left unchanged.

Problem Find the transition matrix for this Markov process.

Solution We use modified tree diagrams in Figure 2 to determine the columns of the transition matrix T. We refer to the three states as the three outcomes:

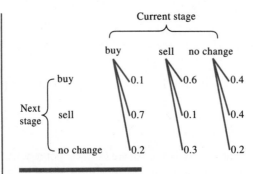

Current stage

FIGURE 2

(1) buy, (2) sell, and (3) no change. Hence the transition matrix T is

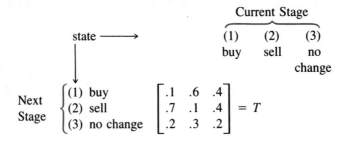

In general, the transition matrix for a process with three outcomes looks like this:

Current Stage

	state 1	state 2	state 3
state 1	p_{11}	p_{12}	p_{13}
state 2	p_{21}	p_{22}	p_{23}
state 3	p_{31}	p_{32}	p_{33}

where p_{ij} is the conditional probability

$$p_{ij} = Pr(\text{process will be in state } i \mid \text{process is in state } j)$$

Distribution Matrices

The transition matrix enables the process to go from one stage of the process to the next. Now we will see how the process gets started. In Example 1 the process starts by looking at one particular election and measuring the percentage of people who voted Democratic and Republican, respectively. It is called a **distribution.** In the 1948 presidential election 80% of Elmira's Christians voted Democratic, while 20% voted Republican. This is called the **initial distribution** and is represented by the column matrix, $V = \begin{bmatrix} .8 \\ .2 \end{bmatrix}$. Let W be the distribution of voters in the next election. Then W is the matrix product TV, where T is the transition matrix, $T = \begin{bmatrix} .9 & .3 \\ .1 & .7 \end{bmatrix}$. This means that to find the distribution of voters in the next election, compute TV. Hence

$$W = TV = \begin{bmatrix} .9 & .3 \\ .1 & .7 \end{bmatrix} \begin{bmatrix} .8 \\ .2 \end{bmatrix} = \begin{bmatrix} .78 \\ .22 \end{bmatrix}$$

To see why $W = TV$, we compute the percentage of (1) the Democratic vote, which is the (1, 1) entry of W, and (2) the Republican vote, which is the (2, 1) entry of W.

1. There are two types of people who will vote Democratic in the next election: (a) those who in the previous election voted Democratic and (b) those who in the previous election voted Republican.
   (a) Of the 80% who voted Democratic in the first election, 90% will vote Democratic in the next election, so .8 × .9 = .72 or 72% will vote Democratic.
   (b) Of the 20% who voted Republican in the first election, 30% will vote Democratic in the next election, so .2 × .3 = .06 or 6% will vote Democratic.
   Therefore in all, .72 + .06 = .78 or 78% will vote Democratic in the next election. This is the (1, 1) entry of TV.

2. How many will vote Republican in the next election?
   (a) Of the 80% who voted Democratic in the first election, 10% will vote Republican in the next election.
   (b) Of the 20% who voted Republican in the first election, 70% will vote Republican in the next election.
   Thus (.8 × .1) + (.2 × .7) = .08 + .14 will vote Republican. Therefore in all, .08 + .14 = .22 or 22% will vote Republican in the next election.

   This shows that the computations used to compute W are the same as those in the matrix product TV. Thus $TV = W$. This says that to find the next distribution from the previous one, take the product of the transition matrix and the distribution matrix.
   Let us apply this technique to the process in Example 3.

EXAMPLE 4

Problem

Given the transition matrix T in Example 3 and the initial distribution

$$V = \begin{bmatrix} .2 \\ .3 \\ .5 \end{bmatrix}$$

find the distribution of the next stage, W.

Solution Compute TV.

$$\begin{bmatrix} .1 & .6 & .4 \\ .7 & .1 & .4 \\ .2 & .3 & .2 \end{bmatrix} \begin{bmatrix} .2 \\ .3 \\ .5 \end{bmatrix} = \begin{bmatrix} .40 \\ .37 \\ .23 \end{bmatrix}$$

Thus if the manager on one day bought shares for 20% of the fund's stocks, sold shares for 30%, and left 50% unchanged, then on the next day 40% would be bought, 37% would be sold, and 23% would be left unchanged.

Further Stages

To go from one distribution V to the next stage's distribution W, we compute $W = TV$. Applying the same reasoning, the following stage's distribution, say, U, is obtained by computing $U = TW$. Substituting for W yields $U = TW = T(TV) = T^2V$, so the distribution after two stages is T^2V. Similarly, the third stage's distribution is T^3V, and, in general, the distribution of n stages after the initial stage is T^nV.

EXAMPLE 5

Problem

Suppose a Markov process has transition matrix $T = \begin{bmatrix} .4 & .2 \\ .6 & .8 \end{bmatrix}$ and initial distribution $V = \begin{bmatrix} .3 \\ .7 \end{bmatrix}$. Find the distributions for the next three stages.

Solution We need to compute TV, T^2V, and T^3V. First

$$TV = \begin{bmatrix} .4 & .2 \\ .6 & .8 \end{bmatrix} \begin{bmatrix} .3 \\ .7 \end{bmatrix} = \begin{bmatrix} .26 \\ .74 \end{bmatrix}$$

Since $T^2V = T(TV)$, the matrix T^2V is obtained by multiplying T by TV.

$$T^2V = T(TV) = \begin{bmatrix} .4 & .2 \\ .6 & .8 \end{bmatrix} \begin{bmatrix} .26 \\ .74 \end{bmatrix} = \begin{bmatrix} .252 \\ .748 \end{bmatrix}$$

Similarly,

$$T^3V = T(TTV) = \begin{bmatrix} .4 & .2 \\ .6 & .8 \end{bmatrix} \begin{bmatrix} .252 \\ .748 \end{bmatrix} = \begin{bmatrix} .2504 \\ .7496 \end{bmatrix}$$

There are two ways to compute successive distributions from the present stage. We can use powers of the transition matrix.

$$TV, T^2V, T^3V, \ldots, T^nV, \ldots$$

The other way is to use the sequence

$$TV, T(TV), T(T(TV)), \ldots$$

The sequences are the same; only the method of computation differs. The latter method is preferred in this section, but in the next section we will use the sequence T^nV to determine the long-term trends of Markov processes.

EXERCISE SET 9–1

In Problems 1 and 2 determine the transition matrix of the Markov process.

1. Researchers determine that of the people who voted Democratic in a given election, 85% voted Democratic in the next election and 15% voted Republican (only those who voted in every election were considered). Of those who voted Republican in the current election, 63% voted Republican in the next election and 37% voted Democratic.

2. The manager of the investment fund has determined that the following strategy will be employed for the 100 blue chip stock portfolio: (1) If a stock issue was purchased in one day's trading, then 20% of the time it will be purchased the next day, 50% of the time it will be sold, and 30% of the time it will be left unchanged. (2) If a stock issue was sold in one day's trading, then 60% of the time it will be purchased the next day, 15% of the time it will be sold, and 25% of the time it will be left unchanged. (3) If a stock issue was left unchanged in one day's trading, then 35% of the time it will be purchased the next day, 45% of the time it will be sold, and 20% of the time it will be left unchanged.

In Problems 3 to 6 determine whether the matrix could be a transition matrix of a Markov process.

3. $\begin{bmatrix} .9 & .2 \\ .1 & .8 \end{bmatrix}$

4. $\begin{bmatrix} .1 & 0 \\ .9 & 1 \end{bmatrix}$

5. $\begin{bmatrix} .9 & .4 & .1 \\ 0 & .3 & .6 \\ .1 & .3 & .3 \end{bmatrix}$

6. $\begin{bmatrix} .3 & .7 & .1 \\ .5 & .1 & .1 \\ .2 & .2 & 1.8 \end{bmatrix}$

In Problems 7 and 8 determine a, b, and c so that the matrix could be a transition matrix.

7. $\begin{bmatrix} .3 & .4 & .5 \\ .2 & .5 & .3 \\ a & b & c \end{bmatrix}$

8. $\begin{bmatrix} .8 & .5 & c \\ .2 & .2 & .4 \\ a & b & 0 \end{bmatrix}$

In Problems 9 and 10 the transition matrix of a Markov process is given. Determine the probability that the process

will be in state 2 in the next stage, given that its current stage is in (a) state 1, (b) state 2.

9. $\begin{bmatrix} .2 & .6 \\ .8 & .4 \end{bmatrix}$

10. $\begin{bmatrix} .1 & .3 & .6 \\ .4 & .2 & .1 \\ .5 & .5 & .3 \end{bmatrix}$

In Problems 11 to 14 determine whether the column matrix could be a probability distribution for a Markov process.

11. $\begin{bmatrix} .3 \\ .7 \end{bmatrix}$

12. $\begin{bmatrix} .4 \\ .2 \end{bmatrix}$

13. $\begin{bmatrix} .2 \\ .7 \\ .2 \end{bmatrix}$

14. $\begin{bmatrix} .4 \\ .6 \\ 0 \end{bmatrix}$

In Problems 15 to 20 let T represent the transition matrix of a Markov process. For the given probability distribution V find the distribution of the next three stages of the process.

$$T = \begin{bmatrix} .1 & .5 \\ .9 & .5 \end{bmatrix}$$

15. $V = \begin{bmatrix} .5 \\ .5 \end{bmatrix}$

16. $V = \begin{bmatrix} .6 \\ .4 \end{bmatrix}$

17. $V = \begin{bmatrix} .2 \\ .8 \end{bmatrix}$

18. $V = \begin{bmatrix} .3 \\ .7 \end{bmatrix}$

19. $V = \begin{bmatrix} .1 \\ .9 \end{bmatrix}$

20. $V = \begin{bmatrix} .4 \\ .6 \end{bmatrix}$

Problems 21 to 26 refer to the following transition matrix T of the Markov process determined by a weather bureau. The first state represents the likelihood of rain in the forecast and the second state represents the likelihood of no rain in the forecast. Each stage represents a daily forecast. That is, the present stage refers to today's weather forecast and the next stage refers to tomorrow's forecast.

$$\begin{array}{c} \\ \text{rain} \\ \text{no rain} \end{array} \quad \begin{array}{cc} \text{rain} & \text{no rain} \end{array}$$
$$T = \begin{bmatrix} .3 & .2 \\ .7 & .8 \end{bmatrix}$$

In Problems 21 to 24 find the conditional probability of the event.

21. It will rain tomorrow given that it is raining today.

22. It will rain tomorrow given that it is not raining today.

23. It will not rain tomorrow given that it is raining today.

24. It will not rain tomorrow given that it is not raining today.

25. Find the probability that it will rain tomorrow if the probability of rain today is .5.

26. Find the probability it will not rain tomorrow if the probability of rain today is .3.

In Problems 27 to 34 a Markov process is described. Find its transition matrix.

27. A manufacturer classifies outstanding debts as high priority (*H*) and low priority (*L*). In any given month 20% of the debts that are high priority will become low priority (because all or most of the debt will have been paid) and 40% of the low priority debts will remain low priority.

28. A city determines that in a given year 10% of the city population moves to the suburbs and 5% of the suburban populace moves to the city.

29. An audio equipment retailer advertises each week in either television, radio, or the newspaper. The retailer's strategy for the next week is influenced primarily by the previous week's choice. There are three possibilities: (1) If in the current week a television ad was used, then for the next week's ad 10% of the time television will be used, 30% of the time the radio will be used, and 60% of the time the newspaper will be used; (2) if in the current week a radio ad was used, then for the next week's ad 45% of the time television will be used, 5% of the time the radio will be used, and 50% of the time the newspaper will be used; (3) if in the current week a newspaper ad was used, for the next week's ad 55% of the time television will be used, 35% of the time the radio will be used, and 10% of the time the newspaper will be used.

30. An audio equipment retailer advertises each week in either television, radio, or the newspaper. The retailer is influenced primarily by the previous week's choice, and so if in the current week (1) a television ad was used, for the next week's ad 0% of the time the television will be used, 30% of the time the radio will be used, and 70% of the time the newspaper will be

used; (2) a radio ad was used, for the next week's ad 35% of the time television will be used, 15% of the time the radio will be used, and 50% of the time the newspaper will be used; (3) a newspaper ad was used, for the next week's ad 100% of the time television will be used, 0% of the time the radio will be used, and 0% of the time the newspaper will be used.

31. Three brands of beer, *A*, *B*, and *C*, share a particular market. An advertising agency determines that because of a new ad campaign, Brand *A* beer will gain 20% of Brand *B*'s customers, and 30% of Brand *C*'s customers each month over the next few months while retaining 90% of its own customers. During the same period it is expected that Brand *B* will retain 70% of its customers and Brand *C* will retain 60%. Thus each will lose the usual 10% to each other. Brand *A* will lose only 5% to each of the others.

32. In a manufacturing process the state of the assembly line is monitored hourly as either being ahead of schedule, on schedule, or behind schedule. If the line is on schedule, the probability that it will be ahead of schedule the next hour is .2, on schedule is .7, and behind schedule is .1. If the line is ahead of schedule, the probability that it will be ahead of schedule the next hour is .3, on schedule is .6, and behind schedule is .1. If the line is behind schedule, the probability that it will be ahead of schedule the next hour is .1, on schedule is .5, and behind schedule is .4.

33. Each week the manager of a supply house checks the inventory of stock. The amount ordered depends primarily on how much stock was ordered the previous week. The order is either large, medium, or small. If in the current week the order was large, then in the next week 10% of the time the order will be large, 45% of the time it will be medium, and otherwise it will be small. If in the current week the order was medium, then in the next week 10% of the time the order will be large, and 60% of the time the order will be medium. If in the current week the order was small, then in the next week 65% of the time the order will be large, and 30% of the time the order will be medium.

34. A population study of a state has determined that in any given year the percentage of the population that remains urban, rural, or suburban depends primarily on the previous year's figures. Of the urban population, 85% remains urban, 10% moves to rural areas, and 5% moves to suburbs. Of the rural population,

95% remains rural, 4% moves to urban areas, and 1% moves to suburbs. Of the suburban population, 85% remains suburban, 10% moves to rural areas, and 5% moves to urban areas.

In Problems 35 and 36 let T represent the transition matrix of a Markov process. For the given probability distribution V find the distribution of the next stage of the process.

$$T = \begin{bmatrix} .3 & .1 & .7 \\ .6 & .4 & .2 \\ .1 & .5 & .1 \end{bmatrix}$$

35.
$$V = \begin{bmatrix} .1 \\ .1 \\ .8 \end{bmatrix}$$

36.
$$V = \begin{bmatrix} .3 \\ .3 \\ .4 \end{bmatrix}$$

In Problems 37 to 40 let T represent the transition matrix of the Markov process in Problems 31 to 34, respectively. For the given probability distribution V find the distribution of the next stage of the process.

37.
$$V = \begin{bmatrix} .2 \\ .5 \\ .3 \end{bmatrix}$$

38.
$$V = \begin{bmatrix} .3 \\ .2 \\ .5 \end{bmatrix}$$

39.
$$V = \begin{bmatrix} .3 \\ .3 \\ .4 \end{bmatrix}$$

40.
$$V = \begin{bmatrix} .2 \\ .1 \\ .7 \end{bmatrix}$$

41. Show that $T^2V = TTV$ for

$$T = \begin{bmatrix} .3 & .1 & .7 \\ .6 & .4 & .2 \\ .1 & .5 & .1 \end{bmatrix} \qquad V = \begin{bmatrix} .1 \\ .1 \\ .8 \end{bmatrix}$$

Referenced Exercise Set 9-1

1. In an article* on urban ecology within a large city researchers used Markov processes to analyze community change in Chicago. They used four states based on economic status (ES) and family status (FS); for each community each status was rated as high or low. Thus the four states were HES/HFS (high ES and high FS), HES/LFS (high ES and low FS), LES/HFS, and LES/LFS. They constructed the following transition matrix, where the first stage corresponded to 1950 and the next stage to 1960.

| | | First Stage | | | |
		HES/HFS	HES/LFS	LES/HFS	LES/LFS
Next Stage	HES/HFS	.81	.06	.26	.00
	HES/LFS	.07	.70	.00	.09
	LES/HFS	.12	.00	.68	.35
	LES/LFS	.00	.24	.06	.56

(a) Find the probability that a community went from state HES/LFS to LES/LFS.

(b) Find the next distribution if the initial distribution was

$$\begin{bmatrix} 0.2 \\ 0.3 \\ 0.4 \\ 0.1 \end{bmatrix}$$

2. Consumers switch brands of a product for many reasons. Sometimes a new brand appears on the market or perhaps a different brand is on sale. One of the most important marketing tools used by marketing directors is the analysis of how and why consumers switch brands. In an article† studying purchasing behavior of various types of consumers, Blattberg and Sen classify consumers into four categories: (1) last-purchase-loyal, LPL, which stands for a person who is likely to purchase the same brand for a while but then switches to a new brand, which is purchased for a while, and then switches to still another brand; (2) low-brand-loyal, LBL, which stands for a person who remains loyal to a brand, then switches to another brand but usually returns to the favorite brand; (3) high-brand-loyal, HBL, which stands for a person who is very loyal to the favorite brand, hardly ever purchasing another brand; (4) deal-not-loyal, DNL, which stands for a person who purchases a brand that offers the best deal, such as a brand on sale. The data used are from the *Chicago Tribune's* Panel of Women in the greater Chicago Area. The purchasing patterns of 200 consumers of aluminum foil were studied over a four-year period. The data relating how the consumers changed from one state to another are given by the following transition matrix.

*Albert Hunter, "Community Change: A Stochastic Analysis of Chicago's Local Communities, 1930–1960," *American Journal of Sociology*, Vol. 79, No. 4, pp. 923–947, 1974.

†Robert Blattberg and Sbrata K. Sen, "A Bayesian Technique to Discriminate Between Stochastic Models of Brand Choice," *Management Science*, Vol. 21, pp. 682–696, 1975.

		Current Stage			
		LPL	LBL	HBL	DNL
	LPL	.61	.06	0	.07
Next	LBL	.17	.74	.15	.03
Stage	HBL	.02	.17	.85	0
	DNL	.20	.03	0	.90

(a) Find the probability that a consumer went from the HBL category to the DNL category.

(b) State why the main diagonal of the matrix has the four highest values.

3. It has been predicted that in the 1990s there will be a significant increase in the number of physicians in the United States. The delivery of primary health care is dependent not only on the number of physicians available but also on how those physicians are dispersed within a given geographic region. In a paper modeling the transitions of physicians, researchers use Markov processes to predict the percentages of physicians that will move from one type of practice to another.[‡] They consider six states and two types of geographical areas, urban and rural. Each area has three types of practices: private practice (P), hospital practice (H), and clinics (C), including health maintenance organizations and government-funded clinics. Data from Washington state in 1982 were used to generate the following transition matrix.

			Current Stage					
			urban			rural		
			P	H	C	P	H	C
	urban	P	0	.5	.4	.1	.05	0
		H	.3	0	.2	.1	.05	0
Next		C	.4	.3	0	.2	.1	.2
Stage	rural	P	.2	.05	.05	0	.4	.5
		H	0	.05	.3	.4	0	.3
		C	.1	.1	.05	.2	.4	0

(a) Find the probability that a physician went from an urban hospital practice to a rural private practice.

(b) State why the main diagonal of the matrix has all zeros.

(c) Find the next distribution if the initial distribution is

$$\begin{bmatrix} .4 \\ .4 \\ .2 \\ 0 \\ 0 \\ 0 \end{bmatrix}$$

[‡]Vandan Trivedi et al., "A Semi-Markov Model for Primary Health Care Manpower Supply Predictions," *Management Science*, Vol. 33, pp. 149–160, 1987.

9–2 Regular Markov Processes

In this section we study the long-term trend of Markov processes. The basic question is, what happens to the distribution in the nth stage of the process as n gets larger and larger? Certain transition matrices have a predictable trend but others do not. In this section we study one type that exhibits a stable long-term trend.

Long-Term Trends

In the Elmira voting study the percentage of people who vote Democratic will become stable in the long run, that is, after many elections. The transition matrix and initial probability distribution for the Markov process are

$$T = \begin{bmatrix} .9 & .3 \\ .1 & .7 \end{bmatrix} \quad \text{and} \quad V = \begin{bmatrix} .8 \\ .2 \end{bmatrix}$$

What will the distribution of voters be after four additional elections? From the last section this is done by calculating $T^n V$ for $n = 1$ to 4. (We round off the matrix entries to two decimal places.)

$$TV = \begin{bmatrix} .9 & .3 \\ .1 & .7 \end{bmatrix} \begin{bmatrix} .8 \\ .2 \end{bmatrix} = \begin{bmatrix} .78 \\ .22 \end{bmatrix}$$

$$T^2 V = \begin{bmatrix} .84 & .48 \\ .16 & .52 \end{bmatrix} \begin{bmatrix} .8 \\ .2 \end{bmatrix} = \begin{bmatrix} .768 \\ .232 \end{bmatrix}$$

$$T^3 V = \begin{bmatrix} .80 & .59 \\ .20 & .41 \end{bmatrix} \begin{bmatrix} .8 \\ .2 \end{bmatrix} = \begin{bmatrix} .758 \\ .242 \end{bmatrix}$$

$$T^4 V = \begin{bmatrix} .78 & .65 \\ .22 & .35 \end{bmatrix} \begin{bmatrix} .8 \\ .2 \end{bmatrix} = \begin{bmatrix} .754 \\ .246 \end{bmatrix}$$

The pattern seems to indicate that the percentage that will vote Democratic approaches 75%. As we take higher powers of T, the distribution approaches $\begin{bmatrix} .75 \\ .25 \end{bmatrix}$. This is what is meant by the **long-term trend.** Example 1 explores some further ideas about the long-term trend of Markov processes and determines whether the initial probability distribution affects the long-term trend.

EXAMPLE 1

Problem

Given the transition matrix T in the Elmira voting study, let $V = \begin{bmatrix} .4 \\ .6 \end{bmatrix}$. How is the long-term trend affected?

Solution We can use the same powers of T to get

$$TV = \begin{bmatrix} .9 & .3 \\ .1 & .7 \end{bmatrix} \begin{bmatrix} .4 \\ .6 \end{bmatrix} = \begin{bmatrix} .54 \\ .46 \end{bmatrix}$$

$$T^2 V = \begin{bmatrix} .84 & .48 \\ .16 & .52 \end{bmatrix} \begin{bmatrix} .4 \\ .6 \end{bmatrix} = \begin{bmatrix} .624 \\ .376 \end{bmatrix}$$

$$T^3 V = \begin{bmatrix} .80 & .59 \\ .20 & .41 \end{bmatrix} \begin{bmatrix} .4 \\ .6 \end{bmatrix} = \begin{bmatrix} .674 \\ .326 \end{bmatrix}$$

$$T^4 V = \begin{bmatrix} .78 & .65 \\ .22 & .35 \end{bmatrix} \begin{bmatrix} .4 \\ .6 \end{bmatrix} = \begin{bmatrix} .702 \\ .298 \end{bmatrix}$$

The first term is increasing and the second is decreasing, but the long-term trend is not clear yet because the distribution has not stabilized. We must use higher powers of T. After some tedious labor we see that the powers of T all become equal. For instance, when we round to two significant digits

$$T^{12} = \begin{bmatrix} .75 & .75 \\ .25 & .25 \end{bmatrix}$$

All powers greater than 12 are the same. Hence for n greater than 12

$$T^n V = \begin{bmatrix} .75 & .75 \\ .25 & .25 \end{bmatrix} \begin{bmatrix} .4 \\ .6 \end{bmatrix} = \begin{bmatrix} .75 \\ .25 \end{bmatrix}$$

Thus the long-term trend with this new initial distribution is the same as with the first one—75% vote Democratic.

Example 1 computed the long-term trend for the initial distribution. The same long-term trend will hold for any initial distribution $\begin{bmatrix} a \\ b \end{bmatrix}$ because $a + b = 1$ implies that for all n greater than or equal to 12

$$\begin{bmatrix} .75 & .75 \\ .25 & .25 \end{bmatrix} \begin{bmatrix} a \\ b \end{bmatrix} = \begin{bmatrix} .75(a + b) \\ .25(a + b) \end{bmatrix} = \begin{bmatrix} .75 \\ .25 \end{bmatrix}$$

Regular Matrices

Example 1 indicates that the long-term trend for this particular transition matrix T does not depend on the choice of the initial probability distribution. However, only certain types of matrices have this property. The entries of every transition matrix are either positive or zero. No entry can be negative because it represents a probability. The special property that we focus on now requires a matrix to have all positive entries, that is, no zero entries. But we can relax that requirement a little, as we see in the next definition.

DEFINITION

A transition matrix is **regular** if some power of the matrix contains all positive entries. A Markov process is *regular* if its transition matrix is regular.

EXAMPLE 2

Problem

Which of the following matrices are regular?

(a) $A = \begin{bmatrix} .3 & .4 \\ .7 & .6 \end{bmatrix}$ (b) $B = \begin{bmatrix} .9 & 1 \\ .1 & 0 \end{bmatrix}$ (c) $C = \begin{bmatrix} .8 & 0 \\ .2 & 1 \end{bmatrix}$

Solution (a) A has all positive entries so it is regular.
(b) B has a zero entry so it may not be regular, but we must determine whether its powers have zeros. We compute $B^2 = \begin{bmatrix} .91 & .9 \\ .09 & .1 \end{bmatrix}$. So B is regular since one of its powers has all positive entries.
(c) This looks very similar to B. Compute $C^2 = \begin{bmatrix} .64 & 0 \\ .36 & 1 \end{bmatrix}$. The answer is still

not clear. Compute higher powers of C. The same second column appears in all C^n. Hence every power of C will have a zero in the (1, 2) entry. Thus C is not regular since no power of it will have all positive entries.

The key property of regular Markov processes is that long-range predictions are possible. For instance, in Example 1 the transition matrix is regular, so the long-term trend of the process can be determined. The higher powers of T approach the matrix

$$\begin{bmatrix} .75 & .75 \\ .25 & .25 \end{bmatrix}$$

This matrix is called the *stable matrix* for T. No matter what the initial distribution is, after a sufficient number of stages in the process the distribution approaches the limiting distribution $\begin{bmatrix} .75 \\ .25 \end{bmatrix}$. This is called the *stable distribution*. These are important ideas, so we record them here.

DEFINITION

If T is a regular matrix, then the powers of T will approach a limiting matrix, called the **stable matrix** of T. This means that T^n gets arbitrarily close to the stable matrix as n gets large. Once the process has reached the stage of the stable matrix, any initial distribution will eventually lead to a distribution that remains the same in all succeeding stages. This limiting distribution is called the **stable distribution.**

Computing Stable Matrices

Finding the stable matrix by computing $T^n V$ for large enough n is often time-consuming. A faster method exists. Suppose V is the initial distribution and after n stages the stable distribution W is reached. Thus $T^n V = W$. The next stage's distribution is $TT^n V = TW$. But the definition of the stable distribution implies that further stages' distributions are equal to W. Thus $TW = W$. This means that in order to compute the stable distribution, we solve the equation $TX = X$. For example, start with the matrices $T = \begin{bmatrix} .9 & .3 \\ .1 & .7 \end{bmatrix}$ and $X = \begin{bmatrix} x \\ y \end{bmatrix}$ and solve $TX = X$.

$$TX = \begin{bmatrix} .9 & .3 \\ .1 & .7 \end{bmatrix} \begin{bmatrix} x \\ y \end{bmatrix} = \begin{bmatrix} .9x + .3y \\ .1x + .7y \end{bmatrix} = X = \begin{bmatrix} x \\ y \end{bmatrix}$$

Thus we must solve the system of equations

$$.9x + .3y = x$$

$$.1x + .7y = y$$

Subtract x from each side of the first equation and subtract y from each side of the second equation. This yields

$$-.1x + .3y = 0$$
$$.1x - .3y = 0$$

The equations are multiples of each other, so there is an infinite number of solutions. Solving for x in terms of y yields $x = 3y$. Since x and y represent probabilities, they must add up to 1 because they are the only outcomes. Hence

$$x + y = 1$$

Combining the last equation with $x = 3y$ yields $3y + y = 1$, so $4y = 1$. Thus $y = \frac{1}{4}$ and $x = \frac{3}{4}$. Hence the stable distribution is

$$W = \begin{bmatrix} \frac{3}{4} \\ \frac{1}{4} \end{bmatrix}$$

Here is the procedure for finding the stable distribution of a regular matrix.

Procedure for Finding the Stable Distribution of a Regular Matrix

Let T be a regular $n \times n$ matrix and let X be a variable column matrix with n rows. The system of equations $TX = X$ has n equations in n unknowns. The sum of the entries of X must equal 1 because X is a probability distribution. Since X has n rows the "sum of the entries of $X = 1$" is an equation with n variables. The stable distribution W is the solution of the system of $n + 1$ equations in n unknowns.

$$TX = X$$
$$\text{sum of the entries of } X = 1$$

The next two examples illustrate this procedure.

EXAMPLE 3

A retail store has 55% of the market versus 45% for the lone competitor. An advertisement firm predicts that an ad campaign will draw 20% of the competitor's customers to the store, while the store will lose only 15% of their own customers to the competitor.

Problem

What will be the long-term trend of the ad campaign?

Solution Refer to the store as A and the competitor as C. The results of the ad campaign can be represented as a Markov process whose transition matrix T is

$$T = \begin{matrix} \\ A \\ C \end{matrix} \begin{matrix} A & \quad C \\ \begin{bmatrix} .85 & .2 \\ .15 & .8 \end{bmatrix} \end{matrix}$$

Notice that T is a regular matrix. Finding the long-term trend is the same as finding the stable distribution. Solve $TX = X$ and $x + y = 1$. First write $TX = X$ as a system of equations.

$$TX = \begin{bmatrix} .85 & .2 \\ .15 & .8 \end{bmatrix} \begin{bmatrix} x \\ y \end{bmatrix} = \begin{bmatrix} x \\ y \end{bmatrix}$$

$.85x + .2y = x$ so $-.15x + .2y = 0$

$.15x + .8y = y$ so $.15x - .2y = 0$

Adding these equations eliminates one of them since they are multiples of each other. Combining the remaining equation with $x + y = 1$ yields

$.15x - .2y = 0$

$x + \quad y = 1$

Using elimination, we get $.35x = .2$. Hence $x = .2/.35 = 20/35 = 4/7$. Since $x + y = 1$, we get $y = 3/7$. The long-term trend is $\begin{bmatrix} 4/7 \\ 3/7 \end{bmatrix}$. Since $4/7 \approx 57\%$, the ad campaign would increase the store's advantage by 2 percentage points, from 55% to 57%.

To illustrate further the procedure, consider the Markov process used by the investment manager whose strategy was described by the transition matrix T.

Current Stage

		buy	sell	no change	
Next Stage	buy	.1	.6	.4	
	sell	.7	.1	.4	$= T$
	no change	.2	.3	.2	

EXAMPLE 4

Problem

Find the long-term trend of the investment strategy.

Solution We introduce the matrix X, which has three entries since T has three columns, and then compute $TX = X$.

$$TX = \begin{bmatrix} .1 & .6 & .4 \\ .7 & .1 & .4 \\ .2 & .3 & .2 \end{bmatrix} \begin{bmatrix} x \\ y \\ z \end{bmatrix} = \begin{bmatrix} .1x + .6y + .4z \\ .7x + .1y + .4z \\ .2x + .3y + .2z \end{bmatrix} = \begin{bmatrix} x \\ y \\ z \end{bmatrix}$$

Coupling this with "the sum of the entries of $X = 1$" gives us the system of equations

$.1x + :6y + .4z = x$

$.7x + .1y + .4z = y$

$.2x + .3y + .2z = z$

$x + \quad y + \quad z = 1$

Rewriting the equations after gathering like terms and expressing the system in matrix form as in Chapter 2 yields

$$\begin{array}{rl}
-.9x + .6y + .4z = 0 \\
.7x - .9y + .4z = 0 \\
.2x + .3y - .8z = 0 \\
x + y + z = 1
\end{array}
\qquad
\begin{bmatrix}
1 & 1 & 1 & 1 \\
-.9 & .6 & .4 & 0 \\
.7 & -.9 & .4 & 0 \\
.2 & .3 & -.8 & 0
\end{bmatrix}$$

Using elementary row operations to reduce the matrix yields

$$\begin{bmatrix}
1 & 0 & 0 & 60/163 \\
0 & 1 & 0 & 64/163 \\
0 & 0 & 1 & 39/163 \\
0 & 0 & 0 & 0
\end{bmatrix}$$

If the fractions are converted to decimals and rounded to the nearest percent, we have $x = 37\%$, $y = 39\%$, and $z = 24\%$. This means that over the long haul the manager will buy 37% of the time, sell 39% of the time, and leave the stocks unchanged 24% of the time.

In Chapters 1 and 2 we said that the skill of solving systems of equations is an important tool that is used often in many areas of mathematics. Here we see that it is an indispensable tool in working with Markov processes.

EXERCISE SET 9–2

In Problems 1 to 4 compute $T^n V$ for $n = 1, 2,$ and 3.

1. $T = \begin{bmatrix} .9 & .3 \\ .1 & .7 \end{bmatrix}, \quad V = \begin{bmatrix} .6 \\ .4 \end{bmatrix}$

2. $T = \begin{bmatrix} .9 & .3 \\ .1 & .7 \end{bmatrix}, \quad V = \begin{bmatrix} .1 \\ .9 \end{bmatrix}$

3. $T = \begin{bmatrix} .5 & .2 \\ .5 & .8 \end{bmatrix}, \quad V = \begin{bmatrix} .5 \\ .5 \end{bmatrix}$

4. $T = \begin{bmatrix} .2 & .7 \\ .8 & .3 \end{bmatrix}, \quad V = \begin{bmatrix} .3 \\ .7 \end{bmatrix}$

In Problems 5 to 8 find the long-term trend of the Markov process whose transition matrix is T and initial distribution is V by computing $T^n V$ for $n = 1$ to 6.

5. $T = \begin{bmatrix} .9 & .3 \\ .1 & .7 \end{bmatrix}, \quad V = \begin{bmatrix} .6 \\ .4 \end{bmatrix}$

6. $T = \begin{bmatrix} .9 & .3 \\ .1 & .7 \end{bmatrix}, \quad V = \begin{bmatrix} .1 \\ .9 \end{bmatrix}$

7. $T = \begin{bmatrix} .5 & .2 \\ .5 & .8 \end{bmatrix}, \quad V = \begin{bmatrix} .5 \\ .5 \end{bmatrix}$

8. $T = \begin{bmatrix} .2 & .7 \\ .8 & .3 \end{bmatrix}, \quad V = \begin{bmatrix} .3 \\ .7 \end{bmatrix}$

In Problems 9 to 12 determine whether the matrix is regular.

9. $T = \begin{bmatrix} 0 & .2 & .1 \\ 0 & .8 & .1 \\ 1 & 0 & .8 \end{bmatrix}$

10. $T = \begin{bmatrix} 1 & .2 & .7 \\ 0 & .8 & .2 \\ 0 & 0 & .1 \end{bmatrix}$

11. $T = \begin{bmatrix} .2 & .1 & 0 \\ .8 & .1 & 0 \\ 0 & .8 & 1 \end{bmatrix}$

12. $T = \begin{bmatrix} .1 & .7 & 0 \\ .8 & .2 & 1 \\ .1 & .1 & 0 \end{bmatrix}$

In Problems 13 to 22 find the stable matrix for the regular transition matrix T.

13. $T = \begin{bmatrix} .4 & .5 \\ .6 & .5 \end{bmatrix}$

14. $T = \begin{bmatrix} .7 & .2 \\ .3 & .8 \end{bmatrix}$

15. $T = \begin{bmatrix} .9 & .3 \\ .1 & .7 \end{bmatrix}$

16. $T = \begin{bmatrix} .8 & .3 \\ .2 & .7 \end{bmatrix}$

17. $T = \begin{bmatrix} .5 & .2 \\ .5 & .8 \end{bmatrix}$

18. $T = \begin{bmatrix} .2 & .7 \\ .8 & .3 \end{bmatrix}$

19. $T = \begin{bmatrix} 0 & .2 & .1 \\ 0 & .7 & .1 \\ 1 & .1 & .8 \end{bmatrix}$

20. $T = \begin{bmatrix} .4 & .2 & .7 \\ .3 & .8 & .2 \\ .3 & 0 & .1 \end{bmatrix}$

21. $T = \begin{bmatrix} .2 & .1 & 1 \\ .8 & .1 & 0 \\ 0 & .8 & 0 \end{bmatrix}$

22. $T = \begin{bmatrix} .1 & .7 & .3 \\ .8 & .2 & .2 \\ .1 & .1 & .5 \end{bmatrix}$

In Problems 23 to 40 find the long-term trend of the Markov process.

23. A retail store with one main competitor has 65% of the market as compared to 35% for the competitor. The store is considering running a weekly advertisement campaign claiming that each week 25% of the competitor's customers will switch allegiance to the store. It is estimated that the ads will drive 10% of the store's customers to the competitor.

24. Researchers found that of the people who voted Democratic in a given election, 85% voted Democratic in the next election and 15% voted Republican. Of those who voted Republican in the current election, 63% voted Republican in the next election and 37% voted Democratic.

25. A weather forecaster determines that if it is raining on one day, then there is a 30% chance it will rain the next day, and if it is not raining on one day, there is a 75% chance it will not rain the next day.

26. A weather forecaster determines that if it is raining on one day, then for the following day there is a 30% chance it will rain, a 5% chance it will snow, and a 65% chance of no precipitation. If it is snowing on one day, then for the following day there is a 10% chance it will rain, a 15% chance it will snow, and a 75% chance of no precipitation. If there is no precipitation on one day, then for the following day there is a 10% chance it will rain, a 5% chance it will snow, and an 85% chance of no precipitation.

27. The manager of an investment fund has determined that the following strategy will be employed for the 100 blue chip stock portfolio: (1) If a stock issue was purchased in one day's trading, then 20% of the time it will be purchased the next day, 50% of the time it will be sold, and 30% of the time it will be left unchanged. (2) If a stock issue was sold in one day's trading, then 60% of the time it will be purchased the next day, 15% of the time it will be sold, and 25%

of the time it will be left unchanged. (3) If a stock issue was left unchanged in one day's trading, then 35% of the time it will be purchased the next day, 45% of the time it will be sold, and 20% of the time it will be left unchanged.

28. Each week a manager reports on the work of her staff. Her ratings are poor, average, and excellent. If an individual was rated poor the previous week, the probability that in the next week the individual will be rated poor is .2, average is .7, and excellent is .1. If an individual was rated average the previous week, the probability that in the next week the individual will be rated poor is .3, average is .4, and excellent is .3. If an individual was rated excellent the previous week, the probability that in the next week the individual will be rated poor is .1, average is .7, and excellent is .2. Determine the long-term trend of the percentage of people rated poor, average, and excellent.

29. An audio equipment retailer advertises each week in either television, radio, or the newspaper. The retailer is influenced primarily by the previous week's choice, and so if in the current week (1) a television ad was used, for the next week's ad 10% of the time television will be used, 30% of the time the radio will be used, and 60% of the time the newspaper will be used; (2) a radio ad was used, for the next week's ad 45% of the time television will be used, 5% of the time the radio will be used, and 50% of the time the newspaper will be used; (3) a newspaper ad was used, for the next week's ad 55% of the time television will be used, 35% of the time the radio will be used, and 10% of the time the newspaper will be used.

30. An audio equipment retailer advertises each week in either television, radio, or the newspaper. The retailer is influenced primarily by the previous week's choice, and so if in the current week (1) a television ad was used, for the next week's ad 0% of the time television will be used, 30% of the time the radio will be used, and 70% of the time the newspaper will be used; (2) a radio ad was used, for the next week's ad 35% of the time television will be used, 15% of the time the radio will be used, and 50% of the time the newspaper will be used; (3) a newspaper ad was used, for the next week's ad 100% of the time television will be used, 0% of the time the radio will be used, and 0% of the time the newspaper will be used.

31. A manufacturer classifies outstanding debts as high priority (H) and low priority (L). In any given month 20% of the debts that are high priority will become low priority (because all or most of the debt will have been paid) and 40% of the low priority debts will remain low priority.

32. A city determines that in a given year 10% of the city population moves to the suburbs and 5% of the suburban populace moves to the city.

33. Three brands of beer, A, B, and C, share a particular market. An advertising agency determines that because of a new ad campaign, Brand A beer will gain 20% of Brand B's customers and 30% of Brand C's customers each month over the next few months while retaining 90% of its own customers. During the same period it is expected that Brand B will retain 70% of its customers and Brand C will retain 60%. Thus each will lose the usual 10% to each other. Brand A will lose only 5% to each of the others.

34. In a manufacturing process the state of the assembly line is monitored hourly as either being ahead of schedule, on schedule, or behind schedule. If the line is on schedule, the probability that it will be ahead of schedule the next hour is .2, on schedule is .7, and behind schedule is .1. If the line is ahead of schedule, the probability that it will be ahead of schedule the next hour is .3, on schedule is .6, and behind schedule is .1. If the line is behind schedule, the probability that it will be ahead of schedule the next hour is .1, on schedule is .5, and behind schedule is .4.

35. Each week the manager of a supply house checks the inventory of stock. The amount ordered depends primarily on how much stock was ordered the previous week, and the order is either large, medium, or small. If in the current week the order was large, then in the next week 10% of the time the order will be large, 45% of the time it will be medium, and 45% of the time it will be small. If in the current week the order was medium, then in the next week 10% of the time the order will be large, 60% of the time the order will be medium, and 30% of the time the order will be small. If in the current week the order was small, then in the next week 65% of the time the order will be large, 30% of the time the order will be medium, and 5% of the time the order will be small.

36. A population study of a small midwestern state has found that in any given year the percentage of the population that remains urban, rural, or suburban depends primarily on the previous year's figures. Of the urban population, 85% remains urban, 10% moves to rural areas, and 5% moves to suburbs. Of the rural population, 95% remains rural, 4% moves to urban areas, and 1% moves to suburbs. Of the suburban population, 85% remains suburban, 10% moves to rural areas, and 5% moves to urban areas.

37. A retail store with one main competitor has 65% of the market to 35% for the competitor. The retail store is considering running a weekly advertisement campaign claiming that each week 10% of the competitor's customers will switch allegiance to the store. It estimates that the ads will drive 5% of its own customers to the competitor.

38. Work the previous problem if the ad campaign will cause 20% of the competitor's customers to switch allegiance to the store and the ads will drive 10% of its own customers to the competitor.

39. A manufacturer classifies outstanding debts as high priority (H), medium priority (M), and low priority (L). In any given month (1) of the debts that are high priority, 40% will remain high priority, 40% will become medium priority, and 20% will become low priority; (2) of the debts that are medium priority, 20% will become high priority, 60% will remain medium priority, and 20% will become low priority; (3) of the debts that are low priority, 0% will become high priority, 30% will become medium priority, and 70% will remain low priority.

40. An internal accountant classifies a firm's adherence to auditing procedures as good (G), fair (F), and poor (P). In any given month (1) if the previous month's adherence was good, then 50% of the time the present month's adherence will be good, 30% of the time it will be fair, and 20% of the time it will be poor; (2) if the previous month's adherence was fair, then 30% of the time the present month's adherence will be good, 40% of the time it will be fair, and 30% of the time it will be poor; (3) if the previous month's adherence was poor, then 20% of the time the present month's adherence will be good, 40% of the time it will be fair, and 40% of the time it will be poor.

41. Show that if a transition matrix has all positive entries then every power of the matrix has all positive entries.

42. Suppose a transition matrix for a Markov process is $T = \begin{bmatrix} a & b \\ c & d \end{bmatrix}$. Find a relationship between a and c

and between *b* and *d*. Find the stable matrix for *T*. When will this matrix be regular?

Referenced Exercise Set 9–2

In Problems 1 to 3 a Markov process is described. Find the long-term trend.

1. Most large department stores encourage customers to purchase products with the store's credit card. An important accounting consideration is to estimate the average number of months it takes for certain types of classifications of credit accounts to be paid up. In an article studying how to compute profit from credit card purchases, Wort and Zumwalt* classify credit card accounts in three categories according to the average number of months an account is carried before it is paid up (1) less than one month, (2) during the second month, and (3) more than 2 months. In one part of the article data from the DAK corporation were studied. The findings were†: if an account was less than one month old, there was a 70% chance it would be paid up within the month, a 27% chance it would be paid up within the second month, and a 3% chance it would be paid up after the second month; if an account was between one and two months old, there was a 75% chance it would be paid up within the first month, a 23% chance it would be paid up within the second month, and a 2% chance it would be paid up after the second month; if an account was more than two months old, there was a 90% chance it would be paid up within the first month, a 9% chance it would be paid up within the second month, and a 1% chance it would be paid up after the second month. Assume that all accounts are eventually paid in full. (The Case Study deals with the situation in which an account is never paid up; it becomes a "bad debt.")

2. Most businesses use various means to forecast the market share of their product, where market share is their product's sales as a percentage of the total market sales. In an article studying the likelihood of households continuing to use the same type of heating unit they used in the recent past or switching to another type, Ezzati uses Markov processes to forecast the market share for three types of home heating units: oil burner, gas burner, and electric heat.‡ The data were gathered from the Federal Energy Office. Suppose a household decided to purchase a new heating unit. The data show that: (1) if the household had an oil burner, there is an 85% chance it would purchase an oil burner, a 10% chance it would purchase a gas burner, and a 5% chance it would purchase an electric heater; (2) if the household had a gas burner, there is a 6% chance it would purchase an oil burner, an 88% chance it would purchase a gas burner, and a 6% chance it would purchase an electric heater; (3) if the household had an electric heater, there is a 1% chance it would purchase an oil burner, a 6% chance it would purchase a gas burner, and a 93% chance it would purchase an electric heater.

3. What is a stable culture? How likely is it for a tribe to remain in a certain social and demographic nature? These are important questions raised by anthropologists that apply not only to tribes but to our own culture as well. In an article applying Markov processes to these questions Hoffman studies data from the Galla tribes of Ethiopia.§ In one part of the study a mature male tribe member is classified in one of three states based on his age when his first son is born. The states are the age groups, youth (ages 13–19), adult (ages 20–29), senior (ages 30 or more). The transition from one stage of the process to the next constitutes the probability that a son enters a state (by having his own first son when he is in a particular age group) given that his father is classified in a particular state. The data show that: (1) given that a male's father is classified in the youth state, the probability is .16 that he will be classified in the youth state, .38 that he will be classified

---

*Donald H. Wort and J. Kenton Zumwalt, "Determining the Impact of Alternate Trade Discount Policies," *Decision Sciences*, Vol. 16, pp. 48–56, 1985.

†The "total balance method" of accounting is used here, and it assumes that all dollars in the account balance are put in the age category of the oldest dollar. This means that a current purchase charge by a customer with a two-month past due balance would be classified in category 2 rather than category 1. This accounts for the fact that some dollars can move from category 2 to category 1 or stay in category 2.

‡Ali Ezzati, "Forecasting Market Shares of Alternative Home-heating Units by Markov Processes Using Transition Probabilities Estimated from Aggregate Time Series Data," *Management Science*, Vol. 21, pp. 462–473, 1974.

§Hans Hoffman, "Markov Chains in Ethiopia," in *Explorations in Mathematical Anthropology*, Paul Lay (ed.), Cambridge, MA, MIT Press, pp. 181–190, 1971.

in the adult state, and .46 that he will be classified in the senior state (for example, this means that if a male's father had his first son at age 18, the probability that the son will have his first son at age 30 or more is .46.); (2) given that a male's father is classified in the adult state, the probability is .37 that he will be classified in the youth state, .40 that he will be classified in the adult state, and .23 that he will be classified in the senior state; (3) given that a male's father is classified in the senior state, the probability is .20 that he will be classified in the youth state, .60 that he will be classified in the adult state, and .20 that he will be classified in the senior state.

Cumulative Exercise Set 9–2

1. Determine the transition matrix of the Markov process. A radio dispatcher for a taxi service determined that the following strategy will be employed for dispersing taxis within the three zones of a city: (1) If a taxi is sent to zone 1 in one trip, then 50% of the time it will be sent to zone 1 on the next trip, 40% of the time it will be sent to zone 2, and 10% of the time it will be sent to zone 3. (2) If a taxi is sent to zone 2 in one trip, then 30% of the time it will be sent to zone 1 on the next trip, 45% of the time it will be sent to zone 2, and 25% of the time it will be sent to zone 3. (3) If a taxi is sent to zone 3 in one trip, then 5% of the time it will be sent to zone 1 on the next trip, 35% of the time it will be sent to zone 2, and 60% of the time it will be sent to zone 3.

2. Determine whether each matrix could be a transition matrix of a Markov process.
   (a) $\begin{bmatrix} 0.9 & 0.2 \\ 0.2 & 0.8 \end{bmatrix}$ (b) $\begin{bmatrix} 0.1 & 1 \\ 0.9 & 0 \end{bmatrix}$

3. Refer to the following transition matrix T of the Markov process determined by a weather bureau. The first state represents the likelihood of heavy rain in the forecast, the second state represents the likelihood of light rain in the forecast, and the third state represents the likelihood of no rain in the forecast. Each stage represents a daily forecast. That is, the present stage refers to today's weather forecast and the next stage refers to tomorrow's forecast.

		heavy rain	light rain	no rain
heavy rain		.3	.1	.1
light rain	$T =$	.2	.2	.1
no rain		.5	.7	.8

Find the conditional probability of each of the following events.
   (a) It will rain heavily tomorrow given that it is not raining today.
   (b) It will rain lightly tomorrow given that it is raining heavily today.
   (c) It will not rain tomorrow given that it is raining heavily today.

4. Refer to the transition matrix in Problem 3. Find the probability it will not rain tomorrow if the probability that it will rain heavily today is .1 and the probability it will not rain today is .6.

5. Find the long-term trend of the Markov process whose transition matrix is T and initial distribution is V by computing $T^n V$ for $n = 1$ to 6.

$$T = \begin{bmatrix} 0.8 & 0.3 \\ 0.2 & 0.7 \end{bmatrix} \qquad V = \begin{bmatrix} 0.6 \\ 0.4 \end{bmatrix}$$

6. Find the stable matrix for the regular transition matrix T.

$$T = \begin{bmatrix} 0 & 0.2 & 0.2 \\ 1 & 0.7 & 0.2 \\ 0 & 0.1 & 0.6 \end{bmatrix}$$

In Problems 7 to 9 find the long-term trend of the Markov process.

7. A retail store with one main competitor has 45% of the market to 55% for the competitor. The retail store is considering running a weekly advertisement campaign claiming that each week 15% of the competitor's customers will switch allegiance to the store. It estimates that the ads will drive 10% of its own customers to the competitor.

8. A manufacturer classifies outstanding debts as high priority (H), medium priority (M), and low priority (L). In any given month (1) of the debts that are high priority, 30% will remain high priority, 50% will become medium priority, and 20% will become low priority; (2) of the debts that are medium priority, 20% will become high priority, 50% will remain medium priority, and 30% will become low priority; (3) of the debts that are low priority, 0% will become high priority, 10% will become medium priority, and 90% will remain low priority.

9. An environmental official classifies a city's air quality as good (G), fair (F), and poor (P). In any given day (1) if the previous day's quality was good, then 50%

of the time the present day's quality will be good, 30% of the time it will be fair, and 20% of the time it will be poor; (2) if the previous day's quality was fair, then 30% of the time the present day's quality will be good, 40% of the time it will be fair, and 30% of the time it will be poor; (3) if the previous day's quality was poor, then 20% of the time the present day's quality will be good, 40% of the time it will be fair, and 40% of the time it will be poor.

9–3 Absorbing Markov Processes

Regular transition matrices are important because they exhibit a long-term trend. In this section we study another type of transition matrix that enables the process to stabilize in the long run.

Absorbing Matrices

Suppose a transition matrix has a column with 1 on the main diagonal and 0's elsewhere. When a higher power is computed, that column will stay the same. To see why, let column i of transition matrix T have a 1 in the (i, i) entry and 0's elsewhere. When computing the (i, i) entry of T^2, the only nonzero contribution will be when the two 1's are multiplied together, producing 1. This means that the (i, i) entry is 1. Therefore the other values in column i must be 0 because T^2 is a transition matrix. Let us look at a specific matrix.

Consider the transition matrix T whose states are labeled 1 to 4. Remember that the columns represent the current stage and the rows represent the next stage of the experiment.

$$
\begin{array}{c}
\text{Current Stage} \\
\begin{array}{cc}
\begin{array}{c} \\ \text{state} \\ \begin{array}{c} \text{Next} \\ \text{Stage} \end{array} \left\{ \begin{array}{c} 1 \\ 2 \\ 3 \\ 4 \end{array} \right. \end{array}
&
\begin{array}{c}
\begin{array}{cccc} 1 & 2 & 3 & 4 \end{array} \\
\begin{bmatrix} 1 & .4 & .3 & 0 \\ 0 & .3 & .2 & 0 \\ 0 & .2 & 0 & 0 \\ 0 & .1 & .5 & 1 \end{bmatrix}
\end{array}
\end{array}
\end{array}
$$

It is not regular because the 1's in the 1-1 and 4-4 entries will never change when we take higher powers of T. If a column has an entry that is 1, then the other entries in that column must be 0's. If the 1 is on the main diagonal, as both of the 1's are in T above, then once the process enters that state the process never leaves it. This is because higher powers of T have 1's on the same main diagonals, as described previously.

In the previous matrix T, once state 1 or state 4 is entered the process will always stay in that state. For this reason these states are called **absorbing states.** Notice that state 2 is not absorbing because the process can go from state 2 to another state; in fact, it can go to states 1, 3, or 4 (with corresponding probabilities .4, .2, and .1). This is called a **nonabsorbing state.** State 3 is nonabsorbing because the process can go from state 3 to states 1, 2, or 4. Thus state n is an absorbing state if the (n, n) entry is 1; otherwise it is nonabsorbing.

There is a second property that is important in this section. In the preceding matrix T the process can go from any nonabsorbing state into some absorbing state. For example, state 2, a nonabsorbing state, can go to the absorbing state, state 1, because the (1, 2) entry is not zero. In fact, state 2 can go to state 1 with probability .4 since the (1, 2) entry is .4. State 2 can also go to state 4 with probability .1 because the (4, 2) entry is .1. Similarly, the nonabsorbing state 3 can go to state 1 (with probability .3) or state 4 (with probability .5).

Thus the second property that is important in this section is the ability of each nonabsorbing state to go to some absorbing state. We can now give a formal definition of this type of matrix.

DEFINITION

An **absorbing matrix** is a transition matrix in which

1. There is at least one absorbing state, and
2. It is possible to go from any nonabsorbing state to an absorbing state after a finite number of stages.

An **absorbing Markov process** is one whose transition matrix is absorbing.

It may take more than one step in the process for a nonabsorbing state to reach an absorbing one. Example 1 will help clarify this definition.

EXAMPLE 1

Problem

Which of the matrices are absorbing?

(a)
$$A = \begin{bmatrix} 0 & .5 & .3 \\ 0 & .2 & .1 \\ 1 & .3 & .6 \end{bmatrix}$$

(b)
$$B = \begin{bmatrix} 1 & 0 & .1 & .2 \\ 0 & 0 & .3 & .3 \\ 0 & 0 & .1 & 0 \\ 0 & 1 & .5 & .5 \end{bmatrix}$$

(c)
$$C = \begin{bmatrix} .8 & 0 & .7 & 0 \\ 0 & 1 & 0 & 0 \\ .2 & 0 & .3 & 0 \\ 0 & 0 & 0 & 1 \end{bmatrix}$$

Solution (a) A does not have an absorbing state, so it is not an absorbing matrix. The only 1 in the matrix is in the column corresponding to state 1, but since it is in the third row, the process will go from state 1 to state 3. So state 1 is not an absorbing state. The 1 must be on the main diagonal for the state to be absorbing.

(b) B has one absorbing state, state 1, so the first property of the definition is satisfied. To be an absorbing process, it must be possible to go from each non-absorbing state into state 1. State 3 can go to state 1 directly (with probability .1). State 4 can also go to state 1 directly (with probability .2). To see that state 2 can go to state 1 is a bit more complicated. It cannot go directly because the (1, 2) entry is zero. In fact, if the process is in state 2, it must go to state 4 in the next stage because the (4, 2) entry is 1. But then state 4 can go to state 1 in the following stage. Thus state 2 can go first to state 4 and then to state 1. So state 2 can go to state 1 after two stages. This means that each nonabsorbing state of matrix B can go to an absorbing state. Therefore B is an absorbing matrix.

(c) C has two absorbing states, states 2 and 4. For C to be an absorbing matrix, states 1 and 3 must eventually lead to states 2 or 4. But if the process is in state 1 or state 3 it can go to neither state 2 nor state 4. Hence C is not absorbing.

Long-Term Trends — Stable Matrices

Not all transition matrices have long-term trends (see Problem 34). The importance of absorbing matrices is that they have a limiting matrix. As in the last section, the limiting matrix is called the **stable matrix.** In order to compute the stable matrix, we arrange the transition matrix so that the absorbing states come before the nonabsorbing states. The rows and columns might have to be rearranged, but this does not affect the meaning of the transition matrix. For example, consider the matrix T, introduced in the beginning of this section.

$$T = \begin{bmatrix} 1 & .4 & .3 & 0 \\ 0 & .3 & .2 & 0 \\ 0 & .2 & 0 & 0 \\ 0 & .1 & .5 & 1 \end{bmatrix}$$

To put the absorbing states first, keep column 1 in its place but move column 4 to column 2. Equivalently, state 4 becomes state 2. We also keep column 3 where it is but move column 2 to column 4. (Columns 2 and 3 could just as well have been moved to columns 3 and 4, respectively.) The rows change accordingly. The new matrix, which represents the same Markov process but with the names of the states changed, is

$$T = \begin{bmatrix} 1 & 0 & .3 & .4 \\ 0 & 1 & .5 & .1 \\ 0 & 0 & 0 & .2 \\ 0 & 0 & .2 & .3 \end{bmatrix}$$

In this way every absorbing matrix can be arranged so that the absorbing states are to the left of the nonabsorbing states. When in this form the matrix can be divided (sometimes called **partitioned**) into four submatrices, called I, Q, O, and R, in the following manner:

$$\begin{matrix} \text{Absorbing} & \text{Nonabsorbing} \\ \text{states} & \text{states} \end{matrix}$$

$$\left[\begin{array}{c|c} I & Q \\ \hline O & R \end{array} \right]$$

where I is an identity matrix and O is the zero matrix with the same number of rows as R, and columns as I. For the previous matrix the partition is

$$\left[\begin{array}{cc|cc} 1 & 0 & .3 & .4 \\ 0 & 1 & .5 & .1 \\ \hline 0 & 0 & 0 & .2 \\ 0 & 0 & .2 & .3 \end{array} \right]$$

where

$$I = \begin{bmatrix} 1 & 0 \\ 0 & 1 \end{bmatrix} \quad Q = \begin{bmatrix} .3 & .4 \\ .5 & .1 \end{bmatrix} \quad R = \begin{bmatrix} 0 & .2 \\ .2 & .3 \end{bmatrix} \quad O = \begin{bmatrix} 0 & 0 \\ 0 & 0 \end{bmatrix}$$

Computing the Stable Matrix

One way to determine the long-term trend of absorbing matrices is to compute their powers to find the stable matrix. But as we saw in the previous section, this computation can be laborious and time-consuming. A more direct method involves finding the inverse of the matrix $I - R$, then multiplying by Q, so the matrix to compute is $Q(I - R)^{-1}$. Then the stable matrix S is defined by its partition in this way:

$$S = \left[\begin{array}{c|c} I & Q(I - R)^{-1} \\ \hline O & O \end{array} \right]$$

In the original matrix I and R may be different sizes. This means that in the matrix S the two identity matrices I and the two zero matrices O may be different sizes. As an illustration of the procedure, let us calculate S from this partitioned transition matrix:

$$T = \left[\begin{array}{cc|c} 1 & 0 & .2 \\ 0 & 1 & .7 \\ \hline 0 & 0 & .1 \end{array} \right]$$

where

$$I = \begin{bmatrix} 1 & 0 \\ 0 & 1 \end{bmatrix} \quad Q = \begin{bmatrix} .2 \\ .7 \end{bmatrix} \quad O = [0 \quad 0] \quad R = [.1]$$

Since R is a 1×1 matrix, we form $I - R = [1] - [.1] = [.9]$, where I is the same size as R. Then $(I - R)^{-1} = [.9]^{-1} = [1/.9]$. Therefore

$$Q(I - R)^{-1} = \begin{bmatrix} .2 \\ .7 \end{bmatrix} [1/.9] = \begin{bmatrix} .2/.9 \\ .7/.9 \end{bmatrix} = \begin{bmatrix} 2/9 \\ 7/9 \end{bmatrix}$$

The stable matrix is

$$S = \left[\begin{array}{cc|c} 1 & 0 & 2/9 \\ 0 & 1 & 7/9 \\ \hline 0 & 0 & 0 \end{array} \right]$$

This means that state 3 leads to state 1 two-ninths of the time, while it leads to state 2 seven-ninths of the time. Example 2 illustrates this procedure for a 4×4 matrix.

EXAMPLE 2

Problem

Find the stable matrix of the transition matrix T.

$$T = \left[\begin{array}{cc|cc} 1 & 0 & 0 & .2 \\ 0 & 1 & .2 & 0 \\ \hline 0 & 0 & .5 & 0 \\ 0 & 0 & .3 & .8 \end{array} \right]$$

Solution We have

$$I = \begin{bmatrix} 1 & 0 \\ 0 & 1 \end{bmatrix} \quad Q = \begin{bmatrix} 0 & .2 \\ .2 & 0 \end{bmatrix} \quad R = \begin{bmatrix} .5 & 0 \\ .3 & .8 \end{bmatrix}$$

Compute

$$I - R = \begin{bmatrix} 1 & 0 \\ 0 & 1 \end{bmatrix} - \begin{bmatrix} .5 & 0 \\ .3 & .8 \end{bmatrix} = \begin{bmatrix} .5 & 0 \\ -.3 & .2 \end{bmatrix}$$

Then

$$(I - R)^{-1} = \begin{bmatrix} .2/.1 & -0/.1 \\ .3/.1 & .5/.1 \end{bmatrix} = \begin{bmatrix} 2 & 0 \\ 3 & 5 \end{bmatrix}$$

So

$$Q(I - R)^{-1} = \begin{bmatrix} 0 & .2 \\ .2 & 0 \end{bmatrix} \begin{bmatrix} 2 & 0 \\ 3 & 5 \end{bmatrix}$$

$$= \begin{bmatrix} .6 & 1 \\ .4 & 0 \end{bmatrix}$$

Hence the stable matrix is

$$S = \left[\begin{array}{cc|cc} 1 & 0 & .6 & 1 \\ 0 & 1 & .4 & 0 \\ \hline 0 & 0 & 0 & 0 \\ 0 & 0 & 0 & 0 \end{array}\right]$$

This means that state 4 will eventually always lead to state 1, while state 3 will lead to state 1 60% of the time and to state 2 40% of the time.

The Effect of the Initial Distribution

To compute the long-term trend of regular matrices, we use systems of equations, and to find the long-term trend of absorbing matrices, we use the inverse and multiplication of matrices. Another difference between regular and absorbing matrices is that for regular matrices different initial distributions yield the same stable distribution, but for absorbing matrices different initial distributions yield different distributions in the long run. The next example demonstrates the latter difference.

EXAMPLE 3

Problem

Find the distribution of the long-term trend of the Markov process in Example 2 for the two initial probability distributions

(a)
$$V = \begin{bmatrix} .1 \\ 0 \\ .5 \\ .4 \end{bmatrix}$$
and
(b)
$$W = \begin{bmatrix} .2 \\ .8 \\ 0 \\ 0 \end{bmatrix}$$

Solution In Example 2 the stable matrix of this process was calculated to be

$$S = \begin{bmatrix} 1 & 0 & .6 & 1 \\ 0 & 1 & .4 & 0 \\ 0 & 0 & 0 & 0 \\ 0 & 0 & 0 & 0 \end{bmatrix}$$

Therefore to find the distribution of the long-term trend, we compute SV and SW.

(a)
$$SV = \begin{bmatrix} 1 & 0 & .6 & 1 \\ 0 & 1 & .4 & 0 \\ 0 & 0 & 0 & 0 \\ 0 & 0 & 0 & 0 \end{bmatrix} \begin{bmatrix} .1 \\ 0 \\ .5 \\ .4 \end{bmatrix} = \begin{bmatrix} .8 \\ .2 \\ 0 \\ 0 \end{bmatrix}$$

and

(b)
$$SW = \begin{bmatrix} 1 & 0 & .6 & 1 \\ 0 & 1 & .4 & 0 \\ 0 & 0 & 0 & 0 \\ 0 & 0 & 0 & 0 \end{bmatrix} \begin{bmatrix} .2 \\ .8 \\ 0 \\ 0 \end{bmatrix} = \begin{bmatrix} .2 \\ .8 \\ 0 \\ 0 \end{bmatrix}$$

This means that if the process starts with the initial distribution V, that is, with state 1 comprising 10%, state 2 0%, state 3 50%, and state 4 40%, then in the long run 80% will be in state 1 and 20% will be in state 2. But if the initial distribution is W, these percentages are reversed.

Example 3 shows that the long-term trend of the distribution depends on the initial distribution. Let us summarize the facts that we have presented concerning absorbing Markov processes. The proof of this result is beyond the scope of this text.

Absorbing Markov Processes

1. The powers of the transition matrix approach some particular matrix, called the stable matrix.
2. The long-term trend depends on the initial distribution.
3. In a finite number of steps the process will enter an absorbing state and stay there.
4. If T is the transition matrix for an absorbing Markov process, then the rows and columns of T can be rearranged so that the absorbing states come first and the matrix will have the form

   $$T = \left[\begin{array}{c|c} I & Q \\ \hline O & R \end{array} \right]$$

   where I is an identity matrix (with the same number of rows as absorbing states) and O is a zero matrix. Then the stable matrix S is

   $$S = \left[\begin{array}{c|c} I & Q(I - R)^{-1} \\ \hline O & O \end{array} \right]$$

   where the identity matrix in the expression $Q(I - R)^{-1}$ is the same size as R.

5. The matrix $Q(I - R)^{-1}$ gives the probabilities that the nonabsorbing states will lead to particular absorbing states.

EXERCISE SET 9–3

In Problems 1 to 4 determine the absorbing and nonabsorbing states of each matrix.

1. $\begin{bmatrix} .2 & .3 & 0 \\ .8 & .2 & 1 \\ 0 & .5 & 0 \end{bmatrix}$

2. $\begin{bmatrix} 1 & .3 & 0 \\ 0 & .2 & 0 \\ 0 & .5 & 1 \end{bmatrix}$

3. $\begin{bmatrix} .2 & .1 & 0 & 0 \\ .6 & .5 & 1 & 0 \\ .1 & .2 & 0 & 0 \\ .1 & .2 & 0 & 1 \end{bmatrix}$

4. $\begin{bmatrix} 0 & .1 & 0 & 0 \\ 0 & .5 & 0 & 1 \\ .5 & .2 & 1 & 0 \\ .5 & .2 & 0 & 0 \end{bmatrix}$

In Problems 5 to 12 determine whether the matrix is absorbing.

5. $\begin{bmatrix} 1 & .3 & 0 \\ 0 & .2 & 1 \\ 0 & .5 & 0 \end{bmatrix}$

6. $\begin{bmatrix} 0 & .3 & 1 \\ 1 & .2 & 0 \\ 0 & .5 & 0 \end{bmatrix}$

7. $\begin{bmatrix} 0 & .3 & 0 \\ 1 & .6 & 1 \\ 0 & .1 & 0 \end{bmatrix}$

8. $\begin{bmatrix} 1 & .3 & 0 \\ 0 & .7 & 0 \\ 0 & 0 & 1 \end{bmatrix}$

9. $\begin{bmatrix} .1 & .1 & 0 & 0 \\ .7 & .5 & 0 & 0 \\ .2 & .4 & 0 & 0 \\ 0 & 0 & 1 & 1 \end{bmatrix}$

10. $\begin{bmatrix} 1 & .1 & 0 & 0 \\ 0 & .5 & 0 & 1 \\ 0 & .2 & 1 & 0 \\ 0 & .2 & 0 & 0 \end{bmatrix}$

11. $\begin{bmatrix} 0 & .1 & 1 & 0 \\ 0 & .8 & 0 & 0 \\ 1 & .1 & 0 & 0 \\ 0 & 0 & 0 & 1 \end{bmatrix}$

12. $\begin{bmatrix} 1 & -1 & 0 & 0 \\ 0 & .5 & 0 & 1 \\ 0 & .2 & 1 & 0 \\ 0 & .2 & 0 & 0 \end{bmatrix}$

In Problems 13 and 14 compute T^4 and determine the stable matrix of the Markov process whose transition matrix is T.

13. $T = \begin{bmatrix} 1 & 0 & 0 \\ 0 & 1 & .9 \\ 0 & 0 & .1 \end{bmatrix}$

14. $T = \begin{bmatrix} .1 & 0 & 0 & 0 \\ .7 & 1 & 0 & 0 \\ .2 & 0 & 1 & 0 \\ 0 & 0 & 0 & 1 \end{bmatrix}$

In Problems 15 to 20 arrange the absorbing matrix so that the absorbing states come before the nonabsorbing ones.

15. $\begin{bmatrix} 1 & .3 & 0 \\ 0 & .2 & 1 \\ 0 & .5 & 0 \end{bmatrix}$

16. $\begin{bmatrix} 0 & .3 & 0 \\ 1 & .2 & 0 \\ 0 & .5 & 1 \end{bmatrix}$

17. $\begin{bmatrix} .2 & .3 & 0 & 0 \\ .8 & .2 & 0 & 0 \\ 0 & .4 & 1 & 0 \\ 0 & .1 & 0 & 1 \end{bmatrix}$

18. $\begin{bmatrix} .1 & .1 & 0 & 0 \\ .7 & .5 & 0 & 0 \\ .2 & .4 & 0 & 0 \\ 0 & 0 & 1 & 1 \end{bmatrix}$

19. $\begin{bmatrix} 1 & .1 & 0 & 0 \\ 0 & .5 & 0 & 1 \\ 0 & .2 & 1 & 0 \\ 0 & .2 & 0 & 0 \end{bmatrix}$

20. $\begin{bmatrix} .2 & 0 & .1 & 0 \\ .6 & 1 & .2 & 0 \\ .1 & 0 & .3 & 0 \\ .1 & 0 & .4 & 1 \end{bmatrix}$

In Problems 21 and 22 compute $(I - R)^{-1}$.

21. $\begin{bmatrix} 1 & 0 & .1 & 0 \\ 0 & 1 & .5 & 1 \\ 0 & 0 & .2 & 0 \\ 0 & 0 & .2 & 0 \end{bmatrix}$

22. $\begin{bmatrix} 1 & 0 & .1 & .4 \\ 0 & 1 & .5 & .1 \\ 0 & 0 & .4 & .4 \\ 0 & 0 & 0 & .1 \end{bmatrix}$

In Problems 23 to 26 find the stable matrix.

23. $\begin{bmatrix} 1 & 0 & .3 & .1 \\ 0 & 1 & .2 & 0 \\ 0 & 0 & .5 & .6 \\ 0 & 0 & 0 & .3 \end{bmatrix}$

24. $\begin{bmatrix} 1 & 0 & .2 & .3 \\ 0 & 1 & 0 & .1 \\ 0 & 0 & .8 & .5 \\ 0 & 0 & 0 & .1 \end{bmatrix}$

25. $\begin{bmatrix} 1 & 0 & .1 & 0 \\ 0 & 1 & .5 & 1 \\ 0 & 0 & .2 & 0 \\ 0 & 0 & .2 & 0 \end{bmatrix}$

26. $\begin{bmatrix} 1 & 0 & 0 & .1 & 0 \\ 0 & 1 & 0 & .5 & 0 \\ 0 & 0 & 1 & .2 & .8 \\ 0 & 0 & 0 & .2 & .1 \\ 0 & 0 & 0 & 0 & .1 \end{bmatrix}$

In Problems 27 and 28 show that the probability distribution of the long-term trend depends on the initial distribution by calculating SV and SW, where S is the stable matrix and V and W are two initial distributions.

27. $S = \begin{bmatrix} 1 & 0 & .3 \\ 0 & 1 & .7 \\ 0 & 0 & 0 \end{bmatrix}$, $V = \begin{bmatrix} .2 \\ .3 \\ .5 \end{bmatrix}$,

$W = \begin{bmatrix} .5 \\ .4 \\ .1 \end{bmatrix}$

28. $S = \begin{bmatrix} 1 & 0 & .2 & .1 \\ 0 & 1 & .8 & .9 \\ 0 & 0 & 0 & 0 \\ 0 & 0 & 0 & 0 \end{bmatrix}$, $V = \begin{bmatrix} .5 \\ .2 \\ .2 \\ .1 \end{bmatrix}$, $W = \begin{bmatrix} .1 \\ .1 \\ .5 \\ .3 \end{bmatrix}$

In Problems 29 to 32 a game is given that can be described by a Markov process. Find the transition matrix and its stable matrix, and then answer the given question.

29. Two players, A and B, engage in a coin-tossing game where the coin is not fair and the probability of heads is $\frac{2}{3}$. Player A always bets on heads. If A wins, B gives A $1; and if A loses, A gives B $1. The players have a total of $3 between them and continue to play until one of them has accumulated all the money. What is the probability that A will win $3 if A starts with $2?

30. Two players, A and B, engage in a coin-tossing game where the coin is not fair and the probability of heads is $\frac{2}{5}$. Player A always bets on heads. If A wins, B gives A $2; and if A loses, A gives B $1. The players have a total of $4 between them and continue to play until one of them has accumulated all the money. What is the probability that A will win $4 if A starts with $2?

31. Consider a game where A continually bets $1 on each play. If A wins, A receives $1 in addition to getting back the bet of $1. A starts with $2 and plays until either going broke or winning $3. The probability of winning on each play is $\frac{2}{5}$. What is the probability of A going broke?

32. Consider a game where A continually bets $1 on each play. If A wins, A receives $2 in addition to getting back the bet of $1. A starts with $1 and plays until either going broke or winning $3. The probability of winning on each play is $\frac{2}{5}$. What is the probability of A going broke?

33. Show that an absorbing transition matrix is not regular.

34. Show that the transition matrix $\begin{bmatrix} 0 & 1 \\ 1 & 0 \end{bmatrix}$ does not have a stable matrix.

35. For the transition matrix

$$T = \begin{bmatrix} 1 & 0 & .2 \\ 0 & 1 & .7 \\ 0 & 0 & .1 \end{bmatrix}$$

show that the limiting matrix of T^n is equal to the stable matrix computed in the text.

Referenced Exercise Set 9–3

1. Many public and private enterprises are based on a service system requiring multiple service stations. Examples include banks, grocery stores, and government application offices. In an article describing inventory levels for such systems Serfozo studies the process that the Bell System uses to transmit television programs, such as news, sports, and special events, from local sites without permanent transmission equipment.[*] In this case the server is the transmission equipment and the service station is the local site, say, a speech or a sporting event. Portable transmission equipment is used to relay the program to the regional pool. As an example, data are taken from the regional pool in Chicago. Requests for service come from northern Illinois, northern Indiana, and southern Wisconsin. Transmission of a program from a local site consists of a series of microwave or cable transmission links, usually under 25 miles, that are set up to transmit the program to a nearby entry into the permanent Bell System network. Sometimes the network can handle the demand, and at other times it routes the transmission to the Chicago pool. The Markov process has four states, each corresponding to a geographic area in the network: Chicago, northern Illinois, northern Indiana, and southern Wisconsin. We refer to these states as Chi, Ill, Ind, and Wis. The probabilities in the transition matrix represent the chance that a program originating in the current state (geographical area) will be transmitted to the next state, that is, transmitted to another geographical area. An example of a typical transition matrix is

		Current Stage			
		Chi	Ill	Ind	Wis
Next Stage	Chi	1	.6	.4	.3
	Ill	0	.4	0	0
	Ind	0	0	.6	0
	Wis	0	0	0	.7

(a) Show that this transition matrix is an absorbing transition matrix. (b) Find the absorbing and nonabsorbing states. (c) Find the stable matrix.

*Richard Serfozo, "Allocation of Servers for Stochastic Service Stations with One Overflow Station," *Management Science,* Vol. 31, pp. 1011–1018, 1985.

2. In a paper† that studied the movement of coronary patients within a hospital to obtain length-of-stay and patient-day statistics, the researchers used data from St. Raphael Hospital, New Haven, Conn., to define the following Markov transition matrix:

$$
\begin{array}{c}
\text{CCU} \\
\text{PCCU} \\
\text{ICU} \\
\text{MED} \\
\text{SURG} \\
\text{AMB} \\
\text{ECF} \\
\text{HOME} \\
\text{DIED}
\end{array}
\begin{bmatrix}
0 & .02 & 0 & 0 & 0 & 0 & 0 & 0 & 0 \\
.74 & 0 & .58 & .01 & 0 & 0 & 0 & 0 & 0 \\
.01 & .01 & 0 & .04 & 0 & 0 & 0 & 0 & 0 \\
.12 & .02 & .17 & 0 & 0 & 0 & 0 & 0 & 0 \\
.01 & .01 & .08 & .02 & 0 & 0 & 0 & 0 & 0 \\
.01 & .01 & 0 & .03 & 0 & 0 & 0 & 0 & 0 \\
0 & .06 & 0 & .08 & 0 & 0 & 1 & 0 & 0 \\
.01 & .83 & 0 & .70 & 1 & 1 & 0 & 1 & 0 \\
.10 & .04 & .17 & .12 & 0 & 0 & 0 & 0 & 1
\end{bmatrix}
$$

The abbreviations stand for the following: CCU (coronary care unit), PCCU (postcoronary care unit), ICU (intensive care unit), MED (a medical unit), SURG (surgical unit), AMB (ambulatory unit), ECF (extended care unit), HOME (the patient was discharged), DIED (the patient died).

(a) Find the absorbing states.

(b) Is the matrix an absorbing matrix?

(c) Rearrange the matrix so that the absorbing states are to the left of the nonabsorbing states.

(d) Find the probability that a patient was discharged from a medical unit.

3. Many psychological theories of learning are based on probability theory. One such theory, presented in a book by Wickens on the mathematical study of human behavior, is based on the "all-or-none" type of learning experiment.‡ A subject is asked to learn a list of 100 paired-associate items, such as nonsense syllables paired with digits from 1 to 5. The first part of the pair is called the stimulus, the second is the response. There are 100 different syllables each paired with a digit. For example, suppose one such pair is "poj-4." The subject is given the syllable "poj" and asked for its corresponding digit. The correct response is "4." Incorrect responses would be 1, 2, 3 or 5. The first time the subject goes through the list he or she is guessing which response corresponds with each syllable. After the subject chooses a response, the correct pair is shown, thereby giving feedback and a chance to study the pair.

Then another, usually different, stimulus is given. The experiment continues until all pairs are learned or a specific amount of time has elapsed. There are two states in this model. An item is either learned, so it is in the L state, or the subject must guess, so it is in the G state. Initially, all items are in the G state. An item can change from the G state to the L state, but once it is in the L state it remains there. The transition probabilities are

Pr(error on presentation $k + 1$ given an error on presentation k)

The next refinement of the model separates state G into two states, a guess that is correct, state GC, and a guess that is an error, state GE. For a given item and a given subject an example of a transition matrix is presented below:

Current Stage

$$
\begin{array}{c}
\text{Next} \\
\text{Stage}
\end{array}
\begin{array}{c}
\text{L} \\
\text{GC} \\
\text{GE}
\end{array}
\begin{array}{ccc}
\text{L} & \text{GC} & \text{GE} \\
\end{array}
\begin{bmatrix}
1 & .1 & .2 \\
0 & .6 & .3 \\
0 & .3 & .5
\end{bmatrix}
$$

(a) Show that this transition matrix is an absorbing transition matrix. (b) Find the absorbing and nonabsorbing states. (c) Find the stable matrix.

Cumulative Exercise Set 9–3

1. Determine the transition matrix of the following Markov process:

An environmental protection agency reports the following information about air quality. If the air is clear on one day, then the chances of it being clear, moderate, or poor the next day are 60%, 35%, and 5%, respectively. If the air is moderate on one day, then the chances of it being clear, moderate, or poor the next day are 25%, 50%, and 25%, respectively. If the air is poor on one day, then the chances of it being clear, moderate, or poor the next day are 10%, 25%, and 65%, respectively.

---

†Edward P. C. Kao, "Modeling the Movement of Coronary Patients Within a Hospital by Semi-Markov Processes," *Operations Research,* pp. 683–699, July–August 1974.

‡Thomas D. Wickens, *Models for Behavior: Stochastic Processes in Psychology,* San Francisco, W. H. Freeman and Co., 1982.

2. Determine a, b, c, and d so that the matrix is a transition matrix.

$$\begin{bmatrix} .2 & .5 & .1 & d \\ .1 & b & .4 & .2 \\ a & .1 & .4 & .1 \\ .3 & .2 & c & .1 \end{bmatrix}$$

3. For the transition matrix T and the probability distribution V find the distribution of the next three stages.

$$T = \begin{bmatrix} .3 & .1 & .6 \\ .5 & .4 & .2 \\ .2 & .5 & .3 \end{bmatrix} \quad V = \begin{bmatrix} .15 \\ .4 \\ .45 \end{bmatrix}$$

4. In the following transition matrix the first state represents the likelihood that a stock will rise and the second state represents the likelihood that it will fall. The present stage refers to one day's behavior and the second stage refers to the next day's behavior.

$$\begin{array}{c} \\ \text{rise} \\ \text{fall} \end{array} \begin{array}{cc} \text{rise} & \text{fall} \\ T = \begin{bmatrix} .4 & .3 \\ .6 & .7 \end{bmatrix} \end{array}$$

Find the probability that the stock will rise tomorrow if the probability is .8 that it will rise today.

5. Determine whether the matrix is regular.

$$\begin{bmatrix} 0 & 1 & 0 \\ \frac{1}{2} & 0 & \frac{1}{4} \\ \frac{1}{2} & 0 & \frac{3}{4} \end{bmatrix}$$

6. Let T be the transition matrix and V the initial distribution of a Markov process. Compute T^nV for $n = 1$ to 5. Use this information to find the long-term trend.

$$T = \begin{bmatrix} \frac{1}{3} & \frac{2}{5} \\ \frac{2}{3} & \frac{3}{5} \end{bmatrix} \quad V = \begin{bmatrix} \frac{1}{4} \\ \frac{3}{4} \end{bmatrix}$$

7. Find the stable matrix for the absorbing matrix.

$$T = \begin{bmatrix} .1 & 0 & 0 \\ .7 & 1 & 0 \\ .2 & 0 & 1 \end{bmatrix}$$

8. Find the long-term trend of the following Markov process:

A state's utility commission has revealed the following information about the conversion between energy sources for home heating. Of the homes that use oil one year, half will convert to electricity the next year and half will convert to gas. All the homes that use electricity will convert to oil. Of the homes that use gas one year, 25% will convert to electricity the next year and none will convert to oil.

9. Determine if the following matrix is absorbing. If it is, list the absorbing states.

$$\begin{bmatrix} .5 & 0 & .6 & .4 \\ .2 & 1 & .2 & 0 \\ .2 & 0 & 0 & .4 \\ .1 & 0 & .2 & .2 \end{bmatrix}$$

10. Consider the transition matrix T defined by

$$\begin{bmatrix} 1 & .3 & 0 \\ 0 & .5 & 0 \\ 0 & .2 & 1 \end{bmatrix}$$

(a) Arrange the absorbing matrix so that the absorbing states come before the nonabsorbing states.
(b) Find the stable matrix.

11. Find the stable matrix of the transition matrix T where

$$T = \begin{bmatrix} 1 & 0 & .5 & .3 \\ 0 & 1 & .1 & 0 \\ 0 & 0 & .3 & .5 \\ 0 & 0 & .1 & .2 \end{bmatrix}$$

12. The following game can be described by a Markov process:

Two players, A and B, engage in a game of tossing dice. Player A always bets on doubles. If doubles are rolled, then player B gives $1 to player A; otherwise player A gives $1 to player B. The players have a total of $3 between them and continue to play until one of them has accumulated all the money.
(a) Find the transition matrix.
(b) Find the stable matrix.
(c) What is the probability that A will win $3 if A starts with $2?

CASE STUDY

Allowance for Doubtful Accounts

In retail businesses, especially large firms, a very important quantity called "bad debts" has a significant effect on income. The term "bad debts" refers to the amount of money due that the firm determines will never be collected. The total amount of money owed to the firm is referred to as "accounts receivable." Thus bad debts is that fraction of accounts receivable that will not be collected. These uncollectible debts must be estimated and an allowance made for them in the accounting process. Otherwise estimated income will be much greater than actual income received. When a large portion of a firm's assets is tied up in accounts receivable, an accurate estimation of the bad debt allowance is a crucial accounting technique. Businesses that rely heavily on credit purchases, such as large department stores and automobile dealers, are examples of such firms.

Before the 1960s bad debt allowances were made on the basis of the accounting department's best guess. Then R. M. Cyert* and others started investigating more accurate and scientific methods of estimating the allowances. Their approach treats the process of how some debts eventually become bad debts as a Markov process.

In order to describe their methods, we must define the *allowance for doubtful accounts* that, at any point in time, is the estimated dollar amount of accounts receivable that will prove uncollectible in the future. For example, if a firm has a total amount of $1,000,000 accounts receivable, and it determines that 5% of that amount will never be collected, then the allowance for doubtful accounts is $1,000,000 $\times$ .05 = $50,000. The crucial issue is the accuracy of the estimate. The problem is that the total amount of accounts receivable continually changes with time. In a given time period most debts are paid but some are not. In the next period new debts are added to the accounts receivable, some of which are not paid, and some of the previous period's accounts are paid but some are not. So the allowance for doubtful accounts also changes with time.

Thus the method for determining an accurate amount is a process whose stages are specific time periods. The time periods vary from firm to firm. They might be months, quarters, or even years. For any choice of time periods the dollar amounts of accounts receivable are classified into various age categories. For instance, at a given time there are debts that are due in that time period, some are overdue by one period, some are overdue by two periods, and so on until a time period, say, n, is determined to be that period in which any debt overdue by n periods or more is called a bad debt.

We define for a balance of accounts receivable at a given time

P_0 = % of accounts receivable 0 periods overdue
   (these are the current debts)
P_1 = % of accounts receivable one period overdue
P_2 = % of accounts receivable two periods overdue
P_3 = % of accounts receivable three periods overdue
P_n = % of accounts receivable n periods overdue

*R. M. Cyert et al., "Estimation of the Allowance for Doubtful Accounts by Markov Chains," *Management Science* Vol. 8, pp. 287–303, April 1962.

One category is chosen as the last, and so all dollars in it and those following it are referred to as bad debts. A delinquent account may eventually be repaid, but it is customary in accounting procedure to lump all debts in both this age category and those after it into one bad debt classification.

For example, suppose a firm determines that its time periods will be months and any debt overdue by three or more months is a bad debt. Suppose it has $1,000,000 in accounts receivable, with $700,000 (or 70%) currently due in the present month, $200,000 (or 20%) overdue by one month, $60,000 (or 6%) overdue by two months, and $40,000 (or 4%) overdue by three or more months. Then

$$P_0 = .70, \quad P_1 = .20, \quad P_2 = .06, \quad P_3 = .04$$

Now consider the firm's accounts receivables in the next month. Many of these debts will be paid while some will not, and, of course, new debts will be accrued since the previous period. The P_i for the next period will usually be different from the previous ones. Those accounts that are in the current month's P_0 and will not be paid by the next period will move to the next month's P_1. Similarly, the current month's accounts in P_1 that are not paid will move to the next month's P_2. The firm determines that in a given month the percentage of accounts receivable in each category depends primarily on the previous month's percentages. Thus the process is a Markov process. For this firm there are five categories if we let 0* represent the accounts that were paid during the time period and let 0, 1, 2, and 3 represent the accounts 0 months overdue (current), one month overdue, two months overdue, and three or more months overdue, respectively.

Suppose it is determined that of those accounts receivable during the current month that are in (1) category 0, 10% were paid, 60% stayed in category 0, 30% became one month overdue; (2) category 1, 25% were paid, 35% moved to category 0, 20% stayed in category 1, 20% moved to category 2; (3) category 2, 10% were paid, 10% moved to category 0, 20% moved to category 1, 15% stayed in category 2, 45% moved to category 3.†

Therefore the transition matrix T is

From Category

	0*	0	1	2	3
0*	1	.1	.25	.1	0
0	0	.6	.35	.1	0
1	0	.3	.2	.2	0
2	0	0	.2	.15	0
3	0	0	0	.45	1

To Category

†The "total balance method" of accounting is used here, and it assumes that all dollars in the account balance are put in the age category of the oldest dollar. This means that a current purchase charge by customer with a two-month past due balance would be classified in category 2 rather than in category 0. This accounts for the fact that some dollars can move from category 1 to category 0 or stay in category 1.

Here is the transition matrix that Cyert and his colleagues studied. They prepared the matrix from a random sample of approximately 1000 department stores in order to test their theory.

From Category

months overdue

	0*	0	1	2	3	4	5	6
0*	1	.21	.13	.13	.10	.14	.09	0
0	0	.67	.19	.08	.01	.02	.01	0
1	0	.12	.44	.20	.04	0	.02	0
2	0	0	.24	.36	.17	.09	.01	0
3	0	0	0	.23	.29	.20	.10	0
4	0	0	0	0	.39	.41	.12	0
5	0	0	0	0	0	.14	.47	0
6	0	0	0	0	0	0	.18	1

To Category

Let us turn to a simpler example to see how the theory of Markov processes can help businesses determine an accurate allowance for doubtful accounts. The transition matrix T is the one used by Cyert for illustrative purposes.

EXAMPLE 1

Suppose the time period is months and accounts two months old are bad debts. Let

$$
T = \begin{array}{c} 0^* \\ 2 \\ 0 \\ 1 \end{array}
\begin{array}{cccc} 0^* & 2 & 0 & 1 \\ \left[\begin{array}{cccc} 1 & 0 & .3 & .5 \\ 0 & 1 & 0 & .1 \\ 0 & 0 & .5 & .3 \\ 0 & 0 & .2 & .1 \end{array}\right] \end{array}
$$

We have arranged the rows and columns to put the absorbing states first. Let the distribution matrix be

$$
V = \begin{bmatrix} 0 \\ 0 \\ .7 \\ .3 \end{bmatrix}
$$

Problem

What is the allowance for bad debts?

Solution The transition matrix has the partition

$$
T = \left[\begin{array}{cc|cc} 1 & 0 & .3 & .5 \\ 0 & 1 & 0 & .1 \\ \hline 0 & 0 & .5 & .3 \\ 0 & 0 & .2 & .1 \end{array}\right]
$$

where

$$I = \begin{bmatrix} 1 & 0 \\ 0 & 1 \end{bmatrix}, \quad Q = \begin{bmatrix} .3 & .5 \\ 0 & .1 \end{bmatrix}, \quad R = \begin{bmatrix} .5 & .3 \\ .2 & .1 \end{bmatrix}$$

We compute

$$I - R = \begin{bmatrix} 1 & 0 \\ 0 & 1 \end{bmatrix} - \begin{bmatrix} .5 & .3 \\ .2 & .1 \end{bmatrix} = \begin{bmatrix} .5 & -.3 \\ -.2 & .9 \end{bmatrix}$$

and so

$$(I - R)^{-1} = \begin{bmatrix} .9/.39 & .3/.39 \\ .2/.39 & .5/.39 \end{bmatrix} = \begin{bmatrix} 2.31 & .77 \\ .51 & 1.28 \end{bmatrix}$$

$$Q(I - R)^{-1} = \begin{bmatrix} .3 & .5 \\ 0 & .1 \end{bmatrix} \begin{bmatrix} 2.31 & .77 \\ .51 & 1.28 \end{bmatrix}$$

$$= \begin{bmatrix} .95 & .87 \\ .05 & .13 \end{bmatrix}$$

Hence the stable matrix is

$$S = \left[\begin{array}{cc|cc} 1 & 0 & .95 & .87 \\ 0 & 1 & .05 & .13 \\ \hline 0 & 0 & 0 & 0 \\ 0 & 0 & 0 & 0 \end{array} \right]$$

The long-term trend is SV.

$$SV = \left[\begin{array}{cc|cc} 1 & 0 & .95 & .87 \\ 0 & 1 & .05 & .13 \\ \hline 0 & 0 & 0 & 0 \\ 0 & 0 & 0 & 0 \end{array} \right] \begin{bmatrix} 0 \\ 0 \\ .7 \\ .3 \end{bmatrix}$$

$$= \begin{bmatrix} .926 \\ .074 \\ 0 \\ 0 \end{bmatrix}$$

Thus the allowance for bad debts is 7.4%. This means that for every $100 of accounts receivable the firm should reserve $7.40 for bad debts.

This is very close to the amount Cyert calculated in his study of 1000 retail stores. The allowance for doubtful accounts for most of the firms was about 7%.

Case Study Exercises

In Problems 1 to 4 compute the allowance for bad debts using the transition matrix T in Example 1 and the given distribution matrix V.

1.
$$V = \begin{bmatrix} 0 \\ 0 \\ .6 \\ .4 \end{bmatrix}$$

2.
$$V = \begin{bmatrix} 0 \\ 0 \\ .8 \\ .2 \end{bmatrix}$$

3.
$$V = \begin{bmatrix} 0 \\ 0 \\ .5 \\ .5 \end{bmatrix}$$

4.
$$V = \begin{bmatrix} 0 \\ 0 \\ .1 \\ .9 \end{bmatrix}$$

In Problems 5 to 8 compute the allowance for bad debts using the following transition matrix T and the given distribution matrix V.

$$T = \left[\begin{array}{cc|cc} 1 & 0 & .2 & .3 \\ 0 & 1 & 0 & .1 \\ \hline 0 & 0 & .6 & .4 \\ 0 & 0 & .2 & .2 \end{array} \right]$$

5.
$$V = \begin{bmatrix} 0 \\ 0 \\ .6 \\ .4 \end{bmatrix}$$

6.
$$V = \begin{bmatrix} 0 \\ 0 \\ .8 \\ .2 \end{bmatrix}$$

7.
$$V = \begin{bmatrix} 0 \\ 0 \\ .5 \\ .5 \end{bmatrix}$$

8.
$$V = \begin{bmatrix} 0 \\ 0 \\ .1 \\ .9 \end{bmatrix}$$

CHAPTER REVIEW

Key Terms

9–1 Basic Properties
Markov Process
Stage
Current Stage
Next Stage
States
Initial Stage
Transition Matrix
Distribution
Initial Distribution

9–2 Regular Markov Processes
Long-Term Trend
Regular Matrices
Stable Distribution

9–3 Absorbing Markov Processes
Absorbing States
Nonabsorbing States
Absorbing Matrix
Absorbing Markov Process
Partitioned Matrix
Stable Matrix

Summary of Important Concepts

Stable distribution

Formulas

$$TX = X$$
$$x + y = 1$$

Partitioned Matrix	$$\left[\begin{array}{c|c} I & Q \\ \hline O & R \end{array}\right]$$
Stable Matrix	$$\left[\begin{array}{c|c} I & Q(I - R)^{-1} \\ \hline O & 0 \end{array}\right]$$

REVIEW PROBLEMS

In Problems 1 and 2 determine the transition matrix of the Markov process.

1. Researchers determine that of the people who voted Democratic in a given election, 45% voted Democratic in the next election and 55% voted Republican (only those who voted in every election were considered). Of those who voted Republican in the current election, 60% voted Republican in the next election and 40% voted Democratic.

2. The manager of an investment fund has determined that the following strategy will be employed for the 100 blue chip stock portfolio: (1) If a stock issue was purchased in one day's trading, then 10% of the time it will be purchased the next day, 60% of the time it will be sold, and 30% of the time it will be left unchanged. (2) If a stock issue was sold in one day's trading, then 50% of the time it will be purchased the next day, 35% of the time it will be sold, and 15% of the time it will be left unchanged. (3) If a stock issue was left unchanged in one day's trading, then 35% of the time it will be purchased the next day, 25% of the time it will be sold, and 40% of the time it will be left unchanged.

3. Determine whether the matrix could be a transition matrix of a Markov process.

$$\begin{bmatrix} .3 & .4 & .1 \\ .4 & .3 & .6 \\ .3 & .1 & .3 \end{bmatrix}$$

4. Determine a, b, and c so that the matrix could be a transition matrix.

$$\begin{bmatrix} .1 & .5 & .1 \\ .2 & .5 & .3 \\ a & b & c \end{bmatrix}$$

5. For the transition matrix T and the probability distribution V find the distribution of the next three stages of the process.

$$T = \begin{bmatrix} .1 & .2 & .4 \\ .1 & .3 & .4 \\ .8 & .5 & .2 \end{bmatrix} \qquad V = \begin{bmatrix} .3 \\ .2 \\ .5 \end{bmatrix}$$

Problems 6 and 7 refer to the following transition matrix T of the Markov process determined by a weather bureau. The first state represents the likelihood of snow in the forecast and the second state represents the likelihood of no snow in the forecast. Each stage represents a daily forecast. That is, the present stage refers to today's weather forecast and the next stage refers to tomorrow's forecast.

$$\begin{array}{cc} & \text{snow} \quad \text{no snow} \end{array}$$
$$\begin{array}{c} \text{snow} \\ \text{no snow} \end{array} T = \begin{bmatrix} .1 & .2 \\ .9 & .8 \end{bmatrix}$$

6. Find the probability that it will snow tomorrow if the probability of snow today is .5.

7. Find the probability that it will not snow tomorrow if the probability of snow today is .3.

8. Determine whether the matrix is regular.

$$T = \begin{bmatrix} 0 & .3 & .2 \\ 0 & .7 & .2 \\ 1 & 0 & .6 \end{bmatrix}$$

9. Compute $T^n V$ for $n = 1$, 2, and 3.

$$T = \begin{bmatrix} .1 & .7 \\ .9 & .3 \end{bmatrix} \qquad V = \begin{bmatrix} .6 \\ .4 \end{bmatrix}$$

10. Find the long-term trend of the Markov process whose transition matrix is T and initial distribution is V by computing $T^n V$ for $n = 1$ to 6.

$$T = \begin{bmatrix} .9 & .3 \\ .1 & .7 \end{bmatrix} \qquad V = \begin{bmatrix} .6 \\ .4 \end{bmatrix}$$

In Problems 11 and 12 find the stable matrix for the regular transition matrix T.

11. $T = \begin{bmatrix} .8 & .1 \\ .2 & .9 \end{bmatrix}$

12. $T = \begin{bmatrix} .3 & .8 & .6 \\ .2 & .2 & .3 \\ .5 & 0 & .1 \end{bmatrix}$

In Problems 13 and 14 find the long-term trend of the Markov process.

13. A weather forecasting service has found that (1) if it rains on a given day, then the probability that it will rain the next day is .3 and (2) if it does not rain on a given day, then the probability that it will rain the next day is .4.

14. A weather forecaster determines that if it is raining on one day, then for the following day there is a 20% chance it will rain, a 5% chance it will snow, and a 75% chance of no precipitation. If it is snowing on one day, then for the following day there is a 10% chance it will rain, a 10% chance it will snow, and an 80% chance of no precipitation. If there is no precipitation on one day, then for the following day there is a 5% chance it will rain, a 5% chance it will snow, and a 90% chance of no precipitation.

15. Determine the absorbing and nonabsorbing states of the matrix.

$$\begin{bmatrix} .2 & .1 & 0 & 0 \\ .5 & .4 & 1 & 0 \\ .1 & .2 & 0 & 0 \\ .2 & .3 & 0 & 1 \end{bmatrix}$$

16. Determine if the matrix is absorbing.

$$\begin{bmatrix} 1 & .1 & 0 & 0 \\ 0 & .5 & 0 & 0 \\ 0 & .3 & 1 & 0 \\ 0 & .1 & 0 & 1 \end{bmatrix}$$

17. Find the stable matrix.

$$\begin{bmatrix} 1 & 0 & .5 & 0 \\ 0 & 1 & .1 & 1 \\ 0 & 0 & .3 & 0 \\ 0 & 0 & .1 & 0 \end{bmatrix}$$

18. Arrange the absorbing matrix so that the absorbing states come before the nonabsorbing ones.

$$\begin{bmatrix} .1 & .3 & 0 & 0 \\ .8 & .2 & 0 & 0 \\ .1 & .1 & 1 & 0 \\ 0 & .4 & 0 & 1 \end{bmatrix}$$

19. Compute $(I - R)^{-1}$.

$$\begin{bmatrix} 1 & 0 & .2 & 0 \\ 0 & 1 & .4 & 1 \\ 0 & 0 & .2 & 0 \\ 0 & 0 & .2 & 0 \end{bmatrix}$$

20. Show that the probability distribution of the long-term trend depends on the initial distribution by calculating SV and SW, where S is the stable matrix and V and W are two initial distributions.

$$S = \begin{bmatrix} 1 & 0 & .2 \\ 0 & 1 & .8 \\ 0 & 0 & 0 \end{bmatrix} \qquad V = \begin{bmatrix} .2 \\ .4 \\ .4 \end{bmatrix}$$

$$W = \begin{bmatrix} .5 \\ .4 \\ .1 \end{bmatrix}$$

21. The following game can be described by a Markov process. Find the transition matrix and its stable matrix, and then answer the given question.

Two players, A and B, engage in a coin-tossing game where the coin is not fair and the probability of heads is $\frac{3}{7}$. Player A always bets on heads. If A wins, B gives A \$1; and if A loses, A gives B \$1. The players have a total of \$3 between them and continue to play until one of them has accumulated all the money. What is the probability that A will win \$3 if A starts with \$2?

PROGRAMMABLE CALCULATOR EXPERIMENTS

1. Write a program that computes successive distributions $TV, T^2V, \ldots, T^{10}V$ for the given matrices T and V.

$$T = \begin{bmatrix} .4 & .1 \\ .6 & .9 \end{bmatrix} \qquad V = \begin{bmatrix} .7 \\ .3 \end{bmatrix}$$

2. Do Problem 1 for the given matrices T and V.

$$T = \begin{bmatrix} .4 & .1 & .3 \\ .2 & .8 & .4 \\ .4 & .1 & .3 \end{bmatrix} \qquad V = \begin{bmatrix} .5 \\ .2 \\ .3 \end{bmatrix}$$

3. Write a program that computes the stable matrix of the matrix T in Problem 1 in two ways: (a) by computing T^nV several times until a pattern is discerned and (b) by solving the matrix equation $TX = X$ and the sum of the entries $= 1$.

4. Do Problem 3 for the given matrix T in Problem 2.

5. Write a program that computes the stable matrix of the following absorbing Markov matrix:

$$\begin{bmatrix} 1 & 0 & .2 & 0 \\ 0 & 1 & 0 & .2 \\ 0 & 0 & .3 & .8 \\ 0 & 0 & .5 & 0 \end{bmatrix}$$

6. Write a program that computes the stable matrix of an arbitrary 4×4 absorbing Markov matrix.

Game Theory

Game theory is applicable to the industrial merger mentioned in a case study, and it also has other interesting uses. (A. McWhirter/H. Armstrong Roberts)

Chapter Overview

Game theory's novel mathematical properties were created in the 1930s by John von Neumann and Oskar Morgenstern to solve problems in economics and management. Their book, *The Theory of Games and Economic Behavior,** showed how social events can be described by models of certain types of games.

Game theory is in an active state of development. During 1983–1984 the program at the Institute for Mathematics and its Applications in Minneapolis was devoted to economics, and the primary tool was game theory. Game theory has been applied to many disciplines, from economics to animal conflict, but it has also been applied to some surprising areas such as evolution.

The thrust of the theory is to derive strategies that maximize an individual's expected gain in a broad spectrum of competitive situations. Game theory's approach to explaining social interaction is significantly different from previous approaches. The physical sciences study the behavior of inanimate objects. Their mathematical techniques fall short when applied to the highly complex social behavior. Game theory assumes that human behavior is sometimes cooperative and sometimes combative. The key to game theory's approach is the blend of probability with the theory of matrices.

CASE STUDY PREVIEWS

Game theory has been most fully developed for the two-person situation—when two individuals or groups are in conflict. Since key military decisions fit this model, it is not surprising that game theory can readily be applied to military decisions. But many business conflicts are also two-person or two-group competitions.

The case study following Section 10–1 is taken from a classic paper written by Colonel O. G. Haywood, Jr. in 1954. It shows that game theory can be used to analyze military decisions by centering on a key battle in World War II.

The second case study at the end of the chapter tells the story of the very stormy merger of Norway's three cement corporations in the 1960s. Even though three groups were involved in the negotiation, the conflict was resolved by using a two-person-game-theoretic approach.

10–1 Strictly Determined Games

In his book *Game Theory and Politics*† Steven J. Brams shows how game theory can help politicians make important decisions, not just concerning elections but also in the military, environmental, and judicial arenas. One example deals with local district elections in Illinois. Each district of the Illinois legislature elects three representatives. Each party can nominate one, two, or three candidates in the general election, and the three with the largest number of votes win. The party's decision

*John von Neumann and Oskar Morgenstern, *The Theory of Games and Economic Behavior,* Princeton, NJ, Princeton University Press, 1953.

†Steven J. Brams, *Game Theory and Politics,* New York, The Free Press, 1975.

of how many candidates to run is made under uncertainty because neither party knows what the other will do. The decision is based on many factors including the percentage of vote the party expects to receive and how that percentage will be split among the candidates. For example, suppose in a particular district the party that anticipates getting the majority of votes chooses to run three candidates and the minority party chooses to run two. If in the election the majority party gets 54% of the vote and the three majority party candidates split the vote evenly then each candidate gets 18%. If the minority party's 46% is also split evenly, the two minority candidates would each get 23% of the vote. This would result in both minority candidates having more votes than each of the majority candidates, so the majority party would win only one of the three seats and the minority party would win two seats.

Brams introduces game theory in the following way. Consider each district election a game with two players (the two parties), and each player has three options, to run one, two, or three candidates. The outcome, called the **payoff,** is the number of seats won by the majority party. In most cases one party had a clear majority so that it never ran just one candidate and the minority party never ran three. Brams studied various categories of voting percentages. If the minority party had a significant percentage of the vote, then the four possible outcomes could be expressed in a matrix, called a **payoff matrix.**

		Minority Party Options	
		run 1	run 2
Majority	run 2	2	2
Party Options	run 3	3	1

or simply

$$P = \begin{bmatrix} 2 & 2 \\ 3 & 1 \end{bmatrix}$$

This means that if the majority party runs two candidates, the majority party will win two seats (and so the minority party will win one seat), no matter what strategy the minority party chooses. That is, if the minority party runs one candidate while the majority party runs two candidates, the payoff is the $(1, 1)$ entry, 2, meaning that the majority party wins two seats. Likewise, the $(1, 2)$ entry, 2, means the majority party wins two seats if it runs two candidates and the minority party runs two candidates. But if the majority party runs three candidates, the majority party will either win three seats if the minority party runs one candidate (since the $(2, 1)$ entry is 3), or it will win one seat if the minority party runs two candidates (since the $(2, 2)$ entry is 1).

The $(2, 2)$ entry, 1, is the result previously described when the majority party evenly splits 54% of the vote among three candidates and the minority party evenly splits 46% of the vote between two candidates.

This is a typical example of a mathematical game. The term **game** in game theory can refer to very serious examples of competition and conflict as well as recreations. In such a game there are two or more players and each player is allowed to make a move. As a result of their moves, there is a payoff (positive, negative, or zero) to each player. The mathematical problem of the game is to determine the best strategy for each player in order for each to get the largest possible payoff.

The Fundamental Goal of Game Theory

For each player, determine the strategy that maximizes the player's gain.

Two-Person Games

Throughout this chapter we consider only **two-person games,** meaning that there are only two players. Brams' voting study is an example of a two-person game. In a two-person game the outcomes can always be expressed in a matrix, called the payoff matrix, where one player's decisions are represented by the rows and the other player's decisions are represented by the columns. The payoff matrix consists of rows corresponding to the options for the row player, which we refer to as R, and columns corresponding to the moves for the column player, referred to as C. The entries of the payoff matrix are the amount won (or lost in the case of a negative entry) by the row player, R. Thus if R chooses row 2 and C chooses column 3, the payoff of the game to R is the (2, 3) entry. For example, suppose a game has payoff matrix

$$R \begin{array}{c} \\ \end{array} \begin{array}{c} C \\ \begin{bmatrix} 2 & -1 & 3 \\ -1 & 3 & -2 \end{bmatrix} \end{array}$$

The entries in the matrix are the possible payoffs to R. That is, if the number is positive, R wins that many units and C loses that many, and if the number is negative, R loses that many while C wins that many. For example, if R chooses row 2 and C chooses column 3, the payoff is -2, which means that R loses two units (and C wins two units). Let us look at an example.

EXAMPLE 1

Consider the game with the following payoff matrix:

$$R \begin{array}{c} \\ \end{array} \begin{array}{c} C \\ \begin{bmatrix} 3 & 4 & -7 \\ -6 & 2 & -2 \end{bmatrix} \end{array}$$

Problem

Suppose R chooses row 1. What is the payoff if C chooses (a) column 2, (b) column 3?

Solution R has two choices corresponding to the two rows. C has three choices corresponding to the three columns.

(a) If R chooses row 1 and C chooses column 2, then the payoff is the $(1, 2)$ entry, 4. This means that R wins four units and C loses four units.

(b) The payoff is the $(1, 3)$ entry, -7, meaning that R loses seven units and C wins seven units.

This is an example of a **zero-sum game,** which means that the payoff to one player is the loss of the other. After the game is over the sum of the winnings is zero (where a loss is a negative win). Most poker games, for example, are zero-sum games. In a nonzero-sum game both players may lose simultaneously. For instance, in some union-management negotiations both sides can lose—management from lost profits during a strike and union members from lost wages.

Pure and Mixed Strategies

Some games are played repeatedly. The players adopt **strategies** that allow them to maximize their gains or minimize their losses. The simplest type of strategy consists of choosing the same row (for R, or the same column for C) on every play of the game. This is called a **pure strategy.** When a player chooses various options on successive moves it is called a **mixed strategy.**

Let us look at a simple example of a two-person, zero-sum game in which each player has two options.

EXAMPLE 2

Rose and Collin play a game with cards. Each has an ace and a king. They choose one card and place it on the table. The ace beats the king and it pays $2. If both play aces, then Rose wins $1, and if both play kings, then Collin wins $3. Here is the payoff matrix.

$$
R \begin{cases} \text{ace} \\ \text{king} \end{cases}
\overset{\displaystyle \overset{C}{\overbrace{\text{ace} \quad \text{king}}}}{\begin{bmatrix} 1 & 2 \\ -2 & -3 \end{bmatrix}}
$$

Problem What strategy should each player choose?

Solution If Rose plays the ace, she wins either $1 or $2, but if she plays the king, she loses $2 or $3. Therefore, Rose should always play the ace. The option for Collin that maximizes his gain (or loss) is not so clear. He either (1) loses $1 or wins $2 if he plays the ace or (2) loses $2 or wins $3 if he plays the king. Neither option seems preferable. But look at the game from the following perspective: If Collin determines that Rose will always play the ace, then he should also always play the ace in order to minimize his loss. Hence in this game each player's optimum strategy is a pure strategy: Each should always play the ace.

Let us generalize the reasoning in Example 2. *R* wants to choose the row that will maximize her payoff. But once she chooses a row she recognizes that *C* will want to choose the column that contains the smallest number in that row. *R* wants this number to be as large as possible. Therefore if *R* is looking for a pure strategy that will maximize her gain she will choose her row as follows:

For the pure strategy that maximizes the gain for the row player:

1. Determine the smallest entry in each row and circle it.
2. Determine the largest of the circled entries and choose that row.

The first step for *R* is to identify, for each row, the column that *C* will choose if *R* selects that row, that is, the smallest entry. Once each smallest entry is identified, *R* chooses the row that has the largest such number. Thus *R* wants to choose the maximum of the row minima, which is called the **maximin.** In a similar way, *C* will choose the **minimax.**

How should *C* determine the best pure strategy? *C* wants to find the column that will minimize the payoff to *R*. But once *C* chooses a column, *R* will select the row in which the largest entry of the column lies. *C* wants this number to be as small as possible since that will be the payoff to *R*. We put squares about the numbers that *C* identifies to distinguish them from *R*'s selections.

For the pure strategy that maximizes the gain for the column player:

1. Determine the largest number in each column and put a square about it.
2. Determine the smallest of these numbers and choose that column.

Let us look at an example that illustrates these pure strategies.

EXAMPLE 3

Consider the payoff matrix

$$R \begin{matrix} & C \\ & \begin{bmatrix} 2 & 1 \\ 3 & -2 \\ -3 & 0 \end{bmatrix} \end{matrix}$$

Problem

Choose the optimum pure strategies for *R* and *C*.

Solution First *R* circles the smallest numbers in each row and *C* puts a square about the largest numbers in each column.

Step 1

R's strategy   C's strategy

$$\begin{bmatrix} 2 & \boxed{1} \\ 3 & -2 \\ -3 & 0 \end{bmatrix} \quad \begin{bmatrix} 2 & 1 \\ 3 & -2 \\ -3 & 0 \end{bmatrix}$$

Step 2 R chooses the largest circled number, the (1, 2) entry, 1. Therefore, R's optimum pure strategy is to choose row 1. C chooses the smaller of the numbers in squares, the (1, 2) entry, 1. Therefore, C's optimum pure strategy is to choose column 2.

Saddle Points and Strictly Determined Games

Now view the pure strategies for both R and C in the same matrix. The circled numbers in Example 3 are those that R identified, the smallest in each row, while those with a square about them are the ones that C identified, the largest in each column.

$$R \begin{bmatrix} 2 & 1 \\ 3 & -2 \\ -3 & 0 \end{bmatrix}$$

The optimum choice is the one with both a circle and a square. Notice that the (1, 2) entry, 1, is simultaneously the smallest number in its row and the largest number in its column. Such an entry is called a **saddle point** for the game. Not all games have a saddle point. But if a game does have a saddle point, then the optimum pure strategy for R is to choose the row containing the saddle point. Likewise, C should choose the column containing the saddle point. Hence the payoff will always be the saddle point. For example, in the preceding game R will always gain one unit because it is the saddle point.

A game with a saddle point is called a **strictly determined game** because the saddle point, v, will always be the payoff if the players choose the optimum strategy. The number, v, is called the **value** of the game.

The next example demonstrates how to determine the value of a strictly determined game. In general, you cannot tell whether a payoff matrix yields a strictly determined game until you find the players' optimum strategies.

EXAMPLE 4

Problem

Determine the optimum strategies for the payoff matrix and show that it is a strictly determined game.

$$
\begin{array}{cc}
& C \\
R & \begin{bmatrix} 1 & -4 & 0 \\ -1 & -2 & 1 \end{bmatrix}
\end{array}
$$

Solution We circle the smallest numbers in each row and put a square about the largest numbers in each column.

$$
\begin{array}{cc}
& C \\
R & \begin{bmatrix} \boxed{1} & \widehat{-4} & 0 \\ -1 & \widehat{-2} & \boxed{1} \end{bmatrix}
\end{array}
$$

Note that the (2, 2) entry, -2, is the saddle point of the game. Hence $v = -2$ is the value of the game and the game is strictly determined.

The next example shows how game theory is used in decision making.

EXAMPLE 5

A politician is planning a reelection campaign. There are three types of advertising: radio, newspaper, and television. The campaign committee has raised \$100,000 to be apportioned among the three media. The decision depends on the opponents' strategy. If the opponent decides to attack the incumbent's past record, then 70% of the money should go to television and the remainder divided equally between radio and newspaper advertising. But if the opponent's strategy is to center on future issues, then only 50% should be spent on television, 30% on newspaper ads, and 20% on the radio.

Problem

Express the possible expenditures in a payoff matrix. Is the game strictly determined?

Solution We let the opponent's strategies be represented by the columns and the type of media by the rows. (The numbers are in \$1000.)

	Opponent's strategy	
Media Type	Past	Future
Radio	15	20
Newspaper	15	30
Television	70	50

The optimum pure strategies are row 3 and column 2, so it is a strictly determined game with value 50.

From a game theory point of view Example 5 indicates that the incumbent should spend 50% of the money on television and the opponent should center on future issues. As we will see in the first Case Study, however, people often choose options that are not indicated by game theory analysis, often to their detriment.

EXERCISE SET 10–1

For the payoff matrices in Problems 1 to 4 suppose R chooses row 1 and C chooses column 2. Determine which player wins the game and find the payoff of the game.

1. $\begin{bmatrix} 1 & 2 \\ 3 & -4 \end{bmatrix}$

2. $\begin{bmatrix} 4 & -2 \\ 0 & 4 \\ -1 & 0 \end{bmatrix}$

3. $\begin{bmatrix} 3 & -1 & 2 \\ -2 & 0 & -4 \end{bmatrix}$

4. $\begin{bmatrix} 0 & -1 & 4 \\ 3 & -2 & -3 \\ -4 & 2 & -1 \end{bmatrix}$

For the payoff matrices in Problems 5 to 8 suppose R chooses row 2 and C chooses column 3. Determine which player wins the game and find the payoff of the game.

5. $\begin{bmatrix} 1 & -3 & 6 \\ 0 & -4 & 7 \end{bmatrix}$

6. $\begin{bmatrix} 4 & -2 & -2 \\ 0 & 4 & 1 \\ -1 & 0 & 0 \end{bmatrix}$

7. $\begin{bmatrix} -2 & 1 & -2 \\ 3 & -4 & 4 \end{bmatrix}$

8. $\begin{bmatrix} 2 & -1 & -4 \\ -3 & 5 & 3 \\ 3 & -2 & 1 \end{bmatrix}$

The payoff matrices in Problems 9 to 16 are the matrices of a strictly determined game. Determine the optimum strategies for R and C and the value of the game.

9. $\begin{bmatrix} -5 & 0 \\ 3 & 2 \end{bmatrix}$

10. $\begin{bmatrix} 4 & 2 \\ -1 & -3 \end{bmatrix}$

11. $\begin{bmatrix} 1 & 2 \\ -3 & -2 \\ 0 & 1 \end{bmatrix}$

12. $\begin{bmatrix} 1 & 2 \\ -4 & -1 \\ 4 & 3 \end{bmatrix}$

13. $\begin{bmatrix} 1 & -1 & 0 \\ 2 & 3 & 5 \end{bmatrix}$

14. $\begin{bmatrix} -3 & -2 & 4 \\ 3 & -1 & 1 \end{bmatrix}$

15. $\begin{bmatrix} 1 & 0 & -3 \\ 6 & -3 & -1 \\ 3 & 1 & 2 \end{bmatrix}$

16. $\begin{bmatrix} -3 & 1 & -2 \\ 4 & -2 & -1 \\ 3 & 2 & 1 \end{bmatrix}$

In Problems 17 to 26 determine if the game with the given payoff matrix is a strictly determined game. If so, find its value.

17. $\begin{bmatrix} 1 & -2 \\ -3 & 5 \end{bmatrix}$

18. $\begin{bmatrix} 1 & -2 \\ 3 & -5 \end{bmatrix}$

19. $\begin{bmatrix} 1 & -2 \\ -3 & -2 \\ 0 & 3 \end{bmatrix}$

20. $\begin{bmatrix} 0 & 2 \\ -4 & -1 \\ 5 & 3 \end{bmatrix}$

21. $\begin{bmatrix} 1 & 2 \\ 5 & 1 \\ 6 & 3 \end{bmatrix}$

22. $\begin{bmatrix} -3 & -2 & 4 \\ 0 & 1 & -1 \end{bmatrix}$

23. $\begin{bmatrix} 4 & 2 & 5 \\ 0 & 1 & 1 \end{bmatrix}$

24. $\begin{bmatrix} 2 & 0 & -4 \\ 7 & -3 & -2 \\ 2 & 1 & 2 \end{bmatrix}$

25. $\begin{bmatrix} -5 & 0 & -1 \\ 4 & -3 & -1 \\ 4 & 2 & 1 \end{bmatrix}$

26. $\begin{bmatrix} 0 & 5 & 4 \\ 9 & 2 & 4 \\ 9 & 7 & 5 \end{bmatrix}$

27. A politician is planning a reelection campaign. There are three types of advertising: radio, newspaper, and television. The campaign committee has raised $100,000 to be apportioned among the three media. The decision depends on the opponent's strategy. The opponent has three types of strategies to choose from: (1) to attack the incumbent's past record, (2) to center on future issues, or (3) to attack the incumbent's intention to seek higher office during the present term of office. If the opponent decides to attack the incumbent's past record, then 60% of the money should go to television, 25% to radio, and 15% to newspaper ads. If the opponent's strategy is to center on future issues, then 20% should be spent on television, 50% on newspaper ads, and 30% on the radio. If the opponent's strategy is to attack the incumbent's desire for higher office, then 80% should be spent on television, 15% on newspaper ads, and 5% on the radio. Express the percentages in a payoff matrix and determine if the game is strictly determined.

28. In Problem 27 double the percentage of money spent on radio and divide the remaining amount evenly between television and newspaper ads. Express the percentages in a payoff matrix and determine if the game is strictly determined.

In Problems 29 to 32 determine if the game with the given payoff matrix is a strictly determined game. If so, find its value.

29. $\begin{bmatrix} 0 & 2 & 1 & 3 \\ -4 & -1 & 2 & -1 \\ -5 & 3 & 0 & -2 \end{bmatrix}$

30. $\begin{bmatrix} 1 & 3 & 2 & 4 \\ -3 & 0 & 3 & 0 \\ -4 & 4 & 1 & -1 \\ 2 & 1 & -3 & -1 \end{bmatrix}$

31. $\begin{bmatrix} 1 & -2 & -5 & 0 & -1 \\ 4 & 0 & -3 & 5 & -1 \\ 4 & 2 & 2 & 2 & 1 \end{bmatrix}$

32. $\begin{bmatrix} -2 & 3 & 5 & -1 & -3 \\ 3 & 2 & -3 & -5 & -1 \\ 5 & -2 & 3 & 1 & -1 \\ 2 & 2 & 3 & 1 & 0 \end{bmatrix}$

In Problems 33 to 39 a game is described. Find a payoff matrix for the game and decide if it is a strictly determined game. If it is strictly determined, find the value of the game.

33. Each of two players, A and B, has two cards, a 2 and a 5. Each player selects a card and they simultaneously show their cards. If the cards match, then A pays B $1. If A's card is greater than B's, then A wins $5 from B. If B's card is greater than A's, then B wins $3.

34. Player A has two cards, a 3 and a 5. Player B has three cards, a 2, a 4, and a 7. Each player selects a card and they simultaneously show the cards. If the sum is even, then A pays B $4. If the sum is odd, then B pays A $3.

35. Player A has two cards, a 3 and a 6. Player B has three cards, a 2, a 4, and a 7. Each player selects a card and they simultaneously show the cards. If the sum is even, then A pays B $4. If the sum is odd, then B pays A $5.

36. Each of two players has a stone, a scissors, and a paper. Each player selects one, and they simultaneously show their choice. The rules are: a stone beats a scissors, a scissors beats a paper, and a paper beats a stone. The winner receives $1 from the loser, and the payoff is $0 in case of a tie.

37. Each of two players has a coin. They match their coins by turning up either heads or tails. If the same side appears on each coin, then A pays B $1. If A turns up heads, and B turns up tails, then A pays B $2. If A turns up tails and B turns up heads, then B pays A $3.

38. An automobile agency places an order for new cars and vans quarterly. The number of cars ordered depends on projected sales during the specified period. Also, a key factor is the type of cars ordered. There are four types: large cars, midsized cars, compact cars, and recreational vans. The state of the economy plays a significant role in the decision. If the economy is strong during the period, not only will more total cars and vans be purchased, but more large cars and vans will be sold. On the other hand, if the economy is slow, fewer cars will be sold and most of them will be compacts. Here is the agency's prediction of the number of each type of car and van to be sold depending on the economy. (1) If the economy is strong, then 45 large cars, 20 midsized cars, 25 compact cars, and 15 vans will be sold. (2) If the economy is slow, then 20 large cars, 15 midsized cars, 45 compact cars, and 5 vans will be sold. Given that the profit on each large car is $3000, on each midsized car is $2000, on each compact car is $1000, and on each van is $4000, express the profit as a payoff matrix. Determine if the game is strictly determined.

39. The owner of a small video store stocks two types of blank video tapes: Beta and VHS. In the past the owner has stocked them in equal amounts. The industries that make each type of tape have announced a forthcoming breakthrough that will have a dramatic effect on tape sales. The owner must decide among three strategies: order (1) mostly Beta tapes, (2) mostly VHS tapes, or (3) the same amount of each. The profit from tape sales depends on which industry has the more significant breakthrough. The owner determines that profit on the sales will be the following: Suppose the Beta industry has the superior breakthrough: If mostly Beta tapes are ordered, the profit will be $8000; if mostly VHS tapes are ordered, the profit will be $2000; and if an equal amount of each type of tape is ordered, the profit will be $5000. Suppose the VHS industry has the superior breakthrough: If mostly Beta tapes are ordered, the profit will be $3000; if mostly VHS tapes are ordered, the profit will be $9000; and if an equal amount of each type of tape is ordered, the profit will be $5500. Express the profit in a payoff matrix and determine if the game is strictly determined.

Referenced Exercise Set 10–1

1. Military science is a field that utilizes game theory in a wide variety of settings. In a paper studying the peacetime use of aircraft carriers, Grotte and Brooks* measure the naval presence of large fleets versus small fleets in several areas. They study two forces, force R and force C, which deploy fleets in one or two areas of an ocean. Force R has three small fleets at its disposal, and force C has two large fleets. The fleets are allocated simultaneously into the two areas, with force R deploying 0, 1, 2, or 3 fleets to an area (with the sum of R's fleets being 3) and force C deploying 0, 1, or 2 fleets to an area (with the sum of C's fleets being 2). The outcome in each area is decided by which force has the greater presence. In a given area, force R is awarded 1 point if its presence is greater than C's, − 1 point if its presence is less than C's, and 0 points are awarded if its presence is equal to C's. The payoff of the game for a given choice of options for R and for C is the sum of the outcomes in the two areas. The options are expressed as ordered pairs: the first number represents the number of fleets deployed to the first area, and the second number is the number of fleets deployed to the second area. Thus force R has four options: (3, 0), (2, 1), (1, 2) and (0, 3). Force C has three options: (2, 0), (1, 1) and (0, 2). There are 12 entries in the payoff matrix. The rules of the game for a given area are as follows: if R has 3 fleets, its presence is greater if C has 0 or 1 fleet (so R gets 1 point in that area) and it is a draw if C has 2 fleets (so the outcome is 0 points in that area); if R has 2 fleets, its presence is greater if C has 0 fleets, it is a draw if C has 1 fleet, and C's presence is greater if C has 2 fleets; if R has 1 fleet, its presence is greater if C has 0 fleets and C's presence is greater if C has 1 or 2 fleets; if R has 0 fleets, it is a draw if C has 0 fleets, and C's presence is greater if C has 1 or 2 fleets. Find the payoff matrix and determine if the game is strictly determined.

2. In a study designed to define optimal strategies for various types of voting situations, Gerald Glasser† used game theory to model corporate voting. Voting for corporate directors often pits one faction against another. Each side has a certain number of votes to divide among several candidates. Suppose a firm has three director-ships open and the two opposing factions, A and B, have 16 votes and 8 votes, respectively, to divide among three candidates. A strategy is an ordered triple. For example, faction A could choose the strategy (8, 7, 1), meaning that their first candidate would get 8 votes, the second 7 votes and the third 1 vote. Let the entries in the payoff matrix be the number of directors won by faction A. Find the payoff matrix if A chooses between the two strategies (8, 7, 1) and (6, 5, 5), while B chooses between (6, 1, 1) and (4, 3, 1). Is the game strictly determined?

3. Adam Smith is often referred to as the founder of modern economics. His major book, *The Wealth of Nations*, deals with the problem of how social order and human progress can be possible in a society where individuals follow their own self-interest.‡ Tullock shows how the common game theory problem, the "prisoner's dilemma," can be used to explain some of Smith's theory. In the game two prisoners have committed a crime together and are separated for interrogation. The police claim that they have enough evidence to convict them. Each has the choice of confessing, thereby squealing on the other, or not confessing. If only one confesses, he or she will receive a suspended sentence while the other will receive two years in jail. If both confess or neither confesses they both will receive one year in jail. Tullock compares this game to an ordinary competitive market with two conglomerates deciding whether to confess to having signed secret agreements together. The payoff would be the loss of reputation in the business community, corresponding to the number of years in jail for the prisoners. The corresponding payoff matrix is

	confess	not confess
confess	1	0
not confess	2	1

Is the game strictly determined?

---

*J. H. Grotte and P. S. Brooks, "Measuring Naval Presence Using Blotto Games," *International Journal of Game Theory,* Vol. 12, pp. 225–236, 1983.

†Adapted from Gerald Glasser, "Game Theory and Cumulative Voting for Corporate Directors," *Management Science,* pp. 151–156, January 1959.

‡Gordon Tullock, "Adam Smith and the Prisoner's Dilemma," *Quarterly Journal of Economics,* Vol. 100, pp. 1073–1081, 1985.

CASE STUDY   **Game Theory and the Normandy Invasion**

In August 1944, two months after the Allies first landed at Normandy on D-day, they had amassed a force of one million soldiers. They broke out of their beach-head at Cherbourg on the Cotentin peninsula in Normandy, the northwestern part of France. (See Figure 1.) The German Ninth Army was instructed by Hitler to seal off the peninsula "at all costs," thus halting the Allied advance. Hitler felt this could be done easily because he assumed the invasion of Normandy was a trick designed to mask the bulk of the Allies' invasion of Europe, which would come later in northern France near Calais.

The German line at the mouth of the peninsula held strong until August when the Allies broke through along a narrow gap by the sea at the town of Avranches. This threatened the west flank of the German position. The German commander, General von Kluge, was forced to choose between two options: to attack the advancing Allied front in the west or to retreat and take a fortified defensive position in the east by the Seine River.

General Omar Bradley, commander of the Allies, ordered the U.S. Third Army, under General George Patton, to sweep west and south from the gap at Avranches. The First Army threatened the German Ninth to the east. Bradley's problem was how to deploy his reserve consisting of four divisions. He had three options: (1) he could send them to reinforce the First Army at the gap in case the Germans attacked, (2) he could send them east to try to encircle the Germans, or (3) he could hold them back for 24 hours to see whether the First Army was attacked by the Germans. If so, he could use them to reinforce the gap or, if the gap held, he could then send them east.

In his paper "Military Decision and Game Theory," O. G. Haywood[*] used game theory to analyze the options of the two generals by assigning appropriate numbers to the outcomes. Haywood reasoned that for each of the two German options there were three Allied alternatives—totaling six possibilities altogether. They are listed with their most probable outcomes in Figure 2. The outcomes in the table are expressed in terms of the Allies' point of view. For example, the (1, 1) entry, "win one battle," means that the Allies would win the battle if each general chose the corresponding strategy. The term "draw" means that neither side gains an immediate advantage.

		German Decision	
		attack	retreat
Allied Decision	reinforce position	win one battle	draw
	proceed east	lose two battles	win one battle
	hold in reserve	win two battles	draw

Thus the situation is a two-person, zero-sum game with the Allied commander representing the row player and the German commander representing the

[*]O. G. Haywood, "Military Decision and Game Theory," *Journal of the Operations Research Society of America*, Vol. 2, pp. 365–385, 1954.

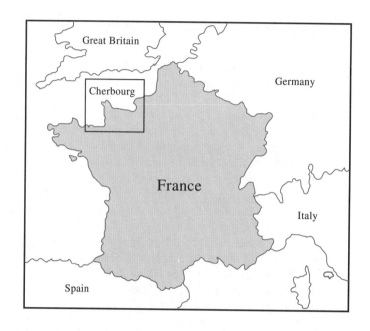

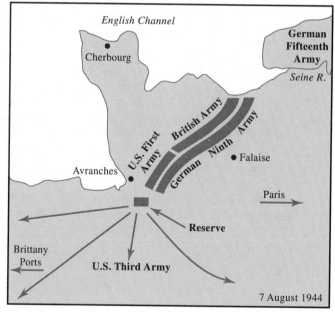

FIGURE 1

column player. There were two possible battles: (1) the Germans attack the Allies at the gap and (2) the Allies harass the Germans in the east. The payoff matrix entries are the number of battles won by the Allies. Thus the (1, 1) entry, corresponding to the Allies reinforcing their position and the Germans attacking, is 1 because the Allies would win the battle. The (2, 1) entry, corresponding to

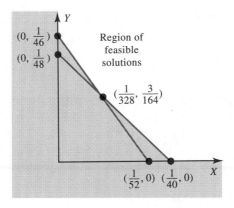

FIGURE 2

the Allies moving their reserves east while the Germans attack, is -2 because the Allies would lose both battles. That is, the Germans would harass the Allies' reserves and defeat the weakened Allied force at the gap. The (3, 1) entry is 2 because the Allies would hold the gap since reinforcements would be available, and the reserve would be free to move east to harass the German rear. In the second column the German retreat would be successful if the Allies reinforced their position with the reserve or if they held the reserve in place. These possibilities would result in a draw, so there is a 0 in the (1, 2) and (3, 2) entries. If the Allies sent the reserve east, they would be able to harass the German retreat. Hence the (2, 2) entry is 1. Here is the payoff matrix.

| | | German Decision | |
		attack	retreat
	reinforce position	1	0
Allied Decision	proceed east	-2	1
	hold in reserve	2	0

or simply

$$\begin{bmatrix} 1 & 0 \\ -2 & 1 \\ 2 & 0 \end{bmatrix}$$

Which strategy do you think each side chose? The Allies chose option 3—to hold back their reserves 24 hours. Von Kluge decided to retreat—his second option. This would have resulted in a relative stalemate corresponding to the entry 0 in the (3, 2) entry of the payoff matrix. The Germans then could have held their position near the gap of the peninsula to harass the Allied advance from their beachhead. At the last moment, however, Hitler, the "Austrian corporal," 300 miles from the battlefield, countermanded von Kluge's decision and ordered him to attack the gap. The Allied forces withstood the attack the first day. Then Bradley deployed his four divisions held in reserve. By the next day the Germans were almost entirely surrounded. Von Kluge was able to escape with only a remnant of his army. A day later he committed suicide.

Case Study Exercises

1. Find the optimum pure strategy for the Allies.
2. Find the optimum pure strategy for the Germans.
3. Is the game defined by the payoff matrix in the Case Study strictly determined?
4. If the (2, 2) entry were changed to -1, would the game be strictly determined?

10–2 Mixed Strategies

In the remainder of this chapter we consider games that are not strictly determined. The players of this type of game usually vary their choice of a row (or column). The mathematics increases in complexity as probability theory plays a key role.

The Even-and-Odd Game

Let us look at the age-old "even-and-odd" child's game. One person places some marbles in his hand and the other player guesses whether there is an even or odd number. If she guesses correctly she wins a marble. Otherwise she gives the other player a marble. If R is the player who guesses and C is the player who puts marbles in his hand, the payoff matrix is

$$
\begin{array}{c}
\qquad\qquad C \\
\qquad\overbrace{\text{even}\quad\text{odd}} \\
R\ \begin{cases}\text{even}\\ \text{odd}\end{cases}
\begin{bmatrix} 1 & -1 \\ -1 & 1 \end{bmatrix}
\end{array}
$$

The game is not strictly determined, so the players cannot have an optimum pure strategy. They must vary their choices if they hope to win. The question is, what is the best way to vary the choices?

In a delightful account of this game Edgar Allen Poe* recounted the story of an 8-year-old with psychic powers. He could read minds and therefore knew precisely how many marbles his opponent had in hand. He won all the marbles in the school. It seemed impossible to win even once against such an invincible foe. Later in the story Poe gave a hint as to how to thwart the "brain-child" as he described one opponent's strategy:

> As poet and mathematician, he would reason well; as mere mathematician, he should not reason at all.

*From Edgar Allen Poe, *Poe's Tales*, New York, 1845.

The key is not to reason at all! Leave the choice to chance. Flip a coin, roll a die, or use some other chance device to determine how many marbles to hold so the brain-child cannot predict your strategy.

Mixed Strategies

When strategies in game theory are determined by chance, that is, by probability methods, they are called **mixed strategies.** Consider the game with payoff matrix

$$P = \begin{bmatrix} 1 & -2 \\ -3 & 5 \end{bmatrix}$$

The game is not strictly determined, so the players must vary their choices. Suppose R decides to choose row 1 30% of the time and row 2 70% of the time. And suppose C decides to favor column 1 over column 2 by choosing it 80% of the time versus 20% of the time for column 2. These mixed strategies are expressed in matrix form as follows: R's strategy is a row matrix, $A = [.3 \quad .7]$ and C's strategy is a column matrix $B = \begin{bmatrix} .8 \\ .2 \end{bmatrix}$. In this way the game can be considered a probability experiment. The sample space consists of the four outcomes $\{1, -2, -3, 5\}$, which are the entries in the payoff matrix P. The probability assigned to each element in the sample space is the frequency with which the players choose that entry. For instance, the $(1, 1)$ entry, 1, occurs when R chooses row 1 and C chooses column 1. These choices are independent, so we multiply their probabilities to find the probability that both happen at the same time. Thus

$$Pr(1) = Pr(R \text{ chooses row 1 and } C \text{ chooses column 1})$$

$$= Pr(\text{row 1}) \times Pr(\text{column 1}) = .3 \times .8 = .24$$

Hence $Pr(1) = .24$. This means that the $(1, 1)$ entry will occur with probability .24. Another way of seeing this is to note that R will choose row 1 30% of the time and, of that 30%, C will choose column 1 80% of the time. So row 1 and column 1 will occur together $30\% \times 80\% = 24\%$ of the time. We similarly compute the other three probabilities where we abbreviate "R chooses row n" by "row n" and "C chooses column n" by "column n."

$$Pr(-2) = Pr(\text{row 1 and column 2}) = .3 \times .2 = .06$$

$$Pr(-3) = Pr(\text{row 2 and column 1}) = .7 \times .8 = .56$$

$$Pr(5) = Pr(\text{row 2 and column 2}) = .7 \times .2 = .14$$

We express these probabilities in a probability distribution.

x	p
1	.24
-2	.06
-3	.56
5	.14

Example 1 illustrates how strategies for R and C produce the probabilities of the entries of a payoff matrix.

EXAMPLE 1

Suppose the strategy for R is $A = [0.3 \quad 0.5 \quad 0.2]$ and the strategy for C is $B = \begin{bmatrix} .6 \\ .4 \end{bmatrix}$.

Problem

Determine the size of the payoff matrix and the probability of each entry.

Solution There are three choices for R and two for C, so the payoff matrix must be 3×2. Hence there are six entries. The following probability distribution gives the probabilities of each entry of the payoff matrix:

x	p
(1, 1) entry	.18
(1, 2) entry	.12
(2, 1) entry	.30
(2, 2) entry	.20
(3, 1) entry	.12
(3, 2) entry	.08

Expected Value

Let us return to the game whose payoff matrix P and probability distribution are

$$P = \begin{bmatrix} 1 & -2 \\ -3 & 5 \end{bmatrix}$$

x	p
1	.24
-2	.06
-3	.56
5	.14

What is the long-term trend of this game? That is, if the game is played many times, which player will win the most? The expected value, E, of a probability distribution answers this question. To compute E, sum the products xp. For the above game the expected value is

$$E = (1)(.24) + (-2)(.06) + (-3)(.56) + 5(.14) \tag{1}$$
$$= -.86$$

This means that R can expect to lose .86 units per play. That is, if the game were played 100 times, R would most likely lose about 86 units.

Another way to compute the expected value of a game is to use matrix multiplication, which is why the strategies are expressed as matrices. Consider the three previous matrices and multiply them as follows:

$$APB = [.3 \quad .7] \begin{bmatrix} 1 & -2 \\ -3 & 5 \end{bmatrix} \begin{bmatrix} .8 \\ .2 \end{bmatrix}$$

$$= [.3 \quad .7] \begin{bmatrix} (1)(.8) + (-2)(.2) \\ (-3)(.8) + (5)(.2) \end{bmatrix}$$

$$= [(.3)(1)(.8) + (.3)(-2)(.2) + (.7)(-3)(.8) + (.7)(5)(.2)] \quad (2)$$

$$= [.24 - .12 - 1.68 + .7] = [-.86]$$

This is the same expected value that was computed in equation 1. Look at the numbers in equations 1 and 2. The numbers are the same, only their order is different. As seen in Chapter 2, matrices are often used to establish a specific order to a set of numbers. By using matrices to define the strategies, we have the numbers in a convenient form for computing the expected value. This computation is a particular instance of an important result that we single out.

The Expected Value of a Game with Mixed Strategies
If a payoff matrix P is an $n \times m$ matrix and the mixed strategy for R is the $1 \times n$ matrix A and the mixed strategy for C is the $m \times 1$ matrix B, then the expected value E of the game is equal to

$$E = APB$$

Example 2 illustrates this result.

EXAMPLE 2

Suppose a game has payoff matrix P and strategies A and B.

$$P = \begin{bmatrix} 1 & 0 & -1 \\ 0 & -2 & 3 \\ -3 & 4 & 0 \end{bmatrix} \quad A = [.1 \quad .5 \quad .4] \quad B = \begin{bmatrix} .7 \\ .1 \\ .2 \end{bmatrix}$$

We compute the expected value as follows:

$$APB = [.1 \quad .5 \quad .4] \begin{bmatrix} 1 & 0 & -1 \\ 0 & -2 & 3 \\ -3 & 4 & 0 \end{bmatrix} \begin{bmatrix} .7 \\ .1 \\ .2 \end{bmatrix}$$

$$= [.1 \quad .5 \quad .4] \begin{bmatrix} .7 \\ .1 \\ .2 \end{bmatrix} = [-.43]$$

Hence the expected value is $-.43$. Therefore, in the long run R can expect to lose .43 units per play with these strategies.

Suppose R learns that C will play the mixed strategy in Example 2. Is there a better strategy for R to use? We could try some different numbers and compute the product of the matrices, but that would be a hit-and-miss approach. Instead, express R's strategy as a row matrix in general form.

$A = [a \quad b \quad c]$

Here a, b, and c are probabilities with $a + b + c = 1$. We compute the expected value and choose the values that maximize E.

$$E = APB = [a \quad b \quad c] \begin{bmatrix} 1 & 0 & -1 \\ 0 & -2 & 3 \\ -3 & 4 & 0 \end{bmatrix} \begin{bmatrix} .7 \\ .1 \\ .2 \end{bmatrix}$$

$$= [.5a + .4b - 1.7c] \tag{3}$$

Since the third term has a negative sign, and since a, b, and c are nonnegative, R must make c as small as possible. R should also choose a and b to be almost equal because their coefficients are close. For instance, if R chooses the strategy $[.5 \quad .4 \quad .1]$, then

$$E = .5(.5) + .4(.4) - 1.7(.1) = .25 + .16 - .17 = .24$$

With this new strategy R changes the losing game of Example 2 into a winning one. So far so good, but is this R's optimum strategy? Example 3 investigates this question further.

EXAMPLE 3

Problem

Given the game with payoff matrix P and strategies A and B above, find values for a, b, and c that maximize E.

Solution From equation 3, $E = .5a + .4b - 1.7c$. Since the coefficient of c is negative, c should be made as small as possible, so let $c = 0$. Since the coefficient of a, .5, is larger than that of b, .4, E will be maximized by letting a be as large as possible. So let $a = 1$ and $b = 0$. Therefore, E is maximized with the strategy $[1 \quad 0 \quad 0]$. Then $E = .5(1) + .4(0) - 1.7(0) = .5$.

The value of E in Example 3 is more than double the previous expected value of .24, but it is a pure strategy and C would easily uncover it. Then C would certainly change strategies accordingly. Thus to maximize R's gain in a nonstrictly determined game, we must keep in mind that C will most likely choose the mixed strategy that minimizes R's chances. This will be the starting point of the next section.

Matrices with Positive Entries

Thus far most of the payoff matrices that were studied have had both positive and negative entries. However, many important results in game theory are derived for matrices with only positive entries. Therefore, in the next two sections where some of these results are presented, we will stipulate that every entry in the payoff matrix must be positive. This seems at first to be a significant restriction because it rules out many possible matrices; in fact, it rules out almost all that we have studied so far. But we are primarily interested in computing the expected values of these matrices, and with this view the stipulation is not restrictive. That is, if a matrix has a zero or negative entry, then we can add a large enough number to each entry to make all the entries positive. The payoff matrix is changed—but how is the expected value changed? If the number n is added to the entries of a payoff matrix, the expected value of the new matrix is equal to the original expected value plus n. Let us look at an example.

EXAMPLE 4

Suppose the payoff matrix is

$$P = \begin{bmatrix} 1 & -3 \\ -4 & 5 \end{bmatrix}$$

Problem

Construct a matrix Q from P by adding the smallest positive integer to all the entries of P so that Q has all positive entries. Then compare the expected values of P and Q.

Solution Since the smallest entry in P is -4, add 5 to all entries to get Q.

$$Q = \begin{bmatrix} 6 & 2 \\ 1 & 10 \end{bmatrix}$$

Compare the expected values of P and Q using some specific strategies. Let $A = [.6 \quad .4]$ and $B = \begin{bmatrix} .2 \\ .8 \end{bmatrix}$. The expected value for P is

$$APB = [.6 \quad .4] \begin{bmatrix} 1 & -3 \\ -4 & 5 \end{bmatrix} \begin{bmatrix} .2 \\ .8 \end{bmatrix} = [-.04]$$

The expected value for Q is

$$AQB = [.6 \quad .4] \begin{bmatrix} 6 & 2 \\ 1 & 10 \end{bmatrix} \begin{bmatrix} .2 \\ .8 \end{bmatrix} = [4.96]$$

Thus the expected value for Q is 5 greater than the one for P since $4.96 = -.04 + 5$.

In the remainder of this chapter we deal only with payoff matrices with positive entries. Example 4 shows that no generality is lost because if a game has a payoff matrix with negative entries, a new payoff matrix can be constructed with all positive entries and their expected values differ by the constant. Section 10–4 also shows that the optimal strategies remain the same.

The general result is given here.

> If P and Q are payoff matrices with $Q = P + c$ for some constant c, then the optimum strategies are the same for P and Q, and for any strategies A and B the expected values for P and Q differ by c; that is,
>
> $$AQB = APB + c$$

EXERCISE SET 10–2

In Problems 1 to 6 determine whether the games are strictly determined.

1. $\begin{bmatrix} 3 & 2 \\ 2 & -1 \end{bmatrix}$

2. $\begin{bmatrix} 0 & 2 \\ 2 & -1 \end{bmatrix}$

3. $\begin{bmatrix} 1 & -1 \\ 2 & 0 \\ 1 & 3 \end{bmatrix}$

4. $\begin{bmatrix} 1 & -1 & 6 \\ -3 & 2 & 1 \end{bmatrix}$

5. $\begin{bmatrix} 1 & 2 & 4 \\ 0 & 1 & 3 \end{bmatrix}$

6. $\begin{bmatrix} -3 & -2 & 1 \\ 2 & -1 & 0 \\ 0 & -3 & 4 \end{bmatrix}$

In Problems 7 to 10 the strategies for R and C are given. Determine the size of the payoff matrix and compute the probability of each entry being played.

7. $[.4 \quad .6], \quad \begin{bmatrix} .5 \\ .5 \end{bmatrix}$

8. $[.5 \quad .5], \quad \begin{bmatrix} .2 \\ .8 \end{bmatrix}$

9. $[.8 \ .1 \ .1], \quad \begin{bmatrix} .4 \\ .6 \end{bmatrix}$

10. $[.6 \quad .4], \quad \begin{bmatrix} .4 \\ .1 \\ .5 \end{bmatrix}$

In Problems 11 to 14 the game has payoff matrix

$$P = \begin{bmatrix} 1 & 2 \\ 2 & -4 \end{bmatrix}$$

Calculate the expected value for each of these strategies.

11. $[.5 \quad .5], \quad \begin{bmatrix} .4 \\ .6 \end{bmatrix}$

12. $[.6 \quad .4], \quad \begin{bmatrix} .7 \\ .3 \end{bmatrix}$

13. $[.1 \quad .9], \quad \begin{bmatrix} .2 \\ .8 \end{bmatrix}$

14. $[.3 \quad .7], \quad \begin{bmatrix} .5 \\ .5 \end{bmatrix}$

In Problems 15 to 18 the game has payoff matrix

$$P = \begin{bmatrix} 1 & -2 \\ 0 & -1 \\ -2 & 4 \end{bmatrix}$$

Compute the expected value of the game with the given strategies.

15. $[.1 \quad .2 \quad .7], \quad \begin{bmatrix} .5 \\ .5 \end{bmatrix}$

16. $[.1 \quad .7 \quad .2], \quad \begin{bmatrix} .5 \\ .5 \end{bmatrix}$

17. $[.1 \quad .1 \quad .8], \quad \begin{bmatrix} .2 \\ .8 \end{bmatrix}$

18. $[.2 \quad .1 \quad .7], \quad \begin{bmatrix} .2 \\ .8 \end{bmatrix}$

In Problems 19 to 22 add the smallest possible integer to the matrix to get a matrix with all positive entries; that is, choose the number such that the smallest entry is 1.

19. $\begin{bmatrix} 3 & -4 \\ -2 & 6 \end{bmatrix}$

20. $\begin{bmatrix} 3 & -4 \\ -2 & 6 \\ -4 & 5 \end{bmatrix}$

21. $\begin{bmatrix} -1 & 4 & -5 \\ 2 & -6 & 7 \end{bmatrix}$

22. $\begin{bmatrix} -7 & 4 & -3 \\ 2 & -6 & 5 \\ 2 & -8 & -7 \end{bmatrix}$

In Problems 23 to 26 the game has payoff matrix

$$P = \begin{bmatrix} 1 & 0 & -2 \\ -1 & 1 & 3 \end{bmatrix}$$

Let C play the given strategy and let R play the strategy $[a \quad b]$. Determine what values of a and b will maximize the payoff to R.

23. $\begin{bmatrix} .7 \\ .2 \\ .1 \end{bmatrix}$

24. $\begin{bmatrix} .2 \\ .4 \\ .4 \end{bmatrix}$

25. $\begin{bmatrix} .3 \\ .2 \\ .5 \end{bmatrix}$

26. $\begin{bmatrix} .4 \\ .3 \\ .3 \end{bmatrix}$

In Problems 27 to 30 the game has payoff matrix

$$P = \begin{bmatrix} -2 & 1 & 2 \\ 1 & -2 & -3 \end{bmatrix}$$

Let C play the given strategy and let R play the strategy $[a \quad b]$. Determine what values of a and b will maximize the payoff to R.

27. $\begin{bmatrix} .1 \\ .8 \\ .1 \end{bmatrix}$

28. $\begin{bmatrix} .4 \\ .2 \\ .4 \end{bmatrix}$

29. $\begin{bmatrix} .4 \\ .1 \\ .5 \end{bmatrix}$

30. $\begin{bmatrix} 0 \\ .3 \\ .7 \end{bmatrix}$

In Problems 31 to 34 add the smallest possible integer to the payoff matrix P to obtain the payoff matrix Q with all positive entries and compute the expected values using the matrices A and B, the strategies for R and C, respectively. How are the two expected values related?

31. $P = \begin{bmatrix} 2 & -1 \\ -2 & 1 \end{bmatrix}$, $A = [.4 \quad .6]$,

$B = \begin{bmatrix} .3 \\ .7 \end{bmatrix}$

32. $P = \begin{bmatrix} 1 & -4 \\ -5 & 2 \end{bmatrix}$, $A = [.6 \quad .4]$,

$B = \begin{bmatrix} .8 \\ .2 \end{bmatrix}$

33. $P = \begin{bmatrix} -1 & 4 & 9 \\ -2 & 5 & -6 \end{bmatrix}$, $A = [.2 \quad .8]$,

$B = \begin{bmatrix} .7 \\ .2 \\ .1 \end{bmatrix}$

34. $P = \begin{bmatrix} -3 & 3 \\ -2 & 4 \\ 1 & -5 \end{bmatrix}$, $A = [.6 \quad .1 \quad .3]$,

$B = \begin{bmatrix} .6 \\ .4 \end{bmatrix}$

In Problems 35 and 36 the game has payoff matrix

$$P = \begin{bmatrix} 2 & 3 & -4 \\ 1 & 0 & -2 \\ -1 & 1 & 3 \end{bmatrix}$$

Compute the expected value of the game with the given strategies.

35. $[.1 \quad .2 \quad .7]$, $\begin{bmatrix} .5 \\ .4 \\ .1 \end{bmatrix}$

36. $[.1 \quad .1 \quad .8]$, $\begin{bmatrix} .2 \\ .5 \\ .3 \end{bmatrix}$

In Problems 37 and 38 the game has payoff matrix

$$P = \begin{bmatrix} 0 & -2 & 3 & 0 \\ 1 & 1 & 1 & 2 \\ -2 & 2 & 1 & 1 \end{bmatrix}$$

Compute the expected value of the game with the given strategies.

37. $[.2 \quad .5 \quad .3]$, $\begin{bmatrix} .4 \\ .4 \\ .1 \\ .1 \end{bmatrix}$

38. $[.1 \quad .1 \quad .8]$, $\begin{bmatrix} .1 \\ .5 \\ .3 \\ .1 \end{bmatrix}$

Problems 39 and 40 refer to two sporting goods manufacturers, Action Sportswear and Bargain Sportswear, that place new products on the market at the same time. Each manufacturer uses newspaper and television (TV) ads to market the product. The payoff matrix shows increases in sales (in millions of dollars) for Action Sportswear depending on which choice each company makes for its ad campaign.

Bargain

newspaper TV

Action $\begin{cases} \text{newspaper} \\ \text{TV} \end{cases}$ $\begin{bmatrix} 3 & -1 \\ -2 & 2 \end{bmatrix}$

39. What is the optimum strategy for Action sportswear if Bargain Sportswear puts 20% of its advertising budget into newspaper ads and 80% into TV ads? What is the value of the game?

40. What is the optimum strategy for Action Sportswear if Bargain Sportswear puts 80% of its advertising budget into newspaper ads and 20% into TV ads? What is the value of the game?

Problems 41 and 42 refer to the town of Springfield, which has two shopping malls that compete for business. Both the East Mall and the West Mall advertise on television (T), on radio (R), in the newspaper (N), and via mail (M). The following payoff matrix indicates the number (in thousands) of shoppers attracted to, or lost by, each mall.

West Mall

$$\text{East Mall} \begin{array}{c} T \\ R \\ N \\ M \end{array} \begin{bmatrix} T & R & N & M \\ 0 & -1 & -1 & 0 \\ 1 & 2 & -1 & -1 \\ 0 & -1 & 0 & 1 \\ -1 & -1 & -1 & 0 \end{bmatrix}$$

41. What is the optimum strategy for the West Mall if the East Mall always places radio ads? What is the value of the game?

42. What is the optimum strategy for the West Mall if the East Mall always places newspaper ads? What is the value of the game?

Referenced Exercise Set 10–2

1. In an article by T. Ingold that studied the problems of reindeer owners in Finland a game theory model was used to help herdsmen decide what to do with reindeer that were expected to calve the following season.* (See matrix that follows.) That is, how many should be slaughtered for food and how many should be left in the wild for possible future profit, either through off-spring or slaughter. The entries of the payoff matrix represented revenue to the owner in dollars per deer. The entries took into account the value of the cow, the value of the calf, and an estimate of the future value of each. There were three possibilities for the column player, which could be regarded as Nature: (1) cow and calf found next season, (2) cow not found but still alive, and (3) cow dies a natural death. The two possibilities for the row player, the owner, were (1) leave the cow alive or (2) slaughter the cow. The payoff matrix was

$$\begin{array}{ccc} & \text{not found} & \text{natural} \\ \text{found} & \text{but alive} & \text{death} \end{array}$$

$$\begin{array}{c} \text{leave alive} \\ \text{slaughter} \end{array} \begin{bmatrix} 300 & 100 & 0 \\ 200 & 200 & 200 \end{bmatrix}$$

Find the expected value of the game with the strategies

$$[.2 \quad .8] \text{ and } \begin{bmatrix} .5 \\ .3 \\ .2 \end{bmatrix}.$$

2. Most observers of the 1980 strike by the Major League Players Association felt that the very unusual "discontinuous" strike threat by the players' union was strange at best and futile at worst. The players voted on April 1, 1980 to strike the remaining games of the exhibition season (for which they do not get paid in any case), but to return to work for the first six weeks of the regular season. Then, if negotiations were still deadlocked on May 23, they would strike again. Few people at the time saw the logic in the players' strategy. The key is that May 23 represented the start of the owners' peak revenue period. A study by two economists, Lawrence DeBrack and Alvin Roth, applied game theory to show that the players' strategy was an optimum threat to the owners.[†] The following game theory model was used to describe the players' union strategy. Let the union be the column player and the owners the row player. The union had to choose between two strategies, a continuous strike from the beginning of the season or a discontinuous strike. The owners had to choose between the threat of a lockout (in which case they would not have to pay those not on strike and save that cost) or not to lockout. The entries in the payoff matrix represent revenue of the owners in relative terms—the larger the number is, the more the revenue will be. The payoff matrix is

---

*T. Ingold, "Statistical Husbandry: Chance, Probability, and Choice in a Reindeer Management Economy," *Science*, Vol. 32, pp. 1029–1038, 1986.

†Lawrence DeBrock and Alvin Roth, "Strike Two: Labor-Management Negotiations in Major League Baseball," *Bell Journal of Economics*, Vol. 16, pp. 413–421, 1985.

	Union continuous strike	discontinuous strike

Owners $\begin{cases} \text{lockout} \\ \text{no lockout} \end{cases}$ $\begin{bmatrix} 4 & 3 \\ 2 & 1 \end{bmatrix}$

Is the game strictly determined? What is the optimum strategy for the players' union?

3. Fellingham and Newman illustrate how the exercise of an accountant auditing a client can be considered a strategic game between the auditor and the client.‡ The authors assume that the auditor is the row player and the client is the column player. Each has two strategies. The auditor can "extend auditing procedures" or "not extend auditing procedures." This means the auditor must decide whether ordinary procedures are enough to generate a complete and accurate audit or whether further study and procedures are necessary. The latter produces a significant increase in cost. The client chooses between "more effort" or "less effort" when preparing for the audit. The latter produces a smaller cost to the client. The payoff is the relative profit of the auditor. It is a "relative" profit because there are certain considerations that are computed in the profit for each choice of strategies that cannot be expressed in real dollars, such as the cost of a poor audit causing loss of further work for the client as well as loss of reputation. For one particular auditor and client, the following payoff matrix is given:

	lower effort	higher effort
extend	47	72
not extend	85	67

What is the optimum strategy for the auditor?

Cumulative Exercise Set 10–2

1. For the given payoff matrices suppose R chooses row 2 and C chooses column 3. Determine which player wins the game and find the payoff of the game.

   (a) $\begin{bmatrix} 3 & 2 & -6 \\ 0 & -4 & 4 \end{bmatrix}$

   (b) $\begin{bmatrix} 4 & -3 & -2 \\ 1 & 5 & 6 \\ -1 & 0 & 2 \end{bmatrix}$

2. The given payoff matrices are the matrices of a strictly determined game. Determine the optimum strategies for R and C and the value of the game.

   (a) $\begin{bmatrix} -6 & 1 \\ 3 & 2 \end{bmatrix}$

   (b) $\begin{bmatrix} 2 & 4 \\ -3 & -2 \\ 0 & -1 \end{bmatrix}$

3. Determine if the game with the given payoff matrix is a strictly determined game. If it is, find its value.

   (a) $\begin{bmatrix} 4 & -2 \\ -1 & 6 \end{bmatrix}$

   (b) $\begin{bmatrix} 2 & -2 \\ -3 & -2 \\ 1 & 0 \end{bmatrix}$

4. Find a payoff matrix for the game and decide if it is a strictly determined game. If so, find its value. Each of two players has a coin. They match their coins by turning up either heads or tails. If the same side appears on each coin, then A pays B \$3. If A turns up heads and B turns up tails, then A pays B \$5. If A turns up tails and B turns up heads, then B pays A \$7.

5. A game has payoff matrix

   $$P = \begin{bmatrix} 0 & 1 \\ 2 & -4 \end{bmatrix}$$

   Calculate the expected value for the strategies

   (a) $[0.5 \quad 0.5]$, $\begin{bmatrix} 0.3 \\ 0.7 \end{bmatrix}$

   (b) $[0.6 \quad 0.4]$, $\begin{bmatrix} 0.8 \\ 0.2 \end{bmatrix}$

6. A game has payoff matrix

   $$P = \begin{bmatrix} 1 & -2 & 3 & 1 \\ 2 & 0 & -1 & 3 \\ -3 & 1 & 3 & 0 \end{bmatrix}$$

   Compute the expected value of the game with the strategies

   $[0.2 \quad 0.5 \quad 0.3]$, $\begin{bmatrix} 0.4 \\ 0.3 \\ 0.2 \\ 0.1 \end{bmatrix}$

‡John C. Fellingham and D. Paul Newman, "Strategic Considerations in Auditing," *The Accounting Review*, Vol. 60, pp. 634–648, 1985.

7. A game has payoff matrix

$$P = \begin{bmatrix} 3 & 2 & -2 \\ -1 & 1 & 4 \end{bmatrix}$$

Let R play the strategy $[a \quad b]$. Determine what values of a and b will maximize the payoff to R if C plays the strategy

$$\begin{bmatrix} 0.1 \\ 0.2 \\ 0.7 \end{bmatrix}$$

8. Add the smallest possible integer to the payoff matrix P to obtain the payoff matrix Q with all positive entries, and compute the expected values using the matrices A and B, the strategies for R and C, respectively. How are the two expected values related?

$$P = \begin{bmatrix} 3 & -2 \\ -1 & 1 \end{bmatrix}, \quad A = [0.7 \quad 0.3], \quad B = \begin{bmatrix} 0.2 \\ 0.8 \end{bmatrix}$$

9. Active Shoes places a new style of running shoe on the market at the same time that its primary competitor, Fleet Shoes, introduces its own new line. Each brand will use radio and television ads to market the new line. The payoff matrix shows the increase in sales (in millions of dollars) for Active depending on which choice each company makes for its ad campaign.

| | Fleet | |
	Radio	TV
Active Radio	2	-1
TV	-3	1

Suppose Active learns that Fleet will put 20% of its

advertising budget into radio ads and 80% into TV ads. Determine the optimum strategy for Active and the value of the game.

10. A union of workers for a clothing manufacturer decides to threaten a "discontinuous" strike by striking for three weeks if a contract is not ratified by the last date that the previous contract is in effect, and then return to work for three weeks to finish manufacturing the summer season line of clothes. However, if negotiations are still deadlocked three weeks later, the union will strike again to threaten the entire fall season line of clothes, which represents the owners' peak revenue period. Show that the union's strategy is an optimum threat to the owners by applying the following game theory model to describe the union's strategy. Let the union be the column player and the owners the row player. The union has to choose between two strategies, a continuous strike from the end of the previous contract or a discontinuous strike. The owners have to choose between the threat of a lockout (in which case they will not have to pay those not on strike and save that cost) or not to lockout. The entries in the payoff matrix represent revenue of the owners in relative terms—the larger the number is, the more revenue there is. The payoff matrix is

| | Union | |
	continuous strike	discontinuous strike
Owners lockout	8	4
no lockout	5	1

Is the game strictly determined? What is the optimum strategy for the workers' union?

10–3 Optimal Mixed Strategies

Section 10–2 showed how closely aligned game theory is with probability. Strategies are expressed in terms of probabilities and the expected value measures whether the game favors one player. This section is based on the fact that **optimal mixed strategies** take into account the opponent's **counterstrategy**. We compute R's optimal strategy given that C chooses the column that minimizes the expected value. C's optimal strategy will be considered in Section 10–4.

Linear Equations

For the time being we consider only 2×2 payoff matrices. Let R's strategy be unknown and let us refer to it by the matrix $[x \quad y]$, where x and y are nonnegative numbers such that $x + y = 1$. We look for the values of x and y that maximize the expected value for different strategies of C. Example 1 demonstrates how to do this.

EXAMPLE 1

Let a game have payoff matrix P and let A be R's strategy where

$$P = \begin{bmatrix} 2 & 1 \\ 1 & 3 \end{bmatrix} \qquad A = [x \quad y]$$

Problem

Find the expected value E of the game if C chooses (a) column 1 and (b) column 2.

Solution (a) Since C chooses column 1, then C's strategy is $B = \begin{bmatrix} 1 \\ 0 \end{bmatrix}$. The expected value is

$$E = APB = [x \quad y] \begin{bmatrix} 2 & 1 \\ 1 & 3 \end{bmatrix} \begin{bmatrix} 1 \\ 0 \end{bmatrix} = [x \quad y] \begin{bmatrix} 2 \\ 1 \end{bmatrix} = [2x + y]$$

Since $x + y = 1$, we have $E = 2x + y = 2x + (1 - x) = x + 1$.
(b) C's strategy is $B = \begin{bmatrix} 0 \\ 1 \end{bmatrix}$, and so the expected value is

$$E = APB = [x \quad y] \begin{bmatrix} 2 & 1 \\ 1 & 3 \end{bmatrix} \begin{bmatrix} 0 \\ 1 \end{bmatrix} = [x \quad y] \begin{bmatrix} 1 \\ 3 \end{bmatrix} = [x + 3y]$$

Since $x + y = 1$, we have $E = x + 3y = x + 3(1 - x) = 3 - 2x$. Hence if C chooses column 1, we have $E = x + 1$, and if C chooses column 2, we have $E = 3 - 2x$.

The expressions for the expected value in Example 1 are linear equations in x. Figures 1 and 2 show the graphs of the equations $E = x + 1$ and $E = 3 - 2x$. Each is a linear segment drawn only between the values $x = 0$ and $x = 1$ because x is a probability and thus is a number between 0 and 1. The graphs help us to visualize R's strategy. Remember that a choice for x determines the value of y since $y = 1 - x$, so it gives a complete strategy for R. For instance, if $x = \frac{1}{4}$, then $y = 1 - \frac{1}{4} = \frac{3}{4}$, so $A = [\frac{1}{4} \quad \frac{3}{4}]$. Table 1 refers to the game in Example 1. It lists some choices for x, the corresponding strategies for R, and the expected values of the game depending on C's choice of column. Notice that as x increases the expected value increases when C chooses column 1, but it decreases when C chooses column 2. C seeks to minimize E, which implies that C will choose the column that produces the smaller value of E. For instance, if R chooses $x = \frac{1}{4}$, then C will choose column 1 because $\frac{5}{4}$ is less than $\frac{5}{2}$. But if R chooses $x = .7$, then C will choose column 2 because 1.6 is smaller than 1.7.

Here is the problem that R faces. Which strategy will maximize E subject to the fact that C can select the smaller of the two values of E? The graphs will help answer this question. In Figure 3 we draw them in the same coordinate system.

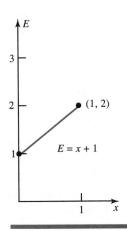

FIGURE 1

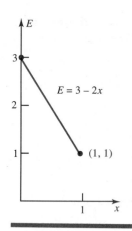

FIGURE 2

Table 1

		C Chooses	
		Column 1	Column 2
x	$A = [x \quad y]$	$E = x + 1$	$E = 3 - 2x$
$\frac{1}{4}$	$[\frac{1}{4} \quad \frac{3}{4}]$	$\frac{5}{4}$	$\frac{5}{2}$
$\frac{1}{3}$	$[\frac{1}{3} \quad \frac{2}{3}]$	$\frac{4}{3}$	$\frac{7}{3}$
.4	$[.4 \quad .6]$	1.4	2.2
.7	$[.7 \quad .3]$	1.7	1.6

Here is how to view C's strategy geometrically. If R chooses $x = \frac{1}{4}$, then C draws a vertical line through $x = \frac{1}{4}$. It intersects both lines corresponding to E. C then chooses the value of E corresponding to the lower line. It corresponds to column 1, so $E = \frac{5}{4}$ since the point of intersection is $(\frac{1}{4}, \frac{5}{4})$. However, if R chooses $x = .7$, the lower line intersecting $x = .7$ corresponds to column 2, so $E = 1.6$ since the point of intersection is $(.7, 1.6)$.

Optimal Strategies

The point of intersection is a key to the optimal strategy. It is obtained by setting the two expressions for E equal to each other.

Set

$$E = x + 1 \qquad \text{equal to} \qquad E = 3 - 2x$$
$$x + 1 = 3 - 2x$$
$$3x = 2$$
$$x = \tfrac{2}{3}$$

Therefore, the point of intersection is $(\frac{2}{3}, \frac{5}{3})$, graphed in Figure 4. Notice that if R chooses a value of x between 0 and $\frac{2}{3}$, then C will choose the line $E = x + 1$ (column 1), because it is always the lower line from $x = 0$ to $\frac{2}{3}$. But if R chooses a value of x between $\frac{2}{3}$ and 1, then C will choose the line $E = 3 - 2x$ (column 2) since in that interval it is the lower line. This describes C's counterstrategy.

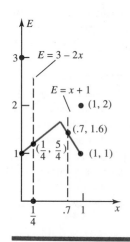

FIGURE 3

C's Optimal Counterstrategy

Choose

column 1, $E = x + 1,$ if $0 \le x \le \frac{2}{3}$
column 2, $E = 3 - 2x,$ if $\frac{2}{3} \le x \le 1$

In Figure 4 the lower lines are shaded; they indicate C's strategy. Since the second coordinates represent the expected value and since R seeks to maximize E, the largest value of E on the shaded lines occurs at the point of intersection.

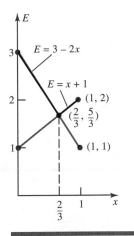

FIGURE 4

Therefore, given that C will choose the lower line for each value of x, R maximizes E by choosing the value of x at the point of intersection. This means that R chooses $x = \frac{2}{3}$, and so R's optimal strategy is $[\frac{2}{3} \quad \frac{1}{3}]$. The expected value computed for R's optimal strategy is called the **payoff of the game.** For this game it is $E = \frac{5}{3}$.

Let us review our procedure.

Procedure for Finding the Optimal Mixed Strategy for R

To find the optimal mixed strategy for R that maximizes the expected value, assuming that C's counterstrategy will always result in the smallest possible expected value, we let P be the playoff matrix and $A = [x \quad y]$ be R's strategy. Then

1. Compute $E = AP \begin{bmatrix} 1 \\ 0 \end{bmatrix}$ and $E = AP \begin{bmatrix} 0 \\ 1 \end{bmatrix}$ corresponding to C's choice of either column 1 or 2. This yields two equations in E, x and y.
2. Substitute $y = 1 - x$ to get two linear equations in E and x.
3. Graph the lines and shade the lower portions.
4. R's optimal strategy is the value of x that produces the highest shaded point. It will occur at the point of intersection of the two lines, or on the E-axis, or on the vertical line $x = 1$.
5. The expected value computed for R's optimal strategy is the payoff of the game.

The next two examples illustrate this procedure.

EXAMPLE 2

Let $P = \begin{bmatrix} 2 & 5 \\ 4 & 3 \end{bmatrix}$ be the payoff matrix of a game.

Problem

Find the optimal mixed strategy for R and the payoff of the game.

Solution Follow the steps in the procedure.

1.
$$E = [x \quad y] \begin{bmatrix} 2 & 5 \\ 4 & 3 \end{bmatrix} \begin{bmatrix} 1 \\ 0 \end{bmatrix} = [x \quad y] \begin{bmatrix} 2 \\ 4 \end{bmatrix} = [2x + 4y]$$

$$E = [x \quad y] \begin{bmatrix} 2 & 5 \\ 4 & 3 \end{bmatrix} \begin{bmatrix} 0 \\ 1 \end{bmatrix} = [x \quad y] \begin{bmatrix} 5 \\ 3 \end{bmatrix} = [5x + 3y]$$

2. Substitute $y = 1 - x$ into each expression.

$$E = 2x + 4y = 2x + 4(1 - x) = 4 - 2x$$

$$E = 5x + 3y = 5x + 3(1 - x) = 2x + 3$$

3. The graph is drawn in Figure 5.

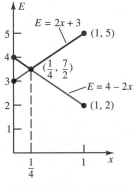

FIGURE 5

4. The highest part of the shaded portion of the graph occurs at the point of intersection.

Set

$$E = 4 - 2x \qquad \text{equal to} \qquad E = 2x + 3$$

$$4 - 2x = 2x + 3$$

$$4x = 1$$

$$x = \tfrac{1}{4}$$

Therefore, the optimal mixed strategy for R is $[\tfrac{1}{4} \quad \tfrac{3}{4}]$.

5. The point of intersection is $(\tfrac{1}{4}, \tfrac{7}{2})$, and so the payoff of the game is $E = \tfrac{7}{2}$.

The next example shows that R's optimal mixed strategy does not always occur at the point of intersection.

EXAMPLE 3

Let $P = \begin{bmatrix} 2 & 1 \\ 3 & 4 \end{bmatrix}$ be the payoff matrix of a game.

Problem

Find the optimal mixed strategy for R and the payoff of the game.

Solution We follow the steps in the procedure.

1.

$$E = [x \quad y] \begin{bmatrix} 2 & 1 \\ 3 & 4 \end{bmatrix} \begin{bmatrix} 1 \\ 0 \end{bmatrix} = [x \quad y] \begin{bmatrix} 2 \\ 3 \end{bmatrix} = [2x + 3y]$$

$$E = [x \quad y] \begin{bmatrix} 2 & 1 \\ 3 & 4 \end{bmatrix} \begin{bmatrix} 0 \\ 1 \end{bmatrix} = [x \quad y] \begin{bmatrix} 1 \\ 4 \end{bmatrix} = [x + 4y]$$

2. Substitute $y = 1 - x$ into each expression.

$$E = 2x + 3y = 2x + 3(1 - x) = 3 - x$$

$$E = x + 4y = x + 4(1 - x) = 4 - 3x$$

3. The graph is drawn in Figure 6.
4. The highest part of the shaded portion of the graph occurs at the point $(0, 3)$. Hence the optimal mixed strategy for R is $[0 \quad 1]$.

5. The payoff of the game is $E = 3 - 0 = 3$.

The solution to Example 3 has an important implication. The fact that R's optimal strategy occurs at $x = 0$ means that R plays the pure strategy $[0 \quad 1]$, always selecting row 2. Certainly C should select the first column, using the strategy $\begin{bmatrix} 1 \\ 0 \end{bmatrix}$. Therefore the game has a saddle point, the $(2, 1)$ entry, 3. Thus the game is strictly determined. This says that the procedure for calculating optimal mixed strategies works for strictly determined games as well.

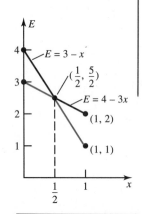

FIGURE 6

EXERCISE SET 10–3

In Problems 1 to 10 the payoff matrix is given. Compute the expected value of the game if R's strategy is $[x \quad y]$ and C chooses (a) column 1 and (b) column 2.

1. $\begin{bmatrix} 0 & 1 \\ 2 & 0 \end{bmatrix}$

2. $\begin{bmatrix} 1 & 2 \\ 0 & 0 \end{bmatrix}$

3. $\begin{bmatrix} 1 & 2 \\ -1 & 0 \end{bmatrix}$

4. $\begin{bmatrix} 2 & 0 \\ -1 & 1 \end{bmatrix}$

5. $\begin{bmatrix} 2 & 1 \\ -1 & -2 \end{bmatrix}$

6. $\begin{bmatrix} 3 & 1 \\ -2 & 1 \end{bmatrix}$

7. $\begin{bmatrix} 2 & -1 \\ 1 & -2 \end{bmatrix}$

8. $\begin{bmatrix} 3 & -1 \\ -2 & 1 \end{bmatrix}$

9. $\begin{bmatrix} 5 & 2 \\ 3 & 4 \end{bmatrix}$

10. $\begin{bmatrix} 3 & 6 \\ 7 & 4 \end{bmatrix}$

In Problems 11 to 16 two expressions for E are given. The first corresponds to C choosing column 1 and the second corresponds to C choosing column 2. Graph each expression and find the point of intersection.

11. $E = x + 2y$ and $E = x + 3y$

12. $E = 3x + y$ and $E = 2x + 4y$

13. $E = 2x + 5y$ and $E = 2x + y$

14. $E = x + 2y$ and $E = 2x + 3y$

15. $E = 6x + 3y$ and $E = 8x + 2y$

16. $E = 4x + 2y$ and $E = 2x + 5y$

In Problems 17 to 32 the payoff matrix is given. Compute the optimal mixed strategy for R that maximizes the expected value, given that C will select the best counterstrategy. Then compute the payoff.

17. $\begin{bmatrix} 1 & 2 \\ 3 & 1 \end{bmatrix}$

18. $\begin{bmatrix} 5 & 2 \\ 3 & 4 \end{bmatrix}$

19. $\begin{bmatrix} 5 & 2 \\ 3 & 1 \end{bmatrix}$

20. $\begin{bmatrix} 10 & 2 \\ 3 & 11 \end{bmatrix}$

21. $\begin{bmatrix} 11 & 22 \\ 23 & 11 \end{bmatrix}$

22. $\begin{bmatrix} 2 & 3 \\ 3 & 1 \end{bmatrix}$

23. $\begin{bmatrix} 2 & 6 \\ 7 & 1 \end{bmatrix}$

24. $\begin{bmatrix} 1 & 3 \\ 7 & 6 \end{bmatrix}$

25. $\begin{bmatrix} 11 & 3 \\ 2 & 16 \end{bmatrix}$

26. $\begin{bmatrix} 21 & 23 \\ 2 & 6 \end{bmatrix}$

27. $\begin{bmatrix} 3 & \frac{1}{2} \\ 2 & 1 \end{bmatrix}$

28. $\begin{bmatrix} 2 & \frac{3}{2} \\ 4 & 1 \end{bmatrix}$

29. $\begin{bmatrix} \frac{2}{3} & 3 \\ 1 & 4 \end{bmatrix}$

30. $\begin{bmatrix} 5 & 3 \\ 1 & \frac{1}{2} \end{bmatrix}$

31. $\begin{bmatrix} \frac{1}{3} & 5 \\ 2 & \frac{2}{3} \end{bmatrix}$

32. $\begin{bmatrix} 5 & \frac{2}{5} \\ \frac{1}{4} & 4 \end{bmatrix}$

In Problems 33 to 38 find the optimum strategy for the game with the given payoff matrix whose entries are not all positive, given that C will select the best counterstrategy. First add a large enough positive integer to each entry.

33. $\begin{bmatrix} 3 & -2 \\ 2 & 1 \end{bmatrix}$

34. $\begin{bmatrix} 2 & 5 \\ 4 & -3 \end{bmatrix}$

35. $\begin{bmatrix} 1 & 3 \\ 5 & -4 \end{bmatrix}$

36. $\begin{bmatrix} -3 & 5 \\ 2 & -3 \end{bmatrix}$

37. $\begin{bmatrix} -7 & 2 \\ 0 & -1 \end{bmatrix}$

38. $\begin{bmatrix} 0 & 3 \\ 1 & -1 \end{bmatrix}$

In Problems 39 to 42 a game is described. Find a payoff matrix for the game and find the optimum strategy for the row player, given that the column player will select the best counterstrategy.

39. Each of two players, A and B, has two cards: a 2 and a 4. Each player selects a card and they simultaneously show their cards. If the cards are both 2's, then A pays B \$1. If they are both 4's, then B pays A \$1. If A's card is greater than B's, then A wins \$5 from B. If B's card is greater than A's, then B wins \$3.

40. Player A has two cards: a 3 and a 5. Player B has two cards: a 2 and a 6. Each player selects a card and they simultaneously show the cards. If the sum is even, then A pays B \$4. If the sum is odd, then B pays A \$3.

41. Each of two players has a coin. They match their coins by turning up either heads or tails. If the same side appears on each coin, then A pays B \$1. If A turns up heads and B turns up tails, then A pays B \$2. If A turns up tails and B turns up heads, then B pays A \$3.

42. Two players simultaneously show one or two fingers. If the number of fingers shown by each player matches, then A wins the amount of dollars equal to the sum of the fingers shown. If the number of fingers shown by each player does not match, then B wins the amount of dollars equal to the number of fingers shown.

Referenced Exercise Set 10–3

1. Contingent valuation (CV) has become a commonly used technique for assigning a value to assets such as natural resources that are not traded in markets. CV's are used in many areas such as tax assessments and insurance rates. A common way of conducting a CV is to interview people and ask their opinion of the value of an asset. Boyle et al.* use a game theory approach to model a CV of the scenic beauty of the Lower Wisconsin River. An interviewer telephones individuals at random and offers an initial bid of the value of the scenic beauty, say, $50. The respondent then gives a counter bid, either higher, lower, or the same. The authors study whether the starting bid of the interviewer affects the final bid of the respondent. They let the interviewer be the row player and the respondent the column player. The interviewer has two strategies: offer either a high initial bid or a low initial bid. The respondents are classified in terms of two strategies, whether they would offer a bid higher than or lower than (including equal to) the interviewer's bid. The goal of the interviewer is to find the percentage of times a high initial bid should be given in order to maximize the payoff, which is the final bid by the respondent. In one part of the study the following payoff matrix is given.

Respondent's Bid

		higher	lower
Interviewer's	high	50	30
Initial Bid	low	40	45

What is the optimum strategy for the interviewer?

2. Game theory is often used to model diverse markets where both buyers and sellers have potential market power. Karp and McCalla† apply these techniques to the world corn market. The seller, represented by the row player, is the United States and the buyer is Japan. In one model the United States must choose between including or not including an export tax in the price. Japan must choose between ordering a large or small amount of corn from the United States. The payoff is the price of corn in dollars per metric ton. The payoff matrix is

Buyer

		large	small
Seller	export tax	124	113
	no export tax	116	118

Find the optimum strategy for the row player.

3. The main objective in cash management is referred to as "collect fast—pay slow." This means speed up collections of accounts receivable and slow down accounts payable. Many firms use lock boxes serviced by local depository banks to reduce the amount of funds tied up in the collection process. Firms also use various disbursement banks to handle accounts payable. Each bank has a different cost for its services. Orgler and Tauman‡ use game theory to model the interaction of a seller and a buyer, the former wanting to collect funds quickly and the latter wanting to delay payment as long as possible. The seller, the row player, has two options corresponding to two depository banks, A and B. The buyer has two options corresponding to two disbursement banks, a and b. The payoff is the relative cost to the buyer to enact the transaction. In one example the payoff matrix is

Buyer

		a	b
Seller	A	100	205
	B	77	270

Find the optimum strategy for both players.

Cumulative Exercise Set 10–3

1. Consider the strictly determined game whose payoff matrix is given below.
   (a) Determine the optimum strategies for R and C.
   (b) Determine the value of the game.

$$\begin{bmatrix} 2 & 3 & 1 \\ -3 & 1 & 0 \\ -1 & 6 & -3 \\ 3 & -4 & 0 \end{bmatrix}$$

*Kevin J. Boyle et al., "Starting Point Bias in Contingent Valuation Bidding Games," *Land Economics*, Vol. 61, pp. 188–194, 1985.

†Larry S. Karp and Alex F. McCalla, "Dynamic Games and International Trade: An Application to the World Corn Market," *American Journal of Agricultural Economics*, Vol. 65, pp. 641–650, 1983.

‡Yair Orgler and Yair Tauman, "A Game Theoretic Approach to Collections and Disbursements," *Management Science*, Vol. 32, pp. 1029–1038, 1986.

2. Determine whether the game with the following payoff matrix is strictly determined. If so, find its value.

$$\begin{bmatrix} -3 & 0 & 1 & 3 \\ 6 & 1 & 3 & -4 \\ -1 & -1 & -2 & 2 \end{bmatrix}$$

3. Each of two players, A and B, has a change purse containing a penny, a nickel, and a dime. Each player selects one coin from his own purse to see if the coins match. If they match and the coins are pennies, then A wins a nickel. If they match and the coins are nickels, then A wins a dime. If they match and the coins are dimes, then A wins a quarter. If the coins do not match, then B wins the difference in the values of the coins. Find a payoff matrix for the game. If it is strictly determined, find its value.

4. Let the strategy for R be [.4 .6] and the strategy for C be

$$\begin{bmatrix} .1 \\ .2 \\ .3 \\ .4 \end{bmatrix}$$

   (a) What is the size of the payoff matrix?
   (b) Compute the probability of each entry.

5. Calculate the expected value of the game with payoff matrix P and the given strategies.

$$P = \begin{bmatrix} 1 & -2 & 0 & -1 \\ -1 & 1 & 2 & -3 \end{bmatrix} \quad [.7 \quad .3] \quad \begin{bmatrix} .2 \\ .3 \\ .2 \\ .3 \end{bmatrix}$$

6. Let the payoff matrix of a game be

$$P = \begin{bmatrix} -1 & -2 & 3 \\ 3 & 2 & -4 \end{bmatrix}$$

and let C's strategy be

$$\begin{bmatrix} .1 \\ .2 \\ .7 \end{bmatrix}$$

If R's strategy is [a b], what values of a and b will maximize the payoff to R?

7. Add the smallest possible integer to the payoff matrix P to obtain the payoff matrix Q with all positive entries and compute the expected value using the matrices A and B, which are the strategies for R and C, respectively. How are the two expected values related?

$$P = \begin{bmatrix} -1 & 0 & -2 \\ 3 & -3 & 2 \end{bmatrix}$$

$$A = [.1 \quad .9] \quad B = \begin{bmatrix} .2 \\ .7 \\ .1 \end{bmatrix}$$

8. Compute the expected value of the game whose payoff matrix is given if C chooses column 2 and the strategy of R is [x y].

$$\begin{bmatrix} 7 & 4 \\ 1 & 5 \end{bmatrix}$$

9. Suppose the expression $E = 2x + 3$ corresponds to C choosing column 1 and the expression $E = 4 - 2x$ corresponds to C choosing column 2. Sketch the graph of each expression and find the point of intersection.

10. (a) For the following payoff matrix compute the optimal mixed strategy for R that maximizes the expected value, given that C will select the best counterstrategy.

   (b) Compute the payoff.

$$\begin{bmatrix} 4 & 2 \\ 1 & 3 \end{bmatrix}$$

10–4 Game Theory and Linear Programming

This section reveals the close link between game theory and linear programming. Because the connection might seem surprising at first, the chapter's epilogue will recount a historical meeting between George Dantzig and John von Neumann that forged the common bond between these two areas.

In this section the players plan their optimal strategies simultaneously. All the game-theoretic problems are reduced to linear programming problems that can be solved by either the geometric method (Chapter 3) or the simplex method (Chapter 4).

R's Optimal Mixed Strategy

Consider the game with this payoff matrix.

$$P = \begin{bmatrix} 3 & 2 \\ 1 & 3 \end{bmatrix}$$

We will show how to write it as a linear programming problem. Let R's optimal strategy be $A = [x \quad y]$ and let C's optimal strategy be $B = \begin{bmatrix} u \\ v \end{bmatrix}$. The expected value of the game is

$$E = APB = [x \quad y] \begin{bmatrix} 3 & 2 \\ 1 & 3 \end{bmatrix} \begin{bmatrix} u \\ v \end{bmatrix}$$

$$= [3x + y \quad 2x + 3y] \begin{bmatrix} u \\ v \end{bmatrix}$$

$$= (3x + y)u + (2x + 3y)v$$

Player C must choose either column 1 or column 2. If the choice is column 1, then $u = 1$ and $v = 0$, so

$$E = 3x + y \tag{1}$$

If the choice is column 2, then $u = 0$ and $v = 1$, so

$$E = 2x + 3y \tag{2}$$

The goal for R is to choose the row that maximizes these two values of E. Let e be the smaller of the two values. Then $E \geq e$, so R's goal will be to maximize e since C will always choose the column producing the smaller value of E. Moreover, $e > 0$ since the entries of P are all positive.

The inequality $E \geq e$ leads to the objective function and constraints of a linear programming problem. Equations 1 and 2 become inequalities.

$$3x + \quad y \geq e$$
$$2x + 3y \geq e$$

Divide each inequality by e.

$$3\left(\frac{x}{e}\right) + \left(\frac{y}{e}\right) \geq 1$$

$$2\left(\frac{x}{e}\right) + 3\left(\frac{y}{e}\right) \geq 1$$

Substitute $X = x/e$ and $Y = y/e$.

$$3X + Y \geq 1 \tag{3}$$
$$2X + 3Y \geq 1 \tag{4}$$

Since R's strategy is $A = [x \quad y]$, it follows that $x \geq 0$ and $y \geq 0$. Then $x/e \geq 0$ and $y/e \geq 0$, hence

$$X \geq 0 \tag{5}$$

$$Y \geq 0 \tag{6}$$

Equations 3 to 6 represent the constraints of a linear programming problem.

Since R's strategy is $[x \quad y]$, it follows that $x + y = 1$. Divide both sides by e, then substitute $X = x/e$ and $Y = y/e$ to obtain the objective function

$$X + Y = 1/e \tag{7}$$

Here is a crucial observation that will be explained in the exercises.

Maximizing e is the same thing as minimizing $1/e$.

R's goal of maximizing e translates into minimizing $1/e$. From equation 7 the objective function is

$$\text{Minimize } X + Y \tag{8}$$

Statements 8 and 3 to 6 define a linear programming problem in two variables, X and Y.

Minimize $X + Y$ subject to

$$3X + Y \geq 1$$
$$2X + 3Y \geq 1$$
$$X, Y \geq 0$$

The problem can be solved by the geometric method of Chapter 3 or the dual method of Chapter 4. Figure 1 shows the solution $X = \frac{2}{7}$ and $Y = \frac{1}{7}$ by the geometric method.

However, R's optimal strategy involves x and y. It follows from equation 7 that $1/e = \frac{2}{7} + \frac{1}{7} = \frac{3}{7}$, so the expected value is $e = \frac{7}{3}$. Since $X = x/e$, it follows that $x = eX = (\frac{7}{3})(\frac{2}{7}) = \frac{2}{3}$. Similarly, $y = eY = \frac{1}{3}$. Therefore, the optimal strategy for R is $A = [\frac{2}{3} \quad \frac{1}{3}]$.

The strategy for R is called the *optimal mixed strategy* because it also takes into account C's optimal strategy. The general procedure is described in the display.

Procedure to Determine the Optimal Mixed Strategy for R

Let the payoff matrix for a game with positive entries be

$$P = \begin{bmatrix} q & r \\ s & t \end{bmatrix}$$

1. Solve the linear programming problem:

   Minimize $X + Y$ subject to

   $qX + sY \geq 1$
   $rX + tY \geq 1$
   $X, Y \geq 0$

2. The expected value is $e = 1/(X + Y)$.
3. The optimal mixed strategy for R is $[x \quad y]$, where $x = eX$ and $y = eY$.

Note that the coefficient matrix in step 1 is the transpose of P, not P. Example 1 illustrates the entire procedure.

EXAMPLE 1

Problem

Compute R's optimal mixed strategy for the game with payoff matrix P, where

$$P = \begin{bmatrix} 4 & 3 \\ 2 & 4 \end{bmatrix}$$

Solution First solve the linear programming problem.

   Minimize $X + Y$ subject to
   $4X + 2Y \geq 1$
   $3X + 4Y \geq 1$
   $X, Y \geq 0$

The solution is $X = \frac{2}{10}$ and $Y = \frac{1}{10}$. Then $X + Y = \frac{3}{10}$, so $e = \frac{10}{3}$. R's optimal mixed strategy is $[x \quad y]$, where $x = eX = (\frac{10}{3})(\frac{2}{10}) = \frac{2}{3}$ and $y = eY = (\frac{10}{3})(\frac{1}{10}) = \frac{1}{3}$.

C's Optimal Mixed Strategy

Consider the same payoff matrix P that began the section, this time from C's viewpoint.

$$P = \begin{bmatrix} 3 & 2 \\ 1 & 3 \end{bmatrix}$$

The expected value is

$$E = (3x + y)u + (2x + 3y)v$$

Player R must choose either row 1 or row 2. If the choice is row 1, then $x = 1$ and $y = 0$, so

$$E = 3u + 2v \qquad (9)$$

If the choice is row 2, then $x = 0$ and $y = 1$, so

$$E = u + 3v \qquad (10)$$

The goal for C is to choose the column that minimizes these two values of E. This time let e be the larger of the two values. Then $E \leq e$, so C's goal is to minimize e. Equations 9 and 10 become

$$3u + 2v \leq e$$
$$u + 3v \leq e$$

Divide by e and substitute $U = u/e$ and $V = v/e$.

$$3U + 2V \leq 1 \qquad (11)$$
$$U + 3V \leq 1 \qquad (12)$$

Since $u \geq 0$, $v \geq 0$, and $e > 0$, it follows that

$$U \geq 0 \qquad (13)$$

$$V \geq 0 \qquad (14)$$

Divide both sides of the formula $u + v = 1$ by e, then substitute $U = u/e$ and $V = v/e$.

$$U + V = \frac{1}{e} \qquad (15)$$

Since minimizing e is the same thing as maximizing $1/e$, C's goal of minimizing e translates into maximizing $1/e$, so the objective function becomes

$$\text{Maximize } U + V \qquad (16)$$

Statements 16 and 11 to 14 define a linear programming problem whose solution is $U = \frac{1}{7}$ and $V = \frac{2}{7}$. It follows from equation 15 that $1/e = \frac{3}{7}$, so $e = \frac{7}{3}$. Then $u = eU = (\frac{7}{3})(\frac{1}{7}) = \frac{1}{3}$ and $v = eV = \frac{2}{3}$. The optimal strategy for C is $\begin{bmatrix} \frac{1}{3} \\ \frac{2}{3} \end{bmatrix}$.

Procedure to Determine the Optimal Mixed Strategy for C

Let the payoff matrix for a game with positive entries be

$$P = \begin{bmatrix} q & r \\ s & t \end{bmatrix}$$

1. Solve the linear programming problem.

   Maximize $U + V$ subject to

   $$qU + rV \leq 1$$
   $$sU + tV \leq 1$$
   $$U, V \geq 0$$

2. The expected value is $e = 1/(U + V)$.

3. The optimal mixed strategy for C is $\begin{bmatrix} u \\ v \end{bmatrix}$, where $u = eU$ and $v = eV$.

The coefficient matrix in the linear programming problem is the payoff matrix P.

E X A M P L E 2

Problem

Compute C's optimal mixed strategy for the game with payoff matrix P, where

$$P = \begin{bmatrix} 4 & 3 \\ 2 & 4 \end{bmatrix}$$

Solution First solve the linear programming problem.

Maximize $U + V$ subject to

$$4U + 3V \leq 1$$
$$2U + 4V \leq 1$$
$$U, V \geq 0$$

The solution is $U = \frac{1}{10}$, $V = \frac{2}{10}$. Then $U + V = \frac{3}{10}$, so $e = \frac{10}{3}$. C's optimal mixed strategy is $\begin{bmatrix} u \\ v \end{bmatrix}$, where $u = eU = (\frac{10}{3})(\frac{1}{10}) = \frac{1}{3}$ and $v = eV = (\frac{10}{3})(\frac{2}{10}) = \frac{2}{3}$.

Combined Strategies

Two players' optimal mixed strategies occur when they carry out their strategies simultaneously. The expected value of the game using R's strategy is the same as the expected value of the game using C's strategy. This common number is called the *value* of the game. The value of the game in Examples 1 and 2 is $\frac{10}{3}$. Example 3 considers the two players' optimal mixed strategies simultaneously.

EXAMPLE 3

Let the payoff matrix for a game be $P = \begin{bmatrix} 2 & 5 \\ 4 & 3 \end{bmatrix}$.

Problem

Find the optimal mixed strategies for R and C, and find the value of the game.

Solution The optimal mixed strategy for R is obtained by solving the linear programming problem.

Minimize $X + Y$ subject to
$2X + 4Y \geq 1$
$5X + 3Y \geq 1$
$\quad X, Y \geq 0$

The solution is $X = \frac{1}{14}$ and $Y = \frac{3}{14}$. The expected value is $e = 1/(X + Y) = \frac{7}{2}$. Thus the value of the game is $\frac{7}{2}$. Then $x = eX = \frac{1}{4}$ and $y = eY = \frac{3}{4}$, so R's optimal mixed strategy is $[\frac{1}{4} \quad \frac{3}{4}]$.

For C solve the linear programming problem.

Maximize $U + V$ subject to

$2U + 5V \leq 1$
$4U + 3V \leq 1$
$\quad U, V \geq 0$

The solution is $U = \frac{1}{7}$ and $V = \frac{1}{7}$. Then $u = eU = (\frac{7}{2})(\frac{1}{7}) = \frac{1}{2}$ and $v = eV = \frac{1}{2}$. The optimal mixed strategy for C is $\begin{bmatrix} \frac{1}{2} \\ \frac{1}{2} \end{bmatrix}$.

All the games in this section have consisted of payoff matrices with positive entries. This assumption was necessary to guarantee that the expected value of the game was positive. What can be done when some entries of the payoff matrix P are negative?

As we saw in Section 10–2, the procedure is to add a sufficiently large number n to the entries of P to produce a payoff matrix Q with positive entries. The strategies and value of the game with payoff matrix Q are solved by the methods of this section. The optimal mixed strategies for a game with payoff matrix P are the same as the optimal mixed strategies for a game with payoff matrix Q. However, the value of the game with payoff matrix P is equal to the value of the game with payoff matrix Q minus n. This procedure is verified in the exercises.

EXERCISE SET 10–4

In Problems 1 to 4 compute APB.

1. $A = [x \quad y], \quad P = \begin{bmatrix} 2 & 3 \\ -1 & 0 \end{bmatrix}, \quad B = \begin{bmatrix} 1 \\ 0 \end{bmatrix}$

2. $A = [1 \quad 0], \quad P = \begin{bmatrix} 2 & 3 \\ -1 & 0 \end{bmatrix}, \quad B = \begin{bmatrix} u \\ v \end{bmatrix}$

3. $A = [x \quad y], \quad P = \begin{bmatrix} 2 & 0 \\ -1 & 4 \end{bmatrix}, \quad B = \begin{bmatrix} u \\ v \end{bmatrix}$

4. $A = [x \quad y], \quad P = \begin{bmatrix} -1 & 2 \\ 3 & -4 \end{bmatrix},$

$B = \begin{bmatrix} u \\ v \end{bmatrix}$

In Problems 5 and 6 substitute $X = x/e$, $Y = y/e$, and $1/e = X + Y$ into the stated linear programming problem to form a linear programming problem in the variables X and Y.

5. Maximize e subject to
   $4x + y \le e$
   $5x + 2y \le e$
   $x, y \ge 0$

6. Maximize e subject to
   $2x + 5y \le e$
   $6x + y \le e$
   $x, y \ge 0$

In Problems 7 to 10 find the optimal mixed strategy for R for the game with the stated payoff matrix.

7. $\begin{bmatrix} 1 & 0 \\ 2 & 4 \end{bmatrix}$ 8. $\begin{bmatrix} 4 & 5 \\ 5 & 2 \end{bmatrix}$

9. $\begin{bmatrix} 2 & 3 \\ 3 & 0 \end{bmatrix}$ 10. $\begin{bmatrix} 4 & 2 \\ 1 & 5 \end{bmatrix}$

In Problems 11 to 14 find the optimal mixed strategy for C for the game with the payoff matrices in Problems 7 to 10.

In Problems 15 to 18 find the optimal mixed strategies for R and C simultaneously for the game with the given payoff matrix.

15. $\begin{bmatrix} 3 & 1 \\ 2 & 4 \end{bmatrix}$ 16. $\begin{bmatrix} 1 & 2 \\ 3 & 1 \end{bmatrix}$

17. $\begin{bmatrix} 5 & 2 \\ 3 & 1 \end{bmatrix}$ 18. $\begin{bmatrix} 5 & 2 \\ 3 & 4 \end{bmatrix}$

In Problems 19 to 24 find the optimal mixed strategies for R and C simultaneously for the game with the given payoff matrix, some of whose entries are negative.

19. $\begin{bmatrix} -2 & 1 \\ 0 & -1 \end{bmatrix}$ 20. $\begin{bmatrix} -4 & -1 \\ -2 & -3 \end{bmatrix}$

21. $\begin{bmatrix} 1 & 2 \\ -1 & 0 \end{bmatrix}$ 22. $\begin{bmatrix} 2 & 0 \\ -1 & 1 \end{bmatrix}$

23. $\begin{bmatrix} 2 & 1 \\ -1 & -2 \end{bmatrix}$ 24. $\begin{bmatrix} 3 & 1 \\ -2 & 1 \end{bmatrix}$

In Problems 25 to 28 verify the stated solution to the given example in the text.

25. In Example 1 the solution is $X = \frac{2}{10}$ and $Y = \frac{1}{10}$.

26. In Example 2 the solution is $U = \frac{1}{10}$ and $V = \frac{2}{10}$.

27. In Example 3 the solution for player R is $X = \frac{1}{14}$ and $Y = \frac{3}{14}$.

28. In Example 3 the solution for player C is $U = \frac{1}{7}$ and $V = \frac{1}{7}$.

Problems 29 to 34 examine the assertion, "maximizing e is the same thing as minimizing $1/e$." Let S be the given set. Find the (a) maximum value of e and (b) the minimum value of $1/e$ for all numbers e in S.

29. $S = \{3, 4, 5, 6, 7\}$ 30. $S = \{10, 100, 100\}$

31. $S = \{\frac{1}{2}, \frac{1}{3}, \frac{1}{4}\}$ 32. $S = \{1.1, 2.2, 3.3\}$

33. $S = \{\frac{1}{2}, \frac{1}{3}, \frac{1}{4}, \ldots\}$ 34. $S = \{\frac{2}{1}, \frac{3}{2}, \frac{4}{3}, \ldots\}$

In Problems 35 to 38 let S be the given set. Find the (a) minimum value of e and (b) the maximum value of $1/e$, for all numbers e in S.

35. $S = \{3, 4, 5, 6, 7\}$

36. $S = \{1, 2, \ldots, 100\}$

37. $S = \{.3, .33, .333, \ldots\}$

38. $S = \{\frac{1}{2}, \frac{2}{3}, \frac{3}{4}, \ldots\}$

Referenced Exercise Set 10–4

1. In an article describing competitive encounters among firms, Teng and Thompson* use a game theoretic approach to model how two firms use level of advertising to attempt to increase market share. The players are two firms, and their strategies involve whether or

*Jinn-tsair Teng and Gerald L. Thompson, "Oligopoly Models for Optimal Advertising When Production Costs Obey a Learning Curve," *Management Science*, Vol. 29, pp. 1087–1100, 1983.

not to increase advertising. In one part of the article they model two firms that manufacture pocket calculators. The payoff is the percent of market gained by firm 1, the row player. A typical payoff matrix is

Firm 2

increase same

$$
\text{Firm 1} \begin{cases} \text{increase advertising} \\ \text{keep the same level} \end{cases} \begin{bmatrix} 40 & 60 \\ 55 & 45 \end{bmatrix}
$$

Find the optimal mixed strategies for each player.

2. An agricultural cooperative represents an attempt by farmers, each of whom has a different set of resources and goals, to compete together in the food market. Staatz[†] uses game theory to illustrate how cooperatives can assess fees equitably, depending on the effort and support given to the cooperative by each farmer. In one example, a cooperative consisting of four farmers must assess a fee to each member. The farmers form two coalitions, A and B, to increase their bargaining power. The members in each coalition will pay the same fee. The coalitions try to influence cost allocations through threats and counterthreats. The ability of a coalition to obtain concessions from other members depends on the costs the coalition could impose on the other members if they were to exit the cooperative or cut back on the level of cooperation and effort extended. Each threat has an implied counterthreat, the cost the threatening coalition imposes on itself by withdrawing partially or totally from the cooperative. Each coalition has two strategies: threaten to withdraw or do not exercise the threat. The payoff in the game is the relative gain to coalition A, which is the relative cost to coalition A. A typical payoff matrix is

B

threat no threat

$$
A \begin{cases} \text{exercise threat} \\ \text{do not threaten} \end{cases} \begin{bmatrix} 5 & 15 \\ 20 & 10 \end{bmatrix}
$$

Find the optimal mixed strategies for each player.

Cumulative Exercise Set 10–4

1. The given payoff matrices are the matrices of a strictly determined game. Determine the optimum strategies for R and C and the value of the game.

(a) $\begin{bmatrix} -7 & 1 \\ 4 & 3 \\ 0 & -2 \end{bmatrix}$

(b) $\begin{bmatrix} 1 & -5 & 0 \\ -4 & -2 & 8 \\ 0 & -1 & 1 \end{bmatrix}$

2. Determine if the game with the given payoff matrix is a strictly determined game. If it is, find its value.

(a) $\begin{bmatrix} 3 & -1 \\ -2 & 8 \end{bmatrix}$

(b) $\begin{bmatrix} 2 & -6 & 1 \\ -5 & -1 & 7 \\ 1 & -3 & 0 \\ 3 & 1 & 6 \end{bmatrix}$

3. Find a payoff matrix for the game and decide if it is a strictly determined game. If so, find its value. Each of two players has three cards, and each card has either the number 1, 2, or 3 on it. They match their cards by each turning up one card and adding the numbers. Player A wins the sum of the numbers if the sum is even, while player B wins the sum of the numbers if the sum is odd.

4. A game has payoff matrix

$$
P = \begin{bmatrix} 0 & -1 & 2 & 0 \\ 0 & 1 & -2 & 2 \\ -1 & 3 & 1 & 0 \end{bmatrix}
$$

Compute the expected value of the game with the following strategies:

$$
[0.1 \quad 0.4 \quad 0.5], \quad \begin{bmatrix} 0.1 \\ 0.6 \\ 0.2 \\ 0.1 \end{bmatrix}
$$

5. A game has payoff matrix

$$
P = \begin{bmatrix} 0 & 1 & -1 \\ -1 & 2 & 0 \end{bmatrix}
$$

Let R play the strategy $[a \quad b]$. Determine what values of a and b will maximize the payoff to R if C plays the strategy

$$
\begin{bmatrix} 0.2 \\ 0.4 \\ 0.4 \end{bmatrix}
$$

†John M. Staatz, "The Cooperative as a Coalition: A Game-theoretic Approach," *American Journal of Agricultural Economics*, Vol. 65, pp. 1084–1089, 1983.

6. Add the smallest possible integer to the payoff matrix P to obtain the payoff matrix Q with all positive entries, and compute the expected values using the matrices A and B, the strategies for R and C, respectively. How are the two expected values related?

$$P = \begin{bmatrix} 6 & -4 \\ -2 & 2 \end{bmatrix}, \quad A = \begin{bmatrix} 0.4 & 0.6 \end{bmatrix}, \quad B = \begin{bmatrix} 0.1 \\ 0.9 \end{bmatrix}$$

7. Two expressions for E are given. The first corresponds to C choosing column 1 and the second corresponds to C choosing column 2. Graph each expression and find the point of intersection.

$$E = x + 3y \quad \text{and} \quad E = x + 2y$$

8. For the given payoff matrix, compute the optimal mixed strategy for R that maximizes the expected value, given that C will select the best counterstrategy. Then compute the payoff.

$$\begin{bmatrix} 3 & 5 \\ 2 & 4 \end{bmatrix}$$

9. Find the optimum strategy for both R and C for the game with the given payoff matrix whose entries are not all positive. First add a large enough positive integer to each entry.

$$\begin{bmatrix} 2 & -1 \\ 4 & -5 \end{bmatrix}$$

10. Substitute $X = x/e$, $Y = y/e$, and $1/e = X + Y$ into the stated linear programming problem to form a linear programming problem in the variables X and Y.

Maximize e subject to

$$3x + 5y \leq e$$

$$2x + 7y \leq e$$

$$x, y \geq 0$$

11. Find the optimal mixed strategy for R for the game with payoff matrix

$$\begin{bmatrix} 2 & 0 \\ 3 & 1 \end{bmatrix}$$

12. Find the optimal mixed strategy for R and C simultaneously for the game with payoff matrix

$$\begin{bmatrix} 2 & 4 \\ 3 & 2 \end{bmatrix}$$

13. Each of two players, A and B, has two cards, a 2 and a 4. Each player selects a card and they simultaneously show their cards. If the cards are both 2's, then A pays B \$1. If they are both 4's, then B pays A \$1. If A's card is greater than B's, then A wins \$5 from B. If B's card is greater than A's, then B wins \$3. Find the optimal mixed strategies for each player.

14. Each of the two players has a coin. They match their coins by turning up either heads or tails. If the same side appears on each coin, then A pays B \$4. If A turns up heads and B turns up tails, then A pays B \$2. If A turns up tails and B turns up heads, then B pays A \$5. Find the optimal mixed strategies for each player.

CASE STUDY   ## Norway Cement Company Merger

Game theory has been applied in many areas involving human competition. One of the most useful applications has been in the resolution of conflicts in various types of corporate mergers. When two companies explore the possibility of a merger it is often the division of the assets—who gets what—that is the major stumbling block.

A classic example is the celebrated case of the merger of Norway's three independent cement companies in the 1960s. The companies, which we will call Alpha, Beta, and Delta, were asked to consider consolidating into one large national company. The central question was how to divide the ownership of the new company. What payoff, that is, what share of the new company, should be assigned to each of the participants?

Another significant consideration was that each company had the right to remain independent or form a coalition with one other company, leaving the third company independent. For example, it was possible that Alpha and Beta would form one company and Delta would remain independent. The latter development would have been devastating for Delta because it was by far the smallest company. However, that was very unlikely to occur, because the owners of Alpha and Beta could not even come close to agreement on the division of the new company. Thus Delta, being much more flexible in its negotiations, had sig-nificant bargaining power by threatening to form a coalition with one company against the other.

Even though it was clear to all the owners that each would gain more from merging all three companies, the negotiations dragged on and almost collapsed because they could not agree upon an ownership split that was acceptable to all. Alpha argued that the present value of future earnings should be the main crite-ria, in which case it should receive 52% of the ownership. Beta claimed that production capacity was the best measure to use to split ownership, in which case it would get 47% (as opposed to 38% under Alpha's plan). Delta wanted the average production capacity over the previous decade to be the primary mea-surement, in which case it would get 13% (versus 8% and 9% under the other plans).

The breakthrough occurred when the government's consultant used a com-plicated formula, called the *Shapley value,* to split the ownership. It used all three measurements. But its real significance was that it took into account each company's bargaining power—the ability to threaten to enter into a coalition with one other company, and thereby potentially affecting the third.

The consultant explained the situation by using the following game-theo-retic approach. Consider the merger negotiation from Alpha's point of view. The owners of Alpha had three options: (1) to remain independent (*I*), (2) to merge with Beta (*M/B*), or (3) to merge with Delta (*M/D*). Let the owners of Alpha be the row player. The column "player" is then Beta and Delta, in the sense that the column player has two choices—to remain independent (*I*) or to merge (*M*). Hence there are six options. They can be expressed in a 3 × 2 payoff matrix, where the entries represent the percentage of business that Alpha would receive. For example, the (1, 1) entry, 52, represents the consultant's calculation that Alpha would get 52% of Norway's cement business if all three remained inde-pendent. If Alpha remained independent but Beta and Delta merged, then Alpha would receive 40% of the business, so the (1, 2) entry is 40.

	Beta and Delta's Decision	
	I	*M*
I	52	40
Alpha's Decision *M/B*	44	48
M/D	46	48

Note that the (2, 2) and (3, 2) entries represent the same eventuality, that all three merge, and Alpha would then get 48% ownership in the company. Given this description of the merger, what should Alpha decide? First notice that Alpha would not even consider the second option, because both entries in the third row are greater than or equal to the entries in the second row. The game-theoretic terminology is that row 3 "dominates" row 2, so if Alpha sought a merger it would be with Delta and not with Beta. Thus we can ignore row 2 and consider the payoff matrix P.

$$\text{Alpha} \begin{cases} I \\ M/D \end{cases} \quad P = \overbrace{\begin{bmatrix} I & M \\ 52 & 40 \\ 46 & 48 \end{bmatrix}}^{\text{Beta and Delta}}$$

We can now solve the problem by using our procedure to find Alpha's optimum mixed strategy, $[x \quad y]$. If one number is significantly greater than the other, then Alpha should choose the option corresponding to the larger number. However, when we translate the problem into a linear programming problem it becomes a minimization problem.

The problem can be expressed as the linear programming problem:

Minimize $X + Y$ subject to

$$52X + 46Y \geq 1$$
$$40X + 48Y \geq 1$$
$$X, Y \geq 0$$

The inequalities are graphed in Figure 1.

We then test the corner points.

Corner Points	$X + Y$
(1/40, 0)	1/40
(1/328, 3/164)	7/328
(0, 1/46)	1/46

The smallest number is 7/328, so the solution is (1/328, 3/164). Remember that we want to maximize the expected value e. The corresponding linear program-

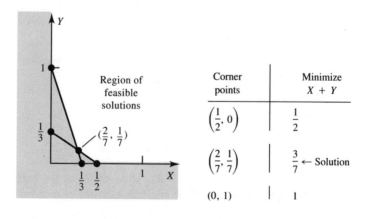

	Corner points	Minimize $X + Y$
	$\left(\dfrac{1}{2}, 0\right)$	$\dfrac{1}{2}$
	$\left(\dfrac{2}{7}, \dfrac{1}{7}\right)$	$\dfrac{3}{7}$ ← Solution
	$(0, 1)$	1

ming problem finds the minimum of $X + Y = 1/e$. Therefore

$$e = \frac{1}{X + Y} = \frac{328}{7}$$

and so

$$x = eX = \left(\frac{328}{7}\right)\left(\frac{1}{328}\right) = \frac{1}{7}$$

$$y = eY = \left(\frac{328}{7}\right)\left(\frac{3}{164}\right) = \frac{6}{7}$$

Alpha's decision was clear-cut. The second option was the most favorable, so they decided to merge. The result was similar from Beta's and Delta's perspectives and the companies formed one national firm. From the Shapley value the owners of Alpha received 48% of the new company. Similar reasoning gave 40% to Beta and 12% to Delta.

Case Study Exercises

In Problems 1 to 4 solve the game theory problem in the case study for the given payoff matrix.

1. $\begin{bmatrix} 50 & 40 \\ 46 & 48 \end{bmatrix}$

2. $\begin{bmatrix} 60 & 40 \\ 46 & 48 \end{bmatrix}$

3. $\begin{bmatrix} 60 & 40 \\ 40 & 50 \end{bmatrix}$

4. $\begin{bmatrix} 60 & 30 \\ 40 & 50 \end{bmatrix}$

EPILOGUE   Dantzig Meets Von Neumann

In a delightful article entitled "Reminiscences About the Origins of Linear Programming," George Dantzig recounted how John von Neumann and he pooled their resources to tackle two intriguing problems.* Dantzig had traveled across the country in the late 1940s eliciting help in solving problems that later became known as linear programming problems. In 1946 he was the Mathematics Adviser to the U.S. Air Force Comptroller, having just completed his Ph.D. He was trying to mechanize the planning process for the Air Force. This led him to formulate a

---

*Arthur Schlissel (ed.), *Essays in the History of Mathematics, Memoirs of the American Mathematical Society*, Vol. 298, pp. 2–16, 1984.

model based on the work of the well-known economist, Wassily Leontief. The solution to the problem eluded him because of the extraordinary number of calculations involved. For example, the simple model of assigning 70 people to 70 jobs involves 70! possible solutions—and 70! is larger than 10^{100}. By mid-1947 the problem had been stated with sufficient clarity that he could seek help from many diverse quarters. But it was his conversation with von Neumann that was particularly fruitful.

Imagine a young mathematician, barely out of graduate school, calling on one of the greatest living scientists. In his own words, here is Dantzig's recollection of the dramatic meeting:

> I decided to consult with the "great" Johnny von Neumann to see what he could suggest in the way of solution techniques. He was considered by many as the leading mathematician in the world. On October 3, 1947, I visited him for the first time at the Institute for Advanced Study at Princeton. I remember trying to describe to von Neumann, as I would an ordinary mortal, the Air Force problem. I began with the formulation of the linear programming model in terms of activities and items, etc. Von Neumann did something which I believe was uncharacteristic of him. "Get to the point," he said impatiently. Having at times a somewhat low kindling-point, I said to myself "O.K., if he wants a quicky, then that's what he'll get." In under one minute I slapped the geometric and the algebraic version of the problem on the blackboard. Von Neumann stood up and said "Oh that!." Then for the next hour and a half he proceeded to give me a lecture on the mathematical theory of linear programs.
>
> At one point, seeing me sitting there with my eyes popping and my mouth open (after all I had searched the literature and found nothing), von Neumann said: "I don't want you to think that I am pulling all this out of my sleeve on the spur of the moment like a magician. I have just recently completed a book with Oscar Morgenstern on the theory of games. What I am doing is conjecturing that the two problems are equivalent. The theory that I am outlining for your problem is an analogue to the one we have developed for games."

CHAPTER REVIEW

Key Terms

10-1 Strictly Determined Games

Payoff	Mixed Strategy
Payoff Matrix	Maximin
Game	Minimax
Two-Person Game	Saddle Point
Zero-Sum Game	Strictly Determined Game
Strategy	Value
Pure Strategy	

10-2 Mixed Strategies
Mixed Strategy

10–3   Optimal Mixed Strategies

Optimal Mixed Strategy   Payoff of the Game
Counterstrategy

Summary of Important Concepts

Fundamental goal of game theory: For each player, determine the strategy that maximizes the player's game.

For the pure strategy that maximizes the gain for the row player:

1. Determine the smallest entry in each row.
2. Choose the largest of these numbers and choose that row.

For the pure strategy that maximizes the gain for the column player:

1. Determine the largest number in each column.
2. Determine the smallest of these numbers and choose that column.

Expected value of a game with mixed strategies: For payoff matrix P and strategy matrices A and B, the expected value is APB.

Procedure for finding the optimal mixed strategy for R:

1. Compute $AP\begin{bmatrix} 1 \\ 0 \end{bmatrix}$ and $AP\begin{bmatrix} 0 \\ 1 \end{bmatrix}$, which yield two equations.
2. Substitute $y = 1 - x$ to get two linear equations in E and x.
3. Graph the lines and shade the lower portions.
4. R's optimal strategy is the value that produces the highest shaded point.
5. The expected value is the payoff of the game.

Procedure to determine the optimal mixed strategy for C:

1. Solve the linear programming problem

   Maximize $U + V$ subject to
   $qU + rV \le 1$
   $sU + tV \le 1$
   $U, V \ge 0$

2. The expected value is $e = 1/(U + V)$.

REVIEW PROBLEMS

1. Determine the optimum strategies for R and C and the value of the strictly determined game whose pay-off matrix is given.

$$\begin{bmatrix} -3 & 1 & 0 \\ -1 & 6 & -3 \\ 2 & 3 & 1 \end{bmatrix}$$

In Problems 2 and 3 determine if the game with the given payoff matrix is a strictly determined game. If so, find its value.

2. $\begin{bmatrix} 0 & 2 & -1 \\ -4 & -1 & -2 \\ 5 & 3 & 0 \end{bmatrix}$

3. $\begin{bmatrix} 1 & -2 & -2 & 4 \\ -3 & 0 & 0 & -1 \\ -4 & -4 & 1 & -1 \\ 3 & 4 & 2 & 3 \end{bmatrix}$

In Problems 4 and 5 a game is described. Find a payoff matrix for the game and decide if it is a strictly determined game. If it is strictly determined, find the value of the game.

4. Each of two players, A and B, has 2 cards, a 2 and a 6. Each player selects a card and they simultaneously show their cards. If the cards match, then A pays B $2. If A's card is greater than B's, then A wins $4 from B. If B's card is greater than A's, then B wins $5 from A.

5. Player A has 2 cards, a 2 and a 5. Player B has 3 cards, a 2, a 5, and a 7. Each player selects a card and they simultaneously show the cards. If the sum is even, then A pays B $3. If the sum is odd, then B pays A $8.

6. The strategies for R and C are given. Determine the size of the payoff matrix and compute the probability of each entry being played.

$$R = [.7 \quad .1 \quad .2], \qquad C = \begin{bmatrix} .4 \\ .6 \end{bmatrix}$$

In Problems 7 and 8 calculate the expected value for the game with payoff matrix P and the given strategies.

7. $P = \begin{bmatrix} 1 & 2 \\ 2 & -4 \end{bmatrix}$ $[.5 \quad .5]$, $\begin{bmatrix} .4 \\ .6 \end{bmatrix}$

8. $P = \begin{bmatrix} -1 & 3 \\ 0 & -1 \\ -1 & 2 \end{bmatrix}$ $[.1 \quad .7 \quad .2]$, $\begin{bmatrix} .5 \\ .5 \end{bmatrix}$

9. A game has payoff matrix

$$P = \begin{bmatrix} 1 & -2 & 3 \\ 0 & -1 & 2 \end{bmatrix}$$

Let C play the given strategy and let R play the strategy $[a \quad b]$. Determine what values of a and b will maximize the payoff to R.

$$\begin{bmatrix} .7 \\ .2 \\ .1 \end{bmatrix}$$

10. Add the smallest possible integer to the matrix to get a matrix with all positive entries; that is, choose the number such that the smallest entry is 1.

$$\begin{bmatrix} -5 & 2 & -1 \\ 1 & -6 & 4 \\ 2 & -9 & -1 \end{bmatrix}$$

11. Add the smallest possible integer to the payoff matrix P to obtain the payoff matrix Q with all positive entries and compute the expected value using the matrices A and B, the strategies for R and C, respectively. How are the two expected values related?

$$P = \begin{bmatrix} -1 & 3 \\ -2 & 2 \\ 3 & -4 \end{bmatrix}, \qquad A = [.5 \quad .2 \quad .3],$$

$$B = \begin{bmatrix} .6 \\ .4 \end{bmatrix}$$

12. The payoff matrix of a game is given. Compute the expected value of the game if C chooses (a) column 1 and (b) column 2.

$$\begin{bmatrix} 3 & 1 \\ 2 & 4 \end{bmatrix}$$

In Problems 13 and 14 two expressions for E are given. The first corresponds to C choosing column 1 and the second corresponds to C choosing column 2. Graph each expression and find the point of intersection.

13. $E = x + 3y$ and $E = x + y$

14. $E = 4x + 2y$ and $E = 3x + 5y$

15. For the given payoff matrix compute the optimal mixed strategy for R that maximizes the expected value, given that C will select the best counterstrategy. Then compute the payoff.

$$\begin{bmatrix} 3 & 1 \\ 1 & 2 \end{bmatrix}$$

16. Find the optimum strategy for both R and C for the game with the given payoff matrix.

$$\begin{bmatrix} 4 & 2 \\ 1 & \frac{1}{2} \end{bmatrix}$$

17. Find a payoff matrix for the game and find the optimum strategy for the row player and the column player. Each of two players has a coin. They match their coins by turning up either heads or tails. If the same side appears on each coin, then A pays B $3. If A turns up heads and B turns up tails, then A pays B $2. If A turns up tails and B turns up heads, then B pays A $1.

18. Substitute $X = x/e$, $Y = y/e$, and $1/e = X + Y$ into the stated linear programming problem to form a linear programming problem in the variables X and Y.

   Maximize e subject to
   $$3x + 4y \leq e$$
   $$2x + 7y \leq e$$
   $$x, y \geq 0$$

19. Find the optimal mixed strategy for R for the game with the stated payoff matrix.

$$\begin{bmatrix} 1 & 4 \\ 2 & 3 \end{bmatrix}$$

20. Find the optimal mixed strategies for R and C for the game with the following payoff matrix by translating the problem into a linear programming problem and solving it.

$$\begin{bmatrix} 4 & 5 \\ 1 & 2 \end{bmatrix}$$

PROGRAMMABLE CALCULATOR EXPERIMENTS

1. Write a program that computes the expected value $E = APB$ for the following matrices:

$$P = \begin{bmatrix} 5 & 1 \\ 3 & 2 \end{bmatrix} \qquad A = [.2 \quad .8] \qquad B = \begin{bmatrix} .7 \\ .3 \end{bmatrix}$$

2. Write a program that computes the expected value $E = APB$ for a given 3×3 payoff matrix P and mixed strategies A and B.

3. Given a 2×2 payoff matrix $\begin{bmatrix} a & b \\ c & d \end{bmatrix}$ write a program that computes the optimal mixed strategy for R that maximizes the expected value, assuming that C's counterstrategy will always result in the smallest possible expected value.

4. Write a program that graphs the two lines corresponding to the two expected values in Problem 3.

5. Write a program that graphs the two lines corresponding to the two expected values in Problem 3 for the matrix $\begin{bmatrix} 3 & 1 \\ 5 & 2 \end{bmatrix}$ and then find the optimal mixed strategy for R graphically by using the zoom feature.

6. Write a program that translates a game theory problem with a 2×2 payoff matrix into a linear programming problem.

7. Use the programs written for Chapter 3 to compute the optimal mixed strategy for R in Problem 5.

Functions

Powerlifting and polynomials: perfect together. (Walt Evans)

Chapter Overview

There are two parts of calculus, differentiation and integration, both of which are based on the concepts of functions and limits. This chapter presents the relevant material on functions, and the next chapter introduces limits.

Section 11–1 marks a return to Sections 1–1 and 1–2 where linear equations were introduced. Here the equations are regarded as functions. Since not all equations are functions, a criterion for distinguishing between them is elaborated. This material is applied to an area called cost analysis.

Section 11–2 introduces power functions, polynomial functions, and rational functions. Two additional kinds of functions are described: those that are defined by different formulas over different intervals and those that are defined implicitly by a graph.

Section 11–3 discusses five operations on functions: the four rational operations and composition. It also introduces quality control. Section 11–4 presents some methods for solving polynomial equations, including a brief review of factoring and the quadratic formula. This section describes how to find the points of intersection of the graphs of functions.

CASE STUDY PREVIEW

"Which one is stronger, an ant or an elephant?" The Case Study will not answer this proverbial question, but it will provide some food for thought by examining an analogous question about human beings.

Powerlifters can lift vast amounts of weight. Their meets are divided into weight classes, and each weight class has a champion. But powerlifting meets also feature a "Champion of Champions" award, and the person who lifts the largest weight may not necessarily win it. The reason is the so-called "Schwartz Formula," a handicapping scheme that distributes points according to a person's bodyweight. The Case Study describes how it is defined in terms of a polynomial function.

11–1   Linear and Quadratic Functions

In Chapter 1 we sketched the graphs of linear equations. Here we introduce the fundamental concept of a function and regard a linear equation as a function. We also introduce quadratic functions and give a criterion for distinguishing between equations and functions.

Linear Functions

The equation $y = 1 - 4x$ has the property that for each value of x there corresponds precisely one value of y. In such a case y is called a *function* of x, written $y = f(x)$ or $f(x) = 1 - 4x$. [$f(x)$ is read "f of x."]

DEFINITION

A **function** f is a rule that assigns to each element x in a set exactly one element $f(x)$ in a (possibly different) set. The element $f(x)$ is called the **image** of x, and x is called the **preimage** of $f(x)$.

Consider the function $f(x) = 1 - 4x$. The **plotting method** for drawing its graph was described in Section 1–1. First, make a table of points satisfying the equation.

x	-1	$-\frac{1}{2}$	0	$\frac{1}{2}$	1
$f(x)$	5	3	1	-1	-3
point	$(-1, 5)$	$(-\frac{1}{2}, 3)$	$(0, 1)$	$(\frac{1}{2}, -1)$	$(1, -3)$

Then plot the points on a Cartesian coordinate system and connect them with the line drawn in Figure 1.

A **linear function** is a function that can be written in the form $f(x) = mx + b$, where m and b are given numbers. We saw in Chapter 1 that the graph of a linear function is a line with slope m and y-intercept b.

The table for the linear function $f(x) = 1 - 4x$ lists the images of five values of x. (Actually, only two images are necessary for drawing the graph of a line.) Since $y = 5$ when $x = -1$, the image of -1 is 5. We write $f(-1) = 5$. This image can be obtained directly by substituting the value of -1 for x wherever x occurs in the equation $f(x) = 1 - 4x$.

$$f(-1) = 1 - 4(-1) = 1 + 4 = 5$$

Example 1 computes some images of another function. Parts (a) and (b) show that it is possible for $f(a) = f(b)$ when $a \neq b$.

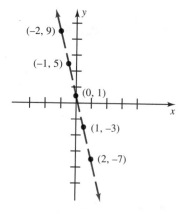

Graph of $y = 1 - 4x$

FIGURE 1

EXAMPLE 1

Let f be the function defined by the equation

$$f(x) = 2x^2 - 4x + 5$$

Problem

Find (a) $f(2)$, (b) $f(0)$, (c) $f(-\frac{1}{2})$.

Solution (a) Since $f(2)$ is the image of 2, it is calculated by substituting 2 for x wherever x occurs in the formula for $f(x)$.

$$f(2) = 2(2)^2 - 4(2) + 5 = 5$$

Thus

$$f(2) = 5$$

(b) $f(0) = 2(0)^2 - 4(0) + 5 = 5$, so

$$f(0) = 5$$

(c) $f(-\frac{1}{2}) = 2(-\frac{1}{2})^2 - 4(-\frac{1}{2}) + 5 = 2(\frac{1}{4}) + 2 + 5 = \frac{1}{2} + 7 = \frac{15}{2}$, so

$$f(-\frac{1}{2}) = \frac{15}{2}$$

Example 2 illustrates the procedure for finding the image of an algebraic expression.

EXAMPLE 2

Let

$$f(x) = 2x^2 - 4x + 5$$

Problem

Find and simplify each of these expressions.

(a) $f(a)$, (b) $f(x + 3)$, (c) $f(x + h)$, (d) $\dfrac{f(x + h) - f(x)}{h}$, for $h \neq 0$.

Solution (a) The value of $f(a)$ is found by substituting a for x wherever x occurs in the formula for $f(x)$.

$$f(x) = 2x^2 - 4x + 5$$
$$f(a) = 2a^2 - 4a + 5$$

(b) To obtain $f(x + 3)$, substitute $x + 3$ for x wherever x occurs in the formula for $f(x)$.

$$\begin{aligned} f(x + 3) &= 2(x + 3)^2 - 4(x + 3) + 5 \\ &= 2x^2 + 12x + 18 - 4x - 12 + 5 \\ &= 2x^2 + 8x + 11 \end{aligned}$$

(c) To obtain $f(x + h)$, substitute $x + h$ for x wherever x occurs in the formula for $f(x)$.

$$\begin{aligned} f(x + h) &= 2(x + h)^2 - 4(x + h) + 5 \\ &= 2x^2 + 4xh + 2h^2 - 4x - 4h + 5 \end{aligned}$$

(d) Using part (c) and the definition of $f(x)$ gives $f(x + h) - f(x)$

$$= 2x^2 + 4xh + 2h^2 - 4x - 4h + 5 - (2x^2 - 4x + 5)$$
$$= 4xh + 2h^2 - 4h = h(4x + 2h - 4)$$

Then

$$\frac{f(x + h) - f(x)}{h} = \frac{h(4x + 2h - 4)}{h}$$

Since $h \neq 0$ the h-factors can be canceled, so

$$\frac{f(x + h) - f(x)}{h} = 4x + 2h - 4$$

In many applications it is advantageous to use letters other than f to represent the function. *Fortune* magazine, which is known for its many rankings of businesses, provides an example. Figure 2 displays the world's 10 biggest industrial corporations in 1986, with the 1985 ranking on the right side of the left-hand column. The list describes a function F (for *Fortune*), whose domain is the set $\{1, 2, \ldots, 10\}$, defined by the rule that $F(n)$ is the sales of the nth biggest corporation in 1986 (in \$thousands). It follows from Figure 2 that $F(1) = 102,813,700$ and $F(8) = 35,211,000$.

Domain and Range

A **function** is a correspondence between two sets in which each element x in one set, called the **domain,** has associated with it precisely one element in another set, called the **range.** Most of the functions in this book will be defined by an equation, with the domain and range being sets of real numbers. Unless otherwise specified,

Rank		Company	Headquarters	Industry	Sales ($ thousands)	Net Income ($ thousands)
1	1	General Motors	Detroit	Motor vehicles and parts	102,813,700	2,944,700
2	2	Exxon	New York	Petroleum refining	69,888,000	5,360,000
3	3	Royal Dutch/Shell Group	The Hague/London	Petroleum refining	64,843,217	3,725,779
4	6	Ford Motor	Dearborn, Mich.	Motor vehicles and parts	62,715,800	3,285,100
5	7	International Business Machines	Armonk, N.Y.	Computers	51,250,000	4,789,000
6	4	Mobil	New York	Petroleum refining	44,866,000	1,407,000
7	5	British Petroleum	London	Petroleum refining	39,855,564	731,954
8	12	General Electric	Fairfield, Conn.	Electronics	35,211,000	2,492,000
9	10	American Tel. & Tel.	New York	Electronics	34,087,000	139,000
10	8	Texaco	White Plains, N.Y.	Petroleum refining	31,613,000	725,000

FIGURE 2   Adapted from ''The World's 50 Biggest Industrial Corporations,'' *Fortune,* August 3, 1987, pp. 23–25.

the domain of a function that is defined by an equation consists of all numbers for which the equation is defined. For example, the domain of a linear function $f(x) = mx + b$ is the set of all real numbers. However, the domain of the function $f(x) = \dfrac{3}{2 - x}$ consists of all numbers different from 2 because division by 0 is undefined. A **function** f is **defined at a number** a if $f(a)$ exists. Notice that the function $f(x) = \dfrac{3}{2 - x}$ is defined at 0 because $f(0) = \dfrac{3}{2}$, but it is not defined at 2.

We are free to choose a value for x and substitute it into the equation to get the corresponding value $f(x)$, provided the equation is defined for that value of x. Thus, we refer to x as the **independent variable** and $f(x)$ as the **dependent variable.** These definitions allow us to give a more rigorous definition of the domain and range of a function.

DEFINITION

The **domain** of a function is the set of values of the independent variable for which the function is defined. The **range** of a function is the set of values of the dependent variable.

The range of a function is the set of images, while the domain of a function is the set of preimages. Example 3 computes the domain and range of another standard type of function.

EXAMPLE 3

Problem

Find the domain and range of the function $f(x)$ defined by the equation $f(x) = \sqrt{1 - x}$.

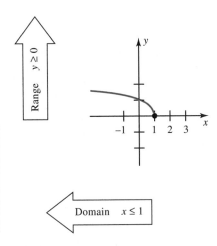

Solution The square root is only defined when the radicand $1 - x$ is nonnegative, that is, when $1 - x \geq 0$. Therefore the domain of $f(x)$ is the set of all numbers x such that $x \leq 1$. As an interval, the domain is $(-\infty, 1]$.

To find the range, observe that $\sqrt{1 - x}$ is never negative, and as x takes on all values in the domain, $f(x)$ takes on all nonnegative numbers. Thus the range is the set of all numbers greater than or equal to 0. As an interval, the range is $[0, \infty)$.

Quadratic Functions

A function defined by the equation

$$f(x) = ax^2 + bx + c$$

where a, b, and c are given real numbers with $a \neq 0$, is called a **quadratic function.** A quadratic function is not a linear function; its graph is a **parabola,** not a line. Chapter 14 will show how to sketch the graphs of quadratic functions by using the techniques of calculus, but for now we rely on the plotting method.

EXAMPLE 4

Problem

Sketch the graph of the quadratic function $f(x) = 2x^2 - 4x + 1$ by finding the images of -1, 0, 1, 2, and 3.

Solution Set $x = -1$. Then $f(-1) = 2(-1)^2 - 4(-1) + 1 = 7$, so the point $(-1, 7)$ lies on the parabola. Similarly, $f(0) = 1$, $f(1) = -1$, $f(2) = 1$, and $f(3) = 7$. Therefore the points $(0, 1)$, $(1, -1)$, $(2, 1)$, and $(3, 7)$ also lie on the parabola. The graph is sketched in Figure 3.

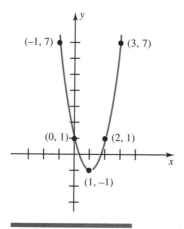

FIGURE 3   Graph of $f(x) = 2x^2 - 4x + 1$.

All parabolas are similar in shape to the one shown in Figure 3. Moreover, the graph of any quadratic function $f(x) = ax^2 + bx + c$ is a parabola that opens upward or downward, has a lowest or highest point (the **vertex**), and has a vertical line of symmetry passing through the vertex. Example 4 shows that the graph of $f(x) = 2x^2 - 4x + 1$ is a parabola that opens upward and has a vertical line of symmetry passing through the vertex (the lowest point on the graph) at $(1, -1)$.

Functions Versus Equations

There is a crucial distinction between functions and equations because not every equation represents a function. Consider the equations $y = x^2$ and $y^2 = x$. Their graphs, which are both parabolas, are drawn in Figure 4. The points $(4, 2)$ and $(4, -2)$ lie on the graph of the equation $y^2 = x$, so if a function $g(x)$ represented this equation, then we would have $g(4) = 2$ and $g(4) = -2$. By definition, a function of x cannot have a value x in the domain with two images. Therefore $y^2 = x$ is an equation that does not represent a function of x. However, for the equation $y = x^2$ each value of x has precisely one image, so it represents the function $f(x) = x^2$.

The **vertical line test** provides a visual means of determining whether a curve represents the graph of a function.

VERTICAL LINE TEST

A curve represents the graph of a function if and only if every vertical line intersects the curve at exactly one point or at no points.

Apply the vertical line test to the curves in Figure 4. In part (a) no vertical lines cross the curve in two or more points, so the curve represents the graph of the function $f(x) = x^2$. Many vertical lines intersect the curve in part (b) in two points,

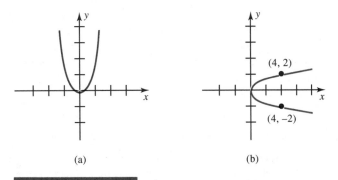

(a)   (b)

FIGURE 4   (a) The equation $y = x^2$ represents a function.
(b) The equation $y^2 = x$ does not represent a function.

so the curve does not represent a function. Example 5 applies the vertical line test to other curves.

EXAMPLE 5

Problem

Which of these curves represent graphs of functions?

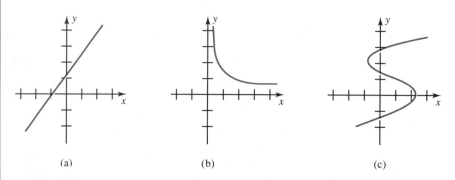

(a)   (b)   (c)

Solution The curve in (a) is a straight line. Each vertical line intersects the curve at precisely one point, so the curve represents the graph of a linear function.

The curve in (b) is drawn in the first quadrant only. The vertical lines through positive values of x meet the curve at one point. The vertical lines through the other values of x do not meet the curve. Thus the curve represents the graph of a function.

Look at the y-axis in (c). It is a vertical line that intersects the curve at more than one point, in fact, at three points. Therefore the curve does not represent the graph of a function.

Cost Analysis

Linear equations can be applied to some manufacturing problems in the following way. Let the cost of producing an item be denoted by C. If C depends on the number x of items produced, then C is called the **cost function,** denoted by $C(x)$. The cost function is composed of two parts, the **fixed cost,** which does not depend on x (such as the manufacturer's insurance), and the **variable cost,** which changes as x changes.

$C(x)$ = variable cost + fixed cost
$C(x)$ = (variable cost per item)x + fixed cost

The **marginal cost** is the change in the cost when one more item is produced. If the variable cost per item is constant, then $C(x)$ is a linear function whose slope is the marginal cost.

As an illustration of these terms consider the cost of receiving cable reception where there is a one-time installation fee of $20 and a charge of $0.30 per day (including fractions of a day). To write the cost function, let x represent the time in days. Since the variable cost is $0.30x and the fixed cost is the flat rate of $20, the cost function is $C(x) = 0.3x + 20$, where $x \geq 0$. The cost for one regular year is $C(365) = 0.3(365) + 20 = 129.5$, or $129.50.

EXAMPLE 6

A printer quotes a price of $5 for producing a 200-page book and $8 for producing a 600-page book.

Problem

Assuming a linear relationship that depends on the number of pages, (a) what is the cost function? and (b) how much will it cost to produce a 300-page book?

Solution (a) Since the cost depends on the number of pages, let x denote the number of pages. There is a linear relationship between $C(x)$ and x, so $C(x) = mx + b$, where m is the variable cost per book and b is the fixed cost. We have $C(200) = 5$ and $C(600) = 8$, so two points on the line are $(200, 5)$ and $(600, 8)$. Therefore the variable cost is $m = (8 - 5)/(600 - 200) = 3/400$. By the point-slope form, using the point $(200, 5)$, we get

$$C(x) - 5 = \left(\frac{3}{400}\right)(x - 200)$$

$$C(x) = \left(\frac{3}{400}\right)x + \left(\frac{7}{2}\right)$$

From this form we see that the fixed cost is $3.50 since $C(0) = 7/2$. (b) Once the cost function is known, the task of finding the cost of producing a book of any number of pages is straightforward. If $x = 300$, then the cost is $5.75 since $C(300) = (3/400)300 + (7/2) = 23/4 = 5.75$. (See Figure 5.)

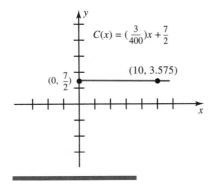

FIGURE 5

EXERCISE SET 11–1

In Problems 1 to 4 compute $f(2)$, $f(0)$, and $f(-\frac{1}{2})$ for the given function.

1. $f(x) = 2x - 3$
2. $f(x) = -3x + 4$
3. $f(x) = -x^2 + x + 3$
4. $f(x) = x^2 + 4x - 5$

In Problems 5 to 8 find and simplify each of these expressions for the given function. (a) $f(a)$, (b) $f(x + 3)$, (c) $f(x + h)$, (d) $\dfrac{f(x + h) - f(x)}{h}$, for $h \neq 0$.

5. $f(x) = 2x - 3$
6. $f(x) = -3x + 4$
7. $f(x) = -x^2 + x + 3$
8. $f(x) = x^2 + 4x - 5$

Problems 9 and 10 refer to Figure 2 in the text, which lists the 10 biggest industrial corporations in the world in 1986. Let F be the function defined by the rule that $F(n)$ was the sales of the nth biggest industrial corporation that year.

9. Compute $F(2)$
10. Compute $F(3)$

In Problems 11 to 14 find the domain of the stated function

11. $f(x) = \sqrt{x + 1}$
12. $f(x) = \sqrt{2x + 1}$
13. $f(x) = \sqrt{1 - 2x}$
14. $f(x) = \sqrt{2 - x}$

In Problems 15 to 18 sketch the graph of the quadratic function by finding the images of -2, -1, 0, 1, and 2.

15. $f(x) = x^2 + 2x - 5$
16. $f(x) = x^2 - 2x + 3$
17. $f(x) = 3x^2 - 5x - 4$
18. $f(x) = 2x^2 - 3x - 5$

In Problems 19 and 20 determine which curves represent the graph of a function.

19.

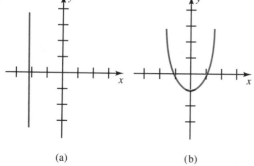

(a)　　　　　　　　(b)

20.

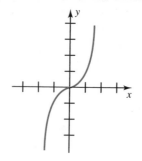

(a)　　　　　　　　(b)

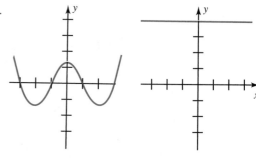

(c)

(c)

In Problems 21 to 24 evaluate $f(x + h) - f(x)$ for the given function.

21. $f(x) = 2x - 3$
22. $f(x) = -3x + 4$
23. $f(x) = -x^2 + x + 3$
24. $f(x) = x^2 + 4x - 5$

In Problems 25 to 28 evaluate $\dfrac{f(x + h) - f(x)}{h}$ for the given function, with $h \neq 0$.

25. $f(x) = 4 - x$
26. $f(x) = -1 - 2x$
27. $f(x) = -2x^2 - 4x$
28. $f(x) = 6x^2 + 3x$

In Problems 29 to 38 find the domain of the stated function.

29. $f(x) = \dfrac{3}{2 - x}$

30. $f(x) = \sqrt{x^2 + 1}$

31. $f(x) = \sqrt{1 - x^2}$

32. $f(x) = \dfrac{1}{x}$

33. $f(x) = \sqrt{x^2 - 1}$

34. $f(x) = \sqrt{x^3 - 1}$

35. $f(x) = \dfrac{1}{x(x - 2)}$

36. $f(x) = \dfrac{x}{x^2 - 9}$

37. $f(x) = \dfrac{\sqrt{x - 3}}{x - 4}$

38. $f(x) = \dfrac{x + 3}{\sqrt{x + 4}}$

In Problems 39 and 40 find all preimages of the given value for the function $f(x) = x^2 + 1$.

39. (a) 5   (b) 0   (c) 1.25

40. (a) 2   (b) -1 (c) 3.25

In Problems 41 and 42 sketch the graph of the equation and determine if it represents a function.

41. $y = \sqrt{x}$

42. $y^2 = 2x$

Referenced Exercise Set 11–1

1. Meteorologists use the equation $d^3 = 216t^2$ as a model to describe the size and intensity of four types of violent storms: tornadoes, thunderstorms, hurricanes, and cyclones.* Here d is the diameter of the storm (in miles) and t is the time (in hours) the storm travels before dissipating.
   (a) Is d a function of t? If so, solve for $d(t)$.
   (b) The world's worst recorded monsoon lasted for 24 hours on November 13–14, 1970, in the Ganges Delta Islands in Bangladesh. More than one million people died. What was the storm's diameter?
   (c) In the United States about 150 tornadoes occur each year, mostly in the central plains states, with the highest frequency in Iowa, Kansas, Arkansas, Oklahoma, and Mississippi. If a tornado's diameter is 2 miles, how long would it be expected to last?

2. Figure 6 lists the top 10 corporations in 1986 located outside the United States.† Let I be the function defined by the rule that $I(n)$ is the sales of the nth biggest corporation that year (in $thousands).
   (a) What is $I(1)$?
   (b) What is the image of 4?
   (c) What is the preimage of 22,668,085?

3. Would the Fortune 500 list of the biggest corporations in 1986 define a function if there is a tie for twelfth place?

Rank 1986	1985	Company	Country	Sales ($ thousands)
1	1	Royal Dutch/Shell Group	Neth./Britain	64,843,217
2	2	British Petroleum	Britain	39,855,564
3	5	IRI	Italy	31,561,709
4	3	Toyota Motor	Japan	31,553,827
5	16	Daimler-Benz	W. Germany	30,168,550
6	7	Matsushita Electric Industrial	Japan	26,459,539
7	6	Unilever	Neth./Britain	25,141,672
8	15	Volkswagen	W. Germany	24,317,154
9	8	Hitachi	Japan	22,668,085
10	4	ENI	Italy	22,549,921

FIGURE 6 The Top 10 Corporations Located Outside the United States, 1986.

*E. E. David and J. G. Truxal, *The Man-Made World*, New York, McGraw-Hill Book Company, 1971, pp. 176–177. The authors thank Glenn Allinger of Montana State University for supplying this excellent source of applications.

†Terence Paré and Wilton Woods, "The International 500: The Fortune Directory of the Biggest Industrial Corporations Outside the U.S.," *Fortune*, August 3, 1987, pp. 214–233.

4. This problem refers to the number of chirps that a cricket makes. Biologists have shown that although different species of crickets produce different numbers of chirps, the number of chirps produced by each species is a linear function of the temperature. For instance, the number of chirps per minute that the cricket *Gryllus pennsylvanicus* makes is given by $C(t) = 4t - 160$, where t is in degrees Fahrenheit.‡

(a) What is the domain of this function? (Make sure to take into account physical considerations.)

(b) How many chirps per minute does this cricket make at 68°?

(c) If this cricket makes 160 chirps per minute, what is the temperature?

5. In 1820 the median age in the United States was 16.7 years. This means that half of the population was younger than that age, and half was older. Table 1 lists the median age in the United States for each year from 1980 to 1987.§

(a) Define a function $y = f(x)$ in which x is a year and y is the corresponding median age.

(b) Evaluate $f(1990)$.

(c) For what value of x is $y = 29.1$?

Year	Median Age
1987	32.1
1986	31.8
1985	31.5
1984	31.2
1983	30.9
1982	30.6
1981	30.3
1980	30.0

‡Philip S. Callahan, *Insect Behavior*, New York, Four Winds Press, 1977.

§Randolf E. Schmid, "Median Age in U.S. Reaches 32.1 Years, Highest on Record," Associated Press, April 6, 1988.

11–2 Polynomial and Other Functions

In this section we define polynomial functions and sketch their graphs by the plotting method. Then we introduce rational functions and sketch their graphs. The concept of a vertical asymptote plays an important role in graphing rational functions. We also introduce functions that are defined by different formulas over different intervals and functions that are defined by a curve instead of a formula.

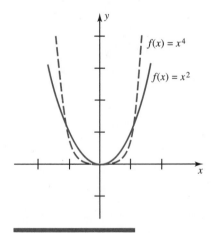

FIGURE 1 Contrast of $f(x) = x^2$ vs. $f(x) = x^4$.

Polynomial Functions

The graph of the quadratic function $f(x) = x^2$ is called a parabola. The graph of the function $f(x) = x^4$ has a very similar shape, although it is slightly different. The two curves are contrasted in Figure 1. The graphs of the functions $f(x) = x^n$ for $n = 6, 8, 10, \ldots$ are similar.

However, the graph of $f(x) = x^n$ is quite different when n is an odd number. For $n = 1$ the graph of $f(x) = x$ is a line. The graph of $f(x) = x^3$ is sketched in Figure 2. The graphs of the functions $f(x) = x^n$ for $n = 5, 7, 9, \ldots$ have similar shapes.

Collectively, functions of the form $f(x) = x^n$, where n can be any real number, are called **power functions.** Since n does not have to be a natural number, some additional examples of power functions are

$$n = 0 \qquad\qquad n = -1 \qquad\qquad n = \frac{1}{2}$$

$$f(x) = x^0 = 1 \qquad f(x) = x^{-1} = \frac{1}{x} \qquad f(x) = x^{\frac{1}{2}} = \sqrt{x}$$

A **polynomial function** of the variable x is defined by the formula

$$f(x) = a_n x^n + a_{n-1} x^{n-1} + \cdots + a_2 x^2 + a_1 x + a_0$$

where n is a nonnegative integer, the coefficients a_i are real numbers, and $a_n \neq 0$. The integer n is called the **degree of the polynomial.** The domain of a polynomial

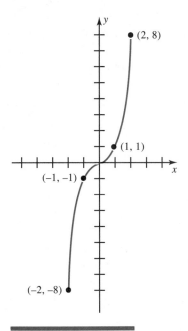

FIGURE 2 Graph of $f(x) = x^3$.

function consists of all real numbers. A polynomial function of degree 2 is a quadratic function, a polynomial function of degree 1 is a linear function, and a polynomial function of degree 0 is a constant function. The function

$$f(x) = x^4 - 8x^3 + 22x^2 + 9$$

is a polynomial function of degree 4 with $a_4 = 1$, $a_3 = -8$, $a_2 = 22$, $a_1 = 0$, and $a_0 = 9$. Some other examples of polynomial functions are

$$g(x) = 12 \qquad F(x) = 4x - 3.5 \qquad G(x) = 3x^2 - 1.5x + 2$$

Some functions which are not polynomial functions are

$$h(x) = \frac{2}{x} \qquad P(x) = \sqrt{x}$$

Example 1 shows how to sketch the graph of a polynomial function by the plotting method.

EXAMPLE 1

Problem

Sketch the graph of the polynomial function

$$f(x) = x^4 - 8x^3 + 22x^2 - 24x + 9$$

Solution Evaluate $f(x)$ at the values $x = -2, -1, 0, 1,$ and 2. The results are displayed in the table.

x	-2	-1	0	1	2
$f(x)$	225	64	9	0	1

These values suggest that the graph is the curve shown in Figure 3a. However, plotting additional points from an expanded table (rounded to one decimal place) leads to the correct graph of the polynomial function shown in Figure 3b.

x	0.5	1.5	2.5	3	3.5	4	5
$f(x)$	1.6	0.6	0.6	0	1.6	9	64

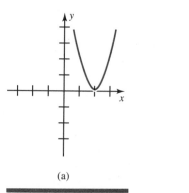

(a)

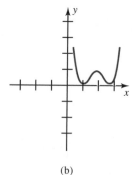

(b)

FIGURE 3 Two views of $f(x) = x^4 - 8x^3 + 22x^2 - 24x + 9$.

Example 1 shows that sometimes it is necessary to plot many points to obtain an accurate graph of a function. Yet the question persists: How do we know that the curve in Figure 3b is the correct graph of the given polynomial? The techniques of calculus will answer this question.

Rational Functions

Consider the power function $f(x)$ defined by the formula

$$f(x) = \frac{1}{x}$$

Its graph, which is called a **hyperbola,** consists of two disconnected branches. (See Figure 4.) The function is not defined at $x = 0$. Notice that as the values of x get closer and closer to 0, the values of $f(x)$ get larger and larger in absolute value. When a number v has the property that the values of a function $f(x)$ get larger and larger in absolute value as the values of x approach v, the line $x = v$ is called a **vertical asymptote** of the graph of $f(x)$. Figure 4 shows that the line $x = 0$ is a vertical asymptote of the function $f(x) = 1/x$.

A **rational function** is the quotient of two polynomial functions. Therefore $f(x)$ is a rational function if it can be written in the form

$$f(x) = \frac{g(x)}{h(x)}$$

for some polynomials $g(x)$ and $h(x)$. The function $f(x) = 1/x$ is a rational function with $g(x) = 1$ and $h(x) = x$.

The domain of a rational function $f(x) = g(x)/h(x)$ consists of all real numbers for which $h(x) \neq 0$. Each number v for which $h(v) = 0$ will produce a vertical asymptote of $f(x)$ at $x = v$ if $g(v) \neq 0$. Example 2 shows how to find vertical asymptotes.

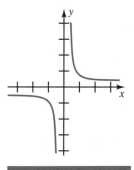

FIGURE 4 Graph of $f(x) = \dfrac{1}{x}$.

EXAMPLE 2

Problem

Find the vertical asymptotes of the function

$$f(x) = \frac{x - 2}{x^2 - 1}$$

Solution Set $g(x) = x - 2$ and $h(x) = x^2 - 1$. If the rational function $f(x)$ has a vertical asymptote, it will occur where $h(x) = 0$. Set $h(x) = x^2 - 1 = 0$. Then $x = 1$ or $x = -1$. Also, $g(1) \neq 0$ and $g(-1) \neq 0$. Therefore the lines $x = 1$ and $x = -1$ are the vertical asymptotes of the graph of $f(x)$.

Consider the rational function $f(x) = \dfrac{x^2 - 4}{x + 2}$. Set $g(x) = x^2 - 4$ and $h(x) = x + 2$. The only possible vertical asymptote will be the line $x = -2$ since -2 is the only number with $h(-2) = 0$. Although $g(-2) = 0$ we cannot necessarily conclude that $x = -2$ is a vertical asymptote. In fact, in this case the line $x = -2$ is not a vertical asymptote of $f(x)$.

A knowledge of the vertical asymptotes of a rational function eases the task of sketching its graph. Example 3 illustrates the procedure.

EXAMPLE 3

Problem

Sketch the graph of the rational function

$$f(x) = \frac{x}{x - 1}$$

Solution Set $g(x) = x$ and $h(x) = x - 1$. The only vertical asymptote of $f(x)$ is the line $x = 1$ since $h(1) = 0$ but $g(1) \neq 0$. Draw it as a dashed line. Construct a table of ordered pairs for $x < 1$.

x	-2	-1	0	0.5	0.9	0.99
$f(x)$	$\frac{2}{3}$	$\frac{1}{2}$	0	-1	-9	-99

Notice that as the values of x approach 1, the images of $f(x)$ get larger and larger in absolute value. This shows the behavior of $f(x)$ to the left of the vertical asymptote. A table for $x > 1$ shows the behavior to the right of the vertical asymptote.

x	4	3	2	1.5	1.1	1.01
$f(x)$	$\frac{4}{3}$	$\frac{3}{2}$	2	3	11	101

Taken together, the graph consists of the two branches that result from plotting the points in these tables. (See Figure 5.)

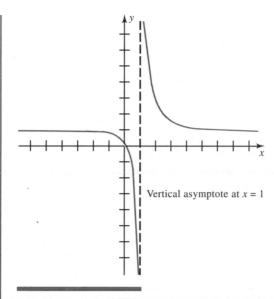

FIGURE 5 Graph of $f(x) = \dfrac{x}{x-1}$.

Next we introduce two additional kinds of functions. The first concerns functions that are defined by different formulas over different intervals.

EXAMPLE 4

Problem

Sketch the graph of the function defined by

$$f(x) = \begin{cases} 2x + 1 & \text{if } x < 2 \\ x^2 - 1 & \text{if } x \geq 2 \end{cases}$$

Solution The notation means that $f(x) = 2x + 1$ for all values of x such that $x < 2$. This is the linear function with slope 2 and y-intercept 1. The graph of $f(x)$ in the interval $(-\infty, 2)$ is part of this line. An open circle at the right end point of the line indicates that this point is not part of the function, but rather, serves as a boundary. The graph of $f(x)$ in the interval $[2, \infty)$ is part of the quadratic function $f(x) = x^2 - 1$. The entire graph consists of the two branches shown in Figure 6.

A function can be defined by a graph as well as by a formula. This sometimes occurs when the equation is either unknown or too complicated to express easily. For each value of x for which there is one point (x, y) on the curve, the image $f(x)$ is defined to be equal to y. This defines the function f implicitly. Example 5 examines one such function.

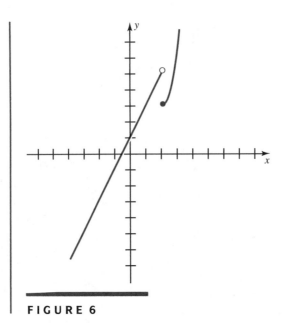

FIGURE 6

EXAMPLE 5

A firm's financial leverage is defined as its debt/assets ratio, or D/A ratio. The D/A ratio is usually given as a percent. There is a relation between the D/A ratio and the firm's expected earnings-per-share (EPS). Figure 7 gives the graph for Santa Clara Industries on January 1, 1985.

Problem

(a) Approximate $f(30)$. (b) What D/A ratio has the maximal expected EPS?

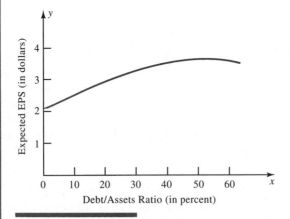

FIGURE 7 Financial Leverage vs. D/A Ratio for Santa Clara Industries. (Adapted from "Financial Leverage," in B. Campsey and E. Brigham, *Introduction to Financial Management,* Hinsdale, IL, The Dryden Press, 1985, pp. 534–541.)

Solution (a) The notation $f(30)$ stands for the image of $x = 30\%$. It corresponds to the value of y from the point $(30, y)$ on the graph. From Figure 7 it appears that $y = 3$, so $f(30) = 3$ and the expected EPS is $3. (According to Santa Clara Industries' tables, $f(30) = \$2.97$. This degree of accuracy cannot be inferred from the graph, however.)

(b) The maximal expected EPS of $3.36 occurs at the point $(50, 3.36)$. Thus $f(50) = 3.36$. The preimage of 3.36 is 50, so the D/A ratio is 50%.

Application

In the previous section we examined linear cost functions. Here we discuss a kind of nonlinear cost function that arises in manufacturing.

EXAMPLE 6

A manufacturer has determined that the cost of producing x items is $C(x) = \dfrac{x^2 + 25}{x}$, where fractional parts of an item can be produced.

Problem

(a) What is the domain of $C(x)$? (b) Sketch the graph of $C(x)$ to determine how many items should be produced in order to minimize cost.

Solution (a) The domain of $C(x)$ does not include $x = 0$ because division by 0 is undefined. In addition, since it makes no sense to talk about a negative num-

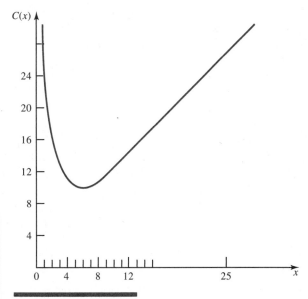

FIGURE 8 The cost function $C(x) = \dfrac{x^2 + 25}{x}$.

ber of items, it is meaningless to let x assume negative values. Therefore the domain of $C(x)$ consists of all positive real numbers, or the interval $(0, \infty)$.

(b) Set $g(x) = x^2 + 25$ and $h(x) = x$. Then $h(x) = 0$ and $g(x) \neq 0$, so the line $x = 0$ is a vertical asymptote of $f(x)$. The relevant graph of $C(x)$ is obtained by the plotting method from the table.

x	1	2	3	4	5	6	7	25
$C(x)$	26	$\frac{29}{2}$	$\frac{34}{3}$	$\frac{41}{4}$	10	$\frac{61}{6}$	$\frac{74}{7}$	26

The graph, which is sketched in Figure 8, indicates that the minimum cost occurs at $x = 5$ items.

In Chapter 14 we will describe how to use calculus to find the minimum cost in Example 6.

EXERCISE SET 11–2

In Problems 1 to 4 sketch the graph of the stated function by the plotting method.

1. $f(x) = 2x^3 - 9x^2 + 12x$

2. $f(x) = x^3 - 9x^2 + 24x - 20$

3. $f(x) = x^4 - 4x^3 + 4x^2 + 1$

4. $f(x) = x^4 - 6x^2$

5. What is the degree of the polynomial in Problem 1?

6. What is the degree of the polynomial in Problem 4?

In Problems 7 to 10 find the vertical asymptotes of each function.

7. $f(x) = \dfrac{x}{x - 5}$

8. $f(x) = \dfrac{3x}{2x - 5}$

9. $f(x) = \dfrac{x - 2}{x^2 - 9}$

10. $f(x) = \dfrac{3 - x}{x^2 - 4}$

In Problems 11 to 16 sketch the graph of the stated function.

11. The function defined in Problem 7.

12. The function defined in Problem 8.

13. $f(x) = \begin{cases} 3x + 2 & \text{if } x \leq 0 \\ x^2 + 2 & \text{if } x > 0 \end{cases}$

14. $f(x) = \begin{cases} 2x + 1 & \text{if } x \geq 3 \\ x^2 - 2 & \text{if } x < 3 \end{cases}$

15. $f(x) = \begin{cases} x^2 - 1 & \text{if } x < 0 \\ 2x - 1 & \text{if } x \geq 0 \end{cases}$

16. $f(x) = \begin{cases} 1 - x^2 & \text{if } x > 2 \\ 1 - 2x & \text{if } x \leq 2 \end{cases}$

Problems 17 to 20 refer to Figure 7, which shows the relation between the D/A ratio and the expected EPS for Santa Clara Industries.

17. Approximate $f(0)$.

18. Approximate $f(60)$.

19. What D/A ratio produces an expected EPS of approximately $3?

20. What D/A ratio produces an expected EPS of approximately $2.75?

In Problems 21 and 22 the cost of producing x items is given. Fractional parts of an item can be produced.
(a) What is the domain of $C(x)$?
(b) Sketch $C(x)$ to determine how many items should be produced in order to minimize cost.

21. $C(x) = \dfrac{x^2 + 16}{x}$

22. $C(x) = \dfrac{x^2 + 81}{x}$

In Problems 23 and 24 sketch the graphs of the two stated functions on the same coordinate axis.

23. $f(x) = x^2$ and $f(x) = x^6$

24. $f(x) = x^3$ and $f(x) = x^5$

In Problems 25 and 26 state the degree of the polynomial.

25. $f(x) = 1 - 2x - 5x^2$

26. $f(x) = 3 - 5x$

In Problems 27 to 34 find the vertical asymptotes, if any, and sketch the graph of the stated function.

27. $f(x) = \dfrac{x - 2}{x^2 - 9}$

28. $f(x) = \dfrac{3 - x}{x^2 - 4}$

29. $f(x) = \dfrac{x}{x^2}$

30. $f(x) = \dfrac{1}{x^2 + 1}$

31. $f(x) = \dfrac{1}{x(x^2 - 1)}$

32. $f(x) = \dfrac{x - 1}{x(x^2 - 4)}$

33. $f(x) = \dfrac{2x}{(3 + 2x)(x - 1)}$

34. $f(x) = \dfrac{x}{(1 + 2x)(x - 2)}$

In Problems 35 to 38 sketch the graph of the stated function.

35. $f(x) = \begin{cases} x & \text{if } x \geq 1 \\ x^2 + 1 & \text{if } x \leq 0 \end{cases}$

36. $f(x) = \begin{cases} 2x - 1 & \text{if } x \geq 3 \\ x^3 & \text{if } x \leq 1 \end{cases}$

37. $f(x) = \begin{cases} 1 - 2x - 5x^2 & \text{if } x \leq 1 \\ x & \text{if } x > 1 \end{cases}$

38. $f(x) = \begin{cases} -x^2 & \text{if } x \geq 0 \\ x^2 & \text{if } x < 0 \end{cases}$

39. Suppose the profit per item is given by the formula $P(x) = 20(3 - x)(x - 25)$, where x is the number of items that are produced. Draw a graph of $P(x)$ to determine the number of items that should be produced to yield the largest profit per item.

40. A store offers this special sale: if 10 compact disks (CDs) are purchased at the full price of $12, then additional CDs can be purchased at half price. There is a limit of 25 CDs per person. Express the cost of the CDs as a function of the number purchased and draw the graph.

In Problems 41 to 44 match the equation with the given graph.

41. $f(x) = \dfrac{x}{x + 1}$

42. $f(x) = \dfrac{x}{1 - x}$

43. $f(x) = \begin{cases} x^2 & x \geq 1 \\ -x^2 & x < 1 \end{cases}$

44. $f(x) = \begin{cases} x^2 & x \geq 0 \\ -x^2 & x < 0 \end{cases}$

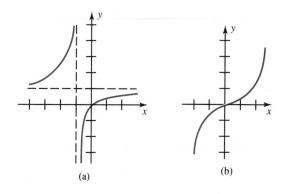

(a)   (b)

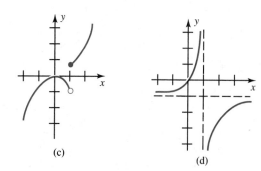

(c)   (d)

In the programming language BASIC there is a function INT that produces the integral part of a number. For instance, $INT(3.14) = 3$. In Problems 45 and 46 sketch the graph of the given function.

45. $f(x) = INT(x)$ for $x \geq 0$

46. $f(x) = INT(x^2)$ for $x \geq 0$

Referenced Exercise Set 11–2

1. The curve in Figure 9 shows the relation between the implied stock price and the D/A ratio for Santa Clara Industries. What percentage yields the maximal D/A ratio?

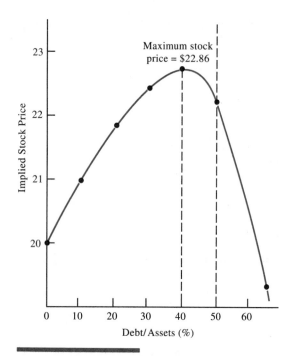

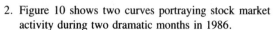

FIGURE 9 Relation Between Implied Stock Price and the D/A Ratio for Santa Clara Industries. (Adapted from B. Campsey and E. Brigham, *Introduction to Financial Management,* Hinsdale, IL, The Dryden Press, 1985, p. 542.)

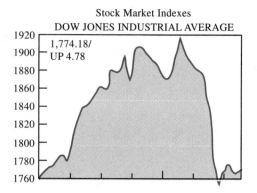

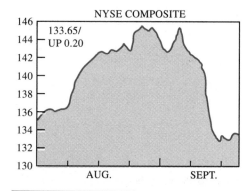

FIGURE 10 Stock market activity during two dramatic months in 1986. (Dow–Jones Index.)

2. Figure 10 shows two curves portraying stock market activity during two dramatic months in 1986.
   (a) What was the maximum value of the Dow Jones Industrial average during this period?
   (b) During which week did the Dow Jones Industrial average tumble precipitously?

3. An article in the *Wall Street Journal* analyzed the federal income tax that was signed into law on October 22, 1986.* Consider this statement as an example of the tax law: "If the income lies in the range from $0 through $29,300 (inclusive) then the tax is 15 percent of the income; if the income exceeds $29,300 then the tax equals $4350 plus 27 percent of the portion of the income above $29,300."

(a) Write this statement as a function $y = f(x)$, where x is the income and y is the corresponding tax.
(b) Sketch the graph of $f(x)$.
(c) Evaluate $f(x)$ for $x = 29,300$.
(d) Evaluate $f(x)$ for $x = 29,301$.
(e) What do you conclude from parts (c) and (d)?

4. Public health officials must often determine the threshold level at which effects on health begin. The threshold level for certain environmental issues occurs when the introduction of one more unit of pollution becomes a detriment to public health. One study† made use of the

*Alan Murray, "Winners? Losers? Estimates Show How Impact of Tax Proposal Varies," *Wall Street Journal,* May 9, 1986, p. 25.

†Louis A. Cox, Jr., "A New Measure of Attributable Risk for Public Health Applications," *Management Science,* July 1985, pp. 800–813.

health function $H(x)$ defined by

$$H(x) = \begin{cases} 0 & \text{if } x < 5 \\ (x - 5)^2 & \text{if } x \geq 5 \end{cases}$$

where x denotes the number of units of pollution.
(a) Sketch the graph of this health function.
(b) At what value of x does the threshold level occur?

5. Human beings have raced over a large span of times and distances. Take women's swimming, for instance. Races have been contested from a scant 50 meters to the 80-kilometer (km) swim across the English Channel. It turns out that when world records for many athletic events are recorded for various distances at any given time, it is possible to derive a power function that relates the time to the distance. The power function for women's swimming up to 1980, with the time t in minutes and the distance x in kilometers (km), is:‡

$$t = 10.578x^{1.03256}$$

Use this equation to calculate the time of the world record for (a) 1500 meters and (b) 80 km.

6. Police departments use a manual when investigating accidents to establish the velocity v (in miles per hour) that the vehicle involved in the accident was traveling. Most manuals use the function

$$v = \sqrt{30\,f(x)}$$

where f is the coefficient of friction (a number that measures the friction between the vehicle's tires and the road's surface, usually between 0 and 1.5) and x is the length of the skid marks in feet.* Set the coefficient of friction equal to 1. (a) If the skid marks are 200 feet, what was the vehicle's velocity? (b) What is the domain of $f(x)$? (c) Sketch the graph of $f(x)$.

Cumulative Exercise Set 11–2

In Problems 1 and 2 sketch the graph of the equation.

1. $y = x^2 + 1$

2. $y = 2 - x^3$

3. Find the slope-intercept form of the line passing through the points $(0,1)$ and $(-1,0)$.

4. Find the equation of the line that passes through the point $(1,0)$ and is (a) vertical, (b) horizontal, (c) parallel to $y = -3x + 1$.

5. Find the domain of the function $f(x) = (x - 4)^{1/2}$.

6. Sketch the graph of the function $f(x) = x^2 - 2x + 1$.

7. Evaluate $\dfrac{f(x + h) - f(x)}{h}$ for $f(x) = x^2 - 3$.

8. Find the domain of the function $f(x) = \dfrac{1}{9 - x^2}$.

9. Sketch the graph of the function $f(x) = x^3 - x^2 + 1$.

10. Sketch the graph of the function

$$f(x) = \begin{cases} 2x & \text{if } x \geq 0 \\ x & \text{if } x < 0 \end{cases}$$

In Problems 11 and 12 find the vertical asymptotes and sketch the graph of the function.

11. $f(x) = \dfrac{x}{x - 4}$.

12. $f(x) = \dfrac{x}{x^2 - 4}$.

‡Peter S. Riegel, "Athletic Records and Human Endurance," *American Scientist*, May-June 1981, pp. 285–290.

*David S. Daniels, "Fast brakes!" *Mathematics Teacher*, February 1989, pp. 104–107 and 111.

11–3 The Algebra of Functions

The four familiar operations on real numbers are addition, subtraction, multiplication, and division. In this section we define these operations on functions and introduce a fifth operation, composition. We then discuss additional types of functions that are defined in terms of these five operations.

Operations on Functions

Consider the polynomial functions f and g defined by

$$f(x) = 2x^2 - 3x - 7 \qquad \text{and} \qquad g(x) = 2x + 3$$

The **sum** of these functions, denoted by $f + g$, is defined by the formula

$$
\begin{aligned}
(f + g)(x) &= f(x) + g(x) \\
&= (2x^2 - 3x - 7) + (2x + 3) \\
&= 2x^2 - x - 4
\end{aligned}
$$

The **difference** $f - g$ is defined similarly.

$$
\begin{aligned}
(f - g)(x) &= (2x^2 - 3x - 7) - (2x + 3) \\
&= 2x^2 - 5x - 10
\end{aligned}
$$

The **product** $f \cdot g$ is defined by carrying out the multiplication of the two polynomials.

$$
\begin{aligned}
(f \cdot g)(x) &= f(x) \cdot g(x) \\
&= (2x^2 - 3x - 7)(2x + 3) \\
&= (2x^2 - 3x - 7)(2x) + (2x^2 - 3x - 7)(3) \\
&= (4x^3 - 6x^2 - 14x) + (6x^2 - 9x - 21) \\
&= 4x^3 - 23x - 21
\end{aligned}
$$

The **quotient** f/g is defined by

$$\left(\frac{f}{g}\right)(x) = \frac{f(x)}{g(x)} = \frac{2x^2 - 3x - 7}{2x + 3}$$

Sometimes the quotient can be left in this form. At other times the division of polynomials must be carried out.

$$
\begin{array}{r}
x - 3 \\
2x + 3 \overline{)\,2x^2 - 3x - 7} \\
2x^2 + 3x \\
\hline
-6x - 7 \\
-6x - 9 \\
\hline
2
\end{array}
$$

Then

$$\left(\frac{f}{g}\right)(x) = x - 3 + \frac{2}{2x + 3}.$$

The operations of $+$, $-$, and $\cdot$ are defined whenever both $f(x)$ and $g(x)$ are defined, so the domain of each one is the intersection of the domain of f and the domain of g. The quotient f/g is somewhat different. It is defined whenever both

$f(x)$ and $g(x)$ are defined except where $g(x) = 0$. In the example above, the domain of f/g consists of all real numbers except $-\frac{3}{2}$ because $g(-\frac{3}{2}) = 0$. The display gives the general definitions.

DEFINITION

If f and g are any functions for which $f(x)$ and $g(x)$ are defined, then

$$(f + g)(x) = f(x) + g(x) \qquad \text{(sum)}$$

$$(f - g)(x) = f(x) - g(x) \qquad \text{(difference)}$$

$$(f \cdot g)(x) = f(x) \cdot g(x) \qquad \text{(product)}$$

$$\left(\frac{f}{g}\right)(x) = \frac{f(x)}{g(x)} \qquad \text{if } g(x) \neq 0 \qquad \text{(quotient)}$$

The four operations $f + g$, $f - g$, $f \cdot g$, and f/g are referred to collectively as the **rational operations**.

EXAMPLE 1

Let f and g be functions defined by

$$f(x) = \frac{x}{x - 1} \qquad \text{and} \qquad g(x) = \frac{x - 3}{x + 1}$$

Problem

Find $f + g$, $f \cdot g$, and f/g. State their domains.

Solution By definition $f + g$ is defined by

$$(f + g)(x) = \frac{x}{x - 1} + \frac{x - 3}{x + 1}$$

The proper form of the sum is obtained by carrying out the addition, which involves finding the least common denominator.

$$(f + g)(x) = \frac{x(x + 1) + (x - 1)(x - 3)}{(x - 1)(x + 1)}$$

$$= \frac{x^2 + x + x^2 - 4x + 3}{x^2 - 1}$$

$$= \frac{2x^2 - 3x + 3}{x^2 - 1}$$

The product $f \cdot g$ is obtained by multiplying the numerators and denominators.

$$(f \cdot g)(x) = \frac{x(x - 3)}{(x - 1)(x + 1)}$$

Often it is convenient to write this expression in a more compact form.

$$(f \cdot g)(x) = \frac{x^2 - 3x}{x^2 - 1}$$

The quotient is obtained from the "invert and multiply" rule

$$\left(\frac{f}{g}\right)(x) = \frac{x}{x - 1} \div \frac{x - 3}{x + 1} = \frac{x}{x - 1} \cdot \frac{x + 1}{x - 3}$$

$$= \frac{x^2 + x}{x^2 - 4x + 3}$$

The domain of f consists of all real numbers except 1 since $f(1)$ is not defined. Similarly, $g(-1)$ is not defined. Therefore the domain of $f + g$ and $f \cdot g$ is the set of all real numbers except 1 and -1. Since $g(3) = 0$, the domain of f/g is the set of all real numbers except 1, -1, and 3.

If a number a lies in the domain of $f + g$, we write $(f + g)(a) = f(a) + g(a)$. The same notation applies to the other three rational operations on two functions. Example 2 shows how to find such images.

EXAMPLE 2

Let f and g be functions defined by

$$f(x) = \frac{x}{x - 1} \quad \text{and} \quad g(x) = \frac{3x - 2}{x + 1}$$

Problem

Compute $(f + g)(\frac{2}{3})$, $(f \cdot g)(\frac{2}{3})$, and $(f/g)(\frac{2}{3})$, if possible.

Solution First $f(\frac{2}{3}) = -2$ and $g(\frac{2}{3}) = 0$. By definition, $(f + g)(\frac{2}{3}) = f(\frac{2}{3}) + g(\frac{2}{3})$ so $(f + g)(\frac{2}{3}) = -2 + 0 = -2$. Similarly, $(f \cdot g)(\frac{2}{3}) = f(\frac{2}{3}) \cdot g(\frac{2}{3}) = (-2) \cdot 0 = 0$. However, $(f/g)(\frac{2}{3})$ is not defined because $g(\frac{2}{3}) = 0$.

Composition of Functions

Consider once again the polynomial functions

$$f(x) = 2x^2 - 3x - 7 \quad \text{and} \quad g(x) = 2x + 3$$

Regard each function as an "input-output machine" such that each input value comes from the domain and the corresponding output lies in the range. Now put the machines in a line with g in front of f, so the output of g becomes the input of f. For example, take $x = 1$. The output of g is $g(1) = 5$. This value becomes the input of f, and since $f(5) = 28$, its output is 28. Altogether the sequence of "machines" has produced an image of 28 from the original input of $x = 1$. This

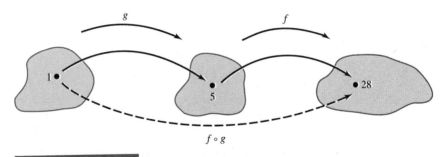

FIGURE 1 The composition of functions.

is shown in Figure 1, where the dashed curve represents the result of the two functions.

The operation of performing one function f on the image of another function g is called **composition** and is denoted by $f \circ g$. It is defined by

$$(f \circ g)(x) = f(g(x))$$

For the polynomials f and g above,

$$(f \circ g)(x) = f(g(x)) = f(2x + 3) = 2[g(x)]^2 - 3g(x) - 7$$

$$= 2(2x + 3)^2 - 3(2x + 3) - 7$$

$$= 2(4x^2 + 12x + 9) - 6x - 9 - 7$$

$$= 8x^2 + 18x + 2$$

The display provides the formal definition.

DEFINITION

If f and g are any functions, then the **composition** of f and g is the function $f \circ g$ defined by

$$(f \circ g)(x) = f(g(x))$$

The *domain* of $f \circ g$ is the set of all real numbers x in the domain of g whose image $g(x)$ lies in the domain of f.

If a is a number in the domain of $f \circ g$, then $(f \circ g)(a) = f(g(a))$. For $f(x) = 2x^2 - 3x - 7$, $g(x) = 2x + 3$, and $a = 1$, we get $(f \circ g)(1) = f(g(1)) = f(5) = 28$. However, $(f \cdot g)(1) = f(1) \cdot g(1) = (-8) \cdot 5 = -40$, so $(f \circ g)(x) \neq (f \cdot g)(x)$.

Warning: Do not confuse $f \circ g$ with $f \cdot g$. The composition of two functions is entirely different from the product of two functions.

EXAMPLE 3

Let f and g be the functions defined by

$$f(x) = \frac{1}{x + 1} \qquad \text{and} \qquad g(x) = \sqrt{x}$$

Problem

Find $f \circ g$ and $g \circ f$.

Solution

$$(f \circ g)(x) = f(g(x))$$

$$= f(\sqrt{x})$$

$$= \frac{1}{\sqrt{x} + 1}$$

Similarly

$$(g \circ f)(x) = g(f(x)) = g\left(\frac{1}{x + 1}\right) = \sqrt{\frac{1}{x + 1}} = \frac{1}{\sqrt{x + 1}}$$

When forming the composition of two functions the order of the functions is important. Notice, for instance, that for the functions f and g in Example 3, $(f \circ g)(1) = 1/2$ but $(g \circ f)(1) = 1/\sqrt{2}$, so $(g \circ f)(1) \neq (f \circ g)(1)$. We single this out as a warning.

☞ *Warning:* In general $f \circ g \neq g \circ f$.

When applying the techniques of calculus, often we find it advantageous to regard a given function as the composition of two functions. For example, consider the function $h(x) = \sqrt{x^2 - 3x + 5}$. Since the polynomial $x^2 - 3x + 5$ is evaluated first, set $g(x) = x^2 - 3x + 5$. Then $h(x) = \sqrt{g(x)}$. Now put $f(x) = \sqrt{x}$. Then $(f \circ g)(x) = f(g(x)) = \sqrt{g(x)} = h(x)$, so $h = f \circ g$. In this way $h(x)$ is written as the composition of the two simpler functions f and g.

EXAMPLE 4

Problem

Write the function

$$k(x) = \frac{(x^2 - 3x + 5)^5 - 1}{(x^2 - 3x + 5)^4 + 1}$$

as a composition of two simpler functions.

Solution The polynomial $x^2 - 3x + 5$ is evaluated first, so set $g(x) = x^2 - 3x + 5$. Then $k(x)$ can be written as

$$k(x) = \frac{[g(x)]^5 - 1}{[g(x)]^4 + 1}$$

Define $f(x)$ by substituting x for $g(x)$ in this expression.

$$f(x) = \frac{x^5 - 1}{x^4 + 1}$$

Then $k = f \circ g$ since

$$(f \circ g)(x) = f(g(x)) = \frac{[g(x)]^5 - 1}{[g(x)]^4 + 1} = k(x)$$

Quality Control

For many companies quality is the most important product. Quality control often depends on the production level; a low production level can cause problems with a worker's concentration and morale, while a high production level can lead to fatigue and nervousness. The usual assembly line offers a prime example.

It is possible to measure quality Q in terms of the production level P, where Q ranges from 0 to 100 (100 being a "perfect" product) and P represents the number of units produced per day where P ranges from 0 units (meaning that the assembly line is shut down) to 18 units (meaning it is operating at peak speed). Production, in turn, often depends on the time of day t, measured in hours. This means that ultimately the quality Q can be regarded as a function of the time of day t. The underlying concept is the composition of functions. Example 5 addresses this issue.

EXAMPLE 5

Let the quality of a product be defined by

$$Q(P) = 50 + 8P - 0.4P^2$$

where P is the number of units produced per day, with $0 \le P \le 18$. Let the daily production be defined by

$$P(t) = 6t - 0.5t^2$$

where t is the time of day in hours, with $0 \le t \le 8$.

Problem

Express the quality Q as a function t.

Solution Write Q as a function of $P(t)$.

$$Q(P(t)) = 50 + 8P(t) - 0.4P(t)^2$$

Substitute the expression for $P(t)$.

$$Q(P(t)) = 50 + 8(6t - 0.5t^2) - 0.4(6t - 0.5t^2)^2$$

Collect similar terms.

$$Q(P(t)) = 50 + 48t - 4t^2 - 0.4(36t^2 - 6t^3 + 0.25t^4)$$
$$= 50 + 48t - 18.4t^2 + 2.4t^3 - 0.1t^4$$

In this way quality Q is defined as a function of time t. (See Figures 2 and 3.)

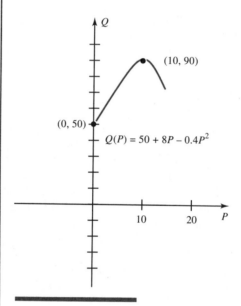

(10, 90)

(0, 50)

$Q(P) = 50 + 8P - 0.4P^2$

10   20 P

FIGURE 2

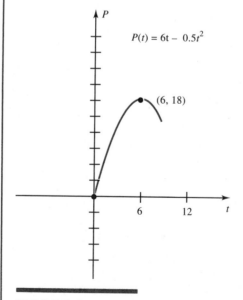

$P(t) = 6t - 0.5t^2$

(6, 18)

6   12 t

FIGURE 3

An important type of relationship between functions is defined by the inverse of functions.

DEFINITION

Two functions $f(x)$ and $g(x)$ are **inverses** of each other if

$$f(g(x)) = x = g(f(x))$$

for each value of the domain of $f(x)$ and $g(x)$.

We denote the inverse of $f(x)$ as $f^{-1}(x)$, read "$f(x)$ inverse." Do not confuse the notation for f inverse with the exponent "-1," meaning $1/f(x)$. By $f^{-1}(x)$ we will always mean $f(x)$ inverse and never the reciprocal of $f(x)$.

EXAMPLE 6

Problem

Show that the two functions $f(x) = \sqrt[3]{2x - 5} = (2x - 5)^{1/3}$ and $g(x) = (x^3 + 5)/2$ are inverses of each other.

Solution Find $f(g(x))$ and $g(f(x))$:

$$f(g(x)) = f[(x^3 + 5)/2]$$

$$= [2(x^3 + 5)/2 - 5]^{1/3}$$

$$= [x^3 + 5 - 5]^{1/3} = (x^3)^{1/3}$$

$$= x$$

$$g(f(x)) = g[(2x - 5)^{1/3}]$$

$$= \{[(2x - 5)^{1/3}]^3 + 5\}/2$$

$$= [(2x - 5) + 5]/2 = 2x/2$$

$$= x$$

Intuitively, the inverse function undoes what the function does. For example, the function $f(x) = x^3$ cubes a number and its inverse function $f^{-1}(x) = x^{1/3}$ takes the cube root of the number. So if you compute the composition of the two functions, you first cube x and then find the cube root of that number, you get back to x, meaning that $f(g(x)) = x$.

To compute the inverse function from a given function, interchange the roles of x and y and solve the equation for x. This reverses each of the operations that constitute the formula for $f(x)$. Then again reverse the roles of x and y. Example 7 illustrates this procedure.

EXAMPLE 7

Problem

Find the inverse of the function $f(x) = \sqrt{3x - 5}$.

Solution Substitute y for $f(x)$ and get $y = (3x - 5)^{1/2}$. To find the inverse function, we solve this equation for x:

$$(3x - 5)^{1/2} = y$$

$$3x - 5 = y^2$$

$$3x = y^2 + 5$$

$$x = (y^2 + 5)/3$$

This means that the inverse function starts with a value y, squares it, adds five to this result, and then divides this number by three. If we write the independent variable as x, we get

$$f^{-1}(x) = (x^2 + 5)/3$$

EXERCISE SET 11-3

In Problems 1 to 4 find $f + g$, $f \cdot g$, and f/g.

1. $f(x) = \dfrac{x + 2}{x - 1}$ and $g(x) = \dfrac{x}{x + 2}$

2. $f(x) = \dfrac{x - 2}{x + 3}$ and $g(x) = \dfrac{x}{x - 3}$

3. $f(x) = \dfrac{x}{x + 1}$ and $g(x) = \dfrac{1}{x}$

4. $f(x) = \dfrac{x}{x - 1}$ and $g(x) = \dfrac{x}{x + 2}$

5. Find the domain of $f + g$ for the functions f and g defined in Problems 1 and 3.

6. Find the domain of $f - g$ for the functions f and g defined in Problems 2 and 4.

7. Find the domain of f/g for the functions f and g defined in Problems 1 and 3.

8. Find the domain of f/g for the functions f and g defined in Problems 2 and 4.

In Problems 9 to 12 compute (a) $(f + g)(a)$, (b) $(f \cdot g)(a)$, and (c) $(f/g)(a)$ for the functions f and g in the stated problem and the given value of a.

9. f and g in Problem 1, $a = 2$

10. f and g in Problem 2, $a = 4$

11. f and g in Problem 3, $a = -2$

12. f and g in Problem 4, $a = 0$

In Problems 13 to 16 find $f \circ g$ and $g \circ f$.

13. $f(x) = \sqrt{x}$ $g(x) = \dfrac{x}{1 - x}$

14. $f(x) = \sqrt{2x}$ $g(x) = \dfrac{1}{1 - x}$

15. $f(x) = 2x$ $g(x) = \dfrac{x - 2}{x + 2}$

16. $f(x) = \dfrac{1}{x}$ $g(x) = \sqrt{x}$

In Problems 17 to 20 write the function as the composition of two functions.

17. $f(x) = \sqrt{x^3 + 2x - 5}$

18. $f(x) = \sqrt{2x + 3}$

19. $f(x) = \dfrac{1}{(x - 2)^2} + (x - 2)^3$

20. $f(x) = \dfrac{1 + (x + 1)^2}{3 + (x + 1)^3}$

In Problems 21 and 22 the quality of a product is defined by $Q(P)$, where P is the number of units produced per day and the daily production is defined by $P(t)$, with t the time of day in hours. Express Q as a function of t.

21. $Q(P) = 20 + 4P - 0.02P^2$
    $P(t) = 8t - 0.5t^2$

22. $Q(P) = 25 + 2P - 0.05P^2$
    $P(t) = 6t - 0.8t^2$

In Problems 23 and 24 let $f(x)$ and $g(x)$ be the functions defined in Problem 1 and $h(x) = \dfrac{1}{x - 2}$.

23. Compute (a) $f + g + h$ (b) $f \cdot g \cdot h$.

24. Compute (a) $(f/g)/h$ (b) $f/(g/h)$
    (c) Is $(f/g)/h = f/(g/h)$?

In Problems 25 and 26 let $f(x)$ and $g(x)$ be the functions defined in Problem 2 and $h(x) = \dfrac{1}{x + 3}$.

25. Compute (a) $f + g + h$ (b) $f \cdot g \cdot h$.

26. Compute (a) $(f/g)/h$ (b) $f/(g/h)$
    (c) Is $(f/g)/h = f/(g/h)$?

27. Evaluate $(f \cdot g \cdot h)(0)$ for the functions in Problem 23.

28. Evaluate $(f \cdot g \cdot h)(0)$, where $f(0)$ and $g(0)$ are defined and $h(0) = 0$.

In Problems 29 to 32 compute $f \circ g$.

29. $f(x) = \dfrac{1}{\sqrt{x}}$ $g(x) = \dfrac{1}{x^2}$

30. $f(x) = 1 - x$ $g(x) = \dfrac{1}{x^3}$

31. $f(x) = \dfrac{x}{x + 1}$ $g(x) = \dfrac{x + 1}{x}$

32. $f(x) = \sqrt{1 - x}$ $g(x) = 1 - x^2$

33. Let $f(x)$ and $g(x)$ be the functions defined in Problem 31. Is it true that $(f \circ g)(1) = (g \circ f)(1)$?

34. Let $f(x)$ and $g(x)$ be the functions defined in Problem 30. Is it true that $(f \circ g)(-1) = (g \circ f)(-1)$?

In Problems 35 and 36 write the given function as the composition of two functions.

35. $f(x) = \sqrt[3]{(x - 1)^2}$

36. $f(x) = \dfrac{5}{(2 - x)^3}$

In Problems 37 and 38 let f and g be the functions defined in Example 1 in the text.

37. Compute $(f + g)(a)$, where a is a given value of x.

38. Compute $(f + g)(x + h)$.

In Problems 39 and 40 let f and g be the functions defined in Example 3 in the text.

39. Compute $(f + g)(a)$, where a is a given value of x.

40. Compute $(f + g)(x + h)$.

41. For the functions f and g in Problem 1, is it true that $f \cdot g = f \circ g$?

42. For the functions f and g in Problem 3, is it true that $f \cdot g = f \circ g$?

43. Suppose the number of hours required to stock x percent of a shipment is $S(x) = g(x)/h(x)$, where $g(x) = 40x$ and $h(x) = 200 - x$.
    (a) What is the domain of the function $S(x)$?
    (b) For what values of x does $S(x)$ have a practical interpretation in this context?
    (c) How many hours were required to stock 40% of the shipment?
    (d) How many hours were required to stock the entire shipment?
    (e) What percentage of the shipment was stocked after 20 hours?

44. Suppose the cost of conducting a political survey of x percent of the eligible voters in a precinct is $C(x) = g(x)/h(x)$ dollars, where $g(x) = 105x$ and $h(x) = 110 - x$.
    (a) What is the domain of the function $C(x)$?
    (b) For what values of x does $C(x)$ have a practical interpretation in this context?
    (c) How much did it cost to poll 40% of the precinct?
    (d) How much did it cost to poll the entire precinct?
    (e) What percentage of the precinct could be polled with a fund of $500?

45. This problem refers to the profit function $P(x)$ on the sales of x units of an item: $P(x) = R(x) - C(x)$, where $R(x) = 480x - 12x^2 - 2100$ is the revenue function and $C(x) = 96x + 204$ is the cost function.
    (a) Draw the graphs of $R(x)$ and $C(x)$ on the same coordinate axis.
    (b) Use the graphs in part (a) to determine the numbers of units of the item that should be sold so the profit is a positive amount.

In Problems 46 and 47 let f, g, and h be the functions

$$f(x) = \frac{1}{x} \quad g(x) = \sqrt{x} \quad h(x) = x + 1$$

46. Compute $\dfrac{f \circ g}{h}$.

TABLE 1

letter	A	B	C	D	...	L	...	X	Y	Z
value	65	66	67	68	...	76	...	88	89	90

47. (a) Compute $(f \circ g) \circ h$.
    (b) Compute $f \circ (g \circ h)$.
    (c) Compute $f \circ g \circ h$.

Referenced Exercise Set 11–3

1. The coding of messages using a computer involves the composition of three functions.* The first function, $h(x)$, assigns a number, called the ASCII value, to each letter in the alphabet. (ASCII stands for the American Standard Code for Information Interchange.) It is partially defined in Table 1.

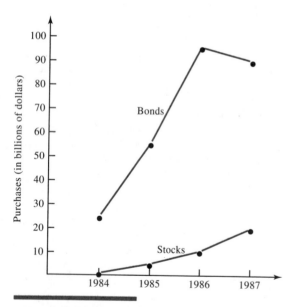

FIGURE 4

The second function, $g(x)$, is called the encoding function. Let $g(x)$ be defined by

$$g(x) = \begin{cases} x + 10 & \text{if } 65 \le x \le 80 \\ x - 16 & \text{if } 81 \le x \le 90 \end{cases}$$

The third function, $f(x)$, uses Table 1 to assign a letter to each ASCII value x, where $65 \le x \le 90$. The code function $c(x)$, which replaces each letter in the alphabet by another letter, is defined by $c = f \circ g \circ h$. For instance, $c(L) = (f \circ g \circ h)(L) = f(g(h(L))) = f(g(76)) = f(86) = V$, so the letter L is always replaced in the code by the letter V. Find the code for each letter in the word FUN.

2. Figure 4 traces Japanese net purchases of foreign stocks and bonds from 1984 through 1987.† Let $B(x)$ denote the purchases of bonds and $S(x)$ denote the purchases of stocks in year x. Write an equation for $B(x) + S(x)$ during the years 1984–1987.

3. When the weather is hot and humid, what is the best time of the day to perform physical activity? Some people think the best time is when the temperature is at its lowest point. Others think it is when the humidity is lowest. Actually the best time to exercise under such conditions is when the sum of the two measures is minimal.‡ The accompanying chart lists hourly measurements of temperature and humidity in Philadelphia on August 27, 1990. The chart shows that the best time to exercise might not occur when the temperature or the humidity is minimal. Most newspapers carry such a chart, so you can verify the same conclusion using records from your own locale.
    (a) If $T(x)$ stands for the temperature at x o'clock, at what value of x is $T(x)$ the lowest?
    (b) If $H(x)$ stands for the humidity at x o'clock, at what value of x is $H(x)$ the lowest?
    (c) At what value of x is $T(x) + H(x)$ the lowest?

---

*James Reagan, "Get the Message? Cryptographs, Mathematics, and Computers," *The Mathematics Teacher,* October 1986, pp. 547–553.

†Mark Beauchamp and John Hines, "We'll Send You VCRs—You Send Us Stocks," *Forbes Magazine,* August 10, 1987, pp. 60–62.

‡You can consult any book on exercise to confirm the results of this problem, for instance, Jim Fixx's *The Complete Book of Running.*

Yesterday in Philadelphia†

A.M.	Temp./Hum.	P.M.	Temp./Hum.
1:00	69/96	1:00	87/62
2:00	70/93	2:00	89/54
3:00	69/100	3:00	90/55
4:00	68/100	4:00	89/53
5:00	68/100	5:00	89/53
6:00	68/100	6:00	87/54
7:00	67/100	7:00	84/62
8:00	71/100	8:00	82/69
9:00	76/90	9:00	79/78
10:00	79/84		
11:00	85/75		
Noon	86/67		

†Source: *The Philadelphia Inquirer*, August 28, 1990.

Cumulative Exercise Set 11–3

In Problems 1 and 2 use the plotting method to sketch the graph of each equation.

1. $3y - 5x = 0$

2. $y = (x - 3)^2$

3. (a) What is the equation of the line that passes through the points $(2, 3)$ and $(2, 4)$?
   (b) What is the y-intercept of the line that passes through the points $(2, 3)$ and $(2, 4)$?

4. Find an equation of the line that passes through the origin and the point $(-4, -2)$.

5. Compute each of the following for the function $f(x) = -2x^2$.
   (a) $f(-2)$ (b) $f(x + h)$
   (c) $f(x + h) - f(x)$ (d) $\dfrac{f(x + h) - f(x)}{h}$
   for $h \neq 0$

6. What is the domain of the function $f(x) = 1 - x^2$?

7. (a) Sketch the graph of $f(x) = 8 - x^3$ by the plotting method.
   (b) What is the degree of $f(x)$?

8. Find the vertical asymptotes of the function
$$f(x) = \frac{2}{4 - x^2}$$

9. Sketch the graph of the function
$$f(x) = \begin{cases} \sqrt{x} & x \geq 0 \\ -x & x < 0 \end{cases}$$

10. Write $f + g$ as a function with denominator $1 - x^2$, where
$$f(x) = \frac{x}{1 - x} \qquad g(x) = \frac{x}{1 + x}$$

11. Find $(f \circ g)(2)$ for the functions f and g in Problem 10.

12. For what function $h(x)$ can $f(x) = (x^2 - 1)^3$ be written in the form $f = g \circ h$, where $g(x) = x^3$?

11–4 Solving Polynomial Equations

Many applications of mathematics require a knowledge of points where the graph of a function $f(x)$ crosses the x-axis. Algebraically this is equivalent to solving the equation $f(x) = 0$.

We begin this section by reviewing quadratic equations and the quadratic formula. Then we apply the method of factoring to solving polynomial equations. We end by using these procedures to find the points of intersection of two graphs.

The Quadratic Formula

The graph of a function $f(x)$ crosses the x-axis at $x = a$ if and only if $f(a) = 0$. Geometrically the point $(a, 0)$ is an x-intercept of the graph. Algebraically the value $x = a$ is called a **solution** (or a **root**) **of the equation** $f(x) = 0$. Therefore solutions of the equation $f(x) = 0$ yield x-intercepts of the graph of f. A **polynomial equation** is an equation $f(x) = 0$, where $f(x)$ is a polynomial function. To "solve a polynomial equation" means to find all of its solutions. Recall that the **degree of a polynomial equation** is the highest power of the polynomial.

A **quadratic equation** is a polynomial equation of degree 2. The general form of a quadratic equation is

$$ax^2 + bx + c = 0$$

where a, b, and c are real numbers with $a \neq 0$. We have seen that the graph of a quadratic function $f(x) = ax^2 + bx + c$ is a parabola, so it has either zero, one, or two x-intercepts. Equivalently, a quadratic equation has either zero, one, or two solutions. The solution of a quadratic equation can be computed from the **quadratic formula.** The quadratic formula also shows when a solution does not exist.

Quadratic Formula
If $ax^2 + bx + c = 0$ and $a \neq 0$, then

$$x = \frac{-b + \sqrt{b^2 - 4ac}}{2a}$$

or

$$x = \frac{-b - \sqrt{b^2 - 4ac}}{2a}$$

The number $D = b^2 - 4ac$, which is called the **discriminant,** determines whether there are zero, one, or two solutions. If $D > 0$, then the quadratic equation has two distinct solutions,

$$x = \frac{-b + \sqrt{D}}{2a} \qquad \text{or} \qquad x = \frac{-b - \sqrt{D}}{2a}$$

If $D = 0$, there is one solution $x = -b/2a$. If $D < 0$, there is no solution because the square root of a negative number is not a real number. Example 1 shows how the three possibilities arise.

EXAMPLE 1

Problem

For each function $f(x)$, solve the quadratic equation $f(x) = 0$ and sketch the graph of $f(x)$.
(a) $f(x) = 2x^2 - x - 10$
(b) $f(x) = x^2 + 4x + 4$
(c) $f(x) = x^2 + 4x + 5$

Solution (a) To use the quadratic formula, set $a = 2$, $b = -1$, and $c = -10$. Then $D = b^2 - 4ac = 81$. Since $D > 0$, the two solutions of the equation $f(x) = 0$ are

$$x = \frac{-(-1) + \sqrt{81}}{2(2)} = \frac{5}{2} \qquad x = \frac{-(-1) - \sqrt{81}}{2(2)} = -2$$

Therefore the graph of $f(x)$ has x-intercepts at the points $(-2, 0)$ and $(\frac{5}{2}, 0)$. Since the graph of a quadratic equation is a parabola that opens upward or downward, it suffices to plot a point (x, y) where $-2 < x < (\frac{5}{2})$. Since $f(0) = -10$, the point $(0, -10)$ lies on the graph so that the parabola opens upward. It is sketched in Figure 1a.

(b) To use the quadratic formula, set $a = 1$, $b = 4$, and $c = 4$. Then $D = 4^2 - 4(1)(4) = 0$, so there is one solution of the equation $f(x) = 0$: $x = -b/2a = -\frac{4}{2} = -2$. This means that the graph of $f(x)$ has one x-intercept, the point $(-2, 0)$, which must be the vertex of the parabola. Since $f(0) = 4$, the point $(0, 4)$ lies on the graph, so the parabola opens upward. It is sketched in Figure 1b.

(c) Set $a = 1$, $b = 4$, and $c = 5$. Then $D = 4^2 - 4(1)(5) = -4$, so there is no solution of the equation $f(x) = 0$. This means that the graph of $f(x)$ has no x-intercepts. The only means at our disposal for sketching the graph is the plot-

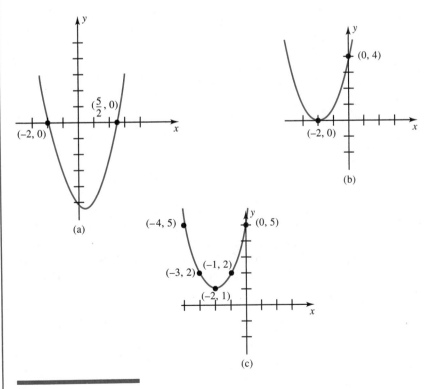

FIGURE 1

ting method using a table like the one here. The resulting graph is sketched in Figure 1c.

x	-4	-3	-2	-1	0
y	5	2	1	2	5

Example 2 illustrates how to use the quadratic formula to determine the domain of a rational function whose denominator is a quadratic function.

EXAMPLE 2

Problem

Determine the domain of the function

$$f(x) = \frac{x}{12x^2 - 11x + 2}$$

Solution The domain consists of all real numbers except where the denominator is 0. Set $12x^2 - 11x + 2 = 0$. Since $D = (-11)^2 - 4(12)(2) = 25$, the values of x where $12x^2 - 11x + 2 = 0$ are

$$x = \frac{-b + \sqrt{D}}{2a} = \frac{11 + 5}{24} = \frac{2}{3}$$

or

$$x = \frac{-b - \sqrt{D}}{2a} = \frac{11 - 5}{24} = \frac{1}{4}.$$

Therefore the domain of $f(x)$ is all real numbers except $\frac{2}{3}$ and $\frac{1}{4}$.

Factoring

A number r is a root of a polynomial equation $f(x) = 0$ if and only if $f(x) = (x - r)g(x)$ for some polynomial $g(x)$. Each of the functions $(x - r)$ and $g(x)$ is called a **factor** of $f(x)$. The method of decomposing a polynomial into a product of polynomials of lesser degree is called **factoring.** Chapter R includes a section on factoring for those students who need a review of this topic.

Consider the general quadratic equation $ax^2 + bx + c = 0$. Since $a \neq 0$, we can divide both sides of the equation by a and get $x^2 + Bx + C = 0$, where $B = b/a$ and $C = c/a$. If r and s are roots of $x^2 + Bx + C = 0$, then

$$x^2 + Bx + C = (x - r)(x - s)$$

Expand the right side of the equation.

$$(x - r)(x - s) = x^2 - (r + s)x + rs$$

Therefore

$$x^2 + Bx + C = x^2 - (r + s)x + rs$$

Set the coefficients equal: $B = -(r + s)$, $C = rs$. The sum of the roots equals $-B$ (since $r + s = -B$) and the product of the roots equals the constant term C. For instance, if $f(x) = x^2 - x - 2$, then $B = -1$ and $C = -2$, so $r + s = 1$ and $rs = -2$. We are searching for two numbers whose sum is 1 and product is -2. The solution by trial and error is $r = 2$ and $s = -1$. Therefore the linear factors of $x^2 - x - 2$ are $(x - 2)$ and $(x + 1)$, so

$$x^2 - x - 2 = (x - 2)(x + 1)$$

Example 3 uses factoring to determine the domain and x-intercepts of a rational function.

EXAMPLE 3

Consider the function

$$f(x) = \frac{x^2 - 4}{x^2 - 2x - 8}$$

Problem

(a) Find the domain of $f(x)$. (b) Find all x-intercepts of the graph of $f(x)$.

Solution (a) The domain of $f(x)$ consists of all real numbers except those where the denominator is 0. Factor the denominator.

$$x^2 - 2x - 8 = (x - 4)(x + 2)$$

The roots are $x = 4$ or $x = -2$. Therefore the domain of $f(x)$ consists of all real numbers except 4 and -2.
(b) The x-intercepts of the graph of $f(x)$ occur where the numerator is 0 and the denominator is not 0. Set $x^2 - 4 = 0$. Since $x^2 - 4 = (x + 2)(x - 2)$, the solutions are $x = 2$ or $x = -2$. However, -2 does not lie in the domain of $f(x)$. Thus the only possible x-intercept occurs at $x = 2$. Since $x^2 - 2x - 8 \neq 0$ when $x = 2$, $f(x)$ has one x-intercept at $(2, 0)$.

The next two examples show other instances where the solution of a polynomial equation is required. Example 4 makes use of the fact that if a and b are numbers with $ab = 0$, then either $a = 0$ or $b = 0$. This fact extends to functions, in which case if $f(x)$ and $g(x)$ are functions with $f(x)g(x) = 0$, then either $f(x) = 0$ or $g(x) = 0$. Therefore if a polynomial has been factored, its roots are found by setting each of the factors equal to 0.

EXAMPLE 4

Problem

Solve $f(x) = 0$, where

$$f(x) = (2x + 3)(x^2 + 3)(x^2 - 3)$$

Solution The solutions to $f(x) = 0$ are found by setting each of the factors equal to 0. The solution to the linear equation $2x + 3 = 0$ is $x = -\frac{3}{2}$. The equation $x^2 = -3$ has no real solutions, so the factor $(x^2 + 3)$ adds no solutions to $f(x) = 0$. However, $x^2 - 3 = (x + \sqrt{3})(x - \sqrt{3})$, so $\sqrt{3}$ and $-\sqrt{3}$ are solutions to $x^2 - 3 = 0$, and hence are solutions to $f(x) = 0$. Altogether the solutions to $f(x) = 0$ are $-\frac{3}{2}$, $\sqrt{3}$, and $-\sqrt{3}$.

Example 5 makes repeated use of one of the basic rules of algebra.

$$a^2 - b^2 = (a + b)(a - b)$$

In particular, $x^2 - 1 = (x + 1)(x - 1)$.

EXAMPLE 5

Problem

Find the x-intercepts of the function

$$f(x) = x^4 - 1$$

Solution The x-intercepts are found by solving $f(x) = 0$. First factor $f(x)$ by making the substitution $y = x^2$.

$$f(x) = x^4 - 1 = y^2 - 1 = (y + 1)(y - 1)$$

Next substitute for y.

$$f(x) = (x^2 + 1)(x^2 - 1)$$

Then factor $x^2 - 1$.

$$f(x) = (x^2 + 1)(x + 1)(x - 1)$$

The roots of $f(x) = 0$ are found by setting each factor equal to 0: this yields $x = 1$, $x = -1$. The factor $x^2 + 1$ is never equal to 0 so it does not yield a root. These are the x-intercepts of $f(x)$.

Points of Intersection

A point at which two graphs meet is called a **point of intersection** of the graphs. We find the intersection of the graphs of two polynomial functions by a method quite different from the one described in Section 1–3 for two linear functions.

Let f and g be any two functions. The intersection of their graphs is obtained by solving the equation $f(x) = g(x)$, or, equivalently, by solving $f(x) - g(x) = 0$. If there are no solutions to this equation, then the graphs have no points in common. Otherwise each solution x, along with the corresponding value of $y = f(x)$ or $y = g(x)$, leads to a point of intersection (x, y).

For example, let $f(x) = 3x$ and $g(x) = 4 - x^2$. Then

$$f(x) - g(x) = 0$$
$$3x - (4 - x^2) = 0$$
$$x^2 + 3x - 4 = 0$$
$$(x + 4)(x - 1) = 0$$

The roots are $x = -4$ and $x = 1$. The corresponding values are $f(-4) = g(-4) = -12$ and $f(1) = g(1) = 3$. Each root leads to a point of intersection, so the points of intersection of $f(x)$ and $g(x)$ are $(-4, -12)$ and $(1, 3)$. The graphs are sketched in Figure 2.

An application of this concept comes from a field called **break-even analysis.** We have already introduced the cost function $C(x)$, where x denotes the number of items being produced. The revenue R is also a function of x, called the **revenue function** and denoted by $R(x)$. The **profit function** $P(x)$ is then defined by

$$P(x) = R(x) - C(x)$$

The **break-even points** occur when cost equals revenue: $R(x) = C(x)$. Much of the information can be conveyed visually by drawing the graphs of $R(x)$ and $C(x)$ on the same coordinate system and finding the points of intersection.

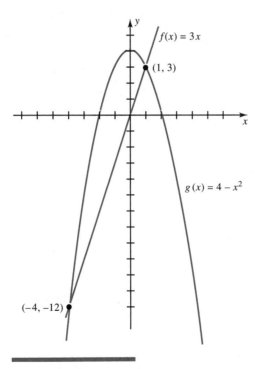

FIGURE 2

EXAMPLE 6

A manufacturer produces compact disks. The variable cost per disk is $2 and the fixed cost is $12. If x disks are manufactured, the revenue function is given by $R(x) = x(10 - x)$.

Problem

What are the break-even points?

Solution The cost function is

$$C(x) = (\text{variable cost per disk})x + (\text{fixed cost})$$

$$C(x) = 2x + 12$$

The revenue function is

$$R(x) = x(10 - x) = 10x - x^2$$

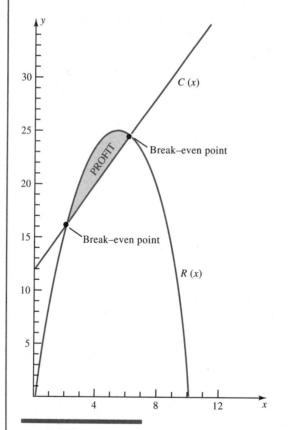

FIGURE 3

The break-even points occur when cost equals revenue.

$$C(x) = R(x)$$

$$2x + 12 = 10x - x^2$$

$$x^2 - 8x + 12 = 0$$

The possible solutions to this quadratic equation are $x = 2$ or $x = 6$, so the break-even points are (2, 16) and (6, 24). (See Figure 3.)

It is informative to view Figure 3 for a deeper understanding of break-even analysis. The functions $C(x)$ and $R(x)$ are sketched on the same coordinate axis. If either two disks or six disks are produced, the manufacturer will break even. A profit is realized if the number of disks produced lies between these numbers. The profit function is

$$P(x) = R(x) - C(x) = x^2 + 8x - 12$$

What production level do you think will maximize profit?

EXERCISE SET 11-4

In Problems 1 to 6 solve the quadratic equation $f(x) = 0$ and sketch the graph of $f(x)$.

1. $f(x) = x^2 + 2x - 8$

2. $f(x) = x^2 - 4x + 4$

3. $f(x) = x^2 + 6x + 9$

4. $f(x) = x^2 - 2x + 5$

5. $f(x) = x^2 + 4$

6. $f(x) = x^2 - 2x - 8$

In Problems 7 to 10 determine the domain and x-intercepts of each function.

7. $f(x) = \dfrac{x - 2}{2x^2 - x - 3}$

8. $f(x) = \dfrac{x - 3}{x^2 + x - 2}$

9. $f(x) = \dfrac{x + 2}{4x^2 + 5x - 6}$

10. $f(x) = \dfrac{x - 3}{5x^2 - 13x - 6}$

In Problems 11 and 12 solve $f(x) = 0$ for the given $f(x)$.

11. $f(x) = (2x - 3)(x^2 - 4)(x^2 + 3)$

12. $f(x) = x(x^2 - 1)(x^2 + 1)$

In Problems 13 to 16 find the x-intercepts of each function.

13. $f(x) = x^4 - 16$

14. $f(x) = x^4 - 81$

15. $f(x) = (x^2 + 3)(x - 4)(x + 3)$

16. $f(x) = (x^2 + 4)(x + 7)(x + 3)$

In Problems 17 to 20 a cost function and a revenue function are given. Find the break-even points.

17. $R(x) = x(10 - x)$ $C(x) = x + 8$

18. $R(x) = x(10 - x)$ $C(x) = 18 - x$

19. $R(x) = x(8 - x)$ $C(x) = 14 - x$

20. $R(x) = x(8 - x)$ $C(x) = x + 6$

In Problems 21 to 24 solve the equation $f(x) = 0$ and sketch the graph of $f(x)$.

21. $f(x) = \dfrac{x - 2}{4x^2 - 12x + 9}$

22. $f(x) = \dfrac{x - 3}{9x^2 + 6x - 1}$

23. $f(x) = \dfrac{x + 1}{x^2 + 1}$

24. $f(x) = \dfrac{x - 1}{5x^2 + 6}$

In Problems 25 to 28 determine the domain and x-intercepts of each function.

25. $f(x) = \dfrac{x - 2}{x^3 - 4x^2 + 4x}$

26. $f(x) = \dfrac{x - 3}{x^3 + 2x^2 + x}$

27. $f(x) = \dfrac{x^2 - x + 2}{x^3 - 4x^2 + 4x}$

28. $f(x) = \dfrac{x^2 - x - 6}{x^3 - 6x^2 + 9x}$

In Problems 29 to 34 find where the graphs of the functions intersect.

29. $f(x) = x + 2$, $g(x) = x^2$

30. $f(x) = 6x - 5$, $g(x) = x^2$

31. $f(x) = x^3$, $g(x) = x^2$

32. $f(x) = x^2$, $g(x) = \sqrt{x}$

33. $f(x) = x^2 + 4$, $g(x) = 22 - x^2$

34. $f(x) = x^3$, $g(x) = x$

In Problems 35 to 38 an object is thrown vertically upward with height $h(t) = 28t - 4t^2$ (in feet) at time t (in seconds). Determine when the object will reach the stated height.

35. 40 feet   36. 48 feet

37. 49 feet   38. 0 feet

39. Let the revenue function on the sales of x units of an item be $R(x) = 480x - 12x^2 - 2100$ and the cost function be $C(x) = 96x + 204$. Determine the break-even points.

In Problems 40 to 43 determine the break-even points and the fixed cost given the graphs of the cost function $C(x)$ and revenue function $R(x)$.

40.

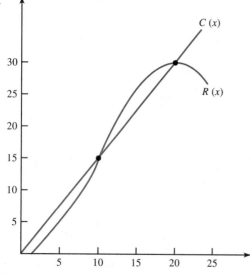

41.

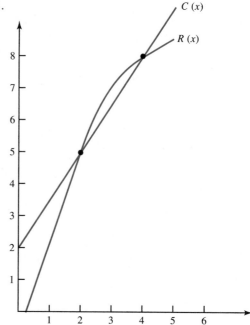

42.

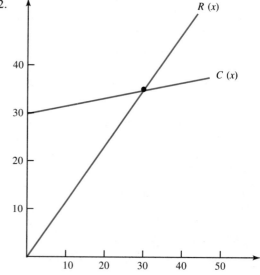

43.

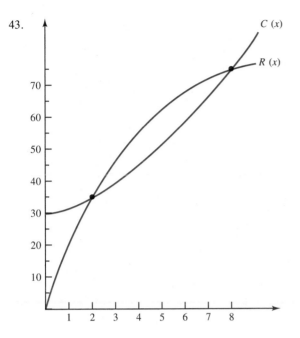

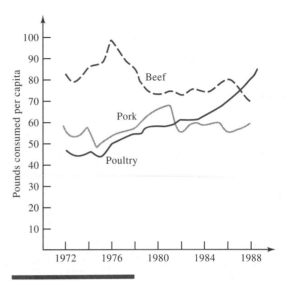

FIGURE 4

Referenced Exercise Set 11–4

Problems 1 and 2 are taken from a lecture that was given by Robert L. Devaney to a sectional meeting of the Mathematical Association of America.* It was accompanied by an exhilarating collection of computer graphics that illustrate the area of mathematics called dynamical systems.

1. Enter any positive number into a calculator. Tap the square-root key. Tap it again. Continue tapping the key about 15 or 20 more times.
   (a) What do you observe?
   (b) Repeat this procedure for several more positive numbers, including numbers between 0 and 1. What general conclusion can be drawn?

2. (a) Find all points of intersection of the graphs of the functions $f(x) = \sqrt{x}$ and $g(x) = x$ for $x > 0$.
   (b) What does Problem 1 have to do with part (a)?

3. Figure 4 illustrates the dramatic shift in the American public's eating habits from red meats to poultry. It has

been drawn from data supplied by the U. S. Department of Agriculture.†
   (a) Interpret the point of intersection of the poultry curve and the pork curve.
   (b) Interpret the point of intersection of the poultry curve and the beef curve.

Cumulative Exercise Set 11–4

1. Find the equation of the line passing through $(-2,0)$ and $(-1,-1)$.

2. Evaluate $f(x + h) - f(x)$ for $f(x) = x^2 + 2x - 1$.

3. Sketch the graph of the function $f(x) = x^2 + 2x - 1$.

4. Find the vertical asymptotes and sketch the graph of the function $f(x) = 1/(1 - x^2)$.

5. Find $f + g$ and $f \cdot g$ for $f(x) = x^2 + x$ and $g(x) = x^3 - 1$.

6. Find $f \circ g$ and $g \circ f$ for $f(x) = 1/x$ and $g(x) = x^3 - x$.

7. Write the function $f(x) = (x^3 + 2)^4$ as the composition of two functions.

*Robert L. Devaney, "Computer Graphics Experiments in Complex Dynamical Systems," Eastern Pennsylvania and Delaware Section of the Mathematical Association of America, Franklin & Marshall College, November 21, 1987.

†Drawn from data extracted from Neill Boroski, "Poultry's Prospects Far from Paltry," *The Philadelphia Inquirer,* June 15, 1987, pp. C1 and C6.

8. Express $Q(P) = 10 + 2P - 3P^2$ as a function of t where $P(t) = 5t - t^2$.

9. Find the domain and the x-intercepts of the function
$$f(x) = x/(x^2 - 4x + 4)$$

10. Solve $f(x) = 0$ for $f(x) = x(x - 3)(x^2 - 9)$.

11. Find the break-even points for the revenue function
$$R(x) = x(8 - x) \text{ and the cost function } C(x) = x + 6.$$

12. Find where the graphs of the functions intersect:
$$f(x) = x^3 - x, \, g(x) = x^2 - x$$

CASE STUDY   **Powerlifting***

Powerlifting is one of the world's youngest sports, having attained international status in 1968 when the strong local programs from the United States and Great Britain joined together with the developing programs in Canada, Japan, and the Scandinavian countries. It consists of three types of lifts: the squat (the bar is set on the back of the neck, and the lifter does a deep knee bend, returning to a standing position); the bench press (the lifter lies on a bench and pushes the weight away from the chest until the arms are locked); and the deadlift (the weight is lifted waist high from the floor, with the arms kept straight). A lifter's total is the sum of the highest weight lifted in each of the three categories.

Surprisingly, a sixth-degree polynomial which plays a crucial role in determining the winner of a competition.

The winner of a powerlifting meet is not necessarily the competitor who has lifted the most weight. In fact, there are many winners. Like other sports in which a competitor's weight is one of the crucial determining factors, powerlifting competitions are divided into several weight classes, and there is a champion in each weight class. However, most meets crown a "Champion of Champions." What criteria should be used to decide this distinction?

Let us consider two lifters, Amos Jones and Bob Brown. Amos weighs 167 pounds (lb) so he competes in the 181-lb class. Bob is a behemoth, who, at 272 lb, competes in the 275-lb class. (There is a super heavyweight class beyond this one.) In a particular meet Amos squatted 710 lb, bench pressed 450 lb, and deadlifted 640 lb, while Bob's best three lifts were 850, 570, and 790 lb, respectively. Therefore Amos's total was 1800 (since $710 + 450 + 640 = 1800$) and Bob's total was 2210. Each lifter was a weight class champion in the meet. Which one should win the Champion of Champions award?

Mathematics to the rescue!

The Schwartz formula   There are several ways to approach the question of who is stronger, Amos or Bob. If only the totals are taken into account, then Bob wins by 410 lb. Yet Amos's total is about 11 times his bodyweight, while Bob's total is about 8 times his bodyweight.

---

*This Case Study was inspired by Joseph A. Gallian's note "Handicapping Bodyweights," which appeared in the August–September 1986 issue of the *American Mathematical Monthly*, pp. 583–584. We are indebted to Lyle Schwartz and Linc Gotshalk for supplying details, and to Walt Evans for supplying the photograph that introduces this chapter.

The International Powerlifting Federation (IPF) sought a handicapping scheme that would compensate for the widely differing body weights of the lifters. In 1976 the IPF chose a scheme that was developed by Lyle Schwartz to fit the empirical data that were available at the time. It has had the effect of taking into account not only body weights and muscular cross-sectional areas, but also body size versus fixed barbell plate, bar diameter, and muscle-ligament attachment leverage.

Dr. Schwartz had developed the scheme in 1967–1968 while he was a Professor of Materials Science at Northwestern University. (Today he is the Director of the Institute for Materials Science and Engineering at the National Bureau of Standards.) His method was published first in the magazine *Muscular Development* in 1970 and adopted by the U.S. Powerlifting Federation in 1975. The method uses a function of the lifter's bodyweight to produce a number that is then multiplied by the total number of pounds lifted to give the lifter's score. The competitor with the highest score wins the Champion of Champions trophy.

The *Schwartz formula* is the following intimidating sixth-degree polynomial function.

$$\begin{aligned}
S(x) = {}& 6.31926 - 1.189995 \cdot 10^{-1} x + 1.052494 \cdot 10^{-3} \cdot x^2 \\
& - 4.850446 \cdot 10^{-6} \cdot x^3 + 1.132899 \cdot 10^{-8} \cdot x^4 \\
& - 1.037119 \cdot 10^{-11} \cdot x^5 - 6.348189 \cdot 10^{-16} \cdot x^6
\end{aligned}$$

Here x is the bodyweight in pounds. The domain of $S(x)$ is $114 \le x \le 275$, so the Schwartz formula only applies to weights in this interval.

Let us evaluate the Schwartz formula for Amos, who weighs 167 lb. The details for computing $S(167)$ are shown in Table 1. They reinforce the fact that all seven terms in $S(x)$ must be included. Since the IPF rounds all numbers to four decimal places, we see that $S(167) = 0.6593$.

The **Schwartz formula total** (denoted SFT) is equal to $S(x)$ when $114 \le x \le 275$. Other methods are used to determine the SFT for lifters who weigh less than 114 lb or more than 275 lb. (See Exercises 9 and 10.) The SFT for Amos is 0.6593. Similar computations show that the SFT for Bob is $S(272) = 0.5232$.

Each lifter's score is the product of the total weight lifted and the SFT. For Amos, the score is $1800 \cdot 0.6593 = 1186.74$. For Bob, the score is $2210 \cdot 0.5232 =$

TABLE 1

Term	Value with $x = 167$	Cumulative Total
6.319260	6.319260	
$-1.189995 \cdot 10^{-1} x$	-19.872917	-13.553657
$1.052494 \cdot 10^{-3} x^2$	29.353005	15.799348
$-4.850446 \cdot 10^{-6} x^3$	-22.590773	-6.791425
$1.132899 \cdot 10^{-8} x^4$	8.811647	2.020222
$-1.037119 \cdot 10^{-11} x^5$	-1.347134	0.673088
$-6.348189 \cdot 10^{-16} x^6$	-0.013770	0.659318

TABLE 2

	Amos	Bob
Body weight x	167	272
$S(x)$	0.6593	0.5232
Total	1800	2210
Score	1186.7	1156.3

1156.272. Therefore Amos is declared the Champion of Champions. Their statistics are summarized in Table 2.

E X A M P L E   1

Jack London won the Pennsylvania state teenage powerlifting championship in 1987. He squatted 701 lb, bench pressed 451 lb, and deadlifted 585 lb. He weighed a mere 275 lb at the time.

Problem

What is his score?

Solution   London's total is $701 + 451 + 585 = 1737$. The Schwartz formula total is $S(275) = 0.5214$. Therefore London's score is $1737 \cdot 0.5214 = 905.67$.

It is not necessary for a powerlifting meet director to employ someone with a calculator to evaluate $S(x)$ for each lifter. Instead, the director uses a table like the one in Figure 1, which also contains the Malone formula total (MFT) for female lifters and the Schwartz master's formula (SMF) for lifters of age 40 or more. However, neither the MFT nor the SMF is defined by a polynomial function, so they will not be described here.

Case Study Exercises

1. Use the Schwartz formula $S(x)$ to determine the Schwartz formula total for a 200-pound (lb) male powerlifter.
2. Use the Schwartz formula $S(x)$ to determine the Schwartz formula total for a 125-lb male powerlifter.
3. Use Figure 1 to verify your answer to Problem 1.
4. Use Figure 1 to verify your answer to Problem 2.
5. Suppose Bob had lost 4 lb and lifted 10 lb less. Would he have won the Champion of Champions award?

Problems 6 to 8 require the use of Figure 1.

6. What is the score of a female powerlifter who weighs 140 lb and lifts 950 lb?
7. Suppose that a 120-lb female powerlifter lifts 640 lb. If she gains 6 lb, how much more weight will she have to lift to maintain the same score?
8. What is the score of a 63-year-old person who lifts 580 lb and weighs 180 lb?

Problems 9 and 10 define the Schwartz Formula in terms of body weight x (in kilograms) for certain male behemoths. Sketch the graph of each formula.

9. $S(x) = \begin{cases} 0.5208 - 0.0012(x - 125) & \text{if } 125 < x \le 135 \\ 0.5088 - 0.0011(x - 135) & \text{if } 135 < x \le 145 \end{cases}$

10. $S(x) = \begin{cases} 0.4978 - 0.0010(x - 145) & \text{if } 145 < x \le 155 \\ 0.4878 - 0.0009(x - 155) & \text{if } 155 < x \le 165 \end{cases}$

Schwartz/Malone Formula (in pounds)*

BWT	Schwartz	Malone	BWT	Schwartz	Malone	BWT	Schwartz	Malone	BWT	Schwartz	Malone	BWT	Schwartz	Malone
90	1.2803	1.1756	123	.8783	.9110	155	.7004	.7565	187	.6077	.6595	219	.5556	.6008
91	1.2627	1.1645	124	.8706	.9086	156	.6967	.7520	188	.6056	.6566	220	.5545	.5993
92	1.2455	1.1557	125	.8630	.9019	157	.6930	.7490	189	.6036	.6543	221	.5535	.5981
93	1.2287	1.1450	126	.8556	.8980	158	.6893	.7453	190	.6014	.6521	222	.5524	.5965
94	1.2124	1.1365	127	.8483	.8902	159	.6857	.7431	191	.5994	.6492	223	.5514	.5953
95	1.1965	1.1261	128	.8412	.8851	160	.6822	.7387	192	.5978	.6464	224	.5504	.5938
96	1.1809	1.1180	129	.8343	.8788	161	.6787	.7358	193	.5954	.6442	225	.5494	.5926
97	1.1657	1.1079	130	.8276	.8738	162	.6753	.7322	194	.5935	.6415	226	.5485	.5911
98	1.1509	1.0980	131	.8210	.8676	163	.6720	.7293	195	.5916	.6387	227	.5476	.5896
99	1.1365	1.0903	132	.8146	.8628	164	.6688	.7258	196	.5897	.6366	228	.5467	.5884
100	1.1223	1.0807	133	.8083	.8568	165	.6656	.7230	197	.5879	.6339	229	.5458	.5869
101	1.1086	1.0732	134	.8022	.8508	166	.6624	.7196	198	.5861	.6317	230	.5449	.5858
102	1.0952	1.0657	135	.7961	.8462	167	.6593	.7168	199	.5843	.6300	231	.5441	.5843
103	1.0821	1.0566	136	.7903	.8401	168	.6563	.7134	200	.5826	.6286	232	.5433	.5831
104	1.0693	1.0494	137	.7846	.8358	169	.6533	.7107	201	.5809	.6269	233	.5426	.5817
105	1.0569	1.0405	138	.7790	.8302	170	.6504	.7074	202	.5792	.6256	234	.5418	.5805
106	1.0448	1.0336	139	.7735	.8257	171	.6475	.7040	203	.5776	.6239	235	.5411	.5791
107	1.0329	1.0250	140	.7682	.8202	172	.6447	.7014	204	.5760	.6226	236	.5405	.5779
108	1.0214	1.0165	141	.7630	.8159	173	.6420	.6981	205	.5744	.6209	237	.5398	.5765
109	1.0101	1.0098	142	.7579	.8105	174	.6392	.6956	206	.5729	.6196	238	.5391	.5754
110	.9991	1.0016	143	.7528	.8052	175	.6365	.6923	207	.5714	.6180	239	.5385	.5740
111	.9884	.9952	144	.7479	.8010	176	.6339	.6898	208	.5700	.6167	240	.5379	.5725
112	.9779	.9872	145	.7432	.7959	177	.6313	.6866	209	.5685	.6151	241	.5373	.5714
113	.9677	.9809	146	.7385	.7918	178	.6288	.6837	210	.5670	.6134	242	.5367	.5700
114	.9578	.9731	147	.7339	.7867	179	.6262	.6810	211	.5657	.6122	243	.5362	.5693
115	.9481	.9670	148	.7294	.7827	180	.6238	.6786	212	.5643	.6109	244	.5357	.5686
116	.9385	.9595	149	.7250	.7769	181	.6214	.6755	213	.5630	.6093	245	.5352	.5681
117	.9293	.9536	150	.7207	.7737	182	.6190	.6731	214	.5617	.6077	246	.5347	.5671
118	.9203	.9462	151	.7165	.7697	183	.6167	.6701	215	.5604	.6064	247	.5342	.5669
119	.9115	.9390	152	.7124	.7666	184	.6144	.6671	216	.5592	.6049	248	.5337	.5662
120	.9029	.9333	153	.7083	.7627	185	.6121	.6639	217	.5580	.6036	249	.5333	.5656
121	.8946	.9263	154	.7044	.7596	186	.6099	.6618	218	.5568	.6021	250	.5328	.5649
122	.8863	.9208												

*Updated (4/84) Schwartz Formula for men, Malone Formula For Women with bodyweights in pounds. To determine the "Best Lifter," multiply each lifter's coefficient (to the right of each body weight listed) by his or her total. The resulting factor is his Schwartz or her Malone formula total (SFT/MFT). The lifter with the highest SFT/MFT is considered the "Best Lifter." The weight class winner with the highest (SFT/MFT) is the Champion of Champions.

FIGURE 1   Schwartz/Malone Formula. (Courtesy of the United States Powerlifting Federation.)

Schwartz Formula (*Continued*)

BWT	Schwartz	BWT	Schwartz	BWT	Schwartz	BWT	Schwartz
251	.5325	279	.5192	307	.5043	335	.4909
252	.5320	280	.5186	308	.5037	336	.4905
253	.5316	281	.5180	309	.5032	337	.4901
254	.5312	282	.5175	310	.5027	338	.4896
255	.5308	283	.5169	311	.5022	339	.4891
256	.5304	284	.5164	312	.5017	340	.4887
257	.5300	285	.5158	313	.5013	341	.4883
258	.5296	286	.5154	314	.5007	342	.4878
259	.5292	287	.5147	315	.5002	343	.4874
260	.5289	288	.5142	316	.4998	344	.4870
261	.5284	289	.5137	317	.4992	345	.4866
262	.5281	290	.5132	318	.4988	346	.4862
263	.5276	291	.5126	319	.4982	347	.4858
264	.5273	292	.5121	320	.4978	348	.4854
265	.5268	293	.5115	321	.4973	349	.4850
266	.5263	294	.5109	322	.4968	350	.4845
267	.5259	295	.5104	323	.4964	351	.4841
268	.5254	296	.5098	324	.4959	352	.4837
269	.5248	297	.5094	325	.4955	353	.4833
270	.5243	298	.5088	326	.4950	354	.4829
271	.5239	299	.5083	327	.4946	355	.4825
272	.5232	300	.5077	328	.4941	356	.4821
273	.5227	301	.5072	329	.4937	357	.4817
274	.5220	302	.5067	330	.4932	358	.4813
275	.5214	303	.5062	331	.4928	359	.4809
276	.5208	304	.5057	332	.4924	360	.4805
277	.5203	305	.5053	333	.4919	361	.4801
278	.5197	306	.5047	334	.4914	362	.4796

Problems 11 to 13 refer to performances at the 1988 National Collegiate Power-lifting Championships.

11. There were four session champions chosen at this meet. For the two lifters listed here, determine their score based on the Schwartz formula. (a) Sheridan Suttles of Middle Tennessee State College, who weighed 129 lb, squatted 440 lb, bench pressed 303 lb, and deadlifted 518 lb. (b) Ty Stapleton Jr. of the University of Oklahoma, who weighed 164 lb, squatted 573 lb, bench pressed 413 lb, and deadlifted 562 lb.

12. Jack London of Temple University weighed 289 lb at this competition, so he entered the superheavyweight class. He squatted 771 lb, bench pressed 458 lb, and deadlifted 661 lb. (a) Use Figure 1 to determine his score. (b) Is this score better than his score in the competition described in Example 1?

13. The Champion of Champions was based on the results of Sheridan Suttles, Ty Stapleton Jr., and Jack London. Who won this prestigious award?

Schwartz Master's Formula*

Age	SMF	Age	SMF
40	1.000	61	1.700
41	1.003	62	1.755
42	1.009	63	1.810
43	1.018	64	1.865
44	1.031	65	1.920
45	1.048	66	1.970
46	1.069	67	2.010
47	1.092	68	2.030
48	1.117	69	2.048
49	1.144	70	2.062
50	1.173	71	2.070
51	1.204	72	2.076
52	1.239	73	2.080
53	1.281	74	2.082
54	1.330	75	2.083
55	1.380	76	2.084
56	1.430	77	2.085
57	1.480	78	2.086
58	1.535	79	2.087
59	1.590	80	2.088
60	1.645		

*Use after the regular formula has been used.

CHAPTER REVIEW

Key Terms

11–1 Linear and Quadratic Functions
Function
Image
Preimage
Plotting Method
Linear Function
Domain
Range
Function Defined at a Number
Dependent Variable

Independent Variable
Quadratic Function
Vertex
Vertical Line Test
Cost Function
Fixed Cost
Variable Cost
Marginal Cost

11–2 Polynomial and Other Functions
Parabola
Power Function
Polynomial Function
Degree of a Polynomial

Hyperbola
Vertical Asymptote
Rational Function

11–3 The Algebra of Functions

Sum of Functions: $f + g$
Difference of Functions: $f - g$
Product of Functions: $f \cdot g$

Quotient of Functions: f/g
Rational Operations
Composition of Functions: $f \circ g$

11–4 Solving Polynomial Equations

Solution of an Equation
Root of an Equation
Polynomial Equation
Degree of a Polynomial Equation
Quadratic Equation
Quadratic Formula
Discriminant

Factor
Factoring
Point of Intersection
Break-Even Analysis
Revenue Function
Profit Function
Break-Even Point

Summary of Important Concepts

Quadratic Formula

If $ax^2 + bx + c = 0$ and $a \neq 0$, then

$$x = \frac{-b + \sqrt{b^2 - 4ac}}{2a}$$

or

$$x = \frac{-b - \sqrt{b^2 - 4ac}}{2a}$$

REVIEW PROBLEMS

Problems 1 to 3 refer to the function $f(x) = x^2 - 4x - 6$.

1. Compute $f(3)$, $f(2)$, $f(1)$, $f(0)$, and $f(-2)$.

2. Draw the graph of f.

3. Evaluate $f(x + h) - f(x)$.

In Problems 4 and 5 determine the domain of each function.

4. $f(x) = \sqrt{2x - 8}$

5. $f(x) = \dfrac{1}{1 - x}$

In Problems 6 and 7 determine whether each curve represents the graph of a function.

6.

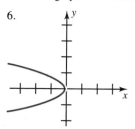

7.

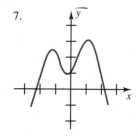

8. Use the plotting method to draw the graph of the function $f(x) = 2x^3 - 3x^2 - 12x + 1$.

9. Find the vertical asymptotes of the graph of the function $f(x) = \dfrac{x}{(2 - x)(x^2 - 9)}$.

10. Draw the graph of the function

$$f(x) = \begin{cases} x^2 + 4 & \text{if } x \geq 1 \\ 3x + 2 & \text{if } x < 1 \end{cases}$$

In Problems 11 to 13 let

$$f(x) = \frac{x + 2}{x - 1} \quad \text{and} \quad g(x) = \frac{x}{x + 2}$$

11. Find (a) $f + g$ (b) $f \cdot g$ (c) f/g.

12. State the domains of (a) $f + g$ (b) $f \cdot g$ (c) f/g.

13. Evaluate (a) $(f + g)(2)$ (b) $(f \cdot g)(2)$ (c) $(f/g)(2)$

14. Find $f \circ g$ and $g \circ f$ for $f(x) = \sqrt{\dfrac{3}{x}}$ and $g(x) = x^2$.

15. Write $f(x)$ as the composition of two functions.

$$f(x) = x + 2 - \sqrt{x + 2}$$

16. Express Q as a function of t, where

$$Q(P) = 40 - 0.1P^2 \quad \text{and} \quad P(t) = 5t - 0.2t^2$$

In Problems 17 to 19 find the x-intercepts of each polynomial function.

17. $f(x) = x^2 + 4x + 5$

18. $f(x) = 6x^2 - 13x + 6$

19. $f(x) = 3(x - 1)(x^2 + 4)(x^2 - 4)$

20. State the domain of the function

$$f(x) = \frac{x - 3}{4x^2 - 12x + 9}$$

21. Determine the domain and x-intercepts of the function

$$f(x) = \frac{x^2 - 1}{x^2 + x - 2}$$

22. Find all points where the graphs of $f(x) = x^4$ and $g(x) = x^2$ intersect.

 PROGRAMMABLE CALCULATOR EXPERIMENTS

1. Write a program that graphs the functions $f(x) = ax^2$ for the following fixed values of a: $a = 1$, $a = 2$, $a = 3$, $a = \frac{1}{2}$, $a = \frac{1}{3}$, $a = .1$.

2. Write a program that graphs the functions $f(x) = -ax^2$ for the following fixed values of a: $a = .01$, $a = .2$, $a = .3$, $a = \frac{1}{2}$, $a = \frac{4}{3}$, $a = 1.1$.

3. Write a program that graphs the functions $f(x) = x^2 + b$ for the following fixed values of b: $b = 1$, $b = 1.1$, $b = 1.5$, $b = -1$, $b = -\frac{1}{3}$, $b = -2$.

4. Write a program that graphs the functions $f(x) = ax^2 + b$ for the following fixed values of a and b: $a = 1$ and $b = 1$, $a = 1$ and $b = -1$, $a = -1$ and $b = 1$, $a = -1$ and $b = -1$.

5. Graph the following functions on the same screen and note the similarities and the differences: $y = x$, $y = x^2$, $y = x^3$, $y = x^4$, $y = x^{1/2}$.

6. Use the zoom feature to approximate the roots of the following equations: (a) $x^2 + x - 1 = 0$, (b) $x^2 - x - 1 = 0$, (c) $x^3 - 3x + 1$.

The Derivative

Economists use differentiation to chart and describe the complexities of worldwide grain production. (H. Armstrong Roberts/M. Roessler)

Chapter Overview

In this chapter we introduce the part of calculus called differentiation. The basic notion is the limit of a function. In Section 12–1 we motivate limits by considering the velocity of a moving object, the rate of change of a function, and lines which are tangent to the graph of a given function. In Section 12–2 we develop limits intuitively in terms of "closer and closer" and "approaching." We also consider limits at infinity.

   The derivative is introduced in Section 12–3, which relies on the properties of functions developed in Chapter 11 and the notion of a limit developed in Sections 12–1 and 12–2. The derivatives of several functions are computed from the definition, then the definition is applied to tangent lines and rates of change. Section 12-4 defines and contrasts continuous functions and differentiable functions.

12–1 Rates of Change and Tangents

There are two broad categories of applications of differential calculus: the rate of change and optimization. In this section the rate of change of a function is introduced by means of the velocity of an object that moves in a straight line. Then the tangent line to the graph of a function is presented by analogy with the rate of change. The tangent forms the basis for many optimization problems. Also discussed is the relationship between average rates of change and secant lines, as well as the relationship between the instantaneous rate of change and the tangent line.

Average Rate of Change

The **velocity** of an object that moves in a straight line is the rate of change of its distance with respect to time. The direction in which the object is traveling must also be taken into account, with a plus sign indicating one direction and a minus sign the opposite direction. We begin by computing the velocity in a specific setting.

   Let $d(t)$ be the distance (in miles) that a car travels due east in time t (in hours after noon). If the car starts at 1 P.M., then $d = 0$ at $t = 1$, so that $d(1) = 0$. If the car travels 100 miles (mi) by 3 P.M., then $d(3) = 100$. The average velocity over the interval of time from 1 to 3 P.M. is 50 miles per hour (mph) since the car traveled 100 mi in 2 hours. This is shown in Figure 1.

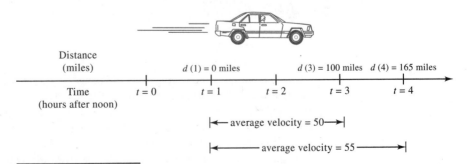

FIGURE 1

Suppose that the car traveled 165 mi by 4 P.M., so that $d(4) = 165$. Then the average velocity over the interval from $t = 1$ to $t = 4$ is 55 mph since

$$\frac{165 - 0}{4 - 1} = 55$$

In general, if a and b denote points in time with $a < b$, then the average velocity over the interval from $t = a$ to $t = b$ is given by the formula

$$\frac{d(b) - d(a)}{b - a}$$

This definition generalizes to the average rate of change of an arbitrary function.

DEFINITION

The **average rate of change** of the function $y = f(x)$ over the interval from $x = a$ to $x = b$, with $a < b$, is

$$\frac{f(b) - f(a)}{b - a} \Rightarrow STOP$$

EXAMPLE 1

Consider the function $f(x) = 3x - 1$.

Problem

(a) What is the average rate of change of $f(x)$ over the interval from $x = -1$ to $x = 4$? (b) What is the average rate of change of $f(x)$ from $x = 1$ to $x = 2$?

Solution (a) First $f(4) = 11$ and $f(-1) = -4$. By definition the average rate of change over the interval from $x = -1$ to $x = 4$ is

$$\frac{f(4) - f(-1)}{4 - (-1)} = \frac{11 - (-4)}{5} = \frac{15}{5} = 3$$

(b) The average rate of change over the interval from $x = 1$ to $x = 2$ is

$$\frac{f(2) - f(1)}{2 - 1} = \frac{5 - 2}{1} = 3$$

In Example 1 the average rate of change of $f(x) = 3x - 1$ over each interval is equal to 3. The exercises will show that the average rate of change of a linear function $f(x) = mx + b$ over any interval is always equal to the slope m. The exercises will also show that linear functions are the only functions whose average rates of change are constant over all intervals. Example 2 indicates that nonlinear functions can have different rates of change over different intervals.

EXAMPLE 2

Problem

Compute the average rate of change of $f(x) = x^3$ over the intervals (a) from 1 to 3 and (b) from 1 to 1.5.

Solution (a) Since $f(3) = 27$ and $f(1) = 1$, the average rate of change over the interval from $x = 1$ to $x = 3$ is

$$\frac{f(3) - f(1)}{3 - 1} = \frac{27 - 1}{2} = 13$$

(b) The average rate of change over the interval from $x = 1$ to $x = 1.5$ is

$$\frac{f(1.5) - f(1)}{1.5 - 1} = \frac{3.375 - 1}{0.5} = 4.75$$

Instantaneous Rate of Change

Table 1 lists the average rates of change of the function $f(x) = x^3$ over the intervals from $x = 1$ to $x = 1 + h$ for various values of h. Example 2 covered $h = 2$ and $h = 0.5$.

Table 2 shows what happens to the average rate of change of $f(x) = x^3$ as the values of h approach 0 for negative values of h.

We can define the instantaneous rate of change of the function at the value $x = 1$ by viewing the numbers in the column for the average rates of change in both Tables 1 and 2. As h approaches 0 from either direction (positive or negative values), the average rates of change approach 3. The number 3 is called the *instantaneous rate of change* of $f(x)$ at $x = 1$, or just the *rate of change* of $f(x)$ at $x = 1$. The general definition is given here.

Table 1

h	Interval	Average Rate of Change
2	from 1 to 3	13
1	from 1 to 2	7
0.5	from 1 to 1.5	4.75
0.1	from 1 to 1.1	3.31
0.01	from 1 to 1.01	3.0301
0.001	from 1 to 1.001	3.003001

Table 2

h	Interval	Average Rate of Change
-0.1	from 0.9 to 1	2.71
-0.01	from 0.99 to 1	2.9701
-0.001	from 0.999 to 1	2.997001

DEFINITION

The **instantaneous rate of change** of the function $y = f(x)$ at a value of x is the number (if it exists) approached by the average rates of change from x to $x + h$ as h approaches 0 and $h \neq 0$.

EXAMPLE 3

Problem

Find the instantaneous rate of change of the function $f(x) = x^2 - 1$ at $x = 2$.

Solution Table 3 shows the average rates of change of $f(x) = x^2 - 1$ over various intervals for $h > 0$ and $h < 0$. Consider, for instance, $h = 0.1$. The interval is from 2 to 2.1, so the average rate of change is

$$\frac{f(2.1) - f(2)}{2.1 - 2} = \frac{(4.41 - 1) - (4 - 1)}{0.1} = \frac{0.41}{0.1} = 4.1$$

Now let h get closer and closer to 0. For $h = 0.01$, the average rate of change is 4.01. For $h = 0.001$, the average rate of change is 4.001. It seems apparent that as h gets closer to 0, the average rate of change approaches 4. Therefore the instantaneous rate of change at $x = 2$ is equal to 4.

Table 3

h	Interval	Average Rate of Change
0.1	from 2 to 2.1	4.1
0.01	from 2 to 2.01	4.01
0.001	from 2 to 2.001	4.001
−0.1	from 1.9 to 2	3.9
−0.01	from 1.99 to 2	3.99
−0.001	from 1.999 to 2	3.999

Secant and Tangent Lines

Thus far we have used algebra to define the instantaneous rate of change in terms of the average rates of change. There is also a geometric analog to this idea.

In general a line that passes through two points on a curve is called a **secant line**. (Recall that the secant line of a circle passes through two points on the circle.) Consider the function $f(x) = x^2$, whose graph is a parabola. The line through the points (2, 4) and (3, 9) is a secant line. (See Figure 2.) By definition the slope of this secant line is equal to

$$m = \frac{9 - 4}{3 - 2} = 5$$

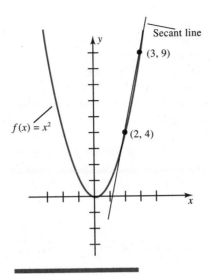

FIGURE 2

Notice that the average rate of change of $f(x) = x^2$ over the interval from $x = 2$ to $x = 3$ is defined similarly.

$$\frac{f(3) - f(2)}{3 - 2} = \frac{9 - 4}{1} = 5$$

These equalities illustrate the fact that *the slope of the secant line is equal to the average rate of change.*

What geometric entity corresponds to the instantaneous rate of change? To answer this question, look at Figure 3, showing some secant lines of the graph of

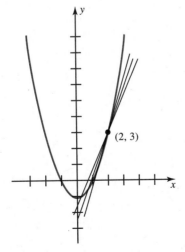

Secant lines of $f(x) = x^2 - 1$

FIGURE 3

$f(x) = x^2 - 1$ from $x = 2$ to $x = 2 + h$ for values of h that approach 0 and $h \neq 0$. Observe that the secant lines get closer and closer to one particular line. This line is called a **tangent line.** (Recall that the tangent to a circle is a line that touches the circle at precisely one point.) Because of this, we define the slope of the line tangent to a given point on a curve to be the number (if it exists) that the slopes of the secant lines approach. From our earlier discussion it follows that *the slope of the line tangent to a point corresponds to the instantaneous rate of change at that point*.

The equation of the tangent line through a given point on the graph of a function can be derived once the slope is known. Example 4 makes use of Example 3 to illustrate the method that is used to derive the equation of the tangent line.

EXAMPLE 4

Problem

Find the slope-intercept form of the line which is tangent to the graph of $f(x) = x^2 - 1$ at $x = 2$.

Solution Example 3 showed that the instantaneous rate of change of $f(x)$ at $x = 2$ is equal to 4. Therefore the slope of the tangent line is 4. Since $f(2) = 3$, the tangent line passes through the point $(2, 3)$. Therefore the equation of the tangent line in point-slope form is $y - 3 = 4(x - 2)$. Solving this equation for y yields the slope-intercept form $y = 4x - 5$. See Figure 4.

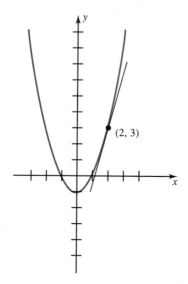

(2, 3)

Tangent line of $f(x) = x^2 - 1$ at $x = 2$

FIGURE 4

The **slope of a function** $f(x)$ at a point on the graph of $f(x)$ is defined to be the slope of the tangent line through that point. The procedure for computing the slope of a function $f(x)$ at any value of x involves three steps:

---

1. Form the **difference quotient:**

$$\frac{f(x + h) - f(x)}{h}$$

2. Simplify the difference quotient, using the fact that $h \neq 0$.
3. The slope of $f(x)$ is the expression that results when h approaches 0 and $h \neq 0$.

EXAMPLE 5

Problem

Find the slope of $f(x) = x^3$ at an arbitrary value of x.

Solution It is instructive to refer to Figure 5 while carrying out the three steps. By step 1 the difference quotient is

$$\frac{f(x + h) - f(x)}{h} = \frac{(x + h)^3 - x^3}{h}$$

By step 2

$$\frac{f(x + h) - f(x)}{h} = \frac{(x^3 + 3x^2h + 3xh^2 + h^3) - x^3}{h}$$

$$= \frac{h(3x^2 + 3xh + h^2)}{h}$$

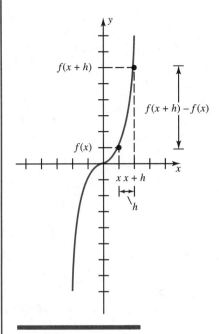

FIGURE 5

The factor h can be canceled since $h \neq 0$, so

$$\frac{f(x + h) - f(x)}{h} = 3x^2 + 3xh + h^2$$

By step 3 let h approach 0. The terms $3xh$ and h^2 also approach 0, so $3x^2 + 3xh + h^2$ approaches $3x^2$. Thus the slope of $f(x) = x^3$ at any value of x is $3x^2$.

The tangent to a curve measures the steepness of the curve. An intuitive way of visualizing this is to put a roller coaster car, with a long board attached to the hubs of the wheels, on the curve. The car is the point, the board is the tangent. (See Figure 6.) As the car moves along the curve from left to right, the board reflects the steepness of the climb and the descent. As the car ascends, the board points upward so the tangent is a positive number. When the car is on the top of the hill, the board is horizontal so the tangent is 0. As the car descends, the board points downward so the tangent is a negative number. We will develop these ideas in greater detail in Chapter 14.

Supply and Demand

The ideas in this section can be applied to the part of economics that deals with supply and demand curves. The **demand** for a product is the number $D(x)$ of items that consumers are willing and able to buy at a given price per unit x. For example, suppose the demand for a new toy called *Cubic* is $D(x) = 5 - x^2$, where x is in dollars per Cubic and $D(x)$ is in thousands of Cubics per week. Notice that the demand slackens as the price rises. For instance, if the price of each Cubic is raised from \$1 to \$2 the average rate of change of the demand is

$$\frac{D(2) - D(1)}{2 - 1} = \frac{1 - 4}{1} = -3$$

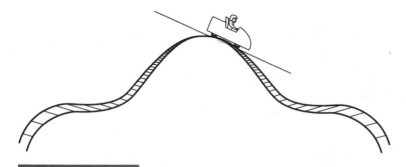

FIGURE 6 The board attached to the roller coaster car approximates the tangent line.

This means that the demand will decrease by 3 (= 3000 Cubics per week) if the price of the Cubics is raised from $1 to $2. The instantaneous rate of change of the demand at any price x is called the **marginal demand.** Example 6 computes the marginal demand of this function.

EXAMPLE 6

Problem

If the demand function is $D(x) = 5 - x^2$, what is the marginal demand?

Solution The marginal demand can be obtained from the three-step procedure. By step 1 the difference quotient is

$$\frac{D(x + h) - D(x)}{h} = \frac{[5 - (x + h)^2] - (5 - x^2)}{h}$$

By step 2

$$\frac{D(x + h) - D(x)}{h} = \frac{5 - x^2 - 2xh - h^2 - (5 - x^2)}{h}$$

$$= \frac{h(-2x - h)}{h}$$

The factor h can be canceled since $h \neq 0$, so

$$\frac{D(x + h) - D(x)}{h} = -2x - h$$

By step 3, as h gets close to 0, the term $-2x - h$ gets close to $-2x$. Thus the marginal demand at an arbitrary price x is equal to $-2x$. This means that the demand for Cubics decreases at a rate that is double the price of each Cubic.

EXERCISE SET 12–1

In Problems 1 to 8 compute the average rate of change of the function $f(x)$ over the stated intervals.

1. $f(x) = 2x - 3$
   (a) from 1 to 2   (b) from -1 to 4

2. $f(x) = 4x - 3$
   (a) from 1 to 2   (b) from -1 to 4

3. $f(x) = 3 - 4x$
   (a) from 0 to 5   (b) from -3 to -2

4. $f(x) = 3 - 2x$
   (a) from 0 to 5   (b) from -3 to -2

5. $f(x) = x^3$
   (a) from 0 to 1   (b) from 0 to 0.1

6. $f(x) = x^3$
   (a) from -1 to 0   (b) from -0.1 to 0

7. $f(x) = x^4$
   (a) from 0 to 1   (b) from 0 to 0.1

8. $f(x) = x^4$
   (a) from -1 to 0   (b) from -0.1 to 0

In Problems 9 to 12 use the method shown in Example 3 to compute the instantaneous rate of change of the function $f(x)$ at the stated value of x.

9. $f(x) = x^2 - 5$ at $x = 1$

10. $f(x) = x^2 - 5$ at $x = -2$

11. $f(x) = 10 - 2x^2$ at $x = -2$

12. $f(x) = 10 - 2x^2$ at $x = 1$

In Problems 13 to 16 find the slope-intercept form of the line tangent to the graph of the function in the stated problem above.

13. Problem 9   14. Problem 10

15. Problem 11   16. Problem 12

In Problems 17 to 20 find the marginal demand at an arbitrary price x, where the demand function is given.

17. $D(x) = 10 - 2x^2$ 18. $D(x) = x^2$

19. $D(x) = x^2 + 2$ 20. $D(x) = 1 - 2x^2$

In Problems 21 to 26 compute the average rate of change of the function $f(x)$ over the stated intervals.

21. $f(x) = x^2 - 5$ (a) from 2 to 2.1 (b) from 2 to 2.01 (c) from 2 to 2.001

22. $f(x) = 10 - x^2$ (a) from 2 to 2.1 (b) from 2 to 2.01 (c) from 2 to 2.001

23. $f(x) = 2x^2 - 4x - 3$ (a) from 3 to 3.1 (b) from 3 to 3.01 (c) from 3 to 3.001

24. $f(x) = 3x^2 + 2x - 29$ (a) from 3 to 3.1 (b) from 3 to 3.01 (c) from 3 to 3.001

25. $f(x) = \sqrt{x}$ (a) from 4 to 4.1 (b) from 4 to 4.01 (c) from 4 to 4.001

26. $f(x) = 1/x$ (a) from 0.5 to 0.51 (b) from 0.5 to 0.501 (c) from 0.5 to 0.5001

In Problems 27 and 28 compute the instantaneous rate of change of the function $f(x)$ at the stated value of x.

27. $f(x) = x^3 + x + 1$ at $x = 1$

28. $f(x) = x^3 + x^2 + 1$ at $x = 1$

In Problems 29 and 30 find the slope-intercept form of the line tangent to the graph of the function in the preceding stated problem.

29. Problem 27   30. Problem 28

In Problems 31 to 34 find the slope of the line tangent to the graph of the function $f(x)$ at an arbitrary value of x.

31. $f(x) = 2x^2 - 4x - 3$

32. $f(x) = 3x^2 + 2x - 29$

33. $f(x) = x^3 + x + 1$

34. $f(x) = x^3 + x^2 + 1$

In Problems 35 to 38 the height of a ball thrown into the air is given by $d(t)$. Find the slope of the line tangent to the graph of the function $d(t)$ at an arbitrary value of t.

35. $d(t) = -16t^2 + 192t$

36. $d(t) = 70t - 16t^2$

37. $d(t) = -16t^2 + 128t + 64$

38. $d(t) = -16t^2 + 32t + 64$

In Problems 39 to 42 choose one of the numbers, -1, 0, 1, 4, that most closely approximates the slope of the tangent line at the point $(a, f(a))$ for the given graph of $y = f(x)$.

39.

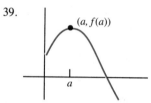

40.

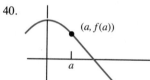

41.

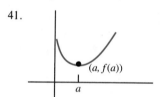

42.

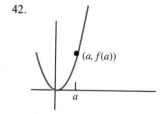

43. Let a, b, and c be arbitrary real numbers. Compute the slope of the line tangent to the graph of $f(x) = ax^2 + bx + c$ at an arbitrary value of x.

44. Prove that if $f(x)$ is a function having the property that the average rate of change is the same number over all intervals, then $f(x) = mx + b$ for some m and b.

45. Prove that if $f(x)$ is a function having the property that the slope of the tangent line at each point on the graph of $f(x)$ is the same number, then $f(x) = mx + b$ for some m and b.

46. Let m, a, b, and c be real numbers and $f(x) = mx + b$.
    (a) Compute the average rate of change of $f(x)$ over the interval from a to c.
    (b) Compute the instantaneous rate of change of $f(x)$ at $x = a$.

Referenced Exercise Set 12–1

Problems 1 and 2 refer to Figure 7, which shows the personal consumption expenditures (in billions of dollars) in the United States on durable goods over the first part of the 1980s.*

1. Estimate the average rate of change of the expenditures.
    (a) from 1980 to 1984
    (b) from 1982 to 1984

2. Estimate the instantaneous rate of change of the expenditures in 1981.

Problems 3 and 4 refer to Figure 8, which shows the U.S. Federal Budget Deficit as a percentage of the Gross National Product (GNP) for the first half of the 1980's.†

3. Estimate the average rate of change of the percentage of the GNP
    (a) from 1981 to 1983
    (b) from 1984 to 1985

4. Estimate the instantaneous rate of change of the percentage of the GNP in 1985.

5. An article by Alan Murray in the *Wall Street Journal* used the fictitious Business Corporation to illustrate the tax law of 1986.‡ The tax law permitted Business Corporation to depreciate its new $6.3 million building over a period of 31.5 years by the "straightline method." What is the rate of depreciation? (The rate of depreciation is defined as the slope of the straight line where the x-axis is time in years and the y-axis is the cost in millions of dollars. The rate is expressed as a percent.)

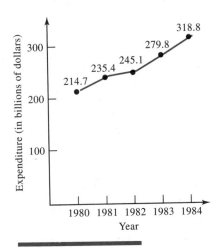

FIGURE 7

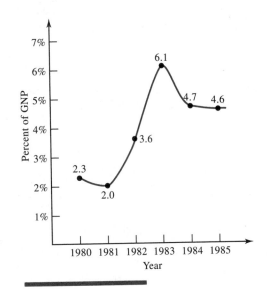

FIGURE 8

Source: Bureau of Economic Analysis, U.S. Commerce Department.

†*The World Almanac and Book of Facts*, New York, Newspaper Enterprise Association Inc., 1986, p. 102.

‡Alan Murray, "Individuals' Top Rate Would Plunge to 28 Percent; Tax-Break Curbs Offset Benefits to Wealthy," *Wall Street Journal*, August 18, 1986, p. 6.

Problems 6 and 7 refer to an editorial by Paul Craig Roberts, who holds the William E. Simon chair of political economy at the Center for Strategic and International Studies in Washington.§ The editorial compared the economy under presidents Carter and Reagan during two 58-month periods using these facts: under President Carter the unemployment rate fell 27% and the CPI (consumer price index) rose 48%, while under President Reagan the unemployment rate fell 45% and the CPI rose 17%.

6. (a) If graphs for the unemployment rate under each administration are sketched with years on the *x*-axis and unemployment rates on the *y*-axis, are the slopes of the functions (which are defined by the graphs) positive or negative?

   (b) Which administration had a better record regarding unemployment?

7. (a) If graphs for the CPI under each administration are sketched with years on the *x*-axis and the CPI on the *y*-axis, are the slopes positive or negative?

   (b) Which administration had a better record regarding the CPI?

§Paul Craig Roberts, "While Critics Carp, the Supply-Side Revolution Sweeps On," *Business Week*, April 25, 1988, p. 15.

12–2 Limits

In the previous section the problem of finding the instantaneous rate of change entailed evaluating the difference quotient as h approached 0. In this section we give a more formal definition of this process. It is called "finding the *limit* of $f(x)$ as x approaches a," which intuitively means that we find the value that $f(x)$ should be when x approaches a. Of course, part of the problem is to discuss what is meant by "should be" and "approaches."

As an example, consider $f(x) = x^2$. As x gets close to 2, $f(x)$ gets close to $2^2 = 4$, so we say that $f(x)$ approaches 4 as x approaches 2. Likewise, as x approaches 3, the function $g(x) = x^3$ approaches $3^3 = 27$. Expressed in terms of limits, we say the limit of $f(x) = x^2$ as x approaches 2 is 4 and the limit of $g(x) = x^3$ as x approaches 3 is 27.

It is easy to find these particular limits because in each case the limit is equal to the functional value, $f(a)$; that is, to find the limit we simply substitute a into the formula of the function. The difficulty arises when the limit is not equal to the functional value. This was the case in the previous section when we had to find the limit of the difference quotient as h approached 0. It could not be done by simply substituting 0 into the expression because the expression was not defined at 0. Another method had to be used. It is this method that we will discuss in a more rigorous way in this section.

The problem is approached from three perspectives. First we look at a *geometric* approach by describing the problem in terms of the graph of $f(x)$. The limit of $f(x)$ as x approaches a, if it exists, is the second coordinate of a point that the function "should" go through. For example, the graph of $f(x) = x^2$ "should" go through (2, 4). That it actually does pass through this point is immaterial in this discussion because we are more concerned with evaluating limits when the function does not go through this point.

Then we will consider an *arithmetic* approach to the problem, where we define "approaches $x = a$" by actually computing functional values of $f(x)$ for numbers within smaller and smaller distances of a. For example, if we want to compute the limit as x approaches 1, we compute functional values at $x = 1.1$, 1.01, and 1.001 and at $x = 0.9, 0.99$, and 0.999, and then see if these functional values are getting close to a specific number. Neither of these two approaches is very rigorous, so, to be more precise, we use a third method, the *algebraic* approach to the problem.

Geometric Approach

The graphs of the two functions $f(x) = x + 2$ and $g(x) = \dfrac{x^2 - 4}{x - 2}$ are identical except for one point: the graph of $f(x)$ includes the point (2, 4) while the graph of $g(x)$ does not include it. This is because $f(2) = 2 + 2 = 4$ while $g(2)$ is $\dfrac{0}{0}$, which is not defined. It is clear that the limit of $f(x)$ as x approaches 2 is 4, because the graph passes through (2, 4) in an unbroken manner. The situation for $g(x)$ is similar, but with a little catch. There is a "hole" in the graph at (2, 4), meaning that the curve would be unbroken if we included (2, 4) in the graph. In an intuitive sense, this is what is meant by saying that $g(x)$ "should" pass through (2, 4), that its graph would be unbroken if that point were included in the graph. For this reason we say the limit of $g(x)$ as x approaches 2 is 4, since 4 is the second coordinate of the point that $g(x)$ should go through to be unbroken.

As another example, consider the function

$$f(x) = \frac{x^2 - 8x + 15}{3 - x} = \frac{(x - 3)(x - 5)}{3 - x}$$

Its graph is the line with the "hole" in it drawn in Figure 1. The open circle indicates that $f(3)$ is not defined. However, even though the function is not defined

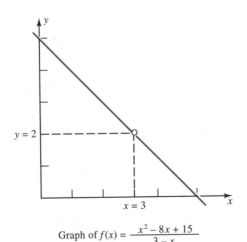

Graph of $f(x) = \dfrac{x^2 - 8x + 15}{3 - x}$

FIGURE 1

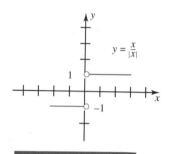

$$y = \frac{x}{|x|}$$

FIGURE 2

at $x = 3$, the limit is defined there because the functional values of $f(x)$ get close to 2 as x gets close to 3. The geometric interpretation of the limit is the height of the graph when the values of x get close to 3.

To see when a function does not have a limit at a particular value of x, consider the function

$$f(x) = \frac{x}{|x|} = \begin{cases} 1 \text{ if } x > 0 \\ -1 \text{ if } x < 0 \end{cases}$$

The function is not defined at $x = 0$. The graph of the $f(x)$ is given in Figure 2. There is a "gap" in the graph at $x = 0$, and so there is no point that we could put into the graph at $x = 0$ so that the function would be unbroken. This means that for $f(x) = \dfrac{x}{|x|}$ **the limit does not exist** at $x = 0$.

Another example of a more familiar function that does not have a limit at particular values of x is the *quality points function,* which many universities use to define grade point average (GPA). If a student's test average is 90 or above, the student gets an A and four quality points are awarded. The chart shows the other values.

Test average	≥ 90	80s	70s	60s	<60
Quality points	4	3	2	1	0

EXAMPLE 1

Let $Q(x)$ denote the quality points function whose graph is drawn in Figure 3. An open circle means that the function is not defined at that end point.

Problem

Explain why the limit of $Q(x)$ as x approaches 80 does not exist.

Solution Note that $Q(80) = 3$ since an average of 80 yields three quality points. In fact, for any number x that is greater than 80 but close to 80, $Q(x) = 3$. However, for values of x that are close to 80 but less than 80, $Q(x) = 2$.

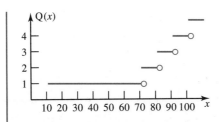

FIGURE 3 Graph of the quality points function $Q(x)$.

As x gets closer to 80, we cannot choose whether $Q(x)$ gets closer to 3 or 2. Since the limit must be a unique number, we conclude that the limit does not exist. For example, $Q(79.9) = 2$ and $Q(80) = 3$.

Arithmetic Approach

In Section 12–1 we derived the instantaneous rate of change of the function $f(x) = x^3$ at $x = 1$ from a table whose essential ingredients are reproduced here (rounded to three places).

h	-0.1	-0.01	-0.001	0.001	0.01	0.1
Average rate	2.71	2.970	2.997	3.003	3.030	3.31

The average rates of change approach 3 as h approaches 0. This leads to an intuitive definition of the limit of a function.

DEFINITION

If there is a unique number L with the property that $f(x)$ approaches L as x approaches the number a and $x \neq a$, then L is called the **limit** of $f(x)$ as x approaches a. In symbols

$$\lim_{x \to a} f(x) = L$$

The notation $\lim_{x \to a} f(x)$ is read "the limit of f of x as x approaches a." If the limit L exists, then it is the only number with the defining property.

To illustrate the definition, consider the function

$$f(x) = x^3 - 3x^2 + 5$$

for values of x that are *near* 1, but are *not equal to* 1. Construct Table 1 with values of x that approach 1 in the top row and the corresponding values of $f(x)$ below them (rounded to three decimal places). It is important to observe that the values of x approach 1 from two directions: from the left (numbers that are less than 1) and from the right (numbers that are greater than 1).

Table 1

	approach 1 from the left				approach 1 from the right		
x	0.9	0.99	0.999	1	1.001	1.01	1.1
$f(x)$	3.299	3.030	3.003		2.997	2.970	2.701

It follows from Table 1 that $f(x)$ approaches 3 as x approaches 1 from both directions, so

$$\lim_{x \to 1} f(x) = 3$$

It is possible to obtain this limit directly by evaluating $f(1)$ since $f(1) = 3$. However, such a substitution is not always possible. For example, consider again

$$f(x) = \frac{x^2 - 8x + 15}{3 - x} = \frac{(x - 3)(x - 5)}{3 - x}$$

What is the behavior of $f(x)$ when x assumes values that are *near* 3 but *not equal to* 3? To answer the question, construct Table 2. Notice that $f(x)$ is not defined at $x = 3$ because both the numerator and denominator are 0.

Table 2

	approach 3 from the left				approach 3 from the right		
x	2.9	2.99	2.999	3	3.001	3.01	3.1
$f(x)$	2.1	2.01	2.001		1.999	1.99	1.9

It follows from Table 2 that $f(x)$ approaches 2 as x approaches 3 from both directions, so

$$\lim_{x \to 3} f(x) = 2$$

EXAMPLE 2

Problem

Compute $\lim_{x \to 1} f(x)$ for $f(x) = \dfrac{4x^2 - 4x}{x - 1} = \dfrac{4x(x - 1)}{x - 1}$.

Solution Construct a table with values of x approaching 1 from both sides in the top row. Compute the corresponding values of $f(x)$ and place them in the bottom row.

	approach 1 from the left				approach 1 from the right		
x	0.9	0.99	0.999	1	1.001	1.01	1.1
$f(x)$	3.6	3.96	3.996		4.004	4.04	4.4

The table shows that $f(x)$ approaches 4 as x approaches 1 from both sides, so

$$\lim_{x \to 1} f(x) = 4$$

In Example 2 the function is defined for values of x that are near 1 but are not equal to 1, yet the limit exists at $x = 1$ (and is equal to 4). Example 3 examines an entirely different situation.

E X A M P L E 3

Problem

Compute $\lim\limits_{x \to 2} f(x)$ where $f(x) = \dfrac{x + 1}{x - 2}$.

Solution The appropriate table is shown here.

	approach 2 from the left				approach 2 from the right		
x	1.9	1.99	1.999	2	2.001	2.01	2.1
$f(x)$	-29	-299	-2999		3001	301	31

The table shows that as x approaches 2 from the left, the values of $f(x)$ get smaller and smaller, and as x approaches 2 from the right, the values of $f(x)$ get larger and larger. Therefore $f(x)$ does not have a limit when x approaches 2, meaning that there is no fixed number that the values of $f(x)$ approach as x approaches 2.

Example 3 illustrates another way in which the limit of a function may fail to exist. Unlike the graph in Example 1, where there was a gap in the graph, the graph of $f(x)$ in Example 3 does not approach any number as x approaches 2 because $f(x)$ **decreases without bound** to the left of $x = 2$ and **increases without bound**

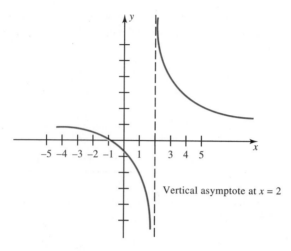

Graph of $f(x) = \dfrac{x+1}{x-2}$

FIGURE 4

to the right of $x = 2$. When a function either increases or decreases without bound near a value $x = a$, we write

$$\lim_{x \to a} f(x) \quad \text{does not exist}$$

Figure 4 shows that the graph of the function $f(x) = \dfrac{x+1}{x-2}$ has a vertical asymptote at $x = 2$. As x approaches 2 from the left, $f(x)$ decreases without bound, and as x approaches 2 from the right, $f(x)$ increases without bound. Therefore

$$\lim_{x \to 2} f(x) \text{ does not exist}$$

In general, whenever a function $f(x)$ has a vertical asymptote at $x = a$ then $\lim_{x \to a} f(x)$ does not exist.

Algebraic Approach

We have evaluated limits by arithmetic and geometric methods. The most rigorous method, however, is to use algebraic simplifications based on the following properties of limits. These properties can be derived from the definition of the limit of a function.

Properties of Limits

Let $f(x)$ and $g(x)$ be functions for which $\lim\limits_{x \to a} f(x)$ and $\lim\limits_{x \to a} g(x)$ exist. Then

1. $\lim\limits_{x \to a} f(x) = \lim\limits_{x \to a} g(x)$ if $f(x) = g(x)$ for all $x \neq a$

2. $\lim\limits_{x \to a} cf(x) = c \lim\limits_{x \to a} f(x)$ for any real number c

3. $\lim\limits_{x \to a} [f(x) \pm g(x)] = \lim\limits_{x \to a} f(x) \pm \lim\limits_{x \to a} g(x)$

4. $\lim\limits_{x \to a} [f(x)g(x)] = \lim\limits_{x \to a} f(x) \lim\limits_{x \to a} g(x)$

5. $\lim\limits_{x \to a} \dfrac{f(x)}{g(x)} = \dfrac{\lim\limits_{x \to a} f(x)}{\lim\limits_{x \to a} g(x)}$,   provided that $\lim\limits_{x \to a} g(x) \neq 0$

6. $\lim\limits_{x \to a} [f(x)]^r = \left[\lim\limits_{x \to a} f(x) \right]^r$ for any real number r

(where x must be restricted to avoid even roots of negative numbers)

The next example makes use of two properties that can be proved from the definition of a limit.

$$\lim_{x \to a} x = a \quad \text{and} \quad \lim_{x \to a} c = c$$

EXAMPLE 4

Problem

Evaluate $\lim\limits_{x \to 0.5} (2x^2 - x - 4)$

Solution It follows from the stated properties that

$$\lim_{x \to 0.5} x = 0.5 \quad \text{and} \quad \lim_{x \to 0.5} 4 = 4$$

By properties 2 and 4

$$\lim_{x \to 0.5} 2x^2 = 2 \left(\lim_{x \to 0.5} x^2 \right) = 2 \left(\lim_{x \to 0.5} x \right)\left(\lim_{x \to 0.5} x \right)$$

$$= 2(0.5)(0.5) = 0.5$$

Using property 3

$$\lim_{x \to 0.5} (2x^2 - x - 4) = \lim_{x \to 0.5} 2x^2 - \lim_{x \to 0.5} x - \lim_{x \to 0.5} 4$$

$$= 0.5 - 0.5 - 4 = -4$$

Property 1 is a powerful tool for evaluating many of the limits that appear in calculus. For instance, let

$$f(x) = \frac{x^2 - 8x + 15}{3 - x}$$

We have seen that $\lim_{x \to 3} f(x) = 2$. The algebraic approach is to reduce $f(x)$ to a function $g(x)$ such that $g(3)$ is defined and $g(x) = f(x)$ for all x except $x = 3$. For this function

$$\frac{x^2 - 8x + 15}{3 - x} = \frac{(x - 3)(x - 5)}{-(x - 3)} = 5 - x$$

The common factor $(x - 3)$ can be canceled because $x \neq 3$. Set $g(x) = 5 - x$. Then $g(x) = f(x)$ for all x except $x = 3$. Since $g(3) = 2$, it follows that $\lim_{x \to 3} f(x) = g(3)$, so

$$\lim_{x \to 3} \frac{x^2 - 8x + 15}{3 - x} = \lim_{x \to 3} (5 - x) = 2$$

Examples 5, 6, and 7 exploit property 1.

EXAMPLE 5

Problem

Evaluate $\lim_{x \to 2} \dfrac{x^2 - 2x}{x - 2}$.

Solution $\lim_{x \to 2} \dfrac{x^2 - 2x}{x - 2} = \lim_{x \to 2} \dfrac{x(x - 2)}{x - 2} = \lim_{x \to 2} x = 2$

EXAMPLE 6

Problem

Evaluate $\lim_{x \to 3} \dfrac{3 - x}{x^2 - 6x + 9}$.

Solution $\lim_{x \to 3} \dfrac{3 - x}{x^2 - 6x + 9} = \lim_{x \to 3} \dfrac{-(x - 3)}{(x - 3)^2} = \lim_{x \to 3} \dfrac{-1}{x - 3}$

This limit does not exist because the function $f(x) = \dfrac{-1}{x - 3}$ has a vertical asymptote at $x = 3$.

Example 7 shows that it is sometimes necessary to rationalize the numerator in order to divide out the factor that results in an expression of the form $\dfrac{0}{0}$.

E X A M P L E 7

Problem

Evaluate $\lim\limits_{x \to 9} \dfrac{\sqrt{x} - 3}{x - 9}$

Solution $\lim\limits_{x \to 9} \dfrac{\sqrt{x} - 3}{x - 9} = \lim\limits_{x \to 9} \dfrac{(\sqrt{x} - 3)(\sqrt{x} + 3)}{(x - 9)(\sqrt{x} + 3)}$

$$= \lim\limits_{x \to 9} \dfrac{x - 9}{(x - 9)(\sqrt{x} + 3)}$$

$$= \lim\limits_{x \to 9} \dfrac{1}{\sqrt{x} + 3} = \dfrac{1}{6}$$

Limits at Infinity

Consider the function $f(x) = \dfrac{1 + 2x}{x}$. Table 3 shows that as x assumes larger and larger values in the positive direction, the corresponding values of $f(x)$ approach the number 2. This is called taking a **limit at infinity.** The notation is

$$\lim\limits_{x \to \infty} f(x) = 2$$

It is read "the limit of $f(x)$ as x approaches infinity is 2," where the symbol for infinity is ∞.

Table 3

x	10	100	1000	10000
$f(x)$	2.1	2.01	2.001	2.0001

Table 4 shows that for values of x that get smaller and smaller in the negative direction, the corresponding values of $f(x)$ also approach the number 2, so

$$\lim\limits_{x \to -\infty} f(x) = 2$$

This is read "the limit of $f(x)$ as x approaches minus infinity is 2."

Table 4

x	-10	-100	-1000	-10000
$f(x)$	1.9	1.99	1.999	1.9999

The display contains the general definitions.

DEFINITION

$\lim\limits_{x\to\infty} f(x) = L$ if there is a unique number L such that the values of $f(x)$ approach L as x assumes larger and larger values. $\lim\limits_{x\to-\infty} f(x) = L$ if there is a unique number L such that the values of $f(x)$ approach L as x assumes smaller and smaller values. If either $\lim\limits_{x\to\infty} f(x) = L$ or $\lim\limits_{x\to-\infty} f(x) = L$, then the horizontal line $y = L$ is called a **horizontal asymptote** of $f(x)$.

Horizontal asymptotes of rational functions can be found by dividing the numerator and the denominator by x^n, where n is the larger of the degrees of the polynomials in the numerator and the denominator. Example 8 illustrates the method.

EXAMPLE 8

Let

$$f(x) = \frac{5x^2 - 1}{2x^2 - x + 5}$$

Problem

(a) Evaluate $\lim\limits_{x\to\infty} f(x)$.   (b) Find all horizontal asymptotes of $f(x)$.

Solution (a) Both the numerator and denominator are polynomials of degree 2. Divide each one by x^2.

$$f(x) = \frac{5x^2 - 1}{2x^2 - x + 5} = \frac{\dfrac{5x^2 - 1}{x^2}}{\dfrac{2x^2 - x + 5}{x^2}} = \frac{\dfrac{5x^2}{x^2} - \dfrac{1}{x^2}}{\dfrac{2x^2}{x^2} - \dfrac{x}{x^2} + \dfrac{5}{x^2}}$$

$$= \frac{5 - \dfrac{1}{x^2}}{2 - \dfrac{1}{x} + \dfrac{5}{x^2}}$$

As the values of x get larger and larger, $\dfrac{1}{x^2}$ approaches 0. Therefore the numerator $5 - \dfrac{1}{x^2}$ approaches 5. Similarly the denominator $2 - \dfrac{1}{x} + \dfrac{5}{x^2}$ approaches 2. Thus the desired limit is $\frac{5}{2}$.

(b) Part (a) shows that $\lim\limits_{x\to\infty} f(x) = \frac{5}{2}$. Similarly $\lim\limits_{x\to-\infty} f(x) = \frac{5}{2}$. Therefore the line $y = \frac{5}{2}$ is the only horizontal asymptote of $f(x)$.

EXERCISE SET 12–2

In Problems 1 to 8 compute the given limits by constructing tables.

1. $\lim\limits_{x \to 1} f(x)$ for $f(x) = \dfrac{3x^2 - 3x}{x - 1}$

2. $\lim\limits_{x \to 1} f(x)$ for $f(x) = \dfrac{2x^2 - 2x}{x - 1}$

3. $\lim\limits_{x \to 3} f(x)$ for $f(x) = \dfrac{2x^2 - 6x}{x - 3}$

4. $\lim\limits_{x \to 3} f(x)$ for $f(x) = \dfrac{3x^2 - 9x}{x - 3}$

5. $\lim\limits_{x \to 2} f(x)$ for $f(x) = \dfrac{x + 1000}{x - 2}$

6. $\lim\limits_{x \to 2} f(x)$ for $f(x) = \dfrac{x + 2}{x - 2}$

7. $\lim\limits_{x \to 3} f(x)$ for $f(x) = \dfrac{3 + x}{3 - x}$

8. $\lim\limits_{x \to 3} f(x)$ for $f(x) = \dfrac{x + 1000}{3 - x}$

In Problems 9 to 12 evaluate the limit, if it exists, of the quality points function $Q(x)$.

9. $\lim\limits_{x \to 70} Q(x)$

10. $\lim\limits_{x \to 90} Q(x)$

11. $\lim\limits_{x \to 81} Q(x)$

12. $\lim\limits_{x \to 79} Q(x)$

In Problems 13 to 26 evaluate each limit, if it exists, by algebraic methods.

13. $\lim\limits_{x \to 2} (x^2 - 2x)$

14. $\lim\limits_{x \to 4} (x^2 - 4x)$

15. $\lim\limits_{x \to 1} (1 - x)$

16. $\lim\limits_{x \to 2} (2 - x)$

17. $\lim\limits_{x \to 2} \dfrac{x^2 - 2x}{x - 2}$

18. $\lim\limits_{x \to 4} \dfrac{x^2 - 4x}{x - 4}$

19. $\lim\limits_{x \to 1} \dfrac{1 - x}{x^2 - 2x + 1}$

20. $\lim\limits_{x \to 2} \dfrac{2 - x}{x^2 - 4x + 4}$

21. $\lim\limits_{x \to -2} \dfrac{x + 2}{x^2 + 4x + 4}$

22. $\lim\limits_{x \to -1} \dfrac{x + 1}{x^2 + 2x + 1}$

23. $\lim\limits_{x \to 1} \dfrac{\sqrt{x} - 1}{x - 1}$

24. $\lim\limits_{x \to 4} \dfrac{\sqrt{x} - 2}{x - 2}$

25. $\lim\limits_{x \to 9} \dfrac{3 - \sqrt{x}}{9 - x}$

26. $\lim\limits_{x \to 1} \dfrac{1 - \sqrt{x}}{1 - x}$

In Problems 27 to 32 evaluate $\lim\limits_{x \to \infty} f(x)$ for each function.

27. $f(x) = \dfrac{5}{x^2}$

28. $f(x) = \dfrac{2}{5x^3}$

29. $f(x) = \dfrac{1 - 4x}{2 + x}$

30. $f(x) = \dfrac{9 + 2x}{1 - 3x}$

31. $f(x) = \dfrac{2x^2 + 3x - 5}{6x^2 - 5}$

32. $f(x) = \dfrac{2x^2 + 12}{5x^3 - 6}$

In Problems 33 to 36 compute each limit, if it exists, by constructing tables.

33. $\lim\limits_{x \to 1} f(x)$ for $f(x) = \dfrac{3x^3 - 3x^2}{x^2 - x}$

34. $\lim\limits_{x \to 1} f(x)$ for $f(x) = \dfrac{2x^3 - 2x^2}{x^2 - x}$

35. $\lim\limits_{x \to 2} f(x)$ for $f(x) = \dfrac{x - 2}{x + 2}$

36. $\lim\limits_{x \to 3} f(x)$ for $f(x) = \dfrac{3 - x}{3 + x}$

In Problems 37 to 40 compute each limit, if it exists, by the geometric approach.

37. $\lim\limits_{x \to 0} |x|$

38. $\lim\limits_{x \to 0} \dfrac{1}{x^2}$

39. $\lim\limits_{x \to 0} f(x)$, where $f(x) = \begin{cases} -x & \text{if } x < 0 \\ 3x & \text{if } x \geq 0 \end{cases}$

40. $\lim\limits_{x \to 1} f(x)$, where $f(x) = \begin{cases} x + 1 & \text{if } x < 1 \\ x & \text{if } x \geq 1 \end{cases}$

In Problems 41 to 52 compute each limit, if it exists, by any method.

41. $\lim\limits_{x \to 1} \dfrac{x^2 - 9}{x - 3}$

42. $\lim\limits_{x \to 1} \dfrac{x^2 - 4}{x - 2}$

43. $\lim\limits_{x \to -2} \dfrac{x^2 + 3x + 2}{x^2 - 4}$

44. $\lim\limits_{x \to 1} \dfrac{x^2 + 3x + 2}{x^2 - 1}$

45. $\lim\limits_{x \to 3} \dfrac{x^3 - 9x}{x - 3}$

46. $\lim\limits_{x \to -3} \dfrac{x^3 - 9x}{x + 3}$

47. $\lim\limits_{x \to 2} \dfrac{x^3 - 4x}{x - 2}$

48. $\lim\limits_{x \to 0} \dfrac{x^2 - 3x}{x}$

49. $\lim\limits_{x \to -2} \dfrac{x^2 + 4x + 4}{x + 2}$

50. $\lim\limits_{x \to -1} \dfrac{x^2 + 2x + 1}{x + 1}$

51. $\lim\limits_{x \to 1} \dfrac{x - 1}{\sqrt{x} - 1}$

52. $\lim\limits_{x \to 4} \dfrac{x - 2}{\sqrt{x} - 2}$

In Problems 53 to 56 evaluate $\lim\limits_{x \to -\infty} f(x)$ for each function.

53. $f(x) = \dfrac{4x + 3}{2x - 6}$

54. $f(x) = \dfrac{x - 3}{5x + 1}$

55. $f(x) = \dfrac{3x^2 - 4x - 5}{x^2 + 5}$

56. $f(x) = \dfrac{7x^2 + 12}{3x^2 - 6}$

In Problems 57 to 60 evaluate $\lim\limits_{x \to \infty} f(x)$ for each function.

57. $f(x) = \dfrac{2 - 3x}{x^3}$

58. $f(x) = \dfrac{x^3}{4 + 5x}$

59. $f(x) = \dfrac{x^2 + 6}{x - 1}$

60. $f(x) = \dfrac{9 + 2x}{x^2 - 3}$

Problems 61 to 64 depict situations where the limit of a function is determined by examining values close to the limiting value.

61. A small midwestern newspaper services 20 small towns within a 20-mile radius of the city where the newspaper is published. A reporter who is charged with recording the daily temperature table notices that 19 of the towns report a noon temperature of 90°F, but the remaining town is missing. What number do you suppose the reporter filled in?

62. The table below lists the population of a bacterium culture, where t is measured in hours and $P(t)$ in millions of bacteria. The measurement is missing at $t = 5$. What value should be inserted?

t	4.6	4.7	4.8	4.9	5	5.1	5.2
$P(t)$	12	14	18	26		74	138

63. A rocket is designed to explode precisely 5 seconds after it has been fired into the air. An instrument measures the distance the rocket travels up to the point of the explosion, but it cannot measure the distance

at the point of explosion. The table gives the distance $s(t)$ traveled in miles after t seconds. How far did the rocket travel before the explosion?

t	4.9	4.99	4.999
$s(t)$	199	199.9	199.99

64. A laboratory tests a new adhesive to see how many pounds it will hold for 1 minute. The exact weight cannot be measured because if the weight falls, it was too heavy, and if the weight does not fall, it was too light. The laboratory compiled the data in the table. What is the maximum weight that the adhesive can hold?

	Held			Did not hold		
Weight (lb)	198	199	199.9	200.1	201	202
Time (min)	60	60	60	59.8	59.5	59

Referenced Exercise Set 12–2

A study by K. Rohrbach proposed a new accounting strategy for auditing called *monetary unit acceptance sampling*.* Figure 5 describes the so-called power functions of the four different statistical distributions which he used. (The nominal power function is based on the binomial distribution.) Problems 1 and 2 refer to Figure 5.

1. Complete the following table for the gamma $- 100$ power function $f(x)$ and use it to evaluate $\lim\limits_{x \to .08} f(x)$.

x	.05	.06	.07	.08
$f(x)$				

2. Repeat Problem 1 for the uniform power function.

3. In an article concerning various demands and price structures of finished goods inventories, F. Arcelus and G. Srinivasan† introduced Figure 6 to show how

*Kermit J. Rohrbach, "Monetary Unit Acceptance Sampling," *Journal of Accounting Research,* Spring 1986, 127–150.

†F. J. Arcelus and G. Srinivasan, "Inventory Policies Under Various Optimizing Criteria and Variable Markup Rates," *Management Science,* June 1987, pp. 756–762.

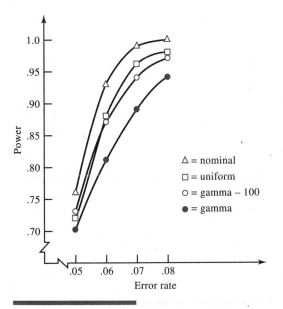

Power (y-axis), Error rate (x-axis)

$\triangle$ = nominal
$\square$ = uniform
$\circ$ = gamma – 100
$\bullet$ = gamma

FIGURE 5

increases in the price elasticity of demand G affect the markup rate m. Use Figure 6 to evaluate $\lim\limits_{-G\to\infty} f(x)$

for each of these functions.
(a) $f(x) = m_1(x)$
(b) $f(x) = m_2(x)$

4. This problem illustrates limits at infinity using concepts from probability. It was suggested by Jerry Johnson of Oklahoma State University.‡
   (a) Suppose you have a huge sock drawer which holds 100 blue socks and 100 brown socks. One morning you are half asleep while dressing for school, and you blindly reach into the drawer to pick out a pair of socks. What is the probability that they match? (*Hint:* Remove one sock. What fraction of the remaining socks match in color?)
   (b) Repeat part (a) for a drawer with n socks of each color.
   (c) Repeat part (a) for a drawer with infinitely many socks of each color.

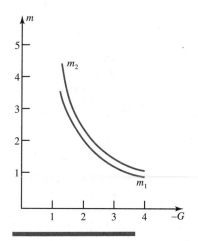

FIGURE 6

Cumulative Exercise Set 12–2

1. Compute the average rate of change for $f(x) = x^2$ from $x = -2$ to $x = -1.99$.

2. Use the method shown in Example 3 of Section 12–1 to compute the instantaneous rate of change of $f(x) = x^2$ at $x = -2$.

3. Use the result of Problem 2 to find the slope-intercept form of the line tangent to the graph of $f(x) = x^2$ at $x = -2$.

4. The height of a ball thrown into the air is given by $d(t) = -16t^2 + 64t + 96$, where t is the time in seconds after the ball is released. What is the instantaneous rate of change of the ball one second after it is released?

5. Compute the limit by constructing a table.

$$\lim_{x\to -1} f(x) \text{ for } f(x) = \frac{2x^2 + 4x + 2}{x + 1}$$

In Problems 6 and 7 evaluate each limit, if it exists, by algebraic methods.

6. $\lim\limits_{x\to -1} \dfrac{x + 1}{2x^2 + 4x + 2}$ 7. $\lim\limits_{x\to -3} \dfrac{(x + 3)^{99}}{98(x + 3)^{98}}$

8. The table below lists the population of a city (in thousands) in the 20th century. If this trend continues, what will the population be in the 21st century?

Year	1900	1910	1920	1930	1940	1950	1960	1970	1980	1990
Pop.	800	480	320	240	200	180	170	165	162.5	161.25

‡Jerry A. Johnson, "An Illustration of Limits," *Mathematics Teacher*, December 1986, pp. 722–723.

12–3 Definition of the Derivative

Section 12–1 provided the motivation and Section 12–2 laid the foundation for the definition of the derivative. Here we define the derivative, show how to compute it, and apply it to graphing techniques and rates of change. Then we discuss continuous and differentiable functions.

Procedure

The methods of computing the instantaneous rate of change of a function $f(x)$ at a value $x = a$ and the line tangent to the graph of $f(x)$ at the point $(a, f(a))$ involve the same procedure: form a difference quotient and evaluate a limit as the denominator approaches 0. This motivates the following fundamental definition.

DEFINITION

The **derivative of a function** $f(x)$ is a function, denoted by $f'(x)$, defined by

$$f'(x) = \lim_{h \to 0} \frac{f(x + h) - f(x)}{h}$$

The domain of $f'(x)$ consists of all x in the domain of $f(x)$ for which the limit exists.

The notation $f'(x)$ is read "f prime of x." The method of computing $f'(x)$ is the same method that was introduced in Section 12–1 to compute the slope of the line tangent to the graph of $f(x)$. The crucial step in evaluating the limit is to perform algebraic simplifications that enable the denominator h to be canceled.

EXAMPLE 1

Problem

Compute $f'(x)$ for $f(x) = x^3 - 2x$.

Solution The difference quotient is

$$\frac{f(x + h) - f(x)}{h} = \frac{[(x + h)^3 - 2(x + h)] - (x^3 - 2x)}{h}$$

$$= \frac{x^3 + 3x^2h + 3xh^2 + h^3 - 2x - 2h - x^3 + 2x}{h}$$

$$= \frac{h(3x^2 + 3xh + h^2 - 2)}{h}$$

Cancel the common factor h.

$$\frac{f(x + h) - f(x)}{h} = 3x^2 + 3xh + h^2 - 2$$

Then let the values of h approach 0:

$$f'(x) = \lim_{h \to 0} (3x^2 + 3xh + h^2 - 2) = 3x^2 - 2$$

The display contains the important fact that the derivative of a function can be interpreted as a rate of change.

> The instantaneous rate of change of a function $f(x)$ at a value $x = a$ is the derivative $f'(a)$, provided that $f'(a)$ exists.

EXAMPLE 2

Problem

Find the instantaneous rate of change of

$$f(x) = \sqrt{x} \quad \text{at } x = 9.$$

Solution First find $f'(x)$. The difference-quotient is

$$\frac{f(x + h) - f(x)}{h} = \frac{\sqrt{x + h} - \sqrt{x}}{h}$$

The algebraic steps that enable h to be canceled are

$$\frac{\sqrt{x + h} - \sqrt{x}}{h} = \frac{(\sqrt{x + h} - \sqrt{x})}{h} \cdot \frac{(\sqrt{x + h} + \sqrt{x})}{(\sqrt{x + h} + \sqrt{x})}$$

$$= \frac{(x + h) - x}{h(\sqrt{x + h} + \sqrt{x})} = \frac{1}{\sqrt{x + h} + \sqrt{x}}$$

Then

$$f'(x) = \lim_{h \to 0} \frac{1}{\sqrt{x + h} + \sqrt{x}} = \frac{1}{\sqrt{x} + \sqrt{x}} = \frac{1}{2\sqrt{x}}$$

Therefore the instantaneous rate of change at $x = 9$ is

$$f'(9) = \frac{1}{2\sqrt{9}} = \frac{1}{6}$$

Tangents

In Section 12–1 the slope of the line tangent to the graph of a function $f(x)$ was shown to correspond to the instantaneous rate of change of $f(x)$. This means that if the derivative $f'(x)$ can be evaluated at $x = a$, then $f'(a)$ is the slope of the line tangent to the curve at that point. The equation of the tangent line can then be written because one point $(a, f(a))$ and the slope $f'(a)$ are known.

PT. SLOPE FORM

> The equation of the line tangent to the graph of a function $f(x)$ at the point $(a, f(a))$ is
>
> $$y - f(a) = f'(a)(x - a)$$
>
> provided that $f'(a)$ exists.

EXAMPLE 3

Problem

Find the slope-intercept form of the line that is tangent to the graph of $f(x) = 2/x$ at $x = -3$.

Solution The approach is to find $f'(x)$ and then to evaluate it at $x = -3$. First find the difference quotient:

$$\frac{f(x + h) - f(x)}{h} = \frac{\dfrac{2}{x + h} - \dfrac{2}{x}}{h}$$

$$= \frac{\dfrac{2x - 2(x + h)}{(x + h)x}}{h}$$

$$= \frac{2x - 2(x + h)}{(x + h)x} \cdot \frac{1}{h}$$

$$= \frac{-2h}{(x + h)xh} = \frac{-2}{(x + h)x}$$

Then find the limit:

$$f'(x) = \lim_{h \to 0} \frac{-2}{(x + h)x} = \frac{-2}{(x)x} = \frac{-2}{x^2}$$

Therefore $f'(-3) = \dfrac{-2}{(-3)^2} = \dfrac{-2}{9}$. The equation of the tangent line is

$$y - f(-3) = f'(-3)(x + 3)$$

$$y - \left(\frac{2}{-3}\right) = \left(\frac{-2}{9}\right)(x + 3)$$

$$y = \frac{-2}{9}x - \frac{4}{3}$$

Velocity

In Section 12–1 the instantaneous velocity of an object moving in a straight line was shown to correspond to the instantaneous rate of change. But the instantaneous rate of change is the derivative. Therefore the derivative of the distance function can also be interpreted as the velocity of a moving object at a given time.

> If $d(t)$ is the distance function of a moving object, then the velocity $v(t)$ is the derivative, provided that $d'(t)$ exists. In symbols, $v(t) = d'(t)$.

If a ball is thrown into the air, it will reach a high point before returning to earth. The ball's velocity is a positive number on the way up, and a negative number on the way down. At the instant when the ball reaches its zenith, the velocity is 0. This observation is crucial for solving problems like Example 4.

EXAMPLE 4

The height of a ball thrown from ground level into the air with an initial velocity of 128 feet per second is given by the equation

$$d(t) = -16t^2 + 128t$$

where d is measured in feet and t in seconds.

Problem

(a) When will the ball reach its maximum height? (b) What is the maximum height?

Solution (a) The velocity at any time t is given by

$$d'(t) = \lim_{h \to 0} \frac{d(t + h) - d(t)}{h}$$

$$= \lim_{h \to 0} \frac{[-16(t + h)^2 + 128(t + h)] - (-16t^2 + 128t)}{h}$$

$$= \lim_{h \to 0} \frac{-16t^2 - 32th - 16h^2 + 128t + 128h + 16t^2 - 128t}{h}$$

$$= \lim_{h \to 0} \frac{h(-32t - 16h + 128)}{h}$$

$$= \lim_{h \to 0} (-32t - 16h + 128)$$

$$= -32t + 128$$

The ball will reach its maximum height when its velocity is 0, so $d'(t) = 0$. Then $-32t + 128 = 0$, so $t = 4$. Therefore the ball will reach its maximum height after 4 seconds. (b) By part (a) the maximum height occurs at $t = 4$. Since $d(4) = -16 \cdot 4^2 + 128 \cdot 4 = 256$, the maximum height is 256 feet. Note that we substitute $t = 4$ into the formula for $d(t)$ to find $d(4)$. Don't be misled into substituting $t = 4$ into the formula for $d'(t)$ to find $d(4)$.

Marginal Cost

Marginal cost functions were introduced in Chapter 11. The **marginal cost** of producing an item is the rate of change in cost when that item is produced. It is an approximation to the change in cost for the next item. For instance, let the cost (in dollars) of producing x liters of distilled water be

$$C(x) = x^2 - 2x + 6 \quad 1 \le x \le 6$$

The additional cost of producing the third liter of distilled water is \$3, the difference between the cost of producing 3 liters and the cost of producing 2 liters, since $C(3) - C(2)/(3 - 2) = 9 - 6 = 3$. This can be seen in Figure 1 as the slope of the secant line from $(2, 6)$ to $(3, 9)$.

However, a liter is an arbitrary unit. If deciliters are used instead, the interval from 2 to 3 liters becomes 20 to 30 deciliters. Then the marginal cost of producing the third liter becomes $C(3) - C(2.9)/(3 - 2.9)$ because the interval runs from 29 to 30 deciliters. If centiliters are used, the marginal cost is $C(3) - C(2.99)/(3 - 2.99)$, while if milliliters are used, the marginal cost is $C(3) - C(2.999)/(3 - 2.999)$. This limit motivates the following definition.

DEFINITION

If $C(x)$ is the cost of producing x units of a product, the **marginal cost** is $C'(x)$. The marginal cost of producing the nth unit is $C'(n)$.

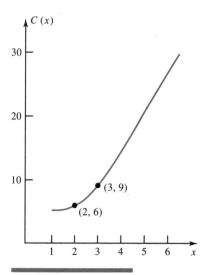

FIGURE 1

EXAMPLE 5

Problem

What is the marginal cost of producing the third unit if the cost function is $C(x) = x^2 - 2x + 6$?

Solution The aim is to compute $C'(x)$ and then to substitute $x = 3$.

$$C'(x) = \lim_{h \to 0} \frac{[(x + h)^2 - 2(x + h) + 6] - (x^2 - 2x + 6)}{h}$$

$$= \lim_{h \to 0} \frac{x^2 + 2xh + h^2 - 2x - 2h + 6 - x^2 + 2x - 6}{h}$$

$$= \lim_{h \to 0} \frac{2xh + h^2 - 2h}{h} = \lim_{h \to 0} \frac{h(2x + h - 2)}{h}$$

$$= \lim_{h \to 0} (2x + h - 2)$$

$$= 2x - 2$$

When $x = 3$, $C'(x) = 4$ so the marginal cost of producing the third item is $4.

EXERCISE SET 12–3

In Problems 1 to 10 compute the derivative of the function.

1. $f(x) = x^2 + 3$

2. $f(x) = x^2 - 5$

3. $f(x) = \sqrt{2x}$

4. $f(x) = -\sqrt{x}$

5. $f(x) = 1 - \sqrt{x}$

6. $f(x) = 1 + \sqrt{x}$

7. $f(x) = 1/x$

8. $f(x) = 3/x$

9. $f(x) = 1 + 1/x$

10. $f(x) = 1 - 1/x$

In Problems 11 to 14 find the equation of the tangent line to the function at the stated value of x.

11. $f(x) = x^2 - 2x + 6$ at $x = 0$

12. $f(x) = x^2 - 2x + 6$ at $x = 1$

13. $f(x) = x^2 + x - 5$ at $x = -1$

14. $f(x) = x^2 + x - 5$ at $x = 1$

In Problems 15 to 18 the height of a ball thrown from ground level into the air is given by $d(t)$, where d is measured in feet and t in seconds. Find (a) the amount of time it takes for the ball to reach its maximum height and (b) the maximum height.

15. $d(t) = -16t^2 + 192t$

16. $d(t) = -16t^2 + 48t$

17. $d(t) = 70t - 16t^2$

18. $d(t) = 96t - 16t^2$

In Problems 19 to 22 determine the marginal cost of producing the third item for the given cost function $C(x)$.

19. $C(x) = x^2 + x + 2$

20. $C(x) = x^2 - x + 5$

21. $C(x) = x^2 - 1$

22. $C(x) = x^2 + 4$

In Problems 23 to 26 compute the derivative of the function.

23. $f(x) = 1/\sqrt{x}$

24. $f(x) = x^3$

25. $f(x) = x^{-2}$

26. $f(x) = x^{-3}$

In Problems 27 to 32 find the equation of the tangent line to the function at the stated value of x.

27. $f(x) = 1/x$ at $x = 2$

28. $f(x) = 6/x$ at $x = -2$

29. $f(x) = 1/x^2$ at $x = -2$

30. $f(x) = \sqrt{x}$ at $x = 4$

31. $f(x) = 3x^2 - 5$ at $x = 0$

32. $f(x) = 4 - 2x^2$ at $x = 1$

In Problems 33 and 34 the height of a ball thrown from the top of a 64-foot building into the air is given by $d(t)$, where d is measured in feet and t in seconds. Find (a) the amount of time it takes for the ball to reach its maximum height and (b) the maximum height.

33. $d(t) = -16t^2 + 128t + 64$

34. $d(t) = -16t^2 + 32t + 64$

35. For how many seconds will the ball in Problem 15 be in the air?

36. For how many seconds will the ball in Problem 18 be in the air?

In Problems 37 and 38 let a, b, and c be real numbers. Find the derivatives of the stated function.

37. $f(x) = ax^2 + bx + c$

38. $f(x) = ax + b$

Referenced Exercise Set 12–3

In the context of income taxes, the *marginal tax rate* is the derivative of the tax function as a percent. Also, *tax brackets* are intervals of income where the derivative is constant. Problems 1 and 2 refer to the following tax function $T(x)$, where x represents a family's income in dollars.*

$T(x) =$

$$\begin{cases} .15x & \text{if } 0 \leq x \leq 29{,}750 \\ .28(x - 29{,}750) + 4463 & \text{if } 29{,}750 < x \leq 71{,}900 \\ .33(x - 71{,}900) + 16{,}265 & \text{if } 71{,}900 < x \leq 227{,}250 \\ .28(x - 227{,}250) + 67{,}531 & \text{if } x > 227{,}250 \end{cases}$$

1. What is the marginal tax rate for a family whose income is
   (a) $25,000   (b) $50,000
   (c) $100,000   (d) $500,000

2. What interval (or intervals) represents each of these tax brackets?
   (a) 15 percent   (b) 28 percent
   (c) 33 percent

The authoritative *Automobile Magazine* declared the Porsche 959 sports car to be the car of the future.† Problems 3 and 4 refer to Figure 2, which gives some indication of the Porsche's astounding power and technology by listing the units for power on the left and the units for torque on the right.

3. (a) At what rpm is the derivative of the power function equal to 0?

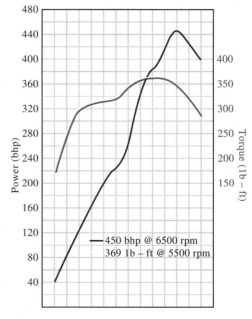

FIGURE 2

(b) What is the power at that point?

4. (a) At what rpm is the derivative of the torque function equal to 0?
   (b) What is the torque at that point?

5. Figure 3 shows the rate of change of certain economic indicators for 1987. Let $f(x)$ denote the function defined by the graph.
   (a) Interpret the meaning of the expression "percent change" in mathematical terms.
   (b) What month had the highest value of the derivative $f'(x)$?
   (c) What month had the lowest value of the derivative $f'(x)$?

Cumulative Exercise Set 12–3

1. Compute the instantaneous rate of change of $f(x) = x^2 + 1$ at $x = 2$.

2. Compute the slope-intercept form of the tangent line to the graph of $f(x) = 2x^2 - 1$ at $x = 1$.

*Alan Murray, "Individuals' Top Rate Would Plunge to 28 Percent; Tax-Break Curbs Offset Benefits to Wealthy," *Wall Street Journal*, August 18, 1986, p. 6.

†Mel Nichols, "Review: Porsche 959," *Automobile Magazine*, November 1987, pp. 60–71.

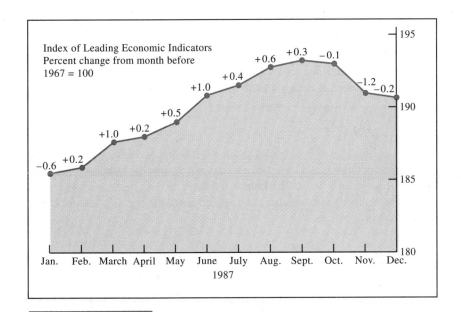

Index of Leading Economic Indicators
Percent change from month before
1967 = 100

FIGURE 3 *Source:* Commerce Department.

3. Compute the average rate of change of $f(x) = x^2 + 1$ from 2 to 2.1 and from 2 to 2.01.

In Problems 4 to 7 find $\lim\limits_{x \to a} f(x)$ for the given function $f(x)$ and $x = a$.

4. $f(x) = \dfrac{x^2 - 16}{x - 4}$, $a = 4$

5. $f(x) = \dfrac{x^2 - 16}{x + 4}$, $a = 4$

6. $f(x) = \dfrac{x^3 - 4x}{2 - x}$, $a = 2$

7. $f(x) = \dfrac{x - 4}{x - 1}$, $a = 1$

8. Compute the slope-intercept form of the tangent line to the graph of $f(x) = x^2 + x - 1$ at $x = 1$.

9. The height of a ball thrown from ground level into the air is given by $d(t) = -16t^2 + 64t$, where d is measured in feet and t in seconds. Find the amount of time it takes for the ball to reach its maximum height.

In Problems 10 to 12 compute the derivative of the function.

10. $f(x) = x^3$

11. $f(x) = x^{-3}$

12. $f(x) = \dfrac{4}{x}$

12–4 Continuity

This section introduces continuous functions by geometric and algebraic means. It also considers differentiable functions and relates them to continuous functions.

Geometric Approach

From a geometrical standpoint, a *function* is *continuous* if a person's pencil need not leave the paper when the graph of the function is being sketched. Consider the graphs of the six functions shown in Figure 1. The graph in part (a) can be sketched without the pencil leaving the paper, so the function $f(x) = 5 - x^2$ is continuous. Similarly, $f(x) = x^3$ is a continuous function as shown in part (b).

However, the other four graphs require the pencil to be lifted from the paper, so they are discontinuous at each value of x where the pencil is lifted. For instance, the function

$$f(x) = \frac{(x - 1)(x + 1)}{x - 1}$$

in part (c) is discontinuous at $x = 1$ because the pencil must be lifted to jump over the hole in the graph at $x = 1$. The hole is indicated by an open circle. Points of discontinuity can also occur at gaps like those in parts (d) and (e), and at jumps like those in part (f).

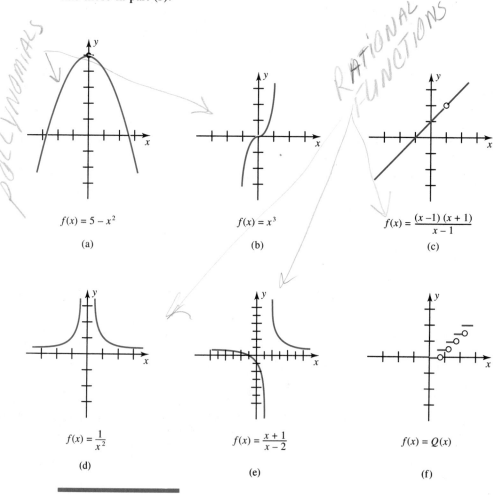

FIGURE 1

Example 1 examines the continuity of functions defined by different expressions on different intervals.

EXAMPLE 1

Problem

Determine whether $f(x)$ is a continuous function, where

$$f(x) = \begin{cases} 5 - x^2 & \text{if } x < 2 \\ x - 1 & \text{if } x \geq 2 \end{cases}$$

Solution The graph of $f(x)$ is sketched in Figure 2. It consists of two sides; the left side is part of a parabola and the right side is part of a line. Because the two sides meet at the point (2, 1) the graph can be sketched without the pencil leaving the paper. Therefore $f(x)$ is a continuous function.

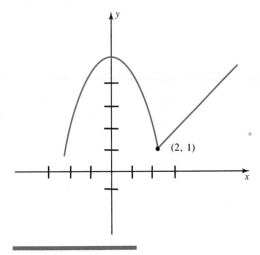

FIGURE 2

Algebraic Approach

The geometric approach to continuous functions leads to the following algebraic definition.

DEFINITION

A function $f(x)$ is **continuous at a value** $x = a$ if

1. $f(x)$ is defined at $x = a$,
2. $\lim\limits_{x \to a} f(x)$ exists, and
3. $\lim\limits_{x \to a} f(x) = f(a)$.

If any one of these conditions is not satisfied then $f(x)$ is a **discontinuous function** at $x = a$. $f(x)$ is **continuous on an interval** if $f(x)$ is continuous at every point in that interval. $f(x)$ is a *continuous function* if $f(x)$ is continuous at every real number.

Although the points of discontinuity of a function can easily be found from its graph, general statements can be made about some special kinds of functions without appealing to their graphs.

1. Every polynomial function is continuous.
2. Every rational function $\dfrac{f(x)}{g(x)}$ is continuous at all values of x for which $g(x) \neq 0$.

Parts (a) and (b) in Figure 1 illustrate the first statement, while parts (c), (d), and (e) illustrate the second statement. Example 2 shows how to determine points of continuity of a function without sketching its graph.

EXAMPLE 2

Problem

Determine all points where the function $f(x)$ is discontinuous:

$$f(x) = \frac{x + 2}{x^2 + 2x}$$

Solution $f(x)$ is a rational function so its continuity can be determined without sketching the graph. According to the second statement in the display, the points of discontinuity occur where $x^2 + 2x = 0$. Since $x^2 + 2x = x(x + 2)$, $f(x)$ is discontinuous at $x = 0$ and at $x = -2$.

Notice that the discontinuity of the rational function $f(x)$ in Example 2 was determined without sketching its graph. The next example shows how both the geometric and algebraic approaches can be utilized in certain cases.

EXAMPLE 3

Problem

Determine where the function $f(x) = \dfrac{x}{|x|}$ is continuous.

Solution The graph of $f(x)$, sketched in Figure 3, shows that $f(x)$ is discontinuous at $x = 0$, and is continuous at every other value of x.

Another way to proceed is to write $f(x)$ as a function defined in different ways on different intervals. Since the absolute value function $|x|$ is equal to x for

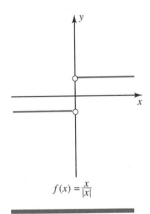

$$f(x) = \frac{x}{|x|}$$

FIGURE 3

$x > 0$, the function $f(x)$ is equal to x/x for $x > 0$. Similarly $|x|$ is equal to $-x$ for $x < 0$, so $f(x)$ is equal to $-x/x$ for $x < 0$. Therefore

$$f(x) = \begin{cases} 1 & \text{if } x > 0 \\ -1 & \text{if } x < 0 \end{cases}$$

Clearly $f(x)$ is continuous for all $x < 0$ and for all $x > 0$. However, when x is negative, $f(x) = -1$, but when x is positive, $f(x) = 1$. So there is a jump from one side of $x = 0$ to the other. This means that $f(x)$ is discontinuous at $x = 0$; $f(x)$ is continuous at all other values of x.

Differentiable Functions

A **function** $f(x)$ is **differentiable** at a value $x = a$ if the derivative exists at $x = a$, that is, if $f'(a)$ exists. The function is **differentiable on an interval** if $f'(x)$ exists at all numbers in that interval, and $f(x)$ is differentiable if $f'(x)$ exists at all real numbers.

Upon first glance the definition of a differentiable function might not seem to carry any geometric connotation. However, the graph of a differentiable function can be viewed as a smooth curve. Historically there was no need to distinguish between continuous functions and differentiable functions until the difference became vital in applications during the first part of the 19th century, when the following fact was proved.

If a function is differentiable then it is also continuous.

The connection between differentiable and continuous functions helps us list some ways in which a function $f(x)$ can fail to be differentiable, because if $f(x)$ is

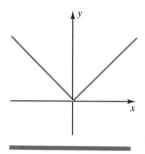

FIGURE 4

not continuous at a point then it is not differentiable there either. Figure 1 thus shows four ways in which functions can fail to be differentiable.

Figure 2 shows another example of a way in which a function can fail to be differentiable. This function also reveals that the converse of the statement above does not hold for all functions, a result worthy of being singled out.

A continuous function need not be differentiable.

The classic example illustrating this fact is the absolute value function $f(x) = |x|$, which is continuous at all values of x but which is not differentiable at $x = 0$. Figure 4 shows that the graph can be sketched without the pencil leaving the paper, so it is continuous, but it is not a smooth curve at $x = 0$ so it is not differentiable there.

Example 4 illustrates one other way in which a function can fail to be differentiable. It uses the criterion of smoothness for determining where a function is differentiable.

EXAMPLE 4

Problem

Determine whether the function $f(x) = \sqrt[3]{x^2}$ is differentiable.

Solution Use the plotting method to sketch the graph of $f(x)$ by first squaring the value of x and then taking the cube root of that value. Thus $f(-8)$ is obtained by squaring -8 to get 64, then taking the cube root of 64 to get 4.

x	-8	-1	0	1	8
$f(x)$	4	1	0	1	4

The graph of $f(x)$ is sketched in Figure 5. It is not smooth at $x = 0$ so $f(x)$ is not differentiable there. However, the graph is smooth at all other values of x, so $f(x)$ is differentiable at all points except $x = 0$.

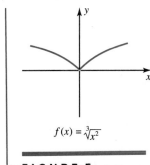

$$f(x) = \sqrt[3]{x^2}$$

FIGURE 5

Example 5 used a geometric way to test for differentiability at the value $x = 0$. There is an algebraic test for this too, but it involves one-sided derivatives, which are not considered in this book.

EXERCISE SET 12–4

The figures in Problems 1 to 10 show the graphs of ten functions. Determine whether each function is continuous.

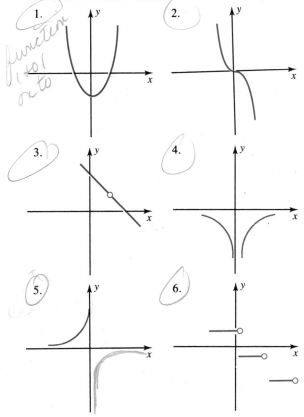

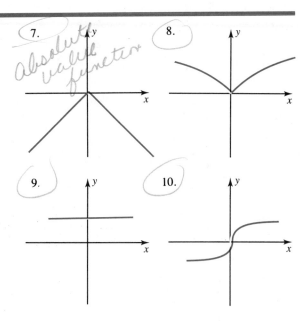

In Problems 11 to 14 determine whether the function $f(x)$ is continuous by sketching its graph.

11. $f(x) = \begin{cases} x^2 & \text{if } x \geq 0 \\ -x & \text{if } x < 0 \end{cases}$

12. $f(x) = \begin{cases} x^3 & \text{if } x \geq 0 \\ -x^2 & \text{if } x < 0 \end{cases}$

13. $f(x) = \begin{cases} x^2 - 5 & \text{if } x \geq 2 \\ x - 1 & \text{if } x < 2 \end{cases}$

14. $f(x) = \begin{cases} x + 1 & \text{if } x \geq 1 \\ -x & \text{if } x < 1 \end{cases}$

In Problems 15 to 24 determine all points where the function $f(x)$ is discontinuous.

15. $f(x) = \dfrac{x - 1}{x^2 - x}$

16. $f(x) = \dfrac{2 - x}{x^2 - 2x}$

17. $f(x) = \dfrac{1}{x^2 - 16}$

18. $f(x) = \dfrac{1}{(x - 2)(x - 3)}$

19. $f(x) = \dfrac{x + 4}{(x + 4)(x - 2)}$

20. $f(x) = \dfrac{x + 2}{x^2 - x - 6}$

21. $f(x) = \dfrac{5}{|x|}$

22. $f(x) = \dfrac{|x|}{x}$

23. $f(x) = \dfrac{x - 3}{x^2 - 3x}$

24. $f(x) = \dfrac{x^2 + 3x}{x + 3}$

In Problems 25 to 34 determine whether the function whose graph is sketched in the stated problem is differentiable.

25. Problem 1   26. Problem 2

27. Problem 3   28. Problem 4

29. Problem 5   30. Problem 6

31. Problem 7   32. Problem 8

33. Problem 9   34. Problem 10

In Problems 35 to 40 determine whether the function $f(x)$ is differentiable by sketching its graph.

35. $f(x) = \sqrt[3]{(1 - x)^2}$

36. $f(x) = \sqrt[3]{(-x)^2}$

37. $f(x) = \sqrt[3]{x^2 - 2x + 1}$

38. $f(x) = \sqrt[3]{4 - 4x + x^2}$

39. $f(x) = 1 - |x|$

40. $f(x) = |1 - x|$

In Problems 41 to 44 determine if the given function is continuous at $x = 2$.

41. $f(x) = x - 2$

42. $f(x) = x^2 - 2$

43. $f(x) = \dfrac{1}{x - 2}$

44. $f(x) = \dfrac{x - 2}{x^2 - 4}$

In Problems 45 to 48 draw the graph of the function and determine whether it is continuous or differentiable.

45. $f(x) = \sqrt{|x|}$

46. $f(x) = \begin{cases} x + 6 & \text{if } x \geq 3 \\ x^2 & \text{if } x < 3 \end{cases}$

47. $f(x) = \dfrac{x^2 - 1}{x - 1}$

48. $f(x) = \begin{cases} x^2 & \text{if } x \leq 0 \\ -x^2 & \text{if } x > 0 \end{cases}$

In Problems 49 to 54 determine whether $f(x)$ is (a) continuous at $x = 2$ and (b) differentiable at $x = 2$, from the given figure.

49.
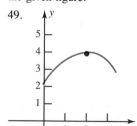

50.
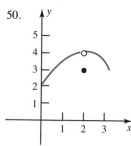

51.
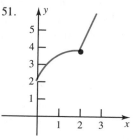

52.
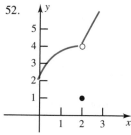

53.
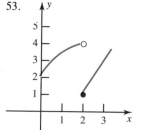

54.
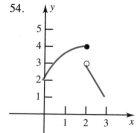

Cumulative Exercise Set 12–4

1. What is the instantaneous rate of change of $f(x) = 1 - x$ at $x = -1$?

2. Find the slope-intercept form of the line tangent to the graph of $f(x) = 1 - x^2$ at $x = -1$.

In Problems 3 and 4 evaluate each limit, if it exists.

3. $\displaystyle\lim_{x \to 1} \dfrac{x^2 - 3x + 2}{x^2 + x - 2}$

4. $\displaystyle\lim_{x \to -1}\left(1 + \dfrac{1}{x}\right)$

In Problems 5 and 6 compute the derivative of the function.

5. $f(x) = 3x^2$

6. $f(x) = x^3$

In Problems 7 and 8 find the equation of the line tangent to the graph of $f(x)$ at the stated value of x.

7. $f(x) = 3x^2$ at $x = -1$

8. $f(x) = x^3$ at $x = -1$

9. Determine whether the function whose graph is shown is continuous or differentiable.

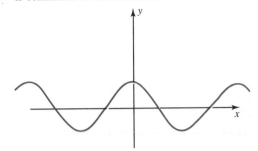

10. Determine whether the function $f(x)$ is continuous by sketching its graph.

$$f(x) = \begin{cases} 2x & \text{if } x \geq 0 \\ x & \text{if } x < 0 \end{cases}$$

11. Determine all points where the function $f(x) = 1 - \dfrac{1}{x}$ is discontinuous.

12. Determine whether the function $f(x) = 1 + \sqrt[3]{x^2}$ is differentiable.

CHAPTER REVIEW

Key Terms

12–1   Rate of Change and Tangents
Velocity
Rate of Change
Average Rate of Change
Instantaneous Rate of Change
Secant Line

Tangent Line
Slope of a Function
Difference Quotient
Demand
Marginal Demand

12–2   Limits
Limit
Limit Does Not Exist
Decreases Without Bound

Increases Without Bound
Limit at Infinity
Horizontal Asymptote

12–3   Definition of the Derivative
Derivative of a Function
Marginal Cost

12–4   Continuity
Continuous Function
Continuous at a Value
Discontinuous Function

Differentiable Function
Differentiable at a Value
Continuous on an Interval

Summary of Important Concepts

Formula

$$f'(x) = \lim_{h \to 0} \frac{f(x + h) - f(x)}{h}$$

Interpretations of the derivative

1. Instantaneous rate of change.
2. Tangent line.
3. Velocity of a moving object.

Steps in computing the derivative of $f(x)$

1. Form the difference quotient $\dfrac{f(x + h) - f(x)}{h}$.
2. Perform algebraic manipulations to factor h from the numerator of the difference quotient.
3. Cancel the h factors.
4. Take the limit as h approaches 0.

REVIEW PROBLEMS

1. Compute the average rate of change of the function $f(x) = x^2 + 1$ over the stated intervals.
   (a) from 1 to 2   (b) from 1 to 1.1

2. Compute the instantaneous rate of change of the function $f(x) = 2x^2 + 1$ at $x = -1$.

3. Find the slope of the tangent line to the graph of the function $f(x) = x^2 - x + 4$ at an arbitrary value of x.

4. Compute the limit by constructing a table.

$$\lim_{x \to 3} f(x) \quad \text{for } f(x) = \frac{x^2 - x - 6}{3 - x}$$

In Problems 5 and 6 evaluate the limit, if it exists, by algebraic methods.

5. $\displaystyle\lim_{x \to 1} \frac{2x^2 - 2x}{x - 1}$

6. $\displaystyle\lim_{x \to 4} \frac{2 - \sqrt{x}}{4 - x}$

In Problems 7 to 10 compute each limit, if it exists, by any method.

7. $\displaystyle\lim_{x \to 1} \frac{x^2 + 2x - 3}{x^2 + x - 2}$

8. $\displaystyle\lim_{x \to 0} \frac{1}{x}$

9. $\displaystyle\lim_{x \to 0} x^2 + x$

10. $\displaystyle\lim_{x \to 1} \frac{x - 1}{(x - 1)^2}$

In Problems 11 and 12 compute the derivative of the function from the definition of the derivative.

11. $f(x) = x^2 - 3x - 7$

12. $f(x) = \dfrac{1}{x - 2}$

13. Find the equation of the tangent line to the function $f(x) = x^2 - 3x - 7$ at the value $x = 4$.

14. The height of a ball thrown from the top of a 64-foot building into the air is given by $d(t) = -16t^2 + 16t + 64$, where d is measured in feet and t in seconds. Find
   (a) the amount of time it takes for the ball to reach its maximum height.
   (b) the maximum height.

PROGRAMMABLE CALCULATOR EXPERIMENTS

In Problems 1 to 4 write a program to evaluate the given function at values $h = 1, .1, .01, .001$ and $.0001$. Use these values to find $\displaystyle\lim_{h \to 0} f(h)$. Then graph $f(h)$ and approximate $f(0)$.

1. $f(h) = \dfrac{2h^2 - 5h + 1}{3h^2 + h - 1}$

2. $f(h) = \dfrac{3h^2 - 7h - 2}{5h^2 + 2h - 1}$

3. $f(h) = \dfrac{4h^3 - 2h - 3}{5h^3 + 3h - 4}$

4. $f(h) = \dfrac{h^4 - h^2 - 3}{h^4 - h^2 + 3h - 4}$

Differentiation Rules

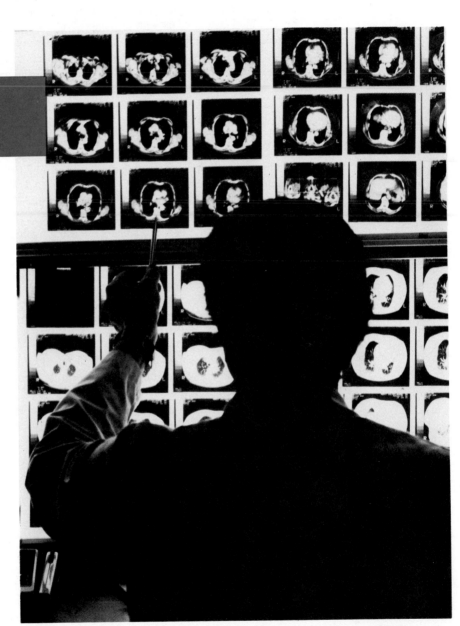

13

The case study discusses mathematical models and cancer therapy. (J. Nettis/H. Armstrong Roberts)

Chapter Overview

Calculus can be applied to a wide variety of real-world problems. Many examples will be given throughout the rest of the book. In each case the application will be described by a function. The problem is solved by analyzing the derivative of the function because the derivative gives information about the original function. Thus we will need to calculate the derivative of many types of functions. This chapter considers various rules for calculating derivatives.

The first section deals with elementary rules that govern functions that can be expressed as the sum of simpler functions or the product of a constant times a function. In the next section more complicated functions are handled, those that can be expressed as the product or quotient of other functions. In the third section the chain rule is presented. The chain rule governs composite functions. The fourth section deals with higher derivatives.

CASE STUDY PREVIEW

The primary treatments for cancer are surgery and radiotherapy. More than half of all tumors can be treated by radiation therapy. The treatment consists of several beams directed at various parts of the tumor. Each beam has a specific intensity depending on the mass of the tumor and the amount of healthy cells the beam will affect. There are many variables that affect the design of an individual patient's treatment. Medical researchers are using mathematical optimization techniques to maximize the amount of cancer cells killed while also minimizing the number of healthy cells affected. The case study shows how the derivative plays an essential role in balancing these two objectives.

13–1 A Few Elementary Rules

We have calculated the derivative of several functions using the definition of the derivative. The definition is cumbersome, so in this section some elementary differentiation rules are developed to simplify the task.

Marginal Product

Figure 1 shows a typical production function from economics. It measures the output of a firm at a given level of input. The input is usually labor or land. For example, the production function given by Samuelson* can be represented by the equation

$$f(x) = -0.16x^2 + 1.6x$$

Here $f(x)$ is the total product, or amount of output, of the firm with a labor force of size x, where x is measured in hundreds of workers and $f(x)$ is measured in units of $10,000. An important economic concept derived from the production function is **marginal product.** It is defined as the rate of increase of production per unit of

*Paul A. Samuelson, *Economics,* New York, McGraw-Hill Book Co., 1987.

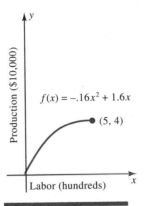

FIGURE 1   *Source:* Paul A. Samuelson, *Economics,* New York, McGraw-Hill Book Co., 1987.

input. Thus marginal product is the derivative of the production function. In this section we study how to find the derivative of functions like $f(x)$ above. At the end of the section we show how marginal product is a key ingredient that economists use to determine a fair wage.

Other Notation for the Derivative

There are several notations for expressing the derivative of $y = f(x)$. The three most common notations are

$$f'(x) \quad D_x f(x) \quad \frac{df}{dx}$$

The last notation can also be written as

$$\frac{dy}{dx} \quad \text{or} \quad \frac{d}{dx} f(x)$$

Each notation has an advantage in a particular situation so that no one notation is preferred over the others in every context. An advantage to the $f'(x)$ notation is that it is more convenient to express the derivative when evaluated at a particular value. For example, it is easier to write $f'(3)$ with this notation than the others.

Using the $\dfrac{dy}{dx}$ and $D_x f(x)$ notations, we express $f'(3)$ as

$$D_x f(x)\big|_{x=3} \quad \frac{d}{dx} f(x)\bigg|_{x=3}$$

The D_x notation is more convenient to use in formulas and when expressing the derivative of more complicated functions in terms of the derivative of simpler functions.

FIGURE 2

Constant Functions

Consider the constant function $f(x) = c$. Its graph is a horizontal straight line. (See Figure 2.) The slope of such a line is 0. Since the derivative measures the slope of the tangent line and the tangent to a line is the line itself, the **derivative of a constant** function is 0. This can be remembered as "the derivative of a constant is 0."

Constant Rule
If $f(x) = c$ for c a real number, then $f'(x) = 0$.
This is also expressed as

$$D_x(c) = 0 \qquad \text{for any constant } c$$

This rule is actually a theorem whose proof is straightforward.

Proof of Constant Rule

If $f(x) = c$ for some real number c, then

$$f'(x) = \lim_{h \to 0} \frac{f(x + h) - f(x)}{h} = \lim_{h \to 0} \frac{c - c}{h}$$

$$= \lim_{h \to 0} \frac{0}{h} = 0$$

Hence $f'(x) = 0$ for every real number x.

Power Rule

The following derivatives were calculated in the previous chapter:

$$D_x(x) = 1 \quad D_x(x^2) = 2x \quad D_x(x^3) = 3x^2 \quad D_x(x^{1/2}) = \tfrac{1}{2}x^{-1/2}$$

Each of these functions has the form $y = x^r$, where r is 1, 2, 3 and $\tfrac{1}{2}$, respectively. There is a pattern to the derivatives. It suggests an important rule called the *power rule*.

Power Rule

$$D_x(x^r) = rx^{r-1} \qquad \text{for any real number } r$$

The exercises indicate how to prove this theorem for r a positive integer. The proof for other real numbers is beyond the scope of this text. However, we emphasize that the power rule works for all real numbers as illustrated by Example 1.

EXAMPLE 1

Problem

Find $f'(x)$ for (a) $f(x) = x^{-5}$, (b) $f(x) = \dfrac{1}{\sqrt{x}}$.

Solution (a) Use the power rule with $r = -5$.

$$f'(x) = D_x[x^{-5}] = -5x^{-5-1} = -5x^{-6}$$

(b) Express the function as $f(x) = x^{-1/2}$. Then use the power rule with $r = -\tfrac{1}{2}$.

$$f'(x) = D_x[x^{-1/2}] = (-\tfrac{1}{2})x^{(-1/2)-1} = (-\tfrac{1}{2})x^{-3/2}$$

Constant Times a Function

From Section 12–3 we compute $D_x[4x^2] = 8x = 4 \cdot 2x = 4 \cdot D_x[x^2]$. The general rule that governs this derivative is "the **derivative of a constant times a function** is the constant times the derivative of the function."

Constant Times a Function Rule

$$D_x[cf(x)] = cD_x[f(x)] \qquad \text{for any real number } c$$

Proof of the Constant Times a Function Rule

If $g(x) = cf(x)$ for some constant c, then

$$g'(x) = \lim_{h \to 0} \frac{g(x + h) - g(x)}{h} = \lim_{h \to 0} \frac{cf(x + h) - cf(x)}{h}$$

$$= \lim_{h \to 0} \frac{c[f(x + h) - f(x)]}{h} = \lim_{h \to 0} c \frac{f(x + h) - f(x)}{h}$$

$$= c \lim_{h \to 0} \frac{f(x + h) - f(x)}{h}$$

$$= cf'(x)$$

EXAMPLE 2

Problem

Find $f'(x)$ for (a) $f(x) = -3x$, (b) $f(x) = 2x^5$.

Solution (a) Use the constant times a function rule with $c = -3$ and $f(x) = x$.

$$f'(x) = D_x[-3x] = -3D_x[x] = -3 \cdot 1 = -3$$

(b)

$$f'(x) = D_x[2x^5] = 2D_x[x^5] = 2 \cdot 5x^4 = 10x^4$$

Derivative of a Sum

The next formula states that "the **derivative of a sum** is the sum of the derivatives." The rule states that this is true for the sum of two functions, but it easily extends to the sum of any finite number of functions.

Sum Rule

$$D_x[f(x) + g(x)] = D_x[f(x)] + D_x[g(x)]$$

Since the difference of two functions, $f(x) - g(x)$, can be regarded as the sum, $f(x) + [-g(x)]$, the derivative of the difference of two functions is the difference of their derivatives.

The exercises will indicate how to prove the sum rule.

EXAMPLE 3

Problem

Find $f'(x)$ for (a) $f(x) = 4x - 3/x$, (b) $f(x) = 8 + 2x^{10} + 3x^{-5} - 7x^{5/3}$.

Solution (a) Recall that $3/x = 3x^{-1}$.

$$f'(x) = D_x(4x - 3x^{-1}) = D_x(4x) + D_x(-3x^{-1})$$

$$= 4D_x(x) - 3D_x(x^{-1}) = 4 \cdot 1 - 3(-1)x^{-2}$$

$$= 4 + \frac{3}{x^2}$$

(b)

$$f'(x) = D_x(8 + 2x^{10} + 3x^{-5} - 7x^{5/3})$$

$$= D_x(8) + D_x(2x^{10}) + D_x(3x^{-5}) + D_x(-7x^{5/3})$$

$$= 0 + 2 \cdot 10x^9 + 3(-5)x^{-6} - 7\left(\frac{5}{3}\right)x^{(5/3 - 1)}$$

$$= 20x^9 - 15x^{-6} - \left(\frac{35}{3}\right)x^{2/3}$$

The next example computes the equation of the tangent line to a curve.

EXAMPLE 4

Problem

Find an equation of the tangent line to the graph of the function $f(x) = x^2 - 5x + 3$ at $x = 1$, that is, at the point $(1, -1)$.

Solution The slope of the tangent line is measured by $f'(1)$. Since $f'(x) = 2x - 5$, $f'(1) = -3$. Using the point-slope form of the equation of the tangent line yields

$$y - (-1) = -3(x - 1)$$

Simplifying gives the equation $y = -3x + 2$.

Applications

The derivative measures instantaneous rate of change. If an object is dropped so that gravity is the only force acting on it, then the relationship between the distance traveled, d, and time, t, is $d(t) = -16t^2$, where d is measured in feet and t in seconds. A negative sign is in the expression because the object is falling and hence is moving in the negative direction. The average velocity is the average rate of change of $d(t)$ with respect to t. The instantaneous velocity measures how fast the object is falling at that instant. It is measured by $d'(t)$. Thus the instantaneous velocity is $v(t) = d'(t) = -32t$. Similarly the instantaneous acceleration of the object is the instantaneous rate of change of its velocity with respect to t. Thus the object's instantaneous acceleration is $a(t) = v'(t) = -32$.

EXAMPLE 5

An object is thrown upward and its position is given by $d(t) = -16t^2 + 64t$.

Problem

Find (a) the velocity and acceleration of the object, (b) the time at which it reaches its maximum height, and (c) how many feet it travels upward.

Solution (a) The velocity is $v(t) = d'(t) = -32t + 64$ and the acceleration is $a(t) = v'(t) = -32$.

(b) The object will reach its maximum height when $v(t) = 0$ because it stops at that instant. Letting $v(t) = 0$ yields $-32t - 64 = 0$, which implies $t = 2$ seconds.

(c) The object travels upward until it stops, at which time it starts downward. Thus it travels upward from $t = 0$ to $t = 2$ seconds. Hence the object travels $d(2)$ feet upward.

$$d(2) = -16(2)^2 + 64(2) = 64$$

The object travels 64 feet upward.

Marginal Product Revisited

The production function from Samuelson given at the beginning of the section is

$$f(x) = -0.16x^2 + 1.6x$$

The marginal product of the firm is the derivative $f'(x)$.

$$f'(x) = -0.32x + 1.6$$

The marginal product measures the rate of change of output as input changes. It makes no sense to let x take on values less than 0, and it is assumed that the largest number of labor units available to the firm is $x = 5$. This function is valid only for values of x in the interval $[0, 5]$. It is a quadratic function whose graph is given in Figure 1. It is typical of production functions because as x increases $f'(x)$ decreases. For example, $f'(0) = 1.6$ and $f'(1) = 1.28$. Another way to see this is to note that the marginal product is a linear function with a negative slope. Thus as x increases $f'(x)$ decreases. Economists refer to this as the "law of diminishing returns." It means that as more labor is used, more output is generated, but the rate of increase of output, the marginal product, decreases.

Economists use marginal product to determine fair wages. The fundamental principle is that *wages = marginal product*. This assumption is derived from the fact that if an extra worker were hired, the worker would increase the firm's productivity only by an amount equal to that worker's marginal product. For instance, consider a particular segment of the work force, so that all of these workers are paid the same wage. If there are 100 workers, thus $x = 1$, then the last one hired, the 100th worker hired, would increase the firm's productivity by the marginal product $f'(1)$. Therefore, in theory, all workers would receive the same fair wage, $f'(1)$, the marginal product of the last worker hired.

EXAMPLE 6

Consider the firm with the production function

$$f(x) = -0.16x^2 + 1.6x$$

Suppose the firm has 100 employees with the same wage.

Problem

Find the theoretical fair wage of the employees.

Solution The theoretical fair wage is the marginal product at $x = 1$ (hundred employees). It is $f'(1)$ where $f'(x) = -0.32x + 1.6$. Thus $f'(1) = -0.32(1) + 1.6 = 1.28$. Each worker is paid $12,800.

Are workers always paid their marginal product? Economist Robert H. Frank answers this question in an article appearing in the *American Economic Review*. His conclusion is that some workers, such as salespersons, do command wages equal to their marginal product, but others, such as college professors, are paid less. Referenced Exercise 3 deals with some of the data from this study.

EXERCISE SET 13–1

In Problems 1 to 26 find $f'(x)$.

1. $f(x) = x^3 + 5$

2. $f(x) = x^6 - 2$

3. $f(x) = 2x^4 - 7$

4. $f(x) = 3x^5 + 1$

5. $f(x) = 5x^{-3} + 5x$

6. $f(x) = 4x^{-5} - x^3 + 1$

7. $f(x) = 2x^{1/3} + 3x^{1/2} - 4x$

8. $f(x) = 7x^{1/5} - x^{2/5} + 2x$

9. $f(x) = 2x^{-1/3} + 7x^{-1/4} - 5$

10. $f(x) = 6x^{-1/6} - 3x^{-3/7} - 6x$

11. $f(x) = x^{0.3} + 3x^{0.4}$

12. $f(x) = 1.2x^{-1.5} - 3.1x^{-2.2}$

13. $f(x) = 1/x^3 + 3x^4$

14. $f(x) = 4x^{0.5/5} - 3x^{2.2/7}$

15. $f(x) = 1/x^{1.5} + 4/x^{1/4} + 4x^{2.4}$

16. $f(x) = 4/x^{2.5} - 2/x^{2/5} - 2x^{2/5}$

17. $f(x) = 2/x^{-3.5} + 4/x^{1.4} + 4x^{-1.3}$

18. $f(x) = 1/x^{-2.5} - 2x^{-3/5} - x^{-2.7/3}$

19. $f(x) = 0.25/x^{-0.5} + 0.48/x^{0.4}$

20. $f(x) = 1.5x^{-0.5} - 2.75x^{-0.05}$

21. $f(x) = 2\sqrt{x} - 2x$

22. $f(x) = \sqrt{x}/9 + 9/x$

23. $f(x) = 5\sqrt{x} - 2/\sqrt{x}$

24. $f(x) = \sqrt{9/x} + 2/\sqrt{7x}$

25. $f(x) = 18/\sqrt{3x} + 2/\sqrt[3]{4x}$

26. $f(x) = 8/\sqrt[4]{3x} + \frac{2}{3}\sqrt[4]{x}$

In Problems 27 to 30 find the equation of the tangent line to $y = f(x)$ at the given value of x.

27. $f(x) = x^2 + 2x$ at $x = 1$

28. $f(x) = x^3 - 4x$ at $x = 2$

29. $f(x) = x^{-2} + 2x^2$ at $x = 1$

30. $f(x) = x^{-3/2} - x$ at $x = 1$

31. An object is thrown upward, and its position is given by $d(t) = -16t^2 + 32t$ with $d(t)$ measured in feet and t in seconds. Find (a) the velocity and acceleration of the object, (b) the time at which it reaches its maximum height, and (c) how many feet it travels upward.

32. An object is thrown upward, and its position is given by $d(t) = -16t^2 + 50t + 6$ with $d(t)$ measured in feet and t in seconds. Find (a) the velocity and accel-

eration of the object, (b) the time at which it reaches its maximum height, and (c) how many feet it travels upward.

33. A firm determines that the relationship of the quantity x that the public is willing to purchase at price p is given by

$$p(x) = 5000 - 300x^{-1}$$

Find the rate of change of $p(x)$ when $x = 5$.

34. The profit P in thousands of dollars from the sales volume of x records is found to be

$$P(x) = 0.2x^2 + 10x - 100$$

Find the marginal profit when $x = 10$.

35. The cost C in thousands of dollars to produce x million batteries is determined to be

$$C(x) = 10 + 20x - 0.1x^2$$

Find the marginal cost when $x = 50$.

36. Business analysts call a product a ''fad'' if its total sales increase rapidly and then level off after a short period of time. Let S represent total sales in thousands of units and let x represent time in weeks. A function that represents the behavior of a fad is

$$S(x) = 100 - 50x^{-1}$$

Show that the rate of change of $S(x)$ with respect to x is decreasing as x increases by computing $S'(x)$ for (a) $x = 1$, (b) $x = 5$, and (c) $x = 10$.

37. The binomial theorem states that for any real numbers p and q and any positive integer n

$$(p + q)^n = p^n + np^{n-1}q + n(n - 1)/2p^{n-2}q^2$$
$$+ \cdots + npq^{n-1} + q^n$$

Use the binomial theorem to prove the Power Rule for positive integers.

38. Use the definition of the derivative to prove the sum rule.

Referenced Exercise Set 13–1

1. In an article illustrating the benefits of quality control to executives, A. W. Whitton, Jr.,* Senior Corporate Quality Consultant for Abbott Laboratories, uses Figure 3 to demonstrate how to reduce total quality cost. Locate point a on the graph of $y = TQC(x)$. It represents the lowest point on the TQC curve because its y-coordinate is the smallest of all points on the graph. The slope of the tangent line at a is 0. From the graph show that the following equations are true:

$$TQC(x) = FC(x) + AP(x) + PC(x)$$
$$TQC'(a) = FC'(a) + AP'(a) + PC'(a)$$

2. In the study of optics the subjective brightness a viewer senses is compared with the actual brightness from the source.† If x measures the actual brightness, then the subjective brightness, S, is a function of x. The functional relationship is usually described by a power function $S(x) = kx^n$, where k and n are constants depending on the light intensity and the type of light the viewer is coming from before the experiment. The *subjective sensitivity* is defined as the rate of change of the subjective brightness. Thus subjective sensitivity is $S'(x)$.

FIGURE 3 *Source:* A. W. Whitton, Jr., ''Methods for Selling Total Quality Cost Systems,'' *American Society of Quality Control — Technical Conference Transactions,* 1972, pp. 365–371.

*A. W. Whitton, Jr., ''Methods for Selling Total Quality Cost Systems,'' *American Society of Quality Control—Technical Conference Transactions,* 1972, pp. 365–371.

†Robert Sekuler et al., ''Structural Modeling of Spatial Vision,'' *Vision Research,* Vol. 24, 1984, pp. 689–700, and Heywood M. Petry et al., ''Spatial Contrast Sensitivity of the Tree Shrew,'' *Vision Research,* Vol. 24, 1984, pp. 1037–1042.

Suppose for those in relative darkness before an experiment that the subjective brightness is given by $S_1(x) = 0.001x^{1/4}$ and for those in light the subjective brightness $S_2(x) = 0.002x^{1/2}$. At what value of x are the subjective sensitivities equal?

3. In an article studying whether certain segments of the work force are paid their theoretical fair wage, defined by economists as the marginal product of the last worker hired, Robert H. Frank computes the following production function for real estate sales agents, where $f(x)$ represents the earnings for a sales force of size x.‡

$$f(x) = -0.024x^2 + 0.575x$$

Compute the marginal product function.

‡Robert H. Frank, ''Are Workers Paid Their Marginal Products?'' *American Economic Review,* Vol. 74, No. 4, 1984, pp. 549–571.

13–2 Product and Quotient Rules

The formulas developed in the previous section are used to compute the derivatives of polynomial functions and functions that are sums of terms of the form ax^r. Some functions do not fit one of these forms. Examples of functions whose derivatives cannot be found directly by these rules are

$$f(x) = x^2(5x + 11)$$

$$h(x) = \frac{x^3}{(4x + 1)}$$

We develop two formulas in this section. The first handles functions that are the product of simpler functions, like $f(x)$, and the second is applied to quotients, like $h(x)$.

The Product Rule

At first it might seem that the derivative of a product is the product of the derivatives, similar to the derivative of a sum. A quick inspection shows that this is not the case. Let $f(x) = x$ and $g(x) = x^2$. Then $f(x)\cdot g(x) = x^3$, so the derivative of the product is $3x^2$. But the product of the derivatives is $f'(x)\cdot g'(x) = 1\cdot 2x = 2x$. Thus the derivative of the product is *not* the product of the derivatives. The formula for the derivative of a product is more complicated.

> **The Product Rule**
> If $f(x) = g(x)\cdot k(x)$, then
>
> $$f'(x) = g(x)k'(x) + k(x)g'(x)$$

The **product rule** states that ''the derivative of a product is the first function times the derivative of the second plus the second times the derivative of the first.'' We prove this theorem at the end of the section.

EXAMPLE 1

Problem

Use the product rule to find $D_x x^2(5x + 11)$.

Solution Let $f(x) = x^2(5x + 11)$ with $g(x) = x^2$ and $k(x) = 5x + 11$. Then $f(x) = g(x) \cdot k(x)$. Since $g'(x) = 2x$ and $k'(x) = 5$, we have

$$f'(x) = g(x)k'(x) + k(x)g'(x) = x^2(5) + (5x + 11)(2x)$$
$$= 5x^2 + 10x^2 + 22x = 15x^2 + 22x$$

The function in Example 1 could be expressed as $f(x) = 5x^3 + 11x^2$ by multiplying through by x^2. Using the rules of the previous section, we have $f'(x) = 15x^2 + 22x$, which is the same derivative computed by the product rule. In all of the examples of this section the functions that are products can be expanded in this way, but in the next section some functions require the product rule to compute their derivatives.

For example, care is needed when computing the derivative of functions that are raised to a power. Consider $f(x) = (x^2 + 1)^2$. It might seem from the power rule that $f'(x) = 2(x^2 + 1)$. But by using the product rule with $g(x) = k(x) = x^2 + 1$, we get

$$f'(x) = (x^2 + 1)2x + (x^2 + 1)2x = 4x(x^2 + 1)$$

In the next section we will learn how to handle power functions that cannot be expanded like this function.

Let us look at one more example of the product rule.

EXAMPLE 2

Problem

Find $D_x(x^2 - 1)(x^3 + x)$.

Solution To apply the product rule let $f(x) = (x^2 - 1)(x^3 + x)$ with $g(x) = x^2 - 1$ and $k(x) = x^3 + x$. Since $g'(x) = 2x$ and $k'(x) = 3x^2 + 1$, we have

$$f'(x) = g(x)k'(x) + k(x)g'(x)$$
$$= (x^2 - 1)(3x^2 + 1) + (x^3 + x)(2x)$$
$$= 3x^4 - 3x^2 + x^2 - 1 + 2x^4 + 2x^2 = 5x^4 - 1$$

The Quotient Rule

Sometimes functions can be expressed as the quotient of simpler functions, such as $h(x) = x^3/(4x + 1)$. To find the derivative of such functions use the following rule.

The Quotient Rule

If $f(x) = \dfrac{g(x)}{k(x)}$, then

$$f'(x) = \frac{k(x)g'(x) - g(x)k'(x)}{[k(x)]^2}$$

The numerator in the quotient rule is similar to the expression in the product rule, but there is a significant difference. The two terms cannot be interchanged as they can in the product rule because the first term has a positive sign and the second has a negative sign. Examples 3 and 4 illustrate how to apply the quotient rule.

EXAMPLE 3

Problem

Find $D_x \dfrac{x^3}{4x + 1}$

Solution Let $f(x) = x^3/(4x + 1)$ with $g(x) = x^3$ and $k(x) = 4x + 1$. Then $f(x) = g(x)/k(x)$. Since $g'(x) = 3x^2$ and $k'(x) = 4$, we have

$$f'(x) = \frac{k(x)g'(x) - g(x)k'(x)}{[k(x)]^2}$$

$$= \frac{(4x + 1)(3x^2) - x^3(4)}{(4x + 1)^2} = \frac{8x^3 + 3x^2}{(4x + 1)^2}$$

EXAMPLE 4

Problem

Find $D_x \dfrac{x^3}{x^2 + 5}$

Solution

$$D_x \frac{x^3}{x^2 + 5} = \frac{(x^2 + 5)D_x x^3 - x^3 D_x(x^2 + 5)}{(x^2 + 5)^2}$$

$$= \frac{(x^2 + 5)3x^2 - x^3(2x)}{(x^2 + 5)^2}$$

$$= \frac{3x^4 + 15x^2 - 2x^4}{(x^2 + 5)^2} = \frac{x^4 + 15x^2}{(x^2 + 5)^2}$$

An Application: The Lerner Index

There are many methods to measure a particular firm's power and influence in the marketplace. One measure, called the **Lerner index,** named after economist A. P. Lerner of England and Michigan State University, depends on the difference between price P and marginal cost, expressed as a proportion of P.* It is defined as

$$\text{Lerner index} = \frac{P - MC}{P}$$

The Lerner index is sometimes referred to as a measure of a firm's monopoly power. If an industry has perfect competition, then $P = MC$, which means that the price each firm establishes for its product is equal to the rate of change of cost, marginal cost. Under perfect competition the Lerner index is 0, meaning the firm has no monopoly power. If a firm has a large amount of monopoly power, it can set a price that is much larger than marginal cost, so the Lerner index will increase. The closer the index is to 1 (its limiting value), the more monopoly power the firm has. Once a firm calculates its price and cost functions it is interested in the rate of change of the Lerner index L, measured by L'.

EXAMPLE 5

A firm determines that its price $P(x)$ and cost $C(x)$ per level of production x are

$$P(x) = x^2 + 4$$
$$C(x) = 3x + 10$$

Problem

Compute the firm's Lerner index $L(x)$ and find the rate of change of $L(x)$.

Solution First calculate the marginal cost function, $MC(x) = C'(x)$.

$$MC(x) = C'(x) = 3$$

Substituting this equation along with the equation of $P(x)$ into the definition of $L(x)$ yields

$$L(x) = \frac{P - MC}{P} = \frac{x^2 + 4 - 3}{x^2 + 4} = \frac{x^2 + 1}{x^2 + 4}$$

Applying the quotient rule gives

$$L'(x) = \frac{2x(x^2 + 4) - (x^2 + 1)(2x)}{(x^2 + 4)^2}$$

$$= \frac{2x^3 + 8x - (2x^3 + 2x)}{(x^2 + 4)^2}$$

$$= \frac{6x}{(x^2 + 4)^2}$$

*Milton H. Spenser, *Contemporary Macroeconomics,* 6th ed., New York, Worth Publishers, 1986.

Proof of the Product Rule

The product rule is not as readily apparent as the derivative of a sum. In fact, intuition leads to an incorrect formula; the derivative of a product looks like it should be the product of the derivatives. Why intuition fails is answered in the proof of the theorem. The proof uses an algebraic ploy—adding and subtracting the same quantity to and from an expression. This results in adding 0 to the expression, so that it remains the same. This is done so the expression can be factored in a certain way. This is why the formula contains two terms, each containing one derivative.

Proof of the Product Rule

If $f(x) = g(x) \cdot k(x)$, then $f(x + h) = g(x + h) \cdot k(x + h)$. From the definition of the derivative of $f(x)$

$$f'(x) = \lim_{h \to 0} \frac{f(x + h) - f(x)}{h}$$

$$= \lim_{h \to 0} \frac{g(x + h)k(x + h) - g(x)k(x)}{h}$$

Here we add and subtract $g(x + h)k(x)$ in the numerator.

$$f'(x) = \lim_{h \to 0} \frac{g(x + h)k(x + h) - g(x + h)k(x) + g(x + h)k(x) - g(x)k(x)}{h}$$

$$= \lim_{h \to 0} \frac{g(x + h)[k(x + h) - k(x)] + k(x)[g(x + h) - g(x)]}{h}$$

$$= \lim_{h \to 0} g(x + h) \left[\frac{k(x + h) - k(x)}{h} \right] + \lim_{h \to 0} k(x) \left[\frac{g(x + h) - g(x)}{h} \right]$$

$$= \lim_{h \to 0} g(x + h) \lim_{h \to 0} \frac{k(x + h) - k(x)}{h} + \lim_{h \to 0} k(x) \lim_{h \to 0} \frac{g(x + h) - g(x)}{h}$$

$$= g(x)k'(x) + k(x)g'(x)$$

provided that $g'(x)$ and $k'(x)$ exist.

EXERCISE SET 13–2

In Problems 1 to 18 find $f'(x)$.

1. $f(x) = x^4(x + 5)$

2. $f(x) = x^6(x - 2)$

3. $f(x) = 2x^3(3x - 7)$

4. $f(x) = 3x^5(2x + 1)$

5. $f(x) = (x^{-3} + 5)(3x^2 - 17)$

6. $f(x) = (x^{-5} - 9)(12x^3 + 1)$

7. $f(x) = (x^{1/3} + x)(3x^{1/2} - 4x)$

8. $f(x) = (x^{1/5} - x)(x^{3/5} + 2x)$

9. $f(x) = (x^{-1/3} + 7x^{-1/4})(4x^2 - 5)$

10. $f(x) = (2x^{-1/6} - x^{-3/7} + 2)(x^{1/2} - 6)$

11. $f(x) = (2x + 1)^2$ 12. $f(x) = (4x - 3)^2$

13. $f(x) = (x^3 + 1)^2$

14. $f(x) = (x^4 - 2)^2$

15. $f(x) = x/(2x + 1)$

16. $f(x) = 2x/(5x - 4)$

17. $f(x) = (x^3 + 5)/3x^2$

18. $f(x) = (x^5 - 11)/x^3$

In Problems 19 to 22 find the equation of the tangent line to $y = f(x)$ at the given point

19. $f(x) = (x^{1/2} - 4)^2$, (1, 9)

20. $f(x) = (x^{4/5} + 1)^3$, (1, 8)

21. $f(x) = (x^3 - 2)/(x^2 - 1)$, (0, 2)

22. $f(x) = (x^{1/2} - 1)/(x^{1/2} + 8)$, (1, 0)

In Problems 23 and 24 a firm determines that its price $P(x)$ and cost $C(x)$ per level of production x are the given functions. Compute the firm's Lerner index $L(x)$ and find the rate of change of $L(x)$.

23. $P(x) = 0.5x^2 + 5$, $C(x) = 2x + 7$

24. $P(x) = 0.22x^2 + 4.1$, $C(x) = 4x + 3$

In Problems 25 to 32 find $f'(x)$.

25. $f(x) = x^2(x^3 + 15)(x^2 - 17)$

26. $f(x) = x^{1/2}(x^{-5} - 9)(12x^3 + 1)$

27. $f(x) = x^{-1}(x^{1/3} + 2)(x^2 - 3)$

28. $f(x) = x^{-1/2}(x^{1.5} - 1)(2x^3 - 3)$

29. $f(x) = x^2(x + 5)/(x^2 - 7)$

30. $f(x) = x^{1/2}(x^2 - 1)/(x^3 + 1)$

31. $f(x) = \dfrac{x^2 + 3}{x(x^2 - 1)}$

32. $f(x) = \dfrac{x^3 - 2}{x^2(x^{-1} + 1)}$

33. The total cost of producing x amount of books is given by $C(x) = 100 + 23x(1 - 2x^2)$. Find the marginal cost.

The *average cost per good* is defined to be the total cost to produce x amount of goods divided by x. If $C(x)$ measures the total cost to produce x amount of goods and $AC(x)$ is the average cost per good, then $AC(x)$ is given by

$$AC(x) = \frac{C(x)}{x}$$

In Problems 34 and 35 find the marginal average cost, $AC'(x)$.

34. $C(x) = 10 + 25x - 2x^2$

35. $C(x) = 100 + 3.5x - 0.4x^2$

36. Use the definition of the derivative and an argument similar to the proof of the product rule to prove the quotient rule.

37. Show why the Lerner index $L = \dfrac{P - MC}{P}$, for price P and marginal cost MC, is never more than 1.

Referenced Exercise Set 13–2

1. With the breakup of AT&T, the transition from monopoly to competition necessitated the use of different tools for business decisions. In an article* describing how GTE Telephone Operations decided on prices for various products, one model centered on the Styleline phone. Of key interest is the total quantity sold q after x years, where q is measured in millions of units. The data can be approximated by

$$q(x) = 0.13x(x - 10)^2$$

GTE was concerned with finding when $q(x)$ would attain its maximum rate of increase. In the next chapter we will show how this is done by using $q'(x)$. Compute $q'(x)$.

2. The Atomic Energy Commission has demanded extensive quality control in nuclear power plants, especially since the Three Mile Island and Chernobyl incidents. In an article describing the significance of AEC regulations, even for "minor" equipment, John B. Silverwood,† Reliability and Assurance Manager for United Engineers and Constructors, outlines various procedures that are necessary in performance testing. In one example he describes the efficiency rating of a centrifugal pump by using a graph (Figure 1) whose equation is approximated by

$$f(x) = \left(\frac{x}{4}\right)\left(5 - \frac{x}{32}\right)$$

$f(x)$ measures the efficiency, in percent, at which the pump is operating when x gallons are being pumped

*Arvind G. Jadhav, "Market Simulation for Telecommunications Services," *Forecasting Public Utilities,* New York, North-Holland Publishing Co., pp. 165–186.

†John B. Silverwood, "Objective Quality Evidence of Product Integrity," *American Society of Quality Control—Technical Conference Transactions,* 1972, pp. 276–283.

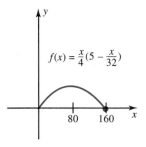

$f(x) = \frac{x}{4}(5 - \frac{x}{32})$

FIGURE 1 *Source:* John B. Silverwood, "Objective Quality Evidence of Product Integrity," *American Society of Quality Control – Technical Conference Transactions,* 1972, pp. 276–283.

per minute. The function is defined in [0, 100]. From the graph it can be seen that as x increases, the efficiency increases for a while and then it starts to decrease. At that point the tangent line has slope 0. Find the point where $f(x)$ changes from increasing to decreasing by finding the value of x that solves $f'(x) = 0$.

3. In a paper‡ studying the effects of private pond fishing on preexisting common property fishing, researchers centered their attention on the crawfish market. The cost function for wild crawfish, those caught in com-

mon property waters, can be approximated by

$$C(x) = \frac{7.12}{13.7x - x^2}$$

Here x is the quantity of crawfish supplied to the market, measured in millions of pounds, and C is measured in millions of dollars. Find the marginal cost $C'(x)$.

Cumulative Exercise Set 13–2

In Problems 1 and 2 find $f'(x)$.

1. $f(x) = 3x^{-2} - x^4 + 5$

2. $f(x) = x^{-1/2} - 2x^{-3}$

3. Find the equation of the tangent line to $f(x) = x^3 - 4x$ at $x = 1$.

4. The profit P in thousands of dollars from the sales volume of x thousands of records is found to be $P(x) = 0.3x^2 + 20x - 200$. Find the marginal profit when $x = 3$.

In Problems 5 to 7 find $f'(x)$.

5. $f(x) = (x^{-5} - 9)(12x^3 + 1)$

6. $f(x) = (x^5 - 1)^2$

7. $f(x) = x^2(x - 5)(x^2 + 7)$

8. The total cost of producing radios is given by $C(x) = 200 + 15x(1 - 3x^2)$. Find the marginal cost.

‡Frederick W. Bell, "Competition from Fish Farming in Influencing Rent Dissipation: The Crawfish Fishery," *American Journal of Agricultural Economics,* Vol. 68, pp. 95–101.

13–3 The Chain Rule

Sometimes more complicated functions can be differentiated by separating the function into two parts, an "inner" and "outer" part, and then computing the derivative of each. The rule that allows us to find a derivative this way is called the chain rule. It has several forms, each governing a different type of function. In this section we study the form of the chain rule called the generalized power rule, so-called because it deals with the derivative of functions raised to a power. In general the chain rule is used to find the derivative of a composite function of the form $f(g(x))$. With this notation, $g(x)$ is the "inner" function and $f(x)$ is the "outer" function. At the end of the section the chain rule is studied in this more general setting so that it can be applied later to other types of functions.

The Generalized Power Rule

The differentiation rules developed so far do not allow us to find the derivative of more complicated functions, such as

$$f(x) = (x^3 + x)^5 \quad s(x) = (x^2 + 1)^{-1} \quad h(x) = (3x^2 + x)^{1/2}$$

Each of these functions is an expression in x raised to a power. The general form of such a function is

$$f(x) = g(x)^n$$

This type of function is called a **general power function.** In each case $g(x)$ is referred to as the ''inside part'' of the function or the inside function. For example, by letting $g(x) = x^3 + x$ and $n = 5$, we can form the previous function $f(x) = (x^3 + x)^5$. Similarly $s(x)$ has $g(x) = x^2 + 1$ and $n = -1$, while $h(x)$ has $g(x) = 3x^2 + x$ and $n = \frac{1}{2}$.

Let us find the derivative of a specific power function.

EXAMPLE 1

Problem

Use the product rule to find the derivative of (a) $f(x) = (x^3 + x)^2$, (b) $g(x) = (x^3 + x)^3$

Solution (a) Consider $f(x)$ as the product of two functions, each being $x^3 + x$. That is, $f(x) = (x^3 + x)(x^3 + x)$. By the product rule

$$\begin{aligned}
f'(x) &= (x^3 + x)D_x(x^3 + x) + (x^3 + x)D_x(x^3 + x) \\
&= 2(x^3 + x)D_x(x^3 + x) \\
&= 2(x^3 + x)(3x^2 + 1)
\end{aligned}$$

(b) Consider $g(x)$ as the product of the two functions $f(x) = (x^3 + x)^2$ from (a) and $x^3 + x$. Thus

$$g(x) = (x^3 + x)^2(x^3 + x)$$

Use the product rule, with $f'(x) = 2(x^3 + x)(3x^2 + 1)$ from (a).

$$\begin{aligned}
g'(x) &= (x^3 + x)^2 D_x(x^3 + x) + (x^3 + x)D_x(x^3 + x)^2 \\
&= (x^3 + x)^2(3x^2 + 1) + (x^3 + x)[(2)(x^3 + x)(3x^2 + 1)]
\end{aligned}$$

Factor out $3x^2 + 1$ and get

$$\begin{aligned}
g'(x) &= [(x^3 + x)^2 + 2(x^3 + x)^2](3x^2 + 1) \\
&= 3(x^3 + x)^2(3x^2 + 1)
\end{aligned}$$

Recall the power rule:

$$D_x x^n = nx^{n-1}$$

It governs only functions of the form $f(x) = x^n$. In Example 1, when finding $f'(x)$, it is tempting to use just the power rule, in which case we would get $2(x^3 + x)$.

While this expression is part of the derivative, it is not precisely the derivative. It varies by a factor of $3x^2 + 1$, which is the derivative of the inside function, $x^3 + x$. The generalized power rule requires that the derivative of a general power function contain the derivative of the inside part of the function as a factor.

Example 1 motivates the chain rule for general power functions, called the **generalized power rule.**

The Generalized Power Rule

$$D_x[g(x)]^n = ng(x)^{n-1}D_x[g(x)]$$

Notice that the power rule is a special case of the generalized power rule when $g(x) = x$. Let us apply the rule to the function in Example 1(a). The function is $f(x) = (x^3 + x)^2$ where $g(x) = x^3 + x$ and $n = 2$. Then

$$f'(x) = 2(x^3 + x)D_x(x^3 + x)$$
$$= 2(x^3 + x)(3x^2 + 1)$$

The next three examples illustrate the generalized power rule for functions that cannot be differentiated by the product rule.

Example 2 illustrates the generalized power rule for a power function whose exponent is a positive fraction.

EXAMPLE 2

Problem

Find $D_x f(x)$ where $f(x) = (x^4 + 2x)^{1/2}$.

Solution In the generalized power rule we have $g(x) = x^4 + 2x$ and $n = \frac{1}{2}$. Therefore

$$D_x(x^4 + 2x)^{1/2} = (\tfrac{1}{2})(x^4 + 2x)^{(1/2)-1}D_x(x^4 + 2x)$$
$$= (\tfrac{1}{2})(x^4 + 2x)^{-1/2}(4x^3 + 2)$$
$$= (2x^3 + 1)(x^4 + 2x)^{-1/2}$$

Notice when simplifying the expression for the derivative, the term corresponding to the derivative of the "inside" part of the function appears on the left of the expression corresponding to the power function.

Example 3 finds the derivative of power function whose exponent is a negative integer.

EXAMPLE 3

Problem

Find $f'(x)$ where $f(x) = (x^3 + 6x^2)^{-4}$.

Solution In the generalized power rule we have $g(x) = x^3 + 6x^2$ and $n = -4$. Therefore

$$f'(x) = -4(x^3 + 6x^2)^{-4-1}D_x(x^3 + 6x^2)$$
$$= -4(x^3 + 6x^2)^{-5}(3x^2 + 12x)$$
$$= -12x(x + 4)(x^3 + 6x^2)^{-5}$$

We look at one more illustration of the generalized power rule in Example 4, this time with the exponent of the power function a negative fraction.

EXAMPLE 4

Problem

Find $f'(x)$ where $f(x) = (x^4 + 12x + 4)^{-1/4}$.

Solution Let $g(x) = x^4 + 12x + 4$ and $n = -\frac{1}{4}$. Then

$$f'(x) = (-\tfrac{1}{4})(x^4 + 12x + 4)^{-5/4}D_x(x^4 + 12x + 4)$$
$$= (-\tfrac{1}{4})(x^4 + 12x + 4)^{-5/4}(4x^3 + 12)$$
$$= -(x^3 + 3)(x^4 + 12x + 4)^{-5/4}$$

Example 5 demonstrates how to use the generalized power rule in conjunction with the product rule. First regard the function as a product, separating the first function from the second. Then express the derivative in terms of the product rule, the first function times the derivative of the second plus the second function times the derivative of the first. Then apply the generalized power rule where appropriate.

EXAMPLE 5

Problem

Compute $f'(x)$ where $f(x) = x^3(x^2 + 5)^{-1}$.

Solution Use the product rule, with the first function being x^3 and the second function being $(x^2 + 5)^{-1}$.

$$f'(x) = x^3 D_x(x^2 + 5)^{-1} + (x^2 + 5)^{-1}D_x x^3$$
$$= x^3(-1)(x^2 + 5)^{-2}(2x) + (x^2 + 5)^{-1}(3x^2)$$
$$= -2x^4(x^2 + 5)^{-2} + 3x^2(x^2 + 5)^{-1}$$
$$= (x^2 + 5)^{-2}[-2x^4 + 3x^2(x^2 + 5)]$$
$$= (x^2 + 5)^{-2}(-2x^4 + 3x^4 + 15x^2)$$
$$= (x^4 + 15x^2)(x^2 + 5)^{-2}$$
$$= x^2(x^2 + 15)(x^2 + 5)^{-2}$$

Notice that the quotient rule could have been used in Example 5 by expressing $f(x)$ as a quotient, $x^3/(x^2 + 5)$.

The General Form of the Chain Rule

The generalized power rule is a special case of the **chain rule.** The general form of the chain rule is stated in terms of the composition of functions.

The Chain Rule
If $f(x) = h(g(x))$, then $f'(x) = h'(g(x))g'(x)$.

To illustrate, suppose $f(x) = (x^2 + 1)^3$. Let $h(x) = x^3$ and $g(x) = x^2 + 1$ so that $f(x) = h(g(x))$. To find $f'(x)$ by the general form of the chain rule, we use $h'(x) = 3x^2$ and $g'(x) = 2x$. Then $h'(g(x)) = 3[g(x)]^2 = 3(x^2 + 1)^2$ so that

$$f'(x) = h'(g(x))g'(x) = 3(x^2 + 1)^2(2x) = 6x(x^2 + 1)^2$$

The chain rule can be expressed in another form, which is easy to memorize. This form uses the $\dfrac{dy}{dx}$ notation of the derivative. Suppose $y = f(g(x))$. Make the substitution of variable $u = g(x)$ so that $y = f(u)$. Then y can be expressed as a function of u or of x. With this notation, we write the derivatives as $g'(x) = \dfrac{du}{dx}$, $f'(u) = \dfrac{dy}{du}$, and $f'(x) = \dfrac{dy}{dx}$. Then the chain rule states

$$\frac{dy}{dx} = \frac{dy}{du}\frac{du}{dx}$$

In the illustration above, $y = f(x) = (x^2 + 1)^3$ and $u = g(x) = x^2 + 1$ so that $y = f(u) = u^3$. Then $\dfrac{dy}{du} = 3u^2$ and $\dfrac{du}{dx} = 2x$. Therefore

$$\frac{dy}{dx} = \frac{dy}{du}\frac{du}{dx} = 3u^2(2x) = 3(x^2 + 1)^2(2x) = 6x(x^2 + 1)^2$$

This agrees with the preceding expression for $f'(x)$.

The generalized power rule is a special case of the chain rule. We now derive the special case, the generalized power rule, from the more general theorem, the chain rule.

Derivation of the Generalized Power Rule from the Chain Rule

Consider the power function $f(x) = g(x)^n$. Let $h(x) = x^n$. Then $f(x) = h(g(x))$ and $h'(x) = nx^{n-1}$, so $h'(g(x)) = ng(x)^{n-1}$. From the general form of the chain rule we have

$$f'(x) = h'(g(x))g'(x) = ng(x)^{n-1}g'(x)$$

This is the statement of the generalized power rule.

An Application

There are many situations in business where one variable depends upon another, and the second variable depends on a third. An example is the important business tool, quality control. Quality $Q(x)$ is measured in terms of level of production x, and x is measured in terms of time t. Q is measured on a scale from 0 to 100 (with 100 being a product with no flaws), x is measured in units of 100, and t is measured in hours. The next example shows how the rate of change of Q can be computed in two ways, by expressing Q in terms of t directly and by using the chain rule.

EXAMPLE 6

A company determines that

$$Q(x) = 90 - 0.4(x - 10)^2 = 50 + 8x - 0.4x^2$$

where $Q(x)$ is defined for $0 < x < 18$. The graph of $Q(x)$ is given in Figure 1. Production level, measured in hours, depends on the time of day, t, and is given by

$$x(t) = 18 - 0.5(t - 6)^2 = 6t - 0.5t^2$$

where $x(t)$ is defined for $0 \le t \le 8$. This means that at time t the factory is producing goods at a rate such that if that rate were held constant all day $100x(t)$ units would be produced. The graph of $x(t)$ is given in Figure 2.

Problem

Find $D_t Q$ by expressing Q as a function of t and differentiating, and also by using the chain rule. Show that the two expressions are equal.

Solution Since Q is a function of x and x is a function of t, Q is a function of t; namely

$$\begin{aligned}
Q(t) &= Q(x(t)) = Q(6t - 0.5t^2) \\
&= 50 + 8(6t - 0.5t^2) - 0.4(6t - 0.5t^2)^2
\end{aligned}$$

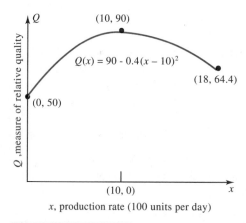

$Q(x) = 90 - 0.4(x - 10)^2$

(10, 90)

(18, 64.4)

(0, 50)

(10, 0)

Q, measure of relative quality

x, production rate (100 units per day)

FIGURE 1

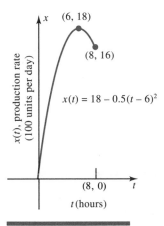

FIGURE 2

To find $Q'(t)$ directly from this formula, differentiate with respect to t.

$$Q'(t) = 8(6 - t) - 0.8(6t - 0.5t^2)(6 - t)$$
$$= 48 - 36.8t + 7.2t^2 - 0.4t^3$$

Using the chain rule to find $Q'(t)$ yields

$$Q'(t) = Q'(x(t))x'(t)$$
$$= D_x[Q(x)]D_t[x(t)]$$
$$= D_x[50 + 8x - 0.4x^2]D_t[6t - 0.5t^2]$$
$$= [8 - 0.8x][6 - t]$$

Now substitute the expression $x(t) = 6t - 0.5t^2$ for x.

$$Q'(t) = [8 - 0.8(6t - 0.5t^2)](6 - t)$$
$$= (8 - 4.8t + 0.4t^2)(6 - t)$$
$$= 48 - 36.8t + 7.2t^2 - 0.4t^3$$

Thus the two expressions for $Q'(t)$ are equal.

The graph of $Q(t)$ is given in Figure 3. In Section 14–2 we will see how to obtain this graph. The quality is lowest at the start of the shift when it is 50; that is, $Q(0) = 50$. The graph of Q then gradually increases until it reaches its largest value, 90, when $t = 2$, meaning that two hours into the shift quality is at its peak. At this point $Q'(t)$ is 0 since the rate of change of Q goes from positive when Q is increasing to negative when it is decreasing. Then Q decreases to the value 64.4 at $t = 6$ hours, at which time it then starts to increase again. Thus when $t = 6$, $Q'(t) = 0$. We will discuss these concepts further in the next chapter.

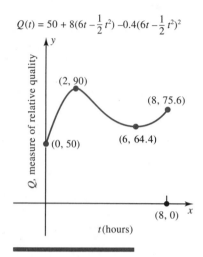

$$Q(t) = 50 + 8(6t - \frac{1}{2}t^2) - 0.4(6t - \frac{1}{2}t^2)^2$$

FIGURE 3

An Application: The Lerner Index

The measure of a firm's monopoly power, called the Lerner index L, was presented in the previous section. It is defined by

$$L = \frac{P - MC}{P}$$

P is the price of the firm's product and MC is the marginal cost. If MC is fixed, then L is a function of P. If P is a function of the units produced x, then L is also a function of x. In the previous section we expressed L as a function of x and then computed $L'(x)$. The next example shows that the chain rule can also be used to find $L'(x)$. The price and cost functions are from Example 5 in the previous section.

EXAMPLE 7

A firm determines that its price $P(x)$ and cost $C(x)$ per level of production x are

$$P(x) = x^2 + 4$$
$$C(x) = 3x + 10$$
$$MC(x) = C'(x) = 3$$

Problem

Use the chain rule to compute $L'(x)$.

Solution Since MC is the constant 3, we can express L as a function of P.

$$L(P) = \frac{P - MC}{P} = \frac{P - 3}{P} = 1 - 3P^{-1}$$

Then we have

$$\frac{dL}{dP} = 3P^{-2}$$

Also,

$$\frac{dP}{dx} = 2x$$

Applying the chain rule yields

$$\frac{dL}{dx} = \frac{dL}{dP}\frac{dP}{dx} = (3P^{-2})(2x) = 3(x^2 + 4)^{-2}(2x)$$

$$= \frac{6x}{(x^2 + 4)^2}$$

This is the same expression for the derivative that was computed in the last section.

EXERCISE SET 13–3

In Problems 1 to 10 find $f'(x)$ by using the generalized power rule.

1. $f(x) = (5x + 2)^3$ 2. $f(x) = (15x + 2)^4$

3. $f(x) = (x^3 + 4)^3$ 4. $f(x) = (5x^2 + 2)^6$

5. $f(x) = (x^3 + 3x^2)^3$ 6. $f(x) = (2x^4 + 3x^3)^6$

7. $f(x) = (x^{3/2} + 4)^{1/3}$ 8. $f(x) = (x^{3/5} - 2)^{2/3}$

9. $f(x) = (x^{1/2} + 2x^{-1})^{-6}$

10. $f(x) = (5x^{1/3} + x^{-2})^{-1/2}$

In Problems 11 to 16 find $f'(x)$ using the generalized power rule in conjunction with the product rule or the quotient rule.

11. $f(x) = x(5x^2 + 2)^2$ 12. $f(x) = 3x(3x^2 + 5)^4$

13. $f(x) = x^2(x^3 + 2x)^2$ 14. $f(x) = 3x^2(x^3 + 3x)^3$

15. $f(x) = \dfrac{x^2}{(x^3 - x)^3}$ 16. $f(x) = 4x^4(x^2 - 2x)^{1/2}$

In Problems 17 to 20 find the equation of the tangent line at the given value of x.

17. $f(x) = (x^2 + 2)^3$ at $x = 0$

18. $f(x) = (x^3 - 1)^2$ at $x = 0$

19. $f(x) = (x^3 + 4x)^{-1}$ at $x = 1$

20. $f(x) = (3x^2 - x - 1)^{-2}$ at $x = 0$

A company determines that its quality control $Q(x)$, measured in terms of level of production x, measured in terms of time t, is given by the functions in Problems 21 and 22. Find D_tQ by expressing Q as a function of t and differentiating and also by using the chain rule. Show that the two expressions are equal.

21. $Q(x) = 95 - 0.2(x - 5)^2$,
    $x(t) = 10 - 0.2(t - 6)^2$

22. $Q(x) = 100 - 0.3(x - 8)^2$,
    $x(t) = 10 - 0.1(t - 6)^2$

A company determines that its price $P(x)$ and cost $C(x)$ per level of production x are given by the functions in Problems 23 and 24. Find the Lerner index $L(x)$ and $L'(x)$.

23. $P(x) = x^2 + 10$, $C(x) = 5x + 20$

24. $P(x) = 100 + 15x - 2x^2$,
    $C(x) = 4x + 25$

In Problems 25 to 34 find $f'(x)$ by using the chain rule together with the rules in the previous section.

25. $f(x) = x^2(x^3 + 2x^2)^4$

26. $f(x) = 3x^4(2x^4 + 3x^3)^6$

27. $f(x) = 2x^{1/2}(x^{3/2} + 4)^{-3}$

28. $f(x) = 4x^{-2}(x^{5/2} + 2x^{-1})^{1/6}$

29. $f(x) = (x^2 + 1)^2(x^3 + 1)^3$

30. $f(x) = (x^2 - 3)^{-1}(x^3 - 1)^2$

31. $f(x) = (x^2 + 1)^{1/2}(x^3 + 3)^{1/3}$

32. $f(x) = (x^{1/2} + 3)^{1/3}(x^2 + 3x)^{-1}$

33. $f(x) = 2x^{1/2}/(x^2 + 2x^{-2})^3$

34. $f(x) = (x^4 + 1)^2/(3x^4 + 3)^{1/3}$

In Problems 35 to 38 find $f'(x)$ by using the chain rule and also by expressing the function as a quotient and using the quotient rule.

35. $f(x) = x(x^2 + 1)^{-1}$ 36. $f(x) = x^2(x^3 - 1)^{-1}$

37. $f(x) = (x - 2)(x^2 + 5)^{-1}$

38. $f(x) = (x + 3)^2(4x - 5)^{-1}$

39. The total profit $p(x)$ for x amount of goods sold is

$$p(x) = 500(200 - x)^2$$

Find the marginal profit.

40. The strength $S(x)$ of an individual's reaction to the quantity x of a drug administered into the bloodstream is given by

$$S(x) = 5x(20 - x/2)^{1/2}$$

Find the rate of change of $S(x)$ with respect to x when $x = 8$.

41. The chain rule can be used in conjunction with the product rule to derive the quotient rule. Suppose $f(x)$ is a quotient function $f(x) = \dfrac{g(x)}{h(x)}$. Express $f(x)$ as the product function $f(x) = g(x)h(x)^{-1}$. First apply the product rule and then the chain rule for the derivative of $h(x)^{-1}$. Factor $h(x)^{-2}$ from each term to get the quotient rule.

Referenced Exercise Set 13–3

1. Many variables affect quality control of production. In describing different quality control procedures for various products, Bartlett and Provost* used as an example the net weight of a #303 can of whole kernel corn.

This type of can should weigh 16 ounces. They show that the acceptance of a lot by a retailer from the producer depends on the average net weight of all cans in the lot, and the average net weight depends on the size of the sample of cans selected from the lot to be inspected. Let P represent the likelihood that a lot will be accepted by the retailer, W the average net weight of all the cans in the lot, and x the sample size. The relationships between the variables can be expressed as follows:

$$P(W) = 0.007W^2 - 0.33W + 100$$
$$W(x) = 16 - 0.41x^{-1/2}$$

Express P as a function of x and find $P'(x)$.

2. Franchised businesses account for over 38% of all retail sales in the United States and originate more than 12% of the gross national product. In an article analyzing methods to optimize franchise contractual agreements, the researchers† derive the following function that relates franchisee's profits P in terms of the quantity Q of goods sold by the equation

$$P = aQ - bQ^2 - F$$

where F is the initial lump-sum franchise fee paid by the franchisee, and a and b are constants whose values depend on the type of industry to which the model is being applied. It is also shown that Q is a function of R where R is the royalty rate charged by the parent company to the franchisee

$$Q(R) = c - dR$$

The constants c and d also depend on the particular industry. Q is measured in thousands of units and P in thousands of dollars. Typical values for the constants in the fast food industry are $a = 0.35$, $b = 0.11$, $c = 0.1$, and Q is measured in 1000 units. Find the marginal profit with respect to R via the chain rule and then express P as a function of R and find the marginal profit with respect to R directly. Show that the two expressions are equal.

3. (a) Find $f'(x)$ for the function

$$f(x) = \sqrt{2x - x^2 - 1}$$

(b) The answer to part (a) is "Does not exist." Use a graphing calculator or a graphing program for a

*Richard P. Bartlett and Lloyd P. Provost, "Tolerances in Standards and Specifications," *Quality Progress*, 1973, pp. 14–18.

†Roger D. Blair and David L. Kaserman, "Optimal Franchising," *Southern Economics Journal*, Vol. 49, 1983, pp. 495–505.

microcomputer to sketch the graph of $f(x)$ to see why the derivative does not exist.‡ [In **MathPath**, the software program that accompanies this text, use the module "Graphing" and define the function $y = $ sqrt $(2x - x^2 - 1)$ between the limits of -5 and 5.]

(c) What is the domain of $f(x)$?

Cumulative Exercise Set 13–3

In Problems 1 to 5 find $f'(x)$.

1. $f(x) = 2x^3 + \sqrt{x} - 7 - \dfrac{1}{x}$

2. $f(x) = \dfrac{3}{\sqrt{2x}}$

3. $f(x) = x(x - 4)$

4. $f(x) = \dfrac{x^2 - 3x + 5}{x^3 - 4x^2 - 9}$

5. $f(x) = (x^2 - 1)^3 - x(x^2 + 1)$

In Problems 6 to 9 find the equation of the line tangent to $y = f(x)$ at the given value of x.

6. $f(x) = x^2 - 4$, $x = -2$

7. $f(x) = \sqrt{9 - x^2}$, $x = 0$

8. $f(x) = \dfrac{x}{1 - x}$, $x = \dfrac{1}{2}$

9. $f(x) = (4 - x)^3$, $x = 2$

10. If the cost C, in hundreds of thousands of dollars, of producing x million batteries is

$$C(x) = 20 + 5.4x - 0.09x^2$$

what is the marginal cost of producing 2 million batteries?

11. Let $f(x) = (x^3 - x - 2)(x^3 - x - 2)$.
    (a) Find $f'(x)$ via the product rule.
    (b) Find $f'(x)$ via the generalized power rule.
    (c) Are the answers in parts (a) and (b) the same?

12. (a) Use the product rule and the generalized power rule to find $f'(x)$ for $f(x) = x(1 - x)^{-1}$
    (b) Use the quotient rule to find $f'(x)$ for
    $$f(x) = \dfrac{x}{1 - x}$$
    (c) Are the answers in parts (a) and (b) the same?

‡Y. L. Cheung, "Examples for graphing calculators," *Mathematics Teacher*, February, 1989, pp. 82–83.

13–4 Higher Derivatives and Other Notation

Just as the derivative $f'(x)$ of a function gives information about $f(x)$, so too the derivative of the derivative gives information about $f'(x)$. The derivative of the derivative is called the "second derivative." This section shows how to compute the second derivative and higher derivatives, as well as discusses a few additional points about alternate notation.

The Second Derivative

The derivative $f'(x)$ of a function is another function that measures the slope of the tangent line of $f(x)$. In the same way the derivative of $f'(x)$ is another function, called the **second derivative** of $f(x)$, and is denoted by $f''(x)$. Using alternate notation, $D_x(D_x f(x)) = f''(x)$, which says that the derivative of the derivative is the second derivative. The first example illustrates how to compute the second derivative.

EXAMPLE 1

Problem

Find $f''(x)$ for (a) $f(x) = 2x^3 + 5x$, (b) $f(x) = \sqrt{x} - 1/x$

Solution (a) $f'(x) = 6x^2 + 5$ and so

$$f''(x) = 12x$$

(b) First write the function as $f(x) = x^{1/2} - x^{-1}$. Then

$$f'(x) = (\tfrac{1}{2})x^{-1/2} - (-x^{-2})$$
$$= (\tfrac{1}{2})x^{-1/2} + x^{-2}$$
$$f''(x) = -(\tfrac{1}{4})x^{-3/2} - 2x^{-3}$$

The second derivative provides important information about $f(x)$. In particular, it helps determine the shape of the curve of $f(x)$ in the vicinity of a given point. This topic will be amplified in the next chapter.

Higher Derivatives

In some applications it is necessary to compute higher derivatives. For example, if $f(x) = x^3$, then $f'(x) = 3x^2$, $f''(x) = 6x$ and the derivative of $f''(x)$, called the **third derivative,** denoted by $f'''(x)$, is the derivative of $6x$. Hence $f'''(x) = 6$. For even **higher derivatives** the notation changes. We write $f^{(n)}(x)$ for $n > 3$ to represent higher derivatives after the third one. In this example $f^{(4)}(x) = 0$ as do all higher derivatives.

EXAMPLE 2

Problem

Find $f^{(n)}(x)$ for $f(x) = x^{-1}$.

Solution Find the first, second, and third derivatives to determine a pattern.

$$f'(x) = -x^{-2}$$
$$f''(x) = -(-2)x^{-3} = 2x^{-3}$$
$$f'''(x) = 2(-3)x^{-4} = -6x^{-4}$$

Successive derivatives change sign so a factor $(-1)^n$ is necessary. The absolute value of the coefficient of each derivative has the form $1 \cdot 2 \cdot 3 \cdot \ldots \cdot n = n!$. The power of x in each derivative is negative and its absolute value is one more than the number of the derivative. For example, the third derivative has the power -4. Hence the nth derivative has the factor x to the power $-(n + 1)$. Therefore the nth derivative is

$$f^{(n)}(x) = (-1)^n n! x^{-(n+1)}$$

In particular, $f^{(4)}(x) = (-1)^4 4! x^{-5} = 24x^{-5}$ and $f^{(5)}(x) = -5! x^{-6} = -120x^{-6}$.

Other Notation

There are several notations for expressing higher derivatives of $y = f(x)$. If the function is $y = f(x)$, then the second derivative is expressed by each of these.

$$f''(x) \quad \frac{d^2y}{dx^2} \quad \frac{d^2f}{dx^2} \quad \frac{d^2}{dx^2} f(x) \quad D_x^2 f(x)$$

These notations might seem to contain powers, but they are merely contractions of more cumbersome notation. Referring to the second derivative as the "derivative of the derivative," the D_x notation is

$$f''(x) = D_x[D_x f(x)]$$

This is shortened to $D_x^2[f(x)]$. The contraction is more obviously preferred when trying to write the third and higher derivatives with this notation. The third derivative is written $D_x^3 f(x)$ instead of $D_x[D_x(D_x f(x))]$. In a similar way the notation $\frac{d^2y}{dx^2}$ is a shortened form of $\frac{d}{dx}\frac{dy}{dx}$. Similarly the notation for the third derivative is $\frac{d^3y}{dx^3}$.

The second derivative of $y = f(x)$ at $x = 3$ is expressed as $f''(3)$. Using the $\frac{dy}{dx}$ and $D_x f(x)$ notations, $f''(3)$ is expressed as

$$\left.\frac{d^2y}{dx^2}\right|_{x=3} \qquad D_x^2[f(x)]\Big|_{x=3}$$

The next example provides practice with these different notations.

EXAMPLE 3

Problem

If $f(x) = x^3 - 2x^2 + 5$, find (a) $\left.\frac{d^2y}{dx^2}\right|_{x=3}$, (b) $D_x^2 f(x)\Big|_{x=4}$

Solution $f'(x) = 3x^2 - 4x$ and $f''(x) = 6x - 4$. Therefore

(a) $\left.\frac{d^2y}{dx^2}\right|_{x=3} = 6(3) - 4 = 14$

(b) $D_x^2 f(x)\Big|_{x=4} = 6(4) - 4 = 20$

Other Variables

In some applications it is convenient to use more appropriate variables than x and y, such as t and v when discussing time and velocity. When the variable is not x, the notation for the derivative is also changed. For instance, if the function is given as $v(t) = t^4$, then we write $v'(t) = D_t v = \frac{dv}{dt} = 4t^3$. The name of the variable does not affect the meaning of the derivative. Thus if $f(x) = 2x$ and $p(s) = 2s$, then $f'(x) = 2$ and $p'(s) = 2$ mean the same thing as far as the derivative is concerned—the slope of the tangent line at any value of the variable is 2. This illustrates that any letter can be used as a variable.

EXAMPLE 4

Problem

Compute the derivative of each function: (a) $G(u) = (3u + 5)^{1/2}$, (b) $z(y) = y^{-3} - y$.

Solution

(a) $G'(u) = (\frac{1}{2})(3u + 5)^{-1/2} \cdot 3$
    $= (\frac{3}{2})(3u + 5)^{-1/2}$
(b) $D_y z(y) = -3y^{-4} - 1$

One of the most important applications of calculus is in the movement of objects. Many modern inventions are based on this application, from radar screens to television broadcasts. In Section 13–1 we considered an object moving in a straight line given by the function $y = f(t)$ at time t. Its velocity, $v(t)$, is given by $v(t) = f'(t)$. The rate of change of velocity is defined to be *acceleration, a(t)*. Thus if $f(t)$ is the position function, then $f'(t) = v(t)$ is the velocity and $f''(t) = a(t)$ is the acceleration. Examples 5 and 6 illustrate how to compute acceleration.

EXAMPLE 5

An object is moving in a straight line with its position at time t given by $f(t) = t^3 - 12t^2 + 36t + 1$.

Problem

Find the value of t where (a) its velocity is 0 and (b) its acceleration is 0.

Solution Find the first and second derivatives.

$f'(t) = 3t^2 - 24t + 36 = 3(t^2 - 8t + 12)$
    $= 3(t - 6)(t - 2)$
$f''(t) = 6t - 24 = 6(t - 4)$

(a) Set $v(t) = f'(t) = 3(t - 6)(t - 2) = 0$. Then $v(t) = 0$ when $t = 6$ and when $t = 2$.
(b) Set $a(t) = 6(t - 4) = 0$. Then $a(t) = 0$ when $t = 4$.

An object comes to rest when $v(t) = 0$. So the object in Example 5 comes to rest at $t = 2$ and $t = 6$. In the next chapter we will learn that when the acceleration is 0, the object's speed changes from slowing down to speeding up.

When studying the velocity and acceleration of an object that is in motion in a vertical direction, usually the positive direction is upward and the negative direction is downward. This convention is used in the next example.

EXAMPLE 6

A ball is thrown straight upward. The ball's vertical distance $d(t)$ feet from the ground after t seconds is measured by the function

$d(t) = -16t^2 + 128t + 6$

Problem

(a) Find the velocity after $t = 1$ and $t = 6$ seconds.
(b) Find the acceleration when $t = 1$ second.
(c) At what value of t is the velocity equal to -32 ft/sec?
(d) When is the acceleration equal to -32 ft/sec^2?
(e) As far as the motion of the ball is concerned, what does it mean to say that the acceleration is a constant function?

Solution (a) First find the velocity function $v(t) = d'(t)$.

$$v(t) = d'(t) = -32t + 128$$

Evaluate $v(t)$ at $t = 1$ and $t = 6$.

$$v(1) = -32 + 128 = 96$$
$$v(6) = -192 + 128 = -64$$

When $t = 1$ second the velocity of the ball is 96 ft/sec. When $t = 6$ seconds the velocity is -64 ft/sec. The negative sign means that the ball is in a downward direction; that is, it is falling.
(b) Compute the acceleration function $a(t) = d''(t)$.

$$a(t) = d''(t) = -32$$

Evaluate $a(t)$ at $t = 1$.

$$a(1) = -32$$

When $t = 1$ second the acceleration is -32 ft/sec^2.
(c) Let $v(t)$ equal -32 and solve for t.

$$v(t) = -32$$
$$-32t + 128 = -32$$
$$-32t = -32 - 128 = -160$$
$$t = 5$$

When $t = 5$ seconds the velocity is -32 ft/sec.
(d) The acceleration function is the constant function $a(t) = -32$. This means that $a(t)$ is -32 ft/sec^2 when evaluated at every value of t.
(e) The second derivative measures the rate of change of the first derivative. Therefore the acceleration measures the rate of change of the velocity. When the ball is thrown in the air the only force acting on it is the force of gravity. It is a constant acceleration of -32 ft/sec^2. Gravity causes the velocity to decrease at this constant rate.

The second derivative is used in business to measure the rate of change of various quantities. For example, the derivative of cost is marginal cost. The rate of change of marginal cost is the derivative of marginal cost, which is the second derivative of cost.

EXAMPLE 7

A firm determines that its cost $C(x)$ of producing x units is given by the function

$$C(x) = 24 - 5x + 0.4x^2 - 0.01x^3$$

Problem

Find the derivative of the marginal cost function.

Solution The marginal cost function $MC(x)$ is

$$MC(x) = C'(x) = -5 + 0.8x - 0.03x^2$$

The derivative of $MC(x)$ is

$$MC'(x) = C''(x) = 0.8 - 0.06x$$

In the next chapter we will see that $MC'(x)$ gives information about $C(x)$.

In an example showing how firms such as fast food chains use price and demand to determine maximum profit, McCarty* uses the second derivative of a profit function. The next example shows how to compute the second derivative of this profit function.

EXAMPLE 8

The profit $P(Q)$ of a firm producing a quantity of Q goods is estimated to be

$$P(Q) = -50 + 390Q - 70.8Q^2$$

Problem

Find the second derivative of $P(Q)$.

Solution Compute $P'(Q)$ and then $P''(Q)$.

$$P'(Q) = 390 - 141.6Q$$
$$P''(Q) = -141.6$$

*Marilu Hurt McCarty, *Managerial Economics*, Glenview, IL, Scott, Foresman and Co., 1986, pp. 350–352.

EXERCISE SET 13–4

In Problems 1 to 14 find $f''(x)$.

1. $f(x) = x^2 + 1$

2. $f(x) = x^3 - 2$

3. $f(x) = 2x^4 - 7$

4. $f(x) = 3x^5 + 1$

5. $f(x) = 4x^{-1} + 5$

6. $f(x) = 2x^{-2} - 1$

7. $f(x) = x^{-3} + 2$

8. $f(x) = 3x^{-4} - 4$

9. $f(x) = x^4 - x^2$

10. $f(x) = x^5 + 3x^2$

11. $f(x) = 2x^{-1/3} + 7x^{-1/4} - 5$

12. $f(x) = 6x^{-1/6} - 3x^{-3/7} - 6x$

13. $f(x) = x^{0.3} + 3x^{0.4}$

14. $f(x) = 1.2x^{-1.5} - 3.1x^{-2.2}$

In Problems 15 to 18 compute $D_x^2 f(x)\big|_{x=2}$.

15. $f(x) = x^2 + 2x^3$

16. $f(x) = 2x^2 + x^4 - 5$

17. $f(x) = x^{-2} - 2$

18. $f(x) = 2x^{-3} + x^{-4}$

In Problems 19 to 22 compute the second derivative.

19. $g(t) = t^2 + 2t^3$

20. $s(u) = 2u^2 + u^3 - 5$

21. $F(g) = g^{1/2} - 2g$ 22. $A(r) = 3r^{-1/3} + r^{-1}$

In Problems 23 to 26 an object is moving in a straight line with its position at time t given by $f(t)$. Find the value of t where (a) its velocity is 0 and (b) its acceleration is 0.

23. $f(t) = t^2 - 2t + 3$ 24. $f(t) = 2t^2 - 4t + 5$

25. $f(t) = t^3 - 2t^2 + t + 5$

26. $f(t) = 3t^3 - t^2 + 4t$

In Problems 27 and 28 a ball is thrown straight upward. The function $d(t)$ is the distance the ball is from the ground after t seconds. Find the acceleration function.

27. $d(t) = -16t^2 + 132t + 10$

28. $d(t) = -16t^2 + 164t + 5$

In Problems 29 and 30 the cost $C(x)$ of a firm producing x units is given. Find the derivative of the marginal cost function.

29. $C(x) = 120 - 22x + 0.9x^2 - 0.003x^3$

30. $C(x) = 233 - 32x + 0.5x^2 - 0.04x^3 + 0.001x^4$

In Problems 31 and 32 the profit $P(x)$ of a firm producing x units is given. Find the second derivative of the profit function.

31. $P(x) = -134 + x(122 - 0.4x)^2$

32. $P(x) = -533 + x(455 - 3.5x)^2$

In Problems 33 to 38 find $f''(x)$.

33. $f(x) = (x^2 + 1)^2$ 34. $f(x) = (x^3 + 4)^2$

35. $f(x) = (2x^2 + 1)^{-1}$ 36. $f(x) = (x^3 + 4)^{-2}$

37. $f(x) = (x^2 + 1)^{1/2}$ 38. $f(x) = (x^3 + 4)^{2/3}$

In Problems 39 to 42 find the values of x for which $f''(x) = 0$.

39. $f(x) = x^4 - 4x^3$ 40. $f(x) = x^4 + 2x^3$

41. $f(x) = x^4 - 6x^2$ 42. $f(x) = x^5 + 10x^3$

In Problems 43 to 46 find $f^{(3)}(x)$ and $f^{(4)}(x)$.

43. $f(x) = x^4 + 3x$

44. $f(x) = 2x^4 - x^3 + x$

45. $f(x) = 2x^{-1} + x^{-2}$

46. $f(x) = (1 + 2x)^{-1}$

In Problems 47 and 48 find $f^{(n)}(x)$.

47. $f(x) = x^{-2}$ 48. $f(x) = (1 - x)^{-1}$

Referenced Exercise Set 13–4

1. When a vendor supplies a business with a large amount of a product, the recipient often selects a small sample to test the quality. If the sample has an acceptable quality level, then the entire lot is accepted. There are two costs involved: the cost of using too many defectives when the lot was accepted but should have been rejected, and the cost of rejecting the lot. The breakeven point is defined to be that fraction of defectives where the two costs are equal. If the fraction of defectives in the sample is greater than the breakeven point, then the lot is rejected. Edgar Dawes,* Director of Quality for the Dictaphone Corporation, illustrates how the breakeven point can be calculated by using the second derivative. In one particular sampling plan he defines the operating characteristic curve $y = f(x)$, where f measures the fraction of times the lot should be accepted when x defectives are found. Thus $f(0) = 1$ implies that when no defectives are found the lot should always be accepted. The function is approximated by

$$f(x) = 960x^3 - 180x^2 + 1$$

The function is defined in $[0, 0.16]$. In the next chapter we will see that the breakeven point occurs when $f''(x) = 0$. Determine the breakeven point for this sampling plan.

2. In an article studying local public goods, such as police and fire protection, Hayes† used empirical data from 90 Illinois municipalities. The cost $C(x)$ of these services can be estimated by the function

$$C(x) = 0.5x^{-0.7}$$

Find the derivative of the marginal cost function.

Cumulative Exercise Set 13–4

1. Find $f'(x)$ for $f(x) = 40 - 3x^{-2} - x^{0.4} + 5x$.

2. Find the equation of the tangent line to $f(x) = (x^3 - 4)^2$ at $x = 1$.

3. Find $f'(x)$ for $f(x) = x^3(2 - x)(1 - x^2)$.

*Edgar W. Dawes, "Optimizing Attribute Sampling Costs—A Case Study," *American Society of Quality Control—Technical Conference Transactions,* 1972, pp. 181–190.

†K. Hayes, "Local Public Goods Demands and Demographic Effects," *Applied Economics,* Vol. 18, 1986, pp. 1039–1045.

4. The total cost of producing stereos is given by $C(x) = 5000 + 25x(1 - 2x^2)$. Find the marginal cost.

In Problems 5 to 8 find $f'(x)$.

5. $f(x) = (3x^4 + x^5)^3$

6. $f(x) = (x^{1/3} + 2x^{-1} + 1)^{-2}$

7. $f(x) = (7x^3 + 2)(1 - 3x)^2$

8. $f(x) = x^3(1 - x^2)^{-1}$

In Problems 9 and 10 find $f''(x)$.

9. $f(x) = x^4 - 3x^{-2}$.

10. $f(x) = (x^4 + x)^2$

11. A ball is thrown straight upward. The function $d(t) = -16t^2 + 164t + 10$ is the distance the ball is from the ground after t seconds. Find the acceleration function.

12. Find $f^{(3)}(x)$ for $f(x) = x^5 - x^{-5}$.

CASE STUDY   Radiotherapy Treatment Design*

This case study describes a new type of weapon used in the treatment of cancer. It discusses the mathematical models developed by David Sonderman and Philip Abrahamson to help cancer therapists devise a comprehensive treatment for their patients.

Cancer therapy   Cancer is second only to heart disease as a cause of human mortality. One-fourth of all deaths of Americans are caused by cancer. After surgery the most important treatment of cancer is radiation therapy. More than half of all tumors can be treated by radiation therapy, which consists of directing ionizing radiation from external beams through a patient's body to the cancer cells. However, the radiation damages healthy cells along with cancerous tissue. Therefore it is necessary to develop a specific strategy of intensity and direction of the radiation for each individual patient in order to minimize the effect of the treatment on healthy cells. The process of selecting the intensity and direction of the beams is called *radiotherapy treatment design*.

It is common practice to develop a treatment design by using a trial-and-error process. The designer studies the patient's anatomy and proposes a treatment by using a series of several beams with various intensities and directions. Computer software is then used to calculate the effect of the radiation at specific points in the patient's body. If the designer is not satisfied with the predicted outcome of the plan, the design is modified or an entirely different treatment is proposed. The new treatment design is then evaluated, and the procedure continues until the predicted outcome is acceptable. This process is time-consuming and offers little guarantee that the design finally implemented will satisfy all the important constraints of the treatment, especially eliminating the cancer cells and minimizing the damage to healthy tissue.

Today many researchers are developing treatment design techniques that use mathematical optimization methods. Mathematical programming and calculus optimization techniques are replacing the trial-and-error process to produce better

*This case study is based on David Sonderman and Philip Abrahamson, "Radiotherapy Treatment Design Using Mathematical Programming Models," *Operations Research*, Vol. 33, No. 4, 1985, pp. 705–725.

treatment designs in much less time. These new rigorous techniques also ensure that all important constraints are satisfied.

Ionizing radiation   Radiation from external beams is administered by sophisticated machines that accurately direct radiation that can vary in energy from several thousand electron volts to more than 40 million electron volts. Radiation dosage is measured in units called rads. While different patients require different dose levels, a common range is 3000 to 7000 rads. The human body cannot withstand such a large dose in one treatment so it is delivered in several treatments of smaller doses over a span of three to six weeks. However, the treatment design is based only on the total dose and its distribution throughout the body. Once the treatment design is formulated, the patient's requirements are determined to decide how often treatments will be given and what fractions of the total dose will be administered at various intervals.

Malignant tissue is slightly more sensitive to radiation than healthy tissue, so the radiation dose at each point of the treatment region must be large enough to kill the cancerous cells but small enough to spare the healthy cells. Tumor cells are usually microscopically interspersed among healthy cells. If the total dose given at a particular point of the tumor is too low, then cancerous tissue will survive and the disease will recur. Too large a dose can cause serious side effects. Therefore the treatment design must include a homogeneous dose distribution over the tumor region.

The total integral dose versus homogeneity   The sum of the individual doses over a finite set of grid points throughout the treatment region is called the *total integral dose* and is denoted by A. The term *homogeneity* is defined to be the percentage of the tumor region that is covered by the radiation beams. If the only concern was to kill the cancerous tissue, then the homogeneity would be set at 100%. But as the homogeneity increases so does the number of healthy cells that are killed. This means that the total integral dose, A, is a function of the homogeneity, u. As u increases so does A.

For almost all patients there is a very gradual increase in A when u ranges between 80% and 85%. However, there is a dramatic increase in A when u approaches 95%. Therefore most treatment designers set u between 80% and 95%. The rule of thumb when designing a treatment is to increase u as much as possible until the increase in A is too great. This means that the rate of increase of A with respect to u, which is measured by the derivative, $A'(u)$, must not be greater than a specific value, usually set at $\frac{2}{3}$ or 0.67. That is, u can be increased as long as $A'(u)$ is less than 0.67.

A specific case   To illustrate their methods, Sonderman and Abrahamson present a specific case of a patient suffering from lung cancer. After studying the patient's anatomy, they used their linear programming techniques to devise a treatment based on maximizing u and minimizing A, subject to various constraints, including the direction and intensity of the beams, the location of the tumor, and the ability of the patient's anatomy to withstand a certain level of

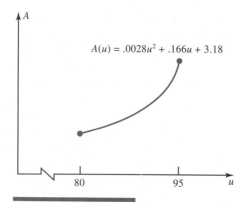

FIGURE 1 *Source:* David Sonderman and Philip Abrahamson, *Operations Research,* Vol. 33, No. 4, 1985, pp. 705–725.

radiation. The final step is to determine the proper value of u. Their model produced the relationship between A and u, graphed in Figure 1, where the domain of A is [80, 100]; that is, u ranges from 80% to 100%. The graph can be approximated in the interval [80, 95] by the function

$$A(u) = 0.0028u^2 + 0.166u + 3.18$$

Figure 1 shows that there is a gradual increase in A when u increases from 80% to 85% but when u is larger than 95% A shows a dramatic rise. This case is typical. To find the value of u for the treatment, first find the derivative of A.

$$A'(u) = 0.0056u + 0.166$$

To demonstrate the behavior of the rate of change of A, we calculate some values of A': $A'(80) = 0.614$, $A'(85) = 0.642$, $A'(90) = 0.670$, and $A'(95) = 0.698$. The rule of thumb shows that the designer should set u at a value of about 90%. At this level of homogeneity the total integral dose of radiation to healthy cells is $A(90) = 40.8$.

The article closes with a comparison of the standard practice method of design treatment versus the mathematical optimization method. The standard practice design would either have yielded a smaller value of u than 90% to produce the same value of A, or, if the same value of u was used, the value of A would have been 49.6. Thus the mathematical model produced a savings of about 22% of A for the same homogeneity.

Case Study Exercises

Suppose a particular patient's profile yielded the stated function of A with respect to u. Find the value of u such that $A'(u) = \frac{2}{3}$.

1. $A(u) = 0.003u^2 + 0.18u + 3.5$
2. $A(u) = 0.003u^2 + 0.15u + 3.1$
3. $A(u) = 0.002u^2 + 0.33u + 2.5$
4. $A(u) = 0.002u^2 + 0.302u + 1.1$

CHAPTER REVIEW

Key Terms

13–1   A Few Elementary Rules

Marginal Product

Derivative of a Constant

Derivative of a Sum

Derivative of a Constant Times a Function

13–2   Product and Quotient Rules

Product Rule

Quotient Rule

Lerner Index

13–3   The Chain Rule

General Power Function

Generalized Power Rule

Chain Rule

13–4   Higher Derivatives

Second Derivative

Third Derivative

Higher Derivatives

Summary of Important Concepts

Formulas

$D_x(c) = 0$ for any constant c

$D_x(x^r) = rx^{r-1}$

$D_x[cf(x)] = cD_x[f(x)]$ for any constant c

$D_x[f(x) + g(x)] = D_x[f(x)] + D_x[g(x)]$

If $f(x) = g(x)k(x)$ then
$f'(x) = g(x)k'(x) + k(x)g'(x)$

If $f(x) = \dfrac{g(x)}{k(x)}$ then

$$f'(x) = \frac{k(x)g'(x) - g(x)k'(x)}{[k(x)]^2}$$

$D_x[g(x)]^n = ng(x)^{n-1} D_x[g(x)]$

REVIEW PROBLEMS

In Problems 1 to 4 find $f'(x)$.

1. $f(x) = 4x^2 + x^3 + 1$

2. $f(x) = 5x^4 - 2x^{-2}$

3. $f(x) = 4x^{-5} - x^{1/3} + 1$

4. $f(x) = \dfrac{5}{x^2} - \dfrac{2}{x^{1/2}}$

5. Find the equation of the tangent line to
   $f(x) = x^{-3/2} - x$ at $x = 1$

In Problems 6 to 16 find $f'(x)$.

6. $f(x) = (x^4 - 7)(11x^3 + 1)$

7. $f(x) = \dfrac{x^3}{x^2 + 2}$

8. $f(x) = \dfrac{x^3 - 2}{x^2 + 2x}$

9. $f(x) = \dfrac{x^{1/3}}{x^{4/5} + 2}$

10. $f(x) = (x^3 + 2)^3$

11. $f(x) = (2x^3 + x)^4$

12. $f(x) = (2 - 3x^{-1})^{-2}$

13. $f(x) = 3x^2 + 3(x^3 + 4x)^3$

14. $f(x) = x^2(x^3 + 2x)^3$

15. $f(x) = (x^2 + 1)^2(x^3 + 1)^3$

16. $f(x) = \dfrac{x^2}{(x^2 + 3x)^2}$

17. Find an equation of the tangent line to
   $f(x) = (x^3 + 3)^2$ at $x = 1$.

In Problems 18 and 19 find $f''(x)$.

18. $f(x) = 6x^{1/2} - 3x^{-3} - 6x$

19. $f(x) = (x^3 + 4)^{-2}$

20. Find the values of x for which $f''(x) = 0$ where
   $f(x) = x^3 - 4x^2$

21. Find $f''(x)$ for $f(x) = x^{-3}$.

PROGRAMMABLE CALCULATOR EXPERIMENTS

In Problems 1 to 4 write a program to compute the derivative of the given type of function.

1. $f(x) = x^n$

2. $f(x) = kx^n + b$

3. $f(x) = (a + bx)^m$

4. $f(x) = x^n(a + bx)^m$

Applications of the Derivative

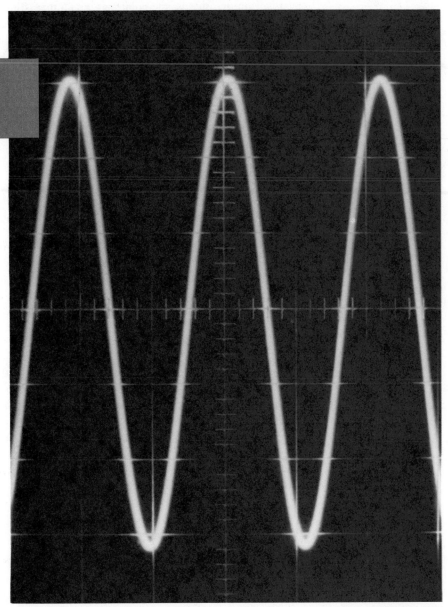

14

Graphs can refer to maximum profits at minimum cost, as described in this chapter.
(Zefa/H. Armstrong Roberts)

Chapter Overview

When a real-world application uses rate of change, the natural tool is the derivative. This chapter shows how the derivative measures various properties about functions that arise in applications. These include rate of inflation, maximum profit, the most efficient route, minimum cost, and minimum traffic noise.

The first section deals with the effect the derivative has in determining where a function is increasing and where it is decreasing. Section 14–2 covers relative extrema, that is, the point on a graph that is higher or lower than any other point on the graph in the vicinity. The derivative is used to determine where a function has a relative maximum and where it has a relative minimum. The second derivative also gives useful information about the function. It helps to determine where a function is concave up or concave down, which means, intuitively, where the function is bending up or bending down.

The next section applies the ideas on relative extrema to problems where an absolute extremum is sought. Section 14–5 deals with problems in which two variables are changing. These are called related rates problems and they occur often in applications.

CASE STUDY PREVIEW

One of the most important considerations of highway operation is safety. Obtaining optimum flow of traffic is also desired, but frequently these two goals work at odds with each other. The case study tells the story of how highway administrators in Houston used calculus to minimize acceleration noise, which is defined as the disturbance of a vehicle's speed from uniform speed. Acceleration noise is a good measure of the threat of hazardous conditions. So minimizing acceleration noise produces safer traffic conditions. Researchers found that the optimum speed for minimizing acceleration noise is very close to the speed that maximizes traffic volume.

14–1 Increasing and Decreasing Functions

One of the most important applications of calculus centers on when a function is increasing and when it is decreasing. A firm wants to know when profit is increasing and when cost is decreasing. A political party studies the latest polls to find out if its percentage of the vote is increasing or decreasing. In this section we study when graphs of functions are increasing and decreasing. The definition of increasing and decreasing functions has a firm geometrical foundation. However, it is also necessary to have an algebraic means for determining where a function is increasing and decreasing. The primary tool is the derivative of the function.

An Application: Laffer's Curve

Let us start by studying a concept from business that depends on when a function is increasing or decreasing.

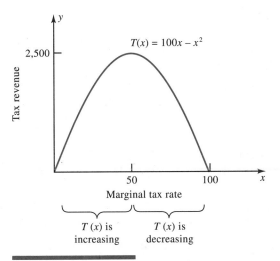

F I G U R E 1 *Source:* Milton H. Spenser, *Contemporary Economics,* 6th ed., New York, Worth Publishers, 1986, p. 293.

Supply-side economics states that governments should stimulate production by giving incentives to business in order to increase economic growth. Other economists believe the best way to increase economic growth is to increase demand by giving incentives to consumers in the way of tax benefits. At the heart of the matter is the concept of *marginal tax rates,* the taxes paid on the last increment of income. Should the government increase or decrease marginal tax rates? Economist Arthur Laffer answers the question with a graph, called the *Laffer curve.** It measures a tax revenue function, $T(x)$, where x is the marginal tax rate measured in percent. A typical Laffer curve is given in Figure 1, where $T(x) = 100x - x^2$. For this curve, tax revenue is highest when $x = 50\%$. From $x = 0\%$ to $x = 50\%$, $T(x)$ is increasing. This means that if the marginal tax rate were at a value from 0% to 50%, an increase in x that is not beyond 50% would generate larger tax revenue. But if x is greater than 50%, an increase in the marginal tax rate would generate less tax revenue. It is this concept of when a function is increasing and decreasing that we study in this section.

Increasing and Decreasing Functions

Consider the graph in Figure 2. Intuitively, the graph is increasing from point A to point B because it goes up as x increases from left to right. The curve is decreasing from B to C because it goes down from left to right. Thus the property of an increasing or decreasing function implies the convention that x always moves from left to right. From the point of view of the x-axis, the curve in Figure 2 is increasing from $x = a$ to $x = b$ and decreasing from $x = b$ to $x = c$.

*Milton H. Spenser, *Contemporary Economics,* 6th ed., New York, Worth Publishers, 1986, p. 293.

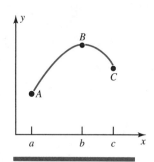

FIGURE 2

DEFINITION

A function $f(x)$ is said to be **increasing** from $x = a$ to $x = b$ if for any x_1 and x_2 in the interval (a, b)

when $x_1 < x_2$, then $f(x_1) < f(x_2)$

A function $f(x)$ is **decreasing** from $x = a$ to $x = b$ if for any x_1 and x_2 in the interval (a, b)

when $x_1 < x_2$, then $f(x_1) > f(x_2)$

A function $f(x)$ is said to be *increasing at* $x = a$ if $f(x)$ is increasing in an interval containing a, and $f(x)$ is said to be *decreasing at* $x = a$ if $f(x)$ is decreasing in an interval containing a.

EXAMPLE 1

Problem

Consider $f(x) = x^2$ whose graph is in Figure 3. Find the intervals where $f(x)$ is increasing and where $f(x)$ is decreasing.

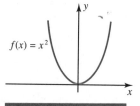

$f(x) = x^2$

FIGURE 3

Solution Because $f(x)$ is going down from left to right for all $x < 0$, $f(x)$ is decreasing in the interval $(-\infty, 0)$. For any x_1 and x_2 in $(-\infty, 0)$, if $x_1 < x_2$, then $f(x_1) > f(x_2)$ because $x_1^2 > x_2^2$. For example, $-2 < -1$ and $f(-2) = (-2)^2 = 4 > 1 = f(-1)$. Similarly $f(x)$ is increasing in $(0, \infty)$.

A Test for Increasing and Decreasing Functions

The primary algebraic tool for determining the intervals where a function is increasing and where it is decreasing is the derivative. For example, if $f(x) = x^2$, then $f'(x) = 2x$. This says that the slope of the tangent line of $f(x)$ is negative for all $x < 0$ and the slope of the tangent line is positive for all $x > 0$. Coupling this fact with Example 1 means that the function $f(x) = x^2$ is increasing when $f'(x) > 0$ and decreasing when $f'(x) < 0$. This is true in general. If the derivative is positive for every value in an interval, then the function is increasing in that interval. A negative derivative means the function is decreasing.

> **Test for Increasing and Decreasing Functions**
> If $f(x)$ is differentiable in an interval (a, b), then
>
> 1. $f(x)$ is increasing in (a, b) if $f'(x) > 0$ for all $x \in (a, b)$
> 2. $f(x)$ is decreasing in (a, b) if $f'(x) < 0$ for all $x \in (a, b)$

If $f'(x) = 0$ at $x = a$, then the slope of the tangent line of $y = f(x)$ at $x = a$ is 0. The tangent line is horizontal at such a point. These are the points that often divide the intervals where $f(x)$ changes from increasing to decreasing.

The function's derivative can also change sign at points where $f'(x)$ does not exist. For example, the derivative of $f(x) = x^{2/3}$ is $f'(x) = \frac{2}{3}x^{-1/3}$, and so $f'(0)$ does not exist. Since $f'(x) < 0$ for all x in $(-\infty, 0)$, $f(x)$ is decreasing to the left of $x = 0$; and since $f'(x) > 0$ in $(0, \infty)$, $f(x)$ is increasing to the right of $x = 0$. Therefore, at $x = 0$, the function's derivative changes sign and the graph changes from decreasing to increasing. The graph of $f(x) = x^{2/3}$ is given in Figure 4. The graph has a "cusp" at $x = 0$. Note that from Figure 4 the tangent line of $f(x) = x^{2/3}$ at $x = 0$ exists—it is the y-axis. But the slope of the tangent line is undefined because the tangent line is a vertical line which has no slope. Algebraically, we have $f'(0) = \dfrac{2}{3 \cdot 0} = \dfrac{2}{0}$, which is undefined.

This suggests how to determine the intervals where $f(x)$ is increasing and intervals where $f(x)$ is decreasing. First find the values $x = a$ where $f'(a) = 0$ or $f'(a)$ does not exist (but $f(a)$ does exist). These values of x are called the **critical**

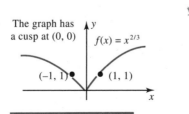

The graph has a cusp at (0, 0) $f(x) = x^{2/3}$
$(-1, 1)$ $(1, 1)$

FIGURE 4

values of $f(x)$. The critical values divide the domain of the function into distinct intervals where $f(x)$ is either increasing or decreasing. To determine the sign of $f'(x)$ in an interval, it is enough to select one value in the interval, say, $x = c$, and then to find $f'(c)$. If $f'(c) > 0$, then $f'(x)$ is positive for all values in the interval. Thus $f(x)$ is increasing in that interval. We treat $f'(c) < 0$ in a similar manner.

Procedure for Determining Where $f(x)$ Is Increasing and Decreasing

1. Find the critical values of $f(x)$, that is, the values $x = a$ where $f'(a) = 0$ or $f'(a)$ does not exist (but $f(a)$ does exist).
2. Determine the distinct intervals where $f(x)$ is either increasing or decreasing using the values in step 1 as end points.
3. For each interval select a value c in the interval and find $f'(c)$.

   i. If $f'(c) > 0$, then $f'(x) > 0$ for all x in that interval.

   ii. If $f'(c) < 0$, then $f'(x) < 0$ for all x in that interval.

4. Apply the test for increasing and decreasing functions.

EXAMPLE 2

Problem

Find the intervals where $f(x) = x^2 - 4x + 3$ is increasing and the intervals where $f(x)$ is decreasing.

Solution Let us follow the steps in the procedure.

1. Compute $f'(x) = 2x - 4$. Solve $f'(x) = 2x - 4 = 0$. This implies $x = 2$. Thus $f'(2) = 0$. There are no values where $f'(x)$ does not exist.
2. This divides the real line into $(-\infty, 2)$ and $(2, \infty)$.
3. Select $0 \in (-\infty, 2)$ and $3 \in (2, \infty)$. (Remember, any suitable values can be chosen.)
   a. for $x = 0$, $f'(0) = -4$, so $f'(x) < 0$ in $(-\infty, 2)$
   b. for $x = 3$, $f'(3) = 2$, so $f'(x) > 0$ in $(2, \infty)$
4. a. $f(x)$ is decreasing in $(-\infty, 2)$ because $f'(x) < 0$ in $(-\infty, 2)$.
   b. $f(x)$ is increasing in $(2, \infty)$ because $f'(x) > 0$ in $(2, \infty)$.

This information is recorded in the chart below.

Test point		
$f'(x) = 0$ at		

Test point   0   3

$f'(x) = 0$ at   2 x

        $f'(0) = -4$ $f'(3) = 2$

Sign of $f'(x)$ $-------$ $+++++++++$

$f(x)$ is   decreasing   increasing

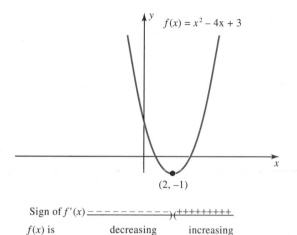

Sign of $f'(x)$ ----------)(++++++++++

$f(x)$ is   decreasing   increasing

FIGURE 5

The graph of $f(x) = x^2 - 4x + 3$ is given in Figure 5. Example 2 shows where $f(x)$ is increasing and where $f(x)$ is decreasing. Also note that there is a horizontal tangent line at $x = 2$ because $f'(2) = 0$. From these facts the graph can be obtained by plotting just a few points. This demonstrates that knowing where a function is increasing and where it is decreasing is a useful tool in graphing the function. In the next few sections we discuss other properties of the graphs of functions.

Example 3 looks at a more complicated function.

EXAMPLE 3

Problem

Find the intervals where $f(x) = x^3 - 6x^2 + 9x$ is increasing and where $f(x)$ is decreasing. Locate the points where the graph has a horizontal tangent line. Graph the function.

Solution

1. Solve $f'(x) = 0$ where

   $f'(x) = 3x^2 - 12x + 9 = 3(x^2 - 4x + 3) = 3(x - 1)(x - 3)$

   Then $3(x - 1)(x - 3) = 0$ implies $x = 1$ or $x = 3$. Therefore $f'(1) = f'(3) = 0$. There are no values where $f'(x)$ does not exist.
2. This divides the real line into the three intervals $(-\infty, 1)$, $(1, 3)$, and $(3, \infty)$.
3. Select $0 \in (-\infty, 1)$, $2 \in (1, 3)$, and $4 \in (3, \infty)$.
   a. for $x = 0$, $f'(0) = 9$, so $f'(x) > 0$ in $(-\infty, 1)$
   b. for $x = 2$, $f'(2) = -3$, so $f'(x) < 0$ in $(1, 3)$
   c. for $x = 4$, $f'(4) = 9$, so $f'(x) > 0$ in $(3, \infty)$
4. a. $f(x)$ is increasing in $(-\infty, 1)$ because $f'(x) > 0$ in $(-\infty, 1)$.
   b. $f(x)$ is decreasing in $(1, 3)$ because $f'(x) < 0$ in $(1, 3)$.
   c. $f(x)$ is increasing in $(3, \infty)$ because $f'(x) > 0$ in $(3, \infty)$.

$f(x)$ has a horizontal tangent line when $f'(x) = 0$. Since $f'(1) = f'(3) = 0$, $f(x)$ has horizontal tangent lines at $x = 1$ and at $x = 3$, that is, at the points $(1, 4)$ and $(3, 0)$. This information is recorded in the following chart.

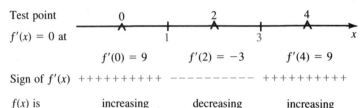

The graph is in Figure 6.

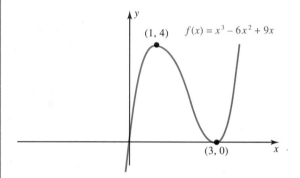

FIGURE 6

EXAMPLE 4

Problem

Let $f(x) = x + \dfrac{4}{x}$. Find the intervals where $f(x)$ is increasing and where $f(x)$ is decreasing. Locate the points where the graph has a horizontal tangent line. Graph the function.

Solution

1. Set $f'(x) = 1 - 4/x^2$ equal to 0. Then $1 = 4/x^2$, so $x^2 = 4$. Hence $x = \pm 2$ are critical values. Also, $f'(x)$ does not exist at $x = 0$, but $x = 0$ is not a critical value because $f(0)$ does not exist.
2. This divides the real line into the four intervals $(-\infty, -2)$, $(-2, 0)$, $(0, 2)$, and $(2, \infty)$.
3. Select $-3 \in (-\infty, -2)$, $-1 \in (-2, 0)$, $1 \in (0, 2)$, and $3 \in (2, \infty)$.
   a. for $x = -3$, $f'(-3) = \frac{5}{9}$, so $f'(x) > 0$ in $(-\infty, -2)$
   b. for $x = -1$, $f'(-1) = -3$, so $f'(x) < 0$ in $(-2, 0)$
   c. for $x = 1$, $f'(1) = -3$, so $f'(x) < 0$ in $(0, 2)$
   d. for $x = 3$, $f'(3) = \frac{5}{9}$, so $f'(x) > 0$ in $(2, \infty)$

4. a. $f(x)$ is increasing in $(-\infty, -2)$ because $f'(x) > 0$ in $(-\infty, -2)$
    b. $f(x)$ is decreasing in $(-2, 0)$ because $f'(x) < 0$ in $(-2, 0)$
    c. $f(x)$ is decreasing in $(0, 2)$ because $f'(x) < 0$ in $(0, 2)$
    d. $f(x)$ is increasing in $(2, \infty)$ because $f'(x) > 0$ in $(2, \infty)$

$f(x)$ has a horizontal tangent line when $f'(x) = 0$. Since $f'(-2) = f'(2) = 0$, $f(x)$ has horizontal tangent lines at $x = -2$ and at $x = 2$; that is, at the points $(-2, -4)$ and $(2, 4)$. The graph approaches the y-axis asymptotically both to the left and right of 0. This information is recorded in the chart below.

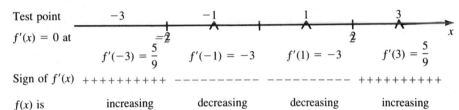

Test point	−3		−1		1		3	
$f'(x) = 0$ at		=−2				=2		
	$f'(-3) = \dfrac{5}{9}$		$f'(-1) = -3$		$f'(1) = -3$		$f'(3) = \dfrac{5}{9}$	
Sign of $f'(x)$	+ + + + + + + + + +		− − − − − − − − − −		− − − − − − − − − −		+ + + + + + + + + +	
$f(x)$ is	increasing		decreasing		decreasing		increasing	

The graph is in Figure 7.

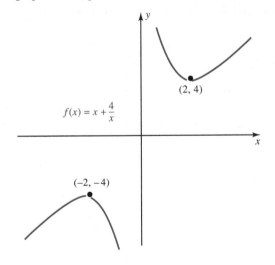

$(2, 4)$

$f(x) = x + \dfrac{4}{x}$

$(-2, -4)$

Sign of $f'(x)$ + + + + + + (− − − − − −)(− − − − − −)(+ + + + + +

$f(x)$ is   increasing decreasing decreasing increasing

FIGURE 7

An Application from Economics

The revenue of a firm depends on many factors, one of which is the quantity q of goods produced. A typical revenue function is dome-shaped as in Figure 8. When q is 0, that is, no goods are produced, the revenue $R(q)$ is 0. As q increases, so

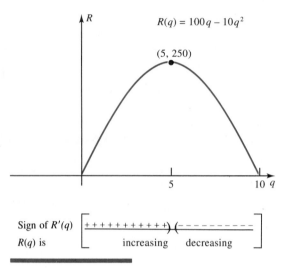

$$R(q) = 100q - 10q^2$$

(5, 250)

Sign of $R'(q)$ $\quad$ $+ + + + + + + + + + +) (- - - - - - - - - - -$

$R(q)$ is $\qquad$ increasing $\quad$ decreasing

FIGURE 8 *Source:* Based on Paul A. Samuelson, *Economics,* 10th ed., New York, McGraw-Hill Book Co., 1985, p. 479.

does $R(q)$ until it reaches a peak where maximum revenue occurs. In Figure 8 this occurs at $q = 5$. Then the values of $R(q)$ decrease for values of q greater than 5. There are several reasons that this decrease in $R(q)$ takes place. One reason is that the firm might not be able to sell such a large amount of goods. The next example computes the intervals where the revenue function in Figure 1 is increasing and where it is decreasing. This function is defined only in the interval [0, 10].

EXAMPLE 5

Problem

Determine the intervals where $R(q)$ is increasing and where $R(q)$ is decreasing.

$$R(q) = 100q - 10q^2, \qquad 0 \leq q \leq 10$$

Solution Compute $R'(q)$.

$$R'(q) = 100 - 20q$$

1. Solve $R'(q) = 0$. Then $100 - 20q = 0$ implies $q = 5$. There are no values where $R'(q)$ does not exist.
2. This divides the domain into [0, 5) and (5, 10].
3. Select $1 \in [0, 5)$ and $6 \in (5, 10]$.
   a. for $q = 1$, $R'(1) = 80$, so $R'(q) > 0$ in [0, 5).
   b. for $q = 6$, $R'(6) = -20$, so $R'(q) < 0$ in (5, 10].
4. a. $R(q)$ is increasing in [0, 5) because $R'(q) > 0$ in [0, 5).
   b. $R(q)$ is decreasing in (5, 10] because $R'(q) < 0$ in (5, 10].

EXERCISE SET 14–1

In Problems 1 to 28 find intervals where $f(x)$ is increasing and where $f(x)$ is decreasing. Find the point(s) where $f(x)$ has a horizontal tangent line. Graph the function.

1. $f(x) = x^2 + 5$

2. $f(x) = 3x^2 - 2$

3. $f(x) = 7 - 2x^2$

4. $f(x) = 1 - 3x^2$

5. $f(x) = x^2 - 9$

6. $f(x) = 4x^2 - 25$

7. $f(x) = x(x - 4)$

8. $f(x) = x(x - 8)$

9. $f(x) = x^2 + 4x - 1$

10. $f(x) = x^2 - 8x + 3$

11. $f(x) = x^3 + 3x^2 - 9x$

12. $f(x) = 2x^3 + 3x^2 - 12x + 1$

13. $f(x) = x^3 - 3x^2 - 9x$

14. $f(x) = 2x^3 + 3x^2 - 36x + 3$

15. $f(x) = -x^3 + 3x^2 + 9x$

16. $f(x) = -x^3 + 6x^2$

17. $f(x) = x + 9/x$

18. $f(x) = 4x + 25/x$

19. $f(x) = (x^2 + 12)/x$

20. $f(x) = (4x^2 + 9)/x$

21. $f(x) = x(x^2 - 4)$

22. $f(x) = x(9 - x^2)$

23. $f(x) = x^2(x - 2)$

24. $f(x) = x^2(3 - x)$

25. $f(x) = x/(x + 1)$

26. $f(x) = x/(x - 4)$

27. $f(x) = (x - 1)/(x + 1)$

28. $f(x) = (x + 3)/(x - 5)$

29. The average cost $AC(x)$ to produce the xth item of production, where $AC(x)$ is measured in dollars and x is measured in hundreds of items, is given by

$$AC(x) = 100 + 48x - 7x^2 + \frac{x^3}{3}$$

If $AC(x)$ is defined for $0 < x < 10$, find the intervals where $AC(x)$ is increasing and where it is decreasing.

30. The amount of glucose $G(t)$ in a patient's bloodstream, measured in tenths of a percent, is measured at time t in minutes during a stress test. The function is

$$G(t) = 12t - 2.1t^2 + 0.1t^3, \text{ where } 0 < t < 12$$

Find the intervals where $G(t)$ is increasing and where $G(t)$ is decreasing.

In Problems 31 to 34 use the information given to sketch the graph of $f(x)$, assuming that $f(x)$ is continuous for all real numbers x.

31. Values where $f'(x) = 0$

Sign of $f'(x)$

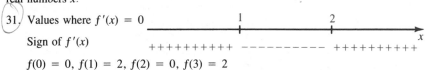

$f(0) = 0$, $f(1) = 2$, $f(2) = 0$, $f(3) = 2$

32. Values where $f'(x) = 0$

Sign of $f'(x)$

$f(-2) = 1$, $f(-1) = 0$, $f(3) = 1$, $f(4) = 0$

33. Values where $f'(x) = 0$

Sign of $f'(x)$

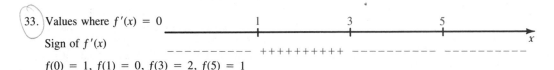

$f(0) = 1$, $f(1) = 0$, $f(3) = 2$, $f(5) = 1$

34. Values where $f'(x) = 0$

Sign of $f'(x)$

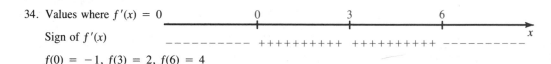

$f(0) = -1$, $f(3) = 2$, $f(6) = 4$

In Problems 35 to 42 find the intervals where $f(x)$ is increasing and where $f(x)$ is decreasing. Graph the functions.

35. $f(x) = x^2/(x - 2)$ 36. $f(x) = x^2/(1 - 2x)$

37. $f(x) = x^2(x^2 - 4)$ 38. $f(x) = x^2(9 - x^2)$

39. $f(x) = (x^2 - 1)(x^2 - 4)$

40. $f(x) = (x^2 - 3)(9 - x^2)$

41. $f(x) = (x - 5)^2(x + 1)^2$

42. $f(x) = (x + 6)^2(x - 4)^2$

In Problems 43 to 46 from the given figure find (a) the intervals where $f(x)$ is increasing, (b) the intervals and where $f(x)$ is decreasing, and (c) the sign of the derivative in each interval.

43.

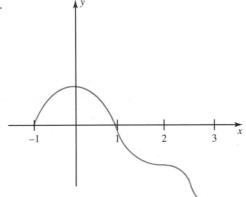

44.

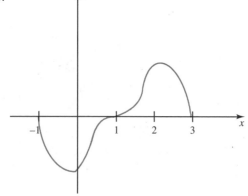

45.

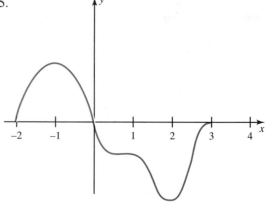

46.

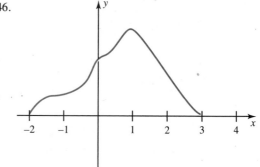

Referenced Exercise Set 14–1

1. In a paper studying the Environmental Protection Agency's motor vehicle emission testing procedures, researchers measured the risk, r, of an incorrect decision on a part versus the percentage, x, of actual deterioration of the part.* That is, laboratory standards were developed for the deterioration level of each part of the emission system. When a part reached the deterioration level where it should fail the EPA test, x was assigned the value 1. For $x < 1$, the part should pass, and it should fail if $x > 1$. (The units of the values assigned to r were arbitrary, but the higher the value the greater the risk, meaning the greater the error.) Field agents were then given the parts to test. There were two types of failure: the part should be passed and the agent failed it, and the part should fail and the agent passed it. The graph in Figure 9 was given measuring the risk r, versus the deterioration x. Explain

*George Miller et al., "Application of Sequential Testing to Motor Vehicle Emission Certification," *Decision Sciences*, Vol. 31, 1985, pp. 249–263.

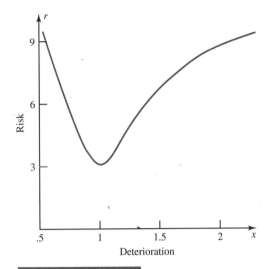

Risk

9

6

3

.5   1   1.5   2 x

Deterioration

FIGURE 9 *Source:* George Miller et al.,
"Application of Sequential Testing to Motor Vehicle
Emission Certification," *Decision Sciences,* Vol. 31,
1985, pp. 249–263.

why the curve is decreasing for $x < 1$ and increasing
for $x > 1$.

2. In the theory of traffic science a "bottleneck" is defined
as a section of roadway with a flow capacity less than
the road ahead. When the volume reaches the capacity
of the bottleneck, the velocity in the bottleneck is much
less than that ahead of the bottleneck. A further increase
in volume creates a queue in back of the bottleneck.
In an article describing why congestion lasts longer
than the interval in which demand exceeds capacity, a
graph measuring the volume during rush-hour traffic

on a large city freeway is studied.[†] The data are approximated by the function $f(t) = -25t^3 + 75t + 150$,
where $f(t)$ is the volume of automobiles passing through
the bottleneck at time t, and where $t = 0$ represents
7:00 A.M. and $t = 1$ represents 7:30 A.M. The function
is defined in the interval [0, 2], that is, from 7:00 A.M.
until 8:00 A.M.
   (a) Find where $f(t)$ is increasing and where $f(t)$ is
   decreasing.
   (b) The capacity of the bottleneck is $f(t) = 170$. From
   the graph estimate when demand exceeds capacity.

3. In an article[‡] appearing in the *Journal of Quality Progress* two administrators in the U.S. Department of Agriculture defined "tolerance" as any limit or allowance
that is specified by a contract to measure quality. One
example referred to the common practice of produce
shippers, such as egg packers, of specifying that "no
more than 20% of the eggs may be below A quality."
The primary question addressed in the article is when
should a shipment with this specification be rejected?
The function used is $y = f(x)$, where f measures the
percent of times the shipment was accepted with $x\%$
of the eggs below A quality. The function is approximated by

$$f(x) = \frac{x^3}{135} - \frac{x^2}{3} + 100$$

   (a) Find where $f(x)$ is increasing and where $f(x)$ is
   decreasing.
   (b) From the definition of $f(x)$, why is it reasonable
   to assume $f(x)$ is decreasing there?
   (c) Find $f(20)$.
   (d) This function models the data only in the interval
   [0, 30]. Give a reason for this.

[†]Karl Moskowitz and Leonard Newman, "Notes on Freeway Capacity," *Highway Resources Board
Record,* Vol. 27, 1963.

[‡]Richard P. Bartlett and Lloyd P. Provost, "Tolerances in Standards and Specifications," *Journal of
Quality Progress,* 1973, pp. 14–18.

14–2 Relative Extrema: The First Derivative Test

A common type of rate of change problem requires that a variable be maximized
or minimized. A business wants to maximize profit; a physician needs to know the
maximum amount of a drug that can be administered safely; a physiologist studies
the minimum rate of stimulus needed to achieve a given response from an organism.
The derivative plays a fundamental role in the solution of such problems.

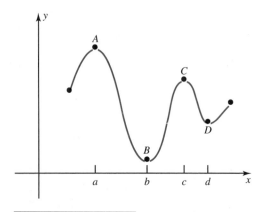

FIGURE 1

Relative Extrema

Consider the graph in Figure 1, which we use to describe the geometric meaning of extrema. Point *A* is the highest point on the curve because its second coordinate, $f(a)$, is greater than all others. We call $f(a)$ an *absolute maximum*. Similarly *B* is the lowest point of the graph so $f(b)$ is called an *absolute minimum*. Point *C* is not the absolute highest point, but it is higher than any of the points in its vicinity. For this reason $f(c)$ is called a *relative maximum*. Similarly $f(d)$ is a *relative minimum*. The term **extremum** refers to either a **maximum** or **minimum.** We also say that $f(x)$ has a relative maximum at $x = c$ and $f(x)$ has a relative minimum at $x = d$. This section deals with relative extrema and the next section considers absolute extrema.

DEFINITIONS

1. $f(a)$ is a **relative maximum** of $f(x)$ if there is an interval (c, d) containing *a* such that

   $$f(x) \leq f(a) \qquad \text{for all } x \text{ in } (c, d)$$

2. $f(a)$ is a **relative minimum** of $f(x)$ if there is an interval (c, d) containing *a* such that

   $$f(x) \geq f(a) \qquad \text{for all } x \text{ in } (c, d)$$

The first example ties together the definition and the geometric meaning of extrema. It considers a function that was graphed in Example 3 of the previous section.

EXAMPLE 1

Problem

Find the extrema of $f(x) = x^3 - 6x^2 + 9x$ by looking at the graph in Figure 2. Then show that the definition is satisfied for each extremum.

Solution The graph has a "peak" at (1, 4) and a "valley" at (3, 0). Thus the graph has a relative maximum of $f(1) = 4$ when $x = 1$ and a relative minimum of $f(3) = 0$ when $x = 3$. To see that $f(1)$ satisfies the definition of a relative maximum, we need an interval containing 1 such that the functional values of all other points in the interval are less than $f(1) = 4$. From Figure 2 it is clear that $f(x) < 4$ for all $x < 3$ (except $x = 1$), so the interval can be chosen to be $(-\infty, 3)$. Any smaller interval containing 1 will also work. (Can a larger interval be selected?) Similarly the largest interval for the relative minimum $f(3) = 0$ is $(0, \infty)$ because the $f(x) > 0$ for all $x > 0$ except $x = 3$.

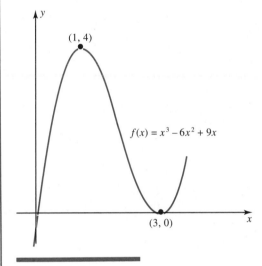

(1, 4)

$f(x) = x^3 - 6x^2 + 9x$

(3, 0)

FIGURE 2

At a relative extremum a curve either has a horizontal tangent line or the derivative does not exist. For instance, in Example 1, $f(x) = x^3 - 6x^2 + 9x$ has horizontal tangent lines at $x = 1$ and at $x = 3$. The graph in Figure 2 has a horizontal tangent line at (1, 4) and (3, 0). In the previous section it was pointed out that the graph of a function has a horizontal tangent line when $f'(x) = 0$. In Example 2, $f'(x) = 3x^2 - 12x + 9 = 3(x - 1)(x - 3)$. By solving $f'(x) = 0$, we get the values $x = 1$ and $x = 3$ where the graph has a horizontal tangent line. This shows that a relative extremum can occur at a value of $f(x)$ where $f'(x) = 0$.

The other type of value $x = a$ that sometimes yields a relative extremum is where $f'(a)$ does not exist but $f(a)$ does exist. Consider the function $f(x) = x^{2/3}$ whose graph is shown in Figure 3. From the graph we see a relative minimum at

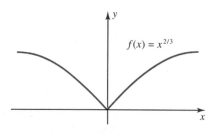

FIGURE 3

$x = 0$, and $f'(x) = \frac{2}{3}x^{-1/3}$ does not exist at $x = 0$. Thus a relative extremum can also occur at a value where $f'(x)$ does not exist.

Recall that the critical values of $f(x)$ are the values $x = a$ in the domain of $f(x)$ where $f'(a) = 0$ or where $f'(a)$ does not exist. This shows that to locate relative extrema, we need to find the critical values of a function.

But a word of caution is in order. It is not true that a relative extremum occurs at every critical value. Consider the function $f(x) = x^3$ whose graph is in Figure 4. To find the critical value(s), set $f'(x) = 3x^2$ equal to 0. This yields the only critical value $x = 0$, since the derivative exists at all values of x. The graph shows that the function has no relative extremum at $x = 0$; in fact the graph has no relative extremum anywhere.

Let us summarize these results.

A **relative extremum** of the function $f(x)$ will occur at a **critical value,** that is, a value $x = a$ in the domain such that $f'(a) = 0$ or $f'(a)$ does not exist. Not all critical values yield relative extrema.

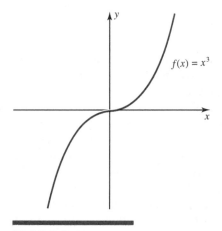

FIGURE 4

The First Derivative Test

The relative maxima and minima of $f(x)$ are computed by finding the critical values and then applying a test to see whether the extremum is (1) a relative maximum, (2) a relative minimum, or (3) neither a maximum nor a minimum. This test involves the derivative of $f(x)$ and uses the properties of $f(x)$ studied in the previous section.

Consider again $f(x) = x^3 - 6x^2 + 9x$. As shown in Figure 2 the graph has a relative maximum at $x = 1$ and a relative minimum at $x = 3$. Also in Figure 2, note that the graph is increasing to the left of $x = 1$ and decreasing to the right of $x = 1$. This is why the critical value $x = 1$ is a relative maximum. Similarly the critical value $x = 3$ is a relative minimum because the graph is decreasing to the left and increasing to the right of $x = 3$.

Recall from the previous section that the intervals where the function is increasing and the intervals where it is decreasing are found by finding where the derivative is greater than 0 and less than 0. This applies to all functions. It is called the **first derivative test.**

The First Derivative Test

To find the relative extrema of $f(x)$

1. Find the critical values of $f(x)$ and the intervals where $f(x)$ is increasing and where $f(x)$ is decreasing.
2. If $x = a$ is a critical value of $f(x)$, then
    i. $f(a)$ is a relative maximum if $f(x)$ is increasing to the left of a and decreasing to the right of a
    ii. $f(a)$ is a relative minimum if $f(x)$ is decreasing to the left of a and increasing to the right of a
    iii. $f(a)$ is not a relative extremum if $f(x)$ is
        (a) decreasing to the left and to the right of a
        (b) increasing to the left and to the right of a

This means that for a critical value $x = a$, $f(a)$ is a relative maximum if $f'(x)$ changes from positive to the left of $x = a$ to negative to the right of $x = a$; $f(a)$ is a relative minimum if $f'(x)$ changes from negative to the left of $x = a$ to positive to the right of $x = a$; $f(a)$ is not a relative extremum if $f'(x)$ does not change sign.

E X A M P L E   2

Problem

Find the relative extrema of $f(x) = (x - 2)^2(x - 5)$. Sketch the graph of the function.

Solution We follow the steps in the first derivative test.

1. Find $f'(x)$.

$$f'(x) = 2(x - 2)(x - 5) + (x - 2)^2 = (x - 2)(2x - 10 + x - 2)$$

$$= 3(x - 2)(x - 4)$$

Solve $f'(x) = 3(x - 2)(x - 4) = 0$. This yields $x = 2$ and $x = 4$. These are the only critical values.

Solve $f'(x) > 0$ and $f'(x) < 0$. The intervals are $(-\infty, 2)$, $(2, 4)$, and $(4, \infty)$. Choose test points, say, 0, 3, and 5. Since $f'(0) = 24$, $f'(3) = -3$ and $f'(5) = 9$, $f(x)$ is increasing in $(-\infty, 2)$ and $(4, \infty)$, while $f(x)$ is decreasing in $(2, 4)$. Record these facts in a chart as follows.

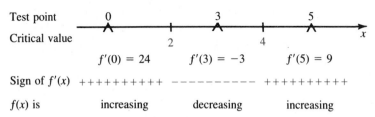

Test point	0		3		5	

Critical value ... 2 ... 4

$f'(0) = 24$ $f'(3) = -3$ $f'(5) = 9$

Sign of $f'(x)$ ++++++++++   ----------   ++++++++++

$f(x)$ is   increasing   decreasing   increasing

2. $f(2) = 0$ is a relative maximum because $f(x)$ is increasing to the left of $x = 2$ and decreasing to the right of $x = 2$.
$f(4) = -4$ is a relative minimum because $f(x)$ is decreasing to the left of $x = 4$ and increasing to the right of $x = 4$.

From this information and by plotting the points $(2, 0)$ and $(4, -4)$, we can sketch the graph as shown in Figure 5.

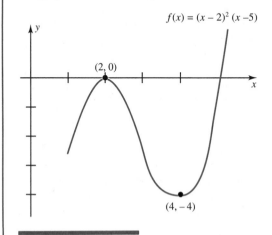

$$f(x) = (x - 2)^2 (x - 5)$$

(2, 0)

(4, -4)

FIGURE 5

Example 3 applies the first derivative test to a fourth degree polynomial. In general, the derivative of a fourth degree polynomial is a third degree polynomial.

It is not always easy to factor a third degree polynomial but the ones given in this text should be fairly straightforward.

EXAMPLE 3

Problem

Find the relative extrema, and sketch the graph of the function

$$f(x) = x^4 - 8x^3 + 16x^2$$

Solution

1. Find $f'(x)$.

$$f'(x) = 4x^3 - 24x^2 + 32x = 4x(x^2 - 6x + 8) = 4x(x - 2)(x - 4)$$

Solve $f'(x) = 0$. This yields $x = 0, 2, 4$. These are the only critical values. Solve $f'(x) > 0$ and $f'(x) < 0$. The intervals are $(-\infty, 0)$, $(0, 2)$, $(2, 4)$, and $(4, \infty)$. Choose test points, say, $-1, 1, 3$, and 5. Evaluate $f'(x)$ at each value and record the results in the following chart.

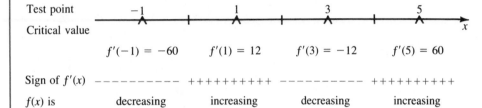

Test point	-1		1		3		5
Critical value							
	$f'(-1) = -60$		$f'(1) = 12$		$f'(3) = -12$		$f'(5) = 60$
Sign of $f'(x)$	$-----------$		$+++++++++++$		$-----------$		$+++++++++++$
$f(x)$ is	decreasing		increasing		decreasing		increasing

2. $f(0) = 0$ is a relative minimum because f is decreasing to the left and increasing to the right of $x = 0$.
$f(2) = 16$ is a relative maximum because f is increasing to the left and decreasing to the right of $x = 2$.
$f(4) = 0$ is a relative minimum because f is decreasing to the left and increasing to the right of $x = 4$.

From this information and by plotting the points $(0, 0)$, $(2, 16)$, and $(4, 0)$, we can sketch the graph as shown in Figure 6.

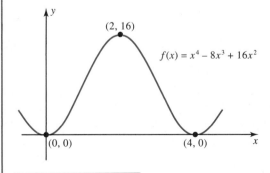

$$f(x) = x^4 - 8x^3 + 16x^2$$

$(2, 16)$
$(0, 0)$
$(4, 0)$

FIGURE 6

The next example shows how to identify when a function has a critical value that does not correspond to an extremum and when a function has an extremum at a point where the derivative does not exist.

EXAMPLE 4

Problem

Find the relative extrema, and sketch the graph of each function.
(a) $g(x) = 2 + x^3$ (b) $f(x) = x - 3x^{1/3}$.

Solution

(a) Compute

$$g'(x) = 3x^2$$

1. Solve $g'(x) = 0$. This yields $x = 0$. Thus the only critical value is $x = 0$. Find the intervals where $g(x)$ is increasing and where $g(x)$ is decreasing by arranging the necessary information in the following chart.

Test point	-1		1	
Critical value		0		x
	$f'(-1) = 3$		$f'(1) = 3$	
Sign of g'	$+++++++++++$		$+++++++++++$	
$g(x)$ is	increasing		increasing	

2. $g(0) = 2$ is neither a relative maximum nor a relative minimum because $g(x)$ is increasing to the left of $x = 0$ and to the right of $x = 0$.

From this information and by plotting the point $(0, 2)$, we can sketch the graph as shown in Figure 7.

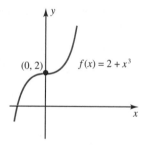

(0, 2) $f(x) = 2 + x^3$

FIGURE 7

(b) Compute

$$f'(x) = 1 - x^{-2/3} = 1 - \frac{1}{x^{2/3}}$$

1. Solve $f'(x) = 0$. This yields $x = \pm 1$. Also, $f'(x)$ does not exist when $x = 0$, but $f(0) = 0$. Thus $x = 0$ is a critical value. The function may change

from increasing to decreasing at $x = 0$, so it is included in the analysis. The critical values are $x = -1$, $x = 0$, and $x = 1$. Find the intervals where $f(x)$ is increasing and where $f(x)$ is decreasing by arranging the necessary information in the following chart.

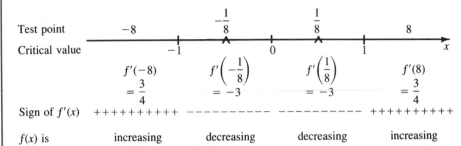

2. $f(-1) = 2$ is a relative maximum because f is increasing to the left and decreasing to the right of $x = -1$.
$f(0) = 0$ is neither a relative maximum nor a relative minimum because f is decreasing to the left and to the right of $x = 0$.
$f(1) = -2$ is a relative minimum because f is decreasing to the left and increasing to the right of $x = 1$.

From this information and by plotting the points $(-1, 2)$, $(0, 0)$, and $(1, -2)$, we can sketch the graph as shown in Figure 8. The graph has a vertical tangent at $(0, 0)$, implying that the derivative does not exist at $x = 0$.

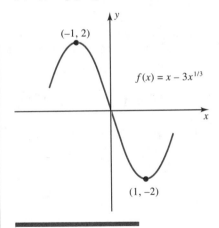

FIGURE 8

An Application: Maximizing Profit

The ultimate goal of most businesses is to maximize profit. A fundamental principle of economics is that a firm will maximize its profit when marginal revenue equals marginal cost.* The first derivative test explains why this is so. The profit $P(x)$ that a firm makes when it sells x units of goods is defined as the difference between the revenue $R(x)$ and cost $C(x)$ functions.

$$P(x) = R(x) - C(x)$$

To maximize $P(x)$, solve $P'(x) = 0$. From the definition of $P(x)$ we have

$$P'(x) = R'(x) - C'(x)$$

Solve $P'(x) = 0$. This yields $R'(x) - C'(x) = 0$. Rearranging the latter equation gives

$$R'(x) = C'(x)$$

This states that marginal revenue $R'(x)$ equals marginal cost $C'(x)$. That is, to maximize profit, find the value of x where marginal revenue equals marginal cost.

The next example is an economic model of a grocery chain.† The model uses the first derivative test to maximize the profit function of a particular product.

EXAMPLE 5

A firm determines that the revenue and cost functions for the volume of sales x for a particular product are

$$R(x) = 20x - 0.1x^2, \qquad x \geq 0$$
$$C(x) = 50 + 9x + 0.1x^2, \qquad x \geq 0$$

Problem

(a) Find the profit function $P(x)$ and the volume of sales x that maximizes profit.
(b) Show that the value of x that maximizes $P(x)$ in part (a) is also the solution of the equation $R'(x) = C'(x)$.

Solution (a) From the definition of the profit function

$$\begin{aligned} P(x) &= R(x) - C(x) \\ &= 20x - 0.1x^2 - (50 + 9x + 0.1x^2) \\ &= -50 + 11x - 0.2x^2 \end{aligned}$$

Solve $P'(x) = 0$.

$$P'(x) = 11 - 0.4x = 0$$

$$x = \frac{11}{0.4} = 27.5$$

The first derivative test shows that $P(x)$ has a relative maximum of $P(27.5) = 101.25$. The graph of $P(x)$ is given in Figure 9.

---

*Paul A. Samuelson, *Economics*, New York, McGraw-Hill Book Co., 1986, p. 491.

†Marilu Hurt McCarty, *Managerial Economics*, Glenview, IL, Scott, Foresman and Co., 1986, pp. 28–29.

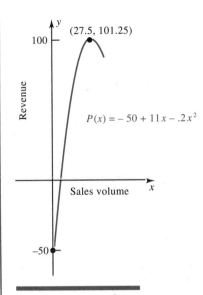

$$P(x) = -50 + 11x - .2x^2$$

(27.5, 101.25)

100

Revenue

Sales volume

−50

FIGURE 9 *Source:* Marilu Hurt McCarty, *Managerial Economics*, Glenview, IL, Scott, Foresman and Co., 1986, pp. 28–29.

(b) Set the marginal revenue equal to the marginal cost.

$$R'(x) = C'(x)$$
$$20 - 0.2x = 9 + 0.2x$$
$$-0.4x = -11$$
$$x = 27.5$$

This states that maximum profit occurs at that volume of sales where marginal revenue equals marginal cost.

EXERCISE SET 14–2 $[-\infty, \infty]$

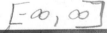 find absolute max/min

In Problems 1 to 30 find the relative extrema and sketch the graph of the functions.

1. $f(x) = x^2 - 4$
2. $f(x) = 2x^2 - 5$
3. $f(x) = 2 - 3x^2$
4. $f(x) = 3 - 4x^2$
5. $f(x) = x^2 - 4x$
6. $f(x) = 3x^2 - 6x$
7. $f(x) = x^2(x + 1)$
8. $f(x) = x^2(x - 3)$
9. $f(x) = x^3 - 6x^2 - 1$
10. $f(x) = x^3 - 9x^2 + 4$
11. $f(x) = x^3 + 3x^2 - 9x$
12. $f(x) = 2x^3 + 3x^2 - 12x + 1$
13. $f(x) = x^3 - 3x^2 - 9x$
14. $f(x) = 2x^3 + 3x^2 - 36x + 3$
15. $f(x) = x^4 - 2x^2$
16. $f(x) = -x^4 + 6x^2$
17. $f(x) = -x^4 + 8x^2$
18. $f(x) = -x^4 + 12x^2 + 1$
19. $f(x) = x + 9/x$
20. $f(x) = 4x + 25/x$
21. $f(x) = (x^2 + 12)/x$
22. $f(x) = (4x^2 + 9)/x$
23. $f(x) = x^3(x - 4)$
24. $f(x) = x^3(8 - x)$
25. $f(x) = x^2(x^2 - 2)$
26. $f(x) = x^2(4 - x^2)$
27. $f(x) = x^2(x^2 + 2)$
28. $f(x) = x^2(4 + x^2)$
29. $f(x) = (x - 1)/(x + 1)$
30. $f(x) = (x + 3)/(x - 5)$

In Problems 31 to 34 find the relative extrema, and sketch the graph of the function from the given information. Assume $f(x)$ is continuous for all real numbers.

31. Critical values

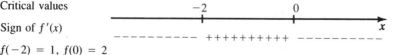

    Sign of $f'(x)$

    $f(0) = 0$, $f(3) = 2$

32. Critical values

    Sign of $f'(x)$

    $f(-2) = 1$, $f(0) = 2$

33. Critical values

    Sign of $f'(x)$

    $f(-1) = 1$, $f(2) = 3$, $f(5) = 4$

34. Critical values

    Sign of $f'(x)$

    $f(2) = -1$, $f(3) = 2$, $f(4) = 4$

35. The profit $P(x)$ in thousands of dollars for selling x television sets in hundreds of sets is

$$P(x) = x^2(x - 8)^2 \qquad \text{for } 0 < x < 6$$

Find the number of television sets that maximizes profit.

36. Psychologists define learning rate $L(x)$ to be the percent of learning that an individual exhibits in x minutes while solving an intricate puzzle. For a particular puzzle the average $L(x)$ is given by

$$L(x) = 100x^2 - 20x^3 + x^4 \text{ for } 0 < x < 12$$

Find where $L(x)$ is a maximum.

In Problems 37 to 44 find the relative extrema, and sketch the graph of the functions.

37. $f(x) = x^3(x^2 - 15)$ 38. $f(x) = x^2(10 - x^3)$

39. $f(x) = (x^2 - 1)(x^2 - 4)$

40. $f(x) = (x^2 - 3)(9 - x^2)$

41. $f(x) = (x - 5)^2(x + 1)^2$

42. $f(x) = (x + 6)^2(x - 4)^2$

43. $f(x) = (x - 5)^{2/3}$ 44. $f(x) = (x + 6)^{2/3}$

Referenced Exercise Set 14–2

1. Nursing is the largest cost item in most hospitals. In a paper* examining decision support models for budgeting nursing work-force requirements in a hospital, a key function is the cost function $C(x)$, where x is the number of nursing hours per week, measured in 1000 hours, and $C(x)$ is the cost of nursing, measured in 100,000 dollars. Several models are developed and applied to data from a large metropolitan hospital in the Sunbelt. For one model the cost function can be described by

$$C(x) = 0.115x^2 - 3.046x + 28.686$$

Use the first derivative test to find the value of x that minimizes C.

2. A national sample of married couples was surveyed to untangle the complex relationship between a person's actual pay and the feeling of being underpaid.† The data show that as income rises, people shift from comparing their income to that which is needed to "get along" to that which is required to "get ahead." One illustrative function measures the actual pay, x, of husbands versus the average amount, $f(x)$, they felt that

*Edward P. Kao and Maurice Queyranne, "Budgeting Costs of Nursing in a Hospital," *Management Science*, Vol. 31, No. 5, pp. 608–621.

†John Mirowsky, "The Psycho-Economics of Feeling Underpaid: Distributive Justice and the Earnings of Husbands and Wives," *American Journal of Sociology*, Vol. 92, No. 6, 1987, pp. 1404–1434.

they were underpaid. The data are approximated by the function

$$f(x) = 0.018x^2 - 0.757x + 9.047$$

This means that, on the average, husbands surveyed earning salary x (in thousands of dollars) felt they were underpaid by $f(x)$ thousands of dollars.

(a) Find the salary where the underpayment is a minimum. The authors refer to this salary as the "optimum pay."
(b) Find the amount of underpayment at the optimum pay.
(c) Graph the function and use it to describe what the authors refer to as a shift in the feelings of an income needed to "get along" versus to "get ahead."

3. Biological cells that are active electrically must function with great precision, making their motion very complicated. A. L. Hodgkin and A. F. Huxley won Nobel Prizes for being the first scientists to construct a model of the squid giant axon, the first such cell to be investigated satisfactorily.‡ An article describing their work states that the model cannot include the extrema of the function $f(x) = 3x - x^3$. What points are thus excluded?

Cumulative Exercise Set 14–2

In Problems 1 to 3, a function $f(x)$ is given.
(a) Find the intervals where $f(x)$ is increasing.
(b) Find the point(s) where $f(x)$ has a horizontal tangent line.
(c) Sketch the graph of the function.

1. $f(x) = 4 - x^2$ 2. $f(x) = x^2(1 - x)$

3. $f(x) = \dfrac{x - 4}{x}$

In Problems 4 and 5 use the information given to sketch the graph of $f(x)$, assuming that $f(x)$ is continuous for all real numbers x.

4. Values where $f'(x) = 0$

Sign of $f'(x)$

$f(-1) = 0,\ f(0) = -2,\ f(1) = 0$

5. Critical values

Sign of $f'(x)$

$f(-4) = 5,\ f(-1) = 3,\ f(3) = 5$

In Problems 6 and 7 find the relative extrema and sketch the graph of $f(x)$.
6. $f(x) = 2 - 8x + x^4$ 7. $f(x) = 8x^2 - x^4$

8. Let $f(x)$ be the function whose graph is shown.
(a) What are the critical values of $f(x)$?
(b) For what intervals is $f'(x) \leq 0$?

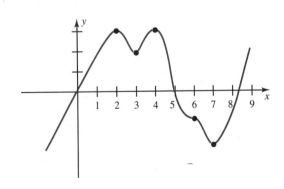

‡Jane Cronin, "Electrically active cells and singular perturbation theory," *The Mathematical Intelligencer*, Vol. 12, No. 4, 1990, pp. 57–64. The function in question is described on page 60.

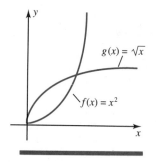

FIGURE 1

14–3 Concavity and the Second Derivative Test

Consider the graphs of the two functions in Figure 1. Each function is increasing in $(0, \infty)$, yet the *rate* of increase is different. The graph of $f(x) = x^2$ is increasing at a much greater rate than that of $g(x) = \sqrt{x}$. To illustrate this, let each function represent the position of a car moving in a straight line, so that at time x the car is at position y. Then the derivative of each function represents its velocity. Each function is increasing, but when x increases from 0 to 4, $f(x)$ moves from position $f(0) = 0$ to $f(4) = 16$, while $g(x)$ moves from $g(0) = 0$ to $g(4) = 2$. This is because the derivative of $f(x)$ is increasing, while the derivative of $g(x)$ is decreasing. In this section we study the rate of increase of the derivative of a function. We learn that the second derivative of the function provides valuable information about the graph of the function.

Concavity

The difference between the two graphs in Figure 1 can be described geometrically by saying that the graph of $f(x) = x^2$ lies above its tangent line at each point in $(0, \infty)$, while the graph of $g(x) = \sqrt{x}$ lies below its tangent line at each point in $(0, \infty)$. This is because $f'(x) = 2x$ is increasing while $g'(x) = 1/(2\sqrt{x})$ is decreasing on $(0, \infty)$. In Figure 2 two graphs are given along with three tangent lines for each: the curve in (a) lies on or above each tangent line and it is said to "open up," while the curve in (b) lies on or below each tangent line and it is said to "open down."

DEFINITION

The graph of a function $y = f(x)$ is **concave up** *on the interval* (a, b) if $f'(x)$ is increasing on (a, b) and is **concave down** *on* (a, b) if $f'(x)$ is decreasing on (a, b).

A function is defined to be *concave up at* $x = a$ if the graph of the function is concave up in some interval containing a and is *concave down at* $x = a$ if the

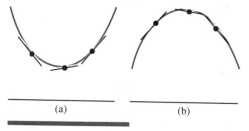

(a)   (b)

FIGURE 2

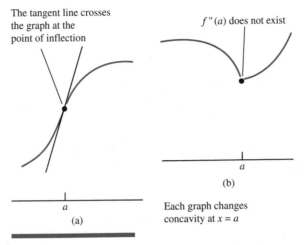

FIGURE 3

graph of the function is concave down in some interval containing a. Geometrically, a function is concave up at $x = a$ *if its graph close to a* lies above its tangent line at a, and it is concave down at $x = a$ if its graph close to a lies below its tangent line at a.

A point on the graph where the graph changes concavity is called a **point of inflection.** Thus at a point of inflection the graph changes from concave up to concave down, or vice versa. At a point of inflection the graph crosses its tangent line or $f''(x)$ does not exist. See Figure 3. Example 1 illustrates these definitions.

EXAMPLE 1

Problem

For the graph in Figure 4 locate the intervals where $f(x)$ is concave up and where $f(x)$ is concave down, and find the points of inflection.

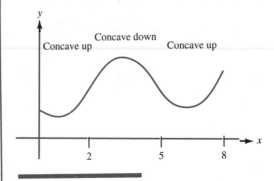

FIGURE 4

Solution $f(x)$ is concave up in $(0, 2)$ and in $(5, 8)$, while it is concave down in $(2, 5)$. There are points of inflection at $x = 2$ and $x = 5$.

It is tempting to try to draw a connection between when a function is concave up and when the function is increasing. But a look at Figure 4 shows that a function can be both increasing and decreasing in an interval where it is concave up: $f(x)$ is decreasing in (0, 1) and increasing in (1, 2), yet $f(x)$ is concave up in (0, 2). Similarly, a function can be increasing or decreasing in an interval where it is concave down. The chart lists the four possibilities.

f(x) is Increasing or Decreasing	f(x) is Concave Up or Concave Down	Graph of f(x)
A. Increasing	concave up	
B. Increasing	concave down	
C. Decreasing	concave up	
D. Decreasing	concave down	

In Figure 5 two functions are graphed along with three tangent lines for each function. The function in (a) is concave up and the function in (b) is concave down. Start at $x = a$ on the left of each graph and move to $x = b$; that is, let x increase from a to b. What can be said about the slopes of the tangent lines? In (a) the slopes of the tangent lines increase; they are negative on the left part of the graph,

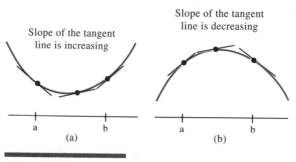

Slope of the tangent line is increasing

Slope of the tangent line is decreasing

(a)   (b)

FIGURE 5

then increase to 0 and then become positive. In (b) the slopes decrease as they start out positive, decrease to 0 and then become negative. This indicates that the derivative increases for a function that is concave up and the derivative decreases for a function that is concave down.

This statement can be coupled with the result from Section 14–1 about increasing functions and decreasing functions to yield a test for concavity. To derive the test, we combine the following two results. Statement I is from the preceding discussion and statement II is from Section 14–1.

I. $f(x)$ is concave up in (a, b) if $f'(x)$ is increasing on (a, b)

II. $f(x)$ is increasing in (a, b) if $f'(x) > 0$ on (a, b)

To tie the two together, replace $f(x)$ by $f'(x)$ in the second statement. We then get

III. $f'(x)$ is increasing in (a, b) if $f''(x) > 0$ on (a, b)

Combining I and III yields the test for concavity.

Procedure for Determining Intervals of Concavity

Assume that the first and second derivatives of $f(x)$ exist for all values of x in the interval (a, b).

1. $f(x)$ is concave up in (a, b) if $f''(x) > 0$ for all x in (a, b)
2. $f(x)$ is concave down in (a, b) if $f''(x) < 0$ for all x in (a, b).

EXAMPLE 2

Problem

Find the intervals of concavity and the point(s) of inflection of $f(x) = x^3 + 3x^2$.

Solution To apply the procedure, first calculate $f'(x)$ and $f''(x)$: $f'(x) = 3x^2 + 6x$ and $f''(x) = 6x + 6 = 6(x + 1)$. Thus $f''(-1) = 0$. Next find where $f''(x)$ is greater than and less than 0. Choose test points to the left and right of $x = -1$, say, $x = -2$ and $x = 0$. This gives $f''(-2) = 6(-1) = -6 < 0$ and $f''(0) = 6 > 0$. Record these facts in the following chart.

Test point		-2		0	
Possible point of inflection			-1		x

Sign of $f''(x)$	$----------$	$++++++++++$
f is	concave down	concave up

From the procedure $f(x)$ is concave up in $(-1, \infty)$ and concave down in $(-\infty, -1)$. Since $f(x)$ changes from concave down to concave up at $x = -1$, $f(x)$ has a point of inflection at $(-1, 2)$. See Figure 6 on the next page.

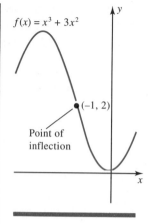

$f(x) = x^3 + 3x^2$

(−1, 2)

Point of inflection

FIGURE 6

If $(a, f(a))$ is a point of inflection, $f''(x)$ changes from positive on one side of $x = a$ to negative on the other. For instance, in Example 2, $f(x)$ has a point of inflection at $(-1, 2)$ and $f''(x)$ is negative to the left of $x = -1$ and positive to the right of $x = -1$. What happens to $f''(x)$ at the point of inflection? In Example 2, $f''(-1) = 0$. But there is another possibility. In Figure 7 the graph of $f(x) = x^{1/3}$ is given. It has a point of inflection at $x = 0$ and $f''(0)$ does not exist. Thus there are two possibilities for $f''(x)$ at a point of inflection.

Second Derivative at Point of Inflection
If $f(x)$ has a point of inflection at $(a, f(a))$, then either $f''(a) = 0$ or $f''(a)$ does not exist.

It is very important to recognize what this result does not say. To find where a function has a point of inflection, we also need to compute the values where $f''(x)$

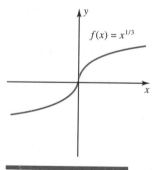

$f(x) = x^{1/3}$

FIGURE 7

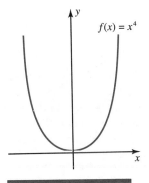

$f(x) = x^4$

FIGURE 8

$= 0$ and where $f''(x)$ does not exist. Recall that it is not enough to compute critical values when trying to locate relative extrema; more information is required about the behavior around the critical value, which is the essence of the first derivative test. So, too, to determine if a value where $f''(x) = 0$ or $f''(x)$ does not exist yields a point of inflection, we need to find where the function is concave up and where it is concave down. For example, $f(x) = x^4$ has $f''(x) = 12x^2$, and $f''(x) = 0$ when $x = 0$. But since $f''(x) > 0$ for all x to the left and to the right of $x = 0$, f is concave up both to the left and to the right of $x = 0$. Thus f does not have a point of inflection at $x = 0$ even though $f''(0) = 0$. See Figure 8.

The Second Derivative Test

The two graphs in Figure 9 illustrate the connection between concavity and relative extrema. Each graph has a horizontal tangent at $x = a$, so $f'(a) = 0$ and the function has a critical value at $x = a$. The graph in (a) is concave down and $f(x)$ has a relative maximum at $x = a$. The graph in (b) is concave up and $f(x)$ has a relative minimum at $x = a$. To summarize, let $x = a$ be a critical value of $f(x)$.

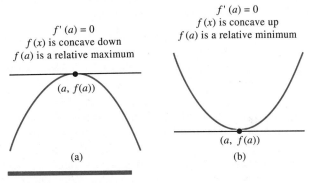

$f'(a) = 0$
$f(x)$ is concave down
$f(a)$ is a relative maximum

$(a, f(a))$

(a)

$f'(a) = 0$
$f(x)$ is concave up
$f(a)$ is a relative minimum

$(a, f(a))$

(b)

FIGURE 9

I. If $f(x)$ is concave down at $x = a$, then $f(x)$ has a relative maximum at $x = a$.

II. If $f(x)$ is concave up at $x = a$, then $f(x)$ has a relative minimum at $x = a$.

Therefore concavity can be used to determine whether a critical value yields a relative maximum or a relative minimum. Since the second derivative measures concavity, the second derivative can be used to determine whether a critical value is a relative maximum or a relative minimum. Statements I and II can be refined by using the test for concavity from the previous section.

Let $x = a$ be a critical value of $f(x)$.

III. If $f''(a) < 0$, then $f(x)$ is concave down at $x = a$, and thus $f(x)$ has a relative maximum at $x = a$.

IV. If $f''(a) > 0$, then $f(x)$ is concave up at $x = a$, and thus $f(x)$ has a relative minimum at $x = a$.

This is the essence of the second derivative test, which, like the first derivative test, determines whether a critical value is a relative maximum or a relative minimum.

The Second Derivative Test

Suppose $f(x)$ is differentiable and $f''(x)$ exists on an interval containing a. Let $x = a$ be a critical value.

1. $f(a)$ is a relative maximum if $f''(x) < 0$ at $x = a$.
2. $f(a)$ is a relative minimum if $f''(x) > 0$ at $x = a$.
3. If $f''(a) = 0$, the test gives no conclusion about a relative extremum at $x = a$; that is, $f(a)$ may be a relative maximum, a relative minimum, or neither.

EXAMPLE 3

Problem

Use the second derivative test to find the relative extrema for
(a) $f(x) = 5x^2 - 10x$ and (b) $f(x) = x^3 - 3x^2$.

Solution (a) Compute $f'(x) = 10x - 10 = 10(x - 1)$. Solve $f'(x) = 0$. The only critical value is $x = 1$. To apply the second derivative test, we compute the second derivative, $f''(x) = 10$. This is a constant function, so

$$f''(1) = 10 > 0$$

Since $f''(1) > 0$, apply part 2 of the second derivative test and conclude that $f(x)$ has a relative minimum at $x = 1$. The relative minimum is $f(1) = -5$.
(b) Compute $f'(x) = 3x^2 - 6x$. Solve $f'(x) = 3x^2 - 6x = 3x(x - 2) = 0$. So $x = 0$ and $x = 2$. The critical values are $x = 0$ and $x = 2$. Compute $f''(x) = 6x - 6$. Substitute each critical value into f''. This yields

$$f''(0) = 0 - 6 = -6 < 0 \qquad f''(2) = 12 - 6 = 6 > 0$$

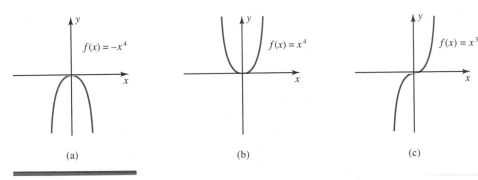

(a)   (b)   (c)

FIGURE 10

Part 1 of the second derivative test states that $f(x)$ has a relative maximum at $x = 0$ because $f''(0) < 0$; part 2 states that $f(x)$ has a relative minimum at $x = 2$ because $f''(2) > 0$. The relative maximum is $f(0) = 0$ and the relative minimum is $f(2) = -4$.

Part 3 of the second derivative test states that the test fails when $f''(a) = 0$ since no conclusion can be drawn about relative extrema. Figure 10 shows why this is the case. Even though each of the three functions graphed in Figure 10 has $f''(0) = 0$, their behavior at $x = 0$ is different. In (a), $f(x) = -x^4$ has a relative maximum at $x = 0$; in (b), $f(x) = x^4$ has a relative minimum at $x = 0$; in (c), $f(x) = x^3$ has neither a relative maximum nor a relative minimum because it has a point of inflection at $x = 0$. This means that if $f''(a) = 0$ for a critical value $x = a$, then the first derivative test must be applied to $x = a$. Also, if $f''(x)$ does not exist, the second derivative test cannot be used so the first derivative test must be applied.

Curve Sketching

The ideas presented thus far in this chapter can be combined to form a handy procedure for sketching graphs. The summary of these various techniques is a process called **curve sketching.**

Curve-Sketching Techniques

1. Compute $f'(x)$.
   a. Find the critical values by solving $f'(x) = 0$ and finding where $f'(x)$ does not exist.
   b. Find the intervals where $f(x)$ is increasing and where $f(x)$ is decreasing by solving $f'(x)$ greater than 0 and less than 0.
2. Compute $f''(x)$.
   a. Find the possible points of inflection by solving $f''(x) = 0$ and locating where $f''(x)$ does not exist.

b. Find the intervals where $f(x)$ is concave up and where $f(x)$ is concave down by setting $f''(x)$ greater than 0 and less than 0. Use this information to locate the points of inflection.
3. Use the first or second derivative test to locate the relative maxima and the relative minima.
4. Plot the relative extrema and the points of inflection. Keep in mind where the function is increasing, decreasing, concave up, and concave down, and plot several additional points that help to get an accurate sketch. For instance, plot the x- and y-intercepts if they exist and are easy to calculate; if the function is defined on a closed or half-open interval, plot the points corresponding to the end points.
5. Draw a sketch of the curve from this information.

Example 4 shows how to use the curve-sketching techniques by applying them to the functions in Example 3.

EXAMPLE 4

Problem

Sketch the graph of the functions (a) $f(x) = 5x^2 - 10x$, (b) $f(x) = x^3 - 3x^2$.

Solution We follow the steps in the procedure even though they do not have to be done precisely in the given order.
(a)
1. $f'(x) = 10x - 10 = 10(x - 1) = 0$ implies $x = 1$.
   a. Since $f'(1) = 0$, the only critical value is $x = 1$.
   b. $f'(x) > 0$ in $(1, \infty)$ and $f'(x) < 0$ in $(-\infty, 1)$, so $f(x)$ is increasing in $(1, \infty)$ and decreasing in $(-\infty, 1)$.
2. $f''(x) = 10$
   a. Since $f''(x)$ is always greater than 0, $f(x)$ is concave up in $(-\infty, \infty)$.
   b. $f''(x)$ is never equal to 0, so there is no point of inflection.

This information is recorded in the following chart.

Critical value		x
Sign of $f'(x)$	$-\,-\,-\,-\,-\,-\,-\,-\,-$	$+\,+\,+\,+\,+\,+\,+\,+\,+\,+$
$f(x)$ is	decreasing	increasing
Sign of $f''(x)$	$+\,+\,+\,+\,+\,+\,+\,+\,+\,+$ $+\,+\,+\,+\,+\,+\,+\,+\,+\,+$	
$f(x)$ is	concave up	

3. The second derivative test shows that $f(x)$ has a relative minimum at $x = 1$ because $f''(1) = 10 > 0$.
4. Plot $(1, -5)$, where $f(x)$ has its relative minimum. To find the y-intercept, let $x = 0$ and get $f(0) = 0$. Since $f(x) = 0$ at $x = 0$ and $x = 2$, plot the x-intercepts $(0, 0)$ and $(2, 0)$.
5. The graph is in Figure 11.

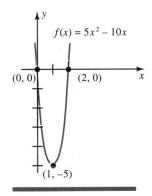

$f(x) = 5x^2 - 10x$

$(0, 0)$ $(2, 0)$ x

$(1, -5)$

FIGURE 11

(b)

1. $f'(x) = 3x^2 - 6x = 3x(x - 2)$
   a. Since $f'(0) = f'(2) = 0$, the critical values occur at $x = 0$ and $x = 2$.
   b. $f'(x) > 0$ in $(-\infty, 0)$ and $(2, \infty)$; $f'(x) < 0$ in $(0, 2)$. Thus $f(x)$ is increasing in $(-\infty, 0)$ and $(2, \infty)$ and $f(x)$ is decreasing in $(0, 2)$.
2. $f''(x) = 6x - 6 = 6(x - 1)$
   a. Since $f''(x) > 0$ in $(1, \infty)$ and $f''(x) < 0$ in $(-\infty, 1)$, $f(x)$ is concave up in $(1, \infty)$ and concave down in $(-\infty, 1)$.
   b. $f''(1) = 0$ and $f(x)$ changes concavity at $x = 1$, so there is a point of inflection at $(1, -2)$.

This information is recorded in the following chart.

Possible point of inflection		1		
Critical values		0	2	x
Sign of $f'(x)$	++++++++++	----------	++++++++++	
$f(x)$ is	increasing	decreasing	increasing	
Sign of $f''(x)$	-------------------		++++++++++++++++	
$f(x)$ is	concave down		concave up	

3. The second derivative test shows that $f(x)$ has a relative maximum at $x = 0$ because $f''(0) = -6 < 0$ and $f(x)$ has a relative minimum at $x = 2$ because $f''(2) = 6 > 0$.
4. Plot $(0, 0)$ and $(2, -4)$, where $f(x)$ has its relative extrema. Plot the point of inflection, $(1, -2)$. To find the y-intercept, let $x = 0$ and get $f(0) = 0$, but $(0, 0)$ is already plotted. Since $f(x) = 0$ at $x = 3$, plot the additional x-intercept $(3, 0)$.
5. The graph is in Figure 12.

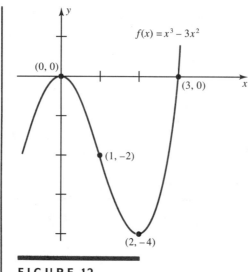

FIGURE 12

An Application: Public Subsidy

In some situations it is not satisfactory to analyze production decisions merely on the basis of cost-effectiveness or profit optimization. Certain not-for-profit institutions provide services to a community at prices below production costs. Total revenue from sales may be insufficient to pay total costs and a subsidy may be required. Examples of such institutions are public transit companies, low-cost higher education institutions, and cultural organizations. To illustrate production decisions when a public subsidy is required, economist Marilu McCarty used as an example a mass transit system in a congested urban area.* The cost $C(x)$ and revenue $R(x)$ functions were estimated for the daily demand x, where $C(x)$ and $R(x)$ are measured in thousands of dollars and x is the number of riders in thousands. If the system expected to make a profit, then it would seek to maximize $R(x) - C(x)$. But the system is designed to absorb a loss. The objective is then to minimize the *economic loss EL(x)*, defined by

$$EL(x) = C(x) - R(x)$$

The goal is to determine the demand x that minimizes $EL(x)$. This minimum value of $EL(x)$ is the subsidy paid by the government to the institution. The next example shows how to solve this problem for cost and revenue functions that are similar to those of the mass transit system. Referenced exercise 2 presents the actual data.

*Marilu Hurt McCarty, *Managerial Economics*, Glenview, IL, Scott, Foresman and Co., 1986, pp. 479–486.

EXAMPLE 5

The cost and revenue functions for the mass transit system are

$$C(x) = 2000 + 140x - 30x^2 + x^3$$

$$R(x) = 20x - 9x^2$$

Problem

Determine the minimum economic loss.

Solution The economic loss function is defined by $EL(x) = C(x) - R(x)$. It is

$$EL(x) = 2000 + 120x - 21x^2 + x^3$$

To find the minimum, use the second derivative test. Compute $EL'(x)$ and $EL''(x)$.

$$EL'(x) = 120 - 42x + 3x^2 = 3(40 - 14x + x^2)$$

$$EL''(x) = -42 + 6x$$

To solve $EL'(x) = 0$, either factor the quadratic or use the quadratic formula.

$$EL'(x) = 3(x - 10)(x - 4) = 0$$

This implies that the critical values are $x = 4$ and $x = 10$. Substituting these values into $EL''(x)$ yields

$$EL''(4) = -18 \quad \text{and} \quad EL''(10) = 18$$

By using the second derivative test, we conclude that the relative minimum of $EL(x)$ occurs at $x = 10$. It is $EL(10) = 2100$. The function is graphed in Figure 13. Notice that there are functional values of $EL(x)$ that are smaller than $EL(10)$; in fact $EL(0) = 2000$. But these values of x are too small to be practical. That is, the transit system would not be properly serving the public if it accommodated so few riders. This means that an implicit assumption in the problem is that the transit system will service a certain minimum number of riders. In the next section we study problems that have natural boundaries like this.

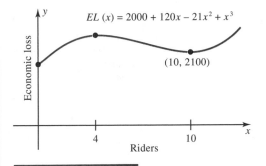

FIGURE 13 *Source:* Marilu Hurt McCarty, *Managerial Economics,* Glenview, IL, Scott, Foresman and Co., 1986, pp. 479–496.

EXERCISE SET 14–3

In Problems 1 to 6 find the largest intervals where the function is concave up and concave down. Find the points of inflection. Assume that the functions are defined on the whole real line.

1.

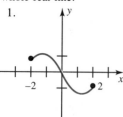

2.

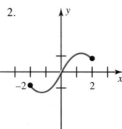

3.

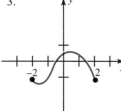

4.

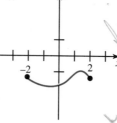

5.

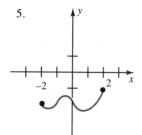

6.

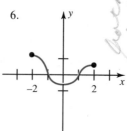

In Problems 7 to 12 find the intervals where the function is concave up and concave down. Find any point of inflection.

7. $f(x) = x^2 - 4$

8. $f(x) = 2x^2 - 5$

9. $f(x) = x^3 - 6x^2 - 2$

10. $f(x) = x^3 - 9x^2$

11. $f(x) = x^3 + 3x^2 - 9x$

12. $f(x) = 2x^3 + 3x^2 - 12x + 1$

In Problems 13 to 18 apply the second derivative test to find the relative extrema.

13. $f(x) = x^3 - 3x^2 - 9x$

14. $f(x) = 2x^3 + 3x^2 - 36x + 3$

15. $f(x) = x^4 - 2x^2$

16. $f(x) = -x^4 + 6x^2$

17. $f(x) = -x^4 + 8x^2$

18. $f(x) = -x^4 + 12x^2 + 1$.

In Problems 19 to 36 use the curve-sketching techniques to graph the function.

19. $f(x) = x^3 + 6x^2$

20. $f(x) = x^3 - x^2$

21. $f(x) = x^3 - 3x^2 - 9x$

22. $f(x) = 2x^3 + 3x^2 - 36x + 3$

23. $f(x) = x^4 - 2x^2$

24. $f(x) = -x^4 + 6x^2$

25. $f(x) = 64x - 2x^4$

26. $f(x) = 2x - 3x^{2/3}$

27. $f(x) = x + 9/x$

28. $f(x) = 4x + 25/x$

29. $f(x) = (x^2 + 12)/x$

30. $f(x) = (4x^2 + 9)/x$

31. $f(x) = x^2(x^2 - 2)$

32. $f(x) = x^2(4 - x^2)$

33. $f(x) = x^2(x^2 + 2)$

34. $f(x) = x^2(4 + x^2)$

35. $f(x) = (x - 1)/(x + 1)$

36. $f(x) = (x + 3)/(x - 5)$

In Problems 37 to 40 use the information given to sketch the graph of $f(x)$. Assume $f(x)$ is continuous for all real numbers.

37. Possible point of inflection

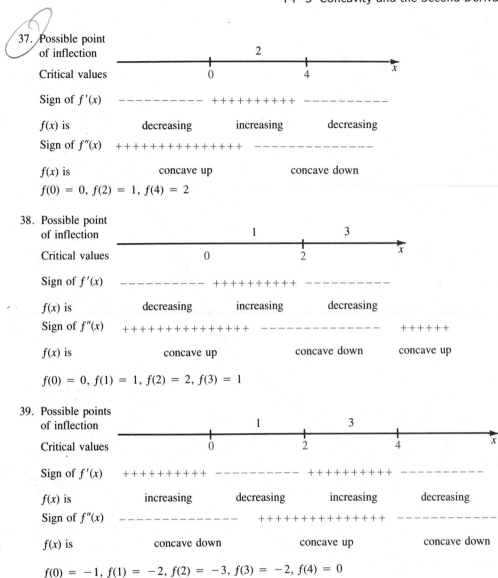

Critical values

Sign of $f'(x)$ – – – – – – – – – + + + + + + + + + + – – – – – – – – –

$f(x)$ is   decreasing   increasing   decreasing

Sign of $f''(x)$ + + + + + + + + + + + + + + + – – – – – – – – – – – –

$f(x)$ is   concave up   concave down

$f(0) = 0,\ f(2) = 1,\ f(4) = 2$

38. Possible point of inflection

Critical values

Sign of $f'(x)$ – – – – – – – – – + + + + + + + + + + – – – – – – – – –

$f(x)$ is   decreasing   increasing   decreasing

Sign of $f''(x)$ + + + + + + + + + + + + + + – – – – – – – – – – – – + + + + + +

$f(x)$ is   concave up   concave down   concave up

$f(0) = 0,\ f(1) = 1,\ f(2) = 2,\ f(3) = 1$

39. Possible points of inflection

Critical values

Sign of $f'(x)$ + + + + + + + + + + – – – – – – – – – – + + + + + + + + + + – – – – – – – – –

$f(x)$ is   increasing   decreasing   increasing   decreasing

Sign of $f''(x)$ – – – – – – – – – – – + + + + + + + + + + + + + – – – – – – – – – – – –

$f(x)$ is   concave down   concave up   concave down

$f(0) = -1,\ f(1) = -2,\ f(2) = -3,\ f(3) = -2,\ f(4) = 0$

40. Possible points of inflection

Critical values

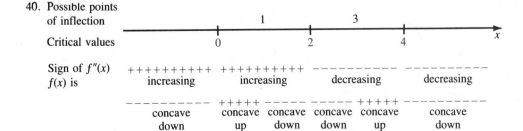

Sign of $f''(x)$ + + + + + + + + + + + + + + + + + + + + – – – – – – – – – – – – – – – – – – – –
$f(x)$ is   increasing   increasing   decreasing   decreasing

– – – – – – – – – – + + + + + – – – – – – – – – – + + + + + – – – – – – – – –
                    concave   concave   concave   concave   concave   concave
                    down   up   down   down   up   down

$f(0) = 0,\ f(1) = 2,\ f(2) = 4,\ f(3) = 2,\ f(4) = 1$

In Problems 41 to 45 use the curve-sketching techniques to graph the function.

41. $f(x) = x^3(x^2 - 15)$ 42. $f(x) = x^2(10 - x^3)$

43. $f(x) = (x^2 - 1)(x^2 - 4)$

44. $f(x) = (x + 6)^2(x - 4)^2$

45. $f(x) = (x + 6)^{2/3}$

In Problems 46 to 49 from the given figure find (a) the intervals where $f(x)$ is increasing, (b) the intervals where $f(x)$ is decreasing, (c) the sign of the derivative in each interval, and (d) the sign of the second derivative in each interval.

46.
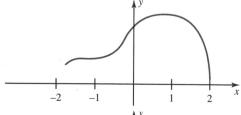

47.
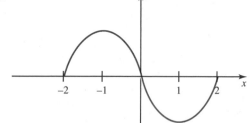

48.
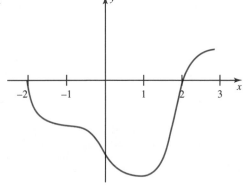

49.
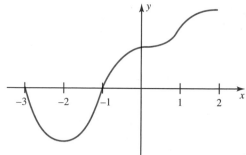

Referenced Exercise Set 14–3

1. In 1970 the U.S. government commissioned the Army to conduct a thorough study of the cost of converting to the metric system of measure. The methodology of the Army study could then be applied by various organizations, which would be similarly affected by the change. In the report* the cumulative costs of metrication for the U.S. Army are estimated by the function

   $$f(x) = -0.02x^3 + 0.12x^2 + 0.72x$$

   where x is measured in units of five-year periods starting with 1972 and $f(x)$ is measured in billions of dollars. For example, $x = 1$ represents 1977 (one five-year period after 1972) and $f(1) = 0.82$ means that the estimated cost of metrication for the Army by 1977 would be $0.82 billion. Find the year when the rate of increase of the budget would be a maximum.

2. For the economic loss function $EL(x)$ given in McCarty† for a mass transit system, find the value of x that minimizes $EL(x)$, assuming that the system must accommodate at least $x = 50$ thousand riders:

   $$EL(x) = 2000 + 72x - 1.1x^2 + 0.005x^3$$

3. Production models in business help a firm decide what is the optimum quantity to produce either to maximize price or profit or to minimize cost. In an article‡ studying agricultural production functions, with an emphasis

*Walter A. Lilius, "The Methodology of the Army Metric Study," *American Society of Quality Control—Technical Conference Transactions*, 1972, pp. 242–249.

†Marilu Hurt McCarty, *Managerial Economics*, Glenview, IL, Scott, Foresman and Co., 1986, pp. 479–486.

‡Jean-Paul Chavas et al., "Modeling Dynamic Agricultural Production Response: The Case of Swine Production," *American Journal of Agricultural Economics*, Vol. 67, 1987, pp. 636–646.

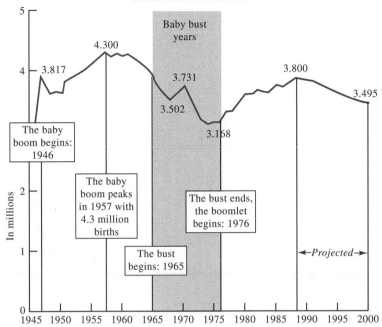

Births in the U.S. since 1945

on the swine industry, the following function relating the price P (dollars) to the weight x (lb) of an individual animal is given:

$$P(x) = 50.536 + 0.0988x - 0.00024x^2$$

Find the value of x that maximizes the price.

Cumulative Exercise Set 14–3

In Problems 1 to 3 find intervals where $f(x)$ is increasing and where $f(x)$ is decreasing. Find the point(s) where $f(x)$ has a horizontal tangent line. Graph the function.

1. $f(x) = x^2 - 12x + 1$

2. $f(x) = x^3 + 6x^2 - 18x$

3. $f(x) = (x^2 + 16)/x$

4. Use the information given to make a sketch of the graph of $f(x)$, assuming that $f(x)$ is continuous for all real numbers.

Values where $f'(x) = 0$ $ -2 3$

Sign of $f'(x)$ $ ----------++++++++++----------$

$f(-3) = 1$, $f(-2) = 0$, $f(3) = 4$, $f(5) = 0$

4. The accompanying graph depicts the number of births in the U.S. from 1945 to 1989 based on figures supplied by American Demographics, Inc. (a) What kind of point on the curve is (1957, 4300)? (b) What kind of point on the curve is (1968, 3502)? (c) The figure describes the year 1965 as the beginning of the "bust." What kind of point on the curve is it?

In Problems 5 and 6 find the relative extrema and sketch the graph of the functions.

5. $f(x) = 2x^3 + 3x^2 - 1$

6. $f(x) = x + 25/x$

7. Find the relative extrema and sketch the graph of the function from the given information. Assume $f(x)$ is continuous for all real numbers.

Critical values -2 2

Sign of $f'(x)$ $- - - - - - - - - -$ $+ + + + + + + + + +$ $- - - - - - - - - -$

$f(-2) = -1, f(0) = 1, f(2) = 3$

8. Find the intervals where the function $f(x) = 2x^3 - 3x^2$ is concave up and where it is concave down. Find any point of inflection.

9. Apply the second derivative test to find the relative extrema of $f(x) = 2x^3 - 3x^2 - 36x$.

In Problems 10 and 11 use the curve-sketching techniques to graph the function.

10. $f(x) = x^4 - 4x^3 - 2$

11. $f(x) = \dfrac{x^2 + 25}{x}$

12. Use the information given to sketch the graph of $f(x)$. Assume $f(x)$ is continuous for all real numbers.

Possible points
of inflection   1   3

Critical values -1 2   5 x

Sign of $f'(x)$ $+ + + + + + + + + +$ $+ + + + + + + + + +$ $- - - - - - - - - -$ $- - - - - - - - - -$

$f(x)$ is   increasing   increasing   decreasing   decreasing

Sign of $f''(x)$ $- - - - - - - - - -$ $+ + + + +$ $- - - - -$ $- - - - -$ $+ + + + +$ $- - - - - - - - - -$

$f(x)$ is   concave   concave   concave   concave   concave   concave
                down   up   down   down   up   down

$f(-1) = 0, f(1) = 2, f(2) = 4, f(3) = 2, f(5) = 1$

14–4 Absolute Extrema and Optimization Problems

A function may have a relative maximum or a relative minimum without being an absolute maximum or an absolute minimum. In fact most of the functions graphed in the previous sections have functional values that are larger or smaller than the relative extrema. In this section we cover absolute extrema. If $f(a)$ is a relative maximum, then $f(a)$ is greater than or equal to the other functional values $f(x)$ for all x restricted to an interval containing a. If $f(a)$ is an absolute maximum, then $f(a)$ is greater than or equal to all other functional values $f(x)$. A similar statement holds for relative minima and absolute minima.

There are several possibilities for absolute extrema. Some functions have neither an absolute maximum nor an absolute minimum. If the function is defined on a closed interval the absolute extremum can occur at an end point. And sometimes an absolute extremum occurs at a relative extremum.

Absolute Extrema

Consider the function $y = f(x)$ graphed in Figure 1. It has a relative minimum at $x = b$ and a relative maximum at $x = c$. But neither is an absolute extremum; $f(a)$ is greater than $f(c)$, so $f(c)$ is not an absolute maximum, and $f(d)$ is less than $f(b)$, so $f(b)$ is not an absolute minimum. In addition, $f(a)$ is greater than all other functional values, so it is the absolute maximum while $f(d)$ is the absolute minimum. Sometimes an absolute extremum occurs at an end point of an interval, as does $f(a)$ in Figure 1, and sometimes it does not, like $f(d)$. This leads to the following definition.

DEFINITION

Let $y = f(x)$ be defined on an interval I, where I can be either an open or closed interval. Suppose a is in I. Then

1. $f(a)$ is the **absolute maximum** of $f(x)$ *on I* if

    $f(a) \geq f(x)$ for all x in I

2. $f(a)$ is the **absolute minimum** of $f(x)$ *on I* if

    $f(a) \leq f(x)$ for all x in I

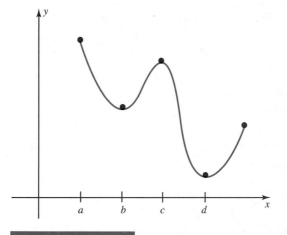

FIGURE 1

The graph in Figure 1 shows that an absolute extremum may occur at an end point of the interval. Of course, this can happen only when the end point is included in the interval, that is when it is closed or half-open (which means one end point is in the interval and the other is not). If the interval is open, then the absolute extrema cannot occur at an end point because the end points are not part of the interval. In fact the function may not have an absolute maximum or an absolute minimum. On the other hand, if the interval is closed, the end points are candidates for the absolute extrema. If an absolute extremum does not occur at an end point, then it occurs at an interior relative extremum. In more advanced courses the following theorem is proved.

Theorem

If $f(x)$ is continuous on the closed interval $[a, b]$, then $f(x)$ has an absolute maximum and an absolute minimum on $[a, b]$.

A nonvertical straight line drawn on an open interval is an example of a function that has neither an absolute maximum nor an absolute minimum on an open interval. If the end points are included, however, meaning the interval is closed, the absolute extrema of a nonvertical straight line occur at the end points.

The procedure to compute absolute extrema is the same as that for finding relative extrema, except that the end points must also be considered. That is, absolute extrema occur either at points that are relative extrema or at end points.

The Procedure for Finding Absolute Extrema for $f(x)$ Defined on a Closed Interval $[a, b]$

1. Calculate the critical values, $c_1, c_2, \ldots, c_n$ that lie in the interval $[a, b]$.
2. Calculate $f(a)$, $f(b)$, $f(c_1)$, $f(c_2)$, $\ldots$, $f(c_n)$.
3. The absolute maximum is the largest value of the numbers in step 2, and the absolute minimum is the smallest value of these numbers.

Example 1 shows how to compute absolute extrema once the relative extrema are known. It uses a function that was defined in Example 2 of Section 14–2 where the relative extrema were computed.

EXAMPLE 1

Let the function $f(x) = (x - 2)^2(x - 5)$ be defined on the interval $[\frac{3}{2}, 6]$.

Problem

Find the absolute extrema of $f(x)$.

Solution In Example 2 of Section 14–2, $f(x)$ was found to have a relative maximum of $f(2) = 0$ and a relative minimum of $f(4) = -4$. See Figure 2 for

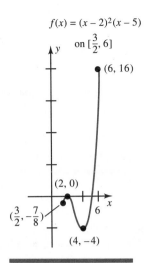

$f(x) = (x - 2)^2(x - 5)$

on $[\frac{3}{2}, 6]$

FIGURE 2

the graph of the function. To see whether the relative extrema are absolute extrema, check the functional values of the end points. The end points occur at $x = \frac{3}{2}$ and $x = 6$. Their functional values are $f(\frac{3}{2}) = -\frac{7}{8}$ and $f(6) = 16$. To find the absolute extrema, find the largest and smallest functional values of the critical values and the end points. We arrange these four functional values in the following table:

	Critical Values		End Points	
x	2	4	$\frac{3}{2}$	6
$f(x)$	0	-4	$-\frac{7}{8}$	16

The largest of the four numbers for $f(x)$ is 16 and the smallest is -4. Therefore the absolute maximum is $f(6) = 16$ and the absolute minimum is $f(4) = -4$. The absolute maximum occurs at the right-hand end-point, while the absolute minimum occurs at an interior relative minimum.

Optimization Problems

Many applications of calculus depend on maximizing or minimizing a certain quantity. Often the variables in such problems have natural boundaries. For example, if x represents the number of items produced by a firm, x must be greater than or equal to 0. If t represents time, measured in hours, during an 8-hour shift, then t is restricted to the interval $[0, 8]$. If an absolute extremum is sought, then the end points must be checked.

An optimization problem is ordinarily a word problem, so it must first be translated into mathematics. Then the relative extrema are computed as well as the functional values of the end points. The extrema are then interpreted in terms of the original problem. Let us demonstrate the method with an example that is made easy because the function is given.

EXAMPLE 2

An object is thrown straight upward so that its height after t seconds is $f(t) = -16t^2 + 64t$, where f is measured in feet.

Problem

How high will the object travel and after how many seconds will it reach its maximum height?

Solution The object will reach its maximum height when $f(t)$ has its absolute maximum.

Are there natural end points for t? It makes no sense for t to be less than 0, and a negative value for $f(t)$ means the object is below ground level. Solving $f(t) < 0$ yields

$$-16t^2 + 64t = -16t(t - 4) < 0$$

This is true when $t < 0$ or when $t > 4$. Thus the natural restriction for $f(t)$ is [0, 4].

There are no values of t where $f'(t)$ fails to exist, so to find the critical values, solve $f'(t) = 0$. Compute $f'(t) = -32t + 64$. Then $32t = 64$ and $t = 2$. The functional values of the end points are $f(0) = 0 = f(4)$. The absolute maximum occurs at the largest of the three values

$$f(0) = 0, \quad f(2) = 64, \quad f(4) = 0$$

Hence the object reaches its maximum height of $f(2) = 64$ at $t = 2$ seconds. The function is graphed in Figure 3.

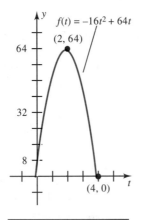

FIGURE 3

Example 3 deals with a familiar problem of building a new facility. Often constraints of cost, location, and size dictate the plans for the building. In this problem the facility is to be constructed utilizing an existing wall, thus allowing a larger area with a fixed cost. The cost is fixed because there is a given amount of fencing to be used.

EXAMPLE 3

A department store wants to build a rectangular outdoor storage facility on the rear wall of its building. An iron fence will enclose the facility on three sides, and the wall of the building will make the fourth side. (See Figure 4.)

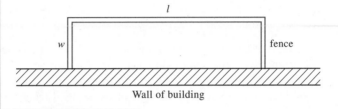

FIGURE 4

Problem

Find the dimensions of the storage facility with the largest area that can be built using 200 feet (ft) of fencing.

Solution From Figure 4 we assign letters to the variable quantities. Let l represent the length and w the width of the storage facility. Then the area, A, is given by the formula

$$A = lw \qquad (1)$$

The variable A is to be maximized since we are asked for the largest storage facility. A must be expressed as a function of l or w alone, so there must be another relationship between w and l. The amount of fencing is fixed at 200 ft, and it is the sum of the three sides, whose lengths are w, w, and l. Therefore

$$200 = 2w + l$$

Solving for l yields

$$l = 200 - 2w \qquad (2)$$

Substitute equation 2 into equation 1, which results in an expression of A as a function of w.

$$A(w) = w(200 - 2w) = 200w - 2w^2$$

To find the end points, note that it makes no sense for A to be negative. So set $A(w) \geq 0$. This yields

$$w(200 - 2w) \geq 0$$

This is satisfied for w in $[0, 100]$. Thus the end points are $w = 0$ and $w = 100$. To find the critical values, solve $A'(w) = 0$.

$$A'(w) = 200 - 4w = 0$$

The solution is $w = 50$. The end points are $w = 0$ and $w = 100$. The three values that are candidates for the absolute maximum are

$A(50) = 5000$
$A(0) = 0$
$A(100) = 0$

The largest value is $A(50) = 5000$. So the storage facility with the largest area has width 50 ft and length $l = 200 - 2w = 200 - 100 = 100$ ft.

In Example 3 the area of the new facility was maximized given that the cost, in terms of a given amount of fencing, was fixed. Another common type of problem is to increase production while keeping other variables fixed, such as cost, labor, and size of production facility. In Example 4 an orchard owner wishes to increase production while keeping the size of the orchard fixed. The object is to maximize the amount of apples produced given a fixed amount of land. The variable that the orchard owner can manipulate is the number of trees planted per acre. It then becomes the independent variable. The number of trees per acre determines how many apples are produced, which is therefore the dependent variable.

EXAMPLE 4

An apple orchard owner has determined that if 40 trees are planted per acre, the yield is 500 apples per tree. For each tree per acre that is added beyond 40, the yield is reduced by 10 apples per tree.

Problem

Find the number of trees per acre that should be planted to yield the maximum crop.

Solution Let x represent the number of trees planted per acre beyond 40. Let Y represent the total crop per acre. Then Y is the number of trees per acre times the number of apples per tree. For example, if $x = 0$, then the number of trees planted per acre is 40 and yield per tree is 500 apples, so $Y = 40 \cdot 500 = 20,000$ apples per acre.

The number of trees per acre is $40 + x$. The yield per tree diminishes by 10 for each tree planted beyond 40 so the yield per tree is $500 - 10x$. Hence the formula for $Y(x)$ is

$$\begin{array}{cc} \text{number} & \text{number of} \\ \text{of trees} & \text{apples per tree} \end{array}$$
$$\begin{aligned} Y(x) &= (40 + x)\ (500 - 10x) \\ &= 20,000 + 100x - 10x^2 \end{aligned}$$

We must find the value of x that maximizes $Y(x)$. First find the critical values. Solve $Y'(x) = 100 - 20x$ equal to 0. This yields

$$Y'(x) = 100 - 20x = 0$$

Thus $x = 5$ is the only critical value. The natural end points are $x = 0$ and $x = 50$. The three values to check are

$Y(5) = (45)(450) = 20{,}250$
$Y(0) = 20{,}000$
$Y(50) = 0$

Therefore the absolute maximum is $Y(5) = 20{,}250$. Hence the maximum yield occurs when 45 trees are planted per acre.

An Application: Dayton Trucking Company

In the mid-1970s the price of gasoline skyrocketed as OPEC (Organization of Petroleum Exporting Countries) forced up the price of crude oil. This caused great hardship for many industries, especially those whose transportation costs were significant. Winger gives an account of how one company, the Dayton Trucking Co., approached the problem.* The price of gasoline had become a critical variable in their cost evaluation for the first time. Truck performance was analyzed. They wanted to determine the speed that produced maximum fuel economy, that is, the speed at which miles per gallon (mpg) was a maximum. The data were summarized in the following function that measured fuel economy $F(x)$ in mpg for a truck operating at speed x, measured in miles per hour (mph).

$F(x) = 0.15x - 0.0015x^2$

The natural bounds for x were $x = 0$, because a negative value for x makes no sense, and $x = 55$, because the government had just imposed the 55 mph speed limit. The next example solves the problem.

EXAMPLE 5

Problem

Find the absolute maximum of the fuel economy function

$F(x) = 0.15x - 0.0015x^2$

where the domain of the function is [0, 55].

Solution Find the critical values by solving $F'(x) = 0$.

$F'(x) = 0.15 - 0.003x = 0$

$x = 50$

Compare the three values $F(0) = 0$, $F(50) = 3.75$, and $F(55) = 3.7125$. The absolute maximum is $F(50) = 3.75$ mpg. Therefore the speed of 50 mph produced the greatest fuel economy.

*Bernard J. Winger, *Cases in Management Economics,* Columbus, OH, Grid Publishing, 1979, pp. 78–81.

EXERCISE SET 14–4

In Problems 1 to 14 find the absolute extrema, if they
exist, for the function defined on the given interval.

1. $f(x) = x^2 + 2x + 1$ on $[-4, 4]$

2. $f(x) = x^2 - 2x + 1$ on $[-4, 4]$ *absolute*
     &
     local

3. $f(x) = 5 - 4x - x^2$ on $[-4, 6]$

4. $f(x) = 8x - 2x^2$ on $[-4, 6]$

5. $f(x) = x^3 + 3x^2 + 1$ on $[-2, 2]$

6. $f(x) = x^3 - 6x^2 + 1$ on $[-1, 5]$

7. $f(x) = x^3 + 3x^2 + 1$ on $[-3, -1]$

8. $f(x) = x^3 - 6x^2 + 1$ on $[-2, 2]$

9. $f(x) = x^4 - 2x^2 + 1$ on $[-2, 2]$

10. $f(x) = x^4 + 4x^3 + 1$ on $[-4, 2]$

11. $f(x) = (x - 2)^2(x - 1)$ on $[-1, 4]$

12. $f(x) = (x + 1)^2(x - 1)$ on $[0, 4]$

13. $f(x) = (x - 3)^2(x - 1)^2$ on $[-1, 4]$

14. $f(x) = (x + 1)^2(x - 1)^2$ on $[0, 4]$

15. An object is thrown straight upward so that its height
after t seconds is $f(t) = -16t^2 + 128t$, where f is
measured in feet. How high will the object travel and
after how many seconds will it reach its maximum
height?

16. An object is thrown straight upward so that its height
after t seconds is $f(t) = -16t^2 + 256t + 6$, where
f is measured in feet. How high will the object travel
and after how many seconds will it reach its maximum
height?

17. A family wants to build a rectangular deck on the
back of their house. It will be enclosed on three sides
by wooden fencing with the back of the house being
the fourth side. If 40 feet of fencing is to be used,
find the dimensions of the deck with the largest area
that can be built.

18. An apple orchard owner has determined that if 30
trees are planted per acre, the yield is 600 apples per
tree. For each tree per acre that is added beyond 30,
the yield is reduced by 15 apples per tree. Find the
number of trees per acre that should be planted to
yield the maximum crop.

19. Divide the number 300 into two parts whose sum is
300 such that the product of one part and the square
of the other is a maximum.

20. Divide the number 120 into two parts whose sum is
120 such that the product of one part and the cube of
the other is a maximum.

21. A farmer wants to enclose with a fence a rectangular
field that is adjacent to a straight river, where no
fencing is required along the river. What are the
dimensions of the field if 4000 feet of fencing is to
be used and the area of the field is to be a maximum?

22. A manufacturing company offers the following discount schedule of prices: $20 for orders of 500 or
fewer, with the cost being reduced 2 cents for each
unit above 500. Find the order that will maximize the
company's revenue.

23. A person is on an island 10 miles from a straight
shore. The person wishes to get to a point 12 miles
up the coast. If the person can row at the rate of 2
mph (miles per hour) and walk at the rate of 4 mph,
to what point on the shore should the person row in
order to minimize the time?

24. A manufacturing company makes an open box from
a 5 by 6-inch rectangular piece of cardboard by cutting
out equal squares at each corner and then folding up
the sides. What are the dimensions of the box with
maximum volume?

25. A manufacturing company wants to package its product in a rectangular box with a square base and a
volume of 32 in.^{3}. The cost of the material used for
the top and bottom is 3 cents per square inch (in.^{2}),
while the cost of the material used for the sides is 6
cents per in.^{2}. What are the dimensions of the box
with the minimum cost?

26. A manufacturing company wants to package its product in a closed can in the shape of a right circular
cylinder. The company has determined that the volume of the can must be 16 in.^{3}. What are the dimensions of the can with the minimum surface area?

27. A manufacturing company wants to package its product in a rectangular box with a square base and a
volume of 32 in.^{3}. The cost of the material used for
the top is 1 cent per in.^{2}, the cost of the material used
for the bottom is 3 cents per in.^{2}, and the cost of the

material used for the sides is 2 cents per in.^{2}. What are the dimensions of the box with the minimum cost?

28. At 1:00 P.M. plane A is 650 miles due west of plane B. If plane A flies at 100 mph due south and plane B flies at 150 mph due west, when will they be nearest each other and how near will they be?

29. U.S. postal regulations state that packages that are sent parcel post must have the sum of the length plus the girth (perimeter of a cross section) no more than 84 in. What are the dimensions of the rectangular package with square ends with the greatest volume that can be sent by parcel post?

30. U.S. postal regulations state that packages that are sent parcel post must have the sum of the length plus the girth (perimeter of a cross section) no more than 84 in. What are the dimensions of the cylindrical package with the greatest volume that can be sent by parcel post?

31. A "Norman window" consists of a rectangle with a semicircular area on top. See the accompanying figure. Find the width of the rectangle if the perimeter of the window is 12 feet and the area is as large as possible.

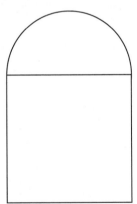

32. A long rectangular sheet of aluminum is to be made into a gutter by turning up equal sides the length of the strip at right angles. How many inches should be turned up on each side to maximize the volume of the gutter?

33. A rectangular pen is to be fenced in, and then an additional fence will be constructed in the center of the pen to make two identical rectangular enclosures. If the area is to be 60 square yards, find the dimensions of the pen with the minimum amount of fencing.

34. Show that among all rectangles having a fixed perimeter, the square has the maximum area.

35. Show that among all rectangles having a fixed area, the square has the minimum perimeter.

36. A principle in economics states that maximum profit is obtained when the marginal revenue, which is the derivative of the revenue function, equals the marginal cost, which is the derivative of the cost function. Explain this principle if $P(x)$ is the profit function, $R(x)$ is the revenue function, and $C(x)$ is the cost function when x goods are produced.

37. A box with no top and a square bottom is to have a volume of 10 in.^{3}. Find the dimensions of the box of least cost if the material of which the sides are to be made costs one-third per square inch as much as the material of which the bottom is to be made.

Referenced Exercise Set 14–4

1. In a study* to increase the productivity of the average personal automobile, Purdue University's Interdisciplinary Engineering Studies researched the feasibility of a plan for members of an enterprise to own and operate jointly a fleet of vehicles. In one model 26 families were monitored by having them keep diaries of when they would use a shared vehicle. By keeping records of cost and availability of the shared vehicles, the researchers generated the following function that relates the number of vehicles, x, to the cost per family per week, $f(x)$:
$$f(x) = 11.8x^3 - 138.2x + 200$$
The function is valid in the interval [0, 4]. Find the value of x that minimizes the cost per family per week.

2. There is extensive literature on the optimal governmental regulation of open-access resources, such as fisheries and watersheds. In a study† of market behavior and exploitation of fisheries an important function is the total revenue R produced that depends on the fishing effort x, measured by the number of days fished

*Jeffery K. Cochran and F. T. Sparrow, "Optimal Management of a Shared Fleet with Peak Demands," *Applications of Management Science*, Vol. 4, 1985, pp. 81–105.

†Lee G. Anderson and Dwight R. Lee, "Optimal Governing Instrument, Operation Level, and Enforcement in Natural Resource Regulation: The Case of the Fishery," *American Journal of Agricultural Economics*, Vol. 68, 1986, pp. 678–690.

by standard boats. One model considered assumes the cost $C(x)$ is a linear function. A typical pair of functions for $R(x)$ and $C(x)$ is

$$R(x) = 0.4x - 0.002x^2$$
and $$C(x) = 0.11x$$

The natural bounds for these functions would be $x \geq 0$ on the left and the natural bound on the right would be that value of x such that $R(x) = C(x)$.

(a) Find the value of x that maximizes $R(x)$.

(b) Define the profit function $P(x) = R(x) - C(x)$ and find the value of x that maximizes $P(x)$.

(c) Graph $R(x)$ and $C(x)$ in the same coordinate system and state why the right-hand bound is a natural boundary for the problem.

3. In an article studying the trade-offs that research and development teams must make, an example from space research was analyzed.[‡] NASA was trying to determine whether it was cost-effective to manufacture silicon cells for use in future applications. The price of silicon when the study was done (1981) was $65 per kilogram (kg). NASA set a goal of producing silicon at a price of $20 per kilogram. They could make silicon at this price, but its quality was not high enough. If they produced silicon at a higher price, what was the likelihood that it would be high quality? The research-

ers generated a function that measured the likelihood $P(x)$, measured in percent, that the silicon produced at price x would be high quality. The function is approximated by

$$P(x) = -0.0001x^3 + 0.0045x^2 - 0.25$$

There were two natural bounds for x. It made no sense to let x take on negative values and the price of $28 per kg was set as the highest price they would be willing to pay. So the domain of $P(x)$ is $[0, 28]$. Find the absolute maximum of $P(x)$ and graph the function.

4. This problem was suggested by John Dawson, our bird-loving friend from Penn State.[§] In the autumn many people put up feeders for wild birds, thereby initiating the annual round of "squirrel wars." Seasoned veterans of the combat have learned to thwart the acrobatic rodents by suspending the feeders from wires in the shape of the letter **Y**. If the wire is suspended between two trees 10 feet apart such that there are two feet below the height at which the wires are attached to the trees, then the length L of the wire is given by the equation

$$L = x + \sqrt{10^2 + 4(2 - x)^2}$$

where x denotes the length of the tail of the **Y**. (See Figure 5.) What value of x will minimize the length of the wire?

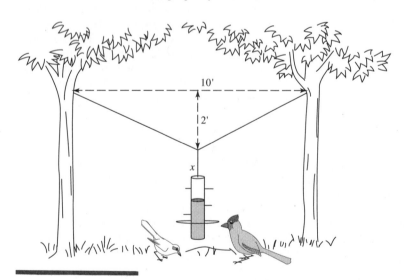

10'

2'

x

FIGURE 5

[‡]Virendra S. Sherlekar and Burton V. Dean, "Assessment of R & D Risk-Cost Trade-Offs," *Applications of Management Science*, Greenwich, CT, JAI Press, 1983, pp. 40–51.

[§]John W. Dawson, Jr., "Hanging a bird feeder: Food for thought," *The College Mathematics Journal*, March, 1990, pp. 129–130.

Cumulative Exercise Set 14–4

In Problems 1 and 2, a function $f(x)$ is given.
(a) Find the intervals where $f(x)$ is increasing.
(b) Find the point(s) where $f(x)$ has a horizontal tangent line.
(c) Sketch the graph of the function.

1. $f(x) = x^3 - 4x^2 - 2$

2. $f(x) = \dfrac{x}{x - 4}$

In Problems 3 to 5 use the information given to sketch the graph of $f(x)$, assuming that $f(x)$ is continuous for all real numbers x.

3. Values where $f'(x) = 0$

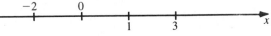

Sign of $f'(x)$ ----   ++++   ----   0000   ++++

$f(-2) = -2$, $f(0) = 2$, $f(1) = 1$, $f(4) = 3$

4. Critical values

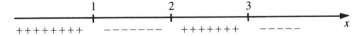

Sign of $f'(x)$ ++++++++   --------   +++++++   -----

$f(1) = 2$, $f(3) = 0$

5. Critical values

Sign of $f'(x)$ ----------   +++++++   +++++++++   +++++

0, $f(0) = 0$

In Problems 6 to 9 find the relative extrema and sketch the graph of $f(x)$.

6. $f(x) = 2 + 4x^3 - x^4$

7. $f(x) = x^5 - x$

8. $f(x) = x(x - 4)^2$

9. $f(x) = (x - 1)^3(x^2 - 1)$

10. Find the absolute extrema for $f(x) = (x^2 - 1)^2$ on the interval $[-2, 3]$.

11. What number exceeds its square by the maximum amount?

12. Let $f(x)$ be the function whose graph is shown.
(a) List the critical values, if there are any.
(b) List the relative extrema, if there are any.
(c) List the points of inflection, if there are any.
(d) List the vertical asymptotes, if there are any.
(e) List the horizontal asymptotes, if there are any.
(f) For what interval(s) is $f' > 0$?
(g) For what interval(s) is $f(x)$ concave upward?

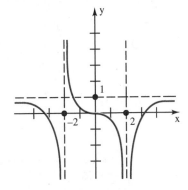

14–5 Implicit Differentiation and Related Rates

Thus far the relationships between the variables have been presented as **explicit functions,** meaning that an equation in the form $y = f(x)$ has been given. Two examples are

$$y = x^2 - 4 \qquad y = (x^3 + 4x)^{1/2}$$

However, often in applications equations do not express y as an explicit function of x. For example, the equation

$$x^2 + y^2 = 1$$

whose graph is a circle of radius 1, is not of the form $y = f(x)$. In this section we study equations that are not expressed as functions. They are called **implicit functions.** We also compute their derivatives. Applications of these types of derivatives, called *related rates,* are also studied.

Implicit Functions

The equation $x^2 + y^2 = 1$ does not express y as a function of x directly because, for every value of x in the interval $(-1, 1)$, there are two corresponding values of y that satisfy the equation. For instance, if $x = 0$, then y can be either 1 or -1; that is, the points $(0, 1)$ and $(0, -1)$ are on the graph. Geometrically, Figure 1 shows that any vertical line $x = a$ for $-1 < a < 1$ intersects the graph in more than one point, so the graph is not a function.

We say that equations such as $x^2 + y^2 = 1$ define y *implicitly* as a function of x. Intuitively this means that even though the equation does not express y in terms of x directly, it is theoretically possible to solve for y. However, this is not always an easy task. For example, given the equation $xy = 1$, it is easy to solve for y to get $y = 1/x$; but it is much more difficult to solve for y given the equation $y^5 - y^3 + xy = 1$. It may be possible to express the equation as several formulas.

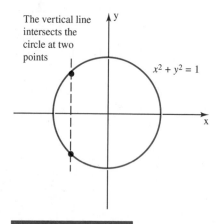

The vertical line intersects the circle at two points

$x^2 + y^2 = 1$

FIGURE 1

For example, $x^2 + y^2 = 1$ can be expressed as the two functions $y = \sqrt{1 - x^2}$ and $y = -\sqrt{1 - x^2}$.

Implicit Differentiation

If an equation expresses y as an implicit function of x, it is possible to compute the derivative of y with respect to x without having to solve for y directly. This technique is called **implicit differentiation.** It assumes that the equation can be expressed in the form $y = f(x)$, and then the derivative of each term of the equation is computed, keeping in mind that y is a function of x. For example, to find y' given the equation $x^2 + y^2 = 1$, find the derivative of each term separately. The derivative of the first term, x^2, is $2x$. However, the derivative of the second term, y^2, is not merely $2y$. The chain rule must be used to compute the derivative of y^2, because it is assumed that $y^2 = [f(x)]^2$. Therefore, using the chain rule, the derivative of y^2 is $2yy'$. The derivative of the constant term, 1, is 0. Putting these steps together and solving for y' yields

$$D_x(x^2 + y^2) = D_x 1$$
$$2x + 2yy' = 0$$
$$2yy' = -2x$$
$$y' = -\frac{2x}{2y}$$
$$y' = -\frac{x}{y}$$

The equation for the derivative contains both x and y, which is usually the case when computing a derivative implicitly. To evaluate the derivative at a particular point both the x and y coordinates must be substituted into the equation.

The important idea to remember when computing implicit derivatives is that the derivative of any term containing y will require the use of the chain rule so that it will have y' as part of its derivative. Also, any term containing a product, such as x^2y^3, will require the product rule as well as the chain rule. For instance, suppose an implicit function contains the term x^2y^3, then the derivative of this term is

$$D_x(x^2y^3) = x^2 D_x(y^3) + y^3 D_x(x^2)$$
$$= x^2 \cdot 3y^2 y' + y^3 \cdot 2x$$
$$= 3x^2y^2y' + 2xy^3$$

Example 1 illustrates how to compute the derivative of an implicit function and how to evaluate the derivative at a point that satisfies the implicit function.

EXAMPLE 1

Problem

Consider the equation

$$x^3 + x^2y - y^4 = 1$$

(a) Find y' using implicit differentiation.
(b) Evaluate y' at (1, 1).

Solution (a) Find the derivative of each term.

$$D_x(x^3 + x^2y - y^4) = D_x(1)$$
$$D_x(x^3) + D_x(x^2y) - D_x(y^4) = 0$$
$$D_x(x^3) + x^2D_x(y) + yD_x(x^2) - D_x(y^4) = 0$$
$$3x^2 + x^2 \cdot y' + y \cdot 2x - 4y^3y' = 0$$
$$3x^2 + x^2y' + 2xy - 4y^3y' = 0$$

Gather terms containing y'.

$$x^2y' - 4y^3y' = -3x^2 - 2xy$$
$$(x^2 - 4y^3)y' = -3x^2 - 2xy$$
$$y' = (-3x^2 - 2xy)/(x^2 - 4y^3)$$

(b) Substitute $x = 1$ and $y = 1$ into the equation for y' in (a).

$$y' = \frac{-3 - 2}{1 - 4} = \frac{-5}{-3} = \frac{5}{3}$$

Often in applications the two variables in an implicit function both depend on a third variable. For example, suppose x and y are related by the equation $x^2 + y^2 = 2500$ where x and y represent distances, and each is changing with respect to time. If t represents time, then x and y are functions of t. By differentiating the equation relating x and y, we get a relationship in the two derivatives D_tx and D_ty. The chain rule must be used on each term, including x^2. That is, the derivative of x^2 with respect to t is not merely $2x$ but $2xx'$, where x' is D_tx. Example 2 shows how to carry this out.

EXAMPLE 2

Let x and y be related by the equation

$$x^2 + y^2 = 2500$$

Suppose x and y change with respect to time, t.

Problem

Find an equation relating D_tx and D_ty. Then evaluate the equation for $x = 40$, $y = 30$, and $D_ty = -4$.

Solution Think of the equation as an implicit function of t, where both x and y are implicit functions of t. Differentiate the equation with respect to t by differentiating each term with respect to t. This yields

$$D_t(x^2 + y^2) = D_t2500$$
$$2x \cdot D_tx + 2y \cdot D_ty = 0$$
$$2xD_tx = -2yD_ty$$
$$D_tx = \frac{-yD_ty}{x}$$

Evaluating the equation for $D_t x$ when $x = 40$, $y = 30$, and $D_t y = -4$ yields

$$D_t x = \frac{-30(-4)}{40} = 3$$

Related Rates

In Example 2 the variables x and y are functions of a third variable t. The rate of change of x is related to the rate of change of y by the formula derived in the example by differentiating the original implicit function with respect to t. Such a problem is called a **related rates** problem. The next example illustrates this type of application.

EXAMPLE 3

A tropical storm is 50 miles offshore, and its path is perpendicular to a straight shoreline. It is approaching the shore at the rate of 4 mph. Meteorologists studying the behavior of the storm from a van on the shore want to stay exactly 50 miles from the storm and remain on the shoreline. They start at the point on the shoreline in the path of the storm.

Problem

Determine a formula for the speed that the truck must maintain to remain 50 miles from the storm. Then find the speed of the truck when the storm is 40 miles from the shore.

Solution Usually the first step in a related rates problem is to make a sketch of the problem. From the sketch the variables can be labeled. See Figure 2. Let the

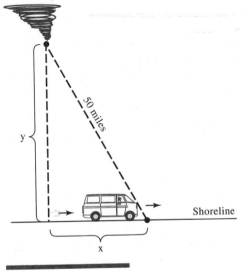

FIGURE 2

distance from the storm to the shore be represented by y and the distance that the van has traveled be represented by x. From the Pythagorean Theorem the relationship between x and y is

$$x^2 + y^2 = 50^2 = 2500$$

The problem asks for $D_t x$, given that $D_t y = -4$, which is negative because y is decreasing. The relationship between $D_t x$ and $D_t y$ is derived by differentiating the formula relating x and y with respect to t. This was done in Example 2. The relationship is

$$D_t x = \frac{-y D_t y}{x}$$

Since $D_t y = -4$, the formula becomes

$$D_t x = \frac{4y}{x}$$

When the storm is 40 mi from the shore, that is, $y = 40$, then from the Pythagorean Theorem $x = 30$. Therefore the speed of the van when $y = 40$ is

$$D_t x = \frac{4(40)}{30} = \frac{16}{3} \text{ mph}$$

Example 3 demonstrates the six steps used in solving related rates problems.

The Key Steps in Solving Related Rates Problems

1. Draw a sketch of the problem if possible.
2. Identify all quantities, both constants and variables, and give labels to the variables.
3. Find a formula relating the variables.
4. Using implicit differentiation, find the derivative with respect to time, t, of each side of the formula. Remember that each variable in the formula is a function of t.
5. The problem asks for the rate of a particular variable. Solve the formula derived in step 4 for this derivative.
6. Substitute the given quantities into the formula for the desired derivative.

It is important to perform steps 4 and 5 before doing the substitution called for in step 6. If the substitution is made beforehand, the variables will not appear in the formula when applying implicit differentiation. Thus the formula derived in step 4 will be incorrect. To illustrate, suppose in Example 3 the value $y = 40$ is substituted into the formula $x^2 + y^2 = 2500$ before the derivative is found. This would result in the formula $x^2 + 1600 = 2500$. Using implicit differentiation on this formula results in $2xx' = 0$, so $x' = 0$ whenever $x \neq 0$, which is certainly incorrect.

The next two examples illustrate how to apply the steps used to solve related rates problems.

EXAMPLE 4

A rock is thrown into a still pond and circular ripples move out. The radius of the disturbed region increases at the rate of 3 ft/sec.

Problem

Find the rate at which the surface area of the disturbed region is increasing when the farthest ripple is 20 feet (ft) from the place where the rock struck the pond.

Solution The sketch of the problem is one of concentric circles given in Figure 3, each circle representing a ripple. The farthest ripple is the largest circle. Let the radius of the farthest ripple be represented by r and the area of the disturbed region by A. The formula relating r and A is the area of a circle.

$$A = \pi r^2$$

It is given that $D_t r = 3$ and we are asked to find $D_t A$ when $r = 20$. Next use implicit differentiation to find $D_t A$ in terms of $D_t r$.

$$D_t A = 2\pi r D_t r$$

Substitute into this formula $r = 20$ and $D_t r = 3$.

$$D_t A = 2\pi(20)(3) = 120\pi \approx 376.99 \text{ ft}^2/\text{sec}$$

Thus when the radius is 20 ft the surface area of the disturbed region is increasing at the rate of about 377 ft^2/sec.

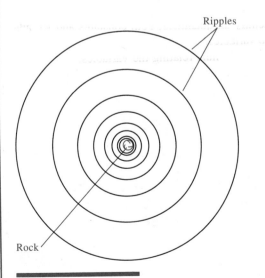

Ripples

Rock

FIGURE 3

EXAMPLE 5

An industrial filter in the shape of a right circular cone is 10 cm (centimeters) high and has a radius of 20 cm at its top. A solution is poured through the filter, and residue gathers in the filter at the rate of 2 cm³/min.

Problem

Find the rate at which the height of the residue is increasing when the height is 2 cm.

Solution The sketch is given in Figure 4. Let the height of the residue be represented by h, the radius by r, and the volume by V. The residue gathers in the shape of a cone. The formula for the volume of a cone is

$$V = \left(\frac{1}{3}\right) \pi r^2 h$$

It is given that $D_t V = 2$ and we are asked to find $D_t h$ when $h = 2$. It is possible to use implicit differentiation to find $D_t h$ in terms of $D_t V$ and $D_t r$, but the value of $D_t r$ is unknown. Since V is expressed as a function of two variables, r and h, and only one value of the two derivatives is given, it is necessary to express V as a function of h alone. The relationship between r and h is given by similar triangles. From Figure 4 the triangle of the cone and the triangle of the residue are similar. The ratios of corresponding sides of similar triangles are equal, which yields

$$\frac{r}{20} = \frac{h}{10}$$

Thus $r = 2h$. Substituting this into the formula for V yields

$$V = \left(\frac{1}{3}\right) \pi (2h)^2 h = \left(\frac{4}{3}\right) \pi h^3$$

Applying implicit differentiation to the formula yields

$$D_t V = \left(\frac{4}{3}\right) \pi \cdot 3h^2 D_t h$$
$$= 4\pi h^2 D_t h$$

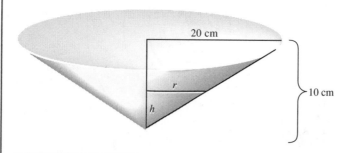

FIGURE 4

Solving for $D_t h$ and making the substitution $D_t V = 2$ gives

$$D_t h = \frac{2}{4\pi h^2}$$

This is the formula for $D_t h$. Next substitute the value $h = 2$ to find the rate at which the height of the residue is increasing when the height is 2 cm.

$$D_t h = \frac{2}{4\pi 2^2} = \frac{1}{8\pi} \simeq 0.04 \text{ cm/min}$$

EXERCISE SET 14-5

In Problems 1 to 8 find $D_x y$ by using implicit differentiation.

1. $x^2 + 3y^2 = 4$

2. $x^2 - 5y^3 = 6$

3. $xy + x^2 y = 2$

4. $xy - x^2 y^2 = 4$

5. $x^3 + xy^3 - y^4 = 1$

6. $x^4 - x^2 y^3 + y^5 = 4$

7. $x^2 + xy^3 - y^{-1} = 1$

8. $x^4 - 2x^2 y + y^{-2} = 0$

In Problems 9 to 12 find $D_x y$ evaluated at the given point.

9. $x^2 + y^3 = 0$; $(1, -1)$

10. $x^3 + y^2 = 2$; $(1, -1)$

11. $3xy + x^2 y = 0$; $(1, 0)$

12. $4xy - x^2 y^2 = 3$; $(1, 1)$

In Problems 13 to 16 x and y are functions of t. Express $D_t y$ in terms of $D_t x$. Then find $D_t y$ when $x = 1$, $y = 1$, and $D_t x = 2$.

13. $x^2 + y^3 = 2$

14. $x^3 + 4y^3 = 5$

15. $3x - x^2 + y^2 = 3$

16. $x^{1/2} + 3y - y^{1/2} = 3$

17. A tropical storm is 100 miles offshore, and its path is perpendicular to a straight shoreline. It is approaching the shore at the rate of 6 mph. Meteorologists studying the behavior of the storm from a van on the shore want to stay exactly 100 miles from the storm and remain on the shoreline. Determine the speed of the truck when the storm is 60 miles from the shore.

18. A rock is thrown into a still pond and circular ripples move out, the radius of the disturbed region increasing at the rate of 2 ft/sec. Find the rate at which the surface area of the disturbed region is increasing when the farthest ripple is 30 ft from the place where the rock struck the pond.

19. A filter in the shape of an inverted cone is 20 cm high and has a radius of 30 cm at its top. A solution is poured through the filter, and residue gathers in the filter at the rate of 5 cm^3/min. Find the rate at which the height of the residue is increasing when the height is 3 cm.

20. A firm determines that output y is related to input x, both measured in hundreds of units, by the formula $y = 3x^{1/3}$. If the demand is increasing at the rate of 20 units per month, at what rate must the input increase to meet this increase in demand if the current input is 2700 units?

21. A ladder 50 ft long is standing against a building. The base of the ladder is moved away from the building at the rate of 3 ft/sec. How fast is the top of the ladder moving down the wall when the base of the ladder is 40 ft from the foot of the building?

22. Water is being pumped into a cylindrical trough at the rate of 10 ft^3/min. If the depth of the trough is 12 ft, the width of the top is 12 ft, and the length of the top is 20 ft, at what rate is the height of the water rising when the height is 6 ft?

23. A 6-ft tall person is walking away from a streetlight at the constant rate of 3 ft/sec. If the streetlight is 20 ft high, how fast is the length of the person's shadow increasing when the person is 30 ft from the streetlight?

24. A water tank is 30 meters (m) wide, 60 m long, and 10 m deep at one end and 15 m deep at the other end, with the depth increasing linearly. Water is pumped in at the rate of 5 m^3/min. How fast is the water level rising when the water level is 3 m?

25. A spherical rubber balloon is being inflated with gas at a constant rate of 15 cm³/sec. What is the rate of change of the radius when the volume is 45 cm³?

26. Water is condensing on the surface of a spherical drop of water so that the water surface remains a sphere and the volume is increasing at the rate of 5 in.³/min. At what rate are the surface area and the radius increasing when the radius is 10 in.?

27. A pulley, mounted on a dock, stands 10 ft above the water and a boat is being towed in toward the dock. The rope is attached to the boat at a point 2 ft above the water. If 0.5 ft of rope is being drawn in each second, at what rate is the boat moving when it is 6 ft from the dock?

28. An off-shore oil drill springs a leak. The oil slick spreads in a circular shape, and the radius of the slick is increasing at the rate of 30 ft/hr. Find the rate that the surface area of the oil slick is increasing when the radius is 100 ft.

29. A balloon is rising vertically at the rate of 15 ft/sec. An observer is standing on the ground 900 ft from the point where the balloon left the ground. Find the rate that the distance between the balloon and the observer is changing when the balloon is 1200 ft high.

30. Two roads intersect at right angles. A truck, traveling at the rate of 30 mph, reaches the intersection 5 minutes (min) before a car that is traveling at the rate of 40 mph on the other road. How fast is the distance between the truck and the car changing 5 min after the car reaches the intersection?

31. One airplane, traveling due north at 200 mph, passes a point 1 hour after another airplane, traveling due east at 150 mph, passes the same point. How fast is the distance between the airplanes increasing 1 hour after the first airplane passed the point?

32. A baseball player, running at the speed of 30 ft/sec, is running in a straight line from second to third base. A fielder, standing at a point 100 ft from third base on the left field foul line (which is at right angles to the path of the runner), throws the ball, at the speed of 150 ft/sec, when the runner is 50 ft from third base.

What is the rate of change of the distance between the ball and the runner 0.5 sec after the fielder throws the ball?

33. When helium expands, its pressure P is related to its volume V by the formula $P = kV^{-5/3}$. When the volume of a helium-filled balloon is 18 m³ the pressure is 0.3 kg/m². What is the rate of change of the volume if the pressure is increasing at the rate of 0.05 N/m²/sec?

34. Liquid is being drawn at the rate of 2 m³/hr from a tank whose shape is a hemisphere with radius 10 m. How fast is the height of the liquid level changing 2 hours after the tank was full?

Referenced Exercise Set 14–5

1. In a paper* studying efficient glass-cutting techniques, various cost-saving measures were explored, including minimizing the amount of waste and the time used in the cutting procedure. Once a firm has an inventory of various sizes of virgin stock, the problem is to determine what sizes of glass should be cut from the different sheets of stock. Solve the following problem, which is representative of those considered in the paper. Two pieces of glass are to be produced by making two cuts in a sheet of stock. A vertical cut will produce one piece and a horizontal cut in the remaining piece of stock will produce the second piece. The piece left over is called the waste and it is to be made square. (See Figure 5.) If the perimeter of the stock plate is 20 in. and the total surface area of two pieces produced by the cuts is 20 in.², find the dimensions of the stock plate that minimizes the total length of the two cuts.

2. Population structure and regeneration patterns of trees are important aspects of forest management. In a paper studying size and age structure of two forested strands in the western Cascade Range in Oregon, Stewart studied four species of trees growing in various types of habitats.† In one part of the study a typical problem involved planting two types of trees, each planted on a different square plot of land. In the first plot seedlings were planted in rows every $1\frac{1}{3}$ meters, and in the other plot the seedlings were planted in rows every 4 meters. Find the minimum area of each plot that would be needed to plant 320 rows of trees.

*Adolf Diegel and Hans Vocker, "Optimal Dimensions of Virgin Stock in Cutting Glass to Order," *Decision Sciences*, Vol. 15, 1984, pp. 260–272.

†Glenn H. Stewart, "Population Dynamics of a Montane Conifer Forest, Western Cascade Range, Oregon, USA," *Ecology*, Vol. 67, 1986, pp. 534–544.

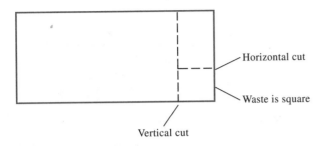

Horizontal cut

Waste is square

Vertical cut

FIGURE 5 *Source:* Adolf Diegel and Hans Vocker, "Optimal Dimensions of Virgin Stock in Cutting Glass to Order," *Decision Sciences,* Vol. 15, 1984, pp. 260–272.

Cumulative Exercise Set 14–5

In Problems 1 and 2 find intervals where $f(x)$ is increasing and where $f(x)$ is decreasing. Find the point(s) where $f(x)$ has a horizontal tangent line. Graph the function.

1. $f(x) = x^3 - 6x^2 - 1$

2. $f(x) = x^4 - 4x^3$

In Problems 3 and 4 find the relative extrema and sketch the graph of the function.

3. $f(x) = 2x^3 + 3x^2 - 12x.$

4. $f(x) = 9x + 4/x$

5. Use the curve-sketching techniques to graph the function

$$f(x) = 3x^4 + 4x^3$$

6. Use the information given to sketch the graph of $f(x)$. Assume $f(x)$ is continuous for all real numbers.

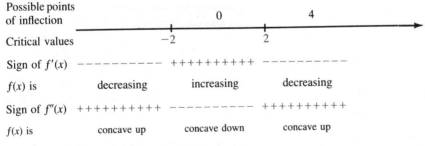

Possible points of inflection ... 0 ... 4

Critical values ... −2 ... 2

Sign of $f'(x)$ — — — — — — — — — + + + + + + + + + + — — — — — — — — —

$f(x)$ is   decreasing   increasing   decreasing

Sign of $f''(x)$ + + + + + + + + + + — — — — — — — — — + + + + + + + + + +

$f(x)$ is   concave up   concave down   concave up

$f(-2) = 0,\ f(0) = 1,\ f(2) = 2,\ f(4) = 1$

7. Find the absolute extrema, if they exist, for the function defined on the given interval.

$$f(x) = x^3 - 6x^2 + 1 \text{ on } [-3, 2]$$

8. Divide the number 240 into two parts whose sum is 240 such that the product of one part and the cube of the other is a maximum.

9. A candy firm makes an open box from a 6-inch by 10-inch rectangular piece of cardboard by cutting out equal squares at each corner and then folding up the sides. What are the dimensions of the box with maximum volume?

10. Find D_{xy} using implicit differentiation:

$$x^3 - 3x^2y + y^3 = 8$$

11. A firm determines that output y is related to input x, both measured in hundreds of units, by the formula $y = 110x^{1/4}$. If the demand is increasing at the rate of 10 units per month, at what rate must the input increase to meet this increase in demand if the current input is 100 units?

12. Two railroad tracks intersect at right angles. One train, traveling at the rate of 50 mph, reaches the intersection 10 minutes before a train that is traveling at the rate of 60 mph on the other track. How fast is the distance between the two trains changing 5 minutes after the second train reaches the intersection?

CASE STUDY   Minimizing Traffic Noise

The private car is by far the most popular means of transportation. However, for many people the car represents more than convenient transportation; their dreams and egos are front seat passengers. It is not surprising that the automobile population rate far outdistances its human counterpart. The role of highway transportation in our lives is profound. Indeed, a significant measure of a country's "development" is its highway system.

The Department of Transportation used to be an agency whose sole responsibility was the structure of roads. Most of our highways were built with static road design as the only critical issue. No concern was given to dynamic factors such as traffic flow, ramp control, merging, and congestion. But today traffic controllers must consider a multitude of factors to minimize traffic snarls and accidents. Estimating traffic demand and capacity has become as much an art as it is a science.

One of the most important considerations of highway operation is acceleration noise control. In the past the sole criterion for improving highway operations was to maximize volume throughout. Little consideration was given to the quality of highway conditions. For example, it is standard practice to reduce highway bottlenecks by ramp control techniques, such as using stoplights to control entrance to the roadway. But nowadays highway administrators give equal concern to the qualitative changes in traffic due to these controls. For instance, a particular control might have reduced overall travel time, but perhaps it created hazardous locations such as rapid decelerations far from the bottlenecks. Measurement of acceleration noise is a valuable tool to identify any hazardous locations. Noise control not only provides more comfortable travel, but it also helps maintain a safer and smoother operation.

One of the first and most extensive traffic flow research projects* was conducted by the Texas Highway Department and the U.S. Bureau of Public Roads in 1963, called the "Gulf Freeway Project." Various surveillance and control devices, including television and noise monitors, were installed along the Gulf Freeway in Houston. The freeway was divided into 56 sections, and each was monitored for several characteristics, including noise level.

Acceleration noise is defined as the disturbance of a vehicle's speed from a uniform speed. It is thus a measure of the deviation from a smooth ride and so it measures the smoothness of traffic. The frequency of violent acceleration and

*Donald R. Drew, *Traffic Flow Theory and Control*, New York, McGraw-Hill Book Co., 1968.

deceleration affects acceleration noise and reflects potentially hazardous conditions.

The primary goal of the Gulf Freeway study was to determine the means to ensure efficient traffic flow. This meant not only maximizing volume but also minimizing the threat of hazardous conditions by minimizing acceleration noise. What range of speeds would best satisfy both of these criteria?

Acceleration noise is a function of speed. Figure 1 is a representation of the relationship of acceleration noise, a, to the average speed of traffic, u, for 1 of the 56 sections of the freeway. The curve that represents the data was given as

$$a = 1.693 - 0.00284u^2 + 0.000045u^3$$

where a is measured in ft/sec^2 and u is measured in miles per hour(mph). This function is valid only in the interval [10, 60]. To see why 10 is a natural endpoint, refer to Figure 1: when $u = 0$, the vehicle is at rest, and $a = 0$. As the vehicle starts to accelerate quickly to merge, acceleration noise increases rapidly. It is represented by the dashed line. As the vehicle enters the traffic pattern, acceleration noise decreases.

The speed that yields minimum acceleration noise is found by calculating the derivative, a', and setting $a' = 0$. The derivative is computed as follows:

$$a' = -2 \cdot 0.00284u + 3 \cdot 0.000045u^2 = -0.00568u + 0.000135u^2$$
$$= u(-0.00568 + 0.000135u)$$

Setting $a' = 0$ yields

$$u(-0.00568 + 0.000135u) = 0$$

This means that either $u = 0$ or $-0.00568 + 0.000135u = 0$. But 0 is not in

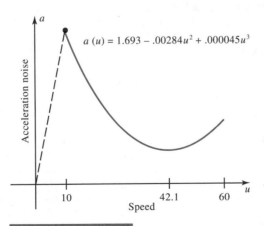

$$a(u) = 1.693 - .00284u^2 + .000045u^3$$

Acceleration noise

Speed

10   42.1   60

FIGURE 1 *Source:* Donald R. Drew, *Traffic Flow Theory and Control,* New York, McGraw-Hill Book Co., 1968.

the domain of the function so the only critical value is the latter one. To find the critical value, we solve for u.

$$-0.00568 + 0.000135u = 0$$
$$0.000135u = 0.00568$$
$$u \simeq 42.1$$

The second derivative is $a'' = -0.00568 + 0.00027u$ and $a''(42.1) > 0$, so a has a minimum at $u = 42.1$. Thus the speed that yields the minimum amount of acceleration noise is 42.1 mph.

In another study the speed that yields the maximum flow of volume was found to be 35 mph. This allowed the highway administrators to establish a target speed interval of 35 to 42 mph. This ensured that while the freeway's average speed fluctuated within this interval, they were either maximizing volume, if the speed was close to 35 mph, or minimizing acceleration noise, if it was close to 42 mph.

Case Study Exercises

In Problems 1 to 4 the acceleration noise levels of several other sections of the Gulf Freeway are given. Find the speed at which the acceleration noise is a minimum.

1. $a = 1.572 - 0.00345u^2 + 0.000078u^3$
2. $a = 1.916 - 0.00402u^2 + 0.000088u^3$
3. $a = 2.006 - 0.00501u^2 + 0.000131u^3$
4. $a = 2.118 - 0.00298u^2 + 0.000073u^3$
5. If an acceleration noise test was done on a highway passing through a large city, would you expect the level of acceleration noise to be greater at 1:00 A.M. or at 5:00 P.M. and why?

CHAPTER REVIEW

Key Terms

14–1   Increasing and Decreasing Functions
Increasing Function   Critical Value
Decreasing Function

14–2   Relative Extrema: The First Derivative Test
Extrema   Relative Minimum
Maximum   Relative Extremum
Minimum   Critical Value
Relative Maximum   First Derivative Test

14–3   Concavity and the Second Derivative Test
Concave Up   Point of Inflection
Concave Down   Curve Sketching

14–4   Absolute Extrema and Optimization Problems
Absolute Maximum   Absolute Extremum
Absolute Minimum

14–5   Implicit Differentiation and Related Rates
Explicit Function   Implicit Differentiation
Implicit Function   Related Rates

Summary of Important Concepts

First Derivative Test: If $f(a)$ is a critical value of $f(x)$, then

1. $f(a)$ is a relative maximum if $f(x)$ is increasing to the left of a and decreasing to the right of a;
2. $f(a)$ is a relative minimum if $f(x)$ is decreasing to the left of a and increasing to the right of a.

The Second Derivative Test: If a is a critical value of $f(x)$, then

1. $f(a)$ is a relative maximum if $f''(x) < 0$ at $x = a$;
2. $f(a)$ is a relative minimum if $f''(x) > 0$ at $x = a$;
3. the test fails if $f''(a) = 0$.

REVIEW PROBLEMS

In Problems 1 to 4 find intervals where $f(x)$ is increasing and decreasing. Find the point(s) where $f(x)$ has a horizontal tangent line. Sketch the graph of the function.

1. $f(x) = 9x^2 - 25$ 2. $f(x) = x^2(4 - x)$

3. $f(x) = x/(3 - x)$ 4. $f(x) = x^2/(4 - x)$

In Problems 5 to 8 find the relative extrema, and sketch the graph of the function.

5. $f(x) = x^2 + 6x - 10$ 6. $f(x) = x^2(x - 4)$

7. $f(x) = x^2(4 - x^2)$ 8. $f(x) = 3x^{2/3} - 2x$

In Problems 9 and 10 use the second derivative test to find the relative extrema.

9. $f(x) = 3x^4 - 4x^3$ 10. $f(x) = x^4 - 2x^2$

In Problems 11 and 12 find the intervals where the function is concave up and concave down, and sketch the graph.

11. $f(x) = 2x^3 - x^4$ 12. $f(x) = x^4 - 6x^2$

In Problems 13 to 15 find the absolute extrema, if they exist.

13. $f(x) = x^3 - 3x^2 + 4$ on $[-2, 2]$

14. $f(x) = (x + 1)^2(x - 1)^2$ on $[0, 4]$

15. $f(x) = x - x^{1/3}$ on $[-1, 8]$

16. An open box is to be constructed from a 10 by 12 inch rectangular piece of cardboard by cutting out equal squares at each corner and then folding up the sides. What are the dimensions of the box with maximum volume?

In Problems 17 and 18 find $D_x y$ by using implicit differentiation.

17. $x^2 + xy + y^2 = 1$

18. $x^3 - 4x^2y + y^{-2} = 1$

19. A water tank is 10 meters (m) wide, 40 m long, and 20 m deep at one end and 30 m deep at the other end, with the depth increasing linearly. Water is pumped in at the rate of 5 m^3/min. How fast is the water level rising when the water level is 3 m?

PROGRAMMABLE CALCULATOR EXPERIMENTS

The problems in the exercises that ask you to sketch the graph of a function by using the first or second derivative tests are designed to enable you to factor the first derivative. In Problems 1 to 6 the derivative of the given function is not easily factored. Use the zoom feature to approximate the roots of the derivative and then use the first derivative test to sketch the graph of the function. Then use the graphing capability of your calculator to evaluate the accuracy of your graph.

1. $f(x) = x^3 + 3x^2 + x$

2. $f(x) = x^3 + 6x^2 + x$

3. $f(x) = x^4 - x$

4. $f(x) = x^4 + 2x$

5. $f(x) = x^4 - 4x^3 + 4x$

6. $f(x) = x^4 - 2x^2 + 4x$

Exponential and Logarithmic Functions

15

The Denver Fire Department used a variety of mathematical techniques to provide the best service at the lowest cost. (H. Armstrong Roberts)

Chapter Overview

Perhaps the most useful and universal functions in the application of calculus are exponential and logarithmic functions. They are used in such diverse areas as the growth of a population, the decay of radioactive material, the growth of investments, and the rate of learning.

The first section defines exponential functions and investigates some of their properties. Some graphs of simple types of exponential functions are sketched. The second section does the same for logarithmic functions. Sections 15–3 and 15–4 cover the derivatives of these functions and show how to use graphing techniques to draw the graphs of more complicated functions.

CASE STUDY PREVIEW

The Frustration of Dieting

Losing weight can be a trying experience. Excuses abound for our inability to diet successfully. This Case Study describes the mathematical foundation for weight control. It offers the frustrated dieter a bit of evidence for the case against short-term dieting.

CASE STUDY PREVIEW

Sears, Clothespins, and Unnatural Logarithms

Sears Roebuck and Co. started its mail-order business at the turn of the 20th century with the same kind of catalogs that are used today. These catalogs are still a major concern for Sears, but they also serve several purposes for economists. The Case Study uses logarithmic functions to explore the U.S. economy and lifestyle by examining automobiles, the Consumer Price Index, and clothespins.

CASE STUDY PREVIEW

Denver Fire Department Deployment Policy Analysis

In 1972 the mayor and fire chief of Denver agreed that a study of the fire department was in order in the face of mounting costs. The challenge was to minimize cost while keeping the same level of fire protection. A research team from the University of Colorado was commissioned to solve the problem. An exponential model was used to predict future trends by means of simulation methods.

15–1   Exponential Functions

In this section we introduce a new kind of function, called an exponential function, in which the exponent is a variable instead of a constant. An exponential function describes an object that "grows exponentially," meaning that it either expands or shrinks at an increasingly fast rate. We draw the graphs of exponential functions, study some of their properties, and define the exponential function involving the number e. The material is applied to an area of psychology that deals with learning curves.

Graphs

The graph of the quadratic function $f(x) = x^2$ is a parabola. It is quite different from the graph of the function $f(x) = 2^x$, which is called an exponential function because the exponent is a variable. Table 1 lists some points on the graph of $f(x) = 2^x$. These points are joined by the curve in Figure 1. In completing the table, remember that negative exponents refer to reciprocals, so that, for instance,

$$2^{-3} = \frac{1}{2^3} = \frac{1}{8} .$$

Table 1

x	-3	-2	-1	0	1	2	3
2^x	$\frac{1}{8}$	$\frac{1}{4}$	$\frac{1}{2}$	1	2	4	8

Table 1 lists only integral exponents, but exponential functions are defined for all real numbers. Rational exponents refer to roots. For instance, $2^{1/2} = \sqrt{2}$,

$2^{5/3} = \sqrt[3]{2^5} = (\sqrt[3]{2})^5$, and $2^{-3/4} = \dfrac{1}{2^{3/4}} = \dfrac{1}{\sqrt[4]{2^3}} = \dfrac{1}{\sqrt[4]{8}}$. The meaning of an

exponential function at an irrational exponent, such as $2^{\sqrt{2}}$, is beyond the scope of this book. However, it can be viewed intuitively from the graph in Figure 1 or derived from a calculator experiment by examining the limit of the numbers $2^{1.4}$, $2^{1.41}$, $2^{1.412}$, ...

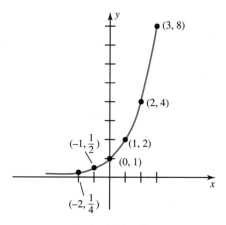

Graph of $f(x) = 2^x$

FIGURE 1

DEFINITION

If b is a positive number with $b \neq 1$, then a function $f(x) = b^x$ is called an **exponential function.** The number b is the **base** of $f(x)$.

The restriction $b \neq 1$ prevents the graph of $f(x) = b^x$ from being the horizontal line $f(x) = 1$. The restriction $b > 0$ ensures that the domain of an exponential function is the set of all real numbers, since otherwise there would be some values of x for which b^x is not a real number. For example, if $b = -1$ and $x = \frac{1}{2}$, then $(-1)^{1/2} = \sqrt{-1}$ is not a real number.

EXAMPLE 1

Problem

Sketch the graph of $f(x) = 3^x$.

Solution Table 2 lists some points on the graph, which is sketched in Figure 2.

Table 2

x	-3	-2	-1	0	1	2	3
3^x	$\frac{1}{27}$	$\frac{1}{9}$	$\frac{1}{3}$	1	3	9	27

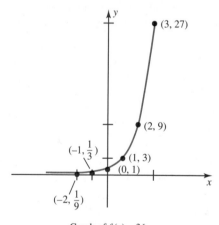

Graph of $f(x) = 3^x$

FIGURE 2

The graphs of $f(x) = 2^x$ and $f(x) = 3^x$ are typical of the graphs of those exponential functions $f(x) = b^x$ with $b > 1$. The graphs pass through the point $(0, 1)$ and increase over the entire domain. As x gets larger and larger the graphs increase without bound, which is expressed compactly by writing $\lim\limits_{x\to\infty} b^x = \infty$. In addition, $\lim\limits_{x\to-\infty} b^x = 0$, so the x-axis is a horizontal asymptote of the graph of $f(x)$.

Example 2 examines a typical exponential function $f(x) = b^x$ for $0 < b < 1$.

EXAMPLE 2

Problem

Sketch the graph of $f(x) = \left(\dfrac{1}{2}\right)^x$.

Solution Write $\left(\dfrac{1}{2}\right)^x$ in the form $\left(\dfrac{1}{2}\right)^x = \dfrac{1}{2^x}$. Then construct Table 3. The graph is drawn in Figure 3.

Table 3

x	-3	-2	-1	0	1	2	3
$(\frac{1}{2})^x$	8	4	2	1	$\frac{1}{2}$	$\frac{1}{4}$	$\frac{1}{8}$

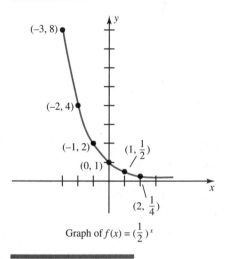

Graph of $f(x) = (\frac{1}{2})^x$

FIGURE 3

Properties

The graph of an exponential function $f(x) = b^x$ shows that if $b^r = b^s$, then $r = s$. This is the first of several properties of exponential functions that we list for reference. The other properties should be familiar from a previous study of algebra.

1. If $b^r = b^s$, then $r = s$
2. $b^r \cdot b^s = b^{r+s}$
3. $\dfrac{b^r}{b^s} = b^r \cdot b^{-s} = b^{r-s}$
4. $(b^r)^s = b^{rs}$

EXAMPLE 3

Problem

Find all values of x for which (a) $2^x = 64$, (b) $9^{(x+3)} = 3^{(3x-9)}$.

Solution (a) Since $2^6 = 64$, one value of x for which $2^x = 64$ is $x = 6$. According to property 1, this is the only value.
(b) Write $9^{(x+3)} = (3^2)^{(x+3)}$. By property 4, $(3^2)^{(x+3)} = 3^{2(x+3)}$. Thus $9^{(x+3)} = 3^{(2x+6)}$. It was given that $9^{(x+3)} = 3^{(3x-9)}$, so $3^{(2x+6)} = 3^{(3x-9)}$. By property 1, $2x + 6 = 3x - 9$. Therefore $x = 15$. This answer can be verified using the properties because $3^{(45-9)} = 3^{36} = (3^2)^{18} = 9^{18} = 9^{(15+3)}$.

THE Exponential Function

Consider the function defined by

$$f(x) = \left(1 + \frac{1}{x}\right)^x$$

Then $f(1) = 2$ and $f(2) = \left(\dfrac{3}{2}\right)^2 = 2.25$. A calculator with a $\boxed{y^x}$ key was used to compute the additional values of $f(x)$ in Table 4. The numbers are rounded to five decimal places. Table 4 indicates that $f(x)$ approaches $2.71828 \ldots$ as x gets larger and larger. The exact number that $f(x)$ approaches is denoted by the letter e.

DEFINITION

$$e = \lim_{x \to \infty} \left(1 + \frac{1}{x}\right)^x$$

Since e lies between 2 and 3, the graph of the function $f(x) = e^x$ is sandwiched between the graphs of $f(x) = 2^x$ and $f(x) = 3^x$, as shown in Figure 4. Although the number e might seem to be contrived, the function $f(x) = e^x$ occurs so often

Table 4

x	f(x)
1	$(2/1)^1 = 2$
2	$(3/2)^2 = 2.25$
3	$(4/3)^3 \approx 2.37037$
4	$(5/4)^4 \approx 2.44141$
5	$(6/5)^5 \approx 2.48832$
6	$(7/6)^6 \approx 2.52163$
7	$(8/7)^7 \approx 2.54650$
8	$(9/8)^8 \approx 2.56578$
9	$(10/9)^9 \approx 2.58117$
10	≈ 2.59374
100	≈ 2.70481
1,000	≈ 2.71692
10,000	≈ 2.71815
100,000	≈ 2.71827
1,000,000	≈ 2.71828
10,000,000	≈ 2.71828

that it is called *the* exponential function. A table in the Appendix lists the values of e^x for various values of x. Many calculators have an $\boxed{e^x}$ key. An approximation of e^k can be obtained by the sequence $\boxed{k}\ \boxed{e^x}$, although a color-coded key sometimes has to be pressed before tapping the $\boxed{e^x}$ key.

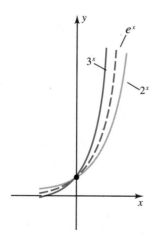

Graphs of three
exponential functions

FIGURE 4

DEFINITION

The function $f(x) = e^x$ is called **the exponential function.** The domain consists of all real numbers and the range is the set of positive numbers.

Example 2 examined the graph of the function $f(x) = \left(\dfrac{1}{2}\right)^x$, which is typical of all exponential functions of the form $f(x) = b^x$ for $0 < b < 1$. Example 4 examines the function $f(x) = e^{-x}$. It is important to remember that

$$e^{-x} = \left(\frac{1}{e}\right)^x = \frac{1}{e^x}$$

EXAMPLE 4

Problem

Sketch the graph of $f(x) = e^{-x}$.

Solution Construct Table 5 by using a calculator or by selecting the appropriate values for x from the table in the Appendix. The graph of $f(x) = e^{-x}$ is sketched by joining these points with the smooth curve shown in Figure 5.

Table 5

x	-2	-1	0	1	2
e^{-x}	7.38906	2.71828	1	0.36788	0.13534

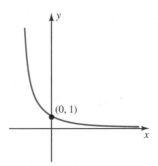

Graph of $f(x) = e^{-x}$

FIGURE 5

The graph of $f(x) = e^{-x}$ is similar to the graph of $f(x) = \left(\dfrac{1}{2}\right)^x$. To see this, compare Figure 5 with Figure 3. Notice that both curves pass through the point $(0, 1)$, decrease for all values of x, have the x-axis as a horizontal asymptote, and increase without bound as the values of x get smaller and smaller. The last two properties can be written $\lim\limits_{x \to \infty} e^{-x} = 0$ and $\lim\limits_{x \to -\infty} e^{-x} = \infty$.

Learning Curves

We have all experienced the range of emotions involved in learning a new skill, be it a physical activity like jumping rope or a mental discipline like differentiating functions. At first we are overwhelmed by the seeming complexity of the new task, but soon progress begins and improvement proceeds apace. However, later we reach a point at which further progress requires increasingly more time and dedication.

Psychologists introduced the name "learning curve" to deal with such situations in which a person's ability to learn a new skill is a function of the time devoted to learning it. A **learning curve** is the graph of an exponential function of the form $f(t) = a - be^{-ct}$, where a, b, and c are positive constants. Example 5 presents a typical setting for a learning curve.

EXAMPLE 5

The rate at which a person can learn to type is given by $f(t) = 70 - 60e^{-0.1t}$, where t is in weeks and $f(t)$ is in words per minute (wpm).

Problem

How many words per minute will the person ultimately be able to type?

Solution We derive the answer from the learning curve. The domain of $f(t)$ is restricted to $t \geq 0$ due to physical considerations. Without any training ($t = 0$), the person can type $f(0) = 70 - 60 = 10$ wpm. After one week the person can type $f(1) = 70 - 60e^{-0.1} = 70 - 54.3 = 15.7$ wpm. Table 6 lists additional time periods. (All wpm are rounded to one decimal place.)

The learning curve is the graph that is obtained from this table. It is sketched in Figure 6 and shows that $\lim\limits_{t \to \infty} f(t) = 70$, so as time goes by, the person's speed approaches 70 wpm.

Table 6

t	0	1	2	4	10	20	30	40
$f(t)$	10	15.7	20.9	29.8	47.9	61.9	67	68.9

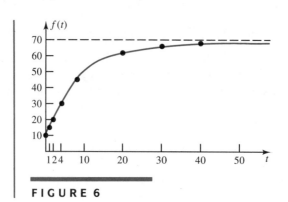

FIGURE 6

EXERCISE SET 15–1

In Problems 1 to 8 sketch the graph of the stated function by the plotting method.

1. $f(x) = 5^x$

2. $f(x) = 4^x$

3. $f(x) = 1.5^x$

4. $f(x) = 2.5^x$

5. $f(x) = (1/3)^x$

6. $f(x) = (1/4)^x$

7. $f(x) = (0.2)^x$

8. $f(x) = (0.1)^x$

In Problems 9 to 14 find all values of x that satisfy the given equality.

9. $2^x = 32$

10. $2^x = 128$

11. $5^x = 625$

12. $7^x = 16,807$

13. $9^{x+6} = 3^{20-2x}$

14. $9^{x+3} = 3^{x+10}$

In Problems 15 to 18 sketch the graph of the given exponential function.

15. $f(x) = e^{x+2}$

16. $f(x) = e^{x-2}$

17. $f(x) = e^{2-x}$

18. $f(x) = e^{-2x}$

In Problems 19 and 20 $f(t)$ represents the rate at which a person can learn to type, where t is in weeks and $f(t)$ is in words per minute (wpm). Find (a) the number of wpm that the person will be able to type after 50 weeks and (b) the number of wpm that the person will ultimately be able to type.

19. $f(t) = 60 - 55e^{-0.1t}$

20. $f(t) = 45 - 40e^{-0.01t}$

In Problems 21 and 22 $f(t)$ represents the number of bricks that an apprentice bricklayer can lay per hour, where t is in weeks. Find (a) the number of bricks that the apprentice will be able to lay per hour after 10 weeks and (b) the

number of bricks that the apprentice will ultimately be able to lay.

21. $f(t) = 28 - 27e^{-0.5t}$

22. $f(t) = 45 - 40e^{-0.2t}$

In Problems 23 to 30 sketch the graph of the stated function by the plotting method.

23. $f(x) = (1.1)^x$

24. $f(x) = (0.9)^x$

25. $f(x) = 6 - 3e^{-x}$

26. $f(x) = 15 - 20e^{-x}$

27. $f(x) = \dfrac{5}{1 + 3e^{-4x}}$

28. $f(x) = \dfrac{6}{1 + 2e^{-3x}}$

29. $f(x) = \dfrac{e^x - e^{-x}}{2}$

30. $f(x) = \dfrac{e^x + e^{-x}}{2}$

In Problems 31 to 34 find all values of x that satisfy the given equality.

31. $2^{2x+4} = 48^{-x}$

32. $2^{2x-3} = 8^{x-4}$

33. $2^x = 6^x$

34. $3^x = 2^x$

35. Sketch the graphs of the exponential functions $f(x)$ and $g(x)$ on the same coordinate axis, where
$$f(x) = (e^x)^2 \quad \text{and} \quad g(x) = e^{(x^2)}$$

36. Sketch the graph of the exponential function
$$f(x) = e^{-x^2}$$

In Problems 37 and 38 let $f(x) = e^{cx}$, for some number c, and let $f(2) = 3$.

37. Use the law of exponents to find $f(8)$.

38. Use the law of exponents to find $f(-8)$.

39. If a thermometer is moved from inside a house where it reads 70°F, to outside the house where it is 10°F, the temperature of the thermometer will read

$$T(t) = 10 + 60e^{-0.46t}$$

where t is in minutes and $T(t)$ is in degrees Fahrenheit.
(a) What temperature does the thermometer read after 5 minutes?
(b) What temperature will the thermometer eventually reach?

40. If $1000 is deposited into an account that pays 5% interest compounded annually, the future amount is given by

$$A(t) = 1000(1.05)^t$$

where t is in years and $A(t)$ is in dollars. What is the future amount after 10 years?

41. It is estimated that the population of a certain city will be given by the equation

$$P(t) = \frac{5}{2 + e^{-0.1t}}$$

where t is in years after 1990 and $P(t)$ is in millions of inhabitants.
(a) Sketch the graph of $P(t)$.
(b) Is the population increasing or decreasing?
(c) What will happen to the population in the distant future?

42. Suppose a disease spreads according to the equation

$$E(t) = \frac{8}{2 + 5e^{-0.1t}}$$

where t is in days after the outbreak of the disease and $E(t)$ is in thousands of people.
(a) Sketch the graph of $E(t)$.
(b) How many people contracted the disease after 10 days?
(c) In the long run, how many people will contract the disease?

43. Let $f(x) = (-1)^x$. Plot the points on the graph of $f(x)$ for x an integer from 0 to 4. Can the entire graph be drawn by connecting these points with a smooth curve?

44. (a) Use a calculator to approximate this infinite sum to five decimal places.

$$1 + \frac{1}{1!} + \frac{1}{2!} + \frac{1}{3!} + \frac{1}{4!} + \cdots$$

(b) What number do you think this sum is equal to?

45. Let $f(x) = e^x$. Evaluate each limit by using a calculator to form a table in which one row contains values of h and the other row contains the corresponding difference quotient.

(a) $\lim\limits_{h \to 0} \dfrac{f(3 + h) - f(3)}{h}$

(b) $\lim\limits_{h \to 0} \dfrac{f(4 + h) - f(4)}{h}$

Referenced Exercise Set 15–1

1. A study of a game called a *lottery* derived a way to distinguish between different types of risk takers.* Consider a given lottery L whose utility is denoted by $U(L) = 0.817$. (*Utility* is a technical term whose meaning is not necessary to define here). A decision maker is said to be *risk-averse* with respect to another lottery G if $U(G) < U(L)$, is *risk-neutral* if $U(G) = U(L)$, and is *risk-prone* if $U(G) > U(L)$. Let G be a lottery in which a player wins $2 90% of the time and loses $2 otherwise. The expected value is $E = (2)(.9) + (-2)(.1) = 1.6$. The utility of G is defined by

$$U(G) = (.9)f(2) + (.1)f(-2)$$

where

$$f(x) = \frac{1 - e^{-E(.1x + .5)}}{1 - e^{-E}}$$

Calculate $U(G)$, and determine whether the decision maker is risk-averse, risk-prone, or risk-neutral with respect to the lottery L.

2. An article on forecasting market development showed that two growth curves could be applied in a wide variety of settings.† A *logistic curve* is of the form

$$l(x) = \frac{a}{1 + ce^{-bx}}$$

*Joao L. Becker and Rakesh K. Sarin, "Lottery Dependent Utility," *Management Science*, November 1987, pp. 1367–1382.
†Nigel Meade, "Forecasting Using Growth Curves—An Adaptive Approach," *Journal of the Operational Research Society*, December 1985, pp. 1103–1115.

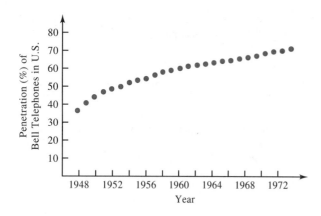

FIGURE 7

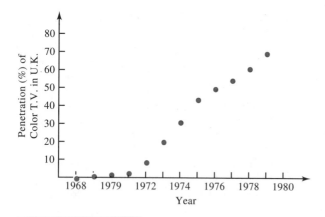

FIGURE 8

and a *Gompertz curve* is of the form

$$g(x) = ae^{-c(e^{-bx})}$$

where a, b, and c are positive numbers. Figure 7 shows the percentage of households in the United States that had Bell telephones in the period 1948–1972, while Figure 8 shows the percentage of households in the United Kingdom that had color television sets in the period 1968–1980. Which figure is modeled by which curve? (*Hint:* set $a = b = c = 1$, and sketch the graphs of the resulting logistic curve and Gompertz curve.)

3. A study of the spread of automatic teller machines (ATMs) in the banking industry found that the proportion of banks that had adopted the innovation could be expressed as the logistic function

$$P(t) = \frac{1}{1 + e^{5.293 - 0.698t}}$$

where t denotes the number of years after the ATMs were first introduced in 1972.‡ Determine the proportion of banks which used ATMs in each listed year by calculating $P(t)$ for the appropriate value of t.
(a) 1975   (b) 1980   (c) 1988

‡Timothy H. Hannan and John M. McDowell, "Market Concentration and the Diffusion of New Technology in the Banking Industry," *The Review of Economics and Statistics*, November 1984, pp. 686–691.

4. A special report to the mathematical community entitled "A Challenge of Numbers" stated that "from entry into the ninth grade through receipt of the doctoral degree, a reduction of one-half per year of the number of students enrolled gives curiously accurate estimates of the numbers of students proceeding toward degrees."§ There were 3,600,000 ninth graders in 1976. How many of them earned a Ph.D. in mathematics by 1990?

§Bernard L. Madison, "A Challenge of Numbers," *Notices of the American Mathematical Society,* May–June 1990, pp. 547–554.

C A S E   S T U D Y   **The Frustration of Dieting***

Who among us has not had to curtail eating habits in order to keep the body as trim as possible? It is true that some people are able to eat as much as they like, and as often as they desire, yet still maintain a slim figure. But others have to work hard to keep their weight down, whether for health or for vanity.

This Case Study examines dieting from a mathematical perspective. It draws one inescapable conclusion: there is a mathematical reason behind the fact that dieting is a long and arduous—in a word, frustrating—ordeal.

Equilibrium weight   A person's weight depends primarily on two factors: energy intake and energy consumption. There are several other important factors, such as age, sex, and metabolic rate, but we restrict our consideration to the two energy factors. We also assume that each of the two main factors remains constant each day, meaning that a diet consists of the same number of calories per day and that the energy consumption is the same number of calories per pound per day.

Most people's energy consumption varies between 15 and 20 calories per pound per day. *We will assume that the energy consumption is 17.5 calories per pound per day.*

Let I denote the daily energy intake (in calories) and C denote the daily energy consumption (in calories per pound). A person who follows such a regimen will maintain a *weight equilibrium* of I/C pounds (lb).

$$w_{eq} = \frac{I}{C}$$

For instance, a person who eats 2625 calories a day will maintain a weight equilibrium of $2625/17.5 = 150$ lb. Since the energy consumption C remains constant, if I increases above 2625 calories, then the person will gain weight, while if I decreases below 2625 calories, then the person will lose weight.

The weight equation   One of the basic assumptions in physiology is that the rate of change of a person's weight w is directly proportional to the difference

*Adapted from Arthur C. Segal, "A Linear Diet Model," *College Mathematics Journal,* January 1987, pp. 44–45.

between the energy intake I and the energy consumption C. Since the rate of change is the derivative $\dfrac{dw}{dt}$, this assumption can be written as an equation.

$$\frac{dw}{dt} = k(I - C)$$

The most commonly used dietetic conversion factor is $k = 1/3500$ lb per calorie, so

$$\frac{dw}{dt} = \frac{1}{3500}(I - C)$$

In Section 15–4 it will be possible to verify that the solution to this equation is

$$w(t) = w_{\text{eq}} + (w_0 - w_{eq})e^{-0.005t}$$

This equation is called the *weight equation,* where $w(t)$ represents the weight after t days and w_0 represents the initial weight. It is an important equation in many people's lives. Let us illustrate the weight equation for a hypothetical person named Rufus.

Rufus presently weighs 200 lb, so $w_0 = 200$. This means that his usual daily energy intake is $I = 200 \cdot 17.5 = 3500$ calories a day. He goes on a 3150 calories-per-day diet in order to lose weight. Now since $I = 3150$, the weight equilibrium will be

$$w_{\text{eq}} = \frac{3150}{17.5} = 180$$

Therefore, by adopting this diet, Rufus will eventually reach a weight of 180 lb and will maintain it.

How long will it take him to achieve this goal? To answer this question, set up the weight equation.

$$\begin{aligned} w(t) &= w_{\text{eq}} + (w_0 - w_{\text{eq}})e^{-0.005t} \\ &= 180 + (200 - 180)e^{-0.005t} \\ &= 180 + 20e^{-0.005t} \end{aligned}$$

Rufus' weight after $t = 7$ days is

$$w(7) = 180 + 20e^{-0.005 \cdot 7}$$

Use a calculator to evaluate this expression: $w(7) = 199.3$ (rounded to one decimal place). Thus Rufus will lose less than 1 lb during the first week of the diet. After 4 weeks Rufus will weigh about 197 lb because

$$w(28) = 180 + 20e^{-0.005 \cdot 28} \approx 197.4$$

Therefore after almost a month Rufus will have lost less than 3 lb. After a year, however, his weight will be $w(365) = 183.2$ lb, which is about 3 lb from his goal.

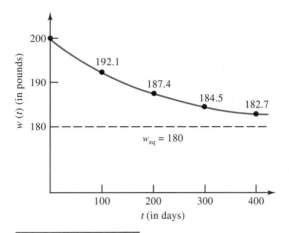

FIGURE 1

Does this mathematical model convince you of the difficulty involved in losing weight strictly by dieting? Poor Rufus stuck to his diet every day for a year, yet he was still 3 lb shy of his goal. Figure 1 shows the asymptotic nature of the weight equation. No wonder that losing weight is a long and arduous struggle!

Example 1 considers another person's valiant fight to remove extra pounds.

EXAMPLE 1

Tina weighs 130 lb. She wants to lose 10 lb.

Problem

(a) What should Tina's daily caloric intake be?
(b) If she maintains the diet for 100 days, how much weight will she lose?

Solution (a) Set $w_0 = 130$. Since Tina wants to lose 10 lb, her equilibrium weight is $w_{eq} = 120$. It is assumed that $C = 17.5$. Since $w_{eq} = I/C$, the daily caloric intake is $I = C \cdot w_{eq} = 17.5 \cdot 120 = 2100$. Therefore Tina must go on a 2100 calories-per-day diet.
(b) Tina's weight equation is

$$w(t) = 120 + 10e^{-0.005t}$$

If she adheres to this diet for $t = 100$ days, her weight will be

$$w(100) = 120 + 10e^{-0.5} = 120 + 10 \cdot 0.6 = 126$$

Since Tina will drop from 130 to 126 lb, she will end up losing 4 lb in 100 days.

Notice that Tina's weight can never reach exactly 120 lb on this diet because the weight equation $w(t) = 120 + 10e^{-0.005t}$ is always greater than 120. (Recall that e^x is positive for all values of x.) However, it is possible to determine when her weight will reach $w(t) = 121$ lb by solving the weight equation for t.

$$121 = 120 + 10e^{-0.005t}$$

The solution requires the logarithmic function, which will be discussed in the next section. By direct substitution into the weight equation, Tina will drop to 121 lb after 460 days. Now that's an ordeal!

Considering the cases of Rufus and Tina, is it any wonder that so many dieters give up in frustration?

Case Study Exercises

1. Rufus' initial weight is 200 pounds (lb). If he goes on a 2500 calories-per-day diet, how much weight will he lose in 100 days?
2. Rufus' initial weight is 200 lb. If he goes on a 2000 calories-per-day diet, how much weight will he lose in 100 days?
3. Tina weighs 120 lb and wants to lose 10 lb.
   (a) What should her daily caloric intake be?
   (b) If she maintains the diet for 100 days, how much weight will she lose?
4. Joan weighs 125 lb and wants to lose 15 lb.
   (a) What should her daily caloric intake be?
   (b) If she maintains the diet for 100 days, how much weight will she lose?
5. For Chris, $w_0 = 150$ and $I = 2275$. How much weight will Chris lose in 100 days?
6. Let your present weight be w_{eq}. Assume an average daily consumption of 17.5 calories per pound.
   (a) What is your daily energy intake I?
   (b) If you reduce I by 500 calories a day, what will your new equilibrium weight be?
7. (a) What is your weight equation for losing 10 lb?
   (b) What will you weigh in 100 days if you stick to the diet?

15–2 Logarithmic Functions

This section is organized like Section 15–1. First a new function, the natural logarithmic function, is introduced in terms of more familiar functions. Then its graph is drawn between the graphs of two similar functions and several of its properties are derived.

This approach is appropriate since the definition of the natural logarithmic function is based on the exponential function $f(x) = e^x$ and the natural logarithmic function is shown to be the inverse of $f(x)$. This inverse relationship is exploited in a discussion of the doubling time of an investment.

Graphs

Logarithmic functions are defined in terms of exponential functions: $y = \log_2 x$ if and only if y is the number such that $2^y = x$. Therefore $3 = \log_2 8$ because $2^3 = 8$. This leads to the definition of a *logarithmic function*.

$$f(x) = \log_2 x \qquad \text{if and only if} \qquad 2^{f(x)} = x$$

(Usually $\log_2 x$ is read "the logarithm of x to the base 2.") Notice that the logarithm of a number is an exponent. The definition yields $f(8) = 3$, so the point $(8, 3)$ lies on the graph of $f(x) = \log_2 x$. Additional points are listed in Table 1. The graph is drawn in Figure 1.

Table 1

x	1	2	4	8	16	32
$\log_2 x$	0	1	2	3	4	5

DEFINITION

Let b be a positive number with $b \neq 1$. The **logarithmic function with base b** is defined by

$$f(x) = \log_b x \qquad \text{if and only if} \qquad b^{f(x)} = x$$

The domain is the set of positive numbers and the range consists of all real numbers.

The next example illustrates that to find points on the graph of a logarithmic function $y = \log_b x$ it is usually easier to start with convenient values of y and to find the corresponding value of x using the equation $x = b^y$.

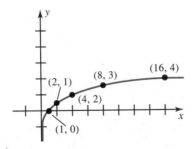

The graph of $f(x) = \log_2 x$

FIGURE 1

EXAMPLE 1

Problem

Sketch the graph of $f(x) = \log_{10} x$.

Solution By definition $y = \log_{10} x$ if $10^y = x$. Substitute some numbers for y and solve for x. For instance, if $y = 2$, then $x = 10^2 = 100$. Thus $(100, 2)$ is a point on the graph. Additional points are listed in the table. The graph is sketched in Figure 2.

x	0.01	0.1	1	10	100	1000
$\log_{10} x$	-2	-1	0	1	2	3

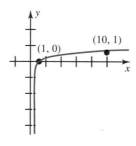

The graph of $f(x) = \log_{10} x$

FIGURE 2

Logarithms to the base 10 are called **common logarithms.** They are usually written $\log x$ instead of $\log_{10} x$. Common logarithms were invented about 1600 as an aid for astronomers in performing involved calculations. Scientific calculators, however, have rendered this use obsolete.

The graphs of $f(x) = \log_2 x$ and $f(x) = \log_{10} x$ are typical of the graphs of logarithmic functions $f(x) = \log_b x$ for $b > 1$. They pass through the point $(1, 0)$, they always increase, they are negative for $0 < x < 1$, and they are positive for $x > 1$.

The Natural Logarithm

Figure 3 shows the graphs of the functions $f(x) = 2^x$ and $f(x) = \log_2 x$ on the same coordinate system. The graphs are symmetric about the dashed line $y = x$, meaning that if one of them is flipped about this line, then it would land precisely on the other. As an example, the point $(3, 8)$ lies on the graph of $f(x) = 2^x$, the point $(8, 3)$ lies on the graph of $f(x) = \log_2 x$, and the distance from $(3, 8)$ to the dashed line is equal to the distance from $(8, 3)$ to the dashed line. When such a relation holds between two functions, each one is called the **inverse function** of the other. Thus $f(x) = \log_2 x$ is the inverse of $f(x) = 2^x$.

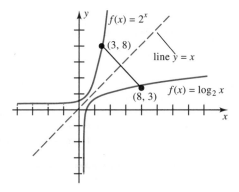

The functions $f(x) = 2^x$ and $f(x) = \log_2 x$
are inverses of each other

FIGURE 3

Figure 4 shows the graph of the exponential function $f(x) = e^x$ and its inverse, which is called the **natural logarithmic function.** The natural logarithm, which is denoted by $\ln x$ instead of $\log_e x$, is defined by

$$y = \ln x \qquad \text{if and only if} \qquad e^y = x$$

Notice that $\ln e = 1$ (since $e^1 = e$) and $\ln 1 = 0$ (since $e^0 = 1$). However, $\ln 0$ is not defined. The graph of $f(x) = \ln x$ is similar to the graph of $f(x) = \log_2 x$. It can be obtained from the graph of $f(x) = e^x$ as in Figure 4, or from Table 2. (In Table 2 the values of $\ln x$ are rounded to two decimal places. Each value of $\ln x$ in the table is an irrational number except for $x = 1$.) A more complete table can be found in the Appendix. Many calculators have $\boxed{\ln}$ and $\boxed{\log}$ keys. The key sequence $\boxed{.5}\ \boxed{\ln}$ produces the value -0.69 in Table 2.

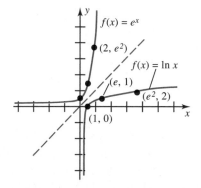

$\ln x$ is the inverse of e^x

FIGURE 4

Table 2

x	0.05	0.1	0.25	0.5	1	2	3	4
$\ln x$	-3.00	-2.30	-1.39	-0.69	0	0.69	1.10	1.39

Example 2 sketches the graph of a function that is defined in terms of the natural logarithmic function. Its domain is the set of all real numbers and its graph is unlike Figures 1 and 2.

EXAMPLE 2

Problem

Sketch the graph of the function

$$f(x) = \ln(x^2 + 1)$$

Solution Construct a table using a calculator or the table of logarithmic functions in the Appendix. Plot the points and draw a smooth curve through them. This is carried out in Figure 5. Notice that $f(x)$ is defined for all real numbers.

x	-3	-2	-1	0	1	2	3
$\ln(x^2 + 1)$	2.30	1.61	0.69	0	0.69	1.61	2.30

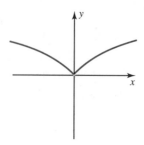

Graph of $f(x) = \ln(x^2 + 1)$

FIGURE 5

Properties

We derive several properties of logarithmic functions from the properties of exponential functions. Although these properties are stated in terms of the natural logarithmic function $f(x) = \ln x$, they hold for all logarithmic functions of the form $f(x) = \log_b x$ as well (for $b > 0$ and $b \neq 1$). In properties 4 to 7 assume that $r > 0$, $s > 0$, and k is any number.

1. $e^{\ln x} = x$ for $x > 0$
2. $\ln e^x = x$ for all x
3. $\ln 1 = 0$
4. If $\ln r = \ln s$, then $r = s$
5. $\ln rs = \ln r + \ln s$
6. $\ln \dfrac{r}{s} = \ln r - \ln s$
7. $\ln r^k = k \cdot \ln r$

The natural logarithm is defined by

$$y = \ln x \qquad \text{if and only if} \qquad e^y = x$$

Property 1 is obtained by substituting $\ln x$ for y in the right-hand equality. Similarly, substituting e^y for x in $y = \ln x$ yields $y = \ln e^y$. Writing this equality in the variable x gives property 2. Property 3 follows by substituting $x = 0$ in property 2. Property 4 can be inferred from the graph of $f(x) = \ln x$. Alternately, if $\ln r = \ln s$, then $e^{\ln r} = e^{\ln s}$. By property 1, $r = e^{\ln r}$ and $s = e^{\ln s}$, so $r = s$.

To prove property 5, use property 1 to write $r = e^{\ln r}$ and $s = e^{\ln s}$. Then $rs = e^{\ln r} \cdot e^{\ln s} = e^{\ln r + \ln s}$. By property 1, $rs = e^{\ln rs}$, so $e^{\ln rs} = e^{\ln r + \ln s}$. Property 5 then follows from property 1 of exponential functions. Properties 6 and 7 can be derived similarly.

Examples 3 and 4 illustrate these properties.

EXAMPLE 3

Problem

Write these expressions in simplified form: (a) $\ln(4x^3)$, (b) $\ln \left(\dfrac{6x^2y}{x + y} \right)$.

Solution (a) Using properties 4 and 6, we obtain

$$\ln(4x^3) = \ln(2^2 x^3) = \ln(2^2) + \ln(x^3) = (2 \ln 2) + (3 \ln x).$$

(b) Again by using properties 4 and 6, we have

$$\ln(6x^2y) = \ln 6 + 2 \ln x + \ln y.$$

Using property 5, we get

$$\ln \left(\frac{6x^2y}{x + y} \right) = \ln(6x^2y) - \ln(x + y)$$

$$= \ln 6 + 2 \ln x + \ln y - \ln(x + y)$$

Warning: It is *not* true that $\ln(x + y) = \ln x + \ln y$. Do not try to use such an equality to simplify Example 3(b) further than it is.

EXAMPLE 4

Problem

Simplify (a) $\ln e^{(3x+4)}$, (b) $e^{\ln(3x+4)}$.

Solution (a) Rewrite property 2 in the variable v, so that $\ln e^v = v$. Now set $v = 3x + 4$. Then $\ln e^{(3x+4)} = 3x + 4$.
(b) Similarly rewrite property 1 in the variable v, so that $e^{\ln v} = v$. Once again set $v = 3x + 4$. Then $e^{\ln(3x+4)} = 3x + 4$.

The same reasoning used in Example 4 leads to a generalization of properties 1 and 2:

$1'.\ e^{\ln f(x)} = f(x) \quad$ for $f(x) > 0$
$2'.\ \ln e^{f(x)} = f(x) \quad$ for all $f(x)$

Doubling Time

One way to measure the effectiveness of an investment is by means of **doubling time,** which is the time required for the initial investment to double. By the Compound Interest Formula, if $1 is invested at an interest rate of $100r\%$ compounded annually (where r is written as a decimal), the formula for the amount after t years is given by

$$A(t) = (1 + r)^t$$

This is an exponential function. The doubling time is computed by solving the equation $A(t) = 2$ for t.

$$2 = (1 + r)^t$$

Since the logarithmic function is the inverse of the exponential function, take the natural logarithm of both sides.

$$\ln 2 = \ln(1 + r)^t$$
$$\ln 2 = t \cdot \ln(1 + r)$$
$$t = \frac{\ln 2}{\ln(1 + r)}$$

For instance, the doubling time of an 8% investment is 9 years because

$$t = \frac{\ln 2}{\ln 1.08} = \frac{0.69315}{0.07696} \approx 9$$

EXAMPLE 5

Problem

What is the doubling time for an investment that earns 16% compounded annually?

Solution By the Compound Interest Formula

$$A(t) = (1 + r)^t$$

Here $A(t) = 2$ and $r = 0.16$, so

$$2 = (1.16)^t$$
$$\ln 2 = t \cdot \ln(1.16)$$
$$t = \frac{\ln 2}{\ln 1.16} = \frac{0.69315}{0.14842} \approx 4.67017$$

Therefore the doubling time is about $4\frac{2}{3}$ years, or 4 years and 8 months.

EXERCISE SET 15–2

In Problems 1 to 4 sketch the graph of the stated function by the plotting method, using the definition of the logarithmic function.

1. $f(x) = \log_{10} 2x$

2. $f(x) = \log_{10} \frac{x}{2}$

3. $f(x) = \log_3 x$

4. $f(x) = \log_{20} x$

In Problems 5 to 8 sketch the graph of the stated function by the plotting method, using the table in the Appendix or a calculator.

5. $f(x) = \ln(3x)$

6. $f(x) = \ln(x/2)$

7. $f(x) = \ln(x^2 + 3)$

8. $f(x) = \ln(x^2 + 4)$

In Problems 9 to 12 write the given expression in simplified form.

9. $\ln(3x^4)$

10. $\ln(2x^5)$

11. $\ln\left(\frac{3xy^2}{x - y}\right)$

12. $\ln\left(\frac{x^3y^2}{x + y}\right)$

In Problems 13 to 16 simplify the expression.

13. (a) $\ln e^{3x}$ (b) $e^{\ln(5 - x)}$

14. (a) $\ln e^{-2}$ (b) $e^{\ln 4}$

15. (a) $\ln e^5$ (b) $e^{\ln 0.5}$

16. (a) $\ln e^{3 - 8x}$ (b) $e^{\ln(2x - 6)}$

In Problems 17 to 20 determine the doubling time for an investment that earns the stated interest compounded annually?

17. 6%   18. 10%   19. 20%   20. 5%

In Problems 21 to 26 sketch the graph of the stated function by the plotting method, using the table in the Appendix or a calculator.

21. $f(x) = \ln \frac{1}{x^2}$

22. $f(x) = \ln\left(\frac{1}{x^2 + 2}\right)$

23. $f(x) = \ln(2x + 3)$

24. $f(x) = \ln(x - 4)$

25. $f(x) = \ln |x|$

26. $f(x) = \ln \sqrt{x}$

In Problems 27 to 30 state the domain of the function in the problem stated above.

27. Problem 21   28. Problem 23

29. Problem 25   30. Problem 26

In Problems 31 to 34 write the given expression in simplified form.

31. $\ln \sqrt{xy}$

32. $\ln \sqrt{x + y}$

33. $\ln \dfrac{1}{\sqrt{x^3y}}$

34. $\ln \dfrac{1}{xy^2}$

In Problems 35 and 36 let $f(x) = \ln x$ and $g(x) = e^x$.

35. Compute $(f \circ g)(x)$.   36. Compute $(g \circ f)(x)$.

In Problems 37 to 44 solve for x.

37. $e^{2x} = 3$

38. $e^{5x} = 1$

39. $18 = e^{-x} - 2$

40. $5 = \dfrac{1}{1 - 2e^{-x}}$

41. $5 \ln x = 1$

42. $4 \ln(1/x) = 1$

43. $3^x = e^2$

44. $2^x = e^5$

In Problems 45 to 48 evaluate the expression without using a calculator.

45. (a) $\ln e$ (b) $e^{\ln 4}$

46. (a) $\ln 1$ (b) $e^{3 \ln 2}$

47. (a) $\ln \sqrt[3]{e}$ (b) $e^{4 \ln 3 - 3 \ln 4}$

48. (a) $\ln(1/\sqrt{e})$ (b) $e^{-\ln 2}$

49. Use a calculator to try to evaluate
    (a) $\ln 0$ (b) $\ln -2$

50. Use a calculator to show that $\ln(2 + 3) \neq \ln 2 + \ln 3$.

51. If a thermometer is moved from inside a house where it reads 70°F, to outside the house where it is 10°F, the temperature of the thermometer will read

$$T(t) = 10 + 60e^{-0.46t}$$

where t is in minutes and $T(t)$ is in degrees Fahrenheit. If the thermometer is digital (meaning that it only displays whole degrees), when will the reading first reach 10°F?

52. Answer true or false.
    (a) $\ln(x^3 + 5x^2) = 3 \ln x + 10 \ln x$
    (b) $\ln(x^5) = (\ln x)^5$

In Problems 53 to 56 modify the argument that derived the doubling time formula to determine how long it will take an investment to increase to the given size at the given rate.

53. To double if it earns 10% compounded annually.

54. To triple if it earns 10% compounded annually.

55. To quadruple if it earns 5% compounded annually.

56. To increase by a factor of 2.5 if it earns 8% compounded annually.

57. Determine how long it will take for an investment that earns 8% interest compounded annually to triple.

58. (a) Prove property 6 of logarithmic functions.
    (b) Prove property 7 of logarithmic functions.

59. The weight equation for a 140-pound (lb) person who wants to lose 10 lb is

$$f(t) = 130 + 10e^{-0.005t}$$

where $f(t)$ is the weight (in lb) after being on the diet for t days. How long will it take the person to lose 5 lb? (This weight equation is derived in the Case Study in this chapter titled "The Frustration of Dieting.")

Referenced Exercise Set 15–2

Consider a job in which a number of workers produce a certain amount of products, say Boeing 707 jets. When the job is performed for the first time it cannot be expected to be completed in an acceptable amount of time. A certain period of time must elapse before the workers become familiar with the tasks involved and learn how to do them in a shorter amount of time. A *learning curve* is the graph of a function which measures the mastery of the job. If one hour is needed to produce the first unit, then the number of hours needed to produce the xth unit is given by

$$Y(x) = x^{\log L/\log 2}$$

where L is the learning rate of the particular job.* Data from the airframe manufacturing industry reveal that $L = 0.8$ is a typical learning rate. Assume such a value in Problems 1 and 2.

1. Compute the time (as a decimal part of an hour) needed to produce (a) the second unit, (b) the third unit, (c) the tenth unit, (d) the one-hundredth unit.

2. How many units have to be produced before a unit can be produced in each of these time periods?
   (a) 30 minutes (0.5 hour)
   (b) 6 minutes (0.1 hour)

Simple products and machine-paced processes were shown to exhibit learning rates between 0.9 and 1.0.†

3. Repeat Problem 1 for a learning rate of $L = 0.95$.

4. Repeat Problem 2 for a learning rate of $L = 0.95$.

In the "classical secretary problem" each applicant for a job is ranked by a manager according to some quality. As each applicant appears, her rank relative to those preceding her is recorded and the manager decides either to select or reject her. The object is to hire the best applicant, and the manager is rewarded accordingly. A Japanese specialist in operations research has solved this problem for various kinds of selections that the manager can make.‡

*Louis E. Yelle, "The Learning Curve: Historical Review and Comprehensive Survey," *Decision Sciences,* April 1979, pp. 302–328.

†Timothy L. Smunt, "A Comparison of Learning Curve Analysis and Moving Average Ratio Analysis for Detailed Operational Planning," *Decision Sciences,* Fall 1986, pp. 475–495.

‡Mitsushi Tamaki, "A Generalized Problem of Optimal Selection and Assignment," *Operations Research,* May–June 1986, pp. 486–493.

Problems 5 and 6 are taken from this study.

5. The reward for the model "without promotion" is $R(x) = x(1 - x)$, where x is a solution to the equation

$$2(1 - x) + \ln x = 0 \qquad 0 < x < 1$$

Solve the equation to one decimal place (by trial-and-error) and determine the reward.

6. The reward for the model "with promotion" is $R(x) = x\left(\dfrac{2}{\sqrt{e}} - x\right)$, where x is a solution to the equation

$$\sqrt{e}(1 + x) - \ln x = 3.5 \qquad 0 < x < 1$$

Solve the equation to one decimal place (by trial-and-error) and determine the reward.

7. In 1987 the world's population reached 5 billion.§ If it continues to grow 2% each year, in what year will it reach 10 billion?

8. Donald Knuth, arguably America's best-known computer scientist, has proved that the number of computer operations needed to alphabetize n words is about $n(\log_2 n)$.‖ About how many computer operations will it take to alphabetize 10,000 words?

Cumulative Exercise Set 15–2

In Problems 1 and 2 sketch the graph of $f(x)$ by the plotting method.

1. $f(x) = (0.2)^x$

2. $f(x) = \dfrac{2}{1 + 2e^{-3x}}$

3. Find all values of x that satisfy the equality $2^{x-2} = 8^{2x+1}$

4. A disease spreads according to the equation

$$E(t) = \frac{5}{2 + 3e^{-0.1t}}$$

where t is in days after the outbreak of the disease and $E(t)$ is in thousands of people. (a) How many people contracted the disease after 10 days? (b) In the long run, how many people will contract the disease?

5. Write the expression in simplified form:

$$\ln\left(\frac{x^2 y^3}{x - y}\right)$$

6. Determine the doubling time for an investment that earns 12% interest compounded annually.

7. Solve for x: $6 \ln(1/x) = 1$

8. Evaluate the expression without using a calculator:

$$e^{2 \ln 3 - 3 \ln 2}$$

CASE STUDY   ### Sears, Clothespins, and Unnatural Logarithms*

Consider the clothespin. No, not the one shown in Figure 1. That one is actually a Claes Oldenberg sculpture that stands in downtown Philadelphia. Instead, regard the usual clothespin, the one that has graced porches and backyards for over a century. It will serve as a central prop in our study of the American economy.

Sears catalogs also play an important role. Sears Roebuck and Co. began its mail-order operation at the turn of the 20th century based on these catalogs. The first ones were distributed in rural areas as a marketing strategy to sell items that were otherwise unavailable. Their overwhelming success spawned a vast

§Dan Shannon, "World Population at 5 Billion," *Los Angeles Times,* July 12, 1987.

‖Donald E. Knuth, *Sorting and Searching: The Art of Computer Programming III,* Addison-Wesley, Reading, MA, 1973.

*Adapted from Elliott W. Montroll, "On the Dynamics and Evolution of Some Sociotechnical Systems," *Bulletin of the American Mathematical Society,* January 1987, pp. 1–46.

F I G U R E 1 A photo of a Claes Oldenberg sculpture of a clothespin that stands in downtown Philadelphia.

nationwide network of over 800 stores which are woven into the lives of millions of people.

In 1987 Sears produced its first videocassette catalog to market some of its products. We will not attempt to predict whether this venture will be as successful in the 21st century as the standard catalogs have been in the 20th century. Instead, we will document several ways in which the standard catalogs have been used to study the American way of life over the past 100 years. We will see that they provide an economic index that measures the cost of living and indicates changes in the American life-style.

Cost of living On a given page in a Sears catalog you might see bicycles, VCRs, and—yes—clothespins. Since the clothespin is an invariant item, the 1988 model being indistinguishable from the 1900 model, its change in price reflects one kind of measure of the cost of living. Another measure results from examining bicycles. Since the models of the 1990s are considerably different from the models from 1900, the variation in price reflects an evolving technol-

Table 1

P	$\log_2 P$	$(\log_2 P - M)^2$
128.00	7	$2.94^2 = 8.6436$
39.40	5.3	$1.24^2 = 1.5376$
16.00	4	$-0.06^2 = 0.0036$
8.00	3	$-1.06^2 = 1.1236$
2.00	1	$-3.06^2 = 9.3636$
Sum	20.3	20.672

$$M = \frac{20.3}{5} = 4.06 \quad S = \sqrt{\frac{20.672}{5}} \approx 2.03$$

ogy and a varying public taste. VCRs, on the other hand, certainly were not featured in the 1900 catalog. Conversely, you are not likely to find buggy whips in the 1992 catalog. Yet in spite of such variations, one constant emerges. We describe it now.

Consider a page in a Sears catalog containing five items whose sales prices are \$128, \$39.40, \$16, \$8, and \$2. It is possible to compute the mean and standard deviation for the items on this page. However, instead of regarding each price P as a data entry, we use its logarithm to the base 2. For $P = \$128$ the data entry becomes $\log_2 P = \log_2 128 = 7$ because $2^7 = 128$. The remaining entries are listed in the second column of Table 1.

The mean of the $\log_2 P$ entries is $20.3/5 = 4.06$. We denote this mean by M, to distinguish it from the mean of the prices P, and refer to it as the *log-mean*. Similarly the standard deviation of the $\log_2 P$ entries is denoted by S and called the *log-standard deviation*. The third column of Table 1 reveals that $S = \sqrt{20.672/5} \approx 2.03$.

The values that we have obtained for this hypothetical page actually correspond to the log-mean M and the log-standard deviation S for all items that were listed in the 1975–1976 Sears catalog. Table 2 contains the values of M and S for selected annual Sears catalogs since 1900.

Table 2

Year	M	S
1900	0.150	2.43
1908	-0.023	2.29
1916	-0.068	2.38
1924–1925	0.422	2.32
1932–1933	0.691	1.91
1939–1940	0.627	2.62
1951–1952	1.785	2.34
1962	2.403	2.24
1975–1976	4.060	2.03

One fact stands out in Table 2. Although M varies considerably, S remains almost the same. For this reason S has been described as an "economic constant of motion" for marketing. For all Sears catalogs, which have ranged over various economic climates and two world wars, the standard deviation of the log-standard deviations is only 0.17. This reflects a remarkable uniformity and a very consistent marketing strategy.

Example 1 provides more practice in computing the log-standard deviation S.

EXAMPLE 1

A page of a Sears catalog lists five items at these prices: $78.25, $3.36, $1.68, $1.06, and $1.04.

Problem

Which annual Sears catalog has a log-mean and a log-standard deviation that are similar to the log-mean and the log-standard deviation of these five items?

Solution The prices P of the items are listed in the left-hand column of Table 3. The associated $\log_2$ prices are listed in column 2. Each number in the second column can be obtained by using the natural logarithm. For instance,

$$\log_2 78.25 = \frac{\ln 78.25}{\ln 2} = \frac{4.3599}{0.6931} = 6.29$$

The remaining calculations, which are shown at the bottom of Table 3, reveal that $M = 1.786$ and $S \approx 2.33$. These numbers approximate the values of M and S in the 1951–1952 Sears catalog.

Table 3

P	$\log_2 P$	$(\log_2 P - M)^2$
78.25	6.29	$4.50^2 \approx 20.2860$
3.36	1.75	$-0.04^2 \approx 0.0013$
1.68	0.75	$-1.04^2 \approx 1.0733$
1.06	0.08	$-1.71^2 \approx 2.9104$
1.04	0.06	$-1.73^2 \approx 2.9791$
Sum	8.93	27.2501

$$M = \frac{8.93}{5} = 1.786 \quad S = \sqrt{\frac{27.25}{5}} \approx 2.33$$

Unnatural logarithms The analysis of Sears catalogs has shown one use of a logarithm other than the natural logarithm. Another "unnatural" use involves the graphing of data. We show two ways in which such graphs arise.

In most graphs if the x-axis and the y-axis have the same units, then these units are marked uniformly along both axes. Sometimes it is more beneficial to scale one of the axes differently. Consider Figure 2, for instance. The y-axis is

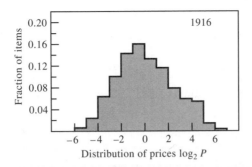

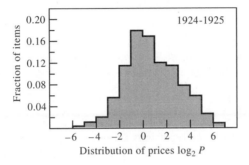

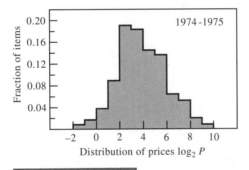

FIGURE 2 Histogram of distribution of prices in Sears Roebuck catalogs for years 1916, 1924–1925, and 1974–1975.

scaled uniformly, with each mark indicating 2% of the items in the 1916 Sears catalog. However, the units along the x-axis represent $\log_2 P$, where P is the price of an item. Therefore the interval from $x = 0$ to $x = 2$ represents prices in the range from $P = \$1$ to $P = \$4$ (since $\log_2 1 = 0$ and $\log_2 4 = 2$), while the interval from $x = 2$ to 4 represents prices in the range from $P = \$4$ to $P = \$16$.

Each histogram in Figure 2 is very close to a normal distribution. The same type of histogram results for every Sears catalog when $\log_2 P$ is used for the x-axis. In their marketing wisdom the Sears Roebuck Company has created catalogs whose price distribution year after year has maximized profit by maximizing a function associated with $\log_2 P$ (the so-called entropy function).

E X A M P L E   2

Problem

Use Figure 2 to determine which price range was listed most frequently in the 1916 catalog.

Solution The maximum interval of the histogram occurs from $x = -1$ to $x = 0$. The associated prices are \$0.50 (since $\log_2 0.5 = -1$) and \$1. Therefore items costing between \$0.50 and \$1 were listed most frequently. (They comprise about 17% of all listings.)

Figure 3 shows the graph of the log-mean M of the Sears catalogs. The years are marked uniformly along the x-axis and the values of $\log_2 M$ are marked along the y-axis. Therefore the point (1975, 4) lies on the graph because the average item in the 1975–1976 Sears catalog listed for about $M = \$16$ and $\log_2 16 = 4$.

The Sears catalogs list many diverse products, so the variation in the annual log-mean price represents an average over numerous technologies. The graph, called *the SR index,* has become a standard for comparison. For example, Figure 3 includes the average factory wholesale price to automobile dealers in

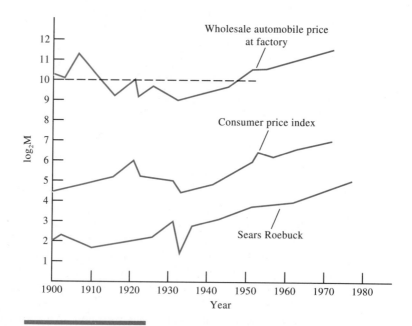

F I G U R E 3 *Source:* From annual volumes, "Automobile Facts and Figures," American Automobile Association.

the United States. The graph is similar to the SR index after the initial drop in prices until about 1935. (The computer industry has also experienced such a period of decrease in prices since its inception.)

EXAMPLE 3

Problem

Use Figure 3 to approximate those years when the average factory wholesale price to automobile dealers was about $1000.

Solution The units along the y-axis of Figure 3 refer to $\log_2 M$, where M is the price to automobile dealers. Since $\log_2 1000 \approx 10$, we can locate the approximate years by drawing a horizontal line at $y = 10$. (It is the dashed line in Figure 3.) The approximate years are 1912, 1920, and 1945.

Case Study Exercises

1. Compute the log-mean and log-standard deviation for the five items whose sales prices are $128, $32, $8, $2, and $1.
2. Which annual Sears catalog (from Table 2) has a log-mean and a log-standard deviation that are similar to the log-mean and the log-standard deviation of these five items: $39.40, $32, $1.62, $1.41, and $1.41?
3. Use Figure 2 to determine which price range was listed most frequently in the 1924–1925 catalog.
4. Use Figure 2 to determine which price range was listed most frequently in the 1974–1975 catalog.

Problems 5 to 8 refer to Figure 3.

5. Approximate the average sales price of an item in the 1970 Sears catalog.
6. Approximate the average factory wholesale price to automobile dealers in 1970.
7. What was the Consumer Price Index in 1970?
8. Approximate the years in which the Consumer Price Index was about $64.

15–3 Differentiation of Logarithmic Functions

The definition of the derivative has been used for two purposes so far. First, in order to demonstrate its meaning, we applied it to compute the derivatives of several simple functions. Then it was used to derive the standard formulas that were applied to a wide range of functions. The derivative of the logarithmic function, however, cannot be derived by these techniques, so in this section we return to the definition. We begin by appealing to the definition to suggest a rule for the derivative of $f(x) = \ln x$. This rule will then be generalized, illustrated, and proved more rigorously.

The domain of the function $\ln x$ is $(0, \infty)$. In order to include all nonzero values of x in the domain, we often use the function $\ln |x|$. Since including absolute value signs complicates the formulas and problems, in the first part of the section we assume that the domain of each function is the largest possible set of real numbers. At the end of the section we handle functions like $\ln |x|$.

The Derivative of $\ln x$

Let us compute the derivative of $f(x) = \ln x$ at $x = a$ using the definition of the derivative. By definition

$$f'(a) = \lim_{h \to 0} \frac{f(a + h) - f(a)}{h}$$

$$= \lim_{h \to 0} \frac{\ln(a + h) - \ln(a)}{h}$$

In previous similar problems the next step is to simplify the difference quotient. However, this quotient does not simplify easily. At the end of the section a rigorous method for calculating this limit is given, but for now we evaluate the limit for two particular values of a and for various choices of h that approach 0. Using a calculator, we obtain the following table of values for $a = 5$ and $a = 8$. They give an indication of the rule for the derivative of $f(x) = \ln x$ (see Table 1).

For $a = 5$, as h approaches 0 the values of the quotient approach 0.2. This suggests that $f'(5) = 0.2$. It is instructive to write

$$f'(5) = \frac{1}{5}$$

In the same way, for $a = 8$, Table 1 suggests that $f'(8) = 0.125$. Once again, write

$$f'(8) = \frac{1}{8}$$

These two results suggest that $f'(a) = 1/a$. This is the rule for the derivative of $f(x) = \ln x$. The formal proof at the end of the section is based in part on this discussion.

Table 1

	$a = 5$		$a = 8$
h	$\dfrac{\ln(5 + h) - \ln 5}{h}$	h	$\dfrac{\ln(8 + h) - \ln 8}{h}$
0.1	0.198026	0.1	0.124225
0.01	0.199800	0.01	0.124922
0.001	0.199980	0.001	0.124992
0.0001	0.199998	0.0001	0.124999

> ### The Derivative of the Logarithmic Function
>
> $$D_x \ln x = \frac{1}{x}$$

EXAMPLE 1

Problem

Find the derivative of $f(x) = x^3 - 5 \ln x$.

Solution Use the rule for the derivative of a sum and the rule for a constant times a function.

$$D_x(x^3 - 5 \ln x) = D_x(x^3) - 5D_x(\ln x) = 3x^2 - 5\left(\frac{1}{x}\right) = 3x^2 - \frac{5}{x}$$

The Logarithmic Function Version of the Chain Rule

Most logarithmic functions found in applications are of the form $\ln u$ where $u = g(x)$ is a function of x. If $g'(x)$ exists, the chain rule can be used to find the derivative of $\ln u$. Recall that the chain rule states that if $y = f(g(x))$, then

$$y' = f'(g(x))g'(x)$$

To compute the derivative of $y = \ln u$, let $u = g(x)$ and $f(x) = \ln x$ in the formula. Then $y = f(g(x)) = f(u) = \ln u = \ln g(x)$. From the rule for the derivative of the logarithmic function, $f'(u) = f'(g(x)) = 1/u = 1/g(x)$. From the chain rule we get

$$y' = f'(g(x))g'(x) = \left(\frac{1}{g(x)}\right)g'(x) = \frac{g'(x)}{g(x)}$$

The form of the chain rule applied to the logarithmic function $f(x) = \ln g(x)$ is the formula

> ### Derivative of Logarithmic Functions
>
> $$D_x \ln g(x) = \frac{g'(x)}{g(x)}$$

EXAMPLE 2

Problem

Find the derivative of (a) $f(x) = \ln(3x^2 + 5)$, (b) $f(x) = \ln(\ln x)$.

Solution (a) Let $g(x) = 3x^2 + 5$. Then $g'(x) = 6x$ and we get

$$D_x \ln(3x^2 + 5) = \frac{g'(x)}{g(x)} = \frac{6x}{3x^2 + 5}$$

(b) Let $g(x) = \ln x$. Then $g'(x) = 1/x$ and we get

$$D_x \ln(\ln x) = \frac{g'(x)}{g(x)} = \frac{(1/x)}{\ln x} = \frac{1}{x \ln x}$$

The next example demonstrates how one derivative can require two versions of the chain rule, the generalized power rule and the logarithmic function version.

EXAMPLE 3

Problem

Find the derivative of

$$f(x) = [2 + \ln(1 + 3x)]^3$$

Solution First apply the generalized power rule. We get

$$f'(x) = 3[2 + \ln(1 + 3x)]^2[D_x(2 + \ln(1 + 3x))]$$

The derivative of $2 + \ln(1 + 3x)$ requires the derivative of a sum with $g(x) = 1 + 3x$. Then $g'(x) = 3$ and $D_x \ln(1 + 3x) = 3/(1 + 3x)$. Then

$$f'(x) = 3[2 + \ln(1 + 3x)]^2[D_x(2 + \ln(1 + 3x))]$$

$$= 3[2 + \ln(1 + 3x)]^2 \left[\frac{3}{1 + 3x}\right]$$

$$= \frac{9[2 + \ln(1 + 3x)]^2}{1 + 3x}$$

Graphing Logarithmic Functions

The curve-sketching techniques of the previous chapter can be used to graph logarithmic functions. The methods are illustrated in the next example.

EXAMPLE 4

Problem

Find the relative extrema and the points of inflection of the function $f(x) = x - \ln x$. Then draw the graph.

Solution Find the extrema via the second derivative test. Compute $f'(x)$ and $f''(x)$.

$$f'(x) = 1 - \frac{1}{x} \qquad f''(x) = \frac{1}{x^2}$$

Solve $f'(x) = 0$; $f'(x) = 1 - 1/x = 0$. Then $x = 1$, so the only critical value is $x = 1$. Although $f'(0)$ does not exist, $x = 0$ is not a critical value because 0 is not in the domain of $f(x)$. This is because $\ln x$ is not defined at $x = 0$.

Since $f''(1) = 1$, $f(x)$ has a relative minimum at $x = 1$. Since $f''(x) > 0$ for all x, $f(x)$ has no point of inflection. As x approaches 0, $\ln x$ approaches $-\infty$, and so $-\ln x$ approaches ∞, meaning that $f(x)$ has a vertical asymptote at $x = 0$. Note that $f(1) = 1 - \ln 1 = 1 - 0 = 1$. These facts can be used to draw the graph of $f(x)$ shown in Figure 1.

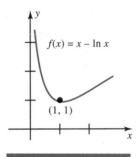

$f(x) = x - \ln x$

$(1, 1)$

FIGURE 1

Consider the function $y = \ln |x|$ whose domain is the set of all real numbers except $x = 0$. The definition of this function is

$$\ln |x| = \begin{cases} \ln x & \text{if } x > 0 \\ \ln(-x) & \text{if } x < 0 \end{cases}$$

If $x > 0$, then $y = \ln x$ and $y' = 1/x$. Suppose $x < 0$. Then $y = \ln(-x)$ and to find y' we use the rule for the derivative of logarithmic functions with $g(x) = -x$. The derivative of $\ln(-x)$ is $-1/-x = 1/x$. Thus the derivative of $\ln(-x)$ is equal to the derivative of $\ln x$. A similar argument shows that the derivative of $\ln g(x)$ is the same as $\ln(-g(x))$.

$$D_x \ln |x| = \frac{1}{x}$$

$$D_x \ln |g(x)| = \frac{g'(x)}{g(x)}$$

EXAMPLE 5

Problem

Find the derivative of (a) $f(x) = \ln |3x - 2|$, (b) $f(x) = x \ln |1 + x^{-3}|$.

Solution (a) Let $g(x) = 3x - 2$. From the formula we get

$$f'(x) = \frac{3}{3x - 2}$$

(b) Let $g(x) = 1 + x^{-3}$ and apply the product rule.

$$f'(x) = xD_x \ln |1 + x^{-3}| + \ln |1 + x^{-3}|D_x x$$

$$= x\left[\frac{-3x^{-4}}{1 + x^{-3}}\right] + \ln |1 + x^{-3}|(1)$$

$$= \frac{-3x^{-3}}{1 + x^{-3}} + \ln |1 + x^{-3}|$$

$$= \frac{-3}{x^3 + 1} + \ln |1 + x^{-3}|$$

Proof of the Rule for the Derivative of the Logarithmic Function

Proof. If $f(x) = \ln x$, then

$$f'(x) = \lim_{h \to 0}\left[\frac{f(x + h) - f(x)}{h}\right] = \lim_{h \to 0}\left[\frac{\ln(x + h) - \ln x}{h}\right]$$

$$= \lim_{h \to 0}\left(\frac{1}{h}\right)\ln\left[\frac{x + h}{x}\right]$$

$$= \lim_{h \to 0}\ln\left[\frac{x + h}{x}\right]^{1/h}$$

We first deal with the case where $h > 0$. The case where $h < 0$ is handled in a similar way. To simplify this expression, let $m = x/h$. In this expression x is being held constant while h varies. As $h \to 0$, note that $m \to \infty$ and vice versa. In the last limit above write $(x + h)/x = 1 + h/x = 1 + 1/m$. The limit then becomes

$$f'(x) = \lim_{m \to \infty}\ln\left[1 + \frac{1}{m}\right]^{m/x} = \lim_{m \to \infty}\ln\left[\left(1 + \frac{1}{m}\right)^m\right]^{1/x}$$

$$= \lim_{m \to \infty}\frac{1}{x} \cdot \ln\left[1 + \frac{1}{m}\right]^m = \frac{1}{x} \cdot \lim_{m \to \infty}\ln\left[1 + \frac{1}{m}\right]^m$$

$$= \frac{1}{x} \cdot \ln\left[\lim_{m \to \infty}\left(1 + \frac{1}{m}\right)^m\right]$$

From the definition of e we have

$$\lim_{m \to \infty}\left(1 + \frac{1}{m}\right)^m = e$$

Therefore

$$f'(x) = \frac{1}{x} \cdot \ln\left[\lim_{m \to \infty}\left(1 + \frac{1}{m}\right)^m\right] = \frac{1}{x}\ln e = \frac{1}{x} \cdot 1 = \frac{1}{x}$$

Thus $D_x \ln x = 1/x$, as desired.

Elasticity

Economists define the **rate of growth** $Rf(x)$ of the function $f(x)$ as the ratio

$$Rf(x) = f'(x)/f(x)$$

It measures the relative rate of change of $f(x)$. For example, suppose an automobile manufacturer sells its car for \$20,000 and another sells its car for \$8000. Suppose both manufacturers increase the price, the first to \$21,000 and the second to \$8,500. So the first manufacturer's increase is \$1000 and the second manufacturer's increase is \$500. Which is the larger relative increase, meaning which is the larger percentage increase? Let the average price per car be represented by $f(x)$ for the first manufacturer and $g(x)$ for the second manufacturer. Let $x = 0$ represent the time when the original prices of \$20,000 and \$8000 were in effect. Then $f(0) = 20{,}000$ and $g(0) = 8000$. The rate of increase for each function is given by $f'(0) = 1000$ and $g'(0) = 500$. The rate-of-growth functions are

$$Rf(0) = f'(0)/f(0) = 1000/20000 = 1/20 = .05$$
$$Rg(0) = g'(0)/g(0) = 500/8000 = 1/16 = .0625$$

This means that the relative rate of increase for the lower priced car is greater than that of the higher priced car, even though the actual size of the increase is only half that of the higher priced car.

Another way to view the rate-of-growth function is to express it in terms of the derivative of the logarithm of $f(x)$:

$$Rf(x) = f'(x)/f(x) = D_x \ln f(x)$$

It is called the *logarithmic derivative of $f(x)$*. Economists use the rate-of-growth function to measure the relative change of various quantities, such as revenue and demand. In this way the relative changes of a product made by different companies can be compared. For instance, suppose two companies have demand curves $y = Q_1(p)$ and $y = Q_2(p)$, whose graphs are given in Figure 2. Because the graph of

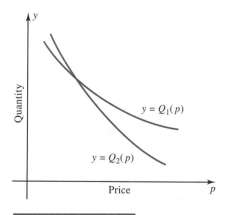

FIGURE 2

$Q_2(p)$ is steeper than the graph of $Q_1(p)$, a small change in the price p will result in a larger change in the quantity $Q_2(p)$ demanded versus the quantity $Q_1(p)$. From an intuitive point of view, economists refer to a steep demand curve such as the graph of $y = Q_2(p)$ as elastic while the graph of $y = Q_1(p)$ is called inelastic.

We now give a concrete definition of the **elasticity** $E(p)$ of a demand curve $y = Q(p)$. It is defined as the negative of the ratio of the rate of growth of $Q(p)$ to the rate of growth of p. That is:

$$E(p) = \frac{-(\text{rate of growth of } Q(p))}{\text{rate of growth of } p} = \frac{-D_p \ln Q(p)}{D_p \ln p} = \frac{-Q'(p)/Q(p)}{1/p} = \frac{-pQ'(p)}{Q(p)}$$

Elasticity is defined with a negative sign so that $E(p)$ is always positive. Notice that p and $Q(p)$ are always positive and $Q'(p)$ is always negative because $y = Q(p)$ is a decreasing function. If $E(p) > 1$ the demand curve is called **elastic** at p and if $E(p) < 1$ it is called **inelastic** at p.

EXAMPLE 6

The demand for oil is given by the formula

$$Q(p) = 60 - 2p$$

where p is the price per barrel and $Q(p)$ is the number of barrels demanded at price p in millions of barrels, for p between \$5 and \$30.

Problem

Find the values of p where $Q(p)$ is (a) elastic (b) inelastic.

Solution First compute $Q'(p) = -2$. Then

$$E(p) = \frac{-p(-2)}{60 - 2p} = \frac{2p}{60 - 2p}$$

Determine the value of p where $E(p) = 1$ by solving

$$E(p) = \frac{2p}{60 - 2p} = 1$$

This simplifies to $2p = 60 - 2p$, and then $4p = 60$, which implies that $p = 15$. $E(p)$ is a continuous function that crosses the line $E = 1$ at $p = 15$. This means that $E(p)$ is either greater than 1 at each point in the interval to the left of 15, or it is less than 1 at each point in the interval. The same is true for the interval greater than 15. This means that we must determine whether $E(p)$ is greater than or less than 1 on the intervals $(5, 15)$ and $(15, 30)$. Choose a test point in each interval, say 10 and 20, respectively: $E(10) = 20/40 = 1/2$ and $E(20) = 40/20 = 2$.
(a) Since 10 is in the interval $(5, 15)$ and $E(10) < 1$, $E(p) < 1$ in $(5, 15)$ and so $Q(p)$ is inelastic there.
(b) Since 20 is in the interval $(15, 30)$ and $E(20) > 1$, $E(p) > 1$ in $(15, 30)$ and so $Q(p)$ is elastic there.

If the demand curve is inelastic, then an increase in p results in an increase in revenue while a decrease in p results in a decrease in revenue. This means that revenue is an increasing function in an interval where the demand is inelastic. If the demand curve is elastic, the opposite is true: an increase in p results in a decrease in revenue, meaning that revenue is decreasing in such an interval. Let us see why this is so using the demand curve in Example 6. If $Q(p) = 60 - 2p$, then the revenue $R(p)$, defined by (price $\times$ quantity), is

$$R(p) = pQ(p) = p(60 - 2p) = 60p - 2p^2$$

The question is to determine when $R(p)$ is increasing, so we compute $R'(p)$ and solve $R'(p) > 0$:

$$R'(p) = 60 - 4p > 0$$
$$4(15 - p) > 0$$

This is true when $p < 15$, meaning that $R(p)$ is increasing in $(5, 15)$. Therefore, from Example 6, in the interval $(5, 15)$, $E(p) < 1$, and thus $Q(p)$ is inelastic in this interval. The previous computation shows that $R(p)$ is increasing in this interval. In a similar way it can be shown that in the interval $(15, 30)$ the demand curve is elastic and $R(p)$ is decreasing.

EXERCISE SET 15-3

In Problems 1 to 24 find $f'(x)$.

1. $f(x) = \ln 6x$

2. $f(x) = \ln 9x$

3. $f(x) = \ln 3x + 5$

4. $f(x) = \ln 5x - 6$

5. $f(x) = \ln(x + 5)$

6. $f(x) = \ln(x - 7)$

7. $f(x) = \ln(x^2 + 3)$

8. $f(x) = \ln(3x^2 - 5)$

9. $f(x) = \ln(1 + 4x^2)$

10. $f(x) = \ln(3 - 5x^2)$

11. $f(x) = \ln(2x^3 + x^{-1})$

12. $f(x) = \ln(4 - x^{1/2} + x)$

13. $f(x) = x \ln 4x$

14. $f(x) = 3x \ln 7x$

15. $f(x) = x^{-3} \ln(x^2 - 1)$

16. $f(x) = 5x^{2/3} \ln\left(2x - \dfrac{3}{2}\right)$

17. $f(x) = (\ln 3x)/(2x - 5)$

18. $f(x) = \dfrac{x^5 - 11}{\ln 3x}$

19. $f(x) = \ln |3 - 5x|$

20. $f(x) = \ln |2x^3 + 1|$

21. $f(x) = \ln |4 - x^{5/2}|$

22. $f(x) = \ln |2x^3 + x|$

23. $f(x) = x \ln |7x|$

24. $f(x) = \dfrac{x^3}{\ln |x^2 - 1|}$

In Problems 25 to 28 find the relative extrema and sketch the graph.

25. $f(x) = x \ln x$

26. $f(x) = x \ln |x|$

27. $f(x) = \dfrac{\ln x}{x}$

28. $f(x) = x - \ln |x|$

In Problems 29 and 30 the demand function $Q(p)$ is defined in the given interval. Find where $Q(p)$ is (a) elastic and (b) inelastic.

29. $Q(p) = 80 - 5p$, $[5, 15]$

30. $Q(p) = 100 - 5p$, $[5, 30]$

In Problems 31 to 36 find $f'(x)$.

31. $f(x) = (x + \ln x)^2$

32. $f(x) = (4x + \ln 3x)^{-3}$

33. $f(x) = (\ln x + x^2)/\ln x$

34. $f(x) = x^2(x - \ln x)^2$

35. $f(x) = \ln(x^2 + \ln x)$

36. $f(x) = \ln(x + x \ln x)$

37. Sound intensity is measured in *decibels*. If $D(x)$ is the number of decibels when x is the power of the sound, measured in watts/cm^2, then the relationship between D and x is given by

$$D(x) = 10 \log_{10}(10^{16}x)$$

Find $D'(x)$.

38. The relationship between cost C and labor x is often given by a logarithmic equation. It is a form of a Cobb–Douglas function. Suppose a firm determines that

$$\ln C(x) = 8.3x^{-0.17}$$

Find $C'(x)$.

39. Economists define the *rate of growth* of a function $f(x)$ as the ratio $F(x)$ given by

$$F(x) = \frac{f'(x)}{f(x)}$$

Show that $F(x) = D_x(\ln f(x))$.

40. A manufacturer has a cost function

$$C(x) = 5000 + 600x - 0.3x^2$$

Find the rate of growth of $C(x)$ when $x = 100$.

41. Let $f(x) = \ln x$. Find a formula for $f^{(n)}(x)$ for n a positive integer by computing the first few derivatives of $f(x)$ and finding a pattern.

42. Find $f'(x)$ for $f(x) = \ln(\ln(\ln x))$.

43. Show that $f(x) = \ln x^2$ is not the same function as $g(x) = (\ln x)^2$ by first showing that their functional values are different for a particular value of x and then showing that their derivatives are different.

Referenced Exercise Set 15–3

1. The relationship between home range and body size of mammals is believed to be as important to carnivores as it is to herbivores. In an article studying this relationship, Lindstedt et al. use a formula that relates home range A, measured in hectares, and body mass m, measured in kilograms.* The relationship is

$$\ln A = 1 + \ln m$$

Express A in terms of m directly, and find $A'(1)$.

2. Accurate leaf-area measurements are critical for estimating fluxes of carbon, solar energy, and water in forests. In an article comparing various methods of estimating leaf areas in fir trees, Marshall and Waring show that the leaf-area index L, which is the projected surface area of foliage per unit ground area, is a function of the irradiance q below the canopy, which is the light level below the tops of the trees.† The relationship for a Douglas-fir forest is estimated as

$$L(q) = 2.5(2q - \ln q)$$

Find the value of q where $L(q)$ is a minimum.

Cumulative Exercise Set 15–3

1. Sketch the graphs of the functions $f(x) = 3^x$ and $g(x) = 3^{-x}$ on the same coordinate axis.

2. Sketch the graphs of the functions $f(x) = e^{x+1}$ and $g(x) = \ln(x + 1)$ on the same coordinate axis.

3. Find all values of x that satisfy the given equality.
   (a) $2^x = 1/4$
   (b) $3^{x-2} = 81$
   (c) $2^{2x-1} = 8^{x-1}$

4. A student's grade after t weeks of a 15-week semester is

$$g(t) = 100 - 60e^{-0.1t}$$

Determine the student's grade (a) at the beginning of the semester, (b) after one week, (c) after 10 weeks, and (d) at the end of the semester. (e) If the semester is extended indefinitely, what will the student's ultimate grade be?

5. Sketch the graphs of the functions $f(x) = \ln x$, $g(x) = \log_2 x$, and $h(x) = \log_3 x$ on the same coordinate axis.

6. Use the table in the Appendix or a calculator to sketch the graph of the function $f(x)$ by the plotting method, where

$$f(x) = e^{x/2} - (\ln x)^4$$

[The command for this function in the graphing module of the software program **MathPath** is $\exp(x/2) - (\ln(x))\wedge 4$.]

7. Solve each equation for x.
   (a) $\ln e^{2x} - e^{\ln 6} = \ln 1$
   (b) $e^{\ln 3x} + 1 = \ln e^4$

---

*Stan L. Lindstedt et al., "Home Range and Body Size in Mammals," *Ecology*, Vol. 67, 1986, pp. 413–418.

†J. D. Marshall and R. H. Waring, "Comparison of Methods of Estimating Leaf-area Index in Old-growth Douglas Fir," *Ecology*, Vol. 67, 1986, pp. 413–418.

8. Suppose the weight equation for a 130-pound person who goes on a diet is given by

$$w(t) = 120 + 10e^{-0.005t}$$

where t is the time in days after the start of the diet. How many days will it take the person to drop to 126 pounds?

In Problems 9 to 11 find $f'(x)$.

9. $f(x) = \ln(4 - x^2)$

10. $f(x) = \ln(e^x) - \dfrac{1}{\ln x}$

11. $f(x) = x(1 - \ln x)^2$

12. The purpose of this problem is to develop a formula for the derivative of logarithmic functions to any base.
(a) Let $y = \log_b x$. Fill in the blank: $b^y =$ ____
(b) Take the natural logarithm of both sides of the equation $b^y = x$ and solve for y.
(c) Show that for any $b > 0$,

$$D_x \log_b x = \frac{1}{\ln b} \frac{1}{x}$$

15–4 Differentiation of Exponential Functions

Recall that the natural logarithm function and the exponential function are inverses of each other. This means that $\ln e^x = e^{\ln x} = x$. This property enables us to derive the derivatives of the functions $y = e^x$ and $y = e^{g(x)}$.

The Derivative of $y = e^x$

Because e^x and $\ln x$ are inverses of each other, the following property holds:

$$\ln e^x = x$$

Take the derivative of each side.

$$D_x(\ln e^x) = D_x(x)$$

On the left-hand side use the rule for the derivative of logarithmic functions, with $g(x) = e^x$.

$$\frac{D_x e^x}{e^x} = 1$$

Multiply by e^x to get

$$D_x e^x = e^x$$

The Derivative of the Exponential Function

$$D_x e^x = e^x$$

EXAMPLE 1

Problem

Find $f'(x)$ for

(a) $f(x) = x^4 e^x - 5e^x$, (b) $f(x) = \dfrac{e^x}{x^2 + 5x}$.

Solution (a) Use the rule for the derivative of a sum, the rule for the derivative of a product, and the rule for a constant times a function.

$$D_x(x^4e^x - 5e^x) = D_x(x^4e^x) - 5D_x(e^x) = x^4D_x(e^x) + e^xD_x(x^4) - 5D_x(e^x)$$
$$= x^4e^x + e^x4x^3 - 5e^x = e^x(x^4 + 4x^3 - 5)$$

(b) Use the rule for the derivative of a quotient.

$$D_x\left(\frac{e^x}{x^2 + 5x}\right) = \frac{(x^2 + 5x)D_x(e^x) - e^xD_x(x^2 + 5x)}{(x^2 + 5x)^2}$$

$$= \frac{(x^2 + 5x)e^x - e^x(2x + 5)}{(x^2 + 5x)^2}$$

$$= \frac{e^x(x^2 + 5x - 2x - 5)}{(x^2 + 5x)^2} = \frac{e^x(x^2 + 3x - 5)}{(x^2 + 5x)^2}$$

The Exponential Function Version of the Chain Rule

The general exponential function is expressed in the form e^u where $u = g(x)$ is a function of x. If $g'(x)$ exists, the chain rule can be used to find the derivative of e^u. Recall that the chain rule states that if $y = f(g(x))$, then

$$y' = f'(g(x))g'(x)$$

To compute the derivative of $y = e^u$, let $u = g(x)$ and $f(x) = e^x$ in the formula. Then $y = f(g(x)) = f(u) = e^u = e^{g(x)}$. From the rule for the derivative of the exponential function $f'(u) = e^u$ or $f'(g(x)) = e^{g(x)}$. From the chain rule we get

$$y' = f'(g(x))g'(x) = e^{g(x)} \cdot g'(x) = g'(x)e^{g(x)}$$

The form of the chain rule applied to the exponential function

$$f(x) = e^{g(x)}$$

is the formula

Derivative of Exponential Functions

$$D_xe^{g(x)} = g'(x)e^{g(x)}$$

Example 2 illustrates how to apply the exponential function version of the chain rule.

E X A M P L E   2

Problem

Find the derivative of (a) $f(x) = e^{2x}$, (b) $f(x) = e^{3x^2+7}$.

Solution (a) Let $g(x) = 2x$. Then $g'(x) = 2$ and

$$D_x e^{2x} = g'(x)e^{g(x)} = 2e^{2x}$$

(b) Let $g(x) = 3x^2 + 7$. Then $g'(x) = 6x$ and

$$D_x e^{(3x^2+7)} = g'(x)e^{g(x)} = 6xe^{3x^2+7}$$

The next example demonstrates how one derivative can require two versions of the chain rule, the generalized power rule, and the exponential function version.

EXAMPLE 3

Problem

Find the derivative of

$$f(x) = (x^{-1} + e^{5x+2})^3$$

Solution First apply the generalized power rule. We get

$$f'(x) = 3(x^{-1} + e^{5x+2})^2 \cdot D_x(x^{-1} + e^{5x+2})$$

The derivative of $x^{-1} + e^{5x+2}$ requires the derivative of a sum. Let $g(x) = 5x + 2$. Then $g'(x) = 5$ and $D_x e^{5x+2} = 5e^{5x+2}$. Thus

$$f'(x) = 3(x^{-1} + e^{5x+2})^2(-x^{-2} + 5e^{5x+2})$$

Graphing Exponential Functions

The curve-sketching techniques of the previous chapter can be used to graph exponential functions. The methods are illustrated in the next two examples.

EXAMPLE 4

Problem

Graph the function $f(x) = \dfrac{1}{1 + e^{-x}}$.

Solution To find $f'(x)$, write $f(x) = (1 + e^{-x})^{-1}$. Then

$$f'(x) = -(1 + e^{-x})^{-2}(-e^{-x}) = e^{-x}(1 + e^{-x})^{-2}$$

$$= \frac{1}{e^x(1 + e^{-x})^2} = \frac{1}{e^x(1 + 2e^{-x} + e^{-2x})} = \frac{1}{2 + e^x + e^{-x}}$$

The numerator is never 0, so there is no value of x such that $f'(x) = 0$. Each term in the denominator is positive so there is no value of x such that $f'(x)$ does not exist. Thus there is no relative extremum. To test for a point of inflection, we calculate $f''(x)$ using the quotient rule.

$$f''(x) = \frac{-(e^x - e^{-x})}{(2 + e^x + e^{-x})^2} = \frac{e^{-x} - e^x}{(2 + e^x + e^{-x})^2}$$

Solving $f''(x) = 0$ yields $e^{-x} - e^x = 0 = e^{-x}(1 - e^{2x})$. This is satisfied when $e^{2x} = 1$, which is true when $x = 0$. Therefore $f(x)$ has a possible point of inflection at $x = 0$. When $x < 0$, $f''(x) > 0$ since $e^{-x} > e^x$. When $x > 0$, $f''(x) < 0$. This means that $f(x)$ is concave up in $(-\infty, 0)$ and concave down in $(0, \infty)$. Thus $f(x)$ has a point of inflection at $x = 0$. Since $f'(x) > 0$ for all values of x, the graph is always increasing. As x gets very large e^{-x} approaches 0, so $f(x)$ approaches $1/(1 + 0) = 1$. Similarly as x approaches $-\infty$, $f(x)$ approaches 0. Putting this information together yields the graph in Figure 1.

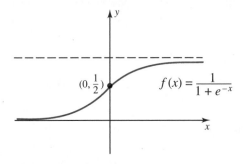

FIGURE 1

EXAMPLE 5

Problem

Graph the function $f(x) = \dfrac{1}{\sqrt{2\pi}} e^{-x^2/2}$.

Solution First find $f'(x)$.

$$f'(x) = \frac{1}{\sqrt{2\pi}} \left(\frac{-2x}{2} \right) e^{-x^2/2} = \left(\frac{-x}{\sqrt{2\pi}} \right) e^{-x^2/2}$$

The only value of x such that $f'(x) = 0$ is $x = 0$ since $e^{-x^2/2}$ is always positive. Thus the only critical value is $x = 0$. Applying the first derivative test shows that $f(x)$ is increasing in $(-\infty, 0)$ and decreasing in $(0, \infty)$ because $e^{-x^2/2}$ is always positive, and so $f'(x)$ is positive and negative when x is negative and positive. Therefore $f(x)$ has a relative maximum of $f(0) = 1/\sqrt{2\pi}$. The points of inflection are found by solving $f''(x) = 0$.

$$f''(x) = \left(\frac{x^2}{\sqrt{2\pi}} \right) e^{-x^2/2} - \frac{1}{\sqrt{2\pi}} e^{-x^2/2} = \left(\frac{1}{\sqrt{2\pi}} \right)(x^2 - 1)e^{-x^2/2}$$

The function has points of inflection at $x = -1$ and $x = 1$ and $f(x)$ is concave up in the intervals $(-\infty, -1)$ and $(1, \infty)$ and concave down in the interval $(-1, 1)$. The graph of $f(x)$ is given in Figure 2.

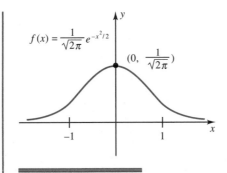

FIGURE 2

The function in Example 5 is the well-known "bell-shaped" curve that is used extensively in statistics. This function is known as the **standard normal probability density function.**

The Logistic Curve

Psychologists use various techniques to model the learning process. One type of model that uses exponential functions of the form $y(x) = a - be^{-ct}$, was introduced in Section 15–1. A more complex model of learning uses an exponential function of the form

$$y(x) = \frac{A}{1 + Be^{-kx}}$$

It is called the general form of the **logistic equation.** An example of a specific logistic equation is the function in Example 4, where $A = B = k = 1$. It is the function

$$f(x) = \frac{1}{1 + e^{-x}}$$

Its graph is given in Figure 1.

The first known application using the logistic equation was a study of world population made by the Belgian mathematician P. Verhulst. In 1840 Verhulst created a model of world population that was off by only 1% a century later.

One of the first applications of the logistic curve in the social sciences was made by Clark L. Hull.* He created a model of the learning process that has had a wide variety of uses in many disciplines.

The graph of a logistic equation is an S-shaped curve, as shown in Figure 3. When $x = 0$ the denominator is $1 + B$, so

$$y(0) = \frac{A}{1 + B}$$

*Clark L. Hull, *A Behavior System*, New Haven, CT, Yale University Press, 1952.

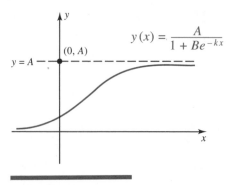

FIGURE 3

As x increases without bound, the denominator approaches 1 since e^{-kx} approaches 0. This means that $y(x)$ approaches the value $A/1$ or simply A. It never reaches A, but as x increases, $y(x)$ gets arbitrarily close to A. When the logistic equation is applied to population growth, the constant A is the maximum level of the population that the environment can sustain.

In Hull's model of learning A represents the maximum amount of learning. For example, suppose a puzzle is given to a large group of individuals. Let $f(x)$ be the average amount of learning, measured in percent, per time x spent solving the puzzle. Then A is 1, or 100%. It represents total mastery of the puzzle. One such puzzle consists of arranging several different cubes with variously colored sides in a specific order. An individual's learning consists in discovering the various characteristics of the cubes. The puzzle is usually solved the first time merely by chance. Little actual learning has taken place. When the cubes are mixed up, it takes the person almost the same amount of time to solve it again. Once some information is acquired about the cubes, say their individual color schemes or which one must come first in the sequence, then learning takes place. Hull found that when solving such tasks, learning takes place very slowly at first. The rate of learning then rises rapidly. When mastery of the task is approached the rate of learning slows again. These are the characteristics of the logistic equation.

The maximum learning rate is the maximum of the derivative of the logistic equation, $y = f(x)$. It occurs at the point of inflection of $f(x)$. The next example looks at a logistic curve similar to one used by Hull.

E X A M P L E   6

A psychologist gives a puzzle to a large group of individuals. In the directions of the puzzle each subject was told 20% of the solution. On the average, after 10 minutes a subject had learned 50% of the solution.

Problem

Find the specific form of the logistic equation that measures the percent of learning $f(x)$ after x minutes of solving the puzzle.

Solution The maximum amount of learning is 100%, so $A = 1$. At $x = 0$ the time when the subject starts to solve the puzzle, 20% of the solution is known, so $f(0) = 0.2$. Therefore

$$f(0) = \frac{1}{1 + Be^0} = 0.2$$

Thus $1 = 0.2 + 0.2B$ and $0.8 = 0.2B$, so $B = 4$. Now solve the equation for k using the fact that $f(10) = 0.5$.

$$f(10) = \frac{1}{1 + 4e^{-10k}} = 0.5$$
$$1 = 0.5 + 2e^{-10k}$$
$$0.5 = 2e^{-10k}$$
$$0.25 = e^{-10k}$$
$$\ln(0.25) = \ln(e^{-10k}) = -10k$$
$$k = \frac{1.39}{10} = 0.139$$

Therefore the equation is

$$f(x) = \frac{1}{1 + 4e^{-0.139x}}$$

The logistic equation is also used in business to model a wide variety of situations. For example, Hannan and McDowell[†] studied how information spreads in an industry, with the banking industry as a specific example. They found that information spreads slowly at first, meaning that the proportion of firms using a new technological process remains small for a period of time. Then as firms using the process enjoy success with it the proportion of firms adopting it increases rapidly. To model the growth of automatic teller machines in the banking industry, they studied 89 local banks and developed the logistic equation

$$f(x) = \frac{1}{1 + 1.97e^{-0.33x}}$$

The logistic equation is used widely in marketing to predict future trends. Suppose a firm produces an innovative product that immediately captures a certain percentage of the market. The firm has allocated a specific amount of money to advertise the new product. The basic question is: how quickly should the advertising budget be spent? Past performance shows that the increased market share of innovative products often follows a logistic growth pattern. The market share increases slowly at first, then gains momentum, and then slows again (see Figure 4). Once the market share increase gains momentum, there is less need for advertising because

[†]Timothy H. Hannan and John M. McDowell, "Market Concentration and the Diffusion of New Technology in the Banking Industry," *Review of Economics and Statistics,* Vol. 66, 1984, pp. 686–691.

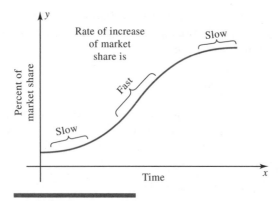

FIGURE 4

information about the product will spread by word of mouth. In addition, as the market becomes saturated there are fewer potential buyers so the market share increase must slow. These factors imply that the advertising budget has the greatest effect in the early stages of the selling period. In fact, it should be entirely spent by the time the market share increase reaches a maximum. If $y = f(x)$ represents the percent of market share of the product at time x, then the rate of increase of market share is $f'(x)$. Thus the rate of increase of market share will attain a maximum when $f'(x)$ is a maximum, that is, when $f''(x) = 0$, or when $f(x)$ has a point of inflection. This means that the advertising budget should be spent by the time $x = a$ when $f(x)$ has a point of inflection (see Figure 5).

In practice, this is a dynamic process for the marketing department. As they spend advertising dollars, they monitor the product's market share. They compare the actual figures with their predictions, alter the predictions, and then make a decision on future advertising expenditures. If it appears that the product is close to its point of inflection on the predicted logistic curve, they may choose to spend more advertising dollars in hopes of altering the actual curve so that they reach the point of inflection sooner (see Figure 6).

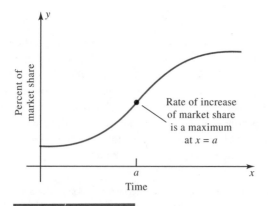

FIGURE 5

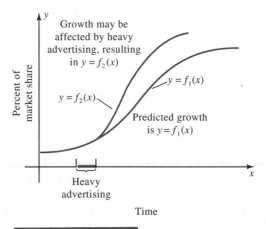

Percent of market share

Growth may be
affected by heavy
advertising, resulting
in $y = f_2(x)$

$y = f_1(x)$

$y = f_2(x)$

Predicted growth
is $y = f_1(x)$

Heavy
advertising

Time

FIGURE 6

Thus in marketing, as in many other fields, the most important point on the curve is the point of inflection, which measures when $f(x)$ attains its maximum rate of growth. The next example computes the point of inflection of a logistic curve.

EXAMPLE 7

Consider the automatic teller machine logistic equation.

$$f(x) = \frac{1}{1 + 1.97e^{-0.33x}}$$

Problem

Find the point of inflection of $f(x)$ and interpret its significance from a marketing point of view.

Solution We compute $f'(x)$ and then $f''(x)$:

$$f'(x) = \frac{1.97(-0.33)e^{-0.33x}}{(1 + 1.97e^{-0.33x})^2} = \frac{-0.65e^{-0.33x}}{(1 + 1.97e^{-0.33x})^2}$$

To find $f''(x)$, let us make some substitutions that ease the computation. Let $a = -0.65$, $B = 1.97$ and $k = -0.33$. Then

$$f'(x) = \frac{ae^{kx}}{(1 + Be^{kx})^2}$$

$$f''(x) = \frac{a[ke^{kx}(1 + Be^{kx})^2 - 2e^{kx}(1 + Be^{kx})Bke^{kx}]}{(1 + Be^{kx})^4}$$

$$= \frac{ake^{kx}(1 + Be^{kx})[1 + Be^{kx} - 2Be^{kx}]}{(1 + Be^{kx})^4}$$

$$= \frac{ake^{kx}(1 - Be^{kx})}{(1 + Be^{kx})^3}$$

Now solve $f''(x) = 0$; $ake^{kx}(1 - Be^{kx}) = 0$ implies that $1 - Be^{kx} = 0$ and so $e^{kx} = 1/B$. Take the natural logarithm of each side and get $\ln e^{kx} = \ln(1/B) = kx$, since $\ln e^{kx} = kx$. Simplify $\ln(1/B) = \ln B^{-1} = -\ln B$. Therefore $x = \dfrac{-\ln B}{k}$. Substituting back for B and k yields

$$x = \frac{-\ln(1.97)}{-0.33} = \frac{0.68}{0.33} \approx 2.06$$

Thus the point of inflection occurs near $x = 2.06$. This means that the rate of increase of the automatic teller machines will be maximum just after two years. This means that advertising expenditures should be spent over two years. In practice, the marketing analysts will monitor the market share to see if the advertising or other factors will alter this predicted logistic growth.

EXERCISE SET 15–4

In Problems 1 to 24 find $f'(x)$.

1. $f(x) = e^{6x}$

2. $f(x) = e^{9x}$

3. $f(x) = e^{3x+5}$

4. $f(x) = e^{5x-6}$

5. $f(x) = 4e^{7x+5}$

6. $f(x) = 6e^{4-7x}$

7. $f(x) = e^{x^2+3}$

8. $f(x) = e^{3x^2-5}$

9. $f(x) = 2e^{1+4x^2}$

10. $f(x) = -3e^{3-5x^2}$

11. $f(x) = 4 - e^{2x^3+x-1}$

12. $f(x) = 3 + e^{4-x^{1/2}+x}$

13. $f(x) = xe^{4x}$

14. $f(x) = 3xe^{7x}$

15. $f(x) = x^{-3}e^{x^2-1}$

16. $f(x) = 5x^{2/3}e^{2x-3/2}$

17. $f(x) = (e^{3x})/(2x - 5)$

18. $f(x) = x^5 - 11/e^{3x}$

19. $f(x) = e^{5x} + e^{-5x}$

20. $f(x) = e^{3x} - 3e^{-3x}$

21. $f(x) = \dfrac{1}{2 + e^{-x}}$

22. $f(x) = \dfrac{1}{1 + e^{-2x}}$

23. $f(x) = \dfrac{1}{1 + e^x + e^{-2x}}$

24. $f(x) = \dfrac{2}{e^{2x} + e^{-2x}}$

In Problems 25 to 28 sketch the graph.

25. $f(x) = xe^x$

26. $f(x) = x + e^x$

27. $f(x) = \dfrac{1}{2 + e^{-2x}}$

28. $f(x) = \dfrac{4}{1 + e^{-3x}}$

In Problems 29 to 36 find $f'(x)$.

29. $f(x) = \ln(x + e^{4x})$

30. $f(x) = \ln(3 + xe^{2x})$

31. $f(x) = (\ln x + e^{5x})^2$

32. $f(x) = (2 \ln x + e^{3x})^{-3}$

33. $f(x) = \dfrac{e^{x^2} + x^2}{e^x}$

34. $f(x) = \dfrac{\ln x}{x - e^{3x}}$

35. $f(x) = e^{x^3} \ln x$

36. $f(x) = e^{3x} \ln(x + e^x)$

37. Let $f(x) = e^{2x}$. Find a formula for $f^{(n)}(x)$ for n a positive integer by computing the first few derivatives of $f(x)$ and finding a pattern.

38. Find $f'(x)$ for $f(x) = xe^x \ln x$.

In Problems 39 to 42 match the equation with the graphs in figures a to d on page 811.

39. $f(x) = \dfrac{1}{1 + e^{-x}}$

40. $f(x) = \dfrac{100}{1 + e^{-x}}$

41. $f(x) = \dfrac{1}{1 + 9e^{-x}}$

42. $f(x) = \dfrac{100}{1 + 9e^{-x}}$

(a)

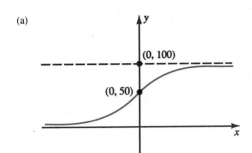

(b)

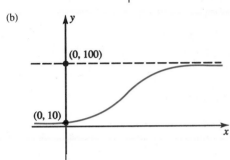

(c)

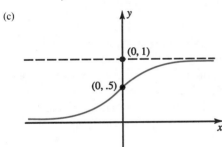

(d)

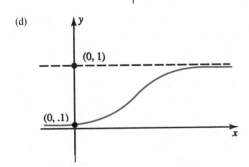

Referenced Exercise Set 15–4

1. In an article* studying the social cost of population growth, researchers used educational tax revenue as a measure of the cost of an expanding population. They focused on California's 1965 tax profile as a baseline for a community's ability to fund fully its educational system. The proportion of the tax paid by various age groups increased from age 16 to 65. Because many people retired at age 65, the proportion of tax paid per age decreased linearly from age 65 to age 75, after which it was always 0. The data are given by the formula $y = f(x)$ where $f(x)$ is the proportion of the total educational tax paid by those of age x. In the interval [16, 65] the data are given as an exponential function, and in the interval (65, 75] as a linear function.

$$f(x) = \begin{cases} 0.03e^{-0.0008(x-65)^2} & x \in [16, 65] \\ -0.003x + 0.225 & x \in (65, 75] \end{cases}$$

Graph $f(x)$ in [16, 75].

2. Thermal recovery of the skin after cooling is not only an important medical treatment but it is also a key indicator of normal versus pathological circulation. In a paper studying thermal recovery, Steketee and Van Der Hoek cooled the skin of the forehead of patients and measured the time it took to reach a steady-state temperature.† The temperature T of the forehead, measured in degrees centigrade, was found to be a function of time t, measured in minutes. For one group of patients, the functional relationship was found to be

$$T(t) = 35 + 2.1e^{-.017t}$$

The steady-state temperature was defined to be that value of t when $T'(t) = -.028$. Find the steady-state temperature for this group of patients.

3. The problem of finding an optimal advertising policy over a period of time is one of the most important questions in the field of marketing. In a paper studying oligopoly models, that is, business problems involving more than one firm, Teng and Thompson studied how firms market new products.‡ The price of the new

*Norman R. Glass et al., "Human Ecology and Educational Crises," in *Is There an Optimum Level of Population?*, Fred Singer (ed.), New York, McGraw-Hill, 1971, pp. 205–218.

†J. Steketee, and M. J. Van Der Hoek, "Thermal Recovery of the Skin after Cooling," *Journal of Physics and Medical Biology*, Vol. 24, 1979, pp. 583–592.

‡Jinn-tsair Teng and Gerald L. Thompson, "Oligopoly Models for Optimal Advertising when Production Costs Obey a Learning Curve," *Management Science*, Vol. 29, 1983, pp. 1087–1104.

product p, measured in dollars, was a function of time t, measured in months. In one market the new product sales price was found to obey the formula

$$p(t) = 100e^{.01t}$$

The object of the firm is to allow the price to increase to the point at which the increase is so great that it influences customers to purchase competitors' products. This point was reached when $p'(t) = 1.02$. Find this value of t.

Cumulative Exercise Set 15–4

1. Sketch the graph of $f(x) = e^{3-x}$.

2. If $2000 is deposited into an account that pays 10% interest compounded annually, the future amount is given by

$$A(t) = 2000(1.1)^t$$

where t is in years and $A(t)$ is in dollars. What is the future amount after 5 years?

3. Solve for x: $e^{\ln(x+2)} = \ln e^{4x}$

4. Use a calculator to show that $\ln(4 - 3)$ is not equal to $\ln 4 - \ln 3$.

In Problems 5 to 12 find $f'(x)$.

5. $f(x) = x^2 \ln (1 - x^3)$

6. $f(x) = (x + \ln 3x)^{-2}$

7. $f(x) = (\ln x)(x - \ln x)^2$

8. $f(x) = \ln(1 + 3x \ln x)$

9. $f(x) = e^{2-5x^2}$

10. $f(x) = x^2 e^{3x^2}$

11. $f(x) = 2/(3 + e^{-5x})$

12. $f(x) = (3 \ln x + e^{5x})^{-2}$

C A S E   S T U D Y   ## Denver Fire Department Deployment Policy Analysis*

In 1972 the mayor and the chief of the Fire Department of Denver, Colorado, agreed that a study of the fire department was in order in the face of mounting costs. They decided to apply modern operations analysis by contracting for a research study with faculty from the Department of Operations Research at the University of Colorado. The study was funded by HUD's Office of Policy Development and Research at a level of $140,000. The challenge given the research team was: Can the fire department provide approximately the same level of fire suppression service at a lower cost?

The most critical cost was maintenance. Denver spent in excess of $250,000 per year to staff each fire engine or ladder truck around the clock. In contrast, the cost of the vehicle itself was about $60,000, and a fire house had a one-time cost of $400,000. Thus the number of vehicles and their location accounted for the bulk of the city's expenditure for fire protection.

In the first phase of the analysis the research team used linear programming to study fire company locations. The objective function minimized cost and the constraints were a set of response time requirements. The latter included the assumptions that every vehicle was always available to respond to a fire incident, and that *alarm rates*, the number of incidents per time period, were less than or equal to a specific threshhold. The conclusion of the study was that the level of service would be unchanged with a reduction of five companies—from 44 to 39 fire stations. This reduction would take place by closing obsolete stations and building new houses at new locations.

*Adapted from D. E. Monarchi et al., "Simulation for Fire Department Deployment Policy Analysis," *Decision Sciences*, Vol. 8, 1977, pp. 211–227.

The recommendations were to take place over a period of seven years. But strong opposition, especially from the fire fighters' union, prevented immediate implementation. The union contended that if alarm rates increased dramatically, suppression service would significantly decline because of the fewer number of stations.

The second challenge for the researchers was to provide evidence supporting their conclusion. They needed to produce realistic data depicting possible future trends of alarm rates. Their approach was to use a dynamic analysis that compared the service provided by the existing fire company locations versus the proposed configuration. The term "dynamic" means that every possible fire incident situation in the city would be considered. For each situation it would be determined whether a vehicle could respond. They first had to study suppression service with the current alarm rate and then determine the effect on service if the alarm rate increased. The important question was whether the new configuration of fire stations recommended by the research team would provide the same service as the current one, both for the current alarm rate and for possible increased rates in the future.

The specific method used for this dynamic analysis was a computer simulation model developed by the New York City–Rand Institute. By using simulation they could explore the effects of increasing alarm rates on fire department performance for each configuration.

The simulation model had three parts. It created an "incident generator," which developed 1000 separate types of incidents, reflecting what, when, and where fires occur and the equipment required to service them. A random number generator was used to simulate the occurrence of the incidents, depending on the probability of occurrence assigned to each incident. The simulator then monitored the equipment—it dispatched the necessary equipment, assigned a time that the equipment would expect to be in service, and then "returned" the vehicle to await the next call.

The third part of the model compared the performance of two fire station configurations, the current one versus the proposed one. The simulation was run for various alarm rates.

One of the key measures of performance is the occurrence of long response times. If a vehicle takes longer than 5 minutes to respond, its performance is poor. For a given average alarm rate the number of responses greater than 5 minutes (for the first vehicle to respond) was measured. The data for the new configuration are shown in Figure 1. As the average alarm rate increases, the number of long responses increases. For example, for the lowest alarm rate $t = 2.5$ (or 2.5 alarms per hour), $N = 3$, meaning there were 3 long responses out of 1000. For $t = 5$, $N = 14$, so when the alarm rate doubles from $t = 2.5$ to $t = 5$, N increases more than fourfold from 3 to 14.

A key point is that N increases at a fast rate from 2.5 to 5, but then the rate of increase of N slows dramatically. That is, while N increases as t increases, the rate of increase, measured by the derivative, decreases when t approaches 5. The data suggest that the best type of function to model the

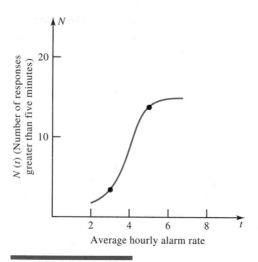

FIGURE 1

behavior of $N(t)$ is a logistic curve, whose general equation is

$$N(t) = \frac{a}{b + e^{-kt}}$$

Analysis of the data from the simulation suggests that the logistic equation of best fit for the proposed configuration is

$$N(t) = \frac{0.375}{0.013 + e^{-0.87t}}$$

This function is used to determine when $N(t)$ starts to level off. At that point, with the new configuration, the fire department could maintain service even as the average alarm rate increases. That is, even though the alarm rate might increase, the number of long responses remains almost constant. The value of t where $N(t)$ starts to level off is the point where the derivative is a maximum, that is, at the point of inflection. It is found by solving $N''(t) = 0$. To compute this number, we use the quotient rule and the chain rule.

$$N'(t) = \frac{0.326e^{-0.87t}}{(0.013 + e^{-0.87t})^2}$$

$$N''(t) = \frac{-0.284e^{-0.87t}(0.013 - e^{-0.87t})}{(0.013 + e^{-0.87t})^3}$$

When we set $N''(t) = 0$ we get $0.013 = e^{-0.87t}$, and so $t = (\ln 0.013)/(-0.87) = 5.0$.

The simulation showed that $N(t)$ leveled off from $t = 5$ to 6, as in the logistic equation model, but when the average alarm rate was increased beyond 6, $N(t)$ began to rise appreciably. Hence the value $t = 5$ is critical. If Denver's average alarm rate ever reaches it, service can be maintained until t reaches 6, but by then the city would have to implement some kind of strategy, such as building new stations, to keep the same level of service. These relatively high alarm rates occur infrequently, however, as Denver's average fire company is available to respond from its quarters about 95 percent of the time.

When the simulation was run for the current configuration the results were almost identical. The only significant discrepancy occurred when t increased beyond 6, for then N increased at a far greater rate for the new configuration than for the current one. The research team recommended that if the new configuration was implemented, then the alarm rate would have to be closely monitored. If and when it reached $t = 5$, the city still had time until it reached $t = 6$ to take further steps to maintain service, like building new stations.

In summary, the research team used linear programming to determine that the fire company's cost would be minimized, subject to the present level of service being maintained, by closing five fire stations. The team used dynamic simulation at first to test the system to see if there were any hidden elements that would cause problems of service, such as poor response time for some combination of incidents. None was discovered. Then, when their conclusions were challenged the fact that the city had been modeled in a dynamic environment gave the team a ''more believable'' presentation. The predictions of the study had more credibility because the simulation effectively replicated the historical fire suppression data of the city.

CHAPTER REVIEW

Key Terms

15–1 Exponential Functions
Exponential Function
Base

e
Learning Curve

15–2 Logarithmic Functions
Logarithmic Function with Base b
Common Logarithms
Inverse Functions

Natural Logarithmic Function
Doubling Time

15–3 Differentiation of Logarithmic Functions
Rate of Growth
Elasticity

15–4 Differentiation of Exponential Functions
Standard Probability Density Function
Logistic Curve

Summary of Important Concepts

Formulas

$$e = \lim_{x \to \infty} \left(1 + \frac{1}{x}\right)^x$$

$$D_x \ln x = \frac{1}{x}$$

$$D_x \ln g(x) = \frac{g'(x)}{g(x)}$$

$$D_x e^x = e^x$$

$$D_x e^{g(x)} = g'(x) e^{g(x)}$$

REVIEW PROBLEMS

In Problems 1 to 5 sketch the graph of the function.

1. $f(x) = 2^{x+3}$

2. $f(x) = 5 + e^{-0.5x}$

3. $f(x) = \frac{1}{\sqrt{6}} e^{-x^2/2}$

4. $f(x) = \log_{1/e}(x - 3)$

5. $f(x) = \ln(4 - x^2)$

6. Evaluate (a) $\ln e^{-6.2}$ (b) $e^{\ln 0.5}$

7. Solve for x.

$$10 = \frac{1}{e^{-x} - 2}$$

8. Evaluate

$$\lim_{h \to 0} \frac{\ln(10 + h) - \ln 10}{h}$$

9. Evaluate

$$\lim_{h \to 0} \frac{e^{2+h} - e^2}{h}$$

In Problems 10 to 12 find the derivative of the function.

10. $f(x) = 3 \ln x - \dfrac{1}{\ln x}$ 11. $f(x) = [\ln(1 + 3x)^4]^5$

12. $f(x) = x \ln |x^2 - 1|$

13. Evaluate the derivative of the function $f(x) = \ln \sqrt{2x}$ in two ways: using the formula and using the properties of logarithmic functions.

14. Find the relative extrema and the points of inflection of the function $f(x) = \dfrac{\ln x}{x}$. Sketch the graph.

In Problems 15 and 16 find the derivative of the function.

15. $f(x) = x^2 e^x - e^{-x}$ 16. $f(x) = (x^2 - e^{2x-1})^4$

17. Find the relative extrema and the points of inflection of the function $f(x) = e^x + e^{-x}$. Sketch the graph.

▩ PROGRAMMABLE CALCULATOR EXPERIMENTS

In Problems 1 to 5 sketch the graphs of the given functions in the same coordinate system. Then answer the accompanying question.

1. $y = 2^x$, $y = e^x$, $y = 3^x$; where will the graph of $y = 4^x$ appear?

2. $y = 2^{-x}$, $y = e^{-x}$, $y = 3^{-x}$; where will the graph of $y = 4^{-x}$ appear?

3. $y = e^x$, $y = e^{-x}$, $y = -e^{-x}$; where will the graph of $y = 1 - e^{-x}$ appear?

4. $y = e^x$, $y = \ln x$; where will the graph of $y = x$ appear?

5. $y = \ln x$, $y = \ln(x - 1)$, $y = -\ln(x - 1)$; where will the graph of $y = 1 - \ln(x - 1)$ appear?

The Integral

The case study connects calculus with fisheries. (H. Armstrong Roberts)

**Chapter
Overview**

Calculus is divided into two branches, differential calculus, which was discussed in the previous four chapters, and integral calculus, which we introduce in this chapter. Just as the derivative of a function $f(x)$ is a new function derived from $f(x)$, so too the indefinite integral of $f(x)$ is a function computed from $f(x)$. The process of "finding the integral" of $f(x)$ is the reverse operation of differentiation. From a geometric point of view the derivative measures the slope of the tangent line, while the definite integral measures the area bounded by $f(x)$.

In the first section we study the reverse process of differentiation called antidifferentiation. A function whose derivative is equal to $f(x)$ is called an anti-derivative of $f(x)$. Section 16–2 uses the limit process to calculate certain types of areas in the plane that are bounded by curves. Section 16–3 shows the close connection between antiderivatives and the areas studied in Section 16–2. The fourth section extends this concept of computing areas to more intricate regions.

**CASE STUDY
PREVIEW**

What happens to a stable industry when a new, innovative production technique is introduced? How can the effect on the industry be measured? One method is called consumers' surplus, which is a measure of the gain to consumers as a result of the new mode of business. The Case Study centers on the effect on the Louisiana crawfish industry when fish farms using private ponds compete with standard fisheries using public watersheds, such as lakes and rivers. We see that the consumer gains via significant price decreases at the expense of the standard fisheries.

16–1 Antiderivatives

In this section several analogs of differentiation rules are presented. Each differentiation rule shows how to start with a function and then find its derivative. This section deals with the reverse operation: start with the derivative and find a function that has it as its derivative. These new rules will be applied to the geometry of curves, the velocity of an object, and carbon-14 dating.

The differentiation rules considered in this section are the power rule, the exponential and logarithmic function rules, the sum of functions rule, and the constant multiple rule. Later sections will look at the analogs of various other differentiation rules.

Antiderivatives in Applications

There are many applications where the rate of change of a function is computed rather than computing the function directly. For instance, it is usually easier to compute the rate of increase of a population than the actual size of the population. In archaeology it is easier to estimate the rate of change of the aging process of items found at an ancient site, rather than computing the age of the site directly.

Suppose a firm computes that the marginal cost $MC(x)$ of a product is given by

$$MC(x) = x + 1$$

The marginal cost is the derivative of the cost function $C(x)$. Can $C(x)$ be computed from $MC(x)$? This is the type of question we answer in this section.

Antiderivatives

We know that the derivative of $f(x) = x^2$ is $2x$. Here we start with x^2 and compute $2x$ via the power rule. The reverse operation starts with the derivative, $2x$, and computes a function, x^2, whose derivative is $2x$. The process is called **antidifferentiation** and the function $f(x) = x^2$ is called an *antiderivative* of $2x$. In general antidifferentiation answers the question: Given a function $y = f(x)$, what is a function whose derivative is $f(x)$?

DEFINITION

Given a function $f(x)$, if $F'(x) = f(x)$, then $F(x)$ is an **antiderivative** of $f(x)$.

Example 1 illustrates not only the definition, but also why we said "an" instead of "the" antiderivative.

EXAMPLE 1

Problem

Verify that three antiderivatives of $f(x) = 3x^2$ are x^3, $x^3 + 1$, and $x^3 - 2$.

Solution The function x^3 is an antiderivative of $3x^2$ because $D_x x^3 = 3x^2$. The two other functions are also antiderivatives because $D_x(x^3 + 1) = D_x(x^3 - 2) = 3x^2$.

Example 1 uses the fact that the derivative of a constant function is 0. The converse of this is true and is a useful result. The proofs of this theorem and the next one are outlined in the exercises. Theorem 1 states that if a function has derivative 0, then it must be a constant function.

Theorem 1

If $F'(x) = 0$ for all x, then $F(x) = C$ for some real number C.

In Example 1 three functions are given whose derivatives are equal. Since the derivative measures the slope of the tangent line, the graphs of the three functions have the same slope at every point. See Figure 1.

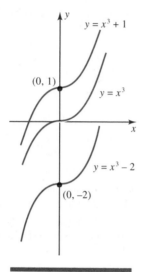

FIGURE 1

The three antiderivatives of $3x^2$ mentioned in Example 1 are similar because each is of the form x^3 plus a constant. In fact any function of the form $x^3 + C$, where C is a constant, is an antiderivative of $3x^2$.

In general if $F(x)$ is an antiderivative of $f(x)$, then $F(x) + C$ is also an antiderivative of $f(x)$. Are there other functions besides these that are antiderivatives? The next theorem shows that all antiderivatives of $f(x)$ are of the form $F(x) + C$. The constant C is called the **constant of integration.** It can be a positive number, a negative number, or zero.

Theorem 2

If $F(x)$ and $G(x)$ are both antiderivatives of $f(x)$, then there is a constant C such that

$$F(x) = G(x) + C$$

Another way to interpret this theorem is to say that any two antiderivatives of the same function differ by a constant.

The Indefinite Integral

Theorem 2 states that once one antiderivative $F(x)$ of $f(x)$ is known, then all antiderivatives are of the form $F(x) + C$. The **integral sign** $\int$ is used to express this.

$$\int f(x)\, dx = F(x) + C$$

This is called the general form of the antiderivative of $f(x)$. It states that the antiderivatives of $f(x)$ are the functions $F(x) + C$. The integral is similar to the derivative, $\dfrac{dy}{dx}$, in that it indicates an operation on the function. The meaning of the notation $\dfrac{dy}{dx}$ is to find the derivative, while the meaning of the notation $\int f(x)\, dx$ is to find all antiderivatives of $f(x)$. The term "dx" serves the same purpose in each notation. It indicates what the variable is. Thus $\int 2x\, dx$ is different from $\int 2x\, dt$. The expression $\int f(x)\, dx$ is called the **indefinite integral,** and $f(x)$ is called the **integrand.** To find the indefinite integral means to find all antiderivatives of the integrand.

Example 2 shows how to use this notation when finding the indefinite integral of another power function.

EXAMPLE 2

Problem

Find the indefinite integral $\displaystyle\int x^3\, dx$.

Solution Work backward as before. The function whose derivative is almost x^3 is x^4. But the derivative of x^4 is $4x^3$; that is, $D_x(x^4) = 4x^3$. Recall that the derivative of a constant times a function is the constant times the derivative. This means that the differentiation statement above can be divided by 4 to get

$$\frac{1}{4} D_x(x^4) = D_x\left(\frac{x^4}{4}\right) = x^3$$

Therefore

$$\int x^3\, dx = \frac{x^4}{4} + C$$

The Power Rule for Antiderivatives

Example 2 illustrates how to compute the indefinite integral of all power functions x^n, except $n = -1$. To find the function whose derivative is x^n, start with x^{n+1}. The derivative of x^{n+1} has a factor of $n + 1$, so divide the function by that factor. Thus an antiderivative of x^n is $\dfrac{x^{n+1}}{n+1}$. This is the power rule for antiderivatives. It holds for all real numbers n except -1, because $n = -1$ produces division by 0.

The Power Rule

$$\int x^n \, dx = \frac{x^{n+1}}{n+1} + C \qquad n \neq -1$$

EXAMPLE 3

Problem

Find the indefinite integrals (a) $\int x^{-4} \, dx$, (b) $\int \sqrt{x} \, dx$, (c) $\int dx$.

Solution (a) Use the power rule with $n = -4$.

$$\int x^{-4} \, dx = \frac{x^{-4+1}}{-4+1} + C = \frac{x^{-3}}{-3} + C = -\frac{1}{3} x^{-3} + C$$

(b) Use the power rule with $n = 1/2$.

$$\int x^{1/2} \, dx = \frac{x^{(1/2)+1}}{1/2+1} + C = \frac{x^{3/2}}{3/2} + C = \frac{2x^{3/2}}{3} + C$$

(c) The integrand is understood to be $1 = x^0$. From the power rule with $n = 0$ we have

$$\int dx = \int x^0 \, dx = \frac{x^{0+1}}{0+1} + C = x + C$$

Exponential and Logarithmic Functions

The derivative rules for exponential and logarithmic functions can also be expressed as indefinite integrals. Recall the rules

$$D_x(\ln |x|) = \frac{1}{x} \qquad D_x e^{rx} = re^{rx}$$

The power rule for antiderivatives excludes $n = -1$, so it does not give the antiderivatives of x^{-1}. However, since $D_x(\ln x) = 1/x = x^{-1}$, the general form of the antiderivative of x^{-1} is expressed in the logarithmic function rule.

Logarithmic Function Rule

Antiderivative of $f(x) = \frac{1}{x} = x^{-1}$:

$$\int x^{-1} \, dx = \ln |x| + C$$

From the formula $D_x(e^{rx}) = re^{rx}$ divide by r to get the exponential function rule.

> **Exponential Function Rule**
> Antiderivative of $f(x) = e^{rx}$:
>
> $$\int e^{rx}\, dx = \frac{e^{rx}}{r} + C$$

EXAMPLE 4

Problem

Find the indefinite integral of $f(x) = e^{5x}$.

Solution From the antiderivative rule of $f(x) = e^{rx}$ with $r = 5$ we have

$$\int e^{5x}\, dx = \frac{e^{5x}}{5} + C$$

Addition and Constant Rules

The first three antidifferentiation rules presented so far can be derived by working backward from their corresponding differentiation rules. In a similar way the formulas that state (a) "the derivative of a sum is the sum of the derivatives" and (b) "the derivative of a constant times a function is the constant times the derivative of the function" can be expressed as indefinite integrals.

> **The Sum Rule**
>
> $$\int [f(x) + g(x)]\, dx = \int f(x)\, dx + \int g(x)\, dx$$

> **The Constant Times a Function Rule**
>
> $$\int kf(x)\, dx = k \int f(x)\, dx \qquad \text{for any constant } k$$

A function with more than one term can be antidifferentiated term by term. It can be extended to a finite number of functions, so that

$$\int [f_1(x) + \cdots + f_n(x)]\, dx = \int f_1(x)\, dx + \cdots + \int f_n(x)\, dx$$

The constant times a function rule states that a constant can be moved through the integral sign in the same way that it can be moved through a derivative sign. The next two examples illustrate these rules.

E X A M P L E 5

Problem

Find the indefinite integrals (a) $\int (x^4 + e^{-x}) \, dx$, (b) $\int \left(3x^{-5} - \dfrac{4}{x} \right) dx$.

Solution

(a) $\displaystyle\int (x^4 + e^{-x}) \, dx = \int x^4 \, dx + \int e^{-x} \, dx = \frac{x^{4+1}}{4+1} + \frac{e^{-x}}{-1} + C$

$\displaystyle\qquad = \frac{x^5}{5} - e^{-x} + C$

(b) $\displaystyle\int \left(3x^{-5} - \frac{4}{x} \right) dx = \int 3x^{-5} \, dx + \int \left(\frac{-4}{x} \right) dx$

$\displaystyle\qquad = 3 \int x^{-5} \, dx + (-4) \int \left(\frac{1}{x} \right) dx$

$\displaystyle\qquad = \frac{3x^{(-5)+1}}{-5+1} - 4 \int \frac{1}{x} \, dx + C$

$\displaystyle\qquad = \frac{-3x^{-4}}{4} - 4 \ln |x| + C$

Sometimes algebraic manipulations can be used to express the integrand in a form that matches the antidifferentiation rules. The next example demonstrates one such manipulation.

E X A M P L E 6

Problem

Find the indefinite integral $\int \sqrt{x}(x^2 - 1) \, dx$.

Solution The integrand does not match any of those in the antidifferentiation rules, so an algebraic simplification is called for. Multiply through the parentheses to get

$\displaystyle\int \sqrt{x}(x^2 - 1) \, dx = \int (\sqrt{x}\, x^2 - \sqrt{x}) \, dx = \int (x^{5/2} - x^{1/2}) \, dx$

$\displaystyle\qquad = \int (x^{5/2}) \, dx - \int (x^{1/2}) \, dx = \frac{x^{7/2}}{7/2} - \frac{x^{3/2}}{3/2} + C$

$\displaystyle\qquad = \frac{2x^{7/2}}{7} - \frac{2x^{3/2}}{3} + C$

EXAMPLE 7

Problem

Find the function whose slope of the tangent line at x is $f'(x) = 3x^2 - 1$ and whose graph passes through the point $(0, 1)$.

Solution The first step is to find the indefinite integral of $f'(x)$. Use the fact that $f(x)$ is the antiderivative of $f'(x)$.

$$f(x) = \int f'(x)\, dx = \int (3x^2 - 1)\, dx = x^3 - x + C$$

This gives all functions whose slope is the given formula. To find which of these functions also passes through $(0, 1)$, substitute $f(0) = 1$ into the formula.

$$1 = f(0) = 0^3 - 0 + C = C$$

The function whose constant of integration is $C = 1$ is the solution, which is $f(x) = x^3 - x + 1$.

Recall that a production function $P(x)$ measures the production level P, or the output, of a firm using x amount of labor. The derivative of $P(x)$ is called marginal product, $MP(x)$. Antiderivatives can be used to compute the production function from the marginal product.

EXAMPLE 8

Problem

A firm determines that its marginal product $MP(x)$ is given by

$$MP(x) = x^2 - 10x + 20$$

where $MP(x)$ is measured in thousands of dollars and x in hundreds of workers. Compute the production function.

Solution Since $P'(x) = MP(x)$, we have

$$P(x) = \int MP(x)\, dx$$

$$= \int (x^2 - 10x + 20)\, dx = \frac{x^3}{3} - 5x^2 + 20x + C$$

Similar statements hold for marginal cost, marginal revenue, and marginal profit. If $R(x)$ is the revenue function of a firm, then $R'(x)$ is the marginal revenue. $R(x)$ can be computed from marginal revenue by the formula

$$R(x) = \int R'(x)\, dx$$

The Growth Equation

Many interesting applications of calculus lead to equations involving derivatives called **differential equations.** In fact whenever an antiderivative is computed we are solving a differential equation of the form $f'(x) = g(x)$, where $g(x)$ is given and $f(x)$ is unknown. A common and powerful type of differential equation is

$$f'(x) = kf(x)$$

for some constant k. It is called the **growth equation.** If k is positive, then k is called the *growth constant*. If k is negative, then k is called the *decay constant*. The next two examples illustrate an application from archaeology, where the growth equation plays a key role.

The rate at which radioactive material changes to lead is proportional to the amount of the material present; that is, if $f(t)$ is the amount present, then $f'(t) = kf(t)$ for some constant k. Hence $f(t)$ is a function whose derivative is a constant times itself. The only type of function with that property is $f(t) = Ce^{kt}$. To check that $f(t)$ satisfies the differential equation, we compute $f'(t) = Cke^{kt}$, which is equal to $kf(t)$ as needed.

EXAMPLE 9

Living plant and animal tissues contain the radioactive substance carbon-14, which has a half-life of 5568 years (at the end of that time one-half of that amount remains). When the organism dies, no new carbon is received.

Problem

Find a formula for the amount of carbon-14 remaining after t years.

Solution From above, $f(t) = Ce^{kt}$. Since $f(0) = Ce^0 = C$, the amount present at time $t = 0$ is $C = f(0)$. When $t = 5568$, $f(0)/2$ is present, and thus

$$f(5568) = f(0)e^{5568k} = \frac{f(0)}{2}$$

Therefore $e^{5568k} = 1/2$, and so $\ln e^{5568k} = \ln 1/2$, which gives $5568k = \ln 1/2$ and $k = (1/5568)\ln 1/2$. Therefore

$$f(t) = Ce^{(1/5568)(\ln 1/2)t}$$

Example 10 illustrates one way that archaeologists use the function in Example 9 to date uncovered ruins.

EXAMPLE 10

Suppose an analysis of ashes found in an unearthed settlement reveals that 1/5 of the carbon-14 present in the original ashes has decomposed, so that 4/5 of $f(0)$ remains.

Problem

How old is the settlement?

Solution The task is to find t such that $f(t) = (4/5)f(0)$. Then

$$(4/5)f(0) = f(0)e^{kt} \qquad \text{implies} \qquad 4/5 = e^{kt}$$

Hence $\ln 4/5 = \ln e^{kt} = kt$. Thus

$$t = \left(\frac{1}{k}\right) \ln 4/5 = \frac{5568(\ln 4/5)}{\ln 1/2} = \frac{5568(\ln 4 - \ln 5)}{\ln 1 - \ln 2}$$

$$\approx \frac{5568(1.386 - 1.609)}{(0 - 0.693)}$$

$$\approx 1792$$

Thus the settlement is approximately 1792 years old.

EXERCISE SET 16–1

In Problems 1 to 24 find the antiderivative.

1. $\int (3x + 5)\, dx$

2. $\int (4x - 2)\, dx$

3. $\int (2x^4 - 7)\, dx$

4. $\int (3x^5 + 1)\, dx$

5. $\int (6x^{-2} + 5x)\, dx$

6. $\int (5x^{-3} - x^3 + 1)\, dx$

7. $\int (2x^{1/2} + 3x^{1/3} - 5x)\, dx$ $3, 4, 2, 5$

8. $\int (5x^{1/5} - x^{2/3} + 7x)\, dx$

9. $\int (3x^{-1/3} + 7x^{-1/4} - 5)\, dx$ $3, 42, 5$

10. $\int (6x^{-1/6} - 3x^{-3/7} - 6x)\, dx$

11. $\int (x^{0.2} + 3x^{1.4})\, dx$

12. $\int (1.2x^{-1.6} - 4.1x^{-2.2})\, dx$

13. $\int (1/x^2 + 5/x^4)\, dx$

14. $\int (2/x^{0.5} - 4/x^{2.2})\, dx$

15. $\int (1/x^7 + 4/x^{1/4} + 4x^{1/4})\, dx$

16. $\int (3/x^{2.2} - 3/x^{2/5} - 2x^{2/5})\, dx$

17. $\int (3e^{2x} + 5/x)\, dx$

18. $\int (5e^{3x} - 2/x)\, dx$

19. $\int (x^3 + 2e^{-x} + x^{1/2} - 5/x)\, dx$

20. $\int (3e^{-4x} - 4x^{-3} - 15/x + x^{-1/2})\, dx$

21. $\int (5\sqrt{x} - 2e^{-4x} + 1/x)\, dx$

22. $\int (8/x + 2/\sqrt{7x} + 3e^{-2x})\, dx$

23. $\int (18/x + 2/\sqrt[3]{4x} - e^{7x})\, dx$ $3, 4$

24. $\int (8/\sqrt[4]{3x} + 2/x - e^{-10x})\, dx$

In Problems 25 to 28 find the function whose slope of the tangent line to the curve $y = f(x)$ is given and whose graph passes through the given point.

25. $f'(x) = x^2 - 2$ at $(1, -1)$

26. $f'(x) = 2x^2 - 1$ at $(1, 1)$

27. $f'(x) = 4x^{-1} - 2x$ at $(2, -2)$

28. $f'(x) = 2x^{-2} - x$ at $(1, 1)$

In Problems 29 to 34 find $\int f(x)\, dx$ by using algebra to express the integrand in a form that fits the integration rules.

29. $\int (2x^4 - 7x)/x^2\, dx$

30. $\int (3x^5 + 10x)/x^3\, dx$

31. $\int x^2(6x^{-2} + 5x)\, dx$

32. $\int x^3(5x^{-3} - x^3 + 1)\, dx$

33. $\int (2x^4 - 7x)^2\, dx$

34. $\int (3x^5 + 10x)^2\, dx$

In Problems 35 and 36 find the position function $s(t)$ given the velocity function $v(t)$ and the initial position $s(0)$.

35. $v(t) = t^2 + 2t$, $s(0) = 0$

36. $v(t) = t^3 + t$, $s(1) = 1$

37. Find the age of a settlement where ashes found at the site are assumed to be 2/3 of the original amount; that is, 1/3 of the original amount has decomposed.

38. Do Problem 37 with the fraction 2/3 replaced by 3/4.

In Problems 39 to 42 determine the decay constant k so that the function $f(x) = Ce^{kx}$ is a solution to the differential equation $f'(x) = kf(x)$ where $f(x)$ is the amount of a radioactive substance remaining after x years and where the half-life of the substance is given.

39. The half-life is 1000 years.

40. The half-life is 100 years.

41. The half-life is 10 years.

42. The half-life is 1 year.

43. Give a reason for each step in the following proof of Theorem 1:
    Proof of Theorem 1: Suppose $F'(x) = 0$ for all x. Then the graph of $F(x)$ has slope 0 at every point. Thus the tangent line is horizontal at every point. This means that the graph of $F(x)$ is a horizontal line, since a horizontal line is the only graph with that property.

44. Give a reason for each step in the following proof of Theorem 2:
    Proof of Theorem 2: Suppose $F(x)$ and $G(x)$ are two antiderivatives of $f(x)$. We want to show that the function $F(x) - G(x)$ is a constant function, since then $F(x) - G(x) = C$ for some C and so $F(x) = G(x) + C$. Consider the function $H(x) = F(x) - G(x)$. Its derivative is
    $$H'(x) = F'(x) - G'(x) = f(x) - f(x) = 0$$
    Therefore $H(x)$ must be a constant function by Theorem 1. Hence $F(x) - G(x) = C$ and so $F(x) = G(x) + C$.

Referenced Exercise Set 16–1

1. In order to illustrate the application of dating methods in archaeology covered in his book, Joseph Michels discusses one of the archaeological expeditions he organized with students from the Pennsylvania State University.* The site was Sheep Rock Shelter in Huntingdon County, PA. For thousands of years, the shelter, located on the Juniata River, provided small platforms for various living mammals. Artifacts were found in the rock strata as deep as 14 feet below the present surface. In one level it was determined that of the original amount of carbon-14 contained in certain artifacts, 42% remained. Find the age of the artifacts.

In Problems 2 and 3 an article is quoted that discusses growth rates. Find the growth equation, $f'(x) = kf(x)$ that is satisfied by the function $f(x)$ given in the article.

2. In an article studying the causes and effects of wind damage to trees such as sugar maples, King investigates the factor influencing wind damage due to the rapid decline in wind speed beneath the canopy, or top, of the forest.† The relationship between the wind speed $V(h)$ and the height h of the wind is given by
   $$V(h) = 22e^{1.4h}$$

3. Because the bald eagle is an endangered species, the study of the growth rates of sibling eagles is very important to conservationists. In an article investigating the growth rates of nestlings, or broods, with one versus more than one young eagle, Bortolotti calculates the following growth equation that measures the difference in the size of two young eagles in the same nestling, born one day apart.‡ Let m be the difference in the mass, measured in percent, at time t, measured in days. The solution of the growth equation is
   $$m(t) = 38.1e^{-.068t}$$

*Joseph W. Michels, *Dating Methods in Archaeology*, New York, Seminar Press, 1973.

†David A. King, "Tree Form, Height Growth, and Susceptibility to Wind Damage in *Acer saccharum*," *Ecology*, Vol. 67, 1986, pp. 980–990.

‡Gary R. Bortolotti, "Evolution of Growth Rates in Eagles: Sibling Competition vs. Energy Considerations," *Ecology*, Vol. 67, 1986, pp. 182–194.

16–2 Area and the Definite Integral

One application of the definite integral is the measure of the area under a curve, which appears in many real-world problems. Areas of certain types of regions can be found using only geometry, such as squares, rectangles, and triangles. The classical Greeks even knew how to find the area of some regions bounded by specific curves. But the general problem of finding the area under an arbitrary continuous curve was not solved until about 1700 with the advent of calculus.

An Application from Economics

The area under a curve has a wide variety of uses in many disciplines, one of which is economics. One application analyzes the distribution of wealth and income in a society. For centuries economists, politicians, and social scientists have debated the question, "What constitutes a fair and equitable distribution of income and wealth?" Wealth is the value of the goods and property people own, while income is the gain derived from the use of human and material resources. For measuring inequalities in the distribution of income and wealth the most commonly used device is a graph called a **Lorentz diagram.** A Lorentz diagram illustrates what percentage of families received what percentage of a nation's total income. It is used to compare the income distribution of different countries as well as to evaluate an individual country's system.

Consider the Lorentz diagram for families in the United States in Figure 1, which is adapted from *Contemporary Economics* by Milton H. Spenser.* There are two curves in a Lorentz diagram. The lower one, called a *Lorentz curve,* measures the percentage of a nation's total income that is shared by those families with a certain income level. As an example of how to interpret the curve, consider $x = 40$ on the horizontal axis. It denotes those families with the 40% lowest family incomes. The vertical axis represents percentages of the total income of the United

*Milton H. Spenser, *Contemporary Economics,* New York, Worth Publishers, Inc., 1986, pp. 53–55.

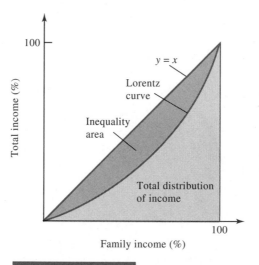

FIGURE 1 *Source:* Milton H. Spenser, *Contemporary Economics,* New York, Worth Publishers, Inc., 1986, pp. 53–55.

States. Corresponding to $x = 40$ on the Lorentz curve is $y = 16$, which means that the families with the lowest 40% of family incomes have 16% of the total income of the United States. The second curve in the diagram is $y = x$. It represents complete equality of distribution of income, because if income were distributed equally, families with the lowest x% of income would have $y = x$% of the total income. Economists call the area between these two curves the *inequality area* because it represents the extent of departure between an equal distribution of income and the actual distribution of income. The area of the region bounded above by the Lorentz curve and below by the x-axis is called the total distribution of income. The area of the region bounded above by $y = x$ and below by the x-axis is the area of complete equality of income distribution.

In this section we study how to compute the latter two areas. Each is the area of a region bounded above by a curve and below by the x-axis. Section 16–4 will show how to evaluate the area between two curves. Then we will compute the inequality income area for the United States.

Area Under a Curve

To find the area under a curve, we approximate the area by using rectangles whose area can be easily computed. The process entails fitting rectangles into the region. The first approximation is often not very close to the actual area. Better approximations are usually found by using more rectangles. The actual area is found by using a limit process.

We start by finding an initial approximation to the area under the curve $f(x) = x^2$ bounded by $x = 0$ to $x = 2$ and above the x-axis. Figure 2 shows this region. A rough approximation uses two rectangles as in Figure 3. Divide the interval $[0, 2]$ into two subintervals of equal width and choose the right-hand end point to calculate the height of each rectangle. Each width is 1 and the two heights are $f(1) = 1^2 = 1$ and $f(2) = 2^2 = 4$. Thus the sum of the areas of the two

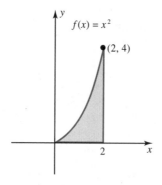

FIGURE 2

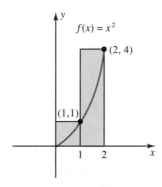

FIGURE 3

rectangles is

$$1 \cdot f(1) + 1 \cdot f(2) = 1 + 4 = 5$$

Figure 3 shows that this approximation is greater than the actual area, but it illustrates how to compute further approximations that are closer. To approximate the area of the region bounded by $y = f(x)$ (for now we must assume that $f(x) \geq 0$), the x-axis, $x = a$, and $x = b$, first divide the interval $[a, b]$ into n equal subintervals, each with width $(b - a)/n$. Evaluate $f(x)$ at the right-hand end point to find the height of each rectangle. Then the approximation is the sum of the areas of these rectangles. Example 1 illustrates this procedure.

EXAMPLE 1

Problem

Find the approximation of the area of the region bounded by $f(x) = x^2$, $x = 0$, $x = 2$, and the x-axis, using four subintervals.

Solution The width of each subinterval is $(2 - 0)/4 = 1/2$. The right-hand end points are $1/2$, 1, $3/2$, and 2. The approximation is the sum of the areas of the rectangles. (See Figure 4.)

$$\frac{1}{2} \cdot f(1/2) + \frac{1}{2} \cdot f(1) + \frac{1}{2} \cdot f(3/2) + \frac{1}{2} \cdot f(2)$$

$$= \frac{1}{2} \cdot \frac{1}{4} + \frac{1}{2} \cdot 1 + \frac{1}{2} \cdot \frac{9}{4} + \frac{1}{2} \cdot 4 = \frac{1}{2}\left(\frac{1}{4} + 1 + \frac{9}{4} + 4\right)$$

$$= \left(\frac{1}{2}\right)\left(\frac{30}{4}\right) = \frac{15}{4}$$

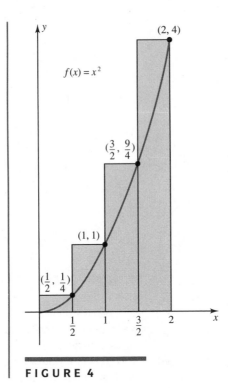

FIGURE 4

Area and the Definite Integral

From Figure 4 it is clear that the approximation of 15/4 is still greater than the area, but it is closer than the previous approximation of 5. To get even closer approximations, use more subintervals with smaller widths. In general the process of approximating the area entails dividing $[0, 2]$ into n subintervals and computing the sum of the areas of the rectangles. Then we let n get large to see if the process has a limiting value. Each subinterval has width $(2 - 0)/n = 2/n$. The right-hand end points are $2/n, 4/n, \ldots , 2$. For example, if $n = 8$, the end points are 1/4, 1/2, 3/4, 1, 5/4, 3/2, 7/4, 2. A computer was used to find the approximations to the area for the following values of n:

n	2	4	8	100	1000	10000	100000
area	5	3.75	3.188	2.707	2.671	2.667	2.6667

The areas of the approximations approach the number 8/3. In fact the larger the number of rectangles that are used, the closer the approximation is to 8/3. This means that as n increases without bound, the limit is equal to 8/3. This is expressed as

$$\lim_{n \to \infty} (\text{sum of the areas of the } n \text{ rectangles}) = \frac{8}{3}$$

Therefore the area of the region bounded by $f(x) = x^2$, the x-axis, $x = 0$, and $x = 2$ is equal to 8/3. This is what is meant by the **definite integral** of $f(x) = x^2$ from $x = 0$ to $x = 2$. It is written as

$$\int_0^2 x^2 \, dx = \frac{8}{3}$$

The function $f(x) = x^2$ is called the *integrand*. The number 2 is called the **upper limit of integration** and the number 0 is called the **lower limit of integration.**

For any continuous function $f(x)$, with $f(x) \geq 0$ in $[a, b]$, as the number n of subintervals of $[a, b]$ gets larger, the corresponding approximations get closer to the area under the curve. This can be seen from Figure 5. The limit of these approximations is the definite integral of $f(x)$ from a to b.

DEFINITION

If $f(x)$ is continuous on the interval $[a, b]$ and $[a, b]$ is divided into n subintervals whose right-hand end points are $x_1, x_2, \ldots, x_n$, then the *definite integral of $f(x)$ from a to b* is

$$\int_a^b f(x) \, dx = \lim_{n \to \infty} \frac{(b - a)}{n} [f(x_1) + \cdots + f(x_n)]$$

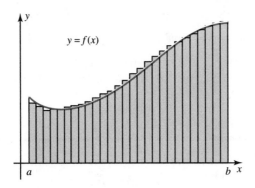

FIGURE 5

The next example shows how this limit behaves with a linear function as the integrand. We use the fact that

$$1 + 2 + \cdots + n = \frac{n(n + 1)}{2}$$

E X A M P L E   2

Problem

Find $\int_0^4 2x \, dx$.

Solution Divide $[0, 4]$ into n subintervals, each of width $4/n$. The right-hand end points are $4/n$, $8/n$, . . . , 4. From the definition with $a = 0$ and $b = 4$ we have

$$\int_0^4 2x \, dx = \lim_{n \to \infty} \frac{(b - a)}{n} [f(x_1) + \cdots + f(x_n)]$$

$$= \lim_{n \to \infty} \frac{(4 - 0)}{n} \left[f\left(\frac{4}{n}\right) + f\left(\frac{8}{n}\right) + \cdots + f(4) \right]$$

$$= \lim_{n \to \infty} \frac{4}{n} \left[\frac{8}{n} + \frac{16}{n} + \cdots + 8 \right] = \lim_{n \to \infty} \frac{4}{n} \left(\frac{8}{n}\right) [1 + 2 + \cdots + n]$$

$$= \lim_{n \to \infty} \frac{32}{n^2} \left[\frac{n(n + 1)}{2} \right] = \lim_{n \to \infty} \frac{16(n + 1)}{n} = \lim_{n \to \infty} \left(16 + \frac{16}{n} \right) = 16$$

Thus $\int_0^4 2x \, dx = 16$.

The definite integral is defined in terms of a limit process. The definition does not stipulate that the function $f(x)$ be nonnegative in the interval, but when the definite integral is linked with the area under a curve this condition is important. When $f(x) \geq 0$ in $[a, b]$, then the area of the region bounded by $f(x)$, the x-axis, $x = a$, and $x = b$ is equal to the definite integral, $\int_a^b f(x) \, dx$. But if $f(x)$ is less than 0 in the interval, so that its graph is below the x-axis somewhere in the interval, then the two are not equal. The next section examines this situation.

D E F I N I T I O N

If $f(x) \geq 0$ in $[a, b]$, then the area bounded by $f(x)$, the x-axis, $x = a$, and $x = b$ is defined by

$$\int_a^b f(x) \, dx$$

A more general definition of the definite integral can be used that does not restrict the subintervals to have equal length. Also, the height of each rectangle can be computed by using any point in the subinterval, rather than the right-hand end point. The definition given here is equivalent to the less restrictive one and is easier to use.

EXAMPLE 3

Problem

Compute $\int_0^1 (3x + 1)\, dx$.

Solution Divide $[0, 1]$ into n subintervals, each of width $1/n$. The right-hand end points are $1/n, 2/n, \ldots, 1$. From the definition with $a = 0$ and $b = 1$ we have

$$\int_0^1 (3x + 1)\, dx = \lim_{n \to \infty} \frac{(b - a)}{n} [f(x_1) + \cdots + f(x_n)]$$

$$= \lim_{n \to \infty} \frac{(1 - 0)}{n} \left[f\left(\frac{1}{n}\right) + f\left(\frac{2}{n}\right) + \cdots + f(1) \right]$$

$$= \lim_{n \to \infty} \frac{1}{n} \left[3\left(\frac{1}{n}\right) + 1 + 3\left(\frac{2}{n}\right) + 1 + \cdots + 3(1) + 1 \right]$$

The term "1" appears n times and we can factor 3 and $1/n$ from each of the remaining terms to get

$$= \lim_{n \to \infty} \frac{1}{n} \left[\frac{3}{n}(1 + 2 + \cdots + n) + n \right]$$

Now use the identity for $1 + 2 + \cdots + n$.

$$= \lim_{n \to \infty} \frac{1}{n} \left[\frac{3}{n} \frac{n(n + 1)}{2} + n \right] = \lim_{n \to \infty} \left[\frac{3(n + 1)}{2n} + \frac{n}{n} \right]$$

$$= \lim_{n \to \infty} \left[\frac{3}{2} \frac{(n + 1)}{n} + 1 \right] = \frac{3}{2} + 1 = \frac{5}{2}$$

Thus

$$\int_0^1 (3x + 1)\, dx = \frac{5}{2}$$

Lorentz Diagrams Revisited

The area of complete equality of income distribution in a Lorentz diagram is the area of the region bounded by $y = x$ and the x-axis. It is the definite integral

$$\int_0^1 x\, dx$$

We use the definition to compute this area. Divide $[0, 1]$ into n subintervals, each of width $1/n$. The right-hand end points are $1/n, 2/n, \ldots, 1$. With $a = 0$, $b = 1$, and $f(x) = x$, we have

$$\int_0^1 x \, dx = \lim_{n \to \infty} \frac{(b - a)}{n} [f(x_1) + \cdots + f(x_n)]$$

$$= \lim_{n \to \infty} \frac{(1 - 0)}{n} \left[f\left(\frac{1}{n}\right) + f\left(\frac{2}{n}\right) + \cdots + f(1) \right]$$

$$= \lim_{n \to \infty} \left(\frac{1}{n}\right) \left[\frac{1}{n} + \frac{2}{n} + \cdots + 1 \right]$$

Factoring $1/n$ from each term, we get

$$= \lim_{n \to \infty} \left(\frac{1}{n^2}\right) [1 + 2 + \cdots + n]$$

Now use the identity for $1 + 2 + \cdots + n$.

$$= \lim_{n \to \infty} \left(\frac{1}{n^2}\right) \left[\frac{n(n + 1)}{2} \right] = \lim_{n \to \infty} \frac{n + 1}{2n} = \frac{1}{2}$$

Thus

$$\int_0^1 x \, dx = \frac{1}{2}$$

This is the area of complete equality of income distribution. A country's actual income distribution is compared with it. For income in the United States the Lorentz curve given in Spenser can be approximated by the curve $y = x^2$. (See Figure 1.) The area of the region bounded by this curve and the x-axis is called the area of total income distribution. For this Lorentz curve it is

$$\int_0^1 x^2 \, dx$$

In the next section we will learn how to compute this area from the fundamental theorem of calculus.

A Few Concluding Remarks

The introduction of this section states that the ancient Greeks knew how to find the area under a curve for several specific curves. One such type of curve is a straight line, the graph of a linear function. This is because the area under a linear function can be divided into triangles and rectangles whose areas are easily computed. For instance, Example 3 shows that the definite integral of the linear function $f(x) = 3x + 1$ from $x = 0$ to $x = 1$ is $5/2$. In the interval $[0, 1]$ the function $f(x) = 3x + 1 > 0$, so the value of the definite integral should be equal to the area under the curve. From Figure 6 the area under the curve can be divided into a square with area 1 and a triangle with area $(1/2)(1)(3) = 3/2$. So the area under the curve

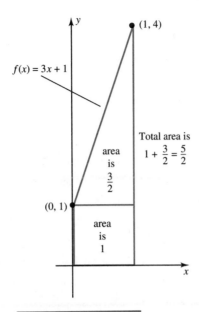

FIGURE 6

is $1 + 3/2 = 5/2$, which is the same as the definite integral computed in the example.

In a similar way consider the Lorentz diagram in Figure 1: the area of total income distribution is the area of the region bounded by the linear function $y = x$. This region is a triangle and its area is $(1/2)(1)(1) = 1/2$, which is the same number computed from the definition. The definition of area is much more difficult to apply for nonlinear functions. The next section will give us an easier way to compute definite integrals.

The notation for the indefinite integral and the definite integral are similar and it is easy to confuse them. In the next section we will discover the close connection between the two concepts—basically the indefinite integral can be used to evaluate the definite integral. Remember that the definite integral $\int_a^b f(x)\,dx$ is a real number, whereas the indefinite integral $\int f(x)\,dx$, whose notation does not include limits of integration, is the set of functions consisting of all antiderivatives, $F(x) + C$ for $F(x)$ a particular antiderivative of $f(x)$.

EXERCISE SET 16–2

In Problems 1 to 6 let $f(x) = x^2$. For the given choice of n, use n rectangles to approximate the area under the curve $y = f(x)$ from $x = a$ to $x = b$.

1. $n = 2$, $a = 1$, $b = 2$

2. $n = 2$, $a = 1$, $b = 4$

3. $n = 4$, $a = 1$, $b = 2$

4. $n = 4$, $a = 1$, $b = 4$

5. $n = 6$, $a = 0$, $b = 3$

6. $n = 8$, $a = 0$, $b = 4$

In Problems 7 to 12 fill in the following table for the given choice of $f(x)$ by dividing $[0, 2]$ into n subintervals and computing the sum of the areas of the rectangles.

n	2	4	8	10
area				

7. $f(x) = 2x + 1$

8. $f(x) = 3x + 1$

9. $f(x) = x^2 + 1$

10. $f(x) = x^2 + 2$

11. $f(x) = 2x^2 + 2$

12. $f(x) = 2 + x^2$

In Problems 13 to 16 use the definition to find the definite integral.

13. $\int_0^4 3x \, dx$

14. $\int_0^8 2x \, dx$

15. $\int_0^4 (2x + 1) \, dx$

16. $\int_0^8 (3x + 1) \, dx$

In Problems 17 to 20 find the area under $f(x)$ from $x = a$ to $x = b$.

17. $f(x) = 2x + 1$, $a = 0$, $b = 2$

18. $f(x) = 3x - 1$, $a = 0$, $b = 1$

19. $f(x) = 3 - 2x$, $a = 0$, $b = 1$

20. $f(x) = 7 - 3x$, $a = 0$, $b = 2$

In Problems 21 to 24 use the definition to find the definite integral.

21. $\int_3^4 3x \, dx$

22. $\int_6^8 2x \, dx$

23. $\int_{-2}^0 (1 - 2x) \, dx$

24. $\int_7^8 (3x - 1) \, dx$

In Problems 25 to 32 use the definition to find the definite integral by using the identity

$$1^2 + 2^2 + \cdots + n^2 = \frac{n(n + 1)(2n + 1)}{6}$$

25. $\int_0^4 x^2 \, dx$

26. $\int_{-2}^2 2x^2 \, dx$

27. $\int_0^4 (2x^2 + 1) \, dx$

28. $\int_1^5 (3x^2 - 1) \, dx$

29. $\int_0^1 (x^2 + x) \, dx$

30. $\int_0^1 (3x^2 + x) \, dx$

31. $\int_1^3 \frac{1}{x} \, dx$

32. $\int_1^4 \frac{2}{x} \, dx$

33. Use the definition to show that $\int_a^b c \, dx = c(b - a)$ for the constant function $f(x) = c$.

34. Verify the result of Example 3 geometrically.

35. In Problem 23 verify that the definite integral is the area under the curve by finding the area geometrically.

Referenced Exercise Set 16–2

1. In an article studying the "well-being" of farm families, Kinsey* quotes one of the original articles† in economics that applied Lorentz curves to income inequality. The latter paper by Carlin and Reinsel constructed a Lorentz curve for the family income of farmers. It can be approximated by

$$y = \frac{x^2}{3} + \frac{2x}{3}.$$

Graph this function using the facts that it passes through $(0, 0)$ and $(1, 1)$ and is similar in shape to $y = x^2$. Shade the inequality area.

2. In an article studying how farmland prices are determined, Brown and Brown discuss the possible options for a prospective seller of a tract of farmland.‡ They define $f(x)$ to be the likelihood that the seller will receive a bid for the land if the price is x thousand dollars. For a particular tract, $f(x) = 1/500$. They define the likelihood that the seller will receive a bid between \$1,900 and \$2,000 to be

$$\int_{1900}^{2000} \frac{1}{500} \, dx$$

Calculate this integral.

*Jean Kinsey, "Measuring the Well-being of Farm Households: Farm, Off-farm and In-kind Sources of Income: Discussion," *American Journal of Agricultural Economics,* Vol. 67, 1985, pp. 1105–1107.

†Thomas A. Carlin and Edward I. Reinsel, "Combining Income and Wealth—An Analysis of Farm Family 'Well-being'," *American Journal of Agricultural Economics,* Vol. 55, 1973, pp. 38–46.

‡Keith C. Brown and Deborah J. Brown, "Heterogenous Expectations and Farmland Prices," *American Journal of Agricultural Economics,* Vol. 66, 1984, pp. 164–169.

Cumulative Exercise Set 16–2

In Problems 1 to 3 find the antiderivative.

1. $\int \left(4x - \dfrac{6}{x^2} \right) dx$ 2. $\int (\sqrt{x} + 7) dx$

3. $\int \left(4e^x - \dfrac{3}{x} \right) dx$

4. Find the function whose slope of the tangent line to the curve is $f'(x) = 1 - 2x + 3x^2$ and whose graph passes through the point $(2, 7)$.

5. If the velocity function of an object is given by

$$v(t) = \frac{4t}{t^2 + 1}$$

what is the distance function $s(t)$ if $s(0) = 4$?

6. (a) Use four rectangles to approximate the area under the curve $y = x^2 + 1$ from $x = 0$ to $x = 1$.

   (b) Use the definition to find the area under the curve $y = x^2 + 1$ from $x = 0$ to $x = 1$.

7. Use four rectangles to approximate the area under the curve $y = e^x$ from $x = 0$ to $x = 1$.

8. Use the definition to evaluate the integral

$$\int_{-1}^{0} (-2x) dx$$

16–3   The Fundamental Theorem of Calculus

The definition of the definite integral involves a somewhat complicated limit process. It would be cumbersome if the definition were necessary to compute each definite integral. However, the **fundamental theorem of calculus** eases the burden by tying together the two branches of calculus, integral and differential calculus. This is done by using antiderivatives, or indefinite integrals, to calculate definite integrals.

The Fundamental Theorem of Calculus

We first present the theorem with an example. Its proof will follow.

The Fundamental Theorem of Calculus

Let $f(x)$ be continuous on the interval $[a, b]$ and let $F(x)$ be any antiderivative of $f(x)$. Then

$$\int_a^b f(x) \, dx = F(b) - F(a)$$

To represent the expression $F(b) - F(a)$ in a more compact form, we use the symbol $F(x) \Big|_a^b$. Thus the fundamental theorem states

$$\int_a^b f(x) \, dx = F(x) \Big|_a^b = F(b) - F(a)$$

Example 1 illustrates how to use the theorem.

E X A M P L E   1

Problem

Use the fundamental theorem of calculus to find the definite integrals

(a) $\int_2^4 2x\, dx$, (b) $\int_{-1}^2 (x^2 + 4x^3)\, dx$.

Solution (a) An antiderivative of $2x$ is x^2, so

$$\int_2^4 2x\, dx = x^2 \Big|_2^4 = 4^2 - 2^2 = 16 - 4 = 12$$

(b) An antiderivative of $x^2 + 4x^3$ is $x^3/3 + x^4$, so

$$\int_{-1}^2 (x^2 + 4x^3)\, dx = (x^3/3 + x^4) \Big|_{-1}^2$$

$$= (8/3 + 16) - (-1/3 + 1) = 18$$

The fundamental theorem states that any antiderivative can be used. To demonstrate what this means consider Example 1(a). The general antiderivative is $x^2 + C$. Using this antiderivative the computation is

$$\int_2^4 2x\, dx = x^2 + C \Big|_2^4 = 4^2 + C - (2^2 + C)$$

$$= (16 + C) - (4 + C) = 12$$

The constant drops out in the calculation, showing that any antiderivative can be used. Also, note that the theorem does not stipulate that $f(x) \geq 0$ in the interval as was required in calculating area in the previous section. Only when we use the definite integral to compute area will this requirement be important.

Verification of the Fundamental Theorem

The verification of the fundamental theorem of calculus provides insight into the connection between area and the definite integral. It starts by looking at areas under the graph of $f(x)$ and quickly shifts to the antiderivative.

Proof of the Fundamental Theorem of Calculus

Proof: Let $f(x) \geq 0$. The proof for $f(x) < 0$ is similar. Define the function $A(x)$ to be the area bounded by $f(x)$ and the x-axis between a and x, where $a \leq x \leq b$. We want to show that $A(x)$ is an antiderivative of $f(x)$, or that $A'(x) = f(x)$. Let h be a small positive number. Then $A(x + h)$ is the area from a to $x + h$. Since $A(x)$ is the area from a to x (Figure 1), $A(x + h) - A(x)$ is the area shaded in Figure 2. This area can be approximated by a rectangle of width h and height $f(x)$, whose area is $h \cdot f(x)$. Thus

$$A(x + h) - A(x) \simeq h \cdot f(x)$$

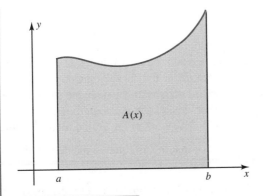

FIGURE 1·

The approximation improves as h gets close to 0. Dividing by h yields

$$\frac{A(x + h) - A(x)}{h} \simeq f(x)$$

Take the limit of both sides as h approaches 0. Since x is not affected by h, the limit of the right-hand side is $f(x)$. Because the approximation improves as h gets close to 0, taking the limit means the approximation sign can be replaced by an equal sign. Thus the limit of the right-hand side is equal to $f(x)$. But the limit of the left-hand side is the definition of the derivative of $A(x)$. Therefore $A'(x) = f(x)$.

There is one more step in the proof. Since x represents any number between a and b, set $x = a$. Then $A(a) = 0$. Also, setting $x = b$ gives $A(b)$, which is the desired area. Since $A(x)$ is an antiderivative of $f(x)$, we have

$$\int_a^b f(x)\, dx = A(x)\,\Big|_a^b = A(b) - A(a) = A(b)$$

Therefore the desired area is the definite integral.

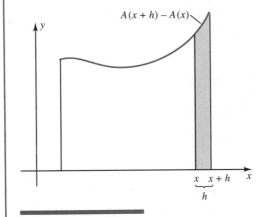

Area

Let us look at an example of the relationship between the definite integral and the area under a curve. Consider the function $f(x) = 2x$ in Example 1(a). Its graph is in Figure 3. The area under the curve bounded by the two lines $x = 2$ and $x = 4$ consists of two areas, the rectangle bounded above by the line $y = 4$ and the triangle above it. The area of the rectangle is $4 \cdot 2 = 8$. The area of the triangle is $1/2$ (base)(height) $= (1/2) \cdot 2 \cdot 4 = 4$. So the total area under $y = 2x$ from $x = 2$ to $x = 4$ is 12. Notice that this is equal to the definite integral $\int_{2}^{4} 2x \, dx$, computed in Example 1(a).

What is the area of the region between the x-axis and $f(x) = 2x$ from $x = -1$ to $x = 2$? It might seem that the answer is $\int_{-1}^{2} 2x \, dx$. However

$$\int_{-1}^{2} 2x \, dx = x^2 \Big|_{-1}^{2} = 4 - 1 = 3$$

From Figure 4 the total area is the sum of the area from $x = -1$ to $x = 0$ and the area from $x = 0$ to $x = 2$; that is, $1 + 4 = 5$.

What went wrong? Why did the definite integral yield an incorrect answer? Because part of the graph is below the x-axis. An area problem must be divided into subproblems, those where the graph is above the x-axis and those where it is below. The definite integral from $x = -1$ to $x = 0$ gives the negative of the area.

$$\int_{-1}^{0} 2x \, dx = x^2 \Big|_{-1}^{0} = 0 - (-1)^2 = -1$$

This leads to the next important property of definite integrals.

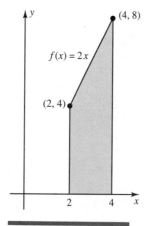

y

(4, 8)

$f(x) = 2x$

(2, 4)

2   4 x

FIGURE 3

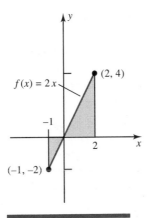

$f(x) = 2x$

$(2, 4)$

-1

2

$(-1, -2)$

FIGURE 4

Areas Under the x-axis
If $f(x) \leq 0$ in $[a, b]$, then the area bounded by $f(x)$, the x-axis, $x = a$, and $x = b$ is given by

$$-\int_a^b f(x)\ dx$$

Therefore if $f(x) \geq 0$, the area and the definite integral are equal, and if $f(x) \leq 0$, the area is the negative of the definite integral. For example, to find the area bounded by $f(x) = 2x$, the x-axis, $x = -1$, and $x = 2$, we need to divide the region into two separate regions: the region from $x = -1$ to $x = 0$ where $f(x) \leq 0$ and the region from $x = 0$ to $x = 2$ where $f(x) \geq 0$. The area of the first region is then the negative of the definite integral from -1 to 0. The area of the second region is the definite integral from 0 to 2. The total area is then

$$-\int_{-1}^0 2x\ dx + \int_0^2 2x\ dx = -(-1) + 4 = 5$$

This means that if the graph of $y = f(x)$ is below the x-axis in the interval over which the area is to be found, then we must find where the function is positive and where it is negative. This is equivalent to evaluating the integral of $|f(x)|$ because if $f(x) < 0$ then $\int_a^b |f(x)|dx = \int_a^b -f(x)dx = -\int_a^b f(x)dx$. Thus

$$\int_a^b |f(x)|dx = \begin{cases} \int_a^b f(x)\ dx & \text{if } f(x) \geq 0 \text{ in } [a, b] \\ -\int_a^b f(x)\ dx & \text{if } f(x) \leq 0 \text{ in } [a, b] \end{cases}$$

For example, to find the area bounded by $f(x) = 2x$, the x-axis, $x = -1$, and $x = 2$, it is necessary to divide the area into two separate regions: the region from $x = -1$ to $x = 0$ where $f(x) \le 0$, and the region from $x = 0$ to $x = 2$ where $f(x) \ge 0$. This means that

$$|f(x)| = |2x| = \begin{cases} -2x \text{ for } x \text{ in } [-1, 0] \\ 2x \text{ for } x \text{ in } [0, 2] \end{cases}$$

The area of the first region is then the negative of the definite integral from -1 to 0. The area of the second region is the definite integral from 0 to 2. The total area is then:

$$\int_{-1}^{2} |2x|dx = -\int_{-1}^{0} 2xdx + \int_{0}^{2} 2xdx = -(-1) + 4 = 5$$

Procedure for Finding the Area Under a Curve

To compute the area of the region bounded by $f(x)$, the x-axis, $x = a$, and $x = b$:

1. From the graph of $f(x)$ find the intervals where $f(x)$ is above the x-axis and where $f(x)$ is below the x-axis by locating the x-intercepts in $[a, b]$.
2. Compute the definite integral of $f(x)$ between each pair of x-intercepts.
3. If $f(x)$ is greater than 0 in an interval, the area is equal to the definite integral. If $f(x)$ is less than 0 in an interval, the area is the negative of the definite integral.
4. The total area is the sum of the areas in step 3.

The next example illustrates the procedure.

EXAMPLE 2

Problem

Find the area bounded by $f(x) = 2 - 2x$, the x-axis, $x = 0$, and $x = 4$.

Solution The graph is in Figure 5. The region consists of two triangles, one above the x-axis and one below it. To find the point of intersection of $f(x)$ and the x-axis, solve $f(x) = 0$; then $2 - 2x = 0$, so $x = 1$. From $x = 0$ to $x = 1$, $f(x)$ is above the x-axis, so the corresponding area is equal to the definite integral

$$\int_{0}^{1} (2 - 2x)\, dx = (2x - x^2) \Big|_{0}^{1} = (2 - 1) - 0 = 1$$

From $x = 1$ to $x = 4$ the graph is below the x-axis, so the area under the curve is equal to the negative of the definite integral.

$$-\int_{1}^{4} (2 - 2x)\, dx = -(2x - x^2) \Big|_{1}^{4} = -[(8 - 16) - (2 - 1)]$$

$$= -(-8 - 1) = -(-9) = 9$$

Therefore the total area is $1 + 9 = 10$.

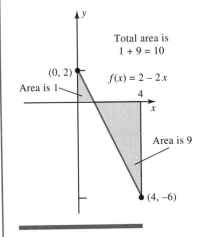

FIGURE 5

Properties of Definite Integrals

Some properties of antiderivatives or indefinite integrals were given in Section 16–1. Similar properties hold for definite integrals. We will list four of them.

Let $f(x)$ and $g(x)$ be functions and a, b, c, and k be real numbers. Then

1. $\displaystyle\int_a^b kf(x)\,dx = k\int_a^b f(x)\,dx$

2. $\displaystyle\int_a^b f(x)\,dx + \int_a^b g(x)\,dx = \int_a^b [f(x) + g(x)]\,dx$

3. $\displaystyle\int_a^a f(x)\,dx = 0$

4. $\displaystyle\int_a^c f(x)\,dx + \int_c^b f(x)\,dx = \int_a^b f(x)\,dx$ for $a \le c \le b$

The first two properties hold for indefinite integrals so they are also valid for definite integrals. Since the distance from a to a is 0, the third property states that the area under the curve between a and a is 0. The fourth property can be illustrated geometrically. The area bounded by the x-axis and $y = f(x)$ from $x = a$ to $x = b$ can be viewed as the sum of the area from $x = a$ to $x = c$ and the area from $x = c$ to $x = b$. This geometric argument works only for $f(x) > 0$, but a similar argument holds for the case when the graph of $f(x)$ is below the x-axis.

The next example shows how to use the properties to find more complicated definite integrals.

EXAMPLE 3

Problem

Find

$$\int_0^1 (4x^3 + 4x - 2e^x)\, dx$$

Solution

$$\int_0^1 (4x^3 + 4x - 2e^x)\, dx = \int_0^1 4x^3\, dx + \int_0^1 4x\, dx - \int_0^1 2e^x\, dx$$

$$= x^4 \Big|_0^1 + 2x^2 \Big|_0^1 - 2e^x \Big|_0^1$$

$$= 1^4 - 0^4 + 2(1^2) - 2(0^2) - 2e^1 + 2e^0$$

$$= 1 + 2 - 2e + 2 = 5 - 2e$$

Wages and Rent

A fundamental principle in economics is that "wages equal marginal product." This means that the employer will pay all employees doing similar work for similar pay, a wage equal to the marginal product of the last worker hired. This is because the last worker's value to the firm is that worker's marginal product, or the amount of increase in production the worker generates.

Along with this idea there are two additional principles. Consider the graph in Figure 6. Its curve is the marginal product function, which is equal to wages.

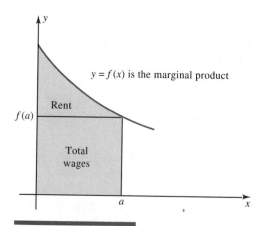

FIGURE 6

That is, $f(x)$ is the wage of each employee when x employees are hired. This curve is decreasing, meaning that as more workers are hired, the marginal product of the last worker decreases, and therefore the wage of the workers also decreases. If the number of employees is $x = a$, then the wage for each employee is $f(a)$. The total wages TW paid to all employees is the area bounded by the constant function $y = f(a)$, which is

$$TW = \int_0^a f(a)\,dx = f(a)x \,\bigg|_0^a = f(a)(a - 0) = af(a)$$

In other words, the region defined by this integral is the rectangle whose length is a and whose height is $f(a)$, so its area is $af(a)$. The **total product** is the total area of the region bounded by $y = f(x)$ and the x-axis from $x = 0$ to $x = a$. Therefore total product TP is defined to be

$$TP = \int_0^a f(x)\,dx$$

There is another region in Figure 6. It is the region bounded by $y = f(x)$ and the horizontal line $y = f(a)$ from $x = 0$ to $x = a$. It represents a savings for the employer because it measures the amount of total product that is not paid out in wages. Economists call the area of this region the employer's *rent*.* From Figure 6 we see that rent is the total product minus total wages. That is

rent $= TP - TW$

When economists study the effects of a new innovation or product on an industry, the concept of rent provides a measure of how much the industry has gained from the innovation. See Referenced Exercise 1 and the Case Study for examples of this type of use of rent.

EXAMPLE 4

A firm determines that the marginal product $MP(x)$ for a level of labor x is

$$MP(x) = 0.03x^2 - 0.6x + 23$$

$MP(x)$ is defined on the interval $[0, 10]$ and is measured in thousands of dollars, while x is measured in hundreds of employees. The number of employees hired is $x = 10$ hundred.

Problem

Find (a) total wages TW, (b) total product TP, and (c) the employer's rent.

Solution The wage paid to each employee is the marginal product of the last employee hired, which is $MP(10) = 3 - 6 + 23 = 20$.
(a) Total wages is defined by

$$TW = \int_0^{10} MP(10)\,dx = \int_0^{10} 20\,dx = 20x \,\bigg|_0^{10} = 200$$

*Paul A. Samuelson, *Economics*, 11th ed., New York, McGraw-Hill Book Co., 1980, pp. 465–467.

(b) The total product *TP* is defined by

$$TP = \int_0^{10} (0.03x^2 - 0.6x + 23)\, dx$$

$$= 0.01x^3 - 0.3x^2 + 23x) \Big|_0^{10} = 10 - 30 + 230 = 210$$

(c) The employer's rent is the area of the region in Figure 7. It is the area of the region under $y = MP(x)$ and above $y = 20$ from $x = 0$ to $x = 10$. This area is *TP* minus *TW*.

$$\text{rent} = TP - TW = 210 - 200 = 10$$

So the employer's rent is $10,000 when 1000 workers are hired at the salary of $20,000.

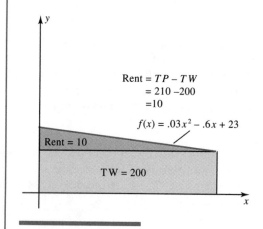

FIGURE 7

EXERCISE SET 16–3

In Problems 1 to 10 find the definite integral by using the fundamental theorem of calculus.

1. $\displaystyle\int_0^1 (3x + 5)\, dx$

2. $\displaystyle\int_1^3 (4x - 2)\, dx$

3. $\displaystyle\int_0^1 (x^2 + 1)\, dx$

4. $\displaystyle\int_{-2}^2 (3x^2 + 1)\, dx$

5. $\displaystyle\int_{-3}^2 (x^2 + 5x)\, dx$

6. $\displaystyle\int_1^2 (x^3 + x)\, dx$

7. $\displaystyle\int_0^8 (-x^2 + \sqrt[3]{x})\, dx$

8. $\displaystyle\int_1^4 (-x^2 + 1/\sqrt{x} + 7e^x)\, dx$

9. $\displaystyle\int_1^4 (-2x^{-3} + 6x^{-4} + 3\sqrt{x})\, dx$

10. $\displaystyle\int_1^4 (x^{-3} - 3x^{-2} - 3\sqrt{x})\, dx$

In Problems 11 and 12 find the definite integral by using geometric properties of the integrand. Verify by using the fundamental theorem of calculus.

11. $\displaystyle\int_{-2}^4 (x - 1)\, dx$

12. $\displaystyle\int_{-1}^4 (2 - x)\, dx$

In Problems 13 to 20 use the definite integral to find the area bounded by $f(x)$, the x-axis, $x = a$, and $x = b$.

13. $f(x) = x - 1$, $a = -3$, $b = 4$

14. $f(x) = 1 - x$, $a = -3$, $b = 3$

15. $f(x) = x^2 - 4$, $a = -2$, $b = 4$

16. $f(x) = 4 - x^2$, $a = -3$, $b = 2$

17. $f(x) = x^2 - x$, $a = 0$, $b = 2$

18. $f(x) = 4x - x^2$, $a = -1$, $b = 2$

19. $f(x) = x^3 - x$, $a = -1$, $b = 1$

20. $f(x) = x^3 - x^2$, $a = 0$, $b = 2$

In Problems 21 to 24 assume $\int_0^1 f(x)\,dx = 10$, $\int_1^2 f(x)\,dx = 15$, and $\int_0^2 g(x)\,dx = -5$. Use the properties of definite integrals to evaluate the integral.

21. $\int_0^2 f(x)\,dx$ 22. $\int_0^2 3f(x)\,dx$

23. $\int_0^2 [f(x) + g(x)]\,dx$

24. $\int_0^2 [2f(x) + 3g(x)]\,dx$

In Problems 25 to 30 find the definite integral.

25. $\int_0^1 (x^{0.2} + 3x^{1.4})\,dx$

26. $\int_0^1 (1.2x^{-1.6} - 4.1x^{-2.2})\,dx$

27. $\int_1^2 (3e^{2x} + 1/x)\,dx$ 28. $\int_1^2 (5e^{3x} - 2/x)\,dx$

29. $\int_1^2 (x^{1.3} - 3e^{1.2x} + 1/2x)\,dx$

30. $\int_1^2 (-e^{1-3x} - e/x)\,dx$

In Problems 31 to 36 use the definite integral to find the area bounded by $f(x)$, the x-axis, $x = a$, and $x = b$.

31. $f(x) = x^3 - x$, $a = -2$, $b = 2$

32. $f(x) = x^3 - x^2$, $a = -1$, $b = 2$

33. $f(x) = x^3 - 2x^2$, $a = 0$, $b = 3$

34. $f(x) = x^3 - 4x$, $a = -3$, $b = 5$

35. $f(x) = 1 - 2/x$, $a = 1$, $b = 3$

36. $f(x) = 2 - 3/x$, $a = 1$, $b = 3$

In Problems 37 to 40 find $\int_1^4 f(x)\,dx$ from the given figure.

37.

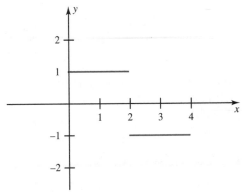

38.

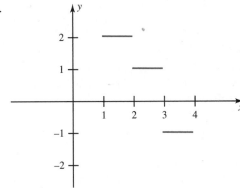

39.

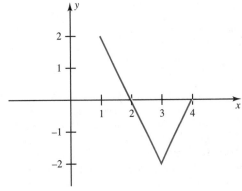

40.

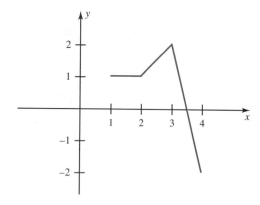

41. Prove the fundamental theorem for $f(x) < 0$.

42. What procedure would you use to find the area of the region bounded by $f(x) = x^2 - 4x + 3$ and the x-axis?

Referenced Exercise Set 16–3

1. Television is an example of an industry whose firms reap benefits from regulation. In an article studying the television industry, Fournier used rents to measure the effects of regulation on the industry.* The marginal cost function $MP(x)$ can be approximated by

$$MP(x) = 3x^2 - 60x + 800$$

where x is the average price of a show. Find the rent when $x = 10$. Here total product is income and total wages is total price.

2. In an article investigating the size of prey versus an optimal diet for carnivorous fish, Bence and Murdoch studied the relationship between the success rate $S(s)$ of the predator catching the prey versus the size s of the prey.† For one type of fish the relationship was calculated to be

$$S(s) = 1 + .861s - 91.57s^2 + 322.7s^3$$

The total success of attacks of prey between a mm (millimeters) and b mm is given by

$$\int_a^b S(s)ds$$

Find the total success for prey between 1 mm and 2 mm.

3. It is often difficult for economists to measure the benefits and costs of the uses of natural resources, such as federally owned lakes. In an article investigating ways to estimate the value of these resources, Seller et al. studied lakes in East Texas.‡ They calculated the demand curve for Lake Livingston to be

$$V(x) = 10.04 - .12x$$

where $V(x)$ is the value of the land when the cost of a family visit is x dollars. One measure of the benefit of the resource involved computing the consumers' surplus, an economic measure defined in Section 16–4. It entails computing the integral of $V(x)$ from $x = 0$ to $x = 2$. Compute this integral.

Cumulative Exercise Set 16–3

In Problems 1 to 4 find the antiderivative.

1. $\int (3x^{-2} - x^4 + 5)dx$

2. $\int (x^{1/5} - x^{5/3} + 8x - 1)dx$

3. $\int (3e^{2x} - 5/x)dx$ \qquad 4. $\int (x^6 + 4x^2)/x^3\, dx$

5. Use four rectangles to approximate the area under the curve $y = f(x) = x^2$ from $x = 0$ to $x = 4$.

In Problems 6 and 7 use the definition to find the definite integral.

6. $\int_0^4 2x\, dx$ \qquad 7. $\int_6^8 3x\, dx$

8. Use the definition to find the definite integral by using the identity

$$1^2 + 2^2 + \cdots + n^2 = \frac{n(n+1)(2n+1)}{6}$$

$$\int_{-2}^2 x^2\, dx.$$

In Problems 9 and 10 find the definite integral by using the fundamental theorem of calculus.

9. $\int_1^2 (x^3 + 2x - 1)dx$

*Gary M. Fournier, "Nonprice Competition and the Dissipation of Rents from Television Regulation," *Southern Economics Journal*, Vol. 51, 1985, pp. 754–765.

†James R. Bence and W. W. Murdoch, "Prey Size Aselection by the Mosquitofish: Relation to Optimal Diet Theory," *Ecology*, Vol. 67, 1986, pp. 324–336.

‡Christine Seller et al., "Validation of Empirical Measures of Welfare Change: A Comparison of Nonmarket Techniques," *Land Economics*, Vol. 61, 1985, pp. 156–175.

10. $\displaystyle\int_{1}^{4} (x^{-2} - 4x^{-2} - 2\sqrt{x})dx$

11. Assume $\displaystyle\int_{1}^{2} f(x)dx = 20$, $\displaystyle\int_{2}^{3} f(x)dx = -1$, and $\displaystyle\int_{1}^{3} g(x)dx = 5$. Use the properties of definite integrals

to evaluate

$$\int_{1}^{3} (f(x) + g(x))dx$$

12. Use the definite integral to find the area bounded by $f(x) = x^3 - x^2$, the x-axis, $x = -1$, and $x = 3$.

16–4   Area Bounded by Curves

Thus far we have considered regions bounded by a single curve and the x-axis. Often in applications a region is bounded both above and below by graphs of functions. In this section the concept of the area between a curve and the x-axis is generalized to include the area between two curves.

Areas Between Two Curves

Consider the area of the region bounded above by $f(x) = x + 2$ and below by $g(x) = x^2$. This region is graphed in Figure 1. View the region in two parts: the first part, region I, is the area below $f(x)$ from $x = -1$ to $x = 2$; the second part, region II, is the area below $g(x)$ from $x = -1$ to $x = 2$. Then the desired region is (region I–region II). (See Figure 2.) Its area is the difference between the two definite integrals.

$$\text{(the desired area)} = \text{region I} \qquad - \text{region II}$$

$$= (\text{area below } f(x)) - (\text{area below } g(x))$$

$$= \int_{-1}^{2} (x + 2)\, dx - \int_{-1}^{2} x^2\, dx$$

$$= \left(\frac{x^2}{2} + 2x\right)\Bigg|_{-1}^{2} - \frac{x^3}{3}\Bigg|_{-1}^{2} = \frac{15}{2} - 3 = \frac{9}{2}$$

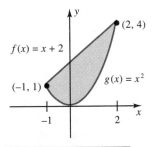

FIGURE 1

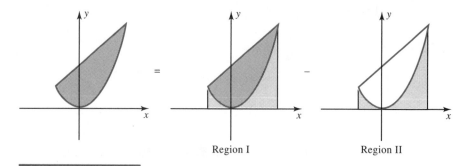

Region I   Region II

FIGURE 2

To compute this area, we can also use the third property from the previous section and get

$$= \int_{-1}^{2} (x + 2)\, dx - \int_{-1}^{2} x^2\, dx = \int_{-1}^{2} (x + 2 - x^2)\, dx$$

$$= \left(\frac{x^2}{2} + 2x - \frac{x^3}{3} \right) \Bigg|_{-1}^{2} = \frac{9}{2}$$

In general the area between two curves is the difference between the two definite integrals. We then use a property of definite integrals to simplify the integration.

The Area Bounded by Two Curves

If a region is bounded above by $f(x)$ and below by $g(x)$ between $x = a$ and $x = b$, then the area of the region is

$$\int_{a}^{b} [f(x) - g(x)]\, dx$$

Lorentz Diagrams

In Section 16–2 we defined a Lorentz diagram. The important part of the diagram is the area between the two curves $y = x$, the curve of complete equality of income distribution, and $y = x^2$, the Lorentz curve of the distribution of income in the United States. This area is called the area of inequality and measures the extent of departure between an equal distribution of income and the actual distribution of income. The area between these two curves is the difference between the area of the region bounded by $y = x$ and the area of the region bounded by $y = x^2$. We can now compute this area. It is

$$\text{area} = \int_0^1 x \, dx - \int_0^1 x^2 \, dx = \int_0^1 (x - x^2) \, dx$$

$$= \left(\frac{x^2}{2} - \frac{x^3}{3} \right) \Big|_0^1 = \frac{1}{2} - \frac{1}{3} = \frac{1}{6}$$

Economists compare this value with the inequality areas of other countries to determine which has a more equitable income distribution.

The formula for the area between two curves does not mention whether the functions are positive or negative. The only stipulation is that $f(x) \geq g(x)$ in $[a, b]$. The next example demonstrates what to do when two curves cross each other in the desired area.

EXAMPLE 1

Problem

Find the area of the region bounded by $f(x) = 3x^2 - 2x$, and $g(x) = x$ from $x = 0$ to $x = 2$.

Solution This region consists of two separate regions as shown in Figure 3. Find the points of intersection by solving $f(x) = g(x)$ to get $x = 0$ and $x = 1$.

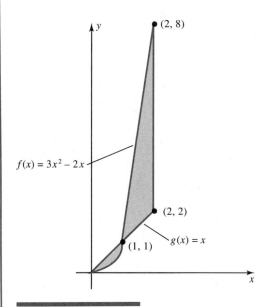

FIGURE 3

From 0 to 1, $g(x) \geq f(x)$, so we integrate $g(x) - f(x)$. From 1 to 2, $f(x) \geq g(x)$, so we integrate $f(x) - g(x)$. The area is the sum of the two integrals.

$$\text{area} = \int_0^1 [x - (3x^2 - 2x)]\, dx + \int_1^2 [(3x^2 - 2x) - x]\, dx$$

$$= \int_0^1 (3x - 3x^2)\, dx + \int_1^2 (3x^2 - 3x)\, dx$$

$$= \left(\frac{3x^2}{2} - x^3 \right) \Big|_0^1 + \left(x^3 - \frac{3x^2}{2} \right) \Big|_1^2$$

$$= \left(\frac{3}{2} - 1 \right) + \left[(8 - 6) - \left(1 - \frac{3}{2} \right) \right] = 3$$

End Points

In some problems the end points of the region are given and in others they are not. If the end points are not specified, they are found by computing the first coordinates of the points of intersection of the two curves. So the first step is to graph the region and locate the points of intersection. The next example shows how to do this.

EXAMPLE 2

Problem

Find the area of the region bounded by $y = x^3 - 2x$ and $y = x^2$.

Solution First sketch the curves. In order to refer to each curve, we name them. Let $f(x) = x^3 - 2x$ and $g(x) = x^2$. To find the points of intersection, solve $f(x) = g(x)$, so $x^3 - 2x = x^2$ or $x^3 - x^2 - 2x = 0$. Factoring yields $x(x - 2)(x + 1) = 0$. Thus the solutions are $x = -1, 0, 2$, and the curves intersect at $(-1, 1)$, $(0, 0)$, and $(2, 4)$. See Figure 4. From $x = -1$ to $x = 0$,

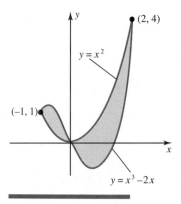

FIGURE 4

$f(x) \geq g(x)$, while from $x = 0$ to $x = 2$, $g(x) \geq f(x)$. Therefore the area of the region bounded by the two curves is

$$\int_{-1}^{0} [(x^3 - 2x) - x^2]\ dx + \int_{0}^{2} [x^2 - (x^3 - 2x)]\ dx$$

$$= \int_{-1}^{0} [x^3 - 2x - x^2]\ dx + \int_{0}^{2} [x^2 - x^3 + 2x]\ dx$$

$$= \left[\frac{x^4}{4} - x^2 - \frac{x^3}{3} \right] \Bigg|_{-1}^{0} + \left[\frac{x^3}{3} - \frac{x^4}{4} + x^2 \right] \Bigg|_{0}^{2}$$

$$= 0 - \left(\frac{1}{4} - 1 + \frac{1}{3} \right) + \frac{8}{3} - 4 + 4 - 0$$

$$= \frac{5}{12} + \frac{8}{3} = \frac{37}{12}$$

Procedure for Finding the Area of a Region Bounded by Two Curves

Find the area of a region bounded by two curves, $y = f(x)$ and $y = g(x)$, with the following steps:

1. Sketch the region. Find the points of intersection.
2. Find the end points, $x = a$ and $x = b$. If they are not given, then they are points of intersection.
3. Find the intervals where $f(x) \geq g(x)$.
   (i)   If $f(x) \geq g(x)$ in $[a, b]$, then the area is

   $$\int_{a}^{b} [f(x) - g(x)]\ dx$$

   (ii) If $g(x) \geq f(x)$ in $[a, b]$, then the area is

   $$\int_{a}^{b} [g(x) - f(x)]\ dx$$

   (iii) If there is a point c between a and b such that $f(x) \geq g(x)$ in $[a, c]$ and $g(x) \geq f(x)$ in $[c, b]$, then the area is

   $$\int_{a}^{c} [f(x) - g(x)]\ dx + \int_{c}^{b} [g(x) - f(x)]\ dx$$

Consumers' Surplus and Producers' Surplus

A pair of economic concepts that are best explained by calculus are *consumers' surplus* and *producers' surplus*. Consider the demand function $y = D(x)$ and the

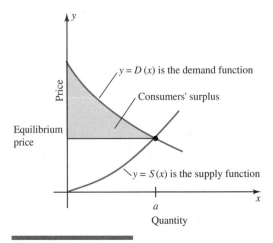

FIGURE 5

supply function $y = S(x)$, where x is the quantity of a product and y is the price. At price $D(x)$ consumers will demand x amount of the product. As x increases, $D(x)$ decreases, meaning that consumers will purchase more of the product at lower prices. This implies that $D(x)$ is a decreasing function. At price $S(x)$ producers are willing to supply x amount of the product. As x increases, $S(x)$ also increases because producers will be willing to supply more of the product at higher prices. Therefore $S(x)$ is an increasing function. In other words, $D(x)$ is the price that consumers are willing to pay for x amount of the product, and producers will supply x amount at price $S(x)$. In a free competitive market the actual selling price for the product will tend to be that price where supply equals demand, that is, where $S(x) = D(x)$. It is called the **equilibrium price,** and it is that quantity $x = a$ where $S(a) = D(a)$.

Figure 5 shows that the demand $D(x)$ is larger than $D(a)$ for values of x less than a. This means that some consumers would be willing to pay a higher price than the equilibrium price. In a certain sense these consumers have gained because the price they pay, the equilibrium price, is lower than what they would have been willing to pay. Economists refer to this gain as **consumers' surplus.** It is defined to be the area of the region bounded above by $y = D(x)$ and below by the constant function $y = D(a)$, from $x = 0$ to $x = a$. This area is

$$\text{consumers' surplus} = \int_0^a [D(x) - D(a)]\, dx$$

EXAMPLE 3

Suppose $D(x) = x^2 - 8x + 20$ for x in $[0, 4]$ and the equilibrium price occurs at $a = 2$.

Problem

Compute the consumers' surplus.

Solution The equilibrium price is $D(2) = 8$. The consumers' surplus is defined to be

$$\int_0^2 [(x^2 - 8x + 20) - 8]\, dx = \frac{x^3}{3} - 4x^2 + 12x \bigg|_0^2 = \frac{8}{3} - 16 + 24 = \frac{32}{3}$$

Similarly the producers who were willing to supply goods at a price smaller than $S(a)$ gain because they would have made less revenue at the lower price. Economists call the sum of their gains the **producers' surplus.** It is defined to be the area of the region bounded above by $y = S(a)$ and below by $y = S(x)$ from $x = 0$ to $x = a$. It is

$$\text{producers' surplus} = \int_0^a [S(a) - S(x)]\, dx$$

EXAMPLE 4

Suppose $S(x) = x^2 + 2x$ for x in $[0, 4]$ and the equilibrium price occurs at $a = 2$.

Problem

Compute the producers' surplus.

Solution The equilibrium price is $S(2) = 8$. The producer's surplus is defined to be

$$\int_0^2 [8 - (x^2 + 2x)]\, dx = \left(8x - \frac{x^3}{3} - x^2 \right) \bigg|_0^2 = 16 - \frac{8}{3} - 4 = \frac{28}{3}$$

Figure 6 shows the supply and demand curves in Examples 3 and 4 as well as the areas defining the consumers' surplus and producers' surplus.

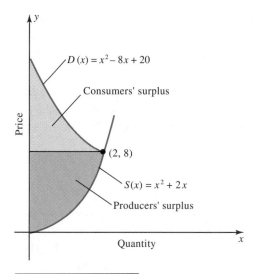

FIGURE 6

The concept of consumers' surplus measures the benefit consumers gain from the power of buying at low prices and not having to pay the higher price a monopolist may insist on. Another use of consumers' surplus is in measuring "social welfare." Sometimes economic value, revenue, or profit cannot be used to account for the value of a good. For instance, the economic value of air is negligible but its social welfare is great. The next example demonstrates how economists use consumers' surplus to help make a decision where the intangible quantity of utility is the only "profit."

E X A M P L E   5

A new community is debating whether to self-impose a tax to install street lights. The community has 3000 homes that will be taxed the same amount for the improvement. Contractors are willing to supply the lights according to the schedule

$$S(x) = x^2 + 9x$$

The community is willing to pay according to the schedule

$$D(x) = x^2 - 18x + 81$$

Price is measured in $100,000 units and x is measured in units of 100 street lights. To find the equilibrium price, solve $S(x) = D(x)$. This yields $a = 3$, so the equilibrium price is $S(3) = D(3) = 36$. Thus each homeowner must pay a tax of $3,600,000/3000 = \$1,200$ for the improvement. Is it worth it to the individual? Since there are no revenues from this project, the only profit is the increased utility achieved. Hence the consumers' surplus gives an indication of the worth of the project.

Problem

Compute the consumers' surplus.

Solution
From the definition the consumers' surplus is

$$\int_0^3 [(x^2 - 18x + 81) - 36]\, dx = \frac{x^3}{3} - 9x^2 + 45x \Big|_0^3 = 9 - 81 + 135 = 63$$

Thus the consumers' surplus is $6,300,000. Each homeowner gets a utility of $6,300,000/3000 = \$2,100$. Therefore the consumers' surplus is far greater than the amount spent on the improvement. The project is worthwhile from this point of view.

EXERCISE SET 16–4

In Problems 1 to 30 find the area between the curves.

1. $x = 0$, $x = 2$,
   $f(x) = x^2 + 1$, $g(x) = 0$

2. $x = 0$, $x = 3$,
   $f(x) = x^2 + 4$, $g(x) = 0$

3. $x = 0$, $x = 2$,
   $f(x) = x^2 + 2x$, $g(x) = 0$

4. $x = 0, \quad x = 3,$
  $f(x) = x^2 + x, \quad g(x) = 0$

5. $x = 0, \quad x = 2,$
  $f(x) = x^3 + 3x, \quad g(x) = 0$

6. $x = 0, \quad x = 4,$
  $f(x) = x^3 + 6x, \quad g(x) = 0$

7. $x = 0, \quad x = 1,$
  $f(x) = x^2, \quad g(x) = x$

8. $x = 0, \quad x = 1,$
  $f(x) = x^2, \quad g(x) = 2x$

9. $x = 0, \quad x = 1,$
  $f(x) = x^2 + x, \quad g(x) = 2x$

10. $x = 0, \quad x = 1,$
  $f(x) = x^2 + 4x, \quad g(x) = 5x$

11. $x = 0, \quad x = 1,$
  $f(x) = 3x - x^2, \quad g(x) = 2x$

12. $x = 0, \quad x = 1,$
  $f(x) = 5x - x^2, \quad g(x) = 6x$

13. $f(x) = x^2 + x, \quad g(x) = 2x$

14. $f(x) = x^2 + 4x, \quad g(x) = 5x$

15. $f(x) = 3x - x^2, \quad g(x) = 2x$

16. $f(x) = 5x - x^2, \quad g(x) = 6x$

17. $f(x) = x^2 - 2x, \quad g(x) = 2x$

18. $f(x) = x^2 - 4x, \quad g(x) = 5x$

19. $f(x) = 2 + x - x^2,$
  $g(x) = 2 - 3x$

20. $f(x) = 3 + 2x - x^2,$
  $g(x) = 3 - 4x$

21. $f(x) = x^3, \quad g(x) = x^2$

22. $f(x) = x^3 - 4x,$
  $g(x) = x^2 - 4x$

23. $x = 0, \quad x = 4, \quad f(x) = x + 2, \quad g(x) = x^2$

24. $x = 0, \quad x = 6, \quad f(x) = 3x + 2,$
  $g(x) = 2x^2$

25. $x = -1, \quad x = 4, \quad f(x) = 2x + 3,$
  $g(x) = x^2$

26. $x = -1, \quad x = 5, \quad f(x) = 3x + 4,$
  $g(x) = x^2 + x + 1$

27. $x = -2, \quad x = 3, \quad f(x) = 1 - 2x,$
  $g(x) = x^2 - 2x - 3$

28. $x = -3, \quad x = 3, \quad f(x) = 3x + 4,$
  $g(x) = x^2 + 3x$

29. $x = -2, \quad x = 2, \quad f(x) = 3x - x^2,$
  $g(x) = x^2 - x$

30. $x = 0, \quad x = 4, \quad f(x) = 4x - x^2,$
  $g(x) = 2x^2 - 5x$

In Problems 31 to 34 compute the consumers' surplus and the producers' surplus.

31. $D(x) = x^2 - 8x + 24, \qquad S(x) = x^2 + 4x$

32. $D(x) = x^2 - 12x + 48, \qquad S(x) = x^2 + 4x$

33. $D(x) = 48 - 10x - x^2, \qquad S(x) = x^2 + 10x$

34. $D(x) = 39 - 12x - x^2, \qquad S(x) = x^2 + 8x$

In Problems 35 to 40 find the area between the curves.

35. $x = 1, \quad x = 2, \quad f(x) = \dfrac{1}{x}, \quad g(x) = 1$

36. $x = 2, \quad x = 3, \quad f(x) = \dfrac{1}{x}, \quad g(x) = x$

37. $x = 1, \quad x = 5, \quad f(x) = \dfrac{4}{x}, \quad g(x) = x$

38. $x = 0, \quad x = 3, \quad f(x) = \dfrac{2}{x + 1},$
  $g(x) = x - 1$

39. $x = 0, \quad x = 3, \quad f(x) = \dfrac{5}{x + 2},$
  $g(x) = x - 2$

40. $x = 1, \quad x = 5, \quad f(x) = \dfrac{8}{x}, \quad g(x) = x^2$

Referenced Exercise Set 16–4

1. A new trash-to-steam plant is proposed for a community to reduce refuse. The community is willing to pay according to the schedule

$$D(x) = 1000 - 50x - x^2$$

Contractors are willing to build the plant according to the schedule

$$S(x) = x^2 + 30x$$

Compute the consumers' surplus and determine if the project is "profitable" in the sense that the consumers' surplus is greater than the cost.

2. In an article studying the adverse effects of ozone and other air pollutants on crop yields, Adams et al. assessed the benefits to agriculture from reductions in these pollutants.* The measurements used were consumers' surplus and producers' surplus. In one model, where a 10% reduction is assumed, the supply and demand curves are approximated by

$$S(x) = 0.03x^2 + 1.2x$$
$$D(x) = 0.03x^2 - 2.7x + 28.2$$

where $S(x)$ and $D(x)$ are measured in billions of dollars. The total benefit was the sum of the consumers' surplus and the producers' surplus. Compute the total benefit.

3. In many areas where soil is good but rainfall is poor, water irrigation projects are developed to support farm products. Often taxpayers are required to subsidize these projects. In 1978, California rice producers were given a subsidy in the form of reduced prices for water to irrigate the rice crop. In an article studying the effect of this subsidy on producers and consumers, Foster et al. used producers' surplus and consumers' surplus to measure the benefit of the subsidy to each group.† The demand curve $D(x)$ and the supply curve $S(x)$ at price x, measured in thousands of dollars, were approximated by

$$D(x) = 34 - 28x$$
$$S(x) = 30 + 2x + 16x^2$$

Compute the producers' surplus and the consumers' surplus.

4. In his book *Applied Mathematical Demography*, Nathan Keyfitz argues that there is an urgent need for haste in lowering the birth rate in less developed countries by showing the lag between the birth rates and death rates of such countries.‡ If the birth rate is represented by $b(t)$ and the death rate by $d(t)$, then the total lag $l(t)$ between the two rates over T years is defined by

$$l(t) = \int_0^T b(t)dt - \int_0^T d(t)dt$$

The birth rate $b(t)$ and the death rate $d(t)$ of one country are approximated by

$$b(t) = .01t^4 - .18t^2 + 1.3$$
$$d(t) = 1.2 - .06t$$

Compute $l(t)$ for $T = 10$.

Cumulative Exercise Set 16–4

In Problems 1 and 2 find the antiderivative.

1. $\int \dfrac{x}{3} \, dx$

2. $\int \dfrac{9}{x^2} \, dx$

3. Find the function whose slope of the tangent line to the curve is $f'(x) = 1/x$ and whose graph passes through the point $(1, 5)$.

4. If the velocity function of an object is given by $v(t) = -32t + 32$, what is the distance function $s(t)$ if $s(0) = 64$?

5. (a) Use four rectangles to approximate the area under the curve $y = 4 - x^2$ from $x = -1$ to $x = 1$.

   (b) Use the definition to find the area under the curve $y = 4 - x^2$ from $x = -1$ to $x = 1$.

6. Use the definition to evaluate the integral

$$\int_0^1 \frac{x}{2} \, dx$$

7. Find the area under the curve $f(x) = x^4$ from $x = -2$ to $x = 2$.

8. Use the fundamental theorem of calculus to evaluate the definite integral

$$\int_1^4 \left(\sqrt{x} - \frac{2}{x} \right) dx$$

9. Use the definite integral to find the area bounded by $f(x) = x^3$ and the x-axis, from $x = -2$ to $x = 2$.

In Problems 10 to 12 find the area between $f(x)$ and $g(x)$.

10. $f(x) = -x$, $g(x) = x^2$

11. $f(x) = x^3$, $g(x) = x$

12. $f(x) = 8x^3$, $g(x) = x^4 + 16x^2$

---

*R. M. Adams et al., "The Benefits of Pollution Control: The Case of Ozone and U.S. Agriculture," *American Journal of Agricultural Economics*, Vol. 68, 1986, pp. 886–893.

†William E. Foster et al., "Distribution Welfare Implications of an Irrigation Water Subsidy," *American Journal of Agricultural Economics*, Vol. 68, 1986, pp. 778–786.

‡Nathan Keyfitz, *Applied Mathematical Demography*, New York, Springer-Verlag, 1985, pp. 1–26.

CASE STUDY   ## Consumers' Surplus and Louisiana Fish Farming

Jim Larkin runs a fleet of small fishing boats in the Atchafalaya River basin in southern Louisiana. Jim's brother Larry owns a small farm nearby. Two brothers with two different careers. But Larry's innovative use of his ponds had a devastating effect on Jim's livelihood in the late 1970s.

The bulk of Jim's catch consists of shellfish, such as shrimp, crabs, and crawfish. Each type of shellfish has a distinct season, depending on its own natural characteristics. For instance, crawfish produce only one brood per year, and young crawfish grow rapidly, doubling in weight each month. Sexual maturity is reached from March to July, and the cycle is repeated. The size of the crawfish population is determined by water levels and temperature. When surface water of their habitat disappears in the fall, crawfish burrow into the ground. In addition, natural predators are reduced in this period. The lower the water level is in the fall, the higher will be the population the next year. This means there will be an abundance of crawfish as the water level rises from January to March. Production of good catches requires that the commercial fishing fleet be in synchronization with water changes and crawfish life cycles.

Rearing crawfish in a small pond is an entirely different story. As early as 1950 a few farmers were harvesting crawfish in ponds, but it was a haphazard effort. The pond-reared crawfish catch amounted to less than 1% of the total crawfish market. By 1980 the total crawfish market had increased from about 1 million pounds in 1950 to about 30 million pounds, a 3000% increase. But here's the catch—the pond production of crawfish accounted for a whopping 85% of the total harvest. Fish farming had become big business at the expense of the commercial fleet. The pond producers could circumvent the environmental barriers facing the commercial fleet. For instance, they could maintain the water level in the ponds year-round as well as supply a constant source of food for the fish.

The innovative aquaculture techniques of the pond producers has had a remarkable impact on the crawfish market, especially on the commercial fleet. But what effect does this have on the public? To what extent has pond-reared crawfish benefited consumers?

Economist Frederick W. Bell describes this phenomenon in an article studying the effects of new and innovative production techniques in the marketplace.* Economists use consumers' surplus to measure the benefit of such techniques to the public. One obvious benefit is that there are more crawfish to consume. As demand increases, the price of crawfish falls. Bell uses 1978 as a representative year in his study. He estimates the demand function in 1978 to be

$$D(x) = -0.0055x + 0.39$$

where x is the quantity of crawfish harvested in millions of pounds and $D(x)$ is dollars per pound. The demand function is graphed in Figure 1. Bell estimates

*Frederick W. Bell, ''Competition from Fish Farming in Influencing Rent Dissipation: The Crawfish Industry,'' *American Journal of Agricultural Economics*, Vol. 68, 1986, pp. 95–101.

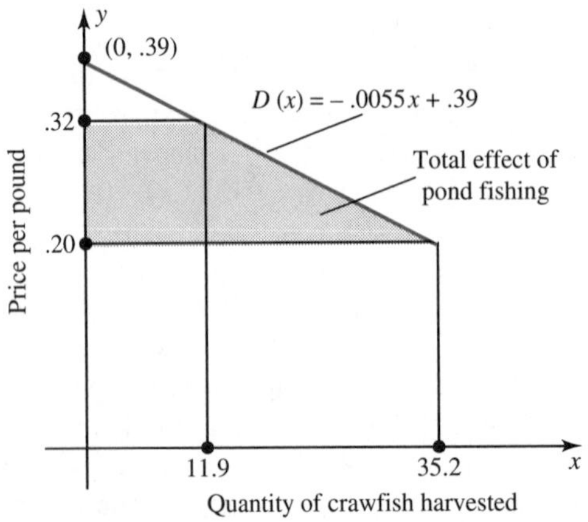

FIGURE 1

that without pond-reared crawfish, only 11.9 pounds would be produced at a price of $0.32. Thus $D(11.9) = 0.32$. With the addition of pond-reared crawfish, 35.2 lb would be harvested at a price of $0.20. Thus $D(35.2) = 0.20$. The total effect of pond-reared crawfish is measured by the difference in the two values of consumers' surplus, with and without pond-reared crawfish.

$$\begin{pmatrix} \text{total effect} \\ \text{of pond fish} \end{pmatrix} = \begin{pmatrix} \text{consumers' surplus} \\ \text{with pond fish} \end{pmatrix} - \begin{pmatrix} \text{consumers' surplus} \\ \text{without pond fish} \end{pmatrix}$$

$$= \int_0^{35.2} (-0.0055x + 0.39)\,dx - 35.2(0.20)$$

$$- \left[\int_0^{11.9} (-0.0055x + 0.39)\,dx - 11.9(0.32) \right]$$

$$= (-0.00275x^2 + 0.39x) \Big|_0^{35.2} - 7.04$$

$$- \left[(-0.00275x^2 + 0.39x) \Big|_0^{11.9} - 3.81 \right]$$

$$= -0.00275(35.2)^2 + 0.39(35.2) - 7.04$$

$$- [-0.00275(11.9)^2 + 0.39(11.9) - 3.81]$$

$$= -3.41 + 13.72 - 7.04 + 0.39 - 4.64 + 3.81$$

$$= 2.83$$

This shows that pond-reared fishing greatly increases the benefit to consumers. The fish farmers have gained income as well as expanding the market for crawfish and increasing demand. On the other hand, commercial fishermen like Jim Larkin have experienced a decline in revenue.

The brothers seldom share crawfish stories.

Case Study Exercises

In Problems 1 to 4 compute the total effect of pond fishing for the given demand function and the quantities of crawfish harvested without pond fishing and those harvested with pond fishing.

1. $D(x) = -0.0055x + 0.30$,
   $x = 12$ and $x = 40$
2. $D(x) = -0.0055x + 0.30$,
   $x = 15$ and $x = 30$
3. $D(x) = -0.011x + 0.40$,
   $x = 10$ and $x = 50$
4. $D(x) = -0.011x + 0.40$,
   $x = 20$ and $x = 40$

In this article Bell computes the marginal cost function to be

$$MC(x) = \frac{7.26}{(13.7 - x)^2}$$

He defines "social welfare loss" by the formula

$$\int_a^b MC(x)\, dx - P(b - a)$$

where a and b are the quantities of crawfish harvested and P is the price per pound. In Problems 5 and 6 compute the social welfare loss for the given values taken from Bell.

5. $a = 7.71$, $b = 10.2$, $P = 0.20$
6. $a = 9.2$, $b = 11.9$, $P = 0.32$

CHAPTER REVIEW

Key Terms

16–1 Antiderivatives
Antidifferentiation
Antiderivative
Constant of Integration
Integral Sign

Indefinite Integral
Integrand
Differential Equation
Growth Equation

16–2 Area and the Definite Integral
Lorentz Diagram
Definite Integral

Limits of Integration
Area under a Curve

16–3 The Fundamental Theorem of Calculus
Fundamental Theorem of Calculus

16–4 Area Bounded by Curves
Area Bounded by Two Curves
Equilibrium Price

Consumers' Surplus
Producers' Surplus

Summary of Important Concepts

Formulas

Growth equation $f'(x) = kf(x)$

The fundamental theorem of calculus $\int_a^b f(x)\, dx = F(b) - F(a)$

REVIEW PROBLEMS

In Problems 1 and 2 find the antiderivative.

1. $\int (5x^{-1/5} - 2x^{-3/5} - 3x)\, dx$

2. $\int \dfrac{x^7 + 3x}{x^3}\, dx$

3. Find the function whose slope of the tangent line is $y' = x^{-2} - x$ and whose graph passes through $(1, 0)$.

4. Use the definition to find the definite integral.

$\int_0^4 (2x + 1)\, dx$

5. Find the area under $f(x)$ from $x = 0$ to $x = 2$ where $f(x) = 7 - 3x$.

6. Use the definition to find the definite integral

$\int_0^2 x^2\, dx$

by using the identity

$1^2 + 2^2 + \cdots + n^2 = \dfrac{n(n + 1)(2n + 1)}{6}$

7. Find the definite integral by using the fundamental theorem of calculus.

$\int_1^4 \left(-x^2 + \dfrac{1}{\sqrt{x}} + 7e^x \right) dx$

8. Use the definite integral to find the area bounded by $f(x) = 4 - x^2$, the x-axis, $x = -4$, and $x = 2$.

9. Find the definite integral.

$\int_1^2 \left(5e^{3x} - \dfrac{2}{x} \right) dx$

In Problems 10 to 12 find the area between the curves.

10. $x = 0$, $x = 1$, $f(x) = x^2 + 4x$, $g(x) = 5x$

11. $f(x) = 5x - x^2$, $g(x) = 6x$

12. $x = 0$, $x = 6$, $f(x) = 3x + 2$, $g(x) = 2x^2$

PROGRAMMABLE CALCULATOR EXPERIMENTS

In Problems 1 to 4 write a program that uses the given number of rectangles n to approximate the area under the curve of an arbitrary function $y = f(x)$ from $x = a$ to $x = b$. Then run the program for the function $f(x) = x^2$ from $x = 0$ to $x = 2$.

1. $n = 4$

2. $n = 6$

3. $n = 8$

4. $n = 10$

5. Write a program to compute the consumers' surplus and producers' surplus for the demand function

$D(x) = x^2 - ax + b$ and supply function $S(x) = x^2 + cx$.

In Problems 6 to 8 run the program in Problem 5 for the given values of a, b, and c and graph the two functions in the same coordinate system.

6. $a = 6$, $b = 24$, $c = 2$

7. $a = 7$, $b = 20$, $c = 3$

8. $a = 2.2$, $b = 6.8$, $c = 1.2$

Techniques of Integration

The social security system may not provide security in the future. (H. Armstrong Roberts)

Chapter Overview

The previous chapter showed how to compute the integral of various simple functions. This chapter shows how to integrate more complicated functions. There is no one method of integration that works for all functions; rather, there are various techniques that are applied on a trial-and-error basis.

The first section deals with the simplest technique. It uses a substitution of a variable to reduce the integrand to a familiar form whose integral can be found by a method that was demonstrated earlier. Section 17–2 considers more difficult functions. It relies on the chain rule. The third section explains how to apply a table of integrals. The table gives several formulas involving a diverse number of integrands. Sometimes an exact integrand cannot be found. In this case it is often preferable to approximate the definite integral by means of one of several techniques. This is called numerical integration, which is covered in Section 17–4.

CASE STUDY PREVIEW

Many people mistakenly believe that the taxes they pay to the social security system each year are put into a savings account to be held until their retirement. In reality, the money paid by the work force goes directly to fund those who are retired. Those in the age range of 20 to 64 support those rẻtired over 64. The plan works well when there is a large work force to support a much smaller retired segment of the population.

The Case Study demonstrates that the birthrate is starting to lag well behind the death rate. And the gap is growing. The demographic consequences show that either the system will prove to be a great burden on the work force or the benefits paid to retirees will be greatly reduced.

17–1 Substitution

The fundamental theorem of calculus states that antiderivatives can be used to compute integrals. Thus far we have integrated only relatively simple functions. Just as there are various rules for finding derivatives that depend on the type of function, so too there are several methods for finding indefinite integrals of different types of functions. In this section we discuss the method of integration called **substitution.** To apply this method the integrand must be in a particular form. Determining whether a function fits the description is part of the technique.

Integration by Substitution

Let us start with an example. By the chain rule for derivatives

$$D_x(x^2 + 1)^4 = 4(x^2 + 1)^3 2x = 8x(x^2 + 1)^3$$

The corresponding indefinite integral is

$$\int 8x(x^2 + 1)^3 \, dx = (x^2 + 1)^4 + C$$

Finding this integral was straightforward because we knew $D_x(x^2 + 1)^4$ beforehand. The method of substitution allows us to perform the integration of such functions without knowing the corresponding derivative beforehand. **Integration by substitution** entails expressing the integrand such as the one above as the product of two functions, one of which is the derivative of the other, or at least almost its derivative. In the integrand above, the function $f(x) = x^2 + 1$ is part of the integrand, and its derivative, $f'(x) = 2x$, is also present. The integrand consists of $f(x)$ raised to the third power times $f'(x)$, with the factor 4 left over. Viewed this way, it is in the proper form for substitution. We write

$$\int 8x(x^2 + 1)^3 \, dx = \int 4 \cdot (x^2 + 1)^3 \cdot 2x \, dx$$

Make the substitution $u = x^2 + 1$. Then $\dfrac{du}{dx} = 2x$. Now formally multiply through by dx to get $du = 2x \, dx$. Then the integrand is written

$$\int 8x(x^2 + 1)^3 \, dx = \int 4 \cdot u^3 \cdot du$$

The new integrand is expressed in terms of the new variable u and is easily integrated.

$$\int 4u^3 \, du = u^4 + C$$

Making the reverse substitution for u in terms of x yields the same integral computed earlier.

$$\int 8x(x^2 + 1)^3 \, dx = (x^2 + 1)^4 + C$$

There are two key steps in applying substitution in this type of problem. First, it must be possible to express the integrand as the product of one function raised to a power and another function that is the derivative of the first function. The entire integrand, with the sole exception of a constant left over, must be described this way. The next step is to let the new variable u equal this function and then compute du, the *differential of u,* which is defined as follows:

DEFINITION

If $u = g(x)$, then the **differential of u** is $du = g'(x) \, dx$.

The differential actually has a deeper meaning and a wider application than is used in this text. Since we use differentials only in this chapter to explain techniques of integration, we will not expand on this definition.

Once the substitution is made, the integrand is expressed in terms of u and constants. Next compute the integral in terms of u if possible. Finally, substitute back for x to find the integral. The result can always be checked by differentiating the result to see if it equals the integrand. Let us look at an example to illustrate the procedure.

EXAMPLE 1

Problem

Compute $\int x(1 - x^2)^5 \, dx$.

Solution Let $u = 1 - x^2$. Then $du = -2x \, dx$. The differential is not present exactly, but it varies only by a constant, -2. In this case the differential can be divided by the constant to get $(-1/2) \, du = x \, dx$. Now the substitution can be made that expresses the integrand entirely in terms of u.

$$\int x(1 - x^2)^5 \, dx = \int (1 - x^2)^5 \, x \, dx = \int u^5 \left(-\frac{1}{2} \right) du = -\frac{1}{2} \int u^5 \, du$$

$$= -\frac{1}{2} \frac{u^6}{6} + C = -\frac{1}{12} (1 - x^2)^6 + C$$

This result can be verified by computing the derivative of $(-1/12)(1 - x^2)^6 + C$ and checking that it equals the original integrand, $x(1 - x^2)^5$.

Notice how the constant $-1/2$ can be taken out of the integrand because of the rule that states that the integral of a constant times a function is the constant times the integral. This is why the expression for du can differ by a constant when deciding what substitution to make for u.

The next example shows how to use substitution to find definite integrals.

EXAMPLE 2

Problem

Find $\int_0^1 (x + 1)(x^2 + 2x + 1)^{-1/2} \, dx$

Solution First find the indefinite integral of $(x + 1)(x^2 + 2x + 1)^{-1/2}$. Let $u = x^2 + 2x + 1$. Then $du = (2x + 2) \, dx = 2(x + 1) \, dx$. The integrand contains only $(x + 1) \, dx$, so we must divide by the constant 2. Dividing by 2 yields $(1/2)du = (x + 1) \, dx$. By the method of substitution

$$\int (x + 1)(x^2 + 2x + 1)^{-1/2} \, dx = \int u^{-1/2} \frac{1}{2} \, du$$

$$= \frac{1}{2} \int u^{-1/2} \, du = \frac{1}{2} \frac{u^{1/2}}{1/2} + C = u^{1/2} + C = (x^2 + 2x + 1)^{1/2} + C$$

Now insert the limits of integration.

$$\int_0^1 (x + 1)(x^2 + 2x + 1)^{-1/2} \, dx = (x^2 + 2x + 1)^{1/2} \Big|_0^1 = 4^{1/2} - 1^{1/2} = 1$$

Sometimes the choice of substitution for u does not work. It might not express the entire integral in terms of u or the resulting integrand may not be easily integrated. Another substitution may be tried. Thus the method often involves trial and error. But not all integrals can be solved by substitution. For example, consider

$$\int (x^2 + 1)^{1/2} \, dx$$

The natural choice would be $u = x^2 + 1$. Then $du = 2x \, dx$. The factor of x is the problem. If the differential differed from the integrand by only a constant, then we could divide through by the constant. But here the differential and the integrand differ by the factor $2x$. If we try to divide through by x, then the left-hand side of the equation for the differential would have two variables, u and x. So the substitution would not entirely replace the variable x by the variable u. Whenever du varies from the integrand by a constant the substitution can be made.

In the examples presented so far the function in the integrand chosen for u has been raised to a power. If we let u be that function, then u' must also be present, at least up to a constant. The general formula governing this type of substitution problem can be expressed as follows, with $u = f(x)$ and $n \neq -1$:

$$\int [f(x)]^n f'(x) \, dx = \int u^n \, du = \frac{u^{n+1}}{n + 1} + C$$

$$= \frac{f(x)^{n+1}}{n + 1} + C, \quad n \neq 1$$

There are other types of functions that can be integrated by substitution. The next type that we consider are certain exponential functions. If e is raised to the power $f(x)$ and $f'(x)$ is also present in the integrand, set $u = f(x)$ and find du to complete the substitution. The general formula, with $u = f(x)$, is

$$\int e^{f(x)} f'(x) \, dx = \int e^u \, du = e^u + C = e^{f(x)} + C$$

The formula can be verified using the definition of the antiderivative.

$$D_x(e^{f(x)} + C) = e^{f(x)} f'(x)$$

EXAMPLE 3

Problem

Find $\int x^2 e^{x^3} \, dx$.

Solution Let $u = x^3$. Then $du = 3x^2\ dx$ and $(1/3)\ du = x^2\ dx$. Thus

$$\int x^2 e^{x^3}\ dx = \int e^{x^3}(x^2\ dx) = \int e^u \frac{1}{3}\ du = \frac{1}{3}\ e^u + C = \frac{1}{3}\ e^{x^3} + C$$

The third form of the substitution method is derived from the derivative formula $D_x \ln |x| = 1/x = x^{-1}$. Thus $\int x^{-1}\ dx = \ln |x| + C$. If a function is raised to the power -1 and its derivative is present in the integrand, then the substitution can be made and the integral involves a logarithmic function. The same is true if the integrand is a rational function whose numerator is a constant multiple of the derivative of the denominator. This means that the denominator can be expressed in the numerator as a function raised to the power -1.

EXAMPLE 4

Problem

Find $\int x(x^2 - 4)^{-1}\ dx$.

Solution Let $u = x^2 - 4$. Then $du = 2x\ dx$ and $(1/2)\ du = x\ dx$. Thus

$$\int x(x^2 - 4)^{-1}\ dx = \int (x^2 - 4)^{-1}(x\ dx)$$

$$= \int u^{-1} \frac{1}{2}\ du$$

$$= \frac{1}{2} \int u^{-1}\ du = \frac{1}{2} \ln |u| + C$$

$$= \frac{1}{2} \ln |x^2 - 4| + C$$

The general formula, with $u = f(x)$, is

$$\int [f(x)]^{-1} f'(x)\ dx = \int \frac{f'(x)}{f(x)}\ dx$$

$$= \int u^{-1}\ du = \int \frac{du}{u}$$

$$= \ln |u| + C = \ln |f(x)| + C$$

The next example illustrates how to use the exponential and logarithmic forms of the substitution method of integration in which the choice for the substitution is not obvious.

EXAMPLE 5

Problem

Find (a) $\int \dfrac{e^{x^{-1}}}{x^2}\,dx$, (b) $\int \dfrac{x^{1/2}}{x^{3/2}-1}\,dx$

Solution (a) Let $u = x^{-1}$. Then $du = -x^{-2}\,dx$ and $-du = x^{-2}\,dx$. Thus

$$\int \frac{e^{x^{-1}}}{x^2}\,dx = \int x^{-2}e^{x^{-1}}\,dx = -\int e^u\,du = -e^u + C = -e^{x^{-1}} + C$$

(b) Let $u = x^{3/2} - 1$. Then $du = (3/2)x^{1/2}\,dx$ and $(2/3)\,du = x^{1/2}\,dx$. Thus

$$\int \frac{x^{1/2}}{x^{3/2}-1}\,dx = \int \frac{1}{u}\left(\frac{2}{3}\right)du = \frac{2}{3}\int \frac{du}{u}$$

$$= \frac{2}{3}\ln|u| + C = \frac{2}{3}\ln|x^{3/2} - 1| + C$$

The method of substitution is versatile. The next example shows how to apply it in a case where the substitution is subtle.

EXAMPLE 6

Problem

Find $\displaystyle\int_0^1 \frac{x\,dx}{x+1}$.

Solution First find the antiderivative and then insert the limits of integration. Let $u = x + 1$. Then $x = u - 1$ and $du = dx$, so

$$\int \frac{x\,dx}{x+1} = \int \frac{u-1}{u}\,du = \int \left(1 - \frac{1}{u}\right)du = u - \ln|u| + C$$

$$= x + 1 - \ln|x+1| + C$$

Insert the limits of integration.

$$\int_0^1 \frac{x\,dx}{x+1} = x + 1 - \ln|x+1| \;\Big|_0^1 = 2 - \ln 2 - (1 - \ln 1)$$

$$= 1 - \ln 2$$

An often troublesome aspect of substitution is the role of the differential, du. The differential is a vital ingredient in putting the integrand in proper form, but when the antiderivative is found it "disappears," in the sense that it merely indicates the variable of integration. That is, once the substitution is made the sole role of du is to label the variable, so it does not appear in the final solution.

EXERCISE SET 17–1

In Problems 1 to 26 find the antiderivative.

1. $\int 2x(x^2 + 5)^3 \, dx$

2. $\int 2x(x^2 - 2)^4 \, dx$

3. $\int 4x^3(x^4 - 7)^2 \, dx$

4. $\int 5x^4(x^5 + 1)^2 \, dx$

5. $\int x(3x^2 + 5)^3 \, dx$

6. $\int x(2x^2 - 3)^5 \, dx$

7. $\int x^3(3x^4 - 5)^3 \, dx$

8. $\int x^4(4x^5 + 1)^4 \, dx$

9. $\int x^5(3x^6 - 5)^{3/2} \, dx$

10. $\int x^3(5x^4 + 1)^{3/4} \, dx$

11. $\int 2x^3(x^4 - 7)^{-3} \, dx$

12. $\int 6x^4(x^5 + 10)^{-4} \, dx$

13. $\int x^{-3}(6x^{-2} + 5)^{-2} \, dx$

14. $\int x^{-4}(5x^{-3} - 1)^{-5} \, dx$

15. $\int (2x + 5)(x^2 + 5x)^{-3} \, dx$

16. $\int (2x - 2)(x^2 - 2x)^{1/4} \, dx$

17. $\int (x^3 - 1)(x^4 - 4x)^{1/2} \, dx$

18. $\int (x^4 + 2x)(x^5 + 5x^2)^{2/5} \, dx$

19. $\int (e^{2x} + 5e^{4x}) \, dx$

20. $\int (5e^{3x} - e^{-3x}) \, dx$

21. $\int (x^3 e^{x^4} + 2e^{-x}) \, dx$

22. $\int (x^{-3} e^{x^{-2}}) \, dx$

23. $\int \frac{2}{3x + 1} \, dx$

24. $\int \frac{x}{x^2 - 1} \, dx$

25. $\int x^2(x^3 - 1)^{-1} \, dx$

26. $\int (x^2 + 1)(x^3 + 3x)^{-1} \, dx$

In Problems 27 to 34 find the definite integral.

27. $\int_0^1 x(x^2 + 1)^2 \, dx$

28. $\int_0^1 x(x^2 - 2)^3 \, dx$

29. $\int_{-1}^1 x^3(x^4 + 1)^2 \, dx$

30. $\int_0^2 x^4(x^5 + 1)^4 \, dx$

31. $\int_0^3 x(x^2 + 1)^{1/2} \, dx$

32. $\int_0^1 x(1 - x^2)^{1/3} \, dx$

33. $\int_1^2 (x^3 + 1)(x^4 + 4x)^{-2} \, dx$

34. $\int_1^2 (x^4 + 5)(x^5 + 5x)^{-2} \, dx$

In Problems 35 to 46 find the antiderivative.

35. $\int (e^x + 1)e^x \, dx$

36. $\int (e^{2x+1} + 10) \, e^{2x} \, dx$

37. $\int \frac{\ln x}{x} \, dx$

38. $\int \frac{\ln x^2}{x} \, dx$

39. $\int x(x - 1)^2 \, dx$

40. $\int x(3x + 4)^4 \, dx$

41. $\int x(1 - 5x)^{1/2} \, dx$

42. $\int 5x(x + 4)^{-4} \, dx$

43. $\int x^3(x^2 - 1)^{2/3} \, dx$

44. $\int x^3(2 - 3x^2)^{-3} \, dx$

45. $\int \frac{x \ln(1 + x^2)}{1 + x^2} \, dx$

46. $\int \frac{e^x \ln(1 + e^x)}{1 + e^x} \, dx$

Referenced Exercise Set 17–1

1. In an article studying the possible environmental impact of uranium mining in northern Australia, Vardavas investigated how contaminants were transported by surface water and ground water.* This study is meant to predict human radiological exposure from contaminants from mine sites. One part of the study requires the calculation of the total water evaporation per day at Manton Dam. The evaporation rate $E(x)$ for a given wind speed x can be approximated by

$$E(x) = \frac{.5x + .8}{x + 1}$$

The total evaporation is then computed by finding the antiderivative of $E(x)$. Find this antiderivative.

---

*Ilias Mihail Vardavas, "Modelling the Seasonal Variation of Net All-wave Radiation Flux and Evaporation in a Tropical Wet-dry Region," *Ecological Modelling*, Vol. 39, 1987, pp. 247–268.

2. Talent searches identifying mathematically gifted high school youth have shown remarkably consistent sex differences. In an article studying the effect of a sex-linked gene that might facilitate high mathematical test score performance, Thomas considered data from the Iowa Test of Basic Skills and the SAT-mathematics test.† One part of the study investigated the properties of the function

$$f(x) = \frac{x}{x^2 + 1}$$

where $f(x)$ is a psychological measure, called the Mills ratio, and x is a percentile. Find the antiderivative of $f(x)$.

†Hoben Thomas, "A Theory of High Mathematical Aptitude," *Journal of Mathematical Psychology,* Vol. 29, 1985, pp. 231–242.

17–2 Integration by Parts

In the previous section we mentioned that substitution is a trial-and-error method that does not work in all cases. In this section we consider another method of integration, integration by parts. To apply the technique, think of the integrand as formed by two distinct functions, or "parts."

Integration by Parts

The product rule for derivatives cannot be transformed directly into an integration formula, but it can be revised to give a powerful rule. Suppose u and v are functions of x. The product rule states that

$$D_x uv = u D_x v + v D_x u$$

Expressing this formula in terms of differentials yields

$$d(uv) = u\, dv + v\, du$$

It is now in a form that is more amenable for integration. But first solve for $u\, dv$.

$$u\, dv = d(uv) - v\, du$$

Taking the integral of both sides yields

$$\int u\, dv = \int d(uv) - \int v\, du = uv - \int v\, du$$

This is the formula for the technique of integration called **integration by parts.**

Integration by Parts
If u and v are differentiable functions, then

$$\int u\, dv = uv - \int v\, du$$

The term "parts" refers to the way that the integrand is separated into two functions, u and dv. Thus there are two parts to the integrand. The underlying idea of the technique is that the integral $\int u\, dv$ is difficult, if not impossible, to calculate. The goal is to choose u and v so that the integral $\int v\, du$ is easier to calculate. The formula for integration by parts shows how to use $\int v\, du$ to compute $\int u\, dv$. The first example illustrates how to use the formula.

EXAMPLE 1

Problem

Compute $\displaystyle\int xe^x\, dx$

Solution Separate the integrand into u and dv as follows:

$$u = x \qquad dv = e^x\, dx$$

Then

$$du = dx \qquad v = \int dv = \int e^x\, dx = e^x$$

Since the constant of integration is usually only included in the final answer, it is not mentioned in the intermediate steps. From the formula

$$\int xe^x\, dx = \int u\, dv = uv - \int v\, du = xe^x - \int e^x\, dx = xe^x - e^x + C$$

It is important to realize that integration by parts is another trial-and-error method. For example, in Example 1 if the choice for u and dv is $u = e^x$ and $dv = x\, dx$, then $du = e^x\, dx$ and $v = \int x\, dx = x^2/2$. Then the formula yields $\int xe^x\, dx = x^2e^x/2 - \int x^2e^x/2\, dx$. The latter integral is more complex than the original one. This choice of u and dv does not work. The object is to choose u and dv so that

1. It is possible to compute $v = \int dv$, and
2. It is possible to compute $\int v\, du$, or $\int v\, du$ is simpler than $\int u\, dv$.

The next example shows how to compute a seemingly simple integral by a judicious choice of u and dv.

EXAMPLE 2

Problem

Compute $\displaystyle\int \ln x\, dx$.

Solution Separate the integrand into u and dv as follows:

$$u = \ln x \qquad dv = dx$$

Then

$$du = \frac{1}{x} dx \qquad v = \int dv = \int dx = x$$

Substituting these expressions in the formula, we get

$$\int \ln x \, dx = \int u \, dv = uv - \int v \, du$$

$$= (\ln x)x - \int x\left(\frac{1}{x}\right) dx = x \ln x - \int dx$$

$$= x \ln x - x + C$$

Sometimes it is necessary to apply the formula more than once to compute the integral, as demonstrated in Example 3.

EXAMPLE 3

Problem

Compute $\int x^2 e^{3x} \, dx$.

Solution The idea is to eliminate the x^2 from the integrand. This is not possible in one application of the technique, but we can replace x^2 by x to simplify the integrand. Separate the integrand into u and dv as follows:

$$u = x^2 \qquad dv = e^{3x} \, dx$$

$$du = 2x \, dx \qquad v = \int dv = \int e^{3x} \, dx = \frac{1}{3} e^{3x}$$

$$\int x^2 e^{3x} \, dx = uv - \int v \, du = x^2 \frac{1}{3} e^{3x} - \int \frac{1}{3} e^{3x} 2x \, dx$$

$$= \frac{1}{3} x^2 e^{3x} - \frac{2}{3} \int x e^{3x} \, dx \qquad (1)$$

The problem has been reduced to finding $\int x e^{3x} \, dx$. To compute the latter integral, use integration by parts again, this time with

$$u = x \qquad dv = e^{3x}$$

$$du = dx \qquad v = \frac{1}{3} e^{3x}$$

This is similar to Example 1. The formula yields

$$\int x e^{3x} \, dx = x \frac{1}{3} e^{3x} - \int \frac{1}{3} e^{3x} \, dx = \frac{1}{3} x e^{3x} - \frac{1}{9} e^{3x}$$

Now substitute this result into equation (1) to get the final answer.

$$\int x^2 e^{3x} \, dx = \frac{1}{3} x^2 e^{3x} - \frac{2}{3} \int x e^{3x} \, dx$$

$$= \frac{1}{3} x^2 e^{3x} - \frac{2}{3} \left[\frac{1}{3} x e^{3x} - \frac{1}{9} e^{3x} \right] + C$$

$$= \frac{1}{3} x^2 e^{3x} - \frac{2}{9} x e^{3x} + \frac{2}{27} e^{3x} + C$$

The next example shows how we sometimes need to employ more than one technique of integration in a problem. It uses substitution and integration by parts.

EXAMPLE 4

Problem

Compute $\int x(2x + 1)^{1/2} \, dx$.

Solution Choose u and dv as follows:

$$u = x \qquad dv = (2x + 1)^{1/2} \, dx$$

$$du = dx \qquad v = \int (2x + 1)^{1/2} \, dx$$

To compute v, we must use substitution. Let $w = 2x + 1$. Then $dw = 2 \, dx$, $(1/2)dw = dx$, and so

$$v = \int (2x + 1)^{1/2} \, dx = \int w^{1/2} \left(\frac{1}{2} \right) dw = \frac{1}{2} \left(\frac{2}{3} \right) w^{3/2} = \frac{1}{3} (2x + 1)^{3/2}$$

Substituting into the formula for integration by parts yields

$$\int x(2x + 1)^{1/2} \, dx = \int u \, dv = uv - \int v \, du$$

$$= x \left(\frac{1}{3} \right) (2x + 1)^{3/2} - \int \frac{1}{3} (2x + 1)^{3/2} \, dx$$

$$= \frac{1}{3} x(2x + 1)^{3/2} - \frac{1}{3} \int (2x + 1)^{3/2} \, dx$$

Again, the substitution $w = 2x + 1$ is needed in the last integral. The final answer is

$$\int x(2x + 1)^{1/2} \, dx = \frac{1}{3} x(2x + 1)^{3/2} - \frac{1}{15} (2x + 1)^{5/2} + C$$

The integral in Example 4 could have been computed by the substitution $u = 2x + 1$. In fact substitution is the first method to try. We used integration by parts to illustrate the method and to show that the two methods can be used on the same problem.

Definite Integrals and Parts

To find a definite integral using integration by parts, first find the antiderivative and then insert the limits of integration. This yields the following formula:

$$\int_a^b u \, dv = uv \Big|_a^b - \int_a^b v \, du$$

The next example shows that first we can calculate $\int v \, du$ and then insert the limits.

EXAMPLE 5

Problem

Compute $\int_0^4 x(2x + 1)^{1/2} \, dx$.

Solution First find the indefinite integral. This was done in Example 4. Then insert the limits of integration.

$$\int_0^4 x(2x + 1)^{1/2} \, dx = \left[\frac{1}{3} x(2x + 1)^{3/2} - \frac{1}{15} (2x + 1)^{5/2} \right] \Big|_0^4$$

$$= \frac{1}{3} 4(9)^{3/2} - \frac{1}{15} (9)^{5/2} - 0 + \frac{1}{15}$$

$$= \frac{108}{3} - \frac{243}{15} + \frac{1}{15} = \frac{298}{15}$$

EXERCISE SET 17-2

In Problems 1 to 20 find the antiderivative.

1. $\int 2xe^x \, dx$

2. $\int 5xe^x \, dx$

3. $\int xe^{3x} \, dx$

4. $\int 5xe^{4x} \, dx$

5. $\int x(3x + 5)^{3/2} \, dx$

6. $\int x(2x - 3)^{5/3} \, dx$

7. $\int 3x(5x - 2)^{-3} \, dx$

8. $\int 2x(4x + 1)^{-4} \, dx$

9. $\int x \ln x \, dx$

10. $\int x \ln 5x \, dx$

11. $\int x^2 \ln x \, dx$

12. $\int x^4 \ln x \, dx$

13. $\int x^2 e^x \, dx$

14. $\int x^2 e^{5x - 3} \, dx$

15. $\int (2x + 5)(x + 5)^{-3} \, dx$

16. $\int (x - 2)(4x - 3)^{-1/2} \, dx$

17. $\int x^2(x - 4)^{1/2} \, dx$

18. $\int x^2(1 - x)^{2/3} \, dx$

19. $\int x^2(1 - 3x)^{-2} \, dx$

20. $\int 4x^2(2 + 3x)^{-3} \, dx$

In Problems 21 to 26 find the definite integral.

21. $\int_0^1 xe^{2x} \, dx$

22. $\int_0^1 xe^{x+2} \, dx$

23. $\int_0^2 x(2x + 5)^{1/2} \, dx$

24. $\int_0^2 x(4x + 1)^{-1/2} \, dx$

25. $\int_0^2 x^2(x + 1)^{-4} \, dx$

26. $\int_0^2 x^2(2x + 3)^{-3} \, dx$

In Problems 27 to 34 find the antiderivative.

27. $\int x^2(x + 1)^{1/2} \, dx$

28. $\int x^2(1 - 3x)^{-1/2} \, dx$

29. $\int x^3(x + 1)^{1/2} \, dx$

30. $\int x^3(1 - 3x)^{-1/2} \, dx$

31. $\int x^3(x^2 + 1)^{1/2} \, dx$

32. $\int x^3(2 - 3x^2)^{-1/2} \, dx$

33. $\int x^3 e^{x^2} \, dx$ 34. $\int x^5 e^{x^3} \, dx$

In Problems 35 and 36 find the definite integral.

35. $\int_0^2 x^5(2x^2 + 1)^{1/2} \, dx$

36. $\int_0^1 x^5(x^2 + 3)^{-4} \, dx$

In Problems 37 to 40 find the antiderivative.

37. $\int (x + 1) \ln x \, dx$

38. $\int (x^2 + 1) \ln x \, dx$

39. $\int x^{-2} \ln x \, dx$

40. $\int x^{-3} \ln x \, dx$

Referenced Exercise Set 17–2

1. In an article studying organic contaminants in natural environments, especially industrial and municipal water wastes, Chang and Rittmann focus attention on the contaminant activated carbon.* Let $q(r)$ be the surface concentration of activated carbon on a circular surface r units from the center of the concentration. The computation of the total concentration involves evaluating the definite integral of $r^2 q(r)$ from $r = 0$ to $r = R$, the radius of the circular region. Compute this integral for $q(r) = e^{-.1r}$ and $R = 1$ mm.

2. In business the term "mark-up" refers to the difference between the selling price to the customer and the cost the retailer paid for the item. The mark-up takes into account the retailer's cost to market and sell the item and the retailer's profit. In an article studying how mark-up varies during various phases of the business cycle, Goldstein applied his model to 20 manufacturing industries from 1949 to 1980.† The goal of each industry is to maximize profit subject to cost restrictions. Goldstein assumed that profit $P(t)$ varied with time t according to the relationship $P'(t) = e^{-rt} f(t)$, where r is the prevailing interest rate and $f(t)$ depended on selling price, market share, and competitors' prices. He defined total profit $P(t)$ as the antiderivative of $P'(t)$. Find $P(t)$ for $f(t) = t$.

Cumulative Exercise Set 17–2

In Problems 1 to 8 find the antiderivative.

1. $\int x^3(2x^4 + 3)^5 \, dx$

2. $\int x^{-2}(3x^{-1} - 1)^{-3} \, dx$

3. $\int (x^3 - 2x)(x^4 - 4x^2)^{2/3} \, dx$

4. $\int x^{-4} e^{x-3} \, dx$

5. $\int 3x e^{2x} \, dx$

6. $\int x(3x - 4)^{1/3} \, dx$

7. $\int x^5 \ln x \, dx$

8. $\int x^3(2 + x)^{1/2} \, dx$

---

*Ted Chang and Bruce E. Rittmann, "Mathematical Modelling of Biofilm on Activated Carbon," *Environmental Science and Technology*, Vol. 21, 1987, pp. 273–280.

†Jonathan Goldstein, "Mark-up Pricing over the Business Cycle: The Microfoundations of the Variable Mark-up," *Southern Economics Journal*, Vol. 53, 1987, pp. 233–246.

17–3 Tables of Integrals

Most of the integrands encountered thus far have required a direct application of one particular technique of integration. Often, however, an integration problem requires lengthy and cumbersome manipulation. For such problems a table of integrals is usually helpful. The table in the Appendix provides a list of antiderivatives that is intended to acquaint the student with such tables. More extensive tables can be found, such as in the *Mathematical Handbook of Formulas and Tables,* which contains more than 500 antiderivatives.

Using a table of integrals is usually straightforward, but there are a few pitfalls that need to be pointed out. Tables of integrals are organized by similarity of the integrands. The first few listings are generally the elementary formulas. These are usually followed by integrands involving $ax + b$ for constants a and b, which in turn are followed by integrands involving $ax^2 + b$, $(ax + b)^2$, and so forth.

Because the table is arranged in this way, it is easy to mistake the correct formula for a similar one. For example, consider the integral

$$\int \frac{dx}{x(3x + 2)}.$$

Referring to the table, we find that formulas 12, 13, and 15 are all similar. Each could be mistaken for the proper formula, namely, formula 14 with $a = 3$ and $b = 2$. Formula 12 is incorrect because there is no x appearing in the denominator of the formula's integrand. Similarly, formulas 13 and 15 are incorrect because $(ax + b)^2$ appears in the integrand rather than $(ax + b)$.

The first example shows how to find an antiderivative when the integrand fits the formula directly. Later examples show how some integrands need to be expressed in a different form to apply a formula.

EXAMPLE 1

Problem

Find $\int \dfrac{dx}{x(3x + 2)^2}$.

Solution The integrand fits formula 15 with $a = 3$ and $b = 2$. Substituting these values into formula 15 yields

$$\int \frac{dx}{x(3x + 2)^2} = \frac{1}{2(3x + 2)} + \frac{1}{4}\ln\left|\frac{x}{3x + 2}\right| + C$$

Often some algebraic manipulation is necessary to see which formula is correct, as illustrated in the next example.

EXAMPLE 2

Problem

Find $\int \dfrac{dx}{1 - 4x^2}$

Solution The integrand fits formula 6, but it cannot be applied immediately because the coefficient of x^2 in the problem is 4, while the coefficient of x^2 in the formula is 1. The coefficient must be factored out of the denominator and the integral sign because the coefficients of x^2 in the problem and the formula must agree. Use the fact that $1 - 4x^2 = 4(1/4 - x^2)$. Hence

$$\int \frac{dx}{1 - 4x^2} = \int \frac{dx}{4(1/4 - x^2)} = \frac{1}{4} \int \frac{dx}{1/4 - x^2}$$

Now formula 6 can be applied with $a = 1/2$.

$$\int \frac{dx}{1 - 4x^2} = \frac{1}{4} \int \frac{dx}{1/4 - x^2} = \frac{1}{4} \ln \left| \frac{1/2 + x}{1/2 - x} \right| + C$$

This formula holds only for $x^2 < a^2$; otherwise the antiderivative is not defined.

Most tables do not include integrals that can be simplified by a simple substitution. Sometimes a simple substitution of variable is required before applying the appropriate formula.

EXAMPLE 3

Problem

Find $\int x(x^4 + 4)^{1/2} \, dx$.

Solution The integrand is similar to formula 16, but it cannot be applied immediately because the problem contains x^4 instead of x^2 in the formula. The x outside the parentheses permits the substitution $u = x^2$. Then $du = 2x \, dx$ and so $(1/2)du = x \, dx$. Since $u^2 = x^4$, this substitution yields

$$\int x(x^4 + 4)^{1/2} \, dx = \frac{1}{2} \int (u^2 + 4)^{1/2} \, du$$

Now apply formula 16 with $a = 2$.

$$\int x(x^4 + 4)^{1/2} \, dx = \frac{1}{2} \int (u^2 + 4)^{1/2} \, du$$

$$= \frac{1}{2} \left[\frac{1}{2} u(u^2 + 4)^{1/2} + \frac{4}{2} \ln |u + (u^2 + 4)^{1/2}| \right] + C$$

$$= \frac{1}{4} x^2(x^4 + 4)^{1/2} + \ln |x^2 + (x^4 + 4)^{1/2}| + C$$

Formula 18 is a recursive formula, meaning that it might have to be applied several times to solve a problem.

EXAMPLE 4

Problem

Compute $\int x^3 e^{2x} \, dx$.

Solution Apply formula 18 with $n = 3$ and $a = 2$.

$$\int x^3 e^{2x}\, dx = \frac{1}{2} x^3 e^{2x} - \frac{3}{2} \int x^2 e^{2x}\, dx \tag{1}$$

To compute the integral on the right, again apply formula 18, this time with $n = 2$ and $a = 2$.

$$\int x^2 e^{2x}\, dx = \frac{1}{2} x^2 e^{2x} - \frac{2}{2} \int x e^{2x}\, dx \tag{2}$$

Once again apply formula 18, now with $n = 1$ and $a = 2$.

$$\int x e^{2x}\, dx = \frac{1}{2} x e^{2x} - \frac{1}{2} \int e^{2x}\, dx$$

$$= \frac{1}{2} x e^{2x} - \frac{1}{4} e^{2x} + C \tag{3}$$

Now put the pieces of the puzzle together. Substitute (3) into (2), and substitute the result into (1). The final answer is

$$\int x^3 e^{2x}\, dx = \frac{1}{2} x^3 e^{2x} - \frac{3}{2} \left[\frac{1}{2} x^2 e^{2x} - \frac{1}{2} x e^{2x} + \frac{1}{4} e^{2x} \right] + C$$

$$= \frac{1}{2} x^3 e^{2x} - \frac{3}{4} x^2 e^{2x} + \frac{3}{4} x e^{2x} - \frac{3}{8} e^{2x} + C$$

To evaluate a definite integral via the table of integrals, first find the indefinite integral and then evaluate the antiderivative at the limits of integration.

EXAMPLE 5

Problem

Find $\displaystyle\int_2^3 \frac{dx}{x(2x - 3)}$.

Solution The integrand fits formula 14 with $a = 2$ and $b = -3$. Substitute these values into the formula without the limits of integration.

$$\int \frac{dx}{x(2x - 3)} = \frac{1}{-3} \ln \left| \frac{x}{(2x - 3)} \right| + C$$

Inserting the limits of integration yields

$$\int_2^3 \frac{dx}{x(2x - 3)} = -\frac{1}{3} \ln \left| \frac{x}{(2x - 3)} \right| \Big|_2^3$$

$$= -\frac{1}{3} \ln \left| \frac{3}{3} \right| + \frac{1}{3} \ln \left| \frac{2}{1} \right| = \frac{1}{3} \ln 2$$

EXERCISE SET 17–3

In Problems 1 to 38 find the integral.

1. $\displaystyle\int \frac{dx}{x(3x + 2)^2}$

2. $\displaystyle\int \frac{dx}{x(5x + 6)^2}$

3. $\displaystyle\int \frac{dx}{x(3x - 4)^2}$

4. $\displaystyle\int \frac{dx}{x(5x - 7)^2}$

5. $\displaystyle\int \frac{dx}{1 - 9x^2}$

6. $\displaystyle\int \frac{dx}{4 - 9x^2}$

7. $\displaystyle\int \frac{dx}{16 - 9x^2}$

8. $\displaystyle\int \frac{dx}{4 - 25x^2}$

9. $\displaystyle\int \frac{dx}{36 - 4x^2}$

10. $\displaystyle\int \frac{dx}{25 - 4x^2}$

11. $\displaystyle\int x(x^4 + 9)^{1/2}\, dx$

12. $\displaystyle\int x(x^4 + 16)^{1/2}\, dx$

13. $\displaystyle\int x(4x^4 + 9)^{1/2}\, dx$

14. $\displaystyle\int x(9x^4 + 16)^{1/2}\, dx$

15. $\displaystyle\int 2x(4x^4 - 25)^{1/2\cdot}\, dx$

16. $\displaystyle\int 3x(9x^4 - 49)^{1/2}\, dx$

17. $\displaystyle\int_2^3 \frac{dx}{x(3x - 5)}$

18. $\displaystyle\int_2^3 \frac{dx}{x(6 - 5x)}$

19. $\displaystyle\int_1^2 \frac{dx}{x(3x + 2)^2}$

20. $\displaystyle\int_1^3 \frac{dx}{x(5x + 6)^2}$

21. $\displaystyle\int_0^1 x(x^4 + 1)^{1/2}\, dx$

22. $\displaystyle\int_0^1 x(x^4 + 4)^{1/2}\, dx$

23. $\displaystyle\int x^3 e^{3x}\, dx$

24. $\displaystyle\int x^4 e^{5x}\, dx$

25. $\displaystyle\int \frac{x\, dx}{(5x + 2)^2}$

26. $\displaystyle\int \frac{x\, dx}{(7x + 6)^2}$

27. $\displaystyle\int \frac{dx}{x(x - 4)}$

28. $\displaystyle\int \frac{x\, dx}{5x - 7}$

29. $\displaystyle\int \frac{dx}{x\sqrt{1 - 9x^2}}$

30. $\displaystyle\int \frac{dx}{\sqrt{4 + 9x^2}}$

31. $\displaystyle\int \frac{x\, dx}{\sqrt{16 + 9x^4}}$

32. $\displaystyle\int \frac{dx}{x\sqrt{4 - x^4}}$

33. $\displaystyle\int \frac{dx}{x\sqrt{9 + 4x^2}}$

34. $\displaystyle\int \frac{x\, dx}{49 - 4x^4}$

35. $\displaystyle\int \frac{x^2\, dx}{9 - 4x^6}$

36. $\displaystyle\int \frac{x^3\, dx}{9 - 4x^4}$

37. $\displaystyle\int x^5 \ln x^2\, dx$

38. $\displaystyle\int x^5 e^{x^2}\, dx$

Referenced Exercise Set 17–3

1. The study of waiting situations, the so-called "queueing problems," usually assumes that people dislike having to wait. In an article studying the resulting psychological cost of waiting, Osuna defines the total stress of waiting $TH(t)$ as the integral of $H(t)$, the stress at time t.* Find the indefinite integral of $H(t)$ for

$$H(t) = \frac{t^2}{10 - t}$$

2. Purchasing land always involves risk, but the purchase of farmland usually involves a greater amount of risk than other types of land because there are so many additional factors. These include complex tax laws and government price supports that affect not only profit but land value. In an article studying farmland pricing strategies, Brown and Brown define the expected earnings per acre $E(a, b)$ of selling a piece of land for a price between $x = a$ and $x = b$ hundred dollars per acre to be

$$E(a, b) = \int_a^b f(x)\, dx$$

where $f(x)$ is the likelihood that the seller will get a bid of price x.† Find $E(1, 1.5)$ for

$$f(x) = \frac{x}{x + 1}$$

Cumulative Exercise Set 17–3

In Problems 1 to 8 find the antiderivative.

1. $\displaystyle\int \sqrt{1 + 5x}\, dx$

2. $\displaystyle\int \frac{6x\, dx}{\sqrt{x^2 + 4}}$

*Edgar Elias Osuna, "The Psychological Cost of Waiting," *Journal of Mathematical Psychology,* Vol. 29, 1985, pp. 82–106.

†Keith C. Brown and Deborah J. Brown, "Heterogenous Expectations and Farmland Prices," *American Journal of Agricultural Economics,* Vol. 66, 1984, pp. 164–169.

3. $\int (3 - x^2)^2 dx$

4. $\int \left(x - \dfrac{3}{x} \right) dx$

5. $\int xe^{-x} dx$

6. $\int x\sqrt{x - 2}\, dx$

7. $\int \dfrac{x\, dx}{2x - 2}$

8. $\int x^3 \ln x\, dx$

In Problems 9 and 10 find the definite integral.

9. $\int_2^8 \dfrac{dx}{x^2}$

10. $\int_0^4 x\sqrt{9 + x^2}\, dx$

In Problems 11 and 12 find the area between the curve $y = f(x)$ and the x-axis from $x = 0$ to $x = 2$.

11. $f(x) = x\sqrt{4 - x^2}$

12. $f(x) = xe^x$

17–4 Numerical Integration

The definite integral was defined by using approximations to the area under a curve that consisted of sums of areas of certain rectangles. A limit process was applied to define the definite integral. The fundamental theorem of calculus permits us to calculate integrals by using antiderivatives, thus avoiding the need to use approximations. But sometimes it is impractical or even impossible to find the correct antiderivative. In this case we need to approximate the area under the curve in order to approximate the definite integral.

There are many ways to approximate the definite integral. They are called **numerical integration** methods because they require several numerical calculations. With the advent of calculators and computers the tedium of these computations has been reduced significantly.

One method of approximation is to use rectangles, as is done when defining the definite integral. In this section we will present another method, called the **trapezoidal rule,** that usually produces a closer approximation than the method using merely rectangles. A third method, **Simpson's rule,** is described in the exercises.

An Example

Rectangles can yield good approximations to the area under a curve, but their upper leg is always horizontal. A trapezoid's upper leg can be chosen so that it usually comes closer to the curve. This is why trapezoids usually produce closer approximations to the area under a curve. In Figure 1 we see that the sum of the areas of the trapezoids is a better approximation to the area under the curve than the sum of the areas of the rectangles. Each trapezoid comes closer to the area under the curve.

Let us look at an example of how to approximate the area under a curve with trapezoids. Then we will develop the formula for approximating $\int_a^b f(x)\, dx$. Consider the area under $f(x) = x^2$ from $x = 0$ to $x = 4$. The first step is to divide the interval $[0, 4]$ into subintervals. We choose a convenient number, say, 4, and select the subintervals so that they have equal length. Then the end points of the subintervals are 0, 1, 2, 3, and 4. Each trapezoid has width 1. The heights of the trapezoids are $f(0) = 0$, $f(1) = 1$, $f(2) = 4$, $f(3) = 9$, and $f(4) = 16$. (See

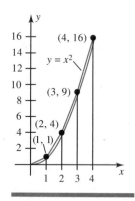

FIGURE 1

Figure 1.) The formula for the area of a trapezoid with heights h and k and width w is

$$\frac{w}{2}(h + k)$$

Let the area of the four trapezoids be A_1, A_2, A_3, and A_4. Then

$$A_1 = \frac{1}{2}[f(0) + f(1)] = \frac{1}{2}[0 + 1] = \frac{1}{2}$$

$$A_2 = \frac{1}{2}[f(1) + f(2)] = \frac{1}{2}[1 + 4] = \frac{5}{2}$$

$$A_3 = \frac{1}{2}[f(2) + f(3)] = \frac{1}{2}[4 + 9] = \frac{13}{2}$$

$$A_4 = \frac{1}{2}[f(3) + f(4)] = \frac{1}{2}[9 + 16] = \frac{25}{2}$$

The approximation is the sum of these areas. It is

$$A_1 + A_2 + A_3 + A_4 = \frac{1}{2} + \frac{5}{2} + \frac{13}{2} + \frac{25}{2} = \frac{44}{2} = 22$$

The actual area can be found using the fundamental theorem. It is

$$\int_0^4 x^2\, dx = \left.\frac{x^3}{3}\right|_0^4 = \frac{64}{3}$$

As a decimal rounded to two places the actual integral is 21.33. This compares favorably with the approximation via the trapezoidal rule with four subintervals, which is 22. A better approximation could be obtained by using more subintervals.

Trapezoidal Rule

We now present the formula for the trapezoidal rule. Assume that $f(x) \geq 0$ in the interval $[a, b]$ so that the area under the curve is equal to the definite integral. To approximate the definite integral $\int_a^b f(x)\, dx$, first divide the interval $[a, b]$ into n equal subintervals, each having width $(b - a)/n$. The elements of the subdivision are

$$a = x_0, x_1, \ldots, x_n = b$$

Observe that $x_k = a + k(b - a)/n$, for $k = 0, 1, \ldots, n$. The area A_k of a trapezoid whose base is $[x_{k-1}, x_k]$ and whose heights are $f(x_{k-1})$ and $f(x_k)$ is given by

$$A_k = \frac{b - a}{2n}[f(x_{k-1}) + f(x_k)]$$

To approximate the area under the curve, we sum all the areas of the trapezoids. Thus

$$\int_a^b f(x)\, dx \simeq A_1 + A_2 + \cdots + A_n$$

$$= \frac{b - a}{2n} [f(x_0) + f(x_1)] + \frac{b - a}{2n} [f(x_1) + f(x_2)]$$

$$+ \cdots + \frac{b - a}{2n} [f(x_{n-1}) + f(x_n)]$$

The expression $(b - a)/2n$ can be factored from each term. Also, each term $f(x_k)$ occurs twice, except the first and last terms, $f(b)$ and $f(a)$. These simplifications result in the following formula, known as the *trapezoidal rule*.

Trapezoidal Rule

Suppose $f(x)$ is a continuous function on the interval $[a, b]$ and $f(x) \geq 0$ on $[a, b]$. To approximate the definite integral of $f(x)$ from a to b by trapezoids, divide $[a, b]$ into n equal subintervals whose end points are $a = x_0, x_1, \ldots, x_n = b$, then

$$\int_a^b f(x)\, dx \simeq \frac{(b - a)}{2n} [f(a) + 2f(x_1)$$

$$+ 2f(x_2) + \cdots + 2f(x_{n-1}) + f(b)]$$

The first example shows that the formula yields the same result as the earlier computations.

EXAMPLE 1

Problem

Use the trapezoidal rule with $n = 4$ to approximate the integral $\int_0^4 x^2\, dx$.

Solution Each subinterval has width 1, and the heights of the trapezoids are $f(0) = 0$, $f(1) = 1$, $f(2) = 4$, $f(3) = 9$, and $f(4) = 16$. The formula yields

$$\int_0^4 x^2\, dx \simeq \frac{4}{8} [f(0) + 2f(1) + 2f(2) + 2f(3) + f(4)]$$

$$= \frac{1}{2} [0 + 2 + 8 + 18 + 16] = \frac{44}{2} = 22$$

This agrees with earlier computations.

The next example shows that better approximations are usually obtained by increasing the number of subintervals.

EXAMPLE 2

Problem

Use the trapezoidal rule with $n = 8$ to approximate the integral $\int_0^4 x^2 \, dx$.

Solution Each subinterval has width $1/2$ since $b = 4$, $a = 0$, and $n = 8$. The heights of the trapezoids are $f(0) = 0$, $f(1/2) = 1/4$, $f(1) = 1$, $f(3/2) = 9/4$, $f(2) = 4$, $f(5/2) = 25/4$, $f(3) = 9$, $f(7/2) = 49/4$, and $f(4) = 16$. The formula yields

$$\int_0^4 x^2 dx \approx \frac{1}{4} \left[f(0) + 2f\left(\frac{1}{2}\right) + 2f(1) + \cdots + f(4) \right]$$

$$= \frac{1}{4} \left[0 + \frac{1}{2} + 2 + \frac{9}{2} + 8 + \frac{25}{2} + 18 + \frac{49}{2} + 16 \right]$$

$$= \frac{1}{4} \left(\frac{172}{2} \right) = \frac{43}{2} = 21.5$$

This is closer to the actual area of $21\frac{1}{3}$ than the earlier approximation using four subintervals.

The next two examples show how to use the trapezoidal rule when no anti-derivative is attainable in a simple form.

EXAMPLE 3

Problem

Use the trapezoidal rule with $n = 4$ to approximate $\int_0^2 (x^3 + 1)^{1/2} \, dx$.

Solution Let $n = 4$, $a = 0$, and $b = 2$. Then

$$f(x_0) = f(0) = 1$$

$$f(x_1) = f\left(\frac{1}{2}\right) = \sqrt{\frac{9}{8}} \approx 1.06$$

$$f(x_2) = f(1) = \sqrt{2} \approx 1.41$$

$$f(x_3) = f\left(\frac{3}{2}\right) = \sqrt{\frac{35}{8}} \approx 2.09$$

$$f(x_4) = f(2) = \sqrt{9} = 3$$

Substituting these values into the formula yields

$$\int_0^2 (x^3 + 1)^{1/2} \, dx \approx \frac{1}{4} \left[f(0) + 2f\left(\frac{1}{2}\right) + 2f(1) + 2f\left(\frac{3}{2}\right) + f(2) \right]$$

$$= \frac{1}{4} (1 + 2.12 + 2.82 + 4.18 + 3) = \frac{1}{4} (13.12) = 3.28$$

EXAMPLE 4

Problem

Use the trapezoidal rule with $n = 5$ to approximate $\int_0^1 e^{-x^2}\,dx$.

Solution Let $n = 5$, $a = 0$, and $b = 1$. Then

$f(x_0) = f(0) = 1$

$f(x_1) = f\left(\dfrac{1}{5}\right) = e^{-1/25} \simeq 0.961$

$f(x_2) = f\left(\dfrac{2}{5}\right) = e^{-4/25} \simeq 0.852$

$f(x_3) = f\left(\dfrac{3}{5}\right) = e^{-9/25} \simeq 0.698$

$f(x_4) = f\left(\dfrac{4}{5}\right) = e^{-16/25} \simeq 0.527$

$f(x_5) = f(1) = e^{-1} \simeq 0.368$

Substituting these values into the formula yields

$$\int_0^1 e^{-x^2}\,dx$$

$$\simeq \frac{1}{10}\left[f(0) + 2f\left(\frac{1}{5}\right) + 2f\left(\frac{2}{5}\right) + 2f\left(\frac{3}{5}\right) + 2f\left(\frac{4}{5}\right) + f(1) \right]$$

$$= \frac{1}{10}(1 + 1.922 + 1.704 + 1.396 + 1.054 + 0.368)$$

$$= \frac{1}{10}(7.444) = 0.7444$$

EXERCISE SET 17-4

In Problems 1 to 20 approximate $\int_a^b f(x)\,dx$ for the given choice of $f(x)$, a, and b, using the trapezoidal rule with n subintervals.

1. $f(x) = x^2$, $a = 0$, $b = 3$, $n = 3$

2. $f(x) = x^2$, $a = 0$, $b = 3$, $n = 6$

3. $f(x) = x^2$, $a = 0$, $b = 6$, $n = 6$

4. $f(x) = x^2$, $a = 0$, $b = 6$, $n = 12$

5. $f(x) = x^3$, $a = 0$, $b = 3$, $n = 3$

6. $f(x) = x^3$, $a = 0$, $b = 3$, $n = 6$

7. $f(x) = x^3$, $a = 0$, $b = 6$, $n = 6$

8. $f(x) = x^3$, $a = 0$, $b = 6$, $n = 12$

9. $f(x) = (x^3 + 1)^{1/2}$, $a = 0$, $b = 5$, $n = 5$

10. $f(x) = (x^3 + 1)^{1/2}$, $a = 0$, $b = 5$, $n = 10$

11. $f(x) = e^{-x^2}$, $a = 0$, $b = 5$, $n = 5$

12. $f(x) = e^{-x^2}$, $a = 0$, $b = 5$, $n = 10$

13. $f(x) = 1/(x + 1)$, $a = 0$, $b = 4$, $n = 4$

14. $f(x) = 1/(x + 1)$, $a = 0$, $b = 4$, $n = 8$

15. $f(x) = 1/(x + 1)$, $a = 0$, $b = 4$, $n = 6$

16. $f(x) = 1/(x + 1)$, $a = 0$, $b = 4$, $n = 10$

17. $f(x) = 1/x^2$, $a = 1$, $b = 5$, $n = 4$

18. $f(x) = 1/x^2$, $a = 1$, $b = 5$, $n = 8$

19. $f(x) = 1/x^2$, $a = 1$, $b = 3$, $n = 8$

20. $f(x) = 1/x^2$, $a = 1$, $b = 5$, $n = 10$

In Problems 21 to 24 approximate the definite integral using the trapezoidal rule with four subintervals.

21. $\int_0^2 (x^2 + x)^{1/2} \, dx$ 22. $\int_0^2 (x^3 + x)^{1/3} \, dx$

23. $\int_0^4 (e^x + x^{1/3}) \, dx$ 24. $\int_1^5 (\ln x + x)^2 \, dx$

In Problems 25 to 28 approximate the definite integral using the trapezoidal rule with eight subintervals.

25. $\int_0^8 (x^2 + 2x)^{1/2} \, dx$ 26. $\int_0^4 (x^3 + x)^{1/4} \, dx$

27. $\int_0^4 (e^{-x} + 2x^{1/2}) \, dx$ 28. $\int_1^5 (x \ln x + x)^2 \, dx$

Simpson's rule is another method of numerical integration that approximates sections of the curve by using segments of parabolas as opposed to line segments in the trapezoidal rule. Usually an approximation by Simpson's rule yields a closer approximation than the trapezoidal rule for the same number of subintervals. It is necessary to have an even number of subintervals for Simpson's rule. The formula for Simpson's rule follows:

Simpson's Rule

Suppose $f(x)$ is a continuous function on the interval $[a, b]$ and suppose $f(x) \geq 0$ on $[a, b]$. To approximate the definite integral of $f(x)$ from a to b by Simpson's rule, divide $[a, b]$ into n (an even number) equal subintervals whose end points are $x_0 = a, x_1, \ldots, x_n = b$, then

$$\int_a^b f(x) \, dx \approx \frac{(b - a)}{3n} [f(a) + 4f(x_1)$$

$$+ 2f(x_2) + 4f(x_3)$$

$$+ 2f(x_4) + 4f(x_5)$$

$$+ \cdots + 2f(x_{n-2})$$

$$+ 4f(x_{n-1}) + f(b)]$$

In Problems 29 to 32 approximate the definite integral by using Simpson's rule with four subintervals.

29. $\int_0^2 (x^2 + x)^{1/2} \, dx$ 30. $\int_0^2 (x^3 + x)^{1/3} \, dx$

31. $\int_0^4 (e^x + x^{1/3}) \, dx$ 32. $\int_1^5 (\ln x + x)^2 \, dx$

In Problems 33 and 34 approximate $\int_0^4 \dfrac{1}{x + 1} \, dx$ using Simpson's rule with n subintervals. Compare this answer with the approximations in Problems 13 and 14. Then compare both approximations with the actual indefinite integral computed by using the fundamental theorem of calculus.

33. $n = 4$ 34. $n = 8$

In Problems 35 to 38 approximate $\int_0^4 \dfrac{1}{1 + x} \, dx$ using Simpson's rule with n subintervals for the given value of n. Compare the approximations with the trapezoidal rule and with the indefinite integral computed by using the fundamental theorem of calculus.

35. $n = 4$ 36. $n = 8$

37. $n = 16$ 38. $n = 2$

Referenced Exercise Set 17–4

1. To ensure that air quality satisfies set standards, air pollution control authorities must determine allowable emission rates for various types of pollutants. An important variable is the cost to treat the pollutants. The total treatment cost $C(x)$ is a function of the emission rate x. In an article investigating how pollution control authorities can establish an optimum scheme for calculating pollution cost functions, Hashimoto and Kimura studied the emission of nitrogen oxide in Tokyo City, where the primary pollutants were utility and industrial boilers and industrial internal combustion engines.[*] They were concerned with the case when the total cost could not be directly assessed, but the marginal cost $g(x)$ could be computed for various zones of the city. Then $C(x)$ is defined to be the integral of $g(x)$, which was computed by numerical integration. In one part of the study seven zones were used. The measurements are given in the table.

zone	1	2	3	4	5	6	7
$g(x)$	101	105	119	163	143	180	105

*Akihiro Hashimoto and Yuri Kimura, "Determining the Optimal Scheme of Zoned Effluent Charges for the Control of Air Pollution," *Socio-Economic Planning Science*, Vol. 14, 1980, pp. 197–208.

Take $g(0) = 100$ and approximate the total cost using the trapezoidal rule.

2. In 1962 Pierre Wenger discovered two persons who were contaminated with radium-226 and strontium-90. They agreed to have their bodily functions closely monitored for 10 years in order to study the long-range effects of contamination. In an article comparing the observed data with the predictions of various models, Wenger and Cosandey centered on the effects of contamination on excretion rates, where the units are in levels of radium-226 excreted per day.[†] They took periodic measurements, and they used numerical integration to compute the sum of the bodily excretion. Their measurements are summarized in the following table.

year	1962	1964	1966	1968	1970	1972
excretion rate	.91	.65	.28	.26	.20	.18

Approximate the sum of the bodily excretion using the trapezoidal rule.

3. In an article investigating the costs and benefits of sexual versus asexual reproduction in various plants in different environments and mating systems, Michaels and Bazzaz studied a perennial herb growing in Kickapoo State Park, Danville, Illinois.[‡] In one part of the study the survival rate was given by $f(t) = 118e^{-.05t^2}$ for $t = 0$ to $t = 5$ months. Approximate the total number of survivors by integrating $f(t)$ from $t = 0$ to $t = 5$ using the trapezoidal rule with 5 intervals.

Cumulative Exercise Set 17–4

In Problems 1 to 8 find the antiderivative.

1. $\int (x^5 + 1)(x^6 + 6x)^{1/3}dx$

2. $\int x(2x - 1)^6 dx$

3. $\int x^{1/4} \ln x \, dx$

4. $\int x^5 e^{x^3} dx$

5. $\int \dfrac{dx}{x(2x + 3)^2}$

6. $\int \dfrac{dx}{9 - 4x^2}$

7. $\int x(25 + x^4)^{1/2} \, dx$

8. $\int \dfrac{x \, dx}{2x - 3}$

In Problems 9 to 11 approximate $\int_a^b f(x)dx$ for the given choice of $f(x)$, a and b, using the trapezoidal rule with n subintervals.

9. $f(x) = x^2$, $a = 0$, $b = 4$, $n = 4$

10. $f(x) = (x^3 + 1)^{1/2}$, $a = 0$, $b = 4$, $n = 4$

11. $f(x) = e^{-x^2}$, $a = 0$, $b = 6$, $n = 6$

12. Approximate the definite integral using the trapezoidal rule with 4 subintervals.

$$\int_0^2 (x^3 + x)^{1/2}dx$$

[†]Pierre Wenger and Maurice Cosandey, "Retention and Excretion of Radium-226 and Strontium-90 in Two Doubly Contaminated Persons," *Health Physics*, Vol. 31, 1976, pp. 225–229.

[‡]H. J. Michaels and F. A. Bazzaz, "Resource Allocation and Demography of Sexual and Apomictic *Antennaria parlinii*," *Ecology*, Vol. 67, 1986, pp. 27–36.

CASE STUDY   ### Demography and the Social Security System

Most people believe that the taxes they pay to the social security system each year are put into a savings account to be held until their retirement. Not so! The U.S. social security system is a pay-as-you-go plan, meaning that the money paid by the work force goes directly to fund those who are retired. In essence, the plan calls for those in the age range of 20 to 64 to support those over 64. The plan has worked well since its inception in 1935 because there has always been a large work force to support a much smaller retired segment of the population.

But times are changing. The present system faces grave consequences in the near future, primarily because both the birthrate and death rate are declining. This means that in a short while fewer people will be entering the work force while a larger number will be retiring. The latter is because of advances in health and medicine.

Demography   Demography answers many age-related questions about various populations, such as what is the effect of a lowered death rate on the retired segment of society, and what is the effect of abortions on the birthrate? The insurance industry was the first industry to use demography extensively, but now many diverse fields, including government, apply demography to a multitude of problems.

In his book on applied demography Nathan Keyfitz* uses calculus to give concise explanations of the theory of demography. As a specific application, Keyfitz describes the ramifications of demographic projections on the U.S. social security system.

Definitions of life table functions   A life table for a given population is a table that contains the percentage of people who are expected to survive to a certain age. These percentages are computed from past records and mathematical projections. The probability of surviving from birth to age x is defined to be $l(x)$. Even though births and deaths occur in single increments, demographers assume $l(x)$ is a continuous function in order to use calculus for their projections. If the population is large, this is a valid assumption.

Consider those who have attained the age of t. The fraction of years survived by these people in the next n years is the "sum" of the probabilities of their surviving at each of these ages from t to $t + n$. Since $l(x)$ is a continuous function, this sum is the integral

$$\int_t^{t+n} l(x) \, dx$$

The definite integral of $l(x)$ from 0 to infinity, or at least to the largest age that anyone lives, say, w, is the sum of all the probabilities, so it is 1.

Example 1 shows how numerical integration is used to compute the number of people in a segment of a population. The example studies the female population of Mexico in 1970. The following partial life table gives the values of $l(x)$ of the female population of Mexico in 1970.

Female Population, Mexico, 1970

Age	Fraction of Those Surviving to a Specific Age, $l(x)$
60	0.008939
62	0.008036
64	0.007071

*Nathan Keyfitz, *Applied Mathematical Demography*, 2nd ed., New York, Springer-Verlag, 1985.

EXAMPLE 1

Problem

Use the trapezoidal rule and the accompanying table to compute the fraction of females in Mexico between the ages of 60 and 64 in 1970.

Solution Find the integral of $l(x)$ from 60 to 64. The trapezoidal rule, with $n = 2$, states

$$\int_{60}^{64} l(x)\, dx \simeq \frac{64 - 60}{4}[l(60) + 2l(62) + l(64)] = l(60) + 2 \cdot l(62) + l(64)$$

$$= 0.008939 + 0.016072 + 0.007071 = 0.032082$$

Therefore the percentage of the female population in Mexico in 1970 between the ages of 60 and 64 was about 3.2%.

Pension cost Consider a pay-as-you-go pension plan in which each person over age 64 is paid the same amount, say, one unit. Also assume that the pension is funded by those aged 20 to 64 and that their salaries are the same amount, one unit. Suppose the total number of individuals in the population is P. Then the amount disbursed to the pensioners is P times the fraction of individuals over 64:

$$P \int_{64}^{w} l(x)\, dx$$

The amount of money paid into the fund will be a fraction of the workers' salaries. Let this fraction be g. Thus g is the premium paid by each worker. The amount of money paid into the fund by the workers is

$$gP \int_{20}^{64} l(x)\, dx$$

Since the fund is a pay-as-you-go plan the amount paid into the fund must equal the amount paid out to the pensioners. This means the two expressions must be equal. Setting them equal and solving for g yields

$$g = \frac{\displaystyle\int_{64}^{w} l(x)\, dx}{\displaystyle\int_{20}^{64} l(x)\, dx}$$

This means that the premium paid by the workers is a percentage of their salary equal to the ratio of the number of individuals over age 64 divided by the number of individuals aged 20 to 64.

The assumptions seem restrictive, but adapting the model to salaries and pension payments that vary with age and even within given ages, as well as other practical considerations—including accounting for some younger persons receiving pensions—complicates the formula for g without altering the principles of the argument. More realistic models have several additional terms that give minor adjustments to the computation of the premium. But the primary conclusion remains: the premium depends on the ratio of those over 64 to those between 20 and 64.

The formula for the premium g illustrates how demographic projections affect pension payments. The premium increases in two ways: if the numerator increases or if the denominator decreases. This means that if the number of pensioners increases, or if the number of individuals in the work force decreases, the premium paid by the workers increases.

Projections for the social security system   The U.S. social security system is a pay-as-you-go plan. The premium paid by the work force is dependent on the ratio of the number of pensioners to the number of individuals in the work force. This type of plan has demographic problems if this ratio has a significant increase. The premium paid by each worker will increase as well. The accompanying table shows the demographic projections of the ratio of pensioners to workers. The values in the table from 1990 on were computed by using numerical integration.

Ratio of Individuals of Pensionable Age Versus Working Age

Year	Age 21–64	Age Over 64	Ratio Over 64/21–64
1950	85,944	12,397	0.1442
1960	92,181	16,675	0.1809
1970	103,939	20,085	0.1932
1980	122,115	24,523	0.2008
1990	137,500	28,933	0.2104
2000	148,589	30,600	0.2059
2025	146,645	45,715	0.3117
2050	147,635	45,805	0.3103

Significant pressure on the social security plan will come after the year 2000. The ratio will increase by more than 50% in the years from 2000 to 2025, from 0.20 to 0.31. The only thing that could avert the swing is a large number of births before the year 2000 that would increase the work force in the early 21st century. This seems unlikely.

The social security system can be understood as a method of borrowing from future generations. When initiated, it was reasonable because there were no start-up funds to pay the retirees during its first few years. In the ensuing years the plan also was viable because the ratio of pensioners to the work force was a relatively small fraction, so that the premium paid by each worker was manageable. In addition, the premium was viewed as insurance for one's own retirement. But the ratio will increase dramatically soon so that either the system will prove to be a great burden on the work force or the benefits paid to retirees will be greatly reduced.

Case Study Exercises

In Problems 1 to 4 use the following table and the trapezoidal rule to find the integral of $l(x)$ from the given ages.

Female Population,
Mexico, 1970

Age	Fraction of Those Surviving to a Specific Age, $l(x)$
60	0.008939
61	0.008570
62	0.008036
63	0.007512
64	0.007071

1. From age 60 to 62
2. From age 61 to 63
3. From age 60 to 63
4. From age 60 to 64
5. Compare the answer in Exercise 4 to Example 1. Which most likely gives the better approximation?
6. Give a reason why the percentage in the ratio of individuals of pensionable age versus working age increases so dramatically from the year 2000 to 2025.

CHAPTER REVIEW

Key Terms

17–1 Substitution
Substitution of a Variable   Differential of u
Integration by Substitution

17–2 Integration by Parts
Integration by Parts

17–4 Numerical Integration
Numerical Integration   Simpson's Rule
Trapezoidal Rule

Summary of Important Concepts

Formulas

If $u = g(x)$ then $du = g'(x)\ dx$

$$\int [f(x)]^n\ f'(x)\ dx = \frac{f(x)^{n+1}}{n+1} + C, \quad n \neq 1$$

$$\int e^{f(x)}\ f'(x)\ dx = e^{f(x)} + C$$

$$\int [f(x)]^{-1}\ f'(x)\ dx = \ln |f(x)| + C$$

$$\int u\ dv = uv - \int v\ du$$

REVIEW PROBLEMS

1. Find $\int (2x + 1)(x^2 + x)^{-3}\ dx$

2. Find $\int (e^{2x} + 3)^{-2} e^{2x}\ dx$

3. Find $\int_0^1 x(x^2 - 2)^3\ dx$

4. Find $\int 3xe^{4x}\ dx$

5. Find $\int x^2(1 - x)^{1/3}\ dx$

6. Find the definite integral $\int_0^2 x(4x + 1)^{-1/2}\ dx$.

In Problems 7 to 9 use the table of integrals to find the integral.

7. $\int \dfrac{dx}{x(5x - 7)^2}$

8. $\int x(9x^4 + 16)^{1/2}\ dx$

9. $\int \dfrac{dx}{14 - 9x^2}$

10. Use the trapezoidal rule with four subintervals to approximate $\int_0^4 x^3\ dx$.

11. Use the trapezoidal rule with six subintervals to approximate $\int_0^3 e^{-x^2}\ dx$.

▦ PROGRAMMABLE CALCULATOR EXPERIMENTS

1. Write a program that approximates the area under the curve of $y = f(x)$ between $x = a$ and $x = b$ using the trapezoidal rule with n intervals.

In Problems 2 to 4 run the program in Problem 1 for the given function and values of a and b.

2. $f(x) = x^2 + 2x$, $a = 0$, $b = 4$

3. $f(x) = x^3 + 2x$, $a = 0$, $b = 4$

4. $f(x) = x^3 + x^2$, $a = 0$, $b = 1$

5. Write a program that approximates the area under the curve of $y = f(x)$ between $x = a$ and $x = b$ using Simpson's rule with n intervals.

In Problems 6 to 8 run the program in Problem 5 for the given function and values of a and b.

6. $f(x) = x^2 + 2x$, $a = 0$, $b = 4$

7. $f(x) = x^3 + 2x$, $a = 0$, $b = 4$

8. $f(x) = x^4 + 2x$, $a = 0$, $b = 1$

9. Compare the approximations to the definite integral of the function in Problems 3 and 7 to the computation of the integral via the fundamental theorem of calculus.

10. Compare the approximations to the definite integral of the function in Problems 2 and 6 to the computation of the integral via the fundamental theorem of calculus.

Functions of Several Variables

18

The costs of high-tech industrial giants can be minimized by using functions of many variables. (H. Armstrong Roberts/Zefa)

Chapter Overview

Many applications of calculus use variable quantities that are functions of more than one variable. For example, profit is a function of cost, price, and sales volume. The total population of a species in a given environment depends on its rate of reproduction, its death rate, the availability of food, and the population of its predators.

In this chapter we study functions of more than one variable. The bulk of the material is on functions of two variables. Such functions are graphed in three dimensions so that it is possible to view them geometrically. In the first section we cover some examples of functions of several variables. The later sections show that many of the concepts that were developed for functions of one variable have analogous statements for functions of several variables.

In Section 18–2 we study the concept of rate of change of a function of several variables. It is an extension of the idea of the derivative of a function of one variable. Likewise in Section 18–3 the idea of relative extrema is covered. In Section 18–4 a constraint is added to the problem of finding relative extrema. These constraints, called Lagrange multipliers, have many applications to business and other disciplines.

The multivariable analog of the integral of a function of one variable is covered in the next two sections. We consider only functions of two variables, so Section 18–5 is called "Double Integrals." In this section we show how such integrals form a natural extension of the ideas developed for integrals of functions of a single variable. Section 18–6 applies the techniques of double integration to volumes of certain kinds of solid figures.

CASE STUDY PREVIEW

Jojoba oil is used for producing amides which, in turn, are an essential ingredient in many chemical processes. The procedure of extracting amides from jojoba oil is time-consuming and expensive. First the oil is refined and purified. Then other chemicals are added and the mixture is heated for several hours. Once the material becomes cakelike the amides can be extracted.

Researchers have tried various methods to maximize the amide yield while conserving cost to produce the cakelike material. The cost is a function of the two variables—temperature and time. The Case Study describes how the chemical industry uses this function of two variables to minimize the cost of producing amides.

18–1 Examples of Functions of Several Variables

Thus far we have considered only functions of one variable. Such functions are of the form $y = f(x)$ where x is the single independent variable. One value of the independent variable is substituted into the formula and a corresponding number is produced. For a function of two variables two numbers, given as an ordered pair, are substituted into a formula and another number is produced. In this section we consider examples of functions of two and more variables.

The Definition of a Function of Two Variables

An example of a function of two variables is

$$z = f(x, y) = 2x + 3y$$

The name of the function is f and z is called the **dependent variable.** The two variables x and y are **independent variables,** meaning that x and y can be chosen to be any numbers, while the corresponding value of z depends on which choice is made for x and y. For instance, let $x = 1$ and $y = -4$, then

$$z = f(1, -4) = 2(1) + 3(-4) = 2 - 12 = -10$$

Thus $z = -10$ when $x = 1$ and $y = -4$. This is succinctly stated by $f(1, -4) = -10$.

This leads to the definition of a function of two variables. The definition of functions of more than two variables is similar.

DEFINITION

The formula $z = f(x, y)$ defines f as a function of the two independent variables x and y if, for each ordered pair of real numbers (x, y) for which the formula is defined, there exists a unique corresponding value for the variable z.

The set of all ordered pairs of numbers for which $f(x, y)$ exists is called the *domain of f*. The set of all values of z that correspond to an ordered pair (x, y) is called the *range of f*. Usually the domain of a given function is not stated, and it is assumed that the domain is the largest set of ordered pairs for which the formula is defined. Occasionally, however, a restricted domain is given.

Sometimes the name of the function is omitted when only one function is being discussed. For example, the function above could have been given as simply $z = 2x + 3y$. If the dependent variable is not specifically mentioned, we assume its name is either z or the name of the function. That is, if the function above would have been given by the formula $f(x, y) = 2x + 3y$, where the dependent variable is not mentioned, it could be named either f or z.

E X A M P L E   1

Problem

If possible, evaluate the given functions at the points $(0, 0)$, $(-1, 2)$, and $(2, -3)$: (a) $f(x, y) = 4x - 5y + 2$, (b) $g(x, y) = 2x^2 + xy - \dfrac{1}{y + 3}$.

Solution (a) $f(0, 0) = 4(0) - 5(0) + 2 = 2$; $f(-1, 2) = 4(-1) - 5(2) + 2 = -12$; $f(2, -3) = 4(2) - 5(-3) + 2 = 8 + 15 + 2 = 25$.
(b) $g(0, 0) = 2(0)^2 + (0)(0) - 1/(0 + 3) = -1/3$; $g(-1, 2) = 2(-1)^2 +$

$(-1)(2) - 1/(2 + 3) = 2 - 2 - 1/5 = -1/5; g(2, -3) = 2(2)^2 + (2)(-3) - 1/(-3 + 3) = 8 - 6 - 1/0$, which is not defined because $1/0$ is not defined, so $(2, -3)$ is not in the domain of g.

A function of three variables is defined similarly to a function of two variables. It differs only in that an ordered triple is substituted into a formula that has three variables. The next example gives an illustration of a function of three variables and one of four variables.

EXAMPLE 2   Define the functions

$$f(x, y, z) = x^2 - 4yz \qquad g(x, y, z, w) = xy + \frac{4z^2}{w}$$

Problem

(a) Find $f(2, 1, -1)$, (b) $g(1, 2, -3, 2)$.

Solution (a) $f(2, 1, -1) = (2)^2 - 4(1)(-1) = 4 + 4 = 8$.
(b) $g(1, 2, -3, 2) = (1)(2) + 4(-3)^2/2 = 2 + 18 = 20$.

Three-Dimensional Coordinate Systems

The graph of a function of two variables requires a three-dimensional coordinate system. This is because a point in the domain (a, b) is plotted in the xy-plane and then its functional value c is plotted on a third axis. Thus the point (a, b, c) is plotted in a three-dimensional system. Figure 1 demonstrates the usual convention for labeling the three axes, called the x-, y-, and z-axes. The z-axis is vertical, the

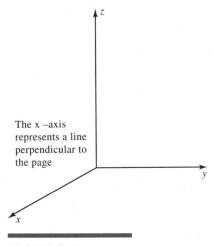

The x –axis
represents a line
perpendicular to
the page

FIGURE 1

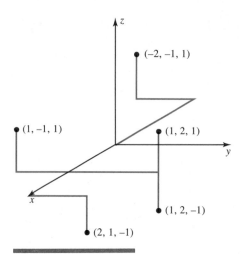

FIGURE 2

y-axis is horizontal, and the x-axis represents a line that is perpendicular to the plane of the paper. Since the figure is a two-dimensional depiction of a three-dimensional object, a perspective drawing is needed. That is why the x-axis is drawn at an angle. Only the positive half of each axis is drawn for clarity.

The point where the three axes intersect is called the *origin*. A point (a, b) in the xy-plane is given the three-dimensional coordinates $(a, b, 0)$. A point (a, b, c) is $|c|$ units above or below the point $(a, b, 0)$, depending on whether c is positive or negative. The coordinate system is divided into eight regions, called *octants*, four above the xy-plane and four below. Only one octant is named. It is the region where all three coordinates are positive, called the *first octant*. Several points in various octants are plotted in Figure 2.

Graphs of Functions of Two Variables

Since the domain of the function $z = f(x, y)$ is a set of ordered pairs (x, y), the graph of the function entails locating the points in the domain in the xy-plane. Then their corresponding functional values are plotted on the z-axis. Thus a point on the graph is an ordered triple (x, y, z). For example, consider the function $z = f(x, y) = 2x + 3y$. To locate the point on the graph corresponding to $(1, 2)$, first plot $(1, 2)$ in the xy-plane and locate $z = f(1, 2) = 8$ on the z-axis. Thus the point on the graph is $(1, 2, 8)$. This point is one unit out on the x-axis, two units to the right on the y-axis and eight units up on the z-axis. This point along with several others is plotted in Figure 3.

There is no easy method for graphing functions of two variables. The problem entails working algebraically with the formula to determine what type of two-dimensional figures arise when the graph is intersected with the graph of vertical and horizontal planes. These intersections are then pieced together to get a perspective figure. The planes used in the perspective drawings are the coordinate

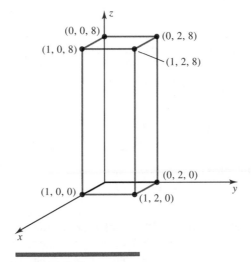

FIGURE 3

planes and those that are parallel to the coordinate planes. For instance, the set of points with z-coordinate 1 is the plane one unit above the xy-plane. Its equation is $z = 1$. Likewise

The graph of $x = a$ is a plane parallel to the yz-plane.
The graph of $y = b$ is a plane parallel to the xz-plane.
The graph of $z = c$ is a plane parallel to the xy-plane.

Figure 4 illustrates these planes.

The intersection of one of the coordinate planes with the graph of the function is called a **trace** of the graph. A **section** of a surface $z = f(x, y)$ is the intersection

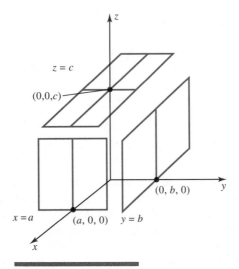

FIGURE 4

of the surface with a plane that is parallel to a coordinate plane. A section of the graph is found by solving simultaneously the two equations, the equation of the function and the equation of the plane. Thus a section is a two-dimensional figure in one of the planes $x = a$, $y = b$, or $z = c$.

The next example illustrates how to use sections to graph functions of two variables.

EXAMPLE 3

Problem

Graph the functions.
(a) $z = f(x, y) = 2x + 3y$, (b) $z = h(x, y) = x^2 + 4y^2$, (c) $z = g(x, y) = y^2 - x^2$.

Solution (a) The sections of the graph corresponding to $x = a$ are the straight lines $z = 2a + 3y$, each in the plane $x = a$, which is parallel to the yz-plane. The sections of the graph corresponding to $y = b$ are the straight lines $z = 2x + 3b$, each in the plane $y = b$, which is parallel to the xz-plane. The sections of the graph corresponding to $z = c$ are the straight lines $c = 2x + 3y$, each in the plane $z = c$, which is parallel to the xy-plane. The graph is the plane sketched in Figure 5.

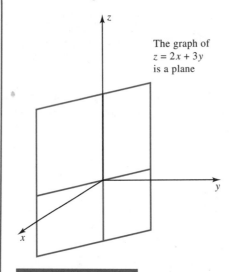

The graph of
$z = 2x + 3y$
is a plane

FIGURE 5

(b) The trace in the yz-plane is found by letting $x = 0$ in the equation. It is the parabola $z = 4y^2$. The trace in the xz-plane is the parabola $z = x^2$. The sections of the graph corresponding to $x = a$ are the parabolas $z = a^2 + 4y^2$. The sections of the graph corresponding to $y = b$ are the parabolas $z = x^2 + 4b^2$. The sections of the graph corresponding to $z = c$ are the ellipses $c = x^2 + 4y^2$. The graph is an infinite bowl-shaped surface, sketched in Figure 6. It is called an elliptic paraboloid.

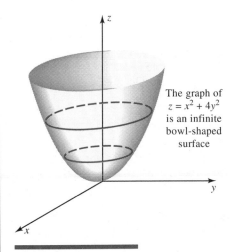

The graph of
$z = x^2 + 4y^2$
is an infinite
bowl-shaped
surface

FIGURE 6

(c) The trace in the yz-plane is the parabola $z = y^2$ (see Figure 7a) and the trace in the xz-plane is the parabola $z = -x^2$ (see Figure 7b). The sections of the graph corresponding to $x = a$ are the parabolas $z = y^2 - a^2$. The sections of

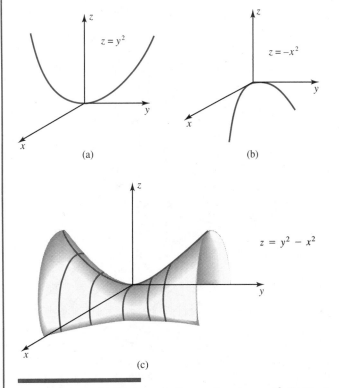

FIGURE 7 (a) The trace in the xy-plane is $z = y^2$. (b) The trace in the xz-plane is $z = -x^2$. (c) The graph of $z = y^2 - x^2$ is a saddle-shaped surface.

the graph corresponding to $y = b$ are the parabolas $z = b^2 - x^2$. The sections of the graph corresponding to $z = c$ are the hyperbolas $c = y^2 - x^2$. The graph is a saddle-shaped surface, sketched in Figure 7c. It is called a hyperbolic paraboloid.

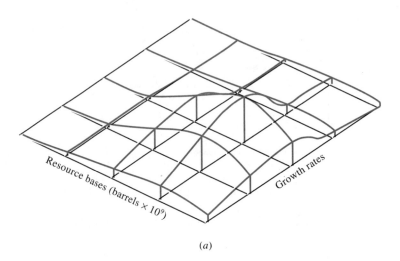

(a)

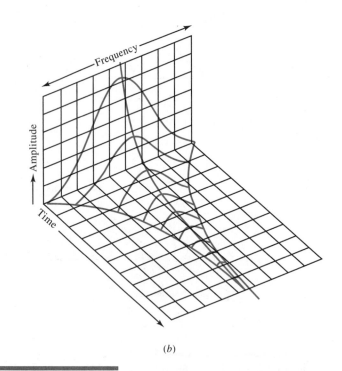

(b)

FIGURE 8

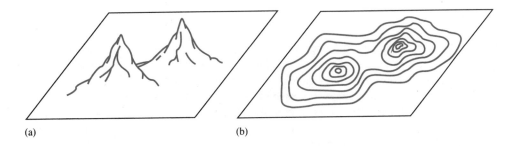

(a)   (b)

FIGURE 9

Applications of Sections

To see how sections are used to give a visual structure to a two-dimensional graph of a three-dimensional surface, consider the two graphs in Figure 8. The first graph appears in a book that investigates the future of the world supply of oil.* The second is taken from a book describing the structure of sound in music.† The surface in Figure 8a describes the relationship between the amount of oil in the ground (resources) and the growth in oil consumption. This affects the likelihood that the world supply of oil will exceed demand. This likelihood depends on the amount of resources available (already taken from the earth) and the growth rate of demand. The surface in Figure 8b describes the relationship among amplitude, frequency, and time of a musical note.

*Peter R. Odell and Kenneth E. Rosing, *The Future of Oil,* New York, Nichols Publishing, 1980, p. 182.
†Robert Erickson, *Sound Structure in Music,* Berkeley, CA, University of California Press, 1975, p. 62.

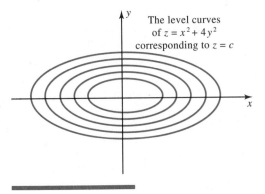

The level curves
of $z = x^2 + 4y^2$
corresponding to $z = c$

FIGURE 10

Level Curves

Another method used to describe the graph of a function of two variables is to sketch the sections $z = c$ in the xy-plane. Each graph is called a **level curve.** Level curves are commonly used on television weather reports and in contour maps, also called topographical maps. For example, Figure 9a illustrates a picture of a mountain range and Figure 9b depicts its corresponding topographical map. The level curves are the intersections of horizontal planes ($z = c$) and the earth, and thus describe the heights of the mountains.

For the function $z = x^2 + 4y^2$, whose graph is sketched in Figure 6, the level curves are the sections corresponding to $z = c$. They are the ellipses graphed in Figure 10.

EXERCISE SET 18–1

In Problems 1 to 4 find the functional value of $f(x, y) = 2x^2 - y^2$ at the given point.

1. $(1, 1)$
2. $(-1, 2)$
3. $(0, -3)$
4. $(-4, 0)$

In Problems 5 to 8 find the functional value of $f(x, y) = y(x^3 + 1)^{1/2} + xy^3$ at the given point.

5. $(0, 2)$
6. $(0, -3)$
7. $(2, 3)$
8. $(2, -3)$

In Problems 9 to 12 find the domain of the function.

9. $f(x, y) = \dfrac{y}{x - 2}$
10. $f(x, y) = \dfrac{2}{x(y^2 - 1)}$

11. $f(x, y) = \dfrac{y}{(x^3 + 1)^{1/2}}$
12. $f(x, y) = \dfrac{2}{x(y - 1)^{1/2}}$

In Problems 13 to 20 graph the functions, giving the traces and at least one section that is parallel to each coordinate plane.

13. $f(x, y) = 3x + 5y$
14. $f(x, y) = y - 4x$

15. $f(x, y) = 2 + 2x - 3y$

16. $f(x, y) = y - 2x - 3$

17. $f(x, y) = 4 + x^2 + 5y^2$

18. $f(x, y) = 1 + 2x^2 + y^2$

19. $f(x, y) = y^2 - 5x^2$
20. $f(x, y) = x^2 - 2y^2$

In Problems 21 to 26 graph the functions.

21. $f(x, y) = 1 - x^2 - y^2$

22. $f(x, y) = 2 - 2x^2 - y^2$

23. $f(x, y) = x^2 + 2x + y^2 + 4y$

24. $f(x, y) = 1 + x^2 + 2x + y^2$

25. $f(x, y) = 2x + x^2$
26. $f(x, y) = y^2 - 4y$

27. The equation of the sphere having a radius of one unit and its center at the origin is $x^2 + y^2 + z^2 = 1$. Is it possible to represent this sphere as a function?

28. Solve the equation $x^2 + y^2 + z^2 = 4$ for z and then keep only the positive square root. Graph the resulting function using the fact about spheres stated in Problem 27.

29. Give a geometrical test using lines perpendicular to the xy-plane to determine whether a given surface is a function $z = f(x, y)$. Recall the geometric test to determine whether a two-dimensional curve is a function.

30. A salesperson hires two employees. One earns $10 per hour and the second earns $12 per hour. Let the number of hours worked per week by the first employee be represented by x and the number of hours worked by the second employee be represented by y. If the salesperson's fixed cost is $300 per week, find a formula for the cost per week.

31. A manufacturing company makes three types of desks, small, medium, and large. If the profit is $40 for each small desk sold, $50 for each medium desk sold, and $60 for each large desk sold, find a formula for the profit by letting x represent the number of small desks sold, y the number of medium desks sold, and z the number of large desks sold.

In Problems 32 to 35 match the equation with the given graph.

32. $f(x, y) = x + y$

33. $f(x, y) = x^2$

34. $f(x, y) = x^2 + y^2$

35. $f(x, y) = x^2 - y^2$

(c)

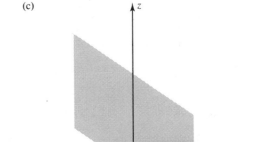

(a)

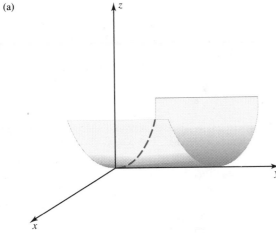

(b)

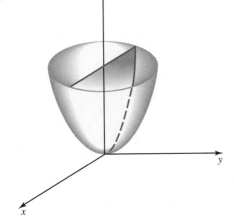

(d)

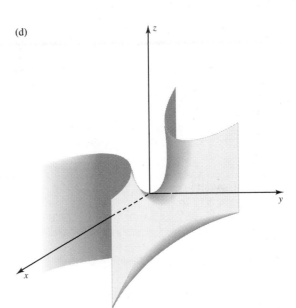

Referenced Exercise Set 18–1

1. Accurate forecasting of ambulance demand directly affects staffing, inventories, and assignment of units within the region served by the ambulance corps. In an article studying ambulance service to four counties in South Carolina, Baker and Fitzpatrick developed a model that accurately forecasted ambulance demand

while minimizing cost.* The cost function to be minimized was

$$C(x, y, z, w) = 103x + 109y + 116z + 117w$$

where the cost per call for an ambulance was $103 in Spartanburg County, $109 in Clarendon County, $116 in Horry County, and $117 in Florence County; and x, y, z and w were the number of calls per day in each county, respectively. For a given week in 1983 in the study, the number of calls in each county was as follows: (a) on Monday, Spartanburg had 670 calls, Clarendon had 370 calls, Horry had 37 calls, Florence had 320 calls; (b) on Tuesday, Spartanburg had 690 calls, Clarendon had 300 calls, Horry had 38 calls, Florence had 290 calls; (c) on Wednesday, Spartanburg had 680 calls, Clarendon had 320 calls, Horry had 40 calls, Florence had 300 calls. Which day produced the least total cost?

2. Most large industries utilize research and development to improve their products and their marketing techniques. In an article evaluating the cost-effectiveness of research in the Florida citrus industry, Stranahan and Shonkwiler centered on the Florida frozen concentrate orange juice market.† The cost C was assumed to be a function of the output Y, the quantity of research R, and the price P. In one part of the study the cost was computed to be

$$C(Y, R, P) = 1.1Y^{.5}R^{.43}P^{-.009}$$

Compute $C(Y, R, P)$ when $Y = 10$ million gallons per day, $R = .2$ million dollars per day, and $P = \$1$ per gallon.

3. Many studies have been done on the harmful effects of large doses of radiation, such as the tragic accident at Chernobyl, U.S.S.R., in 1987. Johnson studied harmful effects of exposure to radiation at very low level dose rates which result from various environmental sources of radiation.‡ These dosages are called high natural radiation background. They occur from diverse sources such as sun rays and wristwatches. The article investigated data from the World Health Organization based on radiation reports from 45 countries. One part of the article studied cancer rates. The number of years required for detection y was found to be a function of the background radiation rate r, measured in rads per year, and the average cancer rate p, measured in cases per million persons per year. Thus

$$y(r, p) = \frac{kp}{r^2}$$

where k is a constant depending upon the country or region being studied. Compute $y(.88, .02)$ for $k = .04$.

*J. R. Baker and K. E. Fitzpatrick, "Determination of an Optimal Forecast Model for Ambulance Demand Using Goal Programming," *Journal of the Operations Research Society,* Vol. 37, 1986, pp. 1047–1059.

†H. A. Stranahan and J. S. Shonkwiler, "Evaluating the Returns to Postharvest Research in the Florida Citrus Processing Subsector," *American Journal of Agricultural Economics,* Vol. 68, 1986, pp. 88–94.

‡Richard E. Johnson, "Problems Involved in Detecting Increased Malignancy Rates in Areas of High Natural Radiation Background," *Health Physics,* Vol. 31, 1976, pp. 148–160.

18–2 Partial Derivatives

One of the most important concepts of calculus is rate of change. For a function of one variable, rate of change is measured by the derivative. For a function of two variables, rate of change can be measured in many ways so that the situation is more complicated. In this section we introduce the concept of a partial derivative, which is the three-dimensional analog of the derivative.

An Application from Business

Rate of change is used extensively in business. The demand function for a particular company or industry often is expressed as a function of more than one variable.

For instance, in an article studying the effect of the "health scare" on the cigarette industry, Bishop and Yoo consider the quantity demanded Q as a function of disposable income D and cigarette advertising expenditures A.* To model the industry's data, the authors use a Cobb–Douglas function of the form

$$Q(D, A) = 231.7D^{0.66}A^{0.07}$$

The article studies the effect on Q when D and A experience increases and decreases. Bishop and Yoo use several types of measures, each of which depends on the partial derivatives of Q. At the end of the section we compute these partial derivatives.

Partial Derivatives: A Geometric Description

The derivative of $y = f(x)$ measures how a change in x affects y. For a function of two variables $z = f(x, y)$ the partial derivatives measure how a change in x or in y affects z. Geometrically the derivative $f'(a)$ is the slope of the tangent line to the curve $y = f(x)$ at $x = a$. But a surface in three dimensions can have a tangent line in any direction, so there are an infinite number of tangent lines. (See Figure 1.) It suffices to study tangent lines in only two directions, the directions parallel to the x-axis and parallel to the y-axis. This is because the tangent lines in every other direction can be expressed in terms of these two.

Consider the direction parallel to the x-axis. How can we interpret the rate of change of $z = f(x, y)$ at the point (a, b) in this direction? The section of the graph in this direction is a two-dimensional curve obtained by taking the intersection of $z = f(x, y)$ with the plane $y = b$. The equation of this curve is $z = f(x, b)$.

*John A. Bishop and Jang H. Yoo, " 'Health Scare,' Excise Taxes and Advertising in the Cigarette Demand and Supply," *Southern Economics Journal*, Vol. 52, 1985, pp. 402–411.

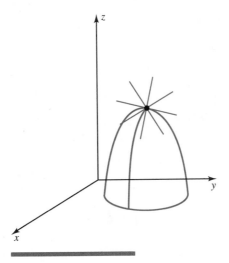

FIGURE 1

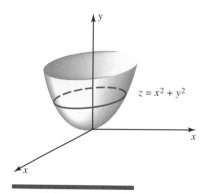

FIGURE 2

Since this curve is two-dimensional, the slope of the tangent line at (a, b) is the derivative of the function of one variable, $z = f(x, b)$. This is called the partial derivative of $z = f(x, y)$ with respect to x. It is found by holding y constant and finding the derivative with respect to x.

For example, consider the function

$$z = f(x, y) = x^2 + y^2$$

Its graph, given in Figure 2, is an infinite bowl-shaped surface. To find the rate of change at $(1, 0)$ in the direction that is parallel to the x-axis, first find the intersection of $y = 0$ and the curve. It is $z = f(x, 0) = x^2 + 0 = x^2$. The derivative of this two-dimensional function is $2x$. This is expressed as

$$f_x(x, y) = 2x$$

It is called the *partial derivative of f with respect to x*. To get the slope of the tangent line at $(1, 0)$, substitute $x = 1$ into the formula for f_x and get $f_x(1, 0) = 2$. From Figure 3 we see that the function $z = f(x, 0)$ is a parabola in the xz plane whose tangent line at $(1, 0)$ is 2.

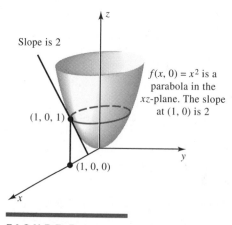

FIGURE 3

Partial Derivatives: Formal Definition

The definition of the partial derivative uses the fact that if one variable is held constant the resulting expression is a function of one variable, and so its derivative can be computed. Thus a partial derivative of $f(x, y)$ is the derivative of a function of one variable that is obtained by holding constant one of the two variables of $f(x, y)$.

DEFINITION

The **partial derivative** of $f(x, y)$ with respect to x is defined as

$$f_x(x, y) = \lim_{h \to 0} \frac{f(x + h, y) - f(x, y)}{h}$$

The partial derivative of $f(x, y)$ with respect to y is defined as

$$f_y(x, y) = \lim_{h \to 0} \frac{f(x, y + h) - f(x, y)}{h}$$

There are several alternate notations for the partial derivatives. If it is clear that f is a function of the two variables x and y, then sometimes $f_x(x, y)$ is expressed as simply f_x. Similarly f_y represents the partial derivative of f with respect to y. The notation for the partial derivative that is similar to the $\frac{dy}{dx}$ notation for the derivative is

$$\frac{\partial f}{\partial x} \quad \text{and} \quad \frac{\partial f}{\partial y}$$

This is read "the partial derivative of $f(x, y)$ with respect to x" and "the partial derivative of $f(x, y)$ with respect to y, respectively. Example 1 illustrates how to compute the partial derivatives of functions of two variables.

EXAMPLE 1

Problem

Compute $f_x(x, y)$ and $f_y(x, y)$ for the functions (a) $f(x, y) = x^3 + xy - y^{-1}$ (b) $f(x, y) = x^2 y^{-3} - e^{2xy} + 3$.

Solution (a) To find f_x treat y as a constant and differentiate with respect to x.

$$f_x(x, y) = 3x^2 + y$$

Since y is held constant, the term y^{-1} is treated as a constant so that the partial derivative of it is 0.

To find f_y, treat x as a constant and differentiate with respect to y.

$$f_y(x, y) = x - (-1)y^{-2} = x + y^{-2}$$

(b) To find f_x, treat y as a constant and differentiate with respect to x.

$$f_x(x, y) = 2xy^{-3} - 2ye^{2xy}$$

To find f_y, treat x as a constant and differentiate with respect to y.

$$f_y(x, y) = -3x^2y^{-4} - 2xe^{2xy}$$

The next example demonstrates how to incorporate the chain rule with partial derivatives.

EXAMPLE 2

Problem

Consider the function $f(x, y) = (x^3 + xy^{-1})^4$. Compute (a) $f_x(x, y)$ (b) $f_y(x, y)$.

Solution (a) Use the extended power rule. Let $u = x^3 + xy^{-1}$ so that the partial derivative of u with respect to x is $3x^2 + y^{-1}$. Treat y as a constant and differentiate with respect to x.

$$f_x(x, y) = 4(x^3 + xy^{-1})^3(3x^2 + y^{-1})$$

(b) Again let $u = x^3 + xy^{-1}$. Then the partial derivative of u with respect to y is $-xy^{-2}$. Treat x as a constant and differentiate with respect to y.

$$f_y(x, y) = 4(x^3 + xy^{-1})^3(-xy^{-2})$$

If the partial derivative with respect to x is to be computed at a particular point (a, b), then the notation used is

$$f_x(a, b)$$

Likewise $f_y(a, b)$ represents the partial derivative with respect to y evaluated at the point (a, b). The next example illustrates this notation.

EXAMPLE 3

Problem

Consider the function $f(x, y) = 2x^3 + xy^2 - y$. Compute (a) $f_x(1, 3)$ (b) $f_y(-1, 2)$.

Solution (a) First find $f_x(x, y)$.

$$f_x(x, y) = 6x^2 + y^2$$

Next substitute the values $x = 1$ and $y = 3$.

$$f_x(1, 3) = 6(1)^2 + (3)^2 = 6 + 9 = 15$$

(b) Find $f_y(x, y)$.

$$f_y(x, y) = 2xy - 1$$

Next substitute the values $x = -1$ and $y = 2$.

$$f_y(-1, 2) = 2(-1)(2) - 1 = -4 - 1 = -5$$

Second-Order Partial Derivatives

Just as we can define higher-order derivatives of functions of one variable, so too we can define higher-order partial derivatives. The partial derivative of a partial derivative is called a second-order partial derivative, or simply a **second partial derivative.** The four second partial derivatives of $f(x, y)$ are

$$f_{xx}(x, y)$$
$$f_{xy}(x, y)$$
$$f_{yx}(x, y)$$
$$f_{yy}(x, y)$$

Just as with first-order partial derivatives, the shortened form of $f_{xx}(x, y)$ is f_{xx}. Also f_{xy}, f_{yx}, and f_{yy} are defined in a similar way. The functions f_{xy} and f_{yx} are called **mixed partial derivatives.** The notation for f_{xy} states that the partial derivative of $f(x, y)$ with respect to x is to be computed first, followed by computing the partial derivative of $f_x(x, y)$ with respect to y; f_{yx} calls for the reverse order of differentiation. For most functions that we will study these two functions are equal. The next example illustrates how to compute second partial derivatives.

EXAMPLE 4

Problem

Consider the function $f(x, y) = x^3y + xy^{-2} - ye^{3x}$. Compute (a) $f_{xx}(x, y)$, (b) $f_{xy}(x, y)$, (c) $f_{yx}(x, y)$, (d) $f_{yy}(x, y)$, (e) $f_{xx}(1, 3)$, (f) $f_{xy}(0, 1)$.

Solution First find $f_x(x, y)$ and $f_y(x, y)$.

$$f_x(x, y) = 3x^2y + y^{-2} - 3ye^{3x}$$
$$f_y(x, y) = x^3 - 2xy^{-3} - e^{3x}$$

(a) To find f_{xx}, compute the partial derivative of f_x with respect to x.

$$f_{xx}(x, y) = \frac{\partial}{\partial x} f_x(x, y) = \frac{\partial}{\partial x}(3x^2y + y^{-2} - 3ye^{3x}) = 6xy - 9ye^{3x}$$

(b) To find f_{xy} compute the partial derivative of f_x with respect to y.

$$f_{xy}(x, y) = \frac{\partial}{\partial y} f_x(x, y) = \frac{\partial}{\partial y}(3x^2y + y^{-2} - 3ye^{3x})$$

$$= 3x^2 - 2y^{-3} - 3e^{3x}$$

(c) To find f_{yx}, compute the partial derivative of f_y with respect to x.

$$f_{yx}(x, y) = \frac{\partial}{\partial x} f_y(x, y) = \frac{\partial}{\partial x}(x^3 - 2xy^{-3} - e^{3x})$$

$$= 3x^2 - 2y^{-3} - 3e^{3x}$$

Notice that $f_{xy} = f_{yx}$.

(d) To find f_{yy}, compute the partial derivative of f_y with respect to y.

$$f_{yy}(x, y) = \frac{\partial}{\partial y} f_y(x, y) = \frac{\partial}{\partial y} (x^3 - 2xy^{-3} - e^{3x}) = 6xy^{-4}$$

(e) To find $f_{xx}(1, 3)$ substitute the values $x = 1$ and $y = 3$ into $f_{xx}(x, y)$.

$$f_{xx}(1, 3) = 6(1)(3) - 9(3)e^{3(1)} = 18 - 27e^3$$

(f) To find $f_{xy}(0, 1)$, substitute the values $x = 0$ and $y = 1$ into $f_{xy}(x, y)$.

$$f_{xy}(0, 1) = 3(0)^2 - 2(1)^{-3} - 3e^{3(0)} = -2 - 3 = -5$$

The Cigarette Industry Revisited

The demand function computed by Bishop and Yoo for the cigarette industry was given in the beginning of the section as

$$Q(D, A) = 231.7D^{0.66}A^{0.07}$$

The analysis of the industry in the article uses several measures of the effect on Q by changes in D and A. These measures are based on the partial derivatives of Q. The investigators reached various conclusions concerning when it would be most advantageous for the industry to increase advertising in order to increase demand. The next example computes these partial derivatives.

EXAMPLE 5

Problem

Find (a) Q_D and (b) Q_A for the above demand function.

Solution (a) Treat A as a constant.

$$Q_D = 231.7(0.66)D^{0.66-1}A^{0.07} = 152.9D^{-0.34}A^{0.07}$$

(b) Treat D as a constant.

$$Q_A = 231.7(0.07)D^{0.66}A^{0.07-1} = 16.22D^{0.66}A^{-0.93}$$

EXERCISE SET 18–2

In Problems 1 to 14 find $f_x(x, y)$ and $f_y(x, y)$.

1. $f(x, y) = 4 + 3x^2 + 5y^{1/2}$

2. $f(x, y) = 1 + 2x^{2/3} + y^{-2}$

3. $f(x, y) = xy^2 - 5x^2$

4. $f(x, y) = x^2 - 2x^3y + y^3$

5. $f(x, y) = xy^2 - 3x^2y$

6. $f(x, y) = x^2y - 2x^2y^3$

7. $f(x, y) = x^{-2} + xy^{-2} - x^2y^3 + 4y$

8. $f(x, y) = x^{1/2} - x^{1/3}y + 3xy^{-3} + y$

9. $f(x, y) = xe^{2y} - \ln(2x + 3y)$

10. $f(x, y) = ye^{3x} + 4e^{(x-y)}$

11. $f(x, y) = (2x + 3y)^4$

12. $f(x, y) = (x^2 + 5y)^3$

13. $f(x, y) = (2x + 4xy)^{-2}$

14. $f(x, y) = (x^2y - 5x)^{1/3}$

In Problems 15 to 18 find $f_x(1, 0)$ and $f_y(2, 1)$.

15. $f(x, y) = 3xy^3 + 5x^{-1}y$

16. $f(x, y) = 4xy^4 - 6x^{-2}y + 3$

17. $f(x, y) = 2x^{1/2} - 3xy^3 + e^y$

18. $f(x, y) = x^3y + 4x^2y^2 + 6e^{2y}$

In Problems 19 to 24 find all second-order partial derivatives.

19. $f(x, y) = 3x^2 + 5y^{-2}$

20. $f(x, y) = 2x^{2/3} + y^{-1/2}$

21. $f(x, y) = 4 + 3x^2y + 5xe^{2y}$

22. $f(x, y) = 1 + 2x^{2/3}y + x \ln(1 + y^{-2})$

23. $f(x, y) = xy^2e^x + e^{2x}$

24. $f(x, y) = 2x^3ye^{3y}$

In Problems 25 and 26 find the values of x and y such that $f_x(x, y) = 0$ and $f_y(x, y) = 0$.

25. $f(x, y) = 1 - x^2 + y^2 + 2xy - 4y$

26. $f(x, y) = 2 - 2x^2 - y^2 + xy + 7x$

In Problems 27 to 30 find f_{xxx}, f_{xxy}, and f_{xyy}.

27. $f(x, y) = x^4 - x^3 + 2x + 2y^5 + 4y^2$

28. $f(x, y) = 1 + x^5 + x^{-1} + y^2$

29. $f(x, y) = 2xy + x^2y - xe^{2y}$

30. $f(x, y) = x^{-2} + xy^{-2} - \ln(y - 1)$

31. Give a definition of the partial derivative with respect to x of a function of three variables $w = f(x, y, z)$.

In Problems 32 and 33 find f_x and f_y by holding constant all variables except the one with which the partial derivative is being taken.

32. $f(x, y, z) = x^2 + x^3z - 2yz^3 + xyz$

33. $f(x, y, z) = 3x^{-2} + x^{1/3}z - x^2z^3 + xyz$

34. Find $f_{xyz}(1, 2, 3)$ for $f(x, y, z) = x^2 - yze^x - 4xyz^2$

35. Show that the function $f(x, y) = x^3 + 5x^2 + e^y$ satisfies the partial differential equation $f_{xy} = 0$.

36. Show that the function $f(x, y) = x^3 - 3xy^2$ satisfies the partial differential equation $f_{xx} + f_{yy} = 0$.

Referenced Exercise Set 18–2

1. In an article studying the economics of urban bus transportation Obeng* gathered data from 62 bus systems. The cost C was treated as a function of demand Q, fuel price F, the price of labor L, and the price of capital P. The data can be described by a Cobb–Douglas function of the form $C = 1.7Q^{1.1}F^{1.5}L^{0.36}P^{0.5}$. Find the partial derivatives of C.

2. In an article studying the optimal use of cobalt beams in treating cancer, Gantchew defines the dose distribution d along a straight line perpendicular to the central ray as a function of the width x of the ray and the length of penetration y as follows:†

$$d(x, y) = 0.93e^{[70x/(60+y)]^{2.7}}$$

Compute d_x.

3. In their article on advertising in the cigarette industry, Bishop and Yoo also expressed price P as a function of disposable income D and cigarette advertising expenditures A.‡ In one part of the article the following function was defined:

$$P(D, A) = 0.125D^{0.56}A^{0.06}$$

Compute (a) P_D and (b) P_A.

Cumulative Exercise Set 18–2

1. Find the functional value of $f(x, y)$ at each point, where

$$f(x, y) = \frac{x^2 - y}{y}$$

(a) $(0.1, 0.0001)$ (b) $(0.1, -0.0001)$

2. Find the domain of the function in Problem 1.

*Kofi Obeng, "The Economics of Bus Transit Operation," *The Logistics and Transportation Review,* Vol. 20, 1986, pp. 45–63.

†M. G. Gantchew, "On the Application of a Simple Model of Cobalt-60 Beams," *Physics in Medicine and Biology,* Vol. 24, 1979, pp. 443–446.

‡John A. Bishop and Jang H. Yoo, " 'Health Scare,' Excise Taxes and Advertising Ban in the Cigarette Demand and Supply," *Southern Economics Journal,* Vol. 52, 1985, pp. 402–411.

In Problems 3 and 4 sketch the graph of the functions, giving the traces and one section that is parallel to each coordinate plane.

3. $f(x, y) = 4 - x - 2y$

4. $f(x, y) = 2 - 3x^2 - y^2$

In Problems 5 and 6 find $f_x(x, y)$ and $f_y(x, y)$.

5. $f(x, y) = \dfrac{x^2 - y}{y}$

6. $f(x, y) = e^{2xy + 3x}$

7. Find $f_x(0, 1)$ and $f_y(0, 1)$ for the function $f(x, y) = x + y + xy$.

8. Find all second-order partial derivatives of the function $f(x, y) = x - y - xy$.

18–3 Relative Extrema

Many applications of functions of more than one variable involve finding a value of a function that is a relative maximum or minimum. We concentrate on functions of two variables because they have a firm geometrical base. The concepts can be extended to functions of more than two variables. The ideas in this section are similar to corresponding results for functions of one variable. For instance, to compute relative extrema we use a theorem with first and second partial derivatives that is similar to the second derivative test for functions of one variable.

Definitions

The definition of a relative maximum and a relative minimum for a function of two variables is similar to the definitions for a function of one variable. The difference is that there are two independent variables instead of one. This means that $f(a, b)$ is a relative maximum if it is larger than all functional values of points (x, y) close to (a, b), whereby "close to" is meant within a circular region about (a, b). Geometrically, a relative maximum resembles a mountain peak or the top of a dome. A relative minimum resembles the bottom of a valley.

DEFINITION

A function $f(x, y)$ has a **relative maximum** at (a, b) if

$$f(a, b) \geq f(x, y)$$

for all (x, y) in some circular region around (a, b); $f(x, y)$ has a **relative minimum** at (a, b) if

$$f(a, b) \leq f(x, y)$$

for all (x, y) in some circular region around (a, b). The term **relative extremum** refers to either a relative maximum or a relative minimum.

EXAMPLE 1

Problem

Determine from the graphs given in Figure 1 if the function has a relative extremum at $(0, 0)$: (a) $f(x, y) = x^2 + 4y^2$, (b) $f(x, y) = y^2 - x^2$.

Solution (a) The graph in Figure 1a shows that all functional values $f(x, y) \geq 0$ for all (x, y), so that any circular region can be chosen to show that $f(x, y)$ has a relative maximum at $(0, 0)$.

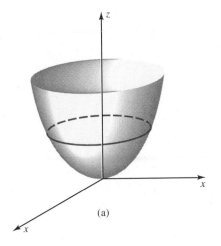

$f(x, y) = x^2 + 4y^2$
has a relative
minimum at $(0, 0)$

(a)

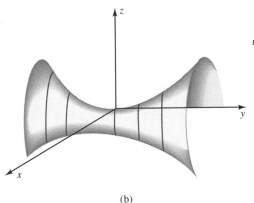

$f(x, y) = y^2 - x^2$
has neither a
relative maximum nor
a relative minimum
at $(0, 0)$

(b)

FIGURE 1

(b) Figure 1b shows that for points $(x, 0)$ on the x-axis, $f(x, 0) = -x^2 \leq 0$, and for points $(0, y)$ on the y-axis, $f(0, y) = y^2 \geq 0$. Any circular region containing $(0, 0)$ must contain points on each axis, so that some functional values will be greater than $f(0, 0) = 0$ and some will be smaller. Therefore $f(x, y)$ has neither a relative maximum nor a relative minimum at $(0, 0)$.

Critical Points

Recall that a function of one variable has a relative extremum at a point only if its derivative is 0 at the point, or if the derivative fails to exist. Figure 2 demonstrates that if a function of two variables has a relative maximum at (a, b), then the tangent lines to the graph at (a, b) must be horizontal, meaning that their slopes are 0. Since the partial derivative measures the slope of the tangent line, $f_x(a, b)$ and $f_y(a, b)$ must be 0. The same holds for relative minima. This is recorded in the following result.

> If $f(x, y)$ has a relative extremum at (a, b) and if both partial derivatives exist, then
>
> $$f_x(a, b) = 0 = f_y(a, b)$$

Extrema can also occur when a partial derivative fails to exist, but we will not consider this case.

A point (a, b) such that $f_x(a, b) = 0 = f_y(a, b)$ is called a **critical point.** The first step in locating relative extrema is to find the critical points. However, just as with functions of one variable, not all critical points correspond to relative extrema. For example, consider the function in Example 1(b), $f(x, y) = y^2 - x^2$. The partial derivatives are $f_x(x, y) = -2x$ and $f_y(x, y) = 2y$, so that each is 0 at $(0, 0)$, but the graph in Figure 1b shows that f has neither a relative maximum nor a relative minimum. This is an example of a **saddle point,** that is, a point where the partial derivatives are 0 but the function does not have a relative extremum at the point.

To find the critical points of a function, we need to solve $f_x = 0$ and $f_y = 0$ simultaneously, meaning that the same point (a, b) must satisfy both equations. This often requires that a system of two equations in two unknowns be solved, as illustrated in the next example.

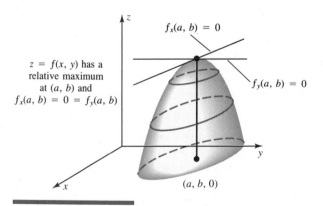

$f_x(a, b) = 0$

$z = f(x, y)$ has a relative maximum at (a, b) and $f_x(a, b) = 0 = f_y(a, b)$

$f_y(a, b) = 0$

$(a, b, 0)$

FIGURE 2

EXAMPLE 2

Problem

Find the critical points of the function

$$f(x, y) = x^2 - 4x + y^2 - 6y$$

Solution First compute the partial derivatives.

$$f_x(x, y) = 2x - 4$$
$$f_y(x, y) = 2y - 6$$

Set each partial derivative equal to 0.

$$f_x(x, y) = 2x - 4 = 0$$
$$f_y(x, y) = 2y - 6 = 0$$

This is a system of two equations in two unknowns. The first equation implies that $x = 2$ and the second implies $y = 3$. Thus $(2, 3)$ is the solution of the system of equations. Therefore $f_x(2, 3) = 0$ and $f_y(2, 3) = 0$, which means that $(2, 3)$ is the only critical point.

The Second Partial Derivative Test

According to the earlier result, if the function $f(x, y)$ in Example 2 has a relative extremum, it must occur at the critical value $(2, 3)$. But this does not mean that there is necessarily a relative extremum at $(2, 3)$; the point could be a saddle point. The following theorem shows how to determine whether critical points are relative extrema. It is referred to as the **second partial derivative test.**

Second Partial Derivative Test

Suppose $z = f(x, y)$ has first and second partial derivatives in a circular region around (a, b) where (a, b) is a critical point; that is

$$f_x(a, b) = 0 = f_y(a, b)$$

To simplify notation, let

$$A = f_{xx}(a, b), \quad B = f_{yy}(a, b) \quad \text{and} \quad C = f_{xy}(a, b)$$

Then

1. If $A > 0$ and $AB - C^2 > 0$, then $f(a, b)$ is a relative minimum.
2. If $A < 0$ and $AB - C^2 > 0$, then $f(a, b)$ is a relative maximum.
3. If $AB - C^2 < 0$, then $f(a, b)$ is not a relative extremum; it is a saddle point.
4. If $AB - C^2 = 0$, then the test yields no conclusion.

The next three examples show how to apply the test.

EXAMPLE 3

Problem

Find the relative extrema of the function $f(x, y) = x^2 - 4x + y^2 - 6y$.

Solution In Example 2 it was shown that the only critical point of $f(x, y)$ is (2, 3). To apply the second partial derivative test, compute $f_{xx}(x, y) = 2$, $f_{yy}(x, y) = 2$, and $f_{xy}(x, y) = 0$. Then evaluate each at the critical point, which is immediate because each partial derivative is a constant.

$$A = f_{xx}(2, 3) = 2$$
$$B = f_{yy}(2, 3) = 2$$
$$C = f_{xy}(2, 3) = 0$$

Since $A > 0$ and $AB - C^2 = 2 \cdot 2 - 0 = 4 > 0$, part 1 of the test applies so that $f(2, 3) = -13$ is a relative minimum.

EXAMPLE 4

Problem

Find the relative extrema of the function $f(x, y) = x^2 + 4x - 2y^3 + 3y^2$.

Solution First find the critical points by setting the partial derivatives equal to 0.

$$f_x(x, y) = 2x + 4$$
$$f_y(x, y) = -6y^2 + 6y$$
$$f_x = 0 = 2x + 4 \tag{1}$$
$$f_y = 0 = -6y^2 + 6y \tag{2}$$

Equation 1 implies that $x = -2$. Equation 2 can be factored into

$$-6y(y - 1) = 0$$

This implies that $y = 0$ and $y = 1$. Thus there are two critical points, $(-2, 0)$ and $(-2, 1)$. We handle each critical point separately after the second partial derivatives are computed.

$$f_{xx}(x, y) = 2$$
$$f_{yy}(x, y) = -12y + 6$$
$$f_{xy}(x, y) = 0$$

(a) Consider the critical point $(-2, 0)$. Then

$$A = f_{xx}(-2, 0) = 2$$
$$B = f_{yy}(-2, 0) = 6$$
$$C = f_{xy}(-2, 0) = 0$$

Since $A > 0$ and $AB - C^2 = 2 \cdot 6 - 0 = 12 > 0$, part 1 of the test applies so that $f(-2, 0) = -4$ is a relative minimum.

(b) Consider the critical point $(-2, 1)$. Then

$$A = f_{xx}(-2, 1) = 2$$
$$B = f_{yy}(-2, 1) = -6$$
$$C = f_{xy}(-2, 1) = 0$$

Then $AB - C^2 = 2(-6) - 0 = -12 < 0$, and so part 3 of the test applies. Thus $f(x, y)$ has a saddle point at $(-2, 1)$.

Sometimes the expressions for f_x and f_y contain both x and y, which makes solving for the critical points a bit more complicated, as illustrated in the next example.

EXAMPLE 5

Problem

Find the relative extrema of the function $f(x, y) = x^2 - 5x - y^2 + xy$.

Solution First find the critical points by setting the partial derivatives equal to 0.

$$f_x(x, y) = 2x - 5 + y = 0$$
$$f_y(x, y) = x - 2y = 0$$

The second equation gives $x = 2y$. Substituting this equation into the equation $f_x = 0$ yields

$$2(2y) - 5 + y = 0$$
$$5y = 5$$
$$y = 1$$

Substituting $y = 1$ into $x = 2y$ yields $x = 2$. Therefore the critical point is $(2, 1)$. Next compute

$$A = f_{xx}(2, 1) = 2$$
$$B = f_{yy}(2, 1) = -2$$
$$C = f_{xy}(2, 1) = 1$$

Then $AB - C^2 = 2(-2) - 1 = -5 < 0$, and so part 3 of the test applies. Thus $f(x, y)$ has a saddle point at $(2, 1)$.

Part 4 of the test needs clarification. If $AB - C^2 = 0$, the second partial derivative test cannot determine whether the critical point is a relative extremum. It is necessary to use another test. For instance, all of the points close to the critical point can be tested. If they are all greater than $f(a, b)$, then $f(a, b)$ is a relative minimum; if they are all less than $f(a, b)$, then $f(a, b)$ is a relative maximum; if there are always some that are greater and some that are less than $f(a, b)$, then $f(x, y)$ has a saddle point at (a, b). This is usually a difficult task, however, and we will not pursue it.

EXERCISE SET 18–3

In Problems 1 to 6 determine whether the graph of the function has a relative extremum at $(0, 0)$.

1.

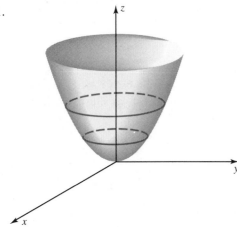

2.

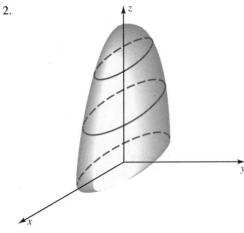

3.

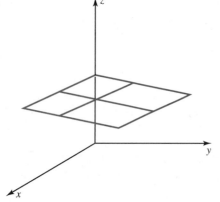

4.

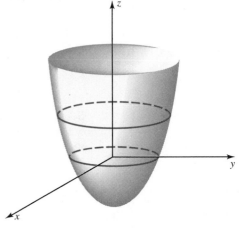

5.

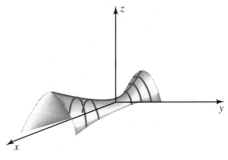

6.

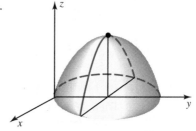

In Problems 7 to 12 find the critical points of the function.

7. $f(x, y) = x^2 + 6x + y^2 + 8y$

8. $f(x, y) = x^2 - 2x + y^2 + 6y$

9. $f(x, y) = 2x^2 + 4xy + y^2 - 6y + 1$

10. $f(x, y) = 3x^2 - 6xy + 2y^2 + 10y - 2$

11. $f(x, y) = x^2 - 2y^3 + 6y$

12. $f(x, y) = x^3 - 3x + 3y^2 + 12y$

In Problems 13 to 24 find the relative extrema and any saddle points of the function.

13. $f(x, y) = x^2 + 4x - 2y^3$

14. $f(x, y) = x^3 + 3y^2 + 12y$

15. $f(x, y) = x^2 - 8x + 2y^3 + 6y^2$

16. $f(x, y) = x^3 - 3x^2 + 3y^2 + 12y$

17. $f(x, y) = x^2 - 2y^3 + 6y$

18. $f(x, y) = x^3 - 3x + 3y^2 + 12y$

19. $f(x, y) = x^2 - 2x - 2xy + 2y^2 - 4y$

20. $f(x, y) = x^2 - 2xy + 3y^2 + 12y$

21. $f(x, y) = x - xy + y^3 - 3y^2$

22. $f(x, y) = x^2 + 4x - 2xy + 12y$

23. $f(x, y) = x^3 - 2x^2 - 2xy + y^2$

24. $f(x, y) = x^2 - 2xy - y^3 + 14y$

In Problems 25 to 30 find the relative extrema of the function.

25. $f(x, y) = x^3 - 2y^3 + 6y$

26. $f(x, y) = x^3 + 3y^3 - 9y$

27. $f(x, y) = x^3 - 6x^2 - 2y^3 + 24y$

28. $f(x, y) = x^3 + 3y^3 - 9y^2$

29. $f(x, y) = x^2 - 8x + 2y^3 + 6y^2 - 3y$

30. $f(x, y) = x^3 - 6x^2 - 8x + 3y^2 + 12y$

31. A firm utilizes x million dollars per year in labor cost and y million dollars per year in production cost. The cost function is

$$C(x, y) = 20 - 24x - 12y + x^2 + 3y^2$$

Find the amount that the firm should spend on labor and production each year to minimize cost.

32. A firm produces two types of radios, AM and AM–FM. If x thousand of AM radios and y thousand AM–FM radios are made per month, the revenue $R(x)$ and cost $C(x)$ functions are given as follows, where $R(x)$ and $C(x)$ are measured in millions of dollars:

$$R(x, y) = 5x + 6y$$
$$C(x, y) = 10 - 5x - 12y + x^2 + 2y^2$$

Find how many radios of each type should be produced to maximize profit.

33. A manufacturing firm wants to package its product in a rectangular box with no top. If the volume of the box is to be 64 cubic inches, find the dimensions that the box should be to use the least amount of material to construct it.

34. A firm wants to package its product in a rectangular box with no top and two intersecting partitions. If the volume of the box is to be 64 cubic inches, find the dimensions that the box should be to use the least amount of material to construct it.

Referenced Exercise Set 18–3

1. Extractive industries are those that extract minerals from the ground, such as metals or petroleum. In an article analyzing environmental controls on extractive industries, Stollery defines a welfare function W in terms of the quantity of minerals extracted x and the quantity of pollutants released y.[*] The welfare function measures the benefit to society based on the positive benefit due to extraction versus the negative benefit due to pollution. Stollery defines the function

$$W(x, y) = ax - bx^2 - cy^2 + d$$

where a, b, c and d are constants depending on the industry and the method of extraction. Compute the values of x and y that maximize $W(x, y)$.

2. The design of an effective ramp is one of the most important aspects of highway design. When a car merges into traffic on a highway from a ramp, the car enters traffic in a gap between two vehicles. The distance between these two vehicles is called the gap size. Pignataro shows that the average gap size g is a function of the angle of approach x of the ramp, measured in degrees, and the length of the ramp y, measured in 100-foot sections.[†] The relationship is

$$g(x, y) = 5.5 - 0.83x + 0.04x^2 - 1.04y + 0.05y^2$$

Find the relative extrema of $g(x, y)$.

[*]Kenneth R. Stollery, "Environmental Controls in Extractive Industries," *Land Economics,* Vol. 61, 1985, pp. 136–144.

[†]Louis J. Pignataro, *Traffic Engineering: Theory and Practice,* Englewood Cliffs, N.J., Prentice-Hall, 1973, p. 169.

Cumulative Exercise Set 18–3

In Problems 1 to 4 graph the functions.

1. $f(x, y) = 2 + 3y - 4x$

2. $f(x, y) = -2 + x^2 + y^2$

3. $f(x, y) = x^2 - 4y^2$

4. $f(x, y) = 1 + x^2 + 2x + y^2$

In Problems 5 to 7 find $f_x(x, y)$ and $f_y(x, y)$.

5. $f(x, y) = x^3 - 2x^2y + y^4$

6. $f(x, y) = x^{-2}y + 6x^{1/2} + 9y^{2/3}$

7. $f(x, y) = (x^3 + 5y)^4$

8. Find all second-order partial derivatives of

$$f(x, y) = x^3 + 4xy^{1/2}$$

In Problems 9 to 12 find the relative extrema and any saddle points of the function.

9. $f(x, y) = x^3 - 3y^2 - 12y$

10. $f(x, y) = x^3 - 3x^2 - 3y^2 + 12y$

11. $f(x, y) = x^2 + 6x + 2xy - 12y$

12. $f(x, y) = x^3 + 3y^3 + 9y^2$

18–4 Lagrange Multipliers

The previous section dealt with finding the relative extrema of a function. This section also deals with optimizing functions, but the types of problems will have a little twist. Often in applications the optimum value of a function is subject to a **constraint.** For example, it might be necessary to find the maximum revenue subject to cost considerations. Another example might be to find the box with minimum surface area where the volume of the box is given.

An Application: Chemical Optimization at Monsanto

"Optimization" is a hallowed term in business. Sometimes a company's primary goal is to maximize profit, but at other times the objective is to minimize cost. As an example of the latter, consider the situation at Monsanto described by a consultant, Raymond Boykin, who was hired to evaluate Monsanto's chemical production methods.* In 1983 Monsanto opened a new chemical plant that caused capacity to exceed demand for many products. This prompted the company to hire a consultant to determine how to optimize production.

Boykin concentrated on one important product, maleic anhydride, which is used to make polyester resins. These resins are used in the manufacture of boat hulls, shower stalls, autobody parts, and counter tops. The chemical is produced by feeding raw material (butane) into a reactor under pressure. Production and cost are dependent on the raw material feed rate, the reactor velocity, and reactor pressure. Hence both production and cost are functions of three variables.

Because Monsanto is such a large producer of maleic anhydride, even small reductions in cost per pound would save large amounts of money. So Boykin recommended that Monsanto concentrate on minimizing cost once the amount of production was determined. This resulted in a constrained optimization problem: minimize cost subject to a certain level of production.

*Raymond F. Boykin, "Optimizing Chemical Production at Monsanto," *Interfaces,* Vol. 15, 1985, pp. 88–95.

In this section we study one method for solving such constrained optimization problems of functions of several variables. At the end of the section we will show how Monsanto solves the problem.

Lagrange Multipliers

Constraine ' optimization problems are often difficult to solve. An 18th-century mathematician, Joseph Louis Lagrange (1736–1813), discovered an ingenious method for solving these types of problems. Let us start with the general problem. Suppose $f(x, y)$ and $g(x, y)$ are functions of two variables. The problem is to find the values of x and y that either maximize or minimize the objective function $f(x, y)$ subject to the constraint $g(x, y) = 0$.

Lagrange's idea is to replace $f(x, y)$ with another function of three variables by introducing a new variable t called a Lagrange multiplier. The new function is

$$F(x, y, t) = f(x, y) + tg(x, y)$$

The variable t is always multiplied by the constraining function $g(x, y)$. The following theorem shows how to use Lagrange's method, called the technique of **Lagrange multipliers.** The proof of the theorem is beyond the scope of this text.

Theorem

If the function $f(x, y)$ has an extremum at (a, b), then there is a value of t, say, $t = c$, such that the partial derivatives of $F(x, y, t)$ each equal 0 at (a, b, c).

The theorem states that to find the extrema, first find the critical values of $F(x, y, t)$ by setting the partial derivatives equal to 0 and solving for x, y, and t. An extremum for $f(x, y)$ subject to $g(x, y) = 0$, if it exists, will be among these critical values. Example 1 illustrates how to apply the theorem.

EXAMPLE 1

Problem

Find the minimum value of $f(x, y) = x^2 + y^2$ subject to $g(x, y) = x + 2y - 10 = 0$.

Solution Form the function

$$F(x, y, t) = x^2 + y^2 + t(x + 2y - 10)$$

Find the partial derivatives of $F(x, y, t)$ and set each equal to 0.

$$F_x(x, y, t) = 2x + t = 0 \tag{1}$$

$$F_y(x, y, t) = 2y + 2t = 0 \tag{2}$$

$$F_t(x, y, t) = x + 2y - 10 = 0 \tag{3}$$

Solving (1) and (2) for t yields $t = -2x$ and $t = -y$. Thus $y = 2x$. Substituting $y = 2x$ into (3) yields

$$x + 2(2x) - 10 = 0$$
$$5x = 10$$
$$x = 2$$

Substitute $x = 2$ into (3) to get $y = 4$. Hence the minimum value of the function $f(x, y) = x^2 + y^2$, subject to $g(x, y) = x + 2y - 10 = 0$, occurs at the point $(2, 4)$. The minimum value is $f(2, 4) = 20$.

We have no easy means to verify that $f(2, 4)$ is actually a minimum rather than a maximum or a saddle point. Often it suffices to substitute into $f(x, y)$ several points close to $(2, 4)$ that also satisfy the constraint. Such points will produce functional values that are larger than 20. For instance, to select a point on $g(x, y) = 0$ close to $(2, 4)$, first choose a value of y that is close to 4, say, $y = 4.1$. Then the corresponding value of x is found by substituting $y = 4.1$ into $g(x, y) = 0$, which yields $x = 1.8$. Then $f(1.8, 4.1) = (1.8)^2 + (4.1)^2 = 20.05$.

The Geometric Interpretation

For functions of two variables the method of Lagrange multipliers has a geometric interpretation. To illustrate, graph both $f(x, y) = x^2 + y^2$ and $g(x, y) = x + 2y - 10 = 0$ in the same coordinate system. Figure 1 shows that $f(x, y)$ is an infinite bowl and $g(x, y) = 0$ is a plane. Their intersection is a two-dimensional curve, a parabola, whose minimum is at $(2, 4)$. Problem 34 will ask you to substitute algebraically $g(x, y) = x + 2y - 10 = 0$ into $f(x, y) = x^2 + y^2$. This produces the equation of the parabola in Figure 1; it is an equation in one variable whose derivative can be computed and whose minimum will occur at $(2, 4)$.

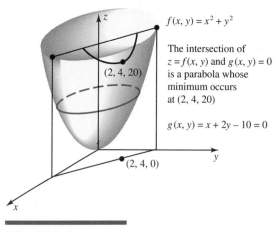

$f(x, y) = x^2 + y^2$

The intersection of $z = f(x, y)$ and $g(x, y) = 0$ is a parabola whose minimum occurs at $(2, 4, 20)$

$g(x, y) = x + 2y - 10 = 0$

$(2, 4, 20)$

$(2, 4, 0)$

FIGURE 1

The Method of Lagrange Multipliers

Before applying the method of Lagrange multipliers to various problems, let us summarize the steps in using the technique.

The Method of Lagrange Multipliers

In order to find the relative extrema of $f(x, y)$ subject to the constraint $g(x, y) = 0$, proceed as follows:

1. Construct the function

$$F(x, y, t) = f(x, y) + tg(x, y)$$

2. Compute the partial derivatives of $F(x, y, t)$ and set each equal to 0.
3. Solve the system of equations formed in part 2.
4. The relative extrema of $f(x, y)$ subject to $g(x, y) = 0$ are among the solutions to the system of equations in step 3.

Example 2 illustrates how to apply the method of Lagrange multipliers to a problem where the system is a bit more complicated than the system in Example 1. This example also shows how to deal with a problem in which the constraint is expressed as an equation rather than as a function.

EXAMPLE 2

Problem

Find the relative maximum value of $f(x, y) = 8x - x^2 + 4y - y^2$ subject to $x + y = 10$.

Solution First express the constraint in the proper form, an expression in x and y, $g(x, y)$, set equal to 0. Subtract 10 from each side of the constraining equation to get

$$g(x, y) = x + y - 10 = 0$$

1. Construct the function

$$F(x, y, t) = 8x - x^2 + 4y - y^2 + t(x + y - 10)$$

2. Compute the partial derivatives of $F(x, y, t)$.

$$F_x(x, y, t) = 8 - 2x + t = 0 \tag{1}$$

$$F_y(x, y, t) = 4 - 2y + t = 0 \tag{2}$$

$$F_t(x, y, t) = x + y - 10 = 0 \tag{3}$$

3. Solving (1) and (2) for t yields

$$t = 2x - 8 \quad \text{and} \quad t = 2y - 4$$

Thus $2x - 8 = 2y - 4$. This simplifies to the equation

$$x - y = 2 \tag{4}$$

Form the system of two equations consisting of (3) and (4).

$$x + y = 10$$
$$x - y = 2$$

Adding these equations yields $2x = 12$, or $x = 6$. Substituting this value into (4) gives $y = 4$.

4. The relative maximum of the value of $f(x, y) = 8x - x^2 + 4y - y^2$ subject to $x + y = 10$ must occur at $(6, 4)$. Testing a few points close to $(6, 4)$ verifies that $f(6, 4) = 12$ is a maximum.

The method of Lagrange multipliers can be applied to functions of more than two variables. Example 3 illustrates how to solve a particular type of word problem by applying Lagrange multipliers to a function of three variables.

EXAMPLE 3

Problem

Find three numbers whose sum is 60 and whose product is a maximum.

Solution First express the problem in a form where Lagrange multipliers can be applied. Let the three numbers be represented by x, y, and z. The problem calls for us to maximize

$$f(x, y, z) = xyz$$

subject to $x + y + z = 60$. Let $g(x, y, z) = x + y + z - 60 = 0$. Apply the steps in the method.

1. Construct the function

$$F(x, y, z, t) = xyz + t(x + y + z - 60)$$

2. Compute the partial derivatives of $F(x, y, z, t)$ and form the system of equations $F_x = 0$, $F_y = 0$, $F_z = 0$, $F_t = 0$.

$$F_x(x, y, z, t) = yz + t = 0 \tag{1}$$

$$F_y(x, y, z, t) = xz + t = 0 \tag{2}$$

$$F_z(x, y, z, t) = xy + t = 0 \tag{3}$$

$$F_t(x, y, z, t) = x + y + z - 60 = 0 \tag{4}$$

3. Solving (1), (2), and (3) for t and combining them yields

$$t = -yz = -xz = -xy \tag{5}$$

Assume that none of the numbers is 0, because then the product would be 0, which is not a maximum. Then (5) simplifies to $x = y = z$. Substituting

these relationships into (4) yields $3x = 60$, so that $x = 20$. This means that $x = y = z = 20$ is the solution.
4. The relative maximum value of $f(x, y, z) = xyz$ subject to $x + y + z = 60$ must occur at $(20, 20, 20)$. Testing a few points close to $(20, 20, 20)$ verifies that the maximum value is $f(20, 20, 20) = 8000$.

It is not always easy to test the points close to the critical value. In this case suppose we considered points of the form $(20, 20 + h, 20 - h)$ for a small number h. These points are close to $(20, 20, 20)$ and satisfy the constraint because they sum to 60. Their product is $f(20, 20 + h, 20 - h) = 20(20 + h)(20 - h) = 20(400 - h^2) = 8000 - 20h^2$. This number is always less than 8000. This does not test all the points close to $(20, 20, 20)$, but it gives a good idea of what is required to prove that the maximum occurs at that point.

The Case Study deals with a common type of function in business called a Cobb–Douglas production function, which is of the form

$$f(x, y) = x^a y^b$$

where f is the production or output of a firm while x and y are the inputs, which could be raw goods or labor. The exponents a and b are constants. Example 4 shows how Lagrange multipliers can be used to solve an optimization problem containing a Cobb–Douglas production function.

EXAMPLE 4

The production capacity of a firm is given by

$$f(x, y) = x^{1/2} y^{1/2}$$

where x is the number of labor-hours available daily and y is the amount of capital invested in machinery. Labor costs are $10 per labor-hour. Assuming that there are 8 working hours in a day and 260 working days in a year, and that the firm allocates $5,000,000 in capital for wages and machinery, the constraining equation is

$$10(8)(260)x + y = 5,000,000$$

or

$$20,800x + y = 5,000,000$$

Problem

Find the values of x and y that maximize production capacity.

Solution Express the constraint in the form

$$g(x, y) = 20,800x + y - 5,000,000.$$

Apply the steps in the method.

1. Construct the function

$$F(x, y, t) = x^{1/2} y^{1/2} + t(20,800x + y - 5,000,000)$$

2. Compute the partial derivatives of $F(x, y, t)$ and form the system of equations $F_x = 0, F_y = 0, F_t = 0$.

$$F_x(x, y, t) = \frac{1}{2} x^{-1/2} y^{1/2} + 20{,}800t = 0 \tag{1}$$

$$F_y(x, y, t) = \frac{1}{2} x^{1/2} y^{-1/2} + t = 0 \tag{2}$$

$$F_t(x, y, t) = 20{,}800x + y - 5{,}000{,}000 = 0 \tag{3}$$

3. Solving (1) and (2) for t and combining them yields

$$-20{,}800t = \frac{1}{2} x^{-1/2} y^{1/2} = \frac{20{,}800}{2} x^{1/2} y^{-1/2}$$

Multiply the equation by $2x^{1/2}y^{1/2}$. This yields

$$y = 20{,}800x$$

Substituting this equation into (3) gives $2y = 5{,}000{,}000$, or $y = 2{,}500{,}000$. Then $x = 2{,}500{,}000/20{,}800 \approx 120.2$.

4. The maximum production occurs at $x \approx 120.2$, or about 120 labor-hours, and $y = \$2{,}500{,}000$.

Let us return to the optimization problem facing Monsanto, which was mentioned at the start of the section. The consultant determined that production P and cost C were functions of three variables. Let x represent the raw material feed rate, y the reactor velocity, and z the reactor pressure. The production function has the form $P = kxy/z$ and the cost function $C = ax + by + c/z$. The constants k, a, b, and c depend on which reactor is used. The object is to minimize cost subject to a given level of production. Boykin developed a computer program for Monsanto, which is run five times a day at each plant. The level of production is set and then the program determines the values of x, y, and z that minimize C. The next example illustrates how the program solves the problem.

EXAMPLE 5

Problem

Let $k = 1$, $a = 0.64$, $b = 4$, and $c = 1.6$ in the preceding production and cost functions. Set the production level at $P = 27$. Find the values of x, y, and z that minimize cost.

Solution The problem is to minimize

$$C = 0.64x + 4y + \frac{1.6}{z}$$

Subject to: $\dfrac{xy}{z} - 27 = 0$

Define

$$C(x, y, z, t) = 0.64x + 4y + \frac{1.6}{z} + t\left(\frac{xy}{z} - 27\right)$$

Find the partial derivatives and form the system of equations.

$$C_x(x, y, z, t) = 0.64 + \frac{ty}{z} = 0$$

$$C_y(x, y, z, t) = 4 + \frac{tx}{z} = 0$$

$$C_z(x, y, z, t) = -1.6z^{-2} - txyz^{-2} = 0$$

$$C_t(x, y, z, t) = \frac{xy}{z} - 27 = 0$$

Solve for $-t$ in the first three equations.

$$-t = \frac{0.64z}{y} = \frac{4z}{x} = \frac{1.6}{xy}$$

$$y = 0.16x, \quad z = \frac{1}{0.4x}$$

Substituting these values into $C_t = 0$ gives

$$x(0.16x)(0.4x) - 27 = 0$$

$$x^3 = \frac{27}{0.064}$$

$$x = \frac{3}{0.4} = 7.5$$

Therefore the solution is $x = 7.5$, $y = 0.16(7.5) = 1.2$, $z = 1/3$.

The next example illustrates another application of Lagrange multipliers for solving real-world optimization problems.

EXAMPLE 6

An aircraft design engineer plans to construct a rectangular storage bin in the space between the rear wall of an aircraft cabin and the fuselage. In the design model of the bin, the floor of the bin is an extension of the floor of the cabin, which is represented by the xy-plane; one side wall of the bin is the rear wall of the cabin, represented by the xz-plane; and the fuselage is represented by the equation $z = 16 - x^2 - y^2$. (See Figure 2.)

Problem

Find the dimensions of the bin with greatest volume that can fit into the given space.

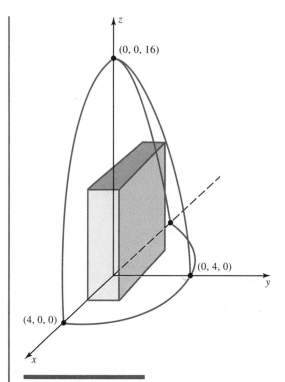

FIGURE 2

Solution The function to be maximized is the volume V of the bin. If we let the length, width, and height of the bin be x, y, and z, respectively, then $V = V(x, y, z) = xyz$. We must have all three variables, x, y, and z, greater than or equal to 0. The constraint is given by $z = 16 - x^2 - y^2$. The problem can be stated in a way to apply Lagrange multipliers as follows:

Maximize $V(x, y, z) = xyz$ subject to $z = 16 - x^2 - y^2$. Let $g(x, y, z) = 16 - x^2 - y^2 - z$. Apply the steps in the method.

1. Construct the function

$$F(x, y, z, t) = xyz + t(16 - x^2 - y^2 - z)$$

2. Compute the partial derivatives of $F(x, y, z, t)$ and form the system of equations $F_x = 0$, $F_y = 0$, $F_z = 0$, $F_t = 0$:

$$F_x(x, y, z, t) = yz - 2xt = 0 \tag{1}$$

$$F_y(x, y, z, t) = xz - 2yt = 0 \tag{2}$$

$$F_z(x, y, z, t) = xy - t = 0 \tag{3}$$

$$F_t(x, y, z, t) = 16 - x^2 - y^2 - z = 0 \tag{4}$$

3. Solving (1), (2), and (3) for $2t$ and combining them yields

$$2t = yz/x = xz/y = 2xy \qquad (5)$$

Assume that none of the numbers is 0, because then the product would be 0, which is not a maximum. Then (5) simplifies to $x^2 = y^2 = z/2$. Substituting these relationships into (4) yields $4x^2 = 16$, or $x^2 = 4$, so $x = 2$. This means that the solution is $x = y = 2$ and $z = 8$.

4. The relative maximum value of $V(x, y, z) = xyz$ subject to $z = 16 - x^2 - y^2$ must occur at $(2, 2, 8)$. Testing a few points close to $(2, 2, 8)$ verifies that the maximum value is $f(2, 2, 8) = 32$.

EXERCISE SET 18-4

In Problems 1 to 18 use the method of Lagrange multipliers to find the extrema of the functions with the given constraints.

1. $f(x, y) = x^2 + y^2$
   subject to $g(x, y) = x + y - 1 = 0$

2. $f(x, y) = x^2 + 2y^2$
   subject to $g(x, y) = x + 2y - 3 = 0$

3. $f(x, y) = x^2 + 2x + y^2 - y$
   subject to $g(x, y) = x + 2y = 0$

4. $f(x, y) = x^2 + 4x + y^2 - 6y$
   subject to $g(x, y) = 3x + y - 1 = 0$

5. $f(x, y) = x^3 + 3x^2 + y^2 - 2y$
   subject to $g(x, y) = 3x + 2y + 3 = 0$

6. $f(x, y) = x^3 + 6x^2 + y^2 - 4y$
   subject to $g(x, y) = x + 3y + 1 = 0$

7. $f(x, y) = 2x^3 - 6x^2 + 2y^2 - 8y$
   subject to $x + 2y - 2 = 0$

8. $f(x, y) = 4x^3 + 6x^2 - y^2 - 2y$
   subject to $2x - 3y + 1 = 0$

9. $f(x, y) = 2x - x^2 + 2y^3 - 6y^2$
   subject to $x - 2y = 10$

10. $f(x, y) = 4x^3 + 6x^2 - y^2 - 2y$
    subject to $2x + y = 2$

11. $f(x, y) = xy$
    subject to $g(x, y) = 2x + y - 1 = 0$

12. $f(x, y) = 2xy$
    subject to $g(x, y) = x + 2y - 3 = 0$

13. $f(x, y) = xy + x^2$
    subject to $2x + 3y = 1$

14. $f(x, y) = xy - y^2$
    subject to $x - 2y = 0$

15. $f(x, y, z) = xyz$
    subject to $g(x, y, z) = 2x + y + z - 2 = 0$

16. $f(x, y, z) = 2xyz$
    subject to $g(x, y, z) = x + 2y - z - 4 = 0$

17. $f(x, y, z) = xyz$
    subject to $x + y + z = 300$

18. $f(x, y, z) = xyz$
    subject to $x + 2y + 3z = 600$

19. Find three numbers that sum to 30 such that their product is a maximum.

20. Find three numbers that sum to 120 such that their product is a maximum.

21. Maximize $f(x, y) = x^{1/2}y^{1/2}$
    subject to $100x + y = 5000$.

22. Maximize $f(x, y) = x^{1/2}y^{1/3}$
    subject to $200x + y = 8000$.

23. Find three numbers such that the sum of two of them and twice the third is equal to 600 and their product is a maximum.

24. Find three numbers such that the sum of two of them and twice the third is equal to 1200 and their product is a maximum.

25. Find the minimum distance from the line $3x + y = 5$ to the origin $(0, 0)$.

26. Find the minimum distance from the line $x - 2y = 4$ to the point $(1, 0)$.

In Problems 27 to 33 use the method of Lagrange multipliers to find the extrema of the functions with the given constraints.

27. $f(x, y) = xy^2$ subject to $2x - y = 1$

28. $f(x, y) = x^2y$ subject to $x + y = 3$

29. $f(x, y, z) = xyz^2$ subject to $2x + 3y + z = 2$

30. $f(x, y, z) = x^2yz$ subject to $x - 2y - 3z = 4$

31. $f(x, y, z) = xy^2 + yz^2$
    subject to $x + y + z = 2$

32. $f(x, y, z) = x^2yz + xz$
    subject to $x - y - z = 4$

33. $f(x, y, z) = xy + xz$
    subject to $xy + yz = 1$

34. Substitute $g(x, y) = x + 2y - 10 = 0$ into $f(x, y) = x^2 + y^2$, and find the extremum of the resulting function.

35. A rectangular box with a top is to have a volume of 16 cubic feet. Find the dimensions that produce the least expensive box if the material for the sides of the box is $2 per square foot and the material for the top and bottom is $4 per square foot.

36. Find the dimensions of a rectangular box with a top that has the least surface area and a volume of 8000 cubic feet.

37. The postal service has a rule that rectangular packages cannot be sent with a specific rate if the length plus girth (the perimeter of the cross section taken perpendicular to the length) is more than 84 inches. Find the dimensions of the rectangular box with largest volume that can be sent with this rate.

38. A soft drink manufacturer sells its product in two markets, retail (in supermarkets) and commercial (in vending machines). The profit from the sale of x gallons of drink in the retail market and y gallons in the commercial market per week is given by

$$f(x, y) = 180x + 60y - 4x^2 - y^2 - xy$$

Find the number of gallons of drink that should be sold in each market per week to maximize profit.

39. A manufacturing company makes two types of suits: wool and polyester. The cost to manufacture each wool suit is $100 and the cost to manufacture each polyester suit is $50. If the manufacturer charges x dollars for each wool suit and y dollars for each polyester suit, the expected revenue per week is determined to be

$$R(x, y) = 140x + 80y - 10x^2 - 8y^2 + 4xy$$

(a) Find the cost function $C(x, y)$ assuming the fixed cost per week is $1000.

(b) Determine the values of x and y that maximize the profit function

$$P(x, y) = R(x, y) - C(x, y).$$

40. A pharmaceutical company makes two types of pills: pain pills and cold pills. The expected revenue $R(x, y)$ and cost $C(x, y)$ functions are determined to be

$$R(x, y) = 60x + 95y - 2x^2 - 2y^2 + 4xy$$
$$C(x, y) = 10 + 20x + 15y$$

Find the profit function $P(x, y) = R(x, y) - C(x, y)$ and determine the values of x and y that maximize profit.

41. A food distributor markets white rice and brown rice. The expected revenue $R(x, y)$ and cost $C(x, y)$ functions are determined to be

$$R(x, y) = 190x + 100y - 20x^2 - 4y^2 - 8xy$$
$$C(x, y) = 15 + 30x + 20y$$

Find the profit function $P(x, y) = R(x, y) - C(x, y)$ and determine the values of x and y that maximize profit.

42. A manufacturing company makes two sizes of its product: large and small. The price at which the company will sell each product, p_l for the large size and p_s for the small size, depends upon the amount of items manufactured. If the company makes x large items and y small items, then it determines that $p_l = 92 - 5x$ and $p_s = 40 - 2y$. The cost function in the given time period is $C(x, y) = 50 + 4x + 8y + xy$. The revenue function $R(x, y)$ is given as $R(x, y) = xp_l + yp_s$. Then the profit function is $P(x, y) = R(x, y) - C(x, y)$. Determine the values of x and y that maximize profit.

43. Find the dimensions of the box with a top with maximum volume if the box is to be made from 216 square inches of material.

44. Find the minimum distance from the plane $6x - y - 2z = -3$ to the point $(1, 0, -1)$.

Referenced Exercise Set 18–4

1. In an article studying the economics of agricultural production, Chavas et al.* consider the swine industry in depth. The study uses constrained optimization to determine the maximum weight that an animal can attain subject to various limitations, such as cost of grain. Solve the following problem via Lagrange multipliers that is similar to the problems presented by Chavas: find the values of x and y that maximize $z = 3x + 4y$ subject to $a = 2x^2 + 3y^2$ where a is the weight of the animal, x is the amount of feed, and y is the amount of veterinary care for the animal.

2. The Swedish roundwood market is a good example of a market in which both buyers and sellers are highly monopolized. This means that various types of simple bargaining models can be tested against the actual results in the marketplace. In one such article, Johannson and Lofgren assume that the owners wish to maximize profit P subject to a utility function u which measures the benefit to the owners when the price of roundwood is greater than the price determined by free competition.† The latter is sometimes possible because of the monopolistic power enjoyed by the sellers. Profit and utility are both functions of the price x and the output y. The two functions are

$$P(x, y) = f(x) - xy$$
$$u(x, y) = x[g(y) - a] - k$$

where $f(x)$ and $g(y)$ are measures of the monopolistic powers of the sellers and the buyers, respectively. Johannson and Lofgren then found conditions when $P(x, y)$ would be maximized subject to the constraint $u(x, y) = 0$ by applying the method of Lagrange multipliers to the function

$$L(x, y, t) = P(x, y) - tu(x, y)$$

Find the maximum of $L(x, y, t)$ for $f(x) = x^2$, $g(y) = y$, $a = 2$ and $k = 1$.

Cumulative Exercise Set 18–4

1. Consider the function

$$f(x, y) = \frac{x^2 - y}{y}$$

(a) Evaluate $f(1, 1)$.
(b) Find the domain of $f(x, y)$.

2. Sketch the graph of the function $f(x, y) = 9 - x^2 - y^2$, giving the traces and one section that is parallel to each coordinate plane.

3. Find $f_x(x, y)$ and $f_y(x, y)$ for the function

$$f(x, y) = \frac{xy}{x - y}$$

4. Find all second-order partial derivatives of the function

$$f(x, y) = x - x^2y + 2y^3.$$

5. Determine whether the graph of the function $f(x, y) = x^2 - y^2$ has a relative extremum at the point $(0, 0)$. (See Figure 3.)

6. Find the critical points of the function
$$f(x, y) = 2x^3 - 6x - y^2.$$

In Problems 7 and 8 find the relative extrema and any saddle points of the given function.

7. $f(x, y) = (x - 4)^2 + (y - 1)^2$

8. $f(x, y) = x^3 + 2y^3 + 3x^2 - 6y^2 - 45x - 48y$

In Problems 9 to 11 use the method of Lagrange multipliers to find extrema of the functions with the given constraints.

9. $f(x, y) = 6x + 20y + 3xy - 2x^2 - 4y^2$ subject to $x + y = 16$.

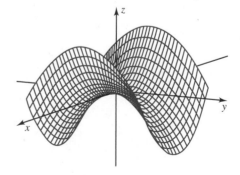

$$f(x, y) = x^2 - y^2$$

FIGURE 3

*Jean-Paul Chavas et al., "Modeling Dynamic Agricultural Production Response: The Case of Swine Production," *American Journal of Agricultural Economics,* Vol. 67, 1985, pp. 636–646.

†Per-Olov Johannson and Karl-Gustaf Lofgren, "A Bargaining Approach to the Modeling of the Swedish Roundwood Market," *Land Economics,* Vol. 61, 1985, pp. 65–75.

10. $f(x, y) = \sqrt{x^2 + y^2}$ subject to $2x + 5y = 14$.

11. $f(x, y, z) = xy^2z^2$ subject to $x + y + z = 15$.

12. Three sides of a box-shaped tree-house need painting: the front, one side, and the roof. The volume of the tree house is 2016 cubic feet. What dimensions of the tree house will minimize the cost of painting if it costs $9 per square foot to paint the face, $7 per square foot to paint the side, and $4 per square foot to paint the top?

18–5 Double Integrals

In the preceding sections of this chapter the concept of the derivative of a function of one variable was generalized to partial derivatives of functions of several variables. In this section we describe how to generalize the integral of a function of one variable to the integral of a function of two variables.

One of the fundamental notions of calculus is that the integral relies on the antiderivative, which is the reverse operation of finding the derivative. In a similar manner the definition of an integral of a function of two variables will also rely on the reverse operation of computing partial derivatives.

Partial Differentiation Reversed

We start with an example of reversing the process of partial differentiation. To find the partial derivative with respect to x of $f(x, y) = 6xy^2 - 5x^4y$, we treat y as a constant and differentiate with respect to x. To reverse the process, hold y constant and find the antiderivative with respect to x. This yields

$$\int (6xy^2 - 5x^4y) \, dx = 6\left(\frac{x^2}{2}\right)y^2 - 5\left(\frac{x^5}{5}\right)y + C(y) = 3x^2y^2 - x^5y + C(y)$$

The constant of integration $C(y)$ represents not just a constant, but an arbitrary function of y. The reason for this can be seen when checking the answer by computing the partial derivative with respect to x of $3x^2y^2 - x^5y + C(y)$ which is $6xy^2 - 5x^4y$. Notice that the partial derivative with respect to x of $C(y)$ is 0.

We write $\int f(x, y) \, dx$ for the operation of computing the partial antiderivative of $f(x, y)$ with respect to x, holding y constant. The first example further illustrates this operation.

EXAMPLE 1

Problem

Compute $\displaystyle\int (6xy^5 + 12x^{-2}y^3 - 7) \, dx$

Solution Hold y constant and find the antiderivative with respect to x.

$$\int (6xy^5 + 12x^{-2}y^3 - 7) \, dx = 6\left(\frac{x^2}{2}\right)y^5 + 12\left(\frac{x^{-1}}{-1}\right)y^3 - 7x + C(y)$$

$$= 3x^2y^5 - 12x^{-1}y^3 - 7x + C(y)$$

The generalization of antidifferentiation to functions of two variables can be used to compute definite integrals of the form

$$\int_a^b f(x, y)\, dx$$

The limits of integration are actually functions. Here they should be viewed as the constant functions $x = a$ and $x = b$.

EXAMPLE 2

Problem

Compute $\int_1^2 (6xy^5 + 12x^{-2}y^3 - 7)\, dx$

Solution The first step is to find the antiderivative with respect to x. This was done in Example 1. Just as with functions of one variable, any antiderivative can be used when computing definite integrals, so we take $C(y) = 0$. Then insert the limits of integration, remembering that they are $x = 1$ and $x = 2$.

$$\int_1^2 (6xy^5 + 12x^{-2}y^3 - 7)\, dx = (3x^2y^5 - 12x^{-1}y^3 - 7x)\, \Big|_{x=1}^{x=2}$$

$$= 3(4)y^5 - 12\left(\frac{1}{2}\right)y^3 - 7(2) - [3(1)y^5 - 12(1)y^3 - 7(1)]$$

$$= 12y^5 - 6y^3 - 14 - 3y^5 + 12y^3 + 7$$

$$= 9y^5 + 6y^3 - 7$$

Notice in Example 2 that the definite integral with respect to x produces a function of y. The variable x does not appear because a number is substituted for x whenever x occurs in the antiderivative. Thus the result of a single integral of $f(x, y)$ is a function of one variable. This function can then be integrated, producing a double integral. Example 3 illustrates this property.

EXAMPLE 3

Problem

Compute $\int_0^2 \left[\int_1^2 (6xy^5 + 12x^{-2}y^3 - 7)\, dx \right] dy$

Solution First work within the brackets. The definite integral in the brackets was computed in Example 2. Substitute its value, $9y^5 + 6y^3 - 7$, for the expression in the brackets to get

$$\int_0^2 \left[\int_1^2 (6xy^5 + 12x^{-2}y^3 - 7)\, dx \right] dy$$

$$= \int_0^2 (9y^5 + 6y^3 - 7)\, dy = \left(\frac{9y^6}{6} + \frac{6y^4}{4} - 7y \right) \Big|_{y=0}^{y=2}$$

$$= \frac{9(64)}{6} + \frac{6(16)}{4} - 7(2) = 96 + 24 - 14 = 106$$

Suppose the order of the integrals in Example 3 was reversed and integration with respect to y was carried out before integration with respect to x. Would the double integral be changed? To answer this question we must examine the processes used in Examples 1, 2, and 3 but with the order reversed. The computations will be carried out in Example 4.

EXAMPLE 4

Problem

Compute

$$\int_1^2 \left[\int_0^2 (6xy^5 + 12x^{-2}y^3 - 7) \, dy \right] dx$$

Solution Consider the integral inside the square brackets. The notation

$$\int f(x, y) \, dy$$

stands for the partial antiderivative of $f(x, y)$ with respect to y. It is evaluated by holding x constant and finding the antiderivative with respect to y.

$$\int (6xy^5 + 12x^{-2}y^3 - 7) \, dy = 6x\left(\frac{y^6}{6}\right) + 12x^{-2}\left(\frac{y^4}{4}\right) - 7y + C(x)$$

$$= xy^6 + 3x^{-2}y^4 - 7y + C(x)$$

Now insert the limits of integration and use the antiderivative with respect to y that was found above (note the $C(x)$ drops out):

$$\int_0^2 (6xy^5 + 12x^{-2}y^3 - 7) \, dy = (xy^6 + 3x^{-2}y^4 - 7y) \Big|_{y=0}^{y=2}$$

$$= x(64) + 3x^{-2}(16) - 7(2) - [x(0) + 3x^{-2}(0) - 7(0)]$$

$$= 64x + 48x^{-2} - 14$$

The definite integral with respect to y produces a function of x that can be integrated to obtain the desired double integral.

$$\int_1^2 \left[\int_0^2 (6xy^5 + 12x^{-2}y^3 - 7) \, dy \right] dx = \int_1^2 (64x + 48x^{-2} - 14) \, dx$$

$$= (32x^2 - 48x^{-1} - 14x) \Big|_{x=1}^{x=2}$$

$$= 32(4) - 48\left(\frac{1}{2}\right) - 14(2) - (32 - 48 - 14) = 76 - (-30) = 106$$

Order of Integrals

Compare the definite integrals in Examples 3 and 4. It is not a coincidence that they are equal. In fact, for most functions $f(x, y)$, if the order of integration is reversed then the double integrals will be the same, that is,

$$\int_a^b \left[\int_c^d f(x, y)\, dx \right] dy = \int_c^d \left[\int_a^b f(x, y)\, dy \right] dx$$

This means that the brackets are not needed, which in turn allows us to define the double integral of $f(x, y)$ without having to specify an order for the integrals.

The limits of integration define a rectangular region R in the plane. Here R is the set of points (x, y) satisfying $c \le x \le d$ and $a \le y \le b$. It is called the **region of integration** and is shown in Figure 1.

The equality of the two orders of integration means that either order of integration can be chosen. The integral is referred to as an **iterated integral** because it is computed by integrating twice. The display gives the formal definition.

DEFINITION

_____

The **double integral** of $f(x, y)$, called the **integrand,** over the rectangular region of integration R, is expressed as

$$\int \int_R f(x, y)\, dx\, dy \qquad \text{or} \qquad \int \int_R f(x, y)\, dy\, dx$$

and is equal to

$$\int_a^b \left[\int_c^d f(x, y)\, dx \right] dy = \int_c^d \left[\int_a^b f(x, y)\, dy \right] dx$$

_____

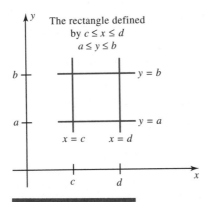

FIGURE 1

As an illustration, consider the region R in the display and let the integrand be the function $f(x, y) = C$ for C a positive number. The double integral of $f(x, y)$ over R is the volume of a box with the lengths of its sides $d - c$ and $b - a$, respectively, and with height C. Then

$$\int \int_R C \, dx \, dy = C[\text{area of } R] = C(d - c)(b - a)$$

The next example shows how to evaluate a double integral when the region is given.

EXAMPLE 5

Problem

Evaluate $\int \int_R x^2(1 - y^3) \, dx \, dy$ over the region R defined by $0 \le x \le 3$ and $-1 \le y \le 1$.

Solution Either order of integration can be selected. We choose to integrate first with respect to x and then with respect to y. The reverse order can be used as a check. The boundaries of the region are $x = 0$ to $x = 3$ and $y = -1$ to $y = 1$. These are the limits and the iterated integral is

$$\int_{-1}^{1} \int_{0}^{3} [x^2(1 - y^3)] \, dx \, dy = \int_{-1}^{1} \left[\left[\frac{x^3}{3}(1 - y^3) \right] \Big|_{x=0}^{x=3} \right] dy$$

$$= \int_{-1}^{1} 9(1 - y^3) \, dy = 9\left(y - \frac{y^4}{4} \right) \Big|_{y=-1}^{y=1}$$

$$= 9\left[\left(1 - \frac{1}{4} \right) - \left(-1 - \frac{1}{4} \right) \right] = 9(2) = 18$$

The final example in this section shows how to evaluate double integrals involving exponential functions.

EXAMPLE 6

Problem

Evaluate the double integral of the function $f(x, y) = ye^{xy}$ over the rectangle $1 \le x \le 2, 0 \le y \le 1$.

Solution We will integrate first with respect to the variable x. Thus our mission is to compute the double integral

$$\int_{0}^{1} \int_{1}^{2} ye^{xy} \, dx \, dy$$

First evaluate the inner integral by treating y as a constant

$$\int ye^{xy} \, dx = \int e^{xy} \, (ydx) = e^{xy} + C(y)$$

Inserting the limits of integration yields

$$\int_1^2 ye^{xy}\, dx = e^{xy}\Big|_{x=1}^{x=2} = e^{2y} - e^y$$

Next,

$$\int_0^1 (e^{2y} - e^y)\, dy = \frac{1}{2}\int_0^1 e^{2y}(2\, dy) - \int_0^1 e^y\, dy$$

$$= .5e^{2y} - e^y\Big|_{y=0}^{y=1}$$

$$= (.5e^2 - e) - (.5 - 1) = .5e^2 - e + .5 \approx 1.4762$$

In Example 6 integration was carried out first with respect to x. To see why this order is preferred, try integrating with respect to y first. We leave this task as an exercise.

EXERCISE SET 18–5

In Problems 1 to 4 compute the antiderivative.

1. $\int (7xy^6 + 6x^3y^{-2} - 5)\, dx$

2. $\int (5x^4y + 6x^2y^{1/2} - 5x + y)\, dx$

3. $\int (8xy^7 - 20x^{-4}y^4 - 3x - y)\, dy$

4. $\int (x^2y^3 - 12x^{1/4}y^{-2} + x - 2y + 3)\, dy$

In Problems 5 to 8 compute the definite integral.

5. $\int_1^2 (5x^4y + 2x^{-1}y^3 - 5)\, dx$

6. $\int_1^3 (4x^3y^2 + x^{-2} - 2y + 1)\, dx$

7. $\int_0^2 (5x^4y + 2x^{-1}y^3 - 5)\, dy$

8. $\int_1^3 (4x^3y^2 + x^{-2} - 2y + 1)\, dy$

In Problems 9 to 12 compute the double integral.

9. $\int_1^3 \left[\int_1^2 (3xy^2 - 12x^{-3}y^2 - 1)\, dx\right] dy$

10. $\int_1^2 \left[\int_{-1}^1 (4xy^3 + 6x^2y^{-1} - y)\, dx\right] dy$

11. $\int_1^2 \left[\int_2^3 (10x^{1/2}y^4 + 12x^{-1/2}y^3 - x)\, dy\right] dx$

12. $\int_0^2 \left[\int_1^3 (3x^{1/3}y^2 + 12x^{-1/3} + x - y)\, dy\right] dx$

In Problems 13 to 16 evaluate the double integral over the rectangular region R.

13. $\iint_R 4x^3(1 - y^2)\, dx\, dy$, where R is defined by $0 \le x \le 2$ and $0 \le y \le 1$.

14. $\iint_R x^2y(2 - 3y^2)\, dx\, dy$, where R is defined by $0 \le x \le 2$ and $1 \le y \le 3$.

15. $\iint_R (x^3y + x - 4y^3)\, dx\, dy$, where R is defined by $1 \le x \le 2$ and $-1 \le y \le 1$.

16. $\iint_R (x^{-2}y + x^{1/2} - 4y)\, dx\, dy$, where R is defined by $1 \le x \le 2$ and $1 \le y \le 3$.

In Problems 17 to 22 compute the antiderivative.

17. $\int x\sqrt{x^2 + 3y}\, dy$

18. $\int \dfrac{3 + 5y}{\sqrt{x}}\, dy$

19. $\displaystyle\int \frac{6x + 2y}{3x^2 + 2xy} \, dx$

20. $\displaystyle\int xe^{x^2 + 9y} \, dy$

21. $\displaystyle\int x\sqrt{x^2 + 3y} \, dx$

22. $\displaystyle\int \frac{10x}{\sqrt{3y + 5x^2}} \, dy$

In Problems 23 to 28 compute the definite integral. Notice that the corresponding indefinite integrals were computed in Problems 17 to 22.

23. $\displaystyle\int_4^5 x\sqrt{x^2 + 3y} \, dy$

24. $\displaystyle\int_2^7 \frac{3 + 5y}{\sqrt{x}} \, dy$

25. $\displaystyle\int_3^5 \frac{6x + 2y}{3x^2 + 2xy} \, dx$

26. $\displaystyle\int_1^6 xe^{x^2 + 9y} \, dy$

27. $\displaystyle\int_0^4 x\sqrt{x^2 + 3y} \, dx$

28. $\displaystyle\int_3^6 \frac{10x}{\sqrt{3y + 5x^2}} \, dy$

In Problems 29 to 32 compute the double integral.

29. $\displaystyle\int_1^3 \left[\int_1^2 \frac{1}{xy} \, dx \right] dy$

30. $\displaystyle\int_1^3 \left[\int_1^2 \left[\frac{x}{1 + y} \right] dy \right] dx$

31. $\displaystyle\int_0^1 \left[\int_0^2 [x(x^2 + y)^{1/2}] \, dx \right] dy$

32. $\displaystyle\int_0^2 \left[\int_1^2 [xy(x^2 + y^2)^{1/2}] \, dy \right] dx$

In Problems 33 to 36 evaluate the double integral over the rectangular region R.

33. $\displaystyle\iint_R ye^{x + y^2} \, dy \, dx$

where R is defined by $-1 \le x \le 1$ and $0 \le y \le 1$.

34. $\displaystyle\iint_R \frac{y}{x^2} e^{y/x} \, dx \, dy$

where R is defined by $1 \le x \le 2$ and $0 \le y \le 1$.

35. $\displaystyle\iint_R (3x^2 + y) \, dx \, dy$

where R is defined by $1 \le x \le 5$ and $0 \le y \le 2$.

36. $\displaystyle\iint_R xe^{x^2 + y} \, dx \, dy$

where R is defined by $0 \le x \le 2$ and $2 \le y \le 3$.

37. Evaluate each integral in parts (a) and (b).

(a) $\displaystyle\int_3^{11} \int_0^4 \sqrt{x} \sqrt{y - 2} \, dx \, dy$

(b) $\displaystyle\int_0^4 \int_3^{11} \sqrt{x} \sqrt{y - 2} \, dy \, dx$

(c) Compare parts (a) and (b). Explain your comparison.

38. Evaluate each integral in parts (a) and (b).

(a) $\displaystyle\int_{-1}^1 \int_0^2 (1 - 6x^2y) \, dx \, dy$

(b) $\displaystyle\int_0^2 \int_{-1}^1 (1 - 6x^2y) \, dy \, dx$

(c) Compare parts (a) and (b). Explain your comparison.

In Problems 39 and 40 compute the double integral carefully because one order of integration is easy while the other is not.

39. $\displaystyle\iint_R xe^{xy} \, dx \, dy$

where R is defined by $0 \le x \le 2$ and $1 \le y \le 3$.

40. $\displaystyle\iint_R \frac{2x + 3x^2y}{1 + 2y + y^2} \, dx \, dy$

where R is defined by $0 \le x \le 1$ and $0 \le y \le 2$.

41. Evalute the double integral in Example 6 in the reverse order by computing

$$\int_1^2 \int_0^1 ye^{xy} \, dy \, dx$$

In Problems 42 to 45 express the limits of integration using both orders of integration, given that a function is to be integrated over the region R in the accompanying graphs.

42.

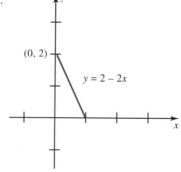

43.

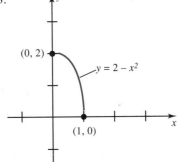

44.

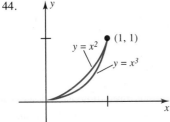

45.

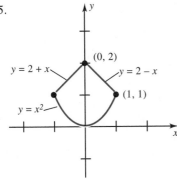

Referenced Exercise Set 18–5

1. In an article investigating the effects of urban renewal on rent increases on the land directly affected, Fishelson and Pines define a transportation cost function $C(x, H)$ that measures the dollars per year the average family spends on transportation when the family lives a distance x miles from the center of the urban renewal

site and H is the yearly housing expenditure of the family.* The authors compute the marginal cost $MC(x, H)$ to be

$$MC(x, H) = \frac{kx^2}{H}$$

where k is a constant depending on the particular city and the location of the urban renewal site in the city. They then calculate the total transportation cost using double integrals. Find the total cost from $x = 0$ to $x = 5$ miles, from $H = 1$ to $H = 10$.

2. Tax reform has become a priority in recent years. Many economists have called for tax revisions for farmers that would replace an income tax with a tax on purchases. In an article investigating this proposal, Chambers and Lopez define a cost of labor function $CL(p, t)$ in terms of profit or income p and income tax t.† They compute the marginal cost function $MCL(p, t)$ to be

$$MCL(p, t) = \frac{kp}{1 - t}$$

where k is a constant depending on the particular financial position of the farm. Then the total cost of labor is the double integral of this function. Find the total cost from $t = 0.25$ to $t = 0.30$ when $k = 1$ and p is

Cumulative Exercise Set 18–5

In Problems 1 and 2 draw the graph of the given function.

1. $f(x, y) = x^2 + 4y^2$

2. $f(x, y) = x^2 + 4x + y^2 - 2y$

3. Find $f_x(x, y)$ and $f_y(x, y)$.

$$f(x, y) = x^2 - 2x^3y + 6y^{1/3}$$

4. Find all second order partial derivatives.

$$f(x, y) = 2x^3 - e^{2x}y^2$$

In Problems 5 and 6 find the relative extrema and any saddle points of the function.

5. $f(x, y) = x^3 - y^2 + 12y$

6. $f(x, y) = x^3 + y^3 + 6y^2$

*Gideon Fishelson and David Pines, "Market Versus Social Valuation of Redevelopment Projects in an Urban Setting," *Socio-Economic Planning*, Vol. 18, 1984, pp. 419–424.

†Robert G. Chambers and Ramon E. Lopez, "Tax Policies and the Financially Constrained Farm Household," *American Journal of Agricultural Economics*, Vol. 69, 1987, pp. 369–377.

In Problems 7 and 8 use the method of Lagrange multipliers to find the extrema of $f(x, y)$ with the given constraints.

7. $f(x, y) = 2xy$ subject to $3x + 2y - 1 = 0$

8. $f(x, y) = x^2y$ subject to $2x + y = 1$

In Problems 9 to 12 compute the double integral.

9. $\int_1^2 \left[\int_{-1}^1 (3xy^2 + 3x^2y - 1)dx \right] dy$

10. $\iint\limits_R (x + 2xy)dx\,dy$ where R is defined by $0 \le x \le 1$ and $0 \le y \le 2$.

11. $\iint\limits_R (x^{-3}y + x^{1/2} - 2y)dx\,dy$ where R is defined by $1 \le x \le 2$ and $1 \le y \le 3$.

12. $\iint\limits_R x^2/(1 + y^2)dx\,dy$ where R is defined by $1 \le x \le 2$ and $1 \le y \le 2$.

18–6 Volume

One measure of a figure in two-dimensional space is its area. We have seen that the definite integral provides a powerful tool for computing area.

The situation is similar when we consider figures in three-dimensional space. There a measure of a figure is its volume. We will see that the definite double integral introduced in the preceding section provides a powerful tool for computing the volume of a solid. First it will be necessary to evaluate double integrals over regions which are not necessarily rectangles.

Double Integrals over Other Regions

In the preceding section we mentioned that the limits of integration of a double integral over a rectangular region are actually functions. However, in that section there was no need to use the fact that they are actually linear functions of the form

$x = $ constant

or

$y = $ constant.

In this section we will show that it is possible to compute double integrals in which the limits of the inner integral are nonlinear functions, provided that the limits of the outer integral are constant functions. Because this remark is so important we single it out. The first example will then illustrate how to compute double integrals with variable limits.

 Warning: In a double integral the limits of the outer integral must always be constant functions. The limits of the inner integral may be variable functions.

EXAMPLE 1

Problem

Compute

$$\int_0^1 \int_{x^2}^x 3xy^2 \, dy \, dx$$

Solution First calculate the inner integral.

$$\int_{x^2}^x 3xy^2 \, dy = xy^3 \Big|_{y=x^2}^{y=x} = xx^3 - x(x^2)^3 = x^4 - x^7$$

Then perform the outer integral of this function.

$$\int_0^1 \int_{x^2}^x 3xy^2 \, dy \, dx = \int_0^1 (x^4 - x^7) \, dx = \left(\frac{x^5}{5} - \frac{x^8}{8}\right)\Big|_0^1 = \frac{1}{5} - \frac{1}{8} = \frac{3}{40}$$

The double integral in Example 1 is defined over a region with variable limits. The shaded region in Figure 1 defines the limits of integration. It is bounded above by the line $y = x$ and below by the parabola $y = x^2$.

There are two different ways to view such a region, and each view corresponds to a different order of integration. Figure 2 shows the two views. In part (a) the

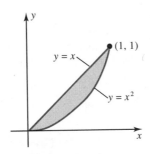

FIGURE 1

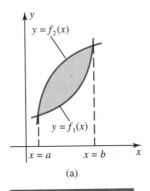

FIGURE 2

region is bounded above by a function of the form $y = f_2(x)$ and below by a function of the form $y = f_1(x)$. This view corresponds to integrating first with respect to y and then with respect to x. The functions $f_1(x)$ and $f_2(x)$ form the limits of the inner integral, while the constant functions $x = a$ and $x = b$ form the limits of the outer integral.

The view in part (b) corresponds to the reverse order of integration, first with respect to x and then with respect to y. Here the boundaries are viewed from a horizontal perspective. The region is bounded on the right by a function of the form $x = g_2(y)$ and on the left by a function of the form $x = g_1(y)$. These functions form the limits of the inner integral, while the constant functions $y = c$ and $y = d$ form the limits of the outer integral. Example 2 evaluates such an integral using both orders of integration.

EXAMPLE 2

Problem

Compute the integral $\displaystyle\iint_R 2xy \, dx \, dy$ over the region defined by $x^2 \le y \le 4$ and $0 \le x \le 2$.

Solution (a) Since the variable boundary of R is given directly as $y = x^2$, the more immediate order of integration is first with respect to y and then x. The limits of the inner integral are $y = x^2$ and $y = 4$, while the limits of the outer integral are $x = 0$ and $x = 2$. See Figure 3(a). The iterated integral is

$$\int_0^2 \int_{x^2}^4 2xy \, dy \, dx = \int_0^2 \left[xy^2 \Big|_{y=x^2}^{y=4} \right] dx = \int_0^2 (16x - x^5) \, dx$$

$$= \left(8x^2 - \frac{x^6}{6} \right) \Big|_{x=0}^{x=2} = \left(32 - \frac{64}{6} \right) - 0 = \frac{64}{3}$$

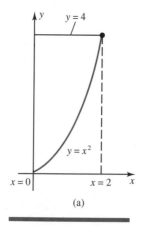

(a)

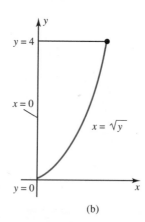

(b)

FIGURE 3

(b) To switch the order of integration, view the region as being bounded by functions of the form $x = g(y)$. [See Figure 3(b).] From this perspective, the formula $y = x^2$ yields the function $x = \sqrt{y}$. For the inner integral, the upper limit is the function $x = \sqrt{y}$ and the lower limit is the function $x = 0$. The limits of the outer integral are $y = 0$ and $y = 4$. The iterated integral is

$$\int_0^4 \int_0^{\sqrt{y}} 2xy \, dx \, dy = \int_0^4 \left[x^2 y \Big|_{x=0}^{x=\sqrt{y}} \right] dy$$

$$= \int_0^4 y^2 \, dy = \frac{y^3}{3} \Big|_0^4 = \frac{64}{3}$$

Notice that the answers in parts (a) and (b) of Example 2 are equal. This is expected because we merely switched the order of integration.

Sometimes computing the integral using one order of integration is easier than the reverse order. Unfortunately there is no general rule to tell which order is better, so instead of worrying about how to start a problem, you are better off attempting one order and, if it does not seem to be working, then try the reverse order.

Volume

One way to compute the area under the graph of a function of one variable is to find the definite integral of the function within the given limits. The initial approach, however, was to form sums of rectangles and then take the limit of the sums.

In a similar way, the volume of a solid in three dimensions can be found by summing the volumes of boxes whose heights are functional values of $f(x, y)$ and whose length and width are small increments of dx and dy. One such box is shown in Figure 4. Taking the analogy with the case in two dimensions one step further,

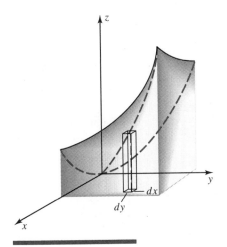

FIGURE 4

the volume of a solid under the graph of a function of two variables can be computed much more easily by finding the definite double integral between appropriate limits.

Look at Figure 5 for instance. It shows the graph of $f(x, y) = x^2 + y^2$ over the region R defined by $0 \le x \le 1$ and $0 \le y \le 2$. The volume of the solid bounded above by $f(x, y)$ and below by R can be found by computing the double integral of $f(x, y)$ over R.

DEFINITION

If $f(x, y) \ge 0$ for all points in a region R, then the **volume** of the solid bounded above by $f(x, y)$ and below by R is equal to

$$\iint\limits_{R} f(x, y) \, dx \, dy$$

Figure 6 portrays this result geometrically. Examples 3 and 4 will show how to compute volumes of solids whose bases are rectangles.

EXAMPLE 3

Problem

Compute the volume of the solid bounded above by $f(x, y) = x^2 + y^2$ and below by the rectangle R defined by $0 \le x \le 1$ and $0 \le y \le 2$. This solid is graphed in Figure 5.

Solution The display states that the volume is the double integral of $f(x, y) = x^2 + y^2$ over R. This is given by

$$\int_0^2 \int_0^1 (x^2 + y^2) \, dx \, dy = \int_0^2 \left[\left(\frac{x^3}{3} + xy^2 \right) \Big|_{x=0}^{x=1} \right] dy$$

$$= \int_0^2 \left(\frac{1}{3} + y^2 \right) dy = \left(\frac{y}{3} + \frac{y^3}{3} \right) \Big|_{y=0}^{y=2}$$

$$= \frac{2}{3} + \frac{8}{3} = \frac{10}{3}$$

In the previous section we saw that the double integral does not depend on the order of the integration. Therefore one way to verify that the volume of the solid in Example 3 is 10/3 is to compute the double integral

$$\int_0^1 \int_0^2 (x^2 + y^2) \, dy \, dx$$

EXAMPLE 4

Problem

Compute the volume of the solid bounded above by $f(x, y) = e^{x+y}$ and below by the rectangle R defined by $0 \le x \le 1$ and $1 \le y \le 2$.

Solution

$$\int_1^2 \int_0^1 e^{x+y}\, dx\, dy = \int_1^2 \left[e^{x+y} \Big|_{x=0}^{x=1} \right] dy = \int_1^2 (e^{1+y} - e^y)\, dy$$

$$= e^{1+y} - e^y \Big|_{y=1}^{y=2} = (e^3 - e^2) - (e^2 - e)$$

$$= e^3 - 2e^2 + e \approx 8.0257.$$

The final example in this section shows how to find the volume of a solid whose base has some boundaries that are not constant functions.

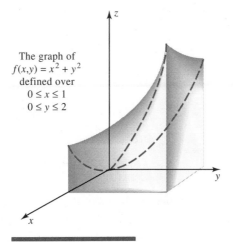

The graph of
$f(x,y) = x^2 + y^2$
defined over
$0 \le x \le 1$
$0 \le y \le 2$

FIGURE 5

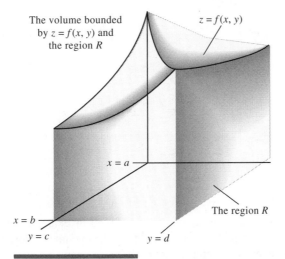

The volume bounded
by $z = f(x, y)$ and
the region R

$z = f(x, y)$

$x = a$

$x = b$

$y = c$

$y = d$

The region R

FIGURE 6

EXAMPLE 5

Problem

Compute the volume of the solid bounded above by the function $f(x, y) = 4 - x - y$ and below by the triangle R formed from the lines $x + y = 4$, $x = 0$, and $y = 0$.

Solution Two issues must be settled before the desired volume can be computed, the order of integration and the limits. We choose to integrate first with respect to y. Thus the limits of the inner integral are of the form $y = f(x)$. It can be seen from Figure 7 that the lower limit is $y = 0$ and the upper limit is $y = 4 - x$.

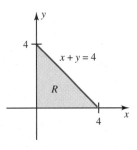

FIGURE 7

Recall the warning that the limits of the outer integral must be constant functions. Here they are $x = 0$ and $x = 4$. Thus the desired volume is

$$\int_0^4 \int_0^{4-x} (4 - x - y)\, dy\, dx = \int_0^4 \left[(4 - x)y - \frac{y^2}{2}\Big|_{y=0}^{y=4-x} \right] dx$$

$$= \int_0^4 \left[(4 - x)^2 - \frac{(4 - x)^2}{2} \right] dx$$

$$= \frac{1}{2} \int_0^4 (4 - x)^2\, dx$$

$$= \frac{1}{2} \frac{(4 - x)^3}{-3}\Big|_{x=0}^{x=4} = \frac{-1}{6} (4 - x)^3\Big|_{x=0}^{x=4}$$

$$= \frac{-1}{6} (0 - 4^3) = \frac{64}{6}$$

The solid whose volume was found in Example 5 is the triangular pyramid shown in Figure 8. It is reminiscent of the pyramids whose volumes were computed by the ancient Egyptians about 4000 years ago.

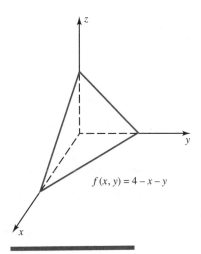

$$f(x, y) = 4 - x - y$$

FIGURE 8

EXERCISE SET 18–6

In Problems 1 to 6 evaluate the double integral.

1. $\int_0^1 \int_{x^2}^x 4xy^3 \, dy \, dx$ 2. $\int_0^2 \int_{x^2}^{2x} 4x^2y^3 \, dy \, dx$

3. $\int_0^1 \int_{y^3}^1 xy^2 \, dx \, dy$ 4. $\int_0^1 \int_{y^3}^y xy^4 \, dx \, dy$

5. $\int_0^2 \int_{-x^2}^{x^2} (x + y^2) \, dy \, dx$

6. $\int_0^2 \int_{1-x^2}^{x^2+1} (x - y) \, dy \, dx$

In Problems 7 to 12 evaluate the double integral over the nonrectangular region R.

7. $\iint_R 4xy \, dx \, dy$, where R is defined by

   $x^2 \le y \le 1$ and $0 \le x \le 1$

8. $\iint_R (x + y) \, dx \, dy$, where R is defined by

   $x^3 \le y \le 1$ and $0 \le x \le 1$.

9. $\iint_R (2x + 3y) \, dx \, dy$, where R is defined by

   $x^3 \le y \le x^2$ and $0 \le x \le 1$.

10. $\iint_R (x - xy) \, dx \, dy$, where R is defined by

    $x^4 \le y \le x^2$ and $0 \le x \le 1$.

11. $\iint_R (x + 2y) \, dx \, dy$, where R is defined by

    $x^2 \le y \le 2x$ and $0 \le x \le 2$.

12. $\iint_R f(x, y) = x + 2y$, where R is defined by

    $0 \le y \le 4$ and $(y/2) \le x \le \sqrt{y}$.

In Problems 13 to 16 compute the volume of the solid bounded above by $f(x, y)$ and below by the rectangle R.

13. $f(x, y) = x^3y + x - 4y^3$, where R is defined by $1 \le x \le 3$ and $0 \le y \le 1$.

14. $f(x, y) = x^{-2}y + x^{1/2} - 4$, where R is defined by $1 \le x \le 3$ and $1 \le y \le 2$.

15. $f(x, y) = xe^{x^2+y}$, where R is defined by $0 \le x \le 2$ and $2 \le y \le 3$.

16. $f(x, y) = y^2e^{2x+y^3}$, where R is defined by $2 \le x \le 3$ and $1 \le y \le 2$.

In Problems 17 to 20 compute the volume of the pyramid bounded above by the given function $f(x, y)$ and below by the triangle R formed from the lines $x = 0$, $y = 0$, and the given line.

17. $f(x, y) = 4 - x - y$, where the line is $x + y = 2$.

18. $f(x, y) = 4 - x - y$, where the line is $2x + y = 4$.

19. $f(x, y) = 6 - x - y$, where the line is $x + y = 6$.

20. $f(x, y) = 6 - x - y$, where the line is $3x + y = 6$.

In Problems 21 to 24 evaluate the double integral.

21. $\int_0^4 \int_0^x \sqrt{xy} \, dy \, dx$

22. $\int_1^4 \int_0^7 (x + 4y) \, dx \, dy$

23. $\int_1^2 \int_y^{3y} \frac{1}{x} \, dx \, dy$ 24. $\int_0^1 \int_{2x}^{3x} e^{x+y} \, dx \, dy$

In Problems 25 to 28 evaluate the double integral over the region R.

25. $\iint_R (\sqrt{x} + y) \, dx \, dy$, where R is defined by

$0 \le y \le 1$ and $0 \le x \le 3y$.

26. $\iint_R \sqrt{1 - x^2} \, dx \, dy$, where R is defined by

$0 \le y \le 1$ and $y \le x \le 1$.

27. $\iint_R \frac{x}{1 + y^2} \, dx \, dy$, where R is defined by

$x^2 \le y \le 4$ and $0 \le x \le 2$.

28. $\iint_R y e^{x^2} \, dx \, dy$, where R is defined by

$0 \le y \le 1$ and $y^2 \le x \le 1$.

In Problems 29 to 32 compute the volume of the solid bounded above by $f(x, y)$ and below by the rectangle R.

29. $f(x, y) = \sqrt{x}$, where R is defined by $0 \le x \le 1$ and $0 \le y \le 4$.

30. $f(x, y) = \sqrt{y}$, where R is defined by $0 \le x \le 4$ and $0 \le y \le 9$.

31. $f(x, y) = e^{x+y}$, where R is defined by $0 \le x \le 2$ and $-1 \le y \le 1$.

32. $f(x, y) = x\sqrt{x^2 + y}$, where R is defined by $0 \le x \le 1$ and $0 \le y \le 1$.

In Problems 33 to 36 compute the volume of the solid bounded above by the given function $f(x, y)$ and below by the given region R.

33. $f(x, y) = x^2 + y^2$, R is the square formed from the lines $x = 0$, $x = 3$, $y = 0$, and $y = 3$.

34. $f(x, y) = x^2 + y^2$, R is the rectangle formed from the lines $x = 0$, $x = 3$, $y = 0$, and $y = 6$.

35. $f(x, y) = 3 - x - y$, and R is the triangle formed from the lines $x = 1$, $y = 0$, and $y = x$.

36. $f(x, y) = 3 - x - y$, and R is the triangle formed from the lines $x = 0$, $y = 0$, and $y = x$.

37. Verify the volume of the solid that was calculated in Example 3 by computing the double integral

$$\int_0^1 \int_0^2 (x^2 + y^2) \, dy \, dx$$

38. Compute the volume of the pyramid in Figure 8 by integrating first with respect to x.

Cumulative Exercise Set 18–6

1. Consider the function

$$f(x, y) = \frac{x^2 + 2xy + y^2}{x + y}$$

(a) Evaluate $f(1, 1)$.
(b) Evaluate $f(1, -1)$.
(c) Find the domain of $f(x, y)$.

2. Sketch the graph of the function $f(x, y) = 4x^2 - y^2$, giving the traces and one section that is parallel to each coordinate plane.

3. Let $f(x, y) = 3x^4 - 4xy^3 - 2y$.
(a) Find $f_x(x, y)$ and $f_y(x, y)$.
(b) Find all second-order partial derivatives of $f(x, y)$.

4. Find the relative extrema and any saddle points of the function $f(x, y) = 12x - 2xy + 4y + y^2$.

5. Use the method of Lagrange multipliers to find extrema of the function $f(x, y) = x^2 + 3y^2 - 12xy$ subject to $x + y = 16$.

6. Find two positive numbers such that the sum of their squares is equal to 50 and their product is a maximum.

7. Compute the antiderivative.

$$\int_{-2}^2 \frac{x}{\sqrt{x^2 + y^2}} \, dx$$

In Problems 8 and 9 compute the double integral.

8. $\int_0^{1/3} \int_{-2}^0 (x + 3y + 2)^4 \, dx \, dy$

9. $\int_0^1 \int_0^{x^3} e^{y/x} \, dx \, dy$

10. Evaluate the double integral over the rectangular region R, where R is defined by $0 \le x \le 3$ and $-1 \le y \le 1$.

$\iint_R (2x + 3y) \, dx \, dy$

11. Compute the volume of the solid bounded above by $f(x, y) = 1 + x^2 + y^2$ and below by the rectangle $0 \le x \le 1$ and $0 \le y \le 4$.

12. Compute the volume of the solid bounded above by $f(x, y) = 6x + 2y + 5$ and below by the square formed from the lines $x = 0$, $x = 2$, $y = 0$, and $y = 2$.

CASE STUDY   **Minimum Cost Production in the Chemical Industry**

Jojoba oil is used for producing amides, which, in turn, are an essential ingredient in many chemical processes. The procedure of extracting amides from jojoba oil is time-consuming and expensive. First the oil is refined and purified. Then other chemicals are added and the mixture is heated for several hours. Once the material becomes cakelike the amides can be extracted.

Researchers have tried various methods to maximize the amide yield while conserving cost to produce the cakelike material. The two most important variables are temperature and time. In an article* discussing various strategies for improving amide yield, researchers discovered that the relationship between the variables is best described by a Cobb–Douglas production function, which is a power function of the form

$$f(x, y) = ax^b y^c$$

Let $f(x, y)$ represent the amount of amide yield (as a percentage of oil weight) at temperature x (measured in degrees Celsius) and time y (measured in hours). Test results showed that proper production function is

$$f(x, y) = 1.3x^{0.5}y^{0.4}$$

This function is useful in planning production. Various constraints on cost are given and the corresponding production yield is computed. A particular firm would decide what the most reasonable limits on cost would be and then use this function to predict whether the amide yield would be great enough. If not, further study of the cost constraints would be required.

For example, cost considerations require that if x increases, then y must decrease. This means that if the processing time increases, then the temperature would have to decrease to keep the production cost the same. One such constraint is

$$x + 2y = 200$$

---

*A. Shani et al., "Synthesis of Jojobamide and Homojojobamide," *Journal of American Oil and Chemical Society*, March 1980, pp. 112–114.

This means that when $x = 100$, $y = 50$, and when $x = 110$, $y = 45$; so that the cost of using a temperature of $100°$ for 50 hours is equal to the cost of using a temperature of $110°$ for 45 hours. The next example shows how Lagrange multipliers are used to determine the maximum amide yield subject to the given cost constraint.

EXAMPLE

Problem

Find the maximum of $f(x, y) = 1.3x^{0.5}y^{0.4}$ subject to the constraint $x + 2y = 200$.

Solution Introduce the Lagrange multiplier t in the function

$$F(x, y, t) = 1.3x^{0.5}y^{0.4} + t(x + 2y - 200)$$

Find the partial derivatives and set them equal to 0.

$$F_x(x, y, t) = 1.3(0.5)x^{-0.5}y^{0.4} + t$$
$$= 0.65x^{-0.5}y^{0.4} + t = 0 \tag{1}$$

$$F_y(x, y, t) = 1.3(0.4)x^{0.5}y^{-0.6} + 2t$$

$$= 0.52x^{0.5}y^{-0.6} + 2t = 0 \tag{2}$$

$$F_t(x, y, t) = x + 2y - 200 = 0 \tag{3}$$

Solve (1) and (2) for $-t$.

$$-t = 0.65x^{-0.5}y^{0.4} = 0.26x^{0.5}y^{-0.6}$$

Disregard $-t$ and multiply the resulting equation by $x^{0.5}y^{0.6}$.

$$0.65y = 0.26x$$

Hence $x = 2.5y$. Substitute this relationship into (3).

$$2.5y + 2y - 200 = 0$$
$$4.5y = 200$$
$$y \approx 44.4$$
$$x \approx 111$$

This means that with the given constraint the maximum amide yield is $f(111, 44.4) \approx 1.3(10.5)(4.56) \approx 62.2\%$ when the material is heated at $111°$ for 44.4 hours.

Case Study Exercises

In Problems 1 to 4 find the maximum amide yield for the given cost function and the production function.

$$f(x, y) = 1.3x^{0.5}y^{0.4}$$

1. $x + 2.4y = 200$ 2. $x + 1.8y = 200$
3. $2x + 4.5y = 440$ 4. $3x + 5.8y = 600$

In Problems 5 and 6 find the maximum amide yield for the given cost function and the given production function.

5. $x + 2.4y = 200,$
   $f(x, y) = 1.3x^{0.6}y^{0.3}$

6. $x + 1.8y = 200,$
   $f(x, y) = 1.7x^{0.55}y^{0.35}$

CHAPTER REVIEW

Key Terms

18–1   Examples of Functions of Several Variables
Dependent Variable
Independent Variable
Function of Two Variables

Trace
Section
Level Curves

18–2   Partial Derivatives
Partial Derivative
Second Partial Derivative

Mixed Partial Derivative

18–3   Relative Extrema
Relative Maximum
Relative Minimum
Relative Extremum

Critical Point
Saddle Point
Second Partial Derivative Test

18–4   Lagrange Multipliers
Constraint

Lagrange Multipliers

18–5   Double Integrals
Region of Integration
Iterated Integral

Double Integral
Integrand

18–6   Volume
Volume

REVIEW PROBLEMS

1. Find the functional value of $f(x, y) = 3x^2 - 4y^2$ at the point $(1, -1)$.

2. Find the domain of the function $f(x, y) = \dfrac{xy}{x^2 - 1}$.

3. Graph the function $f(x, y) = 1 + x^2 + 4y^2$, giving the traces and at least one section parallel to each coordinate plane.

4. Graph the function $f(x, y) = x^2 + 4x + y^2 + 2y$.

5. Find $f_x(x, y)$ and $f_y(x, y)$, where $f(x, y) = xy^3 - (2x + 3y)^4$.

6. Find $f_x(1, 0)$ and $f_y(2, 1)$ for $f(x, y) = 2xy^3 + 3x^{-1}y$.

7. Find all second-order partial derivatives for $f(x, y) = 3x^{2/5} + x^2y^{-2}$.

8. Find the values of x and y such that $f_x(x, y) = 0$ and $f_y(x, y) = 0$ for $f(x, y) = 1 - x^2 + y^2 + 2xy - 4y$.

9. Find the critical points of the function $f(x, y) = 3x^2 - 12xy + y^2 + 10y - 2$.

10. Find the relative extrema and any saddle points of the function $f(x, y) = x - xy + y^3 - 3y^2$.

11. Find the relative extrema of the function $f(x, y) = x^3 - 6x^2 - 2y^3 + 24y$.

12. A firm utilizes x million dollars per year in labor cost and y million dollars per year in production cost. The cost function is

$$C(x, y) = 100 + 24x + 12y + 2x^2 + 3y^2$$

Find the amount that the firm should spend on labor and production each year to minimize cost.

In Problems 13 and 14 use the method of Lagrange multipliers to find the extrema of the function with the given constraints.

13. $f(x, y, z) = xy$ subject to $3x + y - 5 = 0$

14. $f(x, y, z) = xyz$ subject to $x + y + 2z - 1 = 0$

15. Find three numbers that sum to 45 such that their product is a maximum.

16. Maximize $f(x, y) = x^{1/2}y^{1/3}$ subject to $200x + y = 8000$.

17. Compute the antiderivative

$$\int (14xy^6 + 6x^3y^{-2} - y + x - 1)\, dx$$

18. Compute the definite integral.

$$\int_1^2 (10x^4y + 2x^{-3}y^3 - 1)\, dx$$

19. Compute the double integral.

$$\int_1^3 \left[\int_1^2 (6xy^2 - 24x^{-3}y^2 - 1)\, dx \right] dy$$

20. Evaluate the double integral over R:

$$\iint_R 4x^3(1 - y^2)\, dx\, dy, \text{ where } R \text{ is defined by}$$

$0 \le x \le 1$ and $0 \le y \le 1$.

21. Evaluate the double integral over R:

$$\iint_R 4xy\, dx\, dy, \text{ where } R \text{ is defined by}$$

$x^2 \le y \le 1$ and $0 \le x \le 1$.

PROGRAMMABLE CALCULATOR EXPERIMENTS

One way to simulate a three-dimensional coordinate system with a graphics calculator is to sketch the graph of $y = x$ in the standard viewing screen and let it represent the x-axis, then let the x-axis in the screen represent the y-axis of the three-dimensional coordinate system, and let the original y-axis represent the z-axis of the three-dimen-sional system. In Problems 1 to 4 use this depiction of a three-dimensional coordinate system to visualize the given points.

1. $(0, 1, 2)$ 2. $(1, 1, 2)$

3. $(1, 1, -2)$ 4. $(2, 3, 3)$

Appendix A: The Programmable Calculator

The advent of inexpensive handheld programmable calculators might revolutionize the way we teach and learn mathematics. At the very least, it is important for every student to learn how to use these extremely powerful and convenient machines. Their potential is expansive.

At the end of each chapter we have given several problems that invite you to use these calculators to explore the ideas in that chapter more fully. These exercises encourage you to experiment in various ways that will provide a deeper understanding of the material. This appendix is designed to get you started but is not meant to replace your calculator's manual. Once becoming facile with your calculator, you will have fun designing your own experiments. And after you accomplish a few simple tasks with it, you will find the manual less intimidating and foreboding and more pleasant to read.

There are several outstanding calculators on the market, including those by Casio, Texas Instruments, and Hewlett-Packard. As the technology improves and cost plummets, more manufacturers will join the market. Each has advantages and disadvantages, and we do not recommend any one over another. Shop around before buying, talk to friends and find out why they prefer one to another. There are many constraints (you might even want to express them in a linear programming problem). Get the best price and enjoy calculating!

Although all graphing calculators accomplish the same tasks, they have slightly different keyboards and capabilities. To make this appendix as practical as possible, we decided to focus on one particular brand, the Casio fx-7000G. If you have a different one, the examples should be self-explanatory, and with a brief perusal of your manual you should be able to translate the actual key sequences to fit your calculator.

One of the most powerful and compelling capabilities of these calculators is the ability to sketch almost instantaneously the graph of any function. To start, Casio has included the ability to graph several familiar functions immediately. They are called "built-in" functions and you can graph them by simply pressing three keys. There are 20 built-in functions, including the following:

$$x^2, \quad x^{-1}, \quad \sqrt{x}, \quad \sqrt[3]{x}, \quad \ln x, \quad \log x, \quad e^x, \quad 10^x$$

There are also several functions from trigonometry and other special functions. To obtain the graph of one of these built-in functions, press the following key sequence:

Graph		function key		EXE

For example, to sketch the graph $y = x^2$ press the following:

Graph		x^2		EXE

To sketch the graph $y = x^{-1}$, press the following:

Graph		x^{-1}		EXE

It is important to keep in mind that whenever sketching the graph of the function you must set the range on the x-axis and the range on the y-axis. When sketching the graph of one of the built-in functions as before, the calculator uses preset ranges. Only that part of the graph within these ranges appears on the screen. If you want to see any other part of the graph, or to magnify the graph, you must overwrite the preset ranges.

This is not a new idea. When graphing a function on a piece of paper, you determine the ranges. For example, suppose you wanted to sketch the graph of the following system of linear equations:

$$3x + y = 9$$
$$2x + y = 8$$

You would most likely first find the intercepts: the x-intercepts are 3 and 4 and the y-intercepts are 9 and 8. You would then determine how large you want the graph on your paper and then plot the largest intercept, 9, just inside the range you decided to have for the graph. In other words, a little thought and computation goes into the decision of the magnitude of the range of your graph on a piece of paper. The same type of care must be taken when sketching graphs with the calculator.

To set the ranges on the axes, press the Range key and the following appears:

Range
  Xmin: -5
    max:5
    scl:2
  Ymin: -10
    max:5
    scl:2

Xmin represents the leftmost point on the x-axis. Xmax represents the rightmost point on the x-axis, while Ymin and Ymax represent the lower and uppermost points on the y-axis. "scl" represents the scale on each axis.

The position of the minus sign, $-$, before 5 in Xmin will be blinking. The position that is blinking denotes the location of the cursor. The next character you press will appear where the cursor is located. To change the position of the cursor, use the arrow keys ⇨⇦⇧⇩ . To change the values of the ranges, move the cursor to the appropriate value and press the number to which you wish to change it. When you want to enter the appropriate number, press the execute key | EXE | . The cursor will automatically move to the next value. If you do not want to change a value, just press | EXE | . Now press | G⟷T | , which shifts you to the graphing mode (G) of the calculator from the text mode (T) that you were in. The screen will show the coordinate system as you defined it.

If you are satisfied and want to graph a function, press the clear key | AC | . Suppose you want to graph the two preceding linear equations:

$$3x + y = 9$$
$$2x + y = 8$$

The first step is to press | Graph | . On the screen appears

Graph Y = _

You must alter the equations so that they fit the form to which the calculator is set. The equations become

$$y = 9 - 3x$$
$$y = 8 - 2x$$

The variable x is entered by pressing two keys, the red | ALPHA | key and the | + | key. Every key of the calculator has three definitions and the | SHIFT | and | ALPHA | keys give you the alternate definitions. Thus "x" appears in red under the | + | key. The key sequence that sketches the graphs of the first linear equation is

| Graph | 8 | − | 2 | ALPHA X | | EXE |

Appearing on the screen before you press | EXE | is

Graph Y = 8 − 2X

When you press | EXE | the graph of the function appears. To graph the second function, press | AC | or | G⟷T | and enter the following key sequence:

| Graph | 9 | − | 3 | ALPHA X | | EXE |

Both graphs appear on the screen at once. Suppose you wanted to find the point of intersection graphically; you can use the very powerful "zoom-in" feature of the calculator. Of course, it is possible to magnify the graph by changing the ranges, but the zoom-in feature allows you to magnify the graph almost instantaneously. Use the | SHIFT | | * | key sequence to zoom. To find the point of intersection, first press $\boxed{\text{SHIFT}}$ $\boxed{\text{Trace}}$ to access the cursor. It will be blinking and the coordinates of the point in the coordinate system where it is blinking appear on the screen when you press $\boxed{\text{SHIFT}}$ $\boxed{\text{X}\longleftrightarrow\text{Y}}$. Use the arrow keys to move the cursor to the point of intersection. Read the coordinates by pressing $\boxed{\text{SHIFT}}$ $\boxed{\text{X}\longleftrightarrow\text{Y}}$. To get better accuracy, zoom-in and move the cursor to the point of intersection. You can zoom-in as many times as you want, each time obtaining better accuracy.

Statistical Computation

Besides helping you to sketch graphs of functions, programmable calculators can carry out difficult calculations in a breeze. One such capability is in statistical computation. Suppose we have the following data and want to compute the mean and the standard deviation as well as the histogram.

x	1–5	6–10	11–15	16–20	21–25
frequency	7	3	8	4	9

Shift the mode of the calculator to the statistical mode by pressing $\boxed{\text{SHIFT}}$ $\boxed{\text{MODE}}$ $\boxed{\text{X}}$. To clear the screen, press $\boxed{\text{AC}}$. When you press $\boxed{\text{Range}}$ in statistical mode Xmin and Xmax refer to the range of the x values in the table. In this example we enter 0 for Xmin and 25 for Xmax. "scl" refers to the length of the intervals; in this case they are of length 5. Ymin and Ymax refer to the range of values of the frequencies and will be 3 and 9, respectively. "scl" refers to the scale preferred on the vertical axis of the histogram.

Next tell the calculator to reserve 5 memory locations for the 5 intervals by pressing $\boxed{\text{MODE}}$ $\boxed{\text{.}}$ 5 $\boxed{\text{EXE}}$. Next clear the statistical memory by pressing $\boxed{\text{SHIFT}}$ $\boxed{\text{Scl}}$ $\boxed{\text{EXE}}$. This clears a different part of memory from the AC key. Now input the data using the data key $\boxed{\text{DT}}$. This is done by first entering the left (or lower) end point of the interval and then entering its frequency by pressing

$\boxed{\text{SHIFT}}$ $\boxed{\text{;}}$ n $\boxed{\text{DT}}$

where n is the frequency. Thus to enter the first interval and its frequency, press

1 $\boxed{\text{SHIFT}}$ $\boxed{\text{;}}$ 7 $\boxed{\text{DT}}$

Follow this by entering the second interval and its frequency, and so on. Here is the entire key sequence that enters the data.

1 $\boxed{\text{SHIFT}}$ $\boxed{\text{;}}$ 7 $\boxed{\text{DT}}$ 6 $\boxed{\text{SHIFT}}$ $\boxed{\text{;}}$ 3 $\boxed{\text{DT}}$

11 $\boxed{\text{SHIFT}}$ $\boxed{\text{;}}$ 8 $\boxed{\text{DT}}$ 16 $\boxed{\text{SHIFT}}$ $\boxed{\text{;}}$ 4 $\boxed{\text{DT}}$

21 $\boxed{\text{SHIFT}}$ $\boxed{\text{;}}$ 9 $\boxed{\text{DT}}$

To obtain the histogram, press

| Graph | | EXE |

To obtain the sum of the frequencies, press

| ALPHA | | W | | EXE |

To obtain the mean, press

| SHIFT | | x | | EXE |

To obtain the standard deviation, press

| SHIFT | | $x\sigma_{n-1}$ | | EXE |

Appendix B: Tables

Table 1 Compound Interest Table

n	2%	4%	6%	8%	10%	12%	14%
1	1.02000	1.04000	1.06000	1.08000	1.10000	1.12000	1.14000
2	1.04040	1.08160	1.12360	1.16640	1.21000	1.25440	1.29960
3	1.06121	1.12486	1.19102	1.25971	1.33100	1.40493	1.48154
4	1.08243	1.16986	1.26248	1.36049	1.46410	1.57352	1.68896
5	1.10408	1.21665	1.33823	1.46933	1.61051	1.76234	1.92541
6	1.12616	1.26532	1.41852	1.58687	1.77156	1.97382	2.19497
7	1.14869	1.31593	1.50363	1.71382	1.94872	2.21068	2.50227
8	1.17166	1.36857	1.59385	1.85093	2.14359	2.47596	2.85258
9	1.19509	1.42331	1.68948	1.99900	2.35795	2.77308	3.25194
10	1.21900	1.48024	1.79085	2.15892	2.59374	3.10584	3.70722
11	1.24338	1.53945	1.89830	2.33164	2.85312	3.47854	4.22622
12	1.26824	1.60103	2.01220	2.51817	3.13843	3.89597	4.81790
13	1.29361	1.66507	2.13293	2.71962	3.45227	4.36349	5.49240
14	1.31948	1.73167	2.26090	2.93719	3.79750	4.88710	6.26133
15	1.34587	1.80094	2.39656	3.17217	4.17725	5.47355	7.13792
16	1.37279	1.87298	2.54035	3.42594	4.59497	6.13038	8.13723
17	1.40024	1.94790	2.69277	3.70002	5.05447	6.86602	9.27644
18	1.42825	2.02581	2.85434	3.99602	5.55992	7.68995	10.57510
19	1.45681	2.10684	3.02560	4.31570	6.11591	8.61274	12.05570
20	1.48595	2.19112	3.20713	4.66095	6.72750	9.64627	13.74340
21	1.51567	2.27876	3.39956	5.03383	7.40025	10.80380	15.66750
22	1.54598	2.36991	3.60354	5.43654	8.14027	12.10030	17.86100
23	1.57690	2.46471	3.81975	5.87146	8.95430	13.55230	20.36150
24	1.60844	2.56330	4.04893	6.34117	9.84973	15.17860	23.21210
25	1.64061	2.66583	4.29187	6.84847	10.83470	17.00000	26.46180
26	1.67342	2.77246	4.54938	7.39634	11.91820	19.04000	30.16640
27	1.70689	2.88336	4.82234	7.98805	13.11000	21.32480	34.38970
28	1.74103	2.99869	5.11168	8.62709	14.42100	23.88380	39.20430
29	1.77585	3.11864	5.41839	9.31726	15.86310	26.74980	44.69290
30	1.81136	3.24339	5.74349	10.06260	17.44940	29.95980	50.94990
31	1.84759	3.37312	6.08810	10.86770	19.19430	33.55500	58.08290
32	1.88454	3.50804	6.45338	11.73710	21.11380	37.58160	66.21450
33	1.92223	3.64837	6.84059	12.67600	23.22510	42.09130	75.48450
34	1.96068	3.79430	7.25102	13.69010	25.54770	47.14230	86.05230
35	1.99989	3.94607	7.68608	14.78530	28.10240	52.79940	98.09960
36	2.03989	4.10391	8.14725	15.96810	30.91270	59.13530	111.83300
37	2.08069	4.26807	8.63608	17.24560	34.00390	66.23150	127.49000
38	2.12230	4.43879	9.15425	18.62520	37.40430	74.17930	145.33900
39	2.16475	4.61634	9.70350	20.11530	41.14480	83.08080	165.68600
40	2.20804	4.80100	10.28570	21.72450	45.25920	93.05040	188.88200
41	2.25220	4.99304	10.90290	23.46240	49.78520	104.21600	215.32600
42	2.29725	5.19276	11.55700	25.33940	54.76370	116.72200	245.47100
43	2.34319	5.40047	12.25040	27.36660	60.24000	130.72900	279.83700
44	2.39006	5.61648	12.98550	29.55590	66.26410	146.41700	319.01400
45	2.43786	5.84114	13.76460	31.92040	72.89050	163.98700	363.67600
46	2.48662	6.07479	14.59050	34.47400	80.17950	183.66500	414.59100
47	2.53635	6.31778	15.46590	37.23190	88.19750	205.70500	472.63300
48	2.58708	6.57049	16.39390	40.21050	97.01720	230.38900	538.80200
49	2.63882	6.83331	17.37750	43.42730	106.71900	258.03600	614.23400
50	2.69160	7.10664	18.42010	46.90150	117.39100	289.00000	700.22700

Table 2 Present Value I Table
(Compounded Annually)

				r			
n	2%	4%	6%	8%	10%	12%	14%
1	0.980392	0.961539	0.943396	0.925926	0.909091	0.892857	0.877193
2	0.961169	0.924556	0.889996	0.857339	0.826446	0.797194	0.769468
3	0.942322	0.888997	0.839619	0.793832	0.751315	0.711781	0.674972
4	0.923845	0.854805	0.792094	0.735030	0.683014	0.635519	0.592081
5	0.905731	0.821928	0.747258	0.680583	0.620921	0.567427	0.519369
6	0.887971	0.790315	0.704961	0.630170	0.564474	0.506632	0.455587
7	0.870560	0.759918	0.665057	0.583491	0.513158	0.452350	0.399638
8	0.853490	0.730691	0.627413	0.540269	0.466507	0.403884	0.350560
9	0.836755	0.702588	0.591899	0.500249	0.424098	0.360611	0.307508
10	0.820348	0.675565	0.558395	0.463194	0.385543	0.321974	0.269744
11	0.804263	0.649582	0.526788	0.428883	0.350494	0.287477	0.236618
12	0.788493	0.624598	0.496969	0.397114	0.318631	0.256676	0.207560
13	0.773032	0.600575	0.468839	0.367698	0.289664	0.229175	0.182070
14	0.757875	0.577476	0.442301	0.340461	0.263331	0.204620	0.159710
15	0.743014	0.555266	0.417265	0.315242	0.239392	0.182697	0.140097
16	0.728445	0.533909	0.393646	0.291891	0.217629	0.163122	0.122892
17	0.714162	0.513374	0.371365	0.270269	0.197845	0.145645	0.107800
18	0.700159	0.493629	0.350344	0.250249	0.179859	0.130040	0.094561
19	0.686430	0.474644	0.330513	0.231712	0.163508	0.116107	0.082949
20	0.672971	0.456388	0.311805	0.214548	0.148644	0.103667	0.072762
21	0.659775	0.438835	0.294156	0.198656	0.135131	0.092560	0.063826
22	0.646838	0.421957	0.277505	0.183941	0.122846	0.082643	0.055988
23	0.634155	0.405728	0.261797	0.170315	0.111678	0.073788	0.049112
24	0.621721	0.390123	0.246979	0.157700	0.101526	0.065882	0.043081
25	0.609530	0.375118	0.232999	0.146018	0.092296	0.058824	0.037790
26	0.597579	0.360690	0.219810	0.135202	0.083905	0.052521	0.033149
27	0.585861	0.346818	0.207368	0.125187	0.076278	0.046894	0.029078
28	0.574374	0.333479	0.195630	0.115914	0.069343	0.041869	0.025507
29	0.563112	0.320653	0.184557	0.107328	0.063039	0.037383	0.022375
30	0.552070	0.308320	0.174110	0.099378	0.057309	0.033378	0.019627
31	0.541245	0.296461	0.164255	0.092016	0.052099	0.029802	0.017217
32	0.530633	0.285059	0.154957	0.085200	0.047362	0.026609	0.015102
33	0.520228	0.274095	0.146186	0.078889	0.043057	0.023758	0.013248
34	0.510027	0.263553	0.137912	0.073045	0.039143	0.021212	0.011621
35	0.500027	0.253417	0.130105	0.067635	0.035584	0.018940	0.010194
36	0.490222	0.243670	0.122741	0.062625	0.032349	0.016910	0.008942
37	0.480610	0.234298	0.115793	0.057986	0.029408	0.015099	0.007844
38	0.471186	0.225287	0.109239	0.053691	0.026735	0.013481	0.006880
39	0.461947	0.216622	0.103056	0.049714	0.024304	0.012036	0.006036
40	0.452890	0.208290	0.097222	0.046031	0.022095	0.010747	0.005294
41	0.444009	0.200279	0.091719	0.042621	0.020086	0.009595	0.004644
42	0.435303	0.192576	0.086527	0.039464	0.018260	0.008567	0.004074
43	0.426768	0.185169	0.081630	0.036541	0.016600	0.007649	0.003574
44	0.418400	0.178047	0.077009	0.033834	0.015091	0.006830	0.003135
45	0.410196	0.171199	0.072650	0.031328	0.013719	0.006098	0.002750
46	0.402153	0.164615	0.068538	0.029007	0.012472	0.005445	0.002412
47	0.394267	0.158284	0.064658	0.026859	0.011338	0.004861	0.002116
48	0.386537	0.152196	0.060998	0.024869	0.010307	0.004340	0.001856
49	0.378958	0.146342	0.057546	0.023027	0.009370	0.003875	0.001628
50	0.371527	0.140714	0.054288	0.021321	0.008519	0.003460	0.001428

Table 3 Present Value II Table
(Compounded Daily)

				r			
n	2%	4%	6%	8%	10%	12%	14%
1	0.980188	0.960870	0.941776	0.923131	0.904856	0.886942	0.869396
2	0.960768	0.923271	0.886942	0.852171	0.818764	0.786666	0.755849
3	0.941733	0.887143	0.835301	0.786666	0.740863	0.697727	0.657133
4	0.923076	0.852429	0.786666	0.726196	0.670374	0.618844	0.571308
5	0.904788	0.819073	0.740863	0.670374	0.606592	0.548878	0.496693
6	0.886862	0.787022	0.697727	0.618844	0.548878	0.486823	0.431823
7	0.869291	0.756226	0.657103	0.571274	0.496656	0.431784	0.375425
8	0.852069	0.726634	0.618844	0.527361	0.449402	0.382967	0.326393
9	0.835187	0.698201	0.582812	0.486823	0.406644	0.339670	0.283765
10	0.818641	0.670880	0.548878	0.449402	0.367954	0.301267	0.246704
11	0.802422	0.644628	0.516921	0.414857	0.332945	0.267207	0.214484
12	0.786524	0.619404	0.486823	0.382967	0.301267	0.236997	0.186471
13	0.770941	0.595166	0.458478	0.353529	0.272604	0.210203	0.162117
14	0.755667	0.571877	0.431784	0.326354	0.246667	0.186437	0.140944
15	0.740696	0.549500	0.406644	0.301267	0.223198	0.165359	0.122536
16	0.726021	0.527997	0.382967	0.278109	0.201962	0.146664	0.106533
17	0.711637	0.507337	0.360670	0.256732	0.182747	0.130083	0.092619
18	0.697538	0.487485	0.339670	0.236997	0.165359	0.115376	0.080523
19	0.683718	0.468409	0.319893	0.218779	0.149626	0.102331	0.070006
20	0.670172	0.450080	0.301267	0.201962	0.135390	0.090762	0.060863
21	0.656895	0.432468	0.283726	0.186437	0.122509	0.080501	0.052914
22	0.643880	0.415546	0.267207	0.172106	0.110853	0.071399	0.046003
23	0.631124	0.399285	0.251649	0.158877	0.100306	0.063327	0.039995
24	0.618620	0.383661	0.236997	0.146664	0.090762	0.056168	0.034772
25	0.606364	0.368648	0.223198	0.135390	0.082127	0.049817	0.030230
26	0.594350	0.354223	0.210203	0.124983	0.074313	0.044185	0.026282
27	0.582575	0.340362	0.197964	0.115376	0.067242	0.039190	0.022849
28	0.571033	0.327044	0.186437	0.106507	0.060845	0.034759	0.019865
29	0.559719	0.314246	0.175582	0.098320	0.055056	0.030829	0.017271
30	0.548630	0.301950	0.165359	0.090762	0.049817	0.027344	0.015015
31	0.537761	0.290134	0.155731	0.083785	0.045078	0.024252	0.013054
32	0.527106	0.278781	0.146664	0.077345	0.040789	0.021510	0.011349
33	0.516663	0.267873	0.138125	0.071399	0.036908	0.019078	0.009867
34	0.506427	0.257391	0.130083	0.065911	0.033396	0.016921	0.008578
35	0.496394	0.247319	0.122509	0.060845	0.030219	0.015008	0.007458
36	0.486559	0.237641	0.115376	0.056168	0.027344	0.013312	0.006484
37	0.476919	0.228342	0.108658	0.051850	0.024742	0.011807	0.005637
38	0.467471	0.219407	0.102331	0.047864	0.022388	0.010472	0.004901
39	0.458209	0.210822	0.096373	0.044185	0.020258	0.009288	0.004261
40	0.449131	0.202572	0.090762	0.040789	0.018330	0.008238	0.003704
41	0.440233	0.194645	0.085478	0.037653	0.016586	0.007306	0.003221
42	0.431511	0.187029	0.080501	0.034759	0.015008	0.006480	0.002800
43	0.422962	0.179710	0.075814	0.032087	0.013580	0.005748	0.002434
44	0.414582	0.172678	0.071399	0.029621	0.012288	0.005098	0.002116
45	0.406368	0.165921	0.067242	0.027344	0.011119	0.004522	0.001840
46	0.398317	0.159429	0.063327	0.025242	0.010061	0.004010	0.001600
47	0.390425	0.153190	0.059640	0.023301	0.009104	0.003557	0.001391
48	0.382690	0.147196	0.056168	0.021510	0.008238	0.003155	0.001209
49	0.375108	0.141436	0.052897	0.019857	0.007454	0.002798	0.001051
50	0.367677	0.135902	0.049817	0.018330	0.006745	0.002482	0.000914

Table 4 Annuity Table

n	2	4	6	8	10	12
1	1.00000	1.00000	1.00000	1.00000	1.00000	1.00000
2	2.02000	2.04000	2.06000	2.08000	2.10000	2.12000
3	3.06040	3.12160	3.18360	3.24640	3.31000	3.37440
4	4.12161	4.24646	4.37462	4.50611	4.64100	4.77933
5	5.20404	5.41632	5.63709	5.86660	6.10510	6.35285
6	6.30812	6.63297	6.97532	7.33593	7.71561	8.11519
7	7.43428	7.89829	8.39384	8.92280	9.48717	10.08901
8	8.58297	9.21423	9.89747	10.63663	11.43589	12.29969
9	9.75463	10.58279	11.49131	12.48756	13.57948	14.77566
10	10.94972	12.00610	13.18079	14.48657	15.93743	17.54874
11	12.16871	13.48635	14.97164	16.64549	18.53117	20.65458
12	13.41209	15.02580	16.86994	18.97713	21.38429	24.13313
13	14.68033	16.62683	18.88213	21.49530	24.52272	28.02911
14	15.97394	18.29191	21.01506	24.21493	27.97499	32.39260
15	17.29341	20.02358	23.27596	27.15212	31.77249	37.27972
16	18.63928	21.82452	25.67252	30.32429	35.94974	42.75328
17	20.01207	23.69750	28.21287	33.75024	40.54471	48.88368
18	21.41231	25.64540	30.90564	37.45026	45.59918	55.74972
19	22.84055	27.67122	33.75997	41.44628	51.15910	63.43968
20	24.29737	29.77807	36.78557	45.76199	57.27502	72.05245
21	25.78331	31.96919	39.99270	50.42295	64.00252	81.69874
22	27.29898	34.24795	43.39226	55.45678	71.40277	92.50259
23	28.84496	36.61787	46.99579	60.89333	79.54305	104.60290
24	30.42186	39.08258	50.81554	66.76480	88.49736	118.15525
25	32.03029	41.64589	54.86447	73.10599	98.34709	133.33388
26	33.67090	44.31172	59.15633	79.95447	109.18181	150.33395
27	35.34431	47.08419	63.70571	87.35083	121.09999	169.37402
28	37.05120	49.96755	68.52805	95.33890	134.20999	190.69890
29	38.79222	52.96625	73.63973	103.96601	148.63099	214.58277
30	40.56807	56.08490	79.05811	113.28330	164.49410	241.33271
31	42.37943	59.32830	84.80159	123.34597	181.94351	271.29263
32	44.22702	62.70143	90.88968	134.21365	201.13787	304.84775
33	46.11155	66.20948	97.34305	145.95075	222.25166	342.42948
34	48.03379	69.85786	104.18363	158.62682	245.47683	384.52102
35	49.99446	73.65217	111.43464	172.31697	271.02452	431.66354
36	51.99435	77.59825	119.12072	187.10233	299.12698	484.46317
37	54.03423	81.70218	127.26795	203.07053	330.03968	543.59876
38	56.11492	85.97026	135.90402	220.31618	364.04366	609.83061
39	58.23722	90.40907	145.05826	238.94148	401.44803	684.01028
40	60.40196	95.02543	154.76174	259.05681	442.59284	767.09152
41	62.61000	99.82644	165.04744	280.78137	487.85214	860.14251
42	64.86219	104.81950	175.95028	304.24389	537.63736	964.35961
43	67.15944	110.01227	187.50728	329.58341	592.40111	1081.08277
44	69.50263	115.41276	199.75771	356.95010	652.64124	1211.81271
45	71.89268	121.02927	212.74316	386.50612	718.90538	1358.23024
46	74.33053	126.87043	226.50774	418.42663	791.79593	1522.21788
47	76.81714	132.94524	241.09819	452.90078	871.97554	1705.88403
48	79.35348	139.26305	256.56407	490.13286	960.17312	1911.59012
49	81.94055	145.83357	272.95789	530.34351	1057.19045	2141.98094
50	84.57936	152.66690	290.33535	573.77101	1163.90952	2400.01867

Table 5 Amortization Table

| | | | i | | | |
n	2	4	6	8	10	12
1	0.98039	0.96154	0.94340	0.92593	0.90909	0.89286
2	1.94156	1.88609	1.83339	1.78326	1.73554	1.69005
3	2.88388	2.77509	2.67301	2.57710	2.48685	2.40183
4	3.80773	3.62990	3.46511	3.31213	3.16987	3.03735
5	4.71346	4.45182	4.21236	3.99271	3.79079	3.60478
6	5.60143	5.24214	4.91733	4.62288	4.35526	4.11141
7	6.47199	6.00206	5.58238	5.20637	4.86842	4.56376
8	7.32548	6.73275	6.20980	5.74664	5.33493	4.96764
9	8.16224	7.43533	6.80169	6.24689	5.75902	5.32825
10	8.98259	8.11090	7.36009	6.71008	6.14457	5.65022
11	9.78685	8.76048	7.88688	7.13896	6.49506	5.93770
12	10.57534	9.38508	8.38385	7.53608	6.81369	6.19437
13	11.34838	9.98565	8.85269	7.90377	7.10336	6.42355
14	12.10625	10.56313	9.29499	8.24423	7.36669	6.62817
15	12.84927	11.11839	9.71225	8.55948	7.60608	6.81086
16	13.57771	11.65230	10.10590	8.85137	7.82371	6.97399
17	14.29187	12.16567	10.47726	9.12164	8.02155	7.11963
18	14.99203	12.65930	10.82761	9.37188	8.20141	7.24967
19	15.67846	13.13394	11.15812	9.60360	8.36492	7.36578
20	16.35144	13.59033	11.46993	9.81814	8.51356	7.46944
21	17.01121	14.02916	11.76408	10.01680	8.64869	7.56200
22	17.65805	14.45112	12.04159	10.20074	8.77154	7.64465
23	18.29221	14.85685	12.30339	10.37106	8.88322	7.71843
24	18.91393	15.24697	12.55036	10.52875	8.98474	7.78432
25	19.52346	15.62209	12.78336	10.67477	9.07704	7.84314
26	20.12104	15.98278	13.00317	10.80997	9.16094	7.89566
27	20.70690	16.32959	13.21054	10.93516	9.23722	7.94255
28	21.28128	16.66307	13.40617	11.05107	9.30656	7.98442
29	21.84439	16.98372	13.59073	11.15840	9.36960	8.02181
30	22.39646	17.29204	13.76484	11.25778	9.42691	8.05518
31	22.93771	17.58850	13.92909	11.34979	9.47901	8.08499
32	23.46834	17.87356	14.08405	11.43499	9.52637	8.11159
33	23.98857	18.14765	14.23024	11.51388	9.56943	8.13535
34	24.49860	18.41121	14.36815	11.58693	9.60857	8.15656
35	24.99863	18.66462	14.49826	11.65456	9.64416	8.17550
36	25.48885	18.90829	14.62100	11.71719	9.67651	8.19241
37	25.96946	19.14259	14.73679	11.77517	9.70591	8.20751
38	26.44065	19.36787	14.84603	11.82886	9.73265	8.22099
39	26.90260	19.58450	14.94909	11.87858	9.75695	8.23303
40	27.35549	19.79279	15.04631	11.92461	9.77905	8.24378
41	27.79950	19.99306	15.13803	11.96723	9.79913	8.25337
42	28.23480	20.18564	15.22455	12.00669	9.81740	8.26194
43	28.66157	20.37081	15.30618	12.04323	9.83400	8.26959
44	29.07997	20.54885	15.38319	12.07707	9.84909	8.27642
45	29.49017	20.72005	15.45584	12.10840	9.86281	8.28252
46	29.89232	20.88467	15.52438	12.13740	9.87528	8.28796
47	30.28659	21.04295	15.58904	12.16426	9.88662	8.29282
48	30.67313	21.19514	15.65004	12.18913	9.89692	8.29716
49	31.05209	21.34149	15.70758	12.21216	9.90629	8.30104
50	31.42362	21.48220	15.76187	12.23348	9.91481	8.30450

Table 6 Areas Under the Standard Normal Curve

z	A(z)	z	A(z)	z	A(z)
−3.00	.0013	−2.48	.0066	−1.96	.0250
−2.99	.0014	−2.47	.0068	−1.95	.0256
−2.98	.0014	−2.46	.0069	−1.94	.0262
−2.97	.0015	−2.45	.0071	−1.93	.0268
−2.96	.0015	−2.44	.0073	−1.92	.0274
−2.95	.0016	−2.43	.0075	−1.91	.0281
−2.94	.0016	−2.42	.0078	−1.90	.0287
−2.93	.0017	−2.41	.0080	−1.89	.0294
−2.92	.0017	−2.40	.0082	−1.88	.0300
−2.91	.0018	−2.39	.0084	−1.87	.0307
−2.90	.0019	−2.38	.0087	−1.86	.0314
−2.89	.0019	−2.37	.0089	−1.85	.0322
−2.88	.0020	−2.36	.0091	−1.84	.0329
−2.87	.0020	−2.35	.0094	−1.83	.0336
−2.86	.0021	−2.34	.0096	−1.82	.0344
−2.85	.0022	−2.33	.0099	−1.81	.0351
−2.84	.0023	−2.32	.0102	−1.80	.0359
−2.83	.0023	−2.31	.0104	−1.79	.0367
−2.82	.0024	−2.30	.0107	−1.78	.0375
−2.81	.0025	−2.29	.0110	−1.77	.0384
−2.80	.0026	−2.28	.0113	−1.76	.0392
−2.79	.0026	−2.27	.0116	−1.75	.0401
−2.78	.0027	−2.26	.0119	−1.74	.0409
−2.77	.0028	−2.25	.0122	−1.73	.0418
−2.76	.0029	−2.24	.0125	−1.72	.0427
−2.75	.0030	−2.23	.0129	−1.71	.0436
−2.74	.0031	−2.22	.0132	−1.70	.0446
−2.73	.0032	−2.21	.0135	−1.69	.0455
−2.72	.0033	−2.20	.0139	−1.68	.0465
−2.71	.0034	−2.19	.0143	−1.67	.0475
−2.70	.0035	−2.18	.0146	−1.66	.0485
−2.69	.0036	−2.17	.0150	−1.65	.0495
−2.68	.0037	−2.16	.0154	−1.64	.0505
−2.67	.0038	−2.15	.0158	−1.63	.0515
−2.66	.0039	−2.14	.0162	−1.62	.0526
−2.65	.0040	−2.13	.0166	−1.61	.0537
−2.64	.0041	−2.12	.0170	−1.60	.0548
−2.63	.0043	−2.11	.0174	−1.59	.0559
−2.62	.0044	−2.10	.0179	−1.58	.0570
−2.61	.0045	−2.09	.0183	−1.57	.0582
−2.60	.0047	−2.08	.0188	−1.56	.0594
−2.59	.0048	−2.07	.0192	−1.55	.0606
−2.58	.0049	−2.06	.0197	−1.54	.0618
−2.57	.0051	−2.05	.0202	−1.53	.0630
−2.56	.0052	−2.04	.0207	−1.52	.0643
−2.55	.0054	−2.03	.0212	−1.51	.0655
−2.54	.0055	−2.02	.0217	−1.50	.0668
−2.53	.0057	−2.01	.0222	−1.49	.0681
−2.52	.0059	−2.00	.0227	−1.48	.0694
−2.51	.0060	−1.99	.0233	−1.47	.0708
−2.50	.0062	−1.98	.0238	−1.46	.0721
−2.49	.0064	−1.97	.0244	−1.45	.0735

Table 6 Areas Under the Standard Normal Curve
(*continued*)

z	A(z)	z	A(z)	z	A(z)
−1.44	.0749	−0.92	.1788	−0.40	.3446
−1.43	.0764	−0.91	.1814	−0.39	.3483
−1.42	.0778	−0.90	.1841	−0.38	.3520
−1.41	.0793	−0.89	.1867	−0.37	.3557
−1.40	.0808	−0.88	.1894	−0.36	.3594
−1.39	.0823	−0.87	.1921	−0.35	.3632
−1.38	.0838	−0.86	.1949	−0.34	.3669
−1.37	.0853	−0.85	.1977	−0.33	.3707
−1.36	.0869	−0.84	.2004	−0.32	.3745
−1.35	.0885	−0.83	.2033	−0.31	.3783
−1.34	.0901	−0.82	.2061	−0.30	.3821
−1.33	.0918	−0.81	.2090	−0.29	.3859
−1.32	.0934	−0.80	.2119	−0.28	.3897
−1.31	.0951	−0.79	.2148	−0.27	.3936
−1.30	.0968	−0.78	.2177	−0.26	.3974
−1.29	.0985	−0.77	.2206	−0.25	.4013
−1.28	.1003	−0.76	.2236	−0.24	.4052
−1.27	.1020	−0.75	.2266	−0.23	.4090
−1.26	.1038	−0.74	.2296	−0.22	.4129
−1.25	.1056	−0.73	.2327	−0.21	.4168
−1.24	.1075	−0.72	.2358	−0.20	.4207
−1.23	.1093	−0.71	.2388	−0.19	.4246
−1.22	.1112	−0.70	.2420	−0.18	.4286
−1.21	.1131	−0.69	.2451	−0.17	.4325
−1.20	.1151	−0.68	.2482	−0.16	.4364
−1.19	.1170	−0.67	.2514	−0.15	.4404
−1.18	.1190	−0.66	.2546	−0.14	.4443
−1.17	.1210	−0.65	.2578	−0.13	.4483
−1.16	.1230	−0.64	.2611	−0.12	.4522
−1.15	.1251	−0.63	.2643	−0.11	.4562
−1.14	.1271	−0.62	.2676	−0.10	.4602
−1.13	.1292	−0.61	.2709	−0.09	.4641
−1.12	.1314	−0.60	.2742	−0.08	.4681
−1.11	.1335	−0.59	.2776	−0.07	.4721
−1.10	.1357	−0.58	.2810	−0.06	.4761
−1.09	.1379	−0.57	.2843	−0.05	.4801
−1.08	.1401	−0.56	.2877	−0.04	.4840
−1.07	.1423	−0.55	.2912	−0.03	.4880
−1.06	.1446	−0.54	.2946	−0.02	.4920
−1.05	.1469	−0.53	.2981	−0.01	.4960
−1.04	.1492	−0.52	.3015	0.00	.5000
−1.03	.1515	−0.51	.3050	0.01	.5040
−1.02	.1539	−0.50	.3085	0.02	.5080
−1.01	.1562	−0.49	.3121	0.03	.5120
−1.00	.1587	−0.48	.3156	0.04	.5160
−0.99	.1611	−0.47	.3192	0.05	.5199
−0.98	.1635	−0.46	.3228	0.06	.5239
−0.97	.1660	−0.45	.3264	0.07	.5279
−0.96	.1685	−0.44	.3300	0.08	.5319
−0.95	.1711	−0.43	.3336	0.09	.5359
−0.94	.1736	−0.42	.3372	0.10	.5398
−0.93	.1762	−0.41	.3409	0.11	.5438

Table 6 Areas Under the Standard Normal Curve
(*continued*)

z	A(z)	z	A(z)	z	A(z)
0.12	.5478	0.64	.7389	1.16	.8770
0.13	.5517	0.65	.7422	1.17	.8790
0.14	.5557	0.66	.7454	1.18	.8810
0.15	.5596	0.67	.7484	1.19	.8830
0.16	.5636	0.68	.7518	1.20	.8849
0.17	.5675	0.69	.7549	1.21	.8869
0.18	.5714	0.70	.7580	1.22	.8888
0.19	.5754	0.71	.7612	1.23	.8907
0.20	.5793	0.72	.7642	1.24	.8925
0.21	.5832	0.73	.7673	1.25	.8944
0.22	.5871	0.74	.7704	1.26	.8962
0.23	.5910	0.75	.7734	1.27	.8980
0.24	.5948	0.76	.7764	1.28	.8997
0.25	.5987	0.77	.7794	1.29	.9015
0.26	.6026	0.78	.7823	1.30	.9032
0.27	.6064	0.79	.7852	1.31	.9049
0.28	.6103	0.80	.7881	1.32	.9066
0.29	.6141	0.81	.7910	1.33	.9082
0.30	.6179	0.82	.7939	1.34	.9099
0.31	.6217	0.83	.7967	1.35	.9115
0.32	.6255	0.84	.7996	1.36	.9131
0.33	.6293	0.85	.8023	1.37	.9147
0.34	.6331	0.86	.8051	1.38	.9162
0.35	.6368	0.87	.8079	1.39	.9177
0.36	.6406	0.88	.8106	1.40	.9192
0.37	.6443	0.89	.8133	1.41	.9207
0.38	.6480	0.90	.8159	1.42	.9222
0.39	.6517	0.91	.8186	1.43	.9236
0.40	.6554	0.92	.8212	1.44	.9251
0.41	.6591	0.93	.8238	1.45	.9265
0.42	.6628	0.94	.8264	1.46	.9279
0.43	.6664	0.95	.8289	1.47	.9292
0.44	.6700	0.96	.8315	1.48	.9306
0.45	.6736	0.97	.8340	1.49	.9319
0.46	.6772	0.98	.8365	1.50	.9332
0.47	.6808	0.99	.8389	1.51	.9345
0.48	.6844	1.00	.8413	1.52	.9357
0.49	.6879	1.01	.8438	1.53	.9370
0.50	.6915	1.02	.8461	1.54	.9382
0.51	.6950	1.03	.8485	1.55	.9394
0.52	.6985	1.04	.8508	1.56	.9406
0.53	.7019	1.05	.8531	1.57	.9418
0.54	.7054	1.06	.8554	1.58	.9430
0.55	.7088	1.07	.8577	1.59	.9441
0.56	.7123	1.08	.8599	1.60	.9452
0.57	.7157	1.09	.8621	1.61	.9463
0.58	.7190	1.10	.8643	1.62	.9474
0.59	.7224	1.11	.8665	1.63	.9485
0.60	.7258	1.12	.8686	1.64	.9495
0.61	.7291	1.13	.8708	1.65	.9505
0.62	.7324	1.14	.8729	1.66	.9515
0.63	.7357	1.15	.8749	1.67	.9525

Table 6 Areas Under the Standard Normal Curve
(*continued*)

z	A(z)	z	A(z)	z	A(z)
1.68	.9535	2.13	.9834	2.58	.9951
1.69	.9545	2.14	.9838	2.59	.9952
1.70	.9554	2.15	.9842	2.60	.9953
1.71	.9564	2.16	.9846	2.61	.9955
1.72	.9573	2.17	.9850	2.62	.9956
1.73	.9582	2.18	.9854	2.63	.9957
1.74	.9591	2.19	.9857	2.64	.9959
1.75	.9599	2.20	.9861	2.65	.9960
1.76	.9608	2.21	.9865	2.66	.9961
1.77	.9616	2.22	.9868	2.67	.9962
1.78	.9625	2.23	.9871	2.68	.9963
1.79	.9633	2.24	.9875	2.69	.9964
1.80	.9641	2.25	.9878	2.70	.9965
1.81	.9649	2.26	.9881	2.71	.9966
1.82	.9656	2.27	.9884	2.72	.9967
1.83	.9664	2.28	.9887	2.73	.9968
1.84	.9671	2.29	.9890	2.74	.9969
1.85	.9678	2.30	.9893	2.75	.9970
1.86	.9686	2.31	.9896	2.76	.9971
1.87	.9693	2.32	.9898	2.77	.9972
1.88	.9700	2.33	.9901	2.78	.9973
1.89	.9706	2.34	.9904	2.79	.9974
1.90	.9713	2.35	.9906	2.80	.9974
1.91	.9719	2.36	.9909	2.81	.9975
1.92	.9726	2.37	.9911	2.82	.9976
1.93	.9732	2.38	.9913	2.83	.9977
1.94	.9738	2.39	.9916	2.84	.9977
1.95	.9744	2.40	.9918	2.85	.9978
1.96	.9750	2.41	.9920	2.86	.9979
1.97	.9756	2.42	.9922	2.87	.9980
1.98	.9762	2.43	.9925	2.88	.9980
1.99	.9767	2.44	.9927	2.89	.9981
2.00	.9773	2.45	.9929	2.90	.9981
2.01	.9778	2.46	.9931	2.91	.9982
2.02	.9783	2.47	.9932	2.92	.9983
2.03	.9788	2.48	.9934	2.93	.9983
2.04	.9793	2.49	.9936	2.94	.9984
2.05	.9798	2.50	.9938	2.95	.9984
2.06	.9803	2.51	.9940	2.96	.9985
2.07	.9808	2.52	.9941	2.97	.9985
2.08	.9812	2.53	.9943	2.98	.9986
2.09	.9817	2.54	.9945	2.99	.9986
2.10	.9821	2.55	.9946	3.00	.9987
2.11	.9826	2.56	.9948		
2 12	.9830	2.57	.9949		

TABLE 7 Exponentials and Their Reciprocals

x	e^x	e^{-x}
10	22026.46580	0.00005
9	8103.08393	0.00012
8	2980.95799	0.00034
7	1096.63316	0.00091
6	403.42879	0.00248
5	148.41316	0.00674
4	54.59815	0.01832
3	20.08554	0.04979
2	7.38906	0.13534
1.9	6.68589	0.14957
1.8	6.04965	0.16530
1.7	5.47395	0.18268
1.6	4.95303	0.20190
1.5	4.48169	0.22313
1.4	4.05520	0.24660
1.3	3.66930	0.27253
1.2	3.32012	0.30119
1.1	3.00417	0.33287
1.0	2.71828	0.36788
0.9	2.45960	0.40657
0.8	2.22554	0.44933
0.7	2.01375	0.49659
0.6	1.82212	0.54881
0.5	1.64872	0.60653
0.4	1.49183	0.67032
0.3	1.34986	0.74082
0.2	1.22140	0.81873
0.1	1.10517	0.90484
0	1.00000	

TABLE 8 Natural Logarithms

x	ln x	x	ln x	x	ln x	x	ln x
0.05	−2.99573	2	0.69315	25	3.21888	100	4.60517
0.10	−2.30259	3	1.09861	30	3.40120	150	5.01064
0.15	−1.89712	4	1.38629	35	3.55535	200	5.29832
0.20	−1.60944	5	1.60944	40	3.68888	250	5.52146
0.25	−1.38629	6	1.79176	45	3.80666	300	5.70378
0.30	−1.20397	7	1.94591	50	3.91202	350	5.85793
0.35	−1.04982	8	2.07944	55	4.00733	400	5.99146
0.40	−0.91629	9	2.19722	60	4.09434	450	6.10925
0.45	−0.79851	10	2.30259	65	4.17439	500	6.21461
0.50	−0.69315	11	2.39790	70	4.24850	550	6.30992
0.55	−0.59784	12	2.48491	75	4.31749	600	6.39693
0.60	−0.51083	13	2.56495	80	4.38203	650	6.47697
0.65	0.43078	14	2.63906	85	4.44265	700	6.55108
0.70	0.35667	15	2.70805	90	4.49981	750	6.62007
0.75	0.28768	16	2.77259	95	4.55388	800	6.68461
0.80	0.22314	17	2.83321			850	6.74524
0.85	0.16252	18	2.89037			900	6.80239
0.90	0.10536	19	2.94444			950	6.85646
0.95	0.05129	20	2.99572			1000	6.90776
1.00	0.00000						

TABLE 9 Basic Integration Formulas

1. $\displaystyle \int e^x \, dx = e^x$

2. $\displaystyle \int \frac{1}{x} \, dx = \ln |x|$

3. $\displaystyle \int a^x \, dx = \frac{a^x}{\ln a}, \; a > 0 \qquad (a \ne 1)$

4. $\displaystyle \int \frac{1}{\sqrt{a^2 + x^2}} \, dx = \ln \left| x + \sqrt{x^2 + a^2} \right|$

5. $\displaystyle \int \frac{1}{\sqrt{x^2 - a^2}} \, dx = \ln \left| x + \sqrt{x^2 - a^2} \right|$

6. $\displaystyle \int \frac{1}{a^2 - x^2} \, dx = \frac{1}{2a} \ln \left| \frac{a + x}{a - x} \right| \qquad (x^2 < a^2)$

7. $\displaystyle \int \frac{1}{x^2 - a^2} \, dx = -\frac{1}{2a} \ln \left| \frac{x + a}{x - a} \right| \qquad (a^2 < x^2)$

8. $\displaystyle \int \frac{1}{x\sqrt{a^2 - x^2}} \, dx = -\frac{1}{a} \ln \left| \frac{a + \sqrt{a^2 - x^2}}{x} \right| \qquad (0 < x < a)$

9. $\displaystyle \int x^n \ln x \, dx = x^{n+1} \left[\frac{\ln x}{n + 1} - \frac{1}{(n + 1)^2} \right]$

10. $\displaystyle \int \frac{1}{x\sqrt{a^2 + x^2}} \, dx = -\frac{1}{a} \ln \left| \frac{a + \sqrt{a^2 + x^2}}{x} \right|$

11. $\displaystyle \int \ln |x| \, dx = x(\ln |x| - 1)$

12. $\displaystyle \int \frac{x}{ax + b} \, dx = \frac{x}{a} - \frac{b}{a^2} \ln |ax + b|$

13. $\displaystyle \int \frac{x}{(ax + b)^2} \, dx = \frac{b}{a^2(ax + b)} + \frac{1}{a^2} \ln |ax + b|$

14. $\displaystyle \int \frac{1}{x(ax + b)} \, dx = \frac{1}{b} \ln \left| \frac{x}{ax + b} \right|$

15. $\displaystyle \int \frac{1}{x(ax + b)^2} \, dx = \frac{1}{b(ax + b)} + \frac{1}{b^2} \ln \left| \frac{x}{ax + b} \right|$

16. $\displaystyle \int \sqrt{x^2 \pm a^2} \, dx = \tfrac{1}{2}x \sqrt{x^2 \pm a^2} + \frac{a^2}{2} \ln \left| x + \sqrt{x^2 \pm a^2} \right|$

17. $\displaystyle \int x^n \ln x \, dx = x^{n+1} \left[\frac{\ln x}{n + 1} - \frac{1}{(n + 1)^2} \right] \qquad (n \ne -1)$

18. $\displaystyle \int x^n e^{ax} \, dx = \frac{1}{a} x^n e^{ax} - \frac{n}{a} \int x^{n-1} e^{ax} \, dx$

Answers to Odd-Numbered Exercises

CHAPTER R

Exercise Set R–1

1. $x = 7$
5. $s = 12$
9. $x = 2$
13. $y = \frac{30}{46} = \frac{15}{23}$
17. $x = -\frac{18}{29}$
21. $x = -2$
25. $x = 3$
29. $x = 1$
33. $x = 2$
37. $x = 4$
41. $x = 2$

3. $x = \frac{1}{6}$
7. $q = \frac{61}{36}$
11. $x = \frac{4}{3}$
15. $t = -\frac{2}{25}$
19. $x = 11$
23. $x = 1$
27. $x = \frac{4}{9}$
31. $x = \frac{1}{3}$
35. $x = 1$
39. $x = -\frac{1}{2}$
43. $x = 2$

Exercise Set R–2

1. $x < 3$
5. $x > -1$
9. $x > 2$
13. $x > 3$
17. $x \geq \frac{5}{3}$
21. $x > -6$
25. $x < 3$
29. $t \leq -9$
33. $x > \frac{10}{7}$
37. $x < 4$

3. $x > 2$
7. $x \leq 2$
11. $x > -2$
15. $x > -1$
19. $x > 12$
23. $x < 1$
27. $x > -5$
31. $x \leq \frac{7}{2}$
35. $x > 9$
39. $x \leq \frac{13}{2}$

Exercise Set R–3

1. 3^6
5. 2
9. 90
13. $x^{-6}y^6$
17. A^{-2}
21. 7
25. $\frac{1}{6}$
29. $\frac{2}{7}$
33. $\frac{2}{3}$
37. 16
41. -2
45. 25
49. 11
53. 2
57. 42

3. 10^2
7. 6
11. x^{10}
15. x^{-2}
19. x^{-2}
23. 9
27. $\frac{1}{7}$
31. 3
35. $\frac{1}{4}$
39. $\frac{1}{32}$
43. $-\frac{1}{2}$
47. 1
51. $\frac{1}{4}$
55. -2
59. $\frac{81}{2}$

Exercise Set R–4

1. 6
5. -1
9. 1
13. 4
17. 40
21. $\frac{2}{5}$

3. -15
7. 0
11. 7
15. -22
19. $\frac{1}{5}$
23. -5

25. -15

29. 6

33. $\frac{1}{4}$

37. 2

41. $-\frac{73}{9}$

43. $x = -1, \qquad x = \frac{9}{2}$

45. $x = \frac{1}{2}, \qquad x = -\frac{21}{2}$

47. $-1 < x < \frac{9}{2}$

49. $x < -5 \quad$ or $\quad x > -\frac{5}{3}$

51. $-1 < x < \frac{15}{7}$

53. $x < -1 \quad$ or $\quad x > \frac{15}{7}$

27. $\frac{3}{2}$

31. $-\frac{1}{5}$

35. 1

39. -7

Exercise Set R–5

1. $7x^3 + 4x^2 + 3$

3. $3x^4 - 4x^3 - 3x^2$

5. $x^5 + 3x^3 + x^2 + 20x$

7. $3x^6 + 3x^5 - 3x^4 + 6x^2$

9. $2x^7 + 3x^5 + 5x^4$

11. $x^3 + 3x^2 + 2x$

13. $2x^3 + 7x^2 + 5x$

15. $4x^4 + 19x^3 - 2x^2 + 2x + 6$

17. $x^4 + x^3 - 7x^2 + x + 7$

19. $-x^4 + 3x^3 - 5x^2 - 3x + 10$

21. $x^2 - 1$

25. $1 - y^2$

29. $9 + 6x + x^2$

33. $y^2 - 8y + 16$

37. $9 - 6x + x^2$

41. $15x^2 - 22x + 8$

45. $20 + 3x - 9x^2$

49. $9x^2 + 12x + 4$

53. $15x^2 - 2xy - y^2$

55. $27x^3 + 27x^2 + 9x + 1$

57. $8x^3 - 12x^2y + 6xy^2 - y^3$

23. $x^2 - 25$

27. $x^2 + 6x + 9$

31. $t^2 + 16t + 64$

35. $t^2 - 16t + 64$

39. $15x^2 + 22x + 8$

43. $8 - 6y - 9y^2$

47. $16 - 9x^2$

51. $9x^2 + 6xy + y^2$

Exercise Set R–6

1. $x(x - 6)$

5. $x(x^2 - 8x + 1)$

7. $4y(y^2 - 2y + 3)$

9. $(x + 6)(x - 6)$

13. $(1 - 2x)(1 + 2x)$

17. $2(x + 9)(x - 9)$

19. $2x(x + 3)(x - 3)$

21. $x(1 - 2x)(1 + 2x)$

23. $x^2(x + 4)(x - 4)$

25. $6x^2(x + 2)(x - 2)$

27. $(x + 2)^2$

3. $2t(t - 30)$

11. $(7 + t)(7 - t)$

15. $(3 - 2x)(3 + 2x)$

29. $(x - 2)(x + 1)$

31. $(2x + 1)(x + 1)$

35. $(2y + 1)(y - 5)$

39. $(3x + 1)(2x - 3)$

43. $(3 + x)(2 - x)$

33. $(4x + 1)(x + 1)$

37. $(2t + 1)(2t + 3)$

41. $(5 + x)(1 - x)$

45. $x(2x + 1)(x + 2)$

Exercise Set R–7

1. $x = 6 \quad$ and $\quad x = 9$

3. $x = -2 \quad$ and $\quad x = 2$

5. $x = \frac{5}{2} \quad$ and $\quad x = 6$

7. $x = 0 \quad$ and $\quad x = -\frac{16}{3}$

9. $x = -2$

11. $x = -1 \quad$ and $\quad x = 2$

13. $x = -\frac{1}{2} \quad$ and $\quad x = -1$

15. $x = -2 \quad$ and $\quad x = 2$

17. $x = -5 \quad$ and $\quad x = 5$

19. $x = -6 \quad$ and $\quad x = 6$

21. $x = -\frac{2}{3} \quad$ and $\quad x = \frac{2}{3}$

23. $x = -\frac{1}{3} \quad$ and $\quad x = \frac{1}{3}$

25. $x = 0, -4, 4$

27. $x = -\frac{3}{5}, 0, \frac{3}{5}$

29. $x = -3, 0, 3$

31. $x = -5, 0, 5$

33. $x = \dfrac{5 \pm \sqrt{21}}{2}$

35. $x = \dfrac{-5 \pm \sqrt{29}}{2}$

37. $x = \dfrac{-5 \pm \sqrt{33}}{4}$

39. $x = \frac{1}{2} \quad$ and $\quad x = 1$

41. $x = \dfrac{-5 \pm \sqrt{41}}{4} \quad$ and $\quad x = 0$

Exercise Set R–8

1. $\dfrac{2}{2x - 1}$

5. $\dfrac{1}{x - 1}$

9. $x - 4$

11. $2x + 6 = 2(x + 3)$

13. $\dfrac{x^2 - 3}{x - 2}$

17. $4(x - 2)$

21. $\dfrac{x^2 + 1}{x - 2}$

3. $\dfrac{2}{3x - 4}$

7. $\dfrac{1}{x + 2}$

15. $x + 3$

19. $\dfrac{1}{3 - x}$

23. $\dfrac{x - 3}{x^2}$

25. $\dfrac{x + 3}{x^2}$

27. $\dfrac{5x^2 + 1}{(x - 1)(x + 1)}$

29. $-\dfrac{(x + 2)(x + 1)}{x(x - 2)}$

31. $\dfrac{3x}{x - 1}$

33. $\dfrac{2(x + 2)}{x - 1}$

35. $\dfrac{x^2 - x + 1}{x^2(x - 1)}$

37. $\dfrac{5x^3 - x^2 + x + 1}{x(x - 1)(x + 1)}$

39. $\dfrac{-x^3 - 3x^2 - 3x + 2}{x^2(x - 2)}$

41. $2x + h$

43. $\dfrac{-5x^2 + 6x - 5}{(3 - 5x)^2}$

45. $\dfrac{-12x^2 + 6x - 8}{3x^2 - 2}$

Review Exercises

1. $x = -\frac{1}{2}$

3. $x = 21$

5. $x > 1$

7. $x > \frac{2}{3}$

9. x^{-5}

11. $\frac{1}{12}$

13. -8

15. -1

17. $-3x^4 - x^3 - 3x^2 + x$

19. $x^3 - 4x^2 - 4x + 16$

21. $5x^2(x - 2)$

23. $(x - 2)(6x + 1)$

25. $x = 2, \quad x = -\frac{1}{6}$

27. $x = 1 + \left(\dfrac{\sqrt{2}}{2}\right), \quad x = 1 - \left(\dfrac{\sqrt{2}}{2}\right)$

29. $x - 3$

31. $\dfrac{7x^2 - 4x - 1}{x^2 - 1}$

CHAPTER 1

Exercise Set 1–1

1.

x	-2	-1	0	1	2	3
y	-7	-5	-3	-1	1	3
point	$(-2, -7)$	$(-1, -5)$	$(0, -3)$	$(1, -1)$	$(2, 1)$	$(3, 3)$

3.

x	-2	-1	0	1	2	3
y	6	5	4	3	2	1
point	$(-2, 6)$	$(-1, 5)$	$(0, 4)$	$(1, 3)$	$(2, 2)$	$(3, 1)$

5.
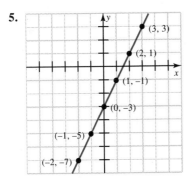

7.
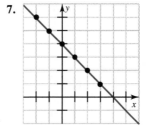

9. $2x + y = 6$ $y = 6 - 2x$

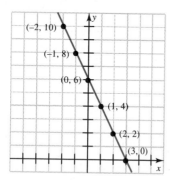

11. $x - y = 3$

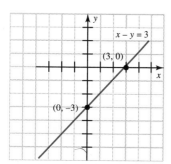

13. $3x - y = -1$

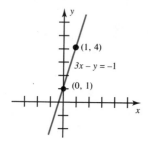

15.

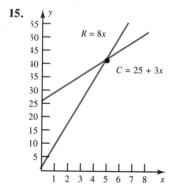

17.

19. (a) \$58,100
(b) 10%

21.

23.

25.

27.

29.

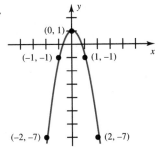

31.

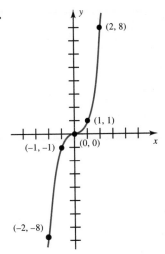

33.

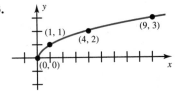

35.

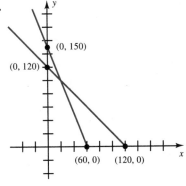

37.

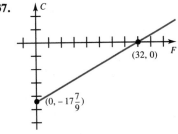

39. 37° Celsius

41. $50,000

43. (a)

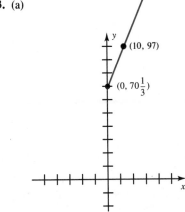

(b) 169 cm
(c) 43 cm

45. This is the break-even point.

47. More than 125 items.

49. 66,706

51. Figure (a)

Referenced Exercise Set 1–1

1. (a) 72.4%
  (b) 1978
3. (a) 2 hours
  (b) 3 hours
  (c) 1 hour, 20 minutes

5. (a) $T = 0.03x$
  (b) $T = 0.05x$

Exercise Set 1–2

1.

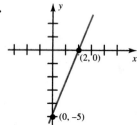

3.

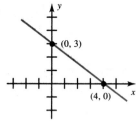

5. $m = 2$
7. $m = -\frac{3}{2}$
9. $m = 2$

11.

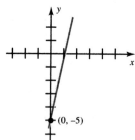

13.

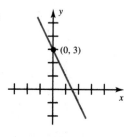

15. not parallel
17. (a) undefined   (b) $m = 0$
19. $y = -7x + 15$

21.

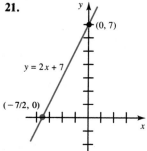

23.

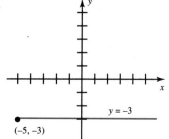

25.

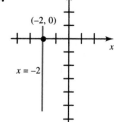

27.

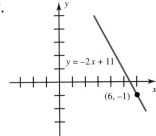

29.

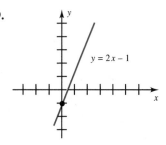

31.

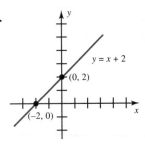

33.

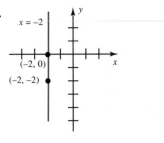

35. (a) positive (b) zero (c) negative

37. (a) $y = -\frac{3}{2}x + 6$ (b) $m = -\frac{3}{2}$ (c) y-intercept $= 6$

39. (a) $y = -\frac{5}{2}x + 5$ (b) $y = -\frac{5}{2}x - 5$

41. $y = -4x + 1$

43. (a)

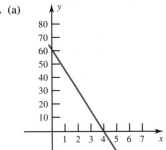

(b) $15,000

(c) The y-intercept represents the worth of the work station when it was purchased.

45. (a) $y = 48643.3x + 17558072$

(b) 18,044,505

47. The population of Iowa decreased from 1980 to 1990.

49. Figure (d)

51. Figure (a)

53. no, for example, $xy = 1$

Referenced Exercise Set 1–2

1. (a) $y = .5x + 7.5$

(b)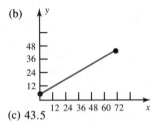

(c) 43.5

3. (a) dead: $y = (1/50)x + 15$
(b) average: $y = (1/100)x + 10$
(c) live: $y = (1/140)x + 20/7$

5. (a) During the first few minutes of running, the percentage of energy supplied to the body by carbohydrates increases.
(b) After a few minutes of running, the percentage of energy supplied to the body by carbohydrates decreases.

Cumulative Exercise Set 1–2

1. $y = 2x - 3$

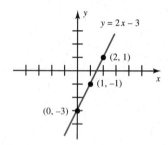

3. $y = -x^2 + 1$

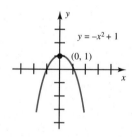

5. $y = -4x$
7. $2x - 4y = -20$
9. \$27,000

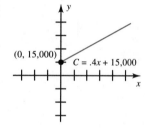

Exercise Set 1–3

1. (4, 1)

3.

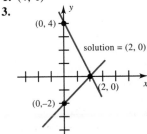

5. (1, 1)

7. (1, −2)

9. ($\frac{7}{5}$, $\frac{3}{5}$)

11. (1, −3)

13. (7/2, 3/2)

15. inconsistent

17. dependent

19. 20,000 boxes of mints

21. ($\frac{2}{3}$, −$\frac{5}{3}$)

23. (4, 2)

25. (20, 100)

27. (2, 0)

29. (100, 20)

31. (0, 0)

33. no solution

35. (10, 4)

37. two 747's and four 707's

39. $1500 in school bonds and $3500 in public utility bonds

41. April, 1983

43. Figure (c)

45. Each adult's ticket costs $5; each child's ticket costs $2.50.

Referenced Exercise Set 1–3

1. (a) 12.2 million cars sold
   (b) 8.9 million cars sold
   (c) 3.3 million cars

3. (a) 9%   (b) 9%

Cumulative Exercise Set 1–3

1. −1

3.

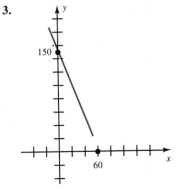

5. $y = -3x + 1$

7. $y = 3x - 7$

9. (−2, 5)

11. inconsistent

13. (a) positive
   (b) positive
   (c) negative

15. (a) $D = .24(x − 1) + .23$
   (b) $S = .11(x − 1) + .10$
   (c) A caller saves $.11 for each additional minute after the first minute.

Exercise Set 1–4

1. $(1, 1, 1)$
3. none
5. $(1, 1, 1)$
7. $(0, 0, 0)$
9. $(2, 1, 1)$
11. $(3, 1, 0)$
13. $(2, 0, 3)$
15. $(0, 1, 2)$

17. 5 small disks, 1 medium disk, and 4 large disks
19. $(1, 2, -1)$
21. $(4, 0, -1)$
23. $(2, 0, 3)$
25. no unique solution
27. $x = 16/9,$ $y = 2/3,$ $z = 10/9$
29. one solo artist and two groups
31. 4 oz fish, 2 oz corn, and 1 oz rice

33. The restated problem is as follows: "A manufacturer makes small, medium, and large tables. The manufacturing costs are $10 for each small table, $15 for each medium table, and $20 for each large table. The shipping costs are $5 for each small table, $8 for each medium table, and $10 for each large table. How many tables of each type should be made per week if the manufacturer must make 100 tables per week, the total manufacturing cost per week is $1700, and the total shipping cost per week is $860?" The solution is for the manufacturer to produce 20 small tables, 20 medium tables, and 60 large tables.

35. (a) $(0, 0, 0)$
(b) no unique solution
(c) $(0, 0, 0)$

37. false
39. true
41. $x = (ed - bf)/(ad - bc), y = (af - ce)/(ad - bc)$

Referenced Exercise Set 1–4

1. $(70, 70, 40)$

Cumulative Exercise Set 1–4

1. $y = 2 - 3x$

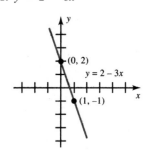

3. $y = -x - 3$
5. $(2, 2)$
7. $(2, -2)$
9. $(2, 1, 1)$
11. $(0, 5, 20)$

Exercise Set 1–5

1. $y = 0.3x + 2$
3. $y = -0.6x + 4$
5. $y = 2x + 2.75$
7. $y = -3.5x + 22.8$
9. $y = 0.316x + 0.163$
11. $y = 2.036x + 6.214$

13. $y = -0.46x - 0.57$
15. $y = \frac{12}{5}x + \frac{92}{3}$
17. $71\frac{1}{2}$ in.
19. Not linear
21. Not linear

Referenced Exercise Set 1–5

1. (a)

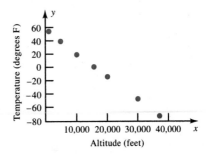

(b) no

(c) $4b + 179m = 100$
$179b + 13899m = 6090$

(d) $y = \dfrac{1292}{4711}x + \dfrac{59,958}{4711}$

(e) $\dfrac{318,358}{4711} \approx 67.6$

(f) $\dfrac{411,142}{1292} \approx 318.2$

3. The points are linear: $y = -0.00355x + 58.56$.

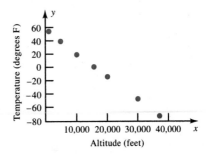

5. (a) least squares   **(b)** $y = 1.25t + 20.5$
(c) 45.5   **(d)** 10
7. (a) $y = .8204x - 1608.0351$ (where x is the year)
(b) 33.59% (or about one third)

Cumulative Exercise Set 1–5

1. -7
3. $y = 3x$
5. $-4/3$
11. No, because three lines in a plane can have at most one point in common.
13. (a) positive   **(b)** positive   **(c)** negative

7. $(-4, -8)$
9. $(1, -2, 1)$

Case Study Exercises

1. $5b + 376m = 1148.4$
$376b + 28380m = 86341.6$
3. $y = -0.173x + 242.65$
5. 3:45.35

7. 2262
9. (a) 3:42   **(b)** 2006
11. (a) $-.32$ **(b)** $\approx \frac{1}{3}$ sec   **(c)** 32 sec

Review Exercises

1.

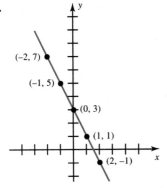

3.

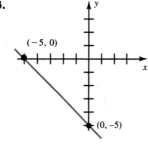

5.

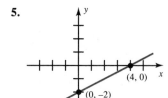

7. $y = 3x - 3$
9. $C = 0.22x + 25$
11. $(-1\frac{2}{3}, 1\frac{1}{3})$
13. $(1, -1), (3, 1)$, other solutions possible
15. $(-3, -1, -2)$
17. $y = 0.3x + 4.6$

CHAPTER 2

Exercise Set 2–1

1. The size of $A = 2 \times 3$; the size of $C = 3 \times 1$; the size of $E = 2 \times 2$.
3. (a) $a_{22} = 4$ (b) $b_{31} = -1$ (c) $d_{13} = 3$ (d) $e_{11} = 1$
5. (a) a_{12} (b) a_{23} (c) a_{22}
7. $x = 5,$ $y = 6,$ $z = 7$

9. (a) $\begin{bmatrix} -2 & 5 \\ 12 & 6 \\ 7 & -3 \end{bmatrix}$ (b) $\begin{bmatrix} 9 & 1 \\ 9 & 16 \\ 2 & -4 \end{bmatrix}$ (c) same as (b)

11. (a) $\begin{bmatrix} 3 & -3 \\ 6 & -6 \\ 9 & -9 \end{bmatrix}$ (b) $\begin{bmatrix} 6 & -12 \\ -20 & -16 \\ -8 & 0 \end{bmatrix}$ (c) $\begin{bmatrix} -11 & 4 \\ 3 & -10 \\ 5 & 1 \end{bmatrix}$

13. (a) $\begin{bmatrix} -7 & 16 \\ 34 & 20 \\ 18 & -6 \end{bmatrix}$ (b) $\begin{bmatrix} 11 & -20 \\ -26 & -28 \\ -6 & -6 \end{bmatrix}$

17. (a) 2×3 (b) 3×3

19. $\begin{bmatrix} -1 & 6 & -7 \\ 2 & -4 & -3 \end{bmatrix}$

21.   March

	cotton	synthetics
Fall River	3	1.5
Passaic	0.5	3
L.A.	5	2

23.   March and April

	cotton	synthetics
Fall River	8	3.5
Passaic	4	5
L.A.	6.5	4.5

25. $x = 6,$ $y = 4$
27. $x = -1,$ $y = -3,$ $z = -8$
29. $x = -3,$ $y = -21,$ $z = -7$
31. $x = 2,$ $y = 3,$ $z = 1$

33. (a) $\begin{bmatrix} -7 & -3 \\ -5 & -20 \\ 4 & -2 \end{bmatrix}$ (b) $\begin{bmatrix} -7 & -3 \\ -5 & -20 \\ 4 & -2 \end{bmatrix}$

35. $R - (S + T) = (R - S) - T$
37. $x = 3,$ $y = 1$
39. Impossible, because the system of equations is inconsistent

41. (a) $K + U = \begin{bmatrix} 165 & 18 \\ 345 & 23 \\ 730 & 28 \end{bmatrix}$ (b) $K - U = \begin{bmatrix} 135 & 6 \\ 305 & 7 \\ 670 & 8 \end{bmatrix}$ (c) $\frac{1}{2}K + 2U = \begin{bmatrix} 105 & 18 \\ 202.5 & 23.5 \\ 410 & 29 \end{bmatrix}$

43. $X = (-1)M$

Referenced Exercise Set 2–1

1. $\begin{bmatrix} 0 & 934 & 3733 \\ 934 & 0 & 4624 \\ 3733 & 4624 & 0 \end{bmatrix}$

3. (a) 34×8 (the "individual" column is not included)

(b) each entry is either 0 or $\frac{1}{99}$

(c) the diagonal entries are $\frac{20}{99}, \frac{17}{99}, \frac{8}{99}, \frac{7}{99}, \frac{16}{99}, \frac{11}{99}, \frac{11}{99}$, and $\frac{9}{99}$; all other entries are 0

(d) the diagonal entries are

$\frac{4}{99}, \frac{2}{99}, \frac{2}{99}, \frac{2}{99}, \frac{3}{99}, \frac{2}{99}, \frac{3}{99}, \frac{4}{99}, \frac{5}{99}, \frac{5}{99}, \frac{3}{99}, \frac{3}{99}, \frac{2}{99}, \frac{4}{99}, \frac{2}{99}, \frac{3}{99}, \frac{2}{99}, \frac{2}{99}, \frac{3}{99}, \frac{2}{99}, \frac{4}{99}, \frac{2}{99}, \frac{2}{99}, \frac{3}{99}, \frac{3}{99}, \frac{3}{99}, \frac{3}{99}, \frac{2}{99}, \frac{3}{99}, \frac{2}{99}, \frac{3}{99}, \frac{3}{99},$
$\frac{3}{99}, \frac{3}{99}, \frac{4}{99};$
all other entries are 0

Exercise Set 2–2

1. $\begin{bmatrix} -2 & -5 & -3 \\ 0 & 60 & 30 \\ 0 & 6 & 1 \end{bmatrix}$

5. $\begin{bmatrix} 1 \\ -10 \\ -3 \end{bmatrix}$

3. $\begin{bmatrix} -1 & -5 & -3 & 3 \\ 20 & 50 & 20 & 30 \\ 6 & 6 & 5 & 4 \end{bmatrix}$

7. $j_{12} = 24$

9. $j_{22} = 10$

11. (a) 4×3 (b) undefined

13. $A = \begin{bmatrix} -1 & 4 & -3 \\ 2 & -2 & 7 \end{bmatrix}$ $X = \begin{bmatrix} x \\ y \\ z \end{bmatrix}$ $C = \begin{bmatrix} 11 \\ -4 \end{bmatrix}$

15. $\begin{aligned} x \quad\quad + 6z &= 0 \\ -2x + 3y + 4z &= 0 \\ x + 2y \quad &= 0 \end{aligned}$

21. No

$AB = \begin{bmatrix} 30 & 6 \\ 2 & -8 \end{bmatrix}$ $BA = \begin{bmatrix} 30 & -4 \\ -3 & -8 \end{bmatrix}$

17. $\begin{aligned} 2x + 6y + 7z &= 0 \\ 3x + 5y - 2z &= 0 \end{aligned}$

19. No

$AB = \begin{bmatrix} 8 & 10 \\ 26 & 22 \end{bmatrix}$ $BA = \begin{bmatrix} 16 & 28 \\ 11 & 14 \end{bmatrix}$

23. $\begin{bmatrix} 2.0 & 4.5 \\ 2.5 & 4.0 \\ 3.5 & 1.5 \end{bmatrix} \begin{bmatrix} 1.0 \\ 1.5 \end{bmatrix} = \begin{bmatrix} 8.75 \\ 8.50 \\ 5.75 \end{bmatrix}$

25. (a) $\begin{bmatrix} -28 & 41 \\ 32 & -52 \end{bmatrix}$ (b) $\begin{bmatrix} -28 & 41 \\ 32 & -52 \end{bmatrix}$ (c) $(AB)C = A(BC)$ This verifies the associative law.

(d) same as (a) and (b)

27. (a) $\begin{bmatrix} -10 & 10 \\ 8 & 1 \end{bmatrix}$ (b) $\begin{bmatrix} -10 & 10 \\ 8 & 1 \end{bmatrix}$ (c) $A(B + C) = AB + AC$ This verifies the distributive law.

29. (a) $\begin{bmatrix} 4 & 5 & 6 \\ 0 & 0 & 0 \\ 0 & 0 & 0 \end{bmatrix}$ (b) $\begin{bmatrix} 0 & 0 & 0 \\ 7 & 8 & 9 \\ 0 & 0 & 0 \end{bmatrix}$

(c) This procedure moves the jth row of A to the ith row. Zeros everywhere else.

31. AB is a 3×4 matrix whose bottom row consists entirely of 0's.

33. How much protein, carbohydrate, and fat is consumed by the three age groups from milk.

35. F1:$3,632,400; F2:$2,701,400; F3:$2,028,600

37. $690

39. $A = \$13.00$, $B = \$10.75$

41. $4,975,000

43. Yes.

45. (a) $AS = SA = \begin{bmatrix} as & bs \\ cs & ds \end{bmatrix}$

(b) Let $B = \begin{bmatrix} u & v \\ w & x \end{bmatrix}$. If $A = \begin{bmatrix} 0 & 1 \\ 1 & 0 \end{bmatrix}$ then $AB = BA$ implies that $u = x$ and $v = w$. If $A = \begin{bmatrix} 0 & 1 \\ 0 & 0 \end{bmatrix}$ then

$AB = BA$ implies that $w = 0$. Therefore $B = \begin{bmatrix} x & 0 \\ 0 & x \end{bmatrix}$.

47. $A = \begin{bmatrix} 0 & 1 \\ 0 & 0 \end{bmatrix}$ and $B = \begin{bmatrix} 1 & 0 \\ 0 & 0 \end{bmatrix}$; other answers are possible.

Referenced Exercise Set 2–2

1. (a) They add up to one because each entry represents one of the three possible options.
(b) E is the most preferable option.
(c) C is the least preferable option.

3. The domestic market D has the higher score, 57 to 35.

5. Common stock financing has the higher score, 18 to 10.

Cumulative Exercise Set 2–2

1. (a) 3×2 (b) -1

3. (a) not defined   (b) $\begin{bmatrix} -3 & 6 \\ 2 & 1 \\ 5 & 7 \end{bmatrix}$ (c) $\begin{bmatrix} 19 & -6 \\ -6 & 23 \\ 15 & 21 \end{bmatrix}$

5. -3

7. The only solution is $x = 0$, $y = 0$.

Exercise Set 2–3

1. $(2, 1, 1)$

3. $(3, 1, 0)$

5. $(-2, 3, 0)$

7. $(3, -1, 6)$

9. $x = 6$, $z = 4$, $y =$ any number

11. $x = -3z - 2$,   if $z = 0$ $(-2, -2, 0)$
$y = -z - 2$,   if $z = 1$ $(-5, -3, 1)$
$z =$ any number,   other solutions possible

13. $x = 4$,   if $z = 0$ $(4, 2, 0)$
$y = z + 2$,   if $z = 1$ $(4, 3, 1)$
$z =$ any number,   other solutions possible

15. $x = -y - 2v + 1$ basic variables: x, u
$y =$ any number   free variables: y, v
$u = -3v$
$v =$ any number

17. $x = -v$ basic variables: x, y, u
     $y = -2v + 5$ free variable: v
     $u = -3v + 4$
     $v =$ any number

19. $x = 4z - 5$ basic variables: x, y
     $y = 6z - 9$
     $z =$ any number

21. $x = 2z + 26$ basic variables: x, y
     $y = -15$
     $z =$ any number

23. two servings of chicken; one serving of rice

25. $(1, 2, -3)$

27. $x = -5 + z$,   if $z = 0$ $(-5, 4, 0)$
     $y = 4 + 2z$,   if $z = 1$ $(-4, 6, 1)$
     $z =$ any number,   other solutions possible

29. $(1, -3, -2)$

31. $x = -2z + 5$,   if $z = 0$ $(5, 4, 0)$
     $y = 4$,   if $z = 1$ $(3, 4, 1)$
     $z =$ any number,   other solutions possible

33. no solution

35. Two servings of chicken and one serving of rice

37. Four different combinations are possible

hard disk drives	6	4	2	0
1.2MB floppy	0	1	2	3
360K floppy	7	7	7	7
Total no. of drives	13	12	11	10

39. Figure (b). Solution $(3, 3)$.

41. Figure (d). No solution.

43. $\begin{bmatrix} 1 & 0 & d/(ad - bc) & -b/(ad - bc) \\ 0 & 1 & -c/(ad - bc) & a/(ad - bc) \end{bmatrix}$

Referenced Exercise Set 2–3

1. $x = 480{,}000/49$, $y = 390{,}000/49$, $z = 1{,}328{,}000/49$, $w = 1{,}171{,}000/49$

Cumulative Exercise Set 2–3

1. (a) $\begin{bmatrix} 3 & 4 & 3 \\ 1 & 7 & 5 \\ 3 & 3 & -1 \\ -3 & 1 & -2 \end{bmatrix}$ (b) $\begin{bmatrix} 3 & 7 & 10 \\ 1 & 8 & 5 \\ 3 & 5 & 1 \\ -2 & 6 & 3 \end{bmatrix}$

3. $(2, 2, 0)$

5. $\begin{bmatrix} -22 & -2 \\ 0 & 21 \end{bmatrix}$

7. $(2, -1, 3)$

9. $x = 2 - z$, $y = 3 - 2z$, $z =$ any number

11. $x = (13/5) + (4/5)z$, $y = (4/5) + (7/5)z$, $z =$ any number

13. $x = 1$ serving of meat, $y = 2$ servings of corn, $z = 1$ serving of rice

Exercise Set 2–4

1. $A^{-1} = Z$ **3.** $A^{-1} = Y$

5. $\begin{bmatrix} -2 & 1 \\ 7 & -3 \end{bmatrix}$

7. not invertible

9. $\frac{1}{4}\begin{bmatrix} 8 & 2 \\ 2 & 1 \end{bmatrix}$

11. $\begin{bmatrix} -1 & 1 & 0 \\ 4 & -1 & -1 \\ -2 & 0 & 1 \end{bmatrix}$

13. $\begin{bmatrix} -3 & 10 & -3 \\ 2 & -4 & 1 \\ -1 & 1 & 0 \end{bmatrix}$

15. $\begin{bmatrix} 1 & 2 & -5 \\ 0 & 1 & -4 \\ 0 & 0 & 1 \end{bmatrix}$

17. $\begin{bmatrix} 0 & \frac{1}{2} & 0 \\ -\frac{1}{3} & 0 & 0 \\ 0 & 0 & \frac{1}{4} \end{bmatrix}$

19. $(AB)^{-1} = -\frac{1}{2}\begin{bmatrix} 0 & 1 \\ 2 & 1 \end{bmatrix} = B^{-1}A^{-1}$

21. $\begin{bmatrix} 4 & 1 \\ -1 & 0 \end{bmatrix}$

23. not invertible

25. $\begin{bmatrix} 1 & 0 & 0 \\ 0 & \frac{1}{3} & 0 \\ 0 & 0 & -1 \end{bmatrix}$ **27.** $\begin{bmatrix} 1 & 2 & -7 & 7 \\ 0 & 1 & -2 & 4 \\ 0 & 0 & 1 & 0 \\ 0 & 0 & 0 & 1 \end{bmatrix}$

29. $A^{-1} = \begin{bmatrix} -2 & 1 \\ 7 & -3 \end{bmatrix}$ $(A^{-1})^{-1} = \begin{bmatrix} 3 & 1 \\ 7 & 2 \end{bmatrix} = A$

31. $A^{-1} = \begin{bmatrix} 2 & \frac{1}{2} \\ \frac{1}{2} & \frac{1}{4} \end{bmatrix}$ $(A^{-1})^{-1} = \begin{bmatrix} 1 & -2 \\ -2 & 8 \end{bmatrix} = A$

33. $(A^{-1})^{-1} = A$

35. $(AB)^{-1} = \begin{bmatrix} 12 & -5 \\ 26 & -11 \end{bmatrix}$ $AB = -\frac{1}{2}\begin{bmatrix} -11 & 5 \\ -26 & 12 \end{bmatrix}$

37. (a) no   (b) $A^{-1} = \frac{1}{2}B$

39. invertible

41. A is not invertible; B is invertible with

$$B^{-1} = \begin{bmatrix} 2995/3 & -5998/3 & 1000 \\ -5996/3 & 11999/3 & -2000 \\ 1000 & -2000 & 1000 \end{bmatrix}$$

43. $B = IB = (CA)B = C(AB) = CI = C$

45. (a) A^{-1} means taking off a sock, B^{-1} means taking off a shoe
   (b) AB means putting on a sock and then a shoe, while $(AB)^{-1}$ means taking off a shoe and a sock
   (c) in order to take off a shoe and a sock, the shoe must be taken off first and the sock second

Referenced Exercise Set 2–4

1. $\begin{bmatrix} 1.0385 & 0.0331 & 0.0788 & 0.1094 & 0.2520 \\ 0.0409 & 1.0378 & 0.1247 & 0.0622 & 0.2531 \\ 0.1353 & 0.1317 & 1.0410 & 0.0722 & 0.2760 \\ 0.0884 & 0.0376 & 0.1225 & 1.0168 & 0.2531 \\ 0.1147 & 0.1137 & 0.0724 & 0.0208 & 1.0643 \end{bmatrix}$

Cumulative Exercise Set 2–4

1. (a) 2×2 (b) -5

3. (a) AC is not defined

(b) $CA = \begin{bmatrix} -25 & 7 \\ 18 & -3 \\ 0 & 0 \end{bmatrix}$

(c) $CAD = \begin{bmatrix} -125 & -14 \\ 90 & 6 \\ 0 & 0 \end{bmatrix}$

5. $x = 0$, $y = 0$ is the only solution

7. $x = -u + v + 3$
   $y = -u \qquad - 1$
   x, y basic variables
   u, v free variables

9. (a)

$$\frac{1}{2}\begin{bmatrix} 11 & -3 \\ -3 & 1 \end{bmatrix}$$

   (b) not invertible

11. $A^{10} = \begin{bmatrix} -1022 & -2046 & 1023 \\ 1023 & 2047 & -1023 \\ 0 & 0 & 1 \end{bmatrix}$

Exercise Set 2–5

1. $(-30, 9, 3)$
3. $(-5, 1, 4)$
5. $(1, 1)$

7. $(1, -2)$
9. $(\frac{7}{5}, \frac{3}{5})$
11. $(5, -3, -4)$

13. $(2, 5, -4)$

15. (a) $(1, 4)$ (b) $(3, -5)$ (c) $(0, 3)$

17. (a) $(-2, 1)$ (b) $(1, 4)$ (c) $(-3, 12)$

19. (a) $(5, 4, 0)$ (b) $(6, 1, -8)$ (c) $(1, 0, -1)$

21. $(2, 3, -2)$

23. $(-3, 5, 4)$

25. $(5, 8, -1)$

27. no solution

29. $(0, 16, -4, 2)$

31. 20 wool suits; 100 blend suits

33. one red pill; three blue pills

35. 105 first-class passengers; 25 executive passengers; 60 economy class passengers

37. $x = -9$, $y = 0$, $z = 6$, $t = 3$

39. $x = 2$, $y = 0$, $z = 5$, $t = 3$, $w = 1$

Referenced Exercise Set 2–5

1. $x = 480{,}000/49$, $y = 390{,}000/49$, $z = 1{,}328{,}000/49$, $w = 1{,}171{,}000/49$

3. (a) $x = \begin{bmatrix} 56{,}327 \\ 84{,}898 \\ 71{,}837 \end{bmatrix}$ (b) $\hat{x} = \begin{bmatrix} 54{,}594.5 \\ 84{,}594.5 \\ 71{,}621.5 \end{bmatrix}$ (c) $\begin{bmatrix} 3.1\% \\ 0.4\% \\ 0.3\% \end{bmatrix}$

Cumulative Exercise Set 2–5

1. (a) $\begin{bmatrix} 2 & 2 & -2 \\ 1 & 5 & -5 \\ 2 & 3 & 1 \end{bmatrix}$ (b) $\begin{bmatrix} -1 & 7 & -6 \\ 0 & 8 & -7 \\ 3 & 11 & -13 \end{bmatrix}$

5. $\begin{bmatrix} 3 & 6 & -2 \\ -1 & -2 & 1 \\ 1 & 3 & -1 \end{bmatrix}$

3. $x = 4 + z$, $y = 5 - z$, $z =$ any number

     if $z = 0$ then $x = 4$, $y = 5$, $z = 0$

     if $z = 1$ then $x = 5$, $y = 4$, $z = 1$

7. $\begin{bmatrix} 3/7 & 2/7 \\ -2/7 & 1/7 \end{bmatrix}$ $(A^{-1})^{-1} = A$

9. $x = 2$, $y = 1$

11. (a) $x = 1$, $y = 1$ (b) $x = 2$, $y = 1$ (c) $x = 3$, $y = 0$

13. 2 regular pills, 1 extra-strength pill, and 1 fortified pill

Exercise Set 2–6

1. $\begin{bmatrix} 65 \\ 40 \end{bmatrix}$

3. $\begin{bmatrix} 85/2 \\ 140/3 \end{bmatrix}$

5. $\begin{bmatrix} 200 \\ 200 \end{bmatrix}$

7. $\begin{bmatrix} 56 \\ 60 \end{bmatrix}$

9. $\begin{bmatrix} 140 \\ 160 \\ 120 \end{bmatrix}$

11. $\begin{bmatrix} 280{,}000 \\ 260{,}000 \end{bmatrix}$

13. $AB = \begin{bmatrix} 1 & 0 & 0 \\ 0 & 1 & 0 \\ 0 & 0 & 1 \end{bmatrix} = I$,   so $A^{-1} = B$ and $B^{-1} = A$

15. $A^{-1} = \dfrac{1}{k} B^{-1}$

Cumulative Exercise Set 2–6

1. They are inverses of each other.

3. $x = z - 1$
$y = 2z + 12$
x, y basic variables
z free variable

5. $\begin{bmatrix} 1 & 2/3 & 11/3 \\ 0 & 1/3 & 5/6 \\ 0 & 0 & -1/2 \end{bmatrix}$

7. $(-1, 2, 1)$

9. (a) $(11, 6)$ (b) $(28, 15)$ (c) $(0, 0)$

11. $(I - T)^{-1}D = \begin{bmatrix} 15/8 & 5/8 \\ 5/12 & 5/4 \end{bmatrix} \begin{bmatrix} 96 \\ 120 \end{bmatrix} = \begin{bmatrix} 255 \\ 190 \end{bmatrix}$

13.

customer	low-risk	high-risk
v	\$60,000	\$40,000
w	\$80,000	\$40,000
x	\$100,000	\$100,000
y	\$200,000	\$300,000
z	\$200,000	\$800,000

Case Study Exercises

1. $\begin{bmatrix} 189,340 \\ 909,920 \\ 1,282,150 \\ 1,624,600 \\ 487,070 \\ 793,060 \end{bmatrix}$

3. $\begin{bmatrix} 399,280 \\ 353,530 \\ 1,121,850 \\ 595,290 \\ 313,920 \\ 1,590,460 \end{bmatrix}$

5. $\begin{bmatrix} 3,827,000,000 \\ 4,929,000,000 \\ 7,577,000,000 \end{bmatrix}$

Review Exercises

1. (a) 5×3 (b) $m_{42} = 5$ (c) $7 = m_{52}$

3. $x = -1, \quad y = 3$

5. $\begin{bmatrix} 29 & 25 & 22 & 23 \\ 0 & 0 & 0 & 0 \\ 33 & 57 & 25 & 53 \end{bmatrix}$

7. $\begin{bmatrix} -6 & 3 \\ 18 & -9 \end{bmatrix}$

9. $\begin{bmatrix} 37 & 54 \\ 81 & 118 \end{bmatrix}$

11.
$$A = \begin{bmatrix} 1 & 1 & -5 & 0 \\ 0 & 3 & 1 & 7 \end{bmatrix} \quad X = \begin{bmatrix} w \\ x \\ y \\ z \end{bmatrix} \quad C = \begin{bmatrix} -1 \\ 8 \end{bmatrix}$$

13. $x = -2 - z,$ if $z = 0$ $(-2, 4, 0)$
$y = \ 4 + z,$ if $z = 1$ $(-3, 5, 1)$
$z = $ any number,   other solutions possible

15. infinitely many solutions
$x = 2y + 26$
$y = $ any number (free variable)
$z = -15$
basic variables: x, z

17. $\frac{1}{3} \begin{bmatrix} 1 & 1 & 1 \\ -2 & 1 & 1 \\ 2 & 2 & -1 \end{bmatrix}$

19. (a) BA^{-1} (b) $CB^{-1}A^{-1}$

21. $A^{-1} = \frac{1}{3}B$

23. $(2, -5, 3)$

25. $\begin{bmatrix} 1300 \\ 1360 \end{bmatrix}$

CHAPTER 3

Exercise Set 3–1

1. $1 + 2 \not> 4$ no; $2 + 2 \not> 4$ no, $2 + 2$ is not *greater than* 4.
3. $-(-1) + 7(1) \geq 8$ yes; $-(0) + 7(0) \not\geq 8$ no
5. $(0, 0)$ is below the line; $(1, 7)$ is above the line.
7. $(1, -1)$ is below the line; $(-1, -1)$ is below the line.

9.
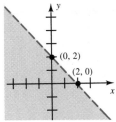

11.
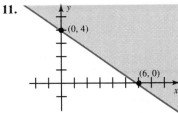

13.
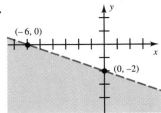

15.
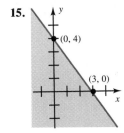

17.
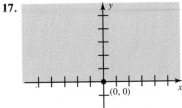

19.
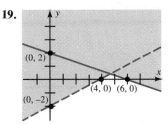

21.
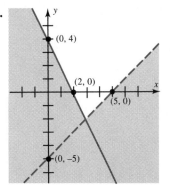

23.

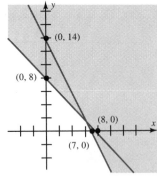

25.

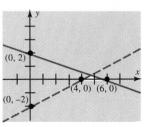

27.

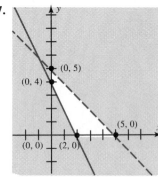

29.

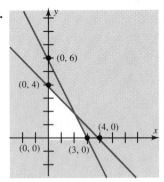

31. $200x + 130y \leq 2600$

33. $40x + 50y \geq 400$

35. $20x + 40y \leq 2400$
$x + y \leq 100$
$x \geq 0, \; y \geq 0$

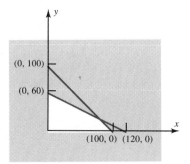

37. $45x + 9y \geq 45$
$9x + 6y \geq 15$
$x \geq 0, \; y \geq 0$

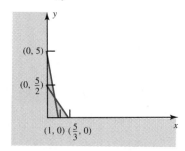

39.

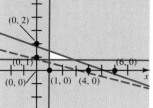

41. no solution

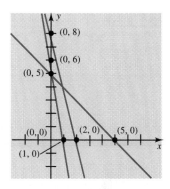

45.

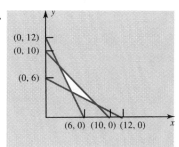

43.

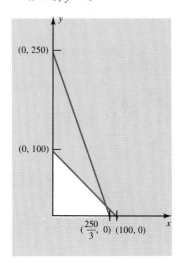

Referenced Exercise Set 3–1

1. $x + \quad y \le \quad 100$
$30x + 10y \le 2500$
$x \ge 0, y \ge 0$

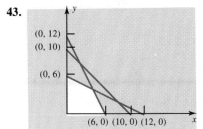

3. $19x + 20y \le 14{,}400$
$\quad x + \quad y \le \quad 750$
$x \ge 0, y \ge 0$

Exercise Set 3–2

1. Maximize: $P = 40x + 30y$
    Subject to: $150x + 100y \leq 12,000$
                  $x + \quad y \leq 100$
                    $x \geq 0, y \geq 0$

3. Minimize: $F = 4x + 2y$
    Subject to: $45x + 9y \geq 45$
                  $9x + 6y \geq 15$
                  $x \geq 0, y \geq 0$

5. Minimize: $C = 132x + 80y$
    Subject to: $3x + 6y \leq 36$
                  $3x + 2y \geq 24$
                  $x \geq 0, y \geq 0$

7. Maximize: $P = 500x + 800y$
    Subject to: $100x + 200y \leq 2000$
                  $3x + \quad 4y \leq 48$
                  $x \geq 0, y \geq 0$

9. Maximize: $P = .5x + .7y$
    Subject to: $.3x + .4y \leq 120$
                  $.2x + .1y \leq \quad 60$
                  $x \geq 0, y \geq 0$

11. Maximize: $G = .15x + .12y$
    Subject to: $x + y \leq 15,000,000$
                        $y \geq 2x$
                  $x \geq 2,000,000$
                  $y \geq 2,000,000$

13. Maximize: $R = 1000x + 1200y$
    Subject to: $8000x + 10,000y \leq 9,000,000$
                  $x + \quad\quad y \leq 1000$
                      $x \geq 200, y \geq 200$

15. Minimize: $C = 20,000x + 15,000y$
    Subject to: $150x + 100y \geq 1000$
                  $50x + \quad 50y \geq \quad 450$
                     $x \geq 0, y \geq 0$

17. Minimize: $C = 9000x + 15,000y$
    Subject to: $x + \quad y \leq \quad 12$
                  $2x + \quad 3y \leq \quad 30$
                  $100x + 300y \geq 1800$
                     $x \geq 0, y \geq 0$

19. Maximize: $P = 600x + 900y$
    Subject to: $x + \quad y \leq \quad 12$
                  $8x + 16y \leq 160$
                  $6x + \quad 3y \leq \quad 66$
                  $x \geq 0, y \geq 0$

21. Maximize: $P = 140x + 90y$
    Subject to: $x + \quad y \leq \quad 100$
                  $60x + 30y \leq 3600$
                  $10x + 15y \leq 1200$
                  $x \geq 0, y \geq 0$

23. Minimize: $C = 10x + 8y$
    Subject to: $x + \quad y \leq \quad 100$
                  $15x + 10y \leq 1050$
                  $x \geq 0, y \geq 0$

25. Maximize: $P = 7x + 5y$
    Subject to: $x + \quad y \geq \quad 100$
                  $15x + \quad 12y \leq 3000$
                  $1.75x + .85y \leq \quad 200$
                  $x \geq 0, y \geq 0$

27. Table b
29. Table a

Referenced Exercise Set 3–2

1. Minimize: $C = .07x + .03y$
    Subject to: $9.2x + 4.6y \geq 165,000,000$
                  $7.6x + 3.1y \leq 230,000,000$
                     $x \geq 0, y \geq 0$

3. Minimize: $P = .34x + .22y$
    Subject to: $x + \quad y \leq 2000$
                  $x - 3y \geq \quad\quad 0$
                    $x \geq 0, y \geq 0$

Cumulative Exercise Set 3–2

1. (a) above the line   (b) on the line   (c) below the line

3.

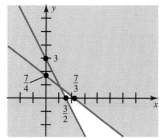

5.

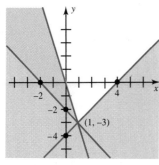

7. (a) Minimize $C = 12{,}000x + 15{,}000y$

(b)
$$\begin{aligned} x + y &\le 15 \\ 150x + 300y &\ge 3000 \\ x &\ge 0 \\ y &\ge 0 \end{aligned}$$

(c)

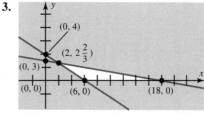

Case Study Exercises

1. Maximize: $N = x + y$
   Subject to: $1200x + 600y \le 35{,}000$
   $42{,}000x + 6{,}000y \ge 672{,}000$
   $x \ge 0,\, y \ge 0,\, x \le 20$

3. Maximize: $N = x + y$
   Subject to: $1400x + 500y \le 35{,}000$
   $40{,}000x + 550y \ge 672{,}000$
   $x \ge 0,\, y \ge 0,\, x \le 20$

Exercise Set 3–3

1.

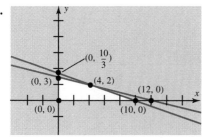

3.

5. no set of feasible solutions; no corner points

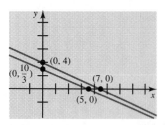

7. (10, 0)
9. $(2, 2\frac{2}{3})$
11. no solution
13. (40, 60)

15. $(\frac{5}{7}, \frac{10}{7})$
17. (6, 3)
19. (8, 6)
21. (0, 300)
23. (5, 10) in millions
25. (500, 500)
27. (2, 7)
29. (0, 6)
31. (3, 1)
33. (0, 8)
35. (4, 4)
37. (2, 8)
39. (40, 40)
41. (600, 200)

Referenced Exercise Set 3–3

1. (0, 90), plant 90 acres of potatoes
3. (200, 0), 200 cattle should be fed the hay ration diet

5. (a)

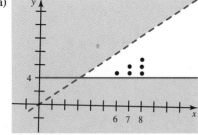

(b) (13, 12) and (20, 6)
(c) The solution (20, 6) was probably rejected because it creates too big a difference between the number of games a team plays against teams in its division vs. games against teams outside its division.

Cumulative Exercise Set 3–3

1.

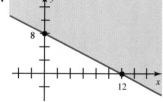

3.

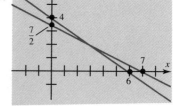

5. Minimize: $z = 900x + 1000y$
 Subject to: $x + \quad y \leq 8$
 $50x + 75y \geq 450$
 $x \geq 0, y \geq 0$

9. 0 type 1, 6 type 2

11. 1500 compact, 500 midsized

7. Maximize: $z = 1000x + 1200y$
 Subject to: $x + \quad\quad y \leq 2{,}000$
 $10{,}000x + 12{,}000y \leq 21{,}000{,}000$
 $x \geq 200, y \geq 200$

Case Study Exercises

1. $x = 0, \quad y = 44$

3. $x = 32, \quad y = 0$

Review Exercises

1.

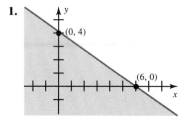

3.

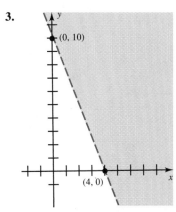

5.

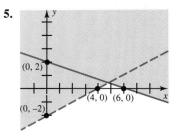

7.

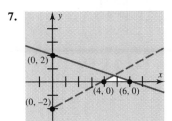

9. $280x + 330y \leq 4800$

11. Minimize: $C = 12{,}000x + 15{,}000y$
 Subject to: $150x + 200y \geq 2500$
 $x + \quad y \leq 15$
 $x \geq 0, y \geq 0$

13. Minimize: $C = 40x + 60y$
 Subject to: $3x + \quad 2y \geq 800$
 $60x + 80y \leq 24{,}000$
 $x \geq 0, y \geq 0$

15.

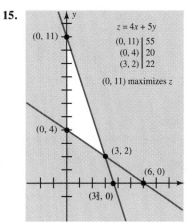

17. $(0, \frac{25}{2})$

19. $(\frac{800}{3}, 0)$

CHAPTER 4

Exercise Set 4–1

1. yes **3.** yes

5. $x + 2y + s_1 \qquad = 4$
$\quad x - y \qquad + s_2 = 5$

7. $-3x - 4y + m = 0$

9.

$$\begin{array}{ccccc} x & y & s_1 & s_2 & m \\ \left[\begin{array}{ccccc|c} 1 & 2 & 1 & 0 & 0 & 10 \\ 3 & 1 & 0 & 1 & 0 & 8 \\ \hline -3 & -4 & 0 & 0 & 1 & 0 \end{array}\right] \end{array}$$

11.

$$\begin{array}{cccccc} x & y & s_1 & s_2 & s_3 & m \\ \left[\begin{array}{cccccc|c} 3 & 8 & 1 & 0 & 0 & 0 & 1 \\ 4 & -5 & 0 & 1 & 0 & 0 & 4 \\ 2 & 7 & 0 & 0 & 1 & 0 & 6 \\ \hline -1 & -5 & 0 & 0 & 0 & 1 & 0 \end{array}\right] \end{array}$$

13.

$$\begin{array}{ccccccc} x & y & z & s_1 & s_2 & s_3 & m \\ \left[\begin{array}{ccccccc|c} 1 & 0 & 1 & 1 & 0 & 0 & 0 & 15 \\ 1 & 2 & 0 & 0 & 1 & 0 & 0 & 8 \\ 0 & 1 & 5 & 0 & 0 & 1 & 0 & 10 \\ \hline -3 & -1 & -5 & 0 & 0 & 0 & 1 & 0 \end{array}\right] \end{array}$$

15. basic: $y, s_1, m,$ nonbasic: x, s_2

17. basic: $x, s_2, m,$ nonbasic: y, s_1

19. basic: $y, z, s_2, m,$ nonbasic: x, s_1, s_3

21. $x = 0, \qquad y = 3, \qquad s_1 = 1, \qquad s_2 = 0, \qquad m = 6$

23. $x = 3, \qquad y = 0, \qquad s_1 = 0, \qquad s_2 = 4, \qquad m = 8$

25. $x = s_1 = s_3 = 0, m = 6, \qquad y = 4, \qquad z = 1, \qquad s_2 = 3$

27. Maximize: $\quad m = 2x + y$
Subject to: $\quad x + 2y \le 3$
$\qquad\qquad 4x + 5y \le 6$
$\qquad\qquad x \ge 0, y \ge 0$

29. Maximize: $\quad m = x + 2y + 5z$
Subject to: $\quad x + 2y + \quad z \le 23$
$\qquad\qquad\qquad 3y + 2z \le 30$
$\qquad\qquad 3x + 4y + 7z \le 41$
$\qquad\qquad x \ge 0, y \ge 0, z \ge 0$

31. $s_1 = 3 \qquad s_2 = 4$

33. $s_1 = 2 \qquad s_2 = 4$

35.

$$\begin{array}{ccccccccc} x & y & z & w & s_1 & s_2 & s_3 & s_4 & m \\ \left[\begin{array}{ccccccccc|c} 1 & 0 & 3 & 1 & 1 & 0 & 0 & 0 & 0 & 10 \\ 0 & 1 & 1 & 2 & 0 & 1 & 0 & 0 & 0 & 13 \\ 0 & 1 & 2 & 3 & 0 & 0 & 1 & 0 & 0 & 8 \\ 0 & 1 & 3 & 2 & 0 & 0 & 0 & 1 & 0 & 12 \\ \hline -1 & -3 & -5 & -6 & 0 & 0 & 0 & 0 & 1 & 0 \end{array}\right] \end{array}$$

37. basic: x, z, s_3, m nonbasic: y, s_1, s_2
basic feasible solution: $\quad x = 8, \qquad y = 0, \qquad z = 2, \qquad s_1 = 0, \qquad s_2 = 0, \qquad s_3 = 5, \qquad m = 9$

39. There is a unique slack variable for each equation.

41.

$$\left[\begin{array}{ccccccc|c} .5 & .25 & .5 & 1 & 0 & 0 & 0 & 40 \\ .5 & .25 & .4 & 0 & 1 & 0 & 0 & 25 \\ 0 & .5 & .1 & 0 & 0 & 1 & 0 & 30 \\ \hline -2 & -1 & -2.5 & 0 & 0 & 0 & 1 & 0 \end{array}\right]$$

Referenced Exercise Set 4–1

1. Maximize: $P = .54x + .44y + .37z$
Subject to: $x + y + z \le 12$
$-\tfrac{1}{2}x + y + z \le 0$
$x \ge 0, y \ge 0, z \ge 0$

$$\begin{bmatrix} 1 & 1 & 1 & 1 & 0 & 0 & | & 12 \\ -\tfrac{1}{2} & 1 & 1 & 0 & 1 & 0 & | & 0 \\ -.54 & -.44 & -.37 & 0 & 0 & 1 & | & 0 \end{bmatrix}$$

in standard form

3. Minimize: $C = 5x + 7y + z$
Subject to: $.0015x + .0015y + .00015z \le .012$
$3x + 4y + 5z \le 30$
$x \ge 0, y \ge 0, z \ge 0$

$$\begin{bmatrix} .0015 & .0015 & .00015 & 1 & 0 & 0 & | & .012 \\ 3 & 4 & 5 & 0 & 1 & 0 & | & 30 \\ -5 & -7 & -1 & 0 & 0 & 1 & | & 0 \end{bmatrix}$$

not in standard form

Exercise Set 4–2

1. no **3.** yes **5.** no
7. $y = s_2 = 0,$ $x = 15,$ $s_1 = 10,$ $m = 20$; no
9. $s_1 = s_2 = s_3 = 0,$ $x = 28$ $y = 34,$ $z = 10,$ $m = 22$; yes
11. the 2nd column; y is the entering variable
13. the 5th column; s_2 is the entering variable
15. the 3rd column; z is the entering variable

17. $x \;\; y \;\; s_1 \;\; s_2 \;\; m$
$$\begin{bmatrix} 0 & 0 & 0 & 1 & 0 & | & 1 \\ 1 & 1 & 1 & 0 & 0 & | & 3 \\ 3 & 0 & 2 & 0 & 1 & | & 17 \end{bmatrix}$$

19. $x \;\;\; y \;\;\; z \;\;\; s_1 \;\; s_2 \;\; s_3 \;\; m$
$$\begin{bmatrix} \tfrac{1}{2} & \tfrac{1}{2} & 0 & 0 & 1 & 0 & 0 & | & \tfrac{3}{2} \\ -\tfrac{1}{2} & -\tfrac{1}{2} & 0 & 1 & 0 & 1 & 0 & | & \tfrac{13}{2} \\ 0 & 0 & 1 & 3 & 0 & 0 & 0 & | & 9 \\ 1 & 3 & 0 & -1 & 0 & 0 & 1 & | & 6 \end{bmatrix}$$

21. $x \;\;\; y \;\;\; z \;\;\; s_1 \;\;\; s_2 \;\;\; s_3 \;\; m$
$$\begin{bmatrix} -\tfrac{1}{3} & 0 & 0 & -\tfrac{1}{3} & -\tfrac{14}{3} & -\tfrac{1}{3} & 0 & | & 7 \\ \tfrac{1}{3} & 0 & 1 & \tfrac{1}{3} & \tfrac{2}{3} & \tfrac{1}{3} & 0 & | & 3 \\ \tfrac{1}{3} & 1 & 0 & \tfrac{1}{3} & \tfrac{5}{3} & \tfrac{4}{3} & 0 & | & 5 \\ 3 & 0 & 0 & 3 & 1 & 6 & 1 & | & 72 \end{bmatrix}$$

23. $x \;\; y \;\; z \;\; s_1 \;\;\; s_2 \;\; s_3 \;\; m$
$$\begin{bmatrix} 0 & 0 & 1 & 0 & -4 & 0 & 0 & | & 10 \\ 1 & 0 & 0 & 1 & 14 & 1 & 0 & | & -21 \\ 0 & 1 & 0 & 0 & -3 & 1 & 0 & | & 12 \\ 0 & 0 & 0 & 0 & -41 & 3 & 1 & | & 135 \end{bmatrix}$$

25. $x \;\;\; y \;\;\;\; z \;\;\; s_1 \;\; s_2 \;\;\; s_3 \;\; m$
$$\begin{bmatrix} 0 & 4 & -3 & 0 & 0 & 4 & 0 & | & 18 \\ 1 & -2 & 5 & 1 & 0 & -1 & 0 & | & 5 \\ 0 & 1 & -1 & 0 & 1 & 1 & 0 & | & 2 \\ 0 & 5 & -14 & 0 & 0 & 8 & 1 & | & 55 \end{bmatrix}$$

27. $x \;\; y \;\;\; z \;\; s_1 \;\;\; s_2 \;\; s_3 \;\; m$
$$\begin{bmatrix} 0 & 0 & 1 & 0 & -4 & 0 & 0 & | & 10 \\ 1 & 0 & 3 & 1 & 2 & 1 & 0 & | & 9 \\ 0 & 1 & -1 & 0 & 1 & 1 & 0 & | & 2 \\ 0 & 0 & -9 & 0 & -5 & 3 & 1 & | & 45 \end{bmatrix}$$

29. $x \;\;\; y \;\; s_1 \;\; s_2 \;\; m$
$$\begin{bmatrix} 1 & 3 & 1 & 0 & 0 & | & 12 \\ 2 & 3 & 0 & 1 & 0 & | & 18 \\ -5 & -7 & 0 & 0 & 1 & | & 0 \end{bmatrix} \qquad \begin{matrix} x \;\;\; y \;\;\; s_1 \;\; s_2 \;\; m \end{matrix} \begin{bmatrix} \tfrac{1}{3} & 1 & \tfrac{1}{3} & 0 & 0 & | & 4 \\ 1 & 0 & -1 & 1 & 0 & | & 6 \\ -\tfrac{8}{3} & 0 & \tfrac{7}{3} & 0 & 1 & | & 28 \end{bmatrix}$$

31. $x \;\;\; y \;\; s_1 \;\; s_2 \;\; m$
$$\begin{bmatrix} 3 & 5 & 1 & 0 & 0 & | & 15 \\ 1 & 5 & 0 & 1 & 0 & | & 10 \\ -4 & -7 & 0 & 0 & 1 & | & 0 \end{bmatrix} \qquad \begin{matrix} x \;\;\; y \;\; s_1 \;\;\; s_2 \;\; m \end{matrix} \begin{bmatrix} 2 & 0 & 1 & -1 & 0 & | & 5 \\ \tfrac{1}{5} & 1 & 0 & \tfrac{1}{5} & 0 & | & 2 \\ -\tfrac{13}{5} & 0 & 0 & \tfrac{7}{5} & 1 & | & 14 \end{bmatrix}$$

33.

x	y	z	s_1	s_2	s_3	m	
1	3	0	1	0	0	0	300
1	0	1	0	1	0	0	150
1	2	1	0	0	1	0	200
-1	-3	-4	0	0	0	1	0

x	y	z	s_1	s_2	s_3	m	
1	3	0	1	0	0	0	300
1	0	1	0	1	0	0	150
0	2	0	0	-1	1	0	50
3	-3	0	0	4	0	1	600

35. Maximize: $m = 2x + 5y + 30z$
Subject to: $3x + 5y + 10z \leq 1500$
$x + y + z \leq 200$
$-x - y + z \leq 0$
$x \geq 0,\ y \geq 0,\ z \geq 0$

x	y	z	s_1	s_2	s_3	m	
1	1	1	1	0	0	0	200
3	5	10	0	1	0	0	1500
-1	-1	1	0	0	1	0	0
-2	-5	-30	0	0	0	1	0

x	y	z	s_1	s_2	s_3	m	
1	1	1	1	0	0	0	200
0	2	7	-3	1	0	0	900
0	0	2	1	0	1	0	200
0	-3	-28	2	0	0	1	400

37. Maximize: $m = 22x + 20y + 15z$
Subject to: $x + y + z \leq 10$
$3x + 6y + 4z \leq 50$
$3x + 2y + z \leq 40$
$x \geq 0,\ y \geq 0,\ z \geq 0$

x	y	z	s_1	s_2	s_3	m	
1	1	1	1	0	0	0	10
3	6	4	0	1	0	0	50
3	2	1	0	0	1	0	40
-22	-20	-15	0	0	0	1	0

1	1	1	1	0	0	0	10
0	3	1	-3	1	0	0	20
0	-1	-2	-3	0	1	0	10
0	2	7	22	0	0	1	220

39. Maximize: $p = .08x + .10y + .12z + .11w$
Subject to: $x + y + z + w \leq 25{,}000$
$x + y \leq 10{,}000$
$x + y + z \leq 15{,}000$
$x + w \leq 20{,}000$
$x \geq 0,\ y \geq 0,\ z \geq 0,\ w \geq 0$

1	1	1	1	1	0	0	0	0	25,000
1	1	0	0	0	1	0	0	0	10,000
1	1	1	0	0	0	1	0	0	15,000
1	0	0	1	0	0	0	1	0	20,000
$-.08$	$-.10$	$-.12$	$-.11$	0	0	0	0	1	0

1	1	1	1	1	0	0	0	0	25,000
0	0	-1	-1	-1	1	0	0	0	$-15,000$
0	0	0	-1	-1	0	1	0	0	$-10,000$
0	-1	-1	0	-1	0	0	1	0	$-5,000$
0	$-.02$	$-.04$	$-.03$	.08	0	0	0	1	2,000

Referenced Exercise Set 4–2

1.
$$\begin{bmatrix} 1 & 1 & 1 & 1 & 0 & 0 & 0 & 60 \\ 0 & 1 & 1 & 0 & 1 & 0 & 0 & 40 \\ 1 & 1 & 0 & 0 & 0 & 1 & 0 & 50 \\ -5 & -6 & -3 & 0 & 0 & 0 & 1 & 0 \end{bmatrix}$$

3. Total capacity for the four machines is 300 units. Machines one and two have a combined capacity of 100 units. Machines two, three, and four have a combined capacity of 230 units. If the profit is \$2 per item for machine one, \$3 per item for machine two, and \$4 per item for machine three, how many items should be made on each machine to maximize profit?

Cumulative Exercise Set 4–2

1.

x	y	s_1	s_2	m	
1	4	1	0	0	10
2	1	0	1	0	6
−6	−7	0	0	1	0

3.

x	y	z	w	s_1	s_2	s_3	s_4	m	
1	0	3	4	1	0	0	0	0	20
1	5	0	1	0	1	0	0	0	6
0	1	2	3	0	0	1	0	0	20
1	0	1	3	0	0	0	1	0	15
−2	−1	−4	−3	0	0	0	0	1	0

5. $x = 25$, $s_1 = 20$, $y = s_2 = 0$, $m = 60$; the test fails

7.

x	y	z	s_1	s_2	s_3	m	
1	−2	1	0	−4	0	0	20
0	5	−2	1	9	0	0	50
0	2	−1	0	4	1	0	5
0	−13	6	0	−25	0	1	205

Exercise Set 4–3

1. Row 1; the pivot is 2
3. Row 2; the pivot is 2
5. Row 3; the pivot is 6
7. The pivot is the (2, 2) entry, 1.

x	y	s_1	s_2	m	
−4	0	1	−2	0	10
1	1	0	1	0	15
1	0	0	1	1	35

9. The pivot is the (1, 1) entry, 2.

x	y	s_1	s_2	m	
1	$\frac{1}{2}$	0	$\frac{1}{2}$	0	5
0	$-\frac{1}{2}$	1	$-\frac{1}{2}$	0	3
0	4	0	0	1	60

11. The pivot is the (3, 5) entry, 1.

x	y	z	s_1	s_2	s_3	m	
1	1	−2	−6	0	0	0	2
0	1	0	1	0	1	0	2
0	1	1	3	1	0	0	9
0	4	3	8	0	0	1	30

13. $x = 9$, $y = 0$, $m = 45$
15. $x = \frac{5}{2}$, $y = \frac{3}{2}$, $m = \frac{41}{2}$
17. $x = 0$, $y = 0$, $z = 30$, $m = 90$
19. $x = 15$, $y = 0$, $z = 0$, $m = 45$
21. $x = 8$, $y = \frac{6}{5}$, $m = \frac{242}{5}$
23. $x = 12$, $y = 12$, $m = 204$
25. $x = 4.8$, $y = 4.8$, $m = 67.2$
27. $x = 0$, $y = 25$, $z = 150$, $m = 675$

29. $x = 0$, $y = 0$, $z = 3$, $w = 2$, $m = 19$
31. $x = 30$, $y = 0$, $z = 0$, $w = 0$, $m = 240$
33. $x = 105$ first class, $y = 25$ executive class, $z = 60$ economy class, $m = \$25{,}200$
35. $x = 71$ deluxe, $y = 0$ regular, $z = 44$ economy, $m = \$13{,}900$
37. $x = 120$ small, $y = 80$ medium, $z = 0$ large, $m = \$15{,}600$
39. Problem 24
41. Problem 23

Referenced Exercise Set 4–3

1. $x = 0$ type 1, $y = 0$ type 2, $z = 72$ type 3, $m = 288$ zones
3. $x = \$20,000$ first payment, $y = \$20,000$ second payment, $z = \$10,000$ third payment

Cumulative Exercise Set 4–3

1. No. The third constraint violates condition (iii).
3. Maximize: $m = 4x + 7y + z$
Subject to: $3x + y + 2z \leq 41$
$4x + 7y + 8z \leq 23$
$7x + 9y + 6z \leq 57$
$x + 5y + 7z \leq 12$
$x, y, z \geq 0$
5. No

7. pivot column: 4
entering variable: s_1

$$\begin{bmatrix} x & y & z & s_1 & s_2 & s_3 & m & \\ \frac{1}{2} & \frac{1}{2} & 0 & 0 & 1 & 0 & 0 & \frac{3}{2} \\ -\frac{1}{2} & -\frac{1}{2} & -\frac{1}{3} & 0 & 0 & 1 & 0 & \frac{7}{2} \\ 0 & 0 & \frac{1}{3} & 1 & 0 & 0 & 0 & 3 \\ 1 & 3 & \frac{1}{3} & 0 & 0 & 0 & 1 & 9 \end{bmatrix}$$

9. $(4, 1)$ entry $= 6$
11. $x = 3, y = 2, m = 22$

Exercise Set 4–4

1. (iii) The sense of the second inequality is reversed.
3. (i) The objective function is minimized. (iii) The sense of the second inequality is reversed.
5. (i) The objective function is minimized. (iii) The sense of the second and third inequalities is reversed.
7. Maximize: $-m = -x - y$
$[1 \ \ 1 \ \ 0 \ \ 0 \ \ 1 \ \ 0]$
9. Maximize: $-m = -x - 3y$
$[1 \ \ 3 \ \ 0 \ \ 0 \ \ 0 \ \ 1 \ \ 0]$
11. Maximize: $-m = -x + y - 2z$
$[1 \ \ -1 \ \ 2 \ \ 0 \ \ 0 \ \ 0 \ \ 1 \ \ 0]$
13. Maximize: $-m = -x - y$
Subject to: $x + 2y \leq 4$
$-x - 3y \leq -1$
$x \geq 0, \ \ y \geq 0$
15. Maximize: $-m = -2x + y$
Subject to: $-2x - y \leq -2$
$-x + y \leq -4$
$x \geq 0, \ \ y \geq 0$

17. Maximize: $-m = -3x - y - z$
Subject to: $x \quad - z \leq -2$
$-x - y \quad \leq -4$
$x \geq 0, \ \ y \geq 0, \ \ z \geq 0$
19. $x = 12, \quad y = 15, \quad m = 60$
21. $x = 0, \quad y = 50, \quad m = 150$
23. $x = 11, \quad y = 0, \quad m = 44$
25. $x = 0, \quad y = 3, \quad m = -12$
27. $x = 0, \quad y = 4, \quad z = 0, \quad m = 0$
29. $x = 0, \quad y = 0, \quad z = 4, \quad m = 4$
31. $x = 0, \quad y = 0, \quad z = 0, \quad m = 0$
33. $x = 0, \quad y = 0, \quad z = 0, \quad u = 0, \quad m = 0$
35. $x = 0$ high quality, $y = 200$ regular quality

37. $x = 10$ chocolate nut bars, $y = 0$ chocolate peanut bars, $z = 30$ chocolate cluster bars

Referenced Exercise Set 4–4

1. $x = 0$ tank 1, $y = 3$ tank 2, $z = 0$ tank 3 (in millions of gallons)

Cumulative Exercise Set 4–4

1. Maximize: $m = 8x + 9y + 3z + 100$

   Subject to: $x + 2z \leq 20$

   $3x + 8y + 3z \leq 35$

   $4x + 2y + 7z \leq 60$

   $x, y, z \geq 0$

3. $x = 250,$ $z = 250,$ $s_1 = 490,$ $y = s_2 = s_3 = 0, m = 1300;$ the test fails.

5.

x	y	z	s_1	s_2	s_3	m	
5	2	3	1	0	0	0	90
2	1	1	0	1	0	0	35
1	2	3	0	0	1	0	60
−2	−5	−6	0	0	0	1	0

4	0	0	1	0	−1	0	30
$\frac{5}{3}$	$\frac{1}{3}$	0	0	1	$-\frac{1}{3}$	0	15
$\frac{1}{3}$	$\frac{2}{3}$	1	0	0	$\frac{1}{3}$	0	20
0	−1	0	0	0	2	1	120

7.

x	y	z	s_1	s_2	s_3	m	
1	0	1	0	1	0	0	20
0	0	−2	1	0	1	0	10
0	1	−3	0	−3	1	0	10
0	0	4	0	1	0	1	85

9. $x = 60,$ $y = 0,$ $z = 10,$ $m = 360$

11. $x = 16,$ $y = 24,$ $z = 0,$ $m = 120$

Exercise Set 4–5

1. Minimize: $m = RX$

   Subject to: $AX \geq C$ and $X \geq 0$, where

   $A = \begin{bmatrix} 3 & 2 \\ 4 & -5 \end{bmatrix}$ $C = \begin{bmatrix} 4 \\ 9 \end{bmatrix}$ $R = \begin{bmatrix} 4 & 2 \end{bmatrix}$ $X = \begin{bmatrix} x \\ y \end{bmatrix}$

3. Minimize: $m = RX$

   Subject to: $AX \geq C$ and $X \geq 0$, where

   $A = \begin{bmatrix} 3 & 5 \\ -1 & 3 \\ 2 & -6 \end{bmatrix}$ $C = \begin{bmatrix} 8 \\ 4 \\ 7 \end{bmatrix}$ $R = \begin{bmatrix} 2 & 4 \end{bmatrix}$ $X = \begin{bmatrix} x \\ y \end{bmatrix}$

5. $\begin{bmatrix} 5 & 2 \\ -1 & -3 \end{bmatrix}$

7. $\begin{bmatrix} -4 & 2 \end{bmatrix}$

9. $\begin{bmatrix} 1 & 6 & -2 \\ -4 & 15 & 1 \\ 11 & 0 & -2 \end{bmatrix}$

11. Maximize: $m = 4u + 9v$

   Subject to: $3u + 4v \leq 4$

   $2u - 5v \leq 2$

   $u \geq 0, \quad v \geq 0$

13. Maximize: $m = 8u + 4v + 7w$

   Subject to: $3u - v + 2w \leq 2$

   $5u + 3v - 6w \leq 4$

   $u \geq 0, \quad v \geq 0, \quad w \geq 0$

15. $x = 5,$ $y = 0,$ $m = 5$

17. $x = 11,$ $y = 0,$ $m = 44$

19. $x = \frac{5}{7}$ serving chicken, $y = \frac{10}{7}$ serving rice

21. $x = 8,$ $y = 0,$ $z = 0,$ $m = 8$

23. $x = 2,$ $y = 0,$ $z = 3,$ $w = 6,$ $m = 30$

25. $x = 2,$ $y = 0,$ $z = 6,$ $m = 8$

27. $x = 10,$ $y = 2,$ $m = 132$

29. $x = 50$ chicken, $y = 40$ rice, $z = 0$ spinach

31. 0 swords, 130 axes, 1300 friends

Referenced Exercise Set 4–5

1. $x = 0$ hopper 1, $y = 100$ hopper 2, $z = 140$ hopper 3

Cumulative Exercise Set 4–5

1. (a) Maximize: $-m = -3x + y - z$

    Subject to: $2x + 3y + z \le 6$

                  $x - 4y - z \le 5$

                    $x, y, z \ge 0$

x	y	z	s_1	s_2	m	
2	3	1	1	0	0	6
1	−4	−1	0	1	0	5
3	−1	1	0	0	1	0

(b) In standard form

x	y	z	s_1	s_2	m	
1	4	2	1	0	0	9
2	3	1	0	1	0	7
−2	−3	−2	0	0	1	0

3. (a) Maximize: $-m = -2x - 3y - 5z$

    (b) $[2 \; 3 \; 5 \; 0 \; 0 \; 0 \; 0 \; 1 \;|\; 0]$

5. (a) Minimize: $m = RX$

    Subject to: $AX \ge C, X \ge 0$

$$A = \begin{bmatrix} 1 & -2 & 1 & 1 \\ 3 & 3 & -1 & 2 \\ 4 & -1 & 0 & 4 \end{bmatrix} \quad X = \begin{bmatrix} x \\ y \\ z \\ t \end{bmatrix} \quad C = \begin{bmatrix} 6 \\ 12 \\ 15 \end{bmatrix} \quad R = [2 \; 3 \; 4 \; 1]$$

(b) Maximize: $m = 6u + 12v + 15w$

    Subject to: $u + 3v + 4w \le 2$

                 $-2u + 3v - w \le 3$

                 $u - v \le 4$

                 $u + 2v + 4w \le 1$

                 $u \ge 0, v \ge 0, w \ge 0$

7. (a) $x = 6,$ $y = 0,$ $m = 24$

    (b) $x = \frac{9}{2},$ $y = 0,$ $z = \frac{3}{2},$ $m = \frac{51}{2}$

9. 6 day shifts, 3 night shifts

11. 6 packages of Cheerios, 3 packages of Wheaties

Case Study Exercise

1. From East Chicago to Niles   None

    From East Chicago to DesPlains   50,000 barrels

    From Hammond to Niles   70,000 barrels

    From Hammond to DesPlains   50,000 barrels

    at a cost of $1590 per mile per 1000 barrels

Review Exercises

1. no

3. $4x + 3y + s_1 = 30$

    $2x + 7y + s_2 = 65$

5.

x	y	z	s_1	s_2	s_3	m	
1	2	5	1	0	0	0	30
2	3	1	0	1	0	0	40
4	3	0	0	0	1	0	20
−1	−4	−2	0	0	0	1	0

7. not maximal

9.

x	y	z	s_1	s_2	s_3	m	
0	0	1	−2	0	−2	0	0
1	0	0	1	2	1	0	10
0	1	0	−1	−2	0	0	2
0	0	0	3	9	4	1	85

11. Maximize: $p = 45x + 9y + 7z$
Subject to: $500x + 150y + 100z \leq 1000$
$4x + 2y + 2z \leq 15$
$5x + 6y + 5z \leq 15$

$$
\begin{array}{ccccccc}
x & y & z & s_1 & s_2 & s_3 & m \\
\end{array}
$$

$$
\left[
\begin{array}{ccccccc|c}
500 & 150 & 100 & 1 & 0 & 0 & 0 & 1000 \\
4 & 2 & 2 & 0 & 1 & 0 & 0 & 15 \\
5 & 6 & 5 & 0 & 0 & 1 & 0 & 15 \\
\hline
-45 & -9 & -7 & 0 & 0 & 0 & 1 & 0 \\
\end{array}
\right]
\qquad
\begin{array}{ccccccc}
x & y & z & s_1 & s_2 & s_3 & m \\
\end{array}
$$

$$
\left[
\begin{array}{ccccccc|c}
1 & \frac{3}{10} & \frac{1}{5} & \frac{1}{500} & 0 & 0 & 0 & 2 \\
0 & \frac{4}{5} & \frac{6}{5} & -\frac{1}{125} & 1 & 0 & 0 & 7 \\
0 & \frac{9}{2} & 4 & -\frac{1}{100} & 0 & 1 & 0 & 5 \\
0 & \frac{9}{2} & 2 & \frac{9}{100} & 0 & 0 & 1 & 90 \\
\end{array}
\right]
$$

13.

$$
\begin{array}{ccccccc}
x & y & z & s_1 & s_2 & s_3 & m \\
\end{array}
$$

$$
\left[
\begin{array}{ccccccc|c}
0 & 0 & 1 & .6 & -.2 & 2.2 & 0 & 7.2 \\
1 & 0 & 0 & .4 & .2 & -.2 & 0 & 4.8 \\
0 & 1 & 0 & -1.6 & .2 & .8 & 0 & 5.8 \\
\hline
0 & 0 & 3 & 1.6 & .8 & -2.8 & 1 & 61.2 \\
\end{array}
\right]
$$

15. $x = 15,$ $s_2 = 5,$ $y = z = s_1 = 0,$ $m = 60$

17. Maximize: $-m = -2x - 5y$
Subject to: $x + 4y \leq 4$
$-x - 2y \leq -7$
$x \geq 0, y \geq 0$

19. $x = \frac{1}{3},$ $y = \frac{5}{3},$ $m = 14$

CHAPTER 5

Exercise Set 5–1

1. (a) $385.00   (b) $32.08   (c) $1.06

3. (a) $3669.85   (b) $494.02   (c) $44.86

5. (a) $869.00   (b) $19.31

7. (a) $1385.00   (b) $1032.08   (c) $1001.06

9. (a) $5657.85   (b) $2482.02   (c) $2032.86

11. (a) $2275.20   (b) $1637.93   (c) $1581.93

13. (a) $2500.00   (b) $4615.38   (c) $4986.34

15. discount = $600.00; proceeds = $1400.00; effective rate = 14.3%

17. discount = $3.29; proceeds = $172.71; effective rate = 11.4%

19. $1079.28

21. (a) $512.50   (b) $584.25

23. (a) $10,512.50   (b) $10,584.25

25. $28,121.48

27. 6.5%

29. 6.25%

31. $8\frac{1}{3}$ years (or 100 months)

33. 6.06 years

35. 30.8%

37. 11.1%

Referenced Exercise Set 5–1

1. (a) $10,000   (b) $5000 at 50%, $7300 at 27%   (c) 46%

Exercise Set 5-2

1. $95.51
3. $1189.10
5. $4808.39
7. $3172.65
9. $3148.91
11. $2208.04
13. $1064.40
15. $709.32
17. $1225.35
19. $1083.11

21. 5.654%
23. $7472.58
25. $573.09
27. $7408.36
29. $2790.87
31. (a) $862.22   (b) $870.45   (c) $874.51
35. $7903.15
37. lump sum of $10,000, by $4.98
39. 4.8%
41. $10,796.76

Referenced Exercise Set 5-2

1. No. The annual yields on 14%, 13.056%, and 9.67% are 15.02%, 13.944%, and 10.15%, respectively.
3. 6.48%
5. $5,080,265.37
7. $1,363,636.36 for 1985
   $1,239,669.42 for 1986
   $1,126,972.20 for 1987

Cumulative Exercise Set 5-2

1. (a) $875   (b) $53.94
3. (a) $8728.30   (b) $8440.25
5. $16.75

7. (a) $58,802.39   (b) $58,093.96
9. $1482

Case Study Exercises

1. $(18.9/6.2) - 1 = 2.05$, so 205%
3. $(12.8/9.1) - 1$ is 41%, $(18.4/12.8) - 1$ is 44%, $(24.2/18.4) - 1$ is 32%, $(31.3/24.2) - 1$ is 29%; the compound growth was 36.2% since $(9.1)(1 + .362)^4 = 31.3$

Exercise Set 5-3

1. $60.55
3. $43,178.50
5. 304,300,859
7. 7.4%
15. (a) 38,608,664   (b) 5,203,219   (c) 3,299,322
17. 29,388.91 cruzeiros
19. 11.9 years
21. $11,706,802.51
23. 281.27 pesos

9. $140,712.62
11. $5.61
13. $3741.66

(d) 7,590,519   (e) 11,512,631   (f) 18,556,165
25. $266,627.43
27. $223.36
29. $27.24

Referenced Exercise Set 5–3

1. 4.32%

Cumulative Exercise Set 5–3

1. $554.40
3. $10,064
5. $2549.99
7. (a) $3876.04   (b) $3914.83   (c) $3933.93

9. $27.59
11. $1615.59
13. 1795.86 pesos

Exercise Set 5–4

1. $9897.47
3. $12,705.69
5. $4291.48
7. $2697.35
9. $4507.74
11. $4377.31
13. $2724.32
15. $1773.96
17. $952.38
19. $223.68
39. (a) 1070 trillion barrels   (b) 1144.9 trillion barrels   (c) 1967.15 trillion barrels   (d) 14,783.6 trillion barrels
41. $37,174.98
43. $2,191.99

21. $223.06
23. $362,164.06
25. $406,140.64
27. $29,390.74
29. $262.26
31. $685.65
33. $3761.40
35. $13,432.70
37. $2.63

Referenced Exercise Set 5–4

1. No, Atlanta's market will be 90,362, and Detroit's market will be 91,802

Cumulative Exercise Set 5–4

1. $24,500
3. $1979.82
5. $3,630.15
7. 6.17%

9. $232,998.63
11. $93.04
13. $2285.85

Exercise Set 5–5

1. $2930.19
3. $385.44
5. $1891.39
7. $5310.85
9. $949.59

11. $964.22
13. $58.75
15. $110.80
17. $539.84
19. $880.52

21.

Payment	Interest	Principal	Balance
0			4000.00
1	240.00	709.59	3290.41
2	197.42	752.17	2538.24
3	152.29	797.30	1740.94
4	104.46	845.13	895.81
5	53.75	895.84	−0.03

23.

Payment	Interest	Principal	Balance
0			60,000.00
1	450.00	89.84	59,910.16
2	449.33	90.51	59,819.65
3	448.65	91.19	59,728.46
4	447.96	91.88	59,636.58

25. $265.57
27. $215.69
29. $210.67
31. (a) $5035.18

(b) Payment	Interest	Principal	Balance
0			600,000.00
1	4500.00	535.18	599,464.82
2	4495.99	539.19	598,925.63
3	4491.94	543.24	598,382.39
4	4487.87	547.31	597,835.08

33.

Payment	Interest	Principal	Balance
0			2500.00
1	12.50	98.30	2401.70
2	12.01	98.79	2302.91
3	11.51	99.29	2203.62
4	11.02	99.78	2103.84
5	10.52	100.28	2003.56
6	10.02	100.78	1902.78
7	9.51	101.29	1801.49
8	9.01	101.79	1699.70
9	8.50	102.30	1597.40
10	7.99	102.81	1494.59
11	7.47	103.33	1391.26
12	6.96	103.84	1287.42
13	6.44	104.36	1183.06
14	5.92	104.88	1078.18
15	5.39	105.41	972.77
16	4.86	105.94	866.83
17	4.33	106.47	760.36
18	3.80	107.00	653.36
19	3.27	107.53	545.83
20	2.73	108.07	437.76
21	2.19	108.61	329.15
22	1.65	109.15	220.00
23	1.10	109.70	110.30
24	0.55	110.25	0.05

35. $6,144,567.11
37. $1755.81
39. $30,578.36
41. $586.51
43. $57,570.67

Cumulative Exercise Set 5–5

1. $2232
3. 16.67%
5. $17,422.13
7. $26,361.59

9. $57,507.39
11. $111.22
13. $75,834.46

Review Exercises

1. $262.50
3. $270.40
5. (a) $7938.32   (b) $7866.49
7. $186.44
9. 92.3%
11. $3353.02
13. $109,729.02
15. $10,046.20

17. (a) $617.17

(b) Payment	Interest	Principal	Balance
0			60,000.00
1	600.00	17.17	59,982.83
2	599.83	17.34	59,965.49
3	599.65	17.52	59,947.97
4	599.48	17.69	59,930.28

CHAPTER 6

Exercise Set 6–1

1. not well-defined
3. well-defined
5. {1, 2, 3, 4, 6, 12}
7. {−3, −2, −1, 0}
9. A: no, B: yes
11. A: yes, B: yes
13. $X' = \{4, 5, 6\}$
15. $X' = \{x \mid x \in N; x > 3\}$
17. $X \cup Y = \{1, 2, 3, 4\}$
19. $X \cup Y = Q$
21. $X \cap Y = \{3, 4\}$
23. $X \cap Y = \{1, 2\}$
25. A: yes B: no
27. A: yes B: yes
29. $X' = \{2, 4, 6, \ldots\}$
31. $X \cup Y = \{-2, -1, 0, 1, 2, 3, \ldots\}$
33. $\{\ldots, -3, -2, -1, 1, 2, 3, \ldots\}$
35. $X \cap Y = \{2, 3, 4, \ldots\}$
37. $X \cap Y = \{1, 2, 3\}$
39. $F \cap X = \{12, 24, 36, 48, \ldots\}$
41. (a) {a, e, i}

(b) {a, e, i, o, u, m, t, h, c, s}
(c) {m, t, h, c, s}
43. (a) {2, 4, 6, 8, 10}
   (b) {1, 2, 3, \ldots, 10, 12, 14, \ldots, 100}
   (c) {1, 3, 5, 7, 9}
45. $\{-2, -1, 0, 1, 2, 3, \ldots\}$
47. (a) true   (b) false   (c) true
49.

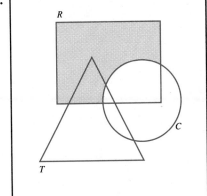

Referenced Exercise Set 6–1

1. {Denver, Phoenix, Columbus}
3. For each of the 10 major cities, either the cost of housing is more than $100 or the cost of miscellaneous goods and services is more than $100
5. (a) $F = $ {Brazil, Canada, China}   (b) $T = $ {Canada, U.S., U.S.S.R.}
7. $\varnothing$ contains no elements, whereas $\{\varnothing\}$ contains one element.
9. (a) $A = $ {Boston, Los Angeles, Miami, New York, Washington}
   (b) $B = $ {Chicago, Los Angeles, New York, San Francisco, Washington}
   (c) Every city having one of the five highest CBD rates would also have to have one of the five highest rates outside CBD's.
   (d) no

Exercise Set 6–2

1. 5
3. 7
5. 3
7. 20
9. 4

11. 65
13. 35,152,000
15. 5,000
17. (a) 2   (b) 6   (c) 8
19. $n(A) = 6$ and $n(B) = 4$

21. 3

23. 20

25.

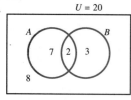

$U = 20$

27. 14

29. yes; if and only if $A = B$

31. (a) 30   (b) 1

33. 102

35. $2^3 = 8$

37. yes; $n(\varnothing) = 0$ and $n(\{\varnothing\}) = 1$

Referenced Exercise Set 6–2

1. 1429

3. 719

5. 30%

7. (a) black

(b) magenta

(c) cyan

(d) black

(e) white

9. $2^6 = 64$ regions; see the reference for the Venn diagram.

Cumulative Exercise Set 6–2

1. $S = \{0, 1, 2, 3, 4, 5, 6, 7, 8\}$

3. (a) $A \cup B = \{1, 2, 3, 4, 5, 6\}$, $A \cap B' = \{1\}$

(b) $A \cup B = \{a, b, c, d, g\}$, $A \cap B' = \{a\}$

5. 32

7. 16

9. 7

Exercise Set 6–3

1. 16 outcomes

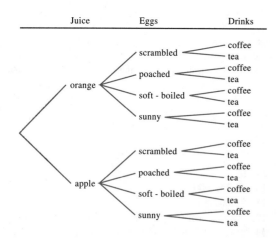

3. four selections

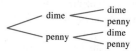

5. 24 choices

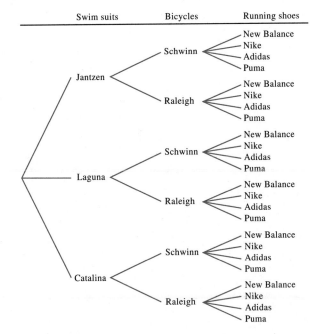

| Swim suits | Bicycles | Running shoes |

7. 95,040
9. 420
11. 15,600
13. 676,000
15. 1,000
17. 27,000
19.

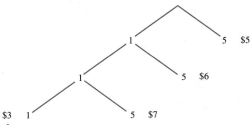

21. $216 = 6^3$
23. $60,466,176 = 6^{10}$
25. $9 = 3^2$
27. $59,049 = 3^{10}$
29. 36^6
31. 8,000,000
33. 24
35. 12 ways: JILL, JLIL, JLLI, IJLL, ILJL, ILLJ, LJIL, LJLI, LIJL, LILJ, LLJI, LLIJ
37. 16
39. 67,600
41. 1,326

43.

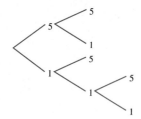

45. 5^{20}
47. 72
49. 1,189,760
51. 8,400
53. 1976 (26 with three alike, and 3·650 with two alike)

Referenced Exercise Set 6–3

1. 3,628,800 = 10·9·8·7·6·5·4·3·2·1
3.

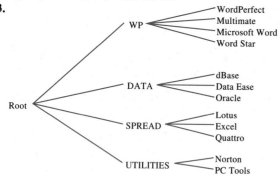

5. 57 minutes

Cumulative Exercise Set 6–3

1. $S = \{2, 4, 6, 8\}$
3. (a) {Jan, Feb, Mar, Apr, May, Jun, Jul, Aug, Sep, Oct, Nov, Dec}
   (b) $A = \{$Jun, Jul, Aug, Sep$\}$
   (c) $B' = \{$Jan, Feb, Mar, May, Jul, Aug, Oct, Dec$\}$
   (d) The set of months not having 30 days.
   (e) {Jun, Sep}
   (f) {Apr, Jun, Jul, Aug, Sep, Nov}
5. $B \subset A$
7. 8
9. (a) 0   (b) 28   (c) 15
11. (a) $26^3 = 17,576$ (b) yes   (c) 26·25·24 = 15,600
13. 72

Exercise Set 6–4

1. (a) 20   (b) 10
3. (a) 30   (b) 45
5. (a) 1   (b) 1
7. (a) 1   (b) 120
9. 90
11. $13{,}800 = 25 \cdot 24 \cdot 23$
13. 12
15. 720
17. 48
19. 15,504
39. 220 (10 with 3 digits the same, 90 with exactly two digits the same, and 120 with three distinct digits)
41. (a) 256   (b) 24
43. (a) 21   (b) 15   (c) 30
45. 72

21. (a) 21   (b) 42
23. 24
25. 144
27. 64
29. (a) 792   (b) 120
31. 1950
33. 845,000
35. 1000
37. 10,000
47. (a) because $5 + 9 = 14$
     (b) because $r + (n - r) = n$

Referenced Exercise Set 6–4

1. $8! = 40{,}320$
3. A list of the $6 = C\,(4,\,2)$ ways of choosing two numbers, without regard to order, from the set $\{1, 2, 3, 4\}$.

Cumulative Exercise Set 6–4

1. (a) yes   (b) no
3. 9
5. $(H, 1)$, $(H, 2)$, $(H, 3)$, $(H, 4)$, $(H, 5)$, $(H, 6)$ $(T, 1)$, $(T, 2)$, $(T, 3)$, $(T, 4)$, $(T, 5)$, $(T, 6)$

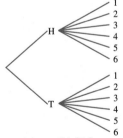

7. (a) 16   (b) 216
9. $10^6 = 1{,}000{,}000$
11. (a) 1320   (b) 220
13. (a) 1   (b)252   (c) 90

Exercise Set 6–5

1. $x^5 + 5x^4y + 10x^3y^2 + 10x^2y^3 + 5xy^4 + y^5$
3. $a^4 + 4a^3b + 6a^2b^2 + 4ab^3 + b^4$
5. $a^6 + 6a^5b + 15a^4b^2 + 20a^3b^3 + 15a^2b^4 + 6ab^5 + b^6$
7. $8x^3 + 12x^2y + 6xy^2 + y^3$

9. $x^9 + 9x^8y + 36x^7y^2 + 84x^6y^3$

11. $a^{10} + 10a^9b + 45a^8b^2 + 120a^7b^3$

13. 36

15. 462

25. $x^{12} + 18x^{10} + 135x^8 + 540x^6 + 1215x^4 + 1458x^2 + 729$

17. 16

19. 127

21. $16x^4 - 96x^3y + 216x^2y^2 - 216xy^3 + 81y^4$

23. $a^8 + 4a^6b + 6a^4b^2 + 4a^2b^3 + b^4$

27. The number of subsets having 0, 1, 2, 3, 4, 5 elements is $\binom{5}{0}, \binom{5}{1}, \binom{5}{2}, \binom{5}{3}, \binom{5}{4}, \binom{5}{5}$, respectively. Therefore

the number of subsets is $\binom{5}{0} + \binom{5}{1} + \binom{5}{2} + \binom{5}{3} + \binom{5}{4} + \binom{5}{5} = 2^5 = 32.$

29. 64

31. 4096

33. 16

35. 64

37. 1

Cumulative Exercise Set 6–5

1. (a) no   (b) no

3. 5

5. $\{(0, 1), (0, 2), (0, 3), (0, 4), (0, 5), (0, 6), (1, 1), (1, 2), (1, 3), (1, 4), (1, 5), (1, 6), (2, 1), (2, 2), (2, 3), (2, 4), (2, 5), (2, 6)\}$

7. (a) 21   (b) $6^4 = 1296$

9. (a) $5^4 = 625$ (b) $5! = 120$

Case Study Exercises

1. 1,352,000

3. 16,208,000

5. 12,000,000

7. no

Review Exercises

1. (a) not well-defined   (b) well-defined

3. (a) $A' = \{f, g, h, i, j, \ldots, z\}$ (b) $V' \cap A' = \{f, g, h, j, k, l, m, n, p, q, r, s, t, v, w, x, y, z\}$
   (c) $(V \cup A)' = V' \cap A'$

5. 10

7. X' is the set of irrational numbers.

9. (a) 3   (b) 17   (c) 13   (d) 22

11. 324

13. (a) 20,736   (b) 11,880

15. 40,320

17. 635,376

19. (a) 6,840   (b) 680

21. 15,600

CHAPTER 7

Exercise Set 7–1

1. $S = \{\text{odd, even}\}$

3. $S = \{p, r, o, b, a, i, l, t, y\}$

5. $S = \{\text{red, black}\}$

7. $S = \{\text{Democratic, Republican, Independent}\}$

9. let g = gained weight; l = lost weight; s = stayed the same
  S = {gg, gl, gs, lg, ll, ls, sg, sl, ss}
11. let i = increases, d = decreases; u = unchanged
  S = {iiii, iiid, iiiu, iidi, iidd, iidu, . . .}
  There are 81 possible combinations in the sample space.
13. let c = corn; p = potatoes; a = alfalfa
  S = {cpa, cap, pca, pac, apc, acp}
15. $E \cup F$ = {1, 2, 3, 4}; $E \cap F$ = {2, 3}; E' = {4, 5, 6}
17. $E \cup F$ = {p, r, o, b, a, i, l, t, y}; $E \cap F$ = $\emptyset$; E' = {p, r, b, l, t, y}
19. $E \cup F$ = {red card}; $E \cap F$ = {heart}; E' = {club, spade, diamond}
21. $E \cup F$ = {2, 3, 4, 5, 6, 7}; $E \cap F$ = {3, 4}; E' = {5, 6, 7, 8, 9, 10, 11, 12}
23. $E \cup F$ = {0 heads, 1 head, 2 heads}; $E \cap F$ = {1 head}; E' = {2 heads}
25. yes
27. no
29. S = {AB, AC, AR, AS, AT, BC, BR, BS, BT, CR, CS, CT, RS, RT, ST};
  E = {AR, AB, AC, AS, AT, BC, BR, BS, BT, CR, CS, CT};   F = {CR, CS, CT, RS, RT, ST};
  G = {AC, AS, BC, BS, CR, CS, CT, RS, ST}
31. (a) and (b) $E \cup F$ = S = {yyy, yyn, yny, ynn, nyy, nyn, nny, nnn}
  (c) $E \cap F$ = {yyn, yny, nyy}
  (d) E' = {ynn, nyn, nny, nnn}
33. (a) $E \cap F$ (b) $E \cap F'$ (c) E' (d) $E \cup F$
35. (a) {0, 1, 2, 3}, E = {2, 3}, F = {0, 2}   (b) {0, 2, 3},   (c) {2},   (d) {0, 1}
37. yes, $E \cap E'$ = $\emptyset$
39. (a) $\emptyset$, {1}, {2}, {3}, {1, 2}, {2, 3}, {1, 3}, S
  (b) $\emptyset$ {1}, {2}, . . . , {n}, {1, 2}, . . . , {1, n}, {2, 3}, . . . , {2, n}, . . . , S

Referenced Exercise Set 7–1

1. The patient had (a) influenza or took aspirin, (b) took aspirin or had an a-d diet, (c) had influenza and an a-d diet, (d) took aspirin and had an a-d diet, (e) did not have influenza, (f) had neither influenza nor took aspirin, (g) had influenza, took aspirin, had an a-d diet.
3. (a) {insects, disease, weather}
  (b) $\frac{1}{3}$
  (c) insects and disease, $\frac{1}{4}$; weather, $\frac{1}{2}$
  (d) insects, $\frac{4}{7}$; disease, $\frac{2}{7}$; weather, $\frac{1}{7}$

Exercise Set 7–2

1. $\frac{1}{6}$
3. $\frac{1}{2}$
5. $\frac{2}{3}$
7. $\frac{1}{2}$
9. $\frac{3}{4}$
11. .43
13. .65
15. $\frac{3}{4}$
17. $\frac{3}{52}$
19. $\frac{25}{26}$

21. yes
23. no, the sum of the probabilities = 1.1; must = 1
25. (a) $\frac{1}{2}$ (b) $\frac{1}{2}$
27. $\frac{9}{19}$
29. $\frac{3}{10}$
31. (a) $\frac{2}{13}$ (b) $\frac{1}{13}$ (c) $\frac{1}{2}$
33. $Pr(E)$ = .83; $Pr(F)$ = .17
35. $Pr(E)$ = .45; $Pr(F)$ = .2, $Pr(G)$ = .35
37. $Pr(E')$ = .55; $Pr(F')$ = .8
39. .7

41. (a) $\frac{10}{50}$ (b) $\frac{16}{50}$ (c) $\frac{40}{50}$

43. $\frac{8}{10}$

45. (a) $\frac{1}{3}$ (b) $\frac{2}{3}$

47. $\frac{1}{2}$

49.

x	a	b	c
p	$\frac{6}{11}$	$\frac{3}{11}$	$\frac{2}{11}$

Referenced Exercise Set 7–2

1. (a) $\dfrac{8442}{12{,}763}$ (b) $\dfrac{4321}{12{,}763}$ (c) $\dfrac{5232}{12{,}763}$ (d) $\dfrac{7531}{12{,}763}$

3. .03, .29

Cumulative Exercise Set 7–2

1. (a) {Sun, Mon, Tue, Wed, Thu, Fri, Sat}
(b) {Jan, Feb, Mar, Apr, May, Jun, Jul, Aug, Sep, Oct, Nov, Dec}
(c) {0, 1, 2, 3, 4}

3. $\frac{1}{2}$

5. $\frac{7}{13}$

7. $\frac{6}{10}$

Exercise Set 7–3

1. $\frac{10}{36} = \frac{5}{18}$ **13.** .75

3. $\frac{2}{3}$ **15.** .33

5. $\frac{8}{13}$ **17.** .11

7. $\frac{4}{13}$ **19.** $\frac{4}{11}$

9. $\frac{2}{3}$ **21.** $\frac{2}{7}$

11. 1

23. odds in favor of E = 2 to 7; odds against F = 4 to 3

25. odds in favor of E = 13 to 37; odds against F = 31 to 19

27. 1   **37.** .78

29. (a) .32   (b) .23   **39.** .3

31. (a) $\frac{9}{19}$ (b) $\frac{9}{19}$ (c) $\frac{18}{19}$ (d) $\frac{10}{19}$ **41.** $\frac{85}{200}$

33. (a) 7 to 13   (b) 7 to 13   **43.** $\frac{65}{200}$

35. .63

Referenced Exercise Set 7–3

1. .68 and .44   **5.** (a) yes
(b) yes

3. .88   (c) Yes, it seems reasonable, *but* . . .
(d) No, the probability of player A winning is $\frac{3}{4}$.

Cumulative Exercise Set 7–3

1. $\{(1, H), (2, H), (3, H), (4, H), (5, H), (6, H), (1, T), (2, T), (3, T), (4, T), (5, T), (6, T)\}$
3. (a) $\{0, 2, 3\}$ (b) $\{2\}$ (c) $\{0, 1\}$ **9.** (a) 0.29   (b) 0.68 and 0.32   (c) 1
5. (a) 0.8   (b) 0.6   (c) 0.3   (d) 1   **11.** (a) 0.63   (b) 0.94
7. (a) $\frac{1}{6}$ (b) $\frac{5}{18}$ (c) $\frac{1}{2}$

Exercise Set 7–4

1. $\frac{6}{64} \simeq .09$
3. $\frac{22}{64} \simeq .34$
5. $\frac{15}{16} \simeq .9375$
7. $\frac{7}{99} \simeq .07$
9. $\frac{42}{99} \simeq .424$
11. $\frac{10}{11} \simeq .91$
13. $\frac{35}{38} \simeq .92$

15. $1/204 \simeq .0049$
17. $226/2118760 \simeq .00011$
19. $12/51 \simeq .24$
21. $22/425 \simeq .052$
23. $62/125 \simeq .496$
25. $2093/5000 \simeq .4186$

27. (a) $15/128 \simeq .117$ (b) $11/64 \simeq .1719$ (c) $121/128 \simeq .9453$
29. (a) $14/323 \simeq .0433$ (b) $94/323 \simeq .291$ **41.** $\frac{32}{56} = \frac{4}{7}$
31. (a) .0002401   (b) .00394   **43.** (a) $\frac{7}{64}$
33. .6032   (b) $\frac{113}{128}$
35. $110/143 \simeq .77$ (c) $\frac{1}{2}\left(\frac{182}{240} + \frac{3192}{4032}\right) = \frac{31}{40}$
37. .856
39. $\frac{3}{8}$ **45.** $\dfrac{60{,}460}{1{,}048{,}576}$

Referenced Exercise Set 7–4

1. $\dfrac{3}{91}$ **3.** $\dfrac{19}{989}$

Cumulative Exercise Set 7–4

1. (a) $\{0, 1, 2\}$ **11.** (a) $\frac{15}{45} = \frac{1}{3}$
     (b) {less than 2, at least 2}   (b) $\frac{48}{90} = \frac{8}{15}$
     (c) $\{0, 1, 2, 3\}$ (c) $\frac{6}{45} = \frac{2}{15}$
3. $\frac{3}{8}$
5. $\frac{1}{13}$ **13.** (a) $1 - \dfrac{19}{20} \cdot \dfrac{18}{20} \cdot \dfrac{17}{20} \cdots \dfrac{11}{20} \simeq 0.9345$
7. $\frac{35}{36}$
9. $\frac{264}{560} = \frac{33}{70}$ (b) $1 - \left(\dfrac{19}{20}\right)^9 \simeq 0.3698$

Exercise Set 7–5

1. $\frac{1}{3}$
3. $\frac{2}{13}$
5. $\frac{7}{25}$
7. $\frac{7}{20}$

9. .9366 . . .
11. $\frac{2}{3}$
13. $\frac{64}{67} \simeq .96$
15. $\frac{19}{83} \simeq .23$

17. not independent, 7/18

19. independent, 1/2

21. .005

23. $Pr(E \mid F') = \frac{1}{3}$; $Pr(E' \mid F') = \frac{2}{3}$

25. $79/180 \approx .44$

35. $Pr(F \mid E) = Pr(F \cap E)/Pr(E) = Pr(F)Pr(E)/Pr(E) = Pr(F)$

37. (b) Use $Pr(E \cap F) + Pr(E' \cap F) = Pr(F)$
(a) and (c), Use Example 1 to disprove.

39. Use the addition principle, the rule for complements and two factorizations.

41. $\frac{12}{40}$

43. $\frac{15}{20}$

27. (a) $\frac{1}{7}$ (b) $\frac{12}{35}$ (c) $\frac{3}{7}$

29. (a) 8%   (b) 36%

31. $\frac{18}{43} \approx .42$

33. $\frac{12}{57} \approx .21$

Referenced Exercise Set 7–5

1. (a) $\frac{3}{13}$ (b) $\frac{4}{26} = \frac{2}{13}$ (c) $\frac{17}{26}$

3. .000000736, $Pr(E) \neq Pr(E \mid C)$, not independent

5. (a) $\frac{234}{365}$ (b) $\left(\frac{234}{365}\right)^3 \approx .26$

Cumulative Exercise Set 7–5

1. (a) {0,2,4,5}   (b) {4}   (c) {0,1,2,3}

3. (a) 0.30   (b) 0.55

5. 0.143

7. (a) $\frac{2}{13}$ (b) $\frac{3}{4}$

9. No, $Pr(E|F) = 12/14$.

11. $\frac{3}{13}$

Case Study Exercises

1. $\frac{19}{33}$

3. $\frac{14}{33}$

5. $\frac{14}{17}$

7. $\frac{5}{70} = \frac{1}{14}$

9. $\frac{20}{30} = \frac{2}{3}$

11. $\frac{10}{15} = \frac{2}{3}$

Exercise Set 7–6

1. $\frac{9}{13} \approx .69$

3. $\frac{2}{23} \approx .087$

5. $\frac{4}{27} \approx .15$

7. $Pr(U \cap E) = .18$; $Pr(H \cap E) = .56$

9. $Pr(U \mid E) = \frac{9}{37} \approx .24$; $Pr(H \mid E) = \frac{28}{37} \approx .76$

11. $Pr(U \cap E) = .18$; $Pr(H \cap E) = .18$; $Pr(W \cap E) = .35$

13. $\frac{18}{71} \approx .25$

15. $\frac{35}{71} \approx .49$

17. $\frac{12}{29} \approx .41$

19. $\frac{5}{13} \approx .38$

21. $\frac{2}{13} \approx .15$

23. $\frac{24}{59} \approx .41$

25. $\frac{20}{73} \approx .27$

27. $25/107 \approx .23$

29. $\frac{3}{11} \approx .27$

31. $\frac{2}{5}$

33. .276

35. $60/166 \approx .36$

37. $\frac{16}{25}$

39. $\frac{7}{23}$

Referenced Exercise Set 7–6

1. $\frac{55}{92}$

3. $\dfrac{51}{157}$

Cumulative Exercise Set 7–6

1. (a) {0, 1, 2, 3}

(b) {yyyy, yyyn, yyny, yynn, ynyy, ynyn, ynny, ynnn, nyyy, nyyn, nyny, nynn, nnyy, nnyn, nnny, nnnn}

3. $\frac{16}{52} = \frac{4}{13}$

5. (a) 0.8   (b) 0.3

7. $\frac{13}{18}$

9. (a) $\frac{2}{46}$ (b) $\frac{6}{50}$

11. $\frac{11}{25}$

13. 0

15. (a) $\frac{1}{3}$ (b) $\frac{11}{24}$ (c) $\frac{127}{660} \simeq 0.1924$

Exercise Set 7–7

1. $\frac{15}{64}$

3. $\frac{22}{64} = \frac{11}{32}$

5. $\frac{28}{256} = \frac{7}{64}$

7. $\frac{15}{1024}$

9. $\frac{19}{4096}$

11. $\frac{4}{625}$

13. $\frac{80}{243}$

15. $\frac{224}{729}$

17. $\frac{11}{243}$

19. $\frac{3072}{6561}$

21. 0.31744

23. $\simeq 0.4059$

25. $\simeq 0.2465$

27. (a) $\simeq 0.2936$ (b) $\simeq 0.7969$

29. (a) $\frac{15}{128}$ (b) $\frac{11}{64}$

31. (a) $\simeq 0.000000001586$ (b) $\simeq 0.000000001605$

33. (a) $\dfrac{1,125}{60,466,176}$ (b) $\dfrac{1,176}{60,466,176}$

35. (a) $\frac{90}{1024}$ (b) $\frac{1008}{1024}$

37. $\dfrac{14,985}{15,625} \simeq 0.959$

39. $\simeq 0.8733$

41. (a) $\dfrac{17,578,125}{60,466,176}$ (b) $\dfrac{46,875,000}{60,466,176}$

43. $\simeq 0.0755$

Referenced Exercise Set 7–7

1. $\simeq 0.0017$

Cumulative Exercise Set 7–7

1. (a) $\frac{5}{9}$ (b) $\frac{1}{18}$

3. $\frac{3}{4}$

5. $\frac{35}{75}$

7. (a) $15,000/6^{10} \simeq 0.00025$

(b) $1 - (1176/6^{10}) \simeq 0.999981$

9. (a) $196,830/1,048,576 \simeq 0.1877$

(b) $551,124/1,048,576 \simeq 0.5256$

11. (a) $\frac{4}{10}$ (b) $\frac{13}{40}$

Exercise Set 7–8

1. 25

3. 2

5. 2.4

7. $\frac{2}{13}$, no

9. $\frac{4}{9}$, no

11. $\frac{1}{3}$, no

13. Change 4 to 2.

15. Site 2

17. Property A

19. Property B

33. $19,250; with averages of 5,000, 15,000, 25,000, and 40,000

35. 1.0

37. .3

21. 3.5

23. .2

25. 1.25

27. 3

29. $-.499$

31. $-.588$

39. .36

41. $46/440 \approx 0.1045$

Referenced Exercise Set 7–8

1. $E(\text{American}) = -\frac{1}{19}$

x	$-.5$	-1	0	1
p	$\dfrac{1}{1369}$	$\dfrac{684}{1369}$	$\dfrac{18}{1369}$	$\dfrac{18}{37}$

$$E\,(\text{European}) = -\frac{18.5}{1369}$$

3. .675, .553

Cumulative Exercise Set 7–8

1. (a) $\frac{15}{36}$ (b) $\frac{5}{36}$

3. $1 - (30/40)(29/39)(28/38) \approx 0.589$

5. $\frac{17}{40}$

7. 50

9. (a) $\frac{45}{1024}$ (b) $\frac{386}{1024}$

11. $17,250, using averages of 5, 15, 25, and 40 thousand dollars

13. (a) $\frac{1}{12}$ (b) $\frac{1}{144}$

Case Study Exercises

1. .7   **3.** .5   **5.** 218.04

Review Exercises

1. {m, a, t, h, e, i, c, s}

3. (a) $E \cup F = S = \{\text{red, blue}\}$ (b) $E \cap F = E = \{\text{blue}\}$ (c) $E' = \{\text{red}\}$

5. (a) $\frac{1}{13}$ (b) $\frac{1}{26}$ (c) $\frac{3}{26}$

7. (a) .1   (b) .9

9. $Pr(E \cap F) = .14$; $Pr(E' \cup F') = .86$

11. $\frac{6}{13} \approx .46$

13. $79/4096 \approx 0.019$

15. $\frac{13}{25} = .52$

17. no

19. $\frac{1}{7}$

21. 75

CHAPTER 8

Exercise Set 8–1

1. .2856

3. .4283

5. $\frac{3}{8}$

7. $\frac{1}{2}$

9.

11.

13.

Heads

Heads

Heads

15.

Nonusers

Users

Blood pressure

Blood pressure

17. There are more lower (100–120) blood pressures among the nonusers group. The pill may elevate blood pressure.

19.

Freq.

18 21 24 27 30

21.

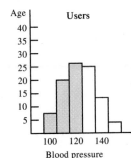

$Pr(X \geq 1) = \frac{15}{16}$

0 1 2 3 4

23.

$Pr(6 \leq X \leq 8) = \frac{4}{9}$

2 3 4 5 6 7 8 9 10 11 12

25.

0 1 2 3

27. $\frac{19}{24} \simeq .79$

29. $\frac{7}{24} \simeq .29$

31.

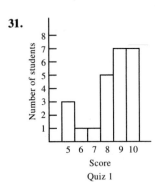
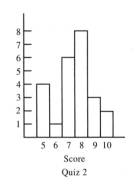

33. 0.7
35. 0.7
37. 0.5
39. 0.3

Referenced Exercise Set 8–1

1. Attendance increased each year until 1980–1981 and has decreased since then. Ticket prices increased throughout the time period.

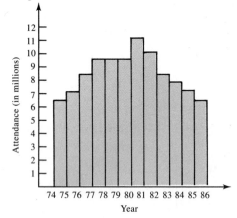

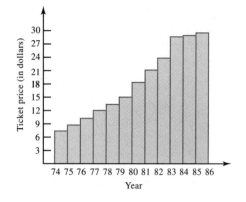

3. Each histogram is close to a normal curve.

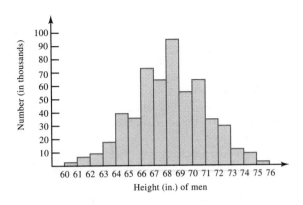

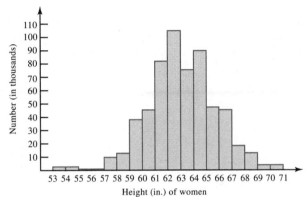

5. Fear of 13
7. no
9. No. The sum of the probabilities is only .95.

Exercise Set 8–2

1. mean $= 161/10 = 16.1$;   median $= 8$
3. mean $= .32/6 \simeq .053$;   median $= 0.055$
5. mean $= \frac{23}{14} \simeq 1.64$;   median $= 1.5$
7. mean $= \frac{47}{7} \simeq 6.7$;   median $= 9$
9. $E(X) = \frac{31}{8}$
11. $E(X) = 3.3$
13. mean $= \frac{37}{22} \simeq 1.68$
15. expected value $= \frac{13}{8} = 1.625$
17. mean $= -.5$;   median $= 0$
19. mean $= 6.695$;   median $= 7.3$

21. mean $= 0$;   median $= 0$
23. $E(X) = -.1$
25. $E(X) = -2$
27. $E(X) = 1.5$
29. mean $= 230.3$ pounds;   median $= 225$ pounds
31. mean $= \frac{43}{13} \simeq 3.3$
33. (a) Each entry is 0   (b) Each entry is 2   (c) no
35. (a) mean $= 4.3/11 \simeq 0.39$ (b) reward
37. mean $= 4.6$;   median $= 4$
39. 1.6

Referenced Exercise Set 8–2

1. (a) 12.09   (b) 8   (c) people argued for median, bank argued for mean
(d) the payment via the median is $(20)(\frac{7}{8})(\$9500) = \$166,250$.
the payment via the mean is $(20)(\frac{7}{12})(\$9500) = \$110,833.33$.
the difference is $55,416.67

3. The expected number is $-.2$ so the expected death month is very close to the birthday.
5. mean is 62.57 inches, median is 62 inches
7. 5.908 (or about 6 games)
9. (a) 1   (b) 1.2   (c) yes

Cumulative Exercise Set 8–2

1.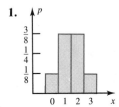

3.

X	2	3	4	5	6	7	8	9	10	11	12
p	$\frac{1}{36}$	$\frac{2}{36}$	$\frac{3}{36}$	$\frac{4}{36}$	$\frac{5}{36}$	$\frac{6}{36}$	$\frac{5}{36}$	$\frac{4}{36}$	$\frac{3}{36}$	$\frac{2}{36}$	$\frac{1}{36}$

$Pr(6 \le X \le 9) = \frac{5}{9}$

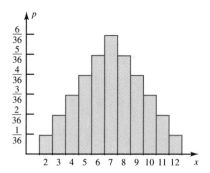

5. 3.2, 2
7. 0.65

Case Study Exercises

1. The mean family size in the Lenin study is $\bar{x} = \dfrac{175.1}{24} \approx 7.3$. The mean family size in the Shcherbina study is $\bar{x} = \dfrac{225,071.3}{33,581} \approx 6.7$.

3. The mean number of cattle owned per household in the Lenin study is $\bar{x} = \dfrac{185.1}{24} \approx 7.7$. The mean number of cattle owned per household in the Shcherbina study is $\bar{x} = \dfrac{147,757.7}{33,581} \approx 4.4$.

5. (a) $\bar{x} = 12.4$ (c) $\bar{x} \approx 8.3$
   (b) $\bar{x} \approx 5.1$ (d) $\bar{x} \approx 5.8$

Exercise Set 8–3

1. $V(X) = 5.05$ $\sigma = 2.25$
3. $V(X) = 11.89$ $\sigma = 3.45$
5. $V(X) = 107.4$ $\sigma = 10.36$
7. $V(X) = 10{,}229/900 = 11.37$ $\sigma = 3.37$
9. $V(X) = 12{,}848/1764 \approx 7.28$ $\sigma = 2.7$
11. $V(X) = 151/64 \approx 2.36$ $\sigma = 1.54$
13. $V(X) = .61$ $\sigma = .78$
15. $\$13{,}800$ $\sigma = 3763$
17. $Pr(40 \le X \le 60) = .75$
19. $V(X) = .108$ $\sigma = .329$
21. $V(X) = .836$ $\sigma = .914$

23. $V(X) = 2883.796$ $\sigma = 53.701$
25. $V(X) = 3.29$ $\sigma = 1.814$
27. $V(X) = 41$ $\sigma = 6.403$
29. $\sigma = 1.26$
31. $\sigma = .992$
33. $\sigma = \sqrt{3}/2 \approx .87$
35. (a) $\mu = 1.87$ (b) $\sigma = 1.26$
37. (a) 2   (b) 2   (c) 2   (d) 2
39. (a) 2   (b) 1
41. $\frac{8}{9}$
45. $\sqrt{n}/2$

Referenced Exercise Set 8–3

1. 3.4

3.

	EF	TS	C	AG
mean	204.1	200.3	198.6	200.0
σ	11.6	8.5	6.4	10.4

This method of comparison is not conclusive; a more exhaustive analysis must be performed.
5. (a) 3.87%   (b) 65.84%   (c) the Standard Line
7. The authors are not convinced.
9. 2

Cumulative Exercise Set 8–3

1. (a)

x	f
0	1
1	4
2	6
3	4
4	1

(b)

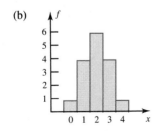

3.

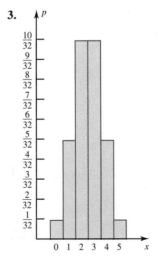

5. (a) $\frac{201}{24} = 8.375$ (b) 9

7. (a) $\frac{2076}{34} \simeq 61.06$ (b) 61

9. 6, 6, 6, 6, 7

11. 2

Exercise Set 8–4

1. about .475

3. about .34

5. about .16

7. about .025

9. .0668

11. .6826

13. 44.35%

15. 11.51%

17. 24.17%

19. 59.87%

21. about .135

23. about .815

25. about .84

27. .6480

29. .6541

31. 27.61%

33. 13.57%

Referenced Exercise Set 8–4

1. 38.21%

3. .3812

5. 0.99% (less than 1%)

7. (a) 9.18%   (b) 84.13%

9. (a) .8413   (b) .7088

Cumulative Exercise Set 8–4

1.

x	0	1	2	3	4	5	6
p	$\frac{1}{64}$	$\frac{6}{64}$	$\frac{15}{64}$	$\frac{20}{64}$	$\frac{15}{64}$	$\frac{6}{64}$	$\frac{1}{64}$

$Pr(0 \leq X \leq 2) = \frac{22}{64}$

3. 6.4

5. $\sigma = \sqrt{6216}/25 \simeq 3.1537$

7. expected value $= 82.5$, $\sigma = \sqrt{18.75} \simeq 4.33$

9. (a) .9573   (b) .9812   (c) .9385

11. 37.38%

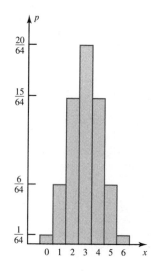

Exercise Set 8–5

1. $\mu = 72$ $\sigma = 6$
3. $\mu = 40.5$ $\sigma = 4.5$
5. .1841
7. .4041
9. .1210
11. .6757
13. .9370
15. .1635
17. .1841
19. .7286
21. .9945
23. .9328

25. .7088
27. .6886
29. .8485
31. .2643
33. .9576
35. .9279
37. .5516
39. virtually zero
41. .1788
43. 0.2420
45. 0.3830

Referenced Exercise Set 8–5

1. (a) .0028   (b) 166

3. (a) .9987   (b) .0013

Cumulative Exercise Set 8–5

1. (a)

X	p
0	$\frac{1}{8}$
1	$\frac{3}{8}$
2	$\frac{3}{8}$
3	$\frac{1}{8}$

(b)

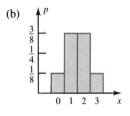

3. expected value $= 0$
5. (a) $\bar{x} = \frac{12}{8}$ (b) $\sigma = \sqrt{0.75}$
7. 0.6247
9. (a) 30   (b) 5
11. 0.1841

Case Study Exercises

1.

	H	M	D	
$\bar{x}$	.44	.73	.80	Not strong enough
σ	.33	.59	.35	

3.

	H	M	D	
$\bar{x}$	7.04	11.52	4	Not strong enough
σ	2.43	3.44	1.73	

5.

	H	M	D	
$\bar{x}$	3.46	7.68	8.5	Madison
σ	1.74	2.70	2.47	

7.

	H	M	D	
$\bar{x}$	7.67	5.96	6.83	Not strong enough
σ	2.13	2.16	2.07	

9.

	H	M	D	
$\bar{x}$	.58	1.11	.77	Hamilton
σ	.43	.83	.45	

Review Exercises

1.

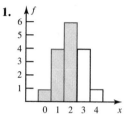

3.

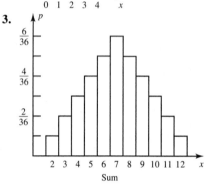

5.

X	probability
3	0.0129
2	0.1376
1	0.4359
0	0.4135

7. .9871

9. mean = 2.4;   median = 2

11. $-\frac{1}{4}$

13. (a) $15,000   (b) $17,916.67

15. average cost = $148.90; $\sigma = 37.31$

17. (a) .6924   (b) .8531

19. .0668

21. $.1294 = C(18, 14)(2/3)^{14}(1/3)^4$

23. $\sigma = 2$

25. 1200

27. .9525

CHAPTER 9

Exercise Set 9–1

1. $T = \begin{bmatrix} .85 & .37 \\ .15 & .63 \end{bmatrix}$

3. This matrix can be a transition matrix.

5. This matrix can be a transition matrix.

7. $a = .5,$ $b = .1,$ $c = .2$

9. (a) .8   (b) .4

11. This could be a probability distribution.

13. This matrix could not be a probability distribution since the sum of the column entries is not 1.

15.

Stage 1	Stage 2	Stage 3
$\begin{bmatrix} .3 \\ .7 \end{bmatrix}$	$\begin{bmatrix} .38 \\ .62 \end{bmatrix}$	$\begin{bmatrix} .348 \\ .652 \end{bmatrix}$

17.

Stage 1	Stage 2	Stage 3
$\begin{bmatrix} .42 \\ .58 \end{bmatrix}$	$\begin{bmatrix} .332 \\ .668 \end{bmatrix}$	$\begin{bmatrix} .3672 \\ .6328 \end{bmatrix}$

19.

Stage 1	Stage 2	Stage 3
$\begin{bmatrix} .46 \\ .54 \end{bmatrix}$	$\begin{bmatrix} .316 \\ .684 \end{bmatrix}$	$\begin{bmatrix} .3736 \\ .6264 \end{bmatrix}$

21. .3

23. .7

25. .25

27. $T = \begin{bmatrix} .8 & .6 \\ .2 & .4 \end{bmatrix}$

29. $T = \begin{bmatrix} .1 & .45 & .55 \\ .3 & .05 & .35 \\ .6 & .50 & .10 \end{bmatrix}$

31. $T = \begin{bmatrix} .90 & .2 & .3 \\ .05 & .7 & .1 \\ .05 & .1 & .6 \end{bmatrix}$

33. $T = \begin{bmatrix} .10 & .10 & .65 \\ .45 & .60 & .30 \\ .45 & .30 & .05 \end{bmatrix}$

35. $\begin{bmatrix} .60 \\ .26 \\ .14 \end{bmatrix}$

37. $\begin{bmatrix} .37 \\ .39 \\ .24 \end{bmatrix}$

39. $\begin{bmatrix} .320 \\ .435 \\ .245 \end{bmatrix}$

41. $T^2 = \begin{bmatrix} .22 & .42 & .30 \\ .44 & .32 & .52 \\ .34 & .26 & .18 \end{bmatrix}$ $T^2V = \begin{bmatrix} .304 \\ .492 \\ .204 \end{bmatrix}$ $TV = \begin{bmatrix} .60 \\ .26 \\ .14 \end{bmatrix}$ $T(TV) = \begin{bmatrix} .304 \\ .492 \\ .204 \end{bmatrix}$

Referenced Exercise Set 9–1

1. (a) .24   (b) $\begin{bmatrix} .284 \\ .233 \\ .331 \\ .152 \end{bmatrix}$

3. (a) .05   (b) Only those who were transferred were studied.   (c) $\begin{bmatrix} .28 \\ .16 \\ .28 \\ .11 \\ .08 \\ .09 \end{bmatrix}$

Exercise Set 9–2

1. $T^1V = \begin{bmatrix} .66 \\ .34 \end{bmatrix}$ $T^2V = \begin{bmatrix} .696 \\ .304 \end{bmatrix}$ $T^3V = \begin{bmatrix} .7176 \\ .2824 \end{bmatrix}$

3. $T^1V = \begin{bmatrix} .35 \\ .65 \end{bmatrix}$ $T^2V = \begin{bmatrix} .305 \\ .695 \end{bmatrix}$ $T^3V = \begin{bmatrix} .2915 \\ .7085 \end{bmatrix}$

5. $T^6V \simeq \begin{bmatrix} .743 \\ .257 \end{bmatrix}$

7. $T^6V \simeq \begin{bmatrix} .286 \\ .714 \end{bmatrix}$

9. T is regular since $T^2 = \begin{bmatrix} .1 & .16 & .10 \\ .1 & .64 & .16 \\ .8 & .20 & .74 \end{bmatrix}$. Thus one of the powers of T contains all positive entries.

11. Not regular, the $(1, 3)$ entry will be 0 in all T^n.

13. $\begin{bmatrix} \frac{5}{11} \\ \frac{6}{11} \end{bmatrix}$

15. $\begin{bmatrix} \frac{3}{4} \\ \frac{1}{4} \end{bmatrix}$

17. $\begin{bmatrix} \frac{2}{7} \\ \frac{5}{7} \end{bmatrix}$

19. $\begin{bmatrix} \frac{1}{9} \\ \frac{2}{9} \\ \frac{2}{3} \end{bmatrix}$

21. $\begin{bmatrix} 45/117 \\ 40/117 \\ 32/117 \end{bmatrix}$

23. The retail store will have $\frac{5}{7}$ or approximately 0.71 of the business, while the competitor will have $\frac{2}{7}$ or approximately .29 of the business. Ad campaign will increase business by approximately 6%.

25. Stable matrix $= \begin{bmatrix} \frac{25}{95} \\ \frac{70}{95} \end{bmatrix}$. So in the long run it will rain about 26% (or $\frac{25}{95}$) of the time and not rain 74% of the time on the next day regardless of the previous day's weather.

27. In the long run a stock will be purchased 38% of the time, sold 36% of the time, and left unchanged 26% of the time.

29. In the long run the retailer will use TV 36% of the time, radio 26% of the time, and newspapers 38% of the time.

31. In the long run 75% of the debts become high priority, while 25% of the debts become low priority.

33. In the long run Brand A has a 71% share of the market, while Brand B has a 16% share and Brand C has a 13% share of the market.

35. In the long run the order will be large 24.8% (116/467) of the time, medium 48.2% (225/467) of the time, and small 27% (126/467) of the time.

37. The retail store will have $\frac{2}{3}$, or 67%, of the business, while the competitor will have $\frac{1}{3}$ of the business. The ad campaign will increase business by approximately 2%.

39. In the long run 15% of the debts become high priority, 45% of the debts become medium priority, and 40% of the debts become low priority.

Referenced Exercise Set 9–2

1. $\begin{bmatrix} \frac{507}{706} \\ \frac{180}{706} \\ \frac{19}{706} \end{bmatrix}$

3. $\begin{bmatrix} .27 \\ .45 \\ .28 \end{bmatrix}$

Cumulative Exercise Set 9–2

1. $\begin{bmatrix} .50 & .30 & .05 \\ .40 & .45 & .35 \\ .10 & .25 & .60 \end{bmatrix}$

3. (a) 0.1   (b) 0.2   (c) 0.5

5. $\begin{bmatrix} .6 \\ .4 \end{bmatrix}$

7. $\begin{bmatrix} .6 \\ .4 \end{bmatrix}$

9. $\begin{bmatrix} \frac{24}{71} \\ \frac{26}{71} \\ \frac{21}{71} \end{bmatrix}$

Exercise Set 9–3

1. All three states are nonabsorbing.

3. State 4 is absorbing; the others are nonabsorbing.

5. absorbing

7. Nonabsorbing, since it does not have an absorbing state.

9. absorbing

11. Nonabsorbing. State 1 can only go to state 3, which can only go to state 1. So state 1 cannot go to state 4, which is the absorbing state.

13. $T^4 = \begin{bmatrix} 1 & 0 & 0 \\ 0 & 1 & .9999 \\ 0 & 0 & .0001 \end{bmatrix}$ Stable matrix $= \begin{bmatrix} 1 & 0 & 0 \\ 0 & 1 & 1 \\ 0 & 0 & 0 \end{bmatrix}$

15. $\begin{bmatrix} 1 & .3 & 0 \\ 0 & .2 & 1 \\ 0 & .5 & 0 \end{bmatrix}$

17.

$$
\begin{array}{c}
 & 3 & 4 & 1 & 2 \\
\begin{matrix} 3 \\ 4 \\ 1 \\ 2 \end{matrix} &
\begin{bmatrix}
1 & 0 & 0 & .4 \\
0 & 1 & 0 & .1 \\
0 & 0 & .2 & .3 \\
0 & 0 & .8 & .2
\end{bmatrix}
\end{array}
$$

19.

$$
\begin{array}{c}
 & 1 & 3 & 2 & 4 \\
\begin{matrix} 1 \\ 3 \\ 2 \\ 4 \end{matrix} &
\begin{bmatrix}
1 & 0 & .1 & 0 \\
0 & 1 & .2 & 0 \\
0 & 0 & .5 & 1 \\
0 & 0 & .2 & 0
\end{bmatrix}
\end{array}
$$

21. $\begin{bmatrix} \frac{5}{4} & 0 \\ \frac{1}{4} & 1 \end{bmatrix}$ or $\begin{bmatrix} 1.25 & 0 \\ .25 & 1 \end{bmatrix}$

23. $\begin{bmatrix} 1 & 0 & \frac{6}{10} & \frac{23}{35} \\ 0 & 1 & \frac{4}{10} & \frac{12}{35} \\ 0 & 0 & 0 & 0 \\ 0 & 0 & 0 & 0 \end{bmatrix}$

25. $\begin{bmatrix} 1 & 0 & .125 & 0 \\ 0 & 1 & .875 & 1 \\ 0 & 0 & 0 & 0 \\ 0 & 0 & 0 & 0 \end{bmatrix}$

27. $SV = \begin{bmatrix} .35 \\ .65 \\ 0 \end{bmatrix}$ $SW = \begin{bmatrix} .53 \\ .47 \\ 0 \end{bmatrix}$

29.

$$
T = \begin{array}{c}
 & \$0 & \$3 & \$1 & \$2 \\
\begin{matrix} \$0 \\ \$3 \\ \$1 \\ \$2 \end{matrix} &
\left[\begin{array}{cc|cc}
1 & 0 & \frac{1}{3} & 0 \\
0 & 1 & 0 & \frac{2}{3} \\
\hline
0 & 0 & 0 & \frac{1}{3} \\
0 & 0 & \frac{2}{3} & 0
\end{array}\right]
\end{array}
\qquad
S = \begin{array}{c}
 & \$0 & \$3 & \$1 & \$2 \\
\begin{matrix} \$0 \\ \$3 \\ \$1 \\ \$2 \end{matrix} &
\left[\begin{array}{cc|cc}
1 & 0 & \frac{3}{7} & \frac{1}{7} \\
0 & 1 & \frac{4}{7} & \frac{6}{7} \\
\hline
0 & 0 & 0 & 0 \\
0 & 0 & 0 & 0
\end{array}\right]
\end{array}
$$

If Player A starts with \$2, he will end up with \$3 with a probability of $\frac{6}{7}$ or .857.

31.

$$
T = \begin{array}{c}
 & \$0 & \$3 & \$1 & \$2 \\
\begin{matrix} \$0 \\ \$3 \\ \$1 \\ \$2 \end{matrix} &
\left[\begin{array}{cc|cc}
1 & 0 & \frac{3}{5} & 0 \\
0 & 1 & 0 & \frac{2}{5} \\
\hline
0 & 0 & 0 & \frac{3}{5} \\
0 & 0 & \frac{2}{5} & 0
\end{array}\right]
\end{array}
\qquad
S = \begin{array}{c}
 & \$0 & \$3 & \$1 & \$2 \\
\begin{matrix} \$0 \\ \$3 \\ \$1 \\ \$2 \end{matrix} &
\left[\begin{array}{cc|cc}
1 & 0 & \frac{15}{19} & \frac{9}{19} \\
0 & 1 & \frac{4}{19} & \frac{10}{19} \\
\hline
0 & 0 & 0 & 0 \\
0 & 0 & 0 & 0
\end{array}\right]
\end{array}
$$

If Player A starts with \$2, he will end up with \$3 with a probability of $\frac{10}{19}$ or .526.

35. $T^n = \begin{bmatrix} 1 & 0 & .222\ldots \\ 0 & 1 & .777\ldots \\ 0 & 0 & 0 \end{bmatrix}$

Referenced Exercise Set 9–3

1. (a) State 1 is an absorbing matrix. States 2, 3, and 4 go to state 1.

(b) Chicago is an absorbing "state." Illinois, Indiana, and Wisconsin are nonabsorbing states.

(c) $S = \begin{bmatrix} 1 & 1 & 1 & 1 \\ 0 & 0 & 0 & 0 \\ 0 & 0 & 0 & 0 \\ 0 & 0 & 0 & 0 \end{bmatrix}$

3. (a) State L is absorbing, and each of the other states can reach L in one transition.

(b) State L is absorbing, states GC and GE are nonabsorbing.

(c) $\begin{bmatrix} 1 & 1 & 1 \\ 0 & 0 & 0 \\ 0 & 0 & 0 \end{bmatrix}$

Cumulative Exercise Set 9–3

1.

$$\begin{array}{c} \\ C \\ M \\ P \end{array}\begin{array}{ccc} C & M & P \\ \begin{bmatrix} .60 & .25 & .10 \\ .35 & .50 & .25 \\ .05 & .25 & .65 \end{bmatrix} \end{array}$$

3. $TV = \begin{bmatrix} .355 \\ .325 \\ .365 \end{bmatrix}$ $T^2V = \begin{bmatrix} .358 \\ .3805 \\ .343 \end{bmatrix}$ $T^3V = \begin{bmatrix} .35125 \\ .3998 \\ .36475 \end{bmatrix}$

5. T is regular because

$$T^3 = \begin{bmatrix} \frac{1}{8} & \frac{1}{2} & \frac{3}{16} \\ \frac{11}{32} & \frac{1}{8} & \frac{17}{64} \\ \frac{17}{32} & \frac{3}{8} & \frac{35}{64} \end{bmatrix}$$

7. $\begin{bmatrix} 0 & 0 & 0 \\ \frac{7}{9} & 1 & 0 \\ \frac{2}{9} & 0 & 1 \end{bmatrix}$

9. Yes, it is absorbing. State 2 is an absorbing state.

11. $\begin{bmatrix} 1 & 0 & \frac{43}{51} & \frac{46}{51} \\ 0 & 1 & \frac{8}{51} & \frac{5}{51} \\ 0 & 0 & 0 & 0 \\ 0 & 0 & 0 & 0 \end{bmatrix}$

Case Study Exercises

1. $\begin{bmatrix} .918 \\ .082 \\ 0 \\ 0 \end{bmatrix}$

3. $\begin{bmatrix} .91 \\ .09 \\ 0 \\ 0 \end{bmatrix}$

5. $SV = \begin{bmatrix} \frac{53}{60} \\ \frac{7}{60} \\ 0 \\ 0 \end{bmatrix}$ Allowance for bad debts $= \frac{7}{60}$ or .1167 or 11.67%.

7. $SV = \begin{bmatrix} \frac{7}{8} \\ \frac{1}{8} \\ 0 \\ 0 \end{bmatrix}$ Allowance for bad debts $= \frac{1}{8}$ or .125 or 12.5%.

Review Exercises

1. $\begin{bmatrix} .45 & .40 \\ .55 & .60 \end{bmatrix}$

3. not a transition matrix

5.

1st stage	2nd stage	3rd stage
$\begin{bmatrix} .27 \\ .29 \\ .44 \end{bmatrix}$	$\begin{bmatrix} .261 \\ .290 \\ .449 \end{bmatrix}$	$\begin{bmatrix} .2637 \\ .2927 \\ .4436 \end{bmatrix}$

7. .83

9. $TV = \begin{bmatrix} .34 \\ .66 \end{bmatrix}$ $T^2V = \begin{bmatrix} .496 \\ .504 \end{bmatrix}$ $T^3V = \begin{bmatrix} .4024 \\ .5976 \end{bmatrix}$

11. $\begin{bmatrix} \frac{1}{3} \\ \frac{2}{3} \end{bmatrix}$

13. $\begin{bmatrix} \frac{4}{11} \\ \frac{7}{11} \end{bmatrix}$ So in the long run it will rain $\frac{4}{11}$ or 36.36% of the time and not rain $\frac{7}{11}$ or 63.64% of the time.

15. State 4 is an absorbing state; states 1, 2, and 3 are nonabsorbing.

17. $\begin{bmatrix} 1 & 0 & \frac{5}{7} & 0 \\ 0 & 1 & \frac{2}{7} & 1 \\ 0 & 0 & 0 & 0 \\ 0 & 0 & 0 & 0 \end{bmatrix}$
19. $\begin{bmatrix} \frac{5}{4} & 0 \\ \frac{1}{4} & 1 \end{bmatrix}$

21.
$T = \begin{array}{c} \\ \$0 \\ \$3 \\ \$1 \\ \$2 \end{array} \begin{array}{c} \$0 \quad \$3 \quad \$1 \quad \$2 \\ \left[\begin{array}{cc|cc} 1 & 0 & \frac{4}{7} & 0 \\ 0 & 1 & 0 & \frac{3}{7} \\ \hline 0 & 0 & 0 & \frac{4}{7} \\ 0 & 0 & \frac{3}{7} & 0 \end{array}\right] \end{array}$
$S = \begin{array}{c} \\ \$0 \\ \$3 \\ \$1 \\ \$2 \end{array} \begin{array}{c} \$0 \quad \$3 \quad \$1 \quad \$2 \\ \left[\begin{array}{cc|cc} 1 & 0 & \frac{28}{37} & \frac{16}{37} \\ 0 & 1 & \frac{9}{37} & \frac{21}{37} \\ \hline 0 & 0 & 0 & 0 \\ 0 & 0 & 0 & 0 \end{array}\right] \end{array}$

If Player A starts with $2, he will end up with $3 with a probability of $\frac{21}{37}$ or .5675.

CHAPTER 10

Exercise Set 10–1

1. R 2

3. C 1

5. R 7

7. R 4

9. R chooses row 2; C chooses column 2; $v = 2$

11. R chooses row 1; C chooses column 1; $v = 1$

13. R chooses row 2; C chooses column 1; $v = 2$

15. R chooses row 3; C chooses column 2; $v = 1$

17. no

19. no

21. yes, $v = 3$

23. yes, $v = 2$

25. yes, $v = 1$

27.

	Past	Future	Higher
TV	60%	20%	80%
Radio	25%	30%	5%
Newspaper	15%	50%	15%

Game is not strictly determined.

29. yes, $v = 0$

31. yes, $v = 1$

33. $\begin{array}{c} \\ 2 \\ 5 \end{array} \begin{array}{c} 2 \quad\quad 5 \\ \begin{bmatrix} -1 & -3 \\ 5 & -1 \end{bmatrix} \end{array}$ yes, $v = -1$

35. $\begin{array}{c} \\ 3 \\ 6 \end{array} \begin{array}{c} 2 \quad\quad 4 \quad\quad 7 \\ \begin{bmatrix} 5 & 5 & -4 \\ -4 & -4 & 5 \end{bmatrix} \end{array}$ no

37. $\begin{array}{c} \\ H \\ T \end{array} \begin{array}{c} H \quad\quad T \\ \begin{bmatrix} -1 & -2 \\ 3 & -1 \end{bmatrix} \end{array}$ yes, $v = -1$

39. $\begin{bmatrix} 8 & 2 & 5 \\ 3 & 9 & 5.5 \end{bmatrix}$ no

Referenced Exercise Set 10–1

1.

	(2, 0)	(1, 1)	(0, 2)
(3, 0)	0	0	0
(2, 1)	0	−1	0
(1, 2)	0	−1	0
(0, 3)	0	0	0

yes, $v = 0$

3. yes, $v = 1$

Case Study Exercises

1. hold in reserve

3. no

Exercise Set 10–2

1. yes

3. no

5. yes

7. 2×2 $\begin{bmatrix} 0.2 & 0.2 \\ 0.3 & 0.3 \end{bmatrix}$

9. 3×2 $\begin{bmatrix} 0.32 & 0.48 \\ 0.04 & 0.06 \\ 0.04 & 0.06 \end{bmatrix}$

11. 0

13. −2.34

15. .55

17. 2.02

19. 5

21. 7

23. $a = 1$, $b = 0$

25. $a = 0$, $b = 1$

27. $a = 1$, $b = 0$

29. $a = 1$, $b = 0$

31. 3.02/payoff with matrix Q is 3 more than the payoff with matrix P (3 is integer that was added to P).

33. 6.4/payoff with matrix Q is 7 more than the payoff with matrix P.

35. .38

37. .51

39. [0, 1], 1.2

41. $\begin{bmatrix} 0 \\ 0 \\ 1 \\ 0 \end{bmatrix}$, −1

Referenced Exercise Set 10–2

1. 196

3. not extend auditing procedures

Cumulative Exercise Set 10–2

1. (a) R wins 4. (b) R wins 6.

3. (a) no (b) yes, 0

5. (a) −0.75 (b) 0.44

7. [0 1]

9. [0 1], 0.2

Exercise Set 10–3

1. $E = 2 - 2x$ $E = x$
3. $E = 2x - 1$ $E = 2x$
5. $E = 3x - 1$ $E = 3x - 2$
7. $E = x + 1$ $E = x - 2$
9. $E = 2x + 3$ $E = 4 - 2x$
11.

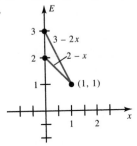

13.

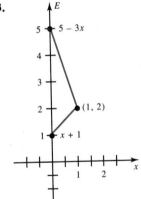

15.

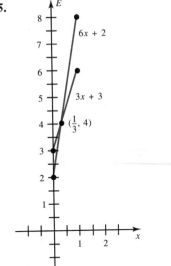

17. $R = [\frac{2}{3}\ \frac{1}{3}]$ $E = \frac{5}{3}$
19. $R = [1\ 0]$ $E = 2$
21. $R = [\frac{12}{23}\ \frac{11}{23}]$ $E = 385/23$
23. $R = [.6\ .4]$ $E = 4$
25. $R = [\frac{7}{11}\ \frac{4}{11}]$ $E = 85/11$
27. $R = [0\ 1]$ $E = 1$
29. $R = [0\ 1]$ $E = 1$
31. $R = [\frac{2}{9}\ \frac{7}{9}]$ $E = 44/27$
33. $R = [0\ 1]$ $E = 1$
35. $R = [\frac{9}{11}\ \frac{2}{11}]$ $E = (74/11) - 5 = \frac{19}{11}$
37. $R = [.1\ .9]$ $E = 7.3 - 8 = -.7$

39. $\begin{bmatrix} -1 & -3 \\ 5 & 1 \end{bmatrix}$ $R = [0\ 1]$

41. $\begin{bmatrix} -1 & -2 \\ 3 & -1 \end{bmatrix}$ $R = [0\ 1]$

Referenced Exercise Set 10–3

1. $R = [.2\ \ .8]$

3. $R = [1\ \ 0],\ C = \begin{bmatrix} 1 \\ 0 \end{bmatrix}$

Cumulative Exercise Set 10–3

1. (a) optimum strategy for R: row 1
   optimum strategy for C: column 3
   (b) value $= 1$

3.   Player B

$$\text{Player A} \quad \begin{array}{c} \\ P \\ N \\ D \end{array} \begin{array}{ccc} P & N & D \\ \begin{bmatrix} 5 & -4 & -9 \\ -4 & 10 & -5 \\ -9 & -5 & 25 \end{bmatrix} \end{array}$$

5. -0.61

7. The expected values for P and Q are -1.21 and 2.79, and $-1.21 + 4 = 2.79$.

9.

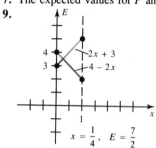

$x = \frac{1}{4}, \quad E = \frac{7}{2}$

Exercise Set 10–4

1. $E = 2x - y$

3. $E = u(2x - y) + 4vy$

5. Minimize $X + Y$: $4X + Y \le 1$

$\qquad\qquad\qquad\qquad 5X + 2Y \le 1$

$\qquad\qquad\qquad\qquad X \ge 0, \quad Y \ge 0$

7. $x = 0, \qquad y = 1$

9. $x = \frac{3}{4}, \qquad y = \frac{1}{4}$

11. $u = 0, \qquad v = 1$

13. $u = \frac{3}{4}, \qquad v = \frac{1}{4}$

15. $x = \frac{1}{2}, \qquad y = \frac{1}{2}, \qquad u = \frac{3}{4}, \qquad v = \frac{1}{4}$

17. $x = 1, \qquad y = 0, \qquad u = 0, \qquad v = 1$

19. $x = \frac{1}{4}, \qquad y = \frac{3}{4}, \qquad u = \frac{1}{2}, \qquad v = \frac{1}{2}$

21. $x = 1, \qquad y = 0, \qquad u = 1, \qquad v = 0$

23. $x = 1, \qquad y = 0, \qquad u = 0, \qquad v = 1$

29. (a) 7   (b) $\frac{1}{7}$

31. (a) $\frac{1}{2}$ (b) 2

33. (a) $\frac{1}{2}$ (b) 2

35. (a) 3   (b) $\frac{1}{3}$

37. (a) .3   (b) $\frac{10}{3}$

Referenced Exercise Set 10–4

1. $[\frac{1}{3}, \frac{2}{3}]$ $[\frac{1}{2}\ \frac{1}{2}]$

7.

(0, 3)
(0, 2)
(1, 1)

Cumulative Exercise Set 10–4

1. (a) row 2, column 2, 3

    (b) row 3, column 2, -1

3. $\begin{bmatrix} 2 & -3 & 4 \\ -3 & 4 & -5 \\ 4 & -5 & 6 \end{bmatrix}$ It is not strictly determined.

5. [0   1]

9. [1   0]   and $\begin{bmatrix} 0 \\ 1 \end{bmatrix}$

11. [0   1]

13. [0   1]   and $\begin{bmatrix} 0 \\ 1 \end{bmatrix}$

Case Study Exercises

1. $x = \frac{1}{6}$; $y = \frac{5}{6}$;   the second option is favorable.

3. $x = \frac{1}{3}$; $y = \frac{2}{3}$;   the second option is favorable.

Review Exercises

1. R chooses row 3; C chooses column 3; $v = 1$

3. yes, $v = 2$

5. $\begin{bmatrix} -3 & 8 & 8 \\ 8 & -3 & -3 \end{bmatrix}$, not strictly determined

7. $E = 0$

9. $a = 1,$ $b = 0$

11. $Q = P + 5$; E (using P) $= .28$; E (using Q) $= 5.28$; E (using P) $+ 5 = E$ (using Q)

13.

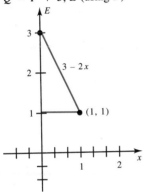

15. $x = \frac{1}{3},$ $y = \frac{2}{3},$ $u = \frac{1}{3},$ $v = \frac{2}{3},$ payoff $= \frac{5}{3}$

17. $\begin{bmatrix} -3 & -2 \\ 1 & -3 \end{bmatrix}$ $R = [\frac{4}{5} \ \frac{1}{5}]$ $C = \begin{bmatrix} \frac{1}{5} \\ \frac{4}{5} \end{bmatrix}$

19. $[0, 1]$

CHAPTER 11

Exercise Set 11–1

1. $1, -3, -4$

3. $1, 3, \frac{9}{4}$

5. (a) $2a - 3$ (b) $2x + 3$ (c) $2x + 2h - 3$
    (d) 2

7. (a) $-a^2 + a + 3$ (b) $-x^2 - 5x - 3$
    (c) $-x^2 - 2xh - h^2 + x + h + 3$
    (d) $-2x + 1 - h$

9. $69,888,000$ **11.** $x \geq -1$

13. $x \leq \frac{1}{2}$

15.

x	-2	-1	0	1	2
$f(x)$	-5	-6	-5	-2	3

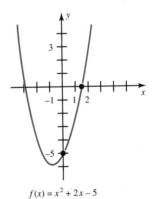

$f(x) = x^2 + 2x - 5$

17.

x	-2	-1	0	1	2
$f(x)$	18	4	-4	-6	-2

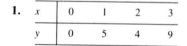

$f(x) = 3x^2 - 5x - 4$

19. (a) not a function **(b)** function **(c)** function
21. $2h$
23. $-2xh - h^2 + h$
25. -1
27. $-4x - 2h - 4$
29. $x \neq 2$
31. $-1 \leq x \leq 1$
33. $x \geq 1$ or $x \leq -1$
35. $x \neq 0$ and $x \neq 2$
37. $x \geq 3$ but $x \neq 4$
39. (a) $\{-2, 2\}$ **(b)** none **(c)** $\{-0.5, 0.5\}$
41. function

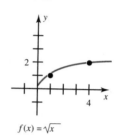

$f(x) = \sqrt{x}$

Referenced Exercise Set 11–1

1. (a) yes, $d = 6t^{2/3}$ or $d = 6\sqrt[3]{t^2}$ **(b)** 49.92 miles **(c)** 0.19 hours   or   11 minutes, 33 seconds
3. yes
5. (a) $f(x) = 0.3x - 564$ **(b)** 33 **(c)** 1977

Exercise Set 11–2

1.

x	0	1	2	3
y	0	5	4	9

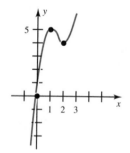

$f(x) = 2x^3 - 9x^2 + 12x$

3.

x	0	1	2	-1
y	1	2	1	10

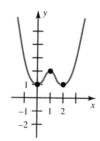

$f(x) = x^4 - 4x^3 + 4x^2 + 1$

5. 3   **7.** the line $x = 5$
9. the lines $x = 3$ and $x = -3$

11.

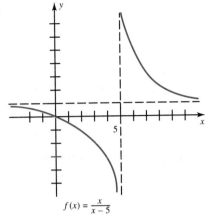

$$f(x) = \frac{x}{x - 5}$$

13.

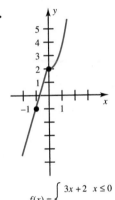

$$f(x) = \begin{cases} 3x + 2 & x \le 0 \\ x^2 + 2 & x > 0 \end{cases}$$

15.

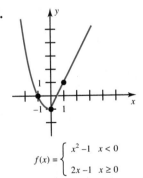

$$f(x) = \begin{cases} x^2 - 1 & x < 0 \\ 2x - 1 & x \ge 0 \end{cases}$$

17. $f(0) = 2$ **19.** 30%

21. (a) $x > 0$
   (b) Four items should be produced to minimize cost.

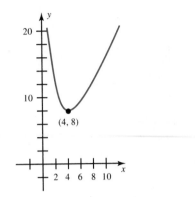

23.

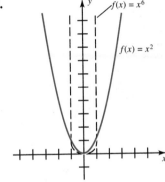

$f(x) = x^6$

$f(x) = x^2$

25. 2
27. the lines $x = 3$ and $x = -3$

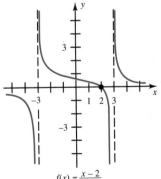

$$f(x) = \frac{x - 2}{x^2 - 9}$$

29. the line $x = 0$

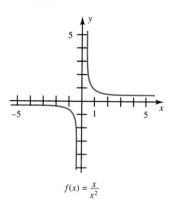

$$f(x) = \frac{x}{x^2}$$

31. the lines $x = 0$, $x = 1$, and $x = -1$

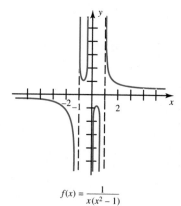

$$f(x) = \frac{1}{x(x^2 - 1)}$$

33. the lines $x = -\frac{3}{2}$ and $x = 1$

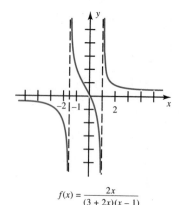

$$f(x) = \frac{2x}{(3 + 2x)(x - 1)}$$

35.

$$f(x) = \begin{cases} x & x \geq 1 \\ x^2 + 1 & x \leq 0 \end{cases}$$

37.

$$f(x) = \begin{cases} 1 - 2x - 5x^2 & x \leq 1 \\ x & x > 1 \end{cases}$$

39.

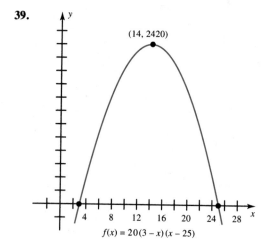

$$f(x) = 20(3 - x)(x - 25)$$

Maximum profit occurs when 14 items are produced.

41. Figure a
43. Figure c

45.

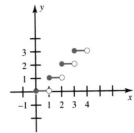

$f(x) = \text{INT } (x)$ for $x \geq 0$

Referenced Exercise Set 11–2

1. 40%

3. (a)

$$f(x) = \begin{cases} .15x & \text{if } 0 \leq x \leq 29{,}300 \\ .27(x - 29{,}300) + 4350 & \text{if } x > 29{,}300 \end{cases}$$

(b)

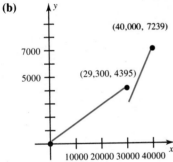

(c) 4395   **(d)** 4350.27
(e) Tax on income of \$29,301 is \$44.73 less than tax on income of \$29,300.
5. (a) 16 minutes and 5 seconds
  (b) 16 hours, 16 minutes and 1 second

Cumulative Exercise Set 11–2

1.

3. $y = x + 1$
5. $x \geq 4$

7. $\dfrac{h(2x + h)}{h} = 2x + h, h \neq 0$

9.

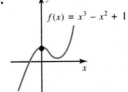

11.

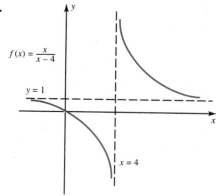

$f(x) = \dfrac{x}{x-4}$

$y = 1$

$x = 4$

Exercise Set 11–3

1. (a) $(f + g)(x) = \dfrac{x + 2}{x - 1} + \dfrac{x}{x + 2}$

$\qquad = \dfrac{2x^2 + 3x + 4}{(x - 1)(x + 2)}$

(b) $(f \cdot g)(x) = \dfrac{(x + 2)}{(x - 1)} \cdot \dfrac{x}{(x + 2)} = \dfrac{x}{x - 1}$

(c) $(f/g)(x) = \dfrac{(x + 2)^2}{x(x - 1)}$

3. (a) $(f + g)(x) = \dfrac{x}{x + 1} + \dfrac{1}{x} = \dfrac{x^2 + x + 1}{x(x + 1)}$

(b) $(f \cdot g)(x) = \dfrac{1}{x + 1}$

(c) $(f/g)(x) = \dfrac{x^2}{x + 1}$

5. (a) all real numbers except $x = 1$ and $x = -2$
(b) all real numbers except $x = -1$ and $x = 0$

7. (a) all real numbers except $x = 1$, $x = -2$, and $x = 0$
(b) all real numbers except $x = -1$ and $x = 0$

9. (a) $\tfrac{9}{2}$ **(b)** 2   **(c)** 8

11. (a) $\tfrac{3}{2}$ **(b)** -1 **(c)** -4

13. (a) $(f \circ g)(x) = \sqrt{\dfrac{x}{1 - x}}$

(b) $(g \circ f)(x) = \dfrac{\sqrt{x}}{1 - \sqrt{x}}$

15. (a) $(f \circ g)(x) = \dfrac{2(x - 2)}{(x + 2)}$

(b) $(g \circ f)(x) = \dfrac{x - 1}{x + 1}$

17. $g(x) = \sqrt{x}$, $k(x) = x^3 + 2x - 5$,
$f(x) = (g \circ k)(x)$

19. $g(x) = \dfrac{1}{x^2} + x^3$, $k(x) = x - 2$,

$\qquad f(x) = (g \circ k)(x)$

21. $Q(t) = -0.005t^4 + 0.16t^3 - 3.28t^2 + 32t + 20$

23. (a) $(f + g + h)(x) = \dfrac{2x^3 - x - 10}{(x - 1)(x + 2)(x - 2)}$

(b) $(f \cdot g \cdot h)(x) = \dfrac{x}{(x - 1)(x - 2)}$

25. (a) $(f + g + h)(x) = \dfrac{2x^2 - x + 3}{(x + 3)(x - 3)}$

(b) $(f \cdot g \cdot h)(x) = \dfrac{x(x - 2)}{(x + 3)^2(x - 3)}$

27. 0

29. $\sqrt{x^2} = |x|$

31. $\dfrac{x + 1}{2x + 1}$

33. no

$\qquad (f \circ g)(1) = \tfrac{2}{3}$
$\qquad (g \circ f)(1) = 3$

35. $g(x) = \sqrt[3]{x^2}$, $k(x) = x - 1$, $f(x) = (g \circ k)(x)$

37. $\dfrac{2a^2 - 3a + 3}{a^2 - 1}$

39. $\dfrac{1}{a + 1} + \sqrt{a}$

41. no

$\qquad (f \cdot g)(x) = \dfrac{x}{x - 1}$

$\qquad (f \circ g)(x) = \dfrac{3x + 4}{-2}$

43. (a) all $x \geq 0$ except $x = 200$ **(b)** $0 \leq x \leq 100$
      (c) 10 hours   **(d)** 40 hours   **(e)** 66.7%

45. (a)

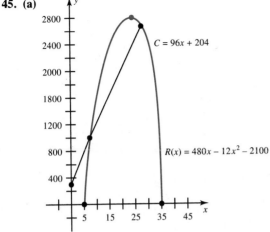

$C = 96x + 204$

$R(x) = 480x - 12x^2 - 2100$

(b) $8 \leq x \leq 24$

47. (a) $\dfrac{1}{\sqrt{x + 1}}$ **(b)** $\dfrac{1}{\sqrt{x + 1}}$ **(c)** $\dfrac{1}{\sqrt{x + 1}}$

Referenced Exercise Set 11–3

1. PEX

3. (a) $x = 7$
   (b) $x = 16$ and $x = 17$ (in a 24-hour clock)
   (c) $x = 18$

Cumulative Exercise Set 11–3

1. $y = (\frac{5}{3})x$

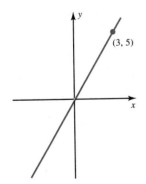

$(3, 5)$

3. (a) The vertical line $x = 2$
   (b) There is no y-intercept.
5. (a) -8
   (b) $-2x^2 - 4xh - 2h^2$
   (c) $h(-4x - 2h)$
   (d) $-4x - 2h$

7. (a)

x	-2	-1	0	1	2
$f(x)$	16	9	8	7	0

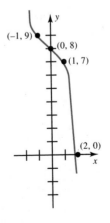

$(-1, 9)$
$(0, 8)$
$(1, 7)$
$(2, 0)$

(b) 3

9.

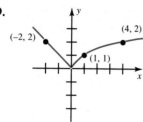

11. 2

Exercise Set 11–4

1. $x = -4$, $x = 2$

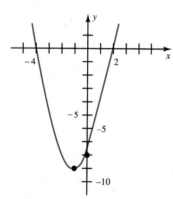

3. $x = -3$

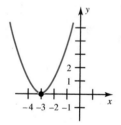

5. no solution

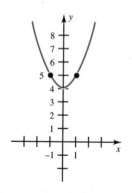

7. domain: real numbers except $x = \frac{3}{2}$, -1
   x-intercept: $x = 2$

9. domain: real numbers except $x = \frac{3}{4}$, -2
   x-intercept: none

11. -2, $\frac{3}{2}$, 2

13. $x = 2$, $x = -2$

15. $x = 4$, $x = -3$

17. (1, 9) and (8, 16)

19. (2, 12) and (7, 7)

21. $x = 2$

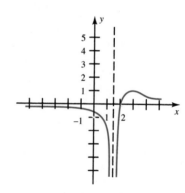

23. $x = -1$

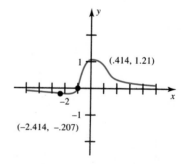

25. domain: all real numbers except $x = 0$ and $x = 2$
    x-intercept: none
27. domain: all real numbers except $x = 0$ and $x = 2$
    x-intercept: none
29. $(-1, 1)$ and $(2, 4)$
31. $(0, 0)$ and $(1, 1)$

33. $(3, 13)$ and $(-3, 13)$
35. 2 seconds and 5 seconds after the object is thrown
37. 3.5 seconds after the object is thrown
39. $(8, 972)$ and $(24, 2508)$
41. $(2, 5)$ and $(4, 8)$; 2
43. $(2, 35)$ and $(8, 75)$; 30

Referenced Exercise Set 11–4

1. $x = 1$
3. (a) Americans have been eating more poultry than pork since 1982.
    (b) Americans have been eating more poultry than beef since 1987.

Cumulative Exercise Set 11–4

1. $y = -x - 2$

3. $f(x) = x^2 + 2x - 1$

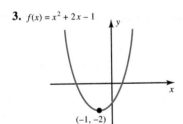

$(-1, -2)$

5. $f + g = x^3 + x^2 + x - 1$
    $f \cdot g = (x^2 + x)(x^3 - 1) = x^5 + x^4 - x^2 - x$
7. $f(x) = h(g(x))$ where $h(x) = x^4$ and $g(x) = x^3 + 2$
9. The domain is all x except $x = 2$; the x-intercept is $x = 0$.
11. $(1, 7)$ and $(6, 12)$

Case Study Exercises

1. 0.58252679
5. no

3. 0.5826
7. 25.16 pounds

11. (a) 1052.05 **(b)** 1035.30
13. Sheridan Shuttles

9.

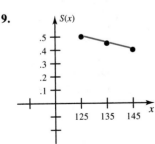

$S(x)$
.5
.4
.3
.2
.1

125   135   145

Review Exercises

1. $-9, -10, -9, -6, 6$

3. $2xh + h^2 - 4h$

5. all real numbers except $x = 1$

7. yes

9. the lines $x = 2, x = 3$, and $x = -3$.

11. (a) $\dfrac{2x^2 + 3x + 4}{(x - 1)(x + 2)}$ (b) $\dfrac{x}{x - 1}$ (c) $\dfrac{(x + 2)^2}{x(x - 1)}$

13. (a) $\frac{9}{2}$ (b) 2 (c) 8

15. $g(x) = x - \sqrt{x}$, $k(x) = x + 2$, $f(x) = (g \circ k)(x)$

17. no x-intercepts

19. $x = 1, 2, -2$

21. domain: all real numbers except -2 and 1
x-intercept: $x = -1$

CHAPTER 12

Exercise Set 12–1

1. (a) 2 (b) 2

3. (a) -4 (b) -4

5. (a) 1 (b) 0.01

7. (a) 1 (b) 0.001

9. 2

11. 8

13. $y = 2x - 6$

15. $y = 8x + 18$

17. $-4x$

19. $2x$

21. (a) 4.1 (b) 4.01 (c) 4.001

23. (a) 8.2 (b) 8.02 (c) 8.002

25. (a) 0.24846 (b) 0.24984 (c) 0.24998

27. 4

29. $y = 4x - 1$

31. $4x - 4$

33. $3x^2 + 1$

35. $-32t + 192$

37. $-32t + 128$

39. 0

41. 0

43. $2ax + b$

Referenced Exercise Set 12–1

1. (a) 26.025 (b) 36.85

3. (a) 2.05% (b) -0.1%

5. 20%

7. (a) positive (b) Reagan

Exercise Set 12–2

1. 3

3. 6

5. no limit

7. no limit

9. no limit

11. 3

13. 0

15. 0

17. 2

19. no limit

21. no limit

23. $\frac{1}{2}$

25. $\frac{1}{6}$

27. 0

29. -4

31. $\frac{1}{3}$

33. 3

35. 0

37. 0

39. 0

41. 4

43. $\frac{1}{4}$

45. 18

47. 8

49. 0

51. 2

53. 2

55. 3

57. 0

59. no limit

61. $90°$

63. 200 miles

Referenced Exercise Set 12–2

1.

x	.05	.06	.07	.08
$f(x)$	.73	.86	.93	.96

$$\lim_{x \to .08} f(x) = .96$$

3. (a) 1   **(b)** 1

Cumulative Exercise Set 12–2

1. -3.99
3. $y = -4x - 4$

5. 0
7. 0

Exercise Set 12–3

1. $2x$

3. $\dfrac{1}{\sqrt{2x}}$

5. $-\dfrac{1}{2\sqrt{x}}$

7. $-\dfrac{1}{x^2}$

9. $-\dfrac{1}{x^2}$

11. $y = -2x + 6$

13. $y = -x - 6$
15. (a) 6 seconds   **(b)** 576 feet
17. (a) $\dfrac{35}{16} = 2.1875$ seconds   **(b)** $\dfrac{1225}{16} = 76.5625$ feet
19. 7

21. 6

23. $-\dfrac{1}{2x\sqrt{x}}$

25. $-\dfrac{2}{x^3}$

27. $y = -\frac{1}{4}x + 1$

29. $y = \frac{1}{4}x + \frac{3}{4}$

31. $y = -5$
33. (a) 4 seconds   **(b)** 320 feet
35. 12 seconds
37. $2ax + b$

Referenced Exercise Set 12–3

1. (a) 15%   **(b)** 28%   **(c)** 33%   **(d)** 28%
3. (a) 6,500 rpm   **(b)** 450 bhp
5. (a) the derivative   **(b)** March and June   **(c)** November

Cumulative Exercise Set 12–3

1. 4
3. 4.1, 4.01
5. 0

7. does not exist
9. $t = 2$ sec
11. $-3x^{-4}$

Exercise Set 12–4

1. continuous
3. not continuous
5. not continuous

7. continuous
9. continuous
11. continuous

13. not continuous
15. $x = 0$ and $x = 1$
17. $x = 4$ and $x = -4$
19. $x = 2$ and $x = -4$
21. $x = 0$
23. $x = 0$ and $x = 3$
25. differentiable
27. not differentiable
29. not differentiable
31. not differentiable
33. differentiable
35. not differentiable
37. not differentiable
39. not differentiable
41. yes
43. not continuous
45. continuous everywhere
differentiable everywhere except $x = 0$

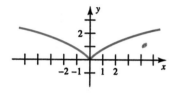

47. continuous everywhere except $x = 1$
differentiable everywhere except $x = 1$

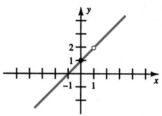

49. continous at $x = 2$, differentiable at $x = 2$
51. continuous at $x = 2$, not differentiable at $x = 2$
53. not continuous at $x = 2$, not differentiable at $x = 2$

Cumulative Exercise Set 12–4

1. -1
3. $-\frac{1}{3}$
5. $6x$

7. $y = -6x - 3$
9. continuous and differentiable
11. $x = 0$

Review Exercises

1. (a) 3   (b) 2.1
5. 2

3. $2x - 1$
7. $\frac{4}{3}$

9. 0
13. $y = 5x - 23$

11. $2x - 3$

CHAPTER 13

Exercise Set 13–1

1. $3x^2$
5. $-15x^{-4} + 5$
7. $\frac{2}{3}x^{-2/3} + \frac{3}{2}x^{-1/2} - 4$
9. $-\frac{2}{3}x^{-4/3} - \frac{7}{4}x^{-5/4}$

3. $8x^3$

11. $0.3x^{-0.7} + 1.2x^{-0.6}$

13. $-\dfrac{3}{x^4} + 12x^3$

15. $-\dfrac{1.5}{x^{2.5}} - \dfrac{1}{x^{5/4}} + 9.6x^{1.4}$

17. $7x^{2.5} - \dfrac{5.6}{x^{2.4}} - \dfrac{5.2}{x^{2.3}}$

19. $\dfrac{0.125}{x^{0.5}} - \dfrac{0.192}{x^{1.4}}$

21. $\dfrac{1}{\sqrt{x}} - 2$

23. $\dfrac{5}{2\sqrt{x}} + \dfrac{1}{\sqrt{x^3}}$

25. $-\dfrac{9}{\sqrt{3}}x^{-3/2} - \dfrac{2}{3\sqrt[3]{4}}x^{-4/3}$

27. $y = 4x - 1$ **29.** $y = 2x + 1$

31. (a) $v = -32t + 32,\ a = -32$

  (b) $t = 1$ second

  (c) 16 feet

33. 12   **35.** 10

Referenced Exercise Set 13–1

3. $-0.048x + 0.575$

Exercise Set 13–2

1. $4x^3(x + 5) + x^4 = 5x^4 + 20x^3$

3. $6x^2(3x - 7) + (3)(2x^3) = 24x^3 - 42x^2$

5. $(-3x^{-4})(3x^2 - 17) + (6x)(x^{-3} + 5)$
$= -3x^{-2} + 51x^{-4} + 30x$

7. $(\tfrac{1}{3}x^{-2/3} + 1)(3x^{1/2} - 4x) + (\tfrac{3}{2}x^{-1/2} - 4)(x^{1/3} + x)$

9. $(-\tfrac{1}{3}x^{-4/3} - \tfrac{7}{4}x^{-5/4})(4x^2 - 5) + (8x)(x^{-1/3} + 7x^{-1/4})$

11. $4(2x + 1)$

13. $6x^2(x^3 + 1)$

15. $\dfrac{1}{(2x + 1)^2}$

17. $\dfrac{x^3 - 10}{3x^3}$

19. $y = -3x + 12$

21. $y = 2$

23. $L = \dfrac{0.5x^2 + 3}{0.5x^2 + 5};\quad L' = \dfrac{2x}{(0.5x^2 + 5)^2}$

25. $2x(x^3 + 15)(x^2 - 17) + 3x^4(x^2 - 17)$
$+ 2x^3(x^3 + 15) = 7x^6 - 85x^4 + 60x^3 - 510x$

27. $-x^{-2}(x^{1/3} + 2)(x^2 - 3)$
$+ \tfrac{1}{3}x^{-5/3}(x^2 - 3) + 2(x^{1/3} + 2)$

29. $\dfrac{x^4 - 21x^2 - 70x}{(x^2 - 7)^2}$

31. $\dfrac{-x^4 - 10x^2 + 3}{x^2(x^2 - 1)^2}$

33. $C'(x) = 23 - 138x^2$

35. $-\dfrac{100}{x^2} - 0.4$

Referenced Exercise Set 13–2

1. $q'(x) = 0.26x(x - 10) + 0.13(x - 10)^2$

3. $C'(x) = \dfrac{-7.12(13.7 - 2x)}{(13.7x - x^2)^2}$

Cumulative Exercise Set 13–2

1. $-6x^{-3} - 4x^3$

3. $y = -x - 2$

5. $36x^2(x^{-5} - 9) - 5x^{-6}(12x^3 + 1)$
$= -5x^{-6} - 24x^{-3} - 324x^2$

7. $2x(x - 5)(x^2 + 7) + x^2(x^2 + 7) + x^2(x - 5)(2x)$
$= 5x^4 - 20x^3 + 21x^2 - 70x$

Exercise Set 13–3

1. $15(5x + 2)^2$ **3.** $9x^2(x^3 + 4)^2$

5. $3(3x^2 + 6x)(x^3 + 3x^2)^2 = 9x(x + 2)(x^3 + 3x^2)^2$

7. $\frac{1}{2}x^{1/2}(x^{3/2} + 4)^{-2/3}$

9. $-6(\frac{1}{2}x^{-1/2} - 2x^{-2})(x^{1/2} + 2x^{-1})^{-7}$

11. $(5x^2 + 2)^2 + 20x^2(5x^2 + 2)$
$$= (5x^2 + 2)(25x^2 + 2)$$

13. $2x(x^3 + 2x)^2 + 2x^2(x^3 + 2x)(3x^2 + 2)$
$$= 8x^3(x^2 + 1)(x^2 + 2)$$

15. $\dfrac{2x(x^3 - x)^3 - 3x^2(x^3 - x)^2(3x^2 - 1)}{(x^3 - x)^6}$
$$= \frac{x^2(-7x^2 + 1)}{(x^3 - x)^4}$$

17. $y = 8$ **19.** $25y + 7x = 12$

21. $0.16[5 - 0.2(t - 6)^2](t - 6)$

23. $L = \dfrac{x^2 + 5}{x^2 + 10}$; $L' = \dfrac{10x}{(x^2 + 10)^2}$

25. $2x^3(x^3 + 2x^2)^3(7x + 10)$

27. $x^{-1/2}(x^{3/2} + 4)^{-3} - 9x(x^{3/2} + 4)^{-4}$

29. $x(x^2 + 1)(x^3 + 1)^2(13x^3 + 9x + 4)$

31. $\dfrac{x(2x^3 + x + 3)}{(x^2 + 1)^{1/2}(x^3 + 3)^{2/3}}$

33. $\dfrac{-11x^{3/2} + 26x^{-5/2}}{(x^2 + 2x^{-2})^4}$

35. $\dfrac{-x^2 + 1}{(x^2 + 1)^2}$

37. $\dfrac{-x^2 + 4x + 5}{(x^2 + 5)^2}$

39. $-1000(200 - x)$

Referenced Exercise Set 13–3

1. $p(x) = 0.007(16 - 0.41x^{-1/2})^2$
$$- 0.33(16 - 0.41x^{-1/2}) + 100$$
$p'(x) = -0.02173x^{-3/2} - 0.00118x^{-2}$

3. (c) The domain is $\{1\}$.

Cumulative Exercise Set 13–3

1. $6x^2 + \dfrac{1}{2\sqrt{x}} + \dfrac{1}{x^2}$

3. $2x - 4$

5. $6x(x^2 - 1)^2 - 3x^2 - 1$

7. $y = 3$

9. $y = -12x + 32$

11. (a) $(x^3 - x - 2)(3x^2 - 1) +$
$$(x^3 - x - 2)(3x^2 - 1)$$
(b) $2(x^3 - x - 2)(3x^2 - 1)$
(c) Yes

Exercise Set 13–4

1. 2 **3.** $24x^2$

5. $\dfrac{8}{x^3} = 8x^{-3}$ **7.** $\dfrac{12}{x^5} = 12x^{-5}$

9. $12x^2 - 2$

11. $\frac{8}{9}x^{-7/3} + \frac{35}{16}x^{-9/4}$

13. $-0.21x^{-1.7} - 0.72x^{-1.6}$

15. 26

17. $\frac{3}{8}$ **19.** $2 + 12t$

21. $-\frac{1}{4}g^{-3/2}$

23. (a) $t = 1$ **(b)** $a(t)$ is never 0

25. (a) $t = \frac{1}{3}$ and $t = 1$ **(b)** $t = \frac{2}{3}$

27. $a(t) = -32$

29. $1.8 - 0.018x$ **31.** $0.96x - 195.2$

33. $12x^2 + 4$ **35.** $\dfrac{24x^2 - 4}{(2x^2 + 1)^3}$

37. $\dfrac{1}{(x^2 + 1)^{3/2}}$ **39.** 0, 2

41. $1, -1$

43. $f^{(3)}(x) = 24x$; $f^{(4)}(x) = 24$

45. $f^{(3)}(x) = -12x^{-4} - 24x^{-5}$;
$f^{(4)}(x) = 48x^{-5} + 120x^{-6}$

47. $f^{(n)}(x) = (-1)^n(n + 1)!\, x^{-(n+2)}$

Referenced Exercise Set 13–4

1. 0.0625

Cumulative Exercise Set 13–4

1. $6x^{-3} - 0.4x^{-0.6} + 5$

3. $6x^2 - 4x^3 - 10x^4 + 6x^5$

5. $3(3x^4 + x^5)^2(12x^3 + 5x^4)$

7. $21x^2(1 - 3x)^2 - 6(7x^3 + 2)(1 - 3x)$

9. $12x^2 - 18x^{-4}$

11. $a(t) = -32$

Case Study Exercises

1. 81.11   **3.** 84.17

Review Exercises

1. $8x + 3x^2$

3. $-20x^{-6} - \frac{1}{3}x^{-2/3}$

5. $5x + 2y = 5$ **7.** $\dfrac{x^4 + 6x^2}{(x^2 + 2)^2}$

9. $\dfrac{\frac{1}{3}(x^{-2/3})(x^{4/5} + 2) - (\frac{4}{5}x^{-1/5})(x^{1/3})}{(x^{4/5} + 2)^2}$

11. $4(2x^3 + x)^3(6x^2 + 1)$

13. $6x + 9(3x^2 + 4)(x^3 + 4x)^2$

15. $x(x^2 + 1)(x^3 + 1)^2(13x^3 + 9x + 4)$

17. $y = 24x - 8$

19. $\dfrac{6x(7x^3 - 8)}{(x^3 + 4)^4}$

21. $\dfrac{(-1)^n(n + 2)! \, x^{-(n+3)}}{2}$

CHAPTER 14

Exercise Set 14–1

1. Increasing on $0 < x < \infty$
Decreasing on $-\infty < x < 0$
Horizontal tangent at $x = 0$

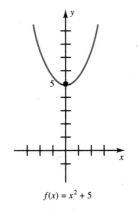

$f(x) = x^2 + 5$

3. Increasing on $-\infty < x < 0$
Decreasing on $0 < x < \infty$
Horizontal tangent at $x = 0$

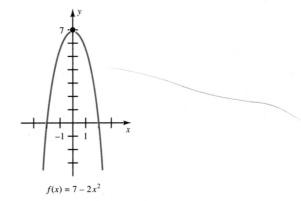

$f(x) = 7 - 2x^2$

5. Increasing on $0 < x < \infty$
Decreasing on $-\infty < x < 0$
Horizontal tangent at $x = 0$

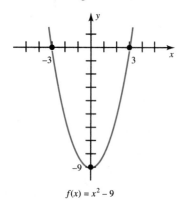

$$f(x) = x^2 - 9$$

7. Increasing on $2 < x < \infty$
Decreasing on $-\infty < x < 2$
Horizontal tangent at $x = 2$

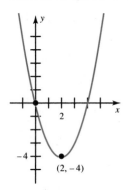

$$f(x) = x(x - 4)$$

9. Increasing on $-2 < x < \infty$
Decreasing on $-\infty < x < -2$
Horizontal tangent at $x = -2$

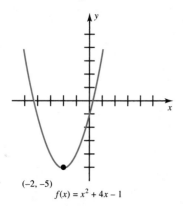

$$f(x) = x^2 + 4x - 1$$

11. Increasing on $-\infty < -3$ and $1 < x < \infty$
Decreasing on $-3 < x < 1$
Horizontal tangents at $x = -3$ and $x = 1$

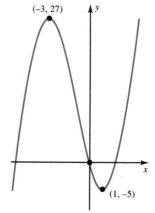

$$f(x) = x^3 + 3x^2 - 9x$$

13. Increasing on $-\infty < x < -1$ and $3 < x < \infty$
Decreasing on $-1 < x < 3$
Horizontal tangents at $x = -1$ and $x = 3$

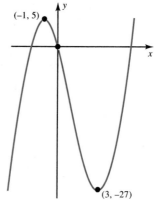

$$f(x) = x^3 - 3x^2 - 9x$$

15. Increasing on $-1 < x < 3$
Decreasing on $-\infty < x < -1$ and $3 < x < \infty$
Horizontal tangents at $x = -1$ and $x = 3$

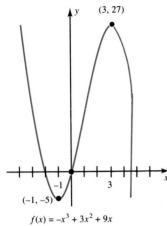

$f(x) = -x^3 + 3x^2 + 9x$

17. Increasing on $-\infty < x < -3$ and $3 < x < \infty$
Decreasing on $-3 < x < 0$ and $0 < x < 3$
Horizontal tangents at $x = -3$ and $x = 3$

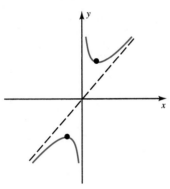

$f(x) = x + 9/x$

19. Increasing on $-\infty < x < -2\sqrt{3}$
and $2\sqrt{3} < x < \infty$
Decreasing on $-2\sqrt{3} < x < 0$ and $0 < x < 2\sqrt{3}$
Horizontal tangents at $x = -2\sqrt{3}$
and $x = 2\sqrt{3}$

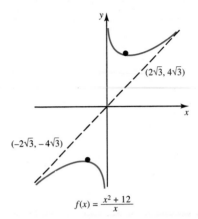

$f(x) = \dfrac{x^2 + 12}{x}$

21. Increasing on $-\infty < x < -\dfrac{2\sqrt{3}}{3}$
and $\dfrac{2\sqrt{3}}{3} < x < \infty$
Decreasing on $-\dfrac{2\sqrt{3}}{3} < x < \dfrac{2\sqrt{3}}{3}$
Horizontal tangents at $x = -\dfrac{2\sqrt{3}}{3}$ and $x = \dfrac{2\sqrt{3}}{3}$

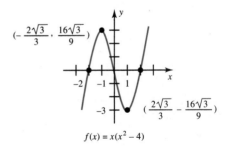

$f(x) = x(x^2 - 4)$

23. Increasing on $-\infty < x < 0$ and $\frac{4}{3} < x < \infty$
Decreasing on $0 < x < \frac{4}{3}$
Horizontal tangents at $x = 0$ and $x = \frac{4}{3}$

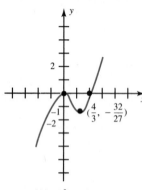

$$f(x) = x^2(x - 2)$$

25. Increasing on $-\infty < x < -1$ and $-1 < x < \infty$
Never decreasing
No horizontal tangents

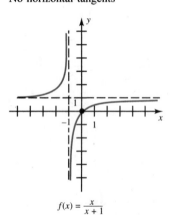

$$f(x) = \frac{x}{x + 1}$$

27. Increasing on $-\infty < x < -1$ and $-1 < x < \infty$
Never decreasing
No horizontal tangents

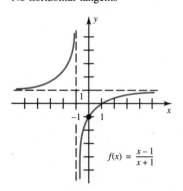

$$f(x) = \frac{x - 1}{x + 1}$$

29. Increasing on $0 < x < 6$ and $8 < x < 10$
Decreasing on $6 < x < 8$

31.

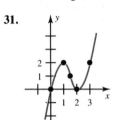

33.

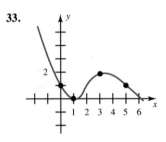

35. Increasing on $-\infty < x < 0$ and $4 < x < \infty$
Decreasing on $0 < x < 2$ and $2 < x < 4$
Horizontal tangents at $x = 0$ and $x = 4$

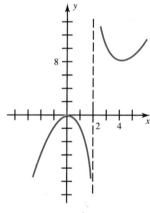

$$f(x) = \frac{x^2}{x - 2}$$

37. Increasing on $-\sqrt{2} < x < 0$ and $\sqrt{2} < x < \infty$
Decreasing on $-\infty < x < -\sqrt{2}$ and $0 < x < \sqrt{2}$
Horizontal tangents at $x = -\sqrt{2}$, $x = 0$, and
$x = \sqrt{2}$

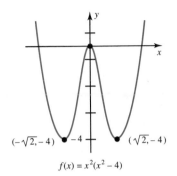

$$f(x) = x^2(x^2 - 4)$$

39. Increasing on $-\dfrac{\sqrt{10}}{2} < x < 0$

and $\dfrac{\sqrt{10}}{2} < x < \infty$

Decreasing on $-\infty < x < -\dfrac{\sqrt{10}}{2}$

and $0 < x < \dfrac{\sqrt{10}}{2}$

Horizontal tangents at $x = -\dfrac{\sqrt{10}}{2}$, $x = 0$,

and $x = \dfrac{\sqrt{10}}{2}$

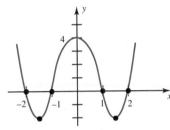

$$\left(\frac{-\sqrt{10}}{2}, -2.25\right) \qquad \left(\frac{\sqrt{10}}{2}, -2.25\right)$$
$$f(x) = (x^2 - 1)(x^2 - 4)$$

41. Increasing on $-1 < x < 2$ and $5 < x < \infty$
Decreasing on $-\infty < x < -1$ and $2 < x < 5$
Horizontal tangents at $x = -1$, $x = 2$, and $x = 5$

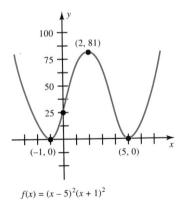

$$f(x) = (x - 5)^2(x + 1)^2$$

43. (a) Increasing in $(-1, 0)$; decreasing in $(0, 3)$
   (b) Positive in $(-1, 0)$; negative in $(0, 3)$
45. (a) Increasing in $(-2, -1)$ and $(2, 3)$; decreasing in $(-1, 1)$, $(1, 2)$ and $(3, 4)$
   (b) Positive in $(-2, -1)$ and $(2, 3)$; negative in $(-1, 1)$, $(1, 2)$ and $(3, 4)$

Referenced Exercise Set 14–1

1. Slope of tangent line is negative for $x < 1$.
Slope of tangent line is positive for $x > 1$.
3. (a) Increasing on $(-\infty, 0)$ and $(30, \infty)$
Decreasing on $(0, 30)$
(b) It represents a percentage, and the percentage should go down.
(c) $f(20) = 25.93$
(d) When $x = 0$, $f(0) = 100$ and when $x = 30$, $f(30) = 0$. Since $f(x)$ represents a percentage, it must vary between 100% and 0%, so x must be in the interval $[0, 30]$; also, no shipments were made with more than 30% of the eggs below A quality.

Exercise Set 14–2

1. Relative minimum at $(0, -4)$

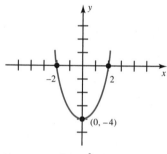

$$f(x) = x^2 - 4$$

3. Relative maximum at $(0, 2)$

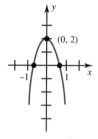

$$f(x) = 2 - 3x^2$$

5. Relative minimum at $(2, -4)$

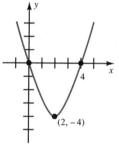

$$f(x) = x^2 - 4x$$

7. Relative maximum at $\left(-\frac{2}{3}, \frac{4}{27}\right)$
Relative minimum at $(0, 0)$

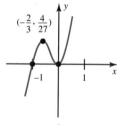

$$f(x) = x^2 (x + 1)$$

9. Relative maximum at $(0, -1)$
Relative minimum at $(4, -33)$

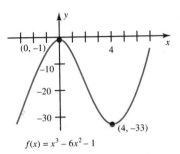

$f(x) = x^3 - 6x^2 - 1$

11. Relative maximum at $(-3, 27)$
Relative minimum at $(1, -5)$

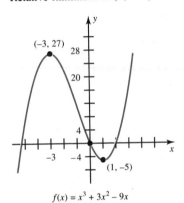

$f(x) = x^3 + 3x^2 - 9x$

13. Relative maximum at $(-1, 5)$
Relative minimum at $(3, -27)$

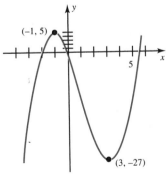

$f(x) = x^3 - 3x^2 - 9x$

15. Relative maximum at $(0, 0)$
Relative minimum at $(-1, -1)$ and $(1, -1)$

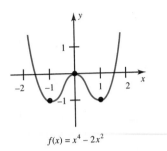

$f(x) = x^4 - 2x^2$

17. Relative maximum at $(2, 16)$ and $(-2, 16)$
Relative minimum at $(0, 0)$

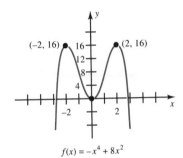

$f(x) = -x^4 + 8x^2$

19. Relative maximum at $(-3, -6)$
Relative minimum at $(3, 6)$

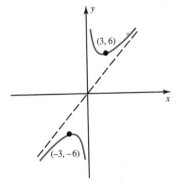

21. Relative maximum at $(-2\sqrt{3}, -4\sqrt{3})$
Relative minimum at $(2\sqrt{3}, -4\sqrt{3})$

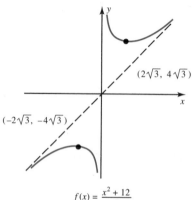

$$f(x) = \frac{x^2 + 12}{x}$$

23. No relative maximum
Relative minimum at $(3, -27)$

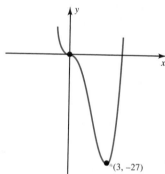

$$f(x) = x^3(x - 4)$$

25. Relative maximum at $(0, 0)$
Relative minimum at $(-1, -1)$ and $(1, -1)$

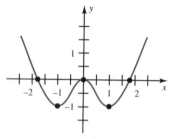

$$f(x) = x^2(x^2 - 2)$$

27. Relative minimum at $(0, 0)$
No relative maximum

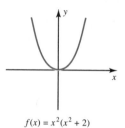

$$f(x) = x^2(x^2 + 2)$$

29. No relative maximum or minimum

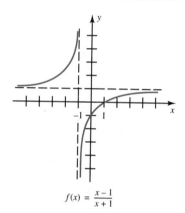

$$f(x) = \frac{x - 1}{x + 1}$$

31. Relative minimum at $(3, 2)$
No relative maximum

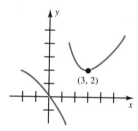

33. Relative maximum at $(5, 4)$
Relative minimum at $(-1, 1)$

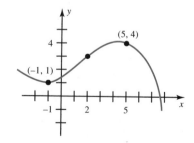

35. $x = 4$

37. Relative maximum at $(-3, 162)$
Relative minimum at $(3, -162)$

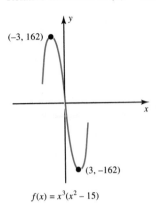

$f(x) = x^3(x^2 - 15)$

39. Relative maximum at $(0, 4)$
Relative minimum at $\left(-\dfrac{\sqrt{10}}{2}, -2.25\right)$
and $\left(\dfrac{\sqrt{10}}{2}, -2.25\right)$

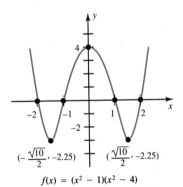

$f(x) = (x^2 - 1)(x^2 - 4)$

41. Relative maximum at $(2, 81)$
Relative minimum at $(-1, 0)$ and $(5, 0)$

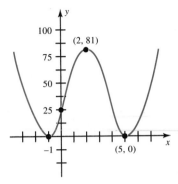

$f(x) = (x - 5)^2(x + 1)^2$

43. Relative minimum at $(5, 0)$
No relative maximum

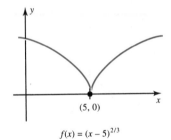

$f(x) = (x - 5)^{2/3}$

Referenced Exercise Set 14–2

1. 13,243 hours

3. $x = 1$ and $x = -1$ are excluded

Cumulative Exercise Set 14–2

1. (a) $(-\infty, 0)$
   (b) $(0, 4)$

   (c)

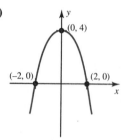

3. (a) $(-\infty, 0)$ and $(0, \infty)$
   (b) none

   (c)

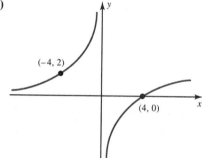

5.

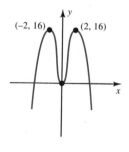

7. relative minimum at $x = 0$
   relative maxima at $x = 2$ and $x = -2$

Exercise Set 14–3

1. Concave down: $(-2, 0)$
   Concave up: $(0, 2)$
   Point of inflection: $(0, 0)$
3. Concave down: $(-1, 2)$
   Concave up: $(-2, -1)$
   Point of inflection: $x = -1$
5. Concave down: $(-1, 0)$
   Concave up: $(-2, -1)$ and $(0, 2)$
   Point of inflection: $x = -1, x = 0$
7. Concave up: $(-\infty, \infty)$
   Concave down: nowhere
   No point of inflection
9. Concave up: $(2, \infty)$
   Concave down: $(-\infty, 2)$
   Point of inflection: $x = 2$
11. Concave up: $(-1, \infty)$
   Concave down: $(-\infty, -1)$
   Point of inflection: $x = -1$

13. Relative maximum at $(-1, 5)$
   Relative minimum at $(3, -27)$
15. Relative maximum at $(0, 0)$
   Relative minimum at $(-1, -1)$ and $(1, -1)$
17. Relative maximum at $(2, 16)$ and $(-2, 16)$
   Relative minimum at $(0, 0)$

19.

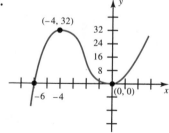

$f(x) = x^3 + 6x^2$

21.

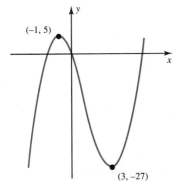

$(-1, 5)$

$(3, -27)$

$$f(x) = x^3 - 3x^2 - 9x$$

23.

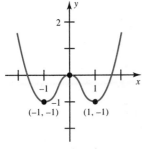

$(-1, -1)$ $(1, -1)$

$$f(x) = x^4 - 2x^2$$

25.

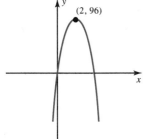

$(2, 96)$

$$f(x) = 64x - 2x^4$$

27.

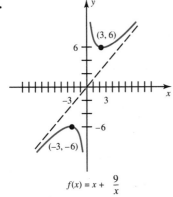

$(3, 6)$

6

-3 3

-6

$(-3, -6)$

$$f(x) = x + \frac{9}{x}$$

29.

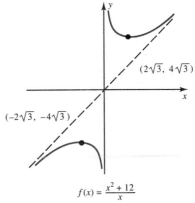

$(2\sqrt{3},\ 4\sqrt{3})$

$(-2\sqrt{3},\ -4\sqrt{3})$

$$f(x) = \frac{x^2 + 12}{x}$$

31.

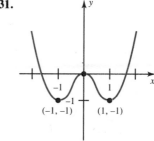

-1 1

-1

$(-1, -1)$ $(1, -1)$

$$f(x) = x^2(x^2 - 2)$$

33.

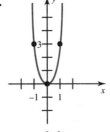

3

-1 1

$$f(x) = x^2(x^2 + 2)$$

35.

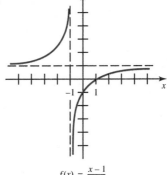

-1 1

$$f(x) = \frac{x - 1}{x + 1}$$

37.

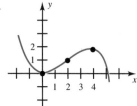

39.

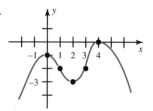

41.

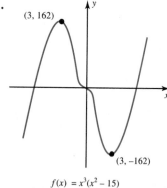

$$f(x) = x^3(x^2 - 15)$$

43.

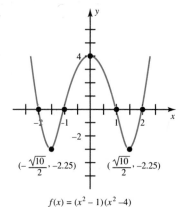

$$\left(-\frac{\sqrt{10}}{2}, -2.25\right) \qquad \left(\frac{\sqrt{10}}{2}, -2.25\right)$$

$$f(x) = (x^2 - 1)(x^2 - 4)$$

45.

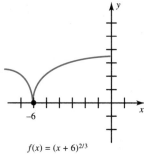

$$f(x) = (x + 6)^{2/3}$$

47. **(a)** Increasing in $(-2, -1)$ and $(1, 2)$; decreasing in $(-1, 1)$
   (b) Positive in $(-2, -1)$ and $(1, 2)$; negative in $(-1, 1)$
   (c) Positive in $(0, 2)$; negative in $(-2, 0)$

49. **(a)** Increasing in $(-2, 0)$ and $(0, 2)$; decreasing in $(-3, -2)$
   (b) Positive in $(-2, 0)$ and $(0, 2)$; negative in $(-3, -2)$
   (c) Positive in $(-3, -\frac{1}{2})$ and $(0, 1)$; negative in $(-\frac{1}{2}, 0)$ and $(1, 2)$

Referenced Exercise Set 14–3

1. 1982
3. 205.83

Cumulative Exercise Set 14–3

1. Increasing on $6 < x < \infty$
   Decreasing on $-\infty < x < 6$
   Horizontal tangent at $x = 6$

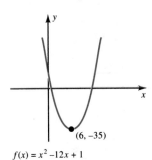

$f(x) = x^2 - 12x + 1$

3. Increasing on $-\infty < x < -4$ and on $4 < x < \infty$
   Decreasing on $-4 < x < 0$ and on $0 < x < 4$
   Horizontal tangents at $x = -4$ and $x = 4$

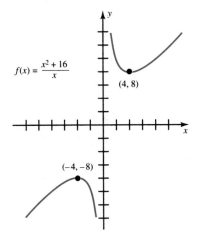

$f(x) = \dfrac{x^2 + 16}{x}$

5. Relative maximum at $x = -1$
   Relative minimum at $x = 0$

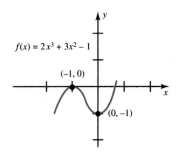

$f(x) = 2x^3 + 3x^2 - 1$

7. Relative maximum at $x = 2$
   Relative minimum at $x = -2$

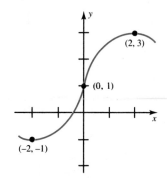

9. Relative maximum at $x = -2$
   Relative minimum at $x = 3$

11.

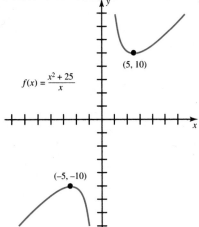

$f(x) = \dfrac{x^2 + 25}{x}$

Exercise Set 14–4

1. Absolute maximum at $f(4) = 25$
   Absolute minimum at $f(-1) = 0$
3. Absolute maximum at $f(-2) = 9$
   Absolute minimum at $f(6) = -55$
5. Absolute maximum at $f(2) = 21$
   Absolute minimum at $f(0) = 1$
7. Absolute maximum at $f(-2) = 5$
   Absolute minimum at $f(-3) = 1$
9. Absolute maximum at $f(-2) = 9$ and $f(2) = 9$
   Absolute minimum at $f(1) = 0$ and $f(-1) = 0$
11. Absolute maximum at $f(4) = 12$
    Absolute minimum at $f(-1) = -18$
13. Absolute maximum at $f(-1) = 64$
    Absolute minimum at $f(1) = 0$ and $f(3) = 0$
15. 256 feet; 4 seconds
17. 10 ft × 20 ft

19. 200 and 100
21. 2000 ft × 1000 ft
23. 6.23 miles from the point he wishes to reach
25. base: 4 inches × 4 inches
    side: 2 inches high
27. base: 3.175 inches × 3.175 inches
    side: 3.175 inches
29. length: 28 inches
    width: 14 inches
31. $\dfrac{6}{1 + \dfrac{\pi}{2}}$ feet = 3.36 feet

33. 6.32 × 9.49
37. base: 1.88 inches × 1.88 inches
    height: 2.82 inches

Referenced Exercise Set 14–4

1. (1.98, 17.96)

3. (28, 1.0828)

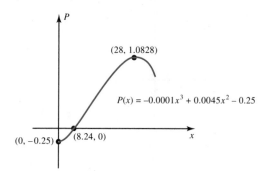

Cumulative Exercise Set 14–4

1. (a) $(-\infty, 0)$ and $(\frac{8}{3}, \infty)$
   (b) $x = 0$ and $x = \frac{8}{3}$
   (c)

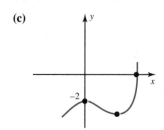

3.

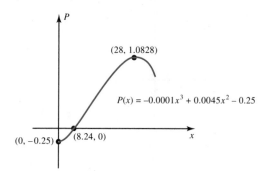

5.

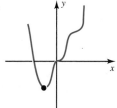

7. relative maximum at $x = -\sqrt[4]{\frac{1}{5}}$
relative minimum at $x = \sqrt[4]{\frac{1}{5}}$

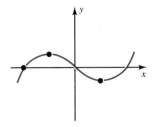

9. relative maximum at $x = -\frac{3}{5}$
relative minimum at $x = 1$

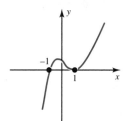

11. $\frac{1}{2}$

Exercise Set 14–5

1. $y' = -\dfrac{x}{3y}$

3. $y' = \dfrac{-y - 2xy}{x + x^2}$

5. $y' = \dfrac{3x^2 + y^3}{-3xy^2 + 4y^3}$

7. $y' = \dfrac{-2x - y^3}{3xy^2 + y^{-2}}$

9. $-\dfrac{2}{3}$

11. 0

13. $D_t y = -\dfrac{2x}{3y^2}; D_t x = -\dfrac{4}{3}$

15. $D_t y = \dfrac{2x - 3}{2y} D_t x; -1$

17. 4.5 mph

19. $\dfrac{dh}{dt} = \dfrac{20}{81\pi} \dfrac{\text{cm}}{\text{min}} = 0.0796 \dfrac{\text{cm}}{\text{min}}$

21. 4 ft/sec

23. $\frac{9}{7}$ ft/sec

25. 0.245 cm/sec

27. $\frac{5}{6}$ ft/sec

29. 12 ft/sec

31. 235.75 mph

33. -1.8 m³/sec

Referenced Exercise Set 14–5

1. 6.58 inches × 3.42 inches

Cumulative Exercise Set 14–5

1. increasing on $-\infty < x < 0$ and on $4 < x < \infty$
decreasing on $0 < x < 4$
horizontal tangents at $x = 0$ and $x = 4$

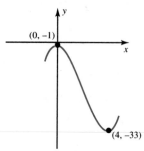

$f(x) = x^3 - 6x^2 - 1$

3.

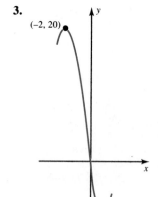

$f(x) = 2x^3 + 3x^2 - 12x$

5.

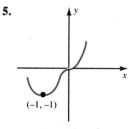

$f(x) = 3x^4 + 4x^3$

7. absolute minimum $f(-3) = -80$
absolute maximum $f(0) = 1$

9. set $x = \dfrac{16 - \sqrt{76}}{6}$; the box is x by $6 - 2x$ by $10 - 2x$

11. $x' = \dfrac{4(100)^{\frac{3}{4}}}{11}$

Case Study Exercises

1. 29.487   **3.** 25.496
5. 5:00 P.M. because it is rush hour

Review Exercises

1. Increasing on $0 < x < \infty$
Decreasing on $-\infty < x < 0$
Horizontal tangent at $x = 0$

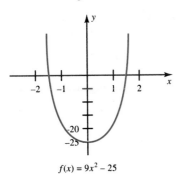

$$f(x) = 9x^2 - 25$$

3. Increasing on $-\infty < x < 3$ and $3 < x < \infty$
No horizontal tangent

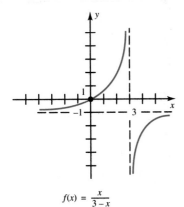

$$f(x) = \frac{x}{3-x}$$

5. Relative minimum at $(-3, -19)$

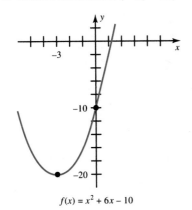

$$f(x) = x^2 + 6x - 10$$

7. Relative maximum at $(\sqrt{2}, 4)$ and $(-\sqrt{2}, 4)$
Relative minimum at $(0, 0)$

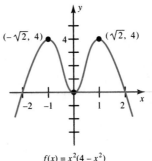

$$f(x) = x^2(4 - x^2)$$

9. Relative minimum at $(1, -1)$
11. Concave up: $0 < x < 1$
Concave down: $-\infty < x < 0$ and $1 < x < \infty$

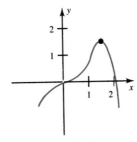

$$f(x) = 2x^3 - x^4$$

13. Absolute maximum of 4 at $x = 0$
Absolute minimum of -16 at $x = -2$
15. Absolute maximum of 6 at $x = 8$
Absolute minimum of -0.3849 at $x = 0.19245$
17. $y' = \dfrac{-2x - y}{x + 2y}$ **19.** $\frac{1}{24}$ m/min

CHAPTER 15

Exercise Set 15–1

1.

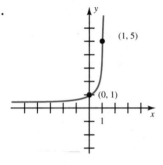

$f(x) = 5^x$

3.

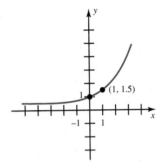

$f(x) = 1.5^x$

5.

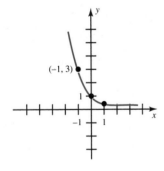

$f(x) = (\frac{1}{3})^x$

7.

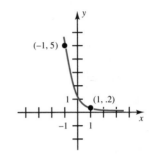

$f(x) = .2^x$

9. 5   **11.** 4   **13.** 2

15.

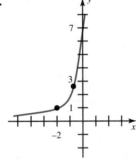

$f(x) = e^{x+2}$

17.

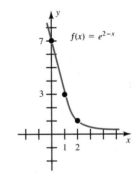

$f(x) = e^{2-x}$

19. (a) 59.6   **(b)** 60
21. (a) 27.8   **(b)** 28

23.

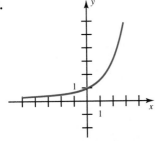

$f(x) = 1.1^x$

25.

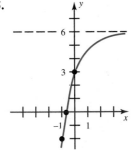

$f(x) = 6 - 3e^{-x}$

27.

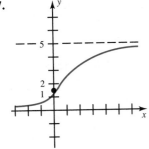

$f(x) = \dfrac{5}{1 + 3e^{-4x}}$

29.

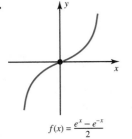

$f(x) = \dfrac{e^x - e^{-x}}{2}$

31. 3
33. 0
35.

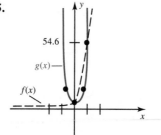

37. 81
39. **(a)** 16°F   **(b)** 10°F

41. **(a)**

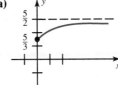

(b) increasing   **(c)** level off and approach 2.5 million
43. no

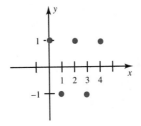

45. **(a)** $e^3 \approx 20.1$ **(b)** $e^4 \approx 54.6$

Referenced Exercise Set 15–1

1. $U(G) \approx 0.808 < 0.817$
risk-averse

3. (a) $t = 3$, $P(t) = 0.039203$
   (b) $t = 8$, $P(t) = 0.5722409$
   (c) $t = 16$, $P(t) = 0.997199$

Case Study Exercises

1. 22 pounds
3. (a) 1925 calories   **(b)** 4 pounds
5. 7.9 pounds

7. Let w be your present weight.
   (a) $w(t) = (w - 10) + 10e^{-0.005t}$
   (b) $(w - 4)$ pounds

Exercise Set 15–2

1.

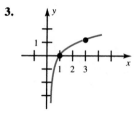

$f(x) = \log_{10} 2x$

3.

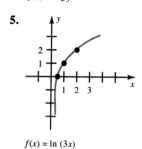

$f(x) = \log_3 x$

5.

$f(x) = \ln (3x)$

7.

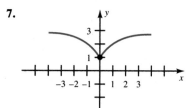

$f(x) = \ln (x^2 + 3)$

9. $\ln 3 + 4 \ln x$
11. $\ln 3 + \ln x + 2 \ln y - \ln(x - y)$
13. (a) $3x$ **(b)** $5 - x$
15. (a) 5   **(b)** 0.5
17. 11.9 years   **19.** 3.8 years

21.

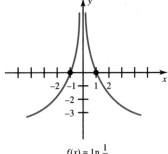

$f(x) = \ln \dfrac{1}{x^2}$

23.

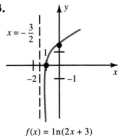

$x = -\frac{3}{2}$

$f(x) = \ln(2x + 3)$

25.

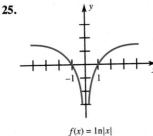

$f(x) = \ln|x|$

27. $x \neq 0$
29. $x \neq 0$
31. $\frac{1}{2}[\ln x + \ln y]$
33. $-\frac{1}{2}[3 \ln x + \ln y]$
35. $\ln(e^x) = x$
37. $\dfrac{\ln 3}{2} \approx 0.5493$ **39.** $-\ln 20$

41. $e^{0.2}$ **43.** $\dfrac{2}{\ln 3}$

45. (a) 1   **(b)** 4
47. (a) $\frac{1}{3}$ **(b)** $\frac{81}{64}$
49. (a) error   **(b)** error
51. about 8.99 minutes
53. $\ln 2/\ln 1.1 \approx 7.3$ years
55. $\ln 4/\ln 1.05 \approx 28.4$ years
57. $t = \dfrac{\ln 3}{\ln 1.08} \approx 14.27$ years

59. $t = \dfrac{\ln(\frac{1}{2})}{-0.005} \approx 138.6$ days

Referenced Exercise Set 15–2

1. (a) 0.8 hour   **(b)** 0.7 hour   **(c)** 0.48 hour
   (d) 0.23 hour
3. (a) .95 hour   **(b)** .92 hour   **(c)** .84 hour
   (d) .71 hour

5. $R(0.2) = 0.16$
7. the year 2022

Cumulative Exercise Set 15–2

1.

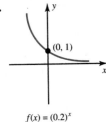

(0, 1)

$f(x) = (0.2)^x$

3. $x = -1$
5. $2 \ln|x| + 3 \ln|y| - \ln|x - y|$
7. $x = e^{-1/6}$

Case Study Exercises

1. $M = 3.2,$ $S \approx 2.56$
3. \$0.50 to \$1.00
5. \$22.63   **7.** \$128

Exercise Set Study 15–3

1. $\dfrac{1}{x}$

3. $\dfrac{1}{x}$

5. $\dfrac{1}{x + 5}$

7. $\dfrac{2x}{x^2 + 3}$

9. $\dfrac{8x}{1 + 4x^2}$

11. $\dfrac{6x^2 - \dfrac{1}{x^2}}{2x^3 + \dfrac{1}{x}} = \dfrac{6x^4 - 1}{2x^5 + x}$

13. $1 + \ln 4x$

15. $-\dfrac{3 \ln(x^2 - 1)}{x^4} + \dfrac{2}{x^2(x^2 - 1)}$

17. $\dfrac{\dfrac{(2x - 5)}{x} - 2 \ln 3x}{(2x - 5)^2} = \dfrac{(2x - 5) - 2x \ln 3x}{x(2x - 5)^2}$

19. $\dfrac{-5}{3 - 5x} = \dfrac{5}{5x - 3}$

21. $\dfrac{-\frac{5}{2}x^{3/2}}{4 - x^{5/2}}$

23. $1 + \ln |7x|$

25. Minimum at $x = e^{-1}$

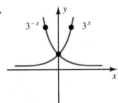

27. Maximum at $x = e$

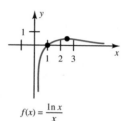

$f(x) = \dfrac{\ln x}{x}$

29. Inelastic in $[5, 8)$, elastic in $(8, 15]$

31. $2(x + \ln x)\left(1 + \dfrac{1}{x}\right)$

33. $\dfrac{2x \ln x - x}{(\ln x)^2}$

35. $\dfrac{2x + 1/x}{x^2 + \ln x} = \dfrac{2x^2 + 1}{x(x^2 + \ln x)}$

37. $\dfrac{10}{x} \cdot \dfrac{1}{\ln 10} = \dfrac{10}{x \ln 10}$

41. $f^n(x) = \dfrac{(-1)^{n+1}(n - 1)!}{x^n}$

43. $f(2) \approx 1.39$ but $g(2) \approx 0.48$;
$f'(2) = 1$ but $g'(2) = \ln 2 \approx 0.69$

Referenced Exercise Set 15–3

1. $A = me$, $A'(1) = e$

Cumulative Exercise Set 15–3

1.

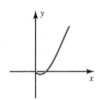

3. (a) $x = -2$ (b) $x = 6$ (c) $x = 2$

5.

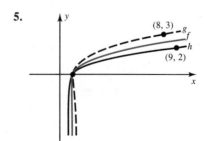

7. (a) $x = 3$ **(b)** $x = 1$

9. $\dfrac{-2x}{4 - x^2}$

11. $(1 - \ln x)^2 - 2(1 - \ln x)$ or $-(1 - \ln x)(1 + \ln x)$

Exercise Set 15–4

1. $6e^{6x}$

3. $3e^{3x+5}$

5. $28e^{7x+5}$

7. $2xe^{x^2+3}$

9. $16xe^{1+4x^2}$

11. $(x^{-2} - 6x^2)e^{2x^3+x^{-1}}$

13. $(4x + 1)e^{4x}$

15. $(2x^{-2} - 3x^{-4})e^{x^2-1}$

17. $\dfrac{(6x - 17)e^{3x}}{(2x - 5)^2}$

19. $5e^{5x} - 5e^{-5x}$

21. $\dfrac{e^{-x}}{(2 + e^{-x})^2}$

23. $\dfrac{2e^{-2x} - e^x}{(1 + e^x + e^{-2x})^2}$

25.

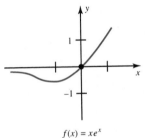

$f(x) = xe^x$

27.

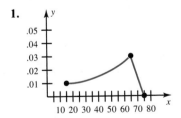

$f(x) = \dfrac{1}{2 + e^{-2x}}$

29. $\dfrac{1 + 4e^{4x}}{x + e^{4x}}$

31. $2(\ln x + e^{5x})\left(\dfrac{1}{x} + 5e^{5x}\right)$

33. $\dfrac{(2x - 1)e^{x^2} + 2x - x^2}{e^x}$

35. $\left(\dfrac{1}{x} + 3x^2 \ln x\right)e^{x^3}$

37. $f''(x) = 2^n e^{2x}$

39. Figure c

41. Figure d

Referenced Exercise Set 15–4

1.

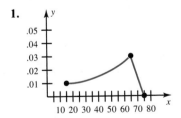

3. 1.98 months

Cumulative Exercise Set 15–4

1.

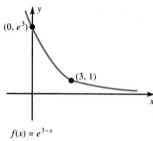

$f(x) = e^{3-x}$

3. $\frac{2}{3}$
5. $2x \ln(1 - x^3) - 3x^4/(1 - x^3)$
7. $(x - \ln x)^2/x + 2(\ln x)(x - \ln x)(1 - 1/x)$
9. $-10xe^{2-5x^2}$
11. $\dfrac{10e^{-5x}}{(3 + e^{-5x})^2}$

Review Exercises

1.

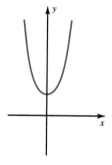

$f(x) = 2^{x+3}$

7. $-\ln(2.1) \approx -0.74$
9. $e^2 \approx 7.4$
11. $\dfrac{60[\ln(1 + 3x)^4]^4}{1 + 3x}$
13. $\dfrac{1}{2x}$
15. $x(x + 2)e^x + e^{-x}$
17. Absolute minimum at $x = 0$
No points of inflection

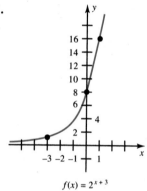

3.

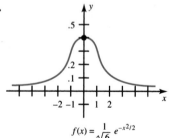

$f(x) = \dfrac{1}{\sqrt{6}} e^{-x^2/2}$

5.

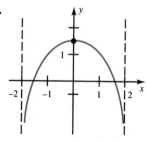

$f(x) = \ln(4 - x^2)$

CHAPTER 16

Exercise Set 16–1

1. $\frac{3}{2}x^2 + 5x + C$
3. $\frac{2}{5}x^5 - 7x + C$
5. $-6x^{-1} + \frac{5}{2}x^2 + C$
7. $\frac{4}{3}x^{3/2} + \frac{9}{4}x^{4/3} - \frac{5}{2}x^2 + C$
9. $\frac{9}{2}x^{2/3} + \frac{28}{3}x^{3/4} - 5x + C$
11. $\frac{x^{1.2}}{1.2} + \frac{3x^{2.4}}{2.4} + C$
13. $-x^{-1} - \frac{5}{3}x^{-3} + C$
15. $-\frac{1}{6x^6} + \frac{16x^{3/4}}{3} + \frac{16x^{5/4}}{5} + C$
17. $\frac{3e^{2x}}{2} + 5 \ln |x| + C$
19. $\frac{1}{4}x^4 - 2e^{-x} + \frac{2}{3}x^{3/2} - 5 \ln |x| + C$

21. $\frac{10}{3}x^{3/2} + \frac{1}{2}e^{-4x} + \ln |x| + C$
23. $18 \ln |x| + \frac{3}{\sqrt[3]{4}}x^{2/3} - \frac{1}{7}e^{7x} + C$
25. $y = \frac{1}{3}x^3 - 2x + \frac{2}{3}$
27. $y = 4 \ln |x| - x^2 + 2 - 4 \ln 2$
29. $\frac{2}{3}x^3 - 7 \ln |x| + C$ **31.** $6x + \frac{5}{4}x^4 + C$
33. $\frac{4}{9}x^9 - \frac{14}{3}x^6 + \frac{49}{3}x^3 + C$
35. $S(t) = \frac{t^3}{3} + t^2$
37. 3257 years
39. $k = -6.93 \times 10^{-4}$
41. $k = -0.0693$

Referenced Exercise Set 16–1

1. 6,969 years   **3.** $m' = -0.068m$

Exercise Set 16–2

1. $3.125 = \frac{25}{8}$ **3.** $2.71875 = \frac{174}{64}$
5. $11.375 = \frac{91}{8}$

7.

n	2	4	8	10
area	8	7	6.5	6.4

9.

n	2	4	8	10
area	7	5.75	5.1875	5.08

11.

n	2	4	8	10
area	14	11.5	10.375	10.16

13. 24   **15.** 20
17. 6   **19.** 2
21. 10.5   **23.** 6
25. $\frac{64}{3}$ **27.** $\frac{140}{3}$
29. $\frac{5}{6}$ **31.** 1.0986

Referenced Exercise Set 16–2

1.

Cumulative Exercise Set 16–2

1. $2x^2 + (6/x) + C$

3. $4e^x - 3(\ln x) + C$

5. $s(t) = 2 \cdot \ln(t^2 + 1) + 4$

7. 1.94

Exercise Set 16–3

1. $\frac{13}{2}$

3. $\frac{4}{3}$

5. $-\frac{5}{6}$

7. $-\dfrac{476}{3}$

9. $\dfrac{481}{32}$

11. 0

13. 12.5

15. $\frac{64}{3}$

17. 1

19. $\frac{1}{2}$

21. 25

23. 20

25. 2.08

27. 71.5068

29. -17.2047

31. 5

33. $\frac{59}{12}$

35. 0.5754

37. -1

39. -1

Referenced Exercise Set 16–3

1. 1,000

3. 19.84

Cumulative Exercise Set 16–3

1. $-3x^{-1} - x^5/5 + 5x + C$

3. $3e^{2x}/2 - 5 \ln x + C$

5. 30

7. 42

9. $\frac{23}{4}$

11. 24

Exercise Set 16–4

1. $\frac{14}{3}$

3. $\frac{20}{3}$

5. 10

7. $\frac{1}{6}$

9. $\frac{1}{6}$

11. $\frac{1}{6}$

13. $\frac{1}{6}$

15. $\frac{1}{6}$

17. $\frac{32}{3}$

19. $\frac{32}{3}$

21. $\frac{1}{12}$

23. 12

25. 13

29. 16

33. C.S. $= \frac{76}{3}$

   P.S. $= \frac{76}{3}$

37. 8.107

27. 13

31. C.S. $= \frac{32}{3}$

   P.S. $= \frac{40}{3}$

35. $1 - \ln 2 \cong 0.3069$

39. 6.08

Referenced Exercise Set 16–4

1. C.S. $= 3166.67$

3. C.S. $= \frac{7}{32}$, P.S. $= \frac{7}{192}$

Cumulative Exercise Set 16–4

1. $(x^2/6) + C$

3. $f(x) = (\ln x) + 5$

5. (a) $\frac{29}{4}$ (b) $\frac{22}{3}$

7. $\frac{64}{5}$

9. 8

11. $\frac{1}{2}$

Case Study Exercises

1. 4.004

3. 13.2

5. 0.36

Review Exercises

1. $\frac{25}{4}x^{4/5} - 5x^{2/5} - \frac{3}{2}x^2 + C$

3. $y = -\frac{1}{x} - \frac{x^2}{2} + \frac{3}{2}$

5. 8

7. 344.16

9. 637.519

11. $\frac{1}{6}$

CHAPTER 17

Exercise Set 17–1

1. $\dfrac{(x^2 + 5)^4}{4} + C$

3. $\dfrac{(x^4 - 7)^3}{3} + C$

5. $\dfrac{(3x^2 + 5)^4}{24} + C$

7. $\dfrac{(3x^4 - 5)^4}{48} + C$

9. $\dfrac{(3x^6 - 5)^{5/2}}{45} + C$

11. $\dfrac{-(x^4 - 7)^{-2}}{4} + C$

13. $\dfrac{(6x^{-2} + 5)^{-1}}{12} + C$

15. $\dfrac{(x^2 + 5x)^{-2}}{-2} + C$

17. $\dfrac{(x^4 - 4x)^{3/2}}{6} + C$

19. $\frac{1}{2}e^{2x} + \frac{5}{4}e^{4x} + C$

21. $\frac{1}{4}e^x - 2e^{-x} + C$

23. $\frac{2}{3}\ln|3x + 1| + C$

25. $\frac{1}{3}\ln|x^3 - 1| + C$

27. $\frac{7}{6}$

29. 0

31. $\frac{1}{3}(10^{3/2} - 1) = 10.2076$

33. $\dfrac{19}{480}$

35. $\dfrac{(e^x + 1)^2}{2} + C$

37. $\dfrac{(\ln x)^2}{2} + C$

39. $\frac{1}{4}x^4 - \frac{2}{3}x^3 + \frac{1}{2}x^2 + C$

41. $\dfrac{-2}{375}(1 - 5x)^{3/2}(15x + 2) + C$

43. $\frac{3}{16}(x^2 - 1)^{8/3} + \frac{3}{10}(x^2 - 1)^{5/3} + C$

45. $\frac{1}{4}[\ln(1 + x^2)]^2 + C$

Referenced Exercise Set 17–1

1. $0.5(x + 1) + 0.3\ln(x + 1) + C$

Exercise Set 17–2

1. $2xe^x - 2e^x + C$

3. $\frac{1}{3}xe^{3x} - \frac{1}{9}e^{3x} + C$

5. $\dfrac{2}{15}x(3x + 5)^{5/2} - \dfrac{4}{315}(3x + 5)^{7/2} + C$

7. $-\frac{3}{10}x(5x - 2)^{-2} - \frac{3}{50}(5x - 2)^{-1} + C$

9. $\frac{1}{2}x^2\ln x - \frac{1}{4}x^2 + C$

11. $\frac{1}{3}x^3\ln x - \frac{1}{9}x^3 + C$

13. $x^2e^x - 2xe^x + 2e^x + C$

15. $-\dfrac{1}{2}(2x + 5)(x + 5)^{-2} - (x + 5)^{-1} + C$

17. $\dfrac{2}{3}x^2(x - 4)^{3/2} - \dfrac{8}{15}(x)(x - 4)^{5/2}$
$+ \dfrac{16}{105}(x - 4)^{7/2} + C$

19. $\frac{1}{3}x^2(1 - 3x)^{-1} - \frac{2}{27}(1 - 3x) + \frac{2}{27}\ln|1 - 3x| + C$

21. $\frac{1}{4}e^2 + \frac{1}{4}$

23. $\dfrac{5\sqrt{5}}{3} + \dfrac{9}{5} \approx 5.5268$

25. $\frac{8}{81}$

27. $\dfrac{2}{3}x^2(x + 1)^{3/2} - \dfrac{8}{15}x(x + 1)^{5/2}$

$$+ \frac{16}{105}(x + 1)^{7/2} + C$$

29. $\dfrac{2}{3}x^3(x + 1)^{3/2} - \dfrac{4}{5}x^2(x + 1)^{5/2}$

$$+ \frac{16}{35}x(x + 1)^{7/2} - \frac{32}{315}(x + 1)^{9/2} + C$$

31. $\frac{1}{3}x^2(x^2 + 1)^{3/2} - \frac{2}{15}(x^2 + 1)^{5/2} + C$

33. $\frac{1}{2}x^2 e^{x^2} - \frac{1}{2}e^{x^2} + C$

35. $\dfrac{2942}{105} \approx 28.0190$

37. $\dfrac{x^2}{2}\ln x - \dfrac{x^2}{4} + x\ln x - x + C$

39. $-\dfrac{\ln x}{x} - \dfrac{1}{x} + C$

Referenced Exercise Set 17–2

1. 0.3093

Cumulative Exercise Set 17–2

1. $\dfrac{(2x^4 + 3)^6}{48} + C$

3. $\dfrac{3(x^4 - 4x^2)^{5/3}}{20} + C$

5. $3xe^{2x}/2 - 3e^{2x}/4 + C$

7. $\dfrac{x^6 \ln x}{6} - \dfrac{x^6}{36} + C$

Exercise Set 17–3

1. $\dfrac{1}{2(3x + 2)} + \dfrac{1}{4}\ln\left|\dfrac{x}{3x + 2}\right| + C$

3. $-\dfrac{1}{4(3x - 4)} + \dfrac{1}{16}\ln\left|\dfrac{x}{3x - 4}\right| + C$

5. $\dfrac{1}{6}\ln\left|\dfrac{1 + 3x}{1 - 3x}\right| + C$

7. $\dfrac{1}{24}\ln\left|\dfrac{4 + 3x}{4 - 3x}\right| + C$

9. $\dfrac{1}{24}\ln\left|\dfrac{3 + x}{3 - x}\right| + C$

11. $\frac{1}{2}[\frac{1}{2}x^2\sqrt{x^4 + 9} + \frac{9}{2}\ln|x^2 + \sqrt{x^4 + 9}|] + C$

13. $\frac{1}{4}[x^2\sqrt{4x^4 + 9} + \frac{9}{2}\ln|2x^2 + \sqrt{4x^4 + 9}|] + C$

15. $\frac{1}{2}[x^2\sqrt{4x^4 - 25} + \frac{25}{2}\ln|2x^2 + \sqrt{4x^4 - 25}|] + C$

17. $-\frac{1}{5}(\ln\frac{3}{4} - \ln 2) = 0.196$

19. $\frac{1}{4}(\ln\frac{1}{4} - \ln\frac{1}{5}) - \frac{3}{80} \approx 0.018$

21. $\dfrac{\sqrt{2}}{4} + \dfrac{1}{4}\ln|1 + \sqrt{2}|$

23. $\frac{1}{3}x^3 e^{3x} - \frac{1}{3}x^2 e^{3x} + \frac{2}{9}xe^{3x} - \frac{2}{27}e^{3x} + C$

25. $\dfrac{2}{25(5x + 2)} + \dfrac{1}{25}\ln|5x + 2| + C$

27. $-\dfrac{1}{4}\ln\left|\dfrac{x}{x - 4}\right| + C$

29. $-\ln\left|\dfrac{1 + \sqrt{1 - 9x^2}}{3x}\right| + C$

31. $\frac{1}{6}\ln|3x^2 + \sqrt{16 + 9x^4}| + C$

33. $-\dfrac{1}{3}\ln\left|\dfrac{3 + \sqrt{9 + 4x^2}}{2x}\right| + C$

35. $\dfrac{1}{36}\ln\left|\dfrac{3 + 2x^3}{3 - 2x^3}\right| + C$

37. $\dfrac{x^6}{2}\left[\dfrac{\ln x^2}{3} - \dfrac{1}{9}\right] + C$

Referenced Exercise Set 17–3

1. $\dfrac{-t^2}{2} - 10t - 100 \ln |t - 10| + C$

Cumulative Exercise Set 17–3

1. $(\frac{2}{15})(1 + 5x)^{3/2} + C$
3. $9x - 2x^3 + (\frac{1}{5})x^5 + C$
5. $-xe^{-x} - e^{-x} + C$

7. $\frac{1}{2} \ln(2x - 2) + \frac{1}{2}x + C$
9. $\frac{3}{8}$
11. $\frac{8}{3}$

Exercise Set 17–4

1. 9.5
5. 22.5
9. 23.88
13. 1.683
17. 0.9436
21. 2.72
25. 38.5861
29. 2.748

3. 73
7. 333
11. 0.8863
15. 1.644
19. 0.6766
23. 62.48
27. 11.5345
31. 58.489

33. $\frac{73}{45}$
35. Simpson's Rule: 1.62
    Trapezoidal Rule: 1.68
    Fundamental Theorem: 1.61
37. Simpson's Rule: 1.609
    Trapezoidal Rule: 1.614
    Fundamental Theorem: 1.609

Referenced Exercise Set 17–4

1. 913.5
3. 413.02

Cumulative Exercise Set 17–4

1. $\dfrac{(x^6 + 6x)^{4/3}}{8} + C$

3. $\dfrac{4x^{5/4} \ln x}{5} - \dfrac{16x^{5/4}}{25} + C$

5. $\dfrac{1}{3(2x + 3)} + \dfrac{1}{9} \ln \left| \dfrac{x}{2x + 3} \right| + C$

7. $\dfrac{x^2}{4} (25 + x^4)^{1/2} + \dfrac{25}{4} \ln (x^2 + (25 + x^4)^{1/2}) + C$
9. 22
11. 0.886

Case Study Exercises

1. 1.7%
3. 2.48%
5. 3.2% (from Exercise 4)

Review Exercises

1. $\dfrac{(x^2 + x)^{-2}}{-2} + C$ **3.** $-\dfrac{15}{8}$

5. $-\dfrac{3}{4}(1 - x)^{4/3} + \dfrac{6}{7}(1 - x)^{7/3} - \dfrac{3}{10}(1 - x)^{10/3} + C$

7. $-\dfrac{1}{7(5x - 7)} + \dfrac{1}{49} \ln \left| \dfrac{x}{5x - 7} \right| + C$

9. $\dfrac{1}{6\sqrt{14}} \ln \left| \dfrac{\sqrt{14} + 3x}{\sqrt{14} - 3x} \right| + C$

11. 0.8862

CHAPTER 18

Exercise Set 18–1

1. 1 **3.** -9

5. 2 **7.** 63

9. All ordered pairs for which $x \neq 2$

11. All ordered pairs for which $x > -1$

13.

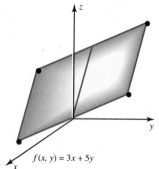

$f(x, y) = 3x + 5y$

15.

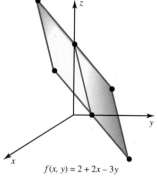

$f(x, y) = 2 + 2x - 3y$

17.

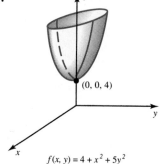

$(0, 0, 4)$

$f(x, y) = 4 + x^2 + 5y^2$

19.

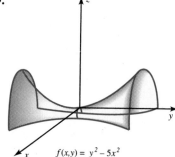

$f(x,y) = y^2 - 5x^2$

21.

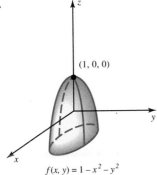

$(1, 0, 0)$

$f(x, y) = 1 - x^2 - y^2$

23.

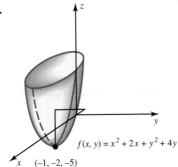

$f(x, y) = x^2 + 2x + y^2 + 4y$

x $(-1, -2, -5)$

25.

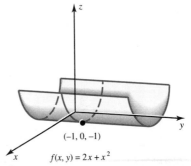

$(-1, 0, -1)$

$f(x, y) = 2x + x^2$

27. no

29. A vertical line will cut the surface in no more than one point.

31. $P = 40x + 50y + 60z$

33. Figure a

35. Figure d

Referenced Exercise Set 18–1

1. Tuesday, \$142,108

3. $y \approx 0.001$

Exercise Set 18–2

1. $f_x(x, y) = 6x$
$f_y(x, y) = \frac{5}{2}y^{-1/2}$

3. $f_x(x, y) = y^2 - 10x$
$f_y(x, y) = 2xy$

5. $f_x(x, y) = y^2 - 6xy$
$f_y(x, y) = 2xy - 3x^2$

7. $f_x(x, y) = -2x^{-3} + y^{-2} - 2xy^3$
$f_y(x, y) = -2xy^{-3} - 3x^2y^2 + 4$

9. $f_x(x, y) = e^{2y} - \dfrac{2}{2x + 3y}$

$f_y(x, y) = 2xe^{2y} - \dfrac{3}{2x + 3y}$

11. $f_x(x, y) = 8(2x + 3y)^3$
$f_y(x, y) = 12(2x + 3y)^3$

13. $f_x(x, y) = -2(2 + 4y)(2x + 4xy)^{-3}$
$f_y(x, y) = -8x(2x + 4xy)^{-3}$

15. $f_x(1, 0) = 0$
$f_y(2, 1) = 20.5$

17. $f_x(1, 0) = 1$
$f_y(2, 1) = -18 + e$

19. $f_{xx}(x, y) = 6$
$f_{xy}(x, y) = 0$
$f_{yy}(x, y) = \dfrac{30}{y^4}$

21. $f_{xx}(x, y) = 6y$
$f_{xy}(x, y) = 6x + 10e^{2y}$
$f_{yy}(x, y) = 20xe^{2y}$

23. $f_{xx} = 2y^2e^x + xy^2e^x + 4e^{2x}$
$f_{yy} = 2xe^x$
$f_{xy} = 2ye^x + 2xye^x$

25. $x = 1$ and $y = 1$

27. $f_{xxx} = 24x - 6$
  $f_{xxy} = 0$
  $f_{xyy} = 0$
29. $f_{xxx} = 0$
  $f_{xxy} = 2$
  $f_{xyy} = -4e^{2y}$

31. $f_x(x, y, z) = \lim\limits_{h \to 0} \dfrac{f(x + h, y, z) - f(x, y, z)}{h}$

33. $f_x = -6x^{-3} + \frac{1}{3}x^{-2/3}z - 2xz^3 + yz$
  $f_y = xz$

Referenced Exercise Set 18–2

1. $\dfrac{\partial C}{\partial Q} = 1.87Q^{0.1}F^{1.5}L^{0.36}P^{0.5}$

$\dfrac{\partial C}{\partial F} = 2.55Q^{1.1}F^{0.5}L^{0.36}P^{0.5}$

$\dfrac{\partial C}{\partial L} = 0.612Q^{1.1}F^{1.5}L^{-0.64}P^{0.5}$

$\dfrac{\partial C}{\partial P} = 0.85Q^{1.1}F^{1.5}L^{0.36}P^{-0.5}$

3. (a) $0.07D^{-0.44}A^{0.06}$ **(b)** $0.0075D^{0.56}A^{-0.94}$

Cumulative Exercise Set 18–2

1. (a) 99 **(b)** -101
3. The plane through the triangle traced in the accompanying figure.

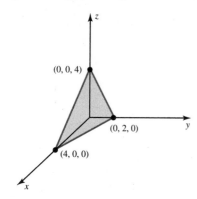

5. $f_x = \dfrac{2x}{y}, f_y = -\dfrac{x^2}{y^2}$

7. $f_x(0, 1) = 2, f_y(0, 1) = 1$

Exercise Set 18–3

1. Relative minimum at $(0, 0)$
3. no
5. no   **7.** $(-3, -4)$
9. $(3, -3)$

11. $(0, +1)$ and $(0, -1)$
13. Saddle point at $(-2, 0)$
  No relative maximum or minimum
15. Relative minimum at $(4, 0)$
  Saddle point at $(4, -2)$

17. Relative minimum at $(0, -1)$
   Saddle point at $(0, 1)$
19. Relative minimum at $(4, 3)$
21. Saddle point at $(-3, 1)$
23. Relative minimum at $(2, 2)$
   Saddle point at $(0, 0)$
25. Saddle point at $(0, 1)$
   Saddle point at $(0, -1)$

27. Saddle point at $(4, 2)$ and $(0, -2)$
   Relative minimum at $(4, -2)$ and relative maximum
   at $(0, 2)$
29. Saddle point at $(4, -2.225)$
   Relative minimum at $(4, 0.225)$
31. $(12, 2)$ is a relative minimum.
33. $l \times w \times h = 4\sqrt[3]{2} \times 4\sqrt[3]{2} \times 2\sqrt[3]{2}$

Referenced Exercise Set 18–3

1. $x = \dfrac{a}{2b}, y = 0$

Cumulative Exercise Set 18–3

1.

$f(x, y) = 2 + 3y - 4x$

3.

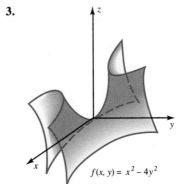

$f(x, y) = x^2 - 4y^2$

5. $f_x(x, y) = 3x^2 - 4xy$
   $f_y(x, y) = -2x^2 + 4y^3$
7. $f_x(x, y) = 12x^2(x^3 + 5y)^3$
   $f_y(x, y) = 20(x^3 + 5y)^3$
9. saddle point at $(0, -2)$
11. saddle point at $(6, -9)$

Exercise Set 18–4

1. $(\frac{1}{2}, \frac{1}{2})$ **3.** $(-1, \frac{1}{2})$
5. $(-1, 0)$ and $(-\frac{5}{2}, \frac{9}{4})$
7. $(1.63, 0.185)$ and $(0.2, 0.9)$
9. $(19.26, 4.63)$ and $(7.4, -1.3)$
11. $(\frac{1}{4}, \frac{1}{2})$
13. $(-\frac{1}{2}, \frac{2}{3})$ **15.** $(\frac{1}{3}, \frac{2}{3}, \frac{2}{3})$
17. $(100, 100, 100)$

19. The numbers are 10, 10, 10.
21. $(25, 2500)$
23. $(200, 200, 100)$
25. $\sqrt{10}/2$ at $(\frac{3}{2}, \frac{1}{2})$
27. $(\frac{1}{2}, 0)$, $(\frac{1}{6}, -\frac{2}{3})$
29. $(\frac{1}{4}, \frac{1}{6}, 1)$ **31.** $(2, 0, 0)$, $(\frac{6}{15}, \frac{16}{15}, \frac{8}{15})$
33. $(1, 1, 0)$, $(-1, -1, 0)$ **35.** $x = y = 2, z = 4$

37. $x = y = 14$, $z = 28$
39. (a) $C(x, y) = 1000 + 100x + 50y$
   (b) $x = y = 2.5$
41. $P(x, y) = 160x + 80y - 20x^2 - 4y^2 - 8xy - 15$; $x = 2.5$, $y = 7.5$
43. $6 \times 6 \times 6$ inches

Referenced Exercise Set 18–4

1. $\left(\sqrt{\dfrac{27a}{118}}, \sqrt{\dfrac{32a}{177}} \right)$

Cumulative Exercise Set 18–4

1. (a) 0   **(b)** $\{(x, y)\,|\,y \neq 0\}$

3. $f_x = \dfrac{-y^2}{(x - y)^2}$

   $f_y = \dfrac{x^2}{(x - y)^2}$

5. no
7. relative minimum at $(4, 1)$
9. $x = 9$, $y = 7$
11. $x = 3$, $y = 6$, $z = 6$

Exercise Set 18–5

1. $\dfrac{7x^2y^6}{2} + \dfrac{3x^4y^{-2}}{2} - 5x + C(y)$

3. $xy^8 - 4x^{-4}y^5 - 3xy - \frac{1}{2}y^2 + C(x)$

5. $31y + 2y^3 \ln 2 - 5$

7. $10x^4 + \dfrac{8}{x} - 10$

9. -2 **11.** 674.44
13. $\frac{32}{3}$ **15.** 3

17. $\dfrac{2x}{9}(x^2 + 3y)^{3/2} + C(x)$

19. $\ln |3x^2 + 2xy| + C(y)$
21. $\frac{1}{3}(x^2 + 3y)^{3/2} + C(y)$

23. $\dfrac{2x}{9}(x^2 + 15)^{3/2} - \dfrac{2x}{9}(x^2 + 12)^{3/2}$

25. $\ln |75 + 10y| - \ln |27 + 6y|$

27. $\frac{1}{9}(16 + 3y)^{3/2} - \frac{1}{9}(3y)^{3/2}$
29. $(\ln 2)(\ln 3)$
31. 3.05

33. $\frac{1}{2}(e^2 - e - 1 + \dfrac{1}{e})$

35. 256
37. (a) $\frac{832}{9}$ **(b)** $\frac{832}{9}$
39. $(\frac{1}{3})e^6 - e^2 + (\frac{2}{3})$

43. $\displaystyle\int_0^1 \int_0^{2-x^2} dy\,dx$ and $\displaystyle\int_0^2 \int_0^{\sqrt{2-y}} dx\,dy$

45. $\displaystyle\int_{-1}^0 \int_{x^2}^{2+x} dy\,dx + \int_0^1 \int_{x^2}^{2-x} dy\,dx$

   and $\displaystyle\int_0^1 \int_{-\sqrt{y}}^{\sqrt{y}} dx\,dy + \int_1^2 \int_{y-2}^{2-y} dx\,dy$

Referenced Exercise Set 18–5

1. 95.9 k

Cumulative Exercise Set 18–5

1.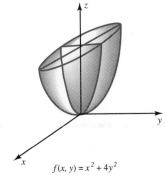

$f(x, y) = x^2 + 4y^2$

3. $f_x(x, y) = 2x - 6x^2y$
  $f_y(x, y) = -2x^3 + 2y^{-2/3}$
5. saddle point at $(0, 6)$
7. $(\frac{1}{6}, \frac{1}{4})$
9. 1
11. -4.062

Exercise Set 18–6

1. 1/15
3. 1/9
5. 424/21
7. 2/3
9. 13/70
11. 28/5
13. 12
15. $\frac{1}{2}(e^7 - e^6 - e^3 + e^2)$
17. 16/3

19. 36
21. 128/9
23. ln 3
25. $(4\sqrt{3}/5) + 1$
27. $\ln(17)/4$
29. 8/3
31. $e^3 - 2e + (1/e)$
33. 54
35. 1

Cumulative Exercise Set 18–6

1. **(a)** 2 **(b)** not defined **(c)** $\{(x, y)|x \neq -y\}$
3. **(a)** $f_x = 12x^3 - 4y^3$, $f_y = -12xy^2 - 2$
  (b) $f_{xy} = f_{yx} = -12y^2$, $f_{xx} = 36x^2$, $f_{yy} = -24xy$
5. $x = 9$, $y = 7$

7. 0
9. $(e/2) - 1$
11. 80/3

Case Study Exercises

1. $x = 111.11$
  $y = 37$
5. $x = 133.33$
  $y = 27.78$

3. $x = 122.22$
  $y = 43.46$

Review Exercises

1. -1

3.

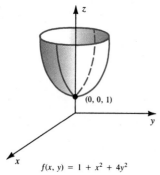

$(0, 0, 1)$

$f(x, y) = 1 + x^2 + 4y^2$

5. $f_x(x, y) = y^3 - 8(2x + 3y)^3$
$f_y(x, y) = 3xy^2 - 12(2x + 3y)^3$

7. $f_{xx} = -\frac{18}{25}x^{-8/5} + 2y^{-2}$
$f_{yy} = 6x^2y^{-4}$
$f_{xy} = f_{yx} = -4xy^{-3}$

9. Critical point $(\frac{10}{11}, \frac{5}{11})$

11. Relative minimum at $(4, -2)$
Relative maximum at $(0, 2)$
Saddle points at $(0, -2)$ and $(4, 2)$

13. $(\frac{5}{6}, \frac{5}{2})$

15. The numbers are 15, 15, and 15.

17. $7x^2y^6 + \dfrac{3x^4y^{-2}}{2} - xy + \dfrac{x^2}{2} - x + C(y)$

19. -2 **21.** $\frac{2}{3}$

Index

CASE STUDY PREVIEWS

Powerlifting

In powerlifting meets, contestants are divided into weight classes, and each class has a champion. But the meet also has a champion of champions, and it is usually not the person who lifts the greatest amount of weight. A handicapping formula, called the Schwartz formula, is used, which is defined in terms of a polynomial function.

Radiotherapy Treatment Design

More than half of all cancerous tumors can be treated by radiation therapy. Medical researchers use mathematical optimization techniques to maximize the amount of cancer cells killed by radiation beams while also minimizing the number of healthy cells affected. How does the derivative help to balance these two objectives?

Minimizing Traffic Noise

Two important goals of highway operation, safety and optimum flow of traffic, often work at odds with each other. Highway administrators in Houston used calculus to determine that the optimum speed for minimizing acceleration noise is very close to the speed that maximizes traffic volume.

The Frustration of Dieting

Losing weight can be a trying experience. Excuses abound for our inability to diet successfully. This case study offers the frustrated dieter a bit of evidence for the case against short-term dieting.

Sears, Clothespins, and Unnatural Logarithms

Sears Roebuck and Co. started its mail-order business at the turn of the 19th century with the same kind of catalogs that are used today. These catalogs are used by economists to explore the U.S. economy and lifestyle by examining automobiles, the Consumer Price Index, and clothespins.